Linear and Nonlinear
Structural Mechanics

Linear and Nonlinear Structural Mechanics

Ali H. Nayfeh
Virginia Polytechnic Institute and State University
Blacksburg, VA

P. Frank Pai
University of Missouri
Columbia, MO

A JOHN WILEY & SONS, INC., PUBLICATION

Copyright © 2004 by John Wiley & Sons, Inc. All rights reserved.

Published by John Wiley & Sons, Inc., Hoboken, New Jersey.
Published simultaneously in Canada.

No part of this publication may be reproduced, stored in a retrieval system, or transmitted in any form or by any means, electronic, mechanical, photocopying, recording, scanning, or otherwise, except as permitted under Section 107 or 108 of the 1976 United States Copyright Act, without either the prior written permission of the Publisher, or authorization through payment of the appropriate per-copy fee to the Copyright Clearance Center, Inc., 222 Rosewood Drive, Danvers, MA 01923, (978) 750-8400, fax (978) 646-8600, or on the web at www.copyright.com. Requests to the Publisher for permission should be addressed to the Permissions Department, John Wiley & Sons, Inc., 111 River Street, Hoboken, NJ 07030, (201) 748-6011, fax (201) 748-6008.

Limit of Liability/Disclaimer of Warranty: While the publisher and author have used their best efforts in preparing this book, they make no representations or warranties with respect to the accuracy or completeness of the contents of this book and specifically disclaim any implied warranties of merchantability or fitness for a particular purpose. No warranty may be created or extended by sales representatives or written sales materials. The advice and strategies contained herein may not be suitable for your situation. You should consult with a professional where appropriate. Neither the publisher nor author shall be liable for any loss of profit or any other commercial damages, including but not limited to special, incidental, consequential, or other damages.

For general information on our other products and services please contact our Customer Care Department within the U.S. at 877-762-2974, outside the U.S. at 317-572-3993 or fax 317-572-4002.

Wiley also publishes its books in a variety of electronic formats. Some content that appears in print, however, may not be available in electronic format.

Library of Congress Cataloging-in-Publication Data is available.

ISBN 0-471-59356-7

Printed in the United States of America.

10 9 8 7 6 5 4 3 2 1

To our wives
Samirah and Lifien

Contents

PREFACE   xv

1   **INTRODUCTION**   1
    1.1   Structural Elements   1
    1.2   Nonlinearities   4
    1.3   Composite Materials   6
    1.4   Damping   7
    1.5   Dynamic Characteristics of Linear Discrete Systems   9
        1.5.1   One-Degree-of-Freedom Systems   9
        1.5.2   Multi-Degree-of-Freedom Systems   14
    1.6   Dynamic Characteristics of Nonlinear Discrete Systems   22
    1.7   Analyses of Linear Continuous Systems   27
        1.7.1   Natural Frequencies and Eigenfunctions   30
        1.7.2   Discretization Using Eigenfunctions   34
        1.7.3   The Ritz Method   35
        1.7.4   Finite Element Method   42
        1.7.5   Weighted Residual Methods   46
        1.7.6   Initial-Value Methods   48
    1.8   Analyses of Nonlinear Continuous Systems   53

	1.8.1	Attacking the Continuous System	57
	1.8.2	Attacking the Discretized System	60
	1.8.3	Time-Averaged Lagrangian	63

2 ELASTICITY — 65
2.1 Principles of Dynamics — 65
- 2.1.1 Newton's Second Law and Energy of a Discrete System — 67
- 2.1.2 Principle of Virtual Work — 69
- 2.1.3 Hamilton's Theories — 70
- 2.1.4 Euler-Lagrange Equations — 72
- 2.1.5 Hamilton's Equations — 74

2.2 Strain-Displacement Relations — 75
2.3 Transformation of Strains and Stresses — 82
2.4 Stress-Strain Relations — 85
- 2.4.1 Anisotropic Materials — 86
- 2.4.2 Orthotropic Materials — 87
- 2.4.3 Isotropic Materials — 90
- 2.4.4 Material Stiffness and Compliance Matrices — 94
- 2.4.5 Fiber-Reinforced Lamina — 96

2.5 Governing Equations — 98
- 2.5.1 Equilibrium Equations — 99
- 2.5.2 Compatibility Conditions — 107
- 2.5.3 Energy Formulation of Structures — 109

3 STRINGS AND CABLES — 111
3.1 Modeling of Taut Strings — 111
- 3.1.1 Exact Equations of Motion — 113
- 3.1.2 Approximate Equations of Motion without Poisson's Effect — 115
- 3.1.3 Approximate Equations of Motion with Poisson's Effect — 117

3.2 Reduction of String Model to Two Equations — 118
- 3.2.1 Attacking the Three-Equation Model — 121
- 3.2.2 Evaluation of the Two-Equation Model — 124
- 3.2.3 Discretized Model — 126

3.3 Nonlinear Response of Strings — 129
- 3.3.1 Frequency-Response Curves — 130
- 3.3.2 Experiments — 132

3.4	Modeling of Cables	136
3.4.1	Exact Equations of Motion	139
3.4.2	Static Deflections	141
3.4.3	Approximate Equations of Motion	145
3.5	Reduction of Cable Model to Two Equations	146
3.6	Natural Frequencies and Modes of Cables	148
3.7	Discretization of the Cable Equations	150
3.8	Single-Mode Response with Direct Approach	152
3.8.1	Primary Resonance of an Inplane Mode	153
3.8.2	Primary Resonance of an out-of-Plane Mode	158
3.9	Single-Mode Response with Discretization Approach	161
3.9.1	The Case of an Inplane Mode	162
3.9.2	The Case of an out-of-Plane Mode	167
3.10	Extensional Bars	168

4 BEAMS — 171

4.1	Introduction	171
4.1.1	Beam Theories	172
4.1.2	Geometric Nonlinearities	176
4.1.3	Shear Deformations, Rotary Inertias, and Gravity	179
4.1.4	Elastic Couplings	180
4.1.5	External and Internal Resonances	180
4.2	Linear Euler-Bernoulli Beam Theory	183
4.3	Linear Shear-Deformable Beam Theories	186
4.3.1	Third-Order Shear-Deformable Theory	189
4.3.2	Timoshenko's Beam Theory	191
4.3.3	Layerwise Shear-Deformable Theory	192
4.4	Mathematics for Nonlinear Modeling	194
4.4.1	Coordinate Transformations and Curvatures	194
4.4.2	Concept of Orthogonal Virtual Rotations	207
4.4.3	Variation of Curvatures	210
4.4.4	Concept of Local Displacements	211
4.5	Nonlinear 2-D Euler-Bernoulli Beam Theory	215
4.5.1	Shortening Effect	221
4.5.2	Stretching Effect	224
4.5.3	Lagrangian and Eulerian Coordinates	225
4.6	Nonlinear 3-D Euler-Bernoulli Beam Theory	226
4.6.1	Isotropic Beams	234

		4.6.2	Composite Beams	235
		4.6.3	Taylor-Series Expansions	235
		4.6.4	Cantilevered Inextensional Beams	240
		4.6.5	Flexural-Flexural Vibration	244
	4.7	Nonlinear 3-D Curved Beam Theory Accounting for Warpings		245
		4.7.1	Inplane and out-of-Plane Warpings	247
		4.7.2	Fully Nonlinear Jaumann Strains	251
		4.7.3	Equations of Motion	254
		4.7.4	Expansions and Simplified Beam Theories	262
		4.7.5	Applications	263
5	DYNAMICS OF BEAMS			267
	5.1	Parametrically Excited Cantilever Beams		267
		5.1.1	Experiments	267
		5.1.2	Principal Parametric Resonance	270
		5.1.3	Combination Parametric Resonance	278
		5.1.4	Nonplanar Dynamics	284
	5.2	Transversely Excited Cantilever Beams		291
		5.2.1	Planar Response to a Primary-Resonance Excitation	291
		5.2.2	External Subcombination Resonance	298
		5.2.3	Nonplanar Dynamics	304
	5.3	Clamped-Clamped Buckled Beams		316
		5.3.1	Buckling Problem	320
		5.3.2	Linear Vibration Problem	321
		5.3.3	Nonlinear Local Vibrations - Direct Approach	324
		5.3.4	Nonlinear Local Vibrations - Discretization Approach	327
		5.3.5	Experiment	331
		5.3.6	Global Dynamics	334
	5.4	Microbeams		341
		5.4.1	Modeling of MEMS Devices	342
		5.4.2	Static Deflection	344
		5.4.3	Linear Mode Shapes and Frequencies	345
		5.4.4	Nonlinear Response to a Primary-Resonance Excitation	346
		5.4.5	Reduced-Order Models of MEMS Devices	351

6 SURFACE ANALYSIS 355
6.1 Initial Curvatures 355
6.2 Inplane Strains and Deformed Curvatures 358
6.3 Orthogonal Virtual Rotations 362
6.3.1 Without Inplane Shear Strains 362
6.3.2 With Inplane Shear Strains 363
6.4 Variation of Curvatures 365
6.5 Local Displacements and Jaumann Strains 366

7 PLATES 371
7.1 Introduction 371
7.1.1 Plate Theories 371
7.1.2 Geometric Nonlinearities 375
7.1.3 Plates with Integrated Piezoelectric Materials 376
7.1.4 Linear Vibrations and Buckling of Plates 377
7.1.5 Nonlinear Analyses of Plates 379
7.2 Linear Classical Plate Theory 382
7.2.1 Rectangular Plates 382
7.2.2 Circular Plates 388
7.2.3 General Plates 392
7.3 Linear Shear-Deformable Plate Theories 396
7.3.1 Formulation for Curvilinear Coordinate Systems 396
7.3.2 Rectangular and Circular Plates 401
7.3.3 Different Shear-Warping Functions 402
7.4 Nonlinear Classical Plate Theory 403
7.4.1 Rectangular Plates 403
7.4.2 von Kármán Plate Theory in Polar Coordinates 408
7.4.3 Thermoelastic Equations in Cartesian Coordinates 412
7.4.4 Thermoelastic Equations in Polar Coordinates 415
7.5 Nonlinear Modeling of Rectangular Surfaces 417
7.5.1 Coordinate Transformation, Inplane Strains, and Curvatures 417
7.5.2 Influence of the Inplane Shear Deformation 420
7.5.3 Variation of the Global Strains 425
7.6 General Nonlinear Classical Plate Theory 426
7.7 Nonlinear Shear-Deformable Plate Theory 435
7.7.1 Equations of Motion 436

xii CONTENTS

       7.7.2   Nonlinear First-Order Theory   446
       7.7.3   Third-Order Theory with von Kármán
              Nonlinearity   446
  7.8   Nonlinear Layerwise Shear-Deformable Plate Theory   446
       7.8.1   Warpings Due to External Loads and Actuators   447
       7.8.2   Equations of Motion   457
       7.8.3   Linear Piezoelectric Plate Theory   465
       7.8.4   Actuator-Induced Loads   467
       7.8.5   Thermal and Moisture Effects   467

8 DYNAMICS OF PLATES   **469**
  8.1   Linear Vibrations of Rectangular Plates   469
       8.1.1   Hinged Edges   470
       8.1.2   Two Hinged Opposite Edges   470
  8.2   Linear Vibrations of Membranes   474
       8.2.1   Circular Membranes   474
       8.2.2   Near Circular Membranes   477
       8.2.3   Elliptic Membranes   480
  8.3   Linear Vibrations of Circular and Annular Plates   481
       8.3.1   Circular Plates   482
       8.3.2   Near Circular and Elliptic Plates   487
       8.3.3   Annular Plates   491
  8.4   Nonlinear Vibrations of Circular Plates   498
       8.4.1   Axisymmetric Vibrations   503
       8.4.2   Asymmetric Vibrations   507
  8.5   Nonlinear Vibrations of Rotating Disks   513
       8.5.1   Static Problem   516
       8.5.2   Natural Frequencies and Mode Shapes   517
       8.5.3   Response to a Primary-Resonance Excitation   522
  8.6   Nonlinear Vibrations of Near-Square Plates   527
  8.7   Micropumps   531
       8.7.1   Annular Plates   532
       8.7.2   Circular Plates   536
  8.8   Thermally Loaded Plates   543
       8.8.1   Linear Natural Frequencies and Mode Shapes   548
       8.8.2   Combination Parametric Resonance of Two
              Axisymmetric Modes   549

9 SHELLS   **559**

9.1	Introduction	559
	9.1.1	Shell Theories	560
	9.1.2	Nonlinear Vibrations of Shells	563
9.2	Linear Classical Shell Theory	566
	9.2.1	Different Shell Geometries	566
	9.2.2	Doubly-Curved Shell Theory	571
	9.2.3	Circular Cylindrical Shell Theory	576
9.3	Linear Shear-Deformable Shell Theories	577
	9.3.1	Formulation for General Shells	577
	9.3.2	Equations of Motion for Different Shells	581
	9.3.3	Shear-Warping Functions	581
9.4	Nonlinear Classical Theory for Doubly-Curved Shells	582
9.5	Nonlinear Shear-Deformable Theories for Circular Cylindrical Shells	588
	9.5.1	Equations of Motion	589
	9.5.2	Simplified Shell Theories	603
	9.5.3	Stiffness Matrices	608
	9.5.4	Classical Linear Theories of Circular Cylindrical Shells	611
9.6	Nonlinear Layerwise Shear-Deformable Shell Theory	615
	9.6.1	Strains and Shear-Warping Functions	615
	9.6.2	Inertia Terms	620
	9.6.3	Structural Terms	622
	9.6.4	Equations of Motion	625
	9.6.5	Shear-Warping Functions	627
9.7	Nonlinear Dynamics of Infinitely Long Circular Cylindrical Shells	630
	9.7.1	Governing Equations	631
	9.7.2	Natural Frequencies and Mode Shapes	635
	9.7.3	Primary Resonance of the Breathing Mode	638
9.8	Nonlinear Dynamics of Axisymmetric Motion of Closed Spherical Shells	641
	9.8.1	Equations of Motion	641
	9.8.2	Natural Frequencies and Mode Shapes	646
	9.8.3	Two-to-One Internal Resonance	649

BIBLIOGRAPHY 654

SUBJECT INDEX 732

PREFACE

In the last few decades engineering materials have gone through different phases and can be categorized into four groups: composite materials, smart materials, micro- and nano-materials, and materials for Gossamer (or ultra-lightweight deployable/inflatable) space structures. The factors that distinguish them are the ability for tailoring; coupling of mechanical, electrical, magnetic, and/or thermal fields; tremendous size decrease; and tremendous increase in size but decrease in mass density. Research in continuum mechanics has followed and advanced through these phases, but the major challenge is still the modeling and analysis of structures built with such materials. Although modeling of such structures built with materials in different groups requires consideration of different effects, all these structures can be modeled as continuum media, especially as cables, beams, plates, and/or shells. However, such structures are usually nonlinear by nature or, in some cases, by design.

For example, thin-walled structures play an important role in the design of aircraft structures because they are often designed to operate in the postbuckling range in order to reduce structural weight. In recent years, the rapid development and use of huge deployable/inflatable structures in aerospace and space exploration has stimulated extensive research into fully nonlinear modeling and analysis, thermal buckling, and control of highly flexible structures. Also, the increasing use of laminated composite materials in modern structures has stimulated the development of refined structural theories that can account for nonclassical effects, such as transverse shear stresses, interlaminar peeling stresses, torsional warping, free-edge effects, and warping restraint effects. Nonlinear problems considered in such structures are mostly those of postbuckling analysis, prediction of stability, and flutter analysis. However, the nonlinear dynamics of such structures have to be ascertained in order to design and control them. Hence, nonlinear modeling and analysis of structures becomes a complex but important step in advancing the design and optimization of modern structural systems.

This book presents mathematically consistent and systematic derivations of comprehensive structural theories developed by the authors as well as well-known linear and nonlinear structural theories in the literature, details the physical meaning of linear and nonlinear structural mechanics, shows how to perform nonlinear structural analysis, points out important nonlinear structural dynamic behaviors, and provides ready-to-use governing equations and boundary conditions, ranging from simple linear ones to complex nonlinear ones, for strings, cables, beams, plates, and shells. The major goal of this book is to close the gap between the practicing engineer and the applied mathematician in the modeling and analysis of geometrically nonlinear structures. This book is written in a common vector-based mathematical language that is understandable by most engineering students. A unique unified approach, more general than those found in most structural mechanics books, is used to model geometric nonlinearities of structures. As a result, the reader can readily extend the methods to formulate and analyze different and/or more complex structures. This book is intended to be a graduate-level text and a reference book for graduate students and structural engineers in mechanical, civil, and aeronautical engineering or in applied mechanics who have had courses in mechanics of materials, ordinary and partial differential equations, and vibrations.

The text is organized into nine chapters. Chapter 1 is essentially an introduction to modeling issues, dynamic characteristics of linear and nonlinear discrete systems, and methods for analyzing linear and nonlinear continuous systems. Chapter 2 presents a self-contained treatment of the basic principles of structural mechanics. Chapter 3 presents linear and geometrically exact formulations, nonlinear analysis, and nonlinear dynamics of taut strings, cables, and bars. Chapter 4 presents linear and geometrically exact formulations of beams. Chapter 5 presents nonlinear analysis and dynamics of different beams, including microbeams for MEMS devices. Chapter 6 presents the mathematics needed for geometrically exact modeling of plates and shells. Chapter 7 presents linear and geometrically exact formulations of plates. Chapter 8 presents nonlinear analysis and dynamics of different plates, including MEMS-based microplates and thermally loaded circular and annular plates. Chapter 9 presents linear and geometrically exact formulations, nonlinear analysis, and nonlinear dynamics of shells. It also includes the nonlinear dynamics of circular cylindrical shells and spherical shells. A long list of references, by no means complete or up-to-date but consisting of most of the important articles in the literature, is provided in the Bibliography at the end of the book.

The authors wish to acknowledge with great appreciation the many valuable suggestions from their colleagues. In particular, the authors thank Dr. Haider Arafat for his valuable comments, thorough proofreading of the entire manuscript, and preparing many of the tables and illustrations. Also, the authors thank Drs. Pramod Malatkar and Eihab Abdel-Rahman for their valuable comments and thorough proofreading of parts of the manuscript. We wish also to thank Mr. Nader Nayfeh for editing and preparing the postscript files for some of the illustrations. Many results presented in this book were obtained under research grants supported by NSF, NASA, ARO, and AFOSR; the support is gratefully acknowledged. Last but not least, the authors want to express their appreciation to Mrs. Sally Shrader for repeatedly typing, correcting, and beautifying the manuscript.

Ali H. Nayfeh
Blacksburg, VA

P. Frank Pai
Columbia, MO

May 2004

1
INTRODUCTION

Mechanics of elastic structures includes linear and nonlinear modeling, statics, dynamics, buckling, postbuckling, flutter, stability, and analyses of stresses, strains, and failure. This book is primarily concerned with the nonlinear modeling and dynamics of elastic structures.

In the literature, there are many different theories of beams, plates, and shells, and the number of structural theories is increasing because of the increase in the use of computers and new structural materials and applications. Moreover, the demand of materials by today's technologies has become so diverse that it often cannot be met by single-component materials. Also, structural engineers are confronted by the challenge of strict requirements of high vehicle performance, less materials, less weight, high safety factors, etc. These requirements cannot be met, in general, except by the use of composites. Although composite materials have characteristics that are better than those of isotropic materials and hence can meet complex design requirements, some nonclassical structural effects, such as transverse shear strains, peeling stresses, free-edge effects, and warping restraint effects, which are usually neglected in isotropic materials, are significant in these materials. Hence, modeling of modern composite structures has become a complex but important step in design, and refined structural theories are needed for the analysis of composite structures.

1.1 STRUCTURAL ELEMENTS

In terms of geometries and loading conditions, structures can be divided into six groups: cables, bars, beams, membranes, plates, and shells. Cables are one-dimensional structures, which can only sustain extensional loads. The buckling loads of cables are zero, and hence they cannot sustain compression loads. Strings are pre-tensioned and initially straight cables.

Bars are one-dimensional structures which can sustain extensional, compressional, and torsional loads. If a bar is only subjected to longitudinal tensile loads, it is usually called a rod. If a bar is only subjected to longitudinal compressive loads, it is usually called a column. Rods and columns are two-force members, and trusses consist of bars. Beams are structures having one dimension much larger than the other two and primarily subjected to lateral loads, resulting in bending of their reference axes. A general beam should be able to sustain extension, compression, bending, transverse shear (flexure), and twisting loads. In other words, cables, strings, bars, rods, and columns are special cases of a general beam, and arches are initially curved beams.

Plates are initially flat structures having two dimensions much larger than the third and can sustain extension, compression, inplane shear, bending, twisting, and transverse shear loads. A membrane is a two-dimensional structure, which can only sustain extensional and inplane shear loads. The membrane stresses of a plate can contribute significantly to its strength. Shells are initially curved structures having two dimensions much larger than the third and can sustain extension, compression, inplane shear, bending, twisting, and transverse shear loads. Shells are the most general engineering structures; they include plates and membranes as special cases. Because the initial curvatures of a shell offer some geometric stiffnesses, the strength of a shell depends on its geometry as well as material.

The complexity of a dynamical system depends on whether the relation between the input and the output is linear or nonlinear, the number of independent and dependent variables used in the system modeling, and the number of parameters. To describe the motion of a discrete dynamical system consisting of N isolated particles, one needs $3N$ dependent variables $u_i(t)$, $v_i(t)$, and $w_i(t)$, $i = 1, 2, \ldots, N$, where u_i, v_i, and w_i are the displacement components of the ith particle along three perpendicular directions and t denotes time. Such a system is called a $3N$-degree-of-freedom system and a $6N$-dimensional system because $3N$ dependent variables are used and second-order time derivatives of all dependent variables are involved.

Any continuous dynamical system can be described by an infinite number of mathematical particles having a volume $dx\,dy\,dz$ if a Cartesian coordinate system xyz is used, and the displacement components of the ith particle are $u(x_i, y_i, z_i, t)$, $v(x_i, y_i, z_i, t)$, and $w(x_i, y_i, z_i, t)$, where (x_i, y_i, z_i) is the location of the ith particle at $t = 0$. For a continuous system without fracture, the distance between two adjacent particles is infinitesimal and particle displacements can only vary continuously from particle to particle. Hence, only three dependent variables u, v, and w are needed but they are continuous functions of three independent spatial variables x, y, and z and time t; that is,

$$u = u(x, y, z, t),\ v = v(x, y, z, t),\ w = w(x, y, z, t) \qquad (1.1.1)$$

If the dependence of u, v, and w on x, y, z, and t can be separated as

$$u = U(x, y, z)q(t),\ v = V(x, y, z)q(t),\ w = W(x, y, z)q(t) \qquad (1.1.2)$$

and if the spatial functions $U, V,$ and W are known, then there is only one unknown dependent variable $q(t)$, which is called a generalized coordinate in structural engineering, and the system has a single degree of freedom; $q(t)$ is governed by a second-order differential equation. However, the spatial functions $U, V,$ and W are load-dependent or unknown or even do not exist (in other words, they are time-dependent functions). Hence, to find the solution of a continuous medium subjected to a general load, engineers usually use an infinite number of assumed, known spatial functions to approximate the solution as

$$u = \sum_{i=1}^{\infty} U_i(x,y,z) q_i(t), \ v = \sum_{i=1}^{\infty} V_i(x,y,z) q_i(t), \ w = \sum_{i=1}^{\infty} W_i(x,y,z) q_i(t)$$
(1.1.3)

In order to have a convergent solution for a general loading condition, one needs to choose the spatial functions from a set of complete functions.

Structures are three-dimensional continuous systems. Structural engineers usually use linear eigenfunctions, which are obtained from the unforced undamped governing equations and are called mode shapes, as the spatial functions. Because an infinite number of dependent variables $q_i(t)$ is involved, any continuous system has infinite degrees of freedom.

Although solving a three-dimensional structural problem may not be impossible, it may require an insurmountable amount of work. Fortunately, most structural elements have one or two of their geometric dimensions much smaller than the others, and hence their motions can be described by the particles on a reference line in the case of cables, bars, and beams or a reference plane in the case of membranes, plates, and shells. Their spatial functions depend on one or two independent variables. However, in addition to the displacements $u, v,$ and w of a general point on the reference line or plane, extra displacement variables are needed in order to describe the motion of a general point that is not on the reference line or plane, which results in different structural theories. Moreover, depending on the type of loading and/or the geometric constraints on the structure, one can adopt some more assumptions about the stress distribution and/or the displacement distribution to simplify the model.

Mathematical modeling is a form of an approximation theory; different models are the result of adopting different assumptions. Simple or rough models are easy to solve, but their accuracy in predicting the system behavior may be poor. To improve the accuracy of structural models, researchers developed refined structural theories by relaxing some of the constraints on the displacement and/or stress field representations. However, relaxing constraints results in an increase in the number and order of the governing equations. In other words, to solve problems modeled by refined theories requires more effort. Consequently, choosing an appropriate model for a specific problem is critical in the analysis.

There are no rules about how to choose a right model, and hence one can only make a decision based on experience, accuracy requirement, and the goal of the analysis. For example, in the dynamic analysis of a pendulum, it can be treated as a rigid body if the forcing frequency is far below its first natural bending frequency. When the forcing frequency is close to or above its first natural bending frequency,

4 INTRODUCTION

the pendulum needs to be treated as an Euler-Bernoulli beam, where transverse shear strains are neglected. When the forcing frequency is higher than the first bending frequency, the shear effect may become important and Timoshenko's beam theory or a higher-order shear-deformation theory is required in order to have accurate results. Moreover, to study the initiation of delamination of laminated composite plates, one needs to choose a plate model that is able to accurately predict the interlaminar shear and peeling stresses because composite laminates are weak in shear.

1.2 NONLINEARITIES

Nonlinear systems are those for which the principle of superposition does not hold. Nature abounds with nonlinear systems; in fact they are the rule rather than the exception. The sources of nonlinearities can be material or constitutive, geometric, inertia, body forces, or friction. The constitutive nonlinearity occurs when the stresses are nonlinear functions of the strains. The geometric nonlinearity is associated with large deformations in solids, such as beams, plates, frames, and shells, resulting in nonlinear strain-displacement relations (e.g., mid-plane stretching, large curvatures of structural elements, large strains, and large rotations of elements). The inertia nonlinearity may be caused by the presence of concentrated or distributed masses; in a Lagrangian formulation, the kinetic energy is a function of the generalized coordinates as well as their rates and, in fluid flow, the acceleration includes a nonlinear convective term. Other examples include Coriolis and centripetal accelerations. The nonlinear body forces are essentially magnetic and electric forces. The friction nonlinearity occurs because the friction force is a nonlinear function of the displacement and velocity, such as dry friction and backlash.

The nonlinearities may appear in the governing partial-differential equations, or the boundary conditions, or both. To some extent, the form of the nonlinearity appearing in the equations and boundary conditions depends on the coordinate system used and the orientation of the body forces, such as gravity. Examples of nonlinear boundary conditions include free surfaces in fluids and deformation-dependent constraints.

The nonlinearities considered in this book are primarily geometric arising from large rotations and displacements and electric arising from the proximity of two plate electrodes to each other. To include the effects of material nonlinearities, one only needs to replace the constant stiffnesses with displacement-dependent ones; they are usually obtained from experiments.

Suppose that a metallic panel is subjected to a longitudinal compressive load. As long as the panel remains flat, it is in equilibrium, and it can only fail by crushing; that is, the compressive stress exceeds the stress that the material can withstand. However, it is well known that panels, whose length is much larger than its thickness, may bend before it fails by crushing. This phenomenon is called buckling or elastic instability, which can occur in bars, beams, plates, and shells. It results in an unproportional increase in the displacement resulting from a small increase in the load. In spite of this, the postbuckling strength of thin-walled structures plays an important role in the design of aircraft structures because conventional aircraft structural elements are often designed to operate in the postbuckling range. To determine the postbuckling behavior, one needs to develop a more inclusive, geometrically nonlinear theory. Hence, nonlinear problems considered in the theory of elastic structures are mostly those of postbuckling analysis, prediction of stability, and nonlinear panel flutter analysis.

In recent years, the rapid developments in aerospace exploration have stimulated extensive research into the dynamics and control of large flexible space structures, such as solar collectors, dish antennas, radar arrays, long truss structures, space telescopes, and space stations. Because these structures are characterized by low flexural rigidities, weak material dampings, and interconnections of rigid and flexible parts, and because there is no air damping in space, maneuvers often lead to destructive large-amplitude vibrations, which introduce excessive material fatigue and affect the operational accuracy of such structures. The tasks of controlling the rotation and high-pointing accuracy and eliminating the structural vibrations in a finite period of time pose difficult control problems, which require theoretical and computational advances. From the dynamic point of view, a great disadvantage of such flexible structures is that their natural frequencies are clustered in very narrow bands, making them more prone to becoming involved in resonant vibrations that cannot be easily controlled. Moreover, flexible structures can undergo large displacements without exceeding the elastic limit. Consequently, the responses of flexible structures exhibit many complicated vibration phenomena, such as multiple solutions, jumps, hysteresis, modal interactions, flutter, chaos, and transfer of energy from high-frequency to low-frequency modes.

To design strategies for the control of large-amplitude structural vibrations, one needs to understand their nonlinear static and dynamic behavior, including modal couplings (transfer of energy among the structure modes) and static and dynamic instabilities. These require accurate nonlinear structural models.

The modeling of structural systems can be divided into three groups: (1) linear modeling, (2) pseudo nonlinear modeling, and (3) nonlinear modeling. In linear modeling, both static and dynamic behaviors of a structure are described by linear models whose static and dynamic solutions are unique. A linear static model can predict the onset of static (or geometric) bifurcation (e.g., buckling), but cannot give the magnitude of buckled displacements. In pseudo nonlinear modeling, the static behavior is described by a nonlinear model, but the dynamic behavior is described by a linear model. A nonlinear static model can predict the magnitudes of buckled displacements of a structure. Then, a linear dynamic model around the static equilibrium position is used to perform dynamic stability analyses and predict the onset of dynamic bifurcations. However, such linear models cannot predict the amplitudes of limit cycles or the presence and character of chaotic attractors, which usually occur after dynamic bifurcations. We note that the parameters of the linear dynamic model will generally depend upon the static (equilibrium) model, and there may be several static equilibria. In nonlinear modeling, both the statics and dynamics are described by nonlinear models. Several distinct possible dynamic equilibria may coexist, and the one observed depends on the static equilibria, the system parameters, and initial conditions.

1.3 COMPOSITE MATERIALS

Because of their high strength-to-weight ratio, long fatigue life, resistance to corrosion, high damping, structural simplicity, and possible use for aeroelastic tailoring, advanced laminated structures made of fiber-reinforced composite materials, such as boron-epoxy, graphite-epoxy, and boron-aluminum, have emerged as primary materials for advanced aerospace vehicle structures, marine structures, large space structures, automotive parts, helicopter rotor blades, turbine blades, and robot manipulators. They show great promise for improved performance. Moreover, the inherent anisotropy is an important property of composite materials and one of the basic reasons for their success because it offers linear, elastic couplings among bending, extension, torsion, and shearing motions, thereby making it possible to satisfy sophisticated design criteria, such as aeroelastic tailoring. For example, the extension-twisting coupling produces different twist distributions along the rotors of a two-speed helicopter when the system rotates at different speeds. Moreover, the bending-twisting coupling produces a pitch-flap stability of helicopter rotor blades.

Flutter of aircraft wings occurs because the speed of flow affects the amplitude ratios and phase shifts between bending and torsional motions of the wing in such a way that energy can be absorbed by the wings from the airstream passing by, resulting in self-excited or self-sustained oscillations. Moreover, experiments (Fung, 1969) on cantilever wings show that the flexural movements at all points across the span are approximately in phase with one another, and likewise the torsional movements are all approximately in phase, but the flexure is considerably out of phase of the torsional movement. This phase difference is apparently the main factor that is responsible for the occurrence of flutter. Hence, the bending-torsion coupling characteristics of composite beams can be used to suppress flutter because bending and torsional vibrations of a composite beam with bending-twisting coupling are forced to be in phase by the fiber-matrix mechanism. One well known example of using the bending-twisting coupling effect is the X-29 demonstrator aircraft; the composite skin of its forward-swept wing has a built-in structural and aerodynamic stability.

However, nonclassical structural effects, such as shear deformations, transverse normal stresses (peeling stresses), warping restraint effects, and boundary-layer effects, can be very significant in composite materials although they are usually negligible in isotropic materials. Because composite structures exhibit relatively weak rigidity in the transverse shear, shear deformations are significant in such materials and need to be included in the study of free vibrations of moderately thick plates, forced vibration amplitudes and stress distributions, high-frequency responses, short-wavelength waves, and localized impacts. Also, peeling stresses can be significant because of non-uniform distributions of Poisson's ratios. Moreover, St. Venant's principle is usually assumed to be valid in the analysis of isotropic structures. This principle states that stresses at a point that is at a sufficient distance from the loading end depend only on the magnitude of the applied load and are practically independent of the manner in which the load is distributed over the end. It also implies that a system of loads having zero resultant forces and moments (i.e., a self-equilibrating stress system) produces a strain field that is negligible at a point that is away from the loading end (Iesan, 1987). But, for highly anisotropic and heterogeneous materials, such a self-equilibrating stress system can result in nontrivial strains with long decay lengths, which are the so-called boundary-layer solutions. To study these nonclassical effects, one requires new, refined structural theories. However, to include these nonclassical effects in the modeling of composite structures is not an easy task, especially if geometric nonlinearities are also involved. Moreover, elastic couplings make it difficult or even impossible to obtain exact linear solutions for some simple structural problems.

In the analysis of composite structures, a macromechanics approach is conventionally used. In this approach perfect bonding between the fibers and matrices is assumed, the material is assumed to be uniform, and the mechanical properties are obtained by taking the average of the properties of the constituent fibers and matrices. However, for real composite structures, there are many problems, which include imperfect bonding, nonuniform distributions of fibers, initially crooked fibers that make the structure behave like a hardening-type material, the existence of gas bubbles at the interfaces of fibers and matrices, local stress concentrations, local elastic-plastic behaviors, cracks, delamination, etc. Hence, one needs to use a micromechanics analysis to obtain valid material and structural properties, or even a statistical approach to account for variations in the many unknown factors and manufacturing processes.

The fundamental mechanics of composite materials can be found in the books by Jones (1975), Christensen (1979), Tsai and Hahn (1980), Whitney (1987), Vinson and Sierakowski (1986), and Reddy (1997, 2003).

1.4 DAMPING

Damping arises from the removal of energy by dissipation or radiation. Dissipative forces in structures can be the result of either internal or external damping. External damping includes aerodynamic and hydrodynamic drag and dissipation in the supports of structures. The drag may be linear or nonlinear. Aerodynamic damping was found to be significant for high-amplitude vibrations of beams of low damping and high modulus of elasticity. Anderson, Nayfeh, and Balachandran (1996) experimentally found the nonlinear aerodynamic damping to be significant for large-amplitude first-mode vibrations of slender parametrically excited beams. Internal damping is usually studied by modeling the mechanisms of energy dissipation in materials. Internal damping mechanisms include thermoelastic, hysteretic, Coulombic, magnetoelastic, and dislocation unpinning and grain boundary relaxation of metals and alloys. For most structural metals, such as steel and aluminum, the energy dissipated per cycle is independent of the frequency over a wide frequency range and is proportional to the square of the amplitude of vibration, and the shape of the hysteretic curve remains unchanged with amplitude and is independent of the strain rate. Internal damping, which fits this classification, is called solid or structural damping and its equivalent linear viscous damping is proportional to the inverse of the frequency of vibration (Thomson, 1981). However, the damping ratios of some structures may exhibit both frequency and amplitude dependence. Moreover, the aerodynamic and hydrodynamic drag is not easily modeled because of structure-fluid interactions.

Although damping forces are small in comparison with the elastic and inertia forces in many applications of structural vibration and wave theory and the influence of damping on structural mode shapes is usually small, damping can be important in controlling the amplitudes of vibration under conditions of steady-state resonance and stationary random excitations. Damping has significant influence on the response amplitudes and phases near resonance and plays a crucial role in fixing the borderline between stability and instability in many dynamical systems. Moreover, damping can significantly affect structural nonlinear responses.

The damping ratios of composite materials, especially nonmetal composites, are much higher than those of structural steels. All damping ratios in the experimental results of Schultz and Tsai (1968) and Ray and Bert (1969) are in the range 0.02% to 2.8%. Using the modified Kennedy-Pancu method (Pendered and Bishop, 1963), Siu and Bert (1974) obtained analytically the damping ratios of laminated boron-epoxy plates with free edges and various orientation angles. The obtained damping ratios are in the range 0.09% to 3.31%. For isotropic materials, the experimental results (e.g., Schultz and Tsai, 1968; Baz and Poh, 1989) usually show that modal dampings decrease with mode number. But for composite beams, the damping ratios may increase with mode number (Schultz and Tsai, 1968). Adams and Bacon (1973) found that the damping ratio of composite materials can be as high as 5.5%. The highest experimentally determined damping ratio of boron-epoxy plates is 5.3% (Clary, 1972). Adams et al. (1969) indicated that the damping capacity of composites under torsion is higher than that under flexure. A comprehensive mathematical technique was developed by Ni and Adams (1984) for predicting the damping of laminated composite beams. They showed that the torsional motion induced by bending-twisting coupling may result in high modal damping ratios for flexural vibrations. Experimental results obtained by Adams and Bacon (1973) show that the damping ratios of composite materials increase with temperature.

Saravanos and Chamis (1990a,b, 1991) showed that damping of composites depends on an array of micromechanics and laminate parameters, including constituent material properties, fiber volume ratios, ply angles, ply thicknesses, ply stacking sequences, temperature, moisture, and existing damage. Damping in composites is also anisotropic, but it exhibits an anisotropy trend that is opposite to that of the stiffness and strength, being minimum in the direction of the fibers and maximum in the transverse direction and in shear. Moreover, metal matrix composites can also undergo energy dissipation at the fiber/matrix interface due to interfacial slip, microplasticity of the matrix, dislocation breakaway, and microcracking at or near the fiber/matrix interface.

The damping mechanisms of most structures are unknown because the sources of energy loss are too complicated. Practicing structural engineers usually use the concepts of modal damping and proportional damping obtained experimentally from modal testing (Ewins, 1984).

1.5 DYNAMIC CHARACTERISTICS OF LINEAR DISCRETE SYSTEMS

Linear systems are those for which the principle of superposition holds. In linear equations of motion, there are no terms containing products of different dependent variables or powers of any dependent variable. Discrete systems are governed by ordinary-differential equations.

A linear single-degree-of-freedom system is characterized by its natural frequency and damping ratio. A linear multi-degree-of-freedom system is characterized by its natural frequencies, modal damping ratios, and mode shapes. Moreover, the response frequency under a single harmonic excitation is the same as the excitation frequency, and the response amplitude is unique and independent of the initial conditions.

1.5.1 One-Degree-of-Freedom Systems

In Figure 1.5.1, we show a typical single-degree-of-freedom spring-mass-damper system, where $x(t)$ denotes the displacement of the mass m from its static equilibrium position, k denotes the spring constant, c denotes the damping coefficient, and F and Ω denote the forcing amplitude and frequency, respectively. Using Newton's second law, we obtain the governing equation as

$$m\ddot{x} + c\dot{x} + kx = F \sin \Omega t \tag{1.5.1}$$

We take the initial conditions in the form

$$x(0) = x_0, \quad \dot{x}(0) = \dot{x}_0 \tag{1.5.2}$$

When m, c, and k are constants, the system is referred to as time-invariant.

The solution of (1.5.1) consists of a particular solution x_p (i.e., the steady-state solution) and a complementary function x_c (i.e., the transient solution), which is the solution of the homogeneous part of (1.5.1). To determine the complementary function $x_c(t)$, we substitute

$$x_c = e^{st} \tag{1.5.3}$$

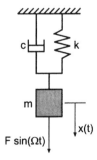

Fig. 1.5.1 A one-degree-of-freedom spring-mass-damper system.

into (1.5.1) with $F = 0$ and obtain

$$(ms^2 + cs + k)e^{st} = 0 \tag{1.5.4}$$

Since e^{st} is time varying, we have

$$ms^2 + cs + k = 0 \tag{1.5.5}$$

which is called the characteristic equation. Hence,

$$s_{1,2} = \frac{-c \pm \sqrt{c^2 - 4mk}}{2m} \tag{1.5.6}$$

We consider the case of positive damping; that is, $c > 0$. There are two types of solutions depending on the sign of the discriminant $D \equiv c^2 - 4mk$. When D is positive, s_1 and s_2 are negative real numbers because m and k are positive, and hence x_c is an exponentially decaying function. In this case, one speaks of an overdamped system. When D is negative, s_1 and s_2 are complex conjugate with negative real part. Hence, x_c oscillates while it decays exponentially. When $D = 0$, $s_1 = s_2 = -c/2m$, and hence x_c decays exponentially, and one speaks of a critically damped system. The value c_c of c that renders $D = 0$ is referred to as the critical damping coefficient. It is given by

$$c_c = 2\sqrt{mk} \tag{1.5.7}$$

Because structural materials usually have small dampings, we only consider the case $D = c^2 - 4mk < 0$. In this case, one speaks of an underdamped system.

Next, we define two linear free-oscillation frequencies: undamped and damped natural frequencies. When $c = 0$, (1.5.6) reduces to

$$s_{1,2} = \pm i\sqrt{\frac{k}{m}}$$

where $i \equiv \sqrt{-1}$. Hence, x_c is a harmonically oscillatory function with the frequency $\sqrt{k/m}$, which is called the undamped natural frequency ω_n, where

$$\omega_n \equiv \sqrt{\frac{k}{m}} \tag{1.5.8a}$$

When $c \neq 0$, we rewrite (1.5.6) as

$$s_{1,2} = -\frac{c}{2m} \pm i\sqrt{\frac{k}{m} - \frac{c^2}{4m^2}} \tag{1.5.8b}$$

The absolute value of the imaginary part of s_1 and s_2 is usually referred to as the damped natural frequency ω_d; that is,

$$\omega_d = \sqrt{\frac{k}{m} - \frac{c^2}{4m^2}} \tag{1.5.8c}$$

Using the damping ratio (damping factor)

$$\xi \equiv \frac{c}{c_c} = \frac{c}{2\sqrt{mk}} \tag{1.5.8d}$$

and the definition of w_n, we rewrite the damped natural frequency as

$$w_d \equiv \sqrt{1-\xi^2}\,w_n \tag{1.5.8e}$$

and the real part of s_1 and s_2 as

$$\frac{c}{2m} = \xi w_n \tag{1.5.8f}$$

Hence,

$$s_{1,2} = -\xi w_n \pm i w_d \tag{1.5.9}$$

Hence, the complementary function x_c is given by

$$x_c = b_1 e^{s_1 t} + b_2 e^{s_2 t} = a e^{-\xi w_n t} \sin(w_d t - \beta) \tag{1.5.10}$$

where b_1 and b_2 are constants and a and β are the free-oscillation amplitude and phase, which are determined by the two initial conditions.

Next, we use the method of undetermined coefficients and seek a particular solution of (1.5.1) in the form

$$x_p = X \sin(\Omega t - \phi) \tag{1.5.11}$$

where X is the amplitude of the steady-state solution and ϕ is the phase lag of the particular solution with respect to the forcing function. Substituting (1.5.11) into (1.5.1), equating each of the coefficients of $\sin \Omega t$ and $\cos \Omega t$ on both sides, and using (1.5.8), we obtain

$$X = \frac{F}{\sqrt{(k-m\Omega^2)^2 + c^2\Omega^2}} = \frac{w_n^2 F}{k\sqrt{(w_n^2 - \Omega^2)^2 + 4\xi^2\Omega^2 w_n^2}} \tag{1.5.12a}$$

$$\phi = \tan^{-1}\left(\frac{c\Omega}{k-m\Omega^2}\right) = \tan^{-1}\frac{2\xi w_n \Omega}{w_n^2 - \Omega^2} \tag{1.5.12b}$$

Because (1.5.1) is linear, the principle of superposition holds and hence its complete solution can be expressed as

$$x(t) = x_p + x_c = X \sin(\Omega t - \phi) + a e^{-\xi w_n t} \sin(w_d t - \beta) \tag{1.5.13}$$

Substituting (1.5.13) into (1.5.2), we obtain

$$a = \sqrt{C_1^2 + C_2^2}, \quad \beta = \tan^{-1}\left(\frac{C_1}{C_2}\right) \tag{1.5.14a}$$

where

$$C_1 \equiv -x_0 - X \sin \phi$$

$$C_2 \equiv \frac{1}{w_d}(\dot{x}_0 + \xi w_n x_0 + \xi w_n X \sin \phi - X\Omega \cos \phi) \tag{1.5.14b}$$

It follows from (1.5.13) that the total response of the system is the sum of two terms: the first term is called the forced component and the second term is called the free-oscillation component.

We note that when $\xi > 0$ the free-oscillation component in (1.5.13) decays with time, and hence the long-time system response is given by the forced component only; that is,

$$x(t) = X \sin(\Omega t - \phi) \qquad (1.5.15a)$$

This solution is usually called the steady-state response. It is periodic, independent of the initial conditions, and unique. The total solution (1.5.13) is usually referred to as the transient response.

When $\xi = 0$ (i.e., undamped system), there are two cases: $\Omega = \omega_n$ and $\Omega \neq \omega_n$. In the latter case, (1.5.13) reduces to

$$x(t) = \frac{F}{m(\omega_n^2 - \Omega^2)} \sin \Omega t + a \sin(\omega_n t - \beta) \qquad (1.5.15b)$$

Clearly, the free-oscillation component does not decay with time. The response consists of two basic frequencies Ω, the excitation frequency, and ω_n, the undamped natural frequency. When Ω/ω_n is a rational number, the response is periodic having the period T, which is the least common multiple of $2\pi/\Omega$ and $2\pi/\omega_n$. When Ω/ω_n is an irrational number, the response is aperiodic and it is usually referred to as two-period quasiperiodic. When $\xi = 0$ and $\Omega = \omega_n$, the solution of (1.5.1) is modified to

$$x(t) = -\frac{F}{2\sqrt{km}} t \cos \omega_n t + a \sin(\omega_n t - \beta) \qquad (1.5.15c)$$

which grows algebraically with time. In this case, one speaks of a primary resonance excitation. As will be discussed later, for primary resonance, Ω need not be exactly equal to ω_n but approximately equal to ω_n.

It is clear from (1.5.12) and (1.5.15a) that the steady-state forced amplitude X and phase ϕ of damped systems are functions of the excitation frequency Ω. To determine the maximum value of the forced amplitude, we differentiate (1.5.12a) with respect to Ω, set the result equal to zero, and obtain

$$1 - \frac{\Omega^2}{\omega_n^2} - 2\xi^2 = 0$$

Hence, the maximum forced amplitude occurs when

$$\Omega = \sqrt{1 - 2\xi^2}\,\omega_n \qquad (1.5.16a)$$

Substituting (1.5.16a) into (1.5.12) yields the maximum forced amplitude

$$X_{max} = \frac{F}{2k\xi\sqrt{1-\xi^2}} \qquad (1.5.16b)$$

and corresponding phase angle

$$\phi = \tan^{-1}\left(\frac{\sqrt{1-2\xi^2}}{\xi}\right) \qquad (1.5.16c)$$

As $\xi \to 0$, it follows from (1.5.16b) and (1.5.16c) that

$$X_{max} \to \infty \text{ and } \phi \to \frac{1}{2}\pi$$

To determine the natural frequency of a system experimentally, one performs what is called a sine-dwell sweep. In such experiments, one excites the system by a harmonic force, gradually sweeps the frequency, locates the maximum steady-state amplitude, and identifies the corresponding frequency as the natural frequency. It is clear from (1.5.16a) that this frequency, which is usually called the resonant frequency, is close to ω_n if ξ is small; that is, if the system is lightly damped.

The frequency-response function can be obtained from the transfer function, which is defined to be the ratio of the Laplace transform $\mathcal{L}[x(t)]$ of the response function $x(t)$ to the Laplace transform $\mathcal{L}[f(t)]$ of the driving function $f(t)$, under the assumption that all initial conditions are zero. Hence, the transfer function of (1.5.1) is

$$H(s) \equiv \frac{\mathcal{L}(x(t))}{\mathcal{L}(f(t))} = \frac{1}{m(s^2 + 2\xi\omega_n s + \omega_n^2)} \qquad (1.5.17)$$

where $f(t) = F\sin\Omega t$. The frequency-response function $H(\Omega)$ is obtained by replacing s with $i\Omega$ in $H(s)$; that is,

$$H(\Omega) \equiv \frac{1}{m(\omega_n^2 - \Omega^2 + 2i\xi\omega_n\Omega)} = \frac{1}{m}\frac{\omega_n^2 - \Omega^2 - 2i\xi\omega_n\Omega}{(\omega_n^2 - \Omega^2)^2 + 4\xi^2\Omega^2\omega_n^2} \qquad (1.5.18)$$

It can be expressed in the polar form as

$$H(\Omega) = |H(\Omega)| e^{-i\phi} \qquad (1.5.19)$$

where it follows from (1.5.12a) that

$$|H(\Omega)| = \frac{\omega_n^2}{k\sqrt{(\omega_n^2 - \Omega^2)^2 + 4\xi^2\Omega^2\omega_n^2}} = \frac{X}{F} \qquad (1.5.20a)$$

$$\phi = \tan^{-1}\left(\frac{2\xi\omega_n\Omega}{\omega_n^2 - \Omega^2}\right) \qquad (1.5.20b)$$

In Figures 1.5.2a and b, we show the magnitude and phase of the frequency-response function. We note that, for lightly damped systems, the response amplitude at $\Omega \simeq \omega_n$ is very large but the peak is always on the left side of ω_n. In fact, the peak occurs at $\Omega = \sqrt{1 - 2\xi^2}\omega_n$ according to (1.5.16a). Also, $x(t)$ and $f(t) = F\cos\Omega t$ are in phase (i.e., $\phi = 0°$) if $\Omega \ll \omega_n$, and out of phase (i.e., $\phi = 180°$) if $\Omega \gg \omega_n$. For lightly damped systems, the phase angle changes by almost 180° as the excitation frequency is swept across the resonance frequency. Experimentalists use this fact in determining the resonance frequency.

14 INTRODUCTION

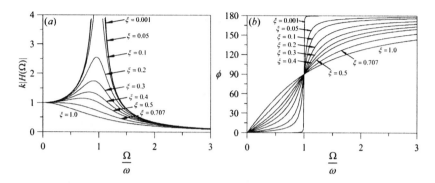

Fig. 1.5.2 The frequency-response function of a linear one-degree-of-freedom system: (a) magnitude and (b) phase angle.

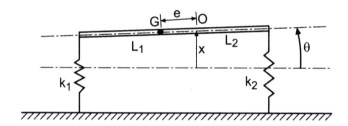

Fig. 1.5.3 A two-degree-of-freedom system.

1.5.2 Multi-Degree-of-Freedom Systems

To show the dynamic characteristics of multiple-degree-of-freedom systems, we consider the two-degree-of-freedom system shown in Figure 1.5.3. Using Newton's second law, we obtain the following equations of motion:

$$m(\ddot{x} - e\ddot{\theta}) = -k_1(x - L_1\theta) - k_2(x + L_2\theta) - c_{11}\dot{x} - c_{12}\dot{\theta} + f_1(t) \quad (1.5.21\text{a})$$

$$J_G\ddot{\theta} - m(\ddot{x} - e\ddot{\theta})e = k_1(x - L_1\theta)L_1 - k_2(x + L_2\theta)L_2 - c_{21}\dot{x} - c_{22}\dot{\theta} + f_2(t) \quad (1.5.21\text{b})$$

where $f_1(t)$ denotes the external force applied at Point O, $f_2(t)$ denotes the external moment acting on the beam, J_G is the rotary inertia with respect to the mass centroid G, and the c_{ij} are damping coefficients. Putting (1.5.21) in matrix form, we obtain

$$\begin{bmatrix} m & -me \\ -me & J \end{bmatrix} \begin{Bmatrix} \ddot{x} \\ \ddot{\theta} \end{Bmatrix} + \begin{bmatrix} c_{11} & c_{12} \\ c_{21} & c_{22} \end{bmatrix} \begin{Bmatrix} \dot{x} \\ \dot{\theta} \end{Bmatrix}$$
$$+ \begin{bmatrix} k_1 + k_2 & k_2 L_2 - k_1 L_1 \\ k_2 L_2 - k_1 L_1 & k_1 L_1^2 + k_2 L_2^2 \end{bmatrix} \begin{Bmatrix} x \\ \theta \end{Bmatrix} = \begin{Bmatrix} f_1 \\ f_2 \end{Bmatrix} \quad (1.5.22)$$

where $J = J_G + me^2$ is the rotary inertia with respect to the reference point O. The matrices

$$\begin{bmatrix} m & -me \\ -me & J \end{bmatrix}, \quad \begin{bmatrix} c_{11} & c_{12} \\ c_{21} & c_{22} \end{bmatrix}, \quad \text{and} \quad \begin{bmatrix} k_1 + k_2 & k_2 L_2 - k_1 L_1 \\ k_2 L_2 - k_1 L_1 & k_1 L_1^2 + k_2 L_2^2 \end{bmatrix}$$

are called the mass, damping, and stiffness matrices, respectively. Clearly, the mass and stiffness matrices are symmetric.

We note that the two degrees of freedom are dynamically coupled by the term $-me$ and are statically coupled by the term $k_2 L_2 - k_1 L_1$. The inertia terms are decoupled if $e = 0$, which can be achieved by choosing the mass centroid to be the reference point. However, to decouple the elastic terms, one needs to choose $k_1/k_2 = L_2/L_1$. In general, one may not be able to simultaneously decouple the inertia and elastic terms by choosing the reference point. However, the system eigenvectors can be used to simultaneously decouple these terms as described next.

We consider the following n-degree-of-freedom system:

$$[M]\{\ddot{x}\} + [C]\{\dot{x}\} + [K]\{x\} = \{f\} \tag{1.5.23}$$

where $\{x\} = \{x_1, x_2, \ldots, x_n\}^T$ and the x_j are physical coordinates, such as θ and x in (1.5.22). Here, we assume that $[M], [C]$, and $[K]$ are real symmetric matrices and that $[M]$ is positive definite. To determine the undamped natural frequencies (i.e., eigenvalues) and eigenvectors of (1.5.23), we let

$$\{x\} = \{X\}e^{i\omega t} \tag{1.5.24}$$

in (1.5.23), assign $[C] = [0]$ and $\{f\} = \{0\}$, and obtain

$$\left[[K] - \omega^2[M]\right] \{X\} = \{0\} \tag{1.5.25}$$

which is a generalized eigenvalue problem.

Because $[M]$ is a positive definite symmetric matrix, it can always be decomposed into

$$[M] = [m]^T[m] \tag{1.5.26}$$

where $[m]$ is a real nonsingular matrix. Substituting (1.5.26) into (1.5.25) and pre-multiplying the result with $[m]^{-T}$, we obtain the following eigenvalue problem in standard form:

$$[A]\{Y\} = \omega^2\{Y\} \tag{1.5.27}$$

where

$$\{Y\} \equiv [m]\{X\}, \quad [A] \equiv [m]^{-T}[K][m]^{-1} \tag{1.5.28}$$

We note that $[A]$ is a real symmetric matrix because $[K]$ is symmetric.

Solving (1.5.25) or (1.5.27) and (1.5.28), we obtain the following eigenvalues and eigenvectors:

$$\omega = \omega_j, \quad \{X\} = \{X\}_j, \quad j = 1, 2, \ldots, n \tag{1.5.29}$$

Next, we show that the ω_j are real and that the eigenvectors are orthogonal with respect to the matrices $[M]$ and $[K]$; that is,

$$\{X\}_m^T[M]\{X\}_n = 0, \text{ and } \{X\}_m^T[K]\{X\}_n = 0, \ m \neq n \qquad (1.5.30)$$

Because $[A]$ is a real matrix, if ω and $\{Y\}$ are solutions of (1.5.27), then their complex conjugates $\bar{\omega}$ and $\{\bar{Y}\}$ are also solutions of (1.5.27); that is,

$$[A]\{\bar{Y}\} = \bar{\omega}^2\{\bar{Y}\} \qquad (1.5.31a)$$

Multiplying (1.5.27) from the left with $\{\bar{Y}\}^T$ yields

$$\{\bar{Y}\}^T[A]\{Y\} = \omega^2\{\bar{Y}\}^T\{Y\} \qquad (1.5.31b)$$

and multiplying (1.5.31a) from the left with $\{Y\}^T$ yields

$$\{Y\}^T[A]\{\bar{Y}\} = \bar{\omega}^2\{Y\}^T\{\bar{Y}\} \qquad (1.5.31c)$$

Taking the transpose of (1.5.31c) and noting that $[A]$ is symmetric, we have

$$\{\bar{Y}\}^T[A]\{Y\} = \bar{\omega}^2\{\bar{Y}\}^T\{Y\} \qquad (1.5.31d)$$

Subtracting (1.5.31d) from (1.5.31b) yields

$$(\omega^2 - \bar{\omega}^2) \parallel Y \parallel = 0 \qquad (1.5.32)$$

where $\parallel Y \parallel \equiv \{\bar{Y}\}^T\{Y\}$ is the norm of $\{Y\}$. For a nontrivial eigenvector $\parallel Y \parallel \neq 0$, (1.5.32) is satisfied if and only if $\omega = \bar{\omega}$; that is, the eigenvalues are real. Because ω, $[A]$, and $[m]$ are real, the eigenvectors $\{Y\}$ and $\{X\}$ are real vectors.

To prove the orthogonality (1.5.30) of the eigenvectors, we specialize (1.5.25) for $\omega = \omega_i$ and ω_j and obtain

$$[K]\{X\}_i - \omega_i^2[M]\{X\}_i = 0 \qquad (1.5.33a)$$

$$[K]\{X\}_j - \omega_j^2[M]\{X\}_j = 0 \qquad (1.5.33b)$$

Multiplying (1.5.33a) and (1.5.33b) from the left with $\{X\}_j^T$ and $\{X\}_i^T$, respectively, yields

$$\{X\}_j^T\left[[K]\{X\}_i - \omega_i^2[M]\{X\}_i\right] = 0 \qquad (1.5.34a)$$

$$\{X\}_i^T\left[[K]\{X\}_j - \omega_j^2[M]\{X\}_j\right] = 0 \qquad (1.5.34b)$$

Taking the transpose of (1.5.34b), recalling that $[M]$ and $[K]$ are symmetric, and subtracting the result from (1.5.34a) yields

$$(\omega_j^2 - \omega_i^2)\{X\}_j^T[M]\{X\}_i = 0 \qquad (1.5.35)$$

Because $\omega_i \neq \omega_j$, (1.5.35) is satisfied if and only if

$$\{X\}_j^T[M]\{X\}_i = 0$$

Then, it follows from (1.5.34a) that

$$\{X\}_j^T[K]\{X\}_i = 0$$

Because the eigenvectors are orthogonal with respect to $[M]$ and because $[M]$ is positive definite, we can normalize them so as to satisfy

$$\{X\}_i^T[M]\{X\}_i = 1 \qquad (1.5.36a)$$

Then, multiplying (1.5.33a) from the left with $\{X\}_i^T$, we obtain

$$\{X\}_i^T[K]\{X\}_i - \omega_i^2\{X\}_i^T[M]\{X\}_i = 0$$

which, upon using (1.5.36a), yields

$$\{X\}_i^T[K]\{X\}_i = \omega_i^2 \qquad (1.5.36b)$$

Using the normalized eigenvectors $\{X\}_i$ and their corresponding eigenvalues ω_i, we decouple the system of equations (1.5.23). To this end, we introduce the linear transformation

$$\{x\} = [X]\{q\} \qquad (1.5.37a)$$

where

$$[X] = [\{X\}_1, \{X\}_2, \ldots, \{X\}_n], \quad \{q\} = \{q_1, q_2, \ldots, q_n\}^T \qquad (1.5.37b)$$

Here, the q_j are the modal coordinates and $[X]$ is the so-called modal matrix. Substituting (1.5.37a) into (1.5.23) and premultiplying the result from the left with $[X]^T$ yields

$$[X]^T[M][X]\{\ddot{q}\} + [X]^T[C][X]\{\dot{q}\} + [X]^T[K][X]\{q\} = [X]^T\{f\} \qquad (1.5.38)$$

It follows from (1.5.36) that

$$[X]^T[M][X] = [I], \quad [X]^T[K][X] = \begin{bmatrix} \omega_1^2 & 0 & 0 & \cdots & 0 \\ 0 & \omega_2^2 & 0 & \cdots & 0 \\ 0 & 0 & \omega_3^2 & \cdots & 0 \\ \vdots & \vdots & \vdots & \ddots & \vdots \\ 0 & 0 & 0 & \cdots & \omega_n^2 \end{bmatrix} \qquad (1.5.39)$$

where $[I]$ is the $n \times n$ identity matrix. If we further assume that the damping is proportional, then

$$[C] = \alpha[M] + \beta[K] \qquad (1.5.40)$$

where α and β are real constants. Hence,

$$[X]^T[C][X] = \begin{bmatrix} \alpha + \beta\omega_1^2 & 0 & \cdots & 0 \\ 0 & \alpha + \beta\omega_2^2 & \cdots & 0 \\ \vdots & \vdots & \ddots & \vdots \\ 0 & 0 & \cdots & \alpha + \beta\omega_n^2 \end{bmatrix} \qquad (1.5.41)$$

18   INTRODUCTION

Using (1.5.39)-(1.5.41), we rewrite (1.5.38) in component form as

$$\ddot{q}_j + 2\xi_j \omega_j \dot{q}_j + \omega_j^2 q_j = \tilde{f}_j, \quad j = 1, 2, \ldots, n \qquad (1.5.42a)$$

where

$$\xi_j \equiv \frac{\alpha + \beta \omega_j^2}{2\omega_j}, \quad \tilde{f}_j = \{X\}_j^T \{f\} \qquad (1.5.42b)$$

Here, ξ_j is the modal damping factor of the jth mode. Using (1.5.37a) and (1.5.39), we express the initial conditions for $\{q\}$ in terms of those for $\{x\}$ as

$$\{q(0)\} = [X]^T[M]\{x(0)\}, \quad \{\dot{q}(0)\} = [X]^T[M]\{\dot{x}(0)\} \qquad (1.5.42c)$$

If $\{f\}$ has the form $\{f\} = \{F\} \sin \Omega t$, then

$$\tilde{f}_j = p_j \sin \Omega t, \quad p_j \equiv \{X\}_j^T \{F\} \qquad (1.5.42d)$$

where p_j is the so-called participation factor of the jth mode in the motion.

As in Section 1.5.1, the solution of (1.5.42a,c,d) is

$$q_j = A_j \sin(\Omega t - \phi_j) + B_j e^{-\xi_j \omega_j t} \sin\left(\sqrt{1 - \xi_j^2} \omega_j t - \beta_j\right) \qquad (1.5.43a)$$

where

$$A_j = \frac{\{X\}_j^T \{F\}}{\sqrt{(\omega_j^2 - \Omega^2)^2 + 4\xi_j^2 \Omega^2 \omega_j^2}}, \quad \phi_j = \tan^{-1}\left(\frac{2\xi_j \omega_j \Omega}{\omega_j^2 - \Omega^2}\right) \qquad (1.5.43b)$$

$$B_i = \sqrt{C_{i1}^2 + C_{i2}^2}, \quad \beta_i = \tan^{-1}\left(\frac{C_{i1}}{C_{i2}}\right)$$

$$C_{i2} \equiv \frac{1}{\sqrt{1-\xi_i^2}\omega_i}(\dot{q}_i(0) + \xi_i \omega_i q_i(0) + \xi_i \omega_i A_i \sin \phi_i - A_i \Omega \cos \phi_i)$$

$$C_{i1} \equiv -q_i(0) - A_i \sin \phi_i \qquad (1.5.43c)$$

Therefore, the solution of (1.5.23) is

$$\{x(t)\} = \sum_{i=1}^{n} \{X\}_i A_i \sin(\Omega t - \phi_i) + \sum_{i=1}^{n} \{X\}_i B_i e^{-\xi_i \omega_i t} \sin\left(\sqrt{1-\xi_i^2}\omega_i t - \beta_i\right)$$

$$(1.5.43d)$$

It is clear from (1.5.43b) and (1.5.43c) that all modes with $p_j = \{X\}_j^T \times \{F\} \neq 0$ contribute to the response. As discussed in Section 1.5.1, the maximum contribution of the jth mode occurs when the excitation frequency Ω is equal to its damped natural frequency $\sqrt{1 - 2\xi_j^2}\omega_j$.

In order that only the jth mode of the system is excited, $\{X\}_i^T \{F\}$ must be zero for all $i \neq j$. This is the case if the forcing vector $\{f\}$ is proportional to the jth eigenvector $\{X\}_j$; that is,

$$\{f\} = (a[M] + b[K])\{X\}_j \sin \Omega t \qquad (1.5.44a)$$

where a and b are real constants. Then,

$$p_j = a + b\omega_j^2 \quad \text{and} \quad p_i = 0 \quad \text{for} \quad i \neq j \tag{1.5.44b}$$

Consequently, $A_i = 0$ for all $i \neq j$. Moreover, it can be seen from (1.5.43c) that, if $\{x(0)\} = c_1\{X\}_j$ and $\{\dot{x}(0)\} = c_2\{X\}_j$, where c_1 and c_2 are constants, then $B_i = 0$ for all $i \neq j$, and hence only the jth mode participates in the motion. Furthermore, (1.5.43a) shows that the steady-state motion, when $A_i = 0$ for all $i \neq j$, is a single-mode vibration and all of the physical coordinates reach the equilibrium position $\{x\} = \{0\}$ at the same time. Such vibrations are called normal-mode vibrations, and the modes are called normal modes; there is no phase difference between its components. We note that Ω need not be close to ω_j for the system to vibrate in the jth mode. However, if Ω is far from ω_j, then the vibration amplitude will be small.

Next, we consider a more general n-degree-of-freedom system governed by (1.5.23) with the damping matrix $[C]$ being nonproportional and/or the matrices $[C]$ and $[K]$ being asymmetric due to gyroscopic and circulatory effects (Meirovitch, 1980). The undamped eigenvectors cannot be used to transform (1.5.23) into a system of decoupled second-order equations. However, one can transform the governing equations into a system of decoupled first-order equations. To accomplish this, we rewrite (1.5.23) as

$$\begin{bmatrix} [M] & [0] \\ [0] & -[K] \end{bmatrix} \begin{Bmatrix} \{\ddot{x}\} \\ \{\dot{x}\} \end{Bmatrix} + \begin{bmatrix} [C] & [K] \\ [K] & [0] \end{bmatrix} \begin{Bmatrix} \{\dot{x}\} \\ \{x\} \end{Bmatrix} = \begin{Bmatrix} \{f\} \\ \{0\} \end{Bmatrix} \tag{1.5.45}$$

We multiply (1.5.45) from the left with the inverse of the first matrix and rewrite it as a system of first-order equations; that is,

$$\{\dot{y}\} = [A]\{y\} + \{\tilde{f}\} \tag{1.5.46a}$$

where

$$\{y\} = \begin{Bmatrix} \{\dot{x}\} \\ \{x\} \end{Bmatrix}, \quad \{\tilde{f}\} = \begin{Bmatrix} [M]^{-1}\{f\} \\ \{0\} \end{Bmatrix}, \quad [A] \equiv \begin{bmatrix} -[M]^{-1}[C] & -[M]^{-1}[K] \\ [I] & [0] \end{bmatrix} \tag{1.5.46b}$$

and $\{y\}$ is the state vector.

To decouple the system (1.5.46a), we first determine the eigenvalues and eigenvectors of its homogeneous part. To this end, we substitute

$$\{y\} = \{Y\}e^{st} = \begin{Bmatrix} \{Y_1\} \\ \{Y_2\} \end{Bmatrix} e^{st} \tag{1.5.47a}$$

into the homogeneous part of (1.5.46a) and obtain

$$[A]\{Y\} = s\{Y\} \tag{1.5.47b}$$

or

$$[[A] - s[I]]\{Y\} = 0 \tag{1.5.47c}$$

Hence, for a nontrivial solution for $\{Y\}$, we demand that

$$|[A] - s[I]| = 0 \tag{1.5.47d}$$

which yields $2n$ eigenvalues $s_1, s_2, \ldots, s_{2n}$. Then, using (1.5.47c), we calculate the $2n$ corresponding eigenvectors $\{Y\}_1, \{Y\}_2, \ldots, \{Y\}_{2n}$, which are also called the right eigenvectors of $[A]$.

Because the matrix $[A]$ is, in general, asymmetric, $[A]^T (\neq [A])$ has eigenvectors different from those of $[A]$. However, their eigenvalues are the same because the determinant of a matrix is the same as the determinant of its transpose. The eigenvectors $\{Z\}_i$ of $[A]^T$ are also called the left eigenvectors of $[A]$ because

$$[A]^T \{Z\}_i = s_i \{Z\}_i \tag{1.5.48a}$$

implies that

$$\{Z\}_i^T [A] = s_i \{Z\}_i^T \tag{1.5.48b}$$

We note that the left and right eigenvectors, $\{Z\}_i$ and $\{Y\}_j$, are orthogonal if $i \neq j$, which can be proved as follows. We specialize (1.5.47b) to $s = s_j$, multiply the result from the left with $\{Z\}_i^T$, and obtain

$$\{Z\}_i^T [A] \{Y\}_j = s_j \{Z\}_i^T \{Y\}_j \tag{1.5.49a}$$

Multiplying (1.5.48b) from the right with $\{Y\}_j$ yields

$$\{Z\}_i^T [A] \{Y\}_j = s_i \{Z\}_i^T \{Y\}_j \tag{1.5.49b}$$

Subtracting (1.5.49b) from (1.5.49a), we have

$$(s_i - s_j) \{Z\}_i^T \{Y\}_j = 0 \tag{1.5.49c}$$

When $s_i \neq s_j$, (1.5.49c) is satisfied if and only if

$$\{Z\}_i^T \{Y\}_j = 0 \tag{1.5.49d}$$

Consequently, the $\{Z\}_i$ and $\{Y\}_j$ can be normalized so that

$$[Z]^T [Y] = [I] \tag{1.5.50a}$$

where $[Y] = [\{Y\}_1, \{Y\}_2, \ldots, \{Y\}_{2n}]$ and $[Z] = [\{Z\}_1, \{Z\}_2, \ldots, \{Z\}_{2n}]$ are the right and left modal matrices. Then, when $i = j$, (1.5.49a) becomes

$$\{Z\}_i^T [A] \{Y\}_i = s_i \{Z\}_i^T \{Y\}_i = s_i \tag{1.5.50b}$$

Hence,

$$[Z]^T [A][Y] = [S] = \begin{bmatrix} s_1 & 0 & 0 & \cdots & 0 \\ 0 & s_2 & 0 & \cdots & 0 \\ 0 & 0 & s_3 & \cdots & 0 \\ \vdots & \vdots & \vdots & \ddots & \vdots \\ 0 & 0 & 0 & \cdots & s_{2n} \end{bmatrix} \tag{1.5.50c}$$

We note that because $[A]$ is, in general, not symmetric, its eigenvalues, in general, are not real, as shown below. Because $[A]$ is real, then if $\{Y\}_i$ is the right eigenvector of $[A]$ corresponding to the eigenvalue s_i, then $\{\bar{Y}\}_i$ is the right eigenvector of $[A]$ corresponding to the eigenvalue $\bar{s}_i$. Hence,

$$[A]\{Y\}_i = s_i\{Y\}_i \qquad (1.5.51a)$$

and

$$[A]\{\bar{Y}\}_i = \bar{s}_i\{\bar{Y}\}_i \qquad (1.5.51b)$$

Consequently,

$$\{\bar{Y}\}_i^T[A]\{Y\}_i = s_i\{\bar{Y}\}_i^T\{Y\}_i \qquad (1.5.51c)$$

and

$$\{Y\}_i^T[A]\{\bar{Y}\}_i = \bar{s}_i\{Y\}_i^T\{\bar{Y}\}_i \qquad (1.5.51d)$$

Subtracting the transpose of (1.5.51d) from (1.5.51c) yields

$$\{\bar{Y}\}_i^T\left[[A] - [A]^T\right]\{Y\}_i = (s_i - \bar{s}_i)\{\bar{Y}\}_i^T\{Y\}_i \qquad (1.5.51e)$$

Because $\{\bar{Y}\}_i\{Y\}_i$, the norm of $\{Y\}_i$, is different from zero and because $[A] \neq [A]^T$, $s_i \neq \bar{s}_i$, and hence s_i is not real.

Using the right and left modal matrices, we decouple the system of $2n$ equations (1.5.46a). To this end, we introduce the linear transformation

$$\{y\} = [Y]\{q\} \qquad (1.5.52a)$$

where the q_j are the modal coordinates, into (1.5.46a) and obtain

$$[Y]\{\dot{q}\} = [A][Y]\{q\} + \{\tilde{f}\} \qquad (1.5.52b)$$

Premultiplying (1.5.52b) with $[Z]^T$ and using (1.5.50a) and (1.5.50c), we have

$$\{\dot{q}\} = [S]\{q\} + \{g\} \qquad (1.5.52c)$$

where

$$\{g\} = [Z]^T\{\tilde{f}\} \qquad (1.5.52d)$$

The element g_i of $\{g\}$ is the modal excitation of the ith mode. Using (1.5.50c), we rewrite (1.5.52c) in component form as

$$\dot{q}_j = s_j q_j + g_j \quad \text{for } j = 1, 2, \ldots, 2n \qquad (1.5.52e)$$

Because, for each j, (1.5.52e) is a first-order linear ordinary-differential equation, it can be exactly solved for any $g_j(t)$. In fact, its general solution is

$$q_j = c_j e^{s_j t} + \int_0^t e^{-s_j(\tau - t)} g_j(\tau) d\tau \qquad (1.5.53a)$$

where $c_j = q_j(0)$ is a constant that depends on the initial conditions. For the case of a harmonic excitation,

$$\{f\} = \text{Real}\left(\{F\}e^{i\Omega t}\right) \tag{1.5.53b}$$

where $\{F\}$ is a complex-valued vector if the components of $\{f\}$ have different phases. Then, using (1.5.46b) and (1.5.52d), we find that

$$g_j = \{Z\}_j^T \{\tilde{f}\} = G_j e^{i\Omega t}, \quad G_j = \{Z\}_j^T \left\{ \begin{array}{c} [M]^{-1}\{F\} \\ 0 \end{array} \right\} \tag{1.5.53c}$$

Consequently, the general solution of (1.5.52e) can be expressed as

$$q_j = \left[q_j(0) - \frac{G_j}{i\Omega - s_j}\right] e^{s_j t} + \frac{G_j}{i\Omega - s_j} e^{i\Omega t} \tag{1.5.54a}$$

Substituting (1.5.54a) into (1.5.52a) and (1.5.46b) yields

$$\{x\} = \text{Real} \sum_{j=1}^{2n} \left[\{Y_2\}_j \frac{G_j}{i\Omega - s_j} e^{i\Omega t} + \{Y_2\}_j \left(q_j(0) - \frac{G_j}{i\Omega - s_j}\right) e^{s_j t}\right] \tag{1.5.54b}$$

For dissipative systems, the real part of each s_j is negative and hence the steady-state response is given by

$$\{x\} = \text{Real} \sum_{j=1}^{2n} \left[\{Y_2\}_j \frac{G_j}{i\Omega - s_j} e^{i\Omega t}\right] \tag{1.5.54c}$$

Frequency-response functions of a multi-degree-of-freedom system can be easily obtained using the Laplace transformation. Taking the Laplace transform of (1.5.23), assuming zero initial conditions, and inverting the obtained matrix, we obtain

$$\{\mathcal{L}(x)\} = [H(s)]\{\mathcal{L}(f)\} \tag{1.5.55a}$$

where the transfer matrix $[H(s)]$ is given by

$$[H(s)] = \left[s^2[M] + s[C] + [K]\right]^{-1} \tag{1.5.55b}$$

Hence, the frequency-response matrix $[H(\Omega)]$ can be obtained as

$$[H(\Omega)] = \left[-\Omega^2[M] + i\Omega[C] + [K]\right]^{-1} \tag{1.5.56}$$

1.6 DYNAMIC CHARACTERISTICS OF NONLINEAR DISCRETE SYSTEMS

Nonlinear systems are those for which the principle of superposition does not hold. In a nonlinear system, the major component of the output to a harmonic input may be at a different frequency: a subharmonic or a superharmonic of the input frequency.

Whereas the steady-state response of a linear system is unique and independent of the initial conditions, the steady-state response of a nonlinear system may not be unique and may depend on the initial conditions. In the absence of an input, some nonlinear systems may possess self-excited (limit-cycle) responses (e.g., the van der Pol and Rayleigh oscillators). The response of a self-excited system to a harmonic input may be a limit-cycle, a quasiperiodic, or a chaotic response. The response of a nonlinear system to a deterministic excitation may be irregular and sensitive to initial conditions, which is called chaos. Chaotic dynamics is characterized by sensitivity to initial conditions.

An autonomous dynamical system having n degrees of freedom is said to have $2n$ dimensions because any second-order differential equation can be transformed into two first-order differential equations and hence the system has $2n$ orthogonal eigendirections in the state space. A single-degree-of-freedom system subject to a periodic force can be rendered autonomous by adding time as a state variable. Then the system can be represented by three first-order equations and hence it is called a third-order system. Symmetry-breaking and period-doubling bifurcations and chaotic motions can only occur in nonlinear systems whose order is no less than three. A dynamical system is said to be fluttering if it has a stable limit cycle. Often flutter is suggested if a system linearized about an equilibrium configuration (fixed point) has two complex conjugate eigenvalues with a positive real part. For a system with two or more degrees of freedom, Hopf bifurcations of a limit cycle (or secondary Hopf bifurcations) are possible.

In deterministically forced single-degree-of-freedom systems, nonlinear phenomena include multiple solutions, jumps, frequency entrainments, natural frequency shifts, subharmonic, superharmonic, combination and ultrasubharmonic resonances, limit cycles, symmetry-breaking and period-multiplying bifurcations, and chaotic motions. In addition to these phenomena, multiple-degree-of-freedom systems may exhibit modal interactions, resulting in energy exchanges among modes.

Most of the available nonlinear dynamic studies involve single- and two-degree-of-freedom systems. The texts by Nayfeh and Mook (1979) and Nayfeh (2000) present comprehensive literature reviews of the nonlinear oscillations of diverse physical systems. They studied and explained complicated nonlinear phenomena in the dynamics of single-degree-of-freedom, multi-degree-of-freedom, and continuous systems. They described the concept of modal interaction, which allows for the reduction of a continuous system to a finite-degree-of-freedom system even when a complex motion is involved. They also argued that, in the presence of damping, modes that are not directly excited through an external resonance or indirectly excited through an internal or autoparametric resonance will decay with time.

A survey paper by Holmes and Moon (1983) contains several examples of diverse physical systems that exhibit complicated dynamics. The examples include experimental studies of postbuckled beams oscillating around their buckled positions; nonlinear circuits (Ueda, 1979); magnetomechanical devices, such as rotating disks in a magnetic field (Robbins, 1977) and torsional oscillations of a compass needle in an oscillating or rotating magnetic field (Croquette and Pointou, 1981); feedback control systems; and chemical reactions, such as the reaction-diffusion system (Rossler,

1976). Moon (1987, 1992) reviewed some of the mathematical models and experiments of physical systems that display chaotic vibrations. He also presented a variety of theoretical and experimental tools to identify chaos.

Nayfeh (1988) discussed several examples of physical systems that exhibit complicated nonlinear behaviors, such as the saturation and jump phenomena, period-multiplying bifurcations, and chaos. He used a numerical-perturbation approach for the analysis of these systems; they include cylindrical shells, experiments with two-beam structures, surface waves in closed cylindrical containers, the subharmonic instability of two-dimensional boundary layers on flat plates, and three-dimensional propagation of sound in partially choked ducts.

Among other books in this field, we quote those of Hale (1963), Urabe (1967), Marsden and McCracken (1976), Jordan and Smith (1977), Iooss and Joseph (1980), Hassard, Kazarinoff, and Wan (1981), Chow and Hale (1982), Vanderbauwhede (1982), Guckenheimer and Holmes (1983), Thompson and Stewart (1986), and Seydel (1988). For the concepts and applications of modern nonlinear dynamics, the readers are referred to the book by Nayfeh and Balachandran (1995).

Single-degree-of-freedom systems with nonlinear characteristics are, in general, easier to study and understand than systems with more degrees of freedom. Therefore, there exists a good and rich amount of literature about the subject. Nayfeh and Khdeir (1986a,b) investigated the nonlinear rolling of ships in regular seas. They modeled the ship as a single-degree-of-freedom system and found out that its response undergoes a cascade of period-doubling bifurcations, culminating in chaos. Dowell and Pezeshki (1986) studied the nonlinear single-mode response of a buckled beam and showed that the response undergoes a sequence of period-doubling bifurcations that culminates in chaos. They reported that their results agree with the experimental results of Moon (1980). Zavodney and Nayfeh (1988) and Zavodney, Nayfeh, and Sanchez (1989, 1990) studied extensively the response of a single-degree-of-freedom system with quadratic and cubic nonlinearities to fundamental and principal parametric resonances. They presented a second-order perturbation solution to the problem and validated the results with analog- and digital-computer simulations. They found period-multiplying and -demultiplying bifurcations as well as chaotic motions. Bolotin (1964), Evan-Iwanowski (1976), Ibrahim (1985), and Schmidt and Tondl (1986) gave comprehensive reviews of parametric excitations.

Kojima et al. (1985) investigated the nonlinear forced vibration of a beam with a mass to parametric excitation and found superharmonic and subharmonic vibrations of order two and one-half, respectively. Gurgoze (1986) studied parametric vibrations of a restrained beam with mass at one end and a displacement excitation at the other end. He used a one-mode Galerkin approximation and reduced the governing partial-differential equation to a Mathieu equation containing cubic nonlinearities. He obtained an approximate solution for the case of a principal parametric resonance. Zavodney and Nayfeh (1989) experimentally and analytically studied the nonlinear response of a beam with tip mass to a principal parametric excitation.

HaQuang (1986) studied the nonlinear response of one- and two-degree-of-freedom systems having quadratic and cubic nonlinearities to external, parametric, and combined external and parametric excitations. He found period-multiplying bifurcations and chaotic responses. Szemplinska-Stupnicka (1978) studied the response of parametrically excited systems. She used the method of harmonic balance and validated her results by using an analog computer. Awrejcewicz (1989) presented an analysis of the transition from regular to chaotic motions in a van der Pol-Duffing's oscillator with time delay after a Hopf bifurcation. He determined conditions for the occurrence of the Hopf bifurcation by means of an approximate method.

Modes may interact in multi-degree-of-freedom systems. Sethna (1965) studied the vibrations of two-degree-of-freedom systems with quadratic nonlinearities. He showed that these systems exhibit amplitude-modulated motions with large modulation periods. Nayfeh, Mook, and Marshall (1973), Mook, Marshall, and Nayfeh (1974), Nayfeh (1988), and Nayfeh and Oh (1995) investigated the coupling between the pitch (heave) and roll modes of a ship. They modeled the ship dynamics with two coupled oscillators with quadratic nonlinearities. They studied the cases of primary and secondary resonances when the linear natural frequency of the pitch mode is approximately twice that of the roll mode (i.e., $\omega_2 \approx 2\omega_1$). They showed that when the excitation frequency is near ω_2, the response exhibits the saturation phenomenon. Furthermore, they showed that when the excitation frequency is near ω_1, there exist conditions under which stable steady-state responses of the ship are not periodic, but rather amplitude- and phase-modulated motions. Later, Yamamoto and Yasuda (1977) applied the method of harmonic balance to study the forced response of systems with quadratic and cubic nonlinearities to a harmonic excitation when one frequency is twice the other. They used analog-computer simulations and obtained amplitude- and phase-modulated steady-state responses for the case of primary resonance of either mode.

Haddow, Barr, and Mook (1984) investigated modal interactions in a two-degree-of-freedom beam structure under a harmonic external excitation. They conducted an experiment and verified the saturation phenomenon. Balachandran and Nayfeh (1990) studied the planar dynamic response of a flexible L-shaped beam-mass structure with a two-to-one internal resonance to a primary resonance. They obtained the equations governing the modulations of the amplitudes and phases by averaging the Lagrangian of the system over the period of the primary oscillation. They experimentally verified the analytical solutions. Hatwal, Mallik, and Ghosh (1983a,b) investigated the forced nonlinear oscillation of a two-degree-of-freedom system. They showed that the motion of the system may be periodic or chaotic in the presence of a two-to-one autoparametric resonance.

Tousi and Bajaj (1985) used the method of averaging and a combination of a shooting technique and a scheme proposed by Chua and Lin (1975) to study in detail the cascade of period-doubling bifurcations in the response of a two-degree-of-freedom mechanical system with a three-to-one autoparametric response. They found out that some of the sequences of period-doubling bifurcations are compatible with the Feigenbaum universality constant (Feigenbaum, 1979).

Nayfeh and Zavodney (1986) studied the response of two-degree-of-freedom systems with quadratic nonlinearities to combination parametric resonances. They presented steady-state solutions and their stability. They also presented limit-cycle solutions and period-doubling bifurcations using numerical techniques. Streit, Bajaj, and

Krousgrill (1988) studied a similar system and found Hopf bifurcations and limit-cycle solutions. They showed that the limit cycles undergo a cascade of period-doubling bifurcations culminating in chaos. Miles (1984a) studied the case of perfectly tuned two-to-one internal resonances in surface waves. He considered the case of principal parametric resonance of the lower mode and found no Hopf bifurcations. He concluded that there are no limit-cycle or chaotic solutions of the modulation equations. Nayfeh (1987a,b) used the method of multiple scales to analyze the problem studied by Miles (1984a, 1985). He relaxed the assumption of perfectly tuned internal resonance and found Hopf bifurcations and period-multiplying bifurcations culminating in chaos. Miles (1984b) used the method of averaging to study the response of two internally resonant coupled oscillators with quadratic nonlinearities to a harmonic excitation. He investigated the stability of the analytical solutions and presented numerical results that demonstrate periodically and chaotically modulated motions when the excitation frequency is near the lower frequency. Nayfeh (1987a) studied a parametrically excited double pendulum in the presence of two-to-one internal resonances. He found out that the response exhibits a Hopf bifurcation leading to amplitude- and phase-modulated motions. Furthermore, he showed that the periodic motions of the averaged equations may undergo period-doubling bifurcations leading to chaos.

Shaw and Shaw (1989) investigated the dynamic response of a two-degree-of-freedom impacting system consisting of an inverted pendulum with motion limiting stops. They attached this system to a sinusoidally excited mass-spring system. They found several types of periodic motions and discussed the stability of two types of motion. They also found chaotic motions and concluded that great care should be taken in the design of an inverted pendulum with an unstable central position as an absorber.

Miles (1984c) studied the case of one-to-one internal resonance between planar and nonplanar motions of spherical pendulums. He showed that a resonating planar motion of the base might excite the nonplanar motion, resulting in a cascade of period-doubling bifurcations and chaos. Maewal (1986a) showed that elastic beams with symmetric cross-sections can exhibit a coupling between their inplane and out-of-plane modes and that a resonance transverse excitation can induce out-of-plane responses. He also showed that the beam may exhibit a chaotically modulated response for certain values of the excitation frequency. Miles (1984a,b) studied the case of one-to-one internal resonance between orthogonal surface-wave modes in a circular cylinder in the case of free and forced oscillations. He concluded that exciting one of these modes may excite the other and the response may be periodically or chaotically modulated. He also showed that this modal coupling does not exist above certain values of the damping and depth-to-radius ratios. Feng and Sethna (1989) studied symmetry-breaking bifurcations of surface waves in nearly square containers subjected to vertical oscillations and in the presence of one-to-one internal resonance. They showed that the system may exhibit periodic, quasiperiodic, or chaotic solutions. Ciliberto and Gollub (1985) conducted experimental investigations into one-to-one internal resonances in surface waves. They examined the case in which the excitation frequency and amplitude are near the intersection of the instability boundary between two degenerate modes and found that they can compete with each other to produce either periodic or chaotic motion. They used the experimental data to reconstruct the attractors in the phase space and calculate their dimensions and Lyapunov exponents. They found that the chaotic attractors have fractal dimensions and at least one positive Lyapunov exponent. Simonelli and Gollub (1989) experimentally studied wave-mode interactions in square and near-square containers. They detected multiple attractors and repellors and studied the involved bifurcations.

1.7 ANALYSES OF LINEAR CONTINUOUS SYSTEMS

A continuous system consists of an infinite number of particles, and hence one needs an infinite number of ordinary-differential equations to describe its motion if it is treated as a particle system. However, because the distance between two adjacent particles is infinitesimal and the displacements must be continuous, the motion of a continuous system is described by a finite number of displacement variables. But, these displacements are functions of spatial coordinates as well as time, and hence the governing equations are partial-differential equations, which need to be supplemented with boundary as well as initial conditions. Hence, solving dynamical problems of continuous systems is more difficult than solving those of discrete systems.

For a discrete system, its eigenvectors are used to decouple the governing ordinary-differential equations. For a continuous system, its eigenfunctions or mode shapes (satisfy the free, undamped governing equations and geometric and natural boundary conditions) or comparison functions (satisfy the geometric and natural boundary conditions) or admissible functions (only satisfy the geometric boundary conditions) are used to discretize the system, and then the method discussed in Section 1.5.2 can be used to decouple the obtained ordinary-differential equations if they are linearly coupled. Eigenfunctions are a natural generalization of the notion of eigenvectors in linear discrete systems.

Dynamics of a continuous system can be expressed in terms of either partial-differential equations with appropriate boundary conditions or a functional. The two main solution approaches are (1) variational methods and (2) weighted-residual methods. Variational methods employ the calculus of variations and involve the determination of the extreme or stationary value of a functional. In structural dynamics, the functional might be the total potential energy and/or complementary energy of the system. The Ritz method is the most widely used variational method in structural engineering, where admissible functions are used. Weighted-residual methods begin with the governing partial-differential equations and boundary conditions. These methods are appropriate for certain problems (e.g., fluid mechanic problems or some aeroelasticity problems) in which a functional does not exist or has not yet been found. Weighted-residual methods include Galerkin, Petrov-Galerkin, least-square, collocation, subdomain, Kantorovich, Trefftz, and moments methods (Meirovitch, 1980). The main difference among these methods is the test functions used. The Galerkin method is the most widely used method in structural engineering, where eigenfunctions or comparison functions are used.

28 INTRODUCTION

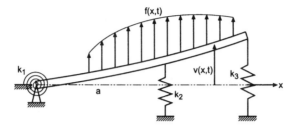

Fig. 1.7.1 A hinged-free beam supported by springs.

There are three variational (energy) principles used in structural mechanics: (a) the theorem of minimum potential energy, (b) the theorem of minimum complementary energy, and (c) Reissner's variational theorem (called stationary principle). Principle (a) is the most widely used one, especially for displacement-based formulations of structural dynamic problems. Principle (b) is used for stress-based formulations, and Principle (c) is used for displacement- and stress-based (or mixed) formulations, which is usually used in finite-element analysis requiring high accuracy in both displacement and stress solutions. We illustrate different variational and weighted-residual methods in the following chapters.

To describe the characteristics of continuous systems and the methods of analyzing them, we consider the beam shown in Figure 1.7.1 and use the extended Hamilton principle

$$\int_{t_1}^{t_2} (\delta T - \delta \Pi + \delta W_{nc})dt = 0 \qquad (1.7.1a)$$

to formulate the problem. The kinetic and elastic energies, T and Π are given by

$$T = \frac{1}{2}\int_0^L (m\dot{v}^2 + j\dot{\theta}^2)dx \qquad (1.7.1b)$$

$$\Pi = \frac{1}{2}\int_0^L EI\kappa^2 dx + \frac{1}{2}k_1\theta^2(0,t) + \frac{1}{2}k_2 v^2(a,t) + \frac{1}{2}k_3 v^2(L,t) \qquad (1.7.1c)$$

where the overdot denotes the derivative with respect to t, m is the mass per unit length, j is the rotary inertia per unit length, L is the beam length, and EI is the bending rigidity. The curvature κ and the rotation angle θ are approximated by their linear parts; that is,

$$\kappa = v'' \text{ and } \theta = v' \qquad (1.7.1d)$$

where the prime denotes the derivative with respect to x. Moreover, the variation δW_{nc} of the non-conservative energy is due to the applied distributed load $f(x,t)$, gravity, and damping; it is given by

$$\delta W_{nc} = \int_0^L (f - mg - c\dot{v})\delta v dx \qquad (1.7.1e)$$

where g is the gravitational acceleration and c is the damping coefficient per unit length.

Substituting (1.7.1d) into (1.7.1b), taking the variation, integrating the result by parts, and assuming that $\delta v(x, t_1) = \delta v(x, t_2) = 0$ and $\delta v'(x, t_1) = \delta v'(x, t_2) = 0$, we obtain

$$\int_{t_1}^{t_2} \delta T dt = \int_{t_1}^{t_2} \int_0^L \left(-m\ddot{v}\delta v - j\ddot{v}'\delta v'\right) dx dt + \int_0^L \left[m\dot{v}\delta v + j\dot{v}'\delta v'\right]_{t_1}^{t_2} dx$$

$$= \int_{t_1}^{t_2} \int_0^L \left[-m\ddot{v} + (j\ddot{v}')'\right] \delta v dx dt - \int_{t_1}^{t_2} \left[j\ddot{v}'\delta v\right]_0^L dt \quad (1.7.2a)$$

Similarly, substituting (1.7.1d) into (1.7.1c), taking the variation, integrating the result by parts, and using the identity $v\delta v \mid_{x=a} = \int_0^L v\hat{\delta}(x-a)\delta v dx$, where $\hat{\delta}(x-a)$ is the spatial Dirac delta function, we obtain

$$\delta \Pi = \int_0^L \left[(EIv'')'' + k_2 v \hat{\delta}(x-a)\right] \delta v dx$$

$$+ k_1 v' \delta v' \mid^0 + k_3 v \delta v \mid^L + \left[EIv'' \delta v' - (EIv'')' \delta v\right]_0^L \quad (1.7.2b)$$

Substituting (1.7.2a), (1.7.2b), and (1.7.1e) into (1.7.1a) yields

$$\int_{t_1}^{t_2} \int_0^L \left[-m\ddot{v} + (j\ddot{v}')' - (EIv'')'' - k_2 v \hat{\delta}(x-a) + f - mg - c\dot{v}\right] \delta v dx dt$$

$$- \int_{t_1}^{t_2} \left\{ (k_1 v' - EIv'') \delta v' - \left[j\ddot{v}' - (EIv'')'\right] \delta v \right\}^0 dt$$

$$- \int_{t_1}^{t_2} \left\{ EIv'' \delta v' + \left[j\ddot{v}' - (EIv'')' + k_3 v\right] \delta v \right\}^L dt = 0 \quad (1.7.3)$$

Because δv is arbitrary, setting the coefficient of δv in the first integral in (1.7.3) equal to zero yields the equation of motion

$$m\ddot{v} - (j\ddot{v}')' + c\dot{v} = -(EIv'')'' - k_2 v \hat{\delta}(x-a) + f - mg \quad (1.7.4a)$$

Moreover, using the fact that $v = 0$ and v' is unknown at $x = 0$ and v and v' are unknown at $x = L$, we obtain from (1.7.3) the following boundary conditions:

$$\begin{aligned} j\ddot{v}' - (EIv'')' \quad \text{unknown} & \qquad v = 0 \\ EIv'' - k_1 v' = 0 & \qquad v' \quad \text{unknown} \end{aligned} \quad (1.7.4b)$$

at $x = 0$ and

$$\begin{aligned} j\ddot{v}' - (EIv'')' + k_3 v = 0 & \qquad v \quad \text{unknown} \\ EIv'' = 0 & \qquad v' \quad \text{unknown} \end{aligned} \quad (1.7.4c)$$

at $x = L$. On the left side of (1.7.4b,c) are the so-called natural or dynamic boundary conditions, and on the right side are the geometric or essential boundary conditions.

1.7.1 Natural Frequencies and Eigenfunctions

Eigenfunctions are free, undamped vibration mode shapes. To obtain the eigenfunctions of the system shown in Figure 1.7.1, we let $f - mg = c = 0$, assume that v varies harmonically with time, and write

$$v(x,t) = V(x)e^{i\omega t} \qquad (1.7.5)$$

The use of separation of variables implies that the motion is synchronous; that is, different points, say the points at $x = x_1$ and x_2, achieve their maxima and minima at the same time although they may have different amplitudes [i.e., $V(x_1), V(x_2)$]. Substituting (1.7.5) into (1.7.4) yields the eigenvalue problem

$$(EIV'')'' + k_2 V \hat{\delta}(x-a) = \omega^2 [mV - (jV')'] \qquad (1.7.6a)$$

$$V = 0, \quad EIV'' - k_1 V' = 0 \text{ at } x = 0 \qquad (1.7.6b)$$

$$k_3 V - (EIV'')' = \omega^2 j V', \quad EIV'' = 0 \text{ at } x = L \qquad (1.7.6c)$$

The eigenvalue problem (1.7.6) yields an infinite number of eigenvalues ω_m, linear undamped natural frequencies, corresponding to an infinite number of eigenfunctions $V_m(x)$, mode shapes. Because the boundary-value problem (1.7.6) is self-adjoint, as shown below, the frequencies ω_m and their corresponding mode shapes $V_m(x)$ are real. Moreover, we show that the mode shapes corresponding to different frequencies satisfy a general orthogonality condition.

First, we show that the boundary-value problem is self-adjoint. To accomplish this, we determine the equation and boundary conditions that govern the adjoint. To this end, we multiply (1.7.6a) with $U(x)$, the adjoint function, integrate the result over the domain $(0 \leq x \leq L)$, and obtain

$$\int_0^L U \left\{ (EIV'')'' + k_2 V \hat{\delta}(x-a) - \omega^2 \left[mV - (jV')' \right] \right\} dx = 0 \qquad (1.7.7a)$$

Integrating (1.7.7a) by parts to transfer the derivatives from V to U and using the boundary conditions (1.7.6b) and (1.7.6c) yields

$$\int_0^L \left\{ (EIU'')'' + k_2 U \hat{\delta}(x-a) - \omega^2 \left[mU - (jU')' \right] \right\} V \, dx$$

$$+ \left\{ \left[k_3 U - (EIU'')' - \omega^2 j U' \right] V + EIU'' V' \right\}^L$$

$$- \left\{ U \left[(EIV'')' + \omega^2 j V' \right] + (EIU'' - k_1 U') V' \right\}^0 = 0 \qquad (1.7.7b)$$

The adjoint equation is obtained by setting the coefficient of V in the integrand in (1.7.7b) equal to zero. The result is

$$(EIU'')'' + k_2 U \hat{\delta}(x-a) = \omega^2 \left[mU - (jU')' \right] \qquad (1.7.8a)$$

The boundary conditions for U are obtained by setting the coefficient of each of the unknown expressions $V'(0)$, $\left[(EIV'')' + \omega^2 jV'\right]^0$, $V(L)$, and $V'(L)$ equal to zero, independently. The result is

$$U = 0, \quad EIU'' - k_1 U' = 0 \text{ at } x = 0 \tag{1.7.8b}$$

$$k_3 U - (EIU'')' = \omega^2 jU', \quad EIU'' = 0 \text{ at } x = L \tag{1.7.8c}$$

Because the adjoint equation (1.7.8a) and associated boundary conditions (1.7.8b) and (1.7.8c) governing U are the same as those governing V, the boundary-value problem governing V is called a self-adjoint problem.

We note that most but not all structural systems are self-adjoint. As an example of systems that are not self-adjoint, we consider a cantilever beam in a supersonic stream. Using the piston theory (Fung, 1969; Dowell, 1975, 1989), the aerodynamic load f is related to the displacement v by $f = \alpha v'$. Hence, the equation governing linear motions of the beam is

$$m\ddot{v} + c\dot{v} = -(EIv'')'' + \alpha v' \tag{1.7.9a}$$

where the rotary inertia and gravity are neglected. The boundary conditions for a cantilever beam are

$$v = 0, \ v' = 0 \text{ at } x = 0 \tag{1.7.9b}$$

$$EIv'' = 0, \ (EIv'')' = 0 \text{ at } x = L \tag{1.7.9c}$$

Substituting (1.7.5) into the undamped (1.7.9) yields

$$(EIV'')'' - \alpha V' = \omega^2 mV \tag{1.7.10a}$$

$$V = 0, \ V' = 0 \text{ at } x = 0 \tag{1.7.10b}$$

$$EIV'' = 0, \ (EIV'')' = 0 \text{ at } x = L \tag{1.7.10c}$$

To determine the adjoint of (1.7.10), we multiply (1.7.10a) with the adjoint function U, integrate the result by parts from $x = 0$ to $x = L$ to transfer the derivatives from V to U, use the boundary conditions (1.7.10b) and (1.7.10c), and obtain

$$\int_0^L \left\{ (EIU'')'' + (\alpha U)' - \omega^2 mU \right\} V \, dx + \left[EIU''V' - ((EIU'')' + \alpha U) V \right]_0^L$$

$$- \left[U(EIV'')' - U'EIV'' \right]^0 = 0 \tag{1.7.11}$$

Setting the coefficient of V in the integrand in (1.7.11) equal to zero yields the adjoint equation

$$(EIU'')'' + (\alpha U)' = \omega^2 mU \tag{1.7.12a}$$

Then, setting each of the coefficients of the unknown expressions $V(L)$, $V'(L)$, $V''(0)$, and $(EIV'')' \mid^0$ equal to zero yields the associated boundary conditions

$$U = 0, \ U' = 0 \text{ at } x = 0 \tag{1.7.12b}$$

$$EIU'' = 0, \quad (EIU'')' + \alpha U = 0 \quad \text{at} \quad x = L \tag{1.7.12c}$$

Clearly, equation (1.7.12a) and its associated boundary conditions (1.7.12b) and (1.7.12c) are different from equation (1.7.10a) and its associated boundary conditions (1.7.10b) and (1.7.10c). Therefore, the problems (1.7.10) and (1.7.12) are not self-adjoint.

To prove that ω is real, we note that if V is a solution of (1.7.6) corresponding to ω, then $\bar{V}$ (the complex conjugate of V) is also a solution of (1.7.6) corresponding to $\bar{\omega}$; that is,

$$(EI\bar{V}'')'' + k_2 \bar{V}\hat{\delta}(x-a) - \bar{\omega}^2\left[m\bar{V} - (j\bar{V}')'\right] = 0 \tag{1.7.13a}$$

$$\bar{V} = 0, \quad EI\bar{V}'' - k_1\bar{V}' = 0 \quad \text{at} \quad x = 0 \tag{1.7.13b}$$

$$k_3\bar{V} - (EI\bar{V}'')' - \bar{\omega}^2 j\bar{V}' = 0, \quad \bar{V}'' = 0 \quad \text{at} \quad x = L \tag{1.7.13c}$$

Multiplying (1.7.6a) by $\bar{V}$, integrating the result by parts, and using the boundary conditions (1.7.6b), (1.7.6c), and (1.7.13b), we obtain

$$\int_0^L EIV''\bar{V}''dx - \omega^2 \int_0^L \left[mV\bar{V} + jV'\bar{V}'\right]dx$$
$$+ k_1 V'\bar{V}' \mid^0 + k_2 V\bar{V} \mid^a + k_3 V\bar{V} \mid^L = 0 \tag{1.7.14}$$

Similarly, multiplying (1.7.13a) by V, integrating the result by parts, and using the boundary conditions (1.7.6b), (1.7.13b), and (1.7.13c), we obtain

$$\int_0^L EIV''\bar{V}''dx - \bar{\omega}^2 \int_0^L \left[mV\bar{V} + jV'\bar{V}'\right]dx$$
$$+ k_1 V'\bar{V}' \mid^0 + k_2 V\bar{V} \mid^a + k_3 V\bar{V} \mid^L = 0 \tag{1.7.15}$$

Subtracting (1.7.14) from (1.7.15) yields

$$(\omega^2 - \bar{\omega}^2) \int_0^L \left[mV\bar{V} + jV'\bar{V}'\right]dx = 0 \tag{1.7.16}$$

Because m and j are positive, (1.7.16) is satisfied if and only if $\omega = \bar{\omega}$; that is, the eigenvalues of (1.7.6) are real. Consequently, the eigenfunctions $V(x)$ of (1.7.6) are also real.

To determine the orthogonality condition of the eigenfunctions of (1.7.6), we consider the rth and sth eigenfunctions $V_r(x)$ and $V_s(x)$ corresponding to the rth and sth eigenvalues ω_r and ω_s. They are governed by

$$(EIV_r''t)'' + k_2 V_r\hat{\delta}(x-a) - \omega_r^2\left[mV_r - (jV_r')'\right] = 0 \tag{1.7.17a}$$

$$V_r = 0, \quad EIV_r'' - k_1 V_r' = 0 \quad \text{at} \quad x = 0 \tag{1.7.17b}$$

$$k_3 V_r - (EIV_r'')' - \omega_r^2 jV_r' = 0, \quad V_r'' = 0 \quad \text{at} \quad x = L \tag{1.7.17c}$$

and

$$(EIV_s'')'' + k_2 V_s\hat{\delta}(x-a) - \omega_s^2\left[mV_s - (jV_s')'\right] = 0 \tag{1.7.18a}$$

$$V_s = 0, \quad EIV_s'' - k_1 V_s' = 0 \quad \text{at} \quad x = 0 \qquad (1.7.18\text{b})$$

$$k_3 V_s - (EIV_s'')' - \omega_s^2 j V_s' = 0, \quad V_s'' = 0 \quad \text{at} \quad x = L \qquad (1.7.18\text{c})$$

Multiplying (1.7.18a) by V_r, integrating the result by parts, and using (1.7.17b), (1.7.18b), and (1.7.18c), we obtain

$$\int_0^L EIV_r'' V_s'' dx + k_1 V_r' V_s' \mid^0 + k_2 V_r V_s \mid^a + k_3 V_r V_s \mid^L$$

$$- \omega_s^2 \int_0^L (mV_r V_s + j V_r' V_s') dx = 0 \qquad (1.7.19)$$

Multiplying (1.7.17a) by V_s, integrating the result by parts, and using (1.7.17b), (1.7.17c), and (1.7.18b), we obtain

$$\int_0^L EIV_r'' V_s'' dx + k_1 V_r' V_s' \mid^0 + k_2 V_r V_s \mid^a + k_3 V_r V_s \mid^L$$

$$- \omega_r^2 \int_0^L (mV_r V_s + j V_r' V_s') dx = 0 \qquad (1.7.20)$$

Subtracting (1.7.20) from (1.7.19) yields

$$(\omega_r^2 - \omega_s^2) \int_0^L (mV_r V_s + j V_r' V_s') dx = 0 \qquad (1.7.21)$$

Therefore, when $\omega_r \neq \omega_s$, (1.7.21) is satisfied if and only if V_r and V_s satisfy the general orthogonality condition

$$\int_0^L (mV_r V_s + j V_r' V_s') dx = 0 \quad \text{for} \quad r \neq s \qquad (1.7.22)$$

Then, it follows from either (1.7.19) or (1.7.20) that

$$\int_0^L EIV_r'' V_s'' dx + k_1 V_r' V_s' \mid^0 + k_2 V_r V_s \mid^a + k_3 V_r V_s \mid^L = 0 \quad \text{for} \quad r \neq s \qquad (1.7.23)$$

When $r = s$, we normalize the eigenfunctions as

$$\int_0^L (mV_r^2 + j V_r'^2) dx = 1 \qquad (1.7.24)$$

Then, it follows from (1.7.20) that

$$\int_0^L EIV_r''^2 dx + k_1 V_r'^2 \mid^0 + k_2 V_r^2 \mid^a + k_3 V_r^2 \mid^L = \omega_r^2 \qquad (1.7.25)$$

1.7.2 Discretization Using Eigenfunctions

To discretize this continuous system by using its first n eigensolutions, we let

$$v = V_0(x,t) + \sum_{s=1}^{n} V_s(x)q_s(t) \tag{1.7.26}$$

where $V_0(x,t)$ is a particular solution that satisfies the nonhomogeneous boundary conditions of the system if any, $q_s(t)$ is the sth modal coordinate, and $V_s(x)$ is the sth eigenfunction of the system with homogeneous boundary conditions. Because the boundary conditions in (1.7.4b) and (1.7.4c) are all homogeneous, $V_0 = 0$ for this problem. Substituting (1.7.26) into (1.7.4a) yields

$$\sum_{s=1}^{n} \left[mV_s\ddot{q}_s - (jV_s')'\ddot{q}_s + cV_s\dot{q}_s + (EIV_s'')''q_s + k_2V_sq_s\hat{\delta}(x-a) \right] - f + mg = 0 \tag{1.7.27}$$

Multiplying (1.7.27) by $V_r(x)$, integrating the result by parts, and using (1.7.6b,c), we obtain

$$\sum_{s=1}^{n} \int_0^L \left\{ (mV_rV_s + jV_r'V_s')\ddot{q}_s + cV_rV_s\dot{q}_s + EIV_r''V_s''q_s \right\} dx - \hat{f}_r$$

$$+ \sum_{s=1}^{n} \left[(k_1V_r'V_s' \mid^0 + k_2V_rV_s \mid^a + k_3V_rV_s \mid^L) \right] q_s = 0 \tag{1.7.28}$$

where (1.7.4b) and (1.7.4c) are used and

$$\hat{f}_r \equiv \int_0^L V_r(f - mg)dx \tag{1.7.29}$$

Using the orthogonality conditions (1.7.22)-(1.7.25), we obtain

$$\ddot{q}_r + \omega_r^2 q_r - \hat{f}_r + \sum_{s=1}^{n} \int_0^L cV_rV_s dx \, \dot{q}_s = 0 \tag{1.7.30}$$

We note that the modal equations (1.7.30) are generally coupled due to the damping term, even though the eigenfunctions were used. Similar to the concept of proportional damping for discrete systems shown in (1.5.40) and (1.5.41), if $c(x)$ is proportional, we have

$$\int_0^L c(x)V_rV_s dx = c_r\delta_{rs} \tag{1.7.31}$$

where the c_r are constants, and (1.7.30) simplifies to

$$\ddot{q}_r + c_r\dot{q}_r + \omega_r^2 q_r = \hat{f}_r \tag{1.7.32}$$

Hence, the vibration modes in this case are decoupled.

Discretization of a non-self-adjoint system results in a discrete system with asymmetric $[K]$ and/or $[M]$ matrices. Similar to the use of right eigenvectors and left eigenvectors for decoupling asymmetric discrete systems (see (1.5.50c)), the eigenfunctions V_i and the adjoint functions U_i need to be used in obtaining modal equations of non-self-adjoint continuous systems.

1.7.3 The Ritz Method

The governing equation of any continuous system can be represented by

$$\mathcal{L}v = f \tag{1.7.33}$$

where f denotes the forcing function and $\mathcal{L}$ is a differential operator, in which the order of the highest derivatives with respect to the spatial coordinates is $2p$ and with respect to time is 2. Closed-form solutions of (1.7.33) subject to appropriate boundary conditions are only possible in relatively few cases, almost invariably in structures with uniform properties. Consequently, solutions are often approximated by using numerical techniques, analytic techniques, and combinations of both. In this section, we describe an analytic approximation technique based on the Ritz method. In Section 1.7.4, we describe methods based on the method of weighted residuals, and in Sections 1.7.5 and 1.7.6, we describe numerical techniques based on finite elements and initial-value methods.

In this section, we approximate the solution of (1.7.33) and its associated boundary conditions by using a finite number n of so-called trial functions V_s as

$$v = \sum_{s=1}^{n} V_s q_s = \{V\}^T \{q\} \tag{1.7.34}$$

which has the same form as (1.7.26) except that V_0 has been dropped in order to simplify the following derivations. This implies that the boundary conditions are homogeneous. To ensure convergence as n increases, one needs to choose the trial functions V_s from a complete set of orthogonal functions. However, the V_s can be the eigenfunctions (functions that satisfy the free, undamped governing equation and all boundary conditions), comparison functions (functions that satisfy the geometric and natural boundary conditions), or admissible functions (functions that satisfy only the geometric boundary conditions), depending on the selected method of solution. For static problems, the q_s are constants. For dynamic problems, the q_s are the generalized coordinates, which are functions of time. In structural dynamics, the V_s are usually chosen to be the eigenfunctions of the system. They are the most convenient and appropriate complete set of orthogonal functions because they can be used to decouple the modal equations, thereby simplifying the analysis. Moreover, when a system is subjected to a low-frequency excitation, high accuracy can be achieved using just a few low-frequency eigenfunctions.

In the absence of a closed-form solution or when the system eigenfunctions do not decouple its modal equations (e.g., (1.7.30)), the most common and systematic way of analyzing the dynamic response of a continuous system is to discretize it first into a finite-degree-of-freedom system in the space of the trial functions V_s, represent the system equations in the form of (1.5.23), and then use the methods discussed in Section 1.5 to decouple the discrete system.

To discretize the system (1.7.33), we seek an approximate solution in the form (1.7.34), premultiply the result by the arbitrary functions ψ_j (the so-called test or weighting functions), integrate the result over the system domain D, and demand that

$$\int_D \psi_j(\mathcal{L}v - f)dD = 0 \text{ for } j = 1, 2, \ldots, n \qquad (1.7.35)$$

We point out that the test functions need not be the adjoint functions. However, because there are n unknowns (i.e., q_s) in (1.7.34), the ψ_j must be linearly independent and preferably orthogonal. Different choices for the trial functions V_s and test functions ψ_j and different approaches of solving for the q_s result in different methods.

For self-adjoint systems, one can use the Ritz method. According to this method, one can use integration by parts to transfer some of the derivatives from v to the test functions ψ_j so that the orders of the spatial derivatives on v and ψ_j are the same, thereby weaken the smoothness requirements on v from $2p$ to p differentiable. As a by-product, the resulting matrices $[M]$, $[C]$, and $[K]$ in the discretized system of equations are symmetric if the problem is self-adjoint. Next, we describe the Ritz method by using the system (1.7.4).

Multiplying (1.7.4a) with ψ_j, integrating the result by parts so that the orders of the spatial derivatives on v and ψ_j are the same, and using the known boundary conditions (both geometric and natural boundary conditions) on v in (1.7.4b,c), we obtain

$$\int_0^L \left[m\psi_j \ddot{v} + j\psi_j' \ddot{v}' + c\psi_j \dot{v} + EI\psi_j'' v'' + k_2 \psi_j v \hat{\delta}(x-a) - f\psi_j + mg\psi_j \right] dx$$
$$+ [k_3 \psi_j v]^L - \left[\psi_j ((EIv'')' - j\ddot{v}') - k_1 \psi_j' v' \right]^0 = 0 \qquad (1.7.36)$$

Approximating v as in (1.7.34) and choosing the ψ_j in such a way that they satisfy the known geometric boundary conditions on v (i.e., $\psi_j = 0$ at $x = 0$), we rewrite (1.7.36) as

$$\sum_{s=1}^n \int_0^L \left\{ \left(m\psi_j V_s + j\psi_j' V_s' \right) \ddot{q}_s + c\psi_j V_s \dot{q}_s + (EI\psi_j'' V_s'' + k_2 \psi_j V_s \hat{\delta}(x-a))q_s \right.$$
$$\left. - f\psi_j + mg\psi_j \right\} dx + \sum_{s=1}^n \left[k_3 \psi_j V_s q_s \mid^L + k_1 \psi_j' V_s' q_s \mid^0 \right] = 0 \quad (1.7.37)$$

for $j = 1, 2, \ldots, n$. We note that whenever both ψ_j and V_s appear in the same term in (1.7.37), they are symmetric because they have the same order of differentiation with respect to x. Moreover, (1.7.37) is linear in both ψ_j and V_s, and hence it has what is called a bilinear form. We note that any self-adjoint system can be put in such a bilinear form.

It is clear from (1.7.37) that ψ_j and V_s are only required to be two times differentiable, whereas, in (1.7.4a), v needs to be four times differentiable. Consequently,

(1.7.37) is called a weak-form formulation. Because the test functions ψ_j need to satisfy only the geometric boundary conditions on v, the system is self-adjoint, and both ψ_j and V_s are only required to be two times differentiable. Therefore, ψ_j and V_s can be chosen from the same set of admissible functions of the system.

Next, we show that system formulations such as (1.7.37) can be put in an energy form. To this end, we multiply (1.7.37) with δq_j, add all equations, and obtain

$$\sum_{j=1}^{n}\sum_{s=1}^{n}\int_{t_1}^{t_2}\left\{\int_0^L\left[\left(m\psi_j V_s + j\psi_j' V_s'\right)\ddot{q}_s\delta q_j + c\psi_j V_s \dot{q}_s \delta q_j + (mg - f)\psi_j \delta q_j\right.\right.$$
$$\left.+ \left(EI\psi_j'' V_s'' + k_2 \psi_j V_s \hat{\delta}(x-a)\right)q_s \delta q_j\right]dx$$
$$\left.+ \left[k_3 \psi_j V_s q_s \delta q_j \mid^L + k_1 \psi_j' V_s' q_s \delta q_j \mid^0\right]\right\}dt = 0 \qquad (1.7.38)$$

Choosing $\psi_k = V_k$ and using (1.7.34), we rewrite (1.7.38) as

$$\int_{t_1}^{t_2}\left\{\int_0^L\left[m\ddot{v}\delta v + j\ddot{v}'\delta v' + EIv''\delta v'' - (f - mg - c\dot{v})\delta v\right]dx\right.$$
$$\left.+ k_1 v'\delta v' \mid^0 + k_2 v \delta v \mid^a + k_3 v \delta v \mid^L \right\}dt = 0 \qquad (1.7.39)$$

Using (1.7.1c), (1.7.1e), and (1.7.2a), we rewrite (1.7.39) as

$$-\int_{t_1}^{t_2}\left(\delta T - \delta \Pi + \delta W_{nc}\right)dt + \int_0^L \left[m\dot{v}\delta v + j\dot{v}'\delta v'\right]_{t_1}^{t_2}dx = 0 \qquad (1.7.40)$$

which, upon using the fact that δv and $\delta v'$ are equal to zero at $t = t_1$ and t_2, becomes

$$\int_{t_1}^{t_2}(\delta T - \delta \Pi + \delta W_{nc})dt = 0 \qquad (1.7.41)$$

Therefore, for a self-adjoint system, the weak-form formulation can always be put in an energy form.

Choosing $\psi_k = V_k$ in (1.7.37), we obtain

$$\sum_{s=1}^{n}(m_{js}\ddot{q}_s + c_{js}\dot{q}_s + k_{js}q_s) = f_j \quad \text{for} \quad j = 1, 2, ..., n \qquad (1.7.42)$$

where

$$m_{js} = \int_0^L (mV_j V_s + jV_j'V_s')dx, \quad c_{js} = \int_0^L cV_j V_s dx$$
$$f_j = \int_0^L (f - mg)V_j dx$$
$$k_{js} = \int_0^L EIV_j''V_s''dx + k_1 V_j'V_s' \mid^0 + k_2 V_j V_s \mid^a + k_3 V_j V_s \mid^L \qquad (1.7.43)$$

38 INTRODUCTION

Equation (1.7.42) can be put in the matrix form (1.5.23) with $[M]$, $[C]$, and $[K]$ being symmetric. These matrices are usually full $n \times n$ matrices. When the number of trial functions increases by r, they become $(n+r) \times (n+r)$ matrices with the m_{ij}, c_{ij}, and k_{ij} $(i,j=1,2,\ldots,n)$ being the same.

For eigenvalue problems and static problems, the Ritz formulation (i.e., (1.7.38) with $\psi_k = V_k$) can be put in an alternate form. For eigenvalue problems, the motion is assumed to be harmonic, free, and undamped, and hence

$$\ddot{q}_s = -\omega^2 q_s, \quad c = 0, \quad mg - f = 0 \tag{1.7.44}$$

Substituting (1.7.44) and $\psi_k = V_k$ into (1.7.38) and using (1.7.34) yields

$$\delta I \equiv \int_0^L \left[-\omega^2(mv\delta v + jv'\delta v') + EIv''\delta v'' \right] dx$$
$$+ k_1 v' \delta v' \big|^0 + k_2 v \delta v \big|^a + k_3 v \delta v \big|^L = 0$$

or

$$\delta I \equiv \sum_{j=1}^n \frac{\partial I}{\partial q_j} \delta q_j = 0 \tag{1.7.45}$$

where $\int_{t_1}^{t_2} dt$ is dropped and

$$I \equiv \frac{1}{2} \int_0^L \left[-\omega^2(mv^2 + jv'^2) + EIv''^2 \right] dx + \frac{1}{2} k_1 v'^2 \big|^0 + \frac{1}{2} k_2 v^2 \big|^a + \frac{1}{2} k_3 v^2 \big|^L$$
$$= \sum_{j=1}^n \sum_{s=1}^n \left\{ \frac{1}{2} \int_0^L \left[-\omega^2(mV_j V_s + jV_j' V_s') + EIV_j'' V_s'' \right] dx \right.$$
$$\left. + \frac{1}{2} k_1 V_j' V_s' \big|^0 + \frac{1}{2} k_2 V_j V_s \big|^a + \frac{1}{2} k_3 V_j V_s \big|^L \right\} q_j q_s \tag{1.7.46}$$

Setting the coefficient of each δq_j in (1.7.45) equal to zero and using (1.7.46) and (1.7.43), we obtain

$$\frac{\partial I}{\partial q_j} = \sum_{s=1}^n \left(-\omega^2 m_{js} + k_{js} \right) q_s = 0 \tag{1.7.47}$$

For static problems, we have $\dot{v} = \ddot{v} = 0$ and hence (1.7.39) becomes

$$\delta I \equiv \int_0^L \left[EIv''\delta v'' + (mg - f)\delta v \right] dx + k_1 v' \delta v' \big|^0 + k_2 v \delta v \big|^a + k_3 v \delta v \big|^L = 0$$

or

$$\delta I \equiv \sum_{j=1}^n \frac{\partial I}{\partial q_j} \delta q_j = 0 \tag{1.7.48}$$

where

$$I \equiv \frac{1}{2}\int_0^L \left[EIv''^2 + 2(mg-f)v\right]dx + \frac{1}{2}k_1 v'^2\big|^0 + \frac{1}{2}k_2 v^2\big|^a + \frac{1}{2}k_3 v^2\big|^L$$

$$= \sum_{j=1}^n \sum_{s=1}^n \left\{\frac{1}{2}\int_0^L \left[EIV_j'' V_s'' q_j q_s + 2(mg-f)V_j q_j\right]dx\right.$$

$$\left. + \frac{1}{2}\left[k_1 V_j' V_s'\big|^0 + k_2 V_j V_s\big|^a + k_3 V_j V_s\big|^L\right] q_j q_s\right\} \tag{1.7.49}$$

Setting the coefficient of each δq_j in (1.7.48) equal to zero and using (1.7.49) and (1.7.43), we obtain

$$\frac{\partial I}{\partial q_j} = \sum_{s=1}^n (k_{js} q_s - f_j) = 0 \tag{1.7.50}$$

It follows from (1.7.47), (1.7.50), (1.7.46), and (1.7.49) that, for eigenvalue and static problems, the Ritz method states that

$$\frac{\partial I}{\partial q_j} = 0 \text{ for } j = 1, 2, \ldots, n \tag{1.7.51}$$

where I is a functional having the form

$$I(v) = \frac{1}{2}B(v,v) + A(v) \tag{1.7.52}$$

with the functional B being bilinear and symmetric and A being linear.

Because the natural boundary conditions on v were used in obtaining (1.7.37), the approximate solution (1.7.34) needs to satisfy the natural boundary conditions on v in order that the bilinear form (1.7.37) be free of errors. Accordingly, a single-term expansion of v cannot be obtained by using admissible functions for V_s, which do not satisfy the natural boundary conditions on v. Even using n admissible functions one cannot satisfy the governing equation and/or the natural boundary conditions by approximating v as in (1.7.34). In other words, n terms ($n \neq 1$) are usually needed and n ordinary-differential equations need to be solved.

Next, we illustrate the application of the Ritz method to the static problem (1.7.4a-c) when $EI = 1$, $mg - f = 1$, $k_1 = k_3 = 1$, $k_2 = 0$, and $L = 1$. The governing equation (1.7.4a) becomes

$$v^{iv} = -1 \tag{1.7.53}$$

and the boundary conditions (1.7.4b,c) become

$$v = 0, \; v'' - v' = 0 \text{ at } x = 0; \text{ and } v - v''' = 0, \; v'' = 0 \text{ at } x = 1 \tag{1.7.54}$$

Integrating (1.7.53) and using the boundary conditions (1.7.54), we obtain the exact solution

$$v = -\frac{13}{56}x - \frac{13}{112}x^2 + \frac{41}{336}x^3 - \frac{1}{24}x^4 \tag{1.7.55}$$

40 INTRODUCTION

To apply the Ritz method, we choose the trial functions $V_s = x^s$, which satisfy the system geometric boundary condition (i.e., $v = 0$ at $x = 0$) but not the natural boundary conditions. We assume an approximate solution of the form

$$v = \sum_{i=1}^{4} x^s q_s \tag{1.7.56}$$

Substituting for the V_s in (1.7.49) and recalling the considered system parameters, we obtain

$$I = \frac{1}{2} \int_0^L \left[EI v''^2 + 2(mg - f)v \right] dx + \frac{1}{2} k_1 v'^2 \Big|^0 + \frac{1}{2} k_3 v^2 \Big|^L$$

$$= \frac{1}{2} \int_0^1 \left[\left(2q_2 + 6q_3 x + 12 q_4 x^2 \right)^2 + 2\left(q_1 x + q_2 x^2 + q_3 x^3 + q_4 x^4 \right) \right] dx$$

$$+ \frac{1}{2} q_1^2 + \frac{1}{2} (q_1 + q_2 + q_3 + q_4)^2 \tag{1.7.57}$$

Substituting (1.7.57) into (1.7.51) yields the four linear algebraic equations

$$2q_1 + q_2 + q_3 + q_4 = -\frac{1}{2}$$

$$q_1 + 5q_2 + 7q_3 + 9q_4 = -\frac{1}{3}$$

$$q_1 + 7q_2 + 13q_3 + 19q_4 = -\frac{1}{4}$$

$$q_1 + 9q_2 + 19q_3 + 1495 q_4 = -\frac{1}{5}$$

Solving this system of equations and substituting the result into (1.7.56), we obtain the exact solution (1.7.55).

If only the trial functions x, x^2, and x^3 were used, then one would obtain

$$v = -\frac{13}{56} x - \frac{25}{336} x^2 + \frac{13}{336} x^3 \tag{1.7.58}$$

If the trial functions x, x^3, and x^4 were used, then one would obtain

$$v = -\frac{39}{160} x - \frac{37}{960} x^3 + \frac{5}{192} x^4 \tag{1.7.59}$$

Neither (1.7.58) nor (1.7.59) satisfies the natural boundary conditions and the governing equation of the system.

Alternatively, again we only use the trial functions x, x^3, and x^4 but insist that (1.7.56) satisfy the governing equation (1.7.53). This fixes q_4 to be $-\frac{1}{24}$. Then, we determine the remaining constants q_1 and q_3 from

$$\frac{\partial I}{\partial q_1} = \frac{\partial I}{\partial q_3} = 0$$

The result would be
$$v = -\frac{13}{50}x + \frac{37}{600}x^3 - \frac{1}{24}x^4 \qquad (1.7.60)$$
which does not satisfy the natural boundary conditions.

Finally, we assume that
$$v = xq_1 + x^2 q_2 + x^3 q_3 + x^5 q_5 \qquad (1.7.61)$$
To determine the four constants, let us demand that v satisfy the natural boundary conditions in (1.7.54) and
$$\frac{\partial I}{\partial q_1} = 0$$
The result would be
$$v = -\frac{227711}{1014216}\left(x + \frac{1}{2}x^2 - \frac{89}{254}x^3 + \frac{7}{127}x^5\right) \qquad (1.7.62)$$
which does not satisfy the governing equation (1.7.53).

It is clear from the above approximate solutions that the Ritz method results in a solution that may not satisfy the system equation and/or natural boundary conditions.

If the selected trial functions are three times differentiable, then one does not need to perform the integration by parts to transfer some of the derivatives from v to the test functions. Instead, one can use the variational
$$\delta I = \int_0^L \left[(EIv'')'' + mg - f\right]\delta v\, dx + (k_1 v' - EIv'')\delta v' \,|^0$$
$$+ \left[k_3 v - (EIv'')'\right]\delta v \,|^L + EIv''\delta v' \,|^L \qquad (1.7.63)$$

We note that the integrand contains the governing equation and the terms outside the integrand contain the natural boundary conditions. Hence, if we can force the approximate solution to satisfy the governing equation, then $\delta I = 0$ will ensure the satisfaction of the natural boundary conditions. This is the essence of the penalty method. In the penalty method, one uses a large specified number (the so-called penalty parameter) λ to define an augmented functional $\tilde{I}$ as
$$\tilde{I} = I + \frac{1}{2}\lambda \int_D (\mathcal{L}v - f)^2 dD \qquad (1.7.64)$$

Then demanding that $\partial \tilde{I}/\partial q_s = 0$ yields the required equations to determine the q_j.

In the Ritz method, because the trial functions need to satisfy the system geometric boundary conditions and be valid for the whole structure, they are usually not polynomial functions and hence the integrand in the functional I is complicated and often the integral needs to be evaluated numerically. Also, if the external loads are highly irregular, many trial functions are needed in order to produce a reasonable accuracy. Moreover, for complex built-up structures, there are no analytical admissible functions. Using the finite element method, one can take full advantage of weak (or energy) formulations, use polynomial trial functions, and integrate exactly the terms in I because the discretization is done for subdomains of the structure rather than for the whole structure.

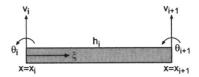

Fig. 1.7.2 A finite beam element.

1.7.4 Finite Element Method

The finite element method makes use of variational methods, very often the Ritz method, to discretize the subdomains (the so-called elements) of a continuous system and uses the displacements at some nodes on the boundary or even in the interior of the subdomains as generalized coordinates. In other words, the finite element method is a piecewise application of a variational method, but only one trial function is used for each element, which consists of interpolation functions (shape functions) and undetermined parameters or generalized coordinates. The interpolation functions are polynomials derived using the concept that any continuous function can be represented by a linear combination of polynomials and the undetermined parameters represent the values of the solution at a finite number of preselected nodes on the boundary and in the interior of the element.

For the one-dimensional problem shown in (1.7.4a,b,c), we choose to use two nodes and four degrees of freedom v_i, θ_i, v_{i+1}, and θ_{i+1} (the so-called primary variables) for the ith element, as shown in Figure 1.7.2. Also, we divide the whole structure into N elements and define a nondimensional local coordinate ξ as

$$\xi = \frac{x - (x_i + x_{i+1})/2}{h_i/2} \qquad (1.7.65)$$

where h_i denotes the length of the ith element. It follows from (1.7.65) that $x = x_i$ corresponds to $\xi = -1$ and $x = x_{i+1}$ corresponds to $\xi = 1$. Because there are four nonhomogeneous geometric boundary conditions, namely,

$$v(x_i) = v_i, \ v(x_{i+1}) = v_{i+1}, \ v'(x_i) = \theta_i, \ v'(x_{i+1}) = \theta_{i+1} \qquad (1.7.66)$$

we approximate v over the ith element by a cubic polynomial as

$$v = c_0 + c_1 \xi + c_2 \xi^2 + c_3 \xi^3 \qquad (1.7.67)$$

It follows from (1.7.66) and (1.7.67) that

$$v = \{N\}^T \{d^{(i)}\} \qquad (1.7.68)$$

where

$$\{d^{(i)}\} \equiv \{v_i, \ \theta_i, \ v_{i+1}, \ \theta_{i+1}\}^T$$

$$\{N\} \equiv \{N_1,\ N_2,\ N_3,\ N_4\}^T$$
$$N_1 = \frac{1}{4}(2 - 3\xi + \xi^3), \quad N_2 = \frac{1}{4}(1 - \xi - \xi^2 + \xi^3)$$
$$N_3 = \frac{1}{4}(2 + 3\xi - \xi^3), \quad N_4 = \frac{1}{4}(-1 - \xi + \xi^2 + \xi^3) \quad (1.7.69)$$

Here, $\{d^{(i)}\}$ is the nodal displacement vector of the ith element and the N_k are usually called shape or interpolation functions. The special shape functions in (1.7.69) are called Hermite cubics. Comparing (1.7.68) with (1.7.34), we note that only one trial function is used in the finite element method. It includes the unknown nodal displacements as coefficients.

It follows from (1.7.65) that

$$dx = \frac{1}{2}h_i d\xi \quad (1.7.70)$$

and it follows from (1.7.39) that

$$\sum_{i=1}^{N} \int_{x_i}^{x_{i+1}} \left[m\ddot{v}\delta v + j\ddot{v}'\delta v' + EIv''\delta v'' - (f - mg - c\dot{v})\delta v \right] dx$$
$$+ k_1 v' \delta v' \big|^{i=1, \xi=-1} + k_2 v \delta v \big|^{i=I, \xi=-1} + k_3 v \delta v \big|^{i=N, \xi=1} = 0 \quad (1.7.71)$$

where I is the number of the element whose left end is at $x = a$. Substituting (1.7.68) and (1.7.70) into (1.7.71) and setting the coefficients of the $\delta\{d^{(i)}\}^T$ equal to zero, we obtain

$$\sum_{i=1}^{N} \{\delta d^{(i)}\}^T \left([m^{(i)}]\{\ddot{d}^{(i)}\} + [c^{(i)}]\{\dot{d}^{(i)}\} + [k^{(i)}]\{d^{(i)}\} - \{f^{(i)}\} \right)$$
$$+ \{\delta d^{(1)}\}^T [\tilde{k}^{(1)}]\{d^{(1)}\} + \{\delta d^{(I)}\}^T [\tilde{k}^{(I)}]\{d^{(I)}\} + \{\delta d^{(N)}\}^T [\tilde{k}^{(N)}]\{d^{(N)}\} = 0 \quad (1.7.72)$$

where

$$[m^{(i)}] = \int_{-1}^{1} \left(\frac{h_i}{2}\{N\}m\{N\}^T + \frac{2}{h_i} \frac{\partial\{N\}}{\partial \xi} j \frac{\partial\{N\}^T}{\partial \xi} \right) d\xi$$

$$[c^{(i)}] = \frac{h_i}{2} \int_{-1}^{1} \{N\}c\{N\}^T d\xi$$

$$[k^{(i)}] = \frac{8}{h_i^3} \int_{-1}^{1} \frac{\partial^2\{N\}}{\partial \xi^2} EI \frac{\partial^2\{N\}^T}{\partial \xi^2} d\xi$$

$$\{f^{(i)}\} = \frac{h_i}{2} \int_{-1}^{1} \{N\}(f - mg) d\xi$$

$$[\tilde{k}^{(1)}] = \frac{4}{h_1^2} \frac{\partial\{N\}}{\partial \xi} k_1 \frac{\partial\{N\}^T}{\partial \xi} \big|_{\xi=-1} = \begin{bmatrix} 0 & 0 & 0 & 0 \\ 0 & 4k_1/h_1^2 & 0 & 0 \\ 0 & 0 & 0 & 0 \\ 0 & 0 & 0 & 0 \end{bmatrix}$$

44 INTRODUCTION

$$\left[\tilde{k}^{(I)}\right] = \{N\}k_2\{N\}^T \mid^{\xi=-1} = \begin{bmatrix} k_2 & 0 & 0 & 0 \\ 0 & 0 & 0 & 0 \\ 0 & 0 & 0 & 0 \\ 0 & 0 & 0 & 0 \end{bmatrix}$$

$$\left[\tilde{k}^{(N)}\right] = \{N\}k_3\{N\}^T \mid^{\xi=1} = \begin{bmatrix} 0 & 0 & 0 & 0 \\ 0 & 0 & 0 & 0 \\ 0 & 0 & k_3 & 0 \\ 0 & 0 & 0 & 0 \end{bmatrix} \qquad (1.7.73)$$

We note that the elemental matrices $[m^{(j)}], [c^{(j)}]$, and $[k^{(j)}]$ are 4×4 symmetric matrices.

Next, we define the global displacement vector as

$$\{D\} \equiv \{v_1, \theta_1, \ldots, v_{N+1}, \theta_{N+1}\}^T \qquad (1.7.74)$$

and rewrite (1.7.72) in terms of the global displacement vector as

$$\{\delta D\}^T ([M]\{\ddot{D}\} + [C]\{\dot{D}\} + [K]\{D\} - \{F\}) = 0$$

where

$$[M] = \begin{bmatrix}
m_{11}^{(1)} & m_{12}^{(1)} & m_{13}^{(1)} & m_{14}^{(1)} & & & & \\
m_{21}^{(1)} & m_{22}^{(1)} & m_{23}^{(1)} & m_{24}^{(1)} & & & & \\
m_{31}^{(1)} & m_{32}^{(1)} & m_{33}^{(1)}+m_{11}^{(2)} & m_{34}^{(1)}+m_{12}^{(2)} & m_{13}^{(2)} & m_{14}^{(2)} & & 0 \\
m_{41}^{(1)} & m_{42}^{(1)} & m_{43}^{(1)}+m_{21}^{(2)} & m_{44}^{(1)}+m_{22}^{(2)} & m_{23}^{(2)} & m_{24}^{(2)} & & \\
 & & m_{31}^{(2)} & m_{32}^{(2)} & m_{33}^{(2)}+m_{11}^{(3)} & m_{34}^{(2)}+m_{12}^{(3)} & & \\
 & & m_{41}^{(2)} & m_{42}^{(2)} & m_{43}^{(2)}+m_{21}^{(3)} & m_{44}^{(2)}+m_{22}^{(3)} & & \\
 & & & & & & & \\
 & & & & 0 & & & \\
\end{bmatrix}$$

$$\begin{bmatrix}
 & & & & & & & & \\
 & & 0 & & & & & & \\
m_{33}^{(N2)}+m_{11}^{(N1)} & m_{34}^{(N2)}+m_{12}^{(N1)} & m_{13}^{(N1)} & m_{14}^{(N1)} & & & \\
m_{43}^{(N2)}+m_{21}^{(N1)} & m_{44}^{(N2)}+m_{22}^{(N1)} & m_{23}^{(N1)} & m_{24}^{(N1)} & & & \\
m_{31}^{(N1)} & m_{32}^{(N1)} & m_{33}^{(N1)}+m_{11}^{(N)} & m_{34}^{(N1)}+m_{12}^{(N)} & m_{13}^{(N)} & m_{14}^{(N)} \\
0 \quad m_{41}^{(N1)} & m_{42}^{(N1)} & m_{43}^{(N1)}+m_{21}^{(N)} & m_{44}^{(N1)}+m_{22}^{(N)} & m_{23}^{(N)} & m_{24}^{(N)} \\
 & & m_{31}^{(N)} & m_{32}^{(N)} & m_{33}^{(N)} & m_{34}^{(N)} \\
 & & m_{41}^{(N)} & m_{42}^{(N)} & m_{43}^{(N)} & m_{44}^{(N)} \\
\end{bmatrix}$$

$$(1.7.75)$$

ANALYSES OF LINEAR CONTINUOUS SYSTEMS

$$[C] = \begin{bmatrix} c_{11}^{(1)} & c_{12}^{(1)} & c_{13}^{(1)} & & c_{14}^{(1)} & & & & \\ c_{21}^{(1)} & c_{22}^{(1)} & c_{23}^{(1)} & & c_{24}^{(1)} & & & & \\ c_{31}^{(1)} & c_{32}^{(1)} & c_{33}^{(1)}+c_{11}^{(2)} & c_{34}^{(1)}+c_{12}^{(2)} & c_{13}^{(2)} & & c_{14}^{(2)} & & 0 \\ c_{41}^{(1)} & c_{42}^{(1)} & c_{43}^{(1)}+c_{21}^{(2)} & c_{44}^{(1)}+c_{22}^{(2)} & c_{23}^{(2)} & & c_{24}^{(2)} & & \\ & & c_{31}^{(2)} & c_{32}^{(2)} & c_{33}^{(2)}+c_{11}^{(3)} & c_{34}^{(2)}+c_{12}^{(3)} & & & \\ & & c_{41}^{(2)} & c_{42}^{(2)} & c_{43}^{(2)}+c_{21}^{(3)} & c_{44}^{(2)}+c_{22}^{(3)} & & & \\ & & & & & & & & \\ & & 0 & & & & & & \end{bmatrix}$$

$$\begin{bmatrix} & & & & & & & 0 & & & \\ c_{33}^{(N2)}+c_{11}^{(N1)} & c_{34}^{(N2)}+c_{12}^{(N1)} & c_{13}^{(N1)} & & c_{14}^{(N1)} & & & \\ c_{43}^{(N2)}+c_{21}^{(N1)} & c_{44}^{(N2)}+c_{22}^{(N1)} & c_{23}^{(N1)} & & c_{24}^{(N1)} & & & \\ c_{31}^{(N1)} & c_{32}^{(N1)} & c_{33}^{(N1)}+c_{11}^{(N)} & c_{34}^{(N1)}+c_{12}^{(N)} & c_{13}^{(N)} & c_{14}^{(N)} \\ 0 \quad c_{41}^{(N1)} & c_{42}^{(N1)} & c_{43}^{(N1)}+c_{21}^{(N)} & c_{44}^{(N1)}+c_{22}^{(N)} & c_{23}^{(N)} & c_{24}^{(N)} \\ & & c_{31}^{(N)} & c_{32}^{(N)} & c_{33}^{(N)} & c_{34}^{(N)} \\ & & c_{41}^{(N)} & c_{42}^{(N)} & c_{43}^{(N)} & c_{44}^{(N)} \end{bmatrix}$$

$$(1.7.76)$$

$$[K] = \begin{bmatrix} k_{11}^{(1)} & k_{12}^{(1)} & k_{13}^{(1)} & k_{14}^{(1)} & & & & \\ k_{21}^{(1)} & k_{22}^{(1)}+4k_1/h_1^2 & k_{23}^{(1)} & k_{24}^{(1)} & & & & \\ k_{31}^{(1)} & k_{32}^{(1)} & k_{33}^{(1)}+k_{11}^{(2)} & k_{34}^{(1)}+k_{12}^{(2)} & k_{13}^{(2)} & k_{14}^{(2)} & & 0 \\ k_{41}^{(1)} & k_{42}^{(1)} & k_{43}^{(1)}+k_{21}^{(2)} & k_{44}^{(1)}+k_{22}^{(2)} & k_{23}^{(2)} & k_{24}^{(2)} & & \\ & & k_{31}^{(2)} & k_{32}^{(2)} & k_{33}^{(2)}+k_{11}^{(3)} & k_{34}^{(2)}+k_{12}^{(3)} & & \\ & & k_{41}^{(2)} & k_{42}^{(2)} & k_{43}^{(2)}+k_{21}^{(3)} & k_{44}^{(2)}+k_{22}^{(3)} & & \\ & & & & & & & \\ & & 0 & & & & & \end{bmatrix}$$

$$\begin{bmatrix} & & & & & & 0 & & \\ k_{33}^{(I2)}+k_{11}^{(I1)} & k_{34}^{(I2)}+k_{12}^{(I1)} & k_{13}^{(I1)} & & k_{14}^{(I1)} & & \\ k_{43}^{(I2)}+k_{21}^{(I1)} & k_{44}^{(I2)}+k_{22}^{(I1)} & k_{23}^{(I1)} & & k_{24}^{(I1)} & & \\ k_{31}^{(I1)} & k_{32}^{(I1)} & k_{33}^{(I1)}+k_{11}^{(I)}+k_2 & k_{34}^{(I1)}+k_{12}^{(I)} \\ 0 \quad k_{41}^{(I1)} & k_{42}^{(I1)} & k_{43}^{(I1)}+k_{21}^{(I)} & k_{44}^{(I1)}+k_{22}^{(I)} \end{bmatrix}$$

$$\begin{bmatrix} k_{33}^{(N2)}+k_{11}^{(N1)} & k_{34}^{(N2)}+k_{12}^{(N1)} & k_{13}^{(N1)} & 0 & & & \\ k_{43}^{(N2)}+k_{21}^{(N1)} & k_{44}^{(N2)}+k_{22}^{(N1)} & k_{23}^{(N1)} & k_{24}^{(N1)} & & & \\ k_{31}^{(N1)} & k_{32}^{(N1)} & k_{33}^{(N1)}+k_{11}^{(N)} & k_{34}^{(N1)}+k_{12}^{(N)} & k_{13}^{(N)} & k_{14}^{(N)} \\ 0 & k_{41}^{(N1)} & k_{42}^{(N1)} & k_{43}^{(N1)}+k_{21}^{(N)} & k_{44}^{(N1)}+k_{22}^{(N)} & k_{23}^{(N)} & k_{24}^{(N)} \\ & & & k_{31}^{(N)} & k_{32}^{(N)} & k_{33}^{(N)}+k_3 & k_{34}^{(N)} \\ & & & k_{41}^{(N)} & k_{42}^{(N)} & k_{43}^{(N)} & k_{44}^{(N)} \end{bmatrix}$$

(1.7.77)

$$\{F\} = \{f_{11}^{(1)},\ f_{12}^{(1)},\ f_{21}^{(1)}+f_{11}^{(2)},\ f_{22}^{(1)}+f_{12}^{(2)},\ \ldots,$$
$$f_{21}^{(N-1)}+f_{11}^{(N)},\ f_{22}^{(N-1)}+f_{12}^{(N)},\ f_{21}^{(N)},\ f_{22}^{(N)}\}^T \quad (1.7.78)$$

and $N1 \equiv N-1$, $N2 \equiv N-2$, $I1 \equiv I-1$, and $I2 \equiv I-2$. Hence the governing equation is

$$[M]\{\ddot{D}\} + [C]\{\dot{D}\} + [K]\{D\} = \{F\} \quad (1.7.79)$$

We note that the global matrices $[M], [C]$, and $[K]$ are band-limited $(2N+2) \times (2N+2)$ matrices and, when the number of nodal displacements (primary variables) increases by r, they become $(2N+2+r) \times (2N+2+r)$ matrices with the m_{ij}, c_{ij}, and k_{ij} ($i,j = 1, 2, \ldots, 2N+2$) being changed, which is different from the Ritz method (see (1.7.43)). Next, we apply the geometric boundary condition

$$v_1 = 0 \quad (1.7.80)$$

and reduce the discretized system of equations (1.7.79) to

$$[\bar{M}]\{\ddot{\bar{D}}\} + [\bar{C}]\{\dot{\bar{D}}\} + [\bar{K}]\{\bar{D}\} = \{\bar{F}\} \quad (1.7.81)$$

where $[\bar{M}], [\bar{C}]$, and $[\bar{K}]$ are symmetric matrices obtained from $[M], [C]$, and $[K]$ by deleting their first rows and columns, and $\{\bar{D}\}$ and $\{\bar{F}\}$ are obtained from $\{D\}$ and $\{F\}$ be deleting their first elements. Then, one can use the methods introduced in Section 1.5.2 to solve (1.7.81).

1.7.5 Weighted Residual Methods

There are many physical systems whose characteristics cannot be put in energy forms. For example, there are no energy forms for a system governed by the Navier-Stokes equations. The method of weighted residuals is appropriate for such systems.

The first step in applying weighted residual methods to (1.7.33) and its associated boundary conditions consists of approximating the solution v in the form (1.7.34), where $V_1, V_2, \ldots, V_n$ are n functions chosen from a complete set of orthonormal functions. These functions are usually called trial or approximating functions. Substituting (1.7.34) into (1.7.33) and assuming that the V_s can be differentiated as needed in the operator $\mathcal{L}$, we obtain the residual (resulting error) R_n as

$$R_n = \sum_{s=1}^{n} \mathcal{L}(V_s q_s) - f \quad (1.7.82)$$

The next step consists of choosing n functions $\psi_1, \psi_2, \ldots, \psi_n$ from another complete set of orthonormal functions and requiring the residual to be orthogonal to the ψ_j for all $j = 1, 2, \ldots, n$. The ψ_j are usually called the weighting or test functions. The result is

$$\int_D \left[\sum_{s=1}^n \mathcal{L}(V_s q_s) - f \right] \psi_j dD = 0 \qquad (1.7.83)$$

which constitutes a system of n equations for the n unknowns q_s.

In weighted residual methods, the test functions ψ_j can be selected, independent of the trial functions V_s, from any orthogonal function set. This flexibility is advantageous in certain nonlinear problems. However, in weighted residual methods, none of the differentiations is transferred from V_s to ψ_j and hence the trial functions V_s need to satisfy both the geometric and natural boundary conditions; that is, the V_s are comparison functions. Moreover, in weighted residual methods, the matrices $[M]$, $[C]$, and $[K]$ of the discretized system of equations (1.5.23) are usually asymmetric.

For example, if the system (1.7.4a-c) is discretized using a weighted residual method, the discretized system is given by (1.5.23) with

$$m_{js} = \int_0^L \psi_j \left[mV_s - (jV_s')' \right] dx, \quad c_{js} = \int_0^L \psi_j c V_s dx$$

$$k_{js} = \int_0^L \psi_j (EIV_s'')'' dx + \psi_j k_2 V_s \big|^a, \quad f_j = \int_0^L \psi_j (f - mg) dx \qquad (1.7.84)$$

We note that $[M]$, $[C]$, and $[K]$ are not symmetric even though the system is self-adjoint.

There are a number of different weighted residual methods, the difference being in the nature of the trial functions V_s and test functions ψ_j. If the trial functions do not satisfy the m boundary conditions, one minimizes the residuals arising from all of the m boundary conditions and uses only the first $n - m$ equations in (1.7.83). This method is called the tau method (Lanczos, 1986; Fox and Parker, 1968); it will not be discussed further because it is rarely used in structural mechanics. Next, we describe five different weighted residual methods.

The Galerkin Method: The Galerkin method uses eigenfunctions or comparison functions as trial functions and the test functions are chosen from the set of trial functions; that is,

$$\psi_j = V_j \qquad (1.7.85)$$

For self-adjoint systems, one can use the system boundary conditions and the V_j being comparison functions to show that the resulting $[M]$, $[C]$, and $[K]$ matrices are symmetric and that the Galerkin method reduces to the Ritz method if comparison functions are used in the latter.

If $\psi_j (\neq V_j)$ are used, the resulting version is called the Petrov-Galerkin method.

The Collocation Method: In this method, the test functions are chosen to be

$$\psi_j = \hat{\delta}(\mathbf{r} - \mathbf{r}_i), \quad j = 1, 2, \ldots, n \qquad (1.7.86)$$

where $\hat{\delta}$ is the spatial Dirac delta function. This method ensures that the residual of the system is equal to zero at the selected n points $\mathbf{r}_1, \mathbf{r}_2, \ldots, \mathbf{r}_n$.

The Subdomain Method: In this method, the system domain is divided into n subdomains and the test functions are chosen to be

$$\psi_j = 1 \text{ if } \mathbf{r} \text{ is within the } j\text{th subdomain}$$
$$\psi_j = 0 \text{ if } \mathbf{r} \text{ is outside the } j\text{th subdomain} \quad (1.7.87)$$

This method ensures that the average residual in each subdomain is zero.

The Least Squares Method: In this method, the test functions are chosen to be

$$\psi_j = \frac{\partial(\mathcal{L}v - f)}{\partial q_j} \quad (1.7.88)$$

which is equivalent to minimizing one-half of the square of the residual over the system domain; that is, minimizing the functional

$$\frac{1}{2} \int_D (\mathcal{L}v - f)^2 dD \quad (1.7.89)$$

The Moments Method: Because $\{x^i \mid i = 0, 1, \ldots, n\}$ is a set of convenient orthogonal functions, one can choose the weighting functions to be

$$\psi_j = x^{j-1} \quad (1.7.90)$$

Of course, one can use the concept of finite elements and apply this method on the subdomains of any continuous system to construct a finite element formulation.

1.7.6 Initial-Value Methods

These methods are well suited for solving for the linear frequencies or buckling loads of structures that can be characterized by a single spatial variable. In describing as well as implementing these methods, one needs to represent the problem as a system of first-order equations; that is, in state-space form. For example, using the variables

$$\zeta_1 = V, \quad \zeta_2 = V', \quad \zeta_3 = EIV'', \quad \text{and} \quad \zeta_4 = (EIV'')' \quad (1.7.91)$$

we rewrite (1.7.6) as

$$\{\zeta\}' = [F(x;\omega)]\{\zeta\} \quad (1.7.92)$$

$$[R_1]\{\zeta\} = \{0\} \text{ at } x = 0 \quad (1.7.93)$$

$$[R_2]\{\zeta\} = \{0\} \text{ at } x = L \quad (1.7.94)$$

where $\{\zeta\} = \{\zeta_1, \zeta_2, \zeta_3, \zeta_4\}^T$, and the matrices $[F]$, $[R_1]$, and $[R_2]$ are given by

$$[F] = \begin{bmatrix} 0 & 1 & 0 & 0 \\ 0 & 0 & (EI)^{-1} & 0 \\ 0 & 0 & 0 & 1 \\ m\omega^2 - k_2\hat{\delta}(x-a) & -j'\omega^2 & -j\omega^2(EI)^{-1} & 0 \end{bmatrix} \quad (1.7.95a)$$

$$[R_1] = \begin{bmatrix} 1 & 0 & 0 & 0 \\ 0 & k_1 & -1 & 0 \end{bmatrix} \quad (1.7.95b)$$

$$[R_2] = \begin{bmatrix} 0 & 0 & 1 & 0 \\ k_3 & -j\omega^2 & 0 & -1 \end{bmatrix} \quad (1.7.95c)$$

As a second example, using the variables in (1.7.91), we rewrite (1.7.10) as in (1.7.92)-(1.7.94) where

$$[F] = \begin{bmatrix} 0 & 1 & 0 & 0 \\ 0 & 0 & (EI)^{-1} & 0 \\ 0 & 0 & 0 & 1 \\ m\omega^2 & \alpha & 0 & 0 \end{bmatrix} \quad (1.7.96a)$$

$$[R_1] = \begin{bmatrix} 1 & 0 & 0 & 0 \\ 0 & 1 & 0 & 0 \end{bmatrix} \quad (1.7.96b)$$

$$[R_2] = \begin{bmatrix} 0 & 0 & 1 & 0 \\ 0 & 0 & 0 & 1 \end{bmatrix} \quad (1.7.96c)$$

Because the problem is linear, the method of superposition holds. To apply this method to (1.7.92)-(1.7.94), we need to calculate four linearly independent solutions of (1.7.92) of the form $\{z\}_1, \{z\}_2, \{z\}_3$, and $\{z\}_4$. Then, the general solution of (1.7.92) can be expressed as

$$\{\zeta(x)\} = c_1\{z(x)\}_1 + c_2\{z(x)\}_2 + c_3\{z(x)\}_3 + c_4\{z(x)\}_4 \quad (1.7.97)$$

In calculating the $\{z\}_n$, one usually starts from one boundary and performs the integration until the second boundary is reached. This is the reason why these methods are called shooting methods. Enforcing the boundary conditions (1.7.93) and (1.7.94) yields four homogeneous algebraic equations for the c_m. The vanishing of the determinant of their coefficient matrix provides the dispersion relation needed to determine ω.

Because the boundary conditions (1.7.93) and (1.7.94) are disjointed (separated), we need only to calculate two linearly independent solutions $\{z\}_1$ and $\{z\}_2$ if their initial conditions are chosen to satisfy the boundary conditions (1.7.93) for integrations starting at $x = 0$. To accomplish this, one uses two linearly independent vector solutions of (1.7.93) as initial conditions for $\{z\}_1$ and $\{z\}_2$. Then, the general solution of (1.7.92) that satisfies (1.7.93) can be expressed as

$$\{\zeta(x)\} = c_1\{z(x)\}_1 + c_2\{z(x)\}_2 \quad (1.7.98)$$

or in matrix form as
$$\{\zeta(x)\} = [Z(x)]\{c\} \tag{1.7.99}$$

where
$$\{c\}^T = \{c_1 \ c_2\} \tag{1.7.100}$$

and the columns of the matrix $[Z]$ are the $\{z\}_m$; that is,
$$[Z(x)] = [\{z(x)\}_1 \ \{z(x)\}_2] \tag{1.7.101}$$

Enforcing the boundary conditions (1.7.94) yields
$$[R_2][Z(L)]\{c\} = \{0\} \tag{1.7.102}$$

Equation (1.7.102) represents a system of two linear homogeneous algebraic equations for the two unknown constants c_1 and c_2. For a nontrivial solution, the determinant $| [R_2][Z(L)] |$ of the coefficient matrix $[R_2][Z(L)]$ must vanish, thereby providing the dispersion relation.

The matrix $[Z]$ is numerically ill-conditioned because the matrix $[F]$ is real and all of its diagonal elements are zero so that its trace is zero. Moreover, the complex eigenvalues of $[F]$ occur in conjugate pairs, and for each real eigenvalue there exists another real eigenvalue of the opposite sign. As L increases, the growth rates of the elements $e^{\lambda_i L}$ (λ_i are the eigenvalues of $[F]$) grow exponentially, while the decay rates of the elements decay exponentially. Hence the calculated solutions lose their numerical independence and the numerical method fails.

As an example illustrating the difficulty arising from the finite word length used by computers, we follow Scott and Watts (1975) and consider the solutions of
$$y'' - y = 0 \tag{1.7.103}$$

Choosing the linearly independent initial conditions (a) $y(0) = 1$ and $\dot{y}(0) = 0$ and (b) $y(0) = 0$ and $\dot{y}(0) = 1$, we find $y_1(x) = \cosh x$ and $y_2(x) = \sinh x$ to be mathematically linearly independent solutions of (1.7.103). However, using a computer that carries six digits, we find that $\cosh x$ and $\sinh x$ are identical for values of x greater than approximately 8. On the other hand, using a computer that carries 14 digits, we find that they are indistinguishable from each other when x exceeds approximately 17.

As an example illustrating the difficulty arising from the presence of rapidly growing solutions, we consider after Bellman and Kalaba (1965) the problem
$$y^{iv} - 24y''' - 169y'' - 324y' - 180y = 0 \tag{1.7.104}$$

whose general solution is
$$y = c_1 e^{-x} + c_2 e^{-2x} + c_3 e^{-3x} + c_4 e^{30x} \tag{1.7.105}$$

We note that the characteristic values associated with the differential operator in (1.7.104) are $(-1, -2, -3, 30)$, which differ greatly in their real parts. Hence, we expect the rapidly growing solution $\exp(30x)$ to dominate the other solutions and hence prevent the accurate numerical calculation of solutions that satisfy the given boundary conditions. To illustrate this point, Bellman and Kalaba (1965) chose the mathematical particular solution

$$y(x) = e^{-x} - 2e^{-2x} + e^{-3x} \tag{1.7.106}$$

which is a special case of (1.7.105) and satisfies the boundary conditions

$$y(0) = 0, y'(0) = 0, y(1) = 0.146996, \text{ and } y'(1) = 0.0241005. \tag{1.7.107}$$

Next, they carried out the above described superposition procedure. They chose the initial conditions

$$y_1(0) = 0, y_1'(0) = 0, y_1''(1) = 1, y_1'''(1) = 0 \tag{1.7.108}$$

and

$$y_2(0) = 0, y_2'(0) = 0, y_2''(1) = 0, y_2'''(1) = 1 \tag{1.7.109}$$

Clearly, the solutions $y_1(x)$ and $y_2(x)$ of (1.7.104) that satisfy (1.7.108) and (1.7.109) also satisfy the boundary condition at $x = 0$ in (1.7.107). Hence

$$y(x) = c_1 y_1(x) + c_2 y_2(x) \tag{1.7.110}$$

Using a Runge-Kutta integration procedure with an integration step of 0.01, they obtained the numerical values

$$y_1(1) = 0.195694 \times 10^{10}, \quad y_1'(1) = 0.587083 \times 10^{11}$$
$$y_2(1) = 0.326157 \times 10^{9}, \quad y_2'(1) = 0.978472 \times 10^{10}$$

Enforcing the boundary conditions (1.7.107) at $x = 1$ and using these values, we obtain

$$c_1 = -0.609136 \times 10^{-2} \text{ and } c_2 = 0.36541 \times 10^{-1}$$

Substituting these values into (1.7.110) and using (1.7.108) and (1.7.109), we find that

$$y''(0) = c_1 = -0.609136 \times 10^{-2}$$
$$y'''(0) = c_2 = 0.3654 \times 10^{-1} \tag{1.7.111}$$

Comparing these derivatives with the exact values $y''(0) = 2$ and $y'''(0) = -12$ obtained from (1.7.106), we clearly see the drawbacks of the straightforward superposition procedure.

Certainly, using a smaller integration step and a more accurate integrator will improve the calculated values of c_1 and c_2. However, due to the finite word length of the computer, it may be difficult to keep the solution vectors linearly independent and hence calculate c_1 and c_2 to a specified accuracy. One way that helps to overcome this difficulty is to orthonormalize the calculated solution vectors whenever an impending loss of independence is detected. Godunov (1961) and Bellman and Kalaba (1965) were the first to propose the use of orthonormalization in implementing the method of superposition. Slightly modified and related procedures were discussed by Conte (1966), Falkenberg (1968), Roberts and Shipman (1972), Davey (1973), and Canosa and Penafiel (1973). Gersting and Jankowski (1972) compared Conte's version of the method of renormalization with several other numerical schemes for solving the Orr-Sommerfeld problem. Scott and Watts (1975, 1977) developed a powerful computer code called SUPORT based on Gudonov's version of the method of orthonormalization that uses a variable-step Runge-Kutta-Fehlberg integration scheme. This code has been extensively used by many researchers. It is robust, easy to use, reliable, and efficient. The disadvantages are the large amount of memory required when too many orthonormalization points are used and lengthy CPU time as compared to other methods. Next, we describe the Gudonov procedure as implemented by Hadian and Nayfeh (1990).

Following the initial-value method in conjunction with continuity, we divide the interval $x = [0, L]$ into $n+1$ nodes at $x_0, x_1, \ldots,$ and x_n, where $x_0 = 0$ and $x_n = L$. From (1.7.93) one can find two orthonormal vectors at $x = 0$ and express the initial value of $\{\zeta\}$ as a linear combination of these vectors as

$$\{\zeta(0)\} = c_1 \{z(0)\}_1^{(1)} + c_2 \{z(0)\}_2^{(1)} \quad (1.7.112)$$

The general solution of (1.7.92) that satisfies (1.7.93) can be expressed in the form

$$\{\zeta(x)\} = c_1 \{z(x)\}_1^{(1)} + c_2 \{z(x)\}_2^{(1)} \quad (1.7.113)$$

These two linearly independent vectors are then evaluated at x_1 and orthonormalized by the modified Gram-Schmidt procedure (Noble and Daniel, 1977). These orthonormalized vectors are used as initial conditions for the second interval $[x_1, x_2]$ and two linearly independent solution vectors are evaluated at x_2. The process is then continued until two linearly independent vectors $\{z(x)\}_1^{(n)}$ and $\{z(x)\}_2^{(n)}$ at x_n are found. The solution at $x = L$ is expressed as a linear combination of these vectors in the form

$$\{\zeta(x)\} = c_1^{(n)} \{z(x)\}_1^{(n)} + c_2^{(n)} \{z(x)\}_2^{(n)} \quad (1.7.114)$$

Substitution of (1.7.114) into (1.7.94) yields

$$[R_2][Z]^{(n)} \{c\}^{(n)} = \{0\} \quad (1.7.115)$$

where $[Z]^{(n)}$ is a 4×2 real matrix whose columns are the $\{z(L)\}_m^{(n)}$ and $\{c\}^{(n)}$ is a constant column vector of length two. For a nontrivial solution,

$$\mid [R_2][Z]^{(n)} \mid = 0 \quad (1.7.116)$$

from which all of the natural frequencies or buckling loads are determined. This method takes a little more computer time due to the orthonormalization process but it yields all of the desired natural frequencies or buckling loads. The total computation time is reduced greatly by increasing the number of nodes even though the number of orthonormalizations is increased, because the absolute value of $\lambda_i L$ is decreased and hence the computer carries smaller numbers.

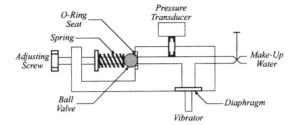

Fig. 1.8.1 An experimental relief valve.

1.8 ANALYSES OF NONLINEAR CONTINUOUS SYSTEMS

There are two general approaches for treating nonlinear continuous systems. In the first approach, which is the most widely used one, the partial-differential equations and boundary conditions are first discretized by using either a variational or weighted residual method to obtain a system of infinite nonlinearly coupled ordinary-differential equations. These equations are usually truncated and then solved either numerically or analytically. In the second approach, the partial-differential equations and boundary conditions are treated directly either analytically or numerically. However, discretizing nonlinear continuous systems first and then treating the resulting nonlinear ordinary-differential equations may lead to erroneous results, especially in the case of quadratic nonlinearities. On the other hand, attacking directly the nonlinear partial-differential equations and boundary conditions avoids the pitfalls associated with the discretization approach (Nayfeh, J. Nayfeh, and Mook, 1992; Nayfeh, S. Nayfeh, and Pakdemirli, 1995; Pakdemirli, S. Nayfeh, and Nayfeh, 1995; Chin and Nayfeh, 1996a,b; Nayfeh, 1997; Nayfeh and Lacarbonara, 1997, 1998; Rega et al., 1997, 1999; Lacarbonara, Nayfeh, and Kreider, 1998; Lacarbonara, 1998; Arafat and Nayfeh, 1999, 2003; Alhazza and Nayfeh, 2001; Nayfeh and Alhazza, 2001; Nayfeh et al., 2002). Nayfeh (1998) developed a method for producing reduced-order models that overcomes the shortcomings of the Galerkin procedure. Treatment of these models yields results that are in agreement with those obtained by treating the governing partial-differential equations and associated boundary conditions. In this section, we illustrate the discrepancy between attacking directly the distributed-parameter system and attacking a discretized version of it using a simple example.

We consider the nonlinear dynamic response of a relief valve used to protect a pneumatic system from overpressure. A schematic representation of the complete valve is shown in Figure 1.8.1. The valve mechanism consists of a ball pressed by a uniform helical spring against a valve seat having nonlinear elastic characteristics. The helical spring is considered to be a distributed-parameter system, and its motion is governed by the wave equation subject to certain boundary conditions. One end of the spring is fixed. In addition to the forces exerted by the spring and seat, the ball is subjected to a pressure when it comes in contact with the fluid. The pressure

54   INTRODUCTION

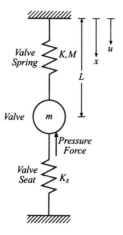

Fig. 1.8.2 A schematic model of a relief valve.

consists of a static component and a sinusoidally varying dynamic component; the latter results from small vibrations in the pneumatic transmission lines.

The relief-valve mechanism is schematically represented in Figure 1.8.2 as a linear helical spring having mass M, length L, stiffness constant K, and viscous damping coefficient μ^*. The displacement at time t^* of the point on the axis of the helical spring originally at the position x^* is given by u^*. The ball has a concentrated mass m and is attached to the helical spring at $x^* = L + u^*(L, t)$. The valve seat restrains the motion of the ball and is modeled as a massless spring having nonlinear characteristics.

We use the extended Hamilton principle to derive the governing equations. The kinetic-energy density of the linear spring is

$$T = \frac{1}{2} \frac{M}{L} \left(\frac{\partial u^*}{\partial t^*} \right)^2 \qquad (1.8.1)$$

Its potential-energy density is

$$\Pi = \frac{1}{2} KL \left(\frac{\partial u^*}{\partial x^*} \right)^2 \qquad (1.8.2)$$

Its Lagrangian function $\tilde{L}$ is

$$\tilde{L} = T - \Pi = \int_0^L \left[\frac{1}{2} \frac{M}{L} \left(\frac{\partial u^*}{\partial t^*} \right)^2 - \frac{1}{2} KL \left(\frac{\partial u^*}{\partial x^*} \right)^2 \right] dx^* \qquad (1.8.3)$$

To account for the mass of the ball and the elasticity of the valve seat at the end $x^* = L$, we introduce the discrete Lagrangian L_0:

$$L_0 = \frac{1}{2} m \left[\frac{\partial u^*}{\partial t^*} (L, t^*) \right]^2 - \frac{1}{2} \alpha \left[u^*(L, t^*) \right]^2 - \frac{1}{4} \beta \left[u^*(L, t^*) \right]^4 \qquad (1.8.4)$$

where the nonlinear character of the restoring force F_s generated by the valve seat is described as follows:

$$F_s = \alpha u^*(L, t^*) + \beta \left[u^*(L, t^*)\right]^3 \tag{1.8.5}$$

The virtual work done by the fluid pressure during a virtual displacement $\delta u^*(x,t)$ is given by

$$\delta \hat{W}_p = -(S^* + F^* \cos \Omega^* t^*) \delta u^*(L, t^*) \tag{1.8.6}$$

where S^*, F^*, and Ω^* are constants. The virtual work done by a linear damping force distributed along the length is given by

$$\delta \hat{W}_D = -\mu^* \int_0^L \frac{\partial u^*}{\partial t}(x^*, t^*) \delta u^*(x^*, t^*) dx^* \tag{1.8.7}$$

where μ^* is a constant. Hamilton's extended principle can be expressed as

$$\int_{t_1}^{t_2} \left[\delta \tilde{L} + \delta \hat{W}_p + \delta \hat{W}_D + \delta L_0\right] dt^* = 0 \tag{1.8.8}$$

where the virtual displacement δu^* is taken to be zero at times t_1 and t_2 along the entire length of the helical spring.

Substituting (1.8.3), (1.8.4), (1.8.6), and (1.8.7) into (1.8.8), integrating the results by parts, and setting the coefficient of δu^* equal to zero, we obtain the following equation of motion and boundary conditions:

$$\frac{\partial^2 u^*}{\partial t^{*2}} + \frac{\mu^*}{M} \frac{\partial u^*}{\partial t^*} = \frac{KL^2}{M} \frac{\partial^2 u^*}{\partial x^{*2}} \tag{1.8.9}$$

$$u^* = 0 \text{ at } x^* = 0 \tag{1.8.10}$$

$$m \frac{\partial^2 u^*}{\partial t^{*2}} + KL \frac{\partial u^*}{\partial x^*} + \alpha u^* + \beta u^{*3} = -(S^* + F^* \cos \Omega^* t^*) \text{ at } x^* = L \tag{1.8.11}$$

Next, we introduce the dimensionless quantities

$$u = \frac{u^*}{L^*}, \quad t = \sqrt{\frac{K}{M}} t^*, \quad x = \frac{x^*}{L}, \quad \text{and} \quad \Omega = \sqrt{\frac{M}{K}} \Omega^* \tag{1.8.12}$$

and rewrite (1.8.9)-(1.8.11) as

$$\frac{\partial^2 u}{\partial x^2} = \frac{\partial^2 u}{\partial t^2} + 2\hat{\mu} \frac{\partial u}{\partial t} \tag{1.8.13}$$

$$u = 0 \text{ at } x = 0 \tag{1.8.14}$$

$$\frac{\partial^2 u}{\partial t^2} + \alpha_1 \frac{\partial u}{\partial x} + ku + \alpha_3 u^3 = -(S + \hat{F} \cos \Omega t) \text{ at } x = 1 \tag{1.8.15}$$

where

$$2\hat{\mu} = \frac{\mu^*}{\sqrt{KM}}, \alpha_1 = \frac{M}{m}, k = \frac{M\alpha}{mK}, \alpha_3 = \frac{\beta L^2 M}{mK}, S = \frac{S^* M}{mKL}, \hat{F} = \frac{F^* M}{mKL}$$

The static equilibrium position u_s of u can be obtained by setting $\hat{F}$ and the time derivatives in (1.8.13)-(1.8.15) equal to zero. Thus, u_s is governed by

$$\frac{d^2 u_s}{dx^2} = 0 \tag{1.8.16a}$$

$$u_s = 0 \text{ at } x = 0 \tag{1.8.16b}$$

$$k u_s + \alpha_1 \frac{du_s}{dx} + \alpha_3 u_s^3 = -S \text{ at } x = 1 \tag{1.8.16c}$$

The solution of (1.8.16a,b,c) is

$$u_s = bx \tag{1.8.16d}$$

where

$$(\alpha_1 + k)b + \alpha_3 b^3 = -S \tag{1.8.16e}$$

To determine the nonlinear vibration of the system around its static equilibrium position, we let

$$u(x,t) = bx + w(x,t) \tag{1.8.17}$$

Substituting (1.8.17) into (1.8.13)-(1.8.15) and using (1.8.16e), we have

$$\frac{\partial^2 w}{\partial x^2} = \frac{\partial^2 w}{\partial t^2} + 2\hat{\mu} \frac{\partial w}{\partial t} \tag{1.8.18}$$

$$w = 0 \text{ at } x = 0 \tag{1.8.19}$$

$$\frac{\partial^2 w}{\partial t^2} + \alpha_1 \frac{\partial w}{\partial x} + \alpha_0 w + \alpha_2 w^2 + \alpha_3 w^3 = -\hat{F} \cos \Omega t \text{ at } x = 1 \tag{1.8.20}$$

where

$$\alpha_0 = k + 3\alpha_3 b^2 \text{ and } \alpha_2 = 3\alpha_3 b \tag{1.8.21}$$

Next, we consider small, but finite, motions around the static equilibrium position; that is, $|w| \ll b$.

One can easily show that (1.8.18) and its associated boundary conditions (1.8.19) and (1.8.20) can be derived from

$$\int_{t_1}^{t_2} \left[\delta \tilde{L} + \delta \hat{W}_p + \delta \hat{W}_D + \delta L_0 \right] dt = 0 \tag{1.8.22}$$

where

$$\tilde{L} = \frac{1}{2} \alpha_1 \int_0^1 \left[\left(\frac{\partial w}{\partial t} \right)^2 - \left(\frac{\partial w}{\partial x} \right)^2 \right] dx \tag{1.8.23}$$

$$L_0 = \frac{1}{2} \left[\frac{\partial w}{\partial t}(1,t) \right]^2 - \frac{1}{2} \alpha_0 [w(1,t)]^2 - \frac{1}{3} \alpha_2 [w(1,t)]^3 - \frac{1}{4} \alpha_3 [w(1,t)]^4 \tag{1.8.24}$$

$$\delta \hat{W}_D = -2\hat{\mu} \alpha_1 \int_0^1 \frac{\partial w}{\partial t}(x,t) \, \delta w(x,t) dx \tag{1.8.25}$$

$$\delta \hat{W}_p = \hat{F} \cos \Omega t \, \delta w(1,t) \tag{1.8.26}$$

1.8.1 Attacking the Continuous System

As discussed earlier, the response of a weakly nonlinear distributed-parameter system can be obtained by either directly treating the partial-differential equations and boundary conditions or treating the discretized form of the system. The latter approach is by far more widely used but, as we show here by means of the simple valve example, it can lead to erroneous results. In this section, we treat the continuous system for the case of primary resonance by using the method of multiple scales, and in Section 1.8.2, we treat the discretized form of the system.

In the case of primary resonance, the excitation frequency Ω is near one of the linear natural frequencies of the system. To determine an approximation to the solution of (1.8.18)-(1.8.20) for small- but finite-amplitude motions, we introduce a small dimensionless measure ϵ of the amplitude of w as a bookkeeping device. Using the method of multiple scales (Nayfeh, 1973b, 1981), we seek an approximation to the solution of (1.8.18)-(1.8.20) in the form (Nayfeh and Bouguerra, 1990)

$$w(x,t;\epsilon) \simeq \epsilon w_1(x,T_0,T_1,T_2) + \epsilon^2 w_2(x,T_0,T_1,T_2) + \epsilon^3 w_3(x,T_0,T_1,T_2) \tag{1.8.27}$$

where $T_0 = t$ is a fast-time scale characterizing changes occurring at the frequencies Ω and ω, $T_1 = \epsilon t$ and $T_2 = \epsilon^2 t$ are slow-time scales characterizing the modulations of the amplitude and phase due to damping, nonlinearity, and possible resonances, and the w_n are $O(1)$ as $\epsilon \to 0$. The damping $\hat{\mu}$ and excitation amplitude $\hat{F}$ are ordered in such a way that they balance the nonlinearity. Thus, we put

$$\hat{\mu} = \epsilon^2 \tilde{\mu} \quad \text{and} \quad \hat{F} = \epsilon^3 F \tag{1.8.28}$$

In terms of the T_n, the first and second time derivatives can be expressed as

$$\frac{\partial}{\partial t} = D_0 + \epsilon D_1 + \epsilon^2 D_2 + \cdots \tag{1.8.29}$$

$$\frac{\partial^2}{\partial t^2} = D_0^2 + 2\epsilon D_0 D_1 + \epsilon^2 (D_1^2 + 2D_0 D_2) + \cdots \tag{1.8.30}$$

where $D_n = \frac{\partial}{\partial T_n}$. Substituting (1.8.27)-(1.8.30) into (1.8.18)-(1.8.20) and equating coefficients of like powers of ϵ, we obtain

Order ϵ:

$$\frac{\partial^2 w_1}{\partial x^2} = D_0^2 w_1 \tag{1.8.31}$$

$$w_1 = 0 \quad \text{at} \quad x = 0 \tag{1.8.32}$$

$$D_0^2 w_1 + \alpha_1 \frac{\partial w_1}{\partial x} + \alpha_0 w_1 = 0 \quad \text{at} \quad x = 1 \tag{1.8.33}$$

Order ϵ^2:

$$\frac{\partial^2 w_2}{\partial x^2} = D_0^2 w_2 + 2 D_0 D_1 w_1 \tag{1.8.34}$$

58 INTRODUCTION

$$w_2 = 0 \text{ at } x = 0 \tag{1.8.35}$$

$$D_0^2 w_2 + \alpha_1 \frac{\partial w_2}{\partial x} + \alpha_0 w_2 = -2D_0 D_1 w_1 - \alpha_2 w_1^2 \text{ at } x = 1 \tag{1.8.36}$$

Order ϵ^3:

$$\frac{\partial^2 w_3}{\partial x^2} = D_0^2 w_3 + 2D_0 D_1 w_2 + (2D_0 D_2 + D_1^2) w_1 + 2\tilde{\mu} D_0 w_1 \tag{1.8.37}$$

$$w_3 = 0 \text{ at } x = 0 \tag{1.8.38}$$

$$D_0^2 w_3 + \alpha_1 \frac{\partial w_3}{\partial x} + \alpha_0 w_3 = -2D_0 D_1 w_2 - (2D_0 D_2 + D_1^2) w_1$$
$$- 2\alpha_2 w_1 w_2 - \alpha_3 w_1^3 - F \cos \Omega T_0 \text{ at } x = 1 \tag{1.8.39}$$

The general solution of (1.8.31)-(1.8.33) can be expressed as

$$w_1(x, T_0, T_1, T_2) = \sum_{m=1}^{\infty} \left[A_m(T_1, T_2) e^{i\omega_m T_0} + cc \right] \frac{\sin \omega_m x}{\sin \omega_m} \tag{1.8.40}$$

where the natural frequencies ω_m are the solutions of

$$\alpha_1 \omega_m + (\alpha_0 - \omega_m^2) \tan \omega_m = 0 \tag{1.8.41}$$

the complex-valued functions A_m are arbitrary at this point, and cc denotes the complex conjugate of the preceding terms. The solution (1.8.40) is a linear combination of all modes. However, in the presence of damping, the modes that are not directly excited by the fluid vibrations or indirectly excited by an internal resonance decay with time. We consider the case in which Ω is near the natural frequency ω_n of the nth mode when this mode is not involved in an internal resonance with any other mode. Hence, the solution of (1.8.31)-(1.8.33) consists of only the mode corresponding to ω_n, given as

$$w_1 = A(T_1, T_2) \frac{\sin \omega x}{\sin \omega} e^{i\omega T_0} + cc \tag{1.8.42}$$

where the subscript n has been dropped for convenience.

Substituting (1.8.42) into (1.8.34)-(1.8.36) yields

$$\frac{\partial^2 w_2}{\partial x^2} = D_0^2 w_2 + 2i\omega D_1 A \frac{\sin \omega x}{\sin \omega} e^{i\omega T_0} + cc \tag{1.8.43}$$

$$w_2 = 0 \text{ at } x = 0 \tag{1.8.44}$$

$$D_0^2 w_2 + \alpha_1 \frac{\partial w_2}{\partial x} + \alpha_0 w_2 = -2i\omega D_1 A e^{i\omega T_0}$$
$$- \alpha_2 \left[A^2 e^{2i\omega T_0} + A\bar{A} \right] + cc \text{ at } x = 1 \tag{1.8.45}$$

The solvability condition (Nayfeh, 1981) for (1.8.43)-(1.8.45) demands that

$$D_1 A = 0 \tag{1.8.46}$$

ANALYSES OF NONLINEAR CONTINUOUS SYSTEMS   59

Substituting (1.8.46) into (1.8.43)-(1.8.45) and solving for w_2 yields

$$w_2 = c_1 A \bar{A} x + c_2 A^2 \frac{\sin 2\omega x}{\sin 2\omega} e^{2i\omega T_0} + cc \tag{1.8.47}$$

where

$$c_1 = -\frac{\alpha_2}{\alpha_1 + \alpha_0} \quad \text{and} \quad c_2 = -\frac{\alpha_2}{\alpha_0 - 4\omega^2 + 2\omega\alpha_1 \cot 2\omega} \tag{1.8.48}$$

Because A is a function of T_2, we express the nearness of Ω to ω by introducing the detuning parameter σ defined by $\Omega = \omega + \epsilon^2 \sigma$. Then, substituting (1.8.42), (1.8.46), and (1.8.47) into (1.8.37)-(1.8.39) yields

$$\frac{\partial^2 w_3}{\partial x^2} = D_0^2 w_3 + 2i\omega(A' + \tilde{\mu} A) \frac{\sin \omega x}{\sin \omega} e^{i\omega T_0} + cc \tag{1.8.49}$$

$$w_3 = 0 \quad \text{at} \quad x = 0 \tag{1.8.50}$$

$$D_0^2 w_3 + \alpha_1 \frac{\partial w_3}{\partial x} + \alpha_0 w_3 = -[2\alpha_2(2c_1 + c_2) + 3\alpha_3] A^2 \bar{A} e^{i\omega T_0} - 2i\omega A' e^{i\omega T_0}$$

$$- \frac{1}{2} F e^{i(\omega T_0 + \sigma T_2)} + NST + cc \quad \text{at} \quad x = 1 \tag{1.8.51}$$

where NST stands for terms that do not produce secular terms and the prime indicates the derivative with respect to T_2. Because the homogeneous parts of equations (1.8.49)-(1.8.51) are the same as (1.8.31)-(1.8.33) and because the latter have a nontrivial solution, the nonhomogeneous equations (1.8.49)-(1.8.51) have a solution only if a certain solvability condition is satisfied (Nayfeh, 1981). To determine this condition, we first write the solution in the form

$$w_3 = \phi(x, T_2) e^{i\omega T_0} + cc + W_3(x, T_0, T_2) \tag{1.8.52}$$

where W_3 is governed by (1.8.49)-(1.8.51) with the terms proportional to $\exp(i\omega T_0)$ being deleted. Therefore, W_3 exists, is unique, and free of secular and so-called small-divisor terms. Substituting (1.8.52) into (1.8.49)-(1.8.51) and equating the coefficients of $\exp(i\omega T_0)$ on both sides of each equation results in

$$\frac{d^2 \phi}{dx^2} + \omega^2 \phi = 2i\omega(A' + \tilde{\mu} A) \frac{\sin \omega x}{\sin \omega} \tag{1.8.53}$$

$$\phi = 0 \quad \text{at} \quad x = 0 \tag{1.8.54}$$

$$\frac{d\phi}{dx} + \alpha \phi = \frac{1}{\alpha_1} \left[-\frac{1}{2} F e^{i\sigma T_2} - 2i\omega A' - 2\alpha_2 B A^2 \bar{A} \right] \quad \text{at} \quad x = 1 \tag{1.8.55}$$

where

$$\alpha = \frac{\alpha_0 - \omega^2}{\alpha_1} \quad \text{and} \quad B = 2c_1 + c_2 + \frac{3\alpha_3}{2\alpha_2} \tag{1.8.56}$$

Hence, determining the solvability condition of (1.8.31)-(1.8.33) has been transformed into determining the solvability condition (1.8.53)-(1.8.55). To determine this solvability condition, we multiply (1.8.53) by $\sin \omega x / \sin \omega$, the solution of the

60 INTRODUCTION

adjoint homogeneous problem (the problem is self-adjoint), integrate the result by parts from $x = 0$ to $x = 1$, use the boundary conditions (1.8.54) and (1.8.55), and obtain

$$2i\omega(A' + \tilde{\mu}A) = -\frac{\Gamma}{\alpha_1}\left[2i\omega A' + 2\alpha_2 BA^2\bar{A} + \frac{1}{2}Fe^{i\sigma T_2}\right] \quad (1.8.57)$$

where

$$\Gamma = \left[\int_0^1 \frac{\sin^2 \omega x}{\sin^2 \omega} dx\right]^{-1} = \frac{4\omega \sin^2 \omega}{2\omega - \sin 2\omega} \quad (1.8.58)$$

Substituting the polar form

$$A = \frac{1}{2}ae^{i(\sigma T_2 - \gamma)} \quad (1.8.59)$$

where a and γ are real functions of T_2, into (1.8.57) and separating the result into real and imaginary parts yields

$$a' = -\mu a - \frac{f}{2\omega}\sin\gamma \quad (1.8.60)$$

$$a\gamma' = \sigma a - \alpha_e a^3 - \frac{f}{2\omega}\cos\gamma \quad (1.8.61)$$

where

$$\mu = \alpha_1\tilde{\mu}/(\alpha_1 + \Gamma) \quad (1.8.62)$$

$$\alpha_e = \frac{\Gamma\alpha_2 B}{4\omega(\alpha_1 + \Gamma)} = \frac{3\alpha}{8\omega} - \frac{\alpha_2\delta}{4\omega}\left[\frac{2}{\alpha_1 + \alpha_0} + \frac{1}{\alpha_0 - 4\omega^2 + 2\omega\alpha_1 \cot 2\omega}\right] \quad (1.8.63)$$

$$(\delta, \alpha, f) = \frac{\Gamma}{\Gamma + \alpha_1}(\alpha_2, \alpha_3, F) = \frac{4\omega \sin^2 \omega}{\alpha_1(2\omega - \sin 2\omega) + 4\omega \sin^2 \omega}(\alpha_2, \alpha_3, F) \quad (1.8.64)$$

Substituting (1.8.42) and (1.8.47) into (1.8.27), recalling that $\Omega = \omega + \epsilon^2\sigma$, and using (1.8.59) and (1.8.48), we obtain the following two-term approximation to the solution:

$$w \simeq \epsilon a \cos(\Omega t - \gamma)\frac{\sin \omega x}{\sin \omega}$$

$$-\frac{1}{2}\epsilon^2\alpha_2 a^2\left[\frac{x}{\alpha_1 + \alpha_0} + \frac{\sin 2\omega x \cos(2\Omega t - 2\gamma)}{(\alpha_0 - 4\omega^2)\sin 2\omega + 2\omega\alpha_1 \cos 2\omega}\right] \quad (1.8.65)$$

where a and γ are defined by (1.8.60) and (1.8.61).

1.8.2 Attacking the Discretized System

One of the most commonly used procedures for obtaining approximate solutions of nonlinear distributed-parameter problems is to discretize the problem first. For the problem at hand, one assumes an approximate solution in the form

$$w(x,t) \simeq \sum_{n=1}^{N} \eta_n(t)\phi_n(x) \qquad (1.8.66)$$

Because (1.8.66) does not satisfy the partial-differential equation (1.8.18) and boundary conditions (1.8.19) and (1.8.20) exactly, one uses a weighted-residual procedure, such as the Galerkin or Ritz procedure, to minimize the errors, thereby obtaining nonlinearly coupled second-order ordinary-differential equations governing the η_n. These equations are called the discretized form of the distributed-parameter problem. Then, one can use either a perturbation or a numerical method to obtain solutions of these discretized equations. In many instances, the $\phi_n(x)$ are chosen to be the mode shapes of the linear problem. Typically, N is taken to be unity and the result is called a single-mode approximation. Thus, for a single-mode approximation of the present problem, we let

$$w(x,t) \simeq \eta(t)\frac{\sin\omega x}{\sin\omega} \qquad (1.8.67)$$

where ω is one of the linear frequencies that can be calculated from the transcendental equation (1.8.41) and $\eta(t)$ is a time-varying function that needs to be determined.

Clearly, there is no two-term function $\eta(t)$ that will make it possible for the assumed approximation in (1.8.67) to match the correct approximation given in (1.8.65). The only possibility is that the first terms in the two expansions match. But, as we show next, even the first terms do not match.

Because one of the boundary conditions is nonlinear, it is more convenient to determine the equation governing $\eta(t)$ by using the Lagrangian. Substituting (1.8.67) into (1.8.22) and integrating the result from $x = 0$ to $x = 1$ yields the following Lagrangian:

$$L = \frac{1}{2}\dot{\eta}^2 - \frac{1}{2}\omega^2\eta^2 - \frac{1}{3}\delta\eta^3 - \frac{1}{4}\alpha\eta^4 - f\eta\cos\Omega t \qquad (1.8.68)$$

where $\delta, \alpha,$ and f are defined in (1.8.64). Writing the Euler-Lagrange equation associated with the Lagrangian in (1.8.68) yields

$$\ddot{\eta} + \omega^2\eta + 2\mu\dot{\eta} + \delta\eta^2 + \alpha\eta^3 = -f\cos\Omega t \qquad (1.8.69)$$

where damping has been added.

Nayfeh (1983) extensively studied the solutions of (1.8.69) for the case of weak nonlinearities by using different perturbation techniques. For the case of primary resonance, we use the method of multiple scales and seek an approximate solution of (1.8.69) in the form

$$\eta(t;\epsilon) = \epsilon\eta_1(T_0,T_2) + \epsilon^2\eta_2(T_0,T_2) + \epsilon^3\eta_3(T_0,T_2) + \cdots \qquad (1.8.70)$$

where ϵ is a small dimensionless parameter that is used as a bookkeeping device and $T_0 = t$ and $T_2 = \epsilon^2 t$. As was evident in the preceding section, secular terms arise first at $O(\epsilon^3)$ and hence there is no dependence on T_1. Using ϵ, we scale μ as $\epsilon^2\mu$ and

f as $\epsilon^3 f$ so that the effects of the damping and nonlinearity balance the effect of the resonance. With this scaling, we rewrite (1.8.69) as

$$\ddot{\eta} + \omega^2 \eta + 2\epsilon^2 \mu \dot{\eta} + \delta\eta^2 + \alpha\eta^3 = -\epsilon^3 f \cos \Omega T_0 \qquad (1.8.71)$$

Substituting (1.8.70) into (1.8.71), using (1.8.29) and (1.8.30), and equating coefficients of like powers of ϵ, we have

Order ϵ:
$$D_0^2 \eta_1 + \omega^2 \eta_1 = 0 \qquad (1.8.72)$$

Order ϵ^2:
$$D_0^2 \eta_2 + \omega^2 \eta_2 = -\delta\eta_1^2 \qquad (1.8.73)$$

Order ϵ^3:
$$D_0^2 \eta_3 + \omega^2 \eta_3 = -2D_0 D_2 \eta_1 - 2\mu D_0 \eta_1 - 2\delta\eta_1 \eta_2 - \alpha\eta_1^3 - f \cos \Omega T_0 \qquad (1.8.74)$$

The solution of (1.8.72) can be expressed as

$$\eta_1 = A(T_2) e^{i\omega T_0} + cc \qquad (1.8.75)$$

Then, (1.8.73) becomes

$$D_0^2 \eta_2 + \omega^2 \eta_2 = -\delta A^2 e^{2i\omega T_0} - \delta A \bar{A} + cc$$

whose solution can be expressed as

$$\eta_2 = \frac{\delta}{3\omega^2} A^2 e^{2i\omega T_0} - \frac{\delta}{\omega^2} A \bar{A} + cc \qquad (1.8.76)$$

Substituting (1.8.75) and (1.8.76) into (1.8.74) yields

$$D_0^2 \eta_3 + \omega^2 \eta_3 = -2i\omega A' e^{i\omega T_0} - \left(3\alpha - \frac{10\delta^2}{3\omega^2}\right) A^2 \bar{A} e^{i\omega T_0}$$
$$- \left(\alpha + \frac{2\delta^2}{3\omega^2}\right) A^3 e^{3i\omega T_0} - \frac{1}{2} f e^{i\Omega T_0} + cc \qquad (1.8.77)$$

For the case of primary resonance, $\Omega = \omega + \epsilon^2 \sigma$, where σ is the detuning parameter. Hence, eliminating the secular terms from (1.8.77) demands that

$$2i\omega A' + 8\omega \hat{\alpha}_e A^2 \bar{A} + \frac{1}{2} f e^{i\sigma T_2} = 0 \qquad (1.8.78)$$

where

$$\hat{\alpha}_e = \frac{3\alpha}{8\omega} - \frac{5\delta^2}{12\omega^3} \qquad (1.8.79)$$

Substituting (1.8.59) into (1.8.78) and separating real and imaginary parts yields

$$a' = -\mu a - \frac{f}{2\omega} \sin \gamma \qquad (1.8.80)$$

$$a\gamma' = \sigma a - \hat{\alpha}_e a^3 - \frac{f}{2\omega} \cos \gamma \qquad (1.8.81)$$

Substituting (1.8.75) and (1.8.76) into (1.8.70) and using the polar form (1.8.59) yields to the second approximation

$$\eta = \epsilon a \cos(\Omega t - \gamma) + \frac{\epsilon^2 \delta a^2}{6\omega^2}\left[\cos(2\Omega t - 2\gamma) - 3\right] + \cdots \tag{1.8.82}$$

where a and γ are governed by (1.8.80) and (1.8.81).

Therefore, to the second approximation,

$$w = \frac{\sin \omega x}{\sin \omega}\left\{ a\cos(\Omega t - \gamma) + \frac{\delta a^2}{6\omega^2}\left[\cos(2\Omega t - 2\gamma) - 3\right] + \ldots \right\} \tag{1.8.83}$$

where a and γ are given by (1.8.80)-(1.8.81). We note that (1.8.80) and (1.8.81) have the same form as (1.8.60) and (1.8.61), but that α_e in (1.8.61) is not the same as $\hat{\alpha}_e$ in (1.8.81). Therefore, we conclude that even the one-term result obtained by the single-mode discretization is wrong; it fails to correctly capture the effect of the nonlinearity, as manifested in the coefficients α_e and $\hat{\alpha}_e$.

1.8.3 Time-Averaged Lagrangian

A quick inspection of the Lagrangian in (1.8.22) shows that the cubic term is the source of the quadratic nonlinearity in the boundary condition (1.8.20). Therefore, this term is responsible for the drift (slowly-varying terms) and the second-harmonic components in the response (1.8.65). Hence, to determine a uniform expansion, we express w to second order in ϵ as

$$w \simeq \epsilon(Ae^{i\omega T_0} + \bar{A}e^{-i\omega T_0})\frac{\sin \omega x}{\sin \omega} + \epsilon^2 \phi_0(x)A\bar{A} + \epsilon^2 \phi_1(x)(A^2 e^{2i\omega T_0} + \bar{A}^2 e^{-2i\omega T_0}) \tag{1.8.84}$$

where $\phi_0(x)$, $\phi_1(x)$, and $A(T_2)$ are to be determined from the analysis. We remark that the determination of ϕ_0 and ϕ_1 may require extensive algebra and numeric calculations in other physical problems when applying the time-averaged Lagrangian method; nevertheless, the neatness and the considerable reduction in the algebra are rather striking in comparison with attacking the nonlinear partial-differential equations and boundary conditions.

Substituting (1.8.84) into (1.8.22), replacing $\hat{F}$ by $\epsilon^3 F$, keeping terms up to $O(\epsilon^4)$, keeping the slowly varying terms only, and recalling that $\Omega = \omega + \epsilon^2 \sigma$, we obtain

$$<L> = i\epsilon^4 \omega \left\{ 1 + \frac{\alpha_1}{\sin^2 \omega} \int_0^1 \sin^2 \omega x\, dx \right\}(A\bar{A}' - \bar{A}A')$$
$$+ \epsilon^4 \left\{ \alpha_1 \int_0^1 \left[4\omega^2 \phi_1^2(x) - \phi_1'^2(x) - \frac{1}{2}\phi_0'^2(x) \right] dx \right.$$
$$- \frac{3}{2}\alpha_3 - 2\alpha_2[\phi_0(1) + \phi_1(1)] + (4\omega^2 - \alpha_0)\phi_1^2(1)$$
$$\left. - \frac{1}{2}\alpha_0 \phi_0^2(1) \right\} A^2 \bar{A}^2 - \frac{1}{2}\epsilon^4 F(Ae^{-i\sigma T_2} + \bar{A}e^{i\sigma T_2}) \tag{1.8.85}$$

To solve for ϕ_0 and ϕ_1, one can either use a finite-element method directly on the Lagrangian or solve the differential equations and boundary conditions governing them. Here, we use the latter approach. To this end, we use the calculus of variations, require that $\delta <L> = 0$ and that $\delta\phi_0$ and $\delta\phi_1$ be independent arbitrary functions, and arrive at the following Euler-Lagrange equations and associated boundary conditions:

$\delta\phi_0$:

$$\alpha_1 \phi_0'' = 0 \text{ for } 0 < x < 1 \tag{1.8.86}$$
$$\phi_0 = 0 \text{ at } x = 0 \tag{1.8.87}$$
$$\alpha_1 \phi_0' + \alpha_0 \phi_0 = -2\alpha_2 \text{ at } x = 1 \tag{1.8.88}$$

$\delta\phi_1$:

$$\phi_1'' + 4\omega^2 \phi_1 = 0 \quad 0 < x < 1 \tag{1.8.89}$$
$$\phi_1 = 0 \text{ at } x = 0 \tag{1.8.90}$$
$$\alpha_1 \phi_1' + (\alpha_0 - 4\omega^2)\phi_1 = -\alpha_2 \text{ at } x = 1 \tag{1.8.91}$$

Solving (1.8.86)-(1.8.91) yields

$$\phi_0 = \frac{-2\alpha_2 x}{\alpha_1 + \alpha_0} \tag{1.8.92}$$

$$\phi_1 = \frac{-\alpha_2 \sin 2\omega x}{(\alpha_0 - 4\omega^2)\sin 2\omega + 2\omega\alpha_1 \cos 2\omega} \tag{1.8.93}$$

which are in full agreement with (1.8.47) and (1.8.48) obtained by attacking the original problem.

Substituting (1.8.92) and (1.8.93) into (1.8.85) and simplifying, we rewrite the time-averaged Lagrangian as

$$<L> = \frac{\Gamma + \alpha_1}{\Gamma} \epsilon^4 \left[i(A\bar{A}' - \bar{A}A') - 4\alpha_e A^2 \bar{A}^2 - \frac{f}{2\omega}\left(Ae^{-i\sigma T_2} + \bar{A}e^{i\sigma T_2}\right) \right] \tag{1.8.94}$$

where α_e is defined in (1.8.63) and Γ is defined in (1.8.58). Introducing the polar form (1.8.59) into (1.8.94) and using (1.8.63), we obtain

$$<L> = \frac{\Gamma + \alpha_1}{\Gamma} \epsilon^4 \left[\frac{1}{2}a^2(\sigma - \gamma') - \frac{1}{4}\alpha_e a^4 - \frac{fa}{2\omega}\cos\gamma \right] \tag{1.8.95}$$

where the prime indicates the derivative with respect to T_2. Writing the Euler-Lagrange equations associated with the Lagrangian (1.8.95), we obtain

$$a' = -\frac{f}{2\omega}\sin\gamma \tag{1.8.96}$$

$$a\gamma' = \sigma a - \alpha_e a^3 - \frac{f}{2\omega}\cos\gamma \tag{1.8.97}$$

where α_e and f are defined in (1.8.63) and (1.8.64). Equations (1.8.96) and (1.8.97) with the proper damping term added to (1.8.96) are in full agreement with (1.8.60) and (1.8.61) obtained by attacking the governing equation and boundary conditions.

2
ELASTICITY

Structures undergoing large deformations but small strains are nonlinearly elastic structures, where nonlinearities are caused by large rigid-body rotations. With respect to a coordinate frame corotated with the rigid-body movement, the relative displacements are small and the problem becomes linearly elastic.

The basic components of linear elasticity are strain-displacement relations, a constitutive law, equilibrium equations, and compatibility conditions. The strain-displacement relations represent the strains in terms of the displacements. The constitutive law represents the stress-strain behavior of the material. The equilibrium equations represent a statement of Newton's laws, which state the balance of the internal forces, body forces, and inertia forces. And, the compatibility conditions are constraints on the strains to ensure the continuity of displacements. In general, these basic components may be collected as a set of differential equations or as an energy principle. Material nonlinearities, which are not considered in this book, enter a structural theory in the constitutive law. Geometric nonlinearities, which are the main concerns of this book, enter a structural theory in the strain-displacement relations and equilibrium equations. In this chapter, we review principles of dynamics and basic components of linear elasticity and derive some identities for future use.

2.1 PRINCIPLES OF DYNAMICS

There are two types of systems: discrete and continuous. Structures are continuous or distributed-parameter systems; particle systems and multi-rigid-body systems are discrete systems. The governing equations of discrete systems are ordinary-differential

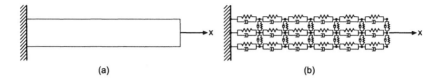

Fig. 2.1.1 (a) A cantilever beam and (b) a discrete model of the beam.

equations, and those of continuous systems are partial-differential equations. A multi-rigid-body system or a continuous system can be represented as a particle system, but there is an infinite number of particles in it and integration is needed, instead of summation, in representing its system energy.

There are two different approaches to the formulation of dynamic problems; one is the vector approach by using Newton's laws, and the other is the energy (variational) approach by applying the calculus of variation to the kinetic energy, work function, and elastic energy of the system. Since Newton's laws are based on vector analysis, a direct geometric consideration is needed, which is difficult for the formulation of a particle system or a multi-body system. Also, the equations of motion obtained by using Newton's laws contain interacting forces between different parts if its a multi-body system, and some extra constraint equations are involved in the modeling when there are elastic components in the system. Moreover, Newton's laws can only be used to derive the governing partial-differential equations of a continuous system, but the corresponding boundary conditions need to be figured out by direct geometric considerations. On the other hand, a variational approach can be used to simultaneously obtain the governing partial-differential equations and the associated boundary conditions of a continuous system. Furthermore, for a multi-rigid-body system, a variational approach leads to unifying formulations, such as Hamilton's principle and Lagrange's equations, which do not depend on the coordinate system used. Hence, variational approaches are commonly used in the formulation of complex mechanical systems and structures. However, because Newton's laws are the basic laws of mechanics, which are based on experimental observations, any other principles of mechanics are just alternate expressions or explanations and can be derived from Newton's laws.

The kinetic energy T and potential energy Π of the spring-mass-damper system shown in Figure 1.5.1 are given by

$$T = \frac{1}{2}m\dot{x}^2 \text{ and } \Pi = \frac{1}{2}kx^2 - mgx \tag{2.1.1}$$

where g is the gravity and the spring is assumed to be unstretched when $x = 0$. Any other discrete or continuous system has similar energy terms as those shown in (2.1.1) because any discrete or continuous elastic system can be ideally modeled as a particle system connected by springs, masses, and dampers. For example, a beam can be represented by a multi-rigid-body system (see Roseau, 1987). Moreover, the cantilever beam shown in Figure 2.1.1a can be ideally modeled as the spring-

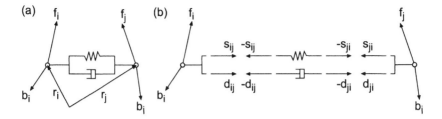

Fig. 2.1.2 (a) A basic element for particle systems and (b) its free-body diagrams.

mass-damper system shown in Figure 2.1.1b. Hence, in this section, we start with reviewing the use of Newton's second law to derive the energy of a discrete system and the equilibrium laws used in statics and dynamics.

2.1.1 Newton's Second Law and Energy of a Discrete System

A particle has no volume, which means that the three perpendicular dimensions of a particle are all infinitesimal. Hence the motion of a particle is only described by its position vector. In other words, there is no need to describe the rotation of a particle and there is no rotary inertia. In Figure 2.1.2a, we show a basic constructing element for particle systems, and in Figure 2.1.2b, we show the free-body diagrams of its components. In Figure 2.1.2b, the ith particle is subject to the force $\mathbf{F}_i$, which is given by

$$\mathbf{F}_i = \mathbf{f}_i + \sum_{j=1,\, j \neq i}^{N} \mathbf{d}_{ij} + \sum_{j=1,\, j \neq i}^{N} \mathbf{s}_{ij} + \mathbf{b}_i \qquad (2.1.2)$$

where bold-face letters represent vectors, N is the total number of particles of the system, $\mathbf{f}_i$ is the externally applied load on the ith particle, $\mathbf{b}_i$ is the body force, $\mathbf{d}_{ij}$ is the internal damping force due to the relative motion of the ith particle with respect to the jth particle, and $\mathbf{s}_{ij}$ is the internal spring force due to the relative motion of the ith particle with respect to the jth particle. Both $\mathbf{d}_{ij}$ and $\mathbf{s}_{ij}$ are defined to be positive if they point from the ith particle to the jth particle. Considering the equilibrium of forces acting on the damper and spring and using the fact that both spring and damper have no mass, we obtain

$$\mathbf{d}_{ij} = -\mathbf{d}_{ji}, \quad \mathbf{s}_{ij} = -\mathbf{s}_{ji} \qquad (2.1.3)$$

Now we treat the N particles as a system, without the springs and dampers. For such a particle system, the work ΔW done on the system by the external loads from time t_1 to time t_2 is

$$\Delta W \equiv \sum_{i=1}^{N} \int_{\mathbf{r}_i(t_1)}^{\mathbf{r}_i(t_2)} \mathbf{F}_i \cdot d\mathbf{r}_i \qquad (2.1.4)$$

where r_i is the position vector of the ith particle. We note that the spring forces s_{ij} and the damping forces d_{ij} are treated as external forces because springs and dampers are not included in the system. For the ith particle, Newton's second law states that

$$\mathbf{F}_i = \frac{d^2}{dt^2}(m_i \mathbf{r}_i) \tag{2.1.5}$$

In this book, we only consider mass-conserved systems and hence

$$\dot{m}_i = 0 \tag{2.1.6}$$

Using the identity $d\mathbf{r}_i = \frac{d\mathbf{r}_i}{dt}dt$ and (2.1.4)-(2.1.6), we obtain

$$\Delta W = \sum_{i=1}^{N} \int_{t_1}^{t_2} \frac{d}{dt}\left(\frac{1}{2}m_i \frac{d\mathbf{r}_i}{dt} \cdot \frac{d\mathbf{r}_i}{dt}\right) dt = T(t_2) - T(t_1) = \Delta T \tag{2.1.7}$$

where the total kinetic energy T of the system is defined as

$$T \equiv \sum_{i=1}^{N} \frac{1}{2} m_i \frac{d\mathbf{r}_i}{dt} \cdot \frac{d\mathbf{r}_i}{dt} \tag{2.1.8}$$

Next, we treat the N particles and all the springs and dampers which connect the particles as a system. Substituting (2.1.2) into (2.1.4) yields

$$\Delta W = \sum_{i=1}^{N} \int_{\mathbf{r}_i(t_1)}^{\mathbf{r}_i(t_2)} \left(\mathbf{f}_i + \sum_{j=1,\, j\neq i}^{N} \mathbf{d}_{ij} + \sum_{j=1,\, j\neq i}^{N} \mathbf{s}_{ij} + \mathbf{b}_i \right) \cdot d\mathbf{r}_i$$
$$= \Delta W_a + \Delta W_d - \Delta \Pi_s - \Delta \Pi_b \tag{2.1.9}$$

where

$$\Delta W_a \equiv \sum_{i=1}^{N} \int_{\mathbf{r}_i(t_1)}^{\mathbf{r}_i(t_2)} \mathbf{f}_i \cdot d\mathbf{r}_i$$

$$\Delta W_d \equiv \sum_{i=1}^{N} \sum_{j=1,\, j\neq i}^{N} \int_{\mathbf{r}_i(t_1)}^{\mathbf{r}_i(t_2)} \mathbf{d}_{ij} \cdot d\mathbf{r}_i$$
$$= \sum_{i=1}^{N} \sum_{j=i+1}^{N} \int_{\mathbf{r}_i(t_1)-\mathbf{r}_j(t_1)}^{\mathbf{r}_i(t_2)-\mathbf{r}_j(t_2)} \mathbf{d}_{ij} \cdot d(\mathbf{r}_i - \mathbf{r}_j)$$

$$\Delta \Pi_s \equiv -\sum_{i=1}^{N} \sum_{j=1,\, j\neq i}^{N} \int_{\mathbf{r}_i(t_1)}^{\mathbf{r}_i(t_2)} \mathbf{s}_{ij} \cdot d\mathbf{r}_i$$
$$= -\sum_{i=1}^{N} \sum_{j=i+1}^{N} \int_{\mathbf{r}_i(t_1)-\mathbf{r}_j(t_1)}^{\mathbf{r}_i(t_2)-\mathbf{r}_j(t_2)} \mathbf{s}_{ij} \cdot d(\mathbf{r}_i - \mathbf{r}_j)$$

$$\Delta \Pi_b \equiv -\sum_{i=1}^{N} \int_{\mathbf{r}_i(t_1)}^{\mathbf{r}_i(t_2)} \mathbf{b}_i \cdot d\mathbf{r}_i \tag{2.1.10}$$

and (2.1.3) was used. Here, ΔW_a is the work done on the system by the applied external loads, ΔW_d is the work done on the system by the dissipative forces ($\Delta W_d \leq 0$ because the direction of $d(\mathbf{r}_i - \mathbf{r}_j)$ is always opposite to that of $\mathbf{d}_{ij}$), $\Delta \Pi_s$ is the increase in the elastic energy in the system ($\Delta \Pi_s \geq 0$ because the direction of $d(\mathbf{r}_i - \mathbf{r}_j)$ is always opposite to that of $\mathbf{s}_{ij}$), and $\Delta \Pi_b$ is the change in the potential energy due to the body forces. We further define the non-conservative work ΔW_{nc} and the total potential energy Π as

$$W_{nc} \equiv W_a + W_d$$
$$= \sum_{i=1}^{N} \int_{\mathbf{r}_i(t_1)}^{\mathbf{r}_i(t_2)} \mathbf{f}_i \cdot d\mathbf{r}_i + \sum_{i=1}^{N} \sum_{j=i+1}^{N} \int_{\mathbf{r}_i(t_1)-\mathbf{r}_j(t_1)}^{\mathbf{r}_i(t_2)-\mathbf{r}_j(t_2)} \mathbf{d}_{ij} \cdot d(\mathbf{r}_i - \mathbf{r}_j)$$
$$\Pi \equiv \Pi_s + \Pi_b$$
$$= -\sum_{i=1}^{N} \sum_{j=i+1}^{N} \int_{\mathbf{r}_i(t_1)-\mathbf{r}_j(t_1)}^{\mathbf{r}_i(t_2)-\mathbf{r}_j(t_2)} \mathbf{s}_{ij} \cdot d(\mathbf{r}_i - \mathbf{r}_j) - \sum_{i=1}^{N} \int_{\mathbf{r}_i(t_1)}^{\mathbf{r}_i(t_2)} \mathbf{b}_i \cdot d\mathbf{r}_i$$

(2.1.11)

Hence, it follows from (2.1.9), (2.1.11), and (2.1.7) that

$$\Delta W = \Delta W_{nc} - \Delta \Pi \tag{2.1.12}$$

or

$$\Delta W_{nc} = \Delta T + \Delta \Pi \tag{2.1.13}$$

Equation (2.1.13) states that the work done on the system by the externally applied loads $\mathbf{f}_i$ and damping forces $\mathbf{d}_{ij}$ is equal to the increase in the kinetic and potential energies, which is the primary principle used in the study of the dynamics of multi-rigid-body systems. If $\Delta T = 0$, (2.1.13) reduces to

$$\Delta W_{nc} = \Delta \Pi \tag{2.1.14}$$

which is the primary principle used in the study of statics.

2.1.2 Principle of Virtual Work

For a particle, D'Alembert's principle states that the applied force $\mathbf{F}_i$ is in equilibrium with the inertia force $-d^2(m_i \mathbf{r}_i)/dt^2$; that is,

$$\mathbf{F}_i - m_i \frac{d^2 \mathbf{r}_i}{dt^2} = 0 \tag{2.1.15}$$

which is equivalent to (2.1.5) when $\dot{m}_i = 0$. Let the variation of $\mathbf{r}_i$ be represented by $\delta \mathbf{r}_i$, which also represents the virtual displacement vector $\delta \mathbf{D}_i$ because

$$\delta \mathbf{r}_i = \delta(\mathbf{R}_i + \mathbf{D}_i) = \delta \mathbf{D}_i$$

where $\mathbf{R}_i$ denotes the initial position vector of the ith particle and $\mathbf{D}_i$ denotes the displacement vector. Virtual displacements are infinitesimal variations from the true position of the system and must be compatible with the constraints of the system. Virtual displacements are not true displacements because there is no time change associated with them.

For a discrete system consisting of N particles, it follows from (2.1.15) that

$$\sum_{i=1}^{N} \left(\mathbf{F}_i - m_i \frac{d^2 \mathbf{r}_i}{dt^2} \right) \cdot \delta \mathbf{r}_i = 0 \qquad (2.1.16)$$

which is the principle of virtual work for a particle system. Using (2.1.2), (2.1.3), and (2.1.11) we express the virtual work done on the particle system by the external loads as

$$\delta W \equiv \sum_{i=1}^{N} \mathbf{F}_i \cdot \delta \mathbf{r}_i = \delta W_a + \delta W_d - \delta \Pi_s - \delta \Pi_b$$
$$= \delta W_{nc} - \delta \Pi \qquad (2.1.17)$$

where

$$\delta W_a \equiv \sum_{i=1}^{N} \mathbf{f}_i \cdot \delta \mathbf{r}_i$$

$$\delta W_d \equiv \sum_{i=1}^{N} \sum_{j=i+1}^{N} \mathbf{d}_{ij} \cdot \delta(\mathbf{r}_i - \mathbf{r}_j)$$

$$\delta \Pi_s \equiv -\sum_{i=1}^{N} \sum_{j=i+1}^{N} \mathbf{s}_{ij} \cdot \delta(\mathbf{r}_i - \mathbf{r}_j)$$

$$\delta \Pi_b \equiv -\sum_{i=1}^{N} \mathbf{b}_i \cdot \delta \mathbf{r}_i \qquad (2.1.18)$$

It follows from (2.1.16) and (2.1.17) that

$$\delta W_{nc} - \delta \Pi - \sum_{i=1}^{N} m_i \frac{d^2 \mathbf{r}_i}{dt^2} \cdot \delta \mathbf{r}_i = 0 \qquad (2.1.19)$$

Equation (2.1.19) states the principle of virtual work for a discrete system consisting of N particles and the springs and dampers which connect these particles.

2.1.3 Hamilton's Theories

Hamilton's Law of Varying Action: Using (2.1.17) and (2.1.8), we rewrite the principle of virtual work (2.1.16) as

$$0 = \sum_{i=1}^{N} \left(\mathbf{F}_i - m_i \frac{d^2 \mathbf{r}_i}{dt^2} \right) \cdot \delta \mathbf{r}_i$$

$$= \delta W - \sum_{i=1}^{N} \frac{d}{dt}\left(m_i \frac{d\mathbf{r}_i}{dt} \cdot \delta \mathbf{r}_i\right) + \frac{1}{2}\delta \sum_{i=1}^{N} m_i \frac{d\mathbf{r}_i}{dt} \cdot \frac{d\mathbf{r}_i}{dt}$$

$$= \delta W - \sum_{i=1}^{N} \frac{d}{dt}(\mathbf{P}_i \cdot \delta \mathbf{r}_i) + \delta T \tag{2.1.20}$$

where $\mathbf{P}_i$ is the momentum of the ith particle and is defined as

$$\mathbf{P}_i \equiv m_i \frac{d\mathbf{r}_i}{dt} \tag{2.1.21}$$

Integrating (2.1.20) from $t = t_1$ to $t = t_2$ yields

$$\int_{t_1}^{t_2} (\delta W + \delta T) dt = \sum_{i=1}^{N} \mathbf{P}_i \cdot \delta \mathbf{r}_i \big|_{t_1}^{t_2} \tag{2.1.22}$$

which is called Hamilton's law of varying action (Hamilton, 1834). The terms on the right-hand side of (2.1.22) are called trailing terms or "boundary terms" at the "boundaries" $t = t_1$ and $t = t_2$.

Extended Hamilton Principle: If it is assumed that the true and varied paths coincide at t_1 and t_2, then $\delta \mathbf{r}_i(t_1) = \delta \mathbf{r}_i(t_2) = 0$ and the trailing terms vanish in (2.1.22) and the result is the extended Hamilton principle:

$$\int_{t_1}^{t_2} (\delta L + \delta W_{nc}) dt = 0 \tag{2.1.23}$$

where (2.1.17) is used and the Lagrangian L of the system is defined as

$$L \equiv T - \Pi \tag{2.1.24}$$

Hamilton's Principle: If there are no non-conservative (path-dependent) forces, then $\delta W_{nc} = 0$ and

$$\int_{t_1}^{t_2} \delta L \, dt = 0 \tag{2.1.25}$$

which is called Hamilton's principle.

Hamilton's law of varying action (2.1.22) is an unconstrained variational law of mechanics, and the extended Hamilton principle (2.1.23) is a valid physical statement of mechanics only when the boundary constraints are such that the trailing terms vanish. But, the extended Hamilton principle has traditionally been used in analytical mechanics as a method of obtaining the equations of motion and boundary conditions for dynamical problems, and the trailing terms are often seen as irrelevant in the derivation of the equations of motion. However, Hamilton's law of varying action is not only a statement of equilibrium, but it is also more easily implementable in a form that can be used to obtain approximate analytical solutions or to construct a time-domain finite-element model, where the trailing terms need to be correctly treated (Bailey, 1975a,b; Hitzl, 1980; Baruch and Riff, 1982; Borri et al., 1985; Borri, 1986; Borri et al., 1988; Peters and Izadpanah, 1988).

2.1.4 Euler-Lagrange Equations

The minimum number of independent coordinates needed to describe the motion of a discrete system is called the number of degrees of freedom of the system. Equation (2.1.22) states Hamilton's law of varying action for a discrete system consisting of particles, springs, and dampers. For such an N-particle system, because three coordinates are used to describe the motion of each particle in three-dimensional space, the number of degrees of freedom is $n = 3N - c$, where c is the number of constraint equations due to the action of the surroundings upon the particles. In other words, there are c constraint equations when there are c dependent coordinates. For a multi-rigid-body system, the displacements of all particles within a rigid body can be described by six coordinates (three translational displacements and three rotational displacements) or less, and some constraint equations may emerge and the number of degrees of freedom is reduced just because some parts of the system are restrained.

If the coordinates used to describe the motion of a system are independent from one another, they are called generalized coordinates. In other words, generalized coordinates, which are denoted by q_i, must be independent of each other. They need not be Newtonian coordinates. They can be, for example, the unknown coefficients in the series expansion of the dependent variables of a system or the modal amplitudes of a continuous system.

Extra coordinates used in the dynamic modeling exceeding the number of degrees of freedom of the system are called superfluous coordinates, and constraint equations equal in number to the superfluous coordinates are necessary for their elimination. Constraints are called holonomic if the extra coordinates can be eliminated through constraint equations. For an n-degree-of-freedom system, holonomic constraints are in the form

$$C(q_1, q_2, ..., q_n) = 0 \quad \text{or} \quad C(q_1, q_2, ..., q_n, t) = 0 \qquad (2.1.26)$$

A system is holonomic if all of the constraint equations are holonomic. On the other hand, if a constraint equation cannot be expressed in the form (2.1.26), it is called nonholonomic. For example,

$$q_1 \dot{q}_1 + q_3 \dot{q}_2 + q_2^2 \dot{q}_3 = 0 \qquad (2.1.27)$$

is a nonholonomic constraint because it cannot be integrated to obtain the form (2.1.26).

For an n-degree-of-freedom holonomic system, the position vector (and displacement components) of a particle can be expressed in terms of the generalized coordinates q_i and time t in the form

$$\mathbf{r}_i = \mathbf{r}_i(q_1, q_2, ..., q_n, t) \qquad (2.1.28)$$

Hence, it follows from (2.1.28), (2.1.8), and (2.1.11) that the kinetic energy T and the potential energy Π can be expressed as

$$T = \frac{1}{2} \sum_{i=1}^{N} m_i \frac{d\mathbf{r}_i}{dt} \cdot \frac{d\mathbf{r}_i}{dt} = T(q_1, q_2, ..., q_n, \dot{q}_1, \dot{q}_2, ..., \dot{q}_n, t) \qquad (2.1.29)$$

$$\Pi = \Pi(q_1, q_2, ..., q_n) \qquad (2.1.30)$$

Hence, it follows from (2.1.22), (2.1.28), (2.1.21), (2.1.24), (2.1.29), and (2.1.30) that Hamilton's law of varying action can be expressed as

$$\int_{t_1}^{t_2} (\delta W + \delta T) dt - \sum_{i=1}^{n} p_i \delta q_i \Big|_{t_1}^{t_2} = 0 \qquad (2.1.31)$$

where p_i is the conjugate momentum of q_i and is defined as

$$p_i \equiv \frac{\partial L}{\partial \dot{q}_i} = \frac{\partial T}{\partial \dot{q}_i} \qquad (2.1.32)$$

Taking the variation of T in (2.1.29) and integrating the result by parts, we rewrite (2.1.31) as

$$\int_{t_1}^{t_2} \left[\sum_{i=1}^{n} Q_i^* \delta q_i - \sum_{i=1}^{n} \frac{d}{dt}\left(\frac{\partial T}{\partial \dot{q}_i}\right) \delta q_i + \sum_{i=1}^{n} \frac{\partial T}{\partial q_i} \delta q_i \right] dt$$
$$+ \sum_{i=1}^{n} \left(\frac{\partial T}{\partial \dot{q}_i} - p_i \right) \delta q_i \Big|_{t_1}^{t_2} = 0 \qquad (2.1.33)$$

where

$$\delta W = \sum_{i=1}^{n} Q_i^* \delta q_i \qquad (2.1.34)$$

and Q_i^* is the generalized force corresponding to q_i, which includes conservative and non-conservative forces. Because of (2.1.32) and the fact that δq_i is arbitrary, it follows from (2.1.33) that

$$\frac{d}{dt}\left(\frac{\partial T}{\partial \dot{q}_i}\right) - \frac{\partial T}{\partial q_i} = Q_i^*, \quad i = 1, 2, ..., n \qquad (2.1.35)$$

which are called the Euler-Lagrange equations. Because any general conservative force can be represented in terms of the potential function Π as $-\partial \Pi / \partial q_i$, we have

$$Q_i^* = -\frac{\partial \Pi}{\partial q_i} + Q_i \qquad (2.1.36a)$$

where the Q_i denote non-conservative forces. It follows from (2.1.34) and (2.1.36a) that

$$\delta W = \sum_{i=1}^{n} \left(Q_i \delta q_i - \frac{\partial \Pi}{\partial q_i} \delta q_i \right) = \delta W_{nc} - \delta \Pi \qquad (2.1.36b)$$

Substituting (2.1.36a) into (2.1.35) and using (2.1.24), we have

$$\frac{d}{dt}\left(\frac{\partial L}{\partial \dot{q}_i}\right) - \frac{\partial L}{\partial q_i} = Q_i, \quad i = 1, 2, ..., n \qquad (2.1.37)$$

This is another form of the Euler-Lagrange equations.

2.1.5 Hamilton's Equations

The Hamiltonian function H of an n-degree-of-freedom system is defined as

$$H \equiv \sum_{i=1}^{n} \dot{q}_i \frac{\partial L}{\partial \dot{q}_i} - L = \sum_{i=1}^{n} \dot{q}_i p_i - L \qquad (2.1.38)$$

where (2.1.32) is used. Consequently,

$$dH = \sum_{i=1}^{n} \dot{q}_i dp_i + \sum_{i=1}^{n} p_i d\dot{q}_i - \sum_{i=1}^{n} \frac{\partial L}{\partial \dot{q}_i} d\dot{q}_i - \sum_{i=1}^{n} \frac{\partial L}{\partial q_i} dq_i - \frac{\partial L}{\partial t} dt$$

$$= \sum_{i=1}^{n} \dot{q}_i dp_i - \sum_{i=1}^{n} \dot{p}_i dq_i + \sum_{i=1}^{n} Q_i dq_i - \frac{\partial L}{\partial t} dt \qquad (2.1.39)$$

where (2.1.32) and (2.1.37) are used. Also, it follows from (2.1.29), (2.1.32), and (2.1.38) that

$$H = H(q_1, q_2, \ldots, q_n, \dot{q}_1, \dot{q}_2, \ldots, \dot{q}_n, t) = H(q_1, q_2, \ldots, q_n, p_1, p_2, \ldots, p_n, t) \qquad (2.1.40)$$

Hence,

$$dH = \sum_{i=1}^{n} \frac{\partial H}{\partial p_i} dp_i + \sum_{i=1}^{n} \frac{\partial H}{\partial q_i} dq_i + \frac{\partial H}{\partial t} dt \qquad (2.1.41)$$

Comparing (2.1.39) and (2.1.41), we conclude that

$$\dot{q}_i = \frac{\partial H}{\partial p_i}, \quad \dot{p}_i = -\frac{\partial H}{\partial q_i} + Q_i, \quad \frac{\partial H}{\partial t} = -\frac{\partial L}{\partial t} \qquad (2.1.42)$$

which are called Hamilton's equations.

Because the generalized coordinates q_i may not be inertial coordinates, the corresponding generalized kinetic energy T consists of three different terms

$$T = \frac{1}{2} \sum_{i=1}^{n} \sum_{j=1}^{n} \theta_{ij} \dot{q}_i \dot{q}_j + \sum_{i=1}^{n} \beta_i \dot{q}_i + T_0 = T_2 + T_1 + T_0 \qquad (2.1.43)$$

where

$$T_2 \equiv \frac{1}{2} \sum_{i=1}^{n} \sum_{j=1}^{n} \theta_{ij} \dot{q}_i \dot{q}_j, \quad T_1 \equiv \sum_{i=1}^{n} \beta_i \dot{q}_i \qquad (2.1.44)$$

and $\theta_{ij} (= \theta_{ji})$, β_i, and T_0 are functions of the q_i and time t. It follows from (2.1.38), (2.1.24), (2.1.32), (2.1.43), and (2.1.44) that

$$H = \sum_{i=1}^{n} \dot{q}_i \frac{\partial T}{\partial \dot{q}_i} - T + \Pi = 2T_2 + T_1 - (T_2 + T_1 + T_0) + \Pi$$

$$= E - 2T_0 - T_1 \qquad (2.1.45)$$

where E is the total energy of the system defined as

$$E \equiv T + \Pi \tag{2.1.46}$$

It follows from (2.1.39) that

$$\begin{aligned}\frac{dH}{dt} &= \sum_{i=1}^{n} \dot{q}_i \dot{p}_i - \sum_{i=1}^{n} \dot{p}_i \dot{q}_i - \frac{\partial L}{\partial t} + \sum_{i=1}^{n} Q_i \dot{q}_i \\ &= -\frac{\partial L}{\partial t} + \sum_{i=1}^{n} Q_i \dot{q}_i \end{aligned} \tag{2.1.47}$$

If $Q_i = 0$ (i.e., $\Delta W_{nc} = 0$), it follows from (2.1.47) and (2.1.42) that

$$\frac{dH}{dt} = -\frac{\partial L}{\partial t} = \frac{\partial H}{\partial t} \tag{2.1.48}$$

If we further assume that L does not include t explicitly, then $\partial L/\partial t = 0$ and

$$\frac{dH}{dt} = 0 \tag{2.1.49}$$

which implies that H is a constant function; that is,

$$T_2 + \Pi - T_0 = \text{constant} \tag{2.1.50}$$

We point out here that there is no T_1 or T_0 in (2.1.1) or (2.1.8) because Newtonian coordinates are used. In other words,

$$T = T_2 \tag{2.1.51}$$

for a system described by Newtonian coordinates.

2.2 STRAIN-DISPLACEMENT RELATIONS

Because structures are continuous systems and external loads on structures may not be uniformly distributed, structural deformations are, in general, functions of spatial coordinates as well as time. Hence, in order to state the balance of forces and moments, one can only treat an infinitesimal structural particle or element. Thus, one needs to introduce stresses and strains in order to relate the external loads to structural displacements. A stress denotes a ratio of an infinitesimal force to an infinitesimal area on which the force is applied, strains can be represented in terms of displacements, and stresses and strains are related by a constitutive law. In other words, external loads and structural displacements are indirectly related in continuous systems.

Because we are concerned in this book with thin-walled structures undergoing large global displacements and rotations but small relative displacements among points in the structure, the strain values are assumed to be small and the stress-strain relations

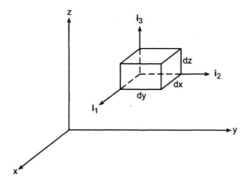

Fig. 2.2.1 A rectangular structural particle.

are assumed to be linear. However, the strain-displacement relations can be nonlinear. Moreover, because engineering strains are defined with respect to undeformed structural dimensions and there are no comoving derivatives involved in the Lagrangian formulation since all variables are represented in terms of the initial undeformed coordinates, we will use Lagrangian coordinates in all the formulations and analyses in this book.

The displacement vector $\mathbf{D}$ of a particle can be referred to any set of right-handed Cartesian or orthogonal curvilinear coordinate system s_1, s_2, and s_3 as

$$\mathbf{D} \equiv u_1 \mathbf{i}_1 + u_2 \mathbf{i}_2 + u_3 \mathbf{i}_3 \tag{2.2.1}$$

where u_1, u_2, and u_3 are the displacements and $\mathbf{i}_1$, $\mathbf{i}_2$, and $\mathbf{i}_3$ are unit vectors along these three axes. The corresponding infinitesimal strain tensor e_{ij} can be represented in the general form

$$e_{ij} \equiv \frac{1}{2}\left(\frac{\partial \mathbf{D}}{\partial s_i} \cdot \mathbf{i}_j + \frac{\partial \mathbf{D}}{\partial s_j} \cdot \mathbf{i}_i\right), \quad i,j = 1,2,3 \tag{2.2.2}$$

It is a second-order tensor requiring two indices to identify its elements or components. A vector is a first-order tensor, and a scalar is a zero-order tensor. It is clear from the definition (2.2.2) of the strain tensor that it is symmetric because $e_{ij} = e_{ji}$. Next, we represent the e_{ij} in terms of the u_i using different coordinate systems.

<u>Cartesian Coordinate System:</u> For a Cartesian coordinate system (see Figure 2.2.1), we have

$$\partial s_1 = \partial x, \quad \partial s_2 = \partial y, \quad \partial s_3 = \partial z$$
$$\frac{\partial \mathbf{i}_m}{\partial x} = \frac{\partial \mathbf{i}_m}{\partial y} = \frac{\partial \mathbf{i}_m}{\partial z} = 0, \quad m = 1,2,3 \tag{2.2.3}$$

Hence,

$$\frac{\partial \mathbf{D}}{\partial x} = \frac{\partial u_1}{\partial x}\mathbf{i}_1 + \frac{\partial u_2}{\partial x}\mathbf{i}_2 + \frac{\partial u_3}{\partial x}\mathbf{i}_3$$

STRAIN-DISPLACEMENT RELATIONS 77

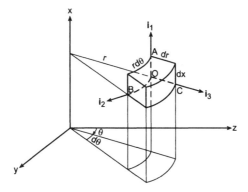

Fig. 2.2.2 A cylindrical structural particle.

$$\frac{\partial \mathbf{D}}{\partial y} = \frac{\partial u_1}{\partial y}\mathbf{i}_1 + \frac{\partial u_2}{\partial y}\mathbf{i}_2 + \frac{\partial u_3}{\partial y}\mathbf{i}_3$$
$$\frac{\partial \mathbf{D}}{\partial z} = \frac{\partial u_1}{\partial z}\mathbf{i}_1 + \frac{\partial u_2}{\partial z}\mathbf{i}_2 + \frac{\partial u_3}{\partial z}\mathbf{i}_3 \qquad (2.2.4)$$

The strains e_{ij} are given by

$$e_{11} = \frac{\partial \mathbf{D}}{\partial x} \cdot \mathbf{i}_1 = \frac{\partial u_1}{\partial x}$$
$$e_{12} = \frac{1}{2}\left(\frac{\partial \mathbf{D}}{\partial y} \cdot \mathbf{i}_1 + \frac{\partial \mathbf{D}}{\partial x} \cdot \mathbf{i}_2\right) = \frac{1}{2}\left(\frac{\partial u_1}{\partial y} + \frac{\partial u_2}{\partial x}\right)$$
$$e_{13} = \frac{1}{2}\left(\frac{\partial \mathbf{D}}{\partial z} \cdot \mathbf{i}_1 + \frac{\partial \mathbf{D}}{\partial x} \cdot \mathbf{i}_3\right) = \frac{1}{2}\left(\frac{\partial u_1}{\partial z} + \frac{\partial u_3}{\partial x}\right)$$
$$e_{22} = \frac{\partial \mathbf{D}}{\partial y} \cdot \mathbf{i}_2 = \frac{\partial u_2}{\partial y}$$
$$e_{23} = \frac{1}{2}\left(\frac{\partial \mathbf{D}}{\partial z} \cdot \mathbf{i}_2 + \frac{\partial \mathbf{D}}{\partial y} \cdot \mathbf{i}_3\right) = \frac{1}{2}\left(\frac{\partial u_2}{\partial z} + \frac{\partial u_3}{\partial y}\right)$$
$$e_{33} = \frac{\partial \mathbf{D}}{\partial z} \cdot \mathbf{i}_3 = \frac{\partial u_3}{\partial z} \qquad (2.2.5)$$

Cylindrical Coordinate System: For a cylindrical coordinate system (see Figure 2.2.2), we have

$$\partial s_1 = \partial x, \quad \partial s_2 = r\partial\theta, \quad \partial s_3 = \partial r \qquad (2.2.6)$$
$$\mathbf{i}_1 = \mathbf{i}_x$$
$$\mathbf{i}_2 = \cos\theta\, \mathbf{i}_y - \sin\theta\, \mathbf{i}_z$$
$$\mathbf{i}_3 = \sin\theta\, \mathbf{i}_y + \cos\theta\, \mathbf{i}_z \qquad (2.2.7)$$

78 ELASTICITY

where $\mathbf{i}_x$, $\mathbf{i}_y$, and $\mathbf{i}_z$ are the base vectors of the reference Cartesian coordinate system xyz. Differentiating (2.2.7) yields

$$\frac{\partial \mathbf{i}_1}{\partial x} = \frac{\partial \mathbf{i}_2}{\partial x} = \frac{\partial \mathbf{i}_3}{\partial x} = \frac{\partial \mathbf{i}_1}{\partial r} = \frac{\partial \mathbf{i}_2}{\partial r} = \frac{\partial \mathbf{i}_3}{\partial r} = 0$$

$$\frac{\partial \mathbf{i}_1}{\partial \theta} = 0, \quad \frac{\partial \mathbf{i}_2}{\partial \theta} = -\mathbf{i}_3, \quad \frac{\partial \mathbf{i}_3}{\partial \theta} = \mathbf{i}_2 \tag{2.2.8}$$

Hence,

$$\frac{\partial \mathbf{D}}{\partial x} = \frac{\partial u_1}{\partial x}\mathbf{i}_1 + \frac{\partial u_2}{\partial x}\mathbf{i}_2 + \frac{\partial u_3}{\partial x}\mathbf{i}_3$$

$$\frac{1}{r}\frac{\partial \mathbf{D}}{\partial \theta} = \frac{1}{r}\frac{\partial u_1}{\partial \theta}\mathbf{i}_1 + \frac{1}{r}\frac{\partial u_2}{\partial \theta}\mathbf{i}_2 + \frac{1}{r}\frac{\partial u_3}{\partial \theta}\mathbf{i}_3 - \frac{u_2}{r}\mathbf{i}_3 + \frac{u_3}{r}\mathbf{i}_2$$

$$\frac{\partial \mathbf{D}}{\partial r} = \frac{\partial u_1}{\partial r}\mathbf{i}_1 + \frac{\partial u_2}{\partial r}\mathbf{i}_2 + \frac{\partial u_3}{\partial r}\mathbf{i}_3 \tag{2.2.9}$$

Consequently, the strain-displacement relations are

$$e_{11} = \frac{\partial \mathbf{D}}{\partial x} \cdot \mathbf{i}_1 = \frac{\partial u_1}{\partial x}$$

$$e_{12} = \frac{1}{2}\left(\frac{1}{r}\frac{\partial \mathbf{D}}{\partial \theta}\cdot \mathbf{i}_1 + \frac{\partial \mathbf{D}}{\partial x}\cdot \mathbf{i}_2\right) = \frac{1}{2}\left(\frac{1}{r}\frac{\partial u_1}{\partial \theta} + \frac{\partial u_2}{\partial x}\right)$$

$$e_{13} = \frac{1}{2}\left(\frac{\partial \mathbf{D}}{\partial r}\cdot \mathbf{i}_1 + \frac{\partial \mathbf{D}}{\partial x}\cdot \mathbf{i}_3\right) = \frac{1}{2}\left(\frac{\partial u_1}{\partial r} + \frac{\partial u_3}{\partial x}\right)$$

$$e_{22} = \frac{1}{r}\frac{\partial \mathbf{D}}{\partial \theta}\cdot \mathbf{i}_2 = \frac{1}{r}\frac{\partial u_2}{\partial \theta} + \frac{u_3}{r}$$

$$e_{23} = \frac{1}{2}\left(\frac{\partial \mathbf{D}}{\partial r}\cdot \mathbf{i}_2 + \frac{1}{r}\frac{\partial \mathbf{D}}{\partial \theta}\cdot \mathbf{i}_3\right) = \frac{1}{2}\left(\frac{\partial u_2}{\partial r} + \frac{1}{r}\frac{\partial u_3}{\partial \theta} - \frac{u_2}{r}\right)$$

$$e_{33} = \frac{\partial \mathbf{D}}{\partial r}\cdot \mathbf{i}_3 = \frac{\partial u_3}{\partial r} \tag{2.2.10}$$

Spherical Coordinate System: For a spherical coordinate system (see Figure 2.2.3), we have

$$\partial s_1 = r\partial \theta, \quad \partial s_2 = r\sin\theta \partial\phi, \quad \partial s_3 = \partial r \tag{2.2.11}$$

$$\mathbf{i}_1 = \cos\theta\cos\phi\,\mathbf{i}_x + \cos\theta\sin\phi\,\mathbf{i}_y - \sin\theta\,\mathbf{i}_z$$

$$\mathbf{i}_2 = -\sin\phi\,\mathbf{i}_x + \cos\phi\,\mathbf{i}_y$$

$$\mathbf{i}_3 = \sin\theta\cos\phi\,\mathbf{i}_x + \sin\theta\sin\phi\,\mathbf{i}_y + \cos\theta\,\mathbf{i}_z \tag{2.2.12}$$

Differentiating (2.2.12) yields

$$\frac{\partial \mathbf{i}_1}{\partial r} = \frac{\partial \mathbf{i}_2}{\partial r} = \frac{\partial \mathbf{i}_3}{\partial r} = 0$$

STRAIN-DISPLACEMENT RELATIONS 79

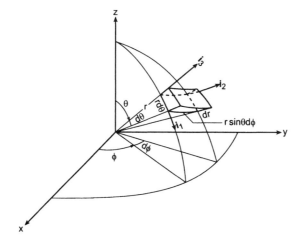

Fig. 2.2.3 A structural particle in a spherical coordinate system.

$$\frac{\partial \mathbf{i}_1}{\partial \theta} = -\mathbf{i}_3, \quad \frac{\partial \mathbf{i}_2}{\partial \theta} = 0, \quad \frac{\partial \mathbf{i}_3}{\partial \theta} = \mathbf{i}_1$$

$$\frac{\partial \mathbf{i}_1}{\partial \phi} = \cos \theta \, \mathbf{i}_2, \quad \frac{\partial \mathbf{i}_2}{\partial \phi} = -\cos \theta \, \mathbf{i}_1 - \sin \theta \, \mathbf{i}_3, \quad \frac{\partial \mathbf{i}_3}{\partial \phi} = \sin \theta \, \mathbf{i}_2 \quad (2.2.13)$$

Hence,

$$\frac{1}{r}\frac{\partial \mathbf{D}}{\partial \theta} = \frac{1}{r}\frac{\partial u_1}{\partial \theta}\mathbf{i}_1 + \frac{1}{r}\frac{\partial u_2}{\partial \theta}\mathbf{i}_2 + \frac{1}{r}\frac{\partial u_3}{\partial \theta}\mathbf{i}_3 + \frac{u_3}{r}\mathbf{i}_1 - \frac{u_1}{r}\mathbf{i}_3$$

$$\frac{1}{r \sin \theta}\frac{\partial \mathbf{D}}{\partial \phi} = \frac{1}{r \sin \theta}\frac{\partial u_1}{\partial \phi}\mathbf{i}_1 + \frac{1}{r \sin \theta}\frac{\partial u_2}{\partial \phi}\mathbf{i}_2 + \frac{1}{r \sin \theta}\frac{\partial u_3}{\partial \phi}\mathbf{i}_3$$

$$- \frac{u_2}{r \tan \theta}\mathbf{i}_1 + \left(\frac{u_1}{r \tan \theta} + \frac{u_3}{r}\right)\mathbf{i}_2 - \frac{u_2}{r}\mathbf{i}_3$$

$$\frac{\partial \mathbf{D}}{\partial r} = \frac{\partial u_1}{\partial r}\mathbf{i}_1 + \frac{\partial u_2}{\partial r}\mathbf{i}_2 + \frac{\partial u_3}{\partial r}\mathbf{i}_3 \quad (2.2.14)$$

Consequently, the strain-displacement relations are

$$e_{11} = \frac{1}{r}\frac{\partial \mathbf{D}}{\partial \theta} \cdot \mathbf{i}_1 = \frac{1}{r}\frac{\partial u_1}{\partial \theta} + \frac{u_3}{r}$$

$$e_{12} = \frac{1}{2}\left(\frac{1}{r \sin \theta}\frac{\partial \mathbf{D}}{\partial \phi} \cdot \mathbf{i}_1 + \frac{1}{r}\frac{\partial \mathbf{D}}{\partial \theta} \cdot \mathbf{i}_2\right) = \frac{1}{2}\left(\frac{1}{r \sin \theta}\frac{\partial u_1}{\partial \phi} + \frac{1}{r}\frac{\partial u_2}{\partial \theta} - \frac{u_2}{r \tan \theta}\right)$$

$$e_{13} = \frac{1}{2}\left(\frac{\partial \mathbf{D}}{\partial r} \cdot \mathbf{i}_1 + \frac{1}{r}\frac{\partial \mathbf{D}}{\partial \theta} \cdot \mathbf{i}_3\right) = \frac{1}{2}\left(\frac{\partial u_1}{\partial r} + \frac{1}{r}\frac{\partial u_3}{\partial \theta} - \frac{u_1}{r}\right)$$

$$e_{22} = \frac{1}{r \sin \theta}\frac{\partial \mathbf{D}}{\partial \phi} \cdot \mathbf{i}_2 = \frac{1}{r \sin \theta}\frac{\partial u_2}{\partial \phi} + \frac{u_1}{r \tan \theta} + \frac{u_3}{r}$$

$$e_{23} = \frac{1}{2}\left(\frac{\partial \mathbf{D}}{\partial r} \cdot \mathbf{i}_2 + \frac{1}{r \sin \theta}\frac{\partial \mathbf{D}}{\partial \phi} \cdot \mathbf{i}_3\right) = \frac{1}{2}\left(\frac{\partial u_2}{\partial r} + \frac{1}{r \sin \theta}\frac{\partial u_3}{\partial \phi} - \frac{u_2}{r}\right)$$

80   ELASTICITY

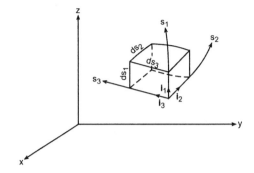

Fig. 2.2.4 A structural particle in a general orthogonal curvilinear coordinate system.

$$e_{33} = \frac{\partial \mathbf{D}}{\partial r} \cdot \mathbf{i}_3 = \frac{\partial u_3}{\partial r} \tag{2.2.15}$$

General Orthogonal Curvilinear Coordinate System: For a general orthogonal curvilinear coordinate system (see Figure 2.2.4), we have

$$\mathbf{i}_j \cdot \mathbf{i}_k = \delta_{jk} \tag{2.2.16}$$

where δ_{jk} is the Kronecker delta. Differentiating (2.2.16) yields the identities

$$\frac{\partial \mathbf{i}_j}{\partial s_m} \cdot \mathbf{i}_j = 0, \quad \frac{\partial \mathbf{i}_j}{\partial s_m} \cdot \mathbf{i}_k = -\frac{\partial \mathbf{i}_k}{\partial s_m} \cdot \mathbf{i}_j \quad \text{for } j, k, m = 1, 2, 3 \tag{2.2.17}$$

Using (2.2.17), we have

$$\frac{\partial}{\partial s_1} \begin{Bmatrix} \mathbf{i}_1 \\ \mathbf{i}_2 \\ \mathbf{i}_3 \end{Bmatrix} = [K_1] \begin{Bmatrix} \mathbf{i}_1 \\ \mathbf{i}_2 \\ \mathbf{i}_3 \end{Bmatrix} \tag{2.2.18}$$

$$\frac{\partial}{\partial s_2} \begin{Bmatrix} \mathbf{i}_1 \\ \mathbf{i}_2 \\ \mathbf{i}_3 \end{Bmatrix} = [K_2] \begin{Bmatrix} \mathbf{i}_1 \\ \mathbf{i}_2 \\ \mathbf{i}_3 \end{Bmatrix} \tag{2.2.19}$$

$$\frac{\partial}{\partial s_3} \begin{Bmatrix} \mathbf{i}_1 \\ \mathbf{i}_2 \\ \mathbf{i}_3 \end{Bmatrix} = [K_3] \begin{Bmatrix} \mathbf{i}_1 \\ \mathbf{i}_2 \\ \mathbf{i}_3 \end{Bmatrix} \tag{2.2.20}$$

where the curvature matrices $[K_j]$ are given by

$$[K_1] \equiv \begin{bmatrix} \mathbf{i}_{1s_1} \cdot \mathbf{i}_1 & \mathbf{i}_{1s_1} \cdot \mathbf{i}_2 & \mathbf{i}_{1s_1} \cdot \mathbf{i}_3 \\ \mathbf{i}_{2s_1} \cdot \mathbf{i}_1 & \mathbf{i}_{2s_1} \cdot \mathbf{i}_2 & \mathbf{i}_{2s_1} \cdot \mathbf{i}_3 \\ \mathbf{i}_{3s_1} \cdot \mathbf{i}_1 & \mathbf{i}_{3s_1} \cdot \mathbf{i}_2 & \mathbf{i}_{3s_1} \cdot \mathbf{i}_3 \end{bmatrix} = \begin{bmatrix} 0 & k_{13} & -k_{12} \\ -k_{13} & 0 & k_{11} \\ k_{12} & -k_{11} & 0 \end{bmatrix} \tag{2.2.21}$$

STRAIN-DISPLACEMENT RELATIONS 81

$$[K_2] \equiv \begin{bmatrix} \mathbf{i}_{1s_2} \cdot \mathbf{i}_1 & \mathbf{i}_{1s_2} \cdot \mathbf{i}_2 & \mathbf{i}_{1s_2} \cdot \mathbf{i}_3 \\ \mathbf{i}_{2s_2} \cdot \mathbf{i}_1 & \mathbf{i}_{2s_2} \cdot \mathbf{i}_2 & \mathbf{i}_{2s_2} \cdot \mathbf{i}_3 \\ \mathbf{i}_{3s_2} \cdot \mathbf{i}_1 & \mathbf{i}_{3s_2} \cdot \mathbf{i}_2 & \mathbf{i}_{3s_2} \cdot \mathbf{i}_3 \end{bmatrix} = \begin{bmatrix} 0 & k_{23} & -k_{22} \\ -k_{23} & 0 & k_{21} \\ k_{22} & -k_{21} & 0 \end{bmatrix} \quad (2.2.22)$$

$$[K_3] \equiv \begin{bmatrix} \mathbf{i}_{1s_3} \cdot \mathbf{i}_1 & \mathbf{i}_{1s_3} \cdot \mathbf{i}_2 & \mathbf{i}_{1s_3} \cdot \mathbf{i}_3 \\ \mathbf{i}_{2s_3} \cdot \mathbf{i}_1 & \mathbf{i}_{2s_3} \cdot \mathbf{i}_2 & \mathbf{i}_{2s_3} \cdot \mathbf{i}_3 \\ \mathbf{i}_{3s_3} \cdot \mathbf{i}_1 & \mathbf{i}_{3s_3} \cdot \mathbf{i}_2 & \mathbf{i}_{3s_3} \cdot \mathbf{i}_3 \end{bmatrix} = \begin{bmatrix} 0 & k_{33} & -k_{32} \\ -k_{33} & 0 & k_{31} \\ k_{32} & -k_{31} & 0 \end{bmatrix} \quad (2.2.23)$$

and $\mathbf{i}_{ms_n} \equiv \partial \mathbf{i}_m / \partial s_n$.

For the cylindrical coordinate system $x\theta r$ shown in Figure 2.2.2, it follows from (2.2.6) and (2.2.8) that

$$[K_1] = [0], \quad [K_2] = \begin{bmatrix} 0 & 0 & 0 \\ 0 & 0 & -1/r \\ 0 & 1/r & 0 \end{bmatrix}, \quad [K_3] = [0] \quad (2.2.24)$$

Moreover, for the spherical coordinate system $\theta\phi r$ shown in Figure 2.2.3, it follows from (2.2.11) and (2.2.13) that

$$[K_1] = \begin{bmatrix} 0 & 0 & -1/r \\ 0 & 0 & 0 \\ 1/r & 0 & 0 \end{bmatrix}, \quad [K_3] = [0],$$

$$[K_2] = \begin{bmatrix} 0 & 1/(r\tan\theta) & 0 \\ -1/(r\tan\theta) & 0 & -1/r \\ 0 & 1/r & 0 \end{bmatrix} \quad (2.2.25)$$

Using (2.2.18)-(2.2.23), we have

$$\frac{\partial \mathbf{D}}{\partial s_1} = \frac{\partial u_1}{\partial s_1}\mathbf{i}_1 + \frac{\partial u_2}{\partial s_1}\mathbf{i}_2 + \frac{\partial u_3}{\partial s_1}\mathbf{i}_3 + \mathbf{i}_1\left(u_3 k_{12} - u_2 k_{13}\right)$$
$$+ \mathbf{i}_2\left(u_1 k_{13} - u_3 k_{11}\right) + \mathbf{i}_3\left(u_2 k_{11} - u_1 k_{12}\right)$$

$$\frac{\partial \mathbf{D}}{\partial s_2} = \frac{\partial u_1}{\partial s_2}\mathbf{i}_1 + \frac{\partial u_2}{\partial s_2}\mathbf{i}_2 + \frac{\partial u_3}{\partial s_2}\mathbf{i}_3 + \mathbf{i}_1\left(u_3 k_{22} - u_2 k_{23}\right)$$
$$+ \mathbf{i}_2\left(u_1 k_{23} - u_3 k_{21}\right) + \mathbf{i}_3\left(u_2 k_{21} - u_1 k_{22}\right)$$

$$\frac{\partial \mathbf{D}}{\partial s_3} = \frac{\partial u_1}{\partial s_3}\mathbf{i}_1 + \frac{\partial u_2}{\partial s_3}\mathbf{i}_2 + \frac{\partial u_3}{\partial s_3}\mathbf{i}_3 + \mathbf{i}_1\left(u_3 k_{32} - u_2 k_{33}\right)$$
$$+ \mathbf{i}_2\left(u_1 k_{33} - u_3 k_{31}\right) + \mathbf{i}_3\left(u_2 k_{31} - u_1 k_{32}\right) \quad (2.2.26)$$

Consequently, the general strain components are

$$e_{11} = \frac{\partial \mathbf{D}}{\partial s_1} \cdot \mathbf{i}_1 = \frac{\partial u_1}{\partial s_1} + u_3 k_{12} - u_2 k_{13}$$

$$e_{12} = \frac{1}{2}\left(\frac{\partial \mathbf{D}}{\partial s_1} \cdot \mathbf{i}_2 + \frac{\partial \mathbf{D}}{\partial s_2} \cdot \mathbf{i}_1\right)$$

$$= \frac{1}{2}\left(\frac{\partial u_2}{\partial s_1} + \frac{\partial u_1}{\partial s_2} + u_1 k_{13} - u_3 k_{11} + u_3 k_{22} - u_2 k_{23}\right)$$

$$e_{13} = \frac{1}{2}\left(\frac{\partial \mathbf{D}}{\partial s_1} \cdot \mathbf{i}_3 + \frac{\partial \mathbf{D}}{\partial s_3} \cdot \mathbf{i}_1\right)$$

$$= \frac{1}{2}\left(\frac{\partial u_3}{\partial s_1} + \frac{\partial u_1}{\partial s_3} + u_2 k_{11} - u_1 k_{12} + u_3 k_{32} - u_2 k_{33}\right)$$

$$e_{22} = \frac{\partial \mathbf{D}}{\partial s_2} \cdot \mathbf{i}_2 = \frac{\partial u_2}{\partial s_2} + u_1 k_{23} - u_3 k_{21}$$

$$e_{23} = \frac{1}{2}\left(\frac{\partial \mathbf{D}}{\partial s_2} \cdot \mathbf{i}_3 + \frac{\partial \mathbf{D}}{\partial s_3} \cdot \mathbf{i}_2\right)$$

$$= \frac{1}{2}\left(\frac{\partial u_3}{\partial s_2} + \frac{\partial u_2}{\partial s_3} + u_2 k_{21} - u_1 k_{22} + u_1 k_{33} - u_3 k_{31}\right)$$

$$e_{33} = \frac{\partial \mathbf{D}}{\partial s_3} \cdot \mathbf{i}_3 = \frac{\partial u_3}{\partial s_3} + u_2 k_{31} - u_1 k_{32} \tag{2.2.27}$$

2.3 TRANSFORMATION OF STRAINS AND STRESSES

Next, we show how to transform strains at a point in an unprimed orthogonal coordinate system $s_1 s_2 s_3$ to those at the same point in another primed orthogonal coordinate system $s_1' s_2' s_3'$. We denote the unit vectors along the axes of these systems by $\mathbf{i}_1, \mathbf{i}_2, \mathbf{i}_3$ and $\mathbf{i}_1', \mathbf{i}_2', \mathbf{i}_3'$, respectively. Moreover, we introduce the direction cosines between the two sets of axes as follows:

$T_{11}, T_{12}, T_{13} =$ direction cosines of the s_1'-axis with respect to the
s_1, s_2, s_3-axes

$T_{21}, T_{22}, T_{23} =$ direction cosines of the s_2'-axis with respect to the
s_1, s_2, s_3-axes

$T_{31}, T_{32}, T_{33} =$ direction cosines of the s_3'-axis with respect to the
s_1, s_2, s_3-axes

Thus,
$$T_{ij} = \mathbf{i}_i' \cdot \mathbf{i}_j \text{ for } i, j = 1, 2, 3 \tag{2.3.1}$$

Consequently, $\mathbf{i}_1', \mathbf{i}_2'$, and $\mathbf{i}_3'$ can be expressed in terms of $\mathbf{i}_1, \mathbf{i}_2$, and $\mathbf{i}_3$ by the relations

$$\mathbf{i}_1' = T_{11}\mathbf{i}_1 + T_{12}\mathbf{i}_2 + T_{13}\mathbf{i}_3$$
$$\mathbf{i}_2' = T_{21}\mathbf{i}_1 + T_{22}\mathbf{i}_2 + T_{23}\mathbf{i}_3$$
$$\mathbf{i}_3' = T_{31}\mathbf{i}_1 + T_{32}\mathbf{i}_2 + T_{33}\mathbf{i}_3 \tag{2.3.2}$$

Using the summation convention, we can express the relations (2.3.2) in compact form as

$$\mathbf{i}_i' = T_{ij}\mathbf{i}_j \tag{2.3.3}$$

where the repeated occurrence of the small Latin-letter subscript j represents a summation over the definition range of j. In this case, $j = 1, 2,$ and 3. In matrix form, (2.3.2) or (2.3.3) can be written as

$$\begin{Bmatrix} \mathbf{i}'_1 \\ \mathbf{i}'_2 \\ \mathbf{i}'_3 \end{Bmatrix} = [T] \begin{Bmatrix} \mathbf{i}_1 \\ \mathbf{i}_2 \\ \mathbf{i}_3 \end{Bmatrix} \qquad (2.3.4)$$

where the elements of $[T]$ are the T_{ij}.

Similarly, one can invert the process and use the direction cosines to express the unprimed unit vectors $\mathbf{i}_j$ in terms of the primed unit vectors $\mathbf{i}'_j$. Thus, we can write

$$\mathbf{i}_j = T_{kj} \mathbf{i}'_k \qquad (2.3.5)$$

or in matrix form as

$$\begin{Bmatrix} \mathbf{i}_1 \\ \mathbf{i}_2 \\ \mathbf{i}_3 \end{Bmatrix} = [T]^T \begin{Bmatrix} \mathbf{i}'_1 \\ \mathbf{i}'_2 \\ \mathbf{i}'_3 \end{Bmatrix} \qquad (2.3.6)$$

where $[T]^T$ is the transpose of $[T]$.

Substituting for the unprimed unit vectors from (2.3.6) into (2.3.4) yields

$$[T][T]^T = [I] \qquad (2.3.7)$$

where $[I]$ is the identity matrix. Thus, $[T]^T$ is equal to the inverse of $[T]$ and hence $[T]$ is called a unitary or orthogonal matrix. The matrix $[T]$ also represents an orthogonal transformation.

Alternatively, substituting for $\mathbf{i}_j$ from (2.3.5) into (2.3.3) yields

$$T_{ij} T_{kj} \mathbf{i}'_k - \mathbf{i}'_i = 0$$

which, upon using the Kronecker delta δ_{ik}, becomes

$$T_{ij} T_{kj} \mathbf{i}'_k - \delta_{ik} \mathbf{i}'_k = (T_{ij} T_{kj} - \delta_{ik}) \mathbf{i}'_k = 0$$

Hence,

$$T_{ij} T_{kj} = \delta_{ik} \qquad (2.3.8)$$

and the columns of the matrix $[T]$ are orthonormal.

Any displacement vector $\mathbf{D}$ can be expressed in terms of the primed and unprimed coordinate systems as

$$\mathbf{D} = u_1 \mathbf{i}_1 + u_2 \mathbf{i}_2 + u_3 \mathbf{i}_3 = u_j \mathbf{i}_j \qquad (2.3.9)$$
$$\mathbf{D} = u'_1 \mathbf{i}'_1 + u'_2 \mathbf{i}'_2 + u'_3 \mathbf{i}'_3 = u'_i \mathbf{i}'_i \qquad (2.3.10)$$

where $u_1, u_2,$ and u_3 are the components of $\mathbf{D}$ along the $s_1, s_2,$ and s_3-axes and $u'_1, u'_2,$ and u'_3 are the components of $\mathbf{D}$ along the $s'_1, s'_2,$ and s'_3-axes. We note that the vector $\mathbf{D}$ is invariant under any coordinate transformation, such as from the primed to the unprimed coordinate system. In other words, the actual vector remains unchanged no matter what coordinate system is used. We note that a vector is a first-order tensor. Dotting (2.3.9) by $\mathbf{i}'_i$, we have

$$\mathbf{D} \cdot \mathbf{i}'_i = u_j \mathbf{i}_j \cdot \mathbf{i}'_i = T_{ij} u_j \tag{2.3.11}$$

according to (2.3.1). But $\mathbf{D} \cdot \mathbf{i}'_i = u'_i$ according to (2.3.10), hence

$$u'_i = T_{ij} u_j \tag{2.3.12}$$

Similarly, one can show that

$$u_j = T_{ij} u'_i \tag{2.3.13}$$

Using arguments similar to those used in (2.3.9)-(2.3.13), one can show that infinitesimal coordinates are related by

$$ds'_i = T_{ij} ds_j \quad \text{or} \quad ds_j = T_{ij} ds'_i \tag{2.3.14}$$

To express the components e'_{ij} of the primed strain tensor in terms of the components e_{ij} of the unprimed strain tensor, we use the definition (2.2.2) and write

$$e'_{ij} = \frac{1}{2}\left(\frac{\partial \mathbf{D}}{\partial s'_i} \cdot \mathbf{i}'_j + \frac{\partial \mathbf{D}}{\partial s'_j} \cdot \mathbf{i}'_i\right) \tag{2.3.15}$$

where $\mathbf{D}$ is given by (2.3.10). Using (2.3.3) and the chain rule in (2.3.15), we obtain

$$e'_{ij} = \frac{1}{2}\left(\frac{\partial \mathbf{D}}{\partial s_k}\frac{\partial s_k}{\partial s'_i} \cdot T_{jm}\mathbf{i}_m + \frac{\partial \mathbf{D}}{\partial s_n}\frac{\partial s_m}{\partial s'_j} \cdot T_{ik}\mathbf{i}_k\right) \tag{2.3.16}$$

which, upon using (2.3.14), becomes

$$e'_{ij} = \frac{1}{2}\left(\frac{\partial \mathbf{D}}{\partial s_k}T_{ik} \cdot T_{jm}\mathbf{i}_m + \frac{\partial \mathbf{D}}{\partial s_m}T_{jm} \cdot T_{ik}\mathbf{i}_k\right)$$
$$= \frac{1}{2}\left(\frac{\partial \mathbf{D}}{\partial s_k} \cdot \mathbf{i}_m + \frac{\partial \mathbf{D}}{\partial s_m} \cdot \mathbf{i}_k\right) T_{ik} T_{jm} \tag{2.3.17}$$

or

$$e'_{ij} = T_{ik} e_{km} T_{jm} \tag{2.3.18}$$

Equation (2.3.18) can be rewritten in matrix form as

$$[e'] = [T][e][T]^T \tag{2.3.19}$$

The relation (2.3.18) can be obtained in an alternate way by using the invariant property of the strain dyadic, which is similar to the invariant property of vectors (i.e., first-order tensors). In other words, the actual strain dyadic $e_{ij}\mathbf{i}_i\mathbf{i}_j$ remains unchanged no matter what coordinate system is used. Therefore, using the invariant property of the strain dyadic, we have

$$e_{ij}\mathbf{i}_i\mathbf{i}_j = e'_{km}\mathbf{i}'_k\mathbf{i}'_m \tag{2.3.20}$$

It follows from (2.3.5) that

$$\mathbf{i}_i = T_{ki}\mathbf{i}'_k \text{ and } \mathbf{i}_j = T_{mj}\mathbf{i}'_m$$

Substituting these into (2.3.20) yields

$$(e'_{km} - e_{ij}T_{ki}T_{mj})\mathbf{i}'_k\mathbf{i}'_m = 0 \tag{2.3.21}$$

Hence,

$$e'_{km} = T_{ki}e_{ij}T_{mj} \text{ or } [e'] = [T][e][T]^T \tag{2.3.22}$$

Using arguments similar to those used for strain tensors, one can relate the stress tensors σ_{ij} and σ'_{ij} in the unprimed and primed coordinate systems by

$$\sigma'_{km} = T_{ki}\sigma_{ij}T_{mj} \text{ or } [\sigma'] = [T][\sigma][T]^T \tag{2.3.23}$$

2.4 STRESS-STRAIN RELATIONS

For a linear anisotropic elastic material, the stresses are related to the strains by the generalized Hooke law

$$\sigma_{ij} = c_{ijkm}e_{km}, \quad i,j,k,m = 1,2,3 \tag{2.4.1}$$

where the summation convention is used, the c_{ijkm} are 81 (=$3 \times 3 \times 3 \times 3$) material parameters, σ_{ij} and e_{ij} are second-order tensors, and c_{ijkl} is a fourth-order tensor. If the parameters c_{ijkl} of a material do not vary from point to point in the body, the material is said to be homogeneous; otherwise it is said to be inhomogeneous (heterogeneous). The material properties of a general anisotropic material vary with direction or orientation of the coordinate axes at any point in the body, and there are no planes of material symmetry. Hence, for a general anisotropic material, the 81 parameters c_{ijkl} are all independent. Fortunately, there is no actual material which has 81 independent elastic parameters.

In the absence of distributed couples (body couples), it follows from the equilibrium of moments for an elastic particle that the stress tensor is symmetric (Malvern, 1969); that is,

$$\sigma_{ij} = \sigma_{ji} \tag{2.4.2}$$

Moreover, it follows from the definition (2.2.2) of the strain tensor that

$$e_{km} = e_{mk} \tag{2.4.3}$$

As a consequence of the symmetries of both σ_{ij} and e_{km}, we have

$$c_{ijkm} = c_{jikm} \text{ and } c_{ijkm} = c_{ijmk} \tag{2.4.4}$$

and there are only 36 (= $(3 \times 3 - 3) \times (3 \times 3 - 3)$) independent parameters. Hence, the constitutive equations (2.4.1) can be rewritten in the compact form

$$\sigma_i = c_{ij}e_j, \quad i,j = 1,2,3,4,5,6 \tag{2.4.5}$$

where the contracted stresses σ_i and contracted strains e_j are defined as

$$\sigma_1 \equiv \sigma_{11}, \ \sigma_2 \equiv \sigma_{22}, \ \sigma_3 \equiv \sigma_{33}, \ \sigma_4 \equiv \sigma_{23}, \ \sigma_5 \equiv \sigma_{13}, \ \sigma_6 \equiv \sigma_{12}$$
$$e_1 \equiv e_{11}, \ e_2 \equiv e_{22}, \ e_3 \equiv e_{33}, \ e_4 \equiv e_{23}, \ e_5 \equiv e_{13}, \ e_6 \equiv e_{12} \quad (2.4.6)$$

We note that the c_{ij} is not a tensor and hence cannot be transformed as such.

2.4.1 Anisotropic Materials

For a general case of loading an elastic material, Green (1837, 1839) postulated the existence of a strain energy function $W(e_i)$ such that

$$\frac{\partial W}{\partial e_i} = \sigma_i \qquad (2.4.7)$$

Because of the stress-strain relation (2.4.5), W can be represented in terms of the e_i as

$$W = \hat{c}_0 + \hat{c}_i e_i + \frac{1}{2}\hat{c}_{ij} e_i e_j, \quad i,j = 1,2,3,4,5,6 \qquad (2.4.8)$$

where the $\hat{c}_0$, $\hat{c}_i$, and $\hat{c}_{ij}$ are constants for a homogeneous solid. Since $W = 0$ when $e_i = 0$, $\hat{c}_0 = 0$. Substituting (2.4.8) into (2.4.7) yields

$$\sigma_k = \hat{c}_i \delta_{ik} + \frac{1}{2}\hat{c}_{ij} e_i \delta_{jk} + \frac{1}{2}\hat{c}_{ij} e_j \delta_{ik} = \hat{c}_k + \frac{1}{2}\hat{c}_{ik} e_i + \frac{1}{2}\hat{c}_{kj} e_j \qquad (2.4.9)$$

because $\partial e_i / \partial e_k = \delta_{ik}$. In the absence of residual stresses (i.e., $\hat{c}_k = 0$), (2.4.9) can be rewritten as

$$\sigma_k = c_{ki} e_i, \quad c_{ki} \equiv \frac{1}{2}(\hat{c}_{ik} + \hat{c}_{ki}) \qquad (2.4.10)$$

It follows from the second relation in (2.4.10) that

$$c_{ik} = c_{ki} \qquad (2.4.11)$$

and hence the matrix $[c]$ whose elements are the c_{ik} is symmetric. This relation reduces the number of elastic parameters from 36 to 21 ($= (6 \times 6 - 6)/2 + 6$) parameters. In other words, there are only 21 independent elastic parameters for elastic materials which possess strain energy functions. Hence, the stress-strain relations are

$$\begin{Bmatrix} \sigma_1 \\ \sigma_2 \\ \sigma_3 \\ \sigma_4 \\ \sigma_5 \\ \sigma_6 \end{Bmatrix} = \begin{bmatrix} c_{11} & c_{12} & c_{13} & c_{14} & c_{15} & c_{16} \\ c_{12} & c_{22} & c_{23} & c_{24} & c_{25} & c_{26} \\ c_{13} & c_{23} & c_{33} & c_{34} & c_{35} & c_{36} \\ c_{14} & c_{24} & c_{34} & c_{44} & c_{45} & c_{46} \\ c_{15} & c_{25} & c_{35} & c_{45} & c_{55} & c_{56} \\ c_{16} & c_{26} & c_{36} & c_{46} & c_{56} & c_{66} \end{bmatrix} \begin{Bmatrix} e_1 \\ e_2 \\ e_3 \\ e_4 \\ e_5 \\ e_6 \end{Bmatrix} \qquad (2.4.12)$$

which are the most general relations within the framework of linear elasticity for general anisotropic materials; that is, solids with no planes of elastic symmetry for the material properties. Such materials are also called triclinic materials.

2.4.2 Orthotropic Materials

The number of independent elastic parameters are less than 21 for solids with planes of elastic symmetry as discussed next. For example, we consider an elastic body whose material properties are symmetric with respect to the $s_1 - s_2$ plane. The resulting symmetry can be expressed by the fact that the c_{ij} are invariant under the transformation $s'_1 = s_1$, $s'_2 = s_2$, and $s'_3 = -s_3$; that is,

$$\sigma_i = c_{ij}e_j, \quad \sigma'_i = c_{ij}e'_j \qquad (2.4.13\text{a,b})$$

where the σ_i and e_i and the σ'_i and e'_j are stresses and strains defined with respect to the unprimed and primed coordinate systems, respectively. The unit vectors $\mathbf{i}'_i$ and $\mathbf{i}_i$ of the primed and unprimed coordinate systems are related by

$$\mathbf{i}'_i = T_{ik}\mathbf{i}_k \qquad (2.4.14\text{a})$$

where the T_{ik} are the elements of the matrix

$$[T] = \begin{bmatrix} 1 & 0 & 0 \\ 0 & 1 & 0 \\ 0 & 0 & -1 \end{bmatrix} \qquad (2.4.14\text{b})$$

Moreover, it follows from (2.3.22) and (2.3.23) that

$$[\sigma'] = [T][\sigma][T]^T, \quad [e'] = [T][e][T]^T \qquad (2.4.15)$$

Substituting (2.4.14b) into (2.4.15) yields

$$\sigma'_1 = \sigma_1, \; \sigma'_2 = \sigma_2, \; \sigma'_3 = \sigma_3, \; \sigma'_4 = -\sigma_4, \; \sigma'_5 = -\sigma_5, \; \sigma'_6 = \sigma_6$$
$$e'_1 = e_1, \; e'_2 = e_2, \; e'_3 = e_3, \; e'_4 = -e_4, \; e'_5 = -e_5, \; e'_6 = e_6 \qquad (2.4.16)$$

Substituting (2.4.16) into (2.4.13b) and comparing the result with (2.4.13a), we conclude that

$$c_{14} = c_{15} = c_{24} = c_{25} = c_{34} = c_{35} = c_{46} = c_{56} = 0 \qquad (2.4.17)$$

Consequently, the number of independent elastic parameters for a material with a single plane of elastic symmetry is 13, and the stress-strain relations are

$$\begin{Bmatrix} \sigma_1 \\ \sigma_2 \\ \sigma_3 \\ \sigma_4 \\ \sigma_5 \\ \sigma_6 \end{Bmatrix} = \begin{bmatrix} c_{11} & c_{12} & c_{13} & 0 & 0 & c_{16} \\ c_{12} & c_{22} & c_{23} & 0 & 0 & c_{26} \\ c_{13} & c_{23} & c_{33} & 0 & 0 & c_{36} \\ 0 & 0 & 0 & c_{44} & c_{45} & 0 \\ 0 & 0 & 0 & c_{45} & c_{55} & 0 \\ c_{16} & c_{26} & c_{36} & 0 & 0 & c_{66} \end{bmatrix} \begin{Bmatrix} e_1 \\ e_2 \\ e_3 \\ e_4 \\ e_5 \\ e_6 \end{Bmatrix} \qquad (2.4.18)$$

Such a material is called monoclinic.

Similarly, if the $s_1 - s_3$ plane is a plane of elastic symmetry, one can show that

$$c_{14} = c_{16} = c_{24} = c_{26} = c_{34} = c_{36} = c_{45} = c_{56} = 0 \qquad (2.4.19)$$

88 ELASTICITY

Consequently, if there are two orthogonal planes of elastic symmetry for a material, the number of independent elastic parameters reduces from 21 to 9.

We should note that if there are two orthogonal planes of elastic symmetry for a material, then symmetry will exist relative to a third mutually orthogonal plane. Thus, in this example, considering the third plane $s_2 - s_3$ to be an elastic plane of symmetry demands that

$$c_{15} = c_{16} = c_{25} = c_{26} = c_{35} = c_{36} = c_{45} = c_{46} = 0$$

These conditions are contained in (2.4.17) and (2.4.19), which imply that if the $s_1 - s_2$ and $s_1 - s_3$ are planes of elastic symmetry, then the $s_2 - s_3$ plane is also a plane of elastic symmetry. In this case, the stress-strain relations are

$$\begin{Bmatrix} \sigma_1 \\ \sigma_2 \\ \sigma_3 \\ \sigma_4 \\ \sigma_5 \\ \sigma_6 \end{Bmatrix} = \begin{bmatrix} c_{11} & c_{12} & c_{13} & 0 & 0 & 0 \\ c_{12} & c_{22} & c_{23} & 0 & 0 & 0 \\ c_{13} & c_{23} & c_{33} & 0 & 0 & 0 \\ 0 & 0 & 0 & c_{44} & 0 & 0 \\ 0 & 0 & 0 & 0 & c_{55} & 0 \\ 0 & 0 & 0 & 0 & 0 & c_{66} \end{bmatrix} \begin{Bmatrix} e_1 \\ e_2 \\ e_3 \\ e_4 \\ e_5 \\ e_6 \end{Bmatrix} \quad (2.4.20a)$$

A material with such elastic symmetry properties is termed an orthotropic material; it has different material properties in three mutually perpendicular directions. There are 9 independent elastic parameters for orthotropic materials. For an orthotropic material whose principal material axes are x, y, z, the thermoelastic constitutive relations can be expressed as

$$\begin{Bmatrix} \sigma_1 \\ \sigma_2 \\ \sigma_3 \\ \sigma_4 \\ \sigma_5 \\ \sigma_6 \end{Bmatrix} = \begin{bmatrix} c_{11} & c_{12} & c_{13} & 0 & 0 & 0 \\ c_{12} & c_{22} & c_{23} & 0 & 0 & 0 \\ c_{13} & c_{23} & c_{33} & 0 & 0 & 0 \\ 0 & 0 & 0 & c_{44} & 0 & 0 \\ 0 & 0 & 0 & 0 & c_{55} & 0 \\ 0 & 0 & 0 & 0 & 0 & c_{66} \end{bmatrix} \begin{Bmatrix} e_1 - \alpha_1 \Delta T \\ e_2 - \alpha_2 \Delta T \\ e_3 - \alpha_3 \Delta T \\ e_4 \\ e_5 \\ e_6 \end{Bmatrix} \quad (2.4.20b)$$

where α_1, α_2, and α_3 are the coefficients of thermal expansion and ΔT is the temperature increment above the stress-free state.

In the literature for orthotropic materials, the material stiffness matrix $[c]$ in (2.4.20a) is replaced with $[Q]$ such that

$$\{\sigma_i\} = [Q]\{e_i\}, \quad [Q] \equiv \begin{bmatrix} Q_{11} & Q_{12} & Q_{13} & 0 & 0 & 0 \\ Q_{12} & Q_{22} & Q_{23} & 0 & 0 & 0 \\ Q_{13} & Q_{23} & Q_{33} & 0 & 0 & 0 \\ 0 & 0 & 0 & 2Q_{44} & 0 & 0 \\ 0 & 0 & 0 & 0 & 2Q_{55} & 0 \\ 0 & 0 & 0 & 0 & 0 & 2Q_{66} \end{bmatrix}$$

(2.4.21a,b)

Solving (2.4.21a) for the strains in terms of the stresses, we obtain

$$\{e_i\} = [S]\{\sigma_i\}, \quad [S] \equiv [Q]^{-1} = \begin{bmatrix} S_{11} & S_{12} & S_{13} & 0 & 0 & 0 \\ S_{12} & S_{22} & S_{23} & 0 & 0 & 0 \\ S_{13} & S_{23} & S_{33} & 0 & 0 & 0 \\ 0 & 0 & 0 & S_{44}/2 & 0 & 0 \\ 0 & 0 & 0 & 0 & S_{55}/2 & 0 \\ 0 & 0 & 0 & 0 & 0 & S_{66}/2 \end{bmatrix}$$
(2.4.22a,b)

where $[S]$ is called the material compliance matrix. Using matrix algebra, we find that the elements of $[S]$ are given by

$$|Q|\,S_{11} = Q_{22}Q_{33} - Q_{23}^2, \quad |Q|\,S_{12} = Q_{13}Q_{23} - Q_{12}Q_{33}$$
$$|Q|\,S_{22} = Q_{33}Q_{11} - Q_{13}^2, \quad |Q|\,S_{13} = Q_{12}Q_{23} - Q_{13}Q_{22}$$
$$|Q|\,S_{33} = Q_{11}Q_{22} - Q_{12}^2, \quad |Q|\,S_{23} = Q_{12}Q_{13} - Q_{23}Q_{11}$$
$$S_{44} = \frac{1}{Q_{44}}, \quad S_{55} = \frac{1}{Q_{55}}, \quad S_{66} = \frac{1}{Q_{66}} \tag{2.4.23}$$

where

$$|Q| = Q_{11}Q_{22}Q_{33} - Q_{11}Q_{23}^2 - Q_{22}Q_{13}^2 - Q_{33}Q_{12}^2 + 2Q_{12}Q_{23}Q_{13} \tag{2.4.24}$$

In (2.4.23) and (2.4.24), one can interchange the symbols Q and S to provide the inverse relations.

For the analysis of classical plates, the three transverse components σ_3, σ_4, and σ_5 are assumed to be zero. Consequently, the stress-strain relations (2.4.21) for orthotropic materials are reduced to

$$\begin{Bmatrix} \sigma_1 \\ \sigma_2 \\ \sigma_6 \end{Bmatrix} = \begin{bmatrix} \tilde{Q}_{11} & \tilde{Q}_{12} & 0 \\ \tilde{Q}_{12} & \tilde{Q}_{22} & 0 \\ 0 & 0 & 2Q_{66} \end{bmatrix} \begin{Bmatrix} e_1 \\ e_2 \\ e_6 \end{Bmatrix} \tag{2.4.25a}$$

where

$$\tilde{Q}_{11} = Q_{11} - \frac{Q_{13}^2}{Q_{33}}, \quad \tilde{Q}_{12} = Q_{12} - \frac{Q_{23}Q_{13}}{Q_{33}}, \quad \tilde{Q}_{22} = Q_{22} - \frac{Q_{23}^2}{Q_{33}} \tag{2.4.25b}$$

And the thermoelastic stress-strain relations (2.4.20b) are reduced to

$$\begin{Bmatrix} \sigma_1 \\ \sigma_2 \\ \sigma_6 \end{Bmatrix} = \begin{bmatrix} \tilde{Q}_{11} & \tilde{Q}_{12} & 0 \\ \tilde{Q}_{12} & \tilde{Q}_{22} & 0 \\ 0 & 0 & 2Q_{66} \end{bmatrix} \begin{Bmatrix} e_1 - \alpha_1 \Delta T \\ e_2 - \alpha_2 \Delta T \\ e_6 \end{Bmatrix} \tag{2.4.25c}$$

Hence, the material compliance matrix $[S]$ is reduced to

$$[S] = \begin{bmatrix} \tilde{S}_{11} & \tilde{S}_{12} & 0 \\ \tilde{S}_{12} & \tilde{S}_{22} & 0 \\ 0 & 0 & S_{66}/2 \end{bmatrix} \tag{2.4.26a}$$

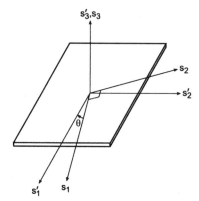

Fig. 2.4.1 Coordinate systems : (a) $s_1 s_2 s_3$ = original coordinate system, and (b) $s'_1 s'_2 s'_3$ = transformed coordinate system.

where

$$\left(\tilde{S}_{11}, \tilde{S}_{12}, \tilde{S}_{22}\right) = \frac{1}{\tilde{Q}_{11}\tilde{Q}_{22} - \tilde{Q}_{12}^2} \left(\tilde{Q}_{22}, -\tilde{Q}_{12}, \tilde{Q}_{11}\right), \quad S_{66} = \frac{1}{Q_{66}} \quad (2.4.26b)$$

2.4.3 Isotropic Materials

If at every point of a material, there is one plane in which the elastic properties are equal in all directions, then the material is termed transversely isotropic. If, for example, the $s_1 - s_2$ plane is a special plane of elastic isotropy, then the stiffness matrix $[Q]$ is invariant under all rotations about the s_3-axis. To determine the resulting simplification to the stiffness matrix, we introduce a new orthogonal set of axes $s'_1 s'_2 s'_3$ by rotating the $s_1 s_2 s_3$ set clockwise about the s_3-axis by an angle θ, as shown in Figure 2.4.1. Thus, the unit vectors along these two sets of axes are related by

$$\begin{Bmatrix} \mathbf{i}'_1 \\ \mathbf{i}'_2 \\ \mathbf{i}'_3 \end{Bmatrix} = \begin{bmatrix} c & -s & 0 \\ s & c & 0 \\ 0 & 0 & 1 \end{bmatrix} \begin{Bmatrix} \mathbf{i}_1 \\ \mathbf{i}_2 \\ \mathbf{i}_3 \end{Bmatrix} = [T] \begin{Bmatrix} \mathbf{i}_1 \\ \mathbf{i}_2 \\ \mathbf{i}_3 \end{Bmatrix} \quad (2.4.27)$$

where $c = \cos\theta$ and $s = \sin\theta$. But the stress tensor σ'_{ij} in the primed system is related to the stress tensor σ_{ij} in the unprimed system by (2.3.23). Hence,

$$\begin{bmatrix} \sigma'_{11} & \sigma'_{12} & \sigma'_{13} \\ \sigma'_{12} & \sigma'_{22} & \sigma'_{23} \\ \sigma'_{13} & \sigma'_{23} & \sigma'_{33} \end{bmatrix} = \begin{bmatrix} c & -s & 0 \\ s & c & 0 \\ 0 & 0 & 1 \end{bmatrix} \begin{bmatrix} \sigma_{11} & \sigma_{12} & \sigma_{13} \\ \sigma_{12} & \sigma_{22} & \sigma_{23} \\ \sigma_{13} & \sigma_{23} & \sigma_{33} \end{bmatrix} \begin{bmatrix} c & s & 0 \\ -s & c & 0 \\ 0 & 0 & 1 \end{bmatrix} \quad (2.4.28)$$

Multiplying the matrices on the right-hand side of (2.4.28) and using the contracted form of the stresses defined in (2.4.6), we obtain

$$\{\sigma_i'\} = [A]\{\sigma_i\} \tag{2.4.29}$$

where

$$[A] = \begin{bmatrix} c^2 & s^2 & 0 & 0 & 0 & -2cs \\ s^2 & c^2 & 0 & 0 & 0 & 2cs \\ 0 & 0 & 1 & 0 & 0 & 0 \\ 0 & 0 & 0 & c & s & 0 \\ 0 & 0 & 0 & -s & c & 0 \\ cs & -cs & 0 & 0 & 0 & c^2 - s^2 \end{bmatrix} \tag{2.4.30}$$

It follows from (2.4.29) that

$$\{\sigma_i\} = [A]^{-1}\{\sigma_i'\} \tag{2.4.31}$$

where $[A]^{-1}$ is the inverse of $[A]$. We note that $[A]^{-1}$ can be obtained from $[A]$ by simply replacing s with $-s$.

According to (2.3.19) the strain tensor can be transformed from the unprimed to the primed system by replacing the stress tensor by the strain tensor in (2.4.29). Hence, in place of (2.4.29), we have

$$\{e_i'\} = [A]\{e_i\} \tag{2.4.32}$$

Thus,

$$\{e_i\} = [A]^{-1}\{e_i'\} \tag{2.4.33}$$

Substituting (2.4.31) and (2.4.33) into (2.4.21a) yields

$$[A]^{-1}\{\sigma_i'\} = [Q][A]^{-1}\{e_i'\}$$

or

$$\{\sigma_i'\} = [A][Q][A]^{-1}\{e_i'\} = [\bar{Q}]\{e_i'\} \tag{2.4.34}$$

where the transformed stiffness matrix $[\bar{Q}]$ is given by

$$[\bar{Q}] = [A][Q][A]^{-1} = \begin{bmatrix} \bar{Q}_{11} & \bar{Q}_{12} & \bar{Q}_{13} & 0 & 0 & 2\bar{Q}_{16} \\ \bar{Q}_{12} & \bar{Q}_{22} & \bar{Q}_{23} & 0 & 0 & 2\bar{Q}_{26} \\ \bar{Q}_{13} & \bar{Q}_{23} & \bar{Q}_{33} & 0 & 0 & 2\bar{Q}_{36} \\ 0 & 0 & 0 & 2\bar{Q}_{44} & 2\bar{Q}_{45} & 0 \\ 0 & 0 & 0 & 2\bar{Q}_{45} & 2\bar{Q}_{55} & 0 \\ \bar{Q}_{16} & \bar{Q}_{26} & \bar{Q}_{36} & 0 & 0 & 2\bar{Q}_{66} \end{bmatrix} \tag{2.4.35}$$

Substituting $[Q]$ and $[A]$ from (2.4.21b) and (2.4.30) into (2.4.35), we express the elements $\bar{Q}_{ij}$ of the transformed stiffness matrix $[\bar{Q}]$ in terms of the Q_{ij} and $\cos\theta$ and $\sin\theta$ as

$$\bar{Q}_{11} = c^4 Q_{11} + 2c^2 s^2 Q_{12} + s^4 Q_{22} + 4c^2 s^2 Q_{66}$$
$$\bar{Q}_{12} = c^2 s^2 (Q_{11} + Q_{22} - 4Q_{66}) + (c^4 + s^4) Q_{12}$$

$$\bar{Q}_{13} = c^2 Q_{13} + s^2 Q_{23}$$
$$\bar{Q}_{16} = cs^3(Q_{12} - Q_{22}) + c^3 s(Q_{11} - Q_{12}) + 2cs(s^2 - c^2)Q_{66}$$
$$\bar{Q}_{22} = s^4 Q_{11} + c^4 Q_{22} + 2c^2 s^2 (Q_{12} + 2Q_{66})$$
$$\bar{Q}_{23} = s^2 Q_{13} + c^2 Q_{23}$$
$$\bar{Q}_{26} = cs^3(Q_{11} - Q_{12}) + c^3 s(Q_{12} - Q_{22}) + 2cs(c^2 - s^2)Q_{66}$$
$$\bar{Q}_{33} = Q_{33}$$
$$\bar{Q}_{36} = cs(Q_{13} - Q_{23})$$
$$\bar{Q}_{44} = c^2 Q_{44} + s^2 Q_{55}$$
$$\bar{Q}_{45} = cs(Q_{55} - Q_{44})$$
$$\bar{Q}_{55} = s^2 Q_{44} + c^2 Q_{55}$$
$$\bar{Q}_{66} = c^2 s^2 (Q_{11} - 2Q_{12} + Q_{22}) + (c^2 - s^2)^2 Q_{66} \qquad (2.4.36)$$

Equations (2.4.34)-(2.4.36) represent a general transformation of the stiffness matrix under a clockwise rotation of the $s_1 s_2 s_3$-axes about the s_3-axis by an angle θ. For a transversely isotropic material in which the $s_1 - s_2$ plane is the special plane of isotropy, $[\bar{Q}]$ must be the same as $[Q]$ for all possible rotation angles θ. In order that $[\bar{Q}]$ defined in (2.4.35) be the same as $[Q]$ defined in (2.4.21b), we demand that

$$\bar{Q}_{16} = \bar{Q}_{26} = \bar{Q}_{36} = \bar{Q}_{45} = 0, \quad \bar{Q}_{11} = Q_{11}, \quad \bar{Q}_{12} = Q_{12}$$

$$\bar{Q}_{13} = Q_{13}, \quad \bar{Q}_{22} = Q_{22}, \quad \bar{Q}_{23} = Q_{23}, \text{ and } \bar{Q}_{66} = Q_{66}$$

Using trigonometric identities, we express $\bar{Q}_{16}$ defined in (2.4.36) as

$$\bar{Q}_{16} = \frac{1}{4}(Q_{11} - Q_{22})\sin 2\theta + \frac{1}{8}(Q_{11} + Q_{22} - 2Q_{12} - 4Q_{66})\sin 4\theta$$

Hence, in order that $\bar{Q}_{16} = 0$ for all angles θ, we have

$$Q_{22} = Q_{11} \text{ and } Q_{66} = \frac{1}{2}(Q_{11} - Q_{12}) \qquad (2.4.37)$$

Using trigonometric identities and (2.4.36), we find that $\bar{Q}_{26} = 0$. In order that $\bar{Q}_{36}$ defined in (2.4.36) be equal to zero for all angles θ, we demand that

$$Q_{13} = Q_{23} \qquad (2.4.38)$$

Moreover, in order that $\bar{Q}_{45}$ defined in (2.4.36) be equal to zero for all θ, we demand that

$$Q_{55} = Q_{44} \qquad (2.4.39)$$

Using (2.4.39), we find that

$$\bar{Q}_{44} = \bar{Q}_{55} = Q_{44} \qquad (2.4.40)$$

Then, one can easily verify that $\bar{Q}_{11} = Q_{11}$, $\bar{Q}_{12} = Q_{12}$, $\bar{Q}_{13} = Q_{13}$, $\bar{Q}_{22} = Q_{22}$, $\bar{Q}_{23} = Q_{23}$, and $\bar{Q}_{66} = Q_{66}$. Therefore, the stress-strain relations for a transversely isotropic material reduces to

$$\begin{Bmatrix} \sigma_1 \\ \sigma_2 \\ \sigma_3 \\ \sigma_4 \\ \sigma_5 \\ \sigma_6 \end{Bmatrix} = \begin{bmatrix} Q_{11} & Q_{12} & Q_{13} & 0 & 0 & 0 \\ Q_{12} & Q_{11} & Q_{13} & 0 & 0 & 0 \\ Q_{13} & Q_{13} & Q_{33} & 0 & 0 & 0 \\ 0 & 0 & 0 & 2Q_{44} & 0 & 0 \\ 0 & 0 & 0 & 0 & 2Q_{44} & 0 \\ 0 & 0 & 0 & 0 & 0 & Q_{11} - Q_{12} \end{bmatrix} \begin{Bmatrix} e_1 \\ e_2 \\ e_3 \\ e_4 \\ e_5 \\ e_6 \end{Bmatrix} \quad (2.4.41a)$$

which has only five independent elastic parameters. And the thermoelastic constitutive relations for a transversely isotropic material are

$$\begin{Bmatrix} \sigma_1 \\ \sigma_2 \\ \sigma_3 \\ \sigma_4 \\ \sigma_5 \\ \sigma_6 \end{Bmatrix} = \begin{bmatrix} Q_{11} & Q_{12} & Q_{13} & 0 & 0 & 0 \\ Q_{12} & Q_{11} & Q_{13} & 0 & 0 & 0 \\ Q_{13} & Q_{13} & Q_{33} & 0 & 0 & 0 \\ 0 & 0 & 0 & 2Q_{44} & 0 & 0 \\ 0 & 0 & 0 & 0 & 2Q_{44} & 0 \\ 0 & 0 & 0 & 0 & 0 & Q_{11} - Q_{12} \end{bmatrix} \begin{Bmatrix} e_1 - \alpha_1 \Delta T \\ e_2 - \alpha_1 \Delta T \\ e_3 - \alpha_3 \Delta T \\ e_4 \\ e_5 \\ e_6 \end{Bmatrix}$$
$$(2.4.41b)$$

If there are an infinite number of planes of elastic symmetry, the material is termed isotropic. To determine the simplification of the stiffness matrix in this case, we assume that the $s_1 - s_3$ plane is a special plane of isotropy. Then, using arguments similar to those used when the $s_1 - s_2$ plane is a special plane of isotropy, we have

$$Q_{33} = Q_{11}, \quad Q_{12} = Q_{23}, \quad Q_{55} = \frac{1}{2}(Q_{11} - Q_{13}), \quad Q_{44} = Q_{66} \quad (2.4.42)$$

Hence, if the $s_1 - s_2$ and $s_1 - s_3$ planes are planes of isotropy, we find from (2.4.37)-(2.4.39) and (2.4.42) that

$$Q_{11} = Q_{22} = Q_{33}, \quad Q_{13} = Q_{23} = Q_{12}, \quad Q_{44} = Q_{55} = Q_{66} = \frac{1}{2}(Q_{11} - Q_{12})$$
$$(2.4.43)$$

Therefore, the stress-strain relations simplify to

$$\begin{Bmatrix} \sigma_1 \\ \sigma_2 \\ \sigma_3 \\ \sigma_4 \\ \sigma_5 \\ \sigma_6 \end{Bmatrix} = \begin{bmatrix} Q_{11} & Q_{12} & Q_{12} & 0 & 0 & 0 \\ Q_{12} & Q_{11} & Q_{12} & 0 & 0 & 0 \\ Q_{12} & Q_{12} & Q_{11} & 0 & 0 & 0 \\ 0 & 0 & 0 & 2Q_{44} & 0 & 0 \\ 0 & 0 & 0 & 0 & 2Q_{44} & 0 \\ 0 & 0 & 0 & 0 & 0 & 2Q_{44} \end{bmatrix} \begin{Bmatrix} e_1 \\ e_2 \\ e_3 \\ e_4 \\ e_5 \\ e_6 \end{Bmatrix} \quad (2.4.44a)$$

$$\begin{Bmatrix} \sigma_1 \\ \sigma_2 \\ \sigma_3 \\ \sigma_4 \\ \sigma_5 \\ \sigma_6 \end{Bmatrix} = \begin{bmatrix} Q_{11} & Q_{12} & Q_{12} & 0 & 0 & 0 \\ Q_{12} & Q_{11} & Q_{12} & 0 & 0 & 0 \\ Q_{12} & Q_{12} & Q_{11} & 0 & 0 & 0 \\ 0 & 0 & 0 & 2Q_{44} & 0 & 0 \\ 0 & 0 & 0 & 0 & 2Q_{44} & 0 \\ 0 & 0 & 0 & 0 & 0 & 2Q_{44} \end{bmatrix} \begin{Bmatrix} e_1 - \alpha\Delta T \\ e_2 - \alpha\Delta T \\ e_3 - \alpha\Delta T \\ e_4 \\ e_5 \\ e_6 \end{Bmatrix}$$
(2.4.44b)

One can easily show that if the $s_1 - s_2$ and $s_1 - s_3$ planes are planes of isotropy, then the $s_2 - s_3$ plane is also a plane of isotropy, and hence the material is isotropic. There are only two independent elastic parameters and a single coefficient of thermal expansion for an isotropic material.

Solving (2.4.44a) for the $\{e_i\}$ in terms of the $\{\sigma_i\}$, we obtain (2.4.22) where the compliance matrix is given by

$$[S] = \begin{bmatrix} S_{11} & S_{12} & S_{12} & 0 & 0 & 0 \\ S_{12} & S_{11} & S_{12} & 0 & 0 & 0 \\ S_{12} & S_{12} & S_{11} & 0 & 0 & 0 \\ 0 & 0 & 0 & S_{11} - S_{12} & 0 & 0 \\ 0 & 0 & 0 & 0 & S_{11} - S_{12} & 0 \\ 0 & 0 & 0 & 0 & 0 & S_{11} - S_{12} \end{bmatrix} \quad (2.4.45)$$

The elements S_{11} and S_{12} of the compliance matrix are related to the elements Q_{11} and Q_{12} of the stiffness matrix by

$$S_{11} = \frac{Q_{11} + Q_{12}}{(Q_{11} - Q_{12})(Q_{11} + 2Q_{12})}, \quad S_{12} = -\frac{Q_{12}}{(Q_{11} - Q_{12})(Q_{11} + 2Q_{12})}$$
(2.4.46)

2.4.4 Material Stiffness and Compliance Matrices

Any orthotropic material has three principal material axes, which are the intersection lines of its three planes of elastic symmetry. To relate the so-far introduced abstract compliance and stiffness matrices to the physical and mechanical properties of an orthotropic material, one usually conducts simple tests, such as uniaxial tension or pure shear tests, along its three principal material directions. For example, conducting a standard tensile test in the principal material direction s_1 of an orthotropic material, one obtains the stress and strain tensors

$$[\sigma] = \begin{bmatrix} \sigma_{11} & 0 & 0 \\ 0 & 0 & 0 \\ 0 & 0 & 0 \end{bmatrix} \quad \text{and} \quad [e] = \begin{bmatrix} e_{11} & 0 & 0 \\ 0 & -\nu_{12}e_{11} & 0 \\ 0 & 0 & -\nu_{13}e_{11} \end{bmatrix} \quad (2.4.47)$$

where the ν_{ij} are Poisson's ratios defined by

$$\nu_{12} \equiv -\frac{e_{22}}{e_{11}} \quad \text{and} \quad \nu_{13} \equiv -\frac{e_{33}}{e_{11}} \quad (2.4.48)$$

The ratios of the strains e_{11}, e_{22}, and e_{33} to the stress σ_{11} are defined to be the compliances S_{i1} and they can be represented in terms of Young's modulus E_{11} and Poisson's ratios as

$$S_{11} \equiv \frac{e_{11}}{\sigma_{11}} \equiv \frac{1}{E_{11}}, \quad S_{21} \equiv \frac{e_{22}}{\sigma_{11}} = -\frac{\nu_{12}}{E_{11}}, \quad S_{31} \equiv \frac{e_{33}}{\sigma_{11}} = -\frac{\nu_{13}}{E_{11}}$$

where (2.4.48) is used.

Performing other tensile and shear tests along other principal material coordinates, one can determine the material compliance matrix for an orthotropic material in terms of the Young's moduli, Poisson's ratios, and shear moduli as

$$[S] \equiv \begin{bmatrix} 1/E_{11} & -\nu_{21}/E_{22} & -\nu_{31}/E_{33} & 0 & 0 & 0 \\ -\nu_{12}/E_{11} & 1/E_{22} & -\nu_{32}/E_{33} & 0 & 0 & 0 \\ -\nu_{13}/E_{11} & -\nu_{23}/E_{22} & 1/E_{33} & 0 & 0 & 0 \\ 0 & 0 & 0 & 1/2G_{23} & 0 & 0 \\ 0 & 0 & 0 & 0 & 1/2G_{13} & 0 \\ 0 & 0 & 0 & 0 & 0 & 1/2G_{12} \end{bmatrix}$$
(2.4.49)

Here, E_{ii} (no summation) is Young's modulus in the i direction, G_{ij} is the shear modulus in the $i-j$ plane, and ν_{ij} is Poisson's ratio of the normal strain in the j direction to the normal strain in the i direction when stressed in the i direction; that is,

$$\nu_{ij} = -\frac{e_{jj}}{e_{ii}} \qquad (2.4.50)$$

for $\sigma_{ii} \neq 0$ and all other stresses are equal to zero. Due to the symmetry of the compliance matrix $[S]$, we have

$$\frac{\nu_{ij}}{E_{ii}} = \frac{\nu_{ji}}{E_{jj}} \qquad (2.4.51)$$

In (2.4.50) and (2.4.51), repeated indices do not imply summation. Using (2.4.51), one can prove that

$$\nu_{12}\nu_{23}\nu_{31} = \nu_{21}\nu_{32}\nu_{13} \qquad (2.4.52)$$

Therefore, the 9 independent elastic constants of an orthotropic material are E_{11}, E_{22}, E_{33}, G_{12}, G_{13}, G_{23}, ν_{12}(or ν_{21}), ν_{13}(or ν_{31}), ν_{23}(or ν_{32}). By inverting the material compliance matrix $[S]$, we obtain the material stiffness matrix $[Q]$. Its elements are given by

$$Q_{11} = E_{11}(1 - \nu_{23}\nu_{32})/\Delta$$
$$Q_{22} = E_{22}(1 - \nu_{31}\nu_{13})/\Delta$$
$$Q_{33} = E_{33}(1 - \nu_{12}\nu_{21})/\Delta$$
$$Q_{12} = E_{22}(\nu_{12} + \nu_{32}\nu_{13})/\Delta$$
$$Q_{13} = E_{33}(\nu_{13} + \nu_{12}\nu_{23})/\Delta$$
$$Q_{23} = E_{33}(\nu_{23} + \nu_{21}\nu_{13})/\Delta$$
$$Q_{44} = G_{23}, \quad Q_{55} = G_{13}, \quad Q_{66} = G_{12}$$
$$\Delta \equiv 1 - \nu_{12}\nu_{21} - \nu_{23}\nu_{32} - \nu_{31}\nu_{13} - 2\nu_{21}\nu_{32}\nu_{13} \qquad (2.4.53)$$

For classical plates, (2.4.49) is reduced to

$$[S] = \begin{bmatrix} 1/E_{11} & -\nu_{21}/E_{22} & 0 \\ -\nu_{12}/E_{11} & 1/E_{22} & 0 \\ 0 & 0 & 1/2G_{12} \end{bmatrix} \quad (2.4.54)$$

Inverting the compliance matrix $[S]$ yields the stiffness matrix $[Q]$ whose elements are

$$(Q_{11}, Q_{12}, Q_{22}) = \frac{1}{1 - \nu_{12}\nu_{21}} (E_{11}, \nu_{12}E_{22} = \nu_{21}E_{11}, E_{22}), \quad Q_{66} = G_{12} \quad (2.4.55)$$

For a transversely isotropic material in which the 1-2 plane is the plane of isotropy, we have

$$E_{11} = E_{22}, \quad G_{13} = G_{23}, \quad G_{12} = \frac{E_{11}}{2(1+\nu_{12})}$$
$$\nu_{12} = \nu_{21}, \quad \nu_{13} = \nu_{23}, \quad \nu_{31} = \nu_{32} \quad (2.4.56)$$

where the last equation is obtained by using (2.4.37). Hence, the number of elastic constants is reduced to 5, which are E_{11} (or G_{12}), E_{33}, G_{13}, ν_{12}, ν_{13}.

For isotropic materials, the material constants are the same in all directions at a point in the body. Hence,

$$E_{11} = E_{22} = E_{33} = E, \quad \nu_{12} = \nu_{13} = \nu_{23} = \nu$$
$$G_{12} = G_{13} = G_{23} = G = \frac{E}{2(1+\nu)}, \quad \Delta = (1+\nu)^2(1-2\nu) \quad (2.4.57)$$

Therefore, there are only two constants E and ν. For thermoelastic plates, the thermoelastic stress-strain relations are

$$\begin{Bmatrix} \sigma_1 \\ \sigma_2 \\ \sigma_6 \end{Bmatrix} = \frac{E}{1-\nu^2} \begin{bmatrix} 1 & \nu & 0 \\ \nu & 1 & 0 \\ 0 & 0 & 1-\nu \end{bmatrix} \begin{Bmatrix} e_1 - \alpha \Delta T \\ e_2 - \alpha \Delta T \\ e_6 \end{Bmatrix} \quad (2.4.58)$$

2.4.5 Fiber-Reinforced Lamina

Next, we consider the stiffness matrix $[Q]$ and compliance matrix $[S]$ of a fiber-reinforced lamina (see Figure 2.4.2), which is an orthotropic material. The stress-strain relations (2.4.21) and the strain-stress relations (2.4.22) are defined with respect to the principal material directions for an orthotropic material. However, these directions do not, in general, coincide with the structural coordinate system used in modeling and analysis. For example, Figure 2.4.2 shows that the material coordinate system 123 is rotated clockwise by an angle θ with respect to the 3-axis to match with the structural coordinate system $s'_1 s'_2 s'_3$, where the straight dashed lines represent fibers in the lamina. Then, the components $\bar{Q}_{ij}$ of the transformed stiffness matrix

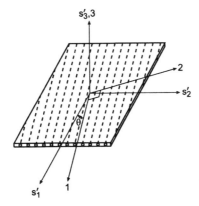

Fig. 2.4.2 Coordinate systems : (a) 123 = material coordinate system of an orthotropic lamina, and (b) $s'_1 s'_2 s'_3$ = transformed or structural coordinate system.

$[\bar{Q}]$ defined with respect to the $s'_1 s'_2 s'_3$-axes are related to the Q_{ij} of the stiffness matrix $[Q]$ defined with respect to the principal material axes by (2.4.36).

Tsai and Pagano (1968) recast (2.4.36) in a form that can be readily used to understand the consequences of rotating a lamina within a laminate. Using trigonometric identities, one can recast (2.4.36) as

$$\bar{Q}_{11} = U_1 + U_2 \cos 2\theta + U_3 \cos 4\theta, \quad \bar{Q}_{13} = U_6 + U_7 \cos 2\theta$$
$$\bar{Q}_{12} = U_4 - U_3 \cos 4\theta, \quad \bar{Q}_{23} = U_6 - U_7 \cos 2\theta$$
$$\bar{Q}_{22} = U_1 - U_2 \cos 2\theta + U_3 \cos 4\theta, \quad \bar{Q}_{44} = U_8 + U_9 \cos 2\theta$$
$$\bar{Q}_{16} = \frac{1}{2} U_2 \sin 2\theta + U_3 \sin 4\theta, \quad \bar{Q}_{55} = U_8 - U_9 \cos 2\theta$$
$$\bar{Q}_{26} = \frac{1}{2} U_2 \sin 2\theta - U_3 \sin 4\theta, \quad \bar{Q}_{45} = -U_9 \sin 2\theta$$
$$\bar{Q}_{66} = U_5 - U_3 \cos 4\theta, \quad \bar{Q}_{36} = U_7 \sin 2\theta \quad (2.4.59)$$

where

$$U_1 = \frac{1}{8}(3Q_{11} + 3Q_{22} + 2Q_{12} + 4Q_{66}), \quad U_6 = \frac{1}{2}(Q_{13} + Q_{23})$$
$$U_2 = \frac{1}{2}(Q_{11} - Q_{22}), \quad U_7 = \frac{1}{2}(Q_{13} - Q_{23})$$
$$U_3 = \frac{1}{8}(Q_{11} + Q_{22} - 2Q_{12} - 4Q_{66}), \quad U_8 = \frac{1}{2}(Q_{44} + Q_{55})$$
$$U_4 = \frac{1}{8}(Q_{11} + Q_{22} + 6Q_{12} - 4Q_{66}), \quad U_9 = \frac{1}{2}(Q_{44} - Q_{55})$$
$$U_5 = \frac{1}{8}(Q_{11} + Q_{22} - 2Q_{12} + 4Q_{66}) \quad (2.4.60)$$

If thermal, hygro-thermal, and piezoelectric effects exist, they can be taken into account by modifying (2.4.21a) as

$$\{\sigma_i\} = [Q](\{e_i\} - \{\tilde{e}_i\}) \tag{2.4.61}$$

The $\tilde{e}_i$ are induced strains due to thermal expansion, moisture absorption, and/or electromechanical effects and are given by

$$\{\tilde{e}_i\}^T = \left\{ \theta_1 T + \beta_1 M + d_{31} E_3,\ \theta_2 T + \beta_2 M + d_{32} E_3, \right.$$
$$\left. \theta_3 T + \beta_3 M + d_{33} E_3,\ 0,\ 0,\ 0 \right\} \tag{2.4.62}$$

where the θ_i, β_i, and d_{3i} denote the thermal expansion coefficients, the hygro-thermal expansion coefficients, and the piezoelectric strain coefficients, respectively; T is the temperature change; M is the increase in moisture measured in percentage weight increase; and E_3 is the electric field intensity applied along the 3-direction. The corresponding form of the transformed constitutive equations (2.4.34) are

$$\{\sigma_i'\} = [\bar{Q}](\{e_i'\} - \{\tilde{e}_i'\}) \tag{2.4.63}$$

where

$$\{\tilde{e}_i'\}^T = \left\{ \bar{\theta}_1 T + \bar{\beta}_1 M + \bar{d}_{31} E_3,\ \bar{\theta}_2 T + \bar{\beta}_2 M + \bar{d}_{32} E_3, \right.$$
$$\left. \bar{\theta}_3 T + \bar{\beta}_3 M + \bar{d}_{33} E_3,\ 0,\ 0,\ \bar{\theta}_6 T + \bar{\beta}_6 M + \bar{d}_{36} E_3 \right\} \tag{2.4.64}$$

and

$$\bar{d}_{31} = d_{31}c^2 + d_{32}s^2,\ \bar{d}_{32} = d_{31}s^2 + d_{32}c^2,\ \bar{d}_{33} = d_{33},\ \bar{d}_{36} = 2(d_{31} - d_{32})cs$$
$$\bar{\beta}_1 = \beta_1 c^2 + \beta_2 s^2,\ \bar{\beta}_2 = \beta_1 s^2 + \beta_2 c^2,\ \bar{\beta}_3 = \beta_3,\ \bar{\beta}_6 = 2(\beta_1 - \beta_2)cs$$
$$\bar{\theta}_1 = \theta_1 c^2 + \theta_2 s^2,\ \bar{\theta}_2 = \theta_1 s^2 + \theta_2 c^2,\ \bar{\theta}_3 = \theta_3,\ \bar{\theta}_6 = 2(\theta_1 - \theta_2)cs \tag{2.4.65}$$

2.5 GOVERNING EQUATIONS

Equations describing the balancing of internal forces, body forces, and inertia forces on a structural particle are called equilibrium equations. Equations describing the relations between displacements and external forces on a structural element are called equations of motion. Because a structural element has at least one of its three dimensions finite, to model it as a particle system, one needs to use a combination of an infinite number of particles connected by a set of springs and dampers, and the spring set must be able to exhibit extensional, bending, twisting, and shear rigidities. In other words, the motion of an element needs to be described by a position vector referred to its reference point (e.g., the mass centroid), a rotation vector which describes the rigid-body rotations, and some other variables which describe the relative motions among the particles in the element. Also, a structural element has nontrivial rotary inertias. On the other hand, a particle has no volume or rotary inertia and its motion is fully described by its position vector.

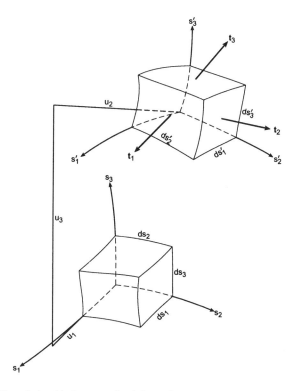

Fig. 2.5.1 The relationship between the deformed configuration of a structural particle and its undeformed configuration in a general orthogonal curvilinear coordinate system.

2.5.1 Equilibrium Equations

Figure 2.5.1 shows a structural particle under the action of traction vectors

$$\mathbf{t}_1 \equiv \sigma'_{11}\mathbf{i}'_1 + \sigma'_{12}\mathbf{i}'_2 + \sigma'_{13}\mathbf{i}'_3 \qquad (2.5.1a)$$

$$\mathbf{t}_2 \equiv \sigma'_{21}\mathbf{i}'_1 + \sigma'_{22}\mathbf{i}'_2 + \sigma'_{23}\mathbf{i}'_3 \qquad (2.5.1b)$$

$$\mathbf{t}_3 \equiv \sigma'_{31}\mathbf{i}'_1 + \sigma'_{32}\mathbf{i}'_2 + \sigma'_{33}\mathbf{i}'_3 \qquad (2.5.1c)$$

where we assume that the deformed coordinate system $s'_1 s'_2 s'_3$ is orthogonal, and the σ'_{ij} are the acting forces per unit of the deformed area; i.e., Cauchy stresses. Considering the force balance for the particle shown in Figure 2.5.1, we have

$$\frac{\partial(\mathbf{t}_1 ds'_2 ds'_3)}{\partial s_1} ds_1 + \frac{\partial(\mathbf{t}_2 ds'_1 ds'_3)}{\partial s_2} ds_2 + \frac{\partial(\mathbf{t}_3 ds'_1 ds'_2)}{\partial s_3} ds_3$$
$$+ (b_1 \mathbf{i}_1 + b_2 \mathbf{i}_2 + b_3 \mathbf{i}_3) ds_1 ds_2 ds_3 = \rho \left(\ddot{u}_1 \mathbf{i}_1 + \ddot{u}_2 \mathbf{i}_2 + \ddot{u}_3 \mathbf{i}_3 \right) ds_1 ds_2 ds_3 \quad (2.5.2)$$

100 ELASTICITY

Note that the body force density b_i and the displacements u_i are defined with respect to the undeformed coordinate system. Defining the "normalized" stresses $\tilde{\sigma}_{ij}$ (acting force per unit of the undeformed area) as

$$\tilde{\sigma}_{ij} \equiv \sigma'_{ij} \frac{ds'_k}{ds_k} \frac{ds'_m}{ds_m}, \qquad i \neq k \neq m \tag{2.5.3}$$

we have

$$\mathbf{t}_1 ds'_2 ds'_3 = (\tilde{\sigma}_{11}\mathbf{i}'_1 + \tilde{\sigma}_{12}\mathbf{i}'_2 + \tilde{\sigma}_{13}\mathbf{i}'_3)\, ds_2 ds_3 \tag{2.5.4a}$$

$$\mathbf{t}_2 ds'_1 ds'_3 = (\tilde{\sigma}_{21}\mathbf{i}'_1 + \tilde{\sigma}_{22}\mathbf{i}'_2 + \tilde{\sigma}_{23}\mathbf{i}'_3)\, ds_1 ds_3 \tag{2.5.4b}$$

$$\mathbf{t}_3 ds'_1 ds'_2 = (\tilde{\sigma}_{31}\mathbf{i}'_1 + \tilde{\sigma}_{32}\mathbf{i}'_2 + \tilde{\sigma}_{33}\mathbf{i}'_3)\, ds_1 ds_2 \tag{2.5.4c}$$

Moreover, if there are no body moments, the equilibrium of moments confirms the mutuality of the Cauchy shear stresses; i.e., $\sigma'_{ij} = \sigma'_{ji}$. However, $\tilde{\sigma}_{ij}$ may not be equal to $\tilde{\sigma}_{ji}$.

Now if the configuration change of the system due to the displacements is neglected, we have

$$ds'_i = ds_i, \quad \mathbf{i}'_j = \mathbf{i}_j, \quad \sigma'_{ij} = \tilde{\sigma}_{ij} = \sigma_{ij}, \quad \sigma_{ij} = \sigma_{ji}, \qquad i,j = 1,2,3 \tag{2.5.5}$$

which are used in linear structural theories. Here, the stresses σ_{ij} denote acting forces per unit of the undeformed area and are defined with respect to the undeformed coordinate system. Substituting (2.5.5) into (2.5.1) and (2.5.2) yields

$$\mathbf{t}_1 = \sigma_{11}\mathbf{i}_1 + \sigma_{12}\mathbf{i}_2 + \sigma_{13}\mathbf{i}_3 \tag{2.5.6a}$$

$$\mathbf{t}_2 = \sigma_{21}\mathbf{i}_1 + \sigma_{22}\mathbf{i}_2 + \sigma_{23}\mathbf{i}_3 \tag{2.5.6b}$$

$$\mathbf{t}_3 = \sigma_{31}\mathbf{i}_1 + \sigma_{32}\mathbf{i}_2 + \sigma_{33}\mathbf{i}_3 \tag{2.5.6c}$$

and

$$\frac{\partial(\mathbf{t}_1 ds_2 ds_3)}{\partial s_1} ds_1 + \frac{\partial(\mathbf{t}_2 ds_1 ds_3)}{\partial s_2} ds_2 + \frac{\partial(\mathbf{t}_3 ds_1 ds_2)}{\partial s_3} ds_3$$
$$+ (b_1\mathbf{i}_1 + b_2\mathbf{i}_2 + b_3\mathbf{i}_3)\, ds_1 ds_2 ds_3 = \rho\,(\ddot{u}_1\mathbf{i}_1 + \ddot{u}_2\mathbf{i}_2 + \ddot{u}_3\mathbf{i}_3)\, ds_1 ds_2 ds_3 \tag{2.5.7}$$

Next, we represent the equilibrium equations of a particle in terms of stresses and displacements for different coordinate systems.

Cartesian Coordinate System: Substituting (2.2.3) into (2.5.7) yields

$$\frac{\partial(\mathbf{t}_1 dydz)}{\partial x}dx + \frac{\partial(\mathbf{t}_2 dxdz)}{\partial y}dy + \frac{\partial(\mathbf{t}_3 dxdy)}{\partial z}dz$$
$$+ (b_1\mathbf{i}_1 + b_2\mathbf{i}_2 + b_3\mathbf{i}_3)\, dxdydz = \rho\,(\ddot{u}_1\mathbf{i}_1 + \ddot{u}_2\mathbf{i}_2 + \ddot{u}_3\mathbf{i}_3)\, dxdydz \tag{2.5.8}$$

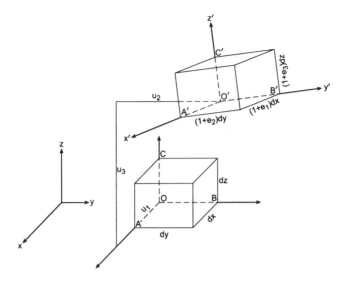

Fig. 2.5.2 The relationship between the deformed configuration of a structural particle and its undeformed configuration in a Cartesian coordinate system.

Substituting (2.5.6) into (2.5.8), dividing by $dxdydz$, and setting each of the coefficients of $\mathbf{i}_1$, $\mathbf{i}_2$, and $\mathbf{i}_3$ equal to zero yields

$$\frac{\partial \sigma_{11}}{\partial x} + \frac{\partial \sigma_{21}}{\partial y} + \frac{\partial \sigma_{31}}{\partial z} + b_1 = \rho \ddot{u}_1 \qquad (2.5.9a)$$

$$\frac{\partial \sigma_{12}}{\partial x} + \frac{\partial \sigma_{22}}{\partial y} + \frac{\partial \sigma_{32}}{\partial z} + b_2 = \rho \ddot{u}_2 \qquad (2.5.9b)$$

$$\frac{\partial \sigma_{13}}{\partial x} + \frac{\partial \sigma_{23}}{\partial y} + \frac{\partial \sigma_{33}}{\partial z} + b_3 = \rho \ddot{u}_3 \qquad (2.5.9c)$$

Next, we derive the nonlinear equilibrium equations for a particle. Figure 2.5.2 shows the relationship between the undeformed and deformed positions of a particle. Although the undeformed coordinate system xyz is Cartesian, the deformed coordinate system $x'y'z'$ is curvilinear; we assume it to be orthogonal. Consequently, the deformation of $\overline{OA}$, $\overline{OB}$, and $\overline{OC}$ can be used to define the transformation matrix $[T]$, which relates the two coordinate systems xyz and $x'y'z'$. In Figure 2.5.2, the coordinates of different points are

$O: (x, y, z)$
$A: (x + dx, y, z)$
$B: (x, y + dy, z)$
$C: (x, y, z + dz)$

O' : $(x+u_1, y+u_2, z+u_3)$
A' : $(x+dx+u_1+u_{1x}dx, y+u_2+u_{2x}dx, z+u_3+u_{3x}dx)$
B' : $(x+u_1+u_{1y}dy, y+dy+u_2+u_{2y}dy, z+u_3+u_{3y}dy)$
C' : $(x+u_1+u_{1z}dz, y+u_2+u_{2z}dz, z+dz+u_3+u_{3z}dz)$ (2.5.10)

It can be seen from Figure 2.5.2 that the axial strains along the x', y', and z' directions, respectively, are

$$e_1 = \frac{\overline{O'A'} - dx}{dx} = \sqrt{(1+u_{1x})^2 + u_{2x}^2 + u_{3x}^2} - 1 \quad (2.5.11)$$

$$e_2 = \frac{\overline{O'B'} - dy}{dy} = \sqrt{u_{1y}^2 + (1+u_{2y})^2 + u_{3y}^2} - 1 \quad (2.5.12)$$

$$e_3 = \frac{\overline{O'C'} - dz}{dz} = \sqrt{u_{1z}^2 + u_{2z}^2 + (1+u_{3z})^2} - 1 \quad (2.5.13)$$

and the base vectors along the x', y', and z' axes are given by

$$\mathbf{i}'_1 = \frac{\mathbf{O'A'}}{(1+e_1)dx} = T_{11}\mathbf{i}_1 + T_{12}\mathbf{i}_2 + T_{13}\mathbf{i}_3$$
$$= \frac{1+u_{1x}}{1+e_1}\mathbf{i}_1 + \frac{u_{2x}}{1+e_1}\mathbf{i}_2 + \frac{u_{3x}}{1+e_1}\mathbf{i}_3 \quad (2.5.14)$$

$$\mathbf{i}'_2 = \frac{\mathbf{O'B'}}{(1+e_2)dx} = T_{21}\mathbf{i}_1 + T_{22}\mathbf{i}_2 + T_{23}\mathbf{i}_3$$
$$= \frac{u_{1y}}{1+e_2}\mathbf{i}_1 + \frac{1+u_{2y}}{1+e_2}\mathbf{i}_2 + \frac{u_{3y}}{1+e_2}\mathbf{i}_3 \quad (2.5.15)$$

$$\mathbf{i}'_3 = \frac{\mathbf{O'C'}}{(1+e_3)dx} = T_{31}\mathbf{i}_1 + T_{32}\mathbf{i}_2 + T_{33}\mathbf{i}_3$$
$$= \frac{u_{1z}}{1+e_3}\mathbf{i}_1 + \frac{u_{2z}}{1+e_3}\mathbf{i}_2 + \frac{1+u_{3z}}{1+e_3}\mathbf{i}_3 \quad (2.5.16)$$

Hence, the transformation that relates the coordinate system xyz to the coordinate system $x'y'z'$ is

$$\begin{Bmatrix} \mathbf{i}'_1 \\ \mathbf{i}'_2 \\ \mathbf{i}'_3 \end{Bmatrix} = [T] \begin{Bmatrix} \mathbf{i}_1 \\ \mathbf{i}_2 \\ \mathbf{i}_3 \end{Bmatrix}, \quad [T] = \begin{bmatrix} T_{11} & T_{12} & T_{13} \\ T_{21} & T_{22} & T_{23} \\ T_{31} & T_{32} & T_{33} \end{bmatrix} \quad (2.5.17)$$

Substituting (2.5.17), (2.5.4), and (2.2.3) into (2.5.2), dividing by $dxdydz$, and setting each of the coefficients of $\mathbf{i}_1$, $\mathbf{i}_2$, and $\mathbf{i}_3$ equal to zero yields

$$\frac{\partial}{\partial x}(\tilde{\sigma}_{11}T_{11} + \tilde{\sigma}_{12}T_{21} + \tilde{\sigma}_{13}T_{31}) + \frac{\partial}{\partial y}(\tilde{\sigma}_{21}T_{11} + \tilde{\sigma}_{22}T_{21} + \tilde{\sigma}_{23}T_{31})$$
$$+ \frac{\partial}{\partial z}(\tilde{\sigma}_{31}T_{11} + \tilde{\sigma}_{32}T_{21} + \tilde{\sigma}_{33}T_{31}) + b_1 = \rho\ddot{u}_1 \quad (2.5.18a)$$

$$\frac{\partial}{\partial x}\left(\tilde{\sigma}_{11}T_{12}+\tilde{\sigma}_{12}T_{22}+\tilde{\sigma}_{13}T_{32}\right)+\frac{\partial}{\partial y}\left(\tilde{\sigma}_{21}T_{12}+\tilde{\sigma}_{22}T_{22}+\tilde{\sigma}_{23}T_{32}\right)$$
$$+\frac{\partial}{\partial z}\left(\tilde{\sigma}_{31}T_{12}+\tilde{\sigma}_{32}T_{22}+\tilde{\sigma}_{33}T_{32}\right)+b_{2}=\rho\ddot{u}_{2} \qquad (2.5.18b)$$
$$\frac{\partial}{\partial x}\left(\tilde{\sigma}_{11}T_{13}+\tilde{\sigma}_{12}T_{23}+\tilde{\sigma}_{13}T_{33}\right)+\frac{\partial}{\partial y}\left(\tilde{\sigma}_{21}T_{13}+\tilde{\sigma}_{22}T_{23}+\tilde{\sigma}_{23}T_{33}\right)$$
$$+\frac{\partial}{\partial z}\left(\tilde{\sigma}_{31}T_{13}+\tilde{\sigma}_{32}T_{23}+\tilde{\sigma}_{33}T_{33}\right)+b_{3}=\rho\ddot{u}_{3} \qquad (2.5.18c)$$

Note that, if the configuration change due to extensionality is negligible (i.e., $e_i \approx 0$), $[T]$ is simplified as

$$[T]=\begin{bmatrix} 1+u_x & v_x & w_x \\ u_y & 1+v_y & w_y \\ u_z & v_z & 1+w_z \end{bmatrix} \qquad (2.5.19)$$

Cylindrical Coordinate System: Substituting (2.2.6) into (2.5.7), we have

$$\frac{\partial(\mathbf{t}_1 r d\theta dr)}{\partial x}dx+\frac{1}{r}\frac{\partial(\mathbf{t}_2 dx dr)}{\partial \theta}rd\theta+\frac{\partial(\mathbf{t}_3 r d\theta dx)}{\partial r}dr$$
$$+\left(b_1\mathbf{i}_1+b_2\mathbf{i}_2+b_3\mathbf{i}_3\right)rdxd\theta dr = \rho\left(\ddot{u}_1\mathbf{i}_1+\ddot{u}_2\mathbf{i}_2+\ddot{u}_3\mathbf{i}_3\right)rdxd\theta dr \qquad (2.5.20)$$

Substituting (2.2.8) and (2.5.6) into (2.5.20), dividing by $rdxd\theta dr$, and setting each of the coefficients of $\mathbf{i}_1$, $\mathbf{i}_2$, and $\mathbf{i}_3$ equal to zero yields

$$\frac{\partial\sigma_{11}}{\partial x}+\frac{1}{r}\frac{\partial\sigma_{21}}{\partial\theta}+\frac{\partial\sigma_{31}}{\partial r}+\frac{\sigma_{31}}{r}+b_1=\rho\ddot{u}_1 \qquad (2.5.21a)$$
$$\frac{\partial\sigma_{12}}{\partial x}+\frac{1}{r}\frac{\partial\sigma_{22}}{\partial\theta}+\frac{\partial\sigma_{32}}{\partial r}+\frac{2\sigma_{32}}{r}+b_2=\rho\ddot{u}_2 \qquad (2.5.21b)$$
$$\frac{\partial\sigma_{13}}{\partial x}+\frac{1}{r}\frac{\partial\sigma_{23}}{\partial\theta}+\frac{\partial\sigma_{33}}{\partial r}+\frac{\sigma_{33}-\sigma_{22}}{r}+b_3=\rho\ddot{u}_3 \qquad (2.5.21c)$$

To obtain a transformation matrix $[T]$ that relates the deformed and undeformed coordinate systems as that shown in (2.5.17), we obtain from Figure 2.2.2 and (2.2.6) and (2.2.8) the displacement vectors $\mathbf{D}$, $\mathbf{D}_a$, $\mathbf{D}_b$, and $\mathbf{D}_c$ of Points O, A, B, and C, respectively, as

$$\mathbf{D}=u_1\mathbf{i}_1+u_2\mathbf{i}_2+u_3\mathbf{i}_3$$
$$\mathbf{D}_a=\mathbf{D}+\frac{\partial\mathbf{D}}{\partial x}dx=\mathbf{D}+\left(u_{1x}\mathbf{i}_1+u_{2x}\mathbf{i}_2+u_{3x}\mathbf{i}_3\right)dx$$
$$\mathbf{D}_b=\mathbf{D}+\frac{\partial\mathbf{D}}{\partial\theta}d\theta=\mathbf{D}+\left(u_{1\theta}\mathbf{i}_1+u_{2\theta}\mathbf{i}_2+u_{3\theta}\mathbf{i}_3+u_3\mathbf{i}_2-u_2\mathbf{i}_3\right)d\theta$$
$$\mathbf{D}_c=\mathbf{D}+\frac{\partial\mathbf{D}}{\partial r}dr=\mathbf{D}+\left(u_{1r}\mathbf{i}_1+u_{2r}\mathbf{i}_2+u_{3r}\mathbf{i}_3\right)dr \qquad (2.5.22)$$

where $u_{1x}\equiv\partial u_1/\partial x$, etc. Hence, the axial strains along the $\mathbf{i}'_1$, $\mathbf{i}'_2$, and $\mathbf{i}'_3$ directions, respectively, are obtained as

$$e_1=\frac{|\mathbf{D}_a+dx\mathbf{i}_1-\mathbf{D}|-dx}{dx}=\sqrt{(1+u_{1x})^2+u_{2x}{}^2+u_{3x}{}^2}-1 \qquad (2.5.23a)$$

104 ELASTICITY

$$e_2 = \frac{|\mathbf{D}_b + rd\theta \mathbf{i}_2 - \mathbf{D}| - rd\theta}{rd\theta}$$

$$= \frac{1}{r}\sqrt{u_{1\theta}^2 + (r + u_3 + u_{2\theta})^2 + (u_{3\theta} - u_2)^2} - 1 \quad (2.5.23b)$$

$$e_3 = \frac{|\mathbf{D}_c + dr\mathbf{i}_3 - \mathbf{D}| - dr}{dr} = \sqrt{u_{1r}^2 + u_{2r}^2 + (1 + u_{3r})^2} - 1 \quad (2.5.23c)$$

and the base vectors of the deformed coordinate system are given by

$$\mathbf{i}'_1 = \frac{\mathbf{D}_a + dx\mathbf{i}_1 - \mathbf{D}}{(1+e_1)dx} = T_{11}\mathbf{i}_1 + T_{12}\mathbf{i}_2 + T_{13}\mathbf{i}_3$$

$$= \frac{1 + u_{1x}}{1 + e_1}\mathbf{i}_1 + \frac{u_{2x}}{1 + e_1}\mathbf{i}_2 + \frac{u_{3x}}{1 + e_1}\mathbf{i}_3 \quad (2.5.24a)$$

$$\mathbf{i}'_2 = \frac{\mathbf{D}_b + rd\theta\mathbf{i}_2 - \mathbf{D}}{(1+e_2)rd\theta} = T_{21}\mathbf{i}_1 + T_{22}\mathbf{i}_2 + T_{23}\mathbf{i}_3$$

$$= \frac{u_{1\theta}}{(1+e_2)r}\mathbf{i}_1 + \frac{r + u_3 + u_{2\theta}}{(1+e_2)r}\mathbf{i}_2 + \frac{u_{3\theta} - u_2}{(1+e_2)r}\mathbf{i}_3 \quad (2.5.24b)$$

$$\mathbf{i}'_3 = \frac{\mathbf{D}_c + dr\mathbf{i}_3 - \mathbf{D}}{(1+e_3)dr} = T_{31}\mathbf{i}_1 + T_{32}\mathbf{i}_2 + T_{33}\mathbf{i}_3$$

$$= \frac{u_{1r}}{1 + e_3}\mathbf{i}_1 + \frac{u_{2r}}{1 + e_3}\mathbf{i}_2 + \frac{1 + u_{3r}}{1 + e_3}\mathbf{i}_3 \quad (2.5.24c)$$

Using (2.5.17), (2.2.18)-(2.2.20), and (2.5.4), we have

$$\frac{\partial(\mathbf{t}_1 rd\theta dr)}{\partial x}dx + \frac{1}{r}\frac{\partial(\mathbf{t}_2 dx dr)}{\partial \theta}rd\theta + \frac{\partial(\mathbf{t}_3 rd\theta dx)}{\partial r}dr$$

$$= rdxd\theta dr\{\mathbf{i}_1, \mathbf{i}_2, \mathbf{i}_3\}\left\{\frac{\partial}{\partial x}\left([T]^T\begin{Bmatrix}\tilde{\sigma}_{11}\\\tilde{\sigma}_{12}\\\tilde{\sigma}_{13}\end{Bmatrix}\right) + \frac{1}{r}\frac{\partial}{\partial \theta}\left([T]^T\begin{Bmatrix}\tilde{\sigma}_{21}\\\tilde{\sigma}_{22}\\\tilde{\sigma}_{23}\end{Bmatrix}\right)\right.$$

$$+ [K_2]^T[T]^T\begin{Bmatrix}\tilde{\sigma}_{21}\\\tilde{\sigma}_{22}\\\tilde{\sigma}_{23}\end{Bmatrix} + \frac{\partial}{\partial r}\left([T]^T\begin{Bmatrix}\tilde{\sigma}_{31}\\\tilde{\sigma}_{32}\\\tilde{\sigma}_{33}\end{Bmatrix}\right) + \frac{1}{r}[T]^T\begin{Bmatrix}\tilde{\sigma}_{31}\\\tilde{\sigma}_{32}\\\tilde{\sigma}_{33}\end{Bmatrix}\right\} \quad (2.5.25)$$

where $[K_2]$ is given by (2.2.24). Substituting (2.5.25) into (2.5.2), dividing by $rdxd\theta dr$, setting the coefficients of $\mathbf{i}_1$, $\mathbf{i}_2$, and $\mathbf{i}_3$ equal to zero, and putting them in matrix form, we obtain

$$\frac{\partial}{\partial x}\left([T]^T\begin{Bmatrix}\tilde{\sigma}_{11}\\\tilde{\sigma}_{12}\\\tilde{\sigma}_{13}\end{Bmatrix}\right) + \frac{1}{r}\frac{\partial}{\partial \theta}\left([T]^T\begin{Bmatrix}\tilde{\sigma}_{21}\\\tilde{\sigma}_{22}\\\tilde{\sigma}_{23}\end{Bmatrix}\right) + [K_2]^T[T]^T\begin{Bmatrix}\tilde{\sigma}_{21}\\\tilde{\sigma}_{22}\\\tilde{\sigma}_{23}\end{Bmatrix}$$

$$+ \frac{\partial}{\partial r}\left([T]^T\begin{Bmatrix}\tilde{\sigma}_{31}\\\tilde{\sigma}_{32}\\\tilde{\sigma}_{33}\end{Bmatrix}\right) + \frac{1}{r}[T]^T\begin{Bmatrix}\tilde{\sigma}_{31}\\\tilde{\sigma}_{32}\\\tilde{\sigma}_{33}\end{Bmatrix} + \begin{Bmatrix}b_1\\b_2\\b_3\end{Bmatrix} = \rho\begin{Bmatrix}\ddot{u}_1\\\ddot{u}_2\\\ddot{u}_3\end{Bmatrix} \quad (2.5.26)$$

GOVERNING EQUATIONS

Spherical Coordinate System: Substituting (2.2.11) into (2.5.7), we have

$$\frac{\partial(\mathbf{t}_1 r \sin\theta d\phi dr)}{r\partial\theta}rd\theta + \frac{\partial(\mathbf{t}_2 r d\theta dr)}{r\sin\theta\partial\phi}r\sin\theta d\phi + \frac{\partial(\mathbf{t}_3 r^2 \sin\theta d\phi d\theta)}{\partial r}dr$$
$$+(b_1\mathbf{i}_1 + b_2\mathbf{i}_2 + b_3\mathbf{i}_3)r^2\sin\theta d\theta d\phi dr = \rho(\ddot{u}_1\mathbf{i}_1 + \ddot{u}_2\mathbf{i}_2 + \ddot{u}_3\mathbf{i}_3)r^2\sin\theta d\theta d\phi dr \quad (2.5.27)$$

Substituting (2.2.13) and (2.5.6) into (2.5.27), dividing by $r^2\sin\theta d\theta d\phi dr$, and setting the coefficients of $\mathbf{i}_1$, $\mathbf{i}_2$, and $\mathbf{i}_3$ equal to zero yield

$$\frac{1}{r\sin\theta}\frac{\partial(\sin\theta\sigma_{11})}{\partial\theta} + \frac{1}{r\sin\theta}\frac{\partial\sigma_{21}}{\partial\phi} + \frac{1}{r^3}\frac{\partial(r^3\sigma_{31})}{\partial r} - \frac{\cot\theta}{r}\sigma_{22} + b_1 = \rho\ddot{u}_1 \quad (2.5.28a)$$

$$\frac{1}{r\sin^2\theta}\frac{\partial(\sin^2\theta\sigma_{12})}{\partial\theta} + \frac{1}{r\sin\theta}\frac{\partial\sigma_{22}}{\partial\phi} + \frac{1}{r^3}\frac{\partial(r^3\sigma_{32})}{\partial r} + b_2 = \rho\ddot{u}_2 \quad (2.5.28b)$$

$$\frac{1}{r\sin\theta}\frac{\partial(\sin\theta\sigma_{13})}{\partial\theta} + \frac{1}{r\sin\theta}\frac{\partial\sigma_{23}}{\partial\phi} + \frac{1}{r^2}\frac{\partial(r^2\sigma_{33})}{\partial r} - \frac{\sigma_{11}+\sigma_{22}}{r} + b_3 = \rho\ddot{u}_3 \quad (2.5.28c)$$

To obtain a transformation matrix $[T]$ that relates the deformed and undeformed coordinate systems as that shown in (2.5.17), we obtain from Figure 2.2.3 and (2.2.11) and (2.2.13) the displacement vectors $\mathbf{D}$, $\mathbf{D}_a$, $\mathbf{D}_b$, and $\mathbf{D}_c$ of Points O, A, B, and C, respectively, as

$$\mathbf{D} = u_1\mathbf{i}_1 + u_2\mathbf{i}_2 + u_3\mathbf{i}_3$$
$$\mathbf{D}_a = \mathbf{D} + \frac{\partial\mathbf{D}}{\partial\theta}d\theta = \mathbf{D} + (u_{1\theta}\mathbf{i}_1 + u_{2\theta}\mathbf{i}_2 + u_{3\theta}\mathbf{i}_3 + u_3\mathbf{i}_1 - u_1\mathbf{i}_3)\,d\theta$$
$$\mathbf{D}_b = \mathbf{D} + \frac{\partial\mathbf{D}}{\partial\phi}d\phi = \mathbf{D} + \big[(u_{1\phi} - u_2\cos\theta)\mathbf{i}_1 + (u_{2\phi} + u_3\sin\theta + u_1\cos\theta)\mathbf{i}_2$$
$$+ (u_{3\phi} - u_2\sin\theta)\mathbf{i}_3\big]d\phi$$
$$\mathbf{D}_c = \mathbf{D} + \frac{\partial\mathbf{D}}{\partial r}dr = \mathbf{D} + (u_{1r}\mathbf{i}_1 + u_{2r}\mathbf{i}_2 + u_{3r}\mathbf{i}_3)\,dr \quad (2.5.29)$$

Hence, the axial strains along the $\mathbf{i}'_1$, $\mathbf{i}'_2$ and $\mathbf{i}'_3$ directions, respectively, are obtained as

$$e_1 = \frac{|\mathbf{D}_a + rd\theta\mathbf{i}_1 - \mathbf{D}| - rd\theta}{rd\theta}$$
$$= \frac{1}{r}\sqrt{(r + u_3 + u_{1\theta})^2 + u_{2\theta}^2 + (u_{3\theta} - u_1)^2} - 1 \quad (2.5.30a)$$

$$e_2 = \frac{|\mathbf{D}_b + r\sin\theta d\phi\mathbf{i}_2 - \mathbf{D}| - r\sin\theta d\phi}{r\sin\theta d\phi}$$
$$= \frac{1}{r\sin\theta}\big[(u_{1\phi} - u_2\cos\theta)^2 + (u_{2\phi} + r\sin\theta + u_3\sin\theta + u_1\cos\theta)^2$$
$$+ (u_{3\phi} - u_2\sin\theta)^2\big]^{1/2} - 1 \quad (2.5.30b)$$

$$e_3 = \frac{|\mathbf{D}_c + dr\mathbf{i}_3 - \mathbf{D}| - dr}{dr} = \sqrt{u_{1r}^2 + u_{2r}^2 + (1+u_{3r})^2} - 1 \qquad (2.5.30c)$$

and the base vectors of the deformed coordinate system are given by

$$\mathbf{i}'_1 = \frac{\mathbf{D}_a + rd\theta\mathbf{i}_1 - \mathbf{D}}{(1+e_1)rd\theta} = T_{11}\mathbf{i}_1 + T_{12}\mathbf{i}_2 + T_{13}\mathbf{i}_3$$

$$= \frac{r + u_3 + u_{1\theta}}{(1+e_1)r}\mathbf{i}_1 + \frac{u_{2\theta}}{(1+e_1)r}\mathbf{i}_2 + \frac{u_{3\theta} - u_1}{(1+e_1)r}\mathbf{i}_3 \qquad (2.5.31a)$$

$$\mathbf{i}'_2 = \frac{\mathbf{D}_b + r\sin\theta d\phi\mathbf{i}_2 - \mathbf{D}}{(1+e_2)r\sin\theta d\phi} = T_{21}\mathbf{i}_1 + T_{22}\mathbf{i}_2 + T_{23}\mathbf{i}_3$$

$$= \frac{u_{1\phi} - u_2\cos\theta}{(1+e_2)r\sin\theta}\mathbf{i}_1 + \frac{r\sin\theta + u_3\sin\theta + u_1\cos\theta + u_{2\phi}}{(1+e_2)r\sin\theta}\mathbf{i}_2 + \frac{u_{3\phi} - u_2\sin\theta}{(1+e_2)r\sin\theta}\mathbf{i}_3 \qquad (2.5.31b)$$

$$\mathbf{i}'_3 = \frac{\mathbf{D}_c + dr\mathbf{i}_3 - \mathbf{D}}{(1+e_3)dr} = T_{31}\mathbf{i}_1 + T_{32}\mathbf{i}_2 + T_{33}\mathbf{i}_3$$

$$= \frac{u_{1r}}{1+e_3}\mathbf{i}_1 + \frac{u_{2r}}{1+e_3}\mathbf{i}_2 + \frac{1+u_{3r}}{1+e_3}\mathbf{i}_3 \qquad (2.5.31c)$$

Using (2.5.17), (2.2.18)-(2.2.20), and (2.5.4), we have

$$\frac{\partial(\mathbf{t}_1 r\sin\theta d\phi dr)}{r\partial\theta}rd\theta + \frac{\partial(\mathbf{t}_2 rd\theta dr)}{r\sin\theta\partial\phi}r\sin\theta d\phi + \frac{\partial(\mathbf{t}_3 r^2\sin\theta d\phi d\theta)}{\partial r}dr$$

$$= r^2\sin\theta d\theta d\phi dr\{\mathbf{i}_1, \mathbf{i}_2, \mathbf{i}_3\}\left\{\frac{1}{r}\frac{\partial}{\partial\theta}\left([T]^T\begin{Bmatrix}\tilde\sigma_{11}\\\tilde\sigma_{12}\\\tilde\sigma_{13}\end{Bmatrix}\right) + [K_1]^T[T]^T\begin{Bmatrix}\tilde\sigma_{11}\\\tilde\sigma_{12}\\\tilde\sigma_{13}\end{Bmatrix}\right.$$

$$+ \frac{1}{\tan\theta}[T]^T\begin{Bmatrix}\tilde\sigma_{11}\\\tilde\sigma_{12}\\\tilde\sigma_{13}\end{Bmatrix} + \frac{1}{r\sin\theta}\frac{\partial}{\partial\phi}\left([T]^T\begin{Bmatrix}\tilde\sigma_{21}\\\tilde\sigma_{22}\\\tilde\sigma_{23}\end{Bmatrix}\right) + [K_2]^T[T]^T\begin{Bmatrix}\tilde\sigma_{21}\\\tilde\sigma_{22}\\\tilde\sigma_{23}\end{Bmatrix}$$

$$\left. + \frac{\partial}{\partial r}\left([T]^T\begin{Bmatrix}\tilde\sigma_{31}\\\tilde\sigma_{32}\\\tilde\sigma_{33}\end{Bmatrix}\right) + \frac{2}{r}[T]^T\begin{Bmatrix}\tilde\sigma_{31}\\\tilde\sigma_{32}\\\tilde\sigma_{33}\end{Bmatrix}\right\} \qquad (2.5.32)$$

where $[K_1]$ and $[K_2]$ are given by (2.2.25). Substituting (2.5.32) into (2.5.2), dividing by $r^2\sin\theta d\theta d\phi dr$, setting the coefficients of $\mathbf{i}_1$, $\mathbf{i}_2$, and $\mathbf{i}_3$ equal to zero, and putting them in matrix form, we obtain

$$\frac{1}{r}\frac{\partial}{\partial\theta}\left([T]^T\begin{Bmatrix}\tilde\sigma_{11}\\\tilde\sigma_{12}\\\tilde\sigma_{13}\end{Bmatrix}\right) + [K_1]^T[T]^T\begin{Bmatrix}\tilde\sigma_{11}\\\tilde\sigma_{12}\\\tilde\sigma_{13}\end{Bmatrix} + \frac{1}{\tan\theta}[T]^T\begin{Bmatrix}\tilde\sigma_{11}\\\tilde\sigma_{12}\\\tilde\sigma_{13}\end{Bmatrix}$$

$$+ \frac{1}{r\sin\theta}\frac{\partial}{\partial\phi}\left([T]^T\begin{Bmatrix}\tilde\sigma_{21}\\\tilde\sigma_{22}\\\tilde\sigma_{23}\end{Bmatrix}\right) + [K_2]^T[T]^T\begin{Bmatrix}\tilde\sigma_{21}\\\tilde\sigma_{22}\\\tilde\sigma_{23}\end{Bmatrix}$$

$$+ \frac{\partial}{\partial r}\left([T]^T\begin{Bmatrix}\tilde\sigma_{31}\\\tilde\sigma_{32}\\\tilde\sigma_{33}\end{Bmatrix}\right) + \frac{2}{r}[T]^T\begin{Bmatrix}\tilde\sigma_{31}\\\tilde\sigma_{32}\\\tilde\sigma_{33}\end{Bmatrix} + \begin{Bmatrix}b_1\\b_2\\b_3\end{Bmatrix} = \rho\begin{Bmatrix}\ddot u_1\\\ddot u_2\\\ddot u_3\end{Bmatrix} \qquad (2.5.33)$$

2.5.2 Compatibility Conditions

For an elasticity boundary-value problem, if the formulation is based on unknown displacements, the governing equations include three equilibrium equations (e.g., (2.5.28a-c) for linear analysis or (2.5.33) for nonlinear analysis), six stress-strain relations (e.g., (2.4.12)), and six strain-displacement relations (e.g., (2.2.27)), and the unknowns include three displacements, six independent stresses, and six independent strains. Hence, the 15 governing equations can be used to solve for the 15 unknowns and obtain a single-valued displacement distribution in a simply connected body. On the other hand, if the formulation is based on unknown stresses (or strains), there are only three equilibrium equations and six stress-strain relations, but the unknowns include six independent stresses and six independent strains. Hence, three more governing equations are needed in order to assure the deformed body is continuous under the obtained stress (or strain) field, and they are the compatibility conditions. Compatibility conditions ensure the integrability of the strain-displacement relations and are necessary and sufficient conditions for the existence of a single-valued displacement distribution in a simply connected body. However, the obtained displacement field may not be unique because of possible arbitrary rigid-body displacements.

The compatibility conditions can be presented in the following form (Malvern, 1969)

$$\nabla \times \mathbf{e} \times \nabla = C_{mn} \mathbf{i}_m \mathbf{i}_n = 0, \quad m, n = 1, 2, 3 \qquad (2.5.34)$$

where

$$\nabla = \frac{\partial}{\partial s_i} \mathbf{i}'_i, \quad \mathbf{e} = e_{mn} \mathbf{i}'_m \mathbf{i}'_n$$

Because e_{mn} is symmetric, C_{mn} is symmetric and (2.5.34) results in only six equations.

For linear problems using a Cartesian coordinate system, $\mathbf{i}'_m = \mathbf{i}_m$ (see Figure 2.5.1) and $\partial \mathbf{i}_m / \partial s_n = 0$. Hence, (2.5.34) results into

$$C_{11} = 2e_{23,23} - e_{22,33} - e_{33,22} = 0$$
$$C_{22} = 2e_{13,13} - e_{11,33} - e_{33,11} = 0$$
$$C_{33} = 2e_{12,12} - e_{11,22} - e_{22,11} = 0$$
$$C_{32} = C_{23} = e_{11,23} + e_{23,11} - e_{13,12} - e_{12,13} = 0$$
$$C_{31} = C_{13} = e_{22,13} - e_{23,12} + e_{13,22} - e_{12,23} = 0$$
$$C_{21} = C_{12} = e_{33,12} - e_{23,13} - e_{13,23} + e_{12,33} = 0 \qquad (2.5.35)$$

where $e_{ij,mn} \equiv \partial^2 e_{ij}/\partial s_m \partial s_n$. One can verify (2.5.35) by substituting for the e_{ij} from (2.2.5). By substituting (2.2.5) into the C_{ij} in (2.5.35) and direct expansion, one can show that

$$C_{11,1} + C_{12,2} + C_{13,3} = 0$$
$$C_{21,1} + C_{22,2} + C_{23,3} = 0$$
$$C_{31,1} + C_{32,2} + C_{33,3} = 0 \qquad (2.5.36)$$

Hence, only three of the six equations in (2.5.35) are independent. However, because the dependency is not direct as shown in (2.5.36), all the six equations are usually needed in actual applications. For the two-dimensional case, $e_{13} = e_{23} = e_{33} = 0$ and $\partial/\partial s_3 = 0$, and hence (2.5.35) reduces to $C_{33} = 0$ or

$$2e_{12,12} - e_{11,22} - e_{22,11} = 0$$

For nonlinear problems using a general orthogonal curvilinear coordinate system, one can use (2.5.17) to present $\mathbf{i}'_m$ in terms of $\mathbf{i}_n$ and use (2.2.18)-(2.2.23) to account for the fact that $\partial \mathbf{i}_m / \partial s_n \neq 0$. The resulting C_{ij} are more complicated than those shown in (2.5.35) because of curvatures.

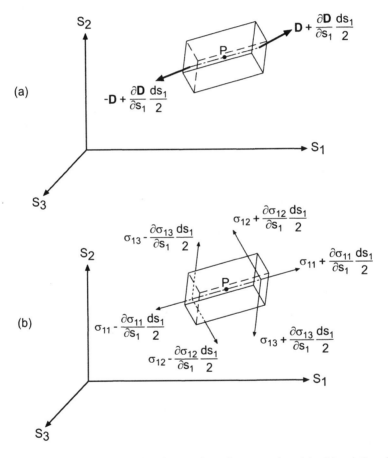

Fig. 2.5.3 The displacements and surface tractions of a structural particle: (a) variation of the displacement vector along the direction s_1 and (b) variation of the stresses along the direction s_1.

2.5.3 Energy Formulation of Structures

The equations of motion for a structural element are different from the equilibrium equations for a particle because an element has finite dimensions and hence it has translational and rotational motions. The general rigid-body motion of an element can be easily modeled by using a vector approach, but both rigid-body motion and relative displacements can occur in an elastic element. Hence, an energy approach is more convenient than a vector approach in the formulation of elastic structures. Moreover, the formulations in Section 2.1 are for discrete systems, and, for distributed or continuous systems, the summations used in Section 2.1 need to be replaced by integrations.

For a structure, the extended Hamilton principle (see (2.1.23)) can be stated as

$$\int_{t_1}^{t_2} (\delta T - \delta \Pi + \delta W_{nc}) dt = 0 \tag{2.5.37}$$

where

$$\delta T = -\int_R \rho \frac{\partial^2 \mathbf{D}}{\partial t^2} \cdot \delta \mathbf{D} \, dR \tag{2.5.38}$$

$$\delta \Pi = \int_R \left\{ \left(\sigma_{11} \delta \frac{\partial \mathbf{D}}{\partial s_1} ds_1 \cdot \mathbf{i}_1 + \sigma_{12} \delta \frac{\partial \mathbf{D}}{\partial s_1} ds_1 \cdot \mathbf{i}_2 + \sigma_{13} \delta \frac{\partial \mathbf{D}}{\partial s_1} ds_1 \cdot \mathbf{i}_3 \right) ds_2 ds_3 \right.$$
$$+ \left(\sigma_{21} \delta \frac{\partial \mathbf{D}}{\partial s_2} ds_2 \cdot \mathbf{i}_1 + \sigma_{22} \delta \frac{\partial \mathbf{D}}{\partial s_2} ds_2 \cdot \mathbf{i}_2 + \sigma_{23} \delta \frac{\partial \mathbf{D}}{\partial s_2} ds_2 \cdot \mathbf{i}_3 \right) ds_1 ds_3$$
$$\left. + \left(\sigma_{31} \delta \frac{\partial \mathbf{D}}{\partial s_3} ds_3 \cdot \mathbf{i}_1 + \sigma_{32} \delta \frac{\partial \mathbf{D}}{\partial s_3} ds_3 \cdot \mathbf{i}_2 + \sigma_{33} \delta \frac{\partial \mathbf{D}}{\partial s_3} ds_3 \cdot \mathbf{i}_3 \right) ds_1 ds_2 \right\} \tag{2.5.39}$$

$$\delta W_{nc} = \int_{S_1} \delta W_{nc1} dS_1 + \delta W_{nc2} \tag{2.5.40}$$

Here, R is the spatial domain of the structure, δW_{nc1} is the virtual work due to applied loads on the structural surface S_1, δW_{nc2} is the virtual work due to forces or moments applied on the boundaries or the motion of the boundaries, ρ denotes the density of the material, and $\mathbf{D}$ is the displacement vector. The variation of the elastic energy $\delta \Pi$ is the sum of the virtual work done by the stresses through the virtual relative displacements along their directions. Figure 2.5.3 shows only the variation of the displacement vector $\mathbf{D}$ and the stresses on the surface $s_2 - s_3$ along the axis s_1. The variations of $\mathbf{D}$ and σ_{ij} along the other axes can be similarly obtained. In (2.5.39), we assume that the stresses are evaluated with respect to the undeformed area, but their directions are defined with respect to the deformed coordinate system. We further assume that $\sigma_{ij} = \sigma_{ji}$ and hence have

$$\delta \Pi = \int_R \left(\sigma_{11} \delta e_{11} + \sigma_{22} \delta e_{22} + \sigma_{33} \delta e_{33} \right.$$
$$\left. + 2\sigma_{23} \delta e_{23} + 2\sigma_{13} \delta e_{13} + 2\sigma_{12} \delta e_{12} \right) ds_1 ds_2 ds_3$$

$$= \int_R (\sigma_{11}\delta\epsilon_{11} + \sigma_{22}\delta\epsilon_{22} + \sigma_{33}\delta\epsilon_{33}$$
$$+ \sigma_{23}\delta\epsilon_{23} + \sigma_{13}\delta\epsilon_{13} + \sigma_{12}\delta\epsilon_{12}) ds_1 ds_2 ds_3 \qquad (2.5.41)$$

where the strains e_{ij} are defined in (2.2.2) and the ϵ_{ij} are the engineering strains given by

$$\epsilon_{11} = e_{11}, \quad \epsilon_{22} = e_{22}, \quad \epsilon_{33} = e_{33}, \quad \epsilon_{23} = 2e_{23}, \quad \epsilon_{13} = 2e_{13}, \quad \epsilon_{12} = 2e_{12} \qquad (2.5.42)$$

For any structure, the displacement vector **D** consists of two parts; that is,

$$\mathbf{D} = \mathbf{U} + \mathbf{B} \qquad (2.5.43)$$

where **B** is due to rigid-body translations and/or rotations and **U** is due to relative displacements among particles in the structure. Because rigid-body motions do not result in any strains or strain energy, the strains can be defined in terms of the relative displacement vector **U** as

$$e_{ij} \equiv \frac{1}{2}\left(\frac{\partial \mathbf{U}}{\partial s_i} \cdot \mathbf{i}_j + \frac{\partial \mathbf{U}}{\partial s_j} \cdot \mathbf{i}_i\right), \qquad i,j = 1,2,3 \qquad (2.5.44)$$

If the relative displacement vector **U** is defined with respect to a coordinate frame corotated with the rigid-body motion, we show in each of the following chapters that one can combine (2.5.44) with an analysis of the reference line (if beams) or the reference surface (if plates or shells) to derive fully nonlinear objective strain-displacement relations without using the complex polar decomposition (Malvern, 1969).

Because the extended Hamilton principle can be derived from the principle of virtual work, as shown in Section 2.1.3, (2.5.37) is also called the dynamic version of the principle of virtual work and is equivalent to the theorem of minimum potential energy. (2.5.37) is a general formulation for any structure. For different structures and/or different assumed displacement (or stress) fields, the final equations of motion obtained from (2.5.37) are different.

Since all structures are essentially three-dimensional bodies, it is difficult to solve a general structural problem. However, any thin-walled structure can be ideally represented by a two-dimensional or a one-dimensional model with an assumed displacement or stress field. Hence, the derivation of the governing equations of a structural element depends on the type of the structure and the assumed displacement (or stress) field, and we treat them separately in different chapters.

3
STRINGS AND CABLES

Cables and strings are one-dimensional structures that can only sustain longitudinal tension loads because they have negligible flexural, torsional, and shear rigidities and have zero buckling loads. Strings are cables with no static sags. The nonlinear static, dynamic, and stability analyses of cables and strings have received considerable attention in recent years because of their wide applications in engineering where light-weight, flexible, or easily deployable structural members or conductors are demanded. Applications of cables and strings as structural members include power cables, information transmission lines, cables for towing and mooring marine vehicles, lines for tethering objects over long distances, cables used in large buildings to provide large column-free areas, tension members of cable-stayed bridges or guy-towers, guy cables for wind turbines, and taut strings for musical instruments. Moreover, cables are very promising structural elements for space applications since they can be packaged easily in a small volume and function as truss members when in a stretched state. Also, cable structures are often used in the offshore industry and oceanographic community. Problems of cables include large static deflections and nonlinear vibrations under ice overloading, wind pressure, and electromagnetic forces due to short circuit currents.

3.1 MODELING OF TAUT STRINGS

Taut strings are initially straight cables subjected to pretension. Nonlinear phenomena in the vibration of strings have been reported for fifty years. Harrison (1948) reported that conditions exist for which a string subjected to a planar force transverse to its length will whirl like a jump rope. Murthy and Ramakrishna (1965) conducted an experimental study of a string subjected to a planar external excitation and observed nonplanar whirling motions and jump phenomena. In the simplest model, it is assumed that the tension in the string is constant. This assumption leads to the linear wave equation in which there is no coupling among modes and the ballooning (whirling) motions cannot be predicted. Therefore, the whirling motions reported by Harrison (1948) and Murthy and Ramakrishna (1965) are a clear indication that the linear equations of motion are invalid for all but the smallest motions of strings.

As a string deflects transversely, its length, and therefore its tension, must increase. Nonlinear terms are required to account for stretching in the equations of motion. This nonlinearity has been considered in a number of recent analyses. Narasimha (1968), Eller (1972), Anand (1973), Lee and Ames (1973), and Miles (1984d) analyzed the coupling between inplane and out-of-plane modes. Narasimha (1968) in particular presented a careful development of the nonlinear equations of motion and a thorough discussion of the various assumptions involved. A detailed analysis of nonplanar string vibrations by the method of multiple scales can be found in Nayfeh and Mook (1979).

Several more modern studies of the nonlinear dynamics of strings have appeared in recent years. Miles (1984d) studied the nonplanar motion of stretched strings. He found Hopf bifurcations and suggested the possibility of strange attractors. He did not investigate limit cycles or chaotic attractors. Yasuda and Torii (1986) used the method of multiple scales to analyze the nonlinear forced response of a string. They studied the multimode responses near the first, third, and fourth primary resonance points. They performed an experiment on a thin steel strip to validate their theoretical results. Tufillaro (1989) predicted various nonlinear phenomena which include quasiperiodic and chaotic motions. Johnson and Bajaj (1989) and Bajaj and Johnson (1992) performed extensive bifurcation analyses and predicted the existence of quasiperiodic torus-doubling bifurcations and chaotic motions as well as the phenomenon of boundary crisis. Molteno and Tufillaro (1990) reported experimental observations of torus-doubling bifurcations and chaotic motions in a string. Johnson and Bajaj (1989) studied the one-to-one internal resonance between the planar and nonplanar responses of strings. They found two branches of periodic solutions between the Hopf bifurcation frequencies. One branch goes through an incomplete sequence of period-doubling bifurcations followed by period-halving bifurcations. The other branch is created by a saddle-node bifurcation and goes through a cascade of period-doubling bifurcations culminating in chaos. Gottlieb (1990) investigated the nonlinear vibrations of a constant-tension string. O'Reilly and Holmes (1992) presented experimental results and investigated them using an integral of the Hamiltonian system with the damping and excitation being set equal to zero.

In all of these studies, the string is excited transversely. Melde (1859) fixed one end of a string and attached the other end to a large tuning fork so that the motion of the tuning fork was parallel to the axis of the string, causing a parametric excitation. He observed that the string could be made to oscillate transversely although the force was along the axis of the string.

S. Nayfeh, Nayfeh, and Mook (1995) studied the nonlinear response of a taut string to an excitation with components both parallel and transverse to its axis, where a model for the transverse vibration of the string was developed by neglecting its longitudinal inertia. The method of multiple scales was applied directly to the obtained governing partial-differential equations and boundary conditions. A continuation method was then employed to determine the equilibrium solutions of the modulation equations and their stability solutions. Resonant responses were predicted to occur simultaneously in as many as three modes. An experimental study was conducted and the results were found to be in good agreement with the theory.

3.1.1 Exact Equations of Motion

In Figure 3.1.1a, we show the deformed configuration of a taut string and the inertial coordinate system $x_1 x_2 x_3$. Point P_0 indicates the position of the observed particle when the string is not loaded, Point $\hat{P}$ indicates the deformed position of P_0 under a static pretension, which changes the undeformed total length L to $\ell = L + \Delta L$, and Point P indicates the deformed position of P_0 under the combined static pretension and dynamic loads. Here, s denotes the undeformed length measured from Point M to the observed particle, x denotes the statically deformed length under pretension, and $\tilde{s}$ represents the dynamically deformed arclength. Also, we let $\rho_0(s)$ denote the mass density of the unloaded string, $A_0(s)$ denote the unloaded cross-section area, and $(\ )' \equiv \partial(\ )/\partial s$. The coordinates of $\hat{P}$ are $(x, 0, 0)$, where $x = s + \int_0^s e_0 ds$, e_0 denotes the static axial strain due to pretension, and $\Delta L = \int_0^L e_0 ds$. We denote the coordinates of P by (x_1, x_2, x_3) and the dynamic displacements of P by $u, v,$ and w along the $x_1, x_2,$ and x_3 axes, respectively. Thus,

$$x_1 = x + u, \quad x_2 = v, \quad x_3 = w \qquad (3.1.1)$$

It follows from (3.1.1) and the free-body diagram shown in Figure 3.1.1b that the axial strain e is given by

$$e = \sqrt{(1 + e_0 + u')^2 + v'^2 + w'^2} - 1 \qquad (3.1.2)$$

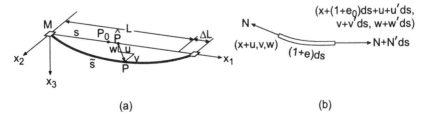

Fig. 3.1.1 String model: (a) the deformed configuration of a taut string, where $x_1 x_2 x_3$ is a Cartesian, inertial coordinate system, and (b) the free-body diagram of a string element.

Because there is in general a transverse contraction when a string is stretched (i.e., Poisson's effect), the deformed cross-section area A is given by

$$A = (1 - \nu e)^2 A_0 \tag{3.1.3}$$

where ν is Poisson's ratio and A_0 is the undeformed cross-section area. We assume that the string is very thin so that the stress can be assumed to be nearly uniform on the cross-action. Moreover, we assume that the string is made up of a linear elastic material. Hence, using Hooke's law and (3.1.3), we obtain the internal tension force N as

$$N(s, t) = EAe = EA_0(1 - \nu e)^2 e \tag{3.1.4}$$

where $E(s)$ is Young's modulus.

It follows from Figure 3.1.1b and Newton's second law that

$$m_0(\ddot{u}\mathbf{i}_1 + \ddot{v}\mathbf{i}_2 + \ddot{w}\mathbf{i}_3) + \tilde{\mu}_1 \dot{u}\mathbf{i}_1 + \tilde{\mu}_2 \dot{v}\mathbf{i}_2 + \tilde{\mu}_3 \dot{w}\mathbf{i}_3 = \mathbf{N}' + \tilde{f}_1 \mathbf{i}_1 + \tilde{f}_2 \mathbf{i}_2 + \tilde{f}_3 \mathbf{i}_3 \tag{3.1.5}$$

where $m_0 \equiv \rho_0 A_0$, the overdot denotes differentiation with respect to time t, the $\mathbf{i}_k$ are the base vectors of the x_k axes, the $\tilde{\mu}_i$ are the damping coefficients per unit of undeformed length along the x_i directions, the $\tilde{f}_i$ denote the distributed dynamic loads per unit of undeformed length along the x_i directions, and

$$\mathbf{N} = \frac{N}{1+e}[(1 + e_0 + u')\mathbf{i}_1 + v'\mathbf{i}_2 + w'\mathbf{i}_3] \tag{3.1.6}$$

Setting each of the coefficients of $\mathbf{i}_k$ in (3.1.5) equal to zero and using (3.1.6), we obtain the equations of motion as

$$m_0 \ddot{u} + \tilde{\mu}_1 \dot{u} = \frac{\partial}{\partial s}\left[\frac{N}{1+e}(1 + e_0 + u')\right] + \tilde{f}_1 \tag{3.1.7}$$

$$m_0 \ddot{v} + \tilde{\mu}_2 \dot{v} = \frac{\partial}{\partial s}\left[\frac{N}{1+e}v'\right] + \tilde{f}_2 \tag{3.1.8}$$

$$m_0 \ddot{w} + \tilde{\mu}_3 \dot{w} = \frac{\partial}{\partial s}\left[\frac{N}{1+e}w'\right] + \tilde{f}_3 \tag{3.1.9}$$

Because either the displacement or the force along the x_i direction must be known at the ends, the boundary conditions are of the form: specify

$$u \quad \text{or} \quad \frac{N(1 + e_0 + u')}{1+e} \tag{3.1.10}$$

$$v \quad \text{or} \quad \frac{Nv'}{1+e} \tag{3.1.11}$$

$$w \quad \text{or} \quad \frac{Nw'}{1+e} \tag{3.1.12}$$

at $s = 0$ and L.

It follows from (3.1.7) that the static equilibrium state (i.e., $u = v = w = 0, \tilde{f}_i = 0$, $e = e_0$) is given by

$$\frac{d}{ds}(N_0) = 0 \qquad (3.1.13)$$

where N_0 denotes the known pretension. Integrating (3.1.13) once and using (3.1.4) yields

$$N_0 \equiv EA_0(1 - \nu e_0)^2 e_0 = \text{constant} \qquad (3.1.14)$$

We note that A_0 and e_0 may be functions of s and that e_0 may be large. Equations (3.1.7)-(3.1.12) are fully nonlinear and account for Poisson's effect and pretension. Next, we develop approximate equations and boundary conditions.

As shown in the next section, the motions in the x_1, x_2, and x_3 directions are linearly uncoupled but nonlinearly coupled.

3.1.2 Approximate Equations of Motion without Poisson's Effect

We note that almost all existing studies of strings do not include the Poisson effect. Neglecting the Poisson effect (i.e., $\nu = 0$), we find from (3.1.4) that $N = EA_0 e$ and hence (3.1.7)-(3.1.9) become

$$m_0 \ddot{u} + \tilde{\mu}_1 \dot{u} = \frac{\partial}{\partial s}\left[\frac{EA_0 e(1 + e_0 + u')}{1+e}\right] + \tilde{f}_1 \qquad (3.1.15)$$

$$m_0 \ddot{v} + \tilde{\mu}_2 \dot{v} = \frac{\partial}{\partial s}\left[\frac{EA_0 e v'}{1+e}\right] + \tilde{f}_2 \qquad (3.1.16)$$

$$m_0 \ddot{w} + \tilde{\mu}_3 \dot{w} = \frac{\partial}{\partial s}\left[\frac{EA_0 e w'}{1+e}\right] + \tilde{f}_3 \qquad (3.1.17)$$

The boundary conditions become: specify

$$u \quad \text{or} \quad \frac{EA_0 e(1 + e_0 + u')}{1+e} \qquad (3.1.18)$$

$$v \quad \text{or} \quad \frac{EA_0 e v'}{1+e} \qquad (3.1.19)$$

$$w \quad \text{or} \quad \frac{EA_0 e w'}{1+e} \qquad (3.1.20)$$

at $s = 0$ and L. It follows from the definition

$$x = s + \int_0^s e_0 ds$$

that

$$dx = (1 + e_0)\, ds = \alpha ds$$

Hence, (3.1.15)-(3.1.17) become

$$m_0 \ddot{u} + \tilde{\mu}_1 \dot{u} = \alpha \frac{\partial}{\partial x}\left[\frac{\alpha EA_0 e}{1+e}\left(1 + \frac{\partial u}{\partial x}\right)\right] + \tilde{f}_1 \qquad (3.1.21)$$

$$m_0\ddot{v} + \tilde{\mu}_2\dot{v} = \alpha\frac{\partial}{\partial x}\left[\frac{\alpha E A_0 e}{1+e}\frac{\partial v}{\partial x}\right] + \tilde{f}_2 \qquad (3.1.22)$$

$$m_0\ddot{w} + \tilde{\mu}_3\dot{w} = \alpha\frac{\partial}{\partial x}\left[\frac{\alpha E A_0 e}{1+e}\frac{\partial w}{\partial x}\right] + \tilde{f}_3 \qquad (3.1.23)$$

Next, we approximate (3.1.21)-(3.1.23) for small but finite amplitudes. It follows from (3.1.2) that

$$e = \alpha\sqrt{(1+u_x)^2 + v_x^2 + w_x^2} - 1 \qquad (3.1.24)$$

Hence,

$$\frac{\alpha e}{1+e} = \alpha - 1 + u_x - u_x^2 + \frac{1}{2}\left(v_x^2 + w_x^2\right) + u_x^3 - \frac{3}{2}u_x\left(v_x^2 + w_x^2\right) + \cdots \qquad (3.1.25)$$

Substituting (3.1.25) into (3.1.21)-(3.1.23) and assuming that α, E, and A_0 are constants, we obtain the approximate equations

$$m_0\ddot{u} + \tilde{\mu}_1\dot{u} - \alpha^2 E A_0 u_{xx} = \alpha E A_0 \frac{\partial}{\partial x}\left[\left(\frac{1}{2} - u_x\right)(v_x^2 + w_x^2)\right] + \tilde{f}_1 \qquad (3.1.26)$$

$$m_0\ddot{v} + \tilde{\mu}_2\dot{v} - \alpha(\alpha-1)EA_0 v_{xx} = \alpha E A_0 \frac{\partial}{\partial x}\left[v_x\left(u_x - u_x^2 + \frac{1}{2}v_x^2 + \frac{1}{2}w_x^2\right)\right]$$
$$+ \tilde{f}_2 \qquad (3.1.27)$$

$$m_0\ddot{w} + \tilde{\mu}_3\dot{w} - \alpha(\alpha-1)EA_0 w_{xx} = \alpha E A_0 \frac{\partial}{\partial x}\left[w_x\left(u_x - u_x^2 + \frac{1}{2}v_x^2 + \frac{1}{2}w_x^2\right)\right]$$
$$+ \tilde{f}_3 \qquad (3.1.28)$$

Dividing (3.1.26)-(3.1.28) by m_0 and using the fact that $m_0 = \rho_0 A_0$, $\rho_0 = \rho(1+e)$, and $N_0 = EA_0 e_0 = EA_0(\alpha - 1)$, we obtain the following final form of the approximate equations:

$$u_{tt} + \mu_1 u_t - c_1^2 u_{xx} = (c_1^2 - c_2^2)\frac{\partial}{\partial x}\left[\left(\frac{1}{2} - u_x\right)(v_x^2 + w_x^2)\right] + f_1 \qquad (3.1.29)$$

$$v_{tt} + \mu_2 v_t - c_2^2 v_{xx} = (c_1^2 - c_2^2)\frac{\partial}{\partial x}\left[v_x\left(u_x - u_x^2 + \frac{1}{2}v_x^2 + \frac{1}{2}w_x^2\right)\right] + f_2 \qquad (3.1.30)$$

$$w_{tt} + \mu_3 w_t - c_2^2 w_{xx} = (c_1^2 - c_2^2)\frac{\partial}{\partial x}\left[w_x\left(u_x - u_x^2 + \frac{1}{2}v_x^2 + \frac{1}{2}w_x^2\right)\right] + f_3 \qquad (3.1.31)$$

where $\mu_n = \tilde{\mu}_n/m_0$, $f_n = \tilde{f}_n/m_0$,

$$c_1^2 = \frac{EA_0\alpha^2}{m_0} = \frac{EA_0\alpha^2}{\rho_0 A_0} = \frac{E\alpha^2}{\rho_0} \qquad (3.1.32)$$

$$c_2^2 = \frac{EA_0\alpha(\alpha-1)}{m_0} = \frac{EA_0\alpha e_0}{\rho_0 A_0} = \frac{N_0\alpha}{\rho_0 A_0} \qquad (3.1.33)$$

Equations (3.1.29)-(3.1.31) are the same as (7.5.9)-(7.5.11) of Nayfeh and Mook (1979). However, the wave speed c_1 in the axial direction given by (3.1.32) is slightly different from (7.5.12) of Nayfeh and Mook. The difference is due to the fact that we defined the strain with respect to the undeformed length s whereas Nayfeh and Mook used the deformed length x. This difference is small for typical metallic strings in the elastic range.

Using (3.1.25), we express the boundary condition (3.1.18)-(3.1.20) in terms of x as: specify

$$u \quad \text{or} \quad EA_0 \left[\alpha u_x + \left(\frac{1}{2} - u_x \right) \left(v_x^2 + w_x^2 \right) \right] \tag{3.1.34}$$

$$v \quad \text{or} \quad EA_0 v_x \left[\alpha - 1 + u_x - u_x^2 + \frac{1}{2} \left(v_x^2 + w_x^2 \right) \right] \tag{3.1.35}$$

$$w \quad \text{or} \quad EA_0 w_x \left[\alpha - 1 + u_x - u_x^2 + \frac{1}{2} \left(v_x^2 + w_x^2 \right) \right] \tag{3.1.36}$$

at $x = 0$ and ℓ.

3.1.3 Approximate Equations of Motion with Poisson's Effect

For small dynamic displacements, we expand (3.1.2) in a Taylor series and obtain

$$e = \alpha - 1 + u' + \frac{v'^2 + w'^2}{2\alpha} - \frac{u'(v'^2 + w'^2)}{2\alpha^2} + \cdots \tag{3.1.37}$$

It follows from (3.1.4) that

$$\frac{N}{1+e} = \frac{EA_0(1 - \nu e)^2 e}{1 + e} \tag{3.1.38}$$

Substituting (3.1.37) into (3.1.38) and expanding the result for small u, v, and w, we obtain

$$\frac{N}{1+e} = EA_0 \left[\Lambda_1 + \alpha \Lambda_2 u' + \Lambda_3 u'^2 + \frac{1}{2} \Lambda_2 (v'^2 + w'^2) + \frac{(1+\nu)^2}{\alpha^4} u'^3 \right.$$
$$\left. + \frac{2\Lambda_3 - \Lambda_2}{2\alpha} u'(v'^2 + w'^2) \right] + \cdots \tag{3.1.39}$$

where

$$\Lambda_1 = (1 + \nu - \nu\alpha)^2 \left(1 - \frac{1}{\alpha} \right) \tag{3.1.40}$$

$$\Lambda_2 = 2\nu^2 - \frac{2\nu + 3\nu^2}{\alpha} + \frac{(1+\nu)^2}{\alpha^3} \tag{3.1.41}$$

$$\Lambda_3 = \nu^2 - \frac{(1+\nu)^2}{\alpha^3} \tag{3.1.42}$$

Substituting (3.1.39) into (3.1.7)-(3.1.9) and using (3.1.40)-(3.1.42), we obtain, to third order,

$$m_0 \ddot{u} + \tilde{\mu}_1 \dot{u} = \frac{\partial}{\partial s} \left\{ EA_0 \left[\left(\Lambda_1 + \alpha^2 \Lambda_2 \right) u' + \alpha \left(\Lambda_2 + \Lambda_3 \right) u'^2 + \nu^2 u'^3 \right. \right.$$

$$+\left(\frac{1}{2}\Lambda_2\alpha + \Lambda_3 u'\right)\left(v'^2 + w'^2\right)\bigg]\bigg\} + \tilde{f}_1 \qquad (3.1.43)$$

$$m_0\ddot{v} + \tilde{\mu}_2\dot{v} = \frac{\partial}{\partial s}\bigg\{EA_0\bigg[\Lambda_1 v' + \alpha\Lambda_2 u'v' + \Lambda_3 u'^2 v'$$
$$+\frac{1}{2}\Lambda_2 v'\left(v'^2 + w'^2\right)\bigg]\bigg\} + \tilde{f}_2 \qquad (3.1.44)$$

$$m_0\ddot{w} + \tilde{\mu}_3\dot{w} = \frac{\partial}{\partial s}\bigg\{EA_0\bigg[\Lambda_1 w' + \alpha\Lambda_2 u'w' + \Lambda_3 u'^2 w'$$
$$+\frac{1}{2}\Lambda_2 w'\left(v'^2 + w'^2\right)\bigg]\bigg\} + \tilde{f}_3 \qquad (3.1.45)$$

Neglecting the nonlinear terms in (3.1.43)-(3.1.45), we obtain the linear equations

$$m_0\ddot{u} + \tilde{\mu}_1\dot{u} = \frac{\partial}{\partial s}\left[EA_0\left(\Lambda_1 + \alpha^2\Lambda_2\right)u'\right] + \tilde{f}_1 \qquad (3.1.46)$$

$$m_0\ddot{v} + \tilde{\mu}_2\dot{v} = \frac{\partial}{\partial s}\left(EA_0\Lambda_1 v'\right) + \tilde{f}_2 \qquad (3.1.47)$$

$$m_0\ddot{w} + \tilde{\mu}_2\dot{w} = \frac{\partial}{\partial s}\left(EA_0\Lambda_1 w'\right) + \tilde{f}_3 \qquad (3.1.48)$$

Neglecting the Poisson effect (i.e., $\nu = 0$) in (3.1.43)-(3.1.45), using the identity $dx = (1 + e_0)ds = \alpha ds$, and assuming that the string is uniform (i.e., A_0, E and α are constants), we obtain (3.1.26)-(3.2.28).

3.2 REDUCTION OF STRING MODEL TO TWO EQUATIONS

In this section, we condense the three-equation model (3.1.29)-(3.1.31) into a two-equation model. To accomplish this, first we consider the linear undamped free-oscillation problem.

Neglecting the damping, forcing, and nonlinear terms in (3.1.29)-(3.1.31), we obtain the following uncoupled equations:

$$u_{tt} - c_1^2 u_{xx} = 0 \qquad (3.2.1)$$
$$v_{tt} - c_2^2 v_{xx} = 0 \qquad (3.2.2)$$
$$w_{tt} - c_2^2 w_{xx} = 0 \qquad (3.2.3)$$

Next, we use the boundary conditions

$$u = v = w = 0 \text{ at } x = 0 \text{ and } \ell \qquad (3.2.4)$$

It follows from (3.2.1) and (3.2.4) that the mode shapes of the axial motion are given by

$$u = \sin\frac{n\pi x}{\ell} \qquad (3.2.5)$$

corresponding to the natural frequencies

$$\omega_n = \frac{n\pi c_1}{\ell} \tag{3.2.6}$$

where n is a positive integer. Similarly, the mode shapes of the transverse motions in the x_2 and x_3 directions are given by

$$v = \sin\frac{m\pi x}{\ell} \quad \text{and} \quad w = \sin\frac{k\pi x}{\ell} \tag{3.2.7}$$

corresponding to the natural frequencies

$$\omega_m = \frac{m\pi c_2}{\ell} \quad \text{and} \quad \omega_k = \frac{k\pi c_2}{\ell} \tag{3.2.8}$$

where m and k are positive integers.

It follows from (3.1.14) with $\nu = 0$, (3.1.32), (3.1.33), and the fact $\alpha = 1 + e_0$ that

$$\frac{c_1}{c_2} = \sqrt{\frac{EA_0\alpha}{N_0}} = \sqrt{\frac{EA_0(1+e_0)}{EA_0 e_0}} = \sqrt{\frac{1+e_0}{e_0}} \tag{3.2.9}$$

which is very large (several hundreds) for typical metals in the elastic range. Hence, for a given frequency order (i.e., $m = k = n$), the transverse frequencies are much less than the longitudinal frequencies according to (3.2.6) and (3.2.8). Consequently, if the excitation frequencies are much smaller than the fundamental longitudinal frequency $\pi c_1/\ell$, the longitudinal inertia in (3.1.29) can be neglected, as shown below. Hence, (3.1.29) can be used to express u in terms of v and w. Substituting the result into (3.1.30) and (3.1.31) yields a two-equation model for the string. Next, we give a formal derivation of the two-equation model.

To analyze the nonlinear response of the string to an excitation having a frequency close to the mth transverse natural frequency applied at the right end, we scale the displacements and time by using ℓ and $\ell/m\pi c_2$ in (3.1.29). Thus we put

$$\xi = \frac{x}{\ell}, \quad \tau = \frac{m\pi c_2 t}{\ell}, \quad U = \frac{u}{\ell}, \quad V = \frac{v}{\ell}, \quad W = \frac{w}{\ell}$$

in (3.1.29) and obtain

$$\frac{m^2\pi^2 c_2^2}{c_1^2}U_{\tau\tau} + \frac{\mu_1 m\pi c_2 \ell}{c_1^2}U_\tau - U_{\xi\xi}$$
$$= \frac{c_1^2 - c_2^2}{c_1^2}\frac{\partial}{\partial \xi}\left[\left(\frac{1}{2} - U_\xi\right)(V_\xi^2 + W_\xi^2)\right] \tag{3.2.10}$$

where the distributed longitudinal excitation $f_1(t)$ is assumed to be zero. For typical metallic strings in the elastic range, $c_2^2/c_1^2 \ll 1$, and hence the terms proportional to $U_{\tau\tau}$ and U_τ can be neglected in (3.2.10). The result is

$$U_{\xi\xi} = -\frac{\partial}{\partial \xi}\left[\left(\frac{1}{2} - U_\xi\right)(V_\xi^2 + W_\xi^2)\right] \tag{3.2.11}$$

Integrating (3.2.11) once with respect to ξ yields

$$U_\xi = -\left(\frac{1}{2} - U_\xi\right)\left(V_\xi^2 + W_\xi^2\right) + b(\tau) \qquad (3.2.12)$$

where $b(\tau)$ depends on the boundary conditions. It follows from (3.2.12) that $U_\xi = O(V_\xi^2) = O(W_\xi^2)$. Hence, to $O(V^2, W^2)$, (3.2.12) becomes

$$U_\xi = -\frac{1}{2}\left(V_\xi^2 + W_\xi^2\right) + b(\tau) \qquad (3.2.13)$$

Because the force is assumed to be imposed at the right end,

$$U(0,\tau) = 0 \quad \text{and} \quad U(1,\tau) = \ell F(\tau) \qquad (3.2.14)$$

Integrating (3.2.13) once and using (3.2.14) yields

$$U = -\frac{1}{2}\int_0^\xi \left(V_\xi^2 + W_\xi^2\right) d\xi + b(\tau)\xi \qquad (3.2.15)$$

where

$$b(\tau) = \frac{1}{2}\int_0^1 \left(V_\xi^2 + W_\xi^2\right) d\xi + \ell F(\tau) \qquad (3.2.16)$$

In terms of dimensional variables, (3.2.13) and (3.2.16) become

$$u_x = -\frac{1}{2}\left(v_x^2 + w_x^2\right) + b(t) \qquad (3.2.17)$$

$$b(t) = \frac{1}{2\ell}\int_0^\ell \left(v_x^2 + w_x^2\right) dx + \ell F(t) \qquad (3.2.18)$$

Hence,

$$u_x + \frac{1}{2}\left(v_x^2 + w_x^2\right) = \ell F(t) + \frac{1}{2\ell}\int_0^\ell \left(v_x^2 + w_x^2\right) dx \qquad (3.2.19)$$

Substituting (3.2.19) into (3.1.30) and (3.1.31), neglecting c_2^2 compared with c_1^2, and keeping terms through cubic in v and w, we obtain the two-equation model

$$v_{tt} + \mu_2 v_t - c_2^2 v_{xx} = h(t)v_{xx} + \frac{c_1^2}{2\ell}v_{xx}\int_0^\ell \left(v_x^2 + w_x^2\right) dx + f_2(x,t) \qquad (3.2.20)$$

$$w_{tt} + \mu_3 w_t - c_2^2 w_{xx} = h(t)w_{xx} + \frac{c_1^2}{2\ell}w_{xx}\int_0^\ell \left(v_x^2 + w_x^2\right) dx + f_3(x,t) \qquad (3.2.21)$$

where $h(t) = c_1^2 \ell F(t)$.

Narasimha (1968) carefully argued that motions in which $u = 0$ when $f_1 = 0$ are generally not possible and the most significant nonlinear problem arises when $u = O(v^2, w^2)$, as employed in deriving (3.2.20) and (3.2.21). Suppose that the assumption $u = 0$ when $f_1 = 0$ was true, then it follows from (3.1.29) that $v_x^2 + w_x^2$

is independent of x. This is possible if the string is straight or it is helical at each instant, including $t = 0$. Clearly such a motion cannot arise except perhaps for special initial conditions or excitations. Therefore, in general $u \neq 0$. Moreover, whenever the motion is planar (i.e., $w = 0$), v_x^2 cannot be independent of x if v is to be zero at the ends unless $v \equiv 0$; that is, the string is straight. However, experiments (Harrison, 1948; Murthy and Ramakrishna, 1965; Molteno and Tufillaro, 1990; S. Nayfeh, Nayfeh, and Mook, 1995) show that, when a string is excited by forces acting in a given plane below a critical frequency, then the motion is also in the same plane and hence u cannot be zero below that frequency. Moreover, there is no reason whatsoever to expect that with or without the onset of whirling motions that u will vanish. In fact, we show in Section 3.2.4 that assuming $u = 0$ leads to erroneous responses of a string to a primary-resonance excitation in a given plane.

3.2.1 Attacking the Three-Equation Model

When c_2^2/c_1^2 is not small, one cannot use (3.2.20) and (3.2.21) and needs to use (3.1.29)-(3.1.31) as described next for the case of primary resonance of the mth mode in a given plane, say the $x_1 - x_2$ plane. Thus, we put $f_1 = f_3 = 0$ and assume $f_2(x, t)$ to be small and vary harmonically with the frequency Ω; that is,

$$f_2(x,t) = \epsilon^3 f(x) \cos \Omega t \qquad (3.2.22)$$

where ϵ is a small dimensionless parameter that is the order of the amplitude of oscillation in the x_2 and x_3 directions. To express the nearness of Ω to $\omega_m = m\pi c_2/\ell$, we introduce the detuning parameter σ defined by

$$\Omega = \frac{m\pi c_2}{\ell} + \epsilon^2 \sigma \qquad (3.2.23)$$

Then, using the method of multiple scales (Nayfeh, 1973b, 1981), we seek an expansion in the form

$$u(x,t) = \epsilon u_1(x, T_0, T_2) + \epsilon^2 u_2(x, T_0, T_2) + \epsilon^3 u_3(x, T_0, T_2) + \cdots \qquad (3.2.24)$$
$$v(x,t) = \epsilon v_1(x, T_0, T_2) + \epsilon^2 v_2(x, T_0, T_2) + \epsilon^3 v_3(x, T_0, T_2) + \cdots \qquad (3.2.25)$$
$$w(x,t) = \epsilon w_1(x, T_0, T_2) + \epsilon^2 w_2(x, T_0, T_2) + \epsilon^3 w_3(x, T_0, T_2) + \cdots \qquad (3.2.26)$$

where $T_0 = t$ and $T_2 = \epsilon^2 t$. Moreover, we order the damping so that its influence balances the influence of the nonlinearities and the primary resonance. Hence, we replace the μ_i with $\epsilon^2 \mu_i$.

Substituting (3.2.22)-(3.2.26) into (3.1.29)-(3.1.31) and (3.2.4) and equating coefficients of like powers of ϵ, we obtain a hierarchy of problems governing the u_n, v_n, and w_n. The first-order problem is

$$D_0^2 u_1 - c_1^2 u_{1xx} = 0 \qquad (3.2.27)$$
$$D_0^2 v_1 - c_2^2 v_{1xx} = 0 \qquad (3.2.28)$$
$$D_0^2 w_1 - c_2^2 w_{1xx} = 0 \qquad (3.2.29)$$
$$u_1 = v_1 = w_1 = 0 \text{ at } x = 0 \text{ and } \ell \qquad (3.2.30)$$

where $D_n = \partial/\partial T_n$. We show in Section 3.2.3 that although the mth transverse modes are commensurate with many other transverse modes, internal resonances with these modes are not activated because the interaction coefficients vanish. Hence, the axial modes and all of the transverse modes except the mth modes will decay with time because of the damping. Hence, the long-time solution of (3.2.27) and (3.2.30) is

$$u_1 = 0 \tag{3.2.31}$$

The solutions of (3.2.28)-(3.2.30) can be expressed as

$$v_1 = A_1(T_2) \sin \frac{m\pi x}{\ell} e^{i\omega_m T_0} + cc \tag{3.2.32}$$

$$w_1 = A_2(T_2) \sin \frac{m\pi x}{\ell} e^{i\omega_m T_0} + cc \tag{3.2.33}$$

where all of the other modes in the transverse directions decay with time due to damping.

The second-order problem is

$$D_0^2 u_2 - c_1^2 u_{2xx} = \frac{1}{2}(c_1^2 - c_2^2)\frac{\partial}{\partial x}\left(v_{1x}^2 + w_{1x}^2\right) \tag{3.2.34}$$

$$D_0^2 v_2 - c_2^2 v_{2xx} = (c_1^2 - c_2^2)\frac{\partial}{\partial x}(u_{1x} v_{1x}) = 0 \tag{3.2.35}$$

$$D_0^2 w_2 - c_2^2 w_{2xx} = (c_1^2 - c_2^2)\frac{\partial}{\partial x}(u_{1x} w_{1x}) = 0 \tag{3.2.36}$$

$$u_2 = v_2 = w_2 = 0 \text{ at } x = 0 \text{ and } \ell \tag{3.2.37}$$

Equations (3.2.35)-(3.2.37) are the same as the first-order problem given by (3.2.28)-(3.2.30). Hence, to specify uniquely the fundamental frequency of the response, we do not include the homogeneous solution of the second-order problem. Consequently, the solutions of (3.2.35)-(3.2.37) are

$$v_2 = w_2 = 0 \tag{3.2.38}$$

Substituting (3.2.32) and (3.2.33) into (3.2.34) yields

$$D_0^2 u_2 - c_1^2 u_{2xx} = -\frac{(c_1^2 - c_2^2)m^3\pi^3}{2\ell^3} \sin \frac{2m\pi x}{\ell}\left[\left(A_1^2 + A_2^2\right) e^{2i\omega_m T_0} + A_1 \bar{A}_1 + A_2 \bar{A}_2\right] + cc \tag{3.2.39}$$

The solution of (3.2.39) and (3.2.37) is

$$u_2 = -\frac{m\pi}{8\ell}\left[\left(A_1^2 + A_2^2\right) e^{2i\omega_m T_0} + \frac{c_1^2 - c_2^2}{c_1^2}\left(A_1 \bar{A}_1 + A_2 \bar{A}_2\right)\right] \sin \frac{2m\pi x}{\ell} + cc \tag{3.2.40}$$

To determine A_1 and A_2, we consider the third-order problem for v and w, namely,

$$D_0^2 v_3 - c_2^2 v_{3xx} = -2D_0 D_2 v_1 - \mu_2 D_0 v_1 + f(x) \cos \Omega T_0$$

$$+(c_1^2 - c_2^2)\frac{\partial}{\partial x}\left[v_{1x}\left(u_{2x} + \frac{1}{2}v_{1x}^2 + \frac{1}{2}w_{1x}^2\right)\right] \quad (3.2.41)$$

$$D_0^2 w_3 - c_2^2 w_{3xx} = -2D_0 D_2 w_1 - \mu_3 D_0 w_1$$

$$+(c_1^2 - c_2^2)\frac{\partial}{\partial x}\left[w_{1x}\left(u_{2x} + \frac{1}{2}v_{1x}^2 + \frac{1}{2}w_{1x}^2\right)\right] \quad (3.2.42)$$

Substituting (3.2.32), (3.2.33), and (3.2.40) into (3.2.41) and (3.2.42) yields

$$D_0^2 v_3 - c_2^2 v_{3xx} = -2i\omega_m A_1' \sin\frac{m\pi x}{\ell}e^{i\omega_m T_0} - i\mu_2 \omega_m A_1 \sin\frac{m\pi x}{\ell}e^{i\omega_m T_0}$$

$$-(c_1^2 - c_2^2)\frac{m^4 \pi^4}{4\ell^4}\sin\frac{m\pi x}{\ell}\left[(A_1^2 + A_2^2)e^{2i\omega_m T_0} + A_1\bar{A}_1 + A_2\bar{A}_2\right]$$

$$\times\left(A_1 e^{i\omega_m T_0} + \bar{A}_1 e^{-i\omega_m T_0}\right)$$

$$-\frac{c_2^2(c_1^2 - c_2^2)}{c_1^2}\frac{m^4\pi^4}{8\ell^4}(A_1\bar{A}_1 + A_2\bar{A}_2)\left(A_1 e^{i\omega_m T_0} + \bar{A}_1 e^{-i\omega_m T_0}\right)$$

$$\times\left(\sin\frac{m\pi x}{\ell} + 3\sin\frac{3m\pi x}{\ell}\right) + \frac{1}{2}f(x)e^{i\Omega T_0} + cc \quad (3.2.43)$$

$$D_0^2 w_3 - c_2^2 w_{3xx} = -2i\omega_m A_2' \sin\frac{m\pi x}{\ell}e^{i\omega_m T_0} - i\mu_3 \omega_m A_2 \sin\frac{m\pi x}{\ell}e^{i\omega_m T_0}$$

$$-(c_1^2 - c_2^2)\frac{m^4\pi^4}{4\ell^4}\sin\frac{m\pi x}{\ell}\left[(A_1^2 + A_2^2)e^{2i\omega_m T_0} + A_1\bar{A}_1 + A_2\bar{A}_2\right]$$

$$\times\left(A_2 e^{i\omega_m T_0} + \bar{A}_2 e^{-i\omega_m T_0}\right)$$

$$-\frac{c_2^2(c_1^2 - c_2^2)}{c_1^2}\frac{m^4\pi^4}{8\ell^4}(A_1\bar{A}_1 + A_2\bar{A}_2)\left(A_2 e^{i\omega_m T_0} + \bar{A}_2 e^{-i\omega_m T_0}\right)$$

$$\times\left(\sin\frac{m\pi x}{\ell} + 3\sin\frac{3m\pi x}{\ell}\right) + cc \quad (3.2.44)$$

Eliminating the terms that produce secular terms from (3.2.43) and (3.2.44), we obtain

$$2i\omega_m(A_1' + \hat{\mu}_2 A_1) + (c_1^2 - c_2^2)\frac{m^4\pi^4}{4\ell^4}\left[(A_1^2 + A_2^2)\bar{A}_1\right.$$

$$\left. + 2\left(1 - \frac{c_2^2}{2c_1^2}\right)(A_1\bar{A}_1 + A_2\bar{A}_2)A_1\right] - Fe^{i\sigma T_2} = 0 \quad (3.2.45)$$

$$2i\omega_m(A_2' + \hat{\mu}_3 A_2) + (c_1^2 - c_2^2)\frac{m^4\pi^4}{4\ell^4}\left[(A_1^2 + A_2^2)\bar{A}_2\right.$$

$$\left. + 2\left(1 - \frac{c_2^2}{2c_1^2}\right)(A_1\bar{A}_1 + A_2\bar{A}_2)A_2\right] = 0 \quad (3.2.46)$$

where the damping is assumed to be modal and

$$\hat{\mu}_n = \frac{1}{\ell}\int_0^\ell \mu_n \sin^2\frac{m\pi x}{\ell}dx \quad\text{and}\quad F = \frac{1}{\ell}\int_0^\ell f(x)\sin\frac{m\pi x}{\ell}dx \quad (3.2.47)$$

3.2.2 Evaluation of the Two-Equation Model

To investigate the accuracy of the two-equation model given by (3.2.20) and (3.2.21), we consider the case of primary resonance of the mth mode in a given plane, say the $x_1 - x_2$ plane. Thus, we put $f_3 = 0$ and assume that f_2 is given by (3.2.22) and (3.2.23). Moreover, we order the damping as in the preceding section and replace μ_i with $\epsilon^2 \mu_i$. Then, we seek a second-order expansion as in (3.2.25) and (3.2.26) with v_2 and w_2 being equal to zero, as shown in (3.2.38) in the preceding section. Substituting (3.2.25), (3.2.26), and (3.2.22) into (3.2.20), (3.2.21), and (3.2.4), setting $h = 0$ (i.e., $f_1 = 0$), and equating coefficients of like powers of ϵ, we obtain

Order ϵ:

$$D_0^2 v_1 - c_2^2 v_{1xx} = 0 \qquad (3.2.48)$$

$$D_0^2 w_1 - c_2^2 w_{1xx} = 0 \qquad (3.2.49)$$

$$v_1 = w_1 = 0 \text{ at } x = 0 \text{ and } \ell \qquad (3.2.50)$$

Order ϵ^3:

$$D_0^2 v_3 - c_2^2 v_{3xx} = -2D_0 D_2 v_1 - \mu_2 D_0 v_1 + \frac{c_1^2}{2\ell} v_{1xx} \int_0^\ell \left(v_{1x}^2 + w_{1x}^2 \right) dx$$
$$+ f(x) \cos(\Omega T_0) \qquad (3.2.51)$$

$$D_0^2 w_3 - c_2^2 w_{3xx} = -2D_0 D_2 w_1 - \mu_3 D_0 w_1 + \frac{c_1^2}{2\ell} w_{1xx} \int_0^\ell \left(v_{1x}^2 + w_{1x}^2 \right) dx \qquad (3.2.52)$$

$$v_3 = w_3 = 0 \text{ at } x = 0 \text{ and } \ell \qquad (3.2.53)$$

Because the mth mode is being directly excited, the solution of the first-order problem (3.2.48)-(3.2.50) can be expressed as

$$v_1 = A_1(T_2) \sin \frac{m\pi x}{\ell} e^{i\omega_m T_0} + cc \qquad (3.2.54)$$

$$w_1 = A_2(T_2) \sin \frac{m\pi x}{\ell} e^{i\omega_m T_0} + cc \qquad (3.2.55)$$

All other modes decay with time (S. Nayfeh, Nayfeh, and Mook, 1995).

Substituting (3.2.54) and (3.2.55) into (3.2.51) and (3.2.52), using (3.2.23), and eliminating the terms that lead to secular terms in v_3 and w_3, we obtain

$$2i\omega_m (A_1' + \hat{\mu}_2 A_1) + \frac{c_1^2 m^4 \pi^4}{4\ell^4} \left[(A_1^2 + A_2^2) \bar{A}_1 + 2A_1 (A_1 \bar{A}_1 + A_2 \bar{A}_2) \right]$$
$$- F e^{i\sigma T_2} = 0 \qquad (3.2.56)$$

$$2i\omega_m (A_2' + \hat{\mu}_3 A_2) + \frac{c_1^2 m^4 \pi^4}{4\ell^4} \left[(A_1^2 + A_2^2) \bar{A}_2 + 2A_2 (A_1 \bar{A}_1 + A_2 \bar{A}_2) \right] = 0 \qquad (3.2.57)$$

where the $\hat{\mu}_n$ and F are defined in (3.2.47). We note that (3.2.56) and (3.2.57) can be obtained from (3.2.45) and (3.2.46) by letting $c_2/c_1 \to 0$. Consequently, the two-equation model (3.2.20) and (3.2.21) is valid as $c_2/c_1 \to 0$.

REDUCTION OF STRING MODEL TO TWO EQUATIONS

Next, we investigate the accuracy of the two-equation model that can be obtained from (3.1.30) and (3.1.31) by putting $u = 0$. The reduced equations are

$$v_{tt} + \mu_2 v_t - c_2^2 v_{xx} = \frac{1}{2}(c_1^2 - c_2^2)\frac{\partial}{\partial x}\left[v_x\left(v_x^2 + w_x^2\right)\right] + f_2 \quad (3.2.58)$$

$$w_{tt} + \mu_3 w_t - c_2^2 w_{xx} = \frac{1}{2}(c_1^2 - c_2^2)\frac{\partial}{\partial x}\left[w_x\left(v_x^2 + w_x^2\right)\right] + f_3 \quad (3.2.59)$$

For the case of primary resonance of the mth mode in the $x_1 - x_2$ direction, $f_3 = 0$ and f_2 is defined in (3.2.22) and (3.2.23). Substituting (3.2.25), (3.2.26), and (3.2.22) into (3.2.58), (3.2.59), and (3.2.4), setting $v_2 = w_2 = 0$, and equating the coefficients of like powers of ϵ, we obtain the first-order problem (3.2.48)-(3.2.50) and the following third-order problem:

$$D_0^2 v_3 - c_2^2 v_{3xx} = -2D_0 D_2 v_1 - \mu_2 D_0 v_1$$
$$+ \frac{1}{2}(c_1^2 - c_2^2)\frac{\partial}{\partial x}\left[v_{1x}\left(v_{1x}^2 + w_{1x}^2\right)\right] + f(x)\cos(\Omega T_0) \quad (3.2.60)$$

$$D_0^2 w_3 - c_2^2 w_{3xx} = -2D_0 D_2 w_1 - \mu_3 D_0 w_1$$
$$+ \frac{1}{2}(c_1^2 - c_2^2)\frac{\partial}{\partial x}\left[w_{1x}\left(v_{1x}^2 + w_{1x}^2\right)\right] \quad (3.2.61)$$

$$v_3 = w_3 = 0 \text{ at } x = 0 \text{ and } \ell \quad (3.2.62)$$

The solution of the first-order problem is given by (3.2.54) and (3.2.55). Substituting this solution into (3.2.60) and (3.2.61), using (3.2.23), and eliminating the terms that produce secular terms in v_3 and w_3, we obtain

$$2i\omega_m(A_1' + \hat{\mu}_2 A_1) + \frac{3m^4\pi^4}{8\ell^4}(c_1^2 - c_2^2)\left[\left(A_1^2 + A_2^2\right)\bar{A}_1 + 2A_1\left(A_1\bar{A}_1 + A_2\bar{A}_2\right)\right]$$
$$- Fe^{i\sigma T_2} = 0 \quad (3.2.63)$$

$$2i\omega_m(A_2' + \hat{\mu}_3 A_2) + \frac{3m^4\pi^4}{8\ell^4}(c_1^2 - c_2^2)\left[\left(A_1^2 + A_2^2\right)\bar{A}_2\right.$$
$$\left. + 2A_2\left(A_1\bar{A}_1 + A_2\bar{A}_2\right)\right] = 0 \quad (3.2.64)$$

where $\hat{\mu}_n$ and F are defined in (3.2.47). Although (3.2.63) and (3.2.64) have the same form as (3.2.45) and (3.2.46), which are obtained by attacking the three-equation model, they do not lead quantitatively to the same description of the nonlinear response of the string because they over predict the coefficient of the effective nonlinearity by 50%. Moreover, it follows from (3.2.25), (3.2.26), (3.2.54), and (3.2.55) that

$$v_x^2 + w_x^2 = \frac{m^2\pi^2}{\ell^2}\left[\left(A_1^2 + A_2^2\right)e^{2i\omega_m t} + A_1\bar{A}_1 + A_2\bar{A}_2\right]\cos^2\frac{m\pi x}{\ell} + cc \quad (3.2.65)$$

which is not independent of x, contrary to the implication of the assumption $u = 0$ used to derive this solution. These results support Narasimha's (1968) statement "there is no reason whatever to expect that with or without the onset of whirling motion u will vanish".

3.2.3 Discretized Model

When the ratio of the speeds of propagation in the longitudinal and transverse directions c_1/c_2 is large, the nonlinear dynamics of strings can be modeled by (3.2.20) and (3.2.21). Moreover, instead of directly attacking the governing partial-differential equations and boundary conditions, we use the Galerkin procedure to discretize these equations into an infinite system of nonlinearly coupled ordinary-differential equations for the case of primary resonance. For fixed ends, the linear mode shapes and corresponding natural frequencies of transverse oscillations are given by (3.2.7) and (3.2.8). Hence, using the Galerkin procedure, we represent the solutions of the nonlinear equations (3.2.20) and (3.2.21) as

$$v(x,t) = \sum_{k=1}^{\infty} \eta_k(t) \sin \frac{k\pi x}{\ell} \qquad (3.2.66)$$

$$w(x,t) = \sum_{k=1}^{\infty} \zeta_k(t) \sin \frac{k\pi x}{\ell} \qquad (3.2.67)$$

which satisfy exactly the boundary conditions given by (3.2.24). Substituting (3.2.66) and (3.2.67) into (3.2.20) and (3.2.21) yields

$$\sum_{k=1}^{\infty} \left[\ddot{\eta}_k + \omega_k^2 \eta_k + \frac{k^2\pi^2}{\ell^2} h \eta_k \right] \sin \frac{k\pi x}{\ell} + \mu_2 \sum_{k=1}^{\infty} \dot{\eta}_k \sin \frac{k\pi x}{\ell}$$

$$= -\sum_{k=1}^{\infty} \frac{c_1^2 k^2 \pi^4}{4\ell^4} \eta_k \sin \frac{k\pi x}{\ell} \sum_{m=1}^{\infty} m^2 \left(\eta_m^2 + \zeta_m^2 \right) + f_2(x,t) \quad (3.2.68)$$

$$\sum_{k=1}^{\infty} \left[\ddot{\zeta}_k + \omega_k^2 \zeta_k + \frac{k^2\pi^2}{\ell^2} h \zeta_k \right] \sin \frac{k\pi x}{\ell} + \mu_3 \sum_{k=1}^{\infty} \dot{\zeta}_k \sin \frac{k\pi x}{\ell}$$

$$= -\sum_{k=1}^{\infty} \frac{c_1^2 k^2 \pi^4}{4\ell^4} \zeta_k \sin \frac{k\pi x}{\ell} \sum_{m=1}^{\infty} m^2 \left(\eta_m^2 + \zeta_m^2 \right) + f_3(x,t) \quad (3.2.69)$$

Multiplying (3.2.68) by $\sin \frac{n\pi x}{\ell}$ and integrating the result from $x=0$ to $x=\ell$, we obtain

$$\ddot{\eta}_n + \omega_n^2 \eta_n + \frac{n^2\pi^2}{\ell^2} h \eta_n + 2\hat{\mu}_{2n} \dot{\eta}_n = -\frac{c_1^2 n^2 \pi^4}{4\ell^4} \eta_n \sum_{m=1}^{\infty} m^2 \left(\eta_m^2 + \zeta_m^2 \right)$$

$$+ F_{2n}(t) \qquad (3.2.70)$$

Similarly, multiplying (3.2.69) by $\sin \frac{n\pi x}{\ell}$ and integrating the result from $x=0$ to $x=\ell$, we obtain

$$\ddot{\zeta}_n + \omega_n^2 \zeta_n + \frac{n^2\pi^2}{\ell^2} h \zeta_n + 2\hat{\mu}_{3n} \dot{\zeta}_n = -\frac{c_1^2 n^2 \pi^4}{4\ell^4} \zeta_n \sum_{m=1}^{\infty} m^2 \left(\eta_m^2 + \zeta_m^2 \right)$$

$$+ F_{3n}(t) \qquad (3.2.71)$$

where the damping is assumed to be modal and

$$\hat{\mu}_{in} = \frac{1}{\ell} \int_0^\ell \mu_i \sin^2 \frac{n\pi x}{\ell} dx \text{ and } F_{in}(x,t) = \frac{2}{\ell} \int_0^\ell f_i(x,t) \sin \frac{n\pi x}{\ell} dx \quad (3.2.72)$$

Therefore, the distributed-parameter problem consisting of (3.2.20), (3.2.21), and (3.2.4) has been transformed into an infinite system of nonlinearly coupled ordinary-differential equations.

For the case of primary resonance of the mth mode in the x_2 direction, $h = 0$ and

$$F_{2n} = \epsilon^3 F_n \cos(\Omega t) \text{ and } \Omega = \omega_m + \epsilon^2 \sigma \quad (3.2.73)$$

To determine a second-order approximate solution of (3.2.70) and (3.2.71) for this case, we use the method of multiple scales and assume that

$$\eta_n = \epsilon \eta_{n1}(T_0, T_2) + \epsilon^3 \eta_{n3}(T_0, T_2) + \cdots \quad (3.2.74)$$

$$\zeta_n = \epsilon \zeta_{n1}(T_0, T_2) + \epsilon^3 \zeta_{n3}(T_0, T_2) + \cdots \quad (3.2.75)$$

Moreover, we order the damping at $O(\epsilon^2)$ so that its influence balances the influence of nonlinearity and the resonance. Thus, we replace $\hat{\mu}_{in}$ by $\epsilon^2 \hat{\mu}_{in}$. Substituting (3.2.74) and (3.2.75) into (3.2.70) and (3.2.71) and equating coefficients of like powers of ϵ, we obtain

Order ϵ:

$$D_0^2 \eta_{n1} + \omega_n^2 \eta_{n1} = 0 \quad (3.2.76)$$
$$D_0^2 \zeta_{n1} + \omega_n^2 \zeta_{n1} = 0 \quad (3.2.77)$$

Order ϵ^3:

$$D_0^2 \eta_{n3} + \omega_n^2 \eta_{n3} = -2D_0 D_2 \eta_{n1} - 2\hat{\mu}_{2n} D_0 \eta_{n1} - \frac{c_1^2 n^2 \pi^4}{4\ell^4} \eta_{n1}$$
$$\times \sum_{k=1}^\infty k^2 \left(\eta_{k1}^2 + \zeta_{k1}^2\right) + F_n \cos(\Omega T_0) \quad (3.2.78)$$

$$D_0^2 \zeta_{n3} + \omega_n^2 \zeta_{n3} = -2D_0 D_2 \zeta_{n1} - 2\hat{\mu}_{3n} D_0 \zeta_{n1} - \frac{c_1^2 n^2 \pi^4}{4\ell^4} \zeta_{n1}$$
$$\times \sum_{k=1}^\infty k^2 \left(\eta_{k1}^2 + \zeta_{k1}^2\right) \quad (3.2.79)$$

The solutions of (3.2.76) and (3.2.77) can be expressed as

$$\eta_{n1} = A_{1n}(T_2) e^{i\omega_n T_0} + cc \quad (3.2.80)$$

$$\zeta_{n1} = A_{2n}(T_2) e^{i\omega_n T_0} + cc \quad (3.2.81)$$

Substituting (3.2.80) and (3.2.81) into (3.2.78) and (3.2.79) yields

$$D_0^2 \eta_{n3} + \omega_n^2 \eta_{n3} = -2i\omega_n(A'_{1n} + \hat{\mu}_{2n} A_{1n})e^{i\omega_n T_0}$$
$$- \frac{c_1^2 n^2 \pi^4}{4\ell^4}\left(A_{1n}e^{i\omega_n T_0} + \bar{A}_{1n}e^{-i\omega_n T_0}\right)\sum_{k=1}^{\infty} k^2 \left[\left(A_{1k}^2 + A_{2k}^2\right) e^{2i\omega_k T_0}\right.$$
$$\left. + A_{1k}\bar{A}_{1k} + A_{2k}\bar{A}_{2k}\right] + \frac{1}{2}F_n e^{i\Omega T_0} + cc \qquad (3.2.82)$$

$$D_0^2 \zeta_{n3} + \omega_n^2 \zeta_{n3} = -2i\omega_n(A'_{2n} + \hat{\mu}_{3n} A_{2n})e^{i\omega_n T_0}$$
$$- \frac{c_1^2 n^2 \pi^4}{4\ell^4}\left(A_{2n}e^{i\omega_n T_0} + \bar{A}_{2n}e^{-i\omega_n T_0}\right)\sum_{k=1}^{\infty} k^2 \left[\left(A_{1k}^2 + A_{2k}^2\right) e^{2i\omega_k T_0}\right.$$
$$\left. + A_{1k}\bar{A}_{1k} + A_{2k}\bar{A}_{2k}\right] + cc \qquad (3.2.83)$$

Using (3.2.73) in eliminating the terms that lead to secular terms in η_{n3} and ζ_{n3}, we obtain

$$2i\omega_n \left(A'_{1n} + \hat{\mu}_{2n} A_{1n}\right) + \frac{c_1^2 n^2 \pi^4}{4\ell^4}\left[2A_{1n}\sum_{k=1}^{\infty} k^2 \left(A_{1k}\bar{A}_{1k} + A_{2k}\bar{A}_{2k}\right)\right.$$
$$\left. + n^2 \left(A_{1n}^2 + A_{2n}^2\right)\bar{A}_{1n}\right] + \frac{1}{2}F_n \delta_{nm}e^{i\sigma T_2} = 0 \quad (3.2.84)$$

$$2i\omega_n \left(A'_{2n} + \hat{\mu}_{3n} A_{2n}\right) + \frac{c_1^2 n^2 \pi^4}{4\ell^4}\left[2A_{2n}\sum_{k=1}^{\infty} k^2 \left(A_{1k}\bar{A}_{1k} + A_{2k}\bar{A}_{2k}\right)\right.$$
$$\left. + n^2 \left(A_{1n}^2 + A_{2n}^2\right)\bar{A}_{2n}\right] = 0 \qquad (3.2.85)$$

Expressing the A_{1n} and A_{2n} in the polar form

$$A_{1n} = \frac{1}{2}a_{1n}e^{i\beta_{1n}} \text{ and } A_{2n} = \frac{1}{2}a_{2n}e^{i\beta_{2n}} \qquad (3.2.86)$$

we separate (3.2.84) and (3.2.85) into real and imaginary parts. Their imaginary parts are given by

$$a'_{1n} = -\hat{\mu}_{2n} a_{1n} + \frac{c_1^2 n^3 \pi^3}{32\ell^3 c_2}a_{1n}a_{2n}^2 \sin(2\beta_{1n} - 2\beta_{2n}) \qquad (3.2.87)$$

$$a'_{2n} = -\hat{\mu}_{3n} a_{2n} + \frac{c_1^2 n^3 \pi^3}{32\ell^3 c_2}a_{1n}^2 a_{2n} \sin(2\beta_{2n} - 2\beta_{1n}) \qquad (3.2.88)$$

when $n \neq m$. Adding a_{1n} times (3.2.87) to a_{2n} times (3.2.88) yields

$$a_{1n}a'_{1n} + a_{2n}a'_{2n} = -\hat{\mu}_{2n}a_{1n}^2 - \hat{\mu}_{3n}a_{2n}^2$$

which can be rewritten as

$$\frac{d}{dT_2}\left(a_{1n}^2 + a_{2n}^2\right) = -2\hat{\mu}_{2n}a_{1n}^2 - 2\hat{\mu}_{3n}a_{2n}^2$$

Because $\hat{\mu}_{2n}$ and $\hat{\mu}_{3n}$ are positive,

$$a_{1n}^2 + a_{2n}^2 \to 0 \text{ as } T_2 \text{ or } t \to \infty$$

Therefore, $a_{1n} \to 0$ and $a_{2n} \to 0$ as $t \to \infty$ for $n \neq m$. Hence, the long-term nonlinear response of the string is governed by (3.2.84) and (3.2.85) with A_{1n} and A_{2n} being equal to zero for $n \neq m$. Thus, the response of the string is governed by

$$2i(A_1' + \mu A_1) + \frac{c_1^2 m^3 \pi^3}{4\ell^3 c_2}\left[2A_1\left(A_1 \bar{A}_1 + A_2 \bar{A}_2\right) + \left(A_1^2 + A_2^2\right)\bar{A}_1\right]$$
$$+ \frac{F_m \ell}{2m\pi c_2} e^{i\sigma T_2} = 0 \qquad (3.2.89)$$

$$2i(A_2' + \mu A_2) + \frac{c_1^2 m^3 \pi^3}{4\ell^3 c_2}\left[2A_2\left(A_1 \bar{A}_1 + A_2 \bar{A}_2\right) + \left(A_1^2 + A_2^2\right)\bar{A}_2\right] = 0 \quad (3.2.90)$$

where the subscript m on A_{1m} and A_{2m} has been dropped and $\hat{\mu}_{2n}$ and $\hat{\mu}_{3n}$ have been replaced with μ for convenience. The damping coefficients in the two transverse directions are assumed to be the same.

Because for the case of primary resonance of the mth mode, a_{1n} and $a_{2n} \to 0$ as $t \to \infty$, η_n and $\zeta_n \to 0$ as $t \to \infty$. Hence, the long-term nonlinear response of the string to a primary resonance of the mth mode is governed by a two-degree-of-freedom model, which can be obtained from (3.2.70) and (3.2.71) by setting η_n and $\zeta_n = 0$ for $n \neq m$. The result is

$$\ddot{\eta}_m + \omega_m^2 \eta_m + 2\mu \dot{\eta}_m = -\frac{c_1^2 m^4 \pi^4}{4\ell^4}\eta_m\left(\eta_m^2 + \zeta_m^2\right) + F_{2m}(x,t) \quad (3.2.91)$$

$$\ddot{\zeta}_m + \omega_m^2 \zeta_m + 2\mu \dot{\zeta}_m = -\frac{c_1^2 m^4 \pi^4}{4\ell^4}\zeta_m\left(\eta_m^2 + \zeta_m^2\right) + F_{3m}(x,t) \quad (3.2.92)$$

where the damping coefficients have been assumed to be equal to μ.

3.3 NONLINEAR RESPONSE OF STRINGS

S. Nayfeh, Nayfeh, and Mook (1995) theoretically and experimentally studied the nonlinear response of a taut string to a primary resonance excitation of the sixth mode. The dynamics is governed by (3.2.89) and (3.2.90). Expressing A_1 and A_2 in the polar form

$$A_1 = \frac{1}{2}a_1 e^{i\beta_1} \text{ and } A_2 = \frac{1}{2}a_2 e^{i\beta_2} \qquad (3.3.1)$$

and separating real and imaginary parts in (3.2.89) and (3.2.90) yields

$$a_1' = -\mu a_1 - \frac{c_1^2 m^3 \pi^3}{32\ell^3 c_2} a_1 a_2^2 \sin\gamma_1 - f\sin\gamma_2 \qquad (3.3.2)$$

$$a_1 \beta_1' = \frac{c_1^2 m^3 \pi^3}{32\ell^3 c_2}\left(3a_1^3 + 2a_1 a_2^2 + a_1 a_2^2 \cos\gamma_1\right) + f\cos\gamma_2 \qquad (3.3.3)$$

$$a_2' = -\mu a_2 + \frac{c_1^2 m^3 \pi^3}{32\ell^3 c_2} a_1^2 a_2 \sin \gamma_1 \qquad (3.3.4)$$

$$a_2 \beta_2' = \frac{c_1^2 m^3 \pi^3}{32\ell^3 c_2} \left(2a_1^2 a_2 + 3a_2^3 + a_1^2 a_2 \cos \gamma_2\right) \qquad (3.3.5)$$

where

$$f = \frac{F_m \ell}{2m\pi c_2}, \quad \gamma_1 = 2\beta_2 - 2\beta_1, \quad \text{and} \quad \gamma_2 = \sigma T_2 - \beta_1 \qquad (3.3.6)$$

3.3.1 Frequency-Response Curves

S. Nayfeh, Nayfeh, and Mook (1995) chose values for the parameters ϵ and μ to match the experiments described in the next section. Their values are

$$\mu = 0.0314\ 1/sec,\ c_1 = 55.7\ m/sec,\ c_2 = 10.5\ m/sec,\ \text{and}\ \epsilon = 0.351 \qquad (3.3.7)$$

They presented results for excitation frequencies in the neighborhood of the sixth natural frequency of the string. They chose to excite the sixth mode rather than a lower-order even mode because better shaker performance is possible at the sixth natural frequency (14.8 Hz) than at lower excitation frequencies. In this case a_1 and a_2, as given in (3.3.2)-(3.3.5), correspond to the response amplitudes of the sixth inplane and sixth out-of-plane modes, respectively.

The obtained frequency-response curves are similar to those obtained by previous researchers. For small enough excitation amplitudes, the response contains only the sixth inplane mode. Thus, $a_2 = 0$ and only a_1 against $\epsilon\sigma$ is plotted. An example of such a condition is shown in Figure 3.3.1 with the external excitation amplitude $f = 3.46 \times 10^{-5}\ m/sec$. In this case, the steady-state response is periodic and in the plane of the excitation.

Increasing the excitation amplitude to $f = 5.39 \times 10^{-5}\ m/sec$, they obtained the frequency-response curves shown in Figure 3.3.2. In this case, as the excitation frequency is increased slightly beyond the natural frequency of the sixth mode, a pitchfork bifurcation occurs and the solution branch with $a_2 = 0$ loses stability to one where a_2 is not zero. Along this branch, one expects periodic whirling motions to occur. Increasing the excitation frequency further, they observed a Hopf bifurcation

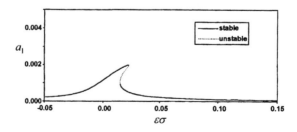

Fig. 3.3.1 Theoretical frequency-response curves when $f = 3.46 \times 10^{-5}\ m/sec$.

and predicted that amplitude- and phase-modulated motions will occur in the neighborhood of this branch. Proceeding further, they observed reverse Hopf and reverse pitchfork bifurcations in succession, leading to planar periodic responses. A slight increase in the excitation frequency causes a jump down to a low-amplitude planar periodic response. In this case, the jump is from a large to a small inplane response, but the out-of-plane response is zero before and after the jump.

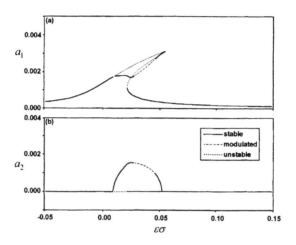

Fig. 3.3.2 Theoretical frequency-response curves when $f = 5.39 \times 10^{-5}\ m/sec$.

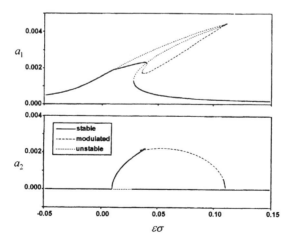

Fig. 3.3.3 Theoretical frequency-response curves when $f = 7.70 \times 10^{-5}\ m/sec$.

At higher excitation amplitudes, periodic and modulated whirling motions are predicted to occur over a broader range of excitation frequencies but the general character of the curves remains the same. Figure 3.3.3 shows the frequency-response curves for an excitation amplitude $f = 7.70 \times 10^{-5}$ m/sec. An added wrinkle here is the jump which occurs along the nontrivial a_2 branch. For more detailed studies of the predicted modulated motions, see Tufillaro (1989), Molteno and Tufillaro (1990), Bajaj and Johnson (1990,1992), and O'Reilly and Holmes (1992).

3.3.2 Experiments

S. Nayfeh, Nayfeh, and Mook (1995) performed experiments on a 2.13 m span of latex tubing with an inner diameter of 1/8 in and outer diameter of 1/4 in. The use of such a soft material for the test specimen has the advantages of allowing a clean excitation to be applied to the string and the motion to be easily discerned by the eye. However, due to the low stiffness of the material, the longitudinal wave speed is not extremely high as assumed in the development. It will be seen that this is significant in determining the validity of the theory.

A schematic diagram of the experimental setup is shown in Figure 3.3.4. A shaker was used to supply the oscillatory boundary condition at one end of the string and the other end was fixed at a rigid wall. They monitored the shaker-head motion by means of an accelerometer mounted to the shaker head. The string was mounted vertically so that gravity does not cause any sag.

To measure the displacement of the string, they employed an optical system modeled after the one that Molteno and Tufillaro (1990) used in their studies, which in turn was modeled after Harrison (1948). The sensor system was mounted close to the fixed end of the string and it returned signals proportional to the inplane and out-of-plane displacements at this point. They decomposed these signals in the frequency domain and computed the modal amplitudes.

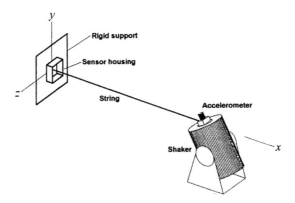

Fig. 3.3.4 Experimental setup for a taut string.

For purposes of comparison with theory, the parameters ϵ and μ as well as the transverse natural frequencies of the string had to be determined. The damping coefficients were determined by the method of logarithmic decrement. Computation of the longitudinal and transverse natural frequencies from the stiffness, density, and geometry of the string is inaccurate because the stiffness of the string is nonlinear and very low. They therefore determined the transverse natural frequencies of the string by examining the frequency spectra of decaying free oscillations. To similarly measure the longitudinal natural frequencies, they attached a piece of thick tape to the string in the vicinity of the optical sensors so that they can detect longitudinal motions. The parameter ϵ can then be determined from the longitudinal and transverse natural frequencies as $\epsilon = \pi^2 c_2^2 / c_1^2$. The parameters identified in this manner are listed in (3.3.7).

S. Nayfeh, Nayfeh, and Mook (1995) performed stationary excitation frequency sweeps by slowly varying the excitation frequency while holding the excitation amplitude constant. At each excitation frequency, they allowed any transients to die out before recording data and continuing the sweep. They presented data corresponding to forward and backward sweeps. In some cases, additional sweeps were necessary in order to obtain responses along isolated branches.

In Figure 3.3.5, the experimentally obtained frequency-response curves are compared with the theoretical results of Figure 3.3.1 obtained for a low-amplitude external excitation. The unstable solutions are not shown for clarity. As predicted by the theory, only the sixth inplane mode is excited. Moreover, the amplitude of the observed motions closely matches the theory. Although the region of overhang is small, the hysteresis phenomenon is clearly discernible in the experiment.

Increasing the excitation amplitude, they obtained the frequency-response curves shown in Figure 3.3.6. As predicted by the theory and shown in Figure 3.3.2, the sixth out-of-plane mode can be excited in this case. The theoretical and experimental results are in good agreement for the planar responses as well as the leading branch of whirling motions. As the excitation frequency was slowly increased beyond the predicted Hopf bifurcation point, modulated motions were observed, but after some time at a constant excitation frequency and amplitude, the response settled to the

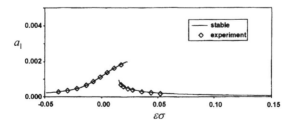

Fig. 3.3.5 Comparison of the theoretical and experimental frequency-response curves when $f = 3.46 \times 10^{-5} \, m/sec$.

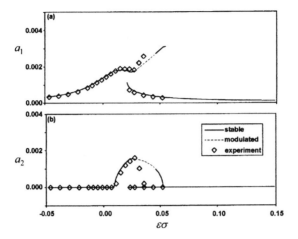

Fig. 3.3.6 Comparison of the theoretical and experimental frequency-response curves when $f = 5.39 \times 10^{-5}$ m/sec.

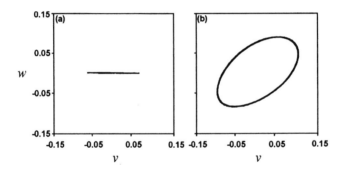

Fig. 3.3.7 Trajectories (in mm) of a point 3 mm from the fixed end when $f = 5.39 \times 10^{-5}$ m/sec: (a) $\epsilon\sigma = 0.0$ and (b) $\epsilon\sigma = 0.025$.

steady-state response whose amplitude is shown in the plot. Thus, where the theory predicts modulated motions, they observed periodic responses with amplitudes far removed from the predicted unstable out-of-plane amplitudes. Trajectories of a point on the string are shown in Figures 3.3.7 for periodic planar and whirling motions.

A further increase in the excitation amplitude yields the frequency-response curves shown in Figure 3.3.8, which correspond to the theoretical results shown in Figure 3.3.3. Good agreement between theory and experiment were obtained for the periodic planar and whirling motions, as shown in the figure. In this case, where the theory predicts a Hopf bifurcation, they observed modulated motions; these are not shown on the plots. The trajectory of a point on the string as it undergoes modulated motions

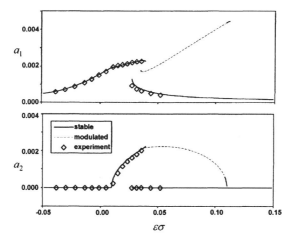

Fig. 3.3.8 Comparison of the theoretical and experimental frequency-response curves when $f = 7.70 \times 10^{-5}\ m/sec$.

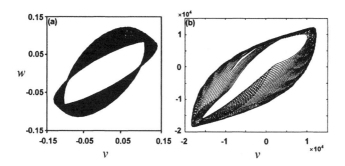

Fig. 3.3.9 Experimentally obtained quasiperiodic motions of a string: (a) S. Nayfeh, Nayfeh, and Mook (1995) and (b) Molteno (1994).

is shown in Figure 3.3.9a. It is qualitatively similar to that obtained experimentally (Figure 3.3.9b) by Molteno (1994). These motions do not persist for the entire length of this branch; instead S. Nayfeh, Nayfeh, and Mook (1995) observed that the response jumps down to the lower planar periodic branch as the excitation frequency was increased beyond a value of $\epsilon\sigma$ of approximately 0.05. This observation is consistent with the work of Bajaj and Johnson (1992) who predicted that a crisis may occur by which the attractor corresponding to modulated motions along this branch will be destroyed.

3.4 MODELING OF CABLES

A cable is a string with initial sag, which makes the modeling and dynamics of cables much more complex than those of strings. Because engineering cables are usually lengthy and flexible, their vibrations involve large displacements and are dominated by geometric nonlinearities. Many large-amplitude vibrations and nonlinear phenomena of cables have been studied analytically (Yamaguchi, Miyata, and Ito, 1981; Beatty and Chow, 1984; Luongo, Rega, and Vestroni, 1984; Rega, Vestroni, and Benedettini, 1984; Al-Noury and Ali, 1985; Pastorel and Beaulieu, 1985; Ali, 1986; Benedettini, Rega, and Vestroni, 1986; Benedettini and Rega, 1987, 1990, 1994; Takahashi and Konishi, 1987a,b; Papazoglou and Mavrakos, 1990; Tadjbakhsh and Wang, 1990; Mitsugi and Yasaka, 1991; Rao and Iyengar, 1991; Takahashi, 1991; Fujino, Warnitchai, and Pacheco, 1993; Luongo and Vestroni, 1994, 1996; Lee and Perkins, 1995b; Pakdemirli, S. Nayfeh, and Nayfeh, 1995; Luongo, Paolone, and Piccardo, 1998; Arafat and Nayfeh, 2003) and experimentally (Shimizu, Yamaguchi, and Ito, 1981; Pastorel and Beaulieu, 1985; Papazoglou and Mavrakos, 1990; Lee and Perkins, 1992, 1995a; Fujino, Warnitchai, and Pacheco, 1993; Benedettini and Rega, 1997; Rega, Alaggio, and Benedettini, 1997; Rega and Alaggio, 2001).

There are almost as many nonlinear cable models as there are researchers because ad hoc assumptions are adopted by different researchers during different steps in the derivation. Because extensional forces are the only elastic loads acting on cables, compressibility and Poisson's effect can be significant. For example, for ideal incompressible, isotropic rubber-like materials, Poisson's effect results in a nonlinear relationship between the tension force and the axial strain (Lawton and King, 1951; King, 1974); they are called neo-Hookean materials. In the literature, most of the cable models do not account for the change in the cross-sectional area due to Poisson's effect and the Poisson ratio ν is absent from the equations of motion. Such theories are valid only for materials with $\nu = 0$. If the Poisson effect is considered but the material is assumed to be incompressible, ν will not appear in the equations of motion (Tadjbakhsh and Wang, 1990); such a theory is valid only for materials with $\nu \simeq 0.5$. We note that the Poisson ratio of most engineering materials is between 0.25 and 0.35 except for rubber and paraffin, for which $\nu \simeq 0.5$.

To account for geometric nonlinearities, some researchers (e.g., Lee and Perkins, 1995b) used the von Kármán type strains or the Lagrangian strains. However, the Lagrangian strains do not fully account for geometric nonlinearities (Pai, Palazotto, and Greer, 1998). Some researchers only account for cubic nonlinearities (Fujino, Warnitchai, and Pacheco, 1993) and some others (Tadjbakhsh and Wang, 1990) only account for quadratic nonlinearities. However, Lee and Perkins (1995b) and Pai and Nayfeh (1991a,b) showed that cubic nonlinearities can greatly change the dynamic responses due to quadratic nonlinearities. Moreover, some theories do not account for initial sag, static loads, and/or extensionality, although they always exist in real cable structures. And, instead of using Cartesian coordinates, some researchers use a curvilinear material (Lagrangian) coordinate, which makes the application awkward.

Since it is well known that the dynamic responses of a nonlinear system are sensitive to small changes in its parameters, an accurate nonlinear model is necessary to predict these responses. Pai and Nayfeh (1992c) used an energy approach to develop a fully nonlinear cable model, which includes the effects of static and dynamic loads, initial sag, compressibility, material nonuniformity, Poisson's effect, and geometric nonlinearities. They used this model to derive nonlinear equations of motion up to cubic nonlinearities for sagged cables, taut strings, and extensional bars. They compared the model with several other models.

The dynamics of suspended cables have received much attention lately due to their importance in many applications in the fields of communications, electricity (e.g., power lines), mooring systems, transportation (cableways), cable-stayed bridges, and crane-operation systems. The linear theory of suspended cables was originally developed by Irvine and Caughey (1974) and Irvine (1981). Triantafyllou (1984) reviewed most of the works dealing with the linear dynamics of cables and chains. The free nonlinear dynamics of suspended cables were investigated by Hagedorn and Schäfer (1980) and Luongo, Rega, and Vestroni (1984), who considered planar motions, and by Benedettini, Rega, and Vestroni (1986) and Takahashi and Konishi (1987a), who considered nonplanar motions.

Benedettini and Rega (1987) investigated the forced planar dynamics of cables near primary resonance and found that the response may be either hardening or softening, depending on the amplitude of oscillation. Takahashi and Konishi (1987b) investigated the forced nonplanar oscillations of cables and determined the regions of instability for several principal parametric and combination parametric resonances. Perkins (1992) considered a suspended cable subjected to a principal parametric resonance in the presence of a two-to-one internal resonance and investigated theoretically and experimentally its nonplanar responses. His experimental results show that large-amplitude nonplanar oscillations may be activated by the internal resonance. Nonplanar oscillations of cables that are excited at primary resonance were investigated by Al-Noury and Ali (1985) and Pakdemirli, S. Nayfeh, and Nayfeh (1995) in the case of a one-to-one internal resonance and by Visweswara Rao and Iyengar (1991) and Lee and Perkins (1992) in the case of a two-to-one internal resonance. The control of nonlinear nonresonant oscillations of cables under inplane excitations was considered by Gattulli, Pasca, and Vestroni (1997). Planar and nonplanar nonlinear dynamics of suspended cables excited by an inplane random loading were investigated by Chang, Ibrahim, and Afaneh (1994, 1996). Moreover, Pilipchuk and Ibrahim (1997) investigated strongly nonlinear stretching-bending interactions in suspended cables with horizontally moving supports. A review of research conducted on the nonlinear dynamics of cables and other mooring systems was presented by Triantafyllou (1987, 1991).

Recent experimental results by Lee and Perkins (1995a), Benedettini and Rega (1997), Rega, Alaggio, and Benedettini (1997), Alaggio and Rega (2000), and Rega and Alaggio (2001) have demonstrated that the response of suspended cables near or away from the first crossover can be large and exhibit very complex behavior due to the simultaneous presence of multiple internal resonances involving several inplane and out-of-plane modes. Examples include the coexistence of different types of periodic motions and the occurrence of quasiperiodic and chaotic oscillations. The latter resulting, in some cases, due to torus breakdown and/or homoclinic bifurcation scenarios.

There are two types of theoretical investigations of multiple internal resonances in cables. First, the excitation is a random loading (Chang and Ibrahim, 1997; Ibrahim and Chang, 1999). Second, the excitation is harmonic at primary resonance (Tadjbakhsh and Wang, 1990; Benedettini and Rega, 1994; Benedettini, Rega, and Alaggio, 1995; Lee and Perkins, 1995b). Moreover, Benedettini and Rega (1994) also considered the case of subharmonic resonance of order one-half. Chang and Ibrahim (1997) and Ibrahim and Chang (1999) used Monte Carlo simulation and found that the response can exhibit an on-off intermittency. Tadjbakhsh and Wang (1990) considered multiple two-to-one internal resonances and used the method of multiple scales (Nayfeh, 1973b, 1981) to determine first-order approximate solutions. They found that the response can exhibit jumps and saturation. Benedettini and Rega (1994), Benedettini, Rega, and Alaggio (1995), and Lee and Perkins (1995b) considered simultaneous two-to-one and one-to-one internal resonances and used the method of multiple scales to determine first-order and second-order expansions of the solutions. Single-mode, two-mode, and three-mode periodic motions of the cable were determined and, in some cases, were found to undergo Hopf bifurcations, thereby leading to quasiperiodic motions.

The analyses of Tadjbakhsh and Wang (1990), Benedettini and Rega (1994), Benedettini, Rega, and Alaggio (1995), and Lee and Perkins (1995b) were conducted by applying the method of multiple scales to a discretized set of ordinary-differential equations resulting from the Galerkin procedure. The resulting modulation equations governing the dynamics of the amplitudes and phases of the interacting modes were then solved for the equilibrium and dynamic solutions and the stability of these solutions was ascertained.

However, as discussed in Section 1.8, attacking the discretized system may lead to quantitative and sometimes qualitative errors (Nayfeh, J. Nayfeh, and Mook, 1992; Nayfeh, S. Nayfeh, and Pakdemirli, 1995; Pakdemirli, S. Nayfeh, and Nayfeh, 1995; Nayfeh, 1997; Nayfeh and Lacarbonara, 1997, 1998; Lacarbonara, Nayfeh, and Kreider, 1998; Arafat and Nayfeh, 1999). Furthermore, Feng and Leal (1995), Arafat, Nayfeh, and Chin (1998), and Arafat and Nayfeh (2000) have demonstrated that, for systems that are conservative in the absence of damping and external forces, the resulting modulation equations must be derivable from a Lagrangian, and hence the coefficients multiplying the nonlinear terms should exhibit certain symmetries.

Accordingly, Rega, Lacarbonara, Nayfeh, and Chin (1997, 1999) investigated the response of suspended cables near the first crossover by applying the method of multiple scales directly to the second-order partial-differential equations of motion and associated boundary conditions. They considered the case of primary resonance of the first symmetric inplane mode and obtained second-order expansions of the solution. They found that, in order to render their reconstituted modulation equations derivable from a Lagrangian, it was necessary to include the homogeneous solutions of the second-order problem. However, their results are inconsistent because some of the coefficients depend on an arbitrary constant. To overcome the inconsistencies, Nayfeh, Arafat, Chin, and Lacarbonara (2002) followed Nayfeh (2000) and

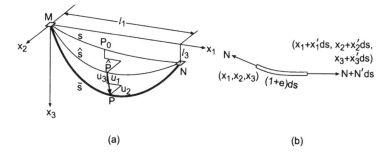

Fig. 3.4.1 Cable model: (a) the deformed configuration of a cable, where $x_1 x_2 x_3$ is a Cartesian, inertial coordinate system, and (b) the free-body diagram of a cable element.

first formulated the two governing partial-differential equations as a system of four first-order (in time) partial-differential equations. Then, they applied the method of multiple scales (Nayfeh, 1973b, 1981, 2000) directly to these equations to obtain second-order uniform asymptotic expansions of the solution. They showed that the resulting reconstituted (Nayfeh, 1985) modulation equations governing the dynamics of the amplitudes and phases of the four interacting modes satisfy all of the symmetry conditions, and therefore they are derivable from a Lagrangian. Furthermore, they showed that all of the coefficients of the nonlinear terms in the modulation equations are determined uniquely, and hence the solution is systematic and consistent.

3.4.1 Exact Equations of Motion

In Figure 3.4.1a, we show the deformed configuration of a cable and the inertial coordinate system $x_1 x_2 x_3$. Point P_0 indicates the position of the observed particle when the cable is not loaded, Point $\hat{P}$ indicates the deformed position of P_0 under static loads, and Point P indicates the deformed position of P_0 under static and dynamic loads. We denote the coordinates of $\hat{P}$ and P by $(\alpha_1, \alpha_2, \alpha_3)$ and (x_1, x_2, x_3), respectively, and denote the dynamic displacements of P along the axes x_1, x_2, and x_3 by u_1, u_2, and u_3, respectively. Thus,

$$x_i = \alpha_i + u_i \quad \text{for} \quad i = 1, 2, 3 \tag{3.4.1}$$

Moreover, s denotes the undeformed arclength measured from Point M to the observed particle at Point P_0; $\hat{s}$ represents the statically deformed arclength; $\tilde{s}$ represents the dynamically deformed arclength; L is the unloaded total length of the cable; $\rho_0(s)$ is the mass density of the unloaded cable; $A_0(s)$ is the cross-section area of the unloaded cable; and $(\ )' \equiv \partial(\ )/\partial s$.

It follows from (3.4.1) and the free-body diagram shown in Figure 3.4.1b that the total axial strain e is given by

$$e = \sqrt{x_1'^2 + x_2'^2 + x_3'^2} - 1 = \sqrt{\eta_0 + 2\eta_1 + \eta_2} - 1 \tag{3.4.2}$$

where

$$\eta_0 \equiv \alpha_1'^2 + \alpha_2'^2 + \alpha_3'^2, \quad \eta_1 \equiv \alpha_1' u_1' + \alpha_2' u_2' + \alpha_3' u_3', \quad \eta_2 \equiv u_1'^2 + u_2'^2 + u_3'^2 \quad (3.4.3)$$

Because of Poisson's effect, the deformed cross-section area A is given by

$$A = (1 - \nu e)^2 A_0 \qquad (3.4.4)$$

where ν is Poisson's ratio. Using Hooke's law and (3.4.4), we obtain

$$N(s,t) = AEe = EA_0(1-\nu e)^2 e \qquad (3.4.5)$$

where N denotes the internal tension force and $E(s)$ is Young's modulus.

The extended Hamilton principle is used to derive the equations of motion, which can be stated as

$$\int_0^t (\delta T - \delta \Pi + \delta W_{nc}) \, dt = 0 \qquad (3.4.6)$$

Here, δT is the variation of the kinetic energy T and is given by

$$\delta T = -\int_0^L m_0 \left(\ddot{u}_1 \delta u_1 + \ddot{u}_2 \delta u_2 + \ddot{u}_3 \delta u_3 \right) ds \qquad (3.4.7)$$

where $m_0 \equiv \rho_0 A_0$ and the overdot denotes differentiation with respect to time t. The variation $\delta \Pi$ of the elastic energy Π can be obtained by using (3.4.2) as

$$\delta \Pi = \int_0^L N \delta e \, ds = \int_0^L \left(\frac{N x_1'}{1+e} \delta x_1' + \frac{N x_2'}{1+e} \delta x_2' + \frac{N x_3'}{1+e} \delta x_3' \right) ds \qquad (3.4.8)$$

Also, δW_{nc} denotes the variation of the non-conservative energy W_{nc}; it is given by

$$\delta W_{nc} = \int_0^L (f_1 \delta x_1 + f_2 \delta x_2 + f_3 \delta x_3 - 2c_1 \dot{u}_1 \delta u_1 - 2c_2 \dot{u}_2 \delta u_2 - 2c_3 \dot{u}_3 \delta u_3) \, ds + \delta W_b \qquad (3.4.9)$$

where f_i denotes the summation of the distributed static $\hat{f}_i$ and dynamic $\tilde{f}_i$ loads per unit of undeformed length along the x_i direction, and c_i is the damping coefficient per unit of the undeformed length along the x_i direction. Moreover, δW_b is the virtual work due to forces applied on the boundaries or due to the motion of the boundaries, which is problem dependent and will not be considered in the derivation. Substituting (3.4.7)-(3.4.9) into (3.4.6), using $\delta \alpha_i' = \delta \alpha_i = 0$ (since the α_i are assumed to be known here), integrating by parts, and then setting each of the coefficients of δu_1, δu_2, and δu_3 equal to zero, we obtain the equations of motion as

$$m_0 \ddot{u}_i + 2c_i \dot{u}_i = \frac{\partial}{\partial s}\left[\frac{N}{1+e}(\alpha_i' + u_i') \right] + f_i \quad \text{for} \quad i = 1, 2, 3 \qquad (3.4.10)$$

and the boundary conditions in the form: specify

$$u_i \quad \text{or} \quad \frac{N}{1+e}(\alpha_i' + u_i') \quad \text{at} \quad s = 0 \quad \text{and} \quad L \quad \text{for} \quad i = 1, 2, 3 \qquad (3.4.11)$$

3.4.2 Static Deflections

To obtain static deflections, we let $u_i = 0$ and $f_i = \hat{f}_i$ in (3.4.10) and obtain

$$\frac{d}{ds}\left[\frac{\hat{N}}{1+\hat{e}}\alpha'_i\right] + \hat{f}_i = 0 \quad \text{for} \quad i = 1, 2, 3 \tag{3.4.12}$$

The boundary conditions are of the form: specify

$$\alpha_i \quad \text{or} \quad \frac{\hat{N}}{1+\hat{e}}\alpha'_i \quad \text{at} \quad s = 0 \text{ and } L \quad \text{for} \quad i = 1, 2, 3 \tag{3.4.13}$$

Here, $\hat{N}$ is the internal extensional force under static loads, $\hat{N} = EA_0(1 - \nu\hat{e})^2\hat{e}$, and $\hat{e}$ is the strain due to static loads, which is obtained from (3.4.2) as

$$\hat{e} = \sqrt{{\alpha'_1}^2 + {\alpha'_2}^2 + {\alpha'_3}^2} - 1 \tag{3.4.14}$$

Equation (3.4.12) shows that $\hat{N}$ is a function of s, which is different from that for strings (see (3.1.14)).

Next, we present a method for solving for the exact large static deflections. We consider the system shown in Figure 3.4.1a. The boundary conditions are

$$\alpha_1 = \alpha_2 = \alpha_3 = 0 \quad \text{at} \quad s = 0 \tag{3.4.15}$$

$$\alpha_1 = l_1, \ \alpha_2 = 0, \ \alpha_3 = l_3 \quad \text{at} \quad s = L \tag{3.4.16}$$

Integrating (3.4.12) once with respect to s yields

$$\frac{\hat{N}}{1+\hat{e}}\alpha'_i = F_i - g_i \quad \text{for } i = 1, 2, 3 \tag{3.4.17}$$

where

$$g_i \equiv \int_0^s \hat{f}_i ds, \quad F_i = \left.\frac{\hat{N}}{1+\hat{e}}\alpha'_i\right|_{s=0} \quad \text{for} \quad i = 1, 2, 3 \tag{3.4.18}$$

and the F_i are constants to be determined in the course of the analysis. Squaring and adding the three equations in (3.4.17) and using (3.4.14), we obtain

$$\hat{N} = \sqrt{(F_1 - g_1)^2 + (F_2 - g_2)^2 + (F_3 - g_3)^2} \tag{3.4.19}$$

For the case of static deflections, (3.4.5) becomes

$$\hat{N} = EA_0(1 - \nu\hat{e})^2\hat{e} \tag{3.4.20}$$

which can be solved for $\hat{e}$ in terms of $\hat{N}$ and hence in terms of the g_i from (3.4.19). Consequently, we rewrite (3.4.17) as

$$\alpha'_i = \frac{(F_i - g_i)(1 + \hat{e})}{\hat{N}} \tag{3.4.21}$$

where the right-hand side depends on the F_i and g_i. Hence, the problem of solving for the exact deflections consists of determining the F_i so that the solution of the initial-value problem consisting of equations (3.4.19)-(3.4.21) and (3.4.15) satisfies the boundary conditions (3.4.16).

To solve for the exact deflection, we start with a guess for the F_i and use a shooting technique to solve (3.4.21) starting from $s = 0$. At each location s, we calculate $\hat{N}$ from (3.4.19), then calculate $\hat{e}$ from (3.4.20), and finally calculate α'_i from (3.4.21). At $s = L$, we check whether the boundary conditions (3.4.16) are satisfied. If not, we use a Newton-Raphson procedure to improve on the initial guess and continue the process until the boundary conditions (3.4.16) are satisfied to within a specified accuracy.

For *incompressible* materials, (3.4.20) can be readily solved for $\hat{e}$ in terms of $\hat{N}$. The incompressibility condition demands that the volume of an infinitesimal element of the cable before and after the deformation be the same; that is,

$$A_0 ds = \hat{A} d\hat{s} = A_0(1 - \nu\hat{e})^2(1 + \hat{e})ds$$

Hence,

$$(1 + \hat{e})(1 - \nu\hat{e})^2 = 1 \qquad (3.4.22)$$

It follows from (3.4.20) and (3.4.22) that

$$\hat{N} = \frac{\hat{e}}{1 + \hat{e}} E A_0 \qquad (3.4.23)$$

Solving (3.4.23) for $1 + \hat{e}$, we have

$$1 + \hat{e} = \frac{E A_0}{E A_0 - \hat{N}} \qquad (3.4.24)$$

where $\hat{N}$ is a function of the $F_i - g_i$ according to (3.4.19). Substituting (3.4.24) into (3.4.21) yields

$$\alpha'_i = \frac{E A_0 (F_i - g_i)}{\hat{N}(E A_0 - \hat{N})} \qquad (3.4.25)$$

Integrating (3.4.25) once and using the boundary conditions (3.4.15), we have

$$\alpha_i = \int_0^s \frac{E A_0 (F_i - g_i)}{\hat{N}(E A_0 - \hat{N})} ds \qquad (3.4.26)$$

where $\hat{N}$ is a function of the F_i, as shown in (3.4.19). Again, because the F_i are unknown, one can use a Newton-Raphson procedure to determine them so that the α_i satisfy the boundary conditions (3.4.16).

We note that the incompressibility assumption is valid for materials with $\nu \approx 0.5$ in the case of infinitesimal strains. To show this, we expand (3.4.22) and obtain

$$(1 - 2\nu)\hat{e} + (\nu^2 - 2\nu)\hat{e}^2 + \nu^2 \hat{e}^3 = 0$$

or
$$(1 - 2\nu) + (\nu^2 - 2\nu)\hat{e} + \nu^2 \hat{e}^2 = 0$$

which, for infinitesimal strains, yields $\nu = 0.5$.

When $\hat{N}/EA_0 \ll 1$, (3.4.20) can be approximately solved for $\hat{e}$ in terms of $\hat{N}/EA_0$. The result is

$$\hat{e} = \frac{\hat{N}}{EA_0} + 2\nu \frac{\hat{N}^2}{E^2 A_0^2} + \cdots \qquad (3.4.27)$$

In the absence of loads, except the cable weight,

$$\alpha_2 = 0, \quad g_1 = 0, \quad g_2 = 0, \quad g_3 = m_0 g$$

where m_0 is the mass density of the cable per unit length. Then, (3.4.19) becomes

$$\hat{N} = \sqrt{F_1^2 + (F_3 - m_0 g s)^2} \qquad (3.4.28)$$

Then, to the first approximation, it follows from (3.4.27) and (3.4.28) that

$$\hat{e} = \frac{\sqrt{F_1^2 + (F_3 - m_0 g s)^2}}{EA_0} \qquad (3.4.29)$$

Consequently, (3.4.17) becomes

$$\alpha_1' = \frac{F_1}{EA_0} + \frac{F_1}{\sqrt{F_1^2 + (F_3 - m_0 g s)^2}} \qquad (3.4.30)$$

$$\alpha_3' = \frac{F_3 - m_0 g s}{EA_0} \left[1 + \frac{EA_0}{\sqrt{F_1^2 + (F_3 - m_0 g s)^2}} \right] \qquad (3.4.31)$$

Integrating (3.4.30) and (3.4.31) and using the boundary conditions (3.4.15), we obtain

$$\alpha_1 = \frac{F_1 s}{EA_0} + \frac{F_1}{m_0 g} \left[\sinh^{-1}\left(\frac{F_3}{F_1}\right) - \sinh^{-1}\left(\frac{F_3 - m_0 g s}{F_1}\right) \right] \qquad (3.4.32)$$

$$\alpha_3 = \frac{2 F_3 s - m_0 g s^2}{2 EA_0} + \frac{1}{m_0 g} \left[\sqrt{F_1^2 + F_3^2} - \sqrt{F_1^2 + (F_3 - m_0 g s)^2} \right] \qquad (3.4.33)$$

Imposing the boundary conditions (3.4.16) yields

$$\ell_1 = \frac{F_1 L}{EA_0} + \frac{F_1 L}{W} \left[\sinh^{-1}\left(\frac{F_3}{F_1}\right) - \sinh^{-1}\left(\frac{F_3 - W}{F_1}\right) \right] \qquad (3.4.34)$$

$$\ell_3 = \frac{2 F_3 L - W L}{2 EA_0} + \frac{L}{W} \left[\sqrt{F_1^2 + F_3^2} - \sqrt{F_1^2 + (F_3 - W)^2} \right] \qquad (3.4.35)$$

where $W = m_0 g L$ is the total weight of the cable. Equations (3.4.34) and (3.4.35) can be solved for F_1 and F_3.

When $\ell_3 = 0$, (3.4.35) yields $F_3 = \frac{1}{2}W$, and hence (3.4.34) becomes

$$\sinh\left[\frac{W\ell_1}{2F_1 L} - \frac{W}{2EA_0}\right] = \frac{W}{2F_1} \qquad (3.4.36)$$

which can be solved for F_1.

When the tension in the cable is much larger than its weight, the cable sag will be small compared to its span (shallow suspended cable) and an approximate expression for the static transverse deflection due to gravity can be obtained as follows. In the absence of loads in the x_2 direction, $\alpha_2' = 0$, and because the tension $\hat{N}$ in the cable is much larger than $m_0 g L$, $\alpha_1' \gg \alpha_3'$. Hence, it follows from (3.4.14) that $\hat{e} = \alpha_1' - 1$. Then, it follows from (3.4.12) for $i = 1$ and $\hat{f}_1 = 0$ that $\hat{N} = N_0 = $ a constant, and hence it follows from (3.4.20) that $\alpha_1' = $ a constant, which is a function of N_0. Consequently, (3.4.12) for $i = 3$ becomes

$$\frac{N_0}{\alpha_1'}\frac{d}{ds}(\alpha_3') = -m_0 g \qquad (3.4.37)$$

When $\alpha_3 = 0$ at $s = 0$ and $s = L$, the solution of (3.4.37) is

$$\alpha_3 = \frac{m_0 g \alpha_1'}{2N_0}s(L - s) \qquad (3.4.38)$$

Alternatively, when the horizontal tension F_1 in the cable is much larger than its total weight W, $F_3 \ll F_1$ and $m_0 g s \ll F_1$. Then, (3.4.30) and (3.4.31) can be approximated by

$$\alpha_1' = \frac{F_1}{EA_0} + 1 + O\left(\frac{W^2}{F_1^2}\right) \qquad (3.4.39)$$

$$\alpha_3' = \frac{F_3 - m_0 g s}{EA_0}\left[1 + \frac{EA_0}{F_1} + O\left(\frac{W^2}{F_1^2}\right)\right] \qquad (3.4.40)$$

Neglecting small terms in (3.4.39) and (3.4.40), integrating the result, and using the boundary conditions (3.4.15), we obtain

$$\alpha_1 = \left(1 + \frac{F_1}{EA_0}\right)s + c_1 \qquad (3.4.41)$$

$$\alpha_3 = \left(1 + \frac{EA_0}{F_1}\right)\frac{2F_3 s - m_0 g s^2}{2EA_0} + c_2 \qquad (3.4.42)$$

where c_1 and c_2 are constants. Imposing the boundary conditions (3.4.16) with $\ell_3 = 0$ yields

$$\alpha_1 = \ell_1 + \left(1 + \frac{F_1}{EA_0}\right)(s - L) \qquad (3.4.43)$$

$$\alpha_3 = \left(1 + \frac{F_1}{EA_0}\right)\frac{m_0 g}{2F_1}(sL - s^2) \qquad (3.4.44)$$

and $F_3 = \frac{1}{2}m_0 gL = \frac{1}{2}W$. Equation (3.4.44) is in full agreement with (3.4.38) because $\alpha_1' = 1 + F_1/EA_0$ and $N_0 = F_1$. Equation (3.4.44) can be rewritten as

$$\alpha_3 = 4b\frac{s}{L}\left(1 - \frac{s}{L}\right) \tag{3.4.45}$$

where

$$b \equiv \left(1 + \frac{F_1}{EA_0}\right)\frac{m_0 gL^2}{8F_1} \approx \frac{m_0 gL^2}{8F_1}$$

or

$$F_1 = \frac{m_0 gL^2}{8b} \tag{3.4.46}$$

3.4.3 Approximate Equations of Motion

For small dynamic displacements u_1, u_2, and u_3, we expand (3.4.2) in a Taylor series and, to third order, obtain

$$e = \sqrt{\eta_0} - 1 + \eta_1 \eta_0^{-\frac{1}{2}} + \frac{1}{2}\left(\eta_2 \eta_0^{-\frac{1}{2}} - \eta_1^2 \eta_0^{-\frac{3}{2}}\right)$$
$$+ \frac{1}{2}\left(\eta_1^3 \eta_0^{-\frac{5}{2}} - \eta_1 \eta_2 \eta_0^{-\frac{3}{2}}\right) + \cdots \tag{3.4.47}$$

Hence, to third order,

$$\frac{N}{1+e} = \frac{EA_0(1-\nu e)^2 e}{1+e}$$
$$= EA_0 \left[\Lambda_1 + \Lambda_2 \eta_1 + \Lambda_3 \eta_1^2 + \frac{1}{2}\Lambda_2 \eta_2 + \Lambda_3 \eta_1 \eta_2 + \Lambda_4 \eta_1^3\right] + \cdots \tag{3.4.48}$$

where

$$\Lambda_1 \equiv \left(1 + \nu - \nu \eta_0^{1/2}\right)^2 \left(1 - \eta_0^{-1/2}\right) \tag{3.4.49}$$

$$\Lambda_2 \equiv 2\nu^2 - \eta_0^{-1/2}\left(2\nu + 3\nu^2\right) + \eta_0^{-3/2}(1+\nu)^2 \tag{3.4.50}$$

$$\Lambda_3 \equiv \eta_0^{-3/2}\left(\nu + \frac{3}{2}\nu^2\right) - \frac{3}{2}\eta_0^{-5/2}(1+\nu)^2 \tag{3.4.51}$$

$$\Lambda_4 \equiv -\eta_0^{-5/2}\left(\nu + \frac{3}{2}\nu^2\right) + \frac{5}{2}\eta_0^{-7/2}(1+\nu)^2 \tag{3.4.52}$$

Substituting (3.4.48) into (3.4.10) and (3.4.11) and using (3.4.12), (3.4.13), and $f_i = \hat{f}_i + \tilde{f}_i$, we obtain the following third-order equations of motion:

$$m_0 \ddot{u}_i + 2c_i \dot{u}_i = \tilde{f}_i + (EA_0 U_i)' \quad \text{for } i = 1, 2, 3 \tag{3.4.53}$$

and the following expanded boundary conditions: specify

$$u_i \quad \text{or} \quad EA_0 U_i \quad \text{at} \quad s = 0 \quad \text{and} \quad L \text{ for } i = 1, 2, 3 \tag{3.4.54}$$

where

$$U_i \equiv \Lambda_1 u_i' + \Lambda_2 \left(\alpha_i' + u_i'\right) \eta_1 + \Lambda_3 \left(\alpha_i' + u_i'\right) \eta_1^2 + \frac{1}{2}\Lambda_2 \left(\alpha_i' + u_i'\right) \eta_2$$
$$+ \Lambda_3 \alpha_i' \eta_1 \eta_2 + \Lambda_4 \alpha_i' \eta_1^3 \qquad (3.4.55)$$

Neglecting the nonlinear terms in (3.4.53)-(3.4.55), we obtain the following linear equations:

$$m_0 \ddot{u}_i + 2c_i \dot{u}_i = \tilde{f}_i + \left[EA_0 \left(\Lambda_1 u_i' + \Lambda_2 \alpha_i' \eta_1\right)\right]' \quad \text{for} \quad i = 1, 2, 3 \qquad (3.4.56)$$

and the boundary conditions are: specify

$$u_i \quad \text{or} \quad EA_0(\Lambda_1 u_i' + \Lambda_2 \alpha_i' \eta_1) \quad \text{at} \quad s = 0 \quad \text{and} \quad L \qquad (3.4.57)$$

for $i = 1, 2, 3$. We note that the $\Lambda_2 \alpha_i' \eta_1$ are linear coupling terms due to the initial sag, and there are no such couplings if two of the α_i' are equal to zero, corresponding to the case of taut strings.

3.5 REDUCTION OF CABLE MODEL TO TWO EQUATIONS

Next, we consider a cable hanging between two horizontal supports under its weight and excited by distributed dynamic loads. When the tension in the cable is much larger than its total weight, $\alpha_3' \ll \alpha_1'$ and the static displacement is given by (3.4.44). For such cables, the longitudinal natural frequencies are much higher than the transverse natural frequencies, and hence the inertia terms in (3.4.53) for $i = 1$ can be neglected, as in the case of strings. In this case, $u_1 = O(u_2^2, u_3^2)$. Then, (3.4.53) can be approximated by

$$\frac{d}{ds}\left\{EA_0\left[\Lambda_1 u_1' + \Lambda_2\left(\alpha_3' u_3' + \frac{1}{2}u_2'^2 + \frac{1}{2}u_3'^2\right)\right]\right\} = 0 \qquad (3.5.1)$$

$$m_0 \ddot{u}_2 + 2c_2 \dot{u}_2 = \frac{d}{ds}\left\{EA_0\left[\Lambda_3 u_2' + \Lambda_2 u_2'\left(u_1' + \alpha_3' u_3' + \frac{1}{2}u_2'^2 + \frac{1}{2}u_3'^2\right)\right]\right\}$$
$$+ \tilde{f}_2 \qquad (3.5.2)$$

$$m_0 \ddot{u}_3 + 2c_3 \dot{u}_3 = \frac{d}{ds}\left\{EA_0\left[\Lambda_3 u_3' + \Lambda_2 \left(\alpha_3' + u_3'\right)\left(u_1' + \alpha_3' u_3'\right.\right.\right.$$
$$\left.\left.\left.+ \frac{1}{2}u_2'^2 + \frac{1}{2}u_3'^2\right)\right]\right\} + \tilde{f}_3 \qquad (3.5.3)$$

where $\alpha_2 = 0$ and

$$\Lambda_1 = (1 + \nu - \nu\alpha_1')(1 + 3\nu - 3\nu\alpha_1') \qquad (3.5.4)$$
$$\Lambda_2 = (1 + \nu - \nu\alpha_1')^3 / \alpha_1' \qquad (3.5.5)$$
$$\Lambda_3 = (\alpha_1' - 1)(1 + \nu - \nu\alpha_1'^2)/\alpha_1' \qquad (3.5.6)$$

Substituting for α_1' from (3.4.39) into (3.5.4)-(3.5.6), we find that

$$\Lambda_1 = 1 + O\left(\frac{F_1}{EA_0}\right), \quad \Lambda_2 = 1 + O\left(\frac{F_1}{EA_0}\right)$$

$$\Lambda_3 = \frac{F_1}{EA_0} + O\left(\frac{F_1^2}{E^2 A_0^2}\right) \tag{3.5.7}$$

For a uniform cable, using (3.5.7) and integrating (3.5.1) once with respect to s yields

$$u_1' + \alpha_3' u_3' + \frac{1}{2}u_2'^2 + \frac{1}{2}u_3'^2 = f(t) \tag{3.5.8}$$

Integrating (3.5.8) once with respect to s and imposing the boundary conditions $u_1 = 0$ at $s = 0$ and $s = L$, we find that

$$f(t) = \frac{1}{L}\int_0^L \left(\alpha_3' u_3' + \frac{1}{2}u_2'^2 + \frac{1}{2}u_3'^2\right) ds \tag{3.5.9}$$

Substituting (3.5.8) into (3.5.2) and (3.5.3) and using (3.5.9), we obtain

$$m_0 \ddot{u}_2 + 2c_2 \dot{u}_2 - F_1 u_2'' = \frac{EA_0}{L} u_2'' \int_0^L \left(\alpha_3' u_3' + \frac{1}{2}u_2'^2 + \frac{1}{2}u_3'^2\right) ds + \tilde{f}_2 \tag{3.5.10}$$

$$m_0 \ddot{u}_3 + 2c_3 \dot{u}_3 - F_1 u_3'' = \frac{EA_0}{L}(\alpha_3'' + u_3'')\int_0^L \left(\alpha_3' u_3' + \frac{1}{2}u_2'^2 + \frac{1}{2}u_3'^2\right) ds + \tilde{f}_3$$

$$\tag{3.5.11}$$

The boundary conditions (3.4.11) yield

$$u_i = 0 \text{ at } s = 0 \text{ and } L \tag{3.5.12}$$

Next, we rewrite (3.5.10)-(3.5.12) in nondimensional form by introducing the nondimensional variables

$$\hat{s} = \frac{s}{L}, \quad \hat{u}_i = \frac{u_i}{L}, \quad \hat{b} = \frac{b}{L}, \quad \hat{\alpha}_3 = \frac{\alpha_3}{b}, \quad \hat{t} = \sqrt{\frac{g}{8b}}t \tag{3.5.13}$$

where $\hat{b}$ is the sag-to-span ratio. Substituting (3.5.13) into (3.5.10)-(3.5.12) yields

$$\frac{\partial^2 \hat{u}_2}{\partial \hat{t}^2} + 2\hat{c}_2 \frac{\partial \hat{u}_2}{\partial \hat{t}^2} - \frac{\partial^2 \hat{u}_2}{\partial \hat{s}^2} - \alpha \frac{\partial^2 \hat{u}_2}{\partial \hat{s}^2} \int_0^1 \left[\hat{b}\frac{d\hat{\alpha}_3}{d\hat{s}}\frac{\partial \hat{u}_3}{\partial \hat{s}}\right.$$

$$\left. + \frac{1}{2}\left(\frac{\partial \hat{u}_2}{\partial \hat{s}}\right)^2 + \frac{1}{2}\left(\frac{\partial \hat{u}_3}{\partial \hat{s}}\right)^2 \right] d\hat{s} = \hat{f}_2 \tag{3.5.14}$$

$$\frac{\partial^2 \hat{u}_3}{\partial \hat{t}^2} + 2\hat{c}_3 \frac{\partial \hat{u}_3}{\partial \hat{t}} - \frac{\partial^2 \hat{u}_3}{\partial \hat{s}^2} - \alpha\left(\frac{\partial^2 \hat{u}_3}{\partial \hat{s}^2} + \hat{b}\frac{d^2 \hat{\alpha}_3}{d\hat{s}^2}\right)$$

$$\times \int_0^1 \left[\hat{b}\frac{d\hat{\alpha}_3}{d\hat{s}}\frac{\partial \hat{u}_3}{\partial \hat{s}} + \frac{1}{2}\left(\frac{\partial \hat{u}_2}{\partial \hat{s}}\right)^2 + \frac{1}{2}\left(\frac{\partial \hat{u}_3}{\partial \hat{s}}\right)^2\right] d\hat{s} = \hat{f}_3 \tag{3.5.15}$$

$$\hat{u}_i = 0 \text{ at } \hat{s} = 0 \text{ and } 1 \tag{3.5.16}$$

where
$$\alpha = \frac{EA_0}{F_1} = \frac{8bEA_0}{m_0 g L^2}, \quad \hat{f}_i = \frac{\tilde{f}_i L}{F_1}, \quad 2\hat{c}_i = \frac{c_i \sqrt{8b}}{m_0 \sqrt{g}} \qquad (3.5.17)$$

3.6 NATURAL FREQUENCIES AND MODES OF CABLES

For convenience of notation, we replace $\hat{u}_3$ with $\hat{u}_1$, $\hat{c}_3$ with $\hat{c}_1$, and $\hat{\alpha}_3$ with $-\hat{y}$, drop the overhat in (3.5.14)-(3.5.16), and rewrite the governing equations as

$$\ddot{u}_1 + 2c_1 \dot{u}_1 - u_1'' = \alpha \left(by'' + u_1'' \right) \int_0^1 \left[by' u_1' + \frac{1}{2} u_1'^2 + \frac{1}{2} u_2'^2 \right] ds + f_1 \qquad (3.6.1)$$

$$\ddot{u}_2 + 2c_2 \dot{u}_2 - u_2'' = \alpha u_2'' \int_0^1 \left[by' u_1' + \frac{1}{2} u_1'^2 + \frac{1}{2} u_2'^2 \right] ds + f_2 \qquad (3.6.2)$$

$$u_i = 0 \text{ at } s = 0 \text{ and } 1 \qquad (3.6.3)$$

$$y = -4s(1-s) \qquad (3.6.4)$$

We will refer, in what follows, to u_1 motions as inplane motions and u_2 motions as out-of-plane motions.

The natural frequencies and mode shapes can be determined by dropping the damping, forcing, and nonlinear terms in (3.6.1) and (3.6.2). The result is

$$\ddot{u}_1 - u_1'' + 32\alpha b^2 \int_0^1 (1 - 2s) u_1' ds = 0 \qquad (3.6.5)$$

$$\ddot{u}_2 - u_2'' = 0 \qquad (3.6.6)$$

subject to the boundary conditions (3.6.3).

It follows from (3.6.6) and (3.6.3) that the normalized out-of-plane mode shapes are given by

$$\chi_n(s) = \sqrt{2} \sin(n\pi s) \qquad (3.6.7)$$

corresponding to the natural frequencies

$$\lambda_n = n\pi \qquad (3.6.8)$$

for $n = 1, 2, 3, \ldots$. The mode shapes are normalized such that $\int_0^1 \chi_n^2(s) ds = 1$. When n is odd, the out-of-plane mode shapes are symmetric; and when n is even, the mode shapes are antisymmetric.

To determine the inplane natural frequencies and mode shapes, we let

$$u_1(s, t) = \phi(s) e^{i\omega t} \qquad (3.6.9)$$

in (3.6.5) and (3.6.3) and obtain the eigenvalue problem

$$\phi'' + \omega^2 \phi = 32\alpha b^2 \int_0^1 (1 - 2s) \phi' ds \qquad (3.6.10)$$

$$\phi = 0 \text{ at } s = 0 \text{ and } 1 \qquad (3.6.11)$$

To solve (3.6.10), we let
$$\Gamma = \int_0^1 (1-2s)\phi' ds \tag{3.6.12}$$
and obtain
$$\phi'' + \omega^2 \phi = 32\alpha b^2 \Gamma \tag{3.6.13}$$
The general solution of (3.6.13) can be expressed as
$$\phi = c_1 \cos(\omega s) + c_2 \sin(\omega s) + \frac{32\alpha b^2}{\omega^2}\Gamma \tag{3.6.14}$$
Imposing the boundary conditions (3.6.11) yields
$$c_1 + \frac{32\alpha b^2}{\omega^2}\Gamma = 0 \tag{3.6.15}$$
$$c_1 \cos\omega + c_2 \sin\omega + \frac{32\alpha b^2}{\omega^2}\Gamma = 0 \tag{3.6.16}$$
Substituting (3.6.14) into (3.6.12) and carrying out the integration, we obtain
$$c_1(2\sin\omega - \omega\cos\omega - \omega) + c_2(2 - 2\cos\omega - \omega\sin\omega) - \omega\Gamma = 0 \tag{3.6.17}$$

There are two possibilities: $\Gamma = 0$ or $\Gamma \ne 0$. When $\Gamma = 0$, it follows from (3.6.15) that $c_1 = 0$, and then it follows from (3.6.16) that $\omega = n\pi$, where n is an integer. In order that (3.6.17) be satisfied when $c_1 = 0$ and $\Gamma = 0$, $\cos\omega$ must be equal to 1, and hence
$$\omega_n = n\pi \text{ for } n = 2, 4, 6, \ldots \tag{3.6.18}$$
Consequently,
$$\phi_n = \sqrt{2}\sin(n\pi s) \text{ for } n = 2, 4, 6, \ldots \tag{3.6.19}$$
where ϕ_n is normalized so that $\int_0^1 \phi_n^2 ds = 1$. These mode shapes are antisymmetric.

When $\Gamma \ne 0$, setting the determinant of the coefficient matrix in (3.6.15)-(3.6.17) equal to zero and after some algebraic manipulations, we obtain
$$\tan\left(\frac{1}{2}\omega_n\right) = \frac{1}{2}\omega_n - \frac{\omega_n^3}{128\alpha b^2} \text{ for } 1, 3, 5, \ldots \tag{3.6.20}$$
Using (3.6.15) and (3.6.16) to eliminate c_1 and c_2 from (3.6.14), we have
$$\phi_n = \kappa_n \left[1 - \cos(\omega_n s) - \tan\left(\frac{1}{2}\omega_n\right)\sin(\omega_n s)\right] \text{ for } n = 1, 3, 5, \ldots \tag{3.6.21}$$
where the κ_n are chosen so that $\int_0^1 \phi_n^2 ds = 1$; that is,
$$\kappa_n^2 = \frac{2\omega_n \cos^2\left(\frac{1}{2}\omega_n\right)}{(2+\cos\omega_n)\omega_n - 3\sin\omega_n} \tag{3.6.22}$$
These mode shapes are symmetric.

150 STRINGS AND CABLES

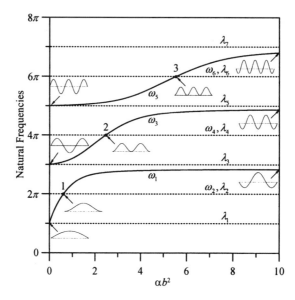

Fig. 3.6.1 Variation of the first few natural frequencies and mode shapes with αb^2.

In Figure 3.6.1, we present the first few natural frequencies as functions of αb^2. We also indicate the first three crossovers: the locations where the first inplane natural frequency crosses the second inplane and out-of-plane natural frequencies, the third inplane natural frequency crosses the fourth inplane and out-of-plane natural frequencies, and the fifth inplane natural frequency crosses the sixth inplane and out-of-plane natural frequencies. At these crossovers, there is the potential of activating two simultaneous one-to-one internal resonances and a two-to-one internal resonance. The dynamics near the first crossover were investigated by Rega et al. (1997, 1999) and Nayfeh et al. (2002).

3.7 DISCRETIZATION OF THE CABLE EQUATIONS

There are two approaches for determining approximate solutions of the cable equations: one based on attacking the integral-partial-differential equations and associated boundary conditions and one based on treating a discretized set of nonlinearly coupled ordinary-differential equations. In this section, we use the Galerkin procedure to discretize the cable equations.

We express the inplane and out-of-plane displacements as infinite series in terms of the linear mode shapes as

$$u_1 = \sum_{m=1}^{\infty} \eta_m(t)\phi_m(s) \qquad (3.7.1)$$

DISCRETIZATION OF THE CABLE EQUATIONS

$$u_2 = \sum_{m=1}^{\infty} \sqrt{2}\zeta_m(t)\sin(m\pi s) \qquad (3.7.2)$$

where the η_m and ζ_m are the generalized coordinates and the inplane mode shapes $\phi_m(s)$ are given by (3.6.19) for even m and by (3.6.21) for odd m. Substituting (3.7.1) and (3.7.2) into (3.6.1) and (3.6.2) yields

$$\sum_m \left[\ddot{\eta}_m\phi_m - \eta_m\phi_m'' + 2c_1\dot{\eta}_m\phi_m + \alpha b^2 y''\Gamma_m\eta_m\right] = \frac{1}{2}\alpha\left(y''b + \sum_m \eta_m\phi_m''\right)$$

$$\times \left[\sum_m m^2\pi^2\zeta_m^2 + \sum_{i,j}\Gamma_{ij}\eta_i\eta_j\right] - \alpha b \sum_i \phi_i''\eta_i \sum_j \Gamma_j\eta_j + f_1 \quad (3.7.3)$$

$$\sqrt{2}\sum_m (\ddot{\zeta}_m + 2c_2\dot{\zeta}_m + m^2\pi^2\zeta_m)\sin(m\pi s) = -\sqrt{2}\alpha \sum_m m^2\pi^2\zeta_m \sin(m\pi s)$$

$$\times \left[\sum_m \left(-b\eta_m\Gamma_m + \frac{1}{2}m^2\pi^2\zeta_m^2\right) + \frac{1}{2}\sum_{i,j}\Gamma_{ij}\eta_i\eta_j\right] + f_2 \quad (3.7.4)$$

where

$$\Gamma_m = \int_0^1 y''\phi_m ds = -\int_0^1 y'\phi_m' ds = \begin{cases} \frac{\kappa_m \omega_m^2}{8\alpha b^2} & \text{for } m = \text{odd} \\ 0 & \text{for } m = \text{even} \end{cases} \quad (3.7.5)$$

$$\Gamma_{ij} = \int_0^1 \phi_i'\phi_j' ds = -\int_0^1 \phi_i''\phi_j ds$$

$$= \begin{cases} -\dfrac{\kappa_i\kappa_j\omega_i^2\omega_j^2}{64\alpha b^2} & \text{for } i,j = \text{odd and } i \neq j \\[4pt] \dfrac{\kappa_k^2\omega_i^2(\omega_i - \sin\omega_i)}{2\cos^2(\frac{1}{2}\omega_i)} & \text{for } i = j = \text{odd} \\[4pt] i^2\pi^2 & \text{for } i = j = \text{even} \\[4pt] 0 & \text{for } i = \text{even}, j = \text{odd, and vice versa, or} \\ & \text{for } i,j = \text{even and } i \neq j \end{cases} \quad (3.7.6)$$

Multiplying (3.7.3) with $\phi_n(s)$ and (3.7.4) with $\sqrt{2}\sin(n\pi s)$, integrating the results with respect to s from $s = 0$ to $s = 1$, and using (3.6.20) and (3.6.21), we obtain

$$\ddot{\eta}_n + 2\mu_n\dot{\eta}_n + \omega_n^2\eta_n = \frac{1}{2}\alpha b \sum_{i,j}(2\Gamma_i\Gamma_{nj} + \Gamma_n\Gamma_{ij})\eta_i\eta_j + \frac{1}{2}\alpha b\Gamma_n \sum_i i^2\pi^2\zeta_i^2$$

$$- \frac{1}{2}\alpha \sum_{i,j,m}\Gamma_{nm}\eta_i\eta_j\eta_m - \frac{1}{2}\alpha \sum_{i,j} i^2\pi^2\Gamma_{nj}\eta_j\zeta_i^2 + f_{1n}(t) \quad (3.7.7)$$

$$\ddot{\zeta}_n + 2\nu_n \dot{\zeta}_n + n^2\pi^2\zeta_n = -\alpha n^2\pi^2\zeta_n \left[\sum_i \left(-b\Gamma_i \eta_i + \frac{1}{2}i^2\pi^2\zeta_i^2 \right) \right.$$
$$\left. + \frac{1}{2}\sum_{i,j} \Gamma_{ij}\eta_i\eta_j \right] + f_{2n}(t) \tag{3.7.8}$$

where

$$f_{1n}(t) = \int_0^1 f_1(s,t)\phi_n(s)ds \tag{3.7.9}$$

$$f_{2n}(t) = \sqrt{2}\int_0^1 f_2(s,t)\sin(n\pi s)ds \tag{3.7.10}$$

$$\mu_n = \int_0^1 c_1\phi_n^2(s)ds \tag{3.7.11}$$

$$\nu_n = 2\int_0^1 c_2 \sin^2(n\pi s)ds \tag{3.7.12}$$

3.8 SINGLE-MODE RESPONSE WITH DIRECT APPROACH

We consider the case of primary resonance of either an inplane or an out-of-plane mode in the absence of internal resonances. We determine a second-order uniform expansion of the cable response by using the method of multiple scales. Thus, we seek an approximate solution of (3.6.1)-(3.6.3) in the form

$$u_1(x,t) = \sum_{n=1}^{3} \epsilon^n u_{1n}(s, T_0, T_2) + \cdots \tag{3.8.1}$$

$$u_2(x,t) = \sum_{n=1}^{3} \epsilon^n u_{2n}(s, T_0, T_2) + \cdots \tag{3.8.2}$$

where $T_0 = t$, $T_2 = \epsilon^2 t$, and ϵ is a small nondimensional bookkeeping parameter. Moreover, we scale the c_i and f_i as $\epsilon^2 c_i$ and $\epsilon^3 f_i$. Then, substituting (3.8.1) and (3.8.2) into (3.6.1)-(3.6.3) and equating coefficients of like powers of ϵ, we obtain

Order ϵ:

$$\mathcal{L}_1(u_{11}) = D_0^2 u_{11} - u_{11}'' - \alpha b^2 y'' \int_0^1 y' u_{11}' ds = 0 \tag{3.8.3}$$

$$\mathcal{L}_2(u_{21}) = D_0^2 u_{21} - u_{21}'' = 0 \tag{3.8.4}$$

$$u_{i1} = 0 \text{ at } s = 0 \text{ and } s = 1 \tag{3.8.5}$$

Order ϵ^2:

$$\mathcal{L}_1(u_{12}) = \alpha b u_{11}'' \int_0^1 y' u_{11}' ds + \frac{1}{2}\alpha b y'' \int_0^1 (u_{11}'^2 + u_{21}'^2) ds \tag{3.8.6}$$

$$\mathcal{L}_2(u_{22}) = \alpha b u_{21}'' \int_0^1 y' u_{11}' \, ds \qquad (3.8.7)$$

$$u_{i2} = 0 \text{ at } s = 0 \text{ and } s = 1 \qquad (3.8.8)$$

Order ϵ^3:

$$\mathcal{L}_1(u_{13}) = -2D_0 D_2 u_{11} - 2c_1 D_0 u_{11} + \alpha b u_{11}'' \int_0^1 y' u_{12}' \, ds$$

$$+ \alpha b u_{12}'' \int_0^1 y' u_{11}' \, ds + \alpha b y'' \int_0^1 (u_{11}' u_{12}' + u_{21}' u_{22}') \, ds$$

$$+ \frac{1}{2} \alpha u_{11}'' \int_0^1 (u_{11}'^2 + u_{21}'^2) \, ds + f_1 \qquad (3.8.9)$$

$$\mathcal{L}_2(u_{23}) = -2D_0 D_2 u_{21} - 2c_2 D_0 u_{21} + \alpha b u_{21}'' \int_0^1 y' u_{12}' \, ds$$

$$+ \alpha b u_{22}'' \int_0^1 y' u_{11}' \, ds + \frac{1}{2} \alpha u_{21}'' \int_0^1 (u_{11}'^2 + u_{21}'^2) \, ds + f_2 \qquad (3.8.10)$$

$$u_{i3} = 0 \text{ at } s = 0 \text{ and } 1 \qquad (3.8.11)$$

Next, we consider the case of primary resonance of either an inplane mode or an out-of-plane mode, starting with the first.

3.8.1 Primary Resonance of an Inplane Mode

To this end, we set $f_2 = 0$ and

$$f_1 = F_1 \cos(\Omega T_0) \qquad (3.8.12)$$

where

$$\Omega = \omega_k + \epsilon^2 \sigma \qquad (3.8.13)$$

and assume that the kth mode is not involved in an internal resonance with any other mode. Consequently, because of the presence of damping, free responses of all of the modes except the kth inplane mode decay with time. Therefore, we express the solutions of (3.8.3)-(3.8.5) as

$$u_{11} = \phi_k(s)\left[A_k(T_2) e^{i\omega_k T_0} + \bar{A}_k(T_2) e^{-i\omega_k T_0}\right] \qquad (3.8.14)$$

$$u_{21} = 0 \qquad (3.8.15)$$

Substituting (3.8.14) and (3.8.15) into (3.8.6) and (3.8.7) yields

$$\mathcal{L}_1(u_{12}) = \alpha b \left(\frac{1}{2} y'' \Gamma_{kk} - \Gamma_k \phi_k''\right) \left(A_k^2 e^{2ikT_0} + 2A_k \bar{A}_k + \bar{A}_k^2 e^{-2ikT_0}\right) \qquad (3.8.16)$$

$$\mathcal{L}_2(u_{22}) = 0 \qquad (3.8.17)$$

The solution of (3.8.17) and (3.8.8) is $u_{22} = 0$ and the solution of (3.8.16) and (3.8.8) can be expressed as

$$u_{12} = \alpha b g_1(s)\left(A_k^2 e^{2i\omega_k T_0} + \bar{A}_k^2 e^{-2i\omega_k T_0}\right) + \alpha b g_2(s) A_k \bar{A}_k \qquad (3.8.18)$$

where

$$g_1'' + 4\omega_k^2 g_1 + \alpha b^2 y'' \int_0^1 y' g_1' ds = \Gamma_k \phi_k'' - \frac{1}{2}\Gamma_{kk} y'' \qquad (3.8.19)$$

$$g_2'' + \alpha b^2 y'' \int_0^1 y' g_2' ds = 2\Gamma_k \phi_k'' - \Gamma_{kk} y'' \qquad (3.8.20)$$

$$g_i = 0 \text{ at } s = 0 \text{ and } 1 \qquad (3.8.21)$$

To determine the solution of (3.8.19) and (3.8.21), we follow a procedure similar to that used in Section 3.6. Thus, we let

$$c_1 = \int_0^1 (1-2s) g_1' ds \qquad (3.8.22)$$

and rewrite (3.8.19) as

$$g_1'' + 4\omega_k^2 g_1 = 32\alpha b^2 c_1 - 4\Gamma_{kk} + \Gamma_k \phi_k'' \qquad (3.8.23)$$

The Case of an Antisymmetric Mode: For an antisymmetric mode, $\omega_k = k\pi$, where k is an even integer. Hence, $\Gamma_k = 0$, $\sin(2\omega_k) = 0$, and $\cos(2\omega_k) = 1$. Then, the general solution of (3.8.19) can be expressed as

$$g_1 = c_2 \sin(2\omega_k s) + c_3 \cos(2\omega_k s) + \frac{8\alpha b^2 c_1 - \Gamma_{kk}}{\omega_k^2} \qquad (3.8.24)$$

where c_1, c_2 and c_3 are constants. Substituting (3.8.24) into (3.8.22) yields

$$c_1 = -2c_3 \qquad (3.8.25)$$

Imposing the condition $g_1(0) = 0$ yields

$$c_3 + \frac{8\alpha b^2}{\omega_k^2} c_1 - \frac{\Gamma_{kk}}{\omega_k^2} = 0 \qquad (3.8.26)$$

Solving (3.8.25) and (3.8.26) for c_1 and c_3, we have

$$c_3 = -\frac{\Gamma_{kk}}{16\alpha b^2 - \omega_k^2} \text{ and } c_1 = \frac{2\Gamma_{kk}}{16\alpha b^2 - \omega_k^2} \qquad (3.8.27)$$

Imposing the boundary condition $g_1(1) = 0$ yields (3.8.26) and hence c_2 is arbitrary. To determine $g_1(s)$ uniquely, we require $g_1(s)$ to be orthogonal to the adjoint of the homogeneous problem governing $g_1(s)$.

SINGLE-MODE RESPONSE WITH DIRECT APPROACH

To determine the adjoint, we multiply the homogeneous part of (3.8.19) with $G_1(s)$, integrate the result from $s = 0$ to 1, and obtain

$$\int_0^1 g_1'' G_1 ds + 4\omega_k^2 \int_0^1 g_1 G_1 ds + \alpha b^2 \left(\int_0^1 y' g_1' ds\right)\left(\int_0^1 y'' G_1 ds\right) = 0 \quad (3.8.28)$$

Integrating the first term in (3.8.28) by parts to transfer the derivatives from g_1 to G_1 and using the boundary conditions (3.8.21), we obtain

$$\int_0^1 g_1'' G_1 ds = g_1' G_1 \Big|_0^1 + \int_0^1 G_1'' g_1 ds \quad (3.8.29)$$

Next, we integrate each of the integrals in the last term in (3.8.28) once by parts, use the boundary conditions (3.8.21), and obtain

$$\int_0^1 y' g_1' ds = -\int_0^1 y'' g_1 ds \quad (3.8.30)$$

$$\int_0^1 y'' G_1 ds = G_1 y' \Big|_0^1 - \int_0^1 y' G_1' ds \quad (3.8.31)$$

Substituting (3.8.29)-(3.8.31) into (3.8.28), we have

$$\int_0^1 \left[G_1'' + 4\omega_k^2 G_1 + \alpha b^2 y'' \int_0^1 y' G_1' ds\right] g_1 ds + g_1' G_1 \Big|_0^1$$
$$- 4[G_1(1) - G_1(0)] \int_0^1 y'' g_1 ds = 0 \quad (3.8.32)$$

The equation governing the adjoint is obtained by setting the coefficient of g_1 in the integrand of the first integral equal to zero. The result is

$$G_1'' + 4\omega_k^2 G_1 + \alpha b^2 y'' \int_0^1 y' G_1' ds = 0 \quad (3.8.33)$$

Then, because g_1' and the integral in the last term are, in general, not zero, we require

$$G_1 = 0 \quad \text{at} \quad s = 0 \text{ and } 1 \quad (3.8.34)$$

Comparing (3.8.33) and (3.8.34) with the homogeneous parts of (3.8.19) and (3.8.21), we find that they are the same, and the problem is self-adjoint.

Following a procedure similar to that used to solve for g_1, we find that $G_1 \propto \sin(2\omega_k s)$. Consequently, requiring (3.8.24) be orthogonal to G_1, we obtain $c_2 = 0$ and hence g_1 becomes

$$g_1(s) = \frac{\Gamma_{kk}}{16\alpha b^2 - \omega_k^2} [1 - \cos(2\omega_k s)] \quad (3.8.35)$$

With $\Gamma_k = 0$, the general solution of (3.8.20) can be expressed as

$$g_2 = 4(4\alpha b^2 c_1 - \Gamma_{kk})s^2 + c_4 s + c_5 \tag{3.8.36}$$

where c_1 is defined in (3.8.22) and c_4 and c_5 are constants. Imposing the boundary conditions (3.8.21) yields $c_5 = 0$ and

$$c_4 + 16\alpha b^2 c_1 - 4\Gamma_{kk} = 0 \tag{3.8.37}$$

Substituting (3.8.36) into (3.8.22) yields

$$c_1 + \frac{16}{3}\alpha b^2 c_1 - \frac{4}{3}\Gamma_{kk} = 0 \tag{3.8.38}$$

Solving (3.8.38) for c_1, we have

$$c_1 = \frac{4\Gamma_{kk}}{3 + 16\alpha b^2} \tag{3.8.39}$$

Then, it follows from (3.8.37) that

$$c_4 = \frac{12\Gamma_{kk}}{3 + 16\alpha b^2} \tag{3.8.40}$$

Finally,

$$g_2(s) = \frac{12\Gamma_{kk}}{3 + 16\alpha b^2}(s - s^2) \tag{3.8.41}$$

The Case of a Symmetric Mode: For a symmetric mode, ϕ_k is given by (3.6.21) and Γ_k and Γ_{kk} are defined in (3.7.5) and (3.7.6). Substituting for ϕ_k in (3.8.23) yields

$$g_1'' + 4\omega_k^2 g_1 = 32\alpha b^2 c_1 - 4\Gamma_{kk} + \Gamma_k \kappa_k \omega_k^2 \left[\cos(\omega_k s) + \tan\left(\frac{1}{2}\omega_k\right)\sin(\omega_k s)\right] \tag{3.8.42}$$

The general solution of (3.8.42) can be expressed as

$$g_1 = c_2 \sin(2\omega_k s) + c_3 \cos(2\omega_k s) + \frac{8\alpha b^2 c_1 - \Gamma_{kk}}{\omega_k^2}$$
$$+ \frac{1}{3}\Gamma_k \kappa_k \left[\cos(\omega_k s) + \tan\left(\frac{1}{2}\omega_k\right)\sin(\omega_k s)\right] \tag{3.8.43}$$

Imposing the boundary conditions (3.8.21) yields

$$c_3 + \frac{8\alpha b^2}{\omega_k^2} c_1 = \frac{\Gamma_{kk}}{\omega_k^2} - \frac{1}{3}\Gamma_k \kappa_k \tag{3.8.44}$$

$$c_2 \sin(2\omega_k) + c_3 \cos(2\omega_k) + \frac{8\alpha b^2}{\omega_k^2} c_1 = \frac{\Gamma_{kk}}{\omega_k^2} - \frac{1}{3}\Gamma_k \kappa_k \left[\cos \omega_k \right.$$
$$\left. + \tan\left(\frac{1}{2}\omega_k\right)\sin \omega_k\right] \tag{3.8.45}$$

Substituting (3.8.43) into (3.8.22) yields

$$c_1 + \left[\sin(2\omega_k) - \frac{1-\cos(2\omega_k)}{\omega_k}\right]c_2 + \left[1 + \cos(2\omega_k) - \frac{\sin(2\omega_k)}{\omega_k}\right]c_3$$
$$= \frac{2\Gamma_k\kappa_k}{3\omega_k}\left[2\tan\left(\frac{1}{2}\omega_k\right) - \omega_k\right] \quad (3.8.46)$$

Solving (3.8.44)-(3.8.46), we have

$$c_1 = \frac{6\Gamma_{kk}(\sin\omega_k - \omega_k\cos\omega_k) - 4\Gamma_k\kappa_k\omega_k^2\tan\left(\frac{1}{2}\omega_k\right)\sin^2\left(\frac{1}{2}\omega_k\right)}{3\cos\omega_k[\omega_k^3 + 16\alpha b^2(\tan\omega_k - \omega_k)]} \quad (3.8.47)$$

$$c_2 = \frac{3\Gamma_{kk}\omega_k + \Gamma_k\kappa_k\left[16\alpha b^2\omega_k - \omega_k^3 - 32\alpha b^2\tan\left(\frac{1}{2}\omega_k\right)\right]}{3\left[\omega_k^3 + 16\alpha b^2(\tan\omega_k - \omega_k)\right]} \quad (3.8.48)$$

$$c_3 = \frac{3\Gamma_{kk}\omega_k\cos\omega_k + 32\alpha b^2\tan\left(\frac{1}{2}\omega_k\right)\sin^2\left(\frac{1}{2}\omega_k\right)}{3\cos\omega_k[\omega_k^3 + 16\alpha b^2(\tan\omega_k - \omega_k)]} \quad (3.8.49)$$

Substituting for ϕ_k and y in (3.8.20) yields

$$g_2'' = 32\alpha b^2 c_1 - 8\Gamma_{kk} + 2\Gamma_k\kappa_k\omega_k^2\left[\cos(\omega_k s) + \tan\left(\frac{1}{2}\omega_k\right)\sin(\omega_k s)\right] \quad (3.8.50)$$

The general solution of (3.8.50) can be expressed as

$$g_2 = 4\left(4\alpha b^2 c_1 - \Gamma_{kk}\right)s^2 + b_1 s + b_2 - 2\Gamma_k\kappa_k\left[\cos(\omega_k s) + \tan\left(\frac{1}{2}\omega_k\right)\sin(\omega_k s)\right] \quad (3.8.51)$$

where b_1 and b_2 are constants. Imposing the boundary conditions (3.8.21) yields

$$b_2 = 2\Gamma_k\kappa_k \quad (3.8.52)$$

$$16\alpha b^2 c_1 + b_1 + b_2 = 4\Gamma_{kk} - 2\Gamma_k\kappa_k\left[\cos\omega_k + \tan\left(\frac{1}{2}\omega_k\right)\sin\omega_k\right] \quad (3.8.53)$$

Substituting (3.8.51) into (3.8.22), we have

$$c_1 = \frac{4\Gamma_{kk}\omega_k + 12\Gamma_k\kappa_k\left[\omega_k - 2\tan\left(\frac{1}{2}\omega_k\right)\right]}{(3 + 16\alpha b^2)\omega_k} \quad (3.8.54)$$

Substituting (3.8.52) and (3.8.54) into (3.8.53), solving for b_1, and using (3.6.20), we obtain

$$b_1 = \frac{3\left(\Gamma_{kk} - \Gamma_k\kappa_k\omega_k^2\right)}{3 + 16\alpha b^2} \quad (3.8.55)$$

Substituting (3.8.14), (3.8.15), and (3.8.18) into (3.8.9) yields

$$\mathcal{L}_1(u_{13}) = -2i\omega_k\left(A_k' + c_1 A_k\right)\phi_k e^{i\omega_k T_0}$$
$$+ \left\{\alpha^2 b^2\left[\phi_k''\int_0^1 y'(g_1' + g_2')\,ds + (g_1'' + g_2'')\int_0^1 y'\phi_k'\,ds\right.\right.$$

$$+y'' \int_0^1 (g_1' + g_2') \phi_k' ds \Big] + \frac{3}{2}\alpha\phi_k'' \int_0^1 \phi_k'^2 ds \Big\} A_k^2 \bar{A}_k e^{i\omega_k T_0}$$
$$+\frac{1}{2}F_1 e^{i\Omega T_0} + cc + NST \qquad (3.8.56)$$

Because the homogeneous problem given by (3.8.56) and (3.8.11) has a nontrivial solution, namely (3.8.14), the nonhomogeneous problem has a solution only if the right-hand side of (3.8.56) is orthogonal to every solution of the adjoint homogeneous problem. Following a procedure similar to that used above, one finds that the homogeneous problem given by (3.8.56) and (3.8.11) is self-adjoint. Hence, the solvability condition of the nonhomogeneous problem demands that the right-hand side of (3.8.56) be orthogonal to $\phi_k \exp(-i\omega_k T_0)$. Using (3.8.13) and enforcing the solvability condition, we obtain

$$2i\omega_k (A_k' + \mu_k A_k) + \alpha_{ek}^* A_k^2 \bar{A}_k - \frac{1}{2}F_k e^{i\sigma T_2} = 0 \qquad (3.8.57)$$

where the effective nonlinearity α_{ek}^* is given by

$$\alpha_{ek}^* = \frac{3}{2}\alpha\Gamma_{kk}^2 + \alpha^2 b^2 \Gamma_{kk} \int_0^1 y' (g_1' + g_2') ds - 2\alpha^2 b^2 \Gamma_k \int_0^1 (g_1' + g_2') \phi_k' ds \qquad (3.8.58)$$

and μ_k and F_k are defined in (3.7.9) and (3.7.11). In (3.8.58), we made use of (3.7.5) and (3.7.6).

3.8.2 Primary Resonance of an out-of-Plane Mode

In this case, $f_1 = 0$ and
$$f_2 = F_2 \cos(\Omega T_0) \qquad (3.8.59)$$

where
$$\Omega = \lambda_m + \epsilon^2 \sigma \qquad (3.8.60)$$

The eigenfunctions are given by (3.6.7) and $\lambda_m = m\pi$, where m is an integer. We assume that the mth out-of-plane mode is not involved in an internal resonance with any inplane or out-of-plane mode.

Because, in the presence of damping, free responses of all of the modes except the mth out-of-plane mode decay with time, the solutions of (3.8.3)-(3.8.5) can be expressed as

$$u_{11} = 0 \qquad (3.8.61)$$
$$u_{21} = \phi_m(s) \left[A_m(T_2) e^{i\lambda_m T_0} + \bar{A}_m(T_2) e^{-i\lambda_m T_0} \right] \qquad (3.8.62)$$

Substituting (3.8.61) and (3.8.62) into (3.8.6) and (3.8.7), we have

$$\mathcal{L}_1(u_{12}) = 4\alpha b m^2 \pi^2 \left[A_m^2 e^{2i\lambda_m T_0} + 2A_m \bar{A}_m + \bar{A}_m^2 e^{-2i\lambda_m T_0} \right] \qquad (3.8.63)$$
$$\mathcal{L}_2(u_{22}) = 0 \qquad (3.8.64)$$

SINGLE-MODE RESPONSE WITH DIRECT APPROACH

The solution of (3.8.64) and (3.8.8) is

$$u_{22} = 0 \qquad (3.8.65)$$

and the solution of (3.8.63) and (3.8.8) can be expressed as

$$u_{12} = \alpha b m^2 \pi^2 h_1(s) \left[A_m^2 e^{2i\lambda_m T_0} + \bar{A}_m^2 e^{-2i\lambda_m T_0} \right] + \alpha b m^2 \pi^2 h_2(s) A_m \bar{A}_m \qquad (3.8.66)$$

where

$$h_1'' + 4m^2\pi^2 h_1 - 32\alpha b^2 \int_0^1 (1-2s) h_1' ds = -4 \qquad (3.8.67)$$

$$h_2'' - 32\alpha b^2 \int_0^1 (1-2s) h_2' ds = -8 \qquad (3.8.68)$$

$$h_i = 0 \text{ at } s = 0 \text{ and } 1 \qquad (3.8.69)$$

To determine the solutions of (3.8.67)-(3.8.69), we let

$$c_1 = \int_0^1 (1-2s) h_1' ds \qquad (3.8.70)$$

Then, the general solution of (3.8.67) can be expressed as

$$h_1(s) = c_2 \sin(2m\pi s) + c_3 \cos(2m\pi s) + \frac{8}{m^2\pi^2} c_1 - \frac{1}{m^2\pi^2} \qquad (3.8.71)$$

Imposing the boundary condition $h_1(0) = 0$ gives

$$c_3 + \frac{8\alpha b^2}{m^2\pi^2} c_1 = \frac{1}{m^2\pi^2} \qquad (3.8.72)$$

Substituting (3.8.71) into (3.8.70) yields

$$2c_3 + c_1 = 0 \qquad (3.8.73)$$

Solving (3.8.72) and (3.8.73) yields

$$c_1 = -\frac{2}{m^2\pi^2 - 16\alpha b^2} \text{ and } c_3 = \frac{1}{m^2\pi^2 - 16\alpha b^2} \qquad (3.8.74)$$

Imposing the boundary condition $h_1(1) = 0$ gives (3.8.72) and hence c_2 is arbitrary. To uniquely determine $h_1(s)$, we require it to be orthogonal to every solution of the adjoint homogeneous problem (3.8.67) and (3.8.69). Following a procedure similar to that used in the preceding section, we find that the solutions of the adjoint problem are $\sin(2m\pi s)$. Consequently, demanding that $h_1(s)$ be orthogonal to $\sin(2m\pi s)$ yields $c_2 = 0$, and hence

$$h_1(s) = \frac{1}{m^2\pi^2 - 16\alpha b^2} [\cos(2m\pi s) - 1] \qquad (3.8.75)$$

The general solution of (3.8.68) is

$$h_2(s) = \left(16\alpha b^2 c_1 - 4\right) s^2 + b_1 s + b_0 \tag{3.8.76}$$

where b_0 and b_1 are constants. Imposing the boundary conditions (3.8.69) yields $b_0 = 0$ and

$$b_1 + 32\alpha b^2 c_1 = 8 \tag{3.8.77}$$

Substituting (3.8.76) into (3.8.70), we have

$$c_1 = \frac{4}{3 + 16\alpha b^2} \tag{3.8.78}$$

Substituting (3.8.78) into (3.8.77) yields

$$b_1 = \frac{12}{3 + 16\alpha b^2} \tag{3.8.79}$$

Therefore,

$$h_2(s) = \frac{12}{3 + 16\alpha b^2} \left(s - s^2\right) \tag{3.8.80}$$

Substituting (3.8.61), (3.8.62), (3.8.65), and (3.8.66) into (3.8.10), we obtain

$$\mathcal{L}_2(u_{23}) = -2i\lambda_m \left(A'_m + c_2 A_m\right) \phi_m e^{i\lambda_m T_0} + \frac{3}{2}\alpha m^2 \pi^2 \phi''_m A_m^2 \bar{A}_m e^{i\lambda_m T_0}$$
$$+ \alpha^2 b^2 m^2 \pi^2 \phi''_m A_m^2 \bar{A}_m \left[\int_0^1 y'\left(h'_1 + h'_2\right) ds\right] e^{i\lambda_m T_0}$$
$$+ \frac{1}{2} F_m e^{i\Omega T_0} + cc + NST \tag{3.8.81}$$

Because the homogeneous problem (3.8.81) and (3.8.69) has a nontrivial solution, the nonhomogeneous problem has a solution only if the right-hand side of (3.8.81) is orthogonal to every solution of the adjoint homogeneous problem, which in this case is $\chi_n(s) \exp(\pm i\lambda_n T_0)$. Using (3.8.60) and imposing the solvability condition, we obtain

$$2i\lambda_m \left(A'_m + \mu_m A_m\right) + \alpha^*_{em} A_m^2 \bar{A}_m - \frac{1}{2} F_m e^{i\sigma T_2} = 0 \tag{3.8.82}$$

where the effective nonlinearity α^*_{em} is given by

$$\alpha^*_{em} = \frac{3}{2}\alpha^4 \pi^4 + \alpha^2 b^2 m^4 \pi^4 \int_0^1 y'\left(h'_1 + h'_2\right) ds \tag{3.8.83}$$

and μ_m and F_m are defined in (3.7.10), (3.7.11), and (3.8.59).

3.9 SINGLE-MODE RESPONSE WITH DISCRETIZATION APPROACH

The forced planar dynamics of suspended cables near primary resonance were investigated by Benedettini and Rega (1987). They used a single-mode Galerkin discretization procedure and then applied the method of multiple scales to calculate a second-order asymptotic expansion of the solution. They found out that the response may be either hardening or softening, depending on the amplitude of oscillation.

In this section, we explore the consequences of using the discretization approach by investigating the response of the cable when only one specific (either inplane or out-of-plane) mode is directly excited by the forcing. We further assume that this excited mode is not involved in an internal resonance with any of the other modes. We apply the method of multiple scales to the discretized set of equations to determine second-order approximate solutions of the response. Then, we compare the solutions obtained when using single-mode and multimode discretizations with each other and with the solutions obtained in Section 3.8 via the direct approach.

We consider the case in which either an inplane mode or an out-of-plane mode is excited by a primary resonance; that is,

$$f_{1k}(t) = F_k \cos(\Omega t) \tag{3.9.1}$$

where F_k is a constant and $\Omega \approx \omega_k$ or $\Omega \approx k\pi = \lambda_k$. And we assume that this mode is not involved in an internal resonance with any other mode. We use the method of multiple scales and seek uniform expansions of the form

$$\eta_n(t;\epsilon) = \sum_{i=1}^{3} \epsilon^i \eta_{ni}(T_0, T_2) + \cdots \tag{3.9.2}$$

$$\zeta_n(t;\epsilon) = \sum_{i=1}^{3} \epsilon^i \zeta_{ni}(T_0, T_2) + \cdots \tag{3.9.3}$$

where $T_0 = t$, $T_2 = \epsilon^2 t$, and ϵ is a small nondimensional bookkeeping parameter. Moreover, we scale μ_n, ν_n, and F_k as $\epsilon^2 \mu_n$, $\epsilon^2 \nu_n$, and $\epsilon^3 F_k$.

Substituting (3.9.1)-(3.9.3) into (3.7.7) and (3.7.8) and equating coefficients of like powers of ϵ, we obtain

Order ϵ:

$$D_0^2 \eta_{n1} + \omega_n^2 \eta_{n1} = 0 \tag{3.9.4}$$

$$D_0^2 \zeta_{n1} + n^2\pi^2 \zeta_{n1} = 0 \tag{3.9.5}$$

Order ϵ^2:

$$D_0^2 \eta_{n2} + \omega_n^2 \eta_{n2} = \frac{1}{2}\alpha b \sum_{i,j} (2\Gamma_i \Gamma_{nj} + \Gamma_n \Gamma_{ij}) \eta_{i1} \eta_{j1}$$

$$+ \frac{1}{2}\alpha b \Gamma_n \sum_i i^2 \pi^2 \zeta_{i1}^2 \tag{3.9.6}$$

$$D_0^2 \zeta_{n2} + n^2\pi^2 \zeta_{n2} = \alpha b n^2 \pi^2 \zeta_{n1} \sum_m \Gamma_m \eta_{m1} \qquad (3.9.7)$$

Order ϵ^3:

$$\begin{aligned} D_0^2 \eta_{n3} + \omega_n^2 \eta_{n3} = & -2D_0 D_2 \eta_{n1} - 2\mu_n D_0 \eta_{n1} \\ & + \frac{1}{2}\alpha b \sum_{i,j} (2\Gamma_i \Gamma_{nj} + \Gamma_n \Gamma_{ij})(\eta_{i1}\eta_{j2} + \eta_{i2}\eta_{j1}) \\ & + \alpha b \Gamma_n \sum_i i^2 \pi^2 \zeta_{i1}\zeta_{i2} - \frac{1}{2}\alpha \sum_{i,j,m} \Gamma_{nm}\Gamma_{ij}\eta_{i1}\eta_{j1}\eta_{m1} \\ & - \frac{1}{2}\alpha \sum_{i,j} i^2 \pi^2 \Gamma_{nj} \eta_{j1} \zeta_{i1}^2 + \delta_{nk} F_k \cos(\Omega T_0) \end{aligned} \qquad (3.9.8)$$

$$\begin{aligned} D_0^2 \zeta_{n3} + n^2\pi^2 \zeta_{n3} = & -2D_0 D_2 \zeta_{n1} - 2\nu_n D_0 \zeta_{n1} \\ & + \alpha b n^2 \pi^2 \zeta_{n1} \sum_m \Gamma_m \eta_{m2} + \alpha b n^2 \pi^2 \zeta_{n2} \sum_m \Gamma_m \eta_{m1} \\ & - \frac{1}{2}\alpha n^2 \pi^2 \zeta_{n1} \left[\sum_i i^2 \pi^2 \zeta_{i1}^2 + \sum_{i,j} \Gamma_{ij} \eta_{i1} \eta_{j1} \right] \\ & + \delta_{nk} F_k \cos(\Omega T_0) \end{aligned} \qquad (3.9.9)$$

3.9.1 The Case of an Inplane Mode

In the presence of damping, the homogeneous solutions of all modes that are not directly or indirectly excited decay with time. Therefore, the solution of the first-order problem (3.9.4) and (3.9.5) can be expressed as

$$\eta_{k1} = A_k(T_2) e^{i\omega_k T_0} + cc \qquad (3.9.10)$$
$$\eta_{n1} = 0 \text{ for } n \neq k \qquad (3.9.11)$$
$$\zeta_{n1} = 0 \text{ for all } n \qquad (3.9.12)$$

Next, we consider two cases: a symmetric inplane mode and an antisymmetric inplane mode. Substituting (3.9.10)-(3.9.12) into (3.9.6) and (3.9.7), we obtain

$$D_0^2 \eta_{n2} + \omega_n^2 \eta_{n2} = \frac{1}{2}\alpha b \left(2\Gamma_k \Gamma_{nk} + \Gamma_n \Gamma_{kk}\right)\left(A_k^2 e^{2i\omega_k T_0} + A_k \bar{A}_k\right) + cc \qquad (3.9.13)$$
$$D_0^2 \zeta_{n2} + n^2\pi^2 \zeta_{n2} = 0 \qquad (3.9.14)$$

for all n. The particular solutions of (3.9.13) and (3.9.14) can be expressed as

$$\eta_{n2} = \frac{1}{2}\alpha b \left(2\Gamma_k \Gamma_{nk} + \Gamma_n \Gamma_{kk}\right)\left(\frac{A_k^2}{\omega_n^2 - 4\omega_k^2} e^{2i\omega_k T_0} + \frac{A_k \bar{A}_k}{\omega_n^2}\right) + cc \qquad (3.9.15)$$
$$\zeta_{n2} = 0 \text{ for all } n \qquad (3.9.16)$$

It is clear from (3.9.15) that the single-mode solution breaks down when $\omega_k \approx \frac{1}{2}\omega_n$ for any n. This is the case of two-to-one internal resonance.

To express the nearness of Ω to ω_k, we introduce the detuning parameter σ defined by

$$\Omega = \omega_k + \epsilon^2 \sigma \tag{3.9.17}$$

Substituting (3.9.11), (3.9.12), and (3.9.16) into (3.9.8) with $n = k$ yields

$$D_0^2 \eta_{k3} + \omega_k^2 \eta_{k3} = -2i\omega_k (A_k' + \mu_k A_k) + \alpha b \sum_n (\Gamma_n \Gamma_{kk} + 2\Gamma_k \Gamma_{nk}) \eta_{k1} \eta_{n2}$$

$$-\frac{1}{2}\alpha \Gamma_{kk}^2 \eta_{k1}^3 + F_k \cos(\Omega T_0) \tag{3.9.18}$$

Substituting (3.9.10), (3.9.15), and (3.9.17) into (3.9.18) and eliminating the terms that lead to secular terms, we obtain

$$2i\omega_k \left(A_k' + \mu_k A_k \right) + \alpha_{ek} A_k^2 \bar{A}_k - \frac{1}{2} F_k e^{i\sigma T_2} = 0 \tag{3.9.19}$$

where the effective nonlinearity coefficient α_{ek} is given by

$$\alpha_{ek} = \frac{3}{2}\alpha \Gamma_{kk}^2 - \frac{1}{2}\alpha^2 b^2 \sum_n \left(\Gamma_n \Gamma_{kk} + 2\Gamma_k \Gamma_{nk} \right)^2 \frac{3\omega_n^2 - 8\omega_k^2}{\omega_n^2 (\omega_k^2 - 4\omega_k^2)} \tag{3.9.20}$$

When $\alpha_{ek} > 0$, the frequency-response curves are bent to the right and the effective nonlinearity is of the hardening type. On the other hand, when $\alpha_{ek} < 0$, the frequency-response curves are bent to the left and the effective nonlinearity is of the softening type.

If the kth mode is symmetric (i.e., k is odd), then substituting for the Γ's from (3.7.5) and (3.7.6) into (3.9.20), we have

$$\alpha_{ek} = \left(\frac{3}{2}\alpha - \frac{5\kappa_k^2 \omega_k^2}{128 b^2} \right) \frac{\kappa_k^4 \omega_k^2 (\omega_k - \sin \omega_k)^2}{4 \cos^4 \left(\frac{1}{2}\omega_k \right)}$$

$$- \sum_{n \neq k} \frac{\kappa_k^4 \kappa_n^2 \omega_k^2 \omega_n^2 (3\omega_n^2 - 8\omega_k^2) \left[32\alpha b^2 (\sin \omega_k - \omega_k) + (1 + \cos \omega_k) \omega_k^3 \right]^2}{524288 \alpha^2 b^6 (\omega_n^2 - 4\omega_k^2) \cos^4 \left(\frac{1}{2}\omega_k \right)} \tag{3.9.21}$$

where n is odd. On the other hand, if the kth mode is antisymmetric (i.e., k is even), then substituting for the Γ's from (3.7.5) and (3.7.6) into (3.9.20) yields

$$\alpha_{ek} = \frac{3}{2}\alpha k^4 \pi^4 - \frac{k^4 \pi^4}{128 \alpha b^2} \sum_{n=\text{odd}} \frac{\kappa_n^2 \omega_n^2 (3\omega_n^2 - 8k^2 \pi^2)}{\omega_n^2 - 4k^2 \pi^2} \tag{3.9.22}$$

Looking at the expressions for the effective nonlinearities presented in (3.9.21) and (3.9.22), we note that they consist of two parts. The first part is the effective nonlinearity coefficient one would obtain if a single-mode discretization is used. The second part is the summation of the contributions of all of the other modes to the effective nonlinearity coefficient of the kth mode. In this case, we note that only the symmetric modes contribute.

As an example, we consider a cable for which $\alpha = 239.16$ and investigate the influence of the number of modes retained in (3.9.21) and (3.9.22) on the value of the effective nonlinearity coefficient α_{ek}. We consider primary resonance of the first symmetric inplane mode (i.e., $k = 1$) and primary resonance of the first antisymmetric inplane mode (i.e., $k = 2$). Because the natural frequencies of the symmetric inplane modes vary as a function of αb^2, we present results for three cases: $\alpha b^2 = \frac{1}{2}$ (left of crossover), $\alpha b^2 = 2$ (right of first crossover), and $\alpha b^2 = 4$ (right of second crossover). The corresponding values of the sag-to-span ratio are approximately $1/21.9$, $1/10.9$, and $1/7.7$, respectively, which are acceptable for "shallow" suspended cable theory (Irvine and Caughey, 1974).

In Figure 3.9.1, we present variation of the effective nonlinearity coefficient α_{ek} of the first symmetric inplane mode when $\alpha b^2 = \frac{1}{2}$ with the number of modes retained in (3.9.21). Using a one-mode discretization, $\alpha_{e1} = -113640.7$. If, instead, both modes 1 and 3 are used in the discretization, α_{e1} increases to -112866.2. As more modes are included in the discretization, α_{e1} monotonously converges onto the correct value. When fifteen modes are retained in the discretization, $\alpha_{e1} = -116871.7$. The value obtained by the direct approach is $\alpha_{e1}^* = -116878.2$. The percent errors in the predicted values of α_{e1} when using one-, two-, three-, and ten-mode discretizations are 2.8%, 3.4%, 0.8%, and 0.019%, respectively.

Choosing $\alpha b^2 = 2$ (right of first crossover), we present in Figure 3.9.2 variation of α_{e1} of the first symmetric inplane mode with the number of modes retained. Switching from a single-mode to a two-mode discretization, we find that α_{e1} jumps from -86661.5 to -137601.1, which is beyond the true value. Relative to the value

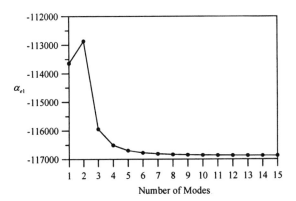

Fig. 3.9.1 The effective nonlinearity coefficient α_{e1} of the first symmetric inplane mode ($k = 1$) as a function of the number of modes considered in (3.9.21) when $\alpha = 239.16$ and $\alpha b^2 = \frac{1}{2}$ (at first crossover), Arafat and Nayfeh (2003).

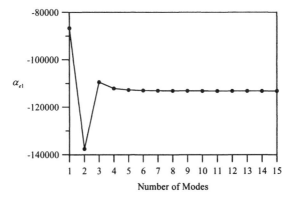

Fig. 3.9.2 The effective nonlinearity coefficient α_{e1} of the first symmetric inplane mode ($k = 1$) as a function of the number of modes considered in (3.9.21) when $\alpha = 239.16$ and $\alpha b^2 = 2$, Arafat and Nayfeh (2003).

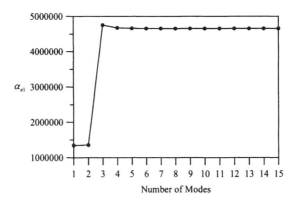

Fig. 3.9.3 The effective nonlinearity coefficient α_{e1} of the first symmetric inplane mode ($k = 1$) as a function of the number of modes considered in (3.9.21) when $\alpha = 239.16$ and $\alpha b^2 = 4$, Arafat and Nayfeh (2003).

obtained from the direct approach ($\alpha_{e1}^* = -113146.7$), the percent errors when using one-, two-, three-, and ten-mode discretizations are 23.4%, 21.6%, 3.3%, and 0.04%, respectively.

From Figures 3.9.1 and 3.9.2, we note that the final value of $\alpha_{e1} < 0$, and therefore the effective nonlinearity for the first symmetric inplane mode for these two cases is softening.

Lastly, we choose $\alpha b^2 = 4$ (right of second crossover) and present in Figure 3.9.3 variation of α_{e1} of the first symmetric inplane mode with the number of modes retained. Using one-mode and two-mode discretizations, we find that the correspond-

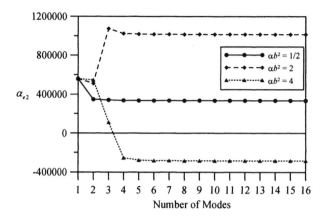

Fig. 3.9.4 The effective nonlinearity coefficient α_{e2} of the first antisymmetric inplane mode ($k = 2$) as a function of the number of modes considered in (3.9.22) when $\alpha = 239.16$ and the cases $\alpha b^2 = \frac{1}{2}$, 2, and 4, Arafat and Nayfeh (2003).

ing values of $\alpha_{e1} = 1345575.9$ and 1356450.4, which are relatively close. However, as we include more modes, a sharp increase in the value of α_{e1} occurs, and after fifteen modes α_{e1} converges to 4654795.8. Relative to the value obtained from the direct approach ($\alpha_{e1}^* = 4654599.6$), the percent errors when using one-, two-, three-, and ten-mode discretizations are 71.1%, 70.9%, 2.1%, and 0.02%, respectively. In addition, we note from Figure 3.9.3 that $\alpha_{e1} > 0$, indicating that, for this case, the effective nonlinearity of the first mode is hardening.

Next, we use (3.9.22) to calculate α_{e2} for the first antisymmetric inplane mode. In Figure 3.9.4, we demonstrate the dependence of α_{e2} on the number of modes (including mode 2) retained for the three cases: $\alpha b^2 = \frac{1}{2}$, $\alpha b^2 = 2$, and $\alpha b^2 = 4$. Because, from (3.9.22), the value of α_{ek} when using a single-mode discretization does not depend on b, in all three cases, α_{e2} begins at $+559112.6$.

For the case $\alpha b^2 = \frac{1}{2}$, α_{e2} quickly converges to $\alpha_{e2}^* = 335392.4$ after using sixteen modes. The relative errors when using one-, two-, three-, and ten-mode discretizations are 66.7%, 3.8%, 1.8%, and 0.013%, respectively. For the case $\alpha b^2 = 2$, the single- and two-mode discretizations give relatively close results, however when using three or more modes, α_{e2} increases and after retaining sixteen modes $\alpha_{e2} = 1015834.36$. The value obtained from the direct approach is $\alpha_{e2}^* = 1015797.4$. In this case, the percent errors when using one-, two-, three-, and ten-mode discretizations are 45.0%, 49.6%, 5.7%, and 0.017%, respectively. In both of these cases, the value of $\alpha_{e2} > 0$, and therefore the effective nonlinearity coefficient of the first antisymmetric inplane mode is hardening.

When $\alpha b^2 = 4$, we find from Figure 3.9.4 that $\alpha_{e2} > 0$ when using one-, two-, and three-mode discretizations. However, using four-mode or higher discretizations, we find that $\alpha_{e2} < 0$, and when we retain sixteen modes $\alpha_{e2} = -283281.9$. The corresponding value of $\alpha_{e2}^* = -283357.1$. Therefore, in this case, the error due to single-mode discretization is qualitative as well as quantitative. That is, for this case, the single-mode discretization predicts the effective nonlinearity of the first antisymmetric inplane mode to be hardening, when in fact it is softening. The percent errors when using one-, two-, three-, four-, and ten-mode discretizations are 297.3%, 293.3%, 138.9%, 10.4%, and 0.13%, respectively.

3.9.2 The Case of an out-of-Plane Mode

In this case, the solution of the first-order problem (3.9.4) and (3.9.5) is expressed as

$$\eta_{n1} = 0 \text{ for all } n \tag{3.9.23}$$

$$\zeta_{k1} = A_k(T_2)e^{ik\pi T_0} + \bar{A}_k(T_2)e^{-ik\pi T_0} \tag{3.9.24}$$

$$\zeta_{n1} = 0 \text{ for all } n \neq k \tag{3.9.25}$$

Substituting (3.9.23)-(3.9.25) into (3.9.6) and (3.9.7) and solving the resulting equations, we obtain

$$\eta_{n2} = \frac{1}{2}\alpha b k^2 \pi^2 \Gamma_n \left(\frac{A_k^2}{\omega_n^2 - 4k^2\pi^2} e^{2ik\pi T_0} + \frac{A_k \bar{A}_k}{\omega_n^2} \right) + cc \text{ for odd } n \tag{3.9.26}$$

$$\zeta_{n2} = 0 \text{ for all } n \tag{3.9.27}$$

It is clear from (3.9.27) that the present expansion breaks down when $\omega_n \approx 2k\pi$, the condition of two-to-one internal resonance.

To express the nearness of Ω to $k\pi$, we introduce the detuning parameter σ defined by

$$\Omega = k\pi + \epsilon^2 \sigma \tag{3.9.28}$$

Substituting (3.9.24), (3.9.26), and (3.9.28) into (3.9.9) with $n = k$, we obtain

$$D_0^2 \zeta_{k3} + k^2 \pi^2 \zeta_{k3} = -2D_0 D_2 \zeta_{k1} - 2\nu_k D_0 \zeta_{k1} + \alpha b k^2 \pi^2 \zeta_{k1} \sum_m \Gamma_m \eta_{m2}$$

$$- \frac{1}{2}\alpha k^4 \pi^4 \zeta_{k1}^3 + F_k \cos(\Omega T_0) \tag{3.9.29}$$

Then, substituting (3.9.25), (3.9.27), and (3.9.28) into (3.9.29) and eliminating the terms that lead to secular terms, we obtain

$$2i\pi k \left(A_k' + \nu_k A_k \right) + \alpha_{ek} A_k^2 \bar{A}_k - \frac{1}{2} F_k e^{i\sigma T_2} = 0 \tag{3.9.30}$$

where the effective nonlinearity coefficient α_{ek} is given by

$$\alpha_{ek} = \frac{3}{2}\alpha k^4 \pi^4 - \frac{1}{2}\alpha^2 b^2 k^4 \pi^4 \sum_n \frac{\Gamma_n^2 \left(3\omega_n^2 - 8k^2\pi^2 \right)}{\omega_n^2 \left(\omega_n^2 - 4k^2\pi^2 \right)} \tag{3.9.31}$$

Substituting for Γ_n from (3.7.5) into (3.9.31) yields the effective nonlinearity coefficient

$$\alpha_{ek} = \frac{3}{2}\alpha k^4 \pi^4 - \frac{k^4 \pi^4}{128 b^2} \sum_{n=\text{odd}} \frac{\kappa_n^2 \omega_n^2 \left(3\omega_n^2 - 8k^2\pi^2 \right)}{\omega_n^2 - 4k^2\pi^2} \tag{3.9.32}$$

The summation term in (3.9.32) is the contribution of the inplane symmetric modes.

3.10 EXTENSIONAL BARS

A bar is a one-dimensional structure that can sustain compression forces and twisting moments as well as tension loads because it has nontrivial compression strength and torsional rigidity, which are different from those of cables. A general composite bar must be treated as a beam because structural couplings, such as extension-bending and torsion-bending couplings, can result in flexural deflections. Moreover, for elastic isotropic bars, there are no geometric nonlinearities and the only nonlinearities are due to the Poisson effect. Hence, we only consider isotropic bars subjected to extensions and compressions in this section.

For an extensional bar, there is no nonlinear geometry involved, but Poisson's effect and a nonlinear stress-strain relation will introduce some nonlinearities. It follows from Figure 3.10.1 that the longitudinal strain e is

$$e \equiv u' \tag{3.10.1}$$

where the prime denotes the partial differentiation with respect to s. When e is large, the stress-strain relation is nonlinear and can usually be formulated as

$$\sigma \equiv \frac{P(s,t)}{A(s,t)} = E(s)e\left[1 + E_1(s)e + E_2(s)e^2 + \cdots\right] \tag{3.10.2}$$

where $E(s)$, $E_1(s)$, and $E_2(s)$ are material parameters, which are constant when the material is uniform, and $A(s,t)$ denotes the instantaneous cross-section area at s. Because of Poisson's effect, the change of the volume ratio is

$$\frac{\Delta V}{A_0 ds} = (1+e)(1-\nu e)^2 - 1 \tag{3.10.3}$$

where ν is the Poisson ratio. Further,

$$\Delta V = A(1+e)ds - A_0 ds \tag{3.10.4}$$

Using (3.10.1), (3.10.3), and (3.10.4), we obtain

$$A = A_0(1-\nu e)^2 = (1-\nu u')^2 A_0 \tag{3.10.5}$$

It follows from the free-body diagram shown in Figure 3.10.1 that the equation of motion is

$$m_0 \ddot{u} + c\dot{u} = P' + f(s,t) \tag{3.10.6}$$

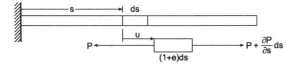

Fig. 3.10.1 A free-body diagram of an extensional bar.

where ρ_0 denotes the material density, $m_0 \equiv \rho_0 A_0$, c is the damping coefficient, and $f(s,t)$ is the distributed load. Substituting (3.10.1), (3.10.2), and (3.10.5) into (3.10.6) yields

$$m_0 \ddot{u} + c\dot{u} = \left\{ EA_0 \left[u' + (E_1 - 2\nu) u'^2 + (E_2 - 2\nu E_1 + \nu^2) u'^3 \right] \right\}' + f(s,t) \tag{3.10.7}$$

The boundary conditions are

$$u = 0 \quad \text{at} \quad s = 0 \tag{3.10.8a}$$

and

$$EA_0 \left[u' + (E_1 - 2\nu) u'^2 + (E_2 - 2\nu E_1 + \nu^2) u'^3 \right] \quad \text{specified at } s = L \tag{3.10.8b}$$

For a bar of uniform material, ν, E, E_1, E_2, and m_0 are constants. If the cross-section is prismatic, A_0 is constant. For a uniform, prismatic bar with e below the proportional limit (i.e., $E_1 = E_2 = 0$), the equation of motion becomes

$$\frac{m_0}{EA_0} \ddot{u} + \frac{c}{EA_0} \dot{u} = u'' \left(1 - 4\nu u' + 3\nu^2 u'^2 \right) \tag{3.10.9}$$

It follows from (3.10.3) that

$$\frac{\Delta V}{A_0 ds} = (1 - 2\nu)e \tag{3.10.10}$$

where higher-order terms are neglected. Hence the material is incompressible if ν is about 0.5. But, the Poisson ratio ν of most engineering materials is between 0.25 and 0.35 except those of rubber and paraffin, for which ν is about 0.5. It follows from (3.10.5) that the cross-section area does not change if $\nu = 0$ (e.g., cork and concrete), and there are no nonlinear terms in (3.10.9).

The cubic nonlinear governing equation (3.10.9) for linearly elastic extensional bars can be deduced from (3.4.53)-(3.4.55), which corresponds to no initial sags or static loads, the string being placed along the x_1 axis, and $u_2 = u_3 = 0$; that is,

$$\alpha_1' = 1, \ \alpha_2' = \alpha_3' = 0, \ \eta_0 = 1, \ \eta_1 = u_1', \ \eta_2 = u_1'^2 \tag{3.10.11}$$

Consequently, (3.4.53) for $i = 1$ reduces to

$$m_0 \ddot{u}_1 + \mu_1 \dot{u}_1 = \tilde{f}_1 + \left[EA_0 \left(u_1' - 2\nu u_1'^2 + \nu^2 u_1'^3 \right) \right]' \tag{3.10.12}$$

and the boundary conditions: specify

$$u_1 \quad \text{or} \quad EA_0 \left(u_1' - 2\nu u_1'^2 + \nu^2 u_1'^3 \right) \quad \text{at } s = 0 \text{ and } L \tag{3.10.13}$$

Equations (3.10.12) and (3.10.13) represent the equation of motion and boundary conditions for a linearly elastic extensional bar without static loads. For extensional bars, there are no geometric nonlinearities involved, and the nonlinear terms in (3.10.12) and (3.10.13) are due to Poisson's effect. We note that nonlinearities due to Poisson's effect can be mistaken for material nonlinearities.

4
BEAMS

Beam theories are one-dimensional models for structures with one dimension being much larger than the other two. Modeling of beams is more difficult than modeling of plates and shells because the former involves the transformation of a three-dimensional problem in nature into a one-dimensional problem, whereas the latter involves the transformation of a three-dimensional problem into a two-dimensional problem. However, solutions of beam problems are easier than solutions of plate and shell problems because the former involve only one independent spatial variable whereas the latter involve two independent spatial variables.

4.1 INTRODUCTION

A beam theory is often used to model helicopter rotor blades, aircraft wings, aviation propeller blades, prop-fan blades, wind-turbine blades, robot manipulators, large space structures, arm-type positioning mechanisms of magnetic disk drives, links of slider-crank mechanisms, etc. In addition to being basic structural elements, beam-type specimens under concentrated loads are also utilized in the experimental characterization of materials. Moreover, beams are major construction components of aircraft, and the postbuckling strength of beams plays an important role in the design of aircraft structures because conventional aircraft structural elements are often designed to operate in the postbuckling range, where nonlinear beam theories are needed. Furthermore, the recent rapid development in aerospace exploration has stimulated extensive use of highly flexible structures that can undergo large elastic deformations, especially beam-type structures. Consequently, development of a general beam theory is a constant research interest, especially for nonlinear composite beams.

4.1.1 Beam Theories

The theory of elasticity has been used to solve some linear, inplane deflection problems for isotropic beams. Chree (1889) derived exact equations in cylindrical coordinates from the general equations of the theory of elasticity, and Cowper (1968) gave a plane stress elasticity solution for the fundamental frequency of a hinged-hinged beam with a rectangular cross section. Moreover, some static, plane-stress beam problems were treated by Saint Venant's semi-inverse method (Timoshenko and Goodier, 1970). However, most beam problems cannot be solved by using the theory of elasticity, especially beams with irregular cross-sections, anisotropic beams with elastic couplings, beams undergoing large deflections in three-dimensional space, or initially curved and twisted beams. Hence, technical or engineering beam theories are commonly used.

Depending on their complexity, beam theories can be divided into the following three groups: (a) Euler-Bernoulli beam theory, (b) shear deformable beam theories (i.e., Timoshenko's theory, third-order shear theory, higher-order shear theory, layer-wise shear theory, etc.), and (c) three-dimensional beam theories, as shown in Figure 4.1.1. In the Euler-Bernoulli beam theory, only the axial strain ϵ_{11} is considered and a plane cross-section perpendicular to the reference axis before deformation is assumed to remain plane and perpendicular to the deformed reference axis after deformation. In other words, the transverse shear-strains ϵ_{12} and ϵ_{13} are neglected. In shear-deformable beam theories, out-of-plane warpings (see Figures 4.1.2d and e) due to ϵ_{12} and ϵ_{13} are taken into account. In three-dimensional beam theories, both out-of-plane and inplane warpings are taken into account. Inplane warpings (see Figures 4.1.2b and c) are due to the transverse normal strains ϵ_{22} and ϵ_{33} and the inplane shear-strain ϵ_{23}, which result in significant transverse normal stresses (i.e., σ_{22} and σ_{33}) and a significant inplane shear stress (i.e., σ_{23}). Hence, effects due to three-dimensional stresses are fully accounted for in three-dimensional beam theories.

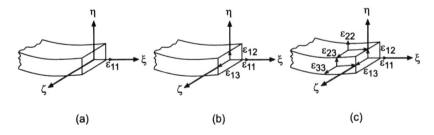

Fig. 4.1.1 Three groups of beam theories: (a) Euler-Bernoulli theory, (b) shear-deformable theories, and (c) theories accounting for three-dimensional stresses.

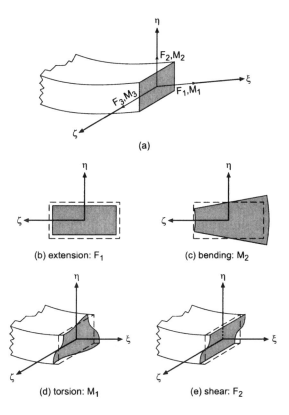

Fig. 4.1.2 Warpings of an isotropic beam: (a) stress resultants, (b) extension-induced inplane warping, (c) bending-induced inplane warping, (d) torsion-induced out-of-plane warping, and (e) shear-induced out-of-plane warping.

From the plane stress solution for a simply-supported beam carrying a uniform load, Shames and Dym (1985) found that the transverse shear and normal stresses are of the order h/L and $(h/L)^2$, respectively, where h and L are the thickness and length of the beam, respectively. Hence, the transverse normal stress is neglected in most beam analyses, but the shear stress is considered for problems with large aspect ratios h/L, short wavelengths, or high frequencies. Moreover, shear effects are significant for composite materials because they are relatively weak in shear. In Timoshenko's beam theory (Timoshenko, 1921, 1922; Thomson, 1981), which is known as the first-order shear theory, the shear-strain is assumed to be uniform over a given cross-section. In this theory, a shear correction coefficient (Timoshenko, 1921; Cowper, 1966) that depends on the shape of the cross-section is introduced to account for the fact that the shear stress and strain are not uniformly distributed over the cross-section. Krishna Murty (1970) included shear deformations in his three linear second-order approximate equations governing the transverse oscillations of uniform short beams. Heyliger and Reddy (1988) presented a higher-order shear-deformation theory in which some quadratic terms are included and no shear correction coefficient is needed. Pai and Schulz (1999) presented a new derivation of shear correction factors for isotropic beams by matching the exact shear-stress resultants and shear-strain energy with those of the equivalent first-order shear deformation theory. The physical meaning of a shear correction factor is shown to be the ratio of the geometric average to the energy average of the transverse shear-strain on a cross-section.

For a compact beam, the warping effect is usually neglected, and its influence on the torsional rigidity is accounted for by using the theory of elasticity (Timoshenko and Goodier, 1970). Bauchau and Hong (1987, 1988), Stemple and Lee (1988), Kane, Ryan, and Banerjee (1987), and Pai and Nayfeh (1992a) accounted for the warpings induced by torsion and/or shears in their equations of motion of composite beams.

In the two-dimensional Euler-Bernoulli theory, there is only one dependent variable. In the two-dimensional Timoshenko beam theory, there are two dependent variables. In higher-order shear deformation theories, there are three (Krishna Murty, 1970; Heyliger and Reddy, 1988) or more (Sheinman and Adan, 1987) dependent variables. Definitely, if inplane warpings, out-of-plane shear deformations, and torsion-related warpings are considered, more dependent variables are required, and the finite-element method is commonly used in these analyses (Bauchau and Hong, 1988; Stemple and Lee, 1988).

For laminated composite beams or box beams, some nonclassical effects, such as transverse shear deformations, torsional warping, inplane warpings due to bending and extension, transverse normal stresses, inplane shear stress, warping restraint effect, and free-edge effect are significant because of the anisotropy, asymmetry of the cross-section, and non-uniformity of Poisson's ratios over the cross-section. These effects are characterized by significant inplane and out-of-plane warping displacements and a three-dimensional stress state. In other words, all three-dimensional stress effects can be important for general anisotropic beams. Hence, a general nonlinear beam theory should include three-dimensional stress effects (due to out-of-plane and inplane warpings) and geometric nonlinearities as well as anisotropy and initial curvatures, which result in linear elastic couplings.

Because of anisotropy and heterogeneity of composite materials and the characteristics of three-dimensional stress states, three-dimensional finite-element methods may be the only way to solve the dynamic problems of general composite beams, which are too expensive for achieving certain accuracy. Although the inertia forces due to inplane warpings and even out-of-plane warpings are negligible because they are relative displacements with respect to the deformed cross-section and are much smaller than the global displacements, these warpings offer extra degrees of freedom for the deformation of the cross-section and hence significantly affect the elastic properties of the cross-section. Moreover, starting from three-dimensional elasticity and using a perturbation analysis with slenderness ratio as the ordering parameter, Parker (1979a,b) showed that a combination of Saint Venant's warping solutions, which are derived from small-displacement linear elasticity, and a one-dimensional nonlinear beam model including large rotations is natural and can account for three-dimensional stress effects. More specifically, Berdichevskii (1981) states that the geometrically nonlinear problem of the three-dimensional beam elasticity can be decoupled into a nonlinear one-dimensional problem and a linear two-dimensional section problem. Consequently, to account for three-dimensional stress effects in a one-dimensional nonlinear beam model, one can neglect the inertia forces due to inplane warpings and even out-of-plane warpings and only include the influence of warpings on the elastic constants. In other words, a one-dimensional nonlinear beam model with stiffness matrices and warping functions determined from a linear, static, two-dimensional sectional analysis is a general and practical approach for solving nonlinear anisotropic beam problems (Parker, 1979a,b; Berdichevskii, 1981; Borri and Merlini, 1986; Hodges, 1990).

To obtain linear extension, torsion, and/or flexure solutions for isotropic beams, Saint Venant's principle is commonly applied. Saint Venant's principle implies that the stresses at a point, which is at a sufficient distance from the loading end, depend only on the magnitude of the applied load and are practically independent of the manner in which the tractions are distributed over the end (Iesan, 1987). The deformations at points away from the ends where nontrivial stress resultants are applied are the so-called Saint Venant's solutions, or central solutions, or particular solutions (Giavotto et al., 1983; Iesan, 1987). On the other hand, Saint Venant's principle also implies that a system of loads having zero resultant forces and moments (i.e., a self-equilibrating stress system) produces a strain field that is negligible at a point which is away from the loading end (Iesan, 1987). But, for highly anisotropic and heterogeneous materials, such a self-equilibrating stress system can result in nontrivial strains with long decay lengths, which are the so-called boundary-layer solutions, or extremity solutions, or eigensolutions, or transitional solutions (Giavotto et al., 1983; Iesan, 1987).

The restraint warping effect is due to the contribution of boundary-layer warping solutions, especially warping solutions which have long decay lengths. If the decay lengths of some boundary-layer warping functions are large, the free warping assumption used in Saint Venant's principle may result in erroneous predictions of the behavior of composite beams when the end cross-sections are restrained against out-of-plane warpings (Oyibo and Berman, 1985; Tsuiji, 1985; Librescu and Simovich, 1988). Moreover, if the distribution of the end loadings are not the same as that of the corresponding Saint Venant's solution, the boundary-layer effect may cause serious error in the prediction by using Saint Venant's solutions only. Hence, boundary-layer solutions are needed in the analysis of composite beams, but they cannot be easily determined by using analytical methods. Moreover, although Saint Venant's solutions for shear and torsional warpings of some prismatic isotropic beams having elliptical, rectangular, or equilateral triangular cross-sections (Love, 1944; Sokolnikoff, 1956; Timoshenko and Goodier, 1970) can be obtained by using the theory of elasticity and Saint Venant's principle, it is almost impossible to obtain analytical Saint Venant's warping functions for a general composite or constructed beam having an arbitrary cross-section. Although approximate analytical Saint Venant's warping functions of a general anisotropic beam can be obtained by using low-order polynomials, substituting a simple geometry for the actual cross-section and/or neglecting the anisotropy and heterogeneity of the beam section (see the review article by Hodges, 1990) yields approximate analytical solutions, which do not satisfy the traction conditions on the lateral boundaries of the beam and the continuity conditions of interlaminar stresses at the interface of two contiguous heterogeneous domains. Hence, a finite-element-based method is a more general and practical approach.

Giavotto et al. (1983) presented a two-dimensional, static, sectional, finite-element analysis of Saint Venant's warping functions and boundary-layer warping solutions for straight beams; however, this formulation is linear and all variables are defined with respect to the undeformed coordinate system. Borri and Merlini (1986) extended the theory of Giavotto et al. (1983) to include geometric nonlinearities by using Green-Lagrange strains. Atilgan and Hodges (1991) and Hodges et al. (1992) presented a systematic, nonlinear formulation of the sectional analysis of straight beams, where geometric nonlinearities are accounted for by using Green-Lagrange strains. However, to obtain the warping functions, one needs to restrain six rigid-body motions (three translations and three rotations) in order to make the problem non-singular. The six constraints used by Hodges et al. (1992) are different from those used by Borri and Merlini (1986). The six constraints used by Borri and Merlini (1986) are based on the assumption that the average work done by the surface traction in producing the warping displacements on the cross-section is zero; and four of the six constraints used by Hodges et al. (1992) are based on the assumption that the average warping displacements and inplane rotation due to warpings on the cross-section are zero and the other two constraints are used to define the deformed, local coordinate system.

Pai and Nayfeh (1994a) used the results of a two-dimensional sectional analysis to model naturally curved and twisted composite rotor blades and account for warpings and three-dimensional stress effects. Pai and Schulz (1999) presented a method for deriving inplane and shear-warping functions from available elasticity solutions. The derived exact warping functions can be used to check the accuracy of a two-dimensional sectional finite-element analysis of central solutions.

4.1.2 Geometric Nonlinearities

When the lateral deflection of a beam is large, the axial force plays a significant role in carrying transverse loads, and geometric nonlinearities couple the equations governing the extension and bending vibrations. Moreover, for an isotropic beam, a linear formulation shows that the vibrations in the two principal planes are independent of each other, and hence a forced motion in one principal plane always remains in that plane. But, when the vibration amplitude is large, the motion may become nonplanar due to inertia and geometric nonlinear coupling between bendings in two principal planes (Pai and Nayfeh, 1990a). Geometric nonlinearities are also important in postbuckling analyses, nonlinear panel flutter, and dynamic stability problems. In the literature, there are many approaches for accounting for geometric nonlinearities. von Kármán strains do not fully account for geometric nonlinearities due to large rotations and hence cannot be used to derive a fully nonlinear beam model (Pai and Nayfeh, 1992a). To fully account for geometric nonlinearities by using the Green-Lagrange strains, one must use the correct conjugate stresses, the second Piola-Kirchhoff stresses and not the Cauchy stresses. However, even if the second Piola-Kirchhoff stresses are used correctly with the Green-Lagrange strains in a formulation, they cannot easily match with real boundary conditions because they are energy measures and are defined with respect to the undeformed system configuration (Pai, Palazotto, and Greer, 1998). Moreover, because the apparent stiffnesses of composite beams are sensitive to fiber directions, the stiffnesses vary with the deflections if the stresses and strains are defined with respect to the undeformed coordinates. Then, geometric nonlinearities can be mistaken as material nonlinearities. Consequently, local stress and strain measures are necessary (Pai and Nayfeh, 1992a). More recently, two or three successive Euler-like rotations are commonly used in obtaining the exact transformation matrix that relates the deformed and undeformed states and hence accounts for large geometric nonlinearities (Hodges and Dowell, 1974; Hodges, 1976; Dowell, Traybar, and Hodges, 1977; Crespo da Silva and Glynn, 1978a; Alkire, 1984; Rosen and Rand, 1986; Bauchau and Hong, 1987; Rosen, Loewy, and Mathew, 1987a; Crespo da Silva, 1988; Minguet and Dugundji, 1990a; Pai and Nayfeh, 1990b, 1992a, 1994a; Banan, Karami, and Farshad, 1991; Simo and Vu-Quoc, 1991). However, because finite rotations are not vector quantities, the variations of three successive Euler-like angles are not unique or independent and not along three perpendicular directions. Moreover, coordinate transformations using three Euler angles result in asymmetric equations of motion and the torsion-related angle does not represent the real twist angle, even if only two Euler angles are used in the transformation (Pai and Nayfeh, 1990b, 1992a). Alkire (1984) showed that the transformation matrix and curvatures obtained by using Euler-like rotations are unique, but the form varies according to the choice of the twist variable. Hence, different sequences of Euler rotations will result in different equations of motion (Pai and Nayfeh, 1990b, 1992a), and the equations are asymmetric because there are implicit asymmetric terms included in the twist variable (Hodges and Dowell, 1974; Rosen, Loewy, and Mathew, 1987b; Hodges, Crespo da Silva, and Peters, 1988; Pai and Nayfeh, 1990b). Hence, the concept of orthogonal virtual rotations (Pai and Nayfeh, 1991c, 1992a) is needed in order to derive geometrically-exact equations governing the motions along three perpendicular directions.

To investigate large amplitude whirling motions of a simply supported beam, Ho, Scott, and Eisley (1975) accounted for large deformations through the use of Green's strain measure in the longitudinal direction. Heyliger and Reddy (1988) and Sheinman and Adan (1987) used von Kármán strains to account for large deflections and rotations. Although Bolotin (1964) and Moody (1967) showed that nonlinear inertia effects are far more significant than nonlinear elasticity effects on the vibrations of thin rods, Crespo da Silva and Glynn (1978a,b) used the extended Hamilton principle and included the nonlinear inertia and the nonlinear bending and twisting curvatures in their derivation of the equations of motion describing the nonplanar, nonlinear dynamics of an inextensional beam. They showed that the generally neglected nonlinear terms arising from the curvature are the same order as the nonlinear terms due to inertia and they might have a significant influence on the response of the system. Nayfeh and Pai (1989) showed that nonlinear structural terms are of the hardening type and dominate low-frequency vibrations, nonlinear inertia terms are of the softening type and dominate high-frequency vibrations, and neglecting geometric nonlinearities leads to erroneous results. An experimental study by Evan-Iwanowski, Sanford, and Kehagioglou (1970) also shows that the nonlinear geometric terms have definite effects on the nonstationary parametric response of a straight column. Friedmann (1977) indicated that rotary-wing aeroelasticity is inherently nonlinear and that the nonlinear terms in the equations of motion are important in the study of aeroelastic stability. Crespo da Silva and Hodges (1986a,b) derived the differential equations of motion for a rotating beam with the objective of retaining all of the possible contributions due to higher-order nonlinearities, and investigated the influence of these terms on the motion of a helicopter rotor blade. They found out that the most significant cubic nonlinear terms are those associated with structural geometric nonlinearities in the torsion equation. Busby and Weingarten (1972) used a combination of the finite-element method and the method of averaging to study the influence of induced stretching on the steady-state response of a beam to periodic excitations.

Hodges and Dowell (1974) used Hamilton's principle and Newton's second law to develop a comprehensive set of equations with quadratic nonlinearities that describe the dynamics of beams. They considered the offsets between the mass centroid and the elastic and area centroid axes and identified several nonlinear terms. Dowell, Traybar, and Hodges (1977) devised a simple, nonrotating beam with a tip weight to evaluate the theory of Hodges and Dowell (1974). The results show that there are systematic differences between theory and experiment when the tip deflections become large. In a follow-on work of Hodges and Dowell (1974), Hodges, Crespo da Silva, and Peters (1988) pointed out that the nonlinear equations of motion governing flexural deformations cannot be consistent unless they include all nonlinear terms up through third-order. Rosen and Friedmann (1979) derived a set of equations more accurate than those of Hodges and Dowell (1974) by including some nonlinear terms with order higher than two. Their numerical results are in good agreement with the experimental data of Dowell, Traybar, and Hodges (1977). Rosen, Loewy, and Mathew (1987a,b) added cubic terms to the curvatures used by Rosen and Friedmann (1979) to improve the modeling.

Hinnant and Hodges (1988) combined multibody and finite-element technology to analyze the nonlinear vibrations of a cantilever beam. Their results are in good agreement with the experimental data of Dowell, Traybar, and Hodges (1977). In the development of nonlinear equations of motion of twisted rotor blades, Houbolt and Brooks (1956) identified many coupling-type terms, which are associated with centrifugal forces. Hodges and Peters (1975) showed that the twisting angle due to flexural displacements must be considered in the derivation of the lateral buckling equation of beams and in the calculation of buckling loads. Peters (1977) pointed out that the classical buckling load formulas that neglect the effect of bending in the plane of greatest flexural rigidity prior to buckling are not always conservative for infinitely slender beams. Rosen, Loewy, and Mathew (1987c) argued that the equations of motion derived by using three Euler-like rotations seem to contain the basis for coupled bending-torsion buckling because the first flapwise frequency tends to zero as the static tip load increases. Minguet and Dugundji (1990a,b) used Newton's second law and Euler rotations to derive the equations of motion for composite beams, and they solved the nonlinear trigonometric equations directly for arbitrarily large static deflections and for the linear natural frequencies.

Sathyamoorthy (1982a) gave a survey of the studies that investigated the effects of geometric nonlinearities due to stretching and curvature and material nonlinearities of the Ramberg-Osgood and Ludwick types in beams. A review of recent developments of the analysis of laminated beams and plates is given by Kapania and Raciti (1989a,b). Sathyamoorthy (1982b) reviewed the use of finite-element methods for analyzing beams.

4.1.3 Shear Deformations, Rotary Inertias, and Gravity

Nayfeh (1973a) investigated the effects of transverse shear and rotary inertia on the nonlinear transverse vibrations of inhomogeneous beams with finite axial restraints. The results show that the natural frequencies increase with the equations governing the amplitude of the response and the axial restraint and that the transverse shear and rotary inertia decrease the natural frequencies. Huang (1961) derived the natural frequencies and mode shapes of uniform Timoshenko beams with simple boundary conditions. It was shown that the shear deformation and rotary inertia have definite influence on the natural frequencies and mode shapes of beams, especially stubby ones. In the case of very large static bending of a cantilever beam, Sinclair (1979) found out that the contributions of the shear and longitudinal deformations to the beam response are of the same order. Rosen, Loewy, and Mathew (1987b) indicated that the influence of rotary inertia on the natural frequencies of the lowest modes is insignificant. Kapania and Raciti (1989c) indicated that, for thin composite beams whose length/thickness is greater than 50, transverse shear effects can be neglected in the large deflection theory. Rao, Raju, and Raju (1976) stated that when $L/r > 100$, where r denotes the radius of gyration and L is the beam length, the shear effect and rotary inertia are negligible. Adams and Bacon (1973) stated that the shear deformation can be neglected for isotropic beams if the aspect ratio $L/h > 20$, where h is the beam thickness; but for unidirectional carbon-fiber-reinforced plastics, an aspect ratio of the order of 100 is necessary if shear effects are to be ignored. In general, shear effects need to be considered for a composite beam unless its aspect ratio is very high.

Sato, Saito, and Otomi (1978) treated parametric vibrations of a simply supported horizontal beam carrying a concentrated mass under the influence of gravity. Their results show that, in addition to parametric resonances, forced resonances occur due to initial static deflections. Zavodney (1987) studied a two-degree-of-freedom system and indicated that the static deflection results in a quadratic term. Crespo da Silva and Zaretzky (1991) studied the equilibrium deflections and natural frequencies of a cantilevered beam with a tip mass. They found out that the effect of the tip mass on the approximation of the deflection is significant when the tip mass is heavy. Shyu, Mook, and Plaut (1993a-c) studied the response of a slender, elastic, cantilevered isotropic beam to transverse harmonic excitations. They considered the static deflection caused by the weight of the beam and found out that the static deflection modifies the natural frequencies and adds quadratic terms to the equations of motion.

4.1.4 Elastic Couplings

One important elastic behavior of composite beams is the coupling among extensional, bending, and torsional stiffnesses, which is desirable in some cases. For example, the extension-twisting coupling produces different twist distributions in a two-speed rotor of a helicopter. Moreover, the bending-twisting coupling produces a pitch-flap stability in helicopter rotor blades. One well-known example of taking advantage of the bending-twisting coupling effect is the X-29 demonstrator aircraft; the composite skin of its forward-swept wing has a built-in structural and aerodynamic stability.

For asymmetric or symmetric angle-ply composite beams, the structural (elastic) couplings introduce other complex features into these vibrations. Moreover, the elastic couplings may induce other nonlinear couplings. Extension-twisting and bending-twisting couplings of a composite beam have been extensively studied (Abarcar and Cunniff, 1972; Bauchau and Hong, 1987, 1988; Whitney, 1987 Kapania and Raciti, 1989c; Minguet and Dugundji, 1990a,b; Pai and Nayfeh, 1991a,b). Krenk (1983) showed that coupling between extension and torsion exists even for pretwisted isotropic beams. Kapania and Raciti (1989c) showed that asymmetrically laminated beams are characterized by a bending-stretching coupling, which induces a quadratic nonlinearity.

4.1.5 External and Internal Resonances

Taking into account the nonlinear inertia terms and considering linear curvature in the differential equations of motion, Haight and King (1971) obtained planar frequency-response curves of a parametrically excited rod by means of an averaging method. They also identified unstable regions in the planar response curves, which correspond to stable motions in the other principal plane. They did not find nonplanar motions. Ho, Scott, and Eisley (1975) showed that the nonlinear terms due to stretching are cubic and cause whirling motions. They concluded that the larger the forcing amplitude is the more likely whirling motions occur. They found out that, in some parameter range, both planar and nonplanar motions are possible. They indicated that the regions in which both planar and nonplanar motions are unstable might correspond to beating motions or to steady-state motions with a more complex modal structure.

For a slender, compact, uniform, initially straight, linearly elastic rod subject to a lateral harmonic excitation in one principal direction, Haight and King (1972) investigated the effect of the nonlinear inertia terms on the response of the first three modes by using the Ritz averaging method. They identified unstable regions in the planar response curves, which correspond to parametrically excited nonplanar motions, without obtaining quantitative results of the resulting nonplanar motions. From their experimental work, they concluded that the nonplanar motions are steady whirling motions with the point on the neutral axis tracing an elliptical path. Using stability analyses, they showed that damping can cause the disappearance of the planar instability phenomenon. Hyer (1979) used Haight and King's (1972) equations to analyze quantitatively the whirling motions. He concluded that all nonplanar vibrations are steady whirling motions. But he indicated that the nonlinear curvature, damping, shear deformation, and rotary inertia might affect the whirling response characteristics. For the free motions of compact beams with fixed ends, Ho, Scott, and Eisley (1976) showed that in resonant cases whirling motions are of the beating type, whereas in non-resonant situations they have a steady-state behavior.

Tezak, Mook, and Nayfeh (1978) studied nonlinear transverse oscillations of parametrically excited columns, including the effects of mid-plane stretching. The results exhibit the jump phenomenon and show that two rather different excitations can produce practically the same response when there is an internal resonance. Nayfeh, Mook, and Sridhar (1974) investigated forced nonlinear oscillations of prismatic beams with mid-plane stretching; they found out that the response to a harmonic excitation can involve strong modal interactions when the natural frequencies are commensurate. In their study of the response of a hinged-clamped beam to a simple harmonic excitation, Nayfeh, Mook, and J. Nayfeh (1987) and Chin and Nayfeh (1997, 1999), considered the effects of geometric (stretching). They showed that multi-mode nonlinear analyses predict complicated responses in case of modal coupling, which cannot be predicted by single-mode nonlinear analyses or by multi-mode linear analyses. Their results also showed that the response may experience period-doubling bifurcations, culminating in a chaotic motion.

Nayfeh (1981, 2000) and Nayfeh and Mook (1979) studied primary, subharmonic, superharmonic, supersubharmonic, principle parametric, fundamental parametric, and combination resonances of many nonlinear systems. They included the motion-induced stretching effect in the governing equations of beams and investigated many cases of primary and secondary resonances. They showed that, in case of modal interactions, if the high-frequency mode is directly excited, the response of the low-frequency mode could be very large due to an energy transfer from the high-frequency mode. But, if the low-frequency mode is directly excited, the response of the high-frequency mode is usually small. They also showed that a small parametric excitation can produce a large response when the excitation frequency is not close to any of the natural frequencies of the system. Furthermore, owing to certain nonlinear coupling terms and the presence of commensurability among the natural frequencies, a parametrically excited mode may in turn indirectly excite other modes through an autoparametric resonance. These phenomena are not disclosed by a linear approximation to these systems. In the case of primary resonance, an external force of order ϵ^2 can result in a response of order ϵ, where ϵ denotes a small parameter. Moreover, modes with different frequencies can be excited simultaneously by a single-frequency harmonic forcing function through an internal or a combination resonance.

Crespo da Silva and Glynn (1979a) used a perturbation method to investigate the inplane and out-of-plane response of a clamped-clamped/sliding beam subject to a planar distributed harmonic excitation. Maganty and Bickford (1987) studied large-amplitude out-of-plane oscillations of a thin circular ring in the case of one-to-one internal resonance and found amplitude-modulated motions. Crespo da Silva and Glynn (1979b) studied the effect of support asymmetry on the free, nonplanar resonant motions of a fixed-free beam with a one-to-one internal resonance. They showed that the torsional-flexural coupling is important when the support conditions are asymmetric. Crespo da Silva and Zaretzky (1988) investigated the influence of a three-to-one modal coupling between the third and second modes of a cantilever beam on its planar response. Modal coupling and the beating phenomenon were observed in the experiments of Ray and Bert (1969), which deal with large-amplitude vibrations of a simply supported flexible beam.

Tso (1968) studied parametrically induced torsional vibrations due to shortening effects in an elastic cantilever beam of rectangular cross-section under a dynamic axial loading. He showed that, when the applied frequency is close to twice one of the longitudinal natural frequencies, the corresponding torsional mode may be excited parametrically. Burton and Kolowith (1988) presented experimental and analytical results for the response of a flexible, parametrically excited, vertically mounted beam. They added nonlinear damping terms to the equations of Crespo da Silva and Glynn (1978a). They concluded that the steady-state amplitudes of the periodic motions of the fourth mode are well predicted by their theory. But they observed chaotic motions at high excitation levels. For the same system, Haddow and Hasan (1988) also observed experimentally chaotic motions, nonplanar motions, and inplane modal interactions. Nayfeh and Pai (1989) analyzed the nonlinear response of an inextensional cantilever beam to a principal parametric base excitation. Their results show that motions in one plane may be destabilized by disturbances in the orthogonal plane, resulting in vibrations in both planes. Moreover, for certain beam and load parameters, nonplanar periodic responses consisting of steady whirling motions are possible. Furthermore, as one of the control parameters varies, the whirling motions may lose stability via a Hopf bifurcation, resulting in quasiperiodic whirling motions whose amplitudes and phases vary with time rather than being constant. For certain beam and load parameters, the whirling response may be a chaotically-modulated motion. The nonlinear planar and nonplanar motions of a long cantilever beam subject to a lateral harmonic base excitation were analyzed by Pai and Nayfeh (1990a). Their results show that inplane vibrations may be destabilized by out-of-plane disturbances, resulting in vibrations in both planes. For certain beam and load parameters, nonplanar responses can be (a) steady whirling motions, (b) unsteady whirling motions (lengths and directions of the axes of the elliptical route continuously change and the motion also shows a low-frequency twisting motion), or (c) chaotically modulated motions. The unsteady whirling motions may be periodic and phase locked if the modulation period is commensurate with that of the base motion. Pai and Nayfeh (1991a,b) investigated the response of an inextensional, symmetric angle-ply graphite-epoxy beam to a harmonic base excitation along either the flapwise direction or the chordwise direction. They studied Hopf bifurcations, symmetry-breaking bifurcations, period-multiplying sequences, and chaotic solutions of the modulation equations. Chaotic solutions are identified from their frequency spectra, Poincaré sections, and Lyapunov's exponents. The results show that the beam motion may be nonplanar although the input force is planar. Nonplanar responses may be periodic, periodically modulated, or chaotically modulated motions.

Zavodney (1987) gave a review of the parametric vibrations of nonlinear systems. Nayfeh and Balachandran (1989) reviewed theoretical and experimental studies of the influence of modal interactions due to one-to-one, two-to-one, three-to-one, and combination internal resonances on the nonlinear responses of harmonically excited structural and dynamical systems. Nayfeh (2000) gave a comprehensive account of nonlinear interactions.

4.2 LINEAR EULER-BERNOULLI BEAM THEORY

For isotropic beams, small vibrations in two principal planes are uncoupled. For anisotropic beams, small vibrations in two principal planes can be linearly coupled due to elastic bending-bending and bending-torsion couplings and hence the motion is three-dimensional. In this and the next section, we consider only isotropic beams undergoing small vibrations in one principal plane.

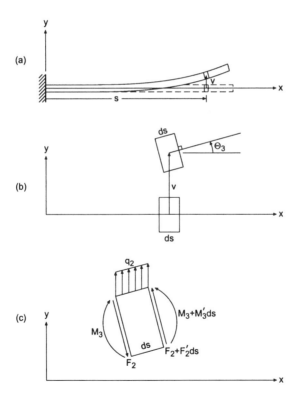

Fig. 4.2.1 The linear Euler-Bernoulli beam theory: (a) the coordinate system x, y, (b) the displacement v and rotation Θ_3, and (c) the free-body diagram.

Figures 4.2.1a-c show the coordinate system, displacements, and a free-body diagram of an Euler-Bernoulli beam element, respectively, where s denotes the undeformed length measured from the root of the beam to the reference point of the observed element. In the Euler-Bernoulli beam theory, a plane cross-section perpendicular to the beam axis before deformation is assumed to remain plane and perpendicular to the deformed axis after deformation, which implies that there are no transverse shear-strains ϵ_{12}, as shown in Figure 4.2.1b. Moreover, the longitudinal displacement u of any point on the beam axis is usually neglected; that is,

$$u = 0 \qquad (4.2.1)$$

Hence, only a single displacement variable $v(s,t)$ is used to represent the deflection of the beam. In other words, we are considering the deflection of a line in a two-dimensional space with each point on the line being able to move in the y direction only. Also, the axial strain e on the beam axis is usually neglected; that is,

$$e = 0 \qquad (4.2.2)$$

Applying Newton's second law of motion to the beam element in Figure 4.2.1c along the y direction yields

$$F_2' ds \cos \Theta_3 + q_2 ds = m \, ds \, \ddot{v} \qquad (4.2.3)$$

where Θ_3 is the rotation angle of the cross-section, q_2 is the distributed external load per unit length along the y direction, the prime denotes the partial derivative with respect to s (i.e., $\partial/\partial s$), the dot denotes the partial derivative with respect to t (i.e., $\partial/\partial t$),

$$F_2 = \int_A \sigma_{12} dA \quad \text{and} \quad m \equiv \int_A \rho \, dA \qquad (4.2.4)$$

are the transverse shear force and mass per unit length, respectively, ρ is the mass density, and A is the area of the cross-section. Similarly, the equilibrium condition of moments along the z direction is

$$M_3' ds + \frac{1}{2} F_2 ds + \frac{1}{2} \left(F_2 + F_2' ds \right) ds = j_3 \, ds \, \ddot{\Theta}_3 \qquad (4.2.5)$$

where the moments are taken with respect to the midpoint between F_2 and $F_2 + F_2' ds$,

$$M_3 = -\int_A y \sigma_{11} dA \qquad (4.2.6)$$

is the bending moment, and

$$j_3 \equiv \int_A \rho y^2 dA \qquad (4.2.7)$$

is the rotary inertia.

For infinitesimal motions, Θ_3 is small and hence

$$\sin \Theta_3 = \Theta_3, \quad \cos \Theta_3 = 1, \quad \Theta_3 = v' \qquad (4.2.8)$$

Using (4.2.8) in (4.2.3) and (4.2.5) and neglecting the higher-order term $\frac{1}{2}F_2'ds^2$, we obtain

$$F_2' + q_2 = m\ddot{v} \qquad (4.2.9)$$

$$M_3' + F_2 = j_3\ddot{v}' \qquad (4.2.10)$$

Eliminating F_2 from (4.2.9) and (4.2.10) yields

$$-M_3'' + q_2 = m\ddot{v} - (j_3\ddot{v}')' \qquad (4.2.11)$$

Next, we relate M_3 to v. To this end, we note from Figure 4.2.1b that the displacements u_1, u_2, and u_3 in the $x, y,$ and z directions of an arbitrary point on the observed cross-section are given, respectively, by

$$u_1 = -y\sin\Theta_3, \quad u_2 = v - y(1 - \cos\Theta_3), \quad u_3 = 0 \qquad (4.2.12)$$

Substituting (4.2.8) into (4.2.12) yields

$$u_1 = -yv', \quad u_2 = v, \quad u_3 = 0 \qquad (4.2.13)$$

Assuming linear strain-displacement relations, we obtain from (4.2.13) the strains as

$$\epsilon_{11} = \frac{\partial u_1}{\partial s} = -yv'', \quad \epsilon_{12} = \frac{\partial u_1}{\partial y} + \frac{\partial u_2}{\partial s} = 0, \quad \epsilon_{22} = \epsilon_{33} = \epsilon_{13} = \epsilon_{23} = 0 \quad (4.2.14)$$

The strain field (4.2.14) has obvious errors even for small deformations. For example, for a thin beam with no transverse loads on its sides, one expects intuitively that

$$\sigma_{22} = \sigma_{33} = 0 \qquad (4.2.15)$$

instead of $\epsilon_{22} = \epsilon_{33} = 0$, which is impossible because of the Poisson effect. Substituting $\epsilon_{22} = \epsilon_{33} = 0$ into (2.4.44a), (2.4.53), and (2.4.57) yields

$$\sigma_{11} = \frac{E(1-\nu)}{1-\nu-2\nu^2}\epsilon_{11} \qquad (4.2.16)$$

which is erroneous. On the other hand, substituting (4.2.15) into (2.4.44a), (2.4.53), and (2.4.57) and using (4.2.14), we obtain

$$\sigma_{11} = E\epsilon_{11} = -Eyv'', \quad \sigma_{12} = 0 \qquad (4.2.17)$$

Substituting for σ_{11} from (4.2.17) into (4.2.6) yields

$$M_3 = \int_A Ev''y^2 dA = EIv'' \qquad (4.2.18)$$

where

$$I \equiv \int_A y^2 dA \qquad (4.2.19)$$

is the area moment of inertia of the cross-section.

Substituting (4.2.18) into (4.2.11) yields the equation of motion

$$-(EIv'')'' + q_2 = m\ddot{v} - (j_3\ddot{v}')' \tag{4.2.20}$$

From mechanics of materials (e.g., Timoshenko and Goodier, 1970), we know that the boundary conditions are to specify

$$\begin{array}{ccc} v & \text{or} & F_2 \\ v' (= \Theta_3) & \text{or} & M_3 \end{array} \tag{4.2.21}$$

at $s = 0$ and L, where L is the length of the beam. However, to obtain such linear boundary conditions or fully nonlinear boundary conditions, it is better to use variational methods in the derivation because the boundary conditions are by-products of such formulations. We use variational methods in the following sections.

Because $\sigma_{12} = 0$ according to (4.2.17), it follows from (4.2.4) that $F_2 = 0$, which is inconsistent with (4.2.10). Substituting (4.2.18) into (4.2.10), we obtain

$$F_2 = -(EIv'')' + j_3\ddot{v}' \tag{4.2.22}$$

This inconsistency is the result of the assumption that plane sections perpendicular to the beam axis before deformation remain plane and perpendicular to this axis after deformation. The shearing force must be associated with shearing stresses, which in turn must be associated with shearing deformations. These deformations vanish at the beam surface and assume their maxima at the reference axis. This warps the cross-section and contradicts the basic assumption of the Euler-Bernoulli beam theory. To overcome this limitation, a number of shear-deformable theories have been proposed, as discussed in the next section.

4.3 LINEAR SHEAR-DEFORMABLE BEAM THEORIES

Figure 4.3.1 shows the displacements and the free-body diagram of a shear-deformable beam element. Again, we assume that the longitudinal displacement and the stretching effects are negligible and hence (4.2.1) and (4.2.2) are valid. However, to account for the shear deformation of the cross-section, we introduce a dependent variable $\gamma_6(s,t)$, which is the shear rotation angle of the observed cross-section at the reference axis, and a shear-warping function $g_3(y)$ in the displacement field defined as

$$u_1 = -y\sin\Theta_3 + \gamma_6 g_3\cos\Theta_3, \quad u_2 = v - y(1-\cos\Theta_3) + \gamma_6 g_3\sin\Theta_3, \quad u_3 = 0 \tag{4.3.1}$$

Substituting (4.2.8) into (4.3.1) and neglecting nonlinear terms, we obtain

$$u_1 = -yv' + g_3\gamma_6, \quad u_2 = v, \quad u_3 = 0 \tag{4.3.2}$$

Using the linear strain-displacement relations, we obtain from (4.3.2) the strains as

$$\epsilon_{11} = \frac{\partial u_1}{\partial s} = -yv'' + g_3\gamma_6', \quad \epsilon_{12} = \frac{\partial u_1}{\partial y} + \frac{\partial u_2}{\partial s} = \frac{dg_3}{dy}\gamma_6$$

$$\epsilon_{22} = \epsilon_{33} = \epsilon_{13} = \epsilon_{23} = 0 \tag{4.3.3}$$

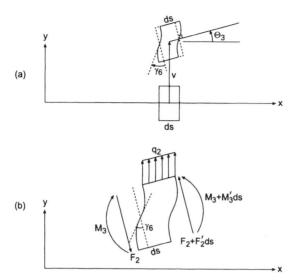

Fig. 4.3.1 A linear shear-deformable beam theory: (a) the displacement v, rotation Θ_3, and shear rotation γ_6 and (b) the free-body diagram.

To derive the equations of motion, we use the extended Hamilton principle, (2.5.37), which states

$$\int_0^t (\delta T - \delta \Pi + \delta W_{nc})\, dt = 0 \qquad (4.3.4)$$

where T is the kinetic energy, Π is the elastic energy, and δW_{nc} is the variation of the nonconservative energy due to external loads and damping. It follows from (4.3.3), (2.5.38), (2.5.39), and (2.5.40) that

$$\delta W_{nc} = \int_0^L q_2 \delta v\, ds \qquad (4.3.5)$$

$$\delta \Pi = \int_0^L \int_A (\sigma_{11}\delta\epsilon_{11} + \sigma_{12}\delta\epsilon_{12})\, dA ds \qquad (4.3.6)$$

$$\delta T = -\int_0^L \int_A \rho \ddot{\mathbf{D}} \cdot \delta \mathbf{D}\, dA ds \qquad (4.3.7)$$

where the displacement vector $\mathbf{D}$ is given by

$$\mathbf{D} = u_1 \mathbf{i}_x + u_2 \mathbf{i}_y + u_3 \mathbf{i}_z = (-yv' + g_3\gamma_6)\,\mathbf{i}_x + v\mathbf{i}_y \qquad (4.3.8)$$

Taking the time derivative and the variation of (4.3.8), we obtain

$$\ddot{\mathbf{D}} = (-y\ddot{v}' + g_3\ddot{\gamma}_6)\,\mathbf{i}_x + \ddot{v}\mathbf{i}_y \qquad (4.3.9)$$

188 BEAMS

$$\delta \mathbf{D} = (-y\delta v' + g_3 \delta\gamma_6)\mathbf{i}_x + \delta v \mathbf{i}_y \qquad (4.3.10)$$

Substituting (4.3.9) and (4.3.10) into (4.3.7), we obtain

$$\delta T = -\int_0^L \int_A \rho \left[\ddot{v}\delta v + (y^2 \ddot{v}' - yg_3\ddot{\gamma}_6)\delta v' + (g_3^2 \ddot{\gamma}_6 - yg_3\ddot{v}')\delta\gamma_6\right] dA\,ds$$

$$= -\int_0^L \left[m\ddot{v}\delta v + (j_3 \ddot{v}' - I_{13}\ddot{\gamma}_6)\delta v' + (I_{33}\ddot{\gamma}_6 - I_{13}\ddot{v}')\delta\gamma_6\right] ds \qquad (4.3.11)$$

where

$$m \equiv \int_A \rho\, dA, \quad j_3 \equiv \int_A \rho y^2\, dA, \quad I_{13} \equiv \int_A \rho y g_3\, dA, \quad I_{33} \equiv \int_A \rho g_3^2\, dA \qquad (4.3.12)$$

Integrating the second term in (4.3.11) by parts yields

$$\delta T = -\int_0^L \left\{\left[m\ddot{v} - (j_3\ddot{v}')' + (I_{13}\ddot{\gamma}_6)'\right]\delta v + (I_{33}\ddot{\gamma}_6 - I_{13}\ddot{v}')\delta\gamma_6\right\} ds$$
$$- \left[(j_3\ddot{v}' - I_{13}\ddot{\gamma}_6)\delta v\right]_0^L \qquad (4.3.13)$$

Substituting (4.3.3) into (4.3.6), we obtain

$$\delta \Pi = \int_0^L \int_A \left(-\sigma_{11} y \delta v'' + \sigma_{11} g_3 \delta\gamma_6' + \sigma_{12} g_{3y} \delta\gamma_6\right) dA\,ds$$

$$= \int_0^L (M_3 \delta v'' + m_3 \delta\gamma_6' + f_2 \delta\gamma_6)\, ds \qquad (4.3.14)$$

where $g_{3y} = \partial g_3/\partial y$ and

$$M_3 \equiv -\int_A \sigma_{11} y\, dA, \quad m_3 \equiv \int_A \sigma_{11} g_3\, dA, \quad f_2 \equiv \int_A \sigma_{12} g_{3y}\, dA \qquad (4.3.15)$$

Integrating the first two terms in (4.3.14) by parts yields

$$\delta \Pi = \int_0^L \left[M_3'' \delta v + (f_2 - m_3')\delta\gamma_6\right] ds + \left[M_3 \delta v' - M_3' \delta v + m_3 \delta\gamma_6\right]_0^L \qquad (4.3.16)$$

Substituting (4.3.5), (4.3.13), and (4.3.16) into (4.3.4), we have

$$\int_0^L \left\{\left[m\ddot{v} - (j_3\ddot{v}')' + (I_{13}\ddot{\gamma}_6)' + M_3'' - q_2\right]\delta v \right.$$
$$\left. + \left[I_{33}\ddot{\gamma}_6 - I_{13}\ddot{v}' - m_3' + f_2\right]\delta\gamma_6\right\} ds$$
$$+ \left[M_3 \delta v' - (M_3' - j_3\ddot{v}' + I_{13}\ddot{\gamma}_6)\delta v + m_3 \delta\gamma_6\right]_0^L = 0 \qquad (4.3.17)$$

Setting each of the coefficients of δv and $\delta\gamma_6$ in the integrand in (4.3.17) equal to zero, we obtain the following equations of motion:

$$-M_3'' + q_2 = m\ddot{v} - (j_3\ddot{v}')' + (I_{13}\ddot{\gamma}_6)' \qquad (4.3.18)$$

$$m_3{}' - f_2 = I_{33}\ddot{\gamma}_6 - I_{13}\ddot{v}' \tag{4.3.19}$$

Then, setting each of the remaining terms in (4.3.17) equal to zero yields the boundary conditions: specify either

$$v \quad \text{or} \quad -M_3' + j_3\ddot{v}' - I_{13}\ddot{\gamma}_6 \tag{4.3.20}$$
$$v' \quad \text{or} \quad M_3 \tag{4.3.21}$$
$$\gamma_6 \quad \text{or} \quad m_3 \tag{4.3.22}$$

at $s = 0$ and L.

Comparing (4.3.18) with (4.2.11), we find that the rotary inertia in the shear-deformable beam theory is $j_3\ddot{v}' - I_{13}\ddot{\gamma}_6$. Using this rotary inertia and the equilibrium condition of moments with respect to the z direction, we obtain from Figure 4.3.1b that

$$M_3' + F_2 = j_3\ddot{v}' - I_{13}\ddot{\gamma}_6 \tag{4.3.23}$$

It follows from (4.3.23) that the natural boundary condition in (4.3.20) is equivalent to specifying F_2, as expected. However, we note that the second term in the rotary inertia $-I_{13}\ddot{\gamma}_6$ is not easily determined by using a vector or Newtonian formulation.

Using the strains (4.3.3) and the stress-strain relations $\sigma_{11} = E\epsilon_{11}$ and $\sigma_{12} = G\epsilon_{12}$, we obtain

$$\sigma_{11} = -E\left(yv'' - g_3\gamma_6'\right) \quad \text{and} \quad \sigma_{12} = Gg_{3y}\gamma_6 \tag{4.3.24}$$

Substituting (4.3.24) into (4.3.15), we have

$$M_3 = EIv'' - F_{13}\gamma_6', \quad m_3 = -F_{13}v'' + F_{33}\gamma_6', \quad f_2 = F_{44}\gamma_6 \tag{4.3.25}$$

where the stiffnesses are defined as

$$I \equiv \int_A y^2 dA, \quad F_{13} \equiv \int_A Eyg_3 dA, \quad F_{33} \equiv \int_A Eg_3^2 dA, \quad F_{44} \equiv \int_A Gg_{3y}^2 dA \tag{4.3.26}$$

Hence, the equations of motion (4.3.18) and (4.3.19) can be rewritten as

$$-(EIv'')'' + (F_{13}\gamma_6')'' + q_2 = m\ddot{v} - (j_3\ddot{v}')' + (I_{13}\ddot{\gamma}_6)' \tag{4.3.27}$$

$$-(F_{13}v'')' + (F_{33}\gamma_6')' - F_{44}\gamma_6 = I_{33}\ddot{\gamma}_6 - I_{13}\ddot{v}' \tag{4.3.28}$$

Equations (4.3.27) and (4.3.28) are linearly coupled by the inertia I_{13} and the stiffness F_{13}. In other words, one needs to solve for v and γ_6 simultaneously.

4.3.1 Third-Order Shear-Deformable Theory

We express g_3 by the following third-degree polynomial

$$g_3 = c_1 y + c_2 y^2 + c_3 y^3 \tag{4.3.29}$$

where the c_i are constants to be determined. Considering isotropic prismatic beams having rectangular cross-sections with thickness h and an axis coinciding with the

190 BEAMS

centroidal axis and assuming that there are no distributed shear loads on the top and bottom surfaces, we have

$$\sigma_{12}\,|_{y=\pm h/2} = 0 \qquad (4.3.30)$$

which, upon using (4.3.24), yields

$$g_{3y}\,|_{y=\pm h/2} = 0 \qquad (4.3.31)$$

Using the definition of γ_6, we have

$$\epsilon_{12}\,|_{y=0} = \gamma_6 \qquad (4.3.32)$$

which, upon using (4.3.3), yields

$$g_{3y}\,|_{y=0} = 1 \qquad (4.3.33)$$

It follows from (4.3.29), (4.3.31), and (4.3.33) that $c_1 = 1$, $c_2 = 0$, and $c_3 = -4/(3h^2)$, and hence

$$g_3 = y - \frac{4}{3h^2}y^3 \qquad (4.3.34)$$

which is the so-called third-order shear deformation function. We note that, if the beam were tapered instead of being prismatic, $\sigma_{12}\,|_{y=\pm h/2} \neq 0$ and (4.3.30) would not hold.

To determine the approximations implied in (4.3.34), we use the linear equilibrium equation (2.5.9a) for an arbitrary elastic particle. Neglecting the inertia $\rho \ddot{u}_1$ and the body force b_1 and using the fact that $\sigma_{31} = 0$, we have

$$\frac{\partial \sigma_{11}}{\partial s} + \frac{\partial \sigma_{12}}{\partial y} = 0 \qquad (4.3.35)$$

Substituting (4.3.24) into (4.3.35) yields

$$G\gamma_6 g_{3yy} + E\gamma_6'' g_3 - Eyv''' = 0 \qquad (4.3.36)$$

In order that the solution of (4.3.36) be a polynomial, γ_6'' must be zero. Then, integrating (4.3.36) yields

$$g_3 = \frac{Ev'''}{G\gamma_6}\left(\frac{1}{6}y^3 + c_1 y + c_2\right) \qquad (4.3.37)$$

Because $g_3(0)$ must be zero so that $u_1(s,0,t) = 0$, (4.3.31) yields $c_1 = -\frac{1}{8}h^2$. Finally, imposing the boundary condition $c_2 = 0$. Then, imposing the condition (4.3.33) demands that

$$Ev'''h^2 = -8G\gamma_6 \qquad (4.3.38)$$

Substituting for c_1 and c_2 in (4.3.37) and using (4.3.38), we obtain (4.3.34).

4.3.2 Timoshenko's Beam Theory

Timoshenko's beam theory (1921) is the so-called first-order shear deformation theory. According to Timoshenko's beam theory, the shear stress resultant F_2 and the shear-strain ϵ_{12} are assumed to be constant and have the forms

$$F_2 \equiv \int_A \sigma_{12} dA = k\bar{\gamma}_6 GA, \quad \epsilon_{12} = \bar{\gamma}_6 \qquad (4.3.39)$$

where $\bar{\gamma}_6$ is a representative shear and k is a correction factor. The shear-strain energy density E_n is derived from (4.3.39) as

$$E_n = \frac{1}{2} F_2 \bar{\gamma}_6 = \frac{1}{2} k \bar{\gamma}_6^2 GA \qquad (4.3.40)$$

Next, we follow Pai and Schulz (1999) to show the physical implications of $\bar{\gamma}_6$ and k.
It follows from (4.3.3) that

$$F_2 \equiv \int_A \sigma_{12} dA = \int_A G g_{3y} \gamma_6 dA = \tilde{\gamma}_6 GA \qquad (4.3.41)$$

$$E_n \equiv \frac{1}{2} \int_A \sigma_{12} \epsilon_{12} dA = \frac{1}{2} \int_A G g_{3y}^2 \gamma_6^2 dA = \frac{1}{2} \hat{\gamma}_6^2 GA \qquad (4.3.42)$$

where $\tilde{\gamma}_6$ is the geometric average of ϵ_{12} and $\hat{\gamma}_6$ is the average of the energy shear-strain. They are defined as

$$\tilde{\gamma}_6 \equiv \frac{\gamma_6 \int_A g_{3y} dA}{A}, \quad \hat{\gamma}_6^2 \equiv \frac{\gamma_6^2 \int_A g_{3y}^2 dA}{A} \qquad (4.3.43)$$

Comparing (4.3.39) with (4.3.41) and (4.3.40) with (4.3.42), we find that

$$\bar{\gamma}_6 = \frac{\hat{\gamma}_6^2}{\tilde{\gamma}_6} = \gamma_6 \frac{\int_A g_{3y}^2 dA}{\int_A g_{3y} dA} \qquad (4.3.44)$$

$$k = \frac{\tilde{\gamma}_6}{\bar{\gamma}_6} = \frac{\tilde{\gamma}_6^2}{\hat{\gamma}_6^2} = \frac{(\int_A g_{3y} dA)^2}{A \int_A g_{3y}^2 dA} \qquad (4.3.45)$$

Equation (4.3.44) indicates that $\bar{\gamma}_6$ is the energy average of ϵ_{12}, and (4.3.45) indicates that k is the ratio of the geometric average shear $\tilde{\gamma}_6$ and the energy average shear $\hat{\gamma}_6$. For isotropic beams with rectangular cross-sections, one can substitute (4.3.34) into (4.3.45) to show that $k = 5/6$.

From (4.3.39) and (4.3.3) one can see that $g_3 = y$ is assumed in Timoshenko's beam theory. Substituting $g_3 = y$ and (4.3.39) into (4.3.12) and (4.3.26) yields

$$I_{13} = I_{33} = j_3, \quad F_{13} = F_{33} = EI, \quad F_{44} = kGA \qquad (4.3.46)$$

Substituting (4.3.46) into (4.3.28) and (4.3.27), replacing γ_6 with $\bar{\gamma}_6$, and assuming the beam to be prismatic, we obtain the equations of motion

$$-EIv''' + EI\bar{\gamma}_6'' - kGA\bar{\gamma}_6 = j_3 \ddot{\bar{\gamma}}_6 - j_3 \ddot{v}' \qquad (4.3.47)$$

$$-EIv^{iv} + EI\bar{\gamma}_6''' + q_2 = m\ddot{v} - j_3\ddot{v}'' + j_3\ddot{\bar{\gamma}}_6' \qquad (4.3.48)$$

Repeating (4.3.47), differentiating it with respect to s, and then subtracting the result from (4.3.48), we obtain

$$-EIv''' + EI\bar{\gamma}_6'' - kGA\bar{\gamma}_6 = j_3\ddot{\bar{\gamma}}_6 - j_3\ddot{v}' \qquad (4.3.49)$$

$$kGA\bar{\gamma}_6' + q_2 = m\ddot{v} \qquad (4.3.50)$$

Equations (4.3.49) and (4.3.50) can be obtained by using a Newtonian approach.

Differentiating (4.3.49) with respect to s and then using (4.3.50) to eliminate $\bar{\gamma}_6$, we obtain the single equation of motion

$$-EIv^{iv} + \frac{EI}{kGA}\left(m\ddot{v}'' - q_2''\right) + q_2 = m\ddot{v} - j_3\ddot{v}'' + \frac{j_3}{kGA}\left(m\frac{\partial^4 v}{\partial t^4} - \ddot{q}_2\right) \qquad (4.3.51)$$

We note that k appears in (4.3.51).

In the literature, the total rotation angle $\psi = \Theta_3 - \bar{\gamma}_6$ of the observed cross-section is often used in the formulation of Timoshenko's beam theory. Using the small angle approximation (4.2.8), we have

$$\bar{\gamma}_6 = v' - \psi \qquad (4.3.52)$$

Using (4.3.52) to eliminate $\bar{\gamma}_6$ from (4.3.49) and (4.3.50), we obtain

$$EI\psi'' + kGA(v' - \psi) = j_3\ddot{\psi} \qquad (4.3.53)$$

$$kGA(v'' - \psi') + q_2 = m\ddot{v} \qquad (4.3.54)$$

which are a set of two coupled equations that need to be solved simultaneously.

We note that the different formulations leading to either the system of equations (4.3.53) and (4.3.54) or the system of equations (4.3.49) and (4.3.50) describe the same dynamic system, and hence they are equivalent.

4.3.3 Layerwise Shear-Deformable Theory

In this section, we determine the shear-warping function g_3 for sandwich beams consisting of N isotropic layers, as shown in Figure 4.3.2. As in (4.3.2), we assume that the displacement field in the ith layer is given by

$$u_1^{(i)} = -yv' + g_3^{(i)}\gamma_6, \quad u_2^{(i)} = v, \quad u_3^{(i)} = 0 \qquad (4.3.55)$$

Using the linear strain-displacement relations, we obtain from (4.3.55) the strains in the ith layer as

$$\epsilon_{11}^{(i)} = \frac{\partial u_1^{(i)}}{\partial s} = -yv'' + g_3^{(i)}\gamma_6', \quad \epsilon_{12}^{(i)} = \frac{\partial u_1^{(i)}}{\partial y} + \frac{\partial u_2^{(i)}}{\partial s} = g_{3y}^{(i)}\gamma_6$$

$$\epsilon_{22}^{(i)} = \epsilon_{33}^{(i)} = \epsilon_{13}^{(i)} = \epsilon_{23}^{(i)} = 0 \qquad (4.3.56)$$

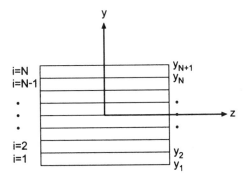

Fig. 4.3.2 The cross-section of a multilayer sandwich beam.

Next, we express the shear-warping function $g_3^{(i)}$ in the ith layer in a third-order polynomial as

$$g_3^{(i)} = y + c_2^{(i)} y^2 + c_3^{(i)} y^3 \qquad (4.3.57)$$

where the $c_j^{(i)}$ are constants to be determined.

We assume that there is no delamination, and hence the displacement u_1 and interlaminar shear stress σ_{12} are continuous across the interface of two contiguous laminae; that is,

$$u_1^{(i)}(s, y_{i+1}, t) - u_1^{(i+1)}(s, y_{i+1}, t) = 0 \text{ for } i = 1, 2, ..., N-1 \qquad (4.3.58)$$

$$\sigma_{12}^{(i)}(s, y_{i+1}, t) - \sigma_{12}^{(i+1)}(s, y_{i+1}, t) = 0 \text{ for } i = 1, 2, ..., N-1 \qquad (4.3.59)$$

where

$$\sigma_{12}^{(i)} = G^{(i)} \epsilon_{12}^{(i)} \qquad (4.3.60)$$

Moreover, we assume that there are no applied shear loads on the bounding surfaces and hence $\sigma_{12} = 0$ at the planes $y = y_1$ and $y = y_{N+1}$; that is,

$$\epsilon_{12}^{(1)}(s, y_1, t) = 0 \qquad (4.3.61)$$

$$\epsilon_{12}^{(N)}(s, y_{N+1}, t) = 0 \qquad (4.3.62)$$

Substituting (4.3.55)-(4.3.57) and (4.3.60) into (4.3.58), (4.3.59), (4.3.61), and (4.3.62) yields

$$y_{i+1}^2 c_2^{(i)} + y_{i+1}^3 c_3^{(i)} - y_{i+1}^2 c_2^{(i+1)} - y_{i+1}^3 c_3^{(i+1)} = 0, \quad i = 1, 2, ..., N-1 \qquad (4.3.63)$$

$$2G^{(i)} y_{i+1} c_2^{(i)} + 3G^{(i)} y_{i+1}^2 c_3^{(i)} - 2G^{(i+1)} y_{i+1} c_2^{(i+1)} - 3G^{(i+1)} y_{i+1}^2 c_3^{(i+1)}$$
$$= G^{(i+1)} - G^{(i)}, \quad i = 1, 2, ..., N-1 \qquad (4.3.64)$$

$$2y_1 c_2^{(1)} + 3y_1^2 c_3^{(1)} = -1 \qquad (4.3.65)$$

$$2y_{N+1} c_2^{(N)} + 3y_{N+1}^2 c_3^{(N)} = -1 \qquad (4.3.66)$$

194 BEAMS

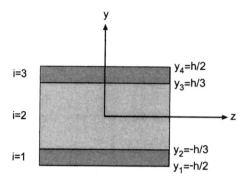

Fig. 4.3.3 The cross-section of a three-layer sandwich beam.

Equations (4.3.63)-(4.3.66) constitute a set of $2N$ linear algebraic equations for the $2N$ unknowns $c_2^{(i)}$ and $c_3^{(i)}$ for $i = 1, 2, ..., N$.

As an example, we consider the three-layer sandwich beam shown in Figure 4.3.3 with $G^{(1)} = G^{(3)} = 10 G^{(2)}$. It follows from (4.3.63)-(4.3.66) that

$$g_3^{(1)} = y + \frac{3}{h}y^2 + \frac{8}{3h^2}y^3, \quad g_3^{(2)} = y - \frac{19}{3h^2}y^3, \quad g_3^{(3)} = y - \frac{3}{h}y^2 + \frac{8}{3h^2}y^3 \quad (4.3.67)$$

To model the beam using Timoshenko's beam theory, one can substitute the obtained shear-warping functions $g_3^{(i)}$ into (4.3.45) and (4.3.44) to obtain the values of k and $\bar{\gamma}_6$ in terms of γ_6.

4.4 MATHEMATICS FOR NONLINEAR MODELING

As pointed out in Section 4.1, local stress and strain measures are needed to fully account for geometric nonlinearities. However, because the local stresses and strains are defined with respect to the deformed system configuration, two coordinate systems are needed: one describes the undeformed system configuration and the other describes the deformed system configuration. The motion of a beam is described by the deformation of its reference line and by inplane and out-of-plane warping displacements with respect to the deformed coordinate system. Different ways of accounting for the warpings result in different beam theories, which are the subjects of Sections 4.5 through 4.7. In this section, the equations describing the deformation of the reference line of a beam are derived.

4.4.1 Coordinate Transformations and Curvatures

In the absence of warpings, a differential beam element is a perfectly rigid thin "plate" having an infinitesimal thickness ds and a fixed finite area. Hence, the motion of such a beam element can be described by three translational displacements and

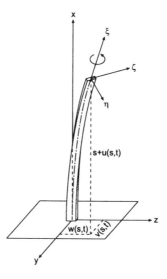

Fig. 4.4.1 Two coordinate systems are used to describe the undeformed and deformed geometries of an initially straight beam, where the xyz system is a Cartesian system describing the undeformed geometry and the $\xi\eta\zeta$ system is a local, orthogonal curvilinear coordinate system describing the deformed geometry.

three rotations. As in rigid-body dynamics, three or two consecutive Euler angles can be used to describe the rotation of such a rigid "plate" from the undeformed position to the deformed position.

A Transformation Using Three Euler Angles: We consider an initially straight beam, as shown in Figure 4.4.1. We introduce two coordinate systems: the Cartesian coordinate system xyz that describes the undeformed geometry, and the orthogonal curvilinear coordinate system $\xi\eta\zeta$ that describes the deformed geometry. Moreover, we let u, v, and w denote the displacements of the reference point on the observed cross-section along the x, y, and z directions, respectively; s denotes the undeformed length from the root of the beam to the reference point; $\mathbf{i}_x$, $\mathbf{i}_y$, and $\mathbf{i}_z$ denote the unit vectors along the x, y, and z axes; and $\mathbf{i}_1$, $\mathbf{i}_2$, and $\mathbf{i}_3$ denote the unit vectors along the ξ, η, and ζ axes. We use three consecutive Euler angles ψ, θ, and φ to describe the rotation from the undeformed position to the deformed one. These angles are executed in the order shown in Figure 4.4.2. First, the xyz system is rotated by an angle ψ about the z axis to an intermediate coordinate system $x_1 y_1 z$; the unit vectors of these two coordinate systems are related by

$$\begin{Bmatrix} \mathbf{i}_{x_1} \\ \mathbf{i}_{y_1} \\ \mathbf{i}_z \end{Bmatrix} = \begin{bmatrix} \cos\psi & \sin\psi & 0 \\ -\sin\psi & \cos\psi & 0 \\ 0 & 0 & 1 \end{bmatrix} \begin{Bmatrix} \mathbf{i}_x \\ \mathbf{i}_y \\ \mathbf{i}_z \end{Bmatrix} \qquad (4.4.1)$$

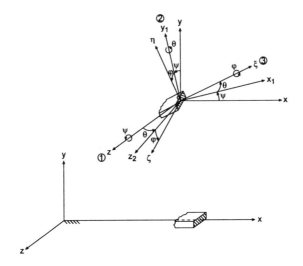

Fig. 4.4.2 Three successive Euler angles describe the rotation of the observed cross-section.

Second, the $x_1 y_1 z$ system is rotated by an angle θ about the y_1 axis to a second intermediate system $\xi y_1 z_2$; the unit vectors of these two systems are related by

$$\left\{ \begin{array}{c} \mathbf{i}_1 \\ \mathbf{i}_{y_1} \\ \mathbf{i}_{z_2} \end{array} \right\} = \left[\begin{array}{ccc} \cos\theta & 0 & -\sin\theta \\ 0 & 1 & 0 \\ \sin\theta & 0 & \cos\theta \end{array} \right] \left\{ \begin{array}{c} \mathbf{i}_{x_1} \\ \mathbf{i}_{y_1} \\ \mathbf{i}_z \end{array} \right\} \qquad (4.4.2)$$

Finally, the $\xi y_1 z_2$ system is rotated by an angle φ about the ξ axis to the system $\xi \eta \zeta$; the unit vectors of these two systems are related by

$$\left\{ \begin{array}{c} \mathbf{i}_1 \\ \mathbf{i}_2 \\ \mathbf{i}_3 \end{array} \right\} = \left[\begin{array}{ccc} 1 & 0 & 0 \\ 0 & \cos\varphi & \sin\varphi \\ 0 & -\sin\varphi & \cos\varphi \end{array} \right] \left\{ \begin{array}{c} \mathbf{i}_1 \\ \mathbf{i}_{y_1} \\ \mathbf{i}_{z_2} \end{array} \right\} \qquad (4.4.3)$$

Therefore, the unit vectors of the two coordinate systems xyz and $\xi \eta \zeta$ are related by

$$\left\{ \begin{array}{c} \mathbf{i}_1 \\ \mathbf{i}_2 \\ \mathbf{i}_3 \end{array} \right\} = [T] \left\{ \begin{array}{c} \mathbf{i}_x \\ \mathbf{i}_y \\ \mathbf{i}_z \end{array} \right\} \qquad (4.4.4)$$

where

$$[T] = \left[\begin{array}{ccc} 1 & 0 & 0 \\ 0 & \cos\varphi & \sin\varphi \\ 0 & -\sin\varphi & \cos\varphi \end{array} \right] \left[\begin{array}{ccc} \cos\theta & 0 & -\sin\theta \\ 0 & 1 & 0 \\ \sin\theta & 0 & \cos\theta \end{array} \right] \left[\begin{array}{ccc} \cos\psi & \sin\psi & 0 \\ -\sin\psi & \cos\psi & 0 \\ 0 & 0 & 1 \end{array} \right]$$

or

$$[T] = \left[\begin{array}{c} \cos\theta \cos\psi \\ -\cos\varphi \sin\psi + \sin\varphi \sin\theta \cos\psi \\ \sin\varphi \sin\psi + \cos\varphi \sin\theta \cos\psi \end{array} \right.$$

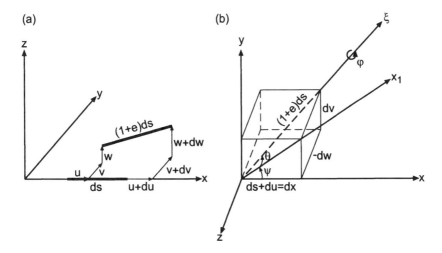

Fig. 4.4.3 Geometric relations between the undeformed and deformed differential elements: (a) displacement relations and (b) relations between displacements and Euler angles.

$$\begin{bmatrix} \cos\theta\sin\psi & & -\sin\theta \\ \cos\varphi\cos\psi + \sin\varphi\sin\theta\sin\psi & & \sin\varphi\cos\theta \\ -\sin\varphi\cos\psi + \cos\varphi\sin\theta\sin\psi & & \cos\varphi\cos\theta \end{bmatrix} \quad (4.4.5)$$

We note that the matrices in (4.4.1)-(4.4.3) are unitary matrices; that is, the columns of each matrix are orthonormal. Consequently, the inverse of each of these matrices is equal to its transpose. Thus, the matrix $[T]$ is unitary and $[T]^{-1} = [T]^T$. Therefore,

$$\begin{Bmatrix} \mathbf{i}_x \\ \mathbf{i}_y \\ \mathbf{i}_z \end{Bmatrix} = [T]^T \begin{Bmatrix} \mathbf{i}_1 \\ \mathbf{i}_2 \\ \mathbf{i}_3 \end{Bmatrix} \quad (4.4.6)$$

It follows from Figure 4.4.3a that

$$(1+e)ds = \sqrt{(ds+du)^2 + dv^2 + dw^2}$$

Hence, the axial strain e on the reference line is

$$e = \sqrt{(1+u')^2 + v'^2 + w'^2} - 1 \quad (4.4.7)$$

where the prime indicates the derivative with respect to s. Also, it follows from Figure 4.4.3b that

$$\cos\psi = \frac{1+u'}{\sqrt{(1+u')^2 + v'^2}}, \quad \sin\psi = \frac{v'}{\sqrt{(1+u')^2 + v'^2}}$$

$$\cos\theta = \frac{\sqrt{(1+u')^2 + v'^2}}{1+e}, \quad \sin\theta = \frac{-w'}{1+e} \quad (4.4.8)$$

198   BEAMS

One can see from (4.4.5) and (4.4.8) that the transformation matrix $[T]$ is a function of four dependent variables u, v, w, and φ.

Since $\mathbf{i}_1, \mathbf{i}_2$, and $\mathbf{i}_3$ are unit vectors along the orthogonal curvilinear coordinate system $\xi\eta\zeta$,

$$\mathbf{i}_j \cdot \mathbf{i}_k = \delta_{jk} \tag{4.4.9}$$

Differentiating (4.4.9) with respect to s yields the identities

$$\mathbf{i}'_j \cdot \mathbf{i}_j = 0, \quad \mathbf{i}'_j \cdot \mathbf{i}_k = -\mathbf{i}'_k \cdot \mathbf{i}_j \quad \text{for } j, k = 1, 2, 3 \tag{4.4.10}$$

Differentiating $\mathbf{i}_1, \mathbf{i}_2$, and $\mathbf{i}_3$ with respect to s and using (4.4.10), we obtain

$$\frac{\partial}{\partial s} \left\{ \begin{array}{c} \mathbf{i}_1 \\ \mathbf{i}_2 \\ \mathbf{i}_3 \end{array} \right\} = [K] \left\{ \begin{array}{c} \mathbf{i}_1 \\ \mathbf{i}_2 \\ \mathbf{i}_3 \end{array} \right\} \tag{4.4.11}$$

where

$$[K] \equiv \begin{bmatrix} \mathbf{i}'_1 \cdot \mathbf{i}_1 & \mathbf{i}'_1 \cdot \mathbf{i}_2 & \mathbf{i}'_1 \cdot \mathbf{i}_3 \\ \mathbf{i}'_2 \cdot \mathbf{i}_1 & \mathbf{i}'_2 \cdot \mathbf{i}_2 & \mathbf{i}'_2 \cdot \mathbf{i}_3 \\ \mathbf{i}'_3 \cdot \mathbf{i}_1 & \mathbf{i}'_3 \cdot \mathbf{i}_2 & \mathbf{i}'_3 \cdot \mathbf{i}_3 \end{bmatrix} = \begin{bmatrix} 0 & \rho_3 & -\rho_2 \\ -\rho_3 & 0 & \rho_1 \\ \rho_2 & -\rho_1 & 0 \end{bmatrix} \tag{4.4.12}$$

$$\rho_1 \equiv \mathbf{i}'_2 \cdot \mathbf{i}_3, \quad \rho_2 \equiv \mathbf{i}'_3 \cdot \mathbf{i}_1, \quad \rho_3 \equiv \mathbf{i}'_1 \cdot \mathbf{i}_2 \tag{4.4.13}$$

Here, $[K]$ is the curvature matrix, ρ_1 is the twisting curvature about the ξ axis, ρ_2 is the bending curvature about the η axis, and ρ_3 is the bending curvature about the ζ axis. Differentiating $\mathbf{i}_2$ in (4.4.4) with respect to s, we have

$$\mathbf{i}'_2 = T'_{21}\mathbf{i}_x + T'_{22}\mathbf{i}_y + T'_{23}\mathbf{i}_z \tag{4.4.14}$$

Substituting for the T_{2i} from (4.4.5) into (4.4.14) yields

$$\begin{aligned}\mathbf{i}'_2 =\ & [(\sin\varphi\sin\psi + \cos\varphi\sin\theta\cos\psi)\varphi' - (\cos\varphi\cos\psi + \sin\varphi\sin\theta\sin\psi)\psi' \\ & + (\sin\varphi\cos\theta\cos\psi)\theta']\mathbf{i}_x + [-(\sin\varphi\cos\psi - \cos\varphi\sin\theta\sin\psi)\varphi' \\ & -(\cos\varphi\sin\psi - \sin\varphi\sin\theta\cos\psi)\psi' + (\sin\varphi\cos\theta\sin\psi)\theta']\mathbf{i}_y \\ & + [(\cos\varphi\cos\theta)\varphi' - (\sin\varphi\sin\theta)\theta']\mathbf{i}_z \end{aligned} \tag{4.4.15}$$

Substituting for $\mathbf{i}_3$ from (4.4.4) and (4.4.5) into the first equation of (4.4.13) and using (4.4.14) and (4.4.15), we obtain

$$\rho_1 = \mathbf{i}'_2 \cdot \mathbf{i}_3 = \sum_{i=1}^{3} T'_{2i} T_{3i} = \varphi' - \psi' \sin\theta \tag{4.4.16}$$

Similarly, it follows from (4.4.13) and (4.4.5) that

$$\rho_2 = \mathbf{i}'_3 \cdot \mathbf{i}_1 = \sum_{i=1}^{3} T'_{3i} T_{1i} = \psi' \cos\theta \sin\varphi + \theta' \cos\varphi \tag{4.4.17}$$

$$\rho_3 = \mathbf{i}'_1 \cdot \mathbf{i}_2 = \sum_{i=1}^{3} T'_{1i}T_{2i} = \psi' \cos\theta \cos\varphi - \theta' \sin\varphi \tag{4.4.18}$$

We note that the ρ_i are normalized curvatures, not the real curvatures, because the differentiation is with respect to the undeformed length ds instead of the actual deformed length $(1+e)ds$.

Alternatively, the curvatures can be determined using Kirchhoff's kinetic analogy (Love, 1944). To accomplish this, we determine the angular velocity $\vec{\omega}$ of the element in Figure 4.4.2; that is,

$$\vec{\omega} = \dot{\varphi}\mathbf{i}_1 + \dot{\theta}\mathbf{i}_{y_1} + \dot{\psi}\mathbf{i}_z \tag{4.4.19}$$

It follows from the inverse of (4.4.3) that

$$\mathbf{i}_{y_1} = \cos\varphi\,\mathbf{i}_2 - \sin\varphi\,\mathbf{i}_3 \tag{4.4.20}$$

Moreover, it follows from (4.4.6) and (4.4.5) that

$$\mathbf{i}_z = \sum_{i=1}^{3} T_{i3}\mathbf{i}_i = -\sin\theta\,\mathbf{i}_1 + \sin\varphi\cos\theta\,\mathbf{i}_2 + \cos\varphi\cos\theta\,\mathbf{i}_3 \tag{4.4.21}$$

Substituting (4.4.20) and (4.4.21) into (4.4.19) yields

$$\vec{\omega} = \omega_1 \mathbf{i}_1 + \omega_2 \mathbf{i}_2 + \omega_3 \mathbf{i}_3 \tag{4.4.22}$$

where

$$\begin{aligned}\omega_1 &= \dot{\varphi} - \dot{\psi}\sin\theta \\ \omega_2 &= \dot{\psi}\sin\varphi\cos\theta + \dot{\theta}\cos\varphi \\ \omega_3 &= \dot{\psi}\cos\varphi\cos\theta - \dot{\theta}\sin\varphi\end{aligned} \tag{4.4.23}$$

Then, according to Kirchhoff's kinetic analogy, the curvatures can be obtained from (4.4.23) by simply replacing the time derivative $\partial/\partial t$ with the spatial derivative $\partial/\partial s$. The result is the same as (4.4.16)-(4.4.18).

Similar to (4.4.11)-(4.4.13), the angular velocity components can be derived as

$$\frac{d}{dt}\begin{Bmatrix}\mathbf{i}_1 \\ \mathbf{i}_2 \\ \mathbf{i}_3\end{Bmatrix} = [\omega]\begin{Bmatrix}\mathbf{i}_1 \\ \mathbf{i}_2 \\ \mathbf{i}_3\end{Bmatrix}, \quad [\omega] \equiv \begin{bmatrix}0 & \omega_3 & -\omega_2 \\ -\omega_3 & 0 & \omega_1 \\ \omega_2 & -\omega_1 & 0\end{bmatrix} \tag{4.4.24}$$

$$\omega_1 \equiv \dot{\mathbf{i}}_2 \cdot \mathbf{i}_3 = \sum_{i=1}^{3}\dot{T}_{2i}T_{3i},\quad \omega_2 \equiv -\dot{\mathbf{i}}_1 \cdot \mathbf{i}_3 = -\sum_{i=1}^{3}\dot{T}_{1i}T_{3i},\quad \omega_3 \equiv \dot{\mathbf{i}}_1 \cdot \mathbf{i}_2 = \sum_{i=1}^{3}\dot{T}_{1i}T_{2i} \tag{4.4.25}$$

Substituting for the T_{ij} from (4.4.5) into (4.4.25) yields (4.4.23).

200 BEAMS

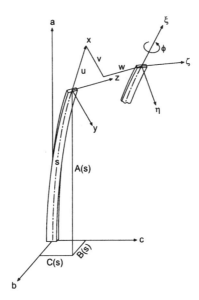

Fig. 4.4.4 Three coordinate systems are used to describe the undeformed and deformed geometries of an initially curved beam, where the abc system is a Cartesian system for reference use, the xyz system is an orthogonal curvilinear system describing the undeformed geometry, and the $\xi\eta\zeta$ system is a local, orthogonal curvilinear coordinate system describing the deformed geometry.

A Transformation Using Two Euler Angles: We consider a naturally curved and twisted beam, as shown in Figure 4.4.4. We employ three coordinate systems. The abc system is a reference Cartesian coordinate system; the xyz system is an orthogonal curvilinear coordinate system in which the x axis is formed by connecting the reference points of all cross-sections of the undeformed beam and is called the reference line; and the $\xi\eta\zeta$ system is a local orthogonal curvilinear coordinate system in which the ξ axis represents the deformed reference line and the η and ζ axes represent the deformed y and z axes in the absence of inplane and out-of-plane warpings. Moreover, we let $\mathbf{i}_a$, $\mathbf{i}_b$, and $\mathbf{i}_c$ denote the unit vectors of the abc coordinate system; $\mathbf{i}_x$, $\mathbf{i}_y$, and $\mathbf{i}_z$ denote the unit vectors of the xyz coordinate system; and $\mathbf{i}_1$, $\mathbf{i}_2$, and $\mathbf{i}_3$ denote the unit vectors of the $\xi\eta\zeta$ coordinate system. Furthermore, we let s denote the undeformed arclength along the x axis from the root of the beam to the observed reference point; t denotes time; and u, v, and w represent the displacement components of the observed reference point with respect to the x, y, and z axes, respectively.

The undeformed position vector $\mathbf{R}$ (see Figure 4.4.4) of the observed reference point is known and given by

$$\mathbf{R} = A(s)\mathbf{i}_a + B(s)\mathbf{i}_b + C(s)\mathbf{i}_c \tag{4.4.26}$$

Also, the angles θ_{21}, θ_{22}, and θ_{23} of the y axis with respect to the abc system are assumed to be known and given by

$$\theta_{21} = \cos^{-1}(\mathbf{i}_y \cdot \mathbf{i}_a), \quad \theta_{22} = \cos^{-1}(\mathbf{i}_y \cdot \mathbf{i}_b), \quad \theta_{23} = \cos^{-1}(\mathbf{i}_y \cdot \mathbf{i}_c) \quad (4.4.27)$$

where the θ_{2i} are functions of s only and $0 \le \theta_{2i} \le 180°$. It follows from (4.4.26) that

$$\mathbf{i}_x = \mathbf{R}' = A'\mathbf{i}_a + B'\mathbf{i}_b + C'\mathbf{i}_c \quad (4.4.28)$$

Using (4.4.27) and (4.4.28) and the identity $\mathbf{i}_z = \mathbf{i}_x \times \mathbf{i}_y$, we obtain

$$\begin{Bmatrix} \mathbf{i}_x \\ \mathbf{i}_y \\ \mathbf{i}_z \end{Bmatrix} = [T^x] \begin{Bmatrix} \mathbf{i}_a \\ \mathbf{i}_b \\ \mathbf{i}_c \end{Bmatrix} \quad (4.4.29)$$

where the transformation matrix $[T^x]$ is given by

$$[T^x] = \begin{bmatrix} A' & B' & C' \\ \cos\theta_{21} & \cos\theta_{22} & \cos\theta_{23} \\ B'\cos\theta_{23} - C'\cos\theta_{22} & C'\cos\theta_{21} - A'\cos\theta_{23} & A'\cos\theta_{22} - B'\cos\theta_{21} \end{bmatrix} \quad (4.4.30)$$

Using (4.4.29) and (4.4.30) and the orthonormality property of $\mathbf{i}_x$, $\mathbf{i}_y$, and $\mathbf{i}_z$ (e.g., $\mathbf{i}'_x \cdot \mathbf{i}_x = 0$, $\mathbf{i}'_x \cdot \mathbf{i}_y = -\mathbf{i}'_y \cdot \mathbf{i}_x$), we obtain

$$\frac{d}{ds}\begin{Bmatrix} \mathbf{i}_x \\ \mathbf{i}_y \\ \mathbf{i}_z \end{Bmatrix} = [k]\begin{Bmatrix} \mathbf{i}_x \\ \mathbf{i}_y \\ \mathbf{i}_z \end{Bmatrix} \quad (4.4.31)$$

where

$$[k] \equiv \begin{bmatrix} \mathbf{i}'_x \cdot \mathbf{i}_x & \mathbf{i}'_x \cdot \mathbf{i}_y & \mathbf{i}'_x \cdot \mathbf{i}_z \\ \mathbf{i}'_y \cdot \mathbf{i}_x & \mathbf{i}'_y \cdot \mathbf{i}_y & \mathbf{i}'_y \cdot \mathbf{i}_z \\ \mathbf{i}'_z \cdot \mathbf{i}_x & \mathbf{i}'_z \cdot \mathbf{i}_y & \mathbf{i}'_z \cdot \mathbf{i}_z \end{bmatrix} = \begin{bmatrix} 0 & k_3 & -k_2 \\ -k_3 & 0 & k_1 \\ k_2 & -k_1 & 0 \end{bmatrix} \quad (4.4.32)$$

$$k_1 = \frac{d\mathbf{i}_y}{\partial s} \cdot \mathbf{i}_z = \sum_{i=1}^{3} \frac{dT^x_{2i}}{ds} T^x_{3i}$$

$$k_2 = -\frac{d\mathbf{i}_x}{\partial s} \cdot \mathbf{i}_z = -\sum_{i=1}^{3} \frac{dT^x_{1i}}{ds} T^x_{3i}$$

$$k_3 = \frac{d\mathbf{i}_x}{\partial s} \cdot \mathbf{i}_y = \sum_{i=1}^{3} \frac{dT^x_{1i}}{ds} T^x_{2i} \quad (4.4.33)$$

Here, k_1, k_2, and k_3 are the initial curvatures with respect to the x, y, and z axes, respectively, and they are functions of s only.

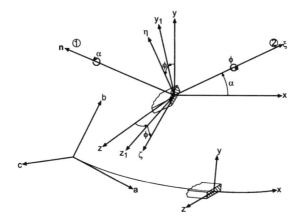

Fig. 4.4.5 Two successive Euler angles describe the rotation of the observed cross-section.

Following Alkire (1984), we use two sequential Euler angles α and ϕ to describe the rotation of the observed element from the undeformed position to the deformed position. The system xyz is rotated by an angle α about the n axis to the intermediate system $\xi y_1 z_1$, as shown in Figure 4.4.5. The n axis is defined later. The transformation relating these two coordinate systems is

$$\begin{Bmatrix} \mathbf{i}_1 \\ \mathbf{i}_{\hat{2}} \\ \mathbf{i}_{\hat{3}} \end{Bmatrix} = [B(\alpha)] \begin{Bmatrix} \mathbf{i}_x \\ \mathbf{i}_y \\ \mathbf{i}_z \end{Bmatrix} \qquad (4.4.34)$$

where $\mathbf{i}_{\hat{2}}$ and $\mathbf{i}_{\hat{3}}$ are unit vectors along the y_1 and z_1 axes, respectively. The transformation matrix $[B(\alpha)]$ is due to the bending rotation α, which rotates the x axis to the ξ axis, the y axis to the y_1 axis, and the z axis to the z_1 axis. We note that the angles between y and y_1 axes and z and z_1 axes are not equal to α because the yy_1 and zz_1 planes are not perpendicular to the n axis.

After the rotation α, the $\xi y_1 z_1$ system is rotated by an angle ϕ about the ξ axis to the $\xi\eta\zeta$ system. The transformation matrix that relates these two coordinate systems is given by

$$\begin{Bmatrix} \mathbf{i}_1 \\ \mathbf{i}_2 \\ \mathbf{i}_3 \end{Bmatrix} = \begin{bmatrix} 1 & 0 & 0 \\ 0 & \cos\phi & \sin\phi \\ 0 & -\sin\phi & \cos\phi \end{bmatrix} \begin{Bmatrix} \mathbf{i}_1 \\ \mathbf{i}_{\hat{2}} \\ \mathbf{i}_{\hat{3}} \end{Bmatrix} \qquad (4.4.35)$$

The second rotation ϕ is related to the torsional motion about the bent reference axis ξ. Hence the transformation which relates the undeformed coordinate system xyz to the deformed coordinate system $\xi\eta\zeta$ is

$$\begin{Bmatrix} \mathbf{i}_1 \\ \mathbf{i}_2 \\ \mathbf{i}_3 \end{Bmatrix} = [T] \begin{Bmatrix} \mathbf{i}_x \\ \mathbf{i}_y \\ \mathbf{i}_z \end{Bmatrix}, \quad [T] = \begin{bmatrix} 1 & 0 & 0 \\ 0 & \cos\phi & \sin\phi \\ 0 & -\sin\phi & \cos\phi \end{bmatrix} [B(\alpha)] \qquad (4.4.36)$$

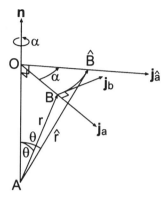

Fig. 4.4.6 The rotational transformation of an arbitrary vector r with respect to n.

Next, we represent $[B(\alpha)]$ in terms of the displacements u, v, and w of the reference point of the observed cross-section. First, we relate $[B(\alpha)]$ to α and the components n_1, n_2, and n_3 of the unit vector $\mathbf{n}$ ($= n_1\mathbf{i}_x + n_2\mathbf{i}_y + n_3\mathbf{i}_z$). To accomplish this, we derive the relationship between an arbitrary vector $\mathbf{r}$ and a vector $\hat{\mathbf{r}}$ obtained by rotating $\mathbf{r}$ by an angle α about an axis n. In Figure 4.4.6, we show a plate OAB rotated by an angle α about $\overline{AO}$. The line $\overline{AO}$ is perpendicular to $\overline{OB}$, and $\mathbf{n}, \mathbf{j}_a, \mathbf{j}_b$, and $\mathbf{j}_{\hat{a}}$ are unit vectors. It follows from Figure 4.4.6 that

$$\mathbf{r} = r\cos\theta\,\mathbf{n} + r\sin\theta\,\mathbf{j}_a \tag{4.4.37}$$

$$\hat{\mathbf{r}} = r\cos\theta\,\mathbf{n} + r\sin\theta\,\mathbf{j}_{\hat{a}} \tag{4.4.38}$$

$$\mathbf{j}_{\hat{a}} = \cos\alpha\,\mathbf{j}_a + \sin\alpha\,\mathbf{j}_b \tag{4.4.39}$$

It follows from (4.4.37) and Figure 4.4.6 that

$$\mathbf{j}_b = \frac{\mathbf{n}\times\mathbf{r}}{|\mathbf{n}\times\mathbf{r}|} = \frac{1}{r\sin\theta}\mathbf{n}\times\mathbf{r} \tag{4.4.40}$$

Moreover, it follows from (4.4.40) and Figure 4.4.6 that

$$\mathbf{j}_a = \mathbf{j}_b \times \mathbf{n} = \frac{1}{r\sin\theta}(\mathbf{n}\times\mathbf{r})\times\mathbf{n}$$

$$= \frac{1}{r\sin\theta}[(\mathbf{n}\cdot\mathbf{n})\mathbf{r} - (\mathbf{r}\cdot\mathbf{n})\mathbf{n}]$$

$$= \frac{1}{r\sin\theta}[\mathbf{r} - (\mathbf{r}\cdot\mathbf{n})\mathbf{n}] \tag{4.4.41}$$

Furthermore, it follows from (4.4.37) that

$$r\cos\theta = \mathbf{r}\cdot\mathbf{n} \tag{4.4.42}$$

204 BEAMS

Substituting (4.4.39)-(4.4.42) into (4.4.38), we obtain

$$\hat{\mathbf{r}} = (1 - \cos\alpha)(\mathbf{r}\cdot\mathbf{n})\mathbf{n} + \cos\alpha\,\mathbf{r} + \sin\alpha\,\mathbf{n}\times\mathbf{r} \tag{4.4.43}$$

which shows the relation between the arbitrary vector $\mathbf{r}$ and its rotated vector $\hat{\mathbf{r}}$.

Because $\mathbf{i}_x$ is transformed into $\mathbf{i}_{\hat{1}}$ in Figure 4.4.5, it follows from (4.4.43) that

$$\mathbf{i}_{\hat{1}} = (1 - \cos\alpha)n_1\mathbf{n} + \cos\alpha\,\mathbf{i}_x + \sin\alpha\,(n_3\mathbf{i}_y - n_2\mathbf{i}_z) \tag{4.4.44}$$

Similarly, because $\mathbf{i}_y$ is transformed into $\mathbf{i}_{\hat{2}}$ (along the y_1 axis) in Figure 4.4.5, it follows from (4.4.43) that

$$\mathbf{i}_{\hat{2}} = (1 - \cos\alpha)n_2\mathbf{n} + \cos\alpha\,\mathbf{i}_y + \sin\alpha\,(-n_3\mathbf{i}_x + n_1\mathbf{i}_z) \tag{4.4.45}$$

Finally, because $\mathbf{i}_z$ is transformed into $\mathbf{i}_{\hat{3}}$ (along the z_1 axis) in Figure 4.4.5, it follows from (4.4.43) that

$$\mathbf{i}_{\hat{3}} = (1 - \cos\alpha)n_3\mathbf{n} + \cos\alpha\,\mathbf{i}_z + \sin\alpha\,(n_2\mathbf{i}_x - n_1\mathbf{i}_y) \tag{4.4.46}$$

Substituting (4.4.44)-(4.4.46) into (4.4.34), we obtain

$$[B(\alpha)] = \begin{bmatrix} n_1^2 & \text{sym.} & \\ n_1n_2 & n_2^2 & \\ n_1n_3 & n_2n_3 & n_3^2 \end{bmatrix}(1 - \cos\alpha) + \begin{bmatrix} 0 & n_3 & -n_2 \\ -n_3 & 0 & n_1 \\ n_2 & -n_1 & 0 \end{bmatrix}\sin\alpha$$
$$+ [I]\cos\alpha \tag{4.4.47}$$

where $[I]$ is a 3×3 identity matrix.

Next, we relate the n_i and α to u, v, and w. To this end, we show in Figure 4.4.7 the relationship between the reference line and the Euler angles α and ϕ. It follows

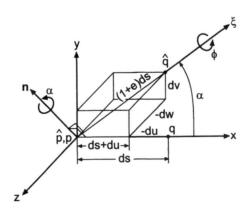

Fig. 4.4.7 The relationship between the reference line and the Euler angles α and ϕ.

from Figure 4.4.7, (4.4.31) and (4.4.32) that the displacement vectors of Points p and q are

$$p: \mathbf{D}_1 = u\mathbf{i}_x + v\mathbf{i}_y + w\mathbf{i}_z \quad (4.4.48)$$

$$\begin{aligned} q: \mathbf{D}_2 &= \mathbf{D}_1 + \frac{\partial \mathbf{D}_1}{\partial s} ds \\ &= \mathbf{D}_1 + \big[(u' - vk_3 + wk_2)\mathbf{i}_x + (v' + uk_3 - wk_1)\mathbf{i}_y \\ &\quad + (w' - uk_2 + vk_1)\mathbf{i}_z\big] ds \end{aligned} \quad (4.4.49)$$

Hence, the vector $\vec{pq}$ is deformed into

$$\begin{aligned} \vec{pq} &= ds\mathbf{i}_x + \mathbf{D}_2 - \mathbf{D}_1 \\ &= \big[(1 + u' - vk_3 + wk_2)\mathbf{i}_x + (v' + uk_3 - wk_1)\mathbf{i}_y \\ &\quad + (w' - uk_2 + vk_1)\mathbf{i}_z\big] ds \end{aligned} \quad (4.4.50)$$

Therefore,

$$\mathbf{i}_1 = \frac{\vec{pq}}{(1+e)ds} = T_{11}\mathbf{i}_x + T_{12}\mathbf{i}_y + T_{13}\mathbf{i}_z \quad (4.4.51)$$

where

$$T_{11} = \frac{1 + u' - vk_3 + wk_2}{1+e}, \quad T_{12} = \frac{v' + uk_3 - wk_1}{1+e}, \quad T_{13} = \frac{w' - uk_2 + vk_1}{1+e} \quad (4.4.52)$$

It follows from (4.4.50) and Figure 4.4.7 that the relationship between the axial strain e and the displacements is

$$\begin{aligned} e &= \frac{\vec{pq} - ds}{ds} \\ &= \sqrt{(1 + u' - vk_3 + wk_2)^2 + (v' + uk_3 - wk_1)^2 + (w' - uk_2 + vk_1)^2} - 1 \end{aligned} \quad (4.4.53)$$

As shown in Figure 4.4.7, we choose the n axis to be

$$\mathbf{n} \equiv \frac{\mathbf{i}_x \times \mathbf{i}_1}{|\mathbf{i}_x \times \mathbf{i}_1|} = n_1\mathbf{i}_x + n_2\mathbf{i}_y + n_3\mathbf{i}_z \quad (4.4.54)$$

Substituting for $\mathbf{i}_1$ from (4.4.51) into (4.4.54), we obtain

$$n_1 = 0, \quad n_2 = -\frac{T_{13}}{\sqrt{T_{12}^2 + T_{13}^2}}, \quad n_3 = \frac{T_{12}}{\sqrt{T_{12}^2 + T_{13}^2}} \quad (4.4.55)$$

Substituting (4.4.55) into (4.4.47), assuming $0 \leq \alpha < 180°$, and using the relationships

$$T_{11}^2 + T_{12}^2 + T_{13}^2 = 1, \quad \cos\alpha = \mathbf{i}_1 \cdot \mathbf{i}_x = T_{11}, \quad \sin\alpha = |\mathbf{i}_1 \times \mathbf{i}_x| = \sqrt{T_{12}^2 + T_{13}^2} \quad (4.4.56)$$

we obtain

$$[B(\alpha)] = \begin{bmatrix} T_{11} & T_{12} & T_{13} \\ -T_{12} & T_{11} + T_{13}^2/(1+T_{11}) & -T_{12}T_{13}/(1+T_{11}) \\ -T_{13} & -T_{12}T_{13}/(1+T_{11}) & T_{11} + T_{12}^2/(1+T_{11}) \end{bmatrix} \quad (4.4.57)$$

We note that $[B(\alpha)]$ is indeterminate when $T_{11} = -1$, which corresponds to $\alpha = 180°$ and $T_{12} = T_{13} = 0$ according to (4.4.56). Hence, it follows from (4.4.55) that $n_1 = 0$ and n_2 and n_3 are indeterminate. In this case, (4.4.47) reduces to

$$[B(\alpha)] = \begin{bmatrix} -1 & 0 & 0 \\ 0 & 2n_2^2 - 1 & 2n_2n_3 \\ 0 & 2n_2n_3 & 2n_3^2 - 1 \end{bmatrix} \quad \text{for} \quad \alpha = 180° \quad (4.4.58)$$

Using the concept of continuity, one can determine the values of n_2, n_3, and $[B(\alpha)]$ at a particular point $s = s_p$ by comparing (4.4.57) with (4.4.58) at adjacent points $s = s_p^-$ or s_p^+. Hence, any arbitrary deformation (i.e., $0 \leq \alpha \leq 180°$) can be modeled.

Differentiating (4.4.36) with respect to s yields

$$\frac{\partial}{\partial s} \begin{Bmatrix} \mathbf{i}_1 \\ \mathbf{i}_2 \\ \mathbf{i}_3 \end{Bmatrix} = [K] \begin{Bmatrix} \mathbf{i}_1 \\ \mathbf{i}_2 \\ \mathbf{i}_3 \end{Bmatrix} = \frac{\partial [T]}{\partial s} \begin{Bmatrix} \mathbf{i}_x \\ \mathbf{i}_y \\ \mathbf{i}_z \end{Bmatrix} + [T] \frac{d}{ds} \begin{Bmatrix} \mathbf{i}_x \\ \mathbf{i}_y \\ \mathbf{i}_z \end{Bmatrix} \quad (4.4.59)$$

where

$$[K] \equiv \begin{bmatrix} \mathbf{i}_1' \cdot \mathbf{i}_1 & \mathbf{i}_1' \cdot \mathbf{i}_2 & \mathbf{i}_1' \cdot \mathbf{i}_3 \\ \mathbf{i}_2' \cdot \mathbf{i}_1 & \mathbf{i}_2' \cdot \mathbf{i}_2 & \mathbf{i}_2' \cdot \mathbf{i}_3 \\ \mathbf{i}_3' \cdot \mathbf{i}_1 & \mathbf{i}_3' \cdot \mathbf{i}_2 & \mathbf{i}_3' \cdot \mathbf{i}_3 \end{bmatrix} = \begin{bmatrix} 0 & \rho_3 & -\rho_2 \\ -\rho_3 & 0 & \rho_1 \\ \rho_2 & -\rho_1 & 0 \end{bmatrix}$$

$$= [T]'[T]^T + [T][k][T]^T \quad (4.4.60)$$

Here, the ρ_i are the deformed curvatures. It follows from (4.4.36), (4.4.59), and (4.4.60) that

$$\rho_1 = \mathbf{i}_2' \cdot \mathbf{i}_3 = \left(T_{21}'\mathbf{i}_x + T_{22}'\mathbf{i}_y + T_{23}'\mathbf{i}_z + T_{21}\mathbf{i}_x' + T_{22}\mathbf{i}_y' + T_{23}\mathbf{i}_z'\right)$$
$$\cdot (T_{31}\mathbf{i}_x + T_{32}\mathbf{i}_y + T_{33}\mathbf{i}_z)$$

which, upon using (4.4.31) and (4.4.32), becomes

$$\rho_1 = \sum_{i=1}^{3} T_{2i}'T_{3i} + (T_{22}T_{33} - T_{23}T_{32})k_1 + (T_{23}T_{31} - T_{21}T_{33})k_2$$
$$+ (T_{21}T_{32} - T_{22}T_{31})k_3 \quad (4.4.61)$$

But

$$[T]^T = [T]^{-1} = \begin{bmatrix} T_{22}T_{33} - T_{23}T_{32} & T_{13}T_{32} - T_{12}T_{33} & T_{12}T_{23} - T_{13}T_{22} \\ T_{23}T_{31} - T_{21}T_{33} & T_{11}T_{33} - T_{13}T_{31} & T_{13}T_{21} - T_{11}T_{23} \\ T_{21}T_{32} - T_{22}T_{31} & T_{12}T_{31} - T_{11}T_{32} & T_{11}T_{22} - T_{12}T_{21} \end{bmatrix}$$
$$(4.4.62)$$

Hence

$$T_{22}T_{33} - T_{23}T_{32} = T_{11}, \quad T_{23}T_{31} - T_{21}T_{33} = T_{12}, \quad T_{21}T_{32} - T_{22}T_{31} = T_{13} \quad (4.4.63)$$

Substituting (4.4.63) into (4.4.61) yields

$$\rho_1 \equiv \mathbf{i_2}' \cdot \mathbf{i_3} = \sum_{i=1}^{3}(T'_{2i}T_{3i} + T_{1i}k_i) \quad (4.4.64)$$

Following a procedure similar to the above, we obtain

$$\rho_2 \equiv -\mathbf{i_1}' \cdot \mathbf{i_3} = \sum_{i=1}^{3}(-T'_{1i}T_{3i} + T_{2i}k_i) \quad (4.4.65)$$

$$\rho_3 \equiv \mathbf{i_1}' \cdot \mathbf{i_2} = \sum_{i=1}^{3}(T'_{1i}T_{2i} + T_{3i}k_i) \quad (4.4.66)$$

We note that, in the absence of elastic deformations, $[T]$ is an identity matrix and $[K] = [k]$.

If the xyz system has an angular velocity $\vec{\omega}^x = \omega_1^x \mathbf{i}_x + \omega_2^x \mathbf{i}_y + \omega_3^x \mathbf{i}_z$, one can follow (4.4.59)-(4.4.66) to show that the angular velocity of the $\xi\eta\zeta$ system is $\vec{\omega} = \omega_1 \mathbf{i_1} + \omega_2 \mathbf{i_2} + \omega_3 \mathbf{i_3}$, where

$$\omega_1 \equiv \frac{d\mathbf{i_2}}{dt} \cdot \mathbf{i_3} = \sum_{i=1}^{3}(\dot{T}_{2i}T_{3i} + T_{1i}\omega_i^x)$$

$$\omega_2 \equiv -\frac{d\mathbf{i_1}}{dt} \cdot \mathbf{i_3} = \sum_{i=1}^{3}(-\dot{T}_{1i}T_{3i} + T_{2i}\omega_i^x)$$

$$\omega_3 \equiv \frac{d\mathbf{i_1}}{dt} \cdot \mathbf{i_2} = \sum_{i=1}^{3}(\dot{T}_{1i}T_{2i} + T_{3i}\omega_i^x)$$

Replacing the time derivative with the spatial derivative and ω_i^x with k_i in ω_i yields the expressions of ρ_i in (4.6.64)-(4.6.66), which reveals that Kirchhoff's kinetic analogy is also applicable to curved beams. If the xyz system is fixed, $\dot{\mathbf{i}}_x = \dot{\mathbf{i}}_y = \dot{\mathbf{i}}_z = \omega_i^x = 0$ and (4.4.25) is also valid for curved beams.

4.4.2 Concept of Orthogonal Virtual Rotations

Variations of the unit vectors $\mathbf{i}_j$ in Figures 4.4.1 and 4.4.4 are due to virtual rigid-body rotations of the coordinate system $\xi\eta\zeta$ and are given by

$$\left\{\begin{array}{c} \delta\mathbf{i_1} \\ \delta\mathbf{i_2} \\ \delta\mathbf{i_3} \end{array}\right\} = \left[\begin{array}{ccc} 0 & \delta\theta_3 & -\delta\theta_2 \\ -\delta\theta_3 & 0 & \delta\theta_1 \\ \delta\theta_2 & -\delta\theta_1 & 0 \end{array}\right] \left\{\begin{array}{c} \mathbf{i_1} \\ \mathbf{i_2} \\ \mathbf{i_3} \end{array}\right\} \quad (4.4.67)$$

208 BEAMS

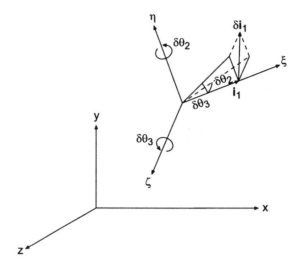

Fig. 4.4.8 The relationship between δi_1 and $\delta\theta_2$ and $\delta\theta_3$.

where $\delta\theta_1$, $\delta\theta_2$, and $\delta\theta_3$ are virtual rigid-body rotations with respect to the ξ, η, and ζ axes, respectively. We note that the $\delta\theta_i$ are infinitesimal rotations and hence they are vector quantities. Moreover, the $\delta\theta_i$ are along three perpendicular directions and hence they are mutually independent. Figure 4.4.8 shows the geometric relationship between δi_1 and $\delta\theta_k$. Next, we develop expressions for $\delta e, \delta\theta_2$, and $\delta\theta_3$ in terms of u, v, and w for initially straight and initially curved beams, respectively.

Initially Straight Beams: For the initially straight beam shown in Figure 4.4.1, the transformation matrix $[T]$ that is obtained using three Euler angles and relates the $\mathbf{i}_k$ to the unit vectors $\mathbf{i}_x, \mathbf{i}_y$, and $\mathbf{i}_z$ is given in (4.4.5), (4.4.7), and (4.4.8). It follows from (4.4.7) that

$$(1+e)^2 = (1+u')^2 + v'^2 + w'^2 \tag{4.4.68}$$

and from (4.4.5) and (4.4.8) that

$$T_{11} = \frac{1+u'}{1+e}, \quad T_{12} = \frac{v'}{1+e}, \quad \text{and } T_{13} = \frac{w'}{1+e} \tag{4.4.69}$$

Moreover, it follows from (4.4.4) and (4.4.69) that

$$(1+e)\mathbf{i}_1 = (1+u')\mathbf{i}_x + v'\mathbf{i}_y + w'\mathbf{i}_z \tag{4.4.70}$$

Taking the variation of (4.4.68), we have

$$(1+e)\delta e = (1+u')\delta u' + v'\delta v' + w'\delta w'$$

which, upon using (4.4.69), becomes

$$\delta e = T_{11}\delta u' + T_{12}\delta v' + T_{13}\delta w' \tag{4.4.71}$$

Taking the variation of (4.4.70) and using the fact that $\delta \mathbf{i}_x = \delta \mathbf{i}_y = \delta \mathbf{i}_z = 0$ yields

$$(1+e)\delta \mathbf{i}_1 = \delta u' \mathbf{i}_x + \delta v' \mathbf{i}_y + \delta w' \mathbf{i}_z - \delta e \mathbf{i}_1 \qquad (4.4.72)$$

Taking the dot product of (4.4.72) with $\mathbf{i}_3$ yields

$$(1+e)\delta \mathbf{i}_1 \cdot \mathbf{i}_3 = \delta u' \mathbf{i}_x \cdot \mathbf{i}_3 + \delta v' \mathbf{i}_y \cdot \mathbf{i}_3 + \delta w' \mathbf{i}_z \cdot \mathbf{i}_3$$

But $\delta \mathbf{i}_1 \cdot \mathbf{i}_3 = -\delta \theta_2$ according to (4.4.67) and

$$\mathbf{i}_x \cdot \mathbf{i}_3 = T_{31}, \quad \mathbf{i}_y \cdot \mathbf{i}_3 = T_{32}, \quad \text{and} \quad \mathbf{i}_z \cdot \mathbf{i}_3 = T_{33}$$

according to (4.4.4). Hence,

$$(1+e)\delta \theta_2 + T_{31}\delta u' + T_{32}\delta v' + T_{33}\delta w' = 0 \qquad (4.4.73)$$

Similarly, taking the dot product of (4.4.72) with $\mathbf{i}_2$, we obtain

$$(1+e)\delta \mathbf{i}_1 \cdot \mathbf{i}_2 = \delta u' \mathbf{i}_x \cdot \mathbf{i}_2 + \delta v' \mathbf{i}_y \cdot \mathbf{i}_2 + \delta w' \mathbf{i}_z \cdot \mathbf{i}_2$$

But $\delta \mathbf{i}_1 \cdot \mathbf{i}_2 = \delta \theta_3$ according to (4.4.67) and $\mathbf{i}_x \cdot \mathbf{i}_2 = T_{21}, \mathbf{i}_y \cdot \mathbf{i}_2 = T_{22}$, and $\mathbf{i}_z \cdot \mathbf{i}_2 = T_{23}$ according to (4.4.4). Hence,

$$-(1+e)\delta \theta_3 + T_{21}\delta u' + T_{22}\delta v' + T_{23}\delta w' = 0 \qquad (4.4.74)$$

We note that (4.4.71), (4.4.73), and (4.4.74) can be put in the matrix form

$$\begin{Bmatrix} \delta e \\ (1+e)\delta \theta_3 \\ -(1+e)\delta \theta_2 \end{Bmatrix} = [T] \begin{Bmatrix} \delta u' \\ \delta v' \\ \delta w' \end{Bmatrix} \qquad (4.4.75)$$

Initially Curved and Twisted Beams: For the initially curved beam shown in Figure 4.4.4, the transformation matrix $[T]$, which is obtained using two Euler angles and relates the $\mathbf{i}_k$ to $\mathbf{i}_x, \mathbf{i}_y$, and $\mathbf{i}_z$, is given in (4.4.36), (4.4.57), and (4.4.52). It follows from (4.4.53) that

$$(1+e)^2 = (1+u'-vk_3+wk_2)^2 + (v'+uk_3-wk_1)^2 + (w'-uk_2+vk_1)^2 \qquad (4.4.76)$$

The variation of (4.4.76) is

$$(1+e)\delta e = (1+u'-vk_3+wk_2)\delta(1+u'-vk_3+wk_2)$$
$$+(v'+uk_3-wk_1)\delta(v'+uk_3-wk_1)$$
$$+(w'-uk_2+vk_1)\delta(w'-uk_2+vk_1)$$

which, upon using (4.4.52), becomes

$$\delta e = T_{11}\delta(1+u'-vk_3+wk_2) + T_{12}\delta(v'+uk_3-wk_1) + T_{13}\delta(w'-uk_2+vk_1) \qquad (4.4.77)$$

It follows from (4.4.51) and (4.4.52) that

$$(1+e)\mathbf{i}_1 = (1 + u' - vk_3 + wk_2)\mathbf{i}_x + (v' + uk_3 - wk_1)\mathbf{i}_y + (w' - uk_2 + vk_1)\mathbf{i}_z \tag{4.4.78}$$

Taking the variation of (4.4.78) and using the fact that $\delta\mathbf{i}_x = \delta\mathbf{i}_y = \delta\mathbf{i}_z = 0$, we obtain

$$(1+e)\delta\mathbf{i}_1 = (\delta u' - k_3\delta v + k_2\delta w)\mathbf{i}_x + (\delta v' + k_3\delta u - k_1\delta w)\mathbf{i}_y$$
$$+ (\delta w' - k_2\delta u + k_1\delta v)\mathbf{i}_z - \delta e\,\mathbf{i}_1 \tag{4.4.79}$$

Taking the dot product of (4.4.79) with $\mathbf{i}_3$ and using (4.4.67) and (4.4.36), we obtain

$$-(1+e)\delta\theta_2 = (1+e)\mathbf{i}_3 \cdot \delta\mathbf{i}_1$$
$$= T_{31}(\delta u' - k_3\delta v + k_2\delta w) + T_{32}(\delta v' + k_3\delta u - k_1\delta w)$$
$$+ T_{33}(\delta w' - k_2\delta u + k_1\delta v) \tag{4.4.80}$$

Similarly, taking the dot product of (4.4.79) with $\mathbf{i}_2$ and using (4.4.67) and (4.4.36), we obtain

$$(1+e)\delta\theta_3 = (1+e)\mathbf{i}_2 \cdot \delta\mathbf{i}_1$$
$$= T_{21}(\delta u' - k_3\delta v + k_2\delta w) + T_{22}(\delta v' + k_3\delta u - k_1\delta w)$$
$$+ T_{23}(\delta w' - k_2\delta u + k_1\delta v) \tag{4.4.81}$$

Using (4.4.32), we put (4.4.77), (4.4.80), and (4.4.81) in matrix form as

$$\left\{ \begin{array}{c} \delta e \\ (1+e)\delta\theta_3 \\ -(1+e)\delta\theta_2 \end{array} \right\} = [T] \left\{ \begin{array}{c} \delta u' \\ \delta v' \\ \delta w' \end{array} \right\} - [T][k] \left\{ \begin{array}{c} \delta u \\ \delta v \\ \delta w \end{array} \right\} \tag{4.4.82}$$

For an initially straight beam (see Figure 4.4.1), $[k] = [0]$ and (4.4.82) reduces to (4.4.75). Hence the virtual rotations $\delta\theta_i$ are independent of the Euler angles used in the transformation.

4.4.3 Variation of Curvatures

It follows from (4.4.16) and (4.4.64) that, for initially straight or curved beams,

$$\int_0^L H\delta\rho_1 ds = \int_0^L H\delta(\mathbf{i}_2' \cdot \mathbf{i}_3)ds = \int_0^L H(\mathbf{i}_2' \cdot \delta\mathbf{i}_3 + \mathbf{i}_3 \cdot \delta\mathbf{i}_2')ds \tag{4.4.83}$$

where H denotes any differential function of s. Integrating the last term in (4.4.83) by parts to transfer the spatial derivative from $\delta\mathbf{i}_2'$ to $H\mathbf{i}_3$ yields

$$\int_0^L H\delta\rho_1 ds = \int_0^L \left(H\mathbf{i}_2' \cdot \delta\mathbf{i}_3 - H\mathbf{i}_3' \cdot \delta\mathbf{i}_2 - H'\mathbf{i}_3 \cdot \delta\mathbf{i}_2 \right) ds + H\mathbf{i}_3 \cdot \delta\mathbf{i}_2 \,|_0^L \tag{4.4.84}$$

Substituting for $\delta \mathbf{i}_2$ and $\delta \mathbf{i}_3$ from (4.4.67) into (4.4.84) yields

$$\int_0^L H\delta\rho_1 ds = \int_0^L \left[H\mathbf{i}_2' \cdot (\delta\theta_2\mathbf{i}_1 - \delta\theta_1\mathbf{i}_2) - H\mathbf{i}_3' \cdot (-\delta\theta_3\mathbf{i}_1 + \delta\theta_1\mathbf{i}_3) \right.$$
$$\left. -H'\mathbf{i}_3 \cdot (-\delta\theta_3\mathbf{i}_1 + \delta\theta_1\mathbf{i}_3)\right] ds + \left[H\mathbf{i}_3 \cdot (-\delta\theta_3\mathbf{i}_1 + \delta\theta_1\mathbf{i}_3)\right]_0^L$$

which, upon using (4.4.64)-(4.4.66) and the fact that $\mathbf{i}_2' \cdot \mathbf{i}_2 = \mathbf{i}_3' \cdot \mathbf{i}_3 = 0$, becomes

$$\int_0^L H\delta\rho_1 ds = \int_0^L (-H'\delta\theta_1 - H\rho_3\delta\theta_2 + H\rho_2\delta\theta_3) ds + H\delta\theta_1 \Big|_0^L \quad (4.4.85)$$

Similarly, we obtain

$$\int_0^L H\delta\rho_2 ds = \int_0^L (H\rho_3\delta\theta_1 - H'\delta\theta_2 - H\rho_1\delta\theta_3) ds + H\delta\theta_2 \Big|_0^L \quad (4.4.86)$$

$$\int_0^L H\delta\rho_3 ds = \int_0^L (-H\rho_2\delta\theta_1 + H\rho_1\delta\theta_2 - H'\delta\theta_3) ds + H\delta\theta_3 \Big|_0^L \quad (4.4.87)$$

We note that (4.4.85)-(4.4.87) can be put in the following matrix form

$$\int_0^L H \begin{Bmatrix} \delta\rho_1 \\ \delta\rho_2 \\ \delta\rho_3 \end{Bmatrix} ds = -\int_0^L (H'[I] + H[K]) \begin{Bmatrix} \delta\theta_1 \\ \delta\theta_2 \\ \delta\theta_3 \end{Bmatrix} ds + \left[H \begin{Bmatrix} \delta\theta_1 \\ \delta\theta_2 \\ \delta\theta_3 \end{Bmatrix} \right]_0^L$$
(4.4.88)

4.4.4 Concept of Local Displacements

In this section, we show that fully nonlinear strains can be obtained by using the concept of local displacements. In Figure 4.4.9, we show the undeformed configuration of a filament $\overline{AB}$ between two cross-sections of the initially curved and twisted beam shown in Figure 4.4.4 and its deformed configuration $\overline{A'B'}$ in the absence of inplane and out-of-plane warpings. The position vector $\mathbf{R}_A$ of Point A is given by

$$\mathbf{R}_A = \mathbf{R}_O + y\mathbf{i}_y + z\mathbf{i}_z$$

Hence, the position vector $\mathbf{R}_B$ of Point B is given by

$$\mathbf{R}_B = \mathbf{R}_A + \frac{\partial \mathbf{R}_A}{\partial s} ds = \mathbf{R}_A + \left(\frac{\partial \mathbf{R}_O}{\partial s} + y\frac{d\mathbf{i}_y}{ds} + z\frac{d\mathbf{i}_z}{ds}\right) ds$$

Using (4.4.31) and (4.4.32) to express $d\mathbf{i}_y/ds$ and $d\mathbf{i}_z/ds$ in terms of k_1, k_2, and k_3 and the fact that $\partial \mathbf{R}_O/ds = \mathbf{i}_x$, we express $\mathbf{R}_B$ as

$$\mathbf{R}_B = \mathbf{R}_A + [(1 - yk_3 + zk_2)\mathbf{i}_x - zk_1\mathbf{i}_y + yk_1\mathbf{i}_z] ds$$

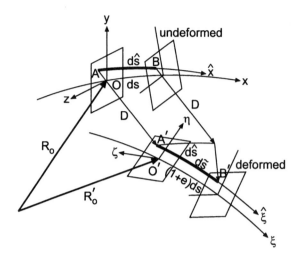

Fig. 4.4.9 The deformation of the filament $\overline{AB}$ in an initially curved and twisted beam.

Hence, the length $d\hat{s}$ of $\overline{AB}$ and the unit vector $\mathbf{i}_{\hat{x}}$ along the $\hat{x}$ axis are given by

$$d\hat{s} = \mid \mathbf{R}_B - \mathbf{R}_A \mid = \tau ds$$

$$\mathbf{i}_{\hat{x}} = \frac{\overrightarrow{AB}}{d\hat{s}} = \frac{1}{\tau}[(1 - yk_3 + zk_2)\mathbf{i}_x - zk_1\mathbf{i}_y + yk_1\mathbf{i}_z]$$

$$\tau \equiv \sqrt{(1 - yk_3 + zk_2)^2 + (zk_1)^2 + (yk_1)^2} \qquad (4.4.89)$$

We note that $\mathbf{i}_x = \mathbf{i}_{\hat{x}}$ only if the initial twisting curvature $k_1 = 0$.

Because the displacement vector of the observed Point O is defined to be $\overrightarrow{OO'} = u\mathbf{i}_x + v\mathbf{i}_y + w\mathbf{i}_z$, the displacement vector $\mathbf{D}$ ($= \overrightarrow{AA'}$) of Point A is given by

$$\mathbf{D} = \overrightarrow{AO} + \overrightarrow{OO'} + \overrightarrow{O'A'}$$
$$= \{u, v, w\}\{\mathbf{i}_{xyz}\} + y\mathbf{i}_2 + z\mathbf{i}_3 - y\mathbf{i}_y - z\mathbf{i}_z \qquad (4.4.90)$$

where $\{\mathbf{i}_{xyz}\} = \{\mathbf{i}_x, \mathbf{i}_y, \mathbf{i}_z\}^T$. Differentiating $\mathbf{D}$ with respect to s yields

$$\frac{\partial \mathbf{D}}{\partial s} = \{u', v', w'\}\{\mathbf{i}_{xyz}\} + \{u, v, w\}\frac{d\{\mathbf{i}_{xyz}\}}{ds}$$
$$+ y\frac{\partial \mathbf{i}_2}{\partial s} + z\frac{\partial \mathbf{i}_3}{\partial s} - y\frac{d\mathbf{i}_y}{ds} - z\frac{d\mathbf{i}_z}{ds}$$

Using (4.4.31) and (4.4.32) to express the derivatives of $\{\mathbf{i}_{xyz}\}$ in terms of the k_i and (4.4.59) and (4.4.60) to express the derivatives of $\mathbf{i}_2$ and $\mathbf{i}_3$ in terms of the ρ_i, we obtain

$$\frac{\partial \mathbf{D}}{\partial s} = \{u', v', w'\}\{\mathbf{i}_{xyz}\} + \{u, v, w\}[k]\{\mathbf{i}_{xyz}\} + y(\rho_1\mathbf{i}_3 - \rho_3\mathbf{i}_1)$$
$$+ z(\rho_2\mathbf{i}_1 - \rho_1\mathbf{i}_2) - y(-k_3\mathbf{i}_x + k_1\mathbf{i}_z) - z(k_2\mathbf{i}_x - k_1\mathbf{i}_y)$$

or
$$\frac{\partial \mathbf{D}}{\partial s} = \{u' - vk_3 + wk_2, v' + uk_3 - wk_1, w' - uk_2 + vk_1\}\{\mathbf{i}_{xyz}\}$$
$$+ (z\rho_2 - y\rho_3)\mathbf{i}_1 - z\rho_1\mathbf{i}_2 + y\rho_1\mathbf{i}_3$$
$$- (zk_2 - yk_3)\mathbf{i}_x + zk_1\mathbf{i}_y - yk_1\mathbf{i}_z \qquad (4.4.91)$$

Using (4.4.52), we rewrite (4.4.91) as
$$\frac{\partial \mathbf{D}}{\partial s} = (1+e)\{T_{11}, T_{12}, T_{13}\}\{\mathbf{i}_{xyz}\} + (z\rho_2 - y\rho_3)\mathbf{i}_1$$
$$- z\rho_1\mathbf{i}_2 + y\rho_1\mathbf{i}_3 - (1 - yk_3 + zk_2)\mathbf{i}_x + zk_1\mathbf{i}_y - yk_1\mathbf{i}_z$$

which, upon using (4.4.89), becomes
$$\frac{\partial \mathbf{D}}{\partial s} = (1+e)\{T_{11}, T_{12}, T_{13}\}\{\mathbf{i}_{xyz}\}$$
$$+ (z\rho_2 - y\rho_3)\mathbf{i}_1 - z\rho_1\mathbf{i}_2 + y\rho_1\mathbf{i}_3 - \tau\mathbf{i}_{\hat{x}} \qquad (4.4.92)$$

Using $\mathbf{i}_1 = \{T_{11}, T_{12}, T_{13}\}\{\mathbf{i}_{xyz}\}$ from (4.4.36), we rewrite (4.4.92) as
$$\frac{\partial \mathbf{D}}{\partial s} = (1 + e + z\rho_2 - y\rho_3)\mathbf{i}_1 - z\rho_1\mathbf{i}_2 + y\rho_1\mathbf{i}_3 - \tau\mathbf{i}_{\hat{x}} \qquad (4.4.93)$$

It follows from Figure 4.4.9 that $\overrightarrow{BB'} = \mathbf{D} + (\partial \mathbf{D}/\partial s)\,ds$ and hence the local axial strain along the $\hat{\xi}$ axis is
$$\epsilon_{11} = \frac{d\tilde{s} - d\hat{s}}{d\hat{s}}$$
$$= \frac{(\frac{\partial \mathbf{D}}{\partial s} ds + d\hat{s}\mathbf{i}_{\hat{x}}) \cdot \mathbf{i}_{\hat{1}} - d\hat{s}}{d\hat{s}}$$
$$= \frac{1}{\tau}\frac{\partial \mathbf{D}}{\partial s} \cdot \mathbf{i}_{\hat{1}} + \mathbf{i}_{\hat{x}} \cdot \mathbf{i}_{\hat{1}} - 1 \qquad (4.4.94)$$

because $d\hat{s} = \tau ds$ from (4.4.89).

Without inplane and out-of-plane warpings, the position vector $\mathbf{R}_{A'}$ of Point A' is given by
$$\mathbf{R}_{A'} = \mathbf{R}_{O'} + y\mathbf{i}_2 + z\mathbf{i}_3$$
Hence, the position vector $\mathbf{R}_{B'}$ of Point B' is given by
$$\mathbf{R}_{B'} = \mathbf{R}_{A'} + \frac{\partial \mathbf{R}_{A'}}{\partial s}ds = \mathbf{R}_{A'} + \left[\frac{\partial \mathbf{R}_{O'}}{\partial s} + y\frac{\partial \mathbf{i}_2}{\partial s} + z\frac{\partial \mathbf{i}_3}{\partial s}\right]ds$$

Using (4.4.11) and (4.4.12) to express the derivatives of $\mathbf{i}_2$ and $\mathbf{i}_3$ in terms of the ρ_i and the fact that $\partial \mathbf{R}_{O'}/\partial s = (1+e)\mathbf{i}_1$, we express $\mathbf{R}_{B'}$ as
$$\mathbf{R}_{B'} = \mathbf{R}_{A'} + [(1 + e - y\rho_3 + z\rho_2)\mathbf{i}_1 - z\rho_1\mathbf{i}_2 + y\rho_1\mathbf{i}_3]ds \qquad (4.4.95)$$

Hence, the unit vector $\mathbf{i}_{\hat{1}}$ along the $\hat{\xi}$ axis is

$$\mathbf{i}_{\hat{1}} = \frac{\mathbf{R}_{B'} - \mathbf{R}_{A'}}{|\mathbf{R}_{B'} - \mathbf{R}_{A'}|}$$

$$= \frac{1}{\hat{\tau}}[(1 + e - y\rho_3 + z\rho_2)\mathbf{i}_1 - z\rho_1\mathbf{i}_2 + y\rho_1\mathbf{i}_3]$$

$$\hat{\tau} \equiv \sqrt{(1 + e - y\rho_3 + z\rho_2)^2 + (z\rho_1)^2 + (y\rho_1)^2} \qquad (4.4.96)$$

We note that, if the influence of local rotations due to the straining of the local displacements is neglected, (4.4.96) reduces to

$$\mathbf{i}_{\hat{1}} = \frac{1}{\tau}[(1 - yk_3 + zk_2)\mathbf{i}_1 - zk_1\mathbf{i}_2 + yk_1\mathbf{i}_3] \qquad (4.4.97)$$

which is a reasonable approximation under the assumption of small strains.

Using (4.4.96), we rewrite (4.4.93) as

$$\frac{\partial \mathbf{D}}{\partial s} = \hat{\tau}\mathbf{i}_{\hat{1}} - \tau\mathbf{i}_{\hat{x}} \qquad (4.4.98)$$

Substituting (4.4.98) into (4.4.94) yields

$$\epsilon_{11} = \frac{\hat{\tau}}{\tau} - 1 \qquad (4.4.99)$$

This is the fully nonlinear strain-displacement relation.

For slender beams, we expand (4.4.99) for small z and y and neglect terms proportional to y^m and z^n, where $m, n \geq 2$, to obtain

$$\epsilon_{11} = e - y[\rho_3 - (1+e)k_3] + z[\rho_2 - (1+e)k_2] \qquad (4.4.100)$$

We note that k_1 does not appear in (4.4.100), which indicates that the influence of k_1 on ϵ_{11} is negligible. If the influence of k_1 is neglected in (4.4.97), $\mathbf{i}_{\hat{1}} = \mathbf{i}_1$. Hence, the axial strain ϵ_{11} in (4.4.100) is along the ξ direction. The presence of the factor $(1+e)$ in (4.4.100) is due to the fact that ρ_2 and ρ_3 are defined using ds, instead of $(1+e)ds$, and are not real curvatures (see (4.4.11)), whereas k_2 and k_3 are real curvatures and ϵ_{11} is a strain defined with respect to the undeformed length. Neglecting ek_2 and ek_3, we rewrite (4.4.100) as

$$\epsilon_{11} = e - y(\rho_3 - k_3) + z(\rho_2 - k_2) \qquad (4.4.101)$$

As pointed out in Section 2.5.3, rigid-body translations and rotations do not produce any strains; strains are due to relative displacements. Hence, one should be able to derive fully nonlinear strains using local displacements defined with respect to the deformed coordinate system $\xi\eta\zeta$. To this end, we assume that the observed cross-section is fixed on the $\eta\zeta$ plane, and the displacements of an arbitrary point on this cross-section can be expressed as

$$u_1(s, y, z, t) = u_1^0(s, t) + z[\theta_2(s, t) - \theta_{20}(s)] - y[\theta_3(s, t) - \theta_{30}(s)]$$
$$u_2(s, y, z, t) = u_2^0(s, t) - z[\theta_1(s, t) - \theta_{10}(s)]$$
$$u_3(s, y, z, t) = u_3^0(s, t) + y[\theta_1(s, t) - \theta_{10}(s)] \qquad (4.4.102)$$

where u_1, u_2, and u_3 are local, strainable displacements with respect to the ξ, η, and ζ axes, respectively; $u_i^0(s,t) \equiv u_i(s,0,0,t)$, $i=1, 2, 3$; θ_1, θ_2, and θ_3 are the rotation angles of the observed cross-section with respect to the ξ, η, and ζ axes, respectively; and θ_{10}, θ_{20}, and θ_{30} are the initial rotation angles of the observed cross-section with respect to the ξ, η, and ζ axes, respectively.

Because the $\xi\eta\zeta$ system is a local coordinate system attached to the observed cross-section and the unit vector $\mathbf{i}_1$ is tangent to the deformed reference axis, we have

$$u_i = u_i^0 = 0, \quad \theta_{i0} = \theta_i = 0, \quad \frac{\partial u_2^0}{\partial s} = \frac{\partial u_3^0}{\partial s} = 0$$

$$e \equiv \frac{\partial u_1^0}{\partial s}, \quad \rho_i \equiv \frac{\partial \theta_i}{\partial s}, \quad k_i \equiv \frac{\partial \theta_{i0}}{\partial s}, \quad i = 1, 2, 3 \qquad (4.4.103)$$

Because all of the variables in (4.4.102) are equal to zero for any point on the cross-section, only the first derivatives of the variables in (4.4.102) with respect to s, y, and z are needed to calculate the strains. The local displacement vector $\mathbf{U}$ is given by

$$\mathbf{U} = u_1 \mathbf{i}_1 + u_2 \mathbf{i}_2 + u_3 \mathbf{i}_3 \qquad (4.4.104)$$

It follows from (4.4.103) that

$$\epsilon_{11} = \frac{\partial \mathbf{U}}{\partial s} \cdot \mathbf{i}_1 = \frac{\partial u_1}{\partial s} = e + z(\rho_2 - k_2) - y(\rho_3 - k_3) \qquad (4.4.105)$$

which is the same as (4.4.101). To obtain (4.4.100), one needs to replace the θ_{i0} with $(1+e)\theta_{i0}$ in (4.4.102).

Since the inplane and out-of-plane warpings are relative displacements with respect to the flat surface on the $\eta\zeta$ plane, they are essentially local displacements. Hence, warping displacements can be superposed on the local displacements shown in (4.4.102), and the above method of deriving local strains is still valid in the presence of warpings, as shown later in Section 4.7.

Equation (4.4.94) shows that the strain ϵ is defined with respect to the deformed coordinate system and it is a geometric strain, not an energy strain. Pai and Nayfeh (1994c) showed that the strains derived using the concept of local displacements are the same as the so-called Jaumann strains. Jaumann strains are defined using the right stretch tensor of the deformation gradient (Malvern, 1969). To separate the right stretch tensor from the deformation gradient requires the use of the complex polar decomposition method. The use of local displacements to derive fully nonlinear strain-displacement relations does not need such polar decomposition.

4.5 NONLINEAR 2-D EULER-BERNOULLI BEAM THEORY

Here, we relax assumptions (4.2.1) and (4.2.2) and derive fully nonlinear equations governing the planar vibration of initially straight Euler-Bernoulli beams. The coordinate systems, displacements, and free-body diagram of an Euler-Bernoulli beam element are shown in Figures 4.5.1a-c. It follows from Figures 4.5.1a and b that

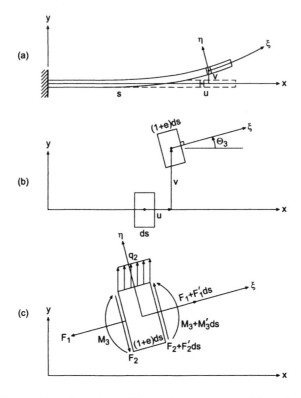

Fig. 4.5.1 The nonlinear Euler-Bernoulli beam theory: (a) the undeformed coordinate system xy and the deformed coordinate system $\xi\eta$, (b) the displacements u and v and rotation Θ_3, and (c) the free-body diagram.

the inertial, Cartesian coordinate system xyz and the local, orthogonal curvilinear coordinate system $\xi\eta\zeta$ are related by the transformation matrix $[T]$ defined by

$$\begin{Bmatrix} \mathbf{i}_1 \\ \mathbf{i}_2 \\ \mathbf{i}_3 \end{Bmatrix} = [T] \begin{Bmatrix} \mathbf{i}_x \\ \mathbf{i}_y \\ \mathbf{i}_z \end{Bmatrix}, \quad [T] = \begin{bmatrix} \cos\Theta_3 & \sin\Theta_3 & 0 \\ -\sin\Theta_3 & \cos\Theta_3 & 0 \\ 0 & 0 & 1 \end{bmatrix} \qquad (4.5.1)$$

where the $\mathbf{i}_j$ are unit vectors along the ξ, η, and ζ axes and $\mathbf{i}_x$, $\mathbf{i}_y$, $\mathbf{i}_z$ are unit vectors along the x, y, and z axes, respectively.

It follows from Figure 4.5.2 that the axial strain e and the rotation angle Θ_3 are related to u' and v' by

$$e = \sqrt{(1+u')^2 + v'^2} - 1, \quad \cos\Theta_3 = \frac{1+u'}{1+e}, \quad \sin\Theta_3 = \frac{v'}{1+e} \qquad (4.5.2)$$

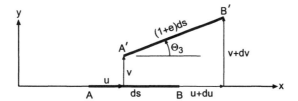

Fig. 4.5.2 The relationship between the displacements and the rotation angle.

It follows from (4.4.12) and (4.5.1) that

$$\begin{aligned}\rho_3 &= \mathbf{i}'_1 \cdot \mathbf{i}_2 \\ &= [(\cos\Theta_3)'\mathbf{i}_x + (\sin\Theta_3)'\mathbf{i}_y] \cdot [-\sin\Theta_3\mathbf{i}_x + \cos\Theta_3\mathbf{i}_y] \\ &= -\sin\Theta_3(\cos\Theta_3)' + \cos\Theta_3(\sin\Theta_3)' \\ &= \Theta'_3 \end{aligned} \quad (4.5.3)$$

and

$$\rho_1 = \mathbf{i}'_2 \cdot \mathbf{i}_3 = 0, \quad \rho_2 = \mathbf{i}'_3 \cdot \mathbf{i}_1 = 0 \quad (4.5.4)$$

The equilibrium of moments with respect to the z direction (see Figure 4.5.1c) yields

$$M'_3 ds + F_2(1+e)ds = j_3 ds \ddot{\Theta}_3$$

or

$$F_2 = -\frac{1}{1+e}(M'_3 - j_3\ddot{\Theta}_3) \quad (4.5.5)$$

where j_3 is the rotary inertia defined in (4.2.7).

Referring to Figure 4.5.1c and using Newton's second law, we obtain

$$\frac{\partial \mathbf{F}}{\partial s} ds + q_2 ds \mathbf{i}_y = m ds\, \mathbf{a} \quad (4.5.6)$$

where

$$\mathbf{F} = F_1 \mathbf{i}_1 + F_2 \mathbf{i}_2 \quad (4.5.7)$$

is the surface traction acting on the cross-section and

$$\mathbf{a} = \ddot{u}\mathbf{i}_x + \ddot{v}\mathbf{i}_y \quad (4.5.8)$$

is the acceleration of the mass centroid of the element, which is assumed to be the same as the area centroid. Substituting (4.5.1), (4.5.7), and (4.5.8) into (4.5.6), we obtain

$$[(F_1\cos\Theta_3)' - (F_2\sin\Theta_3)']\mathbf{i}_x + [(F_1\sin\Theta_3)' + (F_2\cos\Theta_3)']\mathbf{i}_y \\ + q_2\mathbf{i}_y = m\ddot{u}\mathbf{i}_x + m\ddot{v}\mathbf{i}_y$$

Setting each of the coefficients of $\mathbf{i}_x$ and $\mathbf{i}_y$ equal to zero, we obtain the equations of motion as

$$(F_1 \cos \Theta_3)' - (F_2 \sin \Theta_3)' = m\ddot{u} \qquad (4.5.9)$$

$$(F_1 \sin \Theta_3)' + (F_2 \cos \Theta_3)' + q_2 = m\ddot{v} \qquad (4.5.10)$$

Using a variational formulation, as shown later in Section 4.6, we find that the boundary conditions are to specify

$$\begin{array}{ccc} v & \text{or} & F_2 \\ \Theta_3 & \text{or} & M_3 \\ u & \text{or} & F_1 \end{array} \qquad (4.5.11)$$

at $s = 0$ and L.

To complete the problem formulation, we need to relate the stress resultants F_1 and F_2 and the moment M_3 to the displacements u and v. Thus, we have three equations (4.5.5), (4.5.9), and (4.5.10) for the two unknowns u and v. Consequently, the problem is overdetermined. To close the problem, we use (4.5.5) to relate the shear stress resultant F_2 in terms of M_3 and Θ_3, leaving us with the two equations (4.5.9) and (4.5.10) for the two dependent variables u and v. Thus, we only need to relate F_1 and M_3 to u and v.

To fully account for large rotations and displacements, we use a local displacement field to derive fully nonlinear strain-displacement relations. Because according to the Euler-Bernoulli beam theory plane sections perpendicular to the undeformed reference line remain plane and perpendicular to the deformed reference line, we define the local displacement field as

$$u_1 = u_1^0 - y \sin \theta_3, \quad u_2 = u_2^0 - y(1 - \cos \theta_3), \quad u_3 = 0 \qquad (4.5.12)$$

where u_1^0 and u_2^0 are the displacements of the reference point on the observed element with respect to the ξ and η axes and θ_3 denotes the rotation angle of the observed element with respect to the ζ axis. Because the $\xi\eta\zeta$ system is a local coordinate system attached to the observed element, we have

$$u_1^0 = u_2^0 = 0, \quad \theta_3 = \frac{\partial u_2^0}{\partial s} = 0 \qquad (4.5.13)$$

Because u_1 is a local displacement defined with respect to the deformed reference line ξ,

$$\frac{\partial u_1^0}{\partial s} = e \qquad (4.5.14)$$

It follows from the diagram shown in Figure 4.5.3 and the fact that $\rho_3 = \mathbf{i}_1' \cdot \mathbf{i}_2$ that

$$\frac{\partial \theta_3}{\partial s} = \lim_{ds \to 0} \frac{d\mathbf{i}_1 \cdot \mathbf{i}_2}{ds} = \mathbf{i}_1' \cdot \mathbf{i}_2 = \rho_3 \qquad (4.5.15)$$

NONLINEAR 2-D EULER-BERNOULLI BEAM THEORY

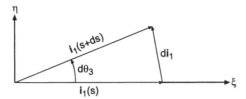

Fig. 4.5.3 Geometric interpretation of the infinitesimal change of a local rotation angle.

We note that ρ_3 is not the real curvature because it is defined with respect to the undeformed length ds. Th real curvature $\tilde{\rho}_3$ should be defined with respect to the deformed length $(1+e)ds$; that is,

$$\tilde{\rho}_3 \equiv \frac{\partial \theta_3}{(1+e)\partial s} = \frac{1}{1+e}\mathbf{i}'_1 \cdot \mathbf{i}_2 \qquad (4.5.16)$$

However, because strains are defined with respect to the undeformed length, it is more convenient to use the normalized curvature ρ_3, rather than the real curvature $\tilde{\rho}_3$.

Differentiating the local displacement vector $\mathbf{U} = u_1\mathbf{i}_1 + u_2\mathbf{i}_2$ with respect to s, we have

$$\frac{\partial \mathbf{U}}{\partial s} = \frac{\partial u_1}{\partial s}\mathbf{i}_1 + \frac{\partial u_2}{\partial s}\mathbf{i}_2 + u_1\frac{\partial \mathbf{i}_1}{\partial s} + u_2\frac{\partial \mathbf{i}_2}{\partial s}$$

Differentiating (4.5.12) with respect to s and using (4.5.13)-(4.5.15), we obtain

$$\frac{\partial u_1}{\partial s} = e - y\cos\theta_3\,\theta'_3 = e - \rho_3 y$$

$$\frac{\partial u_2}{\partial s} = -y\sin\theta_3\,\theta'_3 = 0$$

Substituting (4.5.13) into (4.5.12), we find that the local displacements of the observed element are $u_1 = u_2 = 0$. Therefore,

$$\frac{\partial \mathbf{U}}{\partial s} = (e - y\rho_3)\mathbf{i}_1 \qquad (4.5.17)$$

Differentiating the local displacement vector with respect to y, we have

$$\frac{\partial \mathbf{U}}{\partial y} = \frac{\partial u_1}{\partial y}\mathbf{i}_1 + \frac{\partial u_2}{\partial y}\mathbf{i}_2 + u_1\frac{\partial \mathbf{i}_1}{\partial y} + u_2\frac{\partial \mathbf{i}_2}{\partial y}$$

Differentiating (4.5.12) with respect to y and using (4.5.13) yields

$$\frac{\partial u_1}{\partial y} = -\sin\theta_3 = 0$$

$$\frac{\partial u_2}{\partial y} = -(1 - \cos\theta_3) = 0$$

Therefore,
$$\frac{\partial \mathbf{U}}{\partial y} = 0 \qquad (4.5.18)$$
because the local displacements u_1 and u_2 of the observed element are zero. Similarly,
$$\frac{\partial \mathbf{U}}{\partial z} = \frac{\partial u_1}{\partial z}\mathbf{i}_1 + \frac{\partial u_2}{\partial z}\mathbf{i}_2 + u_1\frac{\partial \mathbf{i}_1}{\partial z} + u_2\frac{\partial \mathbf{i}_2}{\partial z} = 0 \qquad (4.5.19)$$

Using (4.5.17)-(4.5.19), we obtain the strains
$$\epsilon_{11} = \frac{\partial \mathbf{U}}{\partial s} \cdot \mathbf{i}_1 = e - y\rho_3$$
$$\epsilon_{12} = \frac{\partial \mathbf{U}}{\partial s} \cdot \mathbf{i}_2 + \frac{\partial \mathbf{U}}{\partial y} \cdot \mathbf{i}_1 = 0$$
$$\epsilon_{22} = \epsilon_{33} = \epsilon_{13} = \epsilon_{23} = 0 \qquad (4.5.20)$$

Considering isotropic materials for the beam shown in Figure 4.5.1c, assuming that the reference line coincides with the centroidal line, and using (4.5.20), we obtain

$$F_1 \equiv \int_A \sigma_{11} dA = \int_A E(e - y\rho_3) dA = EAe - E\rho_3 \int_A y dA = EAe \qquad (4.5.21)$$

$$M_3 \equiv -\int_A \sigma_{11} y dA = \int_A E(-ye + y^2\rho_3) dA = EI\rho_3 - Ee \int_A y dA = EI\rho_3 \qquad (4.5.22)$$

We note that, if the reference line does not coincide with the centroidal line, $\int_A y dA \neq 0$ and hence the extension and bending are coupled.

Equations (4.5.20)-(4.5.22) can be derived using alternative methods. In this book, we have used the concept of local displacements and local stress and strain measures because it can be easily extended to derive fully nonlinear strains for beams, plates, and shells with inplane and out-of-plane warpings, as shown later.

We note that the Euler-Bernoulli theory has an inconsistency because the shear resultant F_2 should be zero according to (4.5.20), whereas it is nonzero according to (4.5.5).

Substituting (4.5.5) into (4.5.9) and (4.5.10) and using (4.5.21) and (4.5.22), we obtain

$$(EAe\cos\Theta_3)' + \left[\frac{(EI\rho_3)'\sin\Theta_3}{1+e}\right]' = m\ddot{u} + \left[\frac{j_3\ddot{\Theta}_3\sin\Theta_3}{1+e}\right]' \qquad (4.5.23)$$

$$(EAe\sin\Theta_3)' - \left[\frac{(EI\rho_3)'\cos\Theta_3}{1+e}\right]' + q_2 = m\ddot{v} - \left[\frac{j_3\ddot{\Theta}_3\cos\Theta_3}{1+e}\right]' \qquad (4.5.24)$$

where e, Θ_3, and ρ_3 are related to u and v by (4.5.2) and (4.5.3). Assuming that u and v are small but finite, we can expand the expressions in (4.5.2) and obtain

$$e = u' + \frac{1}{2}v'^2 - \frac{1}{2}u'v'^2 + \cdots \qquad (4.5.25)$$

$$\cos\Theta_3 = 1 - \frac{1}{2}v'^2 + u'v'^2 + \cdots \qquad (4.5.26)$$

$$\sin\Theta_3 = v' - u'v' + u'^2v' - \frac{1}{2}v'^3 + \cdots \qquad (4.5.27)$$

$$\Theta_3 = \tan^{-1}\left(\frac{v'}{1+u'}\right) = v' - u'v' + u'^2v' - \frac{1}{3}v'^3 + \cdots \qquad (4.5.28)$$

Substituting (4.5.3) and (4.5.25)-(4.5.28) into (4.5.23) and (4.5.24), expanding the result for small u and v, keeping up to cubic terms, and neglecting nonlinear terms that involve the rotary inertia because j_3 is small, we obtain

$$m\ddot{u} - (EAu')' = \left[EA\left(\frac{1}{2}v'^2 - u'v'^2\right)\right]'$$
$$+ \left\{v'\left[EI(v'-u'v')'\right]' - 2u'v'(EIv'')'\right\} \qquad (4.5.29)$$

$$m\ddot{v} - j_3\ddot{v}'' + (EIv'')'' = \left[EA\left(u'v' - u'^2v' + \frac{1}{2}v'^3\right)\right]'$$
$$+ \left\{\left[EI(u'v')'\right]'(1-u') + (EIv'')'(u' - u'^2 + v'^2)\right.$$
$$\left. - \left[EI\left(u'^2v' - \frac{1}{3}v'^3\right)'\right]'\right\} + q_2 \qquad (4.5.30)$$

For uniform beams, (4.5.29) and (4.5.30) reduce to

$$m\ddot{u} - EAu'' = EA\left(\frac{1}{2}v'^2 - u'v'^2\right)'$$
$$+ EI\left[v'(v''' - u'''v' - 2u''v'' - 3u'v''')\right]' \qquad (4.5.31)$$

$$m\ddot{v} - j_3\ddot{v}'' + EIv^{iv} = EA\left(u'v' - u'^2v' + \frac{1}{2}v'^3\right)'$$
$$+ EI\left[u'v''' + (u'v')'' - (u'^2 - v'^2)v''' - u'(u'v')'' - \left(u'^2v' - \frac{1}{3}v'^3\right)''\right]'$$
$$+ q_2 \qquad (4.5.32)$$

4.5.1 Shortening Effect

For beams with the end at $s = 0$ being fixed or hinged, the other end at $s = L$ being free or sliding, and no external loads acting along the x direction at $s = L$, we have

$$u\,|_{s=0} = 0, \quad F_x\,|_{s=L} = [F_1\cos\Theta_3 - F_2\sin\Theta_3]_{s=L} = 0 \qquad (4.5.33)$$

For such beams, we may assume that the beam is inextensional (i.e., $e = 0$) and obtain from (4.5.2) that

$$u' = \sqrt{1 - v'^2} - 1 \tag{4.5.34}$$

Integrating (4.5.34) once with respect to s and using the boundary condition $u = 0$ at $s = 0$, we have

$$u = \int_0^s [\sqrt{1 - v'^2} - 1] ds \tag{4.5.35}$$

We note that the inextensionality condition reduced the dependent variables from two to one and hence two of the basic equations (4.5.5), (4.5.9), and (4.5.10) will serve as constraints to relate two of the resultant forces F_1 and F_2 and moment M_3 in terms of v. Because the shear-strain ϵ_{12} and axial strain e are neglected, (4.5.5) and (4.5.9) will be used to relate F_1 and F_2 to M_3 and v.

Integrating (4.5.9) once with respect to s and using the boundary condition $F_x = 0$ at $s = L$, we have

$$F_1 \cos \Theta_3 - F_2 \sin \Theta_3 = \int_L^s m \ddot{u} \, ds$$

or

$$F_1 = \frac{1}{\cos \Theta_3} \left[\int_L^s m \ddot{u} \, ds + F_2 \sin \Theta_3 \right] \tag{4.5.36}$$

Differentiating (4.5.35) twice with respect to t and interchanging the order of integration and differentiation, we obtain

$$\ddot{u} = \int_0^s \frac{\partial^2}{\partial t^2} \sqrt{1 - v'^2} \, ds$$

Hence,

$$\int_L^s m \ddot{u} \, ds = \int_L^s m \int_0^s \frac{\partial^2}{\partial t^2} \sqrt{1 - v'^2} \, ds \, ds \tag{4.5.37}$$

Because $e = 0$, it follows from (4.5.2) that

$$\cos \Theta_3 = 1 + u' = \sqrt{1 - v'^2} \quad \text{and} \quad \sin \Theta_3 = v' \tag{4.5.38}$$

Substituting (4.5.37) and (4.5.38) into (4.5.36) yields

$$F_1 = \frac{1}{\sqrt{1 - v'^2}} \left[\int_L^s m \int_0^s \frac{\partial^2}{\partial t^2} \sqrt{1 - v'^2} \, ds \, ds + F_2 v' \right] \tag{4.5.39}$$

We note that this expression for F_1 cannot be obtained from (4.5.21) because we have assumed that $e = 0$.

It follows from (4.5.38) that $\Theta_3 = \sin^{-1} v'$ and hence, it follows from (4.5.3) that

$$\rho_3 = \Theta_3' = (\sin^{-1} v')'$$

Then, it follows from (4.5.22) that

$$M_3 = EI(\sin^{-1} v')'$$

Substituting for Θ_3 and M_3 in (4.5.5) and recalling that $e = 0$ yields

$$F_2 = -[EI(\sin^{-1} v')']' + j_3(\sin^{-1} v')\ddot{} \qquad (4.5.40)$$

Substituting (4.5.39) and (4.5.40) into (4.5.10) and using (4.5.38), we obtain the equation of motion

$$\left[\frac{v'}{\sqrt{1-v'^2}}\int_L^s m \int_0^s \frac{\partial^2}{\partial t^2}\sqrt{1-v'^2}\, ds\, ds\right]' - \left\{\frac{[EI(\sin^{-1} v')']'}{\sqrt{1-v'^2}}\right\}' + q_2$$

$$= m\ddot{v} - \left[\frac{j_3(\sin^{-1} v')\ddot{}}{\sqrt{1-v'^2}}\right]' \qquad (4.5.41)$$

We assume that the beam is prismatic and the nonlinear terms due to rotary inertia j_3 are negligible, expand the terms in (4.5.41) for small v, keep up to cubic terms, and obtain

$$m\ddot{v} + c_1 \dot{v} = -EI\left(v''' + v'v''^2 + v'''v'^2\right)' + j_3 \ddot{v}''$$

$$- \frac{1}{2}\left[v' \int_L^s m \left(\int_0^s v'^2 ds\right)\ddot{}\, ds\right]' + q_2 \qquad (4.5.42)$$

where a linear viscous damping term has been added with c_1 being the damping coefficient. The last two parts of the first term on the right-hand side of (4.5.42) account for the nonlinear curvature (geometry), the second term accounts for the rotary inertia, and the third term accounts for the shortening effect (nonlinear inertia).

In most studies of the nonlinear dynamics of cantilever beams, only the nonlinear inertia was commonly considered (Bolotin, 1964; Evensen and Evan-Iwanowski, 1966; Haight and King, 1971; Nayfeh and Mook, 1979). Bolotin (1964) and Moody (1967) indicated that the effects of nonlinear inertia terms are far more significant than those of nonlinear elasticity terms on the parametric response of thin rods. However, Crespo da Silva and Glynn (1978a,b) showed that the generally neglected nonlinear terms due to curvature are of the same order as the nonlinear terms due to inertia and that the curvature terms may have a significant influence on the response of the system. An experimental study due to Evan-Iwanowski, Sanford, and Kehagioglou (1970) showed that the nonlinear geometric terms have noticeable effects on the nonstationary parametric response of a straight column. Nayfeh and Pai (1989) showed that the often ignored geometric nonlinearities produce a hardening effect, whereas the inertia nonlinearities produce a softening effect. Consequently, the overall effect due to the nonlinearities may be hardening or softening depending on the relative strengths of the inertia and curvature nonlinearities. Further, Nayfeh and Pai showed that for the first mode the effective nonlinearity is of the hardening type, whereas for the second and higher modes the effective nonlinearity is of the softening type. Anderson, Nayfeh, and Balachandran (1996) experimentally and theoretically investigated the first- and second-mode responses of a cantilever beam to principal parametric excitations. Their experimental results verify that the often neglected nonlinear curvature terms play a dominant role in the response of the first mode and that the inertia terms play a dominant role in the response of the second mode.

4.5.2 Stretching Effect

For slender beams, the longitudinal inertia $m\ddot{u}$ in (4.5.23), (4.5.29), and (4.5.31) may be negligible. To show this, we consider the case of a hinged-hinged beam so that the boundary conditions are

$$u(0,t) = 0 \text{ and } u(L,t) = LP(t) \tag{4.5.43}$$

$$v = 0, \ v'' = 0 \text{ at } s = 0 \text{ and } L \tag{4.5.44}$$

Using (4.5.43) and (4.5.44) and neglecting the rotary inertia, we find from (4.5.31) and (4.5.32) that the linear longitudinal and transverse frequencies are given by

$$\lambda_n = \frac{n\pi}{L}\sqrt{\frac{EA}{m}} \text{ and } \omega_k = \frac{k^2\pi^2}{L^2}\sqrt{\frac{EI}{m}}$$

where the subscripts n and k stand for the nth and kth frequencies. Hence,

$$\frac{\lambda_n}{\omega_k} = \frac{nL}{\pi k^2 r}$$

where $r = \sqrt{I/A}$ is the radius of gyration, which is small for slender beams. Consequently, $\lambda_n \gg \omega_k$ if the slenderness ratio $L/r \gg 1$ and $n/\pi k^2$ is not small. In other words, u is mainly induced by the transverse deformation v. In this case, it follows from (4.5.31) that

$$u'' = -\left(\frac{1}{2}v'^2\right)' + \cdots \tag{4.5.45}$$

which means that $u = O(v^2)$. Integrating (4.5.45) twice with respect to s, we have

$$u' = -\frac{1}{2}v'^2 + c_1(t) \tag{4.5.46}$$

$$u = -\frac{1}{2}\int_0^s v'^2 ds + c_1(t)s + c_2(t) \tag{4.5.47}$$

where c_1 and c_2 are functions of time, which can be determined by imposing boundary conditions on u. Imposing the boundary conditions (4.5.43) and (4.5.44) yields $c_2(t) = 0$ and

$$c_1(t) = P(t) + \frac{1}{2L}\int_0^L v'^2 ds \tag{4.5.48}$$

Substituting for u' from (4.5.46) into (4.5.32), recalling that $u = O(v^2)$, and keeping up to cubic terms in v, we have

$$m\ddot{v} + EIv^{iv} = EAc_1(t)v'' + EI\left[2c_1v^{iv} - v''^3 - 2v'v''v'''\right]$$

or

$$m\ddot{v} + EIv^{iv} = \left(EAv'' + 2EIv^{iv}\right)\left[P(t) + \frac{1}{2L}\int_0^L v'^2\,ds\right] - EI(v''^3 + 2v'v''v''') \quad (4.5.49)$$

Because the radius of gyration $r = \sqrt{I/A}$ is usually small, the terms proportional to EI on the right-hand side of (4.5.49) can be neglected. The result is

$$m\ddot{v} - EAP(t)v'' + EIv^{iv} = \frac{EA}{2L}v''\int_0^L v'^2\,ds \quad (4.5.50)$$

4.5.3 Lagrangian and Eulerian Coordinates

To show the difference between the use of Lagrangian and Eulerian coordinates in the nonlinear modeling of structures, we consider the cantilever isotropic beam shown in Figure 4.5.1a. The Eulerian coordinate (i.e., the instantaneous position) of the observed reference point is x and

$$x = s + u, \quad dx = ds + du \quad (4.5.51)$$

We assume the inextensionality condition $e = 0$ to obtain from Figure 4.5.2 that

$$dx^2 + dv^2 = ds^2 \quad (4.5.52)$$

Using (4.5.51) and the boundary condition $u(0,t) = 0$ in (4.5.52) yields

$$u_x = 1 - \sqrt{1 + v_x^2}, \quad u = \int_0^x \left(1 - \sqrt{1 + v_x^2}\right)dx \quad (4.5.53a,b)$$

We note that (4.5.53b) is different from (4.5.35). Moreover, when the Lagrangian coordinate s is used, it follows from (4.5.3) and (4.5.38) that the curvature ρ_3 can be expressed as

$$\rho_3 = \frac{\partial \sin^{-1} v'}{\partial s} = \frac{v''}{\sqrt{1 - v'^2}} = v''\left(1 + \frac{1}{2}v'^2 + \frac{3}{8}v'^4 + \cdots\right) \quad (4.5.54a)$$

On the other hand, when the Eulerian coordinate x is used, it follows from Figure 4.5.2, (4.5.3), (4.5.51) and (4.5.53a) that

$$\rho_3 = \frac{\partial \tan^{-1} v_x}{\partial s} = \frac{\partial \tan^{-1} v_x}{\partial x}\frac{dx}{ds} = \frac{\partial \tan^{-1} v_x}{\partial x}\frac{1}{\sqrt{1 + v_x^2}} = \frac{v_{xx}}{(1 + v_x^2)^{3/2}}$$
$$= v_{xx}\left(1 - \frac{3}{2}v_x^2 + \frac{15}{8}v_x^4 + \cdots\right) \quad (4.5.54b)$$

where

$$v_x \equiv \frac{\partial v}{\partial x} = v'\left(1 + \frac{1}{2}v'^2 + \cdots\right) \quad (4.5.54c)$$

We note that the form of the nonlinear curvature depends on the coordinate being used. Because $|v'| < 1$ for all $|\Theta_3| < 90°$, the series in (4.5.54a) converges for all $|\Theta_3| < 90°$. However, because $|v_x| < 1$ for $|\Theta_3| < 45°$ and $|v_x| > 1$ for $45° < |\Theta_3| < 90°$, the series in (4.5.54b) may diverge.

If the Eulerian coordinate x is used, the co-moving derivative of the displacement v (i.e., the velocity in the y direction) is obtained by using (4.5.51) and (4.5.53b) as

$$\frac{dv}{dt} = \frac{\partial v}{\partial t} + \frac{\partial v}{\partial x}\frac{dx}{dt} = \dot{v} + v_x \frac{du}{dt} = \dot{v} + v_x \frac{d}{dt}\left[\int_0^x \left(1 - \sqrt{1+v_x^2}\right) dx\right] \quad (4.5.55)$$

Similarly, the acceleration of the element is given by

$$\frac{d^2 v}{dt^2} = \ddot{v} + 2\dot{v}_x \frac{du}{dt} + v_{xx}\left(\frac{du}{dt}\right)^2 + v_x \frac{d^2 u}{dt^2} + v_x \frac{du_x}{dt}\frac{du}{dt} \quad (4.5.56)$$

Equations (4.5.55) and (4.5.56) reveal that the Lagrangian coordinate s is more suitable than the Eulerian coordinate x for an inextensional beam because in the latter case the expressions for the velocity and acceleration are more complicated. Moreover, the integration limits of s are 0 and L, but the limits of x are deformation dependent.

4.6 NONLINEAR 3-D EULER-BERNOULLI BEAM THEORY

The main difference between two- and three-dimensional Euler-Bernoulli beam theories is the inclusion of torsional motion. Moreover, elastic couplings due to anisotropy may couple flexural vibrations in two principal directions. Hence the vibration of a general anisotropic beam is three-dimensional. In this section, we develop equations governing the three-dimensional vibration of initially straight beams using the extended Hamilton principle (4.3.4). For the initially straight beam shown in Figure 4.4.1, the deformed coordinate system $\xi\eta\zeta$ can be related to the undeformed coordinate system xyz using three consecutive Euler angles ψ, θ, and φ (see Figure 4.4.2) or two Euler angles α and ϕ (see Figure 4.4.5). However, the transformation using two Euler angles is preferred, as explained later in Section 4.6.3.

Inertia Terms: It follows from Figure 4.4.1 that the deformed position vector $\mathbf{R}$ of an arbitrary point on the observed cross-section is given by

$$\mathbf{R} = (s+u)\mathbf{i}_x + v\mathbf{i}_y + w\mathbf{i}_z + y\mathbf{i}_2 + z\mathbf{i}_3 \quad (4.6.1)$$

Taking the variation of $\mathbf{R}$ and using the fact that $\delta\mathbf{i}_x = \delta\mathbf{i}_y = \delta\mathbf{i}_z = \delta s = 0$, we have

$$\delta\mathbf{R} = \mathbf{i}_x \delta u + \mathbf{i}_y \delta v + \mathbf{i}_z \delta w + y\delta\mathbf{i}_2 + z\delta\mathbf{i}_3$$

Using the concept of orthogonal virtual rotations (4.4.67), we obtain

$$\delta\mathbf{i}_2 = -\delta\theta_3 \mathbf{i}_1 + \delta\theta_1 \mathbf{i}_3 \quad \text{and} \quad \delta\mathbf{i}_3 = \delta\theta_2 \mathbf{i}_1 - \delta\theta_1 \mathbf{i}_2$$

Hence,

$$\delta \mathbf{R} = \mathbf{i}_x \delta u + \mathbf{i}_y \delta v + \mathbf{i}_z \delta w + y(\delta\theta_1 \mathbf{i}_3 - \delta\theta_3 \mathbf{i}_1) + z(\delta\theta_2 \mathbf{i}_1 - \delta\theta_1 \mathbf{i}_2)$$

or

$$\delta \mathbf{R} = \{\mathbf{i}_x,\ \mathbf{i}_y,\ \mathbf{i}_z\} \begin{Bmatrix} \delta u \\ \delta v \\ \delta w \end{Bmatrix} + \{\mathbf{i}_1,\ \mathbf{i}_2,\ \mathbf{i}_3\}[r] \begin{Bmatrix} \delta\theta_1 \\ \delta\theta_2 \\ \delta\theta_3 \end{Bmatrix} \qquad (4.6.2)$$

where

$$[r] \equiv \begin{bmatrix} 0 & z & -y \\ -z & 0 & 0 \\ y & 0 & 0 \end{bmatrix} \qquad (4.6.3)$$

Differentiating (4.6.1) with respect to t and using the following equations:

$$\frac{\partial \vec{\omega}}{\partial t} = \dot{\vec{\omega}} = \dot{\omega}_1 \mathbf{i}_1 + \dot{\omega}_2 \mathbf{i}_2 + \dot{\omega}_3 \mathbf{i}_3, \quad \frac{\partial \mathbf{i}_k}{\partial t} = \vec{\omega} \times \mathbf{i}_k, \quad k = 1, 2, 3$$

we obtain

$$\dot{\mathbf{R}} = \dot{u}\mathbf{i}_x + \dot{v}\mathbf{i}_y + \dot{w}\mathbf{i}_z + \vec{\omega} \times (y\mathbf{i}_2 + z\mathbf{i}_3)$$

The angular velocity $\vec{\omega}$ of the local coordinate system $\xi\eta\zeta$ is given by (4.4.24) and (4.4.25). Differentiating $\dot{\mathbf{R}}$ with respect to t, we have

$$\ddot{\mathbf{R}} = \ddot{u}\mathbf{i}_x + \ddot{v}\mathbf{i}_y + \ddot{w}\mathbf{i}_z + \dot{\vec{\omega}} \times (y\mathbf{i}_2 + z\mathbf{i}_3) + \vec{\omega} \times [\vec{\omega} \times (y\mathbf{i}_2 + z\mathbf{i}_3)]$$

or

$$\begin{aligned}\ddot{\mathbf{R}} =\ & \ddot{u}\mathbf{i}_x + \ddot{v}\mathbf{i}_y + \ddot{w}\mathbf{i}_z + (z\dot{\omega}_2 - y\dot{\omega}_3)\mathbf{i}_1 - z\dot{\omega}_1\mathbf{i}_2 + y\dot{\omega}_1\mathbf{i}_3 \\ &+ (z\omega_1\omega_3 + y\omega_1\omega_2)\mathbf{i}_1 + (-y\omega_1^2 + z\omega_2\omega_3 - y\omega_3^2)\mathbf{i}_2 \\ &+ (-z\omega_1^2 + y\omega_2\omega_3 - z\omega_2^2)\mathbf{i}_3 \end{aligned}$$

or

$$\ddot{\mathbf{R}} = \{\ddot{u}, \ddot{v}, \ddot{w}\} \begin{Bmatrix} \mathbf{i}_x \\ \mathbf{i}_y \\ \mathbf{i}_z \end{Bmatrix} + \{\dot{\omega}_1, \dot{\omega}_2, \dot{\omega}_3\}[r]^T \begin{Bmatrix} \mathbf{i}_1 \\ \mathbf{i}_2 \\ \mathbf{i}_3 \end{Bmatrix} - \{\omega_1, \omega_2, \omega_3\}[r][\omega] \begin{Bmatrix} \mathbf{i}_1 \\ \mathbf{i}_2 \\ \mathbf{i}_3 \end{Bmatrix} \qquad (4.6.4)$$

where

$$[\omega] \equiv \begin{bmatrix} 0 & \omega_3 & -\omega_2 \\ -\omega_3 & 0 & \omega_1 \\ \omega_2 & -\omega_1 & 0 \end{bmatrix} \qquad (4.6.5)$$

In terms of $\ddot{\mathbf{R}}$ and $\delta \mathbf{R}$, variation of the kinetic energy is given by

$$\delta T = -\int_0^L \int_A \rho \ddot{\mathbf{R}} \cdot \delta \mathbf{R}\, dA\, ds$$

Substituting (4.6.2) and (4.6.4) into the expression for δT and using (4.4.4) and (4.4.6), we obtain

$$\delta T = -\int_0^L \int_A \rho \left\{ \left[\begin{Bmatrix} \ddot{u} \\ \ddot{v} \\ \ddot{w} \end{Bmatrix}^T + \begin{Bmatrix} \dot{\omega}_1 \\ \dot{\omega}_2 \\ \dot{\omega}_3 \end{Bmatrix}^T [r]^T [T] - \begin{Bmatrix} \omega_1 \\ \omega_2 \\ \omega_3 \end{Bmatrix}^T [r][\omega][T] \right] \begin{Bmatrix} \delta u \\ \delta v \\ \delta w \end{Bmatrix} \right.$$

$$\left. + \left[\begin{Bmatrix} \dot{\omega}_1 \\ \dot{\omega}_2 \\ \dot{\omega}_3 \end{Bmatrix}^T [r]^T [r] - \begin{Bmatrix} \omega_1 \\ \omega_2 \\ \omega_3 \end{Bmatrix}^T [r][\omega][r] + \begin{Bmatrix} \ddot{u} \\ \ddot{v} \\ \ddot{w} \end{Bmatrix}^T [T]^T [r] \right] \begin{Bmatrix} \delta \theta_1 \\ \delta \theta_2 \\ \delta \theta_3 \end{Bmatrix} \right\} dA\, ds \quad (4.6.6)$$

By direct expansion, one can show that

$$\{\omega_1, \omega_2, \omega_3\}[r][\omega][r] = -\{\omega_1, \omega_2, \omega_3\}[r]^T[r][\omega] \quad (4.6.7)$$

Integrating the right-hand side of (4.6.6) over the area, using (4.6.7), and noting that the ω_i and $[T]$ are independent of y and z, we obtain

$$\delta T = -\int_0^L \left\{ \left[m \begin{Bmatrix} \ddot{u} \\ \ddot{v} \\ \ddot{w} \end{Bmatrix}^T - \begin{Bmatrix} \dot{\omega}_1 \\ \dot{\omega}_2 \\ \dot{\omega}_3 \end{Bmatrix}^T [J_1][T] - \begin{Bmatrix} \omega_1 \\ \omega_2 \\ \omega_3 \end{Bmatrix}^T [J_1][\omega][T] \right] \begin{Bmatrix} \delta u \\ \delta v \\ \delta w \end{Bmatrix} \right.$$

$$\left. + \left[\begin{Bmatrix} \dot{\omega}_1 \\ \dot{\omega}_2 \\ \dot{\omega}_3 \end{Bmatrix}^T [J_2] + \begin{Bmatrix} \omega_1 \\ \omega_2 \\ \omega_3 \end{Bmatrix}^T [J_2][\omega] + \begin{Bmatrix} \ddot{u} \\ \ddot{v} \\ \ddot{w} \end{Bmatrix}^T [T]^T[J_1] \right] \begin{Bmatrix} \delta \theta_1 \\ \delta \theta_2 \\ \delta \theta_3 \end{Bmatrix} \right\} ds \quad (4.6.8)$$

where

$$[J_1] \equiv \int_A \rho [r] dA, \quad m \equiv \int_A \rho\, dA,$$

$$[J_2] \equiv \int_A \rho [r]^T [r] dA = \int_A \rho \begin{bmatrix} y^2 + z^2 & 0 & 0 \\ 0 & z^2 & -yz \\ 0 & -yz & y^2 \end{bmatrix} dA \quad (4.6.9)$$

Because the integrand in (4.6.8) is a scalar, it is equal to its transpose. Then, taking the transpose of this integrand and noting that $[J_1]^T = -[J_1]$ and $[\omega]^T = -[\omega]$, we rewrite (4.6.8) as

$$\delta T = -\int_0^L (A_u \delta u + A_v \delta v + A_w \delta w + A_{\theta_1} \delta \theta_1 + A_{\theta_2} \delta \theta_2 + A_{\theta_3} \delta \theta_3) ds \quad (4.6.10)$$

where

$$\begin{Bmatrix} A_u \\ A_v \\ A_w \end{Bmatrix} \equiv \begin{Bmatrix} m\ddot{u} \\ m\ddot{v} \\ m\ddot{w} \end{Bmatrix} + [T]^T[J_1] \begin{Bmatrix} \dot{\omega}_1 \\ \dot{\omega}_2 \\ \dot{\omega}_3 \end{Bmatrix} - [T]^T[\omega][J_1] \begin{Bmatrix} \omega_1 \\ \omega_2 \\ \omega_3 \end{Bmatrix} \quad (4.6.11)$$

NONLINEAR 3-D EULER-BERNOULLI BEAM THEORY

$$\begin{Bmatrix} A_{\theta_1} \\ A_{\theta_2} \\ A_{\theta_3} \end{Bmatrix} \equiv [J_2] \begin{Bmatrix} \dot{\omega}_1 \\ \dot{\omega}_2 \\ \dot{\omega}_3 \end{Bmatrix} - [\omega][J_2] \begin{Bmatrix} \omega_1 \\ \omega_2 \\ \omega_3 \end{Bmatrix} - [J_1][T] \begin{Bmatrix} \ddot{u} \\ \ddot{v} \\ \ddot{w} \end{Bmatrix} \quad (4.6.12)$$

If the reference point coincides with the mass centroid and η and ζ are the principal axes of the differential beam slice, we have

$$[J_1] = [0], \quad [J_2] = \begin{bmatrix} j_1 & 0 & 0 \\ 0 & j_2 & 0 \\ 0 & 0 & j_3 \end{bmatrix}, \quad j_1 = j_2 + j_3 \quad (4.6.13)$$

$$\begin{Bmatrix} A_u \\ A_v \\ A_w \end{Bmatrix} = \begin{Bmatrix} m\ddot{u} \\ m\ddot{v} \\ m\ddot{w} \end{Bmatrix}, \quad \begin{Bmatrix} A_{\theta_1} \\ A_{\theta_2} \\ A_{\theta_3} \end{Bmatrix} = \begin{Bmatrix} j_1\dot{\omega}_1 - (j_2 - j_3)\omega_2\omega_3 \\ j_2\dot{\omega}_2 - (j_3 - j_1)\omega_1\omega_3 \\ j_3\dot{\omega}_3 - (j_1 - j_2)\omega_1\omega_2 \end{Bmatrix} \quad (4.6.14)$$

Equation (4.6.14) has the same form as the Euler equations used in rigid-body dynamics. However, in rigid-body dynamics, the rotary inertias are obtained by integrating with respect to the system volume instead of the area and $[r]$ is replaced with

$$[r] = \begin{bmatrix} 0 & z & -y \\ -z & 0 & x \\ y & -x & 0 \end{bmatrix}$$

Structural Terms: To fully account for geometric nonlinearities, we use local stress and strain measures. Hence, we introduce the following local displacement field:

$$\mathbf{U} = u_1 \mathbf{i}_1 + u_2 \mathbf{i}_2 + u_3 \mathbf{i}_3$$
$$u_1 = u_1^0 - y\theta_3 + z\theta_2, \quad u_2 = u_2^0 - z\theta_1, \quad u_3 = u_3^0 + y\theta_1 \quad (4.6.15)$$

where

$$u_1^0 = u_2^0 = u_3^0 = 0, \quad \theta_1 = \theta_2 = \theta_3 = 0 \quad (4.6.16)$$

$$\frac{\partial u_2^0}{\partial s} = \frac{\partial u_3^0}{\partial s} = 0, \quad \frac{\partial u_1^0}{\partial s} \equiv e, \quad \frac{\partial \theta_1}{\partial s} \equiv \rho_1, \quad \frac{\partial \theta_2}{\partial s} \equiv \rho_2, \quad \frac{\partial \theta_3}{\partial s} \equiv \rho_3 \quad (4.6.17)$$

It follows from (4.6.15)-(4.6.17) that

$$\frac{\partial \mathbf{U}}{\partial s} = (e - y\rho_3 + z\rho_2)\mathbf{i}_1 - z\rho_1 \mathbf{i}_2 + y\rho_1 \mathbf{i}_3$$

$$\frac{\partial \mathbf{U}}{\partial y} = \frac{\partial \mathbf{U}}{\partial z} = 0$$

Consequently, the local strains are given by

$$\epsilon_{11} = \frac{\partial \mathbf{U}}{\partial s} \cdot \mathbf{i}_1 = e - y\rho_3 + z\rho_2, \quad \epsilon_{12} = \frac{\partial \mathbf{U}}{\partial s} \cdot \mathbf{i}_2 + \frac{\partial \mathbf{U}}{\partial y} \cdot \mathbf{i}_1 = -z\rho_1$$

$$\epsilon_{13} = \frac{\partial \mathbf{U}}{\partial s} \cdot \mathbf{i}_3 + \frac{\partial \mathbf{U}}{\partial z} \cdot \mathbf{i}_1 = y\rho_1, \quad \epsilon_{22} = \epsilon_{23} = \epsilon_{33} = 0 \quad (4.6.18)$$

In terms of the σ_{ij} and ϵ_{ij}, variation of the elastic energy is given by

$$\delta\Pi = \int_0^L \int_A (\sigma_{11}\delta\epsilon_{11} + \sigma_{12}\delta\epsilon_{12} + \sigma_{13}\delta\epsilon_{13} + \sigma_{22}\delta\epsilon_{22} + \sigma_{23}\delta\epsilon_{23} + \sigma_{33}\delta\epsilon_{33})dAds$$

which, upon using the fact that $\epsilon_{22} = \epsilon_{23} = \epsilon_{33} = 0$, reduces to

$$\delta\Pi = \int_0^L \int_A (\sigma_{11}\delta\epsilon_{11} + \sigma_{12}\delta\epsilon_{12} + \sigma_{13}\delta\epsilon_{13})dAds$$

Substituting for the ϵ_{ij} from (4.6.18) into the expression for $\delta\Pi$ yields

$$\delta\Pi = \int_0^L \int_A [\sigma_{11}(\delta e - y\delta\rho_3 + z\delta\rho_2) - \sigma_{12}z\delta\rho_1 + \sigma_{13}y\delta\rho_1] dAds$$

which, upon integration over the area, becomes

$$\delta\Pi = \int_0^L (F_1\delta e + M_2\delta\rho_2 + M_3\delta\rho_3 + M_1\delta\rho_1)ds \qquad (4.6.19)$$

where the stress resultants and moments are defined as

$$F_1 = \int_A \sigma_{11}dA, \quad \begin{Bmatrix} M_1 \\ M_2 \\ M_3 \end{Bmatrix} \equiv \int_A \begin{Bmatrix} \sigma_{13}y - \sigma_{12}z \\ \sigma_{11}z \\ -\sigma_{11}y \end{Bmatrix} dA \qquad (4.6.20)$$

Using (4.4.85)-(4.4.87), we rewrite (4.6.19) as

$$\delta\Pi = \int_0^L \{F_1\delta e - (M_1' + M_3\rho_2 - M_2\rho_3)\,\delta\theta_1 - (M_2' - M_3\rho_1 + M_1\rho_3)\,\delta\theta_2$$
$$- (M_3' + M_2\rho_1 - M_1\rho_2)\,\delta\theta_3\}ds + [M_1\delta\theta_1 + M_2\delta\theta_2 + M_3\delta\theta_3]_0^L$$

which, upon using (4.4.71), becomes

$$\delta\Pi = \int_0^L \{F_1\,(T_{11}\delta u' + T_{12}\delta v' + T_{13}\delta w') - (M_1' + M_3\rho_2 - M_2\rho_3)\,\delta\theta_1$$
$$- (M_2' - M_3\rho_1 + M_1\rho_3)\,\delta\theta_2 - (M_3' + M_2\rho_1 - M_1\rho_2)\,\delta\theta_3\}ds$$
$$+ [M_1\delta\theta_1 + M_2\delta\theta_2 + M_3\delta\theta_3]_0^L \qquad (4.6.21)$$

Nonconservative Terms: Because there are four dependent variables, we assume linear viscous dampings and obtain

$$\delta W_{nc} = \int_0^L (-c_1\dot{u}\delta u - c_2\dot{v}\delta v - c_3\dot{w}\delta w - c_4\dot{\phi}\delta\theta_1$$
$$+ q_1\delta u + q_2\delta v + q_3\delta w + q_4\delta\theta_1)ds \qquad (4.6.22)$$

where the c_i are damping coefficients and the q_i are distributed external loads.

Equations of Motion: To determine the equations of motion, we substitute the expression for $\delta T, \delta \Pi$, and δW_{nc} from (4.6.10), (4.6.21), and (4.6.22) into the statement (4.3.4) of the extended Hamilton principle and obtain

$$\int_0^t \int_0^L \Big[-F_1(T_{11}\delta u' + T_{12}\delta v' + T_{13}\delta w') - (A_u + c_1\dot{u} - q_1)\delta u$$
$$-(A_v + c_2\dot{v} - q_2)\delta v - (A_w + c_3\dot{w} - q_3)\delta w$$
$$+(M_1' + M_3\rho_2 - M_2\rho_3 - A_{\theta_1} - c_4\dot{\phi} + q_4)\delta\theta_1$$
$$+(M_2' - M_3\rho_1 + M_1\rho_3 - A_{\theta_2})\delta\theta_2$$
$$+(M_3' + M_2\rho_1 - M_1\rho_2 - A_{\theta_3})\delta\theta_3 \Big] ds\, dt$$
$$-\int_0^t [M_1\delta\theta_1 + M_2\delta\theta_2 + M_3\delta\theta_3]_0^L \, dt = 0 \qquad (4.6.23)$$

We note that there are four dependent variables, namely, u, v, w, and ϕ, but there are six variation quantities, namely, $\delta u, \delta v, \delta w, \delta\theta_1, \delta\theta_2$, and $\delta\theta_3$. However, the last two variations are related to the first three by (4.4.73) and (4.4.74). To proceed further, we have two alternatives. First, we consider (4.4.73) and (4.4.74) as constraints and increase the dependent variables by two, namely, λ_2 and λ_3. These additional variables are Lagrange multipliers. Hence, we add (4.4.74) times $-\lambda_2$ and (4.4.73) times $-\lambda_3$ to (4.6.23) and obtain

$$\int_0^t \int_0^L \Bigg[-\{F_1, \lambda_2, \lambda_3\}[T] \begin{Bmatrix} \delta u' \\ \delta v' \\ \delta w' \end{Bmatrix} - \begin{Bmatrix} A_u + c_1\dot{u} - q_1 \\ A_v + c_2\dot{v} - q_2 \\ A_w + c_3\dot{w} - q_3 \end{Bmatrix}^T \begin{Bmatrix} \delta u \\ \delta v \\ \delta w \end{Bmatrix}$$
$$+ \Big(M_1' + M_3\rho_2 - M_2\rho_3 - A_{\theta_1} - c_4\dot{\phi} + q_4\Big)\delta\theta_1$$
$$+ \Big[M_2' - M_3\rho_1 + M_1\rho_3 - (1+e)\lambda_3 - A_{\theta_2}\Big]\delta\theta_2$$
$$+ \Big[M_3' + M_2\rho_1 - M_1\rho_2 + (1+e)\lambda_2 - A_{\theta_3}\Big]\delta\theta_3 \Bigg] ds\, dt$$
$$-\int_0^t [M_1\delta\theta_1 + M_2\delta\theta_2 + M_3\delta\theta_3]_0^L \, dt = 0 \qquad (4.6.24)$$

Second, we use (4.4.73) and (4.4.74) to eliminate $\delta\theta_2$ and $\delta\theta_3$ in the integrand in (4.6.23). The result can be expressed as

$$\int_0^t \int_0^L \Bigg[-\{F_1, \lambda_2, \lambda_3\}[T] \begin{Bmatrix} \delta u' \\ \delta v' \\ \delta w' \end{Bmatrix} - \begin{Bmatrix} A_u + c_1\dot{u} - q_1 \\ A_v + c_2\dot{v} - q_2 \\ A_w + c_3\dot{w} - q_3 \end{Bmatrix}^T \begin{Bmatrix} \delta u \\ \delta v \\ \delta w \end{Bmatrix}$$
$$+ \Big(M_1' + M_3\rho_2 - M_2\rho_3 - A_{\theta_1} - c_4\dot{\phi} + q_4\Big)\delta\theta_1 \Bigg] ds\, dt$$
$$-\int_0^t [M_1\delta\theta_1 + M_2\delta\theta_2 + M_3\delta\theta_3]_0^L \, dt = 0 \qquad (4.6.25)$$

where

$$\lambda_2 = \frac{1}{1+e}(-M_3' - M_2\rho_1 + M_1\rho_2 + A_{\theta_3}) \qquad (4.6.26a)$$

$$\lambda_3 = \frac{1}{1+e}(M_2' - M_3\rho_1 + M_1\rho_3 - A_{\theta_2}) \qquad (4.6.26b)$$

Because we added two dependent variables λ_2 and λ_3, $\delta\theta_2$ and $\delta\theta_3$ in (4.6.24) are independent of the other variations. Hence their coefficients must be zero, resulting in the expressions (4.6.26) for λ_2 and λ_3. Thus, the two approaches are equivalent.

Integrating the terms involving $\delta u'$, $\delta v'$, and $\delta w'$ in (4.6.25) by parts yields

$$\int_0^t \int_0^L \left[\frac{\partial}{\partial s}[\{F_1, \lambda_2, \lambda_3\}[T]] \begin{Bmatrix} \delta u \\ \delta v \\ \delta w \end{Bmatrix} - \begin{Bmatrix} A_u + c_1\dot{u} - q_1 \\ A_v + c_2\dot{v} - q_2 \\ A_w + c_3\dot{w} - q_3 \end{Bmatrix}^T \begin{Bmatrix} \delta u \\ \delta v \\ \delta w \end{Bmatrix} \right.$$
$$\left. + \left(M_1' + M_3\rho_2 - M_2\rho_3 - A_{\theta_1} - c_4\dot{\phi} + q_4 \right) \delta\theta_1 \right] ds\, dt$$
$$- \int_0^t \left[M_1\delta\theta_1 + M_2\delta\theta_2 + M_3\delta\theta_3 + \{F_1, \lambda_2, \lambda_3\}[T] \begin{Bmatrix} \delta u \\ \delta v \\ \delta w \end{Bmatrix} \right]_0^L dt = 0$$

(4.6.27)

Setting each of the coefficients of δu, δv, δw, and $\delta\theta_1$ in (4.6.27) equal to zero, we obtain the following equations of motion:

$$[F_1 T_{11} + \lambda_2 T_{21} + \lambda_3 T_{31}]' + q_1 = A_u + c_1\dot{u} \qquad (4.6.28)$$
$$[F_1 T_{12} + \lambda_2 T_{22} + \lambda_3 T_{32}]' + q_2 = A_v + c_2\dot{v} \qquad (4.6.29)$$
$$[F_1 T_{13} + \lambda_2 T_{23} + \lambda_3 T_{33}]' + q_3 = A_w + c_3\dot{w} \qquad (4.6.30)$$
$$M_1' + M_3\rho_2 - M_2\rho_3 + q_4 = A_{\theta_1} + c_4\dot{\phi} \qquad (4.6.31)$$

where λ_2 and λ_3 are given by (4.6.26). The boundary conditions are to specify

$$\delta u = \delta v = \delta w = 0, \ \delta\theta_1 = \delta\theta_2 = \delta\theta_3 = 0 \qquad (4.6.32a)$$

or

$$G_1 = G_2 = G_3 = 0, \ M_1 = M_2 = M_3 = 0 \qquad (4.6.32b)$$

where

$$G_k \equiv \sum_{i=1}^{3} \lambda_i T_{ik}, \ \ k = 1, 2, 3 \qquad (4.6.32c)$$

and $\lambda_1 = F_1$.

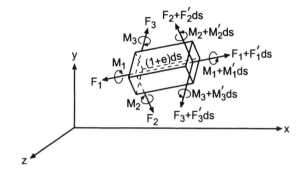

Fig. 4.6.1 The stress resultants on a differential beam element.

Newtonian Formulation: Equations (4.6.28)-(4.6.31) and (4.6.26) can be expressed in matrix form as

$$\frac{\partial}{\partial s}\{[T]^T\{F\}\} + \begin{Bmatrix} q_1 \\ q_2 \\ q_3 \end{Bmatrix} = \begin{Bmatrix} A_u \\ A_v \\ A_w \end{Bmatrix} + \begin{Bmatrix} c_1 \dot{u} \\ c_2 \dot{v} \\ c_3 \dot{w} \end{Bmatrix} \quad (4.6.33)$$

$$\frac{\partial \{M\}}{\partial s} + [K]^T\{M\} + \begin{Bmatrix} 0 \\ -(1+e)\lambda_3 \\ (1+e)\lambda_2 \end{Bmatrix} + \begin{Bmatrix} q_4 \\ 0 \\ 0 \end{Bmatrix} = \begin{Bmatrix} A_{\theta_1} \\ A_{\theta_2} \\ A_{\theta_3} \end{Bmatrix} + \begin{Bmatrix} c_4 \dot{\phi} \\ 0 \\ 0 \end{Bmatrix} \quad (4.6.34)$$

where $\{M\} \equiv \{M_1, M_2, M_3\}^T$ and $\{F\} \equiv \{F_1, \lambda_2, \lambda_3\}^T$. The equivalent vector forms of (4.6.33) and (4.6.34) are

$$\frac{\partial \mathbf{F}}{\partial s} + q_1 \mathbf{i}_x + q_2 \mathbf{i}_y + q_3 \mathbf{i}_z = (A_u + c_1 \dot{u})\mathbf{i}_x + (A_v + c_2 \dot{v})\mathbf{i}_y + (A_w + c_3 \dot{w})\mathbf{i}_z \quad (4.6.35)$$

$$\frac{\partial \mathbf{M}}{\partial s} + (1+e)\mathbf{i}_1 \times \mathbf{F} + q_4 \mathbf{i}_1 = (A_{\theta_1} + c_4 \dot{\phi})\mathbf{i}_1 + A_{\theta_2} \mathbf{i}_2 + A_{\theta_3} \mathbf{i}_3 \quad (4.6.36)$$

where $\mathbf{F} = F_1 \mathbf{i}_1 + \lambda_2 \mathbf{i}_2 + \lambda_3 \mathbf{i}_3$ and $\mathbf{M} = M_1 \mathbf{i}_1 + M_2 \mathbf{i}_2 + M_3 \mathbf{i}_3$. Equations (4.6.35) and (4.6.36) can be directly obtained from a vector approach by using the free-body diagram shown in Figure 4.6.1 (Pai and Nayfeh, 1990b). This shows that the energy formulation (4.6.27) is fully correlated with the Newtonian formulation (4.6.35) and (4.6.36), and the systems of equations obtained from these two approaches are essentially the same. However, the energy formulation is more convenient for accounting for warpings, as shown later.

Because $\mathbf{F} = F_1 \mathbf{i}_1 + \lambda_2 \mathbf{i}_2 + \lambda_3 \mathbf{i}_3$, $G_1 = \sum_{i=1}^{3} \lambda_i T_{i1} = \mathbf{F} \cdot \mathbf{i}_x$. Hence, G_1 is the stress resultant along the x direction. Similarly, G_2 is the stress resultant along the y direction, and G_3 is the stress resultant along the z direction.

4.6.1 Isotropic Beams

For an isotropic beam with no external loads acting on its lateral surfaces, it follows from (2.4.57), (2.4.53), and (4.2.15) that

$$\begin{Bmatrix} \sigma_{11} \\ \sigma_{12} \\ \sigma_{13} \end{Bmatrix} = \begin{bmatrix} E & 0 & 0 \\ 0 & G & 0 \\ 0 & 0 & G \end{bmatrix} \begin{Bmatrix} \epsilon_{11} \\ \epsilon_{12} \\ \epsilon_{13} \end{Bmatrix} \qquad (4.6.37)$$

Using (4.6.18), we express the stresses in terms of the ρ_i and e as

$$\sigma_{11} = E(e - y\rho_3 + z\rho_2), \quad \sigma_{12} = -z\rho_1 G, \quad \sigma_{13} = y\rho_1 G \qquad (4.6.38)$$

Substituting (4.6.38) into (4.6.20) and assuming that the reference line coincides with the centroidal line, we have

$$F_1 = EAe, \quad M_1 = GI_1\rho_1, \quad M_2 = EI_2\rho_2, \quad M_3 = EI_3\rho_3 \qquad (4.6.39)$$

where

$$I_1 = \int_A (y^2 + z^2)\, dA, \quad I_3 = \int_A y^2\, dA, \quad I_2 = \int_A z^2\, dA \qquad (4.6.40)$$

According to (4.6.40), $I_1 = I_2 + I_3$, which is only valid for beams with homogeneous circular cross-sections. For non-circular cross-sections, because of torsional warpings of cross-sections, the torsional rigidity GI_1 is reduced. For example, I_1 for an isotropic rectangular cross-section having a width b and a thin thickness h can be derived using the theory of elasticity to be (Timoshenko and Goodier, 1970)

$$I_1 = \frac{1}{3}bh^3 \left(1 - \frac{192h}{\pi^5 b} \sum_{n=1,3,\ldots}^{\infty} \frac{1}{n^5} \tanh \frac{n\pi b}{2h}\right) \qquad (4.6.41)$$

Using (4.6.39) and (4.6.14), we rewrite (4.6.28)-(4.6.31) as

$$[EAeT_{11} + \lambda_2 T_{21} + \lambda_3 T_{31}]' + q_1 = m\ddot{u} + c_1 \dot{u} \qquad (4.6.42)$$

$$[EAeT_{12} + \lambda_2 T_{22} + \lambda_3 T_{32}]' + q_2 = m\ddot{v} + c_2 \dot{v} \qquad (4.6.43)$$

$$[EAeT_{13} + \lambda_2 T_{23} + \lambda_3 T_{33}]' + q_3 = m\ddot{w} + c_3 \dot{w} \qquad (4.6.44)$$

$$(GI_1\rho_1)' + E(I_3 - I_2)\rho_2\rho_3 + q_4 = j_1\dot{\omega}_1 - (j_2 - j_3)\omega_2\omega_3 + c_4\dot{\phi} \qquad (4.6.45)$$

Substituting (4.6.39) and (4.6.14) into (4.6.26), we have

$$\lambda_2 = \frac{1}{1+e}[-(EI_3\rho_3)' - EI_2\rho_1\rho_2 + GI_1\rho_1\rho_2 + j_3\dot{\omega}_3 - (j_1 - j_2)\omega_1\omega_2]$$

$$\lambda_3 = \frac{1}{1+e}[(EI_2\rho_2)' - EI_3\rho_1\rho_3 + GI_1\rho_1\rho_3 - j_2\dot{\omega}_2 + (j_3 - j_1)\omega_1\omega_3]$$

$$(4.6.46)$$

4.6.2 Composite Beams

For slender composite beams, it is appropriate to assume that

$$\sigma_{22} = \sigma_{33} = \sigma_{23} = 0 \tag{4.6.47}$$

The material stiffness matrix of a general anisotropic material is a full 6×6 matrix, as shown in (2.4.12). Substituting (4.6.47) into (2.4.12) yields

$$\begin{Bmatrix} \sigma_{11} \\ \sigma_{12} \\ \sigma_{13} \end{Bmatrix} = [\hat{Q}] \begin{Bmatrix} \epsilon_{11} \\ \epsilon_{12} \\ \epsilon_{13} \end{Bmatrix}, \quad [\hat{Q}] \equiv \begin{bmatrix} \hat{Q}_{11} & \hat{Q}_{16} & \hat{Q}_{15} \\ \hat{Q}_{16} & \hat{Q}_{66} & \hat{Q}_{65} \\ \hat{Q}_{15} & \hat{Q}_{65} & \hat{Q}_{55} \end{bmatrix} \tag{4.6.48}$$

where $[\hat{Q}]$ is called the reduced stiffness matrix. It follows from (4.6.20), (4.6.48), and (4.6.18) that

$$\begin{Bmatrix} F_1 \\ M_1 \\ M_2 \\ M_3 \end{Bmatrix} = \begin{bmatrix} A_{11} & B_{11} & B_{12} & B_{13} \\ B_{11} & D_{11} & D_{12} & D_{13} \\ B_{12} & D_{12} & D_{22} & D_{23} \\ B_{13} & D_{13} & D_{23} & D_{33} \end{bmatrix} \begin{Bmatrix} e \\ \rho_1 \\ \rho_2 \\ \rho_3 \end{Bmatrix} \tag{4.6.49}$$

where

$$\begin{bmatrix} A_{11} & B_{11} & B_{12} & B_{13} \\ B_{11} & D_{11} & D_{12} & D_{13} \\ B_{12} & D_{12} & D_{22} & D_{23} \\ B_{13} & D_{13} & D_{23} & D_{33} \end{bmatrix} \equiv \int_A [yz]^T [\hat{Q}] [yz] dA$$

$$= \int_A \begin{bmatrix} \hat{Q}_{11} & & & \text{sym.} \\ \hat{Q}_{15}y - \hat{Q}_{16}z & \hat{Q}_{55}y^2 + \hat{Q}_{66}z^2 - 2\hat{Q}_{65}yz & & \\ \hat{Q}_{11}z & \hat{Q}_{15}yz - \hat{Q}_{16}z^2 & \hat{Q}_{11}z^2 & \\ -\hat{Q}_{11}y & \hat{Q}_{16}yz - \hat{Q}_{15}y^2 & -\hat{Q}_{11}yz & \hat{Q}_{11}y^2 \end{bmatrix} dA,$$

$$[yz] \equiv \begin{bmatrix} 1 & 0 & z & -y \\ 0 & -z & 0 & 0 \\ 0 & y & 0 & 0 \end{bmatrix} \tag{4.6.50}$$

Hence, all of the stress resultants are represented in terms of e and the ρ_i. We note that D_{11} is the torsional rigidity and needs to be modified if the cross-section is non-circular and hence torsional warpings are not negligible.

4.6.3 Taylor-Series Expansions

Because the complete nonlinear governing equations (4.6.28)-(4.6.31) are transcendental, closed-form solutions are not available. Moreover, direct numerical procedures suffer from instability and convergence problems. Alternatively, one can use a combination of numerical and perturbation methods to solve them. The first step in such an approach is to expand the nonlinear transcendental terms into polynomials about the static equilibrium position. In what follows, we assume that the equilibrium position is close to the undeformed position.

To obtain the governing equations in polynomial form for small but finite oscillations about the undeformed position, we assume that u, v, w, ϕ (or φ) and their derivatives are $O(\epsilon)$, where ϵ ($\ll 1$) is a small dimensionless parameter that is used as a bookkeeping device. In this section, we expand the e, T_{ij}, and ρ_i in (4.6.28)-(4.6.32) and (4.6.26) in Taylor series by using a symbolic manipulator and keep nonlinear terms up to $O(\epsilon^3)$. We note that u can be $O(\epsilon)$ or $O(\epsilon^2)$, depending on the boundary and loading conditions. When u is $O(\epsilon^2)$, the obtained equations of motion contain extra higher-order terms. Moreover, it follows from (4.6.81) that the rotary inertias in (4.6.13) are $O(\epsilon^2)$.

A Transformation Using Three Euler Angles:
Expanding (4.4.7) and (4.4.8) yields

$$e = u' + \frac{1}{2}(1-u')\left(v'^2 + w'^2\right) + \cdots \qquad (4.6.51)$$

$$\psi = \tan^{-1}\frac{v'}{1+u'} = v' - v'u' + v'u'^2 - \frac{1}{3}v'^3 + \cdots$$

$$\cos\psi = 1 - \frac{1}{2}v'^2 + u'v'^2 + \cdots$$

$$\sin\psi = v' - u'v' + u'^2v' - \frac{1}{2}v'^3 + \cdots \qquad (4.6.52)$$

$$\theta = \tan^{-1}\frac{-w'}{\sqrt{(1+u')^2 + v'^2}}$$

$$= -w' + u'w' - u'^2w' + \frac{1}{2}v'^2w' + \frac{1}{3}w'^3 + \cdots$$

$$\cos\theta = 1 - \frac{1}{2}w'^2 + u'w'^2 + \cdots$$

$$\sin\theta = -w' + u'w' - u'^2w' + \frac{1}{2}v'^2w' + \frac{1}{3}w'^3 + \cdots \qquad (4.6.53)$$

Substituting (4.6.52) and (4.6.53) into (4.4.5), we obtain

$$T_{11} = 1 - \frac{1}{2}v'^2 - \frac{1}{2}w'^2 + u'v'^2 + u'w'^2 + \cdots$$

$$T_{12} = v' - v'u' + v'u'^2 - \frac{1}{2}v'^3 - \frac{1}{2}v'w'^2 + \cdots$$

$$T_{13} = w' - u'w' + u'^2w' - \frac{1}{2}v'^2w' - \frac{1}{2}w'^3 + \cdots$$

$$T_{21} = -v' + u'v' - w'\varphi - u'^2v' + \frac{1}{2}v'^3 + \frac{1}{2}\varphi^2 v' + \varphi u'w' + \cdots$$

$$T_{22} = 1 - \frac{1}{2}v'^2 - \frac{1}{2}\varphi^2 + u'v'^2 - \varphi v'w' + \cdots$$

$$T_{23} = \varphi - \frac{1}{2}\varphi w'^2 - \frac{1}{6}\varphi^3 + \cdots$$

$$T_{31} = -w' + u'w' + \varphi v' - \varphi u'v' - w'u'^2 + v'^2 w' + \frac{1}{2}w'^3 + \frac{1}{2}\varphi^2 w' + \cdots$$

$$T_{32} = -\varphi - v'w' + \frac{1}{2}\varphi v'^2 + \frac{1}{6}\varphi^3 + 2u'v'w' + \cdots$$

$$T_{33} = 1 - \frac{1}{2}w'^2 - \frac{1}{2}\varphi^2 + u'w'^2 + \cdots \tag{4.6.54}$$

Substituting (4.6.52) and (4.6.53) into (4.4.16)-(4.4.18), we obtain

$$\rho_1 = \varphi' + v''w' - 2u'v''w' - u''v'w' + \cdots$$

$$\rho_2 = -w'' + u'w'' + u''w' + v''\varphi - \varphi u'v'' - \varphi u''v' - u'^2 w'' - 2u'u''w'$$
$$+ \frac{1}{2}v'^2 w'' + v'v''w' + w''w'^2 + \frac{1}{2}\varphi^2 w'' + \cdots$$

$$\rho_3 = v'' - u'v'' - u''v' + w''\varphi - \varphi u'w'' - \varphi u''w' + u'^2 v'' + 2u'u''v'$$
$$- \frac{1}{2}w'^2 v'' - v''v'^2 - \frac{1}{2}\varphi^2 v'' + \cdots \tag{4.6.55}$$

Because the j_i are $O(\epsilon^2)$, it follows from (4.6.14), (4.6.52), (4.6.53), and (4.4.23) that

$$A_u = m\ddot{u}, \quad A_v = m\ddot{v}, \quad A_w = m\ddot{w}$$
$$A_{\theta_1} = j_1\ddot{\varphi}, \quad A_{\theta_2} = -j_2\ddot{w}', \quad A_{\theta_3} = j_3\ddot{v}'$$
$$\omega_1 = \dot{\varphi}, \quad \omega_2 = -\dot{w}', \quad \omega_3 = \dot{v}' \tag{4.6.56}$$

Thus, we completed the derivation of nonlinear expressions for the axial strain e, the transformation matrix $[T]$, and the curvatures ρ_i.

In the nonlinear formulation, owing to the use of finite rotations and the contributions from nonlinear bendings, φ does not represent the real twist angle with respect to the ξ axis. To show that a nonzero φ does not necessarily indicate the presence of torsion along the beam, we consider an inextensional uniform isotropic beam (see Figure 4.4.1) with a circular cross-section under a load $\mathbf{Q}$ acting at the tip of the beam along the direction $\mathbf{i}_y + \mathbf{i}_z$. Because in this case $I_2 = I_3$ according to (4.6.40), any direction parallel to the $\eta\zeta$ plane is a principal direction. Hence, the beam will bend in the plane defined by the x axis and $\mathbf{Q}$, the displacements v and w are equal, and torsion is not expected. In other words, the following conditions exist:

$$\rho_1 = 0, \quad \rho_2 = -\rho_3 \tag{4.6.57}$$

Substituting $e = 0$ into (4.6.51) yields

$$u' = -\frac{1}{2}\left(v'^2 + w'^2\right) + \cdots \tag{4.6.58}$$

Substituting (4.6.58) into (4.6.54) and (4.6.55) yields

$$\rho_1 = \varphi' + v''w'$$
$$\rho_2 = -w'' + v''\varphi + \frac{1}{2}w''(\varphi^2 - w'^2)$$

$$\rho_3 = v'' + w''\varphi + \frac{1}{2}v''(v'^2 - \varphi^2) + v'w'w''$$

$$T_{11} = 1 - \frac{1}{2}v'^2 - \frac{1}{2}w'^2, \quad T_{12} = v', \quad T_{13} = w'$$

$$T_{21} = -v' - w'\varphi - \frac{1}{2}v'w'^2 + \frac{1}{2}\varphi^2 v'$$

$$T_{22} = 1 - \frac{1}{2}v'^2 - \frac{1}{2}\varphi^2 - \varphi v'w'$$

$$T_{23} = \varphi - \frac{1}{2}\varphi w'^2 - \frac{1}{6}\varphi^3$$

$$T_{31} = -w' + \varphi v' + \frac{1}{2}v'^2 w' + \frac{1}{2}\varphi^2 w'$$

$$T_{32} = -\varphi - v'w' + \frac{1}{2}\varphi v'^2 + \frac{1}{6}\varphi^3$$

$$T_{33} = 1 - \frac{1}{2}w'^2 - \frac{1}{2}\varphi^2 \qquad (4.6.59)$$

Checking (4.6.59), we see that nontrivial values of φ are needed in order to satisfy the actual conditions (4.6.57). This observation was also discussed by Rosen, Loewy, and Mathew (1987b). We note that the expression for ρ_2 in (4.6.59) cannot be obtained from ρ_3 in (4.6.59) by replacing w by v and v by $-w$ (i.e., rotate the reference coordinate system by 90° with respect to the x axis). Moreover, the expression for T_{22} in (4.6.59) cannot be obtained from T_{33} in (4.6.59) by replacing w by v and v by $-w$. Hence, we conclude that the mathematical model itself is not symmetric and the equations of motion expressed in terms of φ are not symmetric or interchangeable because there are implicit asymmetric terms included in φ.

Moreover, if the rotation sequence used is θ, ψ, $\hat{\phi}$ (see Figure 4.4.2), which implies that the deformation sequence is u, w, v, $\hat{\phi}$, the twisting curvature becomes (Alkire, 1984)

$$\rho_1 = \hat{\phi}' - v'w''$$

Hence, the asymmetry still exists.

A Transformation Using Two Euler Angles: We consider the beam shown in Figure 4.4.4 without initial curvatures and two Euler angles being used for transformation. In other words, the xyz system coincides with the abc system, the beam is initially straight, and

$$A(s) = s, \quad B(s) = C(s) = 0, \quad [k] = [0] \qquad (4.6.60)$$

It follows from (4.4.36), (4.4.52), and (4.4.53) that

$$\mathbf{i}_1 = T_{11}\mathbf{i}_x + T_{12}\mathbf{i}_y + T_{13}\mathbf{i}_z \qquad (4.6.61a)$$

where

$$T_{11} = \frac{1+u'}{1+e}, \quad T_{12} = \frac{v'}{1+e}, \quad T_{13} = \frac{w'}{1+e} \qquad (4.6.61b)$$

$$e = \sqrt{(1+u')^2 + v'^2 + w'^2} - 1 \qquad (4.6.61c)$$

To obtain third-order nonlinear equations of motion, we expand e in (4.6.61c) in a Taylor series and keep nonlinear terms up to $O(\epsilon^3)$ as

$$e = u' + \frac{1}{2}(1 - u')(v'^2 + w'^2) + \cdots \qquad (4.6.62)$$

Substituting (4.6.62) into (4.6.61b), (4.4.36), and (4.4.57), we obtain

$$T_{11} = 1 - \frac{1}{2}v'^2 - \frac{1}{2}w'^2 + u'v'^2 + u'w'^2 + \cdots$$

$$T_{12} = v' - v'u' + v'u'^2 - \frac{1}{2}v'^3 - \frac{1}{2}v'w'^2 + \cdots$$

$$T_{13} = w' - u'w' + u'^2w' - \frac{1}{2}v'^2w' - \frac{1}{2}w'^3 + \cdots$$

$$T_{21} = -v' + u'v' - w'\phi - u'^2v' + \frac{1}{2}v'^3 + \frac{1}{2}\phi^2 v' + \phi u'w' + \frac{1}{2}v'w'^2 + \cdots$$

$$T_{22} = 1 - \frac{1}{2}v'^2 - \frac{1}{2}\phi^2 + u'v'^2 - \frac{1}{2}\phi v'w' + \cdots$$

$$T_{23} = \phi - \frac{1}{2}v'w' - \frac{1}{2}\phi w'^2 - \frac{1}{6}\phi^3 + u'v'w' + \cdots$$

$$T_{31} = -w' + u'w' + \phi v' - \phi u'v' - w'u'^2 + \frac{1}{2}v'^2 w' + \frac{1}{2}w'^3 + \frac{1}{2}\phi^2 w' + \cdots$$

$$T_{32} = -\phi - \frac{1}{2}v'w' + \frac{1}{2}\phi v'^2 + \frac{1}{6}\phi^3 + u'v'w' + \cdots$$

$$T_{33} = 1 - \frac{1}{2}w'^2 - \frac{1}{2}\phi^2 + u'w'^2 + \frac{1}{2}\phi v'w' + \cdots \qquad (4.6.63)$$

Substituting (4.6.63) into (4.4.64)-(4.4.66) and using (4.6.60) yields

$$\rho_1 = \phi' + \frac{1}{2}v''w' - \frac{1}{2}v'w'' - u'v''w' + u'v'w'' + \cdots$$

$$\rho_2 = -w'' + u'w'' + u''w' + v''\phi - \phi u'v'' - \phi u''v' - u'^2 w'' - 2u'u''w'$$

$$+ \frac{1}{2}v'^2 w'' + \frac{1}{2}v'v''w' + w''w'^2 + \frac{1}{2}\phi^2 w'' + \cdots$$

$$\rho_3 = v'' - u'v'' - u''v' + w''\phi - \phi u'w'' - \phi u''w' + u'^2 v'' + 2u'u''v'$$

$$- \frac{1}{2}w'^2 v'' - \frac{1}{2}w'w''v' - v''v'^2 - \frac{1}{2}\phi^2 v'' + \cdots \qquad (4.6.64)$$

Thus, we completed the derivation of the nonlinear expressions for the axial strain e, the transformation matrix $[T]$, and the curvatures ρ_i. The inertias are the same as those in (4.6.56).

Using the transformation

$$\phi \to \phi, \quad u \to u, \quad v \to -w, \quad w \to v$$

which corresponds to the case in which the reference coordinate system xyz is rotated by 90° with respect to the x axis, in (4.6.62)-(4.6.64), we have

$$\begin{array}{llll} & e \to e, & & \\ T_{11} \to T_{11}, & T_{12} \to -T_{13}, & T_{13} \to T_{12} & \\ T_{21} \to -T_{31}, & T_{22} \to T_{33}, & T_{23} \to -T_{32} & \\ T_{31} \to T_{21}, & T_{32} \to -T_{23}, & T_{33} \to T_{22} & \\ \rho_1 \to \rho_1, & \rho_2 \to -\rho_3, & \rho_3 \to \rho_2 & \end{array} \quad (4.6.65)$$

Hence, the expanded form of the governing equations (4.6.28)-(4.6.31) and (4.6.26) remains the same. In other words, the mathematical modeling is symmetric.

To obtain symmetric third-order nonlinear equations of motion, one can substitute (4.6.62)-(4.6.64) and (4.6.39) (or (4.6.49) for composite beams) into (4.6.28)-(4.6.31) and (4.6.26) and keep nonlinear terms up to $O(\epsilon^3)$. We note that the expanded forms of e and the T_{1i} are the same for both transformations because they are not affected by the torsional variable ϕ.

Substituting (4.6.58) into (4.6.64) yields

$$\rho_1 = \phi' + \frac{1}{2}(v''w' - v'w'') + \cdots \quad (4.6.66)$$

For a uniform inextensional isotropic beam having a circular cross-section and being loaded in the $\mathbf{i}_y + 2\mathbf{i}_z$ direction, $\rho_1 = 0$. However, (4.6.66) shows that ϕ' may not be zero. Therefore, it follows from (4.6.59) and (4.6.66) that any twist variable, defined by using a sequence of three Euler-like rotations or even two sequential rotations, is not a real twisting angle because the deformations u, v, w, and ϕ do not occur in sequence as assumed in the mathematical model that uses Euler angles.

4.6.4 Cantilevered Inextensional Beams

Beams under certain loading and boundary conditions can be assumed to be inextensional. However, the assumption of inextensionality is not valid for beams subject to large centrifugal forces due to high-speed rotation (e.g., helicopter rotor blades), beams with fixed boundary conditions, composite beams with elastic twisting-extension and/or bending-extension couplings due to anisotropy, etc. For an inextensional beam, the number of governing equations, (4.6.28)-(4.6.31), can be reduced from four to three. Next, we develop third-order nonlinear partial-differential equations for an initially straight inextensional beam with fixed-free ends, as shown in Figure 4.4.1. It is shown in Section 4.6.4 that a transformation using two Euler angles results in symmetric equations. Hence, we use it hereafter for all derivations.

For an inextensional beam,

$$e = 0 \quad (4.6.67)$$

Substituting (4.6.67) into (4.6.61c), solving for u', and neglecting terms of order higher than three, we obtain

$$u' = -\frac{1}{2}\left(v'^2 + w'^2\right) \quad (4.6.68)$$

Integrating (4.6.68) and using the geometric boundary condition $u(0,t) = 0$ (see Figure 4.4.1), we obtain

$$u = -\frac{1}{2}\int_0^s \left(v'^2 + w'^2\right) ds \tag{4.6.69}$$

which shows that u is a second-order quantity. Next, we need to expand the terms T_{ij}, ρ_i, M_i, and λ_i in (4.6.28)-(4.6.32).

Following the ordering scheme discussed in Section 4.6.4, substituting (4.6.68) into (4.6.63) and (4.6.64), we obtain third-order expansions for the T_{ij} and ρ_i as

$$T_{11} = 1 - \frac{1}{2}v'^2 - \frac{1}{2}w'^2$$
$$T_{12} = v', \qquad T_{13} = w'$$
$$T_{21} = -v' - w'\phi + \frac{1}{2}\phi^2 v'$$
$$T_{22} = 1 - \frac{1}{2}v'^2 - \frac{1}{2}\phi^2 - \frac{1}{2}\phi v'w'$$
$$T_{23} = \phi - \frac{1}{2}v'w' - \frac{1}{2}\phi w'^2 - \frac{1}{6}\phi^3$$
$$T_{31} = -w' + \phi v' + \frac{1}{2}\phi^2 w'$$
$$T_{32} = -\phi - \frac{1}{2}v'w' + \frac{1}{2}\phi v'^2 + \frac{1}{6}\phi^3$$
$$T_{33} = 1 - \frac{1}{2}w'^2 - \frac{1}{2}\phi^2 + \frac{1}{2}\phi v'w' \tag{4.6.70}$$

$$\rho_1 = \phi' + \frac{1}{2}(v''w' - v'w'')$$
$$\rho_2 = -w'' + v''\phi - \frac{1}{2}v'v''w' - \frac{1}{2}w''w'^2 + \frac{1}{2}\phi^2 w''$$
$$\rho_3 = v'' + w''\phi + \frac{1}{2}v''v'^2 + \frac{1}{2}v'w'w'' - \frac{1}{2}\phi^2 v'' \tag{4.6.71}$$

Substituting (4.6.67) into (4.6.49) yields

$$\begin{Bmatrix} M_1 \\ M_2 \\ M_3 \end{Bmatrix} = \begin{bmatrix} D_{11} & D_{12} & D_{13} \\ D_{12} & D_{22} & D_{23} \\ D_{13} & D_{23} & D_{33} \end{bmatrix} \begin{Bmatrix} \rho_1 \\ \rho_2 \\ \rho_3 \end{Bmatrix} \tag{4.6.72}$$

where the ρ_i are expanded in (4.6.71). Because the rotary inertias are $O(\epsilon^2)$, if the reference point coincides with the mass centroid and η and ζ are the principal axes, (4.6.14) is valid and the inertias are the same as those in (4.6.56), except that φ needs to be replaced by ϕ.

It follows from (4.6.26), (4.6.56), and (4.6.67) that

$$\lambda_2 = -M_3' - M_2\rho_1 + M_1\rho_2 + j_3\ddot{v}'$$
$$\lambda_3 = M_2' - M_3\rho_1 + M_1\rho_3 + j_2\ddot{w}' \tag{4.6.73}$$

From Figure 4.4.1 and (4.6.32) we obtain the boundary conditions as

$$v = w = 0, \quad \phi = 0, \quad w_x = v_x = 0 \text{ at } s = 0$$
$$G_2 = G_3 = 0, \quad M_1 = M_2 = M_3 = 0 \text{ at } s = L \quad (4.6.74)$$

Because the axial strain e is constrained to be zero, (4.6.28) needs to be treated as a constraint. Hence, the corresponding stress resultant F_1 (i.e., λ_1) is a Lagrange multiplier and cannot be derived from the constitutive equation (4.6.49). We neglect c_1 and assume $q_1 = 0$ in (4.6.28), integrate from $s = L$ to $s = s$, use the boundary condition $G_1(L,t) = \sum_{i=1}^{3} \lambda_i T_{i1} |^L = 0$, and obtain

$$F_1 = \lambda_1 = \frac{1}{T_{11}} \left[\int_L^s m\ddot{u}\, ds - \lambda_2 T_{21} - \lambda_3 T_{31} \right] \quad (4.6.75)$$

where u is given in (4.6.69). We rewrite (4.6.29)-(4.6.31) as

$$m\ddot{v} + c_2\dot{v} = G_2' + q_2 \quad (4.6.76)$$
$$m\ddot{w} + c_3\dot{w} = G_3' + q_3 \quad (4.6.77)$$
$$j_1\ddot{\phi} + c_4\dot{\phi} = G_4 + q_4 \quad (4.6.78)$$

where

$$G_2 = \lambda_1 T_{12} + \lambda_2 T_{22} + \lambda_3 T_{32}$$
$$G_3 = \lambda_1 T_{13} + \lambda_2 T_{23} + \lambda_3 T_{33}$$
$$G_4 = M_1' + M_3 \rho_2 - M_2 \rho_3 \quad (4.6.79)$$

Substituting (4.6.70)-(4.6.73) and (4.6.75) into (4.6.76)-(4.6.78), one can obtain the governing equations to third-order.

It follows from (4.5.55c) that $v' = 0$ when $v_x = 0$. Moreover, (4.6.71) and (4.6.72) show that $M_1 = M_2 = M_3 = 0$ when $\phi' = 0$ and $w'' = v'' = 0$. Hence the boundary conditions (4.6.74) can be simplified as

$$v = w = 0, \quad \phi = 0, \quad w' = v' = 0 \text{ at } s = 0$$
$$G_2 = G_3 = 0, \quad \phi' = 0, \quad w'' = v'' = 0 \text{ at } s = L \quad (4.6.80)$$

For an isotropic beam, $D_{12} = D_{13} = D_{23} = 0$ in (4.6.72), and $D_{11} = GI_1, D_{22} = EI_2$, and $D_{33} = EI_3$, as shown in (4.6.39). If the cross-section dimensions and material properties of the beam are uniform (i.e., m, D_{11}, D_{22}, D_{33}, j_1, j_2, and j_3 are constant), the equations of motion can be considerably simplified and rewritten in nondimensional form by normalizing the variables using the characteristic length L and the characteristic time $L^2\sqrt{m/D_{22}}$ as

$$s^* = s/L, \quad v^* = v/L, \quad w^* = w/L, \quad t^* = t\sqrt{D_{22}/mL^4}$$
$$c_2^* = c_2 L^2/\sqrt{mD_{22}}, \quad c_3^* = c_3 L^2/\sqrt{mD_{22}}, \quad c_4^* = c_4/\sqrt{mD_{22}j_1^*}$$
$$j_1^* = j_1/mL^2, \quad j_2^* = j_2/mL^2, \quad j_3^* = j_3/mL^2 \quad (4.6.81)$$

Equation (4.6.81) shows that, if v^* and w^* are $O(\epsilon)$, the rotary inertias j_i^* are $O(\epsilon^2)$. For ease of notation, we drop the superscript $*$ and use a prime and an overdot to denote $\partial/\partial s^*$ and $\partial/\partial t^*$, respectively. Letting $\beta_3 \equiv D_{33}/D_{22}$ and $\beta_1 \equiv D_{11}/D_{22}$, we write (4.6.76)-(4.6.78) in the nondimensional form

$$\ddot{v} + c_2\dot{v} + \beta_3 v^{iv} = \beta_1 \left[\left(\frac{1}{2}v'w'' - \frac{1}{2}v''w' - \phi'\right)w''\right]' - \beta_3 \left[v'\left(v'v'' + w'w''\right)'\right]'$$

$$- (1-\beta_3)\left[\left(v''\phi^2 - \phi w'' + \frac{1}{2}v'w'w''\right)' - v''w'w''\right]' + j_3\ddot{v}''$$

$$- \frac{1}{2}\left\{v' \int_1^s \frac{\partial^2}{\partial t^2}\left[\int_0^s \left(v'^2 + w'^2\right) ds\right] ds\right\}' + q_2 \quad (4.6.82)$$

$$\ddot{w} + c_3\dot{w} + w^{iv} = -\beta_1\left[\left(\frac{1}{2}v'w'' - \frac{1}{2}v''w' - \phi'\right)v''\right]' - \left[w'\left(v'v'' + w'w''\right)'\right]'$$

$$+ (1-\beta_3)\left[\left(v''\phi + \frac{1}{2}v'v''w' + w''\phi^2\right)' - v'v''w''\right]' + j_2\ddot{w}''$$

$$- \frac{1}{2}\left\{w' \int_1^s \frac{\partial^2}{\partial t^2}\left[\int_0^s \left(v'^2 + w'^2\right) ds\right] ds\right\}' + q_3 \quad (4.6.83)$$

$$\ddot{\phi} + c_4\dot{\phi} - \frac{\beta_1}{j_1}\phi'' = \frac{\beta_1}{2j_1}\left(v''w' - v'w''\right)'$$

$$+ \frac{\beta_3 - 1}{j_1}\left(v''^2\phi - w''^2\phi - v''w''\right) + q_4 \quad (4.6.84)$$

Because the rotary inertias are small, the boundary conditions can be simplified. Thus, the boundary conditions (4.6.80) become

$$v = w = 0, \quad \phi = 0, \quad w' = v' = 0 \quad \text{at} \quad s = 0$$
$$\beta_3 v''' - j_3\ddot{v}' = w''' - j_2\ddot{w}' = 0, \quad \phi' = 0, \quad w'' = v'' = 0 \quad \text{at} \quad s = 1 \quad (4.6.85)$$

Equations (4.6.82)-(4.6.84) look different from those derived by Crespo da Silva and Glynn (1978a). Crespo da Silva and Glynn (1978a) used three Euler angles for transformation and used ρ_1 defined in (4.6.59) to define a torsional variable γ as

$$\gamma' \equiv \rho_1 = \varphi' + v''w' \quad (4.6.86)$$

Integrating (4.6.86) and using the boundary condition $\varphi = 0$ at $s = 0$ yields

$$\gamma = \varphi + \int_0^s v''w' ds \quad (4.6.87)$$

Comparing the expressions of ρ_1 in (4.6.59) and (4.6.71) we obtain

$$\phi' = \varphi' + \frac{1}{2}(v'w')' \quad (4.6.88)$$

Integrating (4.6.88) and using the boundary conditions $\phi = \varphi = v' = w' = 0$ at $s = 0$ yields

$$\phi = \varphi + \frac{1}{2}v'w' \qquad (4.6.89)$$

Substituting (4.6.89) into (4.6.86) and (4.6.87) yields

$$\phi = \gamma + \frac{1}{2}v'w' - \int_0^s v''w'ds,$$

$$\phi' = \gamma' + \frac{1}{2}v'w'' - \frac{1}{2}v''w' \qquad (4.6.90)$$

Substituting (4.6.88) and (4.6.89) into (4.6.70) and (4.6.71) yields the expressions of ρ_i and T_{ij} shown in (4.6.59), which are derived using three Euler angles. Hence, the governing equations derived using two Euler angles are interchangeable with those derived using three Euler angles. Substituting (4.6.90) into (4.6.82)-(4.6.84) yields

$$\ddot{v} + c_2\dot{v} + \beta_3 v^{iv} = -\beta_1\left(\gamma'w''\right)' - \beta_3\left[v'\left(v'v'' + w'w''\right)'\right]'$$
$$- (1 - \beta_3)\left[\left(v''^2 - w''\gamma\right)' + w'''\int_0^s v''w'ds\right]' + j_3\ddot{v}''$$
$$- \frac{1}{2}\left\{v'\int_1^s \frac{\partial^2}{\partial t^2}\left[\int_0^s \left(v'^2 + w'^2\right)ds\right]ds\right\}' + q_2 \qquad (4.6.91)$$

$$\ddot{w} + c_3\dot{w} + w^{iv} = \beta_1\left(v''\gamma'\right)' - \left[w'\left(v'v'' + w'w''\right)'\right]'$$
$$+ (1 - \beta_3)\left[\left(v''\gamma + w''\gamma^2\right)' + v'''\int_0^s v'w''ds\right]' + j_2\ddot{w}''$$
$$- \frac{1}{2}\left\{w'\int_1^s \frac{\partial^2}{\partial t^2}\left[\int_0^s \left(v'^2 + w'^2\right)ds\right]ds\right\}' + q_3 \qquad (4.6.92)$$

$$\ddot{\gamma} + c_4\dot{\gamma} - \frac{\beta_1}{j_1}\gamma'' = \frac{\beta_3 - 1}{j_1}\left(v''^2\gamma - w''^2\gamma - v''w''\right) + q_4 \qquad (4.6.93)$$

These equations are the same as those of Crespo da Silva and Glynn (1978a). For the case of two-dimensional vibrations of an inextensional cantilever beam, one can obtain (4.5.42) by substituting $w = 0$ and $\gamma = 0$ into (4.6.91) and replacing c_2 by c_1.

4.6.5 Flexural-Flexural Vibration

If the torsional rigidity is relatively high (e.g., a beam with a square cross-section and a long span), the torsional inertia cannot be excited by low-frequency input forces because the twisting natural frequency is relatively high. Hence, in the absence of a torque applied along the beam, the twist is mainly induced by bending deflections. The induced twist angle can be found by setting all terms with time derivatives equal to zero in (4.6.93). The result is

$$\frac{\beta_1}{j_1}\gamma'' + \frac{1-\beta_3}{j_1}\left(v''w'' + w''^2\gamma - v''^2\gamma\right) = 0 \qquad (4.6.94)$$

Using the boundary conditions $\gamma(0,t) = \gamma'(1,t) = 0$ in (4.6.94), we find that the twist angle due to bending is

$$\gamma = \frac{\beta_3 - 1}{\beta_1}\int_0^s\int_1^s v''w''\,ds\,ds + \text{fourth-order quantities} \qquad (4.6.95)$$

Equation (4.6.95) shows that the bending-induced twisting is a nonlinear phenomenon. Using (4.6.95) and neglecting terms of order higher than three, we reduce (4.6.91) and (4.6.92) to

$$\ddot{v} + c_2\dot{v} + \beta_3 v^{iv} = (1-\beta_3)\left[w''\int_1^s v''w''\,ds - w'''\int_0^s v''w'\,ds\right]' + j_3\ddot{v}''$$

$$- \frac{(1-\beta_3)^2}{\beta_1}\left[w''\int_0^s\int_1^s v''w''\,ds\,ds\right]'' - \beta_3\left[v'\left(v'v'' + w'w''\right)'\right]'$$

$$- \frac{1}{2}\left\{v'\int_1^s\frac{\partial^2}{\partial t^2}\left[\int_0^s\left(v'^2 + w'^2\right)ds\right]ds\right\}' + q_2 \qquad (4.6.96)$$

$$\ddot{w} + c_3\dot{w} + w^{iv} = -(1-\beta_3)\left[v''\int_1^s v''w''\,ds - v'''\int_0^s w''v'\,ds\right]' + j_2\ddot{w}''$$

$$- \frac{(1-\beta_3)^2}{\beta_1}\left[v''\int_0^s\int_1^s v''w''\,ds\,ds\right]'' - \left[w'\left(w'w'' + v'v''\right)'\right]'$$

$$- \frac{1}{2}\left\{w'\int_1^s\frac{\partial^2}{\partial t^2}\left[\int_0^s\left(v'^2 + w'^2\right)ds\right]ds\right\}' + q_3 \qquad (4.6.97)$$

The boundary conditions become

$$v = w = 0, \quad w' = v' = 0 \text{ at } s = 0$$
$$\beta_3 v''' - j_3\ddot{v}' = w''' - j_2\ddot{w}' = 0, \quad w'' = v'' = 0 \text{ at } s = 1 \qquad (4.6.98)$$

Equations (4.6.96) and (4.6.97) are the same as those derived by Crespo da Silva and Glynn (1978a).

4.7 NONLINEAR 3-D CURVED BEAM THEORY ACCOUNTING FOR WARPINGS

To show the development of a curved beam model accounting for inplane and out-of-plane warping effects, we consider the helicopter rotor blade shown in Figure 4.7.1,

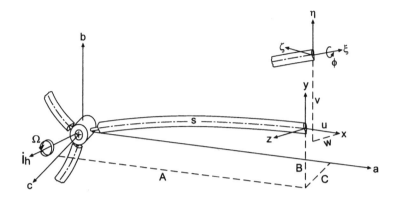

Fig. 4.7.1 Three coordinate systems are used to describe the undeformed and deformed geometries of a helicopter rotor blade, where the abc system is a Cartesian system fixed on the rotating hub, the xyz system is an orthogonal curvilinear frame fixed on the hub and describing the undeformed geometry, and the $\xi\eta\zeta$ system is a local, orthogonal curvilinear coordinate system describing the deformed geometry.

where the three coordinate systems used are the same as those in Figure 4.4.4. The xyz system is an orthogonal curvilinear coordinate system, where the x axis denotes the undeformed reference line of the beam, y and z are rectilinear axes, and s is the undeformed arclength from the root of the beam to the reference point on the observed cross-section. The abc system is a reference Cartesian coordinate system fixed on the hub of the beam, where the c axis is along the center line of the rotor hub and the a axis can be chosen to be tangent to the x axis at the root of the beam. The origin (i.e., the reference point of the observed cross-section) of the xyz system can be at the mass centroid, the area centroid, the shear center, or any other point on the cross-section, and the y and z axes are not necessarily the principal axes of the cross-section. Moreover, the $\xi\eta\zeta$ system is a local orthogonal curvilinear coordinate system, where the ξ axis represents the deformed reference line and the η and ζ are rectilinear axes representing the deformed y and z axes only if there were no shear and torsional warpings. Here, we assume that the cross-section can be of any shape and the beam can be non-prismatic. Because these three coordinate systems are the same as those used in Figure 4.4.4, (4.4.26)-(4.4.66) and Figures 4.4.4-4.4.7 are applicable in this section.

We also consider the case in which the rotor hub and the systems abc and xyz are rotating at an angular speed $\Omega(t)$ with respect to the unit vector $\mathbf{i}_h$, as shown in Figure 4.7.1. The systems abc and xyz at time $t = 0$ are denoted by $\bar{a}\bar{b}\bar{c}$ and $\bar{x}\bar{y}\bar{z}$, as shown in Figure 4.7.2. Moreover, the rotor hub has displacements $U(t)$, $V(t)$, and $W(t)$ along the axes $\bar{a}$, $\bar{b}$, and $\bar{c}$, respectively. These rigid-body motions complicate the derivation of the inertia terms but have no influence on the derivation of the elastic terms.

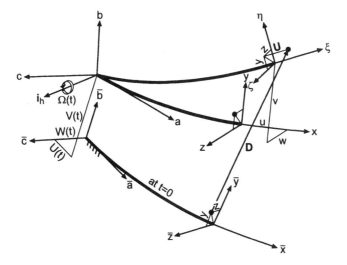

Fig. 4.7.2 The displacements, rotations, and geometric relations of the three coordinate systems.

4.7.1 Inplane and out-of-Plane Warpings

To show cross-sectional warpings of beams, we consider prismatic isotropic beams and use elasticity solutions of stress distributions. For an initially straight beam (see Figure 4.4.1) undergoing small-strain deformations, the displacement field can be represented as (Pai and Schulz, 1999)

$$\begin{aligned}
u_1(x,y,z,t) &= u(x,t) + z\theta_2(x,t) - y\theta_3(x,t) \\
&\quad + \rho_1(x,t)g_{11}(y,z) + \gamma_5(x,t)g_{15}(y,z) + \gamma_6(x,t)g_{16}(y,z) \\
u_2(x,y,z,t) &= v(x,t) - z\theta_1(x,t) \\
&\quad + \rho_2(x,t)g_{22}(y,z) + \rho_3(x,t)g_{23}(y,z) + e(x,t)g_{24}(y,z) \\
u_3(x,y,z,t) &= w(x,t) + y\theta_1(x,t) \\
&\quad + \rho_2(x,t)g_{32}(y,z) + \rho_3(x,t)g_{33}(y,z) + e(x,t)g_{34}(y,z) \quad (4.7.1)
\end{aligned}$$

where u_1, u_2, and u_3 are the displacements of an arbitrary point on the observed cross-section along the axes x, y, and z, respectively, and t is time. Moreover, u, v, and w are the displacements of the reference point on the observed cross-section; θ_1, θ_2, and θ_3 are the rotation angles of the cross-section; ρ_1, ρ_2, and ρ_3 are the deformed curvatures with respect to the axes x, y, and z, respectively; e is the extensional strain of the reference line; γ_5 and γ_6 are the shear rotation angles at the reference point with respect to the axes y and $-z$, respectively; g_{11} is the torsion-induced out-of-plane warping function; g_{15} and g_{16} are shear-induced out-of-plane warping functions; g_{22}, g_{23}, g_{32}, and g_{33} are bending-induced inplane warping functions;

and g_{24} and g_{34} are extension-induced inplane warping functions. Because u, v, and w are defined as $u \equiv u_1(x,0,0,t)$, $v \equiv u_2(x,0,0,t)$, and $w \equiv u_3(x,0,0,t)$, $g_{ij}\,|_{(y,z)=(0,0)} = g_{ij}(0,0) = 0$.

Using $\epsilon_{ii} = \partial u_i/\partial x_i$ and $\epsilon_{ij} = \partial u_i/\partial x_j + \partial u_j/\partial x_i$ ($x_1 \equiv x$, $x_2 \equiv y$, and $x_3 \equiv z$), we find that the engineering strains ϵ_{ij} are

$$\epsilon_{11} = e + z\rho_2 - y\rho_3 + \rho'_1 g_{11} + \gamma'_5 g_{15} + \gamma'_6 g_{16}$$
$$\epsilon_{22} = \rho_2 g_{22y} + \rho_3 g_{23y} + e g_{24y}$$
$$\epsilon_{33} = \rho_2 g_{32z} + \rho_3 g_{33z} + e g_{34z}$$
$$\epsilon_{12} = \rho_1(g_{11y} - z) + \gamma_5 g_{15y} + \gamma_6 g_{16y} + \rho'_2 g_{22} + \rho'_3 g_{23} + e' g_{24}$$
$$\epsilon_{13} = \rho_1(g_{11z} + y) + \gamma_5 g_{15z} + \gamma_6 g_{16z} + \rho'_2 g_{32} + \rho'_3 g_{33} + e' g_{34}$$
$$\epsilon_{23} = \rho_2(g_{22z} + g_{32y}) + \rho_3(g_{23z} + g_{33y}) + e(g_{24z} + g_{34y}) \quad (4.7.2)$$

where $(\ )' \equiv \partial(\ )/\partial x$, $e = u'$, $\theta_2 = -w'$, $\theta_3 = v'$, $\rho_1 = \theta'_1$, $\rho_2 = \theta'_2$, and $\rho_3 = \theta'_3$.

Inplane Warping: To describe a method for obtaining analytical inplane warping functions, we consider isotropic beams with a cross-section that is symmetric with respect to the axes y and z. We also assume that all loads are applied at the ends and hence

$$\rho'_1 = \gamma'_5 = \gamma'_6 = e' = 0 \quad (4.7.3)$$

Using the assumption that $\sigma_{22} = \sigma_{33} = \sigma_{23} = 0$ in the constitutive equation of isotropic materials yields $\sigma_{11} = E\epsilon_{11}$, and

$$\epsilon_{22} = -\nu\epsilon_{11}, \quad \epsilon_{33} = -\nu\epsilon_{11}, \quad \epsilon_{23} = 0 \quad (4.7.4)$$

Here, the σ_{ij} denote engineering stresses, E is Young's modulus, and ν is Poisson's ratio. Substituting (4.7.2) and (4.7.3) into (4.7.4) yields

$$\rho_2(g_{22y} + \nu z) + \rho_3(g_{23y} - \nu y) + e(g_{24y} + \nu) = 0$$
$$\rho_2(g_{32z} + \nu z) + \rho_3(g_{33z} - \nu y) + e(g_{34z} + \nu) = 0$$
$$\rho_2(g_{22z} + g_{32y}) + \rho_3(g_{23z} + g_{33y}) + e(g_{24z} + g_{34y}) = 0 \quad (4.7.5)$$

Since ρ_2, ρ_3, and e are independent of each other, setting their coefficients in (4.7.5) equal to zero yields

$$\begin{array}{lll} g_{22y} + \nu z = 0, & g_{32z} + \nu z = 0, & g_{22z} + g_{32y} = 0 \\ g_{23y} - \nu y = 0, & g_{33z} - \nu y = 0, & g_{23z} + g_{33y} = 0 \\ g_{24y} + \nu = 0, & g_{34z} + \nu = 0, & g_{24z} + g_{34y} = 0 \end{array} \quad (4.7.6)$$

Integrating (4.7.6) and using $g_{ij}(0,0) = 0$, we obtain the inplane warping functions as

$$\begin{array}{lll} g_{22} = -\nu yz, & g_{23} = \tfrac{1}{2}\nu(y^2 - z^2), & g_{24} = -\nu y \\ g_{32} = \tfrac{1}{2}\nu(y^2 - z^2), & g_{33} = \nu yz, & g_{34} = -\nu z \end{array} \quad (4.7.7)$$

Out-of-Plane Shear Warping: To describe a method for deriving out-of-plane warping functions due to transverse shear loading, we consider a prismatic isotropic beam with the reference axis x representing the line through the area centroids of the beam. To avoid complications arising from bending-torsion coupling, we assume that the cross-section and applied static end loads are symmetric with respect to the xz plane and hence the xz plane is the plane of deflection and

$$v = \gamma_6 = \rho_3 = \theta_1 = \rho_1 = 0 \tag{4.7.8a}$$

External loads are assumed to be at the ends only, and hence

$$\gamma_5' = e' = 0 \tag{4.7.8b}$$

Substituting (4.7.8a) and (4.7.8b) into (4.7.2) yields

$$\epsilon_{11} = e + z\rho_2, \quad \epsilon_{12} = \gamma_5 g_{15y} + \rho_2' g_{22}, \quad \epsilon_{13} = \gamma_5 g_{15z} + \rho_2' g_{32} \tag{4.7.9a,b,c}$$

The exact transverse shear stresses in a uniformly loaded beam are the same as those of a tip-loaded cantilever and are given by (Love, 1944)

$$\sigma_{13} = -\frac{F_3}{2(1+\nu)I_2}\left[\frac{\partial \chi}{\partial z} + \frac{1}{2}\nu z^2 + \frac{1}{2}(2-\nu)y^2\right] \tag{4.7.10}$$

$$\sigma_{12} = -\frac{F_3}{2(1+\nu)I_2}\left[\frac{\partial \chi}{\partial y} + (2+\nu)yz\right] \tag{4.7.11}$$

where $\chi(y, z)$ is a harmonic function determined by the shape of the cross-section and $I_2 \equiv \int_A z^2 dA$. Moreover, F_3 is the shear stress resultant, which is equal to the end load in the tip-loaded case and varies linearly in a uniformly-loaded case. Since (4.7.10) and (4.7.11) are exact in the cases of constant and linearly varying F_3, it is expected that they are also valid for cases with F_3 not varying too rapidly along the length of the beam.

For a circular cross-section with a radius a, the function χ can be obtained using elasticity to be (Love, 1944)

$$\chi = -\frac{1}{4}(3+2\nu)a^2 z + \frac{1}{4}(z^3 - 3zy^2) \tag{4.7.12}$$

Substituting (4.7.12) into (4.7.10) and (4.7.11) yields

$$\sigma_{13} = \frac{F_3(3+2\nu)}{8(1+\nu)I_2}\left(a^2 - z^2 - \frac{1-2\nu}{3+2\nu}y^2\right) \tag{4.7.13}$$

$$\sigma_{12} = -\frac{F_3(1+2\nu)}{4(1+\nu)I_2}yz \tag{4.7.14}$$

Since $\sigma_{13}|_{y=z=0} \equiv G\gamma_5$ (G is the shear modulus), it follows from (4.7.13) that

$$G\gamma_5 = \frac{F_3(3+2\nu)}{8(1+\nu)I_2}a^2 \tag{4.7.15}$$

250 BEAMS

It follows from (4.7.9a) that the bending moment $M_2 \equiv \int_A \sigma_{11} z dA = \int_A E\epsilon_{11} z dA = EI_2\rho_2$. Moreover, $M_2 = -\hat{F}_3(L-x)$ (L is the beam length) and $F_3 = \hat{F}_3$ for a cantilever subjected to an end force $\hat{F}_3$, and $M_2 = -q(L-x)^2/2$ and $F_3 = q(L-x)$ for a cantilever subjected to a constant distributed load q. Hence, we obtain, for both cases, that

$$\rho_2' = \frac{M_2'}{EI_2} = \frac{F_3}{EI_2} \tag{4.7.16}$$

Substituting (4.7.13) and (4.7.9c) into the relation $\sigma_{13} = G\epsilon_{13}$ and using (4.7.15), (4.7.16), and $E = 2G(1+\nu)$ yields

$$\frac{F_3(3+2\nu)}{8(1+\nu)I_2}\left(a^2 - z^2 - \frac{1-2\nu}{3+2\nu}y^2\right) = G(\gamma_5 g_{15z} + \rho_2' g_{32})$$

$$= \frac{F_3(3+2\nu)}{8(1+\nu)I_2}a^2\left[g_{15z} + \frac{4}{(3+2\nu)a^2}g_{32}\right] \tag{4.7.17}$$

Substituting g_{32} from (4.7.7) into (4.7.17) yields

$$g_{15z} = -\frac{y^2}{(3+2\nu)a^2} + 1 - \frac{3z^2}{(3+2\nu)a^2} \tag{4.7.18}$$

Hence,

$$g_{15} = -\frac{y^2 z}{(3+2\nu)a^2} + z - \frac{z^3}{(3+2\nu)a^2} + f(y) \tag{4.7.19}$$

Similarly, substituting (4.7.14) and (4.7.9b) into the relation $\sigma_{12} = G\epsilon_{12}$ and using (4.7.15) and (4.7.16) yields

$$-\frac{F_3(1+2\nu)}{4(1+\nu)I_2}yz = G(\gamma_5 g_{15y} + \rho_2' g_{22})$$

$$= \frac{F_3(3+2\nu)}{8(1+\nu)I_2}a^2\left[g_{15y} + \frac{4}{(3+2\nu)a^2}g_{22}\right] \tag{4.7.20}$$

Substituting g_{22} from (4.7.7) into (4.7.20) and integrating the result yields

$$g_{15} = -\frac{y^2 z}{(3+2\nu)a^2} + g(z) \tag{4.7.21}$$

Comparing (4.7.19) and (4.7.21) and using $g_{15}(0,0) = 0$, we find that $f(y) = 0$ and $g(z) = z - z^3/(3+2\nu)a^2$. Hence, the shear-warping function g_{15} is given by

$$g_{15} = -\frac{y^2 z}{(3+2\nu)a^2} + z - \frac{z^3}{(3+2\nu)a^2} \tag{4.7.22}$$

Substituting (4.7.7), (4.7.15), (4.7.16), and (4.7.22) into (4.7.9b) and (4.7.9c) yields

$$\epsilon_{13} = g_{35}\gamma_5, \quad g_{35} \equiv g_{15z} + \frac{\rho_2'}{\gamma_5}g_{32} = 1 - \frac{z^2}{a^2} - \frac{1-2\nu}{(3+2\nu)a^2}y^2 \tag{4.7.23}$$

$$\epsilon_{12} = g_{25}\gamma_5, \quad g_{25} \equiv g_{15y} + \frac{\rho_2'}{\gamma_5}g_{22} = -\frac{2(1+\nu)}{(3+2\nu)a^2}yz \tag{4.7.24}$$

The shear-strain functions g_{35} and g_{25} can be directly obtained from (4.7.13) and (4.7.14) by using the relations $g_{35} = \sigma_{13}/G\gamma_5$ and $g_{25} = \sigma_{12}/G\gamma_5$.

If the beam is subjected to shear loads along both y and z directions,

$$\epsilon_{13} = g_{35}\gamma_5 + g_{36}\gamma_6, \quad \epsilon_{12} = g_{25}\gamma_5 + g_{26}\gamma_6 \qquad (4.7.25)$$

where the shear-strain functions g_{26} and g_{36} can be obtained by following the same steps and considering loads along the y axis to be

$$g_{26} \equiv g_{16y} + \frac{\rho_3'}{\gamma_6}g_{23} = 1 - \frac{y^2}{a^2} - \frac{1-2\nu}{(3+2\nu)a^2}z^2$$

$$g_{36} \equiv g_{16z} + \frac{\rho_3'}{\gamma_6}g_{33} = \frac{2(1+\nu)}{(3+2\nu)a^2}yz \qquad (4.7.26)$$

Out-of-Plane Torsional Warping: The out-of-plane torsional warping of beams is well studied in the elasticity literature. For example, the torsional warping function g_{11} of an isotropic beam with an elliptical cross-section having semi-axes b and c along the y and z directions can be derived using the semi-inverse method to be (Timoshenko and Goodier, 1970)

$$g_{11} = \frac{c^2 - b^2}{c^2 + b^2}yz \qquad (4.7.27)$$

We note that, for a beam with a circular cross-section, $b = c$ and $g_{11} = 0$, which is a well-known phenomenon.

For most of the cross-sections of isotropic and composite beams, analytical inplane and out-of-plane warping functions do not exist. Hence one needs to use, for example, two-dimensional sectional finite-element analysis to calculate the warping functions (Giavotto et al., 1983).

4.7.2 Fully Nonlinear Jaumann Strains

To fully account for geometric nonlinearities, we use Jaumann strains because they are fully nonlinear, objective, and geometric strain measures and their directions are defined with respect to the local coordinate system $\xi\eta\zeta$. The movement of a cross-section consists of two parts. The first part is due to rigid-body displacements u, v, and w of the reference point and the rotation angle ϕ, as shown in Figure 4.4.4. This rigid-body motion rotates the sides dy and dz of the observed cross-section so that they are parallel to the axes η and ζ, respectively. The second part is due to a local, strainable displacement vector $\mathbf{U}$, which consists of relative displacements with respect to the local coordinate system $\xi\eta\zeta$. Because the rigid-body motion does not result in any strain energy, to calculate the elastic energy, we only need to deal with the strainable, local displacement field $\mathbf{U}$. Similar to (4.7.1), the local displacement field can be assumed as

$$\mathbf{U} = u_1\mathbf{i}_1 + u_2\mathbf{i}_2 + u_3\mathbf{i}_3$$

$$u_1(s,y,z,t) = u_1^0(s,t) + z\bar{\theta}_2(s,t) - y\bar{\theta}_3(s,t)$$
$$+ \bar{\rho}_1(s,t)g_{11}(y,z) + \gamma_5(s,t)g_{15}(y,z) + \gamma_6(s,t)g_{16}(y,z)$$
$$u_2(s,y,z,t) = u_2^0(s,t) - z\bar{\theta}_1(s,t)$$
$$+ \bar{\rho}_2(s,t)g_{22}(y,z) + \bar{\rho}_3(s,t)g_{23}(y,z) + e(s,t)g_{24}(y,z)$$
$$u_3(s,y,z,t) = u_3^0(s,t) + y\bar{\theta}_1(s,t)$$
$$+ \bar{\rho}_2(s,t)g_{32}(y,z) + \bar{\rho}_3(s,t)g_{33}(y,z) + e(s,t)g_{34}(y,z) \quad (4.7.28)$$

Here, the Lagrangian coordinates s, y, and z are used to express all functions because Jaumann strains are defined using the undeformed length. Moreover, u_1, u_2, and u_3 are local, strainable displacements with respect to the ξ, η, and ζ axes, respectively; $u_i^0(s,t) \equiv u_i(s,0,0,t)$, $i = 1, 2, 3$; $\bar{\theta}_i \equiv \theta_i - \theta_{i0}$; θ_1, θ_2, and θ_3 are the rotation angles of the observed cross-section with respect to the axes ξ, η, and ζ, respectively; and θ_{10}, θ_{20}, and θ_{30} are the initial rotation angles (after the rigid-body motions u, v, w, and ϕ) of the observed cross-section with respect to the axes ξ, η, and ζ, respectively. Furthermore, $\bar{\rho}_i \equiv \rho_i - k_i$; ρ_1, ρ_2, and ρ_3 are the deformed curvatures with respect to the axes ξ, η, and ζ, respectively; k_1, k_2, and k_3 are the initial curvatures with respect to the axes x, y, and z, respectively. Moreover, γ_5 and γ_6 are the shear rotation angles at the reference point with respect to the axes y and $-z$, respectively.

Because the system $\xi\eta\zeta$ is a local coordinate system attached to the observed cross-section and the unit vector $\mathbf{i}_1$ is tangent to the deformed reference axis, we have

$$u_i^0 = 0, \quad \theta_{i0} = \theta_i = 0, \quad \frac{\partial u_2^0}{\partial s} = \frac{\partial u_3^0}{\partial s} = 0, \quad e = \frac{\partial u_1^0}{\partial s}, \quad \rho_i = \frac{\partial \theta_i}{\partial s}, \quad k_i = \frac{\partial \theta_{i0}}{\partial s}$$
$$(4.7.29)$$

for $i = 1, 2, 3$. It follows from (4.7.28), (4.7.29), and (4.4.59) that

$$\frac{\partial \mathbf{U}}{\partial s} = [e + z\bar{\rho}_2 - y\bar{\rho}_3 + \bar{\rho}_1' g_{11} + \gamma_5' g_{15} + \gamma_6' g_{16}]\mathbf{i}_1$$
$$+ [\rho_2(\bar{\rho}_2 g_{32} + \bar{\rho}_3 g_{33} + e g_{34}) - \rho_3(\bar{\rho}_2 g_{22} + \bar{\rho}_3 g_{23} + e g_{24})]\mathbf{i}_1$$
$$+ [-z\bar{\rho}_1 + \bar{\rho}_2' g_{22} + \bar{\rho}_3' g_{23} + e' g_{24}]\mathbf{i}_2$$
$$+ [\rho_3(\bar{\rho}_1 g_{11} + \gamma_5 g_{15} + \gamma_6 g_{16}) - \rho_1(\bar{\rho}_2 g_{32} + \bar{\rho}_3 g_{33} + e g_{34})]\mathbf{i}_2$$
$$+ [y\bar{\rho}_1 + \bar{\rho}_2' g_{32} + \bar{\rho}_3' g_{33} + e' g_{34}]\mathbf{i}_3$$
$$+ [\rho_1(\bar{\rho}_2 g_{22} + \bar{\rho}_3 g_{23} + e g_{24}) - \rho_2(\bar{\rho}_1 g_{11} + \gamma_5 g_{15} + \gamma_6 g_{16})]\mathbf{i}_3$$
$$\frac{\partial \mathbf{U}}{\partial y} = [\bar{\rho}_1 g_{11y} + \gamma_5 g_{15y} + \gamma_6 g_{16y}]\mathbf{i}_1 + [\bar{\rho}_2 g_{22y} + \bar{\rho}_3 g_{23y} + e g_{24y}]\mathbf{i}_2$$
$$+ [\bar{\rho}_2 g_{32y} + \bar{\rho}_3 g_{33y} + e g_{34y}]\mathbf{i}_3$$
$$\frac{\partial \mathbf{U}}{\partial z} = [\bar{\rho}_1 g_{11z} + \gamma_5 g_{15z} + \gamma_6 g_{16z}]\mathbf{i}_1 + [\bar{\rho}_2 g_{22z} + \bar{\rho}_3 g_{23z} + e g_{24z}]\mathbf{i}_2$$
$$+ [\bar{\rho}_2 g_{32z} + \bar{\rho}_3 g_{33z} + e g_{34z}]\mathbf{i}_3 \quad (4.7.30)$$

Without performing any complex polar decomposition (Malvern, 1969), the Jaumann strains B_{ij} can be derived by using the local displacement field as

$$B_{11} = \frac{\partial \mathbf{U}}{\partial s} \cdot \mathbf{i}_1$$

$$\begin{aligned}
&= e + z\bar{\rho}_2 - y\bar{\rho}_3 + \bar{\rho}'_1 g_{11} + \gamma'_5 g_{15} + \gamma'_6 g_{16} \\
&\quad + \rho_2(\bar{\rho}_2 g_{32} + \bar{\rho}_3 g_{33} + e g_{34}) - \rho_3(\bar{\rho}_2 g_{22} + \bar{\rho}_3 g_{23} + e g_{24})
\end{aligned}$$

$$B_{12} = \frac{\partial \mathbf{U}}{\partial s} \cdot \mathbf{i}_2 + \frac{\partial \mathbf{U}}{\partial y} \cdot \mathbf{i}_1$$

$$\begin{aligned}
&= -z\bar{\rho}_1 + \bar{\rho}'_2 g_{22} + \bar{\rho}'_3 g_{23} + e' g_{24} + \bar{\rho}_1 g_{11y} + \gamma_5 g_{15y} + \gamma_6 g_{16y} \\
&\quad + \rho_3(\bar{\rho}_1 g_{11} + \gamma_5 g_{15} + \gamma_6 g_{16}) - \rho_1(\bar{\rho}_2 g_{32} + \bar{\rho}_3 g_{33} + e g_{34})
\end{aligned}$$

$$B_{13} = \frac{\partial \mathbf{U}}{\partial s} \cdot \mathbf{i}_3 + \frac{\partial \mathbf{U}}{\partial z} \cdot \mathbf{i}_1$$

$$\begin{aligned}
&= y\bar{\rho}_1 + \bar{\rho}'_2 g_{32} + \bar{\rho}'_3 g_{33} + e' g_{34} + \bar{\rho}_1 g_{11z} + \gamma_5 g_{15z} + \gamma_6 g_{16z} \\
&\quad + \rho_1(\bar{\rho}_2 g_{22} + \bar{\rho}_3 g_{23} + e g_{24}) - \rho_2(\bar{\rho}_1 g_{11} + \gamma_5 g_{15} + \gamma_6 g_{16}) \quad (4.7.31)
\end{aligned}$$

The $\bar{\rho}'_1$ and e' in (4.7.31) will be neglected because they are secondary effects and are only important in the study of inplane and torsional warping restraint effects around boundary points. Furthermore, the nonlinear terms $\rho_i \bar{\rho}_j$, $\rho_i \gamma_j$, and $\rho_i e$ are secondary effects due to the coupling of warpings and curvatures. For thick beams, the deformed curvatures ρ_i cannot change very much from the undeformed curvatures k_i before structural failure. For thin beams, warpings are negligible. Hence, the ρ_i will be replaced with k_i in these nonlinear terms. Therefore, without significant loss of accuracy, (4.7.31) can be simplified to

$$\begin{aligned}
B_{11} &= e + z\bar{\rho}_2 - y\bar{\rho}_3 + \gamma'_5 g_{15} + \gamma'_6 g_{16} \\
&\quad + k_2(\bar{\rho}_2 g_{32} + \bar{\rho}_3 g_{33} + e g_{34}) - k_3(\bar{\rho}_2 g_{22} + \bar{\rho}_3 g_{23} + e g_{24}) \\
B_{12} &= -z\bar{\rho}_1 + \bar{\rho}_1 g_{11y} + \gamma_5 g_{25} + \gamma_6 g_{26} \\
&\quad + k_3(\bar{\rho}_1 g_{11} + \gamma_5 g_{15} + \gamma_6 g_{16}) - k_1(\bar{\rho}_2 g_{32} + \bar{\rho}_3 g_{33} + e g_{34}) \\
B_{13} &= y\bar{\rho}_1 + \bar{\rho}_1 g_{11z} + \gamma_5 g_{35} + \gamma_6 g_{36} \\
&\quad + k_1(\bar{\rho}_2 g_{22} + \bar{\rho}_3 g_{23} + e g_{24}) - k_2(\bar{\rho}_1 g_{11} + \gamma_5 g_{15} + \gamma_6 g_{16}) \quad (4.7.32)
\end{aligned}$$

Similar to (4.7.23), (4.7.24), and (4.7.26), the shear-strain functions are defined as

$$g_{25} \equiv g_{15y} + \frac{\bar{\rho}'_2}{\gamma_5} g_{22}, \quad g_{35} \equiv g_{15z} + \frac{\bar{\rho}'_2}{\gamma_5} g_{32}$$

$$g_{26} \equiv g_{16y} + \frac{\bar{\rho}'_3}{\gamma_6} g_{23}, \quad g_{36} \equiv g_{16z} + \frac{\bar{\rho}'_3}{\gamma_6} g_{33} \quad (4.7.33)$$

The strains can be put in matrix form as

$$\{\epsilon\} = [S]\{\psi\} \quad (4.7.34)$$

where

$$\{\epsilon\} \equiv \{B_{11},\ B_{12},\ B_{13}\}^T$$
$$\{\psi\} \equiv \{e,\ \gamma_6,\ \gamma_5,\ \bar{\rho}_1,\ \bar{\rho}_2,\ \bar{\rho}_3,\ \gamma'_5,\ \gamma'_6\}^T$$

$$[S] \equiv \begin{bmatrix} 1 + k_2 g_{34} - k_3 g_{24} & -k_1 g_{34} & k_1 g_{24} \\ 0 & g_{26} + k_3 g_{16} & g_{36} - k_2 g_{16} \\ 0 & g_{25} + k_3 g_{15} & g_{35} - k_2 g_{15} \\ 0 & g_{11y} - z + k_3 g_{11} & g_{11z} + y - k_2 g_{11} \\ z + k_2 g_{32} - k_3 g_{22} & -k_1 g_{32} & k_1 g_{22} \\ -y + k_2 g_{33} - k_3 g_{23} & -k_1 g_{33} & k_1 g_{23} \\ g_{15} & 0 & 0 \\ g_{16} & 0 & 0 \end{bmatrix}^T$$

(4.7.35)

Assuming that the Jaumann stresses J_{22}, J_{33}, and J_{23} are zero, the material stiffness matrix of a general anisotropic material is reduced to the $[\hat{Q}]$ shown in (4.6.48) and the constitutive equation becomes

$$\{\sigma\} = [\hat{Q}]\{\epsilon\} \qquad (4.7.36)$$

where

$$\{\sigma\} \equiv \{J_{11},\ J_{12},\ J_{13}\}^T$$

4.7.3 Equations of Motion

The extended Hamilton principle (4.3.4) is used to derive the equations of motion and is repeated below

$$\int_0^t (\delta T - \delta \Pi + \delta W_{nc}) dt = 0 \qquad (4.7.37)$$

Structural Terms: The variation of the elastic energy Π is obtained using (4.7.34) and (4.7.36) as

$$\delta \Pi = \int_0^L \int_A \{\delta\epsilon\}^T \{\sigma\} dA dx = \int_0^L \int_A \{\delta\psi\}^T [S]^T [\hat{Q}][S]\{\psi\} dA dx$$
$$= \int_0^L \{\delta\psi\}^T \{\hat{F}\} dx \qquad (4.7.38)$$

where

$$\{\hat{F}\} \equiv \{F_1,\ F_2,\ F_3,\ M_1,\ M_2,\ M_3,\ m_2,\ m_3\}^T = [D]\{\psi\}$$
$$[D] \equiv \int_A [S]^T [\hat{Q}][S] dA \qquad (4.7.39)$$

Here, $[D]$ is an 8×8 stiffness matrix. Because e, γ_5, and γ_6 in $\{\psi\}$ correspond to the stress resultants F_1, F_2, and F_3, they are also called force strains. Because the ρ_i in $\{\psi\}$ correspond to the moments M_1, M_2, and M_3, they are also called moment strains. Using (4.4.85)-(4.4.87) and $\delta\bar{\rho}_i = \delta\rho_i$ because the k_i are known initial curvatures, we rewrite (4.7.38) as

$$\delta\Pi = \int_0^L \Big[F_1\delta e + F_2\delta\gamma_6 + F_3\delta\gamma_5 + m_2\delta\gamma_5' + m_3\delta\gamma_6'$$
$$+ M_1\delta\bar{\rho}_1 + M_2\delta\bar{\rho}_2 + M_3\delta\bar{\rho}_3\Big]dx$$
$$= \int_0^L \Big[F_1\delta e + F_2\delta\gamma_6 + F_3\delta\gamma_5 + m_2\delta\gamma_5' + m_3\delta\gamma_6'$$
$$-(M_1' + M_3\rho_2 - M_2\rho_3)\delta\theta_1 - (M_2' - M_3\rho_1 + M_1\rho_3)\delta\theta_2$$
$$-(M_3' + M_2\rho_1 - M_1\rho_2)\delta\theta_3\Big]ds + [M_1\delta\theta_1 + M_2\delta\theta_2 + M_3\delta\theta_3]_0^L$$

which, upon using (4.4.82), becomes

$$\delta\Pi = \int_0^L \Big[\{F_1,\ 0,\ 0\}([T]\{\delta uvw\}' - [T][k]\{\delta uvw\})$$
$$+ F_2\delta\gamma_6 + F_3\delta\gamma_5 + m_2\delta\gamma_5' + m_3\delta\gamma_6' - (M_1' + M_3\rho_2 - M_2\rho_3)\delta\theta_1$$
$$-(M_2' - M_3\rho_1 + M_1\rho_3)\delta\theta_2 - (M_3' + M_2\rho_1 - M_1\rho_2)\delta\theta_3\Big]ds$$
$$+ [M_1\delta\theta_1 + M_2\delta\theta_2 + M_3\delta\theta_3]_0^L \qquad (4.7.40)$$

where $\{\delta uvw\} \equiv \{\delta u,\ \delta v,\ \delta w\}^T$.

Inertia Terms: To make the formulations concise and clear, we use the following notations:

$$\{\mathbf{i}_{xyz}\} \equiv \{\mathbf{i}_x,\ \mathbf{i}_y,\ \mathbf{i}_z\}^T$$
$$\{\mathbf{i}_{123}\} \equiv \{\mathbf{i}_1,\ \mathbf{i}_2,\ \mathbf{i}_3\}^T$$

If the $\mathbf{I}_j$ denote the base vectors of an arbitrary orthogonal coordinate system that has an angular velocity $\vec{\omega}^I$ given by

$$\vec{\omega}^I = \omega_1^I \mathbf{I}_1 + \omega_2^I \mathbf{I}_2 + \omega_3^I \mathbf{I}_3$$

and **X** and **Y** are two arbitrary vectors given by

$$\mathbf{X} = X_1\mathbf{I}_1 + X_2\mathbf{I}_2 + X_3\mathbf{I}_3$$
$$\mathbf{Y} = Y_1\mathbf{I}_1 + Y_2\mathbf{I}_2 + Y_3\mathbf{I}_3$$

we obtain the following identities:

$$\mathbf{X} \times \mathbf{Y} = \{\mathbf{I}_{123}\}^T [P(X)]^T \begin{Bmatrix} Y_1 \\ Y_2 \\ Y_3 \end{Bmatrix} = \{\mathbf{I}_{123}\}^T [P(Y)] \begin{Bmatrix} X_1 \\ X_2 \\ X_3 \end{Bmatrix} \qquad (4.7.41a)$$

$$X_1\dot{\mathbf{i}}_1 + X_2\dot{\mathbf{i}}_2 + X_3\dot{\mathbf{i}}_3 = \vec{\omega}^I \times \mathbf{X} = \{\omega^I\}^T [P(X)]^T \{\mathbf{I}_{123}\} \quad (4.7.41\text{b})$$

$$X_1\ddot{\mathbf{i}}_1 + X_2\ddot{\mathbf{i}}_2 + X_3\ddot{\mathbf{i}}_3 = \dot{\vec{\omega}}^I \times \mathbf{X} + \vec{\omega}^I \times (\vec{\omega}^I \times \mathbf{X})$$
$$= -\{\dot{\omega}^I\}^T [P(X)]\{\mathbf{I}_{123}\} - \{\omega^I\}^T [P(X)][P(\omega^I)]\{\mathbf{I}_{123}\} \quad (4.7.41\text{c})$$

where

$$[P(X)] \equiv [P(X_1 X_2 X_3)] \equiv \begin{bmatrix} 0 & X_3 & -X_2 \\ -X_3 & 0 & X_1 \\ X_2 & -X_1 & 0 \end{bmatrix} \quad (4.7.41\text{d})$$

It follows from Figures 4.7.1 and 4.7.2 that the absolute displacement vector **D** of a generic point on the observed cross-section is given by

$$\begin{aligned}\mathbf{D} = {}& U(t)\mathbf{i}_{\bar{a}} + V(t)\mathbf{i}_{\bar{b}} + W(t)\mathbf{i}_{\bar{c}} + A\mathbf{i}_a + B\mathbf{i}_b + C\mathbf{i}_c \\ & + u\mathbf{i}_x + v\mathbf{i}_y + w\mathbf{i}_z + y\mathbf{i}_2 + z\mathbf{i}_3 \\ & - A\mathbf{i}_{\bar{a}} - B\mathbf{i}_{\bar{b}} - C\mathbf{i}_{\bar{c}} - y\mathbf{i}_{\bar{y}} - z\mathbf{i}_{\bar{z}} \end{aligned} \quad (4.7.42)$$

Although the cross-section warpings are important in obtaining accurate structural stiffnesses, they do not have significant influence on the inertia forces, especially for slender beams. Hence, we neglect the strainable local displacement vector **U** in **D**.

Taking the variation of (4.7.42) and substituting for $\delta\mathbf{i}_2$ and $\delta\mathbf{i}_3$ from (4.4.67) yields

$$\begin{aligned}\delta\mathbf{D} &= \mathbf{i}_x\delta u + \mathbf{i}_y\delta v + \mathbf{i}_z\delta w + y\delta\mathbf{i}_2 + z\delta\mathbf{i}_3 \\ &= \{\mathbf{i}_{xyz}\}^T\{\delta uvw\} + \{\mathbf{i}_{123}\}^T[r]\{\delta\theta_{123}\}\end{aligned} \quad (4.7.43)$$

where U, V, and W are assumed to be known, $\{\delta\theta_{123}\} \equiv \{\delta\theta_1, \delta\theta_2, \delta\theta_3\}^T$, and $[r]$ is defined in (4.6.3) and is repeated below

$$[r] \equiv \begin{bmatrix} 0 & z & -y \\ -z & 0 & 0 \\ y & 0 & 0 \end{bmatrix} \quad (4.7.44)$$

To determine $\ddot{\mathbf{D}}$, we need the angular velocities and rigid-body transformation matrices of the three coordinate systems. Equations (4.4.29) and (4.4.36) relate the three coordinate systems and are repeated below

$$\{\mathbf{i}_{123}\} = [T]\{\mathbf{i}_{xyz}\}, \quad \{\mathbf{i}_{xyz}\} = [T^x]\{\mathbf{i}_{abc}\} \quad (4.7.45)$$

Moreover, the systems $\bar{a}\bar{b}\bar{c}$ and abc are related by

$$\{\mathbf{i}_{abc}\} = [T^a]\{\mathbf{i}_{\bar{a}\bar{b}\bar{c}}\} \quad (4.7.46)$$

where $[T^a]$ is due to rigid-body rotations, as shown in Figure 4.7.2. We assume that the rigid-body motions of the reference frame abc are known and hence $[T^a]$ is a function of three known rotations $\Theta_1(t)$, $\Theta_2(t)$, and $\Theta_3(t)$. For example, if these three angles are consecutive Euler angles (first, Θ_1 around the $\bar{a}$ axis; second, Θ_2 around the rotated $\bar{b}$ axis; and last, Θ_3 around the rotated $\bar{c}$ axis), then

$$[T^a] = \begin{bmatrix} c\Theta_3 & s\Theta_3 & 0 \\ -s\Theta_3 & c\Theta_3 & 0 \\ 0 & 0 & 1 \end{bmatrix} \begin{bmatrix} c\Theta_2 & 0 & -s\Theta_2 \\ 0 & 1 & 0 \\ s\Theta_2 & 0 & c\Theta_2 \end{bmatrix} \begin{bmatrix} 1 & 0 & 0 \\ 0 & c\Theta_1 & s\Theta_1 \\ 0 & s\Theta_1 & c\Theta_1 \end{bmatrix} \tag{4.7.47}$$

where $c\Theta_i \equiv \cos\Theta_i$ and $s\Theta_i \equiv \sin\Theta_i$. We note that, if the governing equations of the rigid-body motions of the reference frame abc (e.g., fuselage) are to be determined as part of the solution, then δU, δV, δW, $\delta\Theta_1$, $\delta\Theta_2$, and $\delta\Theta_3$ are not zero.

Here we define the angular velocity $\vec{\omega}$ of the $\xi\eta\zeta$ frame as

$$\vec{\omega} = \omega_1 \mathbf{i}_1 + \omega_2 \mathbf{i}_2 + \omega_3 \mathbf{i}_3 \tag{4.7.48a}$$

the angular velocity $\vec{\omega}^x$ of the xyz frame as

$$\vec{\omega}^x = \omega_1^x \mathbf{i}_x + \omega_2^x \mathbf{i}_y + \omega_3^x \mathbf{i}_z \tag{4.7.48b}$$

and the angular velocity $\vec{\omega}^a$ of the abc frame as

$$\vec{\omega}^a = \omega_1^a \mathbf{i}_a + \omega_2^a \mathbf{i}_b + \omega_3^a \mathbf{i}_c \tag{4.7.48c}$$

Moreover, Figure 4.7.2 shows that

$$\vec{\omega}^a = \vec{\omega}^x = \Omega \mathbf{i}_h \tag{4.7.48d}$$

Taking the time derivative of (4.7.46) and using $[T^a]^{-1} = [T^a]^T$ yields

$$\frac{d}{dt}\{\mathbf{i}_{abc}\} = [\dot{T}^a][T^a]^T \{\mathbf{i}_{abc}\} \tag{4.7.49a}$$

Also, it follows from (4.7.41a) and (4.7.48c) that

$$\frac{d}{dt}\{\mathbf{i}_{abc}\} = \vec{\omega}^a \times \{\mathbf{i}_{abc}\} = [P(\omega^a)]\{\mathbf{i}_{abc}\} \tag{4.7.49b}$$

Hence, we obtain from (4.7.49a,b) that

$$\omega_1^a = \sum_{i=1}^{3} \dot{T}_{2i}^a T_{3i}^a, \quad \omega_2^a = \sum_{i=1}^{3} \dot{T}_{3i}^a T_{1i}^a, \quad \omega_3^a = \sum_{i=1}^{3} \dot{T}_{1i}^a T_{2i}^a \tag{4.7.50}$$

Moreover, it follows from (4.7.48b,c) and (4.7.45) that

$$\omega_i^x = \sum_{j=1}^{3} \omega_j^a T_{ij}^x, \quad i = 1, 2, 3 \tag{4.7.51}$$

It follows from (4.7.41a) and (4.7.48a,b) that

$$\frac{d}{dt}\{\mathbf{i}_{123}\} = \vec{\omega} \times \{\mathbf{i}_{123}\} = [P(\omega)]\{\mathbf{i}_{123}\}$$

$$\frac{d}{dt}\{\mathbf{i}_{xyz}\} = \vec{\omega}^x \times \{\mathbf{i}_{xyz}\} = [P(\omega^x)]\{\mathbf{i}_{xyz}\} \qquad (4.7.52a)$$

Also it follows from (4.7.45), (4.7.52a), and $[T]^{-1} = [T]^T$ that

$$\frac{d}{dt}\{\mathbf{i}_{123}\} = ([\dot{T}][T]^T + [T][P(\omega^x)][T]^T)\{\mathbf{i}_{123}\} \qquad (4.7.52b)$$

Using (4.7.52a,b) and the orthonormality property of the $\mathbf{i}_j$, we obtain

$$\omega_1 = \sum_{j=1}^{3}(\dot{T}_{2j}T_{3j} + \omega_j^x T_{1j})$$

$$\omega_2 = \sum_{j=1}^{3}(\dot{T}_{3j}T_{1j} + \omega_j^x T_{2j})$$

$$\omega_3 = \sum_{j=1}^{3}(\dot{T}_{1j}T_{2j} + \omega_j^x T_{3j}) \qquad (4.7.53)$$

Equations (4.7.50) and (4.7.51) express the components of $\vec{\omega}^a$ and $\vec{\omega}^x$ in terms of the known Θ_i in (4.7.47), and (4.7.53) expresses the components of $\vec{\omega}$ in terms of the Θ_i, u, v, w, ϕ, and their derivatives.

Taking time derivatives of (4.7.42) and using (4.7.41b,c) we obtain

$$\begin{aligned}\ddot{\mathbf{D}} &= \ddot{U}\mathbf{i}_{\bar{a}} + \ddot{V}\mathbf{i}_{\bar{b}} + \ddot{W}\mathbf{i}_{\bar{c}} + A\ddot{\mathbf{i}}_a + B\ddot{\mathbf{i}}_b + C\ddot{\mathbf{i}}_c + \ddot{u}\mathbf{i}_x + \ddot{v}\mathbf{i}_y + \ddot{w}\mathbf{i}_z \\ &\quad + 2(\dot{u}\dot{\mathbf{i}}_x + \dot{v}\dot{\mathbf{i}}_y + \dot{w}\dot{\mathbf{i}}_z) + u\ddot{\mathbf{i}}_x + v\ddot{\mathbf{i}}_y + w\ddot{\mathbf{i}}_z + y\ddot{\mathbf{i}}_2 + z\ddot{\mathbf{i}}_3 \\ &= \{\ddot{U}, \ddot{V}, \ddot{W}\}\{\mathbf{i}_{\bar{a}\bar{b}\bar{c}}\} \\ &\quad - \left(\{\dot{\omega}^a\}^T[P(ABC)] + \{\omega^a\}^T[P(ABC)][P(\omega^a)]\right)\{\mathbf{i}_{abc}\} \\ &\quad + (\{\ddot{u}, \ddot{v}, \ddot{w}\} - 2\{\omega^x\}^T[P(\dot{u}\dot{v}\dot{w})])\{\mathbf{i}_{xyz}\} \\ &\quad - \left(\{\dot{\omega}^x\}^T[P(uvw)] + \{\omega^x\}^T[P(uvw)][P(\omega^x)]\right)\{\mathbf{i}_{xyz}\} \\ &\quad - \left(\{\dot{\omega}\}^T[r] + \{\omega\}^T[r][P(\omega)]\right)\{\mathbf{i}_{123}\} \\ &= \{Q_1\}^T\{\mathbf{i}_{abc}\} + \{Q_2\}^T\{\mathbf{i}_{xyz}\} \\ &\quad - \left(\{\dot{\omega}\}^T[r] + \{\omega\}^T[r][P(\omega)]\right)\{\mathbf{i}_{123}\} \qquad (4.7.54)\end{aligned}$$

where

$$\begin{aligned}\{Q_1\} &\equiv [T^a]\{\ddot{U},\ \ddot{V},\ \ddot{W}\}^T + [P(ABC)]\{\dot{\omega}^a\} - [P(\omega^a)][P(ABC)]\{\omega^a\} \\ \{Q_2\} &\equiv \{\ddot{u},\ \ddot{v},\ \ddot{w}\}^T + 2[P(\dot{u}\dot{v}\dot{w})]\{\omega^x\} \\ &\quad + [P(uvw)]\{\dot{\omega}^x\} - [P(\omega^x)][P(uvw)]\{\omega^x\} \qquad (4.7.55)\end{aligned}$$

It follows from (4.7.45) and (4.7.46) that

$$\{\mathbf{i}_{abc}\} \cdot \{\mathbf{i}_{xyz}\}^T = [T^x]^T, \quad \{\mathbf{i}_{abc}\} \cdot \{\mathbf{i}_{123}\}^T = [T^x]^T[T]^T,$$
$$\{\mathbf{i}_{xyz}\} \cdot \{\mathbf{i}_{123}\}^T = [T]^T, \quad \{\mathbf{i}_{123}\} \cdot \{\mathbf{i}_{123}\}^T = [I] \quad (4.7.56)$$

Using (4.7.43), (4.7.54), and (4.7.56), we obtain the variation of the kinetic energy T as

$$\delta T = -\int_0^L \int_A \rho \ddot{\mathbf{D}} \cdot \delta \mathbf{D} \, dA \, ds$$
$$= -\int_0^L \left(\{A_u, A_v, A_w\} \begin{Bmatrix} \delta u \\ \delta v \\ \delta w \end{Bmatrix} + \{A_{\theta_1}, A_{\theta_2}, A_{\theta_3}\} \begin{Bmatrix} \delta \theta_1 \\ \delta \theta_2 \\ \delta \theta_3 \end{Bmatrix} \right) ds$$
(4.7.57)

where

$$\{A_u, A_v, A_w\} \equiv \left(\{Q_1\}^T [T^x]^T + \{Q_2\}^T \right) m$$
$$- \left(\{\dot{\omega}\}^T [J_1] + \{\omega\}^T [J_1][P(\omega)] \right) [T]$$
$$\{A_{\theta_1}, A_{\theta_2}, A_{\theta_3}\} \equiv \{\ddot{\omega}\}^T [J_2] + \{\omega\}^T [J_2][P(\omega)]$$
$$+ \left(\{Q_1\}^T [T^x]^T + \{Q_2\}^T \right) [T]^T[J_1]$$
$$m \equiv \int_A \rho \, dA, \quad [J_1] \equiv \int_A \rho[r] \, dA, \quad [J_2] \equiv \int_A \rho[r]^T[r] \, dA \quad (4.7.58)$$

Here, we used the identity

$$\{\omega\}^T [r]^T [P(\omega)][r] = \{\omega\}^T [r]^T[r][P(\omega)] \quad (4.7.59)$$

which can be proved by direct expansion. We note that the components of $\{Q_1\}$ represent the accelerations due to rigid-body translations and rotations of the reference frame abc, and the components of $\{Q_2\}$ represent the accelerations due to the flexural displacements and the rotation of the hub. If ρ is constant, the x axis represents the centroidal line, the y and z axes are principal axes of the cross-section, and $[J_1]$ and $[J_2]$ reduce to (4.6.13).

Nonconservative Terms: Because there are six dependent variables, we assume linear viscous dampings and obtain

$$\delta W_{nc} = \int_0^L \bigl(-c_1 \dot{u} \delta u - c_2 \dot{v} \delta v - c_3 \dot{w} \delta w - c_4 \dot{\phi} \delta \theta_1 - c_5 \dot{\gamma}_5 \delta \gamma_5 - c_6 \dot{\gamma}_6 \delta \gamma_6$$
$$+ q_1 \delta u + q_2 \delta v + q_3 \delta w + q_4 \delta \theta_1 + q_5 \delta \gamma_5 + q_6 \delta \gamma_6 \bigr) ds \quad (4.7.60)$$

where the μ_i are damping coefficients and the q_i are distributed external loads.

Equations of Motion: To determine the equations of motion, we substitute the expressions for $\delta\Pi, \delta T$, and δW_{nc} from (4.7.40), (4.7.57), and (4.7.60) into the extended Hamilton principle (4.7.37) and obtain

$$\int_0^t \int_0^L \Bigg[-\{F_1,\, 0,\, 0\} \left([T]\{\delta uvw\}' - [T][k]\{\delta uvw\} \right)$$

$$- \left\{ \begin{array}{c} A_u + c_1\dot{u} - q_1 \\ A_v + c_2\dot{v} - q_2 \\ A_w + c_3\dot{w} - q_3 \end{array} \right\}^T \{\delta uvw\}$$

$$+ \left(M_1' + M_3\rho_2 - M_2\rho_3 - A_{\theta 1} - c_4\dot{\phi} + q_4 \right) \delta\theta_1$$

$$- \left(F_2 + c_6\dot{\gamma}_6 - q_6 \right) \delta\gamma_6 - \left(F_3 + c_5\dot{\gamma}_5 - q_5 \right) \delta\gamma_5$$

$$- m_2 \delta\gamma_5' - m_3 \delta\gamma_6'$$

$$+ (M_2' - M_3\rho_1 + M_1\rho_3 - A_{\theta 2})\, \delta\theta_2$$

$$+ (M_3' + M_2\rho_1 - M_1\rho_2 - A_{\theta 3})\, \delta\theta_3 \Bigg] ds\, dt$$

$$- \int_0^t [M_1\delta\theta_1 + M_2\delta\theta_2 + M_3\delta\theta_3]_0^L\, dt = 0 \qquad (4.7.61)$$

We note that there are six dependent variables, namely, u, v, w, ϕ, γ_5, and γ_6, but there are eight variation quantities, namely, $\delta u, \delta v, \delta w, \delta\theta_1, \delta\theta_2, \delta\theta_3, \delta\gamma_5$, and $\delta\gamma_6$. However, $\delta\theta_2$ and $\delta\theta_3$ are related to δu, δv, and δw by (4.4.80) and (4.4.81). Substituting for $\delta\theta_2$ and $\delta\theta_3$ from (4.4.82) into (4.7.61) yields

$$\int_0^t \int_0^L \Bigg[-\{F_1,\, \lambda_2,\, \lambda_3\} \left([T]\{\delta uvw\}' - [T][k]\{\delta uvw\} \right)$$

$$- \left\{ \begin{array}{c} A_u + c_1\dot{u} - q_1 \\ A_v + c_2\dot{v} - q_2 \\ A_w + c_3\dot{w} - q_3 \end{array} \right\}^T \{\delta uvw\}$$

$$+ \left(M_1' + M_3\rho_2 - M_2\rho_3 - A_{\theta 1} - c_4\dot{\phi} + q_4 \right) \delta\theta_1$$

$$- \left(F_2 + c_6\dot{\gamma}_6 - q_6 \right) \delta\gamma_6 - \left(F_3 + c_5\dot{\gamma}_5 - q_5 \right) \delta\gamma_5$$

$$- m_2 \delta\gamma_5' - m_3 \delta\gamma_6' \Bigg] ds\, dt$$

$$- \int_0^t [M_1\delta\theta_1 + M_2\delta\theta_2 + M_3\delta\theta_3]_0^L\, dt = 0 \qquad (4.7.62)$$

where

$$\lambda_2 = \frac{1}{1+e}\left(-M_3' - M_2\rho_1 + M_1\rho_2 + A_{\theta 3} \right)$$

$$\lambda_3 = \frac{1}{1+e}\left(M_2' - M_3\rho_1 + M_1\rho_3 - A_{\theta 2} \right) \qquad (4.7.63)$$

Integrating the terms involving $\delta u'$, $\delta v'$, $\delta w'$, $\delta \gamma_5'$, and $\delta \gamma_6'$ in (4.7.62) by parts yields

$$\int_0^t \int_0^L \left[\frac{\partial}{\partial s}(\{F_1, \lambda_2, \lambda_3\}[T])\{\delta uvw\} + \{F_1, \lambda_2, \lambda_3\}[T][k]\{\delta uvw\} \right.$$

$$- \left\{ \begin{array}{c} A_u + c_1\dot{u} - q_1 \\ A_v + c_2\dot{v} - q_2 \\ A_w + c_3\dot{w} - q_3 \end{array} \right\}^T \{\delta uvw\}$$

$$+ \left(M_1' + M_3\rho_2 - M_2\rho_3 - A_{\theta 1} - c_4\dot{\phi} + q_4 \right) \delta\theta_1$$

$$\left. - (-m_3' + F_2 + c_6\dot{\gamma}_6 - q_6)\delta\gamma_6 - (-m_2' + F_3 + c_5\dot{\gamma}_5 - q_5)\delta\gamma_5 \right] ds\, dt$$

$$- \int_0^t \left[\left\{ \begin{array}{c} M_1 \\ M_2 \\ M_3 \end{array} \right\}^T \left\{ \begin{array}{c} \delta\theta_1 \\ \delta\theta_2 \\ \delta\theta_3 \end{array} \right\} + \left\{ \begin{array}{c} F_1 \\ \lambda_2 \\ \lambda_3 \end{array} \right\}^T [T] \left\{ \begin{array}{c} \delta u \\ \delta v \\ \delta w \end{array} \right\} \right.$$

$$\left. + m_2\delta\gamma_5 + m_3\delta\gamma_6 \right]_0^L dt = 0 \qquad (4.7.64)$$

Setting each of the coefficients of δu, δv, δw, $\delta\theta_1$, $\delta\gamma_5$, and $\delta\gamma_6$ in (4.7.64) equal to zero, we obtain the following equations of motion:

$$\frac{\partial}{\partial s}\left([T]^T \left\{ \begin{array}{c} F_1 \\ \lambda_2 \\ \lambda_3 \end{array} \right\}\right) - [k][T]^T \left\{ \begin{array}{c} F_1 \\ \lambda_2 \\ \lambda_3 \end{array} \right\} = \left\{ \begin{array}{c} A_u + c_1\dot{u} - q_1 \\ A_v + c_2\dot{v} - q_2 \\ A_w + c_3\dot{w} - q_3 \end{array} \right\} \qquad (4.7.65)$$

$$M_1' + M_3\rho_2 - M_2\rho_3 + q_4 = A_{\theta 1} - c_4\dot{\phi} \qquad (4.7.66)$$

$$m_3' - F_2 + q_6 = j_3\ddot{\gamma}_6 + c_6\dot{\gamma}_6 \qquad (4.7.67)$$

$$m_2' - F_3 + q_5 = j_2\ddot{\gamma}_5 + c_5\dot{\gamma}_5 \qquad (4.7.68)$$

where λ_2 and λ_3 are given by (4.7.63). The shear inertias $j_3\ddot{\gamma}_6$ and $j_2\ddot{\gamma}_5$ are added here because warpings are neglected in the derivation of the inertia terms in (4.7.57).

The boundary conditions are to specify

$$\delta u = \delta v = \delta w = 0, \quad \delta\theta_1 = \delta\theta_2 = \delta\theta_3 = \delta\gamma_5 = \delta\gamma_6 = 0 \qquad (4.7.69a)$$

or

$$G_1 = G_2 = G_3 = 0, \quad M_1 = M_2 = M_3 = 0, \quad m_2 = m_3 = 0 \qquad (4.7.69b)$$

where

$$G_k \equiv \sum_{i=1}^{3} \lambda_i T_{ik}, \quad k = 1, 2, 3 \qquad (4.7.69c)$$

and $\lambda_1 = F_1$.

Newtonian Formulation: Equations (4.7.65), (4.7.66), and (4.7.63) can be expressed in matrix form as

$$\frac{\partial}{\partial s}\left([T]^T\{F\}\right) + [k]^T[T]^T\{F\} + \begin{Bmatrix} q_1 \\ q_2 \\ q_3 \end{Bmatrix} = \begin{Bmatrix} A_u \\ A_v \\ A_w \end{Bmatrix} + \begin{Bmatrix} c_1\dot{u} \\ c_2\dot{v} \\ c_3\dot{w} \end{Bmatrix} \quad (4.7.70)$$

$$\frac{\partial\{M\}}{\partial s} + [K]^T\{M\} + \begin{Bmatrix} 0 \\ -(1+e)\lambda_3 \\ (1+e)\lambda_2 \end{Bmatrix} + \begin{Bmatrix} q_4 \\ 0 \\ 0 \end{Bmatrix} = \begin{Bmatrix} A_{\theta_1} \\ A_{\theta_2} \\ A_{\theta_3} \end{Bmatrix} + \begin{Bmatrix} c_4\dot{\phi} \\ 0 \\ 0 \end{Bmatrix} \quad (4.7.71)$$

where $\{M\} \equiv \{M_1, M_2, M_3\}^T$ and $\{F\} \equiv \{F_1, \lambda_2, \lambda_3\}^T$. The equivalent vector form of (4.7.70) and (4.7.71) are

$$\frac{\partial \mathbf{F}}{\partial s} + q_1\mathbf{i}_x + q_2\mathbf{i}_y + q_3\mathbf{i}_z = (A_u + c_1\dot{u})\mathbf{i}_x + (A_v + c_2\dot{v})\mathbf{i}_y + (A_w + c_3\dot{w})\mathbf{i}_z \quad (4.7.72)$$

$$\frac{\partial \mathbf{M}}{\partial s} + (1+e)\mathbf{i}_1 \times \mathbf{F} + q_4\mathbf{i}_1 = (A_{\theta_1} + c_4\dot{\phi})\mathbf{i}_1 + A_{\theta_2}\mathbf{i}_2 + A_{\theta_3}\mathbf{i}_3 \quad (4.7.73)$$

where $\mathbf{F} = F_1\mathbf{i}_1 + \lambda_2\mathbf{i}_2 + \lambda_3\mathbf{i}_3$ and $\mathbf{M} = M_1\mathbf{i}_1 + M_2\mathbf{i}_2 + M_3\mathbf{i}_3$. Equations (4.7.72) and (4.7.73) can be directly obtained from a vector approach by using the free-body diagram shown in Figure 4.6.1 and considering the initial curvatures of the xyz system shown in (4.4.31). This shows that the energy formulation (4.7.64) is fully correlated with the Newtonian formulation (4.7.72) and (4.7.73), and the system of equations obtained from these two approaches are essentially the same. However, the energy formulation is more convenient for accounting for warpings and deriving the equations governing γ_5 and γ_6, as shown in (4.7.67) and (4.7.68).

Because $\mathbf{F} = F_1\mathbf{i}_1 + \lambda_2\mathbf{i}_2 + \lambda_3\mathbf{i}_3$, $G_1 = \sum_{i=1}^{3}\lambda_i T_{i1} = \mathbf{F} \cdot \mathbf{i}_x$. Hence, G_1 is the stress resultant along the x direction. Similarly, G_2 is the stress resultant along the y direction, and G_3 is the stress resultant along the z direction.

4.7.4 Expansions and Simplified Beam Theories

Taylor-Series Expansions: Substituting (4.4.52), (4.4.53), and (4.4.57) into (4.4.36), one can obtain exact expressions for the T_{ij} in terms of u, v, w, ϕ, and their derivatives. Substituting these expressions into (4.4.64)-(4.4.66), one can obtain exact expressions for the ρ_i in terms of u, v, w, ϕ, and their derivatives. Moreover, the inertia terms in (4.7.58) are exact, except that the inertias due to warpings are neglected.

To express the governing equations in polynomial form, one can expand the exact expressions of the T_{ij}, ρ_i, and e into polynomials and then use (4.7.39), (4.7.63), and (4.7.58) to expand (4.7.65)-(4.7.68) into six polynomial partial-differential equations.

Simplified Inertias: For the hover and straight forward flight along the b direction in Figure 4.7.2, the rotation axis is always along the c axis, $\mathbf{i}_h = \mathbf{i}_c = \mathbf{i}_{\bar{c}}$, and the transformation matrix $[T^a]$ in (4.7.47) reduces to

$$[T^a] = \begin{bmatrix} \cos\Theta & \sin\Theta & 0 \\ -\sin\Theta & \cos\Theta & 0 \\ 0 & 0 & 1 \end{bmatrix}, \quad \Theta = \int_0^t \Omega(t)dt \qquad (4.7.74)$$

If it is further assumed that the reference frame abc is fixed (i.e., no rigid-body motions), and the reference point coincides with the mass centroid and η and ζ are the principal axes of the differential beam slice, then $[J_1]$ and $[J_2]$ are shown in (4.6.13) and

$$[T^a] = [I], \quad U = V = W = 0, \quad \Omega = \vec{\omega}^a = \vec{\omega}^x = 0 \qquad (4.7.75a)$$

Hence, (4.7.55) becomes $\{Q_1\} = \{0\}$ and $\{Q_2\} = \{\ddot{u},\ \ddot{v},\ \ddot{w}\}^T$ and (4.7.58) reduces to

$$\begin{Bmatrix} A_u \\ A_v \\ A_w \end{Bmatrix} = m\{Q_2\} = m \begin{Bmatrix} \ddot{u} \\ \ddot{v} \\ \ddot{w} \end{Bmatrix} \qquad (4.7.75b)$$

$$\begin{Bmatrix} A_{\theta_1} \\ A_{\theta_2} \\ A_{\theta_3} \end{Bmatrix} = [J_2]\{\dot{\omega}\} - [P(\omega)][J_2]\{\omega\} = \begin{Bmatrix} j_1\dot{\omega}_1 - (j_2 - j_3)\omega_2\omega_3 \\ j_2\dot{\omega}_2 - (j_3 - j_1)\omega_1\omega_3 \\ j_3\dot{\omega}_3 - (j_1 - j_2)\omega_1\omega_2 \end{Bmatrix} \qquad (4.7.75c)$$

which is the same as (4.6.14).

Reducing the Number of Equations: If the beam is not vibrating at high frequencies, the transverse shears γ_5 and γ_6 are mainly induced by flexural deformations. Neglecting the inertia terms and external loads in (4.7.67) and (4.7.68) yields

$$m_3' = F_2 \quad \text{and} \quad m_2' = F_3 \qquad (4.7.76)$$

Equations (4.7.76) can be used to derive expressions for γ_5 and γ_6 in terms of the other strains in $\{\psi\}$ by using (4.7.39) and the boundary conditions. Hence the number of equations can be reduced by two.

If the beam can be assumed to be inextensional due to specific loading and boundary conditions, the number of equations of motion can be reduced one more by using the approach discussed in Section 4.6.4. If the torsional deformation is mainly induced by bendings, one can reduce one more equation by using the approach discussed in Section 4.6.5.

4.7.5 Applications

Certain small offsets of the reference axis of a helicopter rotor blade are often provided to reduce steady blade-bending stresses, to improve rotorcraft flying qualities, or to enhance rotor blade aeroelastic stability (Hodges, 1976). All of the initial bending and twisting curvatures, precone, droop, sweep, torque offset, and blade root offset can be included in the model by properly choosing the functions $A(s), B(s), C(s)$, $\theta_{21}(s), \theta_{22}(s)$, and $\theta_{23}(s)$, which are shown in Figure 4.4.4.

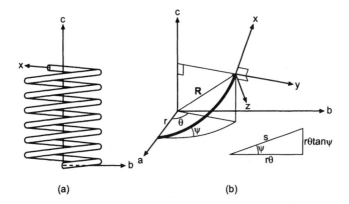

Fig. 4.7.3 A curved beam model of helical springs: (a) the geometry of a helical spring and (b) the geometric relationship between the undeformed system xyz and the reference system abc.

If there are J rotor blades on the rotor hub, one can substitute

$$u = u_j, \ v = v_j, \ w = w_j, \ \phi = \phi_j, \ \gamma_5 = \gamma_{5j}, \ \gamma_6 = \gamma_{6j}$$
$$A(s) = A_j(s), \ B(s) = B_j(s), \ C(s) = C_j(s) \qquad (4.7.77)$$

for $j = 1, 2, \ldots, J$, in (4.7.65)-(4.7.69) to obtain $6J$ equations of motion and their corresponding boundary conditions.

The present curved and twisted beam theory can be used to model helical springs, as shown in Figure 4.7.3. It follows from Figure 4.7.3 that the unloaded position vector $\mathbf{R}$ of the reference point on the observed cross-section is

$$\mathbf{R} = r\cos\theta \mathbf{i}_a + r\sin\theta \mathbf{i}_b + r\theta \tan\psi \mathbf{i}_c \qquad (4.7.78)$$

where r is the radius of the projection of the reference line onto the ab plane and ψ is the pitch angle; both r and ψ are assumed to be constant. Hence, we have

$$\mathbf{i}_x = \frac{d\mathbf{R}}{ds} = -r\theta'\sin\theta \mathbf{i}_a + r\theta'\cos\theta \mathbf{i}_b + r\theta'\tan\psi \mathbf{i}_c \qquad (4.7.79a)$$

$$\mathbf{i}_y = \cos\theta \mathbf{i}_a + \sin\theta \mathbf{i}_b \qquad (4.7.79b)$$

Using (4.7.79a,b) and $\mathbf{i}_z = \mathbf{i}_x \times \mathbf{i}_y$ in (4.7.45), we obtain that

$$[T^x] = \begin{bmatrix} -r\theta'\sin\theta & r\theta'\cos\theta & r\theta'\tan\psi \\ \cos\theta & \sin\theta & 0 \\ -r\theta'\sin\theta\tan\psi & r\theta'\cos\theta\tan\psi & -r\theta' \end{bmatrix} \qquad (4.7.80)$$

Using (4.7.80) and (4.4.33) and the identity $s\cos\psi = r\theta$, we find that the initial curvatures are

$$k_1 = \frac{1}{r}\cos\psi\sin\psi, \quad k_2 = 0, \quad k_3 = -\frac{1}{r}\cos^2\psi \qquad (4.7.81)$$

For circular rings, $\psi = 0$ and the initial curvatures are obtained from (4.7.81) as

$$k_1 = k_2 = 0, \quad k_3 = -\frac{1}{r} \qquad (4.7.82)$$

For straight beams, the initial curvatures k_i are zero.

5
DYNAMICS OF BEAMS

In this chapter, we use the theories developed in the preceding chapter to investigate some aspects of the linear and nonlinear dynamics of beams. In Section 5.1, we treat parametrically excited cantilever beams; in Section 5.2, we treat transversely excited cantilever beams; in Section 5.3, we treat buckled beams; and in Section 5.4, we treat microbeams actuated with electrical forces.

5.1 PARAMETRICALLY EXCITED CANTILEVER BEAMS

In this section, we consider the nonlinear responses of cantilever beams to pulsating axial loads. In Section 5.1.1, we describe experiments in which parametric resonances were observed in the responses of cantilever beams to harmonic excitations. In Sections 5.1.2 and 5.1.3, we treat the cases of principal and combination parametric resonances, respectively. And in Section 5.1.4, we treat nonplanar dynamics.

5.1.1 Experiments

Anderson, Balachandran, and Nayfeh (1992, 1994) conducted an experiment on a vertically mounted cantilever carbon steel beam (see Figure 5.1.1) of dimensions 33.56 *in* × 0.75 *in* × 0.032 *in*. The first four natural frequencies of the beam are 0.65, 5.65, 16.19, and 31.91 Hz. The beam was clamped to a 250-*lb* modal shaker with a custom-table and suspension to allow base excitations of the beam (Zavodney, 1987). The base motion was along the vertical direction. There was a slight bend in the beam with about a 0.25 *in* tip deflection in the static configuration. The base

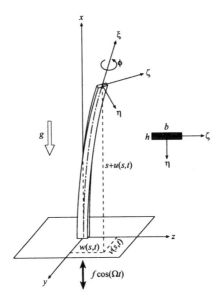

Fig. 5.1.1 A schematic of a cantilever beam under parametric excitation. The xyz is an inertial reference frame and the $\xi\eta\zeta$ are the principal axis of the beam's cross-section at position s, which are fixed on the cross-section.

motion produced a parametric excitation of the beam. In addition, the bend caused the base motion to produce a "small" external excitation.

The base motion was monitored with an accelerometer. The beam response was monitored with two strain gages: one located at $x/L = 0.06$ and the other located at $x/L = 0.25$, where x is the distance along the undeformed beam measured from the base and L is the length of the beam.

The accelerometer and strain gage spectra were monitored as the excitation frequency was varied. For the spectral analyses, there were 1280 lines of resolution in a 40 Hz baseband. A flat-top window was used during periodic excitations, and a Hanning window with thirty overlap averages was used during random excitations. The two strain-gage signals were plotted against each other on the digital oscilloscope, thereby producing a pseudo-state plane. For the Poincaré sections (Moon, 1987, 1992; Nayfeh and Balachandran, 1995), one-half of the excitation frequency was used as the clock frequency. Fourier spectra, pseudo-state planes, autocorrelation functions, and dimension calculations were used to analyze the different motions.

The excitation frequency was chosen to be the control parameter, and the base acceleration was held constant at 0.85 g rms, where g is the acceleration due to gravity and rms denotes root mean square. Initially when the excitation frequency Ω was 33.5 Hz, there was a peak in the response spectrum at the excitation frequency. This peak is due to a primary resonance of the fourth mode. As the excitation frequency was

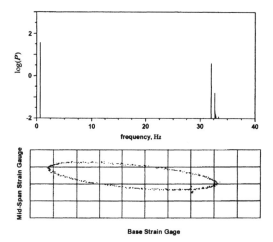

Fig. 5.1.2 Combination parametric resonance of the additive type in the response of a cantilever metallic beam: (a) power spectrum of the response and (b) two-dimensional projection of a Poincaré map of the response.

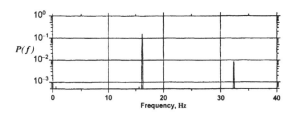

Fig. 5.1.3 Power spectrum of the response of a metallic cantilever beam to a principal parametric excitation of the third mode.

gradually decreased, a combination resonance involving the first and fourth modes was observed between 33.00 Hz and 32.74 Hz. As a result, a typical power spectrum of the response (Figure 5.1.2a) consisted of three peaks: a peak at the excitation frequency, a peak corresponding to the fourth mode, and a peak corresponding to the first mode. In Figure 5.1.2b, we show a two-dimensional projection of a Poincaré section of the attractor. It consists of a closed loop of points, indicating that the motion is two-period quasiperiodic.

As the excitation frequency was gradually decreased further, a response due to a principal parametric resonance of the third mode and a primary resonance of the fourth mode was observed between 32.74 Hz and 32.40 Hz. A spectrum of this type of response is shown in Figure 5.1.3. The third-mode component dominates the response.

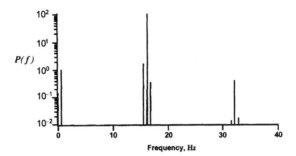

Fig. 5.1.4 Power spectrum of the modulated response showing a transfer of energy from a low-amplitude high-frequency excitation to a high-amplitude low-frequency mode in the response of a parametrically excited cantilever beam (Anderson, Nayfeh, and Balachandran, 1996).

As the excitation frequency was decreased to 32.40 Hz, the periodic response consisting of the third and fourth modes lost stability, and a modulated motion developed in which the amplitudes and phases varied with time. The spectrum of such a response is shown in Figure 5.1.4 with the sidebands around the carrier frequency at $\frac{1}{2}\Omega$ indicating a modulated motion. The sideband spacing f_m is 0.65 Hz, which is close to the first natural frequency of the beam.

Once the modulated motions developed, Anderson, Balachandran, and Nayfeh (1992, 1994) observed that the contribution of the first mode to the response became large. During the experiments, the presence of the first mode was very apparent visually. These results suggest that the activation of the first mode is the result of an interaction of the dynamics of the first mode (which is slow) with the slow dynamics of the third mode, which is represented by the dynamics of its amplitude and phase. Using this observation, S. Nayfeh and Nayfeh (1993) and Nayfeh and Chin (1995) developed a paradigm for explaining this phenomenon.

5.1.2 Principal Parametric Resonance

We consider the nonlinear planar response of a long slender inextensional metallic beam to a principal parametric excitation. Also, we restrict our attention to beams of uniform cross-section and neglect the effects of rotary inertia and shear deformations. In addition, we assume that the damping is composed of linear viscous and quadratic terms. Neglecting the rotary inertia term in (4.5.42) and adding a quadratic damping term, we write the governing equation as

$$m\ddot{v} + c_1\dot{v} + c_2\dot{v}\,|\,\dot{v}\,| + EI\left(v''' + v'v''^2 + v'''v'^2\right)'$$
$$+ \frac{1}{2}\left[v'\int_L^s m\left(\frac{\partial^2}{\partial t^2}\int_0^s v'^2 ds\right) ds\right]' = q_2 \quad (5.1.1)$$

where
$$q_2 = -m\left[v''(s-L) + v'\right]\left[f\cos\Omega t - g\right] \tag{5.1.2}$$

with f being the excitation displacement amplitude and g being the acceleration due to gravity. The associated boundary conditions are

$$v = 0 \quad \text{and} \quad v' = 0 \quad \text{at} \quad s = 0 \tag{5.1.3a}$$

$$v'' = 0 \quad \text{and} \quad v''' = 0 \quad \text{at} \quad s = L \tag{5.1.3b}$$

Next, we introduce nondimensional variables denoted by asterisks as follows:

$$s^* = \frac{s}{L}, \quad v^* = \frac{v}{L}, \quad t^* = t\sqrt{\frac{EI}{mL^4}}, \quad \Omega^* = \Omega\sqrt{\frac{mL^4}{EI}}$$

in (5.1.1)-(5.1.3), drop the asterisks from v, s, t, and Ω, and obtain

$$\ddot{v} + v^{iv} = -\left(v'v''^2 + v'''v'^2\right)' - \frac{1}{2}\left[\int_1^s v'\frac{\partial^2}{\partial t^2}\int_0^s v'^2\,ds\,ds\right]'$$
$$-c_1^*\dot{v} - c_2^*\dot{v}\,|\dot{v}| - [v''(s-1) + v'](f^*\cos\Omega t - g^*) \tag{5.1.4}$$

$$v = 0 \quad \text{and} \quad v' = 0 \quad \text{at} \quad s = 0 \tag{5.1.5a}$$

$$v'' = 0 \quad \text{and} \quad v''' = 0 \quad \text{at} \quad s = 1 \tag{5.1.5b}$$

where

$$c_1^* = \frac{c_1 L^2}{\sqrt{mEI}}, \quad c_2^* = \frac{c_2 L}{m}, \quad f^* = \frac{mfL^3}{EI}, \quad g^* = \frac{mgL^3}{EI}$$

To analyze (5.1.4) and (5.1.5) for weak damping and nonlinearity, we employ the method of multiple scales (Nayfeh, 1973b, 1981) and seek a second-order uniform expansion in the form

$$v(s,t;\epsilon) = \epsilon v_1(s,T_0,T_2) + \epsilon^3 v_3(s,T_0,T_2) + \cdots \tag{5.1.6}$$

where $T_0 = t$ is a fast scale characterizing motions at the frequencies ω_n and Ω and $T_2 = \epsilon^2 t$ is a slow scale characterizing the time variations of the amplitude and phase. Moreover, we scale the damping coefficients c_1^* and c_2^* and the forcing coefficients f^* and g^* as $\epsilon^2 c_1^*$, ϵc_2^*, $\epsilon^2 f^*$, and $\epsilon^2 g^*$. Substituting (5.1.6) into (5.1.4) and (5.1.5) and equating coefficients of like powers of ϵ, we obtain

Order ϵ:

$$D_0^2 v_1 + v_1^{iv} = 0 \tag{5.1.7}$$

$$v_1 = 0 \quad \text{and} \quad v_1' = 0 \quad \text{at} \quad s = 0 \tag{5.1.8a}$$

$$v_1'' = 0 \quad \text{and} \quad v_1''' = 0 \quad \text{at} \quad s = 1 \tag{5.1.8b}$$

Order ϵ^3:

$$D_0^2 v_3 + v_3^{iv} = -2D_0 D_2 v_1 - c_1^* D_0 v_1 - c_2^* D_0 v_1 \mid D_0 v_1 \mid - \left[v_1' \left(v_1' v_1''\right)'\right]'$$

$$-\frac{1}{2}\left[v_1' \int_1^s D_0^2 \left(\int_0^s v_1'^2 ds\right) ds\right]'$$

$$- [v_1''(s-1) + v_1'][f^* \cos(\Omega T_0) - g^*] \qquad (5.1.9)$$

$$v_3 = 0 \text{ and } v_3' = 0 \text{ at } s = 0 \qquad (5.1.10\text{a})$$
$$v_3'' = 0 \text{ and } v_3''' = 0 \text{ at } s = 1 \qquad (5.1.10\text{b})$$

where $D_k = \partial/\partial T_k$.

The general solution of (5.1.7) and (5.1.8) can be expressed as

$$v_1(s, T_0, T_2) = \sum_{m=1}^{\infty} \Phi_m(s) A_m(T_2) e^{i\omega_m T_0} + cc \qquad (5.1.11)$$

where cc stands for the complex conjugate of the preceding terms, $\omega_m = z_m^2$, z_m is the mth root of $\cos z \cosh z + 1 = 0$, and

$$\Phi_m(s) = \kappa_m \left\{ \cosh(z_m s) - \cos(z_m s) + \frac{\cos z_m + \cosh z_m}{\sin z_m + \sinh z_m} [\sin(z_m s) - \sinh(z_m s)] \right\} \qquad (5.1.12)$$

The eigenfunctions are normalized so that $\int_0^1 \Phi_j(s)\Phi_k(s)ds = \delta_{jk}$. The A_m are unknown functions of T_2 at this order of approximation and are found by imposing the solvability condition at the next order of approximation.

The general solution consists of an infinite number of modes corresponding to an infinite number of frequencies. However, any mode that is not directly or indirectly excited will decay to zero with time due to the presence of damping (Nayfeh and Mook, 1979). In this section, we restrict our attention to the case where only one mode is directly excited and assume that this mode is not excited by any internal resonance; that is, we consider a single-mode approximation. Thus, the nondecaying first-order solution can be expressed as

$$v_1(s, T_0, T_2) = \Phi_n(s) \left[A_n(T_2) e^{i\omega_n T_0} + \bar{A}_n(T_2) e^{-i\omega_n T_0}\right] \qquad (5.1.13)$$

where the overbar denotes the complex conjugate and the index n corresponds to the mode being excited.

Substituting (5.1.13) into (5.1.9) yields

$$D_0^2 v_3 + v_3^{iv} = \left\{ -2i\omega_n \Phi_n D_2 A_n e^{i\omega_n T_0} + [\Phi_n''(s-1) + \Phi_n'] g^* A_n e^{i\omega_n T_0} \right.$$
$$\left. - i\omega_n c_1^* \Phi_n A_n e^{i\omega_n T_0} - \left[\Phi_n'(\Phi_n' \Phi_n'')'\right]' \left(A_n^3 e^{3i\omega_n T_0} + 3A_n^2 \bar{A}_n e^{i\omega_n T_0}\right) \right.$$

$$+ 2\omega_n^2 \left(\Phi_n' \int_1^s \int_0^s \Phi_n'^2 ds ds \right)' \left(A_n^3 e^{3i\omega_n T_0} + A_n^2 \bar{A}_n e^{i\omega_n T_0} \right)$$

$$- \frac{1}{2} [\Phi_n''(s-1) + \Phi_n'] f^* \left[A_n e^{i(\Omega+\omega_n)T_0} + \bar{A}_n e^{i(\Omega-\omega_n)T_0} \right] + cc \Big\}$$

$$- c_2^* \omega_n^2 \Phi_n \left(iA_n e^{i\omega_n T_0} - i\bar{A}_n e^{-i\omega_n T_0} \right) \left| \Phi_n \left(iA_n e^{i\omega_n T_0} - i\bar{A}_n e^{-i\omega_n T_0} \right) \right|$$

(5.1.14)

Here, we restrict our discussion to the case of principal parametric resonance of the nth mode (i.e., $\Omega \approx 2\omega_n$). To express the nearness of this resonance, we introduce the detuning parameter σ defined by $\Omega = 2\omega_n + \epsilon^2 \sigma$. Because the homogeneous part of (5.1.14) and (5.1.10) has a nontrivial solution, the nonhomogeneous problem has a solution only if a solvability condition is satisfied (Nayfeh, 1981). To determine this solvability condition, we seek a particular solution free of secular terms in the form

$$v_3 = V_3(s, T_2) e^{i\omega_n T_0} + cc \qquad (5.1.15)$$

Substituting (5.1.15) into (5.1.14) and (5.1.10), multiplying the result by $e^{-i\omega_n T_0}$, and integrating the outcome over the interval $T_0 = 0$ to $2\pi/\omega_n$, we obtain

$$-\omega_n^2 V_3 + V_3^{iv} = -2i\omega_n \Phi_n D_2 A_n - ic_1^* \omega_n \Phi_n A_n + [\Phi_n''(s-1) + \Phi_n'] g^* A_n$$
$$- \frac{1}{2} [\Phi_n''(s-1) + \Phi_n'] f^* \bar{A}_n e^{i\sigma T_2} - 3 \left[\Phi_n' (\Phi_n' \Phi_n'')' \right]' A_n^2 \bar{A}_n$$
$$+ 2\omega_n^2 \left(\Phi_n' \int_1^s \int_0^s \Phi_n'^2 ds ds \right)' A_n^2 \bar{A}_n - \Gamma_n c_2^* \omega_n \Phi_n \mid \Phi_n \mid \quad (5.1.16)$$

$$V_3 = 0 \text{ and } V_3' = 0 \text{ at } s = 0 \qquad (5.1.17a)$$

$$V_3'' = 0 \text{ and } V_3''' = 0 \text{ at } s = 1 \qquad (5.1.17b)$$

where

$$\Gamma_n = \frac{1}{2\pi} \int_0^{2\pi} i \left(A_n e^{i\tau} - \bar{A}_n e^{-i\tau} \right) \mid A_n e^{i\tau} - \bar{A}_n e^{-i\tau} \mid e^{-i\tau} d\tau \qquad (5.1.18)$$

and $\tau = \omega_n T_0$.

To determine the solvability condition, we multiply (5.1.16) by the adjoint Ψ, integrate the result over the interval $s = 0$ to 1, and obtain

$$\int_0^1 \Psi \left(-\omega_n^2 V_3 + V_3^{iv} \right) ds = \int_0^1 \Psi H ds \qquad (5.1.19)$$

where H, which is the right-hand side of (5.1.16), is a function of A_n and Φ_n. Next, we integrate the left-hand side of (5.1.19) by parts to transfer the derivatives from V_3 to Ψ and obtain

$$[\Psi V_3''' - \Psi' V_3'' + \Psi'' V_3' - \Psi''' V_3]_0^1 + \int_0^1 V_3 \left(\Psi^{iv} - \omega_n^2 \Psi \right) ds = \int_0^1 \Psi H ds$$

(5.1.20)

The equation governing the adjoint Ψ is obtained by setting the coefficient of V_3 in the integrand in (5.1.20) equal to zero. The result is

$$\Psi^{iv} - \omega_n^2 \Psi = 0 \tag{5.1.21}$$

Using (5.1.21) and the boundary conditions given in (5.1.17), we obtain

$$[\Psi''V_3' - \Psi'''V_3]^1 - [\Psi V_3''' - \Psi'V_3'']^0 = \int_0^1 \Psi H ds \tag{5.1.22}$$

To determine the boundary conditions needed to define Ψ, we consider the homogeneous problem in which $H = 0$. Then, (5.1.22) becomes

$$[\Psi''V_3' - \Psi'''V_3]^1 - [\Psi V_3''' - \Psi'V_3'']^0 = 0 \tag{5.1.23}$$

We choose the adjoint boundary conditions such that the coefficients of $V_3(1)$, $V_3'(1)$, $V_3''(0)$, and $V_3'''(0)$ in (5.1.23) vanish independently. This results in the following boundary conditions:

$$\Psi = 0 \text{ and } \Psi' = 0 \text{ at } s = 0 \tag{5.1.24a}$$

$$\Psi'' = 0 \text{ and } \Psi''' = 0 \text{ at } s = 1 \tag{5.1.24b}$$

We note that (5.1.21) and (5.1.24) are the same as the spatial part of (5.1.7) and (5.1.8). Hence, the system is self-adjoint. So the solution to the adjoint system, (5.1.21) and (5.1.24), is $\Psi = \Phi_n$.

Having defined the adjoint, we let $\Psi = \Phi_n$ in (5.1.22) and obtain the solvability condition

$$\int_0^1 \Phi_n H ds = 0 \tag{5.1.25}$$

Using the definition of H, we find that the solvability condition yields

$$2i\omega_n D_2 A_n + 2i\omega_n \mu_n A_n + S_{nn} A_n^2 \bar{A}_n + \frac{1}{2}\delta_{nn} f^* \bar{A}_n e^{i\sigma T_2} - \delta_{nn} g^* A_n$$

$$+ \Gamma_n \omega_n \int_0^1 c_2^* \Phi_n^2 |\Phi_n| ds = 0 \tag{5.1.26}$$

where

$$S_{nn} = 3 \int_0^1 \Phi_n \left[\Phi_n' (\Phi_n' \Phi_n'')' \right]' ds - 2\omega_n^2 \int_0^1 \Phi_n \left[\Phi_n' \int_1^s \int_0^s \Phi_n'^2 ds ds \right]' ds$$

$$= 6 \int_0^1 \Phi_n'^2 \Phi_n''^2 ds - 2\omega_n^2 \int_0^1 \left[\int_0^s \Phi_n'^2 ds \right]^2 ds$$

$$\delta_{nn} = \int_0^1 \Phi_n [\Phi_n''(s-1) + \Phi_n'] ds$$

$$2\mu_n = \int_0^1 c_1^* \Phi_n^2 ds \tag{5.1.27}$$

The numerical values for the S_{nn} are $S_{11} = 7.6680$, $S_{22} = -1.0028 \times 10^6$, $S_{33} = -6.8189 \times 10^7$, and $S_{44} = -1.1042 \times 10^8$, and the numerical values for the δ_{nn} are $\delta_{11} = 1.5709$, $\delta_{22} = 8.6471$, $\delta_{33} = 24.952$, and $\delta_{44} = 51.459$. Because $S_{11} > 0$, the effective nonlinearity of the first mode is hardening and the curvature nonlinearity dominates the inertia nonlinearity. On the other hand, $S_{nn} < 0$ for all $n > 1$, and hence the effective nonlinearity of all modes except the first is softening.

Substituting the polar form

$$A_n = \frac{1}{2} a_n e^{i\beta_n} \tag{5.1.28}$$

into (5.1.26), multiplying the result by $e^{-i\beta_n}$, and separating real and imaginary parts, we obtain

$$a'_n = -\mu_n a_n - \frac{\delta_{nn} f^*}{4\omega_n} a_n \sin \gamma_n - C_n a_n^2 \tag{5.1.29}$$

$$a_n \gamma'_n = \left(\sigma + \frac{\delta_{nn} g^*}{\omega_n} \right) a_n - \frac{S_{nn}}{4\omega_n} a_n^3 - \frac{\delta_{nn} f^*}{2\omega_n} a_n \cos \gamma_n \tag{5.1.30}$$

where

$$\gamma_n = \sigma T_2 - 2\beta_n \tag{5.1.31}$$

and

$$C_n = \frac{4\omega_n^2}{3\pi} \int_0^1 c_2^* \mid \Phi_n \mid \Phi_n^2 ds \tag{5.1.32}$$

Substituting (5.1.28) into (5.1.13) and then substituting (5.1.13) into (5.1.6) and using (5.1.31), we find that the response of the beam is given by

$$v(s,t;\epsilon) = \epsilon a_n \Phi_n(s) \cos \left(\frac{1}{2} \Omega t - \frac{1}{2} \gamma_n \right) + \cdots \tag{5.1.33}$$

where a_n and γ_n are given by (5.1.29) and (5.1.30).

It follows from (5.1.33) that periodic motions of the beam correspond to the fixed points or equilibrium solutions of (5.1.29) and (5.1.30). Setting $a'_n = 0$ and $\gamma'_n = 0$ in (5.1.29) and (5.1.30) yields the following two algebraic equations for the fixed points:

$$-\mu_n a_n - \frac{\delta_{nn} f^*}{4\omega_n} a_n \sin \gamma_n - C_n a_n^2 = 0 \tag{5.1.34}$$

$$\left(\sigma + \frac{\delta_{nn} g^*}{\omega_n} \right) a_n - \frac{S_{nn}}{4\omega_n} a_n^3 - \frac{\delta_{nn} f^*}{2\omega_n} a_n \cos \gamma_n = 0 \tag{5.1.35}$$

There are two possibilities. First, $a_n = 0$ and the beam response is trivial. Second, $a_n \neq 0$ and the beam response is at one-half the excitation frequency. In the latter case, dividing (5.1.34) and (5.1.35) by a_n and then eliminating γ_n from the resulting equations yields the frequency-response equation

$$4 \left(\mu_n + C_n a_n \right)^2 + \left[\sigma + \frac{\delta_{nn} g^*}{\omega_n} - \frac{S_{nn}}{4\omega_n} a_n^2 \right]^2 = \frac{\delta_{nn}^2 f^{*2}}{4\omega_n^2} \tag{5.1.36}$$

276 DYNAMICS OF BEAMS

Equation (5.1.36) is a quartic equation in a_n. When $C_n = 0$, it can be solved explicitly for a_n^2. For a given f^*, (5.1.36) can be solved explicitly for σ in terms of a_n to generate the frequency-response curves.

The stability of the nontrivial fixed points can be ascertained by investigating the eigenvalues of the Jacobian matrix of (5.1.29) and (5.1.30) evaluated at the fixed point. However, the stability of the trivial fixed point cannot be ascertained by these eigenvalues because one needs to divide (5.1.30) by a_n. To determine the stability of the trivial fixed points, we go back to (5.1.26), drop the nonlinear terms, and obtain

$$2iD_2 A_n + 2i\mu_n A_n - \frac{\delta_{nn} g^*}{\omega_n} A_n + \frac{\delta_{nn} f^*}{2\omega_n} \bar{A}_n e^{i\sigma T_2} = 0 \tag{5.1.37}$$

To solve (5.1.37), we let

$$A_n = \frac{1}{2}(p_n - iq_n) e^{\frac{1}{2} i\sigma T_2}$$

separate real and imaginary parts, and obtain

$$p'_n = -\mu_n p_n - \frac{1}{2}\left(\sigma + \frac{\delta_{nn} g^*}{\omega_n}\right) q_n - \frac{\delta_{nn} f^*}{4\omega_n} q_n \tag{5.1.38}$$

$$q'_n = -\mu_n q_n + \frac{1}{2}\left(\sigma + \frac{\delta_{nn} g^*}{\omega_n}\right) p_n - \frac{\delta_{nn} f^*}{4\omega_n} p_n \tag{5.1.39}$$

Therefore, the stability of the trivial fixed point can be ascertained by investigating the eigenvalues of the coefficient matrix in (5.1.38) and (5.1.39). They are given by

$$\lambda = -\mu_n \pm \frac{1}{2}\sqrt{\frac{\delta_{nn}^2 f^{*2}}{4\omega_n^2} - \left(\sigma + \frac{\delta_{nn} g^*}{\omega_n}\right)^2}$$

Therefore, a trivial fixed point is stable if and only if

$$\frac{\delta_{nn}^2 f^{*2}}{4\omega_n^2} \leq 4\mu_n^2 + \left(\sigma + \frac{\delta_{nn} g^*}{\omega_n}\right)^2$$

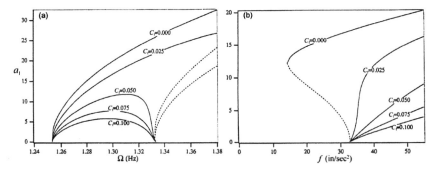

Fig. 5.1.5 Theoretical frequency-response (a) and force-response (b) curves for the first mode for $f = 46.53 \ in/s^2$ and various values of C_1.

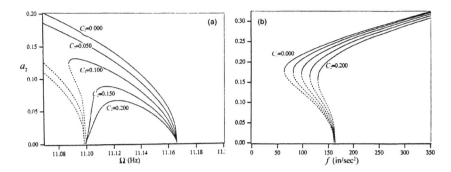

Fig. 5.1.6 Theoretical frequency-response (a) and force-response (b) curves for the second mode for $f = 61.78 \ in/s^2$ and various values of C_2.

In Figure 5.1.5a, we show the influence of the quadratic damping on the frequency-response curves of the first mode where solid lines indicate stable solutions and broken lines indicate unstable solutions. For small C_1, the frequency-response curves are bent to the right, indicating that the nonlinear curvature terms dominate the nonlinear inertia terms and hence the nonlinearity is of the hardening type. For $C_1 = 0$, the unstable and stable branches do not meet as the excitation frequency is increased. For small values of C_1, the stable and unstable branches meet at a turning point or a saddle-node bifurcation. For larger values of C_1, the unstable branch and the saddle-node bifurcation disappear. These results are in qualitative agreement with the experimental results of Anderson, Nayfeh, and Balachandran (1996).

In Figure 5.1.5b, we show the influence of the quadratic damping on the force-response curves for the first mode. For $C_1 = 0$, there is an overhang associated with a subcritical pitchfork bifurcation of the trivial solution. This bifurcation results in a jump from the trivial state to a nontrivial state as the excitation amplitude is increased past the bifurcation point. However, the inclusion of quadratic damping with C_1 greater than a critical value appears to change the subcritical pitchfork bifurcation to a transcritical bifurcation (Nayfeh and Balachandran, 1995). These results are similar to the experimental results of Anderson, Nayfeh, and Balachandran (1996).

In Figure 5.1.6a, we show the influence of the quadratic damping on the frequency-response curves of the second mode. Again, in the absence of the quadratic damping (i.e., $C_2 = 0$), the stable and unstable branches do not meet and the frequency-response curves are bent to the left, indicating a softening nonlinearity and the dominance of the inertia nonlinearity over the curvature nonlinearity. When $C_2 \neq 0$, the stable and unstable branches meet at a turning point or a saddle-node bifurcation. The larger C_2 is, the smaller is the peak amplitude. These results are qualitatively similar to the experimental results of Anderson, Nayfeh, and Balachandran (1996).

In Figure 5.1.6b, we show force-response curves for the second mode for five values of C_2. The larger the value of C_2 is, the smaller the range of excitation amplitudes over which multivalued responses occurs. Again, these results are qualitatively in agreement with the experimental results of Anderson, Nayfeh, and Balachandran (1996).

5.1.3 Combination Parametric Resonance

Dugundji and Mukhopadhyay (1973) investigated the response of a thin cantilever metallic beam to combination parametric resonances involving the first bending and torsional modes (i.e., $\Omega \approx \omega_{B1} + \omega_{T1}$) in one case and the second bending and first torsional modes (i.e., $\Omega \approx \omega_{B2} + \omega_{T1}$) in another. Their experimental results show that the beam exhibits significant oscillations both in bending and in torsion. In addition, at large excitation amplitudes they observed the beam snapping-through and whipping around. Cartmell and Roberts (1987) theoretically and experimentally investigated the stability of a cantilever beam-mass system possessing the two simultaneous combination parametric resonances $\Omega \approx \omega_{B1} + \omega_{T1} \approx \omega_{B2} - \omega_{T1}$. They found good agreement between theory and experiment within certain ranges of the excitation frequency. However, in other regions, where periodic modulations can occur, the correlation was not satisfactory because the theoretical solution could not predict nonstationary responses.

Kar and Sujata (1990) investigated the instability of an elastically restrained cantilever beam subjected to uniaxial and follower forces. They found out that combination parametric resonances of the difference type do not occur when the force is uniaxial or supertangential, but that they are predominant when the force is tangential or subtangential. Kar and Sujata (1992) also investigated the instability of a rotating, pretwisted, and preconed cantilever beam, taking into consideration the Coriolis effects. They found out that the Coriolis force may increase the instability regions in the case of combination parametric resonances.

The experimental results of Dugundji and Mukhopadhyay (1973) and Anderson, Balachandran, and Nayfeh (1994) confirm the occurrence of such resonances in structures. More importantly, their results demonstrate that such resonances can be a mechanism where a high-frequency excitation can activate low-frequency large-amplitude modes. For example, the ratio of the excitation frequency to the natural frequency of the lowest mode excited was approximately 18:1 in the experiments of Dugundji and Mukhopadhyay (1973) and 49:1 in the experiments of Anderson, Balachandran, and Nayfeh (1994). The analyses of Cartmell and Roberts (1987) and Kar and Sujata (1990, 1992) did not take into consideration the effect of nonlinearities inherent in their systems.

Arafat and Nayfeh (1999) investigated the nonlinear response of a beam to a combination parametric resonance of the first bending and torsional modes. Nayfeh and Arafat (1998) investigated the nonlinear response of a uniform thin metallic cantilever beam to either a combination parametric resonance or a subcombination resonance of two modes. They also accounted for both geometric and inertia nonlinearities. They used the method of multiple scales to determine two sets of four first-order nonlinear ordinary-differential equations governing the modulation of the amplitudes and phases of the two interacting modes. Then, they used these modulation equations to generate frequency- and force-response curves.

In this section, we consider the planar response of a long slender inextensional metallic beam to a combination parametric resonance involving the mth and nth modes. The problem is governed by (5.1.4) and (5.1.5) with c_2^* being zero. We seek a second-order uniform expansion as in (5.1.6) and obtain (5.1.7)-(5.1.10). We consider the case

$$\Omega = \omega_m + \omega_n + \epsilon^2 \sigma \qquad (5.1.40)$$

Because in the presence of damping all modes, which are not directly or indirectly excited, decay with time, the solution of (5.1.7) and (5.1.8) can be expressed as

$$v_1 = \Phi_m(s) A_m(T_2) e^{i\omega_m T_0} + \Phi_n(s) A_n(T_2) e^{i\omega_n T_0} + cc \qquad (5.1.41)$$

Substituting (5.1.41) into (5.1.9) with c_2^* being zero yields

$$\begin{aligned}
D_0^2 v_3 + v_3^{iv} = & -\left\{ 2i\omega_n \left(A_n' + \frac{1}{2} c_1^* A_n \right) \Phi_n + 3 A_n^2 \bar{A}_n \left[\Phi_n' (\Phi_n' \Phi_n'')' \right]' \right. \\
& + 2 A_m \bar{A}_m A_n \left[\Phi_m' (\Phi_m' \Phi_n'' + \Phi_n' \Phi_m'')' + \Phi_n' (\Phi_m' \Phi_m'')' \right]' \\
& + 2\omega_n^2 \left[\Phi_n' \int_1^s \int_0^s \Phi_n'^2 ds\, ds \right]' A_n^2 \bar{A}_n - [\Phi_n''(s-1) + \Phi_n'] g^* A_n \\
& \left. + 2\left(\omega_m^2 + \omega_n^2\right) \left[\Phi_m' \int_1^s \int_0^s \Phi_m' \Phi_n' ds\, ds \right]' A_n A_m \bar{A}_m \right\} e^{i\omega_n T_0} \\
& -\left\{ 2i\omega_m \left(A_m' + \frac{1}{2} c_1^* A_m \right) \Phi_m + 3 A_m^2 \bar{A}_m \left[\Phi_m' (\Phi_m' \Phi_m'')' \right]' \right. \\
& + 2 A_n \bar{A}_n A_m \left[\Phi_n' (\Phi_n' \Phi_m'' + \Phi_m' \Phi_n'') + \Phi_m' (\Phi_n' \Phi_n'')' \right]' \\
& + 2\omega_m^2 \left[\Phi_m' \int_1^s \int_0^s \Phi_m'^2 ds\, ds \right]' A_m^2 \bar{A}_m - [\Phi_m''(s-1) + \Phi_m'] g^* A_m \\
& \left. + 2\left(\omega_m^2 + \omega_n^2\right) \left[\Phi_n' \int_1^s \int_0^s \Phi_m' \Phi_n' ds\, ds \right]' A_m A_n \bar{A}_n \right\} e^{i\omega_m T_0} \\
& + \frac{1}{2} [\Phi_m''(s-1) + \Phi_m'] f^* \bar{A}_m e^{i(\Omega - \omega_m) T_0} \\
& + \frac{1}{2} [\Phi_n''(s-1) + \Phi_n'] f^* \bar{A}_n e^{i(\Omega - \omega_n) T_0} + cc + NST \qquad (5.1.42)
\end{aligned}$$

As in the last section, using (5.1.40) and imposing the solvability conditions yields

$$2i\omega_m (A_m' + \mu_m A_m) + S_{mm} A_m^2 \bar{A}_m + S_{mn} A_m A_n \bar{A}_n \\
- \delta_{mm} g^* A_m + \frac{1}{2} \delta_{mn} f^* \bar{A}_n e^{i\sigma T_2} = 0 \qquad (5.1.43)$$

$$2i\omega_n (A_n' + \mu_n A_n) + S_{nn} A_n^2 \bar{A}_n + S_{nm} A_n A_m \bar{A}_m \\
- \delta_{nn} g^* A_n + \frac{1}{2} \delta_{nm} f^* \bar{A}_m e^{i\sigma T_2} = 0 \qquad (5.1.44)$$

where the μ_i, S_{ii}, and δ_{ii} are defined in (5.1.27) and

$$S_{mn} = S_{nm} = 2\int_0^1 \Phi_n \left[\Phi_m' \left(\Phi_m'\Phi_n'' + \Phi_n'\Phi_m''\right)' + \Phi_n' \left(\Phi_m'\Phi_m''\right)'\right] ds$$

$$- 2\left(\omega_m^2 + \omega_n^2\right) \int_0^1 \Phi_n \left[\Phi_m' \int_1^s \int_0^s \Phi_m'\Phi_n' ds ds\right]' ds$$

$$= 2\int_0^1 \left[\Phi_m'^2 \Phi_n''^2 + 4\Phi_m'\Phi_m''\Phi_n'\Phi_n'' + \Phi_m''^2\Phi_n'^2\right] ds$$

$$- 2\left(\omega_m^2 + \omega_n^2\right) \int_0^1 \left(\int_0^s \Phi_m'\Phi_n' ds\right)^2 ds \qquad (5.1.45)$$

$$\delta_{ij} = \int_0^1 \left[\Phi_j''(s-1) + \Phi_j'\right] \Phi_i ds \qquad (5.1.46)$$

The numerical values for the S_{mn} and δ_{ij} are $S_{12} = -1.1222 \times 10^3$, $S_{13} = -1.5929 \times 10^4$, $S_{14} = -3.4595 \times 10^4$, $S_{23} = -7.5835 \times 10^4$, $S_{24} = -9.2741 \times 10^5$, $S_{34} = -4.0175 \times 10^6$, $\delta_{12} = -0.4223$, $\delta_{13} = -1.0721$, $\delta_{14} = -0.8731$, $\delta_{23} = 1.8901$, $\delta_{24} = -3.6434$, and $\delta_{34} = 8.3383$. Next, we transform the complex-valued equations (5.1.43) and (5.1.44) into real-valued equations by introducing the transformation

$$A_m = \frac{1}{2}a_m e^{i\beta_m} \quad \text{and} \quad A_n = \frac{1}{2}a_n e^{i\beta_n} \qquad (5.1.47)$$

into (5.1.43) and (5.1.44), separating real and imaginary parts, and obtaining

$$a_m' = -\mu_m a_m - \frac{\delta_{mn}f^*}{4\omega_m} a_n \sin\gamma \qquad (5.1.48)$$

$$a_m\beta_m' = -\frac{\delta_{mm}g^*}{2\omega_n}a_m + \frac{S_{mm}}{8\omega_m}a_m^3 + \frac{S_{mn}}{8\omega_m}a_m a_n^2 + \frac{\delta_{mn}f^*}{4\omega_m}a_n \cos\gamma \qquad (5.1.49)$$

$$a_n' = -\mu_n a_n - \frac{\delta_{nm}f^*}{4\omega_n} a_m \sin\gamma \qquad (5.1.50)$$

$$a_n\beta_n' = -\frac{\delta_{nn}g^*}{2\omega_n}a_n + \frac{S_{nm}}{8\omega_n}a_m^2 a_n + \frac{S_{nn}}{8\omega_n}a_n^3 + \frac{\delta_{nm}f^*}{4\omega_n}a_m \cos\gamma \qquad (5.1.51)$$

$$\gamma \equiv \sigma T_2 - \beta_m - \beta_n \qquad (5.1.52)$$

Substituting (5.1.47) into (5.1.41) and then substituting the result into (5.1.6), we find that the beam response is given by

$$v(s, t; \epsilon) \approx \epsilon \left[a_m \Phi_m(s) \cos(\omega_m t + \beta_m) + a_n \Phi_n(s) \cos(\omega_n t + \beta_n)\right] \qquad (5.1.53)$$

where the a_i and β_i are given by (5.1.48)-(5.1.52). Using (5.1.40) and (5.1.52) to eliminate ω_n and β_n from (5.1.53), we have

$$v(s, t) \approx \epsilon \{a_m \Phi_m(s) \cos(\omega_m t + \beta_m) + a_n \Phi_n(s) \cos[(\Omega - \omega_m)t - \beta_m - \gamma]\} \qquad (5.1.54)$$

In what follows, we consider the case $g^* = 0$.

The equilibrium solutions or fixed points of (5.1.48)-(5.1.52) correspond to $a'_m = 0$, $a'_n = 0$, and $\gamma' = 0$, which in turn correspond to two-period quasiperiodic responses of the beam according to (5.1.54). There are two possible equilibrium solutions: (a) $a_m = 0$ and $a_n = 0$ and the parametric resonance is not activated, and (b) $a_m \neq 0$ and $a_n \neq 0$ and the beam response is quasiperiodic. In the latter case, (5.1.49), (5.1.51), and (5.1.52) can be used to eliminate β_m and β_n to obtain the following equation for γ:

$$\gamma' = \sigma - \left(\frac{S_{mm}}{8\omega_m} + \frac{S_{nm}}{8\omega_n}\right) a_m^2 - \left(\frac{S_{mn}}{8\omega_m} + \frac{S_{nn}}{8\omega_n}\right) a_n^2$$
$$- \left(\frac{a_n \delta_{mn}}{4 a_m \omega_m} + \frac{a_m \delta_{nm}}{4 a_n \omega_n}\right) f^* \cos\gamma \quad (5.1.55)$$

Thus, for nontrivial solutions, the modulation equations are reduced from four to three first-order differential equations. For equilibrium solutions, we set the time derivatives in (5.1.48), (5.1.50), and (5.1.55) equal to zero and solve for a_m, a_n, and γ, yielding the following closed-form solution:

$$\alpha_e a_m^2 = \sigma \pm \frac{\mu_m + \mu_n}{\sqrt{\mu_m \mu_n}} \sqrt{\frac{\delta_{mn} \delta_{nm} f^{*2}}{16 \omega_m \omega_n} - \mu_m \mu_n} \quad (5.1.56)$$

$$a_n^2 = \frac{\mu_m \omega_m \delta_{nm} f^*}{\mu_n \omega_n \delta_{mn}} a_m^2 \quad (5.1.57)$$

$$\sin\gamma = -\frac{4\mu_m \omega_m}{\delta_{mn} f^*} \frac{a_m}{a_n} = -\frac{4\mu_n \omega_n}{\delta_{nm} f^*} \frac{a_n}{a_m} = \pm 4\sqrt{\frac{\mu_m \mu_n \omega_m \omega_n}{\delta_{mn} \delta_{nm} f^{*2}}} \quad (5.1.58)$$

where

$$\alpha_e = \frac{1}{8}\left[\frac{S_{mm}}{\omega_m} + \frac{S_{nm}}{\omega_n} + \left(\frac{S_{mn}}{\omega_m} + \frac{S_{nn}}{\omega_n}\right)\frac{\mu_m \omega_m \delta_{nm}}{\mu_n \omega_n \delta_{mn}}\right] \quad (5.1.59)$$

The stability of a nontrivial equilibrium solution can then be studied by calculating the eigenvalues of the Jacobian matrix of (5.1.48), (5.1.50), and (5.1.55) evaluated at this equilibrium solution.

To determine the stability of the trivial equilibrium solution, we study the stability of the linearized complex-valued modulation equations (5.1.43) and (5.1.44). To this end, we let

$$A_m = c_m e^{\lambda T_2 + i\sigma T_2} \quad \text{and} \quad A_n = c_n e^{\bar{\lambda} T_2} \quad (5.1.60)$$

in the linearized (5.1.43) and (5.1.44) and obtain

$$2i\omega_m (\lambda + i\sigma + \mu_m) c_m + \frac{1}{2}\delta_{mn} f^* \bar{c}_n = 0 \quad (5.1.61)$$

$$2i\omega_n (\bar{\lambda} + \mu_n) c_n + \frac{1}{2}\delta_{nm} f^* \bar{c}_m = 0 \quad (5.1.62)$$

Hence,

$$\lambda = -\frac{1}{2}(\mu_m + \mu_n + i\sigma) \pm \sqrt{\frac{1}{4}(\mu_m + \mu_n + i\sigma)^2 - \mu_n (\mu_m + i\sigma) + \frac{\delta_{mn} \delta_{nm}}{16 \omega_m \omega_n} f^{*2}} \quad (5.1.63)$$

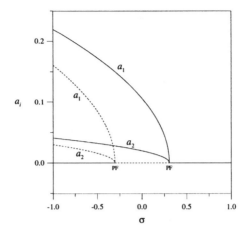

Fig. 5.1.7 Frequency-response curves for a combination parametric resonance of the additive type involving modes 1 and 2 for $f^* = 10$, $\mu_1 = 0.0137$, and $\mu_2 = 0.0635$. Solid lines (———) denote stable fixed points and dashed lines (– – –) denote unstable fixed points.

It follows from (5.1.63) that the trivial solution is stable if the real parts of both λ's are negative.

In Figure 5.1.7, we show typical frequency-response curves for a combination parametric resonance of the additive type of the first two modes when the excitation amplitude is $f^* = 10$. Clearly, the first mode dominates the response. Although the nonlinearity is hardening for the first mode and softening for the second mode, the frequency-response curves are bent to the left, indicating a softening behavior for both modes. This is so because

$$\beta_1' = \frac{S_{11}}{8\omega_1}a_1^2 + \frac{S_{12}}{8\omega_1}a_2^2 + \frac{\delta_{12}f^*a_2}{4\omega_1 a_1}\cos\gamma \tag{5.1.64}$$

according to (5.1.49). Although S_{11} is positive, S_{12} is negative and its magnitude is much larger than S_{11}. Hence, the nonlinearity decreases the frequency of the first mode, and hence bends the frequency-response curves to the left. It follows from Figure 5.1.7 that, depending on how σ is varied, the trivial solution loses stability via either a subcritical or a supercritical pitchfork bifurcation.

In Figure 5.1.8, we show force-response curves for a combination parametric resonance of the additive type of the first two modes. In part (a), the frequency detuning parameter $\sigma = -1$, and in part (b), $\sigma = 1$. When $\sigma = -1$, there are two branches of nontrivial fixed point solutions, one stable and the other unstable. As f^* is increased away from zero, the trivial solution loses stability via a subcritical pitchfork bifurcation, causing the response to jump up to the stable branch of nontrivial solutions. Similarly, a fixed point on the stable nontrivial branch loses stability via a saddle-node bifurcation as f^* is decreased, resulting in a jump down to the trivial branch. When $\sigma \geq 0$, there are only branches of stable nontrivial fixed points, as

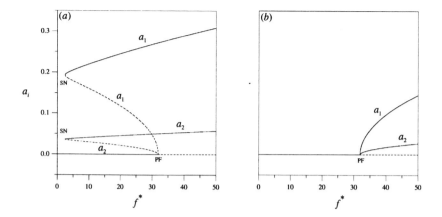

Fig. 5.1.8 Force-response curves for a combination parametric resonance of the additive type involving modes 1 and 2 when $\mu_1 = 0.0137$ and $\mu_2 = 0.0635$: (a) $\sigma = -1$ and (b) $\sigma = 1$. Solid lines (———) denote stable fixed points and dashed lines (– – –) denote unstable fixed points.

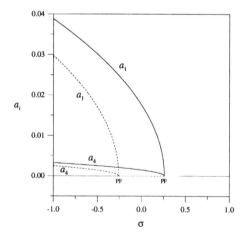

Fig. 5.1.9 Frequency-response curves for a combination parametric resonance of the additive type involving modes 1 and 4 for $f^* = 10$, $\mu_1 = 0.0137$, and $\mu_4 = 0.0573$. Solid lines (———) denote stable fixed points and dashed lines (– – –) denote unstable fixed points.

shown in part (b). The nontrivial solution is activated gradually as the trivial solution undergoes a supercritical pitchfork bifurcation.

In Figures 5.1.9 and 5.1.10, the frequency- and force-response curves are presented when the first and fourth modes are activated by the combination parametric resonance. The forcing amplitude in Figure 5.1.9 is $f^* = 10$ and the detuning parameter

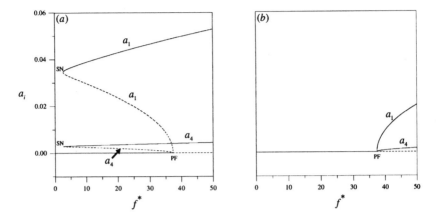

Fig. 5.1.10 Force-response curves for a combination parametric resonance of the additive type involving modes 1 and 4 when $\mu_1 = 0.0137$ and $\mu_4 = 0.0573$. In part (a), $\sigma = -1$ and in part (b), $\sigma = 1$. Solid lines (———) denote stable fixed points and dashed lines (– – –) denote unstable fixed points.

$\sigma = -1$ in Figure 5.1.10a and $\sigma = 1$ in Figure 5.1.10b. We note that the behaviors in Figures 5.1.9 and 5.1.10 are similar to those in Figures 5.1.7 and 5.1.8. However, the amplitudes when modes 1 and 4 are excited are about an order of magnitude smaller than those when modes 1 and 2 are excited.

Results for the case of a combination parametric resonance of the difference type can be obtained by replacing ω_m by $-\omega_m$ and β_m by $-\beta_m$ in (5.1.48)-(5.1.58). However, it can be seen from (5.1.57) that this resonance cannot be activated in this system.

5.1.4 Nonplanar Dynamics

Nayfeh and Pai (1989) used (4.6.96)-(4.6.98) to investigate nonplanar responses of cantilevered rectangular and near-square beams to principal parametric excitations. They used the method of multiple scales to derive four nonlinear first-order ordinary-differential equations governing the modulation of the amplitudes and phases of two interacting modes, one in each plane. The coefficients of these equations are given in quadrature in terms of the linear undamped mode shapes. They evaluated these coefficients numerically. By analytically manipulating the coefficients in the modulation equations of Nayfeh and Pai (1989), Feng and Leal (1995) showed that some of them are related, which produced symmetry in these equations.

Arafat, Nayfeh, and Chin (1998) showed that (4.6.96)-(4.6.98) can be derived by using Hamilton's extended principle from a Lagrangian and a virtual work term. Then, they applied the method of multiple scales directly to the Lagrangian and virtual work term to derive the modulation equations governing the amplitudes and phases of the two interacting modes. These equations exhibit the symmetry conditions found by Feng and Leal (1995). Arafat, Nayfeh, and Chin investigated the effect of the cross-section detuning and forcing frequency detuning on the static and dynamic bifurcations the beam undergoes and found codimension-2 bifurcations. They found that the beam's response can be very complex – including periodic motions, bubble structures, periodically and chaotically modulated motions, attractor-merging crises, and boundary crises.

In this section, we follow Arafat, Nayfeh, and Chin (1998) and consider the nonplanar response of an inextensional cantilever beam with near-square cross-section to a principal parametric resonance excitation of one of its modes. We assume that the beam's torsional frequencies of oscillations are much higher than the frequencies of the directly excited flexural modes and that the mass moments of inertia are very small in comparison with the translational inertia and flexural stiffnesses. Using these conditions and the nondimensional quantities

$$s^* = \frac{s}{L}, \quad t^* = \sqrt{\frac{D_\eta}{mL^4}} t, \quad v^* = \frac{v}{L}, \quad w^* = \frac{w}{L}, \quad \beta_y = \frac{D_\varsigma}{D_\eta}, \quad \beta_\gamma = \frac{D_\xi}{D_\eta} \quad (5.1.65)$$

we rewrite (4.6.96)-(4.6.98) in nondimensional form as

$$\ddot{v} + c\dot{v} + \beta_y v^{iv} = (1 - \beta_y)\left[w'' \int_1^s v''w''ds - w''' \int_0^s v''w'ds\right]'$$
$$- \frac{(1-\beta_y)^2}{\beta_\gamma}\left[w'' \int_0^s \int_1^s v''w''dsds\right]'' - \beta_y \left[v'\left(v'v'' + w'w''\right)'\right]'$$
$$- \frac{1}{2}\left\{v' \int_1^s \frac{\partial^2}{\partial t^2}\left[\int_0^s (v'^2 + w'^2)\,ds\right]ds\right\}'$$
$$- [v''(s-1) + v']F\Omega^2 \cos(\Omega t) \qquad (5.1.66)$$

$$\ddot{w} + c\dot{w} + w^{iv} = -(1 - \beta_y)\left[v'' \int_1^s v''w''ds - v''' \int_0^s v'w''ds\right]'$$
$$- \frac{(1-\beta_y)^2}{\beta_\gamma}\left[v'' \int_0^s \int_1^s v''w''dsds\right]'' - \left[w'\left(v'v'' + w'w''\right)'\right]'$$
$$- \frac{1}{2}\left\{w' \int_1^s \frac{\partial^2}{\partial t^2}\left[\int_0^s (v'^2 + w'^2)\,ds\right]ds\right\}'$$
$$- [w''(s-1) + w']F\Omega^2 \cos(\Omega t) \qquad (5.1.67)$$

along with the boundary conditions

$$v = w = 0 \text{ and } v' = w' = 0 \text{ at } s = 0 \qquad (5.1.68)$$
$$v'' = w'' = 0 \text{ and } v''' = w''' = 0 \text{ at } s = 1 \qquad (5.1.69)$$

Equations (5.1.66)-(5.1.69) can be derived by using Hamilton's extended principle from the following Lagrangian and virtual work term:

$$\mathcal{L} = \int_0^1 \left\{ \frac{1}{2}\left[\frac{\partial}{\partial t}\int_0^s \frac{1}{2}(v'^2 + w'^2)\,ds\right]^2 + \frac{1}{2}\dot{v}^2 + \frac{1}{2}\dot{w}^2 \right.$$

$$-(1-\beta_y)\left(v''w''\int_0^s v''w'ds\right) - \frac{1}{2}\left(w''^2 + w'^2 w''^2\right)$$

$$-\frac{(1-\beta_y)^2}{2\beta_\gamma}\left[\left(\int_1^s v''w''ds\right)^2 + \left(2v''w''\int_0^s\int_1^s v''w''ds\,ds\right)\right]$$

$$\left.-\frac{1}{2}\beta_y\left(v''^2 + v'^2 v''^2 + 2v'v''w'w''\right)\right\}ds \tag{5.1.70}$$

$$\delta W = \int_0^1 (Q_v^*\delta v + Q_w^*\delta w)\,ds$$

$$= -\int_0^1 \left([v''(s-1) + v']F\Omega^2 \cos(\Omega t) + c\dot{v}\right)\delta v\,ds$$

$$-\int_0^1 \left([w''(s-1) + w']F\Omega^2 \cos(\Omega t) + c\dot{w}\right)\delta w\,ds \tag{5.1.71}$$

We consider the case of one-to-one internal resonance between the mth mode in the y direction and the nth mode in the z direction; that is, $\omega_{1m} \approx \omega_{2n}$, where ω_{1m} is the natural frequency in the y direction. To express the nearness of these frequencies quantitatively, we let

$$\beta_y = 1 + \delta_0 + \epsilon^2 \delta_2 \tag{5.1.72}$$

so that

$$\omega_{1m} = z_m^2\sqrt{1+\delta_0} = z_n^2 = \omega_{2n} \tag{5.1.73}$$

where ϵ is a small nondimensional bookkeeping parameter, which can be set equal to unity in the final results. We note that $\delta_0 = 0$ for a near-square cross-section. For the case of principal parametric resonance of the nth mode in the z direction, we put

$$\Omega = 2\omega_{2n}\left(1 + \epsilon^2 \sigma\right) = 2\omega_{1m}\left(1 + \epsilon^2 \sigma\right) \tag{5.1.74}$$

where σ is a detuning parameter. We also set $F = \epsilon^2 f$.

To apply the method of time-averaged Lagrangian, we let

$$v(s, T_0, T_2) = \epsilon\Phi_m(s)A_1(T_2)e^{i\omega_{1m}T_0} + cc + \cdots \tag{5.1.75}$$
$$w(s, T_0, T_2) = \epsilon\Phi_n(s)A_2(T_2)e^{i\omega_{2n}T_0} + cc + \cdots \tag{5.1.76}$$

Substituting (5.1.75) and (5.1.76) into the Lagrangian (5.1.70) and virtual work (5.1.71), performing the spatial integrations, and keeping the slowly varying terms only, we obtain

$$\frac{<\mathcal{L}>}{\epsilon^4} = -\delta_2 z_m^2 A_1\bar{A}_1 - i\omega_{1m}\left(A_1'\bar{A}_1 - A_1\bar{A}_1'\right) - i\omega_{2n}\left(A_2'\bar{A}_2 - A_2\bar{A}_2'\right)$$

$$-\frac{1}{2}S_{mm}A_1^2\bar{A}_1^2 - \frac{1}{2}S_{nn}A_2^2\bar{A}_2^2 - 4S_{mn}A_1\bar{A}_1 A_2\bar{A}_2$$

$$-\Lambda_{mn}\left(A_1^2\bar{A}_1^2 + \bar{A}_1^2 A_2^2\right) + cc \tag{5.1.77}$$

$$\frac{<\delta W>}{\epsilon^4} = Q_1^*\delta A_1 + Q_2^*\delta A_2 + cc$$

$$= -\left[2f\omega_{1m}^2\delta_{mm}A_1e^{-2i\omega_{1m}\sigma T_2} - 2i\omega_{1m}\mu_1\bar{A}_1\right]\delta A_1$$
$$- \left[2f\omega_{2n}^2\delta_{nn}A_2e^{-2i\omega_{2n}\sigma T_2} - 2i\omega_{2n}\mu_2\bar{A}_2\right]\delta A_2 + cc \quad (5.1.78)$$

where

$$S_{nn} = 6\Gamma_{2nn} - 2\omega_{2n}^2\Gamma_{1nn}, \qquad S_{mm} = 6(1+\delta_0)\Gamma_{2mm} - 2\omega_{1m}^2\Gamma_{1mm},$$

$$S_{mn} = (1+\delta_0)\Gamma_{3mn} - \delta_0\Gamma_{4mn} + \frac{\delta_0^2}{2\beta_\gamma}\Gamma_{5mn}, \quad \Lambda_{mn} = S_{mn} - \omega_{1m}\omega_{2n}\Gamma_{6mn},$$

$$\Gamma_{1ii} = \int_0^1 \left(\int_0^s \Phi_i'^2 ds\right)^2 ds, \qquad \Gamma_{4mn} = \int_0^1 \Phi_m''\Phi_n''\left(\int_0^s \Phi_m''\Phi_n' ds\right) ds,$$

$$\Gamma_{2ii} = \int_0^1 \Phi_i'^2\Phi_i''^2 ds, \qquad \Gamma_{3mn} = \int_0^1 \Phi_m'\Phi_n''\Phi_n'\Phi_n'' ds, \qquad \epsilon^2\mu_i = \frac{1}{2}\int_0^1 c\Phi_i^2 ds,$$

$$\Gamma_{5mn} = \int_0^1 \left(\int_0^s \Phi_m''\Phi_n'' ds\right)^2 ds + 2\int_0^1 \Phi_m''\Phi_n''\left(\int_0^s\int_1^s \Phi_m''\Phi_n'' ds ds\right) ds,$$

$$\Gamma_{6mn} = \int_0^1 \left(\int_0^s \Phi_m'^2 ds\right)\left(\int_0^s \Phi_n'^2 ds\right) ds, \qquad \delta_{ii} = \int_0^1 \Phi_i\left[\Phi_i''(s-1) + \Phi_i'\right] ds$$

Arafat, Nayfeh, and Chin (1998) calculated the S_{ij}, Λ_{ij}, and δ_{ii} for combinations of the first, second, and third modes in the two directions. For the case of a near-square beam, $\delta_0 = 0$ and

$$\Gamma_{111} = \Gamma_{611} = 4.597, \quad \Gamma_{122} = \Gamma_{622} = 144.7, \quad \Gamma_{133} = \Gamma_{633} = 999.9$$
$$\Gamma_{211} = \Gamma_{311} = 20.22, \quad \Gamma_{222} = \Gamma_{322} = 6.709 \times 10^3$$
$$\Gamma_{233} = \Gamma_{333} = 1.322 \times 10^5, \quad \delta_{11} = 1.571, \quad \delta_{22} = 8.647, \quad \delta_{33} = 24.95$$

Applying the extended Hamilton principle

$$\frac{d}{dT_2}\left(\frac{\partial <\mathcal{L}>}{\partial \bar{A}_i'}\right) - \frac{\partial <\mathcal{L}>}{\partial \bar{A}_i} = \bar{Q}_i^*$$

and using (5.1.77) and (5.1.78), we obtain

$$2i\omega_{1m}(A_1' + \mu_1 A_1) + \delta_2 z_m^2 A_1 + S_{mm}A_1^2\bar{A}_1 + 4S_{mn}A_1A_2\bar{A}_2 + 2\Lambda_{mn}A_2^2\bar{A}_1$$
$$+ 2f\omega_{1m}^2\delta_{mm}\bar{A}_1 e^{2i\omega_{1m}\sigma T_2} = 0 \quad (5.1.79)$$
$$2i\omega_{2n}(A_2' + \mu_2 A_2) + S_{nn}A_2^2\bar{A}_2 + 4S_{mn}A_2A_1\bar{A}_1$$
$$+ 2\Lambda_{mn}A_1^2\bar{A}_2 + 2f\omega_{2n}^2\delta_{nn}\bar{A}_2 e^{2i\omega_{2n}\sigma T_2} = 0 \quad (5.1.80)$$

Next, we express the A_k in the polar form

$$A_1 = \frac{1}{2}a_1 e^{i(\omega_{1m}\sigma T_2 - \gamma_1)} \quad \text{and} \quad A_2 = \frac{1}{2}a_2 e^{i(\omega_{2n}\sigma T_2 - \gamma_2)}$$

separate the real and imaginary parts in (5.1.79) and (5.1.80), and obtain

$$a_1' = -\mu_1 a_1 - \frac{\Lambda_{mn}}{4\omega_{1m}}a_1 a_2^2 \sin[2(\gamma_1 - \gamma_2)] - f\omega_{1m}\delta_{mm}a_1 \sin(2\gamma_1) \quad (5.1.81)$$

$$a_1\gamma_1' = \left[\omega_{1m}\sigma - \frac{\delta_2 z_m^2}{2\omega_{1m}}\right]a_1 - \frac{S_{mm}}{8\omega_{1m}}a_1^3 - \frac{S_{mn}}{2\omega_{1m}}a_1 a_2^2$$
$$- \frac{\Lambda_{mn}}{4\omega_{1m}}a_1 a_2^2 \cos[2(\gamma_1 - \gamma_2)] - f\omega_{1m}\delta_{mm}a_1 \cos(2\gamma_1) \quad (5.1.82)$$

$$a_2' = -\mu_2 a_2 + \frac{\Lambda_{mn}}{4\omega_{2n}}a_1^2 a_2 \sin[2(\gamma_1 - \gamma_2)] - f\omega_{2n}\delta_{nn}a_2 \sin(2\gamma_2) \quad (5.1.83)$$

$$a_2\gamma_2' = \omega_{2n}\sigma a_2 - \frac{S_{nn}}{8\omega_{2n}}a_2^3 - \frac{S_{mn}}{2\omega_{2n}}a_1^2 a_2 - \frac{\Lambda_{mn}}{4\omega_{2n}}a_1^2 a_2 \cos[2(\gamma_1 - \gamma_2)]$$
$$- f\omega_{2n}\delta_{nn}a_2 \cos(2\gamma_2) \quad (5.1.84)$$

Alternatively, we express the A_k in the Cartesian form

$$A_1 = \frac{1}{2}(p_1 - iq_1)e^{i\omega_{1m}\sigma T_2} \text{ and } A_2 = \frac{1}{2}(p_2 - iq_2)e^{i\omega_{2n}\sigma T_2}$$

separate the real and imaginary parts in (5.1.79) and (5.1.80), and obtain

$$p_1' = -\mu_1 p_1 - \left[\omega_{1m}\sigma - \frac{\delta_2 z_m^2}{2\omega_{1m}}\right]q_1 - \frac{S_{mm}}{8\omega_{1m}}(p_1^2 + q_1^2)q_1 + \frac{S_{mn}}{2\omega_{1m}}(p_2^2 + q_2^2)q_1$$
$$- \frac{\Lambda_{mn}}{4\omega_{1m}}\left[(p_2^2 - q_2^2)q_1 - 2p_1 p_2 q_2\right] - f\omega_{1m}\delta_{mm}q_1 \quad (5.1.85)$$

$$q_1' = -\mu_1 q_1 + \left[\omega_{1m}\sigma - \frac{\delta_2 z_m^2}{2\omega_{1m}}\right]p_1 + \frac{S_{mm}}{8\omega_{1m}}(p_1^2 + q_1^2)p_1 - \frac{S_{mn}}{2\omega_{1m}}(p_2^2 + q_2^2)p_1$$
$$- \frac{\Lambda_{mn}}{4\omega_{1m}}\left[(p_2^2 - q_2^2)p_1 + 2q_1 p_2 q_2\right] - f\omega_{1m}\delta_{mm}p_1 \quad (5.1.86)$$

$$p_2' = -\mu_2 p_2 - \omega_{2n}\sigma q_2 - \frac{S_{nn}}{8\omega_{2n}}(p_2^2 + q_2^2)q_2 + \frac{S_{mn}}{2\omega_{2n}}(p_1^2 + q_1^2)q_2$$
$$- \frac{\Lambda_{mn}}{4\omega_{2n}}\left[(p_1^2 - q_1^2)q_2 - 2p_1 q_1 p_2\right] - f\omega_{2n}\delta_{nn}q_2 \quad (5.1.87)$$

$$q_2' = -\mu_2 q_2 + \omega_{2n}\sigma p_2 + \frac{S_{nn}}{8\omega_{2n}}(p_2^2 + q_2^2)p_2 - \frac{S_{mn}}{2\omega_{2n}}(p_1^2 + q_1^2)p_2$$
$$- \frac{\Lambda_{mn}}{4\omega_{2n}}\left[(p_1^2 - q_1^2)p_2 + 2p_1 q_1 q_2\right] - f\omega_{2n}\delta_{nn}p_2 \quad (5.1.88)$$

We note that the system (5.1.85)-(5.1.88) is invariant under the following transformations: $(p_1, q_1, p_2, q_2) \iff (-p_1, -q_1, p_2, q_2) \iff (p_1, q_1, -p_2, -q_2) \iff (-p_1, -q_1, -p_2, -q_2)$. Therefore, for any asymmetric solution found, three other solutions can be obtained using the above transformations.

We set the time derivatives in (5.1.85)-(5.1.88) equal to zero and solve the resulting system of algebraic equations for p_1, q_1, p_2, and q_2 for a specified value of the forcing detuning parameter σ. A pseudo arclength scheme is used to trace branches of the equilibrium solutions (Seydel, 1994; Nayfeh and Balachandran, 1995). Then, the amplitudes a_1 and a_2 are calculated from $a_i = \sqrt{p_i^2 + q_i^2}$. The stability of a fixed point is determined by investigating the eigenvalues of the Jacobian matrix of the right-hand sides of (5.1.85)-(5.1.88). When $\delta_2 = -0.05$, we show in Figure 5.1.11

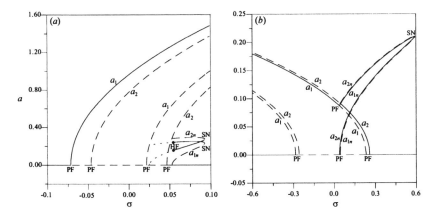

Fig. 5.1.11 Frequency-response curves when $\delta_2 = -0.05$, $\mu_1 = \mu_2 = 0.025$, $f = 0.03$, and $\beta_\gamma = 0.6489$: (a) mode (1,1) and (b) mode (2,2). The planar response amplitudes are denoted by a_1 and a_2 and the nonplanar response amplitudes are denoted by a_{1n} and a_{2n}: (———) stable solution, (– – –) saddles, ($\cdots$) unstable foci, PF = pitchfork bifurcation, SN = saddle-node bifurcation, and HF = Hopf bifurcation.

typical frequency-response curves for modes (1,1) and (2,2). For mode (1,1), the planar response curves are bent to the right and hence the effective nonlinearity is of the hardening type. Therefore, the nonlinear geometric terms are dominant for this mode. In part (b), the planar response curves are bent to the left and hence the effective nonlinearity is of the softening type for mode (2,2). Therefore, the nonlinear inertia terms are dominant for this mode. These results agree with those of Nayfeh and Pai (1989). The nonplanar response of mode (1,1) undergoes a Hopf bifurcation at $\sigma = 0.0566242$, resulting in the creation of limit cycles. The corresponding fixed point at the Hopf bifurcation point is $(p_1, q_1, p_2, q_2) = (0.241107, 0.0240642, -0.0929129, 0.131068)$. On the other hand, the nonplanar response of mode (2,2) does not undergo Hopf bifurcations for the parameters used.

Changing the geometric detuning parameter from $\delta_2 = -0.5$ to 0.5, they obtained the frequency-response curves shown in Figure 5.1.12 for mode (1,1). For the most part, the general characteristics of these curves are qualitatively similar to those in Figure 5.1.11a.

The normal form of any autonomous system near a generic Hopf bifurcation point is (Seydel, 1994; Nayfeh and Balachandran, 1995)

$$\dot{r} = \epsilon_1(\sigma - \sigma_c)r + \alpha_r r^3$$
$$\dot{\theta} = \epsilon_2(\sigma - \sigma_c) + \alpha_i r^2$$

where r is a measure of the amplitude of the created limit cycle. The bifurcation is generic if $\epsilon_1 \neq 0$ and subcritical if $\alpha_r > 0$ or supercritical if $\alpha_r < 0$. The created limit cycle is stable if the bifurcation is supercritical and unstable if the

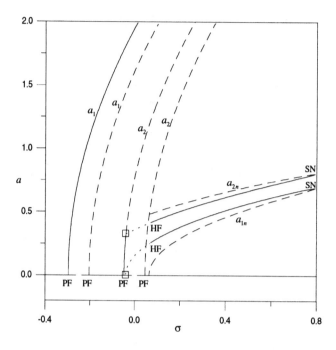

Fig. 5.1.12 Frequency-response curves for mode (1,1) when $\delta_2 = 0.5$, $\mu_1 = \mu_2 = 0.025$, $f = 0.03$, and $\beta_\gamma = 0.6489$. The planar response amplitudes are denoted by a_1 and a_2 and the nonplanar response amplitudes are denoted by a_{1n} and a_{2n}: (———) stable solution, (- - -) saddles, (· · ·) unstable foci, PF = pitchfork bifurcation, SN = saddle-node bifurcation, and HF = Hopf bifurcation.

bifurcation is subcritical. The amplitude of the created limit cycle is given by $r = \sqrt{-\epsilon_1(\sigma - \sigma_c)/\alpha_r}$.

A *Mathematica* code developed by Nayfeh and Chin (1999b) was used to calculate the coefficients in the normal form of the Hopf bifurcations. For the Hopf bifurcation at $\sigma = 0.0694703$, the corresponding fixed point is $(p_1, q_1, p_2, q_2) = (0.254868, 0.0200404, -0.0560086, 0.411458)$. For this point, $\epsilon_1 = -39.1347$ and $\alpha_r = 1.41995$, indicating that the bifurcation is generic and subcritical. Hence, the created limit cycles are unstable. The two-mode solutions occur over a longer range of σ because increasing the magnitude of the geometric detuning δ_2 creates stronger coupling between the inplane and out-of-plane modes. An important difference in this case is the occurrence of a second Hopf bifurcation point at $\sigma = -0.0383338$ and $(p_1, q_1, p_2, q_2) = (0.005642228, 0.0001068165, -0.02463237, 0.324553)$. The corresponding values for ϵ_1 and α_r are 1049.01 and -2844.36, respectively, indicating a generic supercritical Hopf bifurcation with the unstable foci to its right. In Figure 5.1.13, we zoom on this Hopf bifurcation point, which is very close to the pitchfork bifurcation that occurs at $\sigma = -0.0383862$.

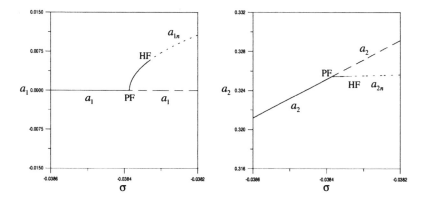

Fig. 5.1.13 Enlargements of the blocked areas shown in Figure 5.1.12. The planar response amplitudes are denoted by a_1 and a_2 and the nonplanar response amplitudes are denoted by a_{1n} and a_{2n}: (——) stable solution, (– – –) saddles, ($\cdots$) unstable foci, PF = pitchfork bifurcation, and HF = Hopf bifurcation.

5.2 TRANSVERSELY EXCITED CANTILEVER BEAMS

In this section, we consider the nonlinear responses of cantilever beams to transverse harmonic excitations. In Section 5.2.1, we treat the planar response to a primary resonance of the nth mode. In Section 5.2.2, we treat the planar response to a sub-combination resonant excitation. And in Section 5.2.3, we treat the case of nonplanar response of a beam with a near-square cross-section to a primary resonance of one of its modes.

5.2.1 Planar Response to a Primary-Resonance Excitation

The nonlinear planar response of a long slender inextensional metallic beam to a primary resonance is governed by (5.1.1) and (5.1.3) where

$$q_2 = mF \cos \Omega t \tag{5.2.1}$$

We introduce the nondimensional variables denoted by asterisks as follows:

$$s^* = \frac{s}{L}, \quad v^* = \frac{v}{L}, \quad t^* = t\sqrt{\frac{EI}{mL^4}}, \quad \Omega^* = \Omega\sqrt{\frac{mL^4}{EI}}$$

in (5.1.1), (5.1.3), and (5.2.1), drop the asterisks from v, s, t, and Ω, and obtain

$$\ddot{v} + v^{iv} = -\left(v'v''^2 + v'''v'^2\right)' - \frac{1}{2}\left[v' \int_1^s \frac{\partial^2}{\partial t^2}\left(\int_0^s v'^2 ds\right) ds\right]'$$
$$-c_1^*\dot{v} - c_2^*\dot{v}\,|\,\dot{v}\,| + F^* \cos \Omega t \tag{5.2.2}$$

$$v = 0 \text{ and } v' = 0 \text{ at } s = 0 \qquad (5.2.3a)$$
$$v'' = 0 \text{ and } v''' = 0 \text{ at } s = 1 \qquad (5.2.3b)$$

where

$$c_1^* = \frac{c_1 L^2}{\sqrt{mEI}}, \quad c_2^* = \frac{c_2 L}{m}, \quad F^* = \frac{mFL^3}{EI}$$

Moreover, we scale the damping coefficients c_1^* and c_2^* and the forcing coefficient F^* as $\epsilon^2 c_1^*, \epsilon c_2^*$, and $\epsilon^3 F^*$. Substituting (5.1.6) into (5.2.2) and (5.2.3) and equating coefficients of like powers of ϵ, we obtain

Order ϵ:

$$D_0^2 v_1 + v_1^{iv} = 0 \qquad (5.2.4)$$
$$v_1 = 0 \text{ and } v_1' = 0 \text{ at } s = 0 \qquad (5.2.5a)$$
$$v_1'' = 0 \text{ and } v_1''' = 0 \text{ at } s = 1 \qquad (5.2.5b)$$

Order ϵ^3:

$$D_0^2 v_3 + v_3^{iv} = -2D_0 D_2 v_1 - c_1^* D_0 v_1 - c_2^* D_0 v_1 \mid D_0 v_1 \mid - \left[v_1' \left(v_1' v_1'' \right)' \right]'$$
$$- \frac{1}{2} \left[v_1' \int_1^s D_0^2 \left(\int_0^s v_1'^2 ds \right) ds \right]' + F^* \cos(\Omega T_0) \qquad (5.2.6)$$
$$v_3 = 0 \text{ and } v_3' = 0 \text{ at } s = 0 \qquad (5.2.7a)$$
$$v_3'' = 0 \text{ and } v_3''' = 0 \text{ at } s = 1 \qquad (5.2.7b)$$

where $D_k = \partial/\partial T_k$.

As discussed in Section 5.1.2, the general solution of (5.2.4) and (5.2.5) consists of an infinite number of modes corresponding to an infinite number of frequencies. However, any mode that is not directly or indirectly excited will decay to zero with time due to the presence of damping (Nayfeh and Mook, 1979). In this section, we restrict our attention to the case where only the nth mode is directly excited and assume that this mode is not excited by any internal resonance. Thus, we express the solution of (5.2.4) and (5.2.5) as in (5.1.13).

Substituting (5.1.13) into (5.2.6) yields

$$D_0^2 v_3 + v_3^{iv} = -2i\omega_n \Phi_n D_2 A_n e^{i\omega_n T_0} - i\omega_n c_1^* \Phi_n A_n e^{i\omega_n T_0} + \frac{1}{2} F^* e^{i\Omega T_0}$$
$$- \left[\Phi_n' \left(\Phi_n' \Phi_n'' \right)' \right]' \left(A_n^3 e^{3i\omega_n T_0} + 3 A_n^2 \bar{A}_n e^{i\omega_n T_0} \right)$$
$$+ 2\omega_n^2 \left(\Phi_n' \int_1^s \int_0^s \Phi_n'^2 ds ds \right)' \left(A_n^3 e^{3i\omega_n T_0} + A_n^2 \bar{A}_n e^{i\omega_n T_0} \right) + cc$$
$$- c_2^* \omega_n^2 \Phi_n \left(i A_n e^{i\omega_n T_0} - i \bar{A}_n e^{-i\omega_n T_0} \right) \mid \Phi_n \left(i A_n e^{i\omega_n T_0} - i \bar{A}_n e^{-i\omega_n T_0} \right) \mid$$
$$\qquad (5.2.8)$$

Here, we restrict our discussion to the case of primary resonance of the nth mode (i.e., $\Omega \approx \omega_n$). To express the nearness of this resonance, we introduce the detuning parameter σ defined by $\Omega = \omega_n + \epsilon^2 \sigma$. As discussed in Section 5.1.2, the solvability condition demands that the right-hand side of (5.2.8) be orthogonal to every solution of the adjoint homogeneous problem. Therefore, demanding that the right-hand side of (5.2.8) be orthogonal to $\Phi_n(s)\exp(-i\omega_n T_0)$, we obtain

$$-2i\omega_n (D_2 A_n + \mu_n A_n) - S_{nn} A_n^2 \bar{A}_n + \frac{1}{2} f e^{i\sigma T_2} - \Gamma_n \int_0^1 c_2^* \Phi_n^2 \mid \Phi_n \mid ds = 0 \tag{5.2.9}$$

where S_{nn} and μ_n are defined in (5.1.27) and $f = \int_0^1 F^* \Phi_n ds$. The numerical values for the S_{nn} are given in Section 5.1.2.

Substituting the polar form

$$A_n = \frac{1}{2} a_n e^{i(\sigma T_2 - \gamma_n)} \tag{5.2.10}$$

into (5.2.9), multiplying the result by $\exp\left[i\left(\gamma_n - \sigma T_2\right)\right]$, and separating real and imaginary parts, we obtain

$$a'_n = -\mu_n a_n - C_n a_n^2 + \frac{f}{2\omega_n} \sin \gamma_n \tag{5.2.11}$$

$$a_n \gamma'_n = \sigma a_n - \frac{S_{nn}}{8\omega_n} a_n^3 + \frac{f}{2\omega_n} \cos \gamma_n \tag{5.2.12}$$

where C_n is defined in (5.1.32). Substituting (5.2.10) into (5.1.13) and then substituting (5.1.13) into (5.1.6), we find that the beam response is given by

$$v(s, t; \epsilon) = \epsilon a_n \Phi_n(s) \cos\left(\Omega t - \gamma_n\right) + \cdots \tag{5.2.13}$$

where a_n and γ_n are given by (5.2.11) and (5.2.12).

It follows from (5.2.13) that periodic motions of the beam correspond to the fixed points or equilibrium solutions of (5.2.11) and (5.2.12). Setting $a'_n = 0$ and $\gamma'_n = 0$ in (5.2.11) and (5.2.12) yields the following algebraic equations for the fixed points:

$$-\mu_n a_n - C_n a_n^2 + \frac{f}{2\omega_n} \sin \gamma_n = 0 \tag{5.2.14}$$

$$\sigma a_n - \frac{S_{nn}}{8\omega_n} a_n^3 + \frac{f}{2\omega_n} \cos \gamma_n = 0 \tag{5.2.15}$$

Eliminating γ_n from (5.2.14) and (5.2.15) yields

$$f = 2\omega_n a_n \sqrt{(\mu_n + C_n a_n)^2 + \left(\sigma - \frac{S_{nn}}{8\omega_n} a_n^2\right)^2} \tag{5.2.16}$$

or

$$\sigma = \frac{S_{nn}}{8\omega_n} a_n^2 \pm \sqrt{\frac{f^2}{4\omega_n^2 a_n^2} - (\mu_n + C_n a_n)^2} \tag{5.2.17}$$

Malatkar (2003) and Malatkar and Nayfeh (2003) conducted an experiment on a transversely excited beam and used the above results to identify the effective nonlinearity S_{nn} and the damping coefficients μ_n and C_n for the third mode.

They excited a steel beam with the dimensions $19.085\ in \times \frac{1}{2}\ in \times \frac{1}{32}\ in$ by a base excitation. The density and Young's modulus of the beam were taken as $7810\ kg/m^3$ and $207\ GPa$, respectively. The beam was mounted vertically on a steel clamping fixture attached to a $100\text{-}lb$ shaker. The output of the shaker was measured using an accelerometer placed on the clamping fixture, and the response of the cantilever beam was measured with a strain gage mounted approximately $33\ mm$ from the fixed end.

The experiment included four test sequences, each of which was run on a separate day. In three of these test sequences, the frequency was swept while the excitation amplitude was held constant, though the excitation amplitude itself was different for each sequence. In the fourth test sequence, the excitation amplitude was varied while the excitation frequency was held constant. They waited for a long time to ensure steady-state response behavior before taking any measurement.

Linear Natural Frequencies

The natural frequencies of the beam were determined using the frequency-response function of the signal analyzer. The beam was excited by a 50% burst-chirp low-amplitude excitation, and a uniform window was used to analyze the power spectra of the accelerometer and strain-gage signals. Peaks in the amplitude portion of the frequency-response function give the linear natural frequencies of the beam. To increase confidence in the experimentally obtained linear natural frequencies, they measured the frequency-response functions at several low excitation levels. No noticeable shifts in the peaks were observed and the coherence was close to unity at the corresponding peaks. Also, a periodic checking of the natural frequencies of the beam was done to detect any fatigue damage.

Before the beginning of each test sequence, they measured the natural frequency of the corresponding mode. The experimentally determined third-mode natural frequencies were 49.078 Hz for the $F = 0.10$ g test sequence, 49.094 Hz for the $F = 0.15$ g test sequence, and 49.094 Hz for the $F = 0.20$ g test sequence. They noted that the variation in the third-mode natural frequency was equal to the frequency resolution of the signal analyzer used to make the measurements. The experimentally measured value of the fourth-mode natural frequency was 96.117 Hz. To match the theoretical frequency-response curves with the experimental ones, they had to shift the latter by about 0.05 Hz. The most probable reason for this discrepancy is the assumption of an ideal-clamp boundary condition (Tabaddor, 2000). In what follows, the theoretically obtained natural frequencies are lowered by 0.05 Hz.

Determination of the Beam Displacement

The strain gage essentially measures the strain at the location where it is mounted on the beam. To measure the beam displacement at that point, one needs to convert the strain into displacement. Their procedure to convert the strain-gage reading into displacement is as follows.

From beam theory, we know that the strain experienced by the strain gage is given by

$$e = \frac{y}{\rho} \tag{5.2.18}$$

where e denotes the strain, ρ denotes the radius of curvature at the location of the strain gage, and y denotes the distance of the strain gage from the beam's neutral axis; that is, $y = \frac{1}{2}b$ where b is the beam thickness. It follows from (5.2.13) that

$$v(s) = \epsilon a_n \Phi_n(s) \tag{5.2.19}$$

The curvature at a distance s from the fixed end is given by

$$\frac{1}{\rho} = \frac{\partial^2 v}{\partial s^2} \left[1 + \left(\frac{\partial v}{\partial s} \right)^2 \right]^{-\frac{3}{2}} \tag{5.2.20}$$

For the displacement amplitudes observed in the test sequences, it was found that the nonlinear expression for the curvature was not necessary. The linear part by itself determines the displacement amplitude to a sufficient degree of accuracy. Hence, they used the following linear expression for the curvature:

$$\frac{1}{\rho} \approx \frac{\partial^2 v}{\partial s^2} \tag{5.2.21}$$

Let l_{sg} denote the distance of the strain-gage center from the fixed end. Using (5.2.19) in (5.2.21), we obtain

$$\left. \frac{1}{\rho} \right|_{s=l_{sg}} = a_n \Phi_n''(l_{sg}) \tag{5.2.22}$$

Consequently, one can determine the strain from the strain-gage reading as follows:

$$e = \frac{4 V_{out}}{V_{excite} G K_g} \tag{5.2.23}$$

where V_{out} denotes the strain-gage amplifier output in volts, V_{excite} denotes the bridge excitation voltage in volts, G denotes the gain of the strain-gage signal conditioner, and K_g denotes the gage factor of the strain gage. Substituting (5.2.22) and (5.2.23) into (5.2.18), we obtain

$$a_n = \frac{8}{V_{excite} G K_g b \Phi_n''(l_{sg})} V_{out} \tag{5.2.24}$$

which when substituted into (5.2.19) gives

$$v(s) = \frac{8 \Phi_n(s)}{V_{excite} G K_g b \Phi_n''(l_{sg})} V_{out} \tag{5.2.25}$$

Therefore, one only needs to multiply the strain-gage reading V_{out} by a constant to obtain the displacement amplitude v at a given s. The values of the constants appearing in (5.2.24) are $V_{excite} = 10\ V$, $G = 1000$, $K_g = 2.095$, $b = 0.794\ mm$, and $l_{sg} = 32.56\ mm$.

Parameter Estimation

Malatkar and Nayfeh (2003) estimated the parameters (S_{nn}, μ_n, and C_n) from the experimental frequency-response results. For a given excitation level, the location of the peak of the corresponding frequency-response curve depends on the damping values (Nayfeh and Mook, 1979). At the peak of the frequency-response curve, it follows from (5.2.17) that

$$\mu_n + C_n a_p = \frac{f}{2\omega_n a_p} \qquad (5.2.26)$$

where a_p denotes the value of a_n at the peak. Further, the effect of the nonlinearity is essentially to shift the peak away from the natural frequency ω_n. At the peak of the frequency-response curve, it follows from (5.2.17) that the shift in frequency is given by

$$\sigma_p = \frac{S_{nn}}{8\omega_n} a_p^2 \qquad (5.2.27)$$

For a system with hardening nonlinearity, the peak is shifted to the right, and in the case of a softening nonlinearity it is shifted to the left. The magnitude of the shift depends on the strength of the nonlinearity. Thus, knowing the location of the peak, one can estimate the damping coefficients and the effective nonlinearity of the system. The coordinates of the peaks of the experimentally obtained third-mode frequency-response curves for three base acceleration levels are

$$F = 0.10 \, g: \quad \Omega_p = 48.963 \text{ Hz}, \quad v_p^* = 2.415 \, mm$$
$$F = 0.15 \, g: \quad \Omega_p = 48.917 \text{ Hz}, \quad v_p^* = 3.220 \, mm$$
$$F = 0.20 \, g: \quad \Omega_p = 48.885 \text{ Hz}, \quad v_p^* = 3.876 \, mm$$

where v_p^* denotes the peak of the displacement at the tip of the beam.

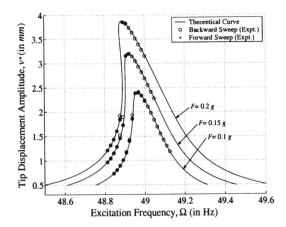

Fig. 5.2.1 Experimentally and theoretically obtained third-mode frequency-response curves for $F = 0.1 \, g, 0.15 \, g$, and $0.2 \, g$.

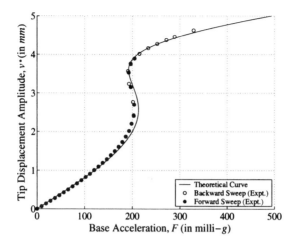

Fig. 5.2.2 Experimentally and theoretically obtained third-mode force-response curves for $\Omega = 48.891$ Hz.

It follows from (5.2.26) that there are two unknowns (μ_n and C_n) and just one equation. Therefore, they used results of two test sequences to determine the two damping coefficients. Using (5.2.26) for two test sequences, denoted by subscripts 1 and 2, we have

$$C_n = \frac{1}{2(a_{2p} - a_{1p})} \left(\frac{f_2}{\omega_n a_{2p}} - \frac{f_1}{\omega_n a_{1p}} \right) \tag{5.2.28}$$

$$\mu_n = \frac{f_2}{2\omega_n a_{2p}} - C_n a_{2p} \tag{5.2.29}$$

To estimate the effective nonlinearity, we find from (5.2.27) that

$$S_{nn} = \frac{8\omega_n \sigma_p}{a_p^2} \tag{5.2.30}$$

where σ_p denotes the value of σ at the peak. Their estimated parameter values for the third mode are

$F = 0.10\ g$: $\quad \mu_3 = 0.0356, \quad C_3 = 0.0533, \quad S_{33} = -1.578 \times 10^9$
$F = 0.15\ g$: $\quad \mu_3 = 0.0338, \quad C_3 = 0.0577, \quad S_{33} = -1.442 \times 10^9$
$F = 0.20\ g$: $\quad \mu_3 = 0.0304, \quad C_3 = 0.0631, \quad S_{33} = -1.447 \times 10^9$

Next, we compare the experimentally and theoretically obtained third-mode force- and frequency-response curves. The theoretical curves are obtained using (5.2.16) and (5.2.17) and the parameter values estimated from the experimental results. Figure 5.2.1 compares the experimental and theoretical frequency-response curves, and Figure 5.2.2 compares the experimental force-response curve with that obtained theoretically. The agreement between the experimental and theoretical results is very good over all of the experimental data points.

5.2.2 External Subcombination Resonance

We consider, after Nayfeh and Arafat (1998), the case of external subcombination resonance $\Omega \approx \frac{1}{2}(\omega_n + \omega_m)$ in systems with cubic inertia and geometric nonlinearities. We assume that modes m and n are not involved in internal resonances with any of the other modes of the system. Because of damping, all of the other modes decay with time, and the dynamics of the system can be obtained by a two-mode truncation involving the mth and nth modes. As in the preceding section, in the presence of linear damping, the planar response of a cantilever inextensional beam to a transverse harmonic excitation is governed by the nondimensional equation

$$\ddot{v} + v^{iv} + c\dot{v} = -\left[v'\left(v'v''\right)'\right]' - \frac{1}{2}\left[v'\int_1^s \frac{\partial^2}{\partial t^2}\left(\int_0^s v'^2 ds\right)ds\right]' + F(s)\cos(\Omega t) \tag{5.2.31}$$

subject to the boundary conditions

$$v = 0 \text{ and } v' = 0 \text{ at } s = 0 \tag{5.2.32a}$$

$$v'' = 0 \text{ and } v''' = 0 \text{ at } s = 1 \tag{5.2.32b}$$

Using Hamilton's extended principle, one can derive (5.2.31) and (5.2.32) from the Lagrangian

$$\mathcal{L} = \int_0^1 \left[\frac{1}{2}\dot{v}^2 + \frac{1}{8}\left(\frac{\partial}{\partial t}\int_0^s v'^2 ds\right)^2 - \frac{1}{2}\left(v''^2 + v'^2 v''^2\right)\right]ds - \cos(\Omega t)\int_0^1 vF(s)ds \tag{5.2.33}$$

and the virtual work term

$$\delta W = -\int_0^1 c\dot{v}\delta v ds \tag{5.2.34}$$

The linear forced undamped response of the beam is given by

$$\ddot{v} + v^{iv} = F(s)\cos(\Omega t) \tag{5.2.35}$$

subject to the boundary conditions (5.2.32). A particular solution of this problem can be expressed as

$$v(s,t) = 2\Phi(s)\cos(\Omega t) \tag{5.2.36}$$

where $\Phi(s)$ is governed by the boundary-value problem

$$\Phi^{iv} - \Omega^2\Phi = \frac{1}{2}F(s) \tag{5.2.37}$$

$$\Phi = 0 \text{ and } \Phi' = 0 \text{ at } s = 0 \tag{5.2.38}$$

$$\Phi'' = 0 \text{ and } \Phi''' = 0 \text{ at } s = 1 \tag{5.2.39}$$

When $F(s)$ is constant, the solution of (5.2.37)-(5.2.39) can be expressed as

$$\Phi(s) = \frac{F}{4\Omega^2}\left[c_1\sin(\sqrt{\Omega}s) + c_2\cos(\sqrt{\Omega}s) + c_3\sinh(\sqrt{\Omega}s) \right.$$
$$\left. + c_4\cosh(\sqrt{\Omega}s) - 2\right] \tag{5.2.40}$$

where

$$c_1 = -c_3 = \frac{\cosh\sqrt{\Omega}\sin\sqrt{\Omega} + \sinh\sqrt{\Omega}\cos\sqrt{\Omega}}{1 + \cosh\sqrt{\Omega}\cos\sqrt{\Omega}} \quad (5.2.41)$$

$$c_2 = \frac{1 - \sinh\sqrt{\Omega}\sin\sqrt{\Omega} + \cosh\sqrt{\Omega}\cos\sqrt{\Omega}}{1 + \cosh\sqrt{\Omega}\cos\sqrt{\Omega}} \quad (5.2.42)$$

$$c_4 = \frac{1 + \sinh\sqrt{\Omega}\sin\sqrt{\Omega} + \cosh\sqrt{\Omega}\cos\sqrt{\Omega}}{1 + \cosh\sqrt{\Omega}\cos\sqrt{\Omega}} \quad (5.2.43)$$

Because the characteristic equation is

$$1 + \cos\sqrt{\omega_n}\cosh\sqrt{\omega_n} = 0$$

we find that (5.2.40)-(5.2.43) break down when Ω is near any of the natural frequencies of the beam. In the present case, $\Omega \approx \frac{1}{2}(\omega_n \pm \omega_m)$, which is away from any of the natural frequencies.

Modulation Equations

To describe the nearness of Ω to $\frac{1}{2}(\omega_n + \omega_m)$, we introduce a detuning parameter σ defined by

$$\Omega = \frac{1}{2}(\omega_n + \omega_m) + \epsilon\sigma \quad (5.2.44)$$

where ϵ is a small nondimensional bookkeeping parameter. In order that the effects of the nonlinearities and damping balance the effect of the resonance, we scale v as $\epsilon^{1/2}v$, the damping coefficient c as ϵc, and the forcing F as $\epsilon^{1/2}F$. Then, we let

$$v(s,t) = \epsilon^{1/2}\left[A_m(T_1)\Phi_m(s)e^{i\omega_m T_0} + A_n(T_1)\Phi_n(s)e^{i\omega_n T_0}\right.$$
$$\left. + \Phi(s)e^{i\Omega T_0}\right] + cc + \cdots \quad (5.2.45)$$

Substituting (5.2.45) into (5.2.33) and (5.2.34), integrating over s, and keeping the slowly varying terms, we obtain

$$<\mathcal{L}> = \epsilon^2\left[i\omega_m\left(A_m\bar{A}'_m - \bar{A}_m A'_m\right) + i\omega_n\left(A_n\bar{A}'_n - \bar{A}_n A'_n\right) - \frac{1}{2}S_{nn}A_n^2\bar{A}_n^2\right.$$
$$-\frac{1}{2}S_{mm}A_m^2\bar{A}_m^2 - S_{mn}A_m\bar{A}_m A_n\bar{A}_n - \Gamma_m A_m\bar{A}_m - \Gamma_n A_n\bar{A}_n$$
$$\left. - \Lambda A_n A_m e^{-2i\sigma T_1} - \Lambda \bar{A}_n\bar{A}_m e^{2i\sigma T_1} + \text{constant}\right] + \cdots \quad (5.2.46)$$

$$<\delta W> = \epsilon^2\left[2i\mu_m\bar{A}_m\delta A_m + 2i\mu_n\bar{A}_n\delta A_n\right] + cc + \cdots \quad (5.2.47)$$

where the S_{ij} are defined in (5.1.27) and (5.1.45) and

$$\Gamma_i = \int_0^1\left[2\Phi_i''^2\Phi'^2 + 8\Phi_i'\Phi_i''\Phi'\Phi'' + 2\Phi_i'^2\Phi''^2\right.$$

$$-2\left(\omega_i^2+\Omega^2\right)\left(\int_0^s \Phi_i'\Phi' ds\right)^2\bigg]ds \qquad (5.2.48)$$

$$\Lambda = \int_0^1 \bigg[2\Phi_m''\Phi_n'\Phi'\Phi'' + \Phi_m'\Phi_n'\Phi''^2 + \Phi_m''\Phi_n''\Phi'^2 + 2\Phi_m'\Phi_n''\Phi'\Phi''$$
$$-\left(\omega_m\Omega+\omega_n\Omega-\Omega^2-\omega_m\omega_n\right)\left(\int_0^s \Phi_m'\Phi' ds\right)\left(\int_0^s \Phi_n'\Phi' ds\right)$$
$$-\Omega\left(\omega_m+\omega_n\right)\left(\int_0^s \Phi_m'\Phi_n' ds\right)\left(\int_0^s \Phi'^2 ds\right)\bigg]ds \qquad (5.2.49)$$

$$2\mu_j = \int_0^1 c\,\Phi_j^2 ds \qquad (5.2.50)$$

Applying Hamilton's principle to the Lagrangian (5.2.46) and the virtual work (5.2.47), we obtain the modulation equations

$$-2i\omega_m\left(A_m'+\mu_m A_m\right) = \Gamma_m A_m + S_{mm}A_m^2\bar{A}_m + S_{mn}A_m A_n\bar{A}_n + \Lambda\bar{A}_n e^{2i\sigma T_1} \qquad (5.2.51)$$

$$-2i\omega_n\left(A_n'+\mu_n A_n\right) = \Gamma_n A_n + S_{nm}A_m\bar{A}_m A_n + S_{nn}A_n^2\bar{A}_n + \Lambda\bar{A}_m e^{2i\sigma T_1} \qquad (5.2.52)$$

In Table 5.1, we present values of the coefficients in these modulation equations.
Substituting the polar transformation

$$A_m = \frac{1}{2}a_m e^{i\beta_m} \quad \text{and} \quad A_n = \frac{1}{2}a_n e^{i\beta_n} \qquad (5.2.53)$$

Table 5.1 Values of the coefficients S_{ij}, Γ_m, Γ_n, and Λ for different combinations of the first four modes.

Modes m & n	Γ_m	Γ_n	Λ
1 & 2	$-0.0107F^2$	$-0.1268F^2$	$0.0130F^2$
1 & 3	$-0.0005F^2$	$-0.0420F^2$	$0.0051F^2$
2 & 3	$-0.0111F^2$	$-0.0531F^2$	$0.0006F^2$
1 & 4	$-0.0588F^2$	$-15.3290F^2$	$0.9893F^2$
2 & 4	$-0.0045F^2$	$-0.0298F^2$	$0.0161F^2$
3 & 4	$-0.0135F^2$	$-0.0420F^2$	$-0.0041F^2$

Modes m & n	S_{mm}	$S_{mn} = S_{nm}$	S_{nn}
1 & 2	7.6680	-1122.2408	-100279.6731
1 & 3	7.6680	-15928.9047	-6818871.8310
2 & 3	-100279.6731	-75835.2975	-6818871.8310
1 & 4	7.6680	-34594.5220	-1.1042×10^8
2 & 4	-100279.6731	-927411.9209	-1.1042×10^8
3 & 4	-6818871.8310	-4017517.1670	-1.1042×10^8

into (5.2.51) and (5.2.52) and separating real and imaginary parts, we obtain the real-valued modulation equations

$$a'_m = -\mu_m a_m - \frac{\Lambda}{2\omega_m} a_n \sin\gamma \tag{5.2.54}$$

$$a_m \beta'_m = \frac{\Gamma_m}{2\omega_m} a_m + \frac{S_{mm}}{8\omega_m} a_m^3 + \frac{S_{mn}}{8\omega_m} a_m a_n^2 + \frac{\Lambda}{2\omega_m} a_n \cos\gamma \tag{5.2.55}$$

$$a'_n = -\mu_n a_n - \frac{\Lambda}{2\omega_n} a_m \sin\gamma \tag{5.2.56}$$

$$a_n \beta'_n = \frac{\Gamma_n}{2\omega_n} a_n + \frac{S_{nm}}{8\omega_n} a_m^2 a_n + \frac{S_{nn}}{8\omega_n} a_n^3 + \frac{\Lambda}{2\omega_n} a_m \cos\gamma \tag{5.2.57}$$

where

$$\gamma = 2\sigma T_1 - \beta_m - \beta_n \tag{5.2.58}$$

Substituting (5.2.53) into (5.2.45), we find that, to the first approximation, the response of the beam is given by

$$v(s,t;\epsilon) = \epsilon^{1/2} \big[a_n \Phi_n(s) \cos(\omega_n t + \beta_n) + a_m \Phi_m(s) \cos(\omega_m t + \beta_m) \\ + 2\Phi(s) \cos(\Omega t) \big] + \cdots \tag{5.2.59}$$

where the a_i and β_i are given by (5.2.54)-(5.2.58). Using (5.2.44) and (5.2.58) to eliminate β_n from (5.2.59), we obtain

$$v(s,t;\epsilon) = \epsilon^{1/2} \big\{ a_n \Phi_n(s) \cos[(2\Omega - \omega_m) t - \beta_m - \gamma] \\ + a_m \Phi_m(s) \cos(\omega_m t + \beta_m) + 2\Phi(s) \cos(\Omega t) \big\} + \cdots \tag{5.2.60}$$

Fixed Points and Their Stability

The fixed points of (5.2.54)-(5.2.58) correspond to $a'_m = a'_n = 0$ and $\gamma' = 0$. There are two possibilities: (a) $a_m = a_n = 0$ and (b) $a_m \neq 0$ and $a_n \neq 0$. In the first case, it follows from (5.2.59) that the beam response is given by

$$v(s,t;\epsilon) = 2\epsilon^{1/2} \Phi(s) \cos(\Omega t) + \cdots \tag{5.2.61}$$

which is periodic having the same period as that of the excitation. In the second case, the response is two-period quasiperiodic because Ω and $\omega_m + \beta'_m$ are, in general, incommensurate.

For nontrivial solutions, we use (5.2.55), (5.2.57), and (5.2.58) to eliminate β_n and β_m and obtain

$$\gamma' = 2\sigma - \frac{1}{2}\left(\frac{\Gamma_m}{\omega_m} + \frac{\Gamma_n}{\omega_n}\right) - \frac{1}{8}\left(\frac{S_{mm}}{\omega_m} + \frac{S_{nm}}{\omega_n}\right) a_m^2 - \frac{1}{8}\left(\frac{S_{mn}}{\omega_m} + \frac{S_{nn}}{\omega_n}\right) a_n^2 \\ - \frac{\Lambda}{2}\left(\frac{a_n}{\omega_m a_m} + \frac{a_m}{\omega_n a_n}\right) \cos\gamma \tag{5.2.62}$$

To determine the nontrivial fixed points, we put $a'_m = 0$, $a'_n = 0$, and $\gamma' = 0$ into (5.2.54), (5.2.56), and (5.2.62). The solution of the resulting algebraic equations can be expressed as

$$\alpha_e a_m^2 = 2\sigma - \frac{1}{2}\left(\frac{\Gamma_m}{\omega_m} + \frac{\Gamma_n}{\omega_n}\right) \pm \frac{\mu_m + \mu_n}{\sqrt{\mu_m \mu_n}}\sqrt{\frac{\Lambda^2}{4\omega_m \omega_n} - \mu_m \mu_n} \quad (5.2.63)$$

$$a_n^2 = \frac{\mu_m \omega_m}{\mu_n \omega_n} a_m^2 \quad (5.2.64)$$

$$\sin\gamma = -\frac{2\mu_m \omega_m}{\Lambda}\frac{a_m}{a_n} = -\frac{2\mu_n \omega_n}{\Lambda}\frac{a_n}{a_m} \quad (5.2.65)$$

where

$$\alpha_e = \frac{1}{8}\left[\frac{S_{mm}}{\omega_m} + \frac{S_{nm}}{\omega_n} + \left(\frac{S_{mn}}{\omega_m} + \frac{S_{nn}}{\omega_n}\right)\frac{\mu_m \omega_m}{\mu_n \omega_n}\right] \quad (5.2.66)$$

The case $\Omega \approx \frac{1}{2}(\omega_n - \omega_m)$ can be obtained from the above results by changing the sign of ω_m. However, it follows from (5.2.64) that this resonance cannot be activated in the cantilever beam.

The stability of the trivial solution can be analyzed by investigating the solution of the linearized complex-valued modulation equations (5.2.51) and (5.2.52). To this end, we let

$$A_m = c_m e^{\lambda T_1 + 2i\sigma T_1} \quad \text{and} \quad A_n = c_n e^{\bar{\lambda} T_1} \quad (5.2.67)$$

in the linearized equations (5.2.51) and (5.2.52) and obtain

$$[2i\omega_m(\mu_m + \lambda + 2i\sigma) + \Gamma_m]c_m + \Lambda \bar{c}_n = 0 \quad (5.2.68)$$

$$\Lambda \bar{c}_m + [2i\omega_n(\mu_n + \bar{\lambda}) + \Gamma_n]c_n = 0 \quad (5.2.69)$$

For trivial solutions, we set the determinant of the coefficient matrix in (5.2.68) and (5.2.69) equal to zero. The result is

$$\lambda^2 + \left[(\mu_m + \mu_n) + i\left(2\sigma - \frac{\Gamma_m}{2\omega_m} + \frac{\Gamma_n}{2\omega_n}\right)\right]\lambda + \left[\left(\mu_m \mu_n - \frac{\sigma \Gamma_n}{\omega_n}\right)\right.$$
$$\left. + \frac{\Gamma_m \Gamma_n - \Lambda^2}{4\omega_m \omega_n} + i\left(2\sigma\mu_n - \frac{\Gamma_m \mu_n}{2\omega_m} + \frac{\Gamma_n \mu_m}{2\omega_n}\right)\right] = 0 \quad (5.2.70)$$

It follows from (5.2.70) that the trivial solution loses stability as one of the λ's crosses the imaginary axis along the real axis from the left-half to the right-half of the complex plane. The stability of a nontrivial fixed point can be ascertained by investigating the eigenvalues of the Jacobian matrix of (5.2.54), (5.2.56), and (5.2.62) evaluated at this fixed point.

In Figure 5.2.3a, we show typical force-response curves for subcombination resonance of the first two modes when $\sigma = -1$. The trivial solution loses stability via a subcritical pitchfork bifurcation as the forcing amplitude is increased, resulting in a jump in the response amplitudes. On the other hand, as the forcing amplitude is decreased from a large value, the trivial solution loses stability through a supercritical

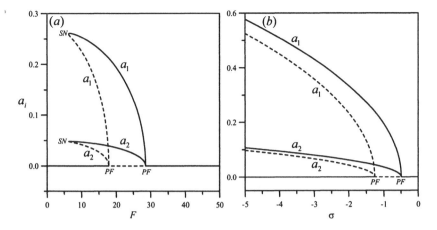

Fig. 5.2.3 Force- and Frequency-response curves for an additive-type subcombination resonance involving modes 1 and 2 when $\mu_1 = 0.0137$ and $\mu_2 = 0.0635$: (a) $\sigma = -1$ and (b) $F = 20$. Solid lines (———) denote stable fixed points and dashed lines (– – –) denote unstable fixed points.

pitchfork bifurcation, resulting in a gradual increase in the response amplitudes. In either case, the nontrivial solution loses stability as F is decreased via a saddle-node bifurcation.

In Figure 5.2.3b, we show typical frequency-response curves for the same resonance when $F = 20$. As in the case of combination parametric resonance, the curves are bent to the left indicating a softening-type nonlinearity. Because Γ_1 and Γ_2 are negative and proportional to F^2, the linear natural frequencies decrease appreciably with increasing F. Consequently, for $F = 20$, unlike the combination parametric resonance, subcombination resonance is activated only for negative values of σ. We also note that increasing the forcing amplitude causes both the stable and unstable branches to shift to the left, the latter more than the former.

In Figure 5.2.4a, we show the force-response curves for a subcombination resonance of the first and third modes when $\sigma = -1$. Comparing Figures 5.2.3a and 5.2.4a, we note that the force-response curves for the subcombination resonance of modes 1 and 3 are qualitatively different from the force-response curves for the subcombination resonance of modes 1 and 2. As in Figure 5.2.3a, the trivial solution in Figure 5.2.4a loses stability via a subcritical pitchfork bifurcation as F is increased, resulting in a jump in the response amplitudes. However, the nontrivial amplitudes in Figure 5.2.4a increase as F is increased. This is in contrast to Figure 5.2.3a, where the nontrivial amplitudes decrease as F is increased.

In Figure 5.2.4b, we show frequency-response curves for subcombination resonance of modes 1 and 3 when $F = 50$. Again the curves are bent to the left, indicating that the nonlinearity and the linear shift $\Gamma_1/2\omega_1$ decrease the frequency of the domi-

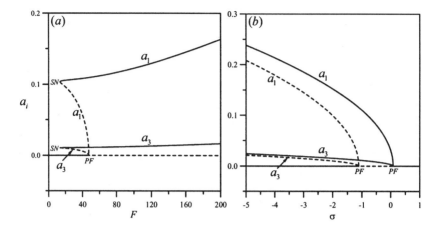

Fig. 5.2.4 Force- and Frequency-response curves for an additive-type subcombination resonance involving modes 1 and 3 when $\mu_1 = 0.0137$ and $\mu_3 = 0.076$: (a) $\sigma = -1$ and (b) $F = 50$. Solid lines (——) denote stable fixed points and dashed lines (- - -) denote unstable fixed points.

nant first mode. Furthermore, this case of subcombination resonance may be activated for positive as well as negative values of σ.

Of all of the cases mentioned in Table 5.1, subcombination resonance of modes 2 and 4 is qualitatively similar to subcombination resonance of modes 1 and 3, whereas the remaining cases are qualitatively similar to subcombination resonance of modes 1 and 2.

5.2.3 Nonplanar Dynamics

We consider the nonlinear response of a cantilever beam to a transverse harmonic excitation that is restricted to a plane, as shown in Figure 5.2.5, when the excitation frequency Ω^* is near the linear natural frequencies ω_{2n}^* and ω_{1m}^* of the nth inplane bending mode and the mth out-of-plane bending mode, respectively. The nondimensional governing equations and boundary conditions are given by (4.6.96)-(4.6.98) with $q_2 = F(s,t) = e\Omega^2 \cos(\Omega t)$. When the rotary inertia is negligible, these equations and boundary conditions can be derived from the following Lagrangian and virtual work term (Arafat, Nayfeh, and Chin, 1998):

$$\mathcal{L} = \int_0^1 \left\{ \frac{1}{2} \left[\frac{\partial}{\partial t} \int_0^s \frac{1}{2} \left(v'^2 + w'^2 \right) ds \right]^2 + \frac{1}{2} \dot{v}^2 + \frac{1}{2} \dot{w}^2 \right. \\
\left. - (1 - \beta_y) \left(v'' w'' \int_0^s v'' w' ds \right) - \frac{1}{2} \left(w''^2 + w'^2 w''^2 \right) \right.$$

TRANSVERSELY EXCITED CANTILEVER BEAMS

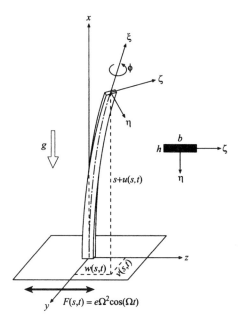

Fig. 5.2.5 A transversely excited cantilever beam. The xyz is an inertial reference frame, and the ξ, η, and ζ are the principal axes of the beam cross-section at position s, which are fixed on the cross-section.

$$-\frac{1}{2}\beta_y \left(v''^2 + v'^2 v''^2 + 2v'v''w'w''\right) - \frac{(1-\beta_y)^2}{2\beta_\gamma}\left[\left(\int_1^s v''w''ds\right)^2\right.$$

$$\left.+ \left(2v''w''\int_0^s \int_1^s v''w''dsds\right)\right]\bigg\}ds \tag{5.2.71}$$

$$\delta W = -\int_0^1 c\dot{v}\delta v ds - \int_0^1 [c\dot{w} - F(s,t)]\delta w ds \tag{5.2.72}$$

It follows from (4.6.96)-(4.6.98) that the nondimensional linear undamped inplane and out-of-plane natural frequencies ω_{2n} and ω_{1m} are given by

$$\omega_{2n} = z_n^2 \quad \text{and} \quad \omega_{1m} = z_m^2 \sqrt{\beta_y} \tag{5.2.73}$$

where the z_j are the roots of

$$\cos z \cosh z + 1 = 0 \tag{5.2.74}$$

The lowest five roots are 1.8751, 4.6941, 7.8548, 10.9955, and 14.1372. The corresponding mode shapes are given in (5.1.12). These modes are orthonormal so that

$$\int_0^1 \Phi_j(s)\Phi_k(s)ds = \delta_{jk} \tag{5.2.75}$$

Modulation Equations

We consider the case for which $\Omega \approx \omega_{2n} \approx \omega_{1m}$. To describe the nearness of Ω to ω_{2n}, we introduce the detuning parameter σ defined by

$$\Omega = \omega_{2n}\left(1 + \epsilon^2 \sigma\right) \tag{5.2.76}$$

To describe the nearness of ω_{1m} to ω_{2n}, we let

$$\beta_y = 1 + \delta_0 + \epsilon^2 \delta_2 \tag{5.2.77}$$

where

$$z_m^2 \sqrt{1 + \delta_0} = \omega_{2n} = z_n^2 \tag{5.2.78}$$

For a near-square cross-section, $\delta_0 = 0$ and hence $\omega_{1m} \approx \omega_{2m}$ for every m. Next, we determine a first-order uniform expansion of the solution of (4.6.96)-(4.6.98) by using the method of time-averaged Lagrangian and virtual work.

Because in the presence of damping all of the modes that are not directly or indirectly excited decay with time, the long-time dynamic behavior of the system can be described with the nth inplane and mth out-of-plane bending modes. Therefore, we seek a first-order uniform expansion in the form

$$v(s, T_0, T_2) = \epsilon \Phi_m(s) A_1(T_2) e^{i\omega_{1m} T_0} + cc + \cdots \tag{5.2.79}$$

$$w(s, T_0, T_2) = \epsilon \Phi_n(s) A_2(T_2) e^{i\omega_{2n} T_0} + cc + \cdots \tag{5.2.80}$$

and scale e as $\epsilon^3 e$ and c as $\epsilon^2 c$. This scaling reflects the fact that a very small primary-resonance excitation produces a large response. We substitute (5.2.79) and (5.2.80) into the Lagrangian (5.2.71) and virtual work (5.2.72), perform the spatial integrations, keep the slowly varying terms only, and obtain

$$\frac{<\mathcal{L}>}{\epsilon^4} = -\delta_2 z_m^2 A_1 \bar{A}_1 - i\omega_{1m}\left(A_1' \bar{A}_1 - A_1 \bar{A}_1'\right) - i\omega_{2n}\left(A_2' \bar{A}_2 - A_2 \bar{A}_2'\right)$$

$$-\frac{1}{2} S_{mm} A_1^2 \bar{A}_1^2 - \frac{1}{2} S_{nn} A_2^2 \bar{A}_2^2 - 4 S_{mn} A_1 \bar{A}_1 A_2 \bar{A}_2$$

$$-\Lambda_{mn}\left(A_1^2 \bar{A}_1^2 + \bar{A}_1^2 A_2^2\right) + \cdots \tag{5.2.81}$$

$$\frac{<\delta W>}{\epsilon^4} = Q_1^* \delta \bar{A}_1 + Q_2^* \delta \bar{A}_2 + cc$$

$$= -2i\omega_{2n}\mu_1 A_1 \delta \bar{A}_1 - \left(2i\omega_{2n}\mu_2 A_2 - f\omega_{2n}^2 e^{i\sigma\omega_{2n} T_0}\right)\delta \bar{A}_2 + cc \tag{5.2.82}$$

where the μ_i, S_{ij}, and Λ_{ij} are defined in Section 5.1.4 and

$$f = \frac{1}{2} e \Omega^2 \int_0^1 \Phi_n(s) ds \tag{5.2.83}$$

The numerical values of the coefficients in the averaged Lagrangian are given in Section 5.1.4 for combinations of the first three modes.

Using (5.2.81) and (5.2.82) in the extended Hamilton principle

$$\frac{d}{dT_2}\left(\frac{\partial <\mathcal{L}>}{\partial \bar{A}'_i}\right) - \frac{\partial <\mathcal{L}>}{\partial \bar{A}_i} = \bar{Q}^*_i$$

we obtain

$$2i\omega_{1m}(A'_1 + \mu_1 A_1) + \delta_2 z_m^2 A_1 + S_{mm}A_1^2 \bar{A}_1 + 4S_{mn}A_1 A_2 \bar{A}_2$$
$$+ 2\Lambda_{mn}A_2^2 \bar{A}_1 = 0 \qquad (5.2.84)$$
$$2i\omega_{2n}(A'_2 + \mu_2 A_2) + S_{nn}A_2^2 \bar{A}_2 + 4S_{mn}A_2 A_1 \bar{A}_1 + 2\Lambda_{mn}A_1^2 \bar{A}_2$$
$$+ f\omega_{2n}^2 e^{i\omega_{2n}\sigma T_2} = 0 \qquad (5.2.85)$$

Next, we express the A_k in the polar form

$$A_k = \frac{1}{2}a_k e^{i(\beta_k + \sigma\omega_{2n}T_2)} \qquad (5.2.86)$$

separate the real and imaginary parts in (5.2.84) and (5.2.85), and obtain

$$a'_1 = -\mu_1 a_1 + \frac{\Lambda_{mn}}{4\omega_{1m}}a_1 a_2^2 \sin[2(\beta_1 - \beta_2)] \qquad (5.2.87)$$

$$a_1\beta'_1 = -\left(\omega_{2n}\sigma - \frac{\delta_2 z_m^2}{2\omega_{1m}}\right)a_1 + \frac{S_{mm}}{8\omega_{1m}}a_1^3 + \frac{S_{mn}}{2\omega_{1m}}a_1 a_2^2$$
$$+ \frac{\Lambda_{mn}}{4\omega_{1m}}a_1 a_2^2 \cos[2(\beta_1 - \beta_2)] \qquad (5.2.88)$$

$$a'_2 = -\mu_2 a_2 - \frac{\Lambda_{mn}}{4\omega_{2n}}a_1^2 a_2 \sin[2(\beta_1 - \beta_2)] - f\omega_{2n}\sin(\beta_2) \qquad (5.2.89)$$

$$a_2\beta'_2 = -\omega_{2n}\sigma a_2 + \frac{S_{nn}}{8\omega_{2n}}a_2^3 + \frac{S_{mn}}{2\omega_{2n}}a_1^2 a_2 + \frac{\Lambda_{mn}}{4\omega_{2n}}a_1^2 a_2 \cos[2(\beta_1 - \beta_2)]$$
$$- f\omega_{2n}\cos(2\beta_2) \qquad (5.2.90)$$

Substituting (5.2.86) into (5.2.79) and (5.2.80) and using (5.2.76), we find, to the first approximation, that

$$v(s,t;\epsilon) = \epsilon a_1 \Phi_m(s) \cos(\Omega t + \beta_1) + \cdots \qquad (5.2.91)$$
$$w(s,t;\epsilon) = \epsilon a_2 \Phi_m(s) \cos(\Omega t + \beta_2) + \cdots \qquad (5.2.92)$$

where the a_i and β_i are given by (5.2.87)-(5.2.90). Except for the difference in notation, the modulation equations (5.2.87)-(5.2.90) are the same as those obtained by Crespo da Silva and Glynn (1978b) and Pai and Nayfeh (1990a).

Alternatively, we express the A_k in the Cartesian form

$$A_k = \frac{1}{2}(p_k - iq_k)e^{i\sigma\omega_{2n}T_2} \qquad (5.2.93)$$

separate the real and imaginary parts in (5.2.84) and (5.2.85), and obtain

$$p'_1 = -\mu_1 p_1 - \left(\omega_{2n}\sigma - \frac{\delta_2 z_m^2}{2\omega_{1m}}\right)q_1 - \frac{S_{mm}}{8\omega_{1m}}(p_1^2 + q_1^2)q_1 + \frac{S_{mn}}{2\omega_{1m}}(p_2^2 + q_2^2)q_1$$

$$-\frac{\Lambda_{mn}}{4\omega_{1m}}\left[\left(p_2^2-q_2^2\right)q_1-2p_1p_2q_2\right] \tag{5.2.94}$$

$$q_1' = -\mu_1 q_1 + \left(\omega_{2n}\sigma - \frac{\delta_2 z_m^2}{2\omega_{1m}}\right)p_1 + \frac{S_{mm}}{8\omega_{1m}}\left(p_1^2+q_1^2\right)p_1 - \frac{S_{mn}}{2\omega_{1m}}\left(p_2^2+q_2^2\right)p_1$$

$$-\frac{\Lambda_{mn}}{4\omega_{1m}}\left[\left(p_2^2-q_2^2\right)p_1+2q_1p_2q_2\right] \tag{5.2.95}$$

$$p_2' = -\mu_2 p_2 - \omega_{2n}\sigma q_2 - \frac{S_{nn}}{8\omega_{1n}}\left(p_2^2+q_2^2\right)q_2 + \frac{S_{mn}}{2\omega_{1n}}\left(p_1^2+q_1^2\right)q_2$$

$$-\frac{\Lambda_{mn}}{4\omega_{1n}}\left[\left(p_1^2-q_1^2\right)q_2-2p_1q_1p_2\right] \tag{5.2.96}$$

$$q_2' = -\mu_2 q_2 + \omega_{2n}\sigma p_2 + \frac{S_{nn}}{8\omega_{1n}}\left(p_2^2+q_2^2\right)p_2 - \frac{S_{mn}}{2\omega_{1n}}\left(p_1^2+q_1^2\right)p_2$$

$$-\frac{\Lambda_{mn}}{4\omega_{1n}}\left[\left(p_1^2-q_1^2\right)p_2+2p_1q_1q_2\right] - f\omega_{2n} \tag{5.2.97}$$

Except for the difference in notation, (5.2.94)-(5.2.97) are the same as those derived by Pai and Nayfeh (1990a).

Theoretical Results

We investigate the response of an almost square cross-section beam, which implies that the natural frequency of a given inplane mode is close to the natural frequency of

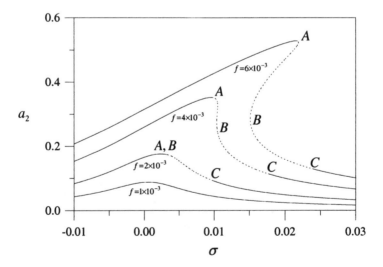

Fig. 5.2.6 Planar frequency-response curves of the first mode for a beam with an aspect ratio $b/h \approx 1.0$: mode (1,1), $\omega_{11} = \omega_{21}, \delta_2 = 0.0, \mu_j = 0.02$, and $\beta_\gamma = 0.6489$; (———) stable and (- - - -) unstable with at least one eigenvalue being positive.

the same out-of-plane mode. For all of the following results, we set Poisson's ratio $\nu = 0.3$ and assume that the cross-section is rectangular with width b and height h. From material mechanics and the theory of elasticity, we know that

$$D_\eta = \frac{1}{12}b^3 hE, \quad D_\zeta = \frac{1}{12}bh^3 E, \quad D_\xi = \frac{k_1 bh^3 E}{2(1+\nu)} \qquad (5.2.98)$$

so that

$$\beta_y = \frac{D_\zeta}{D_\eta} = \left(\frac{h}{b}\right)^2 \approx 1 + \delta_0, \quad \beta_\gamma = \frac{D_\xi}{D_\eta} = \frac{6k_1}{1+\nu}\left(\frac{h}{b}\right)^2 \qquad (5.2.99)$$

and k_1 can be calculated using equations (168) and (171) of Timoshenko and Goodier (1970).

It follows from (5.2.84) and (5.2.85) that the out-of-plane response $v(s, t)$ can only be excited indirectly through the autoparametric resonance and that the only directly excited motion is a planar motion in the z direction. The stability is investigated by assuming disturbances in both the y and z directions. For an almost square cross-section, $\delta_0 = 0.0$, $\beta_\gamma = 0.6489$, and $S_{mm} = S_{nn}$. Next, we present some of the results obtained by Pai and Nayfeh (1990a) and recalculated by Arafat (1999) for the first and second modes assuming that $\mu_1 = \mu_2 = 0.02$.

The planar frequency-response curves of the first mode with different amplitudes of the base excitation are shown in Figure 5.2.6. We note that when the amplitude of the base motion f is small, the response amplitude is small, single-valued, and stable, which corresponds to the linear solution. The phase angles are not constant. If the input frequency is away from the perfect resonant point $\sigma = 0$, β_2 is close to 0° for $\sigma < 0$ and 180° for $\sigma > 0$, as predicted by linear theory. As f increases, the response curves are bent to the right, implying that the geometric nonlinearity dominates the response and produces an effective hardening nonlinearity. It is well-known (Nayfeh and Mook, 1979) that, if there are only inplane disturbances, the branch AB is unstable but the branch BC is stable. Owing to out-of-plane disturbances (i.e., disturbances perpendicular to the direction of motion), the branch BC is unstable. Hence, the notion that nonlinearities are only important for large excitation levels is not true.

When we decrease the geometric detuning δ_2 from 0.0 to -0.01, the unstable frequency range of the planar response curve, which corresponds to nonplanar motions, is moved from AB in Figure 5.2.6 to AB in Figure 5.2.7. The reason is that, when $\delta_2 < 0.0$, $h < b$ and the beam is more flexible in the y direction than in the z direction. So the motion in the y direction can be easily excited autoparametrically, and the unstable region of the planar response curve shifts to the left and enlarges. When δ_2 is positive, the unstable region AB shifts to the right and narrows down because the beam is stiffer in the y direction than in the z direction. It follows from (5.2.89) and (5.2.90) that, when δ_2 changes, the magnitude of the planar response amplitude a_2 does not change because δ_2 appears only in (5.2.88).

As σ is increased from -0.01, the response is planar. As σ exceeds Point A, the planar responses undergoes a supercritical pitchfork bifurcation, resulting in a nonplanar response. As σ increases beyond C, the nonplanar response undergoes a saddle-node bifurcation, resulting in a jump into a planar response. As σ is decreased from 0.03, the response is planar. As σ decreases past B, the planar response undergoes a subcritical pitchfork bifurcation, resulting in a jump to a nonplanar response. As σ is decreased below A, the nonplanar response undergoes a reverse supercritical pitchfork bifurcation, resulting in a smooth transition to a planar response.

In Figure 5.2.8, we show the frequency-response curves of the second mode for three excitation levels. Clearly, the single-mode response curve is bent to the left, and hence the nonlinear inertia dominates the response and produces an effective softening nonlinearity. Again the branch BC is unstable due to out-of-plane disturbances, and the branch AB is unstable due to inplane disturbances. Furthermore, away from the perfect resonant point $\sigma = 0$, the phase angles of the left and right branches are close to $0°$ and $180°$, respectively.

In Figure 5.2.9, we show the frequency-response curves of the second mode when $\delta_2 = 0.002$ and $f = 0.00006$. In contrast with the results for the case $\delta_2 = 0$ in Figure 5.2.8, the response when $\delta_2 > 0$ undergoes Hopf bifurcations at H_1 and H_2, leading to the birth of small limit cycles. Moreover, there is a region where no stable equilibrium solutions exist. In this region, the beam response is expected to be either a periodically or a chaotically modulated motion.

Arafat (1999) studied in detail the beam dynamics in the region between H_1 and H_2 using long-time integration, a combination of a two-point boundary-value scheme and Newton's method, and Floquet theory. He found eleven branches of dynamic solutions.

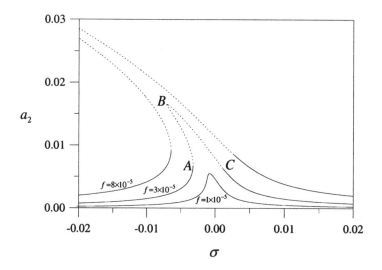

Fig. 5.2.7 Planar frequency-response curves of the second mode for a beam with an aspect ratio $b/h \approx 1.0$: mode (2,2), $\omega_{12} = \omega_{22}, \delta_2 = 0.0, \mu_j = 0.02$, and $\beta_\gamma = 0.6489$; (———) stable and (– – –) unstable with at least one eigenvalue being positive.

TRANSVERSELY EXCITED CANTILEVER BEAMS   311

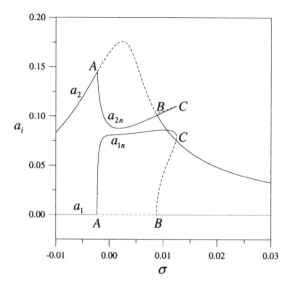

Fig. 5.2.8 Frequency-response curves of the first mode for a beam with an aspect ratio $b/h \approx 1.0$: mode (1,1), $\omega_{11} = \omega_{21}, \delta_2 = -0.01, \mu_j = 0.02, f = 0.002,$ and $\beta_\gamma = 0.6489$; a_2 corresponds to planar response amplitude; a_{1n} and a_{2n} correspond to nonplanar response amplitudes; (———) stable and (– – –) unstable with at least one eigenvalue being positive.

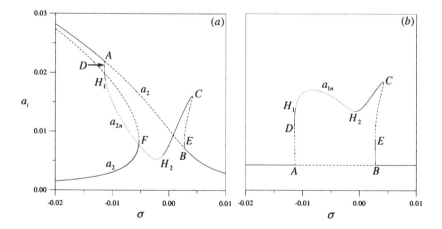

Fig. 5.2.9 Frequency-response curves of the second mode for a beam with an aspect ratio $b/h \approx 1.0$: mode (2,2), $\omega_{12} = \omega_{22}, \delta_2 = 0.002, \mu_j = 0.02, f = 0.00006,$ and $\beta_\gamma = 0.6489$; a_2 corresponds to planar response amplitudes; a_{1n} and a_{2n} correspond to nonplanar response amplitudes; (———) stable and (– – –) unstable with at least one eigenvalue being positive, and (· · ·) unstable with the real part of a complex conjugate pair of eigenvalues being positive.

Fig. 5.2.10 The shaker and base assembly (Zaretzky and Crespo da Silva, 1994).

Experiments

Zaretzky and Crespo da Silva (1994) conducted an experiment to determine the response of aluminum beams to primary-resonance base excitations. The experiments were conducted on four beams: two have rectangular cross-sections and two, except for imperfections, have square cross-sections. Here, we present some of their results on one of the beams with a square cross-section ($\delta_0 = 0$), namely, the beam with 70.5 in length and 1/4 in × 1/4 in cross-section. The signals were generated and monitored with a dynamic signal analyzer. The generated signals were amplified and fed to a long-stroke electrodynamic table-shaker, as shown in Figure 5.2.10. They used a clamp assembly in which the direction of the shaker travel coincides with a principal axis of the undeformed beam.

The motions of the beam and the shaker assembly were measured using three piezoelectric accelerometers. One was attached to the base assembly, as shown in Figure 5.2.10. The other two were mounted at a span location of 7 in from the beam base (i.e., $s = 7/70.5 \approx 0.099$), their locations are indicated by the points P_1 and P_2 in Figure 5.2.11. When the beam was stationary, the measurement axes of the accelerometers coincided with the two principal axes of the beam cross-section. Inspection of Figure 5.2.10 shows that the deflection $v(s,t)$ in the theory is the inplane deflection relative to the base, whereas the inplane accelerometer measures the total acceleration. Therefore, the signal from the base accelerometer was subtracted from that of the inplane accelerometer. Then, this signal as well as the signals from the base

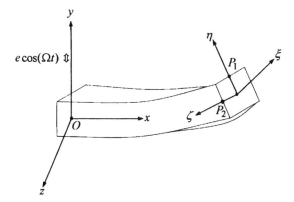

Fig. 5.2.11 Coordinate systems used in the formulation of Zaretzky and Crespo da Silva (1994).

accelerometer and the out-of-plane accelerometer were sent to the dynamic analyzer, which in turn calculated their power spectra.

As a first step, Zaretzky and Crespo da Silva (1994) estimated the beam natural frequencies and their corresponding damping coefficients. For a specific mode, they adjusted the shaker frequency until the beam deflection resembled the desired mode. Then, they turned off the shaker and secured by hand the shaker stops to prevent any interaction between the beam and the shaker. The signal from the inplane accelerometer was then recorded by the signal analyzer. Then, the linear damping coefficient was determined by using the method of logarithmic decrement, and the natural frequency was determined by counting the number of cycles and dividing it with the time interval. This procedure allowed the determination of the nondimensionalization factor $\sqrt{D_\eta/(mL^4)}$ used in the theory to be 2.79. Then, the dimensional natural frequencies of all of the other beam modes were estimated from

$$\omega_j^* = z_j^2 \sqrt{\frac{D_\eta}{mL^4}} (rad/sec) = \frac{z_j^2}{2\pi} \sqrt{\frac{D_\eta}{mL^4}} (\text{Hz}) \qquad (5.2.100)$$

The estimated natural frequencies of the second and third modes were 9.77 Hz and 27.4 Hz and their estimated linear damping coefficients were 0.09 and 0.11, respectively.

For a given base travel amplitude e^* (in mm), Zaretzky and Crespo da Silva (1994) generated frequency-response curves by sweeping forward and backward in a quasi-stationary manner the excitation frequency around the natural frequency of the mode of interest. Each time the excitation frequency was varied, they waited long enough for the beam motion to reach steady state. If the steady-state motion was periodic, the analyzer was used to calculate the FFTs of the inplane and out-of-plane readings and display the power spectra of the results. To maintain the base amplitude constant during the sweeps, they adjusted the output voltage of the dynamic analyzer to achieve

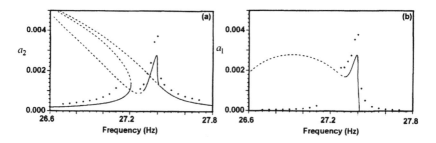

Fig. 5.2.12 The experimental and analytical frequency-response curves of the third mode for $2\hat{f} = 0.081$ ($e^* = 0.15\ mm$), $c = 0.11$, and $\delta_2 = 0$ (Zaretzky and Crespo da Silva, 1994).

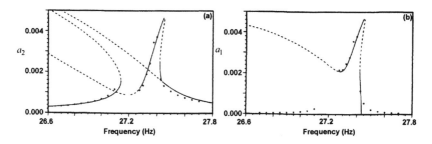

Fig. 5.2.13 The experimental and analytical frequency-response curves of the third mode for $2\hat{f} = 0.137$ ($e^* = 0.15\ mm$), $c = 0.11$, and $\delta_2 = 0$ (Zaretzky and Crespo da Silva, 1994).

a reading for the base accelerometer given by

$$V_p = 20 \log \left\{ \left[(2\pi\Omega^*)^2 e^* / 9810 \right] \times \text{cal} \right\} \tag{5.2.101}$$

where Ω^* is the excitation frequency in Hz and cal is the calibration factor in V_p/g of the base accelerometer.

Initially, they investigated the response of the third mode for a base oscillation amplitude e^* of $0.15\ mm$, which corresponds to $2f = 0.081$. It follows from (5.2.91) and (5.2.92) that

$$\ddot{v} \approx -\epsilon\Omega^2 a_1 \Phi_m(s) \cos(\Omega t + \beta_1) \tag{5.2.102}$$
$$\ddot{w} \approx -\epsilon\Omega^2 a_2 \Phi_n(s) \cos(\Omega t + \beta_2) \tag{5.2.103}$$

Hence, the amplitudes a_2 and a_1 are obtained by dividing the measured peak values of $\ddot{v}$ and $\ddot{w}$ by $\Omega^2 \Phi_n(0.099)$. In Figure 5.2.12, we display the theoretically and experimentally obtained inplane and out-of-plane response amplitudes defined by

$$\hat{a}_j = a_j \Phi_n(0.099) \tag{5.2.104}$$

for periodic motions as functions of the excitation frequency in Hz. The solid and dashed lines represent the stable and unstable theoretically obtained values and the

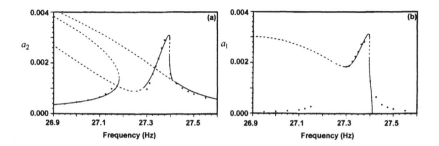

Fig. 5.2.14 The experimental and analytical frequency-response curves of the third mode for $2\hat{f} = 0.091$ ($e^* = 0.1\ mm$), $c = 0.11$, and $\delta_2 = 0$ (Zaretzky and Crespo da Silva, 1994).

solid circles represent the test results. It follows from Figure 5.2.12 that the theoretical results, although indicate the correct trends, underestimate the experimental results even for values of the excitation frequency where the linear theory is expected to hold. Zaretzky and Crespo da Silva (1994) found out that the culprit for this disagreement was the flexibility in the attachment of the base to the tracks, which resulted in a small rocking of the base about an axis normal to the translation axis. They modeled this phenomenon by assuming that the base was attached to a torsional spring of large stiffness K and developed the following expression for the effective excitation amplitude $\hat{f}$:

$$2\hat{f} \approx e\Omega^2 \left[\int_0^1 \Phi_n(s)ds + \frac{1}{2}K^{-1}\Omega^2 \int_0^1 s\Phi_n(s)ds \right]$$
$$\approx e\left(211 + \frac{10700}{K}\right) \quad \text{for mode 2}$$
$$\approx e\left(968 + \frac{235344}{K}\right) \quad \text{for mode 3} \qquad (5.2.105)$$

Using the experimentally obtained data in Figure 5.2.12, they estimated K to be 350, which they used in all of their subsequent correlations. For $e^* = 0.15\ mm$, they found that $2\hat{f} = 0.137$ and that the theoretical and experimental results are in good agreement over the whole frequency range, as shown in Figure 5.2.13.

Next, they decreased e^* to $0.1\ mm$ (corresponding to $2\hat{f} = 0.091$ with $K = 350$) and obtained the results shown in Figure 5.2.14. The agreement is again very good.

The theoretical results displayed in Figures 5.2.13 and 5.2.14 are for $\delta_2 = 0$. It is clear from Figures 5.2.6-5.2.9 that the frequency-response curves are very sensitive to deviations from squareness of the cross-section of the beam. In the present case, the frequency-response curves obtained for $\delta_2 = 0$ are the closest to the experimental results. This is not always the case, and one needs to choose the value of δ_2 that yields the best fit.

In Figure 5.2.15, we compare theoretically and experimentally obtained frequency-response curves for the second mode for a base travel amplitude $e^* = 0.7\ mm$ (which

316 DYNAMICS OF BEAMS

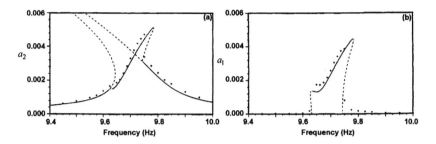

Fig. 5.2.15 The experimental and analytical frequency-response curves of the third mode for $2\hat{f} = 0.094$ ($e^* = 0.7\ mm$), $c = 0.09$, and $\delta_2 = 0.012$ (Zaretzky and Crespo da Silva, 1994).

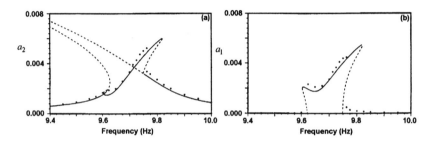

Fig. 5.2.16 The experimental and analytical frequency-response curves of the third mode for $2\hat{f} = 0.115$ ($e^* = 0.85\ mm$), $c = 0.09$, and $\delta_2 = 0.012$ (Zaretzky and Crespo da Silva, 1994).

corresponds to $2\hat{f} = 0.094$) and in Figure 5.2.16 for a base travel amplitude $e^* = 0.85\ mm$ (corresponding to $2\hat{f} = 0.115$). In both cases, Zaretzky and Crespo da Silva (1994) found that the best fit can be obtained for $\delta_2 = 0.012$. The theoretical and experimental results are in good agreement.

5.3 CLAMPED-CLAMPED BUCKLED BEAMS

In this section, we investigate the nonlinear transverse planar vibrations of a clamped-clamped buckled beam subject to a longitudinal compressive static load and transverse harmonic excitations. These vibrations are governed by (Nayfeh and Mook, 1979; Nayfeh, 2000)

$$m\frac{\partial^2 \hat{w}}{\partial \hat{t}^2} + \hat{P}\frac{\partial^2 \hat{w}}{\partial \hat{x}^2} + EI\frac{\partial^4 \hat{w}}{\partial \hat{x}^4} + \hat{c}\frac{\partial \hat{w}}{\partial \hat{t}} = \frac{EA}{2\ell}\frac{\partial^2 \hat{w}}{\partial \hat{x}^2}\int_0^L \left(\frac{\partial \hat{w}}{\partial \hat{x}}\right)^2 d\hat{x} + \hat{F}(\hat{x})\cos\hat{\Omega}\hat{t}$$

$$\hat{w} = 0 \text{ and } \frac{\partial \hat{w}}{\partial \hat{x}} = 0 \text{ at } \hat{x} = 0$$

$$\hat{w} = 0 \text{ and } \frac{\partial \hat{w}}{\partial \hat{x}} = 0 \text{ at } \hat{x} = L$$

where $\hat{w}(\hat{x}, \hat{t})$ is the transverse displacement, $\hat{P}$ is the axial load, and $\hat{F}$ is the spatial distribution of the transverse load. Using the nondimensional quantities

$$w = \frac{\hat{w}}{r}, \quad x = \frac{\hat{x}}{L}, \quad t = \hat{t}\sqrt{\frac{EI}{mL^4}}, \quad \Omega = \hat{\Omega}\sqrt{\frac{mL^4}{EI}}$$

where $r = \sqrt{I/A}$ is the radius of gyration of the cross-section, we rewrite the problem as

$$\ddot{w} + w^{iv} + Pw'' + c\dot{w} - \frac{1}{2}w''\int_0^1 w'^2 dx = F(x)\cos\Omega t \quad (5.3.1)$$

$$w = 0 \text{ and } w' = 0 \text{ at } x = 0 \quad (5.3.2)$$

$$w = 0 \text{ and } w' = 0 \text{ at } x = 1 \quad (5.3.3)$$

where

$$P = \frac{\hat{P}L^2}{EI}, \quad c = \frac{\hat{c}L^2}{\sqrt{mEI}}, \quad F = \frac{\hat{F}L^3}{EIr}$$

There are two approaches used in the literature to investigate the nonlinear vibrations of buckled beams. In the first approach, one attacks directly the equation governing large-amplitude transverse vibrations, but small curvatures, of a straight beam, which is subjected simultaneously to axial and lateral loads. This equation exhibits only a cubic nonlinearity. In the other approach, which is used in this chapter, one applies an axial load first until the beam buckles, and then studies the transverse vibrations around the buckled configurations. The equation governing the nonlinear vibrations of such a beam exhibits quadratic and cubic nonlinearities. A few researchers used the second approach.

The majority of the investigations use a single-mode approximation to discretize the equation of motion and associated boundary conditions. A few researchers investigated the effect of other modes on the total response. Burgreen (1951) investigated the free vibrations of a simply supported buckled beam using a single-mode discretization. He pointed out that the natural frequencies of the buckled beam depend on the amplitude of vibration. He obtained experimentally fairly good results to validate the theory. McDonald (1955) investigated the free vibrations of a simply supported buckled beam whose ends are axially constrained. He represented the response by a series of sinusoidal functions. He reported that the natural frequencies of the buckled beam change with the amplitude of vibration.

Eisley (1964a,b) used a one-mode discretization to investigate the forced vibrations of buckled beams and plates. He considered both simply supported and clamped-clamped boundary conditions. For a clamped-clamped buckled beam, he used the first buckling mode in the discretization process. He obtained similar forms of the governing equations for simply supported and clamped-clamped buckled beams. However, we show in this section that the equation of motion governing the nonlinear vibrations around the buckled configurations of a clamped-clamped beam, unlike that of a simply supported beam, exhibits a quadratic nonlinearity, which cannot be eliminated from the equation.

Holmes (1979) investigated the lateral vibrations around the buckled configuration of a simply supported beam. Using a Lyapunov function, he proved global stability of the response to small and large excitation amplitudes. He generalized his model to apply for any situation in which the undamped system possesses two stable equilibria separated by a saddle. Moon (1980) and Holmes and Moon (1983) investigated the chaotic motions of buckled beams under external harmonic excitations. They used a one-mode approximation to predict the onset of these chaotic motions. Abou-Rayan, Nayfeh, and Mook (1993) and Ramu, Sankar, and Ganesan (1994) investigated the nonlinear dynamics of a simply supported buckled beam using a single-mode approximation. They obtained period-doubling bifurcations and chaos. Abhyankar, Hall, and Hanagud (1993) investigated the nonlinear vibrations of a simply supported buckled beam under lateral harmonic excitation using a finite-difference method and a single-mode Galerkin discretization approach. They obtained a series of period-doubling bifurcations leading to chaos in both approaches.

Eisley and Bennett (1970) investigated the validity of using a single-mode approximation for a simply supported buckled beam by forcing one mode and determining the stability of the other modes in both the prebuckling and postbuckling regimes. They used the method of harmonic balance to determine the amplitude-frequency relations. They concluded that using a single-mode approximation is not valid in the region where higher modes are unstable. Min and Eisley (1972) investigated the free and forced vibrations of simply supported, axially restrained, buckled beams. They used a multimode discretization to investigate the stability of all modes retained in the discretization under the excitation of one mode. They considered two types of motions: one-side motions and snap-through motions.

Tang and Dowell (1988) extended the analysis of Moon (1980) and Holmes and Moon (1983). They investigated the effect of higher modes on the chaotic oscillations of a buckled beam under external excitations. They reported that it is generally necessary to consider the effects of higher modes on the response. They obtained good quantitative agreement between theory and experiment when a sufficient number of modes is included in the theoretical model. Smelova-Reynolds and Dowell (1996a,b) studied the chaotic motion of a simply supported buckled beam under a harmonic excitation using a multimode Galerkin discretization. They noted the importance of higher modes in determining the onset of chaos. For the range they considered, the contributions of the fifth and higher modes are insignificant.

Lestari and Hanagud (2001) studied the nonlinear vibrations of buckled beams with elastic end constraints. The beam is subjected simultaneously to axial and lateral loads without first statically buckling the beam. They used a single-mode approximation and obtained a closed-form solution of the free vibration problem in terms of elliptic functions. They also analyzed the nonlinear vibrations of a buckled beam with elastic constraints using a Fourier sine series.

Tseng and Dugundji (1971) considered a two-mode approximation for a clamped-clamped beam using a linear combination of the first two linear buckled modes. It is important to note that the second mode, which is antisymmetric, does not contribute to the response unless it is parametrically excited by the first mode through an internal resonance. Away from the crossover region, at which the two frequencies of the first and second modes are close to each other, the result is similar to that obtained using a single-mode approximation.

Ji and Hansen (2000) investigated experimentally the postbuckling behavior of a clamped-sliding beam subjected to an axial harmonic excitation. They used a one-mode approximation in their model and obtained a nonlinear oscillator with parametric excitation. Both fundamental and subharmonic resonances were considered. In the case of subharmonic resonance, they obtained a series of period-doubling bifurcations leading to chaos. In the case of primary resonance, they observed windows of period-three and period-six motions embedded within the chaotic region. Furthermore, they observed period-demultiplying bifurcations coming out of chaos.

Afaneh and Ibrahim (1992) investigated analytically and experimentally the nonlinear response of a clamped-clamped buckled beam near a one-to-one internal resonance. They used the method of multiple scales to derive the equations governing the amplitudes and phases of the response. They reported an energy transfer from the first mode, which is externally excited, to the second mode, which is indirectly excited via the internal resonance.

Kreider (1995) and Kreider and Nayfeh (1998) investigated the nonlinear vibrations of a clamped-clamped buckled beam experimentally and theoretically. They used a single-mode approximation in the theory. In one set of the experiments, they observed a sequence of period-doubling bifurcations of the local attractors (limit cycles) culminating in chaos before snapping through. These responses can be predicted with a single-mode discretization. In another set, they observed responses whose FFTs correspond to period-three and period-five motions. Because these responses do not exist within narrow bands embedded within chaos, they cannot be predicted with a single-mode discretization. In a third set, they observed responses whose FFTs consist of peaks corresponding to the excitation frequency and its harmonics as well as sidebands around these peaks, which seem to be incommensurate with the excitation frequency; because these sidebands cannot be predicted by their single-mode discretization, they called them *unexplained sidebands*. In this section, we follow Emam (2002) and Emam and Nayfeh (2004a,b) and show that the latter responses represent quasiperiodic motions, which result from a secondary Hopf bifurcation that cannot be predicted using a single-mode approximation.

In this section, we study the local and global dynamics using both the direct and discretization approaches. The Galerkin procedure is used to discretize the governing integral-partial-differential equation and associated boundary conditions using the symmetric linear vibration modes of the buckled beam as trial functions. As a result, we obtain a set of nonlinearly coupled second-order ordinary-differential equations in time only, which possess quadratic and cubic nonlinearities. Emam and Nayfeh (2002, 2004a) solved for the fixed points of the discretized set. They found out that a single-mode approximation is valid only for limited buckling levels, and that at least a four-mode approximation is required to accurately predict the buckling configurations and the dynamic responses. We treat the buckling problem in Section 5.3.1, the linear vibration modes and frequencies in Section 5.3.2, the nonlinear local dynamics of the buckled beam to a primary-resonance excitation in Sections 5.3.3-5.3.5, and the global dynamics in Section 5.3.6.

5.3.1 Buckling Problem

The buckling problem can be obtained from (5.3.1)-(5.3.3) by dropping the time derivatives and the dynamic load. The result is

$$\psi^{iv} + P\psi'' - \frac{1}{2}\psi'' \int_0^1 \psi'^2 dx = 0 \tag{5.3.4}$$

subject to the boundary conditions

$$\psi = 0 \text{ and } \psi' = 0 \text{ at } x = 0 \tag{5.3.5}$$
$$\psi = 0 \text{ and } \psi' = 0 \text{ at } x = 1 \tag{5.3.6}$$

To determine the exact solution to the nonlinear buckling problem, we note that the integral in (5.3.4) is a constant for a given value of ψ, and hence we let

$$Q = \int_0^1 \psi'^2 dx \tag{5.3.7}$$

where Q is a constant. As a result, (5.3.4) becomes

$$\psi^{iv} + \lambda \psi'' = 0 \tag{5.3.8}$$

where $\lambda = P - \frac{1}{2}Q$. The general solution of (5.3.8) is given by

$$\psi = c_1 + c_2 x + c_3 \cos \sqrt{\lambda} x + c_4 \sin \sqrt{\lambda} x \tag{5.3.9}$$

where the c_i are constants. Substituting (5.3.9) into (5.3.5) and (5.3.6) yields four algebraic equations in the c_i as follows:

$$c_1 + c_3 = 0 \tag{5.3.10}$$
$$c_2 + \sqrt{\lambda} c_4 = 0 \tag{5.3.11}$$
$$c_1 + c_2 + c_3 \cos \sqrt{\lambda} + c_4 \sin \sqrt{\lambda} = 0 \tag{5.3.12}$$
$$c_2 - \sqrt{\lambda} c_3 \sin \sqrt{\lambda} + c_4 \sqrt{\lambda} \cos \sqrt{\lambda} = 0 \tag{5.3.13}$$

This system of equations represents an eigenvalue problem for λ. Equating the determinant of the coefficient matrix of these equations to zero yields the characteristic equation

$$2 - 2\cos \sqrt{\lambda} - \sqrt{\lambda} \sin \sqrt{\lambda} = 0 \tag{5.3.14}$$

For asymmetric modes, the first three roots of (5.3.14) are 80.763, 238.718, and 355.306. For symmetric modes, the solutions of (5.3.14) are $\lambda = 4n^2\pi^2$, where n is an integer. Then, it follows from (5.3.10)-(5.3.13) that $c_2 = c_4 = 0$ and $c_1 = -c_3 = \frac{1}{2}b_n$, where b_n is the rise at the midsection of the nth buckling mode. Hence, for symmetric modes, (5.3.9) becomes

$$\psi = \frac{1}{2}b_n[1 - \cos(2n\pi x)] \tag{5.3.15}$$

To determine b_n for a given load P, we substitute (5.3.15) into (5.3.7) and obtain

$$Q = \frac{1}{2}b_n^2 n^2 \pi^2 \qquad (5.3.16)$$

Recalling that $\lambda = P - \frac{1}{2}Q$ and $\lambda = 4n^2\pi^2$, we relate the buckling rise to the axial load as

$$b_n^2 = 4(P - 4n^2\pi^2)/n^2\pi^2 \qquad (5.3.17)$$

To determine the problem governing nonlinear vibrations around the first buckled configuration, we let

$$w(x,t) = \frac{1}{2}b[1 - \cos(2\pi x)] + u(x,t) \qquad (5.3.18)$$

in (5.3.1)-(5.3.3) and obtain

$$\ddot{u} + u^{iv} + 4\pi^2 u'' - 2b^2\pi^3 \cos(2\pi x) \int_0^1 u' \sin(2\pi x)dx$$
$$= -c\dot{u} + b\pi u'' \int_0^1 u' \sin(2\pi x)dx + b\pi^2 \cos(2\pi x) \int_0^1 u'^2 dx$$
$$+ \frac{1}{2}u'' \int_0^1 u'^2 dx + F(x)\cos(\Omega t) \qquad (5.3.19)$$

$$u = 0 \text{ and } u' = 0 \text{ at } x = 0 \text{ and } x = 1 \qquad (5.3.20)$$

5.3.2 Linear Vibration Problem

We follow Nayfeh, Kreider, and Anderson (1995) and obtain an exact solution for the natural frequencies and corresponding mode shapes of the linear vibrations of a buckled beam. The natural frequencies and mode shapes can be obtained by dropping the nonlinear, damping, and forcing terms from (5.3.19); that is,

$$\ddot{u} + u^{iv} + 4\pi^2 u'' - 2b^2\pi^3 \cos(2\pi x) \int_0^1 u' \sin(2\pi x)dx = 0 \qquad (5.3.21)$$

Next, we let

$$u(x,t) = \phi(x)e^{i\omega t} \qquad (5.3.22)$$

in (5.3.20) and (5.3.21) and obtain

$$\phi^{iv} + 4\pi^2 \phi'' - \omega^2 \phi - 2b^2\pi^3 \cos(2\pi x) \int_0^1 \phi' \sin(2\pi x)dx = 0 \qquad (5.3.23)$$

$$\phi = 0 \text{ and } \phi' = 0 \text{ at } x = 0 \text{ and } x = 1 \qquad (5.3.24)$$

Equation (5.3.23) can be rewritten as

$$\phi^{iv} + 4\pi^2 \phi'' - \omega^2 \phi = 2\Gamma b^2 \pi^3 \cos(2\pi x) \qquad (5.3.25)$$

where
$$\Gamma = \int_0^1 \phi' \sin(2\pi x) dx \tag{5.3.26}$$
is a constant for a given ϕ.

The general solution of (5.3.25) is the superposition of a particular solution ϕ_p and a homogeneous solution ϕ_h; that is,
$$\phi = \phi_h + \phi_p \tag{5.3.27}$$
The homogenous solution of (5.3.25) is given by
$$\phi_h = c_1 \sin s_1 x + c_2 \cos s_1 x + c_3 \sinh s_2 x + c_4 \cosh s_2 x \tag{5.3.28}$$
where the c_i are constants and
$$s_{1,2} = \left(\pm 2\pi^2 + \sqrt{4\pi^4 + \omega^2} \right)^{\frac{1}{2}} \tag{5.3.29}$$
And a particular solution of (5.3.25) has the form
$$\phi_p = c_5 \cos(2\pi x) \tag{5.3.30}$$
Substituting (5.3.27) into (5.3.23) and using (5.3.30) yields
$$(2b^2 \pi^4 - \omega^2) c_5 = 2b^2 \pi^3 \int_0^1 \phi_h'(\sin 2\pi x) \, dx \tag{5.3.31}$$

There are two possibilities: either the integrand in (5.3.31) is equal or not equal to zero. When it is zero,
$$(2b^2 \pi^4 - \omega^2) c_5 = 0 \tag{5.3.32}$$
which implies that $c_5 = 0$ since, in general, $\omega^2 \neq 2b^2 \pi^4$, and hence the mode shape is given by the homogeneous solution. This means that these mode shapes and corresponding natural frequencies do not depend on the buckling level b, they are antisymmetric modes.

When the integrand is different from zero, the general solution of (5.3.23) is given by
$$\phi = c_1 \sin s_1 x + c_2 \cos s_1 x + c_3 \sinh s_2 x + c_4 \cosh s_2 x + c_5 \cos 2\pi x \tag{5.3.33}$$
Substituting (5.3.33) into (5.3.23) and (5.3.24), we obtain
$$c_2 + c_4 + c_5 = 0 \tag{5.3.34}$$
$$s_1 c_1 + s_2 c_3 = 0 \tag{5.3.35}$$
$$c_1 \sin s_1 + c_2 \cos s_1 + c_3 \sinh s_2 + c_4 \cosh s_2 = 0 \tag{5.3.36}$$
$$c_1 s_1 \sin s_1 + c_2 s_1 \cos s_1 + c_3 s_2 \sinh s_2 + c_4 s_2 \cosh s_2 = 0 \tag{5.3.37}$$
$$\frac{s_1(\cos s_1 - 1)}{s_1^2 - 4\pi^2} c_1 - \frac{s_1 \sin s_1}{s_1^2 - 4\pi^2} c_2 + \frac{s_2(1 - \cosh s_2)}{s_2^2 + 4\pi^2} c_3 - \frac{s_2 \sinh s_2}{s_2^2 + 4\pi^2} c_4 + \frac{\omega^2 - 2b^2 \pi^4}{4b^2 \pi^4} c_5 = 0 \tag{5.3.38}$$

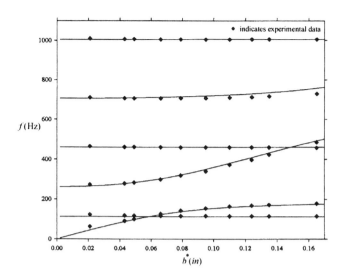

Fig. 5.3.1 A comparison of the analytically and experimentally obtained natural frequencies for clamped-clamped buckled beams (Nayfeh, Kreider, and Anderson, 1995).

Equations (5.3.34)-(5.3.38) represent a system of five algebraic equations in the c_i and the natural frequency ω. Setting the determinant of the coefficient matrix of these equations equal to zero yields an equation for the natural frequency ω. Solving this equation, we obtain the natural frequencies and corresponding mode shapes of the symmetric modes of the linear vibrations of a buckled beam for any buckling level b.

In Figure 5.3.1, we show variation of the analytically and experimentally obtained lowest six natural frequencies (Nayfeh, Kreider, and Anderson, 1995) with the dimensional buckling level b^* when $L = 10.991\ in$, $E = 30 \times 10^6\ psi$, $d = 0.75\ in$, $\rho = 0.282\ lbm/in^3$, and thickness $= 0.0327\ in$. Clearly, the obtained natural frequencies are in excellent agreement with the experimental results. Inspecting Figure 5.3.1, we note that the buckled beam possesses a few internal resonances. Although the second natural frequency is approximately equal to twice the first natural frequency when $b^* = b_1^* \approx 0.027\ in$, a two-to-one internal resonance cannot be activated because the interaction coefficient identically vanish (Chin, Nayfeh, and Lacarbonara, 1997). One-to-one internal resonances between the first and second modes and between the third and fourth modes may be activated when b^* is near $b_2^* \approx 0.059\ in$ and $b_4^* \approx 0.149\ in$, respectively. A three-to-one internal resonance between the third and first modes may be activated when b^* is near $b_5^* \approx 0.189\ in$. Although the frequencies of the third and second modes are approximately in the ratio of three-to-one when $b_3^* \approx 0.087\ in$, a three-to-one internal resonance cannot be activated because the interaction coefficients identically vanish (Nayfeh, Lacarbonara, and Chin, 1999). Also, the fourth and second modes may be activated simultaneously for all buckling levels when the excitation frequency $\Omega \approx 2\omega_2$ because $\omega_4 \approx 4\omega_2$; the second mode will be activated by a subharmonic resonance of order one-half while the fourth mode will be activated by a superharmonic resonance of order two. Nayfeh (2000) considered three-to-one internal resonances when one of the interacting modes is excited by a primary resonance. In the next section, we consider the nonlinear response of buckled beams to primary-resonance excitations in the absence of internal resonances.

5.3.3 Nonlinear Local Vibrations - Direct Approach

Lacarbonara, Nayfeh, and Kreider (1998) used two approaches for calculating the frequency-response curves of a buckled beam in the case of primary resonance of its fundamental vibration mode. In the first approach (direct approach), they attacked directly (5.3.19) and (5.3.20) by using the method of multiple scales. In the second approach (discretization approach), they used the Galerkin method to reduce (5.3.19) and (5.3.20) into a single nonlinear second-order ordinary-differential equation. Then, they found an approximate solution of this equation by using the method of multiple scales. They found qualitative and quantitative discrepancies between the results obtained with both approaches. They also determined experimentally frequency-response curves and found them to be in agreement with those obtained with the direct approach. Lacarbonara (1998) determined the frequency-response curves using a multimode Galerkin discretization and found out that the results converge to those obtained with the direct approach as the number of modes retained in the approximation increases.

We use the method of multiple scales to determine a second-order uniform solution of (5.3.19) and (5.3.20) when the nth mode is excited with a primary-resonance excitation and is not involved in any internal resonance with any other mode. We seek a second-order uniform expansion in the form

$$u(x, t; \epsilon) = \epsilon u_1(x, T_0, T_2) + \epsilon^2 u_2(x, T_0, T_2) + \epsilon^3 u_3(x, T_0, T_2) + \cdots \quad (5.3.39)$$

We note that $u(x, t)$ does not depend on T_1 because secular terms arise at $O(\epsilon^3)$. In order that the nonlinearity balance the effects of damping and excitation, we scale them so that they appear in the same modulation equations. Thus, we replace c and F with $\epsilon^2 c$ and $\epsilon^3 F$, respectively.

Substituting (5.3.39) into (5.3.19) and (5.3.20) and equating coefficients of like powers of ϵ, we obtain

Order ϵ:

$$\mathcal{L}(u_1) = D_0^2 u_1 + u_1^{iv} + 4\pi^2 u_1'' - 2b^2 \pi^3 \cos(2\pi x) \int_0^1 u_1' \sin(2\pi x) dx = 0 \quad (5.3.40)$$

Order ϵ^2:

$$\mathcal{L}(u_2) = b\pi u_1'' \int_0^1 u_1' \sin(2\pi x) dx + b\pi^2 \cos(2\pi x) \int_0^1 u_1'^2 dx \quad (5.3.41)$$

Order ϵ^3:

$$\mathcal{L}(u_3) = -2D_0 D_2 u_1 - c D_0 u_1 + 2b\pi^2 \cos(2\pi x) \int_0^1 u_1' u_2' dx$$
$$+ b\pi \left[u_1'' \int_0^1 u_2' \sin(2\pi x) dx + u_2'' \int_0^1 u_1' \sin(2\pi x) dx \right]$$
$$+ \frac{1}{2} u_1'' \int_0^1 u_1'^2 dx + F(x) \cos(\Omega t) \quad (5.3.42)$$

$$u_i = 0 \text{ and } u_i' = 0 \text{ at } x = 0 \text{ and } x = 1 \text{ for } i = 1, 2, 3 \quad (5.3.43)$$

Because in the presence of damping all modes that are not directly or indirectly excited decay with time (Nayfeh and Mook, 1979), the solution of the first-order problem is assumed to consist of the nth mode; that is,

$$u_1 = A_n(T_2) \phi_n(x) e^{i\omega_n T_0} + cc \quad (5.3.44)$$

Substituting (5.3.44) into (5.3.41) yields

$$\mathcal{L}(u_2) = bh(x) A_n^2 e^{2i\omega_n T_0} + bh(x) A_n \bar{A}_n + cc \quad (5.3.45)$$

where

$$h(x) = \pi \phi_n'' \int_0^1 \phi_n' \sin(2\pi x) dx + \pi^2 (\cos 2\pi x) \int_0^1 \phi_n'^2 dx \quad (5.3.46)$$

A particular solution of (5.3.45) and (5.3.43) can be expressed as

$$u_2 = b \Psi_2(x) A_n^2 e^{2i\omega_n T_0} + b \Psi_1(x) A_n \bar{A}_n + cc \quad (5.3.47)$$

where the Ψ_i are given by

$$\Psi_1^{iv} + 4\pi^2 \Psi_1'' - 2b^2 \pi^3 \cos(2\pi x) \int_0^1 \Psi_1' \sin(2\pi x) dx = h(x) \quad (5.3.48)$$

$$\Psi_2^{iv} + 4\pi^2 \Psi_2'' - 4\omega_n^2 \Psi_2 - 2b^2 \pi^3 \cos(2\pi x) \int_0^1 \Psi_2' \sin(2\pi x) dx = h(x) \quad (5.3.49)$$

The boundary conditions on the Ψ_i are

$$\Psi_i = 0 \text{ and } \Psi_i' = 0 \text{ at } x = 0 \text{ and } x = 1 \quad (5.3.50)$$

Substituting (5.3.44) and (5.3.47) into (5.3.42) yields

$$\mathcal{L}(u_3) = \left[-i\omega_n (2 A_n' + c A_n) \phi_n + \chi A_n^2 \bar{A}_n \right] e^{i\omega_n T_0} + \frac{1}{2} F e^{i\Omega T_0} + cc + NST \quad (5.3.51)$$

where the prime indicates the derivative with respect to T_2 and

$$\chi = b^2\pi\left[(\Psi_2'' + 2\Psi_1'')\int_0^1 \phi_n'\sin(2\pi x)dx + \phi_n''\int_0^1 (\Psi_2' + 2\Psi_1')\sin(2\pi x)dx\right]$$
$$+2b^2\pi^2\cos(2\pi x)\int_0^1 (\Psi_2' + 2\Psi_1')\phi_n' dx + \frac{3}{2}\phi_n''\int_0^1 \phi_n'^2 dx \qquad (5.3.52)$$

To describe the nearness of Ω to ω_n, we introduce the detuning parameter σ defined by

$$\Omega = \omega_n + \epsilon^2\sigma \qquad (5.3.53)$$

Because the homogeneous problem governing u_3 has a nontrivial solution, the corresponding nonhomogeneous problem has a solution only if the right-hand side of (5.3.51) is orthogonal to every solution of the adjoint homogeneous problem governing u_3. Since the latter problem is self-adjoint, the adjoints are given by $\phi_j(x)\exp(\pm i\omega_j T_0)$. Demanding that the right-hand side of (5.3.51) be orthogonal to $\phi_n(x)\exp(-i\omega_n T_0)$ and using equations (5.3.53), we obtain the solvability condition

$$2i\omega_n\left(A_n' + \mu_n A_n\right) + 8\omega_n\alpha_e A_n^2 \bar{A}_n - \frac{1}{2}f_n e^{i\sigma T_2} = 0 \qquad (5.3.54)$$

where

$$f_n = \int_0^1 F(x)\phi_n(x)dx, \quad 2\mu_n = \int_0^1 c\phi_n^2(x)dx \qquad (5.3.55)$$

$$8\omega_n\alpha_e = -\int_0^1 \chi\phi_n dx$$
$$= \frac{3}{2}\left(\int_0^1 \phi_n'^2 dx\right)^2 - b^2\pi\left(\int_0^1 \phi_n'^2 dx\right)\left[\int_0^1 (\Psi_2' + 2\Psi_1')\sin(2\pi x)dx\right]$$
$$-2b^2\pi\left[\int_0^1 \phi_n'\sin(2\pi x)dx\right]\left[\int_0^1 (\Psi_2' + 2\Psi_1')\phi_n' dx\right] \qquad (5.3.56)$$

Substituting the polar form

$$A_n = \frac{1}{2}a_n e^{i(\sigma T_2 - \gamma_n)} \qquad (5.3.57)$$

into (5.3.54), multiplying the result by $\exp\left[i\left(\gamma_n - \sigma T_2\right)\right]$, and separating real and imaginary parts, we obtain

$$a_n' = -\mu_n a_n + \frac{f_n}{2\omega_n}\sin\gamma_n \qquad (5.3.58)$$

$$a_n\gamma_n' = \sigma a_n - \alpha_e a_n^3 + \frac{f_n}{2\omega_n}\cos\gamma_n \qquad (5.3.59)$$

Substituting (5.3.57) into (5.3.44) and (5.3.47) and then substituting the results into (5.3.39), we find that the beam response, to second order, is given by

$$u(x,t;\epsilon) \approx \epsilon a_n \phi_n(x) \cos{(\Omega t - \gamma_n)} + \frac{1}{2}\epsilon^2 b a_n^2 \left[\Psi_2(x) \cos{(2\Omega t - 2\gamma_n)} + \Psi_1(x)\right]$$
(5.3.60)

where a_n and γ_n are given by (5.3.58) and (5.3.59).

5.3.4 Nonlinear Local Vibrations - Discretization Approach

We start by using the Galerkin procedure to discretize (5.3.19) and (5.3.20) into a set of nonlinearly coupled second-order ordinary-differential equations in time. To this end, we assume that

$$u(x,t) = \sum_{n=1}^{\infty} \phi_n(x) q_n(t)$$
(5.3.61)

where the $q_n(t)$ are generalized coordinates and the $\phi_n(x)$ are the linear vibration mode shapes, which are normalized such that

$$\int_0^1 \phi_i(x)\phi_j(x)\, dx = \delta_{ij}$$
(5.3.62)

Substituting (5.3.61) into (5.3.19), multiplying the result by $\phi_m(x)$, then integrating the outcome over the domain, and using the orthonormality condition (5.3.62) yields

$$\ddot{q}_m + \omega_m^2 q_m = -2\mu_m \dot{q}_m + b\sum_{i,j}^{\infty} \Gamma_{mij} q_i q_j + \sum_{i,j,k}^{\infty} \Lambda_{mijk} q_i q_j q_k$$
$$+ f_m \cos(\Omega t), \qquad m = 1, 2, 3, \ldots$$
(5.3.63)

where $2\mu_m = \int_0^1 c\phi_m^2 dx$,

$$\Gamma_{mij} = \pi^2 \left(\int_0^1 \cos(2\pi x)\phi_m\, dx\right)\left(\int_0^1 \phi_i'\phi_j'\, dx\right)$$
$$+ \pi \left(\int_0^1 \sin(2\pi x)\phi_j'\, dx\right)\left(\int_0^1 \phi_i''\phi_m\, dx\right)$$
(5.3.64)

$$\Lambda_{mijk} = \frac{1}{2}\left(\int_0^1 \phi_i''\phi_m\, dx\right)\left(\int_0^1 \phi_j'\phi_k'\, dx\right)$$
(5.3.65)

are the coefficients of the quadratic and cubic nonlinearities in the discretized equations, and

$$f_m = \int_0^1 F(x)\phi_m(x)\, dx$$
(5.3.66)

is the projection of the distributed force $F(x)$ onto the mth mode.

328 DYNAMICS OF BEAMS

Next, we use these discretized equations and the method of multiple scales to investigate the nonlinear response of a buckled beam to a primary-resonance excitation of its nth vibration mode. We seek an approximate solution of (5.3.63) in the form

$$q_n(t;\epsilon) = \epsilon q_{n1}(T_0, T_2) + \epsilon^2 q_{n2}(T_0, T_2) + \epsilon^3 q_{n3}(T_0, T_2) + \cdots \quad (5.3.67)$$

$$q_m(t;\epsilon) = \epsilon^2 q_{m2}(T_0, T_2) + \epsilon^3 q_{m3}(T_0, T_2) + \cdots \quad \text{for} \quad m \neq n \quad (5.3.68)$$

Since the beam is subjected to a primary resonance of its nth vibration mode, we assume that this mode is of lower order than the other modes. Substituting (5.3.67) and (5.3.68) into (5.3.63) and equating the coefficients of like powers of ϵ, we obtain

Order ϵ:

$$D_0^2 q_{n1} + \omega_1^2 q_{n1} = 0 \quad (5.3.69)$$

Order ϵ^2:

$$D_0^2 q_{m2} + \omega_m^2 q_{m2} = b\Gamma_{mnn} q_{n1}^2 \quad (5.3.70)$$

Order ϵ^3:

$$D_0^2 q_{m3} + \omega_m^2 q_{m3} = -2D_0 D_2 q_{n1} - 2\mu_n D_0 q_{n1} + b \sum_{i=1}^{\infty} \Gamma_{mni} q_{n1} q_{i2}$$

$$+ b \sum_{j=1}^{\infty} \Gamma_{mjn} q_{n1} q_{j2} + \Lambda_{mnnn} q_{n1}^3 + f_m \cos \Omega T_0 \quad (5.3.71)$$

The solution of (5.3.69) can be expressed as

$$q_{n1} = A_n(T_2) e^{i\omega_n T_0} + cc \quad (5.3.72)$$

Substituting (5.3.72) into (5.3.70) yields

$$D_0^2 q_{m2} + \omega_m^2 q_{m2} = b\Gamma_{mnn}(A_n^2 e^{2i\omega_n T_0} + A_n \bar{A}_n) + cc \quad (5.3.73)$$

A particular solution of (5.3.73) can be expressed as

$$q_{m2} = b\Gamma_{mnn}\left(\frac{A_n \bar{A}_n}{\omega_m^2} + \frac{A_n^2}{\omega_m^2 - 4\omega_n^2} e^{2i\omega_n T_0} + cc\right) \quad (5.3.74)$$

Substituting (5.3.72) and (5.3.74) into (5.3.71), we obtain

$$D_0^2 q_{m3} + \omega_m^2 q_{m3} = -2i\omega_n \left(A_n' + \mu_n A_n\right) e^{i\omega_n T_0} + 3\Lambda_{mnnn} A_n^2 \bar{A}_n e^{i\omega_n T_0}$$

$$+ b^2 A_n^2 \bar{A}_n \sum_{i=1}^{\infty} \Gamma_{mni} \Gamma_{inn} \left[\frac{2}{\omega_i^2} + \frac{1}{\omega_i^2 - 4\omega_n^2}\right] e^{i\omega_n T_0}$$

$$+ b^2 A_n^2 \bar{A}_n \sum_{j=1}^{\infty} \Gamma_{mjn} \Gamma_{jnn} \left[\frac{2}{\omega_j^2} + \frac{1}{\omega_j^2 - 4\omega_n^2}\right] e^{i\omega_n T_0}$$

$$+ \frac{1}{2} f_m e^{i\Omega T_0} + cc + NST \quad (5.3.75)$$

Using (5.3.53) and eliminating the secular producing terms in (5.3.75), we obtain

$$2i\omega_n(A'_n + \mu_n A_n) + 8\omega_n \alpha_e A_n^2 \bar{A}_n - \frac{1}{2}f_n e^{i\sigma T_2} = 0 \quad (5.3.76)$$

where the effective nonlinearity α_e is given by

$$8\omega_n \alpha_e = -3\Lambda_{nnnn} - \frac{10b^2\Gamma_{nnn}^2}{3\omega_n^2} - b^2 \sum_{i \neq n}^{\infty} \Gamma_{nni}\Gamma_{inn}\left[\frac{2}{\omega_i^2} + \frac{1}{\omega_i^2 - 4\omega_n^2}\right]$$

$$-b^2 \sum_{j \neq n}^{\infty} \Gamma_{njn}\Gamma_{jnn}\left[\frac{2}{\omega_j^2} + \frac{1}{\omega_j^2 - 4\omega_n^2}\right] \quad (5.3.77)$$

and the summation terms account for the contribution of all of the modes, which are not directly excited, to the effective nonlinearity. In a single-mode Galerkin discretization, these summation terms disappear and (5.3.77) reduces to

$$8\omega_n \alpha_e = -3\Lambda_{nnnn} - \frac{10b^2\Gamma_{nnn}^2}{3\omega_n^2} \quad (5.3.78)$$

Substituting the polar form (5.3.57) into (5.3.76) and separating real and imaginary parts, we obtain (5.3.58) and (5.3.59). Substituting (5.3.57) into (5.3.72) and (5.3.74) and then substituting the outcome into (5.3.67) and (5.3.68), we find, to the second approximation, that

$$q_n(t;\epsilon) \approx \epsilon a_n \cos(\Omega t - \gamma_n) + \frac{\epsilon^2 b a_n^2 \Gamma_{nnn}}{6\omega_n^2}\left[3 - \cos(2\Omega t - 2\gamma_n)\right] \quad (5.3.79)$$

$$q_m(t;\epsilon) \approx \frac{1}{2}\epsilon^2 b a_n^2 \sum_{m \neq n}^{\infty} \Gamma_{mnn}\left[\frac{1}{\omega_n^2} + \frac{\cos(2\Omega t - 2\gamma_n)}{\omega_m^2 - 4\omega_n^2}\right] \text{ for } m \neq n \quad (5.3.80)$$

where a_n and γ_n are given by (5.3.58) and (5.3.59). Substituting (5.3.79) and (5.3.80) into (5.3.61), we find that the beam response, to second order, is given by

$$u(x,t;\epsilon) = \left\{\epsilon a_n \cos(\Omega t - \gamma_n) + \frac{\epsilon^2 b a_n^2 \Gamma_{nnn}}{6\omega_n^2}\left[3 - \cos(2\Omega t - 2\gamma_n)\right]\right\}\phi_n(x)$$

$$+\frac{1}{2}\epsilon^2 b a_n^2 \sum_{m \neq n}^{\infty} \Gamma_{mnn}\left[\frac{1}{\omega_n^2} + \frac{\cos(2\Omega t - 2\gamma_n)}{\omega_m^2 - 4\omega_n^2}\right]\phi_m(x) + \cdots \quad (5.3.81)$$

In the single-mode discretization, the summation term in (5.3.81) disappears and it reduces to

$$u(x,t;\epsilon) = \left\{\epsilon a_n \cos(\Omega t - \gamma_n) + \frac{\epsilon^2 b a_n^2 \Gamma_{nnn}}{6\omega_n^2}\left[3 - \cos(2\Omega t - 2\gamma_n)\right]\right\}\phi_n(x)$$
$$\quad (5.3.82)$$

It follows from (5.3.78) and (5.3.82) that, in the one-degree-of-freedom Galerkin model, the quadratic nonlinearity affects the amplitude and phase of the motion through the effective nonlinearity α_e and produces a drift and a second-harmonic term, which multiply the eigenmode in question. On the other hand, in the full-basis Galerkin reduced model or the direct model, the second-order corrections are both temporal and spatial. With the direct approach, it follows from (5.3.56) and (5.3.60) that the quadratic nonlinearities produce a drift term proportional to $a_n^2 \Psi_1(x)$ and a second-harmonic term proportional to $a_n^2 \cos(2\Omega t - 2\gamma_n) \Psi_2(x)$, while retaining the full spectrum of the eigenmodes of the system in the functions Ψ_1 and Ψ_2.

The effective nonlinearity coefficient calculated by either an infinite- or finite-dimensional model, (5.3.77) or (5.3.78), consists of two terms: a term proportional to b^2 and the term $-3\Lambda_{nnnn}/8\omega_n$, which does not depend on the buckling level. The first term is independent, to third order, of the approach (an infinite- or a finite-dimensional discretization or the direct method) used to calculate it. It is positive and due to the cubic nonlinearity; hence, it produces a hardening behavior. On the other hand, the second term is caused by the quadratic nonlinearities, depends on the order of the reduced model, and is negative; hence, it produces a softening behavior. The effective nonlinearity coefficient α_e depends on the relative magnitudes of the hardening (cubic) and softening (quadratic) terms. Consequently, there is a possibility of a sign difference in this coefficient, depending on the order of the model. For example, it follows from (5.3.78) that the effective nonlinearity coefficient calculated with a one-mode discretization vanishes when

$$b = b_c = \frac{3\omega_n}{\sqrt{10}\Gamma_{nnn}} \sqrt{-\Lambda_{nnnn}} \qquad (5.3.83)$$

Next, we summarize the convergence results of Lacarbonara (1998) for the effective nonlinearity coefficient of the first mode. It follows from (5.3.64) that Γ_{11k}, Γ_{k11}, and Γ_{1k1} vanish when k is even, which corresponds to antisymmetric modes. In other words, antisymmetric modes do not contribute to the effective nonlinearity coefficient of the first mode. In Figure 5.3.2, we compare variation of α_e with the buckling level and the number N of the symmetric modes retained in the discretization. Also, we show variation of the value of α_e calculated using the direct approach with the buckling level. We note that, for low buckling levels, the results obtained with reduced-order models, including the one-mode model, are in qualitative as well as, to a certain extent, quantitative agreement with the results obtained with the direct approach. We also note that the effective nonlinearity coefficient obtained with the direct approach is negative for all buckling levels of interest (i.e., the effective nonlinearity of the first mode is always of the softening type). On the other hand, the effective nonlinearity coefficient computed with the one-mode model changes from negative (softening behavior) to positive (hardening behavior) as b increases above the critical buckling level $b_c = 15.20$. At this critical value, the first mode is predicted to behave linearly to second order. The critical value is close but below the second crossover buckling level (i.e., the buckling level at which the natural frequencies of the third and fourth modes are equal). Moreover, above $b \approx 17.5$, the two-mode model predicts a hardening-type effective nonlinearity. However, the three- and four-mode models predict a softening-type effective nonlinearity for all buckling levels of interest, in qualitative agreement with the direct approach model. We conclude that

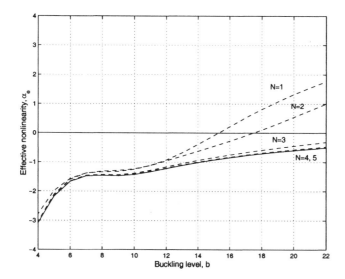

Fig. 5.3.2 Variation of the effective nonlinearity coefficient of the first mode with the buckling level. The coefficient has been computed with the direct model and with the one- through four-mode Galerkin reduced models.

a model consisting of the three lowest symmetric modes is the least reduced-order model that can predict qualitatively primary resonances of the first mode.

A more in-depth analysis of the convergence of α_e as the number of modes retained in the discretization increases can be found in Lacarbonara (1998), Emam (2002), and Emam and Nayfeh (2004a).

5.3.5 Experiment

We present some of the theoretically and experimentally obtained frequency-response curves by Lacarbonara, Nayfeh, and Kreider (1998) for the case of primary resonance of the first mode for buckling levels where qualitative discrepancies between the discretization and direct approaches exist. In Figure 5.3.3, we show the frequency-response curve obtained in the range 165 Hz to 178 Hz for a beam of length 279.2 mm, width 19.05 mm, thickness 0.813 mm, density 7.810 kg/m^3, Young's modulus 207 GPa, and hardness 183 Vickers. The buckling level was 5.842 mm (the nondimensional buckling level and the rise-to-span ratio were 24.36 and 0.0209, respectively), the measured first natural frequency before conducting the experiment was 174 Hz, and the base acceleration was set at 0.332 g. For this buckling level and these excitation frequencies, internal resonances were not activated.

For this buckling level, the single-mode discretization predicts a hardening behavior, whereas the direct approach predicts a softening behavior, as shown in Figure 5.3.4. In the forward sweep, the response was initially linear and increased gradually

332 DYNAMICS OF BEAMS

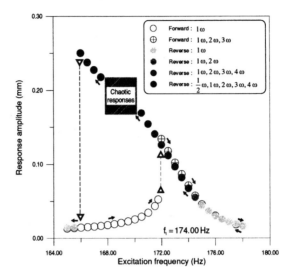

Fig. 5.3.3 Experimental frequency-response curve for a buckled beam when $\hat{b}$ is 5.842 *mm* and the excitation amplitude is 0.332 *g* (Lacarbonara, Nayfeh, and Kreider, 1998).

as the excitation frequency increased. Above approximately 171.5 Hz, the response jumped up to a motion which contained the fundamental frequency and its second and third harmonics, indicating that both quadratic and cubic nonlinearities were activated. The different fillings of the circles in Figure 5.3.3 indicate different frequency contents for the steady-state responses. The arrow indicates the jump direction.

In the reverse sweep, the observed motions were the same as those observed in the forward sweep up to 171.5 Hz. Below this frequency, the response displayed higher-order harmonics and underwent a period-doubling bifurcation. The period-doubled motion persisted over a certain range and then became chaotic. Lacarbonara, Nayfeh, and Kreider (1998) conjectured that the route to chaos is a sequence of period-doubling bifurcations, even though they could not capture experimentally higher-order period-doubling bifurcations. The chaotic response persisted over an extended frequency range, as indicated by the dashed area in Figure 5.3.3. Then, they observed again a period-doubled motion. This motion persisted until it jumped down at approximately 165.6 Hz to a low-amplitude motion. The experimentally obtained frequency-response curve is bent to the left and the up and down jumps occurred below the first natural frequency, and hence the frequency-response curve is of the softening type.

In Figure 5.3.5, we show a frequency-response curve obtained for the same buckling level, but at the lower excitation amplitude 0.209 *g*. The measured first natural frequency was 174.27 Hz. The signatures of the motions are similar to those observed at the higher excitation level. However, the jumps occurred at slightly different frequencies: the jump up occurred at approximately 172.70 Hz and the jump down occurred at 165.25 Hz.

In summary, the results obtained with the direct approach are in qualitative agreement with those obtained experimentally, whereas the single-mode discretization leads to erroneous results for high buckling levels.

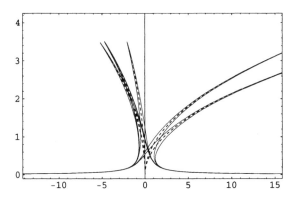

Fig. 5.3.4 Frequency-response curves obtained with one- through four-mode discretizations (from right to left) and with the direct approach when $\hat{b} = 5.842$ mm, $f = 50$, and $\mu = 0.1$.

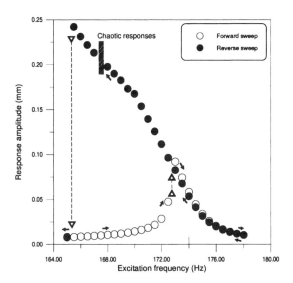

Fig. 5.3.5 Experimental frequency-response curve for a buckled beam when $\hat{b}$ is 5.842 mm and the excitation amplitude is 0.209 g (Lacarbonara, Nayfeh, and Kreider, 1998).

334 DYNAMICS OF BEAMS

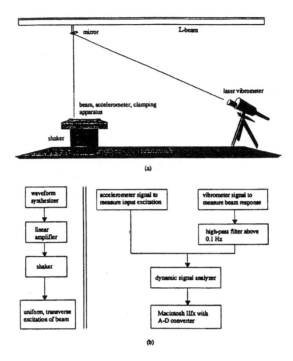

Fig. 5.3.6 A schematic of the setup of the experiment.

5.3.6 Global Dynamics

Experiments

Kreider and Nayfeh (1998) investigated experimentally the nonlinear response of a clamped-clamped buckled beam to a harmonic force that is constant over the length of the beam. To simulate these conditions, they mounted a clamping apparatus to an aluminum slab, which is then attached to an electrodynamic shaker. This arrangement is depicted in Figure 5.3.6.

They considered a spring steel beam with a thickness of 0.032 in, a width of 0.622 in, and a hardness of nearly 45 on the Rockwell C scale (441 Vickers). One part of the clamping apparatus is the aluminum slab to which the clamps are mounted. The dimensions of this slab are as follows: 3 in thick, 15 in wide, and 20 in long. There are three aspects involved in the design of this slab. First, the length and width are chosen to allow a beam of about 11 in to be clamped on the surface. Second, the thickness is required to prevent the corners that hang over the edges of the shaker table from 'flapping' around. Last, aluminum is used to provide enough mass to minimize feedback to the shaker without severely limiting the output amplitude capabilities of the shaker.

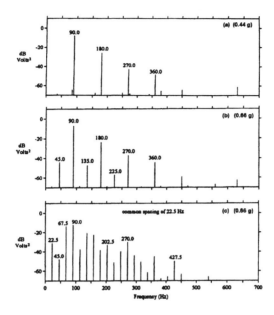

Fig. 5.3.7 Power spectra of the responses for three excitation levels obtained in run #4 of Kreider and Nayfeh (1998). They show period-one, period-two, and period-four motions.

The other part of the clamping apparatus is the clamps themselves. At this point, it is helpful to define the physical requirements of successful clamps. Successful clamps generally provide boundary conditions that enable a meaningful comparison of experimental data to a theoretical model. For a vibrating beam with fixed ends, physical clamps allow some nonzero slopes as well as axial and transverse deflections at the boundaries of the beam. Small nonzero slopes and small transverse deflections at the boundaries are acceptable. Axial deflections due to the elastic behavior of the clamps are also acceptable. Other axial motions of the clamps, such as slipping, change the dynamics of the system. The reason is that the natural frequencies of the buckled beam depend on the static deflection and the beam length, which change as a result of the slipping.

Six runs were conducted in this experiment and in each case the natural frequencies were measured before and after the experiment to ensure that they did not change during the run. In two of these runs, run #1 and run #4, they obtained classical supercritical period-doubling bifurcations of limit cycles. The power spectra of the responses obtained for three excitation levels during run #4 are shown in Figure 5.3.7. It shows period-one, period-two, and period-four motions. These motions can be predicted using a single-mode discretization. In run #3 and run #2, they obtained sidebands, which have a common spacing of one-third and one-fifth of the excitation frequency, respectively, as shown in Figures 5.3.8 and 5.3.9. Consequently, they concluded that

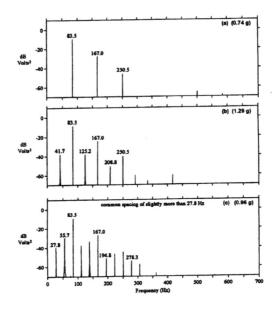

Fig. 5.3.8 Power spectra of the responses for three excitation levels obtained in run #3 of Kreider and Nayfeh (1998). They show period-one, period-two, and period-three motions.

these motions are period-three and period-five, respectively. Since these motions do not exist in narrow windows embedded within chaotic regions, these motions cannot be predicted using a single-mode discretization. In runs #5 and #6, period-one, period-two, and quasiperiodic motions were observed, as shown in Figure 5.3.10. The period-one and period-two motions can be predicted using a single-mode discretization. However, the quasiperiodic motion, consisting of peaks at the excitation frequency and its harmonics and sidebands whose frequencies are incommensurate with the excitation frequency, cannot be predicted using a single-mode discretization. Therefore, they called the sidebands in Figure 5.3.10 *unexplained sidebands*.

Using a multimode Galerkin discretization, Emam and Nayfeh (2002, 2004a) were able to explain the mechanism by which these sidebands are created. By using two or more modes in the discretization, they obtained a classical supercritical period-doubling bifurcation followed by a secondary Hopf bifurcation, resulting in a quasiperiodic motion. This secondary Hopf bifurcation creates a new frequency, which is, in general, incommensurate with the excitation frequency. The power spectrum of such a motion is similar to those obtained by Kreider and Nayfeh (1998) in run #5, as shown in Figure 5.3.10. If the new frequency created by the secondary Hopf bifurcation is commensurate with the excitation frequency, a phase locking of the motion takes place, resulting in a large-period periodic motion. In run #2, the new frequency seems to be one-fifth of the excitation frequency, and as a result a period-five motion was created. Similarly, in run #3, the new frequency seems to be one-third of the excitation frequency, and as a result a period-three motion was created. In the next section, we present numerical results supporting this mechanism.

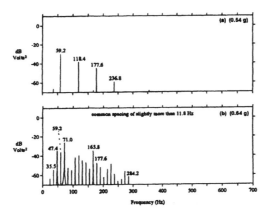

Fig. 5.3.9 Power spectra of the responses for two excitation levels obtained in run #2 of Kreider and Nayfeh (1998). They show period-one and period-five motions.

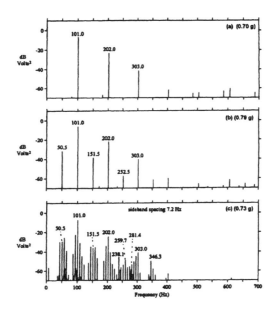

Fig. 5.3.10 Power spectra of the responses for three excitation levels obtained in run #5 of Kreider and Nayfeh (1998). They show period-one, period-two, and quasiperiodic motions.

338 DYNAMICS OF BEAMS

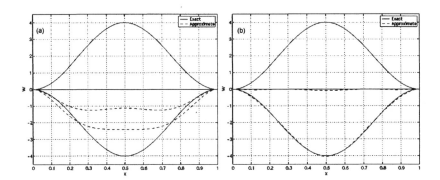

Fig. 5.3.11 Comparison of the buckled configurations obtained with (a) single-mode (- - -) and (b) two-mode (- - -) discretizations with those obtained by solving the exact nonlinear buckling problem (———).

Numerical Results

Emam and Nayfeh (2002, 2004a) calculated the equilibrium solutions (fixed points) of (5.3.63) using single- and multimode approximations to determine the number of modes required to obtain accurate buckled configurations under different buckling levels. The displacement is measured from the upper position of the buckled beam. Figure 5.3.11a shows the approximate buckled configurations obtained using a single-mode approximation compared with the exact configurations obtained by solving the nonlinear static buckling problem. At this buckling level, the single-mode solution does not give a good approximation to two of the three buckling configurations. Consequently, discrepancy in the dynamics of the beam is expected. Adding more modes in the discretization improves the solution; for example, using two modes provides a fairly accurate result, as shown in Figure 5.3.11b.

To investigate the nonlinear vibrations of the buckled beam for different buckling rises, one needs to determine how many modes are needed to provide a good approximation to the periodic responses. One measure of the accuracy is the symmetry of the predicted periodic responses around the two buckled configurations. In other words, the dynamics that takes place around the upper buckled position should be the same as the dynamics that takes place around the lower buckled position before snapping through. To this end, Emam and Nayfeh (2002, 2004a) calculated the motion around the buckled positions using an increasing number of modes. We note from Figure 5.3.12a that using a single-mode approximation results in unsymmetric limit cycles around the buckled positions, which themselves are unsymmetric as predicted by the static analysis. Using a two-mode discretization improves significantly the dynamic solutions, as shown in Figure 5.3.12b. Adding more modes gives the required symmetry of the limit cycles around the buckled positions, as shown in Figure 5.3.12c. Therefore, they used a four-mode discretization to investigate the nonlinear vibrations of the beam.

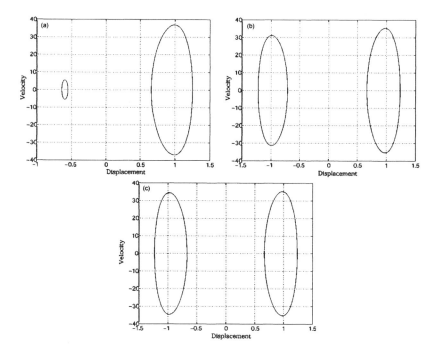

Fig. 5.3.12 The periodic orbits obtained with (a) one-mode, (b) two-mode, and (c) four-mode approximations.

A periodic solution, which is a period-one motion, is shown in Figure 5.3.13a for $F = 41$. As the excitation amplitude is increased, a supercritical period-doubling bifurcation occurs at $F = 440$; the period-two motion obtained for $F = 441$ is shown in Figure 5.3.13b. As the excitation amplitude is increased further, a sequence of supercritical period-doubling bifurcations takes place. Figures 5.3.13c and 5.3.13d show period-four and period-eight motions at $F = 475.1$ and $F = 479.3$, respectively. This sequence of supercritical period-doubling bifurcations of the local attractors culminates in chaos.

The global attractor, which coexists with the local attractors at $F = 41$ becomes the only attractor; it represents a snapping-through motion, as shown in Figure 5.3.14a for $F = 6164$. Increasing the excitation amplitude at the same frequency results in an interesting dynamics corresponding to a quasiperiodic motion at $F = 6165$ where two complex-conjugate Floquet multipliers exit the unit circle away from the real axis, resulting in a Hopf bifurcation. This quasiperiodic motion cannot be obtained by using a single-mode discretization. We note that at least two modes are required to obtain this type of motion. Emam and Nayfeh (2002, 2004a) examined the robustness of this quasiperiodic motion qualitatively as the number of modes retained in the approximation is increased to five. They obtained qualitatively the same results with

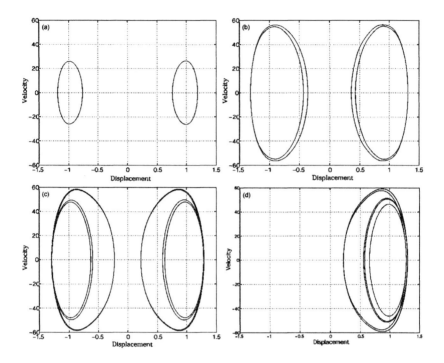

Fig. 5.3.13 (a) A period-one motion obtained at $F = 41$, (b) a period-two motion obtained at $F = 441$, (c) a period-four motion obtained at $F = 475.1$, and (d) a period-eight motion obtained at $F = 479.3$ using a four-mode discretization.

minor quantitative differences. The Hopf bifurcation creates a new frequency, which is, in general, incommensurate with the excitation frequency. The result is a two-period quasiperiodic motion (two-torus). If this new frequency is commensurate with the excitation frequency, a phase locking of the motion occurs, resulting in a larger-period periodic motion. These results are qualitatively in agreement with all of the responses observed by Kreider and Nayfeh (1998). Their unexplained sidebands correspond to two-period quasiperiodic motions, which result from a Hopf bifurcation where the newly created frequency is incommensurate with the excitation frequency. Their apparent "period-three" and "period-five" motions are again created by a Hopf bifurcation where the newly created frequency is commensurate with the excitation frequency.

A Poincaré map of the torus is shown in Figure 5.3.14b. As the excitation amplitude is increased, the torus deforms, as shown in Figure 5.3.14c. A torus-breakdown bifurcation occurs at $F = 6186$, as shown in Figure 5.3.14d, leading to chaos.

Emam and Nayfeh (2002) concluded that a single-mode approximation yields quantitatively as well as qualitatively erroneous static and dynamic results for rela-

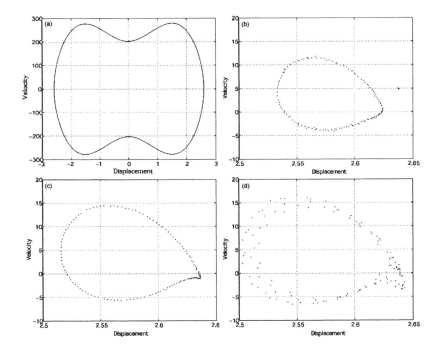

Fig. 5.3.14 (a) A phase portrait of the global attractor, period-one motion, obtained at $F = 6164$, (b) a Poincaré map of the global attractor obtained at $F = 6165$, (c) a Poincaré map of the global attractor obtained at $F = 6175$, and (d) a Poincaré map of the global attractor obtained at $F = 6186$.

tively high buckling levels. Moreover, at least a four-mode approximation is required to accurately predict the buckling configurations and the dynamic responses.

5.4 MICROBEAMS

Microelectromechanical systems (MEMS) and devices drew attention in the eighties as sensors and actuators. The fact that they could be manufactured using existing manufacturing techniques and infrastructure of the semiconductor industry meant that they could be produced at low cost and in large volumes, making their commercialization quite attractive. Their light weight, small size, low-energy consumption, and durability made them even more attractive.

There are two approaches for the analysis of MEMS devices: finite element and reduced-order modes. There are FEM codes dedicated to MEMS, such as MEMCAD (Senturia et al., 1992), CAEMEMS (Crary and Zhang, 1990), IntelliCAD (Yie et al., 1996), SESES (Korvink et al., 1994), and ANSYS. Some researchers, such as Senturia et al. (1992), Zook et al. (1992), Yao et al. (1999), and Chan, Garikipati, and Dutton

(1999), have used FEM models to study the mechanical behavior of various MEMS devices.

However, the use of these codes in the simulation of MEMS dynamics has proved to be cumbersome and impractical. On the other hand, reduced-order models lend themselves very well to dynamic simulations. Models constructed using this approach seek to capture the most significant characteristics of a device behavior in a few variables governed by a few ordinary-differential equations of motion. The resulting system is typically easy to simulate as a standalone model or integrate into system-level simulations.

Turner and Andrews (1995), Zavracky, Majumder, and McGruer (1997), Seeger and Crary (1997), Ayela and Fournier (1998), Castañer and Senturia (1999), Veijola et al. (2000), and Ahn, Guckel, and Zook (2001) used lumped-mass models to study the mechanical behavior of various MEMS devices. Tilmans, Elwespoek, and Fluitman (1992), Ijntema and Tilmans (1992), and Tilmans and Legtenberg (1994) used the method of weighted residuals in a one-mode approximation to study the mechanical behavior of microbeam resonators. Grétillat et al. (1997), Hung and Senturia (1999), Gabbay, Mehner, and Senturia (2000), Mehner, Gabbay, and Senturia (2000), and Jiang, Ng, and Lam (2000) used the method of weighted residuals in a multimode approximation to study the mechanical behavior of various MEMS devices.

5.4.1 Modeling of MEMS Devices

Abdel-Rahman, Younis, and Nayfeh (2002) presented nonlinear models of electrically actuated microbeams, which form the heart of many MEMS devices, such as capacitive microswitches and resonant microsensors. The models account for mid-plane stretching, an applied axial load, a DC electrostatic force, and an AC harmonic force. They developed an approach for calculating the static deflections of microbeams and their natural frequencies and mode shapes around the static deflections. In this section, we treat the modeling and analysis of microbeams.

We consider a resonant microbeam actuated by an electric load composed of a DC component (polarization voltage) V_p and an AC component $v(t)$ and subject to viscous damping per unit length of coefficient $\hat{c}$. We consider the case of a clamped-clamped beam, as shown in Figure 5.4.1. The equation governing the dynamics of microbeams can be expressed as

$$\rho b h \frac{\partial^2 w}{\partial t^2} + EI \frac{\partial^4 w}{\partial x^4} + \hat{c}\frac{\partial w}{\partial t} = \left[\frac{EA}{2L} \int_0^L \left(\frac{\partial w}{\partial x}\right)^2 dx + \hat{N}\right] \frac{\partial^2 w}{\partial x^2}$$
$$+ \frac{1}{2}\epsilon b \frac{[V_p + v(t)]^2}{(d-w)^2} \quad (5.4.1)$$

where t is time, x is the position along the microbeam, ρ is the material density, A and I are the area and moment of inertia of the cross-section, h is the microbeam thickness, E is Young's modulus, d is the capacitor gap width, ϵ is the dielectric constant of the medium, and $\hat{N}$ is the applied axial force. The last term in (5.4.1)

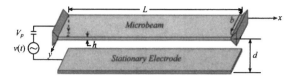

Fig. 5.4.1 A schematic drawing of a typical structural element under parallel-plate electric actuation.

represents the parallel-plate electric forces when neglecting the fringing effect of the electric field (Griffiths, 1981). The corresponding boundary conditions are

$$w = 0 \text{ and } w' = 0 \text{ at } x = 0 \text{ and } x = L \quad (5.4.2)$$

For convenience, we introduce the nondimensional variables (denoted by hats)

$$\hat{w} = \frac{w}{d}, \quad \hat{x} = \frac{x}{L}, \quad \hat{t} = \frac{t}{T} \quad (5.4.3)$$

where T is a time scale, which is defined below. Substituting (5.4.3) into (5.4.1) and (5.4.2) and dropping the hats, we obtain

$$\frac{\partial^2 w}{\partial t^2} + \frac{\partial^4 w}{\partial x^4} + c\frac{\partial w}{\partial t} = [\alpha_1 \Gamma(w,w) + N]\frac{\partial^2 w}{\partial x^2} + \alpha_2 \frac{[V_p + v(t)]^2}{(1-w)^2} \quad (5.4.4)$$

$$w = 0 \text{ and } w' = 0 \text{ at } x = 0 \text{ and } x = 1 \quad (5.4.5)$$

where

$$\Gamma[f_1(x,t), f_2(x,t)] = \int_0^1 \frac{\partial f_1}{\partial x}\frac{\partial f_2}{\partial x} dx \quad (5.4.6)$$

Equation (5.4.4) is a nondimensional integral-partial-differential equation with linear and nonlinear terms. The nondimensional parameters appearing in (5.4.4) are

$$c = \frac{\hat{c}L^4}{EIT}, \quad \alpha_1 = 6\left(\frac{d}{h}\right)^2, \quad N = \frac{\hat{N}L^2}{EI}, \quad \alpha_2 = \frac{6\epsilon L^4}{Eh^3 d^3} \quad (5.4.7)$$

and T is chosen as $T = \sqrt{\rho b h L^4 EI}$.

The microbeam deflection under an electric force is composed of a static component due to the DC voltage, denoted by $w_s(x)$, and a dynamic component due to the AC voltage, denoted by $u(x,t)$; that is,

$$w(x,t) = w_s(x) + u(x,t) \quad (5.4.8)$$

To calculate the static deflection of the microbeam, we set the time derivatives and the AC forcing term in (5.4.4) equal to zero and obtain

$$\frac{d^4 w_s}{dx^4} = [\alpha_1 \Gamma(w_s, w_s) + N]\frac{d^2 w_s}{dx^2} + \frac{\alpha_2 V_p^2}{(1-w_s)^2} \quad (5.4.9)$$

It follows from (5.4.5) and (5.4.8) that the boundary conditions are

$$w_s = 0 \text{ and } w'_s = 0 \text{ at } x = 0 \text{ and } x = 1 \qquad (5.4.10)$$

We generate the problem governing the dynamic behavior of the microbeam around the deflected shape by substituting (5.4.8) into (5.4.4) and (5.4.5) and using (5.4.9) and (5.4.10) to eliminate the terms representing the equilibrium position. To third-order in u, the result is

$$\frac{\partial^2 u}{\partial t^2} + c\frac{\partial u}{\partial t} + \frac{\partial^4 u}{\partial x^4} = [\alpha_1 \Gamma(w_s, w_s) + N]\frac{\partial^2 u}{\partial x^2} + 2\alpha_1 \Gamma(w_s, u)\frac{d^2 w_s}{dx^2}$$

$$+ \frac{2\alpha_2 V_p^2}{(1-w_s)^3}u + \alpha_1 \Gamma(u,u)\frac{d^2 w_s}{dx^2} + 2\alpha_1 \Gamma(w_s, u)\frac{\partial^2 u}{\partial x^2}$$

$$+ \frac{3\alpha_2 V_p^2}{(1-w_s)^4}u^2 + \alpha_1 \Gamma(u,u)\frac{\partial^2 u}{\partial x^2} + \frac{4\alpha_2 V_p^2}{(1-w_s)^5}u^3$$

$$+ \frac{2\alpha_2 V_p}{(1-w_s)^2}v(t) + \frac{4\alpha_2 V_p}{(1-w_s)^3}v(t)u + \cdots \qquad (5.4.11)$$

$$u = 0 \text{ and } u' = 0 \text{ at } x = 0 \text{ and } x = 1 \qquad (5.4.12)$$

5.4.2 Static Deflection

Abdel-Rahman, Younis, and Nayfeh (2002) used a shooting method to solve (5.4.9) and (5.4.10) for $w_s(x)$. They assumed an initial deflection distribution $w_s^0(x)$ and used it to evaluate $\Gamma\left(w_s^0, w_s^0\right)$. In each iteration, they substituted the iterated value of the integral Γ^i into (5.4.9), solved the nonlinear boundary-value problem

$$\frac{d^4 w_s^{i-1}}{dx^4} = (\alpha_1 \Gamma^i + N)\frac{d^2 w_s^{i-1}}{dx^2} + \frac{\alpha_2 V_p^2}{\left(1-w_s^{i-1}\right)^2} \qquad (5.4.13)$$

$$w_s^{i-1} = 0, \quad \frac{dw_s^{i-1}}{dx} = 0 \text{ at } x = 0 \text{ and } x = 1 \qquad (5.4.14)$$

numerically for $w_s^i(x)$, and then calculated a new $\Gamma^{i+1} = \Gamma\left(w_s^i, w_s^i\right)$. They continued to iterate (shoot) on the integral $\Gamma(w_s, w_s)$ until it converged to a value within a predefined tolerance.

There are two problems associated with this numerical method. The first is the sensitivity of the solution to the initial guess as pull-in is approached. We note that the domain of attraction of the solution narrows down as $\alpha_2 V_p^2$ increases and becomes very narrow close to pull-in. Hence, it becomes very difficult to find a suitable initial guess that makes the numerical scheme converge. The second is related to the stiffness of the problem. As the values of α_1 and $\alpha_2 V_p^2$ increase, the boundary-value problem becomes stiffer. Accordingly, at high α_1, choosing initial conditions that lead to convergence becomes difficult.

Numerical solutions of (5.4.9) and (5.4.10) reveal the existence of two static (fixed point) solutions (upper branch and lower branch) for each value of V_p. The lower branch is stable and the upper branch is unstable. Both branches seem to coalesce at the pull-in voltage. However, as aforementioned, it becomes difficult using the above numerical method to calculate the static deflection very close to pull-in and hence to determine the exact location of the pull-in voltage. Using the reduced-order model described in Section 5.4.5, Abdel-Rahman, Younis, and Nayfeh (2002) were able to determine the initial conditions needed near the pull-in voltage.

5.4.3 Linear Mode Shapes and Frequencies

To obtain the microbeam linear mode shapes and the corresponding natural frequencies, we drop the nonlinear, forcing, and damping terms in (5.4.11) and (5.4.12) and obtain the linear-undamped eigenvalue problem

$$\frac{\partial^2 u}{\partial t^2} + \frac{\partial^4 u}{\partial x^4} - (\alpha_1 \Gamma_1 + N)\frac{\partial^2 u}{\partial x^2} - \frac{2\alpha_2 V_p^2}{(1-w_s)^3}u - 2\alpha_1 \Gamma(w_s, u)\frac{d^2 w_s}{dx^2} = 0 \quad (5.4.15)$$

$$u = 0 \text{ and } u' = 0 \text{ at } x = 0 \text{ and } x = 1 \quad (5.4.16)$$

where $\Gamma_1 = \Gamma(w_s, w_s)$. We solve (5.4.15) and (5.4.16) for the undamped mode shapes and natural frequencies under the static deflection distribution $w_s(x)$. Assuming a harmonic motion in the nth mode in the form

$$u(x,t) = \phi_n(x)e^{i\omega_n t} \quad (5.4.17)$$

where $\phi_n(x)$ is the nth mode shape and ω_n is the nth nondimensional natural frequency, we reduce (5.4.15) and (5.4.16) to

$$\phi_n^{iv} - (\alpha_1 \Gamma_1 + N)\phi_n'' - \frac{2\alpha_2 V_p^2}{(1-w_s)^3}\phi_n - 2\alpha_1 \Gamma_2 w_s'' = \omega_n^2 \phi_n \quad (5.4.18)$$

$$\phi_n = 0 \text{ and } \phi_n' = 0 \text{ at } x = 0 \text{ and } x = 1 \quad (5.4.19)$$

where $\Gamma_2 = \Gamma(w_s, \phi_n)$ and the prime denotes the derivative with respect to x. Abdel-Rahman, Younis, and Nayfeh (2002) first solved (5.4.9) and (5.4.10) for the static deflection w_s. Then, they solved (5.4.18) and (5.4.19) for each mode shape ϕ_n and its associated natural frequency ω_n using a combination of a shooting method and an iteration on ω_n and Γ_2 until they converged to within predefined tolerances.

Figure 5.4.2 compares the normalized fundamental natural frequency ω_1/ω_0 calculated using the model of Abdel-Rahman, Younis, and Nayfeh (2002) with the theoretical and experimental results of Tilmans and Legtenberg (1994) for two microbeams with the lengths 210 μm and 510 μm and subject to an axial load of 0.0009 N. Here, ω_1 is the first natural frequency of the deflected beam and ω_0 is the first natural frequency of the straight beam. Hence, $\omega_1 \to \omega_0$ as $V_p \to 0$. There is good agreement between the theoretical and experimental results, even as the microbeams approach their stability limits. Further, the results of Abdel-Rahman, Younis, and Nayfeh show that failure to account for the effect of mid-plane stretching in the microbeam restoring force, as is commonly done in the literature, leads to an under-estimation of the

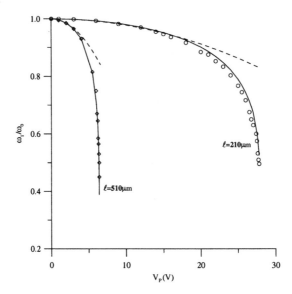

Fig. 5.4.2 Comparison of the fundamental natural frequency calculated using the model of Abdel-Rahman, Younis, and Nayfeh (2002) (———) with that obtained theoretically (– – –) and experimentally by Tilmans and Legtenberg (1994) for two microbeams of lengths 210 μm (∘ ∘ ∘) and 510 μm (⋄ ⋄ ⋄).

stability limits of the resonator. They found out that the ratio of the air-gap width to the microbeam thickness can be tuned so as to extend the domain of the linear relationship between the DC voltage and the fundamental natural frequency. This fact and the ability of the nonlinear model to predict the natural frequencies accurately for any DC voltage allow designers to use a wider range of DC voltages in resonator operations.

5.4.4 Nonlinear Response to a Primary-Resonance Excitation

In this section, we investigate the dynamic response of a microbeam to a primary-resonance electric actuation of its nth mode by applying directly the method of multiple scales to (5.4.11) and (5.4.12). To this end, we seek a second-order uniform solution of (5.4.11) and (5.4.12) in the form

$$u(x,t;\epsilon) = \epsilon u_1(x,T_0,T_2) + \epsilon^2 u_2(x,T_0,T_2) + \epsilon^3 u_3(x,T_0,T_2) + \cdots \quad (5.4.20)$$

We note that $u(x,t)$ does not depend on $T_1 = \epsilon t$ because, as shown below, secular terms arise at $O(\epsilon^3)$. In order that the nonlinearity balances the effects of the damping c and excitation $v(t)$, we scale them so that they appear together in the modulation equations as follows: $\epsilon^2 c$ and $\epsilon^3 V_{AC}\cos(\Omega t)$ where V_{AC} and Ω are the magnitude and frequency of the applied AC voltage. Substituting (5.4.20) into (5.4.11) and equating coefficients of like powers of ϵ, we obtain

Order ϵ:

$$\mathcal{L}(u_1) = D_0^2 u_1 + u_1^{iv} - \alpha_1 \Gamma_1 u_1'' - N u_1'' - 2\alpha_1 \Gamma(w_s, u_1) w_s'' - \frac{2\alpha_2 V_p^2}{(1-w_s)^3} u_1 = 0$$
(5.4.21)

Order ϵ^2:

$$\mathcal{L}(u_2) = \alpha_1 w_s'' \Gamma(u_1, u_1) + 2\alpha_1 \Gamma(w_s, u_1) u_1'' + \frac{3\alpha_2 V_p^2}{(1-w_s)^4} u_1^2 \quad (5.4.22)$$

Order ϵ^3:

$$\mathcal{L}(u_3) = -2D_0 D_2 u_1 - c D_0 u_1 + 2\alpha_1 w_s'' \Gamma(u_1, u_2) + 2\alpha_1 \Gamma(w_s, u_2) u_1''$$
$$+ 2\alpha_1 \Gamma(w_s, u_1) u_2'' + \alpha_1 \Gamma(u_1, u_1) u_1'' + F(x) \cos(\Omega T_0)$$
$$+ \frac{6\alpha_2 V_p^2}{(1-w_s)^4} u_1 u_2 + \frac{4\alpha_2 V_p^2}{(1-w_s)^5} u_1^3 \quad (5.4.23)$$

where

$$F(x) = \frac{2\alpha_2 V_p V_{AC}}{(1-w_s)^2} \quad (5.4.24)$$

The boundary conditions at all orders are given by

$$u_j = 0 \text{ and } u_j' = 0 \quad \text{at } x = 0 \text{ and } x = 1, \ j = 1, 2, 3 \quad (5.4.25)$$

The first-order problem given by (5.4.21) and (5.4.25) is identical to the linear eigenvalue problem, (5.4.15) and (5.4.16). Because, in the presence of damping, the homogeneous solutions corresponding to all modes that are not directly or indirectly excited decay with time, the solution of (5.4.21) and (5.4.25) is assumed to consist of only the directly excited mode. Accordingly, we express u_1 as

$$u_1 = \left[A_n(T_2) e^{i\omega_n T_0} + \bar{A}_n(T_2) e^{-i\omega_n T_0} \right] \phi_n(x) \quad (5.4.26)$$

where ω_n and $\phi_n(x)$ are the natural frequency and corresponding eigenfunction of the nth mode, respectively. The eigenfunction ϕ_n is normalized such that $\int_0^1 \phi_n^2 dx = 1$.
Substituting (5.4.26) into (5.4.22), we obtain

$$\mathcal{L}(u_2) = \left(A_n^2 e^{2i\omega_n T_0} + 2 A_n \bar{A}_n + \bar{A}_n^2 e^{-2i\omega_n T_0} \right) h(x) \quad (5.4.27)$$

where

$$h(x) = \alpha_1 \Gamma(\phi_n, \phi_n) w_s'' + 2\alpha_1 \phi_n'' \Gamma(w_s, \phi_n) + \frac{3\alpha_2 V_p^2}{(1-w_s)^4} \phi_n^2 \quad (5.4.28)$$

348 DYNAMICS OF BEAMS

The solution of (5.4.27) and (5.4.25) can be expressed as

$$u_2 = \Psi_2(x)A_n^2 e^{2i\omega_n T_0} + 2\Psi_1(x)A_n\bar{A}_n + \Psi_2(x)\bar{A}_n^2 e^{-2i\omega_n T_0} \quad (5.4.29)$$

where Ψ_1 and Ψ_2 are the solutions of the boundary-value problems

$$\Psi_1^{iv} - \alpha_1\Gamma_1\Psi_1'' - N\Psi_1'' - 2\alpha_1\Gamma(w_s, \Psi_1)w_s'' - \frac{2\alpha_2 V_p^2}{(1-w_s)^3}\Psi_1 = h(x) \quad (5.4.30)$$

$$\Psi_2^{iv} - 4\omega_n^2\Psi_2 - \alpha_1\Gamma_1\Psi_2'' - N\Psi_2'' - 2\alpha_1\Gamma(w_s, \Psi_2)w_s'' - \frac{2\alpha_2 V_p^2}{(1-w_s)^3}\Psi_2 = h(x)$$
$$(5.4.31)$$

$$\Psi_j = 0 \text{ and } \Psi_j' = 0 \text{ at } x = 0 \text{ and } x = 1, \quad j = 1, 2 \quad (5.4.32)$$

To describe the nearness of the excitation frequency Ω to the fundamental natural frequency ω_n, we introduce the detuning parameter σ defined by

$$\Omega = \omega_n + \epsilon^2 \sigma \quad (5.4.33)$$

Substituting (5.4.26), (5.4.29), and (5.4.33) into (5.4.23) and keeping the terms that produce secular terms only, we obtain

$$\mathcal{L}(u_3) = \left[-i\omega_n(2A_n' + cA_n)\phi_n(x) + \chi(x)A_n^2\bar{A}_n + \frac{1}{2}F(x)e^{i\sigma T_2} \right] e^{i\omega_n T_0}$$
$$+ cc + NST \quad (5.4.34)$$

where $\chi(x)$ is defined as

$$\chi(x) = 2\alpha_1 w_s''\Gamma(\Psi_2, \phi_n) + 4\alpha_1 w_s''\Gamma(\Psi_1, \phi_n) + (2\alpha_1\Psi_2'' + 4\alpha_1\Psi_1'')\Gamma(w_s, \phi_n)$$
$$+ 2\phi_n''\alpha_1\Gamma(w_s, \Psi_2) + 4\phi_n''\alpha_1\Gamma(w_s, \Psi_1) + 3\phi_n''\alpha_1\Gamma(\phi_n, \phi_n)$$
$$+ \frac{6\alpha_2 V_p^2}{(1-w_s)^4}(2\phi_n\Psi_1 + \phi_n\Psi_2) + \frac{12\alpha_2 V_p^2}{(1-w_s)^5}\phi_n^3 \quad (5.4.35)$$

For a uniform second-order approximation, we do not need to solve for u_3. We only need to impose the solvability condition, which provides an equation for the function A_n. Because the homogeneous problem governing u_3 has a nontrivial solution, the corresponding nonhomogeneous problem has a solution only if the right-hand side of (5.4.34) is orthogonal to every solution of the adjoint homogeneous problem governing u_3. Since the latter problem is self-adjoint, the adjoints are given by $\phi_j(x)\exp(\pm i\omega_j T_0)$. Demanding that the right-hand side of (5.4.34) be orthogonal to $\phi_n(x)\exp(-i\omega_n T_0)$ and using equation (5.4.33), we obtain the solvability condition

$$2i\omega_n(A_n' + \mu_n A_n) + 8\omega_n\alpha_e A_n^2\bar{A}_n - \frac{1}{2}f_n e^{i\sigma T_2} = 0 \quad (5.4.36)$$

where

$$8\omega_n\alpha_e = -\int_0^1 \chi\phi_n dx, \quad f_n = \int_0^1 F(x)\phi_n(x)dx, \quad 2\mu_n = \int_0^1 c\phi_n^2(x)dx$$
$$(5.4.37)$$

Substituting the polar form

$$A_n = \frac{1}{2}a_n e^{i(\sigma T_2 - \gamma_n)} \qquad (5.4.38)$$

into (5.4.36), multiplying the result by $\exp[i(\gamma_n - \sigma T_2)]$, and separating real and imaginary parts, we obtain

$$a'_n = -\mu_n a_n + \frac{f_n}{2\omega_n}\sin\gamma_n \qquad (5.4.39)$$

$$a_n \gamma'_n = \sigma a_n - \alpha_e a_n^3 + \frac{f_n}{2\omega_n}\cos\gamma_n \qquad (5.4.40)$$

Substituting (5.4.38) into (5.4.26) and (5.4.29) and then substituting the results into (5.4.20), we find that the beam response, to second order, is given by

$$u(x,t;\epsilon) \approx \epsilon a_n \phi_n(x)\cos(\Omega t - \gamma_n) + \frac{1}{2}\epsilon^2 a_n^2 [\Psi_2(x)\cos(2\Omega t - 2\gamma_n) + \Psi_1(x)] \qquad (5.4.41)$$

where a_n and γ_n are given by (5.4.39) and (5.4.40).

It follows from (5.4.41) that periodic solutions of (5.4.11) and (5.4.12) correspond to constant a_n and γ_n; that is, the fixed points of (5.4.39) and (5.4.40). Thus, letting $a'_n = 0$ and $\gamma'_n = 0$ in (5.4.39) and (5.4.40) and eliminating γ_n from the resulting equations, we obtain the following frequency-response equation:

$$a_n^2\left[\mu_n^2 + (\sigma - \alpha_e a_n^2)^2\right] = \frac{f_n^2}{4\omega_n^2} \qquad (5.4.42)$$

Equation (5.4.42) is an implicit equation for the amplitude a_n of the periodic response as a function of the detuning parameter σ (which is a representation of the excitation frequency), the effective nonlinearity coefficient α_e, the damping coefficient μ_n, and the amplitude of excitation f_n. Solving (5.4.42) for σ, we obtain

$$\sigma = \pm\frac{1}{a_n}\left[\frac{f_n^2}{4\omega_n^2} - \mu_n^2 a_n^2\right]^{\frac{1}{2}} + \alpha_e a_n^2 \qquad (5.4.43)$$

Recalling that $\sigma = \Omega - \epsilon^2\omega$, setting $\epsilon = 1$, and noting that the amplitude a_n is maximum when the expression inside the square root vanishes, we obtain the following equation for the nonlinear resonance frequency:

$$\Omega_r = \omega_n + \frac{\alpha_e f_n^2}{4\omega_n^2 \mu_n^2} \qquad (5.4.44)$$

The parameters μ_n is related to the quality factor Q by

$$\mu_n = \frac{\omega_n}{2Q} \qquad (5.4.45)$$

Substituting (5.4.45) into (5.4.44), we obtain

$$\Omega_r = \omega_n + \frac{\alpha_e Q^2}{\omega_n^4} f_n^2 \qquad (5.4.46)$$

Equation (5.4.46) relates the nonlinear resonance frequency Ω_r to the effective nonlinearity α_e of the system, the amplitude f_n of the AC forcing, and the quality factor Q.

The stability of the fixed points is determined by examining the eigenvalues of the Jacobian matrix of (5.4.39) and (5.4.40) evaluated at the corresponding fixed point. The characteristic equation is

$$\lambda^2 + 2\mu_n \lambda + \left[\mu^2 + (\sigma - 3\alpha_e a_n^2)(\sigma - \alpha_e a_n^2)\right] = 0 \qquad (5.4.47)$$

For asymptotically stable solutions, all of the eigenvalues must be in the left-half plane.

To describe the dynamic response of the microbeam, one needs to determine the natural frequency ω_n, the excitation amplitude f_n, the effective nonlinearity of the system α_e, and either the damping coefficient μ_n or the quality factor Q.

In Figure 5.4.3, we compare the results obtained by Younis and Nayfeh (2003) for the normalized nonlinear resonance frequency Ω_r/ω_1 with those obtained theoretically and experimentally by Tilmans and Legtenberg (1994) for two microbeams of lengths 210 μm and 310 μm and subject to an axial load of 0.0009 N. We show two sets of calculated results for each microbeam obtained using the model of Younis and Nayfeh. The first set, shown as thin solid lines, uses the same values of the quality factor Q used by Tilmans and Legtenberg. These are $Q = 592$ and 151 for the 210 μm and 310 μm microbeams, respectively. As seen in Figure 5.4.3, these values give poor results. Legtenberg and Tilmans (1994) mentioned that the quality factor for the design of their device varies across the wafer due to variations in the sealing pressure of the microbeam encapsulation. Because such variations can lead to an incorrect value of Q and due to the difficulty of measuring the system damping, in general, Younis and Nayfeh determined the quality factors by matching Ω_r/ω_1 obtained using their model at a single voltage, namely $V_{AC} = 0.6\ V$, to the experimental value of Tilmans and Legtenberg. They obtained $Q = 816.6$ and 197 for the 210 μm and 310 μm microbeams, respectively. The results obtained using these values, shown as thick solid lines, are in excellent agreement with the experimental results.

As expected, the device dimensions and the DC voltage V_p affect the qualitative and quantitative nature of the frequency-response curves (Younis and Nayfeh, 2003). For large V_p, they found that the sign of α_e changes from positive, corresponding to a hardening behavior, to negative, corresponding to a softening behavior. This is because the contribution of the electric quadratic nonlinearity to the effective nonlinearity is negative and drastically decreases in magnitude as V_p increases, eventually overcoming the positive contribution of the geometric nonlinearity. Therefore, assuming the nonlinearity of the system to be solely cubic cannot predict the change in the character of the effective nonlinearity (from hardening to softening) as the DC voltage increases. Moreover, neglecting the quadratic nonlinearity precludes the activation of subharmonic resonances of order one-half and superharmonic resonances

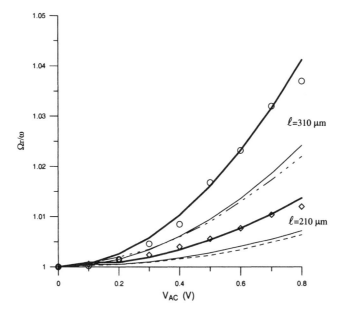

Fig. 5.4.3 Comparison of the normalized nonlinear resonance frequency Ω_r/ω_1 calculated using Younis and Nayfeh (2003) (solid lines) with those obtained theoretically (dashed lines) and experimentally by Tilmans and Legtenberg (1994) for two microbeams of lengths 210 μm (diamonds) and 310 μm (circles). The thin solid lines are based on the quality factors of Tilmans and Legtenburg and the thick solid lines are based on quality factors estimated using our model.

of order two. Further, the assumption of a constant effective nonlinearity, when it is in fact a strong function of the electrostatic force, may lead to more errors in the prediction of the system behavior.

5.4.5 Reduced-Order Models of MEMS Devices

Younis, Abdel-Rahman, and Nayfeh (2003) have developed an approach to create an accurate reduced-order model of a microbeam under parallel-plate electric actuation. The nondimensional equation of motion can be written as

$$\frac{\partial^2 w}{\partial t^2} + c\frac{\partial w}{\partial t} + \frac{\partial^4 w}{\partial x^4} = \left[\alpha_1 \int_0^1 \left(\frac{\partial w}{\partial x}\right)^2 dx + N\right] \frac{\partial^2 w}{\partial x^2} + \alpha_2 \frac{V^2}{(1-w)^2} \quad (5.4.48)$$

where α_1 and α_2 are nondimensional parameters defined in (5.4.7), which are functions of the device dimensions and material properties. The electric forcing term in the equation of motion represents a strong nonlinearity. Traditionally, this nonlinearity is expanded first in a Taylor-series expansion to obtain

$$\frac{\partial^2 w}{\partial t^2} + c\frac{\partial w}{\partial t} + \frac{\partial^4 w}{\partial x^4} = \left[\alpha_1 \int_0^1 \left(\frac{\partial w}{\partial x}\right)^2 dx + N\right] \frac{\partial^2 w}{\partial x^2}$$
$$+ \alpha_2 V^2 \left(1 + 2w + 3w^2 + 4w^3 + 5w^4 + 6w^5\right) \quad (5.4.49)$$

which is then discretized. However, as the deflection of the microbeam increases, the importance of the truncated higher-order terms increases and the quality of the approximation degrades.

Alternatively, Younis, Abdel-Rahman, and Nayfeh (2003) have developed a new approach where both sides of (5.4.48) are multiplied by the term $(1 - w)^2$, thus producing

$$(1-w)^2 \frac{\partial^2 w}{\partial t^2} + c(1-w)^2 \frac{\partial w}{\partial t} + (1-w)^2 \frac{\partial^4 w}{\partial x^4}$$
$$= \left[\alpha_1 \int_0^1 \left(\frac{\partial w}{\partial x}\right)^2 dx + N\right](1-w)^2 \frac{\partial^2 w}{\partial x^2} + \alpha_2 V^2 \quad (5.4.50)$$

which is then discretized. This form of the equation of motion enables one to model the electric force exactly in the discretized equations.

Figure 5.4.4 compares the nondimensional maximum deflection W_{Max} in terms of V_p obtained by solving the boundary-value problem with the solutions obtained from the discretization of (5.4.50) employing the first five symmetric modes. The configuration of the microbeam used in this study yields a pull-in voltage of 8.75 V. Younis, Abdel-Rahman, and Nayfeh (2003) found that, whereas the shooting

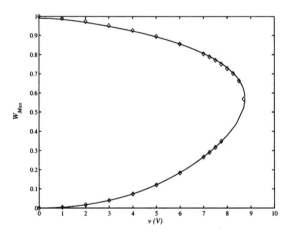

Fig. 5.4.4 Variation of W_{max} calculated using five symmetric modes. The discrete points are results obtained by solving the static boundary-value problem using a shooting method.

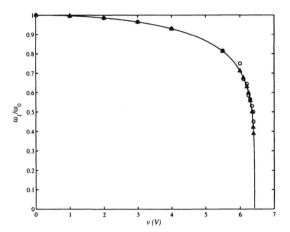

Fig. 5.4.5 Comparison of the normalized fundamental natural frequency calculated using the reduced-order model and employing five symmetric modes in the discretization (solid line) with results obtained (triangles) and the experimental results (circles) obtained by Tilmans and Legtenberg (1994).

method used to solve the BVP fails to converge as the pull-in voltage is approached, the reduced-order model is robust up to the pull-in instability.

Next, they used the reduced-order model to calculate the natural frequencies of a resonant microsensor. In Figure 5.4.5, we compare the normalized fundamental natural frequency calculated using the reduced-order model with five symmetric modes in the discretization (solid line) with results obtained by solving the eigenvalue problem of the distributed-parameter system (triangles) using a shooting method and the experimental results (circles) obtained by Tilmans and Legtenberg (1994) for a resonator with the specifications $L = 210\ \mu m$, $h = 1.5\ \mu m$, $b = 100\ \mu m$, $d = 1.18\ \mu m$, $E = 166\ GPa$, and $\hat{N} = 0.0009\ N$. There is excellent agreement among the results. The reduced-order model is capable of predicting the natural frequency over the whole range even as the microbeam approaches its stability limit where the frequency approaches zero.

6
SURFACE ANALYSIS

Plates and shells are surface structures, whose deformations are mainly described by their reference surfaces. However, shells are initially curved structures and hence curvilinear coordinate systems are always needed in the modeling and analysis. On the other hand, plates have no initial curvatures and may be described by rectilinear coordinates. However, for non-rectangular plates, curvilinear coordinate systems need to be used to describe their boundaries. For example, the polar coordinate system (i.e., a cylindrical coordinate system without the longitudinal coordinate, see Figure 2.2.2) is used in the modeling of circular plates. Hence, the modeling and analysis of plates are almost the same as those of shells.

In this chapter, we formulate initial curvatures, derive fully nonlinear expressions for the inplane strains, curvatures, and a transformation matrix which relates the deformed coordinate system to the undeformed one, relate variations of the curvatures to the orthogonal virtual rotations, and show how to account for the inplane shear strains in the nonlinear modeling. To unify and simplify the formulation, instead of using Lamé parameters or measure numbers and principal curvatures, we use general curvatures in the derivations.

6.1 INITIAL CURVATURES

Similar to beam problems, two coordinate systems are needed in the modeling of a shell structure in order to fully account for geometric nonlinearities. One system describes the undeformed reference surface and the other describes the deformed reference surface. Figure 6.1.1 shows a general shell reference surface before and

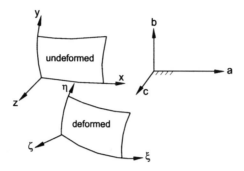

Fig. 6.1.1 Coordinate systems.

after deformation. The xyz system is an orthogonal curvilinear coordinate system with the curvilinear axes x and y being on the undeformed reference surface and the z axis being a rectilinear axis. The $\xi\eta\zeta$ system is an orthogonal curvilinear coordinate system with the curvilinear axes ξ and η being on the deformed reference surface and the ζ axis being a rectilinear axis. Also, a body-fixed Cartesian coordinate system abc is used, for reference purpose, in the calculation of the initial curvatures. We let $\mathbf{j}_1$, $\mathbf{j}_2$, and $\mathbf{j}_3$ denote the unit vectors along the axes x, y, and z; $\mathbf{i}_1$, $\mathbf{i}_2$, and $\mathbf{i}_3$ denote the unit vectors along the axes ξ, η, and ζ; and $\mathbf{i}_a$, $\mathbf{i}_b$, and $\mathbf{i}_c$ denote the unit vectors along the axes a, b, and c, respectively.

Figure 6.1.2 shows an infinitesimal element of a general shell reference surface before and after deformation. The undeformed position vector $\mathbf{P}$ of the observed Point A in Figure 6.1.2 is assumed to be known and given by

$$\mathbf{P} = p_1(x,y)\mathbf{i}_a + p_2(x,y)\mathbf{i}_b + p_3(x,y)\mathbf{i}_c \tag{6.1.1}$$

Taking the first-order derivatives of (6.1.1) with respect to x and y yields

$$\mathbf{j}_1 = \frac{\partial \mathbf{P}}{\partial x} = p_{1x}\mathbf{i}_a + p_{2x}\mathbf{i}_b + p_{3x}\mathbf{i}_c$$

$$\mathbf{j}_2 = \frac{\partial \mathbf{P}}{\partial y} = p_{1y}\mathbf{i}_a + p_{2y}\mathbf{i}_b + p_{3y}\mathbf{i}_c$$

$$\mathbf{j}_3 = \mathbf{j}_1 \times \mathbf{j}_2 = (p_{2x}p_{3y} - p_{3x}p_{2y})\mathbf{i}_a + (p_{3x}p_{1y} - p_{1x}p_{3y})\mathbf{i}_b$$
$$+ (p_{1x}p_{2y} - p_{2x}p_{1y})\mathbf{i}_c \tag{6.1.2}$$

where $(\ )_x \equiv \partial(\ )/\partial x$ and $(\ )_y \equiv \partial(\ )/\partial y$. Hence, the base vectors of the xyz system are related to the base vectors of the abc system by the known transformation matrix $[T^x]$ as

$$\{\mathbf{j}_{123}\} = [T^x]\{\mathbf{i}_{abc}\} \tag{6.1.3}$$

INITIAL CURVATURES

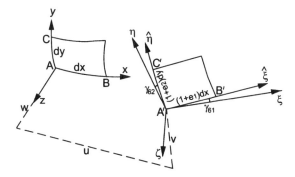

Fig. 6.1.2 The undeformed and deformed geometries of a differential element of a shell reference surface.

where $\{\mathbf{j}_{123}\} \equiv \{\mathbf{j}_1, \mathbf{j}_2, \mathbf{j}_3\}^T$, $\{\mathbf{i}_{abc}\} \equiv \{\mathbf{i}_a, \mathbf{i}_b, \mathbf{i}_c\}^T$, and

$$[T^x] = \begin{bmatrix} p_{1x} & p_{2x} & p_{3x} \\ p_{1y} & p_{2y} & p_{3y} \\ p_{2x}p_{3y} - p_{3x}p_{2y} & p_{3x}p_{1y} - p_{1x}p_{3y} & p_{1x}p_{2y} - p_{2x}p_{1y} \end{bmatrix} \quad (6.1.4)$$

Using (6.1.3) and the identities $[T^x]^{-1} = [T^x]^T$ and

$$\frac{\partial \mathbf{j}_m}{\partial x} \cdot \mathbf{j}_m = \frac{\partial \mathbf{j}_m}{\partial y} \cdot \mathbf{j}_m = 0, \quad \frac{\partial \mathbf{j}_m}{\partial x} \cdot \mathbf{j}_n = -\frac{\partial \mathbf{j}_n}{\partial x} \cdot \mathbf{j}_m, \quad \frac{\partial \mathbf{j}_m}{\partial y} \cdot \mathbf{j}_n = -\frac{\partial \mathbf{j}_n}{\partial y} \cdot \mathbf{j}_m \quad (6.1.5)$$

we obtain

$$\frac{\partial \{\mathbf{j}_{123}\}}{\partial x} = [K_1^0]\{\mathbf{j}_{123}\}, \quad \frac{\partial \{\mathbf{j}_{123}\}}{\partial y} = [K_2^0]\{\mathbf{j}_{123}\}, \quad \frac{\partial \{\mathbf{j}_{123}\}}{\partial z} = 0 \quad (6.1.6)$$

$$[K_1^0] \equiv \begin{bmatrix} \mathbf{j}_{1x} \cdot \mathbf{j}_1 & \mathbf{j}_{1x} \cdot \mathbf{j}_2 & \mathbf{j}_{1x} \cdot \mathbf{j}_3 \\ \mathbf{j}_{2x} \cdot \mathbf{j}_1 & \mathbf{j}_{2x} \cdot \mathbf{j}_2 & \mathbf{j}_{2x} \cdot \mathbf{j}_3 \\ \mathbf{j}_{3x} \cdot \mathbf{j}_1 & \mathbf{j}_{3x} \cdot \mathbf{j}_2 & \mathbf{j}_{3x} \cdot \mathbf{j}_3 \end{bmatrix} = \begin{bmatrix} 0 & k_5^0 & -k_1^0 \\ -k_5^0 & 0 & -k_{61}^0 \\ k_1^0 & k_{61}^0 & 0 \end{bmatrix} = \frac{\partial [T^x]}{\partial x}[T^x]^T$$

$$[K_2^0] \equiv \begin{bmatrix} \mathbf{j}_{1y} \cdot \mathbf{j}_1 & \mathbf{j}_{1y} \cdot \mathbf{j}_2 & \mathbf{j}_{1y} \cdot \mathbf{j}_3 \\ \mathbf{j}_{2y} \cdot \mathbf{j}_1 & \mathbf{j}_{2y} \cdot \mathbf{j}_2 & \mathbf{j}_{2y} \cdot \mathbf{j}_3 \\ \mathbf{j}_{3y} \cdot \mathbf{j}_1 & \mathbf{j}_{3y} \cdot \mathbf{j}_2 & \mathbf{j}_{3y} \cdot \mathbf{j}_3 \end{bmatrix} = \begin{bmatrix} 0 & k_4^0 & -k_{62}^0 \\ -k_4^0 & 0 & -k_2^0 \\ k_{62}^0 & k_2^0 & 0 \end{bmatrix} = \frac{\partial [T^x]}{\partial y}[T^x]^T$$

where $[K_1^0]$ and $[K_2^0]$ are the initial curvature matrices. The initial curvatures are given by

$$k_1^0 \equiv \frac{\partial \mathbf{j}_3}{\partial x} \cdot \mathbf{j}_1 = -\frac{\partial \mathbf{j}_1}{\partial x} \cdot \mathbf{j}_3 = -\sum_{j=1}^{3} \frac{\partial T_{1j}^x}{\partial x} T_{3j}^x$$

$$= -p_{1xx}(p_{2x}p_{3y} - p_{3x}p_{2y}) - p_{2xx}(p_{3x}p_{1y} - p_{1x}p_{3y})$$
$$- p_{3xx}(p_{1x}p_{2y} - p_{2x}p_{1y})$$

$$k_2^0 \equiv \frac{\partial \mathbf{j}_3}{\partial y} \cdot \mathbf{j}_2 = -\frac{\partial \mathbf{j}_2}{\partial y} \cdot \mathbf{j}_3 = -\sum_{j=1}^{3} \frac{\partial T_{2j}^x}{\partial y} T_{3j}^x$$

$$k_{61}^0 \equiv \frac{\partial \mathbf{j}_3}{\partial x} \cdot \mathbf{j}_2 = -\frac{\partial \mathbf{j}_2}{\partial x} \cdot \mathbf{j}_3 = -\sum_{j=1}^{3} \frac{\partial T_{2j}^x}{\partial x} T_{3j}^x$$

$$k_{62}^0 \equiv \frac{\partial \mathbf{j}_3}{\partial y} \cdot \mathbf{j}_1 = -\frac{\partial \mathbf{j}_1}{\partial y} \cdot \mathbf{j}_3 = -\sum_{j=1}^{3} \frac{\partial T_{1j}^x}{\partial y} T_{3j}^x$$

$$k_5^0 \equiv -\frac{\partial \mathbf{j}_2}{\partial x} \cdot \mathbf{j}_1 = \frac{\partial \mathbf{j}_1}{\partial x} \cdot \mathbf{j}_2 = \sum_{j=1}^{3} \frac{\partial T_{1j}^x}{\partial x} T_{2j}^x$$

$$k_4^0 \equiv -\frac{\partial \mathbf{j}_2}{\partial y} \cdot \mathbf{j}_1 = \frac{\partial \mathbf{j}_1}{\partial y} \cdot \mathbf{j}_2 = \sum_{j=1}^{3} \frac{\partial T_{1j}^x}{\partial y} T_{2j}^x \quad (6.1.7)$$

Here, k_{61}^0, k_1^0, and k_5^0 are the initial twisting curvature of the x axis with respect to $-x$, the spiral curvature with respect to y, and the bending curvature with respect to z, respectively; and k_{62}^0, k_2^0, and k_4^0 are the initial twisting curvature of the y axis with respect to y, the spiral curvature with respect to $-x$, and the bending curvature with respect to z, respectively. If $k_{61}^0 = k_{62}^0 = k_4^0 = k_5^0 = 0$, x and y are the principal orthogonal curvilinear coordinates and k_1^0 and k_2^0 are the so-called principal curvatures. We note that $\partial \{\mathbf{j}_{123}\}/\partial z = 0$ is always true in plate and shell analyses because the z axis is rectilinear and perpendicular to the reference surface. Because $\mathbf{P}$ is the undeformed position vector of the observed reference Point A, if the undeformed reference surface is smooth,

$$\mathbf{P}_{xy} = \mathbf{P}_{yx} \quad (6.1.8a)$$

Because $\mathbf{j}_1 = \partial \mathbf{P}/\partial x = \mathbf{P}_x$ and $\mathbf{j}_2 = \mathbf{P}_y$,

$$k_{61}^0 = -\mathbf{P}_{yx} \cdot \mathbf{j}_3 = -\mathbf{P}_{xy} \cdot \mathbf{j}_3 = k_{62}^0 \quad (6.1.8b)$$

6.2 INPLANE STRAINS AND DEFORMED CURVATURES

Figure 6.1.2 shows an infinitesimal element of the reference surface shown in Figure 6.1.1 before and after deformation, where the $\hat{\xi}$ and $\hat{\eta}$ represent the deformed curves of the axes x and y and $\mathbf{i}_{\hat{1}}$ and $\mathbf{i}_{\hat{2}}$ are unit vectors along the $\hat{\xi}$ and $\hat{\eta}$ axes, respectively. Moreover, γ_6 ($\equiv \gamma_{61} + \gamma_{62}$) is the inplane shear deformation, and the ξ and η axes coincide with $\hat{\xi}$ and $\hat{\eta}$ only if the inplane shear deformation γ_6 is zero. In Figure 6.1.2, the displacement vectors $\mathbf{D}_i$ of the vertices of the observed element are

$$A: \mathbf{D}_1 = u\mathbf{j}_1 + v\mathbf{j}_2 + w\mathbf{j}_3$$
$$B: \mathbf{D}_2 = \mathbf{D}_1 + \frac{\partial \mathbf{D}_1}{\partial x} dx$$
$$= \mathbf{D}_1 + [(u_x - vk_5^0 + wk_1^0)\mathbf{j}_1 + (v_x + uk_5^0 + wk_{61}^0)\mathbf{j}_2$$

$$C: \mathbf{D}_3 = \mathbf{D}_1 + \frac{+(w_x - uk_1^0 - vk_{61}^0)\mathbf{j}_3]\,dx}{\partial y}\,dy$$

$$= \mathbf{D}_1 + [(u_y - vk_4^0 + wk_{62}^0)\mathbf{j}_1 + (v_y + uk_4^0 + wk_2^0)\mathbf{j}_2$$
$$+(w_y - uk_{62}^0 - vk_2^0)\mathbf{j}_3]\,dy \qquad (6.2.1)$$

on account of (6.1.6). Here, u, v, and w are the displacements of the observed reference Point A with respect to the axes x, y, and z, respectively. Hence, we obtain from Figure 6.1.2 and (6.2.1) that

$$\overrightarrow{A'B'} = dx\mathbf{j}_1 + \mathbf{D}_2 - \mathbf{D}_1$$
$$= [(1 + u_x - vk_5^0 + wk_1^0)\mathbf{j}_1 + (v_x + uk_5^0 + wk_{61}^0)\mathbf{j}_2$$
$$+(w_x - uk_1^0 - vk_{61}^0)\mathbf{j}_3]\,dx \qquad (6.2.2a)$$

$$\overrightarrow{A'C'} = dy\mathbf{j}_2 + \mathbf{D}_3 - \mathbf{D}_1$$
$$= [(u_y - vk_4^0 + wk_{62}^0)\mathbf{j}_1 + (1 + v_y + uk_4^0 + wk_2^0)\mathbf{j}_2$$
$$+(w_y - uk_{62}^0 - vk_2^0)\mathbf{j}_3]\,dy \qquad (6.2.2b)$$

Therefore, the axial strains along the axes $\hat{\xi}$ and $\hat{\eta}$ are e_1 and e_2 and are given by

$$e_1 = \frac{\overline{A'B'} - dx}{dx} = -1$$
$$+ \sqrt{(1 + u_x - vk_5^0 + wk_1^0)^2 + (v_x + uk_5^0 + wk_{61}^0)^2 + (w_x - uk_1^0 - vk_{61}^0)^2}$$
$$\qquad (6.2.3)$$

$$e_2 = \frac{\overline{A'C'} - dy}{dy} = -1$$
$$+ \sqrt{(u_y - vk_4^0 + wk_{62}^0)^2 + (1 + v_y + uk_4^0 + wk_2^0)^2 + (w_y - uk_{62}^0 - vk_2^0)^2}$$
$$\qquad (6.2.4)$$

The unit vectors along the $\hat{\xi}$ and $\hat{\eta}$ directions are obtained from (6.2.2a,b) as

$$\mathbf{i}_{\hat{1}} = \frac{\overrightarrow{A'B'}}{(1+e_1)dx} = \hat{T}_{11}\mathbf{j}_1 + \hat{T}_{12}\mathbf{j}_2 + \hat{T}_{13}\mathbf{j}_3 \qquad (6.2.5)$$

$$\mathbf{i}_{\hat{2}} = \frac{\overrightarrow{A'C'}}{(1+e_2)dy} = \hat{T}_{21}\mathbf{j}_1 + \hat{T}_{22}\mathbf{j}_2 + \hat{T}_{23}\mathbf{j}_3 \qquad (6.2.6)$$

where

$$\hat{T}_{11} = \frac{1 + u_x - vk_5^0 + wk_1^0}{1 + e_1}, \qquad \hat{T}_{12} = \frac{v_x + uk_5^0 + wk_{61}^0}{1 + e_1}$$

$$\hat{T}_{13} = \frac{w_x - uk_1^0 - vk_{61}^0}{1 + e_1}, \qquad \hat{T}_{21} = \frac{u_y - vk_4^0 + wk_{62}^0}{1 + e_2}$$

$$\hat{T}_{22} = \frac{1 + v_y + uk_4^0 + wk_2^0}{1 + e_2}, \qquad \hat{T}_{23} = \frac{w_y - uk_{62}^0 - vk_2^0}{1 + e_2} \qquad (6.2.7)$$

Using (6.2.5) and (6.2.6), we obtain

$$\gamma_6 \equiv \gamma_{61} + \gamma_{62} = \sin^{-1}(\mathbf{i}_{\hat{1}} \cdot \mathbf{i}_{\hat{2}})$$
$$= \sin^{-1}(\hat{T}_{11}\hat{T}_{21} + \hat{T}_{12}\hat{T}_{22} + \hat{T}_{13}\hat{T}_{23}) \quad (6.2.8a)$$

Hence, γ_6 can be represented in terms of u, v, and w by substituting (6.2.7) into (6.2.8a). To determine the unique expressions of γ_{61} and γ_{62}, one needs to use the symmetry of the inplane shear strain B_{12} (i.e., $B_{12} = B_{21}$). Because $B_{12} = (1+e_1)dx\sin\gamma_{61}/dx$ and $B_{21} = (1+e_2)dy\sin\gamma_{62}/dy$,

$$(1+e_1)\sin\gamma_{61} = (1+e_2)\sin\gamma_{62} \quad (6.2.8b)$$

The unit vector normal to the deformed reference plane is defined as

$$\mathbf{i}_3 = \frac{\mathbf{i}_{\hat{1}} \times \mathbf{i}_{\hat{2}}}{|\mathbf{i}_{\hat{1}} \times \mathbf{i}_{\hat{2}}|} = T_{31}\mathbf{j}_1 + T_{32}\mathbf{j}_2 + T_{33}\mathbf{j}_3 \quad (6.2.9)$$

where

$$T_{31} = (\hat{T}_{12}\hat{T}_{23} - \hat{T}_{13}\hat{T}_{22})/R_0, \; T_{32} = (\hat{T}_{13}\hat{T}_{21} - \hat{T}_{11}\hat{T}_{23})/R_0$$
$$T_{33} = (\hat{T}_{11}\hat{T}_{22} - \hat{T}_{12}\hat{T}_{21})/R_0$$
$$R_0 \equiv \sqrt{(\hat{T}_{12}\hat{T}_{23} - \hat{T}_{13}\hat{T}_{22})^2 + (\hat{T}_{13}\hat{T}_{21} - \hat{T}_{11}\hat{T}_{23})^2 + (\hat{T}_{11}\hat{T}_{22} - \hat{T}_{12}\hat{T}_{21})^2}$$
$$= |\mathbf{i}_{\hat{1}} \times \mathbf{i}_{\hat{2}}| = |\cos\gamma_6| \quad (6.2.10)$$

Using (6.2.5), (6.2.6), (6.2.9), and Figure 6.1.2, we obtain the following transformation relating the undeformed coordinate system xyz to the deformed coordinate system $\xi\eta\zeta$:

$$\{\mathbf{i}_{123}\} = [T]\{\mathbf{j}_{123}\}, \quad [T] \equiv [\Gamma]\begin{bmatrix} \hat{T}_{11} & \hat{T}_{12} & \hat{T}_{13} \\ \hat{T}_{21} & \hat{T}_{22} & \hat{T}_{23} \\ \hat{T}_{31} & \hat{T}_{32} & \hat{T}_{33} \end{bmatrix} \quad (6.2.11a)$$

where $\{\mathbf{i}_{123}\} \equiv \{\mathbf{i}_1, \mathbf{i}_2, \mathbf{i}_3\}^T$,

$$\{\mathbf{i}_{123}\} = [\Gamma]\{\mathbf{i}_{\hat{1}\hat{2}3}\}$$

$$[\Gamma] \equiv \begin{bmatrix} \cos\gamma_{61} & \sin\gamma_{61} & 0 \\ \sin\gamma_{62} & \cos\gamma_{62} & 0 \\ 0 & 0 & 1 \end{bmatrix}^{-1} = \frac{1}{\cos\gamma_6}\begin{bmatrix} \cos\gamma_{62} & -\sin\gamma_{61} & 0 \\ -\sin\gamma_{62} & \cos\gamma_{61} & 0 \\ 0 & 0 & \cos\gamma_6 \end{bmatrix}$$
$$(6.2.11b)$$

and $\{\mathbf{i}_{\hat{1}\hat{2}3}\} \equiv \{\mathbf{i}_{\hat{1}}, \mathbf{i}_{\hat{2}}, \mathbf{i}_3\}^T$.

Using (6.2.11a), (6.1.6), $[T]^{-1} = [T]^T$, and the identities

$$\frac{\partial \mathbf{i}_j}{\partial x} \cdot \mathbf{i}_j = \frac{\partial \mathbf{i}_j}{\partial y} \cdot \mathbf{i}_j = 0, \quad \frac{\partial \mathbf{i}_j}{\partial x} \cdot \mathbf{i}_k = -\frac{\partial \mathbf{i}_k}{\partial x} \cdot \mathbf{i}_j, \quad \frac{\partial \mathbf{i}_j}{\partial y} \cdot \mathbf{i}_k = -\frac{\partial \mathbf{i}_k}{\partial y} \cdot \mathbf{i}_j \quad (6.2.12)$$

we obtain

$$\frac{\partial}{\partial x}\{i_{123}\} = [K_1]\{i_{123}\}$$

$$[K_1] \equiv \begin{bmatrix} i_{1x} \cdot i_1 & i_{1x} \cdot i_2 & i_{1x} \cdot i_3 \\ i_{2x} \cdot i_1 & i_{2x} \cdot i_2 & i_{2x} \cdot i_3 \\ i_{3x} \cdot i_1 & i_{3x} \cdot i_2 & i_{3x} \cdot i_3 \end{bmatrix} = \begin{bmatrix} 0 & k_5 & -k_1 \\ -k_5 & 0 & -k_{61} \\ k_1 & k_{61} & 0 \end{bmatrix}$$

$$= \frac{\partial [T]}{\partial x}[T]^T + [T][K_1^0][T]^T \tag{6.2.13}$$

$$\frac{\partial}{\partial y}\{i_{123}\} = [K_2]\{i_{123}\}$$

$$[K_2] \equiv \begin{bmatrix} i_{1y} \cdot i_1 & i_{1y} \cdot i_2 & i_{1y} \cdot i_3 \\ i_{2y} \cdot i_1 & i_{2y} \cdot i_2 & i_{2y} \cdot i_3 \\ i_{3y} \cdot i_1 & i_{3y} \cdot i_2 & i_{3y} \cdot i_3 \end{bmatrix} = \begin{bmatrix} 0 & k_4 & -k_{62} \\ -k_4 & 0 & -k_2 \\ k_{62} & k_2 & 0 \end{bmatrix}$$

$$= \frac{\partial [T]}{\partial y}[T]^T + [T][K_2^0][T]^T \tag{6.2.14}$$

where $[K_1]$ and $[K_2]$ are the deformed curvature matrices. From (6.2.13) and (6.2.14), we obtain the deformed curvatures as

$$k_1 \equiv -\frac{\partial i_1}{\partial x} \cdot i_3 = -\sum_{m=1}^{3} T_{1mx}T_{3m} - T_{21}k_{61}^0 + T_{22}k_1^0 + T_{23}k_5^0$$

$$k_2 \equiv -\frac{\partial i_2}{\partial y} \cdot i_3 = -\sum_{m=1}^{3} T_{2my}T_{3m} + T_{11}k_2^0 - T_{12}k_{62}^0 - T_{13}k_4^0$$

$$k_{61} \equiv -\frac{\partial i_2}{\partial x} \cdot i_3 = -\sum_{m=1}^{3} T_{2mx}T_{3m} + T_{11}k_{61}^0 - T_{12}k_1^0 - T_{13}k_5^0$$

$$k_{62} \equiv -\frac{\partial i_1}{\partial y} \cdot i_3 = -\sum_{m=1}^{3} T_{1my}T_{3m} - T_{21}k_2^0 + T_{22}k_{62}^0 + T_{23}k_4^0$$

$$k_5 \equiv \frac{\partial i_1}{\partial x} \cdot i_2 = \sum_{m=1}^{3} T_{1mx}T_{2m} - T_{31}k_{61}^0 + T_{32}k_1^0 + T_{33}k_5^0$$

$$k_4 \equiv -\frac{\partial i_2}{\partial y} \cdot i_1 = -\sum_{m=1}^{3} T_{2my}T_{1m} - T_{31}k_2^0 + T_{32}k_{62}^0 + T_{33}k_4^0 \tag{6.2.15}$$

Here, (6.2.11a) and $i_1 = i_2 \times i_3$, $i_2 = i_3 \times i_1$, $i_3 = i_1 \times i_2$ (see (4.4.62) and (4.4.63)) are used. We note that the curvatures in (6.2.13) and (6.2.14) do not represent real curvatures because the differentiations are taken with respect to the undeformed lengths dx and dy, not the deformed lengths along the axes ξ and η. When $\gamma_{61} = \gamma_{62} = e_1 = e_2 = 0$, k_1 and k_2 denote the bending curvatures with respect to the axes η and $-\xi$, respectively; k_{61} and k_{62} are the twisting curvatures with respect to the axes $-\xi$ and η, respectively; and k_4 and k_5 are the "drilling" or "spiral" curvatures, with respect to the axis ζ, of the axes η and ξ, respectively.

6.3 ORTHOGONAL VIRTUAL ROTATIONS

The variations of the unit vectors $\mathbf{i}_j$ of the $\xi\eta\zeta$ system are due to the virtual rigid-body rotations $\delta\theta_i$ of the $\xi\eta\zeta$ frame and are given by

$$\left\{\begin{array}{c}\delta\mathbf{i}_1\\ \delta\mathbf{i}_2\\ \delta\mathbf{i}_3\end{array}\right\} = \left[\begin{array}{ccc}0 & \delta\theta_3 & -\delta\theta_2\\ -\delta\theta_3 & 0 & \delta\theta_1\\ \delta\theta_2 & -\delta\theta_1 & 0\end{array}\right]\left\{\begin{array}{c}\mathbf{i}_1\\ \mathbf{i}_2\\ \mathbf{i}_3\end{array}\right\} \quad (6.3.1\mathrm{a})$$

Substituting for the $\mathbf{i}_j$ from (6.2.11a) and knowing that the $\delta\mathbf{j}_m = 0$, we obtain from (6.3.1a) that

$$\delta\theta_1 = \mathbf{i}_3 \cdot \delta\mathbf{i}_2 = -\mathbf{i}_2 \cdot \delta\mathbf{i}_3 = T_{31}\delta T_{21} + T_{32}\delta T_{22} + T_{33}\delta T_{23}$$
$$\delta\theta_2 = \mathbf{i}_1 \cdot \delta\mathbf{i}_3 = -\mathbf{i}_3 \cdot \delta\mathbf{i}_1 = T_{11}\delta T_{31} + T_{12}\delta T_{32} + T_{13}\delta T_{33}$$
$$\delta\theta_3 = \mathbf{i}_2 \cdot \delta\mathbf{i}_1 = -\mathbf{i}_1 \cdot \delta\mathbf{i}_2 = T_{21}\delta T_{11} + T_{22}\delta T_{12} + T_{23}\delta T_{13} \quad (6.3.1\mathrm{b})$$

6.3.1 Without Inplane Shear Strains

For thin-walled structures, deformations are mainly due to transverse displacements and bending rotations, and the inplane shear strains γ_{61}, γ_{62}, and γ_6 are usually small and negligible. Assuming $\gamma_{61} = \gamma_{62} = \gamma_6 = 0$ yields

$$[\Gamma] = [I], \quad \hat{T}_{1i} = T_{1i}, \quad \hat{T}_{2i} = T_{2i}, \quad \mathbf{i}_1 = \mathbf{i}_{\hat{1}}, \quad \mathbf{i}_2 = \mathbf{i}_{\hat{2}} \quad (6.3.2)$$

where $[I]$ is an identity matrix. Substituting (6.2.7) and (6.3.2) into the identities $T_{11}^2 + T_{12}^2 + T_{13}^2 = T_{21}^2 + T_{22}^2 + T_{13}^2 = 1$ and taking the variation of the results, we obtain

$$\delta e_1 = T_{11}\delta(u_x - vk_5^0 + wk_1^0) + T_{12}\delta(v_x + uk_5^0 + wk_{61}^0)$$
$$+ T_{13}\delta(w_x - uk_1^0 - vk_{61}^0) \quad (6.3.3\mathrm{a})$$

$$\delta e_2 = T_{21}\delta(u_y - vk_4^0 + wk_{62}^0) + T_{22}\delta(v_y + uk_4^0 + wk_2^0)$$
$$+ T_{23}\delta(w_y - uk_{62}^0 - vk_2^0) \quad (6.3.3\mathrm{b})$$

Taking the variations of $\mathbf{i}_1$ and $\mathbf{i}_2$ in (6.2.11a) and using (6.2.7) and (6.3.2), we obtain

$$\delta\mathbf{i}_1 = \mathbf{j}_1\delta T_{11} + \mathbf{j}_2\delta T_{12} + \mathbf{j}_3\delta T_{13}$$
$$= [(\delta u_x - k_5^0\delta v + k_1^0\delta w)\mathbf{j}_1 + (\delta v_x + k_5^0\delta u + k_{61}^0\delta w)\mathbf{j}_2$$
$$+ (\delta w_x - k_1^0\delta u - k_{61}^0\delta v)\mathbf{j}_3]\frac{1}{1+e_1} - \frac{\delta e_1}{1+e_1}\mathbf{i}_1 \quad (6.3.4\mathrm{a})$$

$$\delta\mathbf{i}_2 = \mathbf{j}_1\delta T_{21} + \mathbf{j}_2\delta T_{22} + \mathbf{j}_3\delta T_{23}$$
$$= [(\delta u_y - k_4^0\delta v + k_{62}^0\delta w)\mathbf{j}_1 + (\delta v_y + k_4^0\delta u + k_2^0\delta w)\mathbf{j}_2$$
$$+ (\delta w_y - k_{62}^0\delta u - k_2^0\delta v)\mathbf{j}_3]\frac{1}{1+e_2} - \frac{\delta e_2}{1+e_2}\mathbf{i}_2 \quad (6.3.4\mathrm{b})$$

Substituting (6.3.4a,b) into $\delta\theta_1 = \mathbf{i}_3 \cdot \delta\mathbf{i}_2$ and $\delta\theta_2 = -\mathbf{i}_3 \cdot \delta\mathbf{i}_1$ from (6.3.1a) and using (6.2.11a) yields

$$(1 + e_1)\delta\theta_2 + T_{31}(\delta u_x - k_5^0 \delta v + k_1^0 \delta w) + T_{32}(\delta v_x + k_5^0 \delta u + k_{61}^0 \delta w)$$
$$+ T_{33}(\delta w_x - k_1^0 \delta u - k_{61}^0 \delta v) = 0 \qquad (6.3.5a)$$

$$-(1 + e_2)\delta\theta_1 + T_{31}(\delta u_y - k_4^0 \delta v + k_{62}^0 \delta w) + T_{32}(\delta v_y + k_4^0 \delta u + k_2^0 \delta w)$$
$$+ T_{33}(\delta w_y - k_{62}^0 \delta u - k_2^0 \delta v) = 0 \qquad (6.3.5b)$$

Also, it can be seen from Figure 6.1.2 that

$$\delta\gamma_{61} = \delta\mathbf{i}_{\hat{1}} \cdot \mathbf{i}_2, \quad \delta\gamma_{62} = \delta\mathbf{i}_{\hat{2}} \cdot \mathbf{i}_1 \qquad (6.3.6)$$

Substituting (6.2.5)-(6.2.7) into (6.3.6) and using (6.2.11a), we obtain

$$\delta\gamma_{61} = \frac{1}{1 + e_1} \left[T_{21}(\delta u_x - k_5^0 \delta v + k_1^0 \delta w) + T_{22}(\delta v_x + k_5^0 \delta u + k_{61}^0 \delta w) \right.$$
$$\left. + T_{23}(\delta w_x - k_1^0 \delta u - k_{61}^0 \delta v) \right] \qquad (6.3.7a)$$

$$\delta\gamma_{62} = \frac{1}{1 + e_2} \left[T_{11}(\delta u_y - k_4^0 \delta v + k_{62}^0 \delta w) + T_{12}(\delta v_y + k_4^0 \delta u + k_2^0 \delta w) \right.$$
$$\left. + T_{13}(\delta w_y - k_{62}^0 \delta u - k_2^0 \delta v) \right] \qquad (6.3.7b)$$

6.3.2 With Inplane Shear Strains

If the inplane shear strain γ_6 is not neglected, then the derivation is complicated. Using (6.2.7) and the fact that $\hat{T}_{11}^2 + \hat{T}_{12}^2 + \hat{T}_{13}^2 = \hat{T}_{21}^2 + \hat{T}_{22}^2 + \hat{T}_{13}^2 = 1$ and taking their variations, we obtain the variations of the inplane strains e_1 and e_2 as

$$\delta e_1 = \hat{T}_{11} \delta t_{11} + \hat{T}_{12} \delta t_{12} + \hat{T}_{13} \delta t_{13} \qquad (6.3.8)$$

$$\delta e_2 = \hat{T}_{21} \delta t_{21} + \hat{T}_{22} \delta t_{22} + \hat{T}_{23} \delta t_{23} \qquad (6.3.9)$$

where

$$\delta t_{11} = \delta(1 + u_x - vk_5^0 + wk_1^0) = \delta u_x - k_5^0 \delta v + k_1^0 \delta w$$
$$\delta t_{12} = \delta(v_x + uk_5^0 + wk_{61}^0) = \delta v_x + k_5^0 \delta u + k_{61}^0 \delta w$$
$$\delta t_{13} = \delta(w_x - uk_1^0 - vk_{61}^0) = \delta w_x - k_1^0 \delta u - k_{61}^0 \delta v$$
$$\delta t_{21} = \delta(u_y - vk_4^0 + wk_{62}^0) = \delta u_y - k_4^0 \delta v + k_{62}^0 \delta w$$
$$\delta t_{22} = \delta(1 + v_y + uk_4^0 + wk_2^0) = \delta v_y + k_4^0 \delta u + k_2^0 \delta w$$
$$\delta t_{23} = \delta(w_y - uk_{62}^0 - vk_2^0) = \delta w_y - k_{62}^0 \delta u - k_2^0 \delta v \qquad (6.3.10)$$

We note that (6.3.8) and (6.3.9) are the same as (6.3.3a) and (6.3.3b). Taking the variations of (6.2.5) and (6.2.6) and using (6.2.7) yields

$$\delta\mathbf{i}_{\hat{1}} = \frac{1}{1 + e_1} \left(\mathbf{j}_1 \delta t_{11} + \mathbf{j}_2 \delta t_{12} + \mathbf{j}_3 \delta t_{13} - \mathbf{i}_{\hat{1}} \delta e_1 \right) \qquad (6.3.11a)$$

$$\delta i_{\hat{2}} = \frac{1}{1+e_2}(j_1\delta t_{21} + j_2\delta t_{22} + j_3\delta t_{23} - i_{\hat{2}}\delta e_2) \tag{6.3.11b}$$

It follows from (6.2.8a) that $\sin\gamma_6 = i_{\hat{1}} \cdot i_{\hat{2}}$. Hence,

$$\delta\gamma_6 = (\delta i_{\hat{1}} \cdot i_{\hat{2}} + i_{\hat{1}} \cdot \delta i_{\hat{2}})/\cos\gamma_6$$

Substituting (6.3.11a,b), (6.2.5), and (6.2.6) into this equation and using (6.3.8) and (6.3.9) yields

$$\delta\gamma_6 = \frac{(\hat{T}_{21} - \sin\gamma_6\hat{T}_{11})\delta t_{11} + (\hat{T}_{22} - \sin\gamma_6\hat{T}_{12})\delta t_{12} + (\hat{T}_{23} - \sin\gamma_6\hat{T}_{13})\delta t_{13}}{\cos\gamma_6(1+e_1)}$$
$$+ \frac{(\hat{T}_{11} - \sin\gamma_6\hat{T}_{21})\delta t_{21} + (\hat{T}_{12} - \sin\gamma_6\hat{T}_{22})\delta t_{22} + (\hat{T}_{13} - \sin\gamma_6\hat{T}_{23})\delta t_{23}}{\cos\gamma_6(1+e_2)}$$
$$\tag{6.3.12}$$

Taking the variation of (6.2.8b) and using the fact that $\delta\gamma_6 = \delta\gamma_{61} + \delta\gamma_{62}$, we obtain

$$\delta\gamma_{61} = \frac{(1+e_2)\cos\gamma_{62}\delta\gamma_6 - \sin\gamma_{61}\delta e_1 + \sin\gamma_{62}\delta e_2}{(1+e_1)\cos\gamma_{61} + (1+e_2)\cos\gamma_{62}} \tag{6.3.13a}$$

$$\delta\gamma_{62} = \frac{(1+e_1)\cos\gamma_{61}\delta\gamma_6 + \sin\gamma_{61}\delta e_1 - \sin\gamma_{62}\delta e_2}{(1+e_1)\cos\gamma_{61} + (1+e_2)\cos\gamma_{62}} \tag{6.3.13b}$$

Using (6.3.1a), (6.2.11b), and (6.3.11a,b) and the fact that $i_3 \cdot i_{\hat{1}} = i_3 \cdot i_{\hat{2}} = 0$, we obtain

$$\delta\theta_1 = \delta i_2 \cdot i_3 = \frac{\cos\gamma_{61}}{\cos\gamma_6}\delta i_{\hat{2}} \cdot i_3 - \frac{\sin\gamma_{62}}{\cos\gamma_6}\delta i_{\hat{1}} \cdot i_3$$
$$= \frac{\cos\gamma_{61}}{\cos\gamma_6(1+e_2)}(T_{31}\delta t_{21} + T_{32}\delta t_{22} + T_{33}\delta t_{23})$$
$$- \frac{\sin\gamma_{62}}{\cos\gamma_6(1+e_1)}(T_{31}\delta t_{11} + T_{32}\delta t_{12} + T_{33}\delta t_{13}) \tag{6.3.14}$$

$$\delta\theta_2 = -\delta i_1 \cdot i_3 = \frac{\sin\gamma_{61}}{\cos\gamma_6}\delta i_{\hat{2}} \cdot i_3 - \frac{\cos\gamma_{62}}{\cos\gamma_6}\delta i_{\hat{1}} \cdot i_3$$
$$= \frac{\sin\gamma_{61}}{\cos\gamma_6(1+e_2)}(T_{31}\delta t_{21} + T_{32}\delta t_{22} + T_{33}\delta t_{23})$$
$$- \frac{\cos\gamma_{62}}{\cos\gamma_6(1+e_1)}(T_{31}\delta t_{11} + T_{32}\delta t_{12} + T_{33}\delta t_{13}) \tag{6.3.15}$$

Using (6.3.1a), (6.2.11b), (6.3.11a,b), (6.3.8), and (6.3.9), we obtain

$$\delta\theta_3 = \frac{1}{2}(\delta i_1 \cdot i_2 - \delta i_2 \cdot i_1)$$
$$= \delta\gamma_{62} - \delta\gamma_{61} + \frac{1}{2\cos\gamma_6}(\delta i_{\hat{1}} \cdot i_{\hat{2}} - \delta i_{\hat{2}} \cdot i_{\hat{1}})$$
$$= \frac{(\hat{T}_{21} - \sin\gamma_6\hat{T}_{11})\delta t_{11} + (\hat{T}_{22} - \sin\gamma_6\hat{T}_{12})\delta t_{12} + (\hat{T}_{23} - \sin\gamma_6\hat{T}_{13})\delta t_{13}}{2\cos\gamma_6(1+e_1)}$$

$$-\frac{(\hat{T}_{11} - \sin\gamma_6 \hat{T}_{21})\delta t_{21} + (\hat{T}_{12} - \sin\gamma_6 \hat{T}_{22})\delta t_{22} + (\hat{T}_{13} - \sin\gamma_6 \hat{T}_{23})\delta t_{23}}{2\cos\gamma_6(1+e_2)}$$
$$+\delta\gamma_{62} - \delta\gamma_{61} \tag{6.3.16}$$

Hence, δe_1, δe_2, $\delta\gamma_6$, $\delta\gamma_{61}$, $\delta\gamma_{62}$, $\delta\theta_1$, $\delta\theta_2$, and $\delta\theta_3$ can be represented in terms of δu, δv, δw, δu_x, δv_x, δw_x, δu_y, δv_y, and δw_y.

6.4 VARIATION OF CURVATURES

Taking the variations of the deformed curvatures defined in (6.2.15), integrating by parts, using (6.2.13), (6.2.14), and (6.3.1a), we obtain

$$\int_A m\delta k_1 dxdy = \int_A m\left(-\mathbf{i}_{1x}\cdot\delta\mathbf{i}_3 - \mathbf{i}_3\cdot\delta\mathbf{i}_{1x}\right)dxdy$$
$$= \int_A \left(-m\mathbf{i}_{1x}\cdot\delta\mathbf{i}_3 + \frac{\partial m}{\partial x}\mathbf{i}_3\cdot\delta\mathbf{i}_1 + m\mathbf{i}_{3x}\cdot\delta\mathbf{i}_1\right)dxdy - \int_y m\mathbf{i}_3\cdot\delta\mathbf{i}_1\Big|_{x=0}^{x=X}dy$$
$$= \int_A \left(mk_5\delta\theta_1 - \frac{\partial m}{\partial x}\delta\theta_2 + mk_{61}\delta\theta_3\right)dxdy + \int_y m\delta\theta_2\Big|_{x=0}^{x=X}dy \tag{6.4.1a}$$

$$\int_A m\delta k_2 dxdy = \int_A m\left(-\mathbf{i}_{2y}\cdot\delta\mathbf{i}_3 - \mathbf{i}_3\cdot\delta\mathbf{i}_{2y}\right)dxdy$$
$$= \int_A \left(-m\mathbf{i}_{2y}\cdot\delta\mathbf{i}_3 + \frac{\partial m}{\partial y}\mathbf{i}_3\cdot\delta\mathbf{i}_2 + m\mathbf{i}_{3y}\cdot\delta\mathbf{i}_2\right)dxdy - \int_x m\mathbf{i}_3\cdot\delta\mathbf{i}_2\Big|_{y=0}^{y=Y}dx$$
$$= \int_A \left(\frac{\partial m}{\partial y}\delta\theta_1 + mk_4\delta\theta_2 - mk_{62}\delta\theta_3\right)dxdy - \int_x m\delta\theta_1\Big|_{y=0}^{y=Y}dx \tag{6.4.1b}$$

$$\int_A m\delta k_{61} dxdy = \int_A \left(\frac{\partial m}{\partial x}\delta\theta_1 + mk_5\delta\theta_2 - mk_1\delta\theta_3\right)dxdy - \int_y m\delta\theta_1\Big|_{x=0}^{x=X}dy \tag{6.4.1c}$$

$$\int_A m\delta k_{62} dxdy = \int_A \left(mk_4\delta\theta_1 - \frac{\partial m}{\partial y}\delta\theta_2 + mk_2\delta\theta_3\right)dxdy + \int_x m\delta\theta_2\Big|_{y=0}^{y=Y}dx \tag{6.4.1d}$$

$$\int_A m\delta k_4 dxdy = \int_A \left(-mk_{62}\delta\theta_1 - mk_2\delta\theta_2 - \frac{\partial m}{\partial y}\delta\theta_3\right)dxdy + \int_x m\delta\theta_3\Big|_{y=0}^{y=Y}dx \tag{6.4.1e}$$

$$\int_A m\delta k_5 dxdy = \int_A \left(-mk_1\delta\theta_1 - mk_{61}\delta\theta_2 - \frac{\partial m}{\partial x}\delta\theta_3\right)dxdy + \int_y m\delta\theta_3\Big|_{x=0}^{x=X}dy \tag{6.4.1f}$$

where A denotes the undeformed area of the reference plane, m represents a moment, and X and Y are the boundary values of x and y, respectively.

We note that (6.4.1c,a,f) and (6.4.1b,d,e) can be put in the following matrix forms:

$$\int_A m \begin{Bmatrix} -\delta k_{61} \\ \delta k_1 \\ \delta k_5 \end{Bmatrix} dxdy = -\int_A \left(\frac{\partial m}{\partial x} + m[K_1]\right) \begin{Bmatrix} \delta\theta_1 \\ \delta\theta_2 \\ \delta\theta_3 \end{Bmatrix} dxdy$$

$$+ \int_y m \left\{ \begin{array}{c} \delta\theta_1 \\ \delta\theta_2 \\ \delta\theta_3 \end{array} \right\}_{x=0}^{x=X} dy \qquad (6.4.2a)$$

$$\int_A m \left\{ \begin{array}{c} -\delta k_2 \\ \delta k_{62} \\ \delta k_4 \end{array} \right\} dxdy = -\int_A \left(\frac{\partial m}{\partial y} + m[K_2] \right) \left\{ \begin{array}{c} \delta\theta_1 \\ \delta\theta_2 \\ \delta\theta_3 \end{array} \right\} dxdy$$

$$+ \int_x m \left\{ \begin{array}{c} \delta\theta_1 \\ \delta\theta_2 \\ \delta\theta_3 \end{array} \right\}_{y=0}^{y=Y} dx \qquad (6.4.2b)$$

Integrating (6.4.2a,b) by parts yields

$$\left\{ \begin{array}{c} -\delta k_{61} \\ \delta k_1 \\ \delta k_5 \end{array} \right\} = \frac{\partial}{\partial x} \left\{ \begin{array}{c} \delta\theta_1 \\ \delta\theta_2 \\ \delta\theta_3 \end{array} \right\} - [K_1] \left\{ \begin{array}{c} \delta\theta_1 \\ \delta\theta_2 \\ \delta\theta_3 \end{array} \right\} \qquad (6.4.3a)$$

$$\left\{ \begin{array}{c} -\delta k_2 \\ \delta k_{62} \\ \delta k_4 \end{array} \right\} = \frac{\partial}{\partial y} \left\{ \begin{array}{c} \delta\theta_1 \\ \delta\theta_2 \\ \delta\theta_3 \end{array} \right\} - [K_2] \left\{ \begin{array}{c} \delta\theta_1 \\ \delta\theta_2 \\ \delta\theta_3 \end{array} \right\} \qquad (6.4.3b)$$

Hence, the variations of the curvatures δk_j can be represented in terms of δu, δv, δw, δu_x, δv_x, δw_x, δu_y, δv_y, δw_y, δu_{xx}, δv_{xx}, δw_{xx}, δu_{yy}, δv_{yy}, δw_{yy}, δu_{xy}, δv_{xy}, and δw_{xy}.

6.5 LOCAL DISPLACEMENTS AND JAUMANN STRAINS

Here, we show that fully nonlinear Jaumann strains can be derived by using the concept of local displacements without performing polar decomposition. Figure 6.5.1 shows the undeformed configuration of a filament $\overline{AB}$ and its deformed configuration $\overline{A'B'}$ in the absence of transverse and inplane shear deformations. The position vectors of Points A and B are given by

$$\mathbf{R}_A = \mathbf{R}_o + z\mathbf{j}_3$$
$$\mathbf{R}_B = \mathbf{R}_A + \frac{\partial \mathbf{R}_A}{\partial x} dx = \mathbf{R}_A + \left[(1 + zk_1^0)\mathbf{j}_1 + zk_{61}^0 \mathbf{j}_2\right] dx \qquad (6.5.1)$$

where $\partial \mathbf{R}_o/\partial x = \mathbf{j}_1$ and (6.1.6) are used. Hence, the length $d\tilde{x}$ of $\overline{AB}$ and the unit vector $\mathbf{j}_{\tilde{1}}$ along the $\tilde{x}$ axis are obtained as

$$d\tilde{x} = |\mathbf{R}_B - \mathbf{R}_A| = \tau dx, \quad \mathbf{j}_{\tilde{1}} = \frac{\overrightarrow{AB}}{d\tilde{x}} = \frac{1}{\tau}[(1 + zk_1^0)\mathbf{j}_1 + zk_{61}^0 \mathbf{j}_2] \qquad (6.5.2)$$

where $\tau \equiv \sqrt{(1 + zk_1^0)^2 + (zk_{61}^0)^2}$. We note that $\mathbf{j}_1 = \mathbf{j}_{\tilde{1}}$ only if the initial twisting curvature $k_{61}^0 = 0$. The displacement vector $\mathbf{D}$ of Point A is

$$\mathbf{D} = \{u, \ v, \ w\}\{\mathbf{j}_{123}\} + z\mathbf{i}_3 - z\mathbf{j}_3 \qquad (6.5.3)$$

LOCAL DISPLACEMENTS AND JAUMANN STRAINS

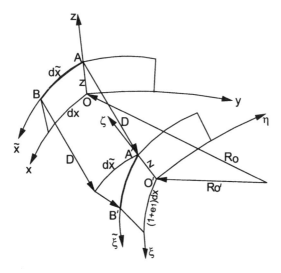

Fig. 6.5.1 The deformation of a filament in a shell.

Taking the derivative of (6.5.3) and using (6.1.6), (6.2.13), and (6.5.2), we obtain

$$\frac{\partial \mathbf{D}}{\partial x} = \{u_x, v_x, w_x\} \{\mathbf{j}_{123}\} + \{u, v, w\} [K_1^0] \{\mathbf{j}_{123}\} + z(k_1\mathbf{i}_1 + k_{61}\mathbf{i}_2) + \mathbf{j}_1 - \tau \mathbf{j}_{\tilde{1}} \quad (6.5.4)$$

It follows from Figure 6.5.1 and (6.5.2) that the local axial strain along the $\tilde{\xi}$ axis is

$$\epsilon_{11} = \frac{(\mathbf{D} + \frac{\partial \mathbf{D}}{\partial x} dx + d\tilde{x}\mathbf{j}_{\tilde{1}} - \mathbf{D}) \cdot \mathbf{i}_{\tilde{1}} - d\tilde{x}}{d\tilde{x}} = \frac{1}{\tau}\frac{\partial \mathbf{D}}{\partial x} \cdot \mathbf{i}_{\tilde{1}} + \mathbf{j}_{\tilde{1}} \cdot \mathbf{i}_{\tilde{1}} - 1 \quad (6.5.5)$$

where $\mathbf{i}_{\tilde{1}}$ is the unit vector along the $\tilde{\xi}$ axis. In the absence of transverse and inplane shears, the position vectors of Points A' and B' are $\mathbf{R}_{A'}$ and $\mathbf{R}_{B'}$, which are given by

$$\mathbf{R}_{A'} = \mathbf{R}_{o'} + z\mathbf{i}_3$$
$$\mathbf{R}_{B'} = \mathbf{R}_{A'} + \frac{\partial \mathbf{R}_{A'}}{\partial x} dx = \mathbf{R}_{A'} + \left[(1 + e_1 + zk_1)\mathbf{i}_1 + zk_{61}\mathbf{i}_2\right] dx \quad (6.5.6)$$

where $\partial \mathbf{R}_{o'}/\partial x = (1 + e_1)\mathbf{i}_1$ and (6.2.13) is used. Hence, the unit vector $\mathbf{i}_{\tilde{1}}$ along the $\tilde{\xi}$ axis is

$$\mathbf{i}_{\tilde{1}} = \frac{\mathbf{R}_{B'} - \mathbf{R}_{A'}}{A'B'} = \frac{1}{\tilde{\tau}}\left[(1 + e_1 + zk_1)\mathbf{i}_1 + zk_{61}\mathbf{i}_2\right] \quad (6.5.7a)$$

where $\tilde{\tau} \equiv \sqrt{(1 + e_1 + zk_1)^2 + (zk_{61})^2}$. We note that, if the influence of local rotations due to strainable local displacements is neglected, (6.5.7a) reduces to

$$\mathbf{i}_{\tilde{1}} = \frac{1}{\tau}\left[(1 + zk_1^0)\mathbf{i}_1 + zk_{61}^0\mathbf{i}_2\right] \quad (6.5.7b)$$

which is an approximation under the assumption of small strains. Substituting (6.5.7a) and (6.5.4) into (6.5.5) and using (6.2.8), we obtain

$$\epsilon_{11} = \frac{1 + e_1 + zk_1}{\tau\tilde{\tau}}[(u_x - vk_5^0 + wk_1^0)T_{11} + (v_x + uk_5^0 + wk_{61}^0)T_{12}$$
$$+ (w_x - uk_1^0 - vk_{61}^0)T_{13} + zk_1 + T_{11}] + \frac{zk_{61}}{\tau\tilde{\tau}}[(u_x - vk_5^0 + wk_1^0)T_{21}$$
$$+ (v_x + uk_5^0 + wk_{61}^0)T_{22} + (w_x - uk_1^0 - vk_{61}^0)T_{23} + zk_{61} + T_{21}]$$
$$- \mathbf{i}_{\tilde{1}} \cdot \mathbf{j}_{\tilde{1}} + \mathbf{j}_{\tilde{1}} \cdot \mathbf{i}_{\tilde{1}} - 1$$
$$= \frac{1 + e_1 + zk_1}{\tau\tilde{\tau}}[1 + e_1 + zk_1] + \frac{zk_{61}}{\tau\tilde{\tau}}[(1 + e_1)\mathbf{i}_1 \cdot \mathbf{i}_2 + zk_{61}] - 1$$
$$= \frac{\tilde{\tau}}{\tau} - 1 \quad (6.5.8)$$

Expanding (6.5.8) and neglecting terms proportional to z^n, $n \geq 2$, we obtain

$$\epsilon_{11} = e_1 + z[k_1 - (1 + e_1)k_1^0] \quad (6.5.9a)$$

We note that k_{61}^0 does not appear in (6.5.9a) and hence the expression of the axial strain along the ξ axis is the same as that along the $\tilde{\xi}$ axis (i.e., (6.5.9a)) because $\mathbf{i}_{\tilde{1}} = \mathbf{i}_1$ if the influence of k_{61}^0 is neglected (see (6.5.7b)). The factor $(1 + e_1)$ in (6.5.9a) is due to the fact that k_1 is not a real curvature but k_1^0 is a real curvature and ϵ_{11} is a strain defined with respect to the undeformed length. Assuming $1 + e_1 \simeq 1$, we rewrite (6.5.9a) as

$$\epsilon_{11} = e_1 + z(k_1 - k_1^0) \quad (6.5.9b)$$

Because rigid-body translations and rotations produce no strains and the strains are due to relative displacements, one can choose the observed ζ axis to be fixed and hence the displacements of an arbitrary point, which is very close to the ζ axis, can be expressed as

$$u_1(x, y, z, t) = u_1^0(x, y, t) + z[\theta_2(x, y, t) - \theta_{20}(x, y)] \quad (6.5.10a)$$
$$u_2(x, y, z, t) = u_2^0(x, y, t) - z[\theta_1(x, y, t) - \theta_{10}(x, y)] \quad (6.5.10b)$$
$$u_3(x, y, z, t) = u_3^0(x, y, t) \quad (6.5.10c)$$

where the $u_i^0 (i = 1, 2, 3)$ are the displacements (with respect to the local coordinate system) of the observed reference point (i.e., the point which is on the reference surface), θ_1 and θ_2 are the rotation angles of the observed shell element with respect to the ξ and η axes, and θ_{10} and θ_{20} are the initial rotation angles of the observed shell element with respect to the ξ and η axes, respectively. Because the $\xi\eta\zeta$ frame is a local coordinate system attached to the observed shell element and the $\xi\eta$ plane is tangent to the deformed reference surface, we have

$$u_1^0 = u_2^0 = u_3^0 = \theta_1 = \theta_2 = \theta_{10} = \theta_{20} = \partial u_3^0/\partial x = \partial u_3^0/\partial y = 0 \quad (6.5.11a)$$

$$e_1 \equiv \frac{\partial u_1^0}{\partial x}, \quad e_2 \equiv \frac{\partial u_2^0}{\partial y}, \quad \gamma_6 \equiv \frac{\partial u_1^0}{\partial y} + \frac{\partial u_2^0}{\partial x} \quad (6.5.11b)$$

$$k_1 \equiv \frac{\partial \theta_2}{\partial x}, \quad k_2 \equiv -\frac{\partial \theta_1}{\partial y}, \quad k_{61} \equiv -\frac{\partial \theta_1}{\partial x}, \quad k_{62} \equiv \frac{\partial \theta_2}{\partial y}, \quad k_6 \equiv k_{61} + k_{62} \quad (6.5.11c)$$

$$k_1^0 \equiv \frac{\partial \theta_{20}}{\partial x}, \quad k_2^0 \equiv -\frac{\partial \theta_{10}}{\partial y}, \quad k_{61}^0 \equiv -\frac{\partial \theta_{10}}{\partial x}, \quad k_{62}^0 \equiv \frac{\partial \theta_{20}}{\partial y}, \quad k_6^0 \equiv k_{61}^0 + k_{62}^0$$
$$(6.5.11d)$$

Because all of the variables in (6.5.10a-c) are zero for any point on the observed shell element and only the first derivatives of (6.5.10a-c) with respect to $x, y,$ and z are needed to calculate the strains, nonlinear terms are not needed in (6.5.10a-c). Moreover, because the local displacement vector $\mathbf{U}$, which is given by

$$\mathbf{U} = u_1 \mathbf{i}_1 + u_2 \mathbf{i}_2 + u_3 \mathbf{i}_3 \quad (6.5.12)$$

is an infinitesimal vector defined with respect to the deformed coordinate system, it follows from (6.5.11a-d) that the Jaumann strain B_{11} is given by

$$B_{11} = \frac{\partial \mathbf{U}}{\partial x} \cdot \mathbf{i}_1 = e_1 + z(k_1 - k_1^0) \quad (6.5.13)$$

which is the same as (6.5.9b). To obtain (6.5.9a), one needs to substitute $(1+e_1)\theta_{20}$ for θ_{20} and $(1+e_2)\theta_{10}$ for θ_{10} in (6.5.10a-c).

Since warpings are relative displacements with respect to the ζ axis, they are essentially local displacements. Hence, warping displacements can be superposed on the local displacements shown in (6.5.10a-c), and the above method of deriving local strains is still valid even with nontrivial warping displacements.

7
PLATES

Plates are important construction elements for structures and are the most studied elements in structural engineering. Plate theories are two-dimensional (2-D) models of initially flat structures with one dimension being smaller than the other two characteristic dimensions. Solving plate problems is more difficult than solving beam problems because two, instead of one, independent spatial variables are involved.

7.1 INTRODUCTION

Because both plates and shells are 2-D structures, they have similar characteristics in their modeling, mechanics, and dynamic responses. Consequently, most of the discussions and characteristics of plates introduced in this section are also applicable to shells and will not be repeated in Chapter 9.

7.1.1 Plate Theories

In the classical plate theory, the midplane of a plate is chosen as the reference plane and the Kirchhoff hypothesis is used. The Kirchhoff hypothesis assumes that (1) the deflection of the midplane is smaller than the plate thickness; (2) the midplane remains unstrained and neutral during bending; (3) plane sections initially normal to the midplane remain plane and normal to the midplane after bending, which implies that transverse shear strains are neglected; and (4) the transverse normal stress is negligible compared to the other stress components. However, experimental results showed that the classical plate theory underpredicts the deflections and overpredicts the natural frequencies due to the omission of the transverse shear strains, especially when the structure vibrates at high frequencies. Moreover, the shear strains may be significant in composite materials because of their relatively low shear strength.

Srinivas, Joga Rao, and Rao (1970) showed that the vibration analysis of homogeneous thick laminated plates carried out without considering transverse shear deformations is highly inaccurate. Pagano (1969, 1970) demonstrated the deficiencies of the classical plate theory for thick laminated plates by comparing his exact solutions for cylindrical bending of a simply-supported rectangular plate subjected to a distributed harmonic transverse load with the results obtained using the classical plate theory. Hence, an adequate plate theory must account for transverse shear stresses, especially for composite plates.

To fully account for effects due to transverse shear and normal stresses in laminated composite plates, Seide (1975), Srinivas and Rao (1970), Srinivas, Joga Rao, and Rao (1970), Pagano (1970), and Pagano and Hatfield (1972) proposed 3-D plate theories. These theories can give a better estimation of the complex stress states of composite plates than the classical plate theory. However, when the number of layers increases, the number of governing equations needed in these theories becomes large and the problem becomes intractable. Hence, lamination theories are commonly used in the dynamic analysis of laminated composite plates (Lekhnitski,1968; Jones, 1975; Vinson and Sierakowski, 1986; Whitney,1987; Reddy, 1997, 2003; Ye, 2003; Qatu, 2004), where all layers are assumed to be perfectly bonded to one another and the laminate is treated as an anisotropic layer. Koiter (1960, 1967) and John (1965) pointed out that the transverse shear stresses of plates and shells are the order $E\epsilon\Theta$ and the transverse normal stress is the order $E\epsilon\Theta^2$, where E is the modulus of elasticity, $\Theta = \max(h/D, \sqrt{h/R}, \sqrt{\epsilon})$, h is the thickness, D is the distance from the point on the middle surface under consideration to the edge of the shell, R is the smallest principal radius of curvature of the undeformed middle surface, and ϵ is the largest extension of the middle surface. Hence, the transverse normal stress is generally smaller than the transverse shear stresses and is very often neglected in plate and shell analyses.

Reissner (1944, 1945, 1947) presented a stress-based shear-deformation plate theory in which the transverse shear stresses are determined using the 3-D theory of elasticity. Basset (1890) presented a displacement-based theory for shells, in which he gave geometric constraints on the shear warping of the cross sections by assuming that the displacements can be expanded in power series of the thickness coordinate. Following the idea of Basset (1890), Hildebrand, Reissner, and Thomas (1949), Hencky (1947), and Mindlin (1951) presented the so-called first-order shear-deformation theory. It results in five equations of motion governing the translational displacements u, v, and w and the rotational displacements ψ_1 and ψ_2, where ψ_1 and ψ_2 are the rotation angles of the normal of the undeformed midplane.

Application of the first-order shear-deformation theory to isotropic plates yields a system of six partial-differential equations that allow the specification of three boundary conditions at each edge. Medwadowski (1958) extended Reissner's theory (1945) to orthotropic plates. He used the von Kármán nonlinearity and studied the effects of transverse shear and rotary inertia. Stavsky (1965) applied the first-order shear-deformation theory to laminated plates. Yang, Norris, and Stavsky (1966) generalized the Reissner-Mindlin theory for homogeneous isotropic plates to arbitrary laminated anisotropic plates by including the effects of shear strains and rotary inertia. Mau (1973), Pagano (1969, 1970), Pagano and Hatfield (1972), Srinivas and Rao (1970), Whitney (1969a,b), and Sun and Whitney (1973) are among the researchers who applied the Yang, Norris, and Stavsky (YNS) theory (1966) to laminated anisotropic plates. They showed that the YNS theory is adequate for predicting the deflections and natural frequencies of the first few modes of laminated anisotropic plates. Reddy (1980) used the YNS theory to develop a simple finite-element model to analyze the linear transient responses of laminated composite plates.

Brunelle and Robertson (1976) and Roufaeil and Dawe (1982) employed the Mindlin-type shear-deformation theory to study linear vibrations of plates. Noor and Hartley (1977) included the effect of transverse shear and used a finite-element method to investigate the nonlinear responses of composite plates. Reddy and Chao (1981) formulated a quadratic finite element to study large-amplitude free vibrations of laminated composite plates. Rajgopal et al. (1986) extended the formulation of Reddy and Chao to study the nonlinear vibrations of sandwich plates. Among the many other works on the nonlinear analysis of plates accounting for transverse shear, we quote those of Schmidt (1977), Wu and Vinson (1969a,b), Singh, Sundararajan, and Das (1974), Sathyamoorthy (1978, 1979, 1984a), Sathyamoorthy and Chia (1980), and Prathap and Pandalai (1979). Wu and Vinson (1971) used Berger's approach and discussed the effect of transverse shear deformations on the nonlinear responses of cross-ply laminated plates. Sathyamoorthy (1981) suggested an improved version of Berger's theory for the dynamic analysis of rectilinearly orthotropic moderately thick circular plates. Sathyamoorthy (1984b) used Berger's approach to investigate the effect of transverse shear deformations, the geometrical nonlinearity, and modal interactions on the dynamic behavior of a moderately thick orthotropic circular plate clamped along the boundary. He compared these results with those he obtained using the von Kármán nonlinearity and validated Berger's approximation for moderately thick plates. Sathyamoorthy and Prasad (1983) considered the multimode nonlinear analysis of isotropic circular plates and showed the significance of modal interactions, especially for moderately thick plates.

The Reissner-Mindlin plate theory assumes a linear variation of the inplane displacements with the thickness coordinate. Hence it does not account for the nonuniform distribution of transverse shear stresses and does not satisfy the condition of zero-shear stresses on the top and bottom surfaces of the plates, which result in the need of a shear-correction factor. Unfortunately, it is difficult to determine shear-correction factors for composite laminates because they are functions of the material and geometric properties and the lamination sequence (Pai, 1995). Therefore, higher-order shear-deformation theories are desirable, especially for the analysis of thick plates and composite laminates.

A number of higher-order shear-deformation theories have been proposed by various authors. Green and Naghdi (1981) developed a dynamic theory of orthotropic laminated plates and investigated wave propagation in three layered plates. Adnan Nayfeh (1995) documented his extensive work on wave propagation in layered anisotropic media, and Taylor and Adnan Nayfeh (1997) investigated damping characteristics of laminated thick plates. Khoroshun and Ivanov (1981) applied the theory developed by Khoroshun (1978) to wave propagation in two-layer laminated plates. Sun and Tan (1981) used the Laplace transform in time and the Fourier transform in space to analyze the response of graphite-epoxy laminates to a localized impact loading. Lo, Christensen, and Wu (1977a,b) derived a higher-order plate theory in which the transverse and inplane displacements were expanded as quadratic and cubic functions of the transverse coordinate, respectively. Levinson (1980) considered similar expansions for the inplane displacements.

Librescu (1966, 1969), Nelson and Lorch (1974), Whitney and Sun (1973), Levinson (1980), Murthy (1981), Schmidt (1977), and Lo, Christensen, and Wu (1977a,b) presented some higher-order displacement-based shear-deformation theories. Some of these theories used variationally inconsistent equilibrium equations, and therefore the strain energy associated with the assumed displacement field is not correctly accounted for. Moreover, these higher-order shear-deformation theories require the use of shear-correction factors, which are unknown. Furthermore, these higher-order shear-deformation theories are usually computationally expensive, and the assumed transverse shear stresses do not vanish on the top and bottom surfaces of the plates. Reddy (1984b), Bhimaraddi and Stevens (1984), and Librescu and Reddy (1989) presented third-order shear-deformation theories in which the five governing differential equations were derived using variational principles. The third-order shear-deformation theory of Reddy (1984a,b) neglects the transverse normal strain and satisfies zero-shear stresses on the top and bottom surfaces of the plate, and its expanded displacement field is similar to that of Levinson (1980) and Murthy (1981). The third-order theory accounts for a cubic variation of the inplane displacements through the plate thickness and the vanishing of the shear stress on the top and bottom surfaces, and therefore no shear-correction factors are needed. A similar third-order shear-deformation theory was also proposed by Bhimaraddi and Stevens (1984) in which the generalized variables are slightly different from those of Reddy (1984b). The higher-order shear-deformation theories of Nelson and Lorch (1974) and Lo et al. (1977a) have nine and eleven dependent variables, respectively. The third-order shear-deformation theories of Reddy (1984b) and Bhimaraddi and Stevens (1984) have five dependent variables. A review and comparison of all third-order shear-deformation theories of plates was presented by Reddy (1990). It was shown in the works of Reddy (1984c, 1985), Librescu, Khdeir, and Reddy (1987), Librescu and Khdeir (1988), and Khdeir and Librescu (1988) that higher-order shear-deformation theories provide accurate global estimates of the deflections, frequencies, and buckling loads for laminated plates and are efficient for problems not having regions of acute discontinuities.

However, most of the shear-deformation theories, including the third-order theory, do not account for the continuity of interlaminar shear stresses and the elastic coupling between two transverse shear strains of laminated composite plates (Pai, 1995). Di Sciuva (1987) and Librescu and Schmidt (1991) used a piecewise linear interpolation function to satisfy the continuity of interlaminar shear stresses. Pai et al. (1993) extended the piecewise linear displacement field used by Di Sciuva (1987) by adding quadratic and cubic interpolation functions to satisfy the continuity of interlaminar shear stresses, to accommodate free shear-stress conditions on the top and bottom surfaces, and to account for nonuniform distributions of the transverse shear stresses in each layer.

7.1.2 Geometric Nonlinearities

Large elastic displacements and rotations of a thin-walled structure introduce geometric nonlinearities in its governing equations. Geometric nonlinearities play important roles in the postbuckling characteristics, stability, and nonlinear flutter. However, modeling of geometric nonlinearities is a challenging task, especially for highly flexible structures. Moreover, solving structural problems with geometric nonlinearities is not an easy task.

When the transverse deflection of a plate is large, the midplane is strained and assumptions (1) and (2) discussed at the beginning of Section 7.1.1 are no longer applicable. A nonlinear plate theory derived by assuming small strains and rotations but accounting for large deflections through the use of induced inplane force resultants can be variationally inconsistent. To systematically account for moderate geometric nonlinearities of thin plates, von Kármán strains were introduced (von Kármán, 1910). Berger (1955) used a variational principle and ignored the second invariant in the energy term to obtain simplified governing equations for plates. Nash and Modeer (1959) followed Berger's approach and derived the dynamic analog of Berger's equations. They showed that the application of these equations to simply supported plates leads to results that are in excellent agreement with those obtained by using the dynamic analogue of von Kármán equations. Wah (1963) extended Berger's approach and reduced the coupled fourth-order partial-differential equations of von Kármán to a single fourth-order partial-differential equation. Huang and Al-Khattat (1977) considered the vibration of a circular plate and showed that solutions based on Berger's approach are accurate at low amplitudes of vibration but the accuracy decreases as the amplitude increases. They also showed that Berger's approach is entirely unsuitable for plates with movable edges. Nowinski and Ohnabe (1972) also pointed out that Berger's approach may lead to inaccurate and even meaningless results for plates without inplane restraints.

Most nonlinear plate theories account for geometric nonlinearities by using von Kármán strains in conjunction with the classical plate theory, the first-order shear-deformation theory, or a higher-order shear-deformation theory (e.g., Reissner, 1948, 1953; Schmidt, 1977; Reddy, 1984a; Bhimaraddi, 1987c). All of these theories use linear curvatures, and the configuration change in the curvature-displacement relation is neglected, which is also variationally inconsistent. Although von Kármán strains result in some quadratic and cubic terms, they do not fully account for all of the geometric nonlinearities (Pai and Nayfeh, 1991c). For more historic reviews of nonlinear plate theories, the reader is referred to the articles by Reddy (1984a, 1985).

In the literature, geometric nonlinearities of plates have not received the attention they deserve, and the configuration change due to inplane extension and shear deformations is usually neglected. In contrast, geometric nonlinearities of beams have received considerable attention, and two or three Euler rotations are commonly used to relate the deformed and undeformed configurations (Hodges and Dowell, 1974; Dowell, Traybar, and Hodges, 1977; Crespo da Silva and Glynn, 1978a; Alkire, 1984; Hodges, Crespo da Silva, and Peters, 1988; Pai and Nayfeh, 1990b). Pai and Nayfeh (1991c) used new concepts of local engineering stresses and strains and orthogonal virtual rotations and a coordinate transformation to derive a set of consistent nonlinear equations governing the motions of flexible laminated plates. The resulting nonlinear curvatures and midplane strains are combined with the third-order shear-deformation theory (Bhimaraddi and Stevens, 1984; Reddy, 1984b) to derive five variationally consistent, third-order shear-deformable, fully nonlinear equations of motion. Moreover, Pai and Palazotto (1995) showed that the local engineering strains used by Pai and Nayfeh (1991c) are the same as the Jaumann strains (Malvern, 1969). They also showed that the Jaumann strains are fully nonlinear strains and can be derived using the concept of local displacements without performing the complex polar decomposition operation.

7.1.3 Plates with Integrated Piezoelectric Materials

Lead zirconate titanate (PZT) patches and polyvinylidene fluoride (PVDF) films can be integrated into structures for sensing and actuation and for the design of active structures. However, modeling of such structures is challenging. Models of plates with integrated PZT patches and PVDF films are very limited in the literature because most of the researchers concentrated their efforts on the implementation of control algorithms. Most of the models in the literature (e.g., Im and Atluri, 1989; Lazarus and Crawley, 1989; Lee and Moon, 1989; Tzou and Gadre, 1989; Lee, 1990) are based on the Kirchhoff hypothesis and hence neglect transverse shear deformations. However, shear effects are significant for composite plates because the ratios of the inplane Young's moduli E_α ($\alpha = 1, 2$) to the transverse shear moduli G_{α_3} are much larger than those of isotropic materials. As discussed in Section 7.1.1, there are several refined shear-deformable plate theories with the third-order shear theory (Bhimaraddi and Stevens, 1984; Reddy, 1984a) being the most recommended theory. But most of the shear deformation theories, including the third-order theory, do not account for the continuity of interlaminar shear stresses and the elastic coupling between two transverse shear deformations. Pai et al. (1993) developed a layerwise higher-order shear-deformation theory, which satisfies the continuity of interlaminar shear stresses, accommodates free shear-stress conditions on the bounding surfaces, and accounts for non-uniform distributions of the transverse shear stresses in each layer.

Geometric nonlinearities are either totally ignored or considered in an incomplete manner in most of piezoelectric plate models. Geometric nonlinearities introduce nonlinear dynamic responses, such as flutter and chaotic vibrations, into a plate system under steady external excitations. Furthermore, in a nonlinear system, an unstable "steady" solution with small amplitude can be sustained for a long time before it diverges, and hence it may be mistaken for a linear stable solution (Fujino, Warnitchai, and Pacheco, 1993). Although the motion of a structure subject to control actions is a transient vibration, Balachandran et al. (1994) and Nayfeh (2000) showed that modal interactions can produce small-amplitude transient vibrations in nonlinear systems, which pose difficulties in their identification using linear identification methods (e.g., moving-block analysis or time domain techniques). For example, modal interactions can cause the identified damping coefficients to be oscillatory and to assume negative values. Hence, geometric nonlinearities need to be accurately modeled in an adequate piezoelectric plate theory.

Pai et al. (1993) extended the basic idea underlying the development of the nonlinear plate theory of Pai and Nayfeh (1991c) to derive a set of mathematically consistent, nonlinear equations governing the motions of laminated piezoelectric plates. The surface analysis was performed by using a vector approach, and the resulting expressions for the nonlinear curvatures and midplane strains were combined with a layerwise higher-order shear-deformation theory and the extended Hamilton principle to derive variationally consistent, shear-deformable, nonlinear equations of motion. The theory fully accounts for geometric nonlinearities by using local stress and strain measures and an exact coordinate transformation, which result in nonlinear strain-displacement relations that contain the von Kármán strains as a special case. Moreover, the model accounts for the continuity of interlaminar shear stresses, the elastic shear coupling effects, extensionality, orthotropic properties of piezoelectric materials, integrated actuators and sensors at various orientations, and actuator-induced local actuating forces and moments. Moreover, the piezoelectric actuator-induced warping was addressed and comparisons with other simplified models and nonlinear theories were made.

7.1.4 Linear Vibrations and Buckling of Plates

Bhimaraddi and Stevens (1984) studied the vibrations of thick laminated composite plates by using a higher-order shear-deformation theory. Stein and Jegley (1987) investigated the effects of transverse shear deformations on the cylindrical bending and vibration of laminated plates. Bowlns, Palazotto, and Whitney (1987) studied the effects of transverse shear deformations and rotary inertia on the vibration of symmetrically laminated rectangular plates by using the Galerkin method. Sivakumaran (1987) considered the vibrations of thick symmetrically laminated rectangular plates having free-edge conditions.

Recent research on the free-vibration analysis of rectangular plates by using higher-order shear-deformation theories include the studies by Reddy (1983, 1984a), Di Sciuva (1986), Bhimaraddi and Stevens (1984), and Doong and Chen (1987). Bert and Mayberry (1969) conducted a free-vibration analysis of unsymmetrically laminated plates with clamped edges. Whitney (1969b, 1970) studied the effects of boundary conditions on the bending vibration and buckling of asymmetric laminates. Ashton (1969) and Jones (1973) also studied the response of asymmetric laminates. Baharlou and Leissa (1987) proposed a general method for the analysis of generally laminated thin plates with different boundary conditions by using a technique based on the Ritz method. Khdeir (1988b) analyzed the free vibrations of antisymmetric angle-ply laminates with various boundary conditions. He used a generalized Levy-type solution (Reddy, Khdeir, and Librescu, 1987) in conjunction with the state-space concept

(Librescu, Khdeir, and Reddy, 1987). Kamal and Durvasula (1986) studied the free vibrations of asymmetric laminates considering all possible boundary conditions by using a shear-deformation plate theory.

The study of free vibrations of laminated plates was initiated by Ashton (1969), Bert and Mayberry (1969), and Whitney and Leissa (1969). Chandra and Raju (1975b) and Chandra (1976) investigated the free vibrations of cross-ply laminated rectangular plates. Mohan and Kingsbury (1971) analyzed the free vibrations of generally laminated orthotropic plates and Noor (1973a) used the first-order shear-deformation theory to analyze the free vibrations of cross-ply laminated plates. Fortier and Rossettos (1973) analyzed the free vibrations of thick asymmetric cross-ply laminated rectangular plates. Whitney and Pagano (1970) presented closed-form solutions for the free vibrations of antisymmetric angle-ply rectangular plates. Bert and Chen (1978) presented closed-form solutions for the free vibrations of simply supported antisymmetric angle-ply laminated plates. Whitney and Leissa (1969) developed a Navier solution for the classical laminate theory, which was extended by Whitney (1969b) by including the first-order shear-deformation theory and by Reddy (1984a,b) through use of a refined shear-deformation theory.

Reddy, Khdeir, and Librescu (1987) developed a Levy-type solution for symmetric cross-ply laminated plates. They used the first-order shear-deformation theory and applied the state-space concept to solve the resulting equations for a variety of boundary conditions. Lekhnitski (1968) and Leissa (1981a) surveyed the literature on the vibration behavior of anisotropic plates. Bert (1985) reviewed the dynamic response of laminated composites. Reddy (1985) reviewed the literature concerning the application of the finite-element method to the vibrations of plates.

More recently, free vibrations of composite plates have been studied by Anderson and Nayfeh (1996), Chai (1996), Cunha (1997), and De Almeida and Hansen (1997). Fan and Ye (1990) developed an exact solution for laminated plates. Qatu and Leissa (1991) used the Ritz method to determine the vibration frequencies of completely free laminated plates and shallow shells. They showed that, for the same accuracy, the Ritz method requires fewer degrees of freedom when compared with the finite-element method. Anderson and Nayfeh (1996) built and tested composite plates in cantilever, free-free, and clamped configurations. The experimental results are in agreement with their finite-element results. Rajalingham, Bhat, and Xistris (1996) studied the vibration of rectangular plates using the plate characteristic functions as the deflection shapes in the Rayleigh-Ritz method. Lee, Chung, and Chung (1997) analyzed the free vibrations of symmetrically laminated composite rectangular plates using the Rayleigh-Ritz and Kantorovich methods. Chandrashekhara and Kolli (1997) used the finite-element method to study the free vibration characteristics of laminated plates. For a detailed account of the linear vibrations of composite plates, we refer the reader to the book of Qatu (2004).

The solution of a flat-plate stability problem was first given by Bryan (1891). A great many publications on the subject have appeared since then. Srinivas and Rao (1970) used the three-dimensional theory of elasticity to carry out extensive studies on the effects of plate thickness on the vibration and buckling of homogeneous and laminated plates. Jones (1973) presented an exact theory and numerical results for the buckling of asymmetric cross-ply laminated rectangular plates. Noor (1975) investigated the stability of multilayered plates. He used the classical plate theory and the Mindlin-Reissner type shear-deformation theory and compared his results with the results obtained using the three-dimensional theory of elasticity (Noor, 1973b). Reddy and Phan (1985) used the higher-order shear-deformation theory of Reddy (1984b) to investigate the stability of laminated plates. Khdeir (1988a, 1989) and Reddy and Khdeir (1989) used the third-order shear theory of Reddy (1984b) to determine Levy-type solutions for the problems of free vibration and buckling of cross-ply laminated plates by using the state-space concept. However, their method of solution is limited because it yields numerically ill-conditioned problems for thin and many moderately thick plates. Asfar, Masad, and Nayfeh (1990) encountered a similar stiffness problem in treating the stability of boundary layers and proposed a method based on continuity to overcome the problem. Hadian and Nayfeh (1993) used the continuity-based method proposed by Asfar, Masad, and Nayfeh to overcome the problem of stiffness. Pai and Nayfeh (1991a,b) also encountered a similar stiffness problem in investigating the response of an inextensional, symmetric angle ply graphite epoxy beam to a harmonic base excitation along the flapwise direction; they used the fundamental matrix method to find the natural frequencies of bending-torsional coupling modes. Another alternative is the method of orthonormalization in conjunction with continuity, which is one of the most efficient methods to deal with problems of this nature. Extensive treatments of flat-plate stability analyses are given by Bleich (1952) and Bulson (1969). Some other textbooks on buckling are Timoshenko and Gere (1961), Cox (1963), and Brush and Almroth (1975).

7.1.5 Nonlinear Analyses of Plates

The dynamic analogue of the von Kármán equations, which take into account the stretching of the midsurface, were used by several investigators to study the response of plates to harmonic excitations. Some early experimental results were obtained by Tobias (1958) for main-resonant vibrations of free circular plates. Yamaki (1961) obtained approximate solutions for rectangular and circular plates with various boundary conditions. He used a single-term Galerkin expansion. Kung and Pao (1972) used a combination of the Galerkin method and the method of harmonic balance to analyze axisymmetric vibrations of circular plates. Rehfield (1974) proposed a method for investigating single-mode main resonances of structures in which Hamilton's principle is combined with a perturbation procedure; he applied the method to beams and rectangular plates. Eisley (1964a) used the Galerkin method in combination with the Lindstédt-Poincaré perturbation technique to investigate the main-resonant vibrations of buckled rectangular plates. Huang and Sandman (1971) used the Kantorovich method to study axisymmetric vibrations of circular plates. They used a numerical technique to solve the resulting two-point boundary-value problems. Huang (1973) used the same approach to study the main-resonant vibration of an axisymmetric orthotropic circular plate. Young and Chen (1995) examined the nonlinear vibrations of cantilever skew plates subjected to aerodynamic and inplane exciting forces.

Ramesh and Krishnamoorthy (1995) investigated the use of dynamic relaxation for the analysis of geometrically nonlinear plates and shallow shells.

Whitney and Leissa (1969) formulated the equations of motion for the large-deflection behavior of laminated anisotropic plates that account for the von Kármán geometric nonlinearity. Bennett (1971) considered the response of simply supported rectangular laminated plates. He used a four-term Galerkin expansion to approximate solutions of the nonlinear coupled equations for angle-ply rectangular composite plates having simply supported edge conditions. He reported that the bending-stretching coupling appears only in the linear terms, which indirectly affects the nonlinear term in the frequency equation. Wu and Vinson (1971) used Berger's approximation to study the nonlinear vibrations of laminated plates with clamped and simply supported edges. Employing the Galerkin procedure, Bert (1973) studied the vibrations of clamped laminated composite plates. Chandra and Raju (1975a, 1975b) investigated the nonlinear vibrations of angle-ply and cross-ply plates by using the Galerkin method and a perturbation technique. They assumed the plate edges to be clamped and simply supported for out-of-plane boundary conditions and to be movable and immovable for inplane boundary conditions. They presented a two-term (multimode) solution for the case of angle-ply plates. Chandra (1976) extended this analysis to study nonlinear oscillations of simply supported clamped cross-ply plates. Chia and Prabhakara (1976, 1978) studied the nonlinear responses of cross-ply and angle-ply laminated rectangular plates using the approach of Chandra (1976). Chia (1982) presented a single-mode solution for the large-amplitude vibrations of rigidly clamped antisymmetric cross-ply plates. He expressed the dynamic von Kármán type nonlinear equations in terms of three displacement components.

Using the finite-element method, Reddy and Chao (1981, 1982) and Reddy (1982) considered nonlinear oscillations of laminated plates with all edges clamped and all edges simply supported. Sivakumaran and Chia (1984, 1985) used Hamilton's principle and Reissner's variational principle to determine single-mode vibrations of laminated plates with all edges being clamped, all edges being simply supported, and two parallel edges clamped and the other edges simply supported. Reddy and Chao (1981, 1982), Reddy (1982), and Sivakumaran and Chia (1984, 1985) included the effects of transverse shear and rotary inertia. Reddy (1982) and Sivakumaran and Chia (1985) included rectangular cutouts and transverse normal stresses, respectively. Chia (1985) investigated the amplitude and frequency responses of angle-ply plates on a Winkler-Pasternak elastic foundation having the edges nonuniformly restrained against rotation by using a single-mode solution to the von Kármán nonlinear equations. Manoj et al. (2000) analyzed the nonlinear vibrations of thin laminated rectangular plates on elastic foundations. Tenneti and Chandrashekhara (1994) used a refined shear flexible element to analyze nonlinear vibrations of laminated plates. Ganapathi, Varadan, and Sarma (1991) considered nonlinear flexural vibrations of laminated orthotropic plates and Kuo and Yang (1989) investigated the generic nonlinear behavior of antisymmetric angle-ply laminated plates. Singh, Rao, and Iyengar (1995) used finite elements to analyze the nonlinear vibrations of moderately thick unsymmetrically laminated composite plates and Ribeiro and Petyt (1999) used the hierarchical FEM to study the nonlinear vibrations of composite laminated plates.

Kim and Kim (2002) investigated the nonlinear vibrations of viscoelastic laminated composite plates and Chen (2000) analyzed the nonlinear transient behavior of rectangular composite laminated plates.

Hui (1985a,b) investigated the effects of geometric imperfections and inplane edge compressive loads on the large-amplitude vibrations of simply supported antisymmetric laminates using a perturbation method and the Galerkin procedure in conjunction with the dynamic von Kármán-type plate equations. Large-amplitude vibrations of imperfect plates were also studied by Bhimaraddi (1990,1993).

Eslami and Kandil (1989a) used the method of multiple scales to study the large-amplitude responses of special orthotropic plates to harmonic excitations. They used a single-mode Galerkin discretization to transform the nonlinear coupled partial-differential equations into a Duffing-type equation. They studied primary, subharmonic, and superharmonic resonances. Eslami and Kandil (1989b) also studied multimode nonlinear vibrations of orthotropic panels by using the method of multiple scales in conjunction with the Galerkin procedure. They discussed different resonances and compared the solution in the case of primary resonance with that obtained by using the method of harmonic balance and found good agreement. Abe, Kobayashi, and Yamada (1998a) analyzed subharmonic resonance of moderately thick antisymmetric angle-ply laminated plates by using the method of multiple scales, whereas Birman and Zahed (1989) considered nonlinear problems of parametric vibrations of imperfect laminated plates. Harichandran and Naja (1997) treated the case of random vibrations of laminated composite plates with material nonlinearity.

Kapania and Raciti (1989a) reviewed recent developments in the analyses of laminated beams and plates with an emphasis on shear effects and buckling. They presented a discussion of various shear-deformation theories for plates and beams and a review of the recently developed finite-element method for the analysis of thin and thick laminated beams and plates. Moreover, they included recent studies on the buckling and postbuckling behavior of perfect and geometrically imperfect plates and the delamination buckling and growth in beams and plates. Kapania and Raciti (1989b) presented a summary of recent advances in the analysis of laminated beams and plates with an emphasis on vibrations and wave propagation. They reviewed the free-vibration analyses of symmetrically laminated plates for various geometric shapes and edge conditions. They also reviewed various developments in the analysis of unsymmetrically laminated beams and plates. Moreover, they presented a survey of the nonlinear vibrations of perfect and geometrically imperfect laminated plates. They pointed out that, due to the bending-stretching coupling, the nonlinear behavior of unsymmetrically laminated perfect and imperfect plates may be of the hardening- or softening-type, depending upon the boundary conditions. In addition, they reviewed wave propagation in and the linear and nonlinear transient response of laminated materials.

Sridhar, Mook, and Nayfeh (1975, 1978) and Lobitz, Nayfeh, and Mook (1977) investigated primary resonant responses of a circular isotropic plate. They considered the interaction of modes and the possibility of multimode solutions. Multimode nonlinear vibrations of angle-ply and cross-ply plates were investigated by Chia and Prabhakara (1978) for all clamped and all simply supported stress-free edges.

382 PLATES

Among the books with approximate solutions for nonlinear plates of simple regular shapes, we quote those of Timoshenko (1959), Sokolnikoff and Redheffer (1966), Brush and Almroth (1975), Nayfeh and Mook (1979), Chia (1980), Sathyamoorthy (1997), and Nayfeh (2000). A comprehensive literature review of the nonlinear analyses of plates can be found in the text by Chia (1980).

7.2 LINEAR CLASSICAL PLATE THEORY

The derivations in Chapter 4 demonstrate that the energy approach using the extended Hamilton principle is more straightforward than the Newtonian approach, especially for obtaining boundary conditions and governing equations for shear rotations. Hence, we use the energy approach throughout all of the following chapters and correlate the energy and Newtonian approaches.

For plates, the extended Hamilton principle can be stated, in the absence of body forces, as

$$0 = \int_0^t (\delta T - \delta \Pi + \delta W_{nc}) dt \tag{7.2.1}$$

where δW_{nc} denotes variation of the nonconservative energy W_{nc}, which is problem dependent, and variations of the kinetic and elastic energies T and Π are given by

$$\delta T = -\int_z \int_A \rho \ddot{\mathbf{D}} \cdot \delta \mathbf{D} \, dA \, dz \tag{7.2.2a}$$

$$\delta \Pi = \int_z \int_A (\sigma_{11} \delta \epsilon_{11} + \sigma_{22} \delta \epsilon_{22} + \sigma_{33} \delta \epsilon_{33} + \sigma_{23} \delta \epsilon_{23} + \sigma_{13} \delta \epsilon_{13} + \sigma_{12} \delta \epsilon_{12}) \, dA \, dz \tag{7.2.2b}$$

Here, ρ is the mass density, A denotes the undeformed area of the reference plane, $\mathbf{D}$ denotes the displacement vector of an arbitrary point of the differential plate element under observation, the σ_{ij} are the Jaumann stresses, and the ϵ_{ij} are the Jaumann strains.

7.2.1 Rectangular Plates

The classical plate theory is based on the Kirchhoff hypothesis (1850), which assumes that the cross-sections are flat and perpendicular to the reference plane before and after deformation. In other words, the shear warpings are assumed to be zero in the classical plate theory.

To derive the governing equations of rectangular plates, we consider a rectangular plate over the domain $0 \leq x \leq a$ and $0 \leq y \leq b$. Figure 7.2.1 shows the undeformed and deformed configurations of a differential plate element. We note that a differential plate element without shear warpings is a rigid straight filament with dimensions $dx \times dy \times h$, where h is the plate thickness. From Figure 7.2.1, one can see that the small displacements u_i of an arbitrary point on the observed plate element are

$$u_1 = u + z\theta_2 = u - zw_x, \quad u_2 = v - z\theta_1 = v - zw_y, \quad u_3 = w \tag{7.2.3}$$

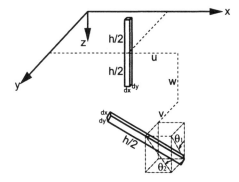

Fig. 7.2.1 A differential plate element without shear deformations.

where u, v, and w are the displacement components of the observed reference point with respect to the xyz coordinate system, and they are independent of z. Using the linear strain-displacement relations, we obtain from (7.2.3) the strains as

$$\epsilon_{11} = \frac{\partial u_1}{\partial x} = u_x - zw_{xx}, \quad \epsilon_{22} = \frac{\partial u_2}{\partial y} = v_y - zw_{yy}$$

$$\epsilon_{12} = \frac{\partial u_1}{\partial y} + \frac{\partial u_2}{\partial x} = u_y + v_x - 2zw_{xy}, \quad \epsilon_{33} = \epsilon_{13} = \epsilon_{23} = 0 \quad (7.2.4)$$

The displacement vector $\mathbf{D}$ is given by

$$\mathbf{D} = u_1 \mathbf{j}_1 + u_2 \mathbf{j}_2 + u_3 \mathbf{j}_3 \quad (7.2.5)$$

where $\mathbf{j}_1$, $\mathbf{j}_2$, and $\mathbf{j}_3$ denote the unit vectors along the axes x, y, and z, respectively. Taking the time derivatives and variation of the displacement vector $\mathbf{D}$, we obtain

$$\ddot{\mathbf{D}} = (\ddot{u} - z\ddot{w}_x)\mathbf{j}_1 + (\ddot{v} - z\ddot{w}_y)\mathbf{j}_2 + \ddot{w}\mathbf{j}_3 \quad (7.2.6a)$$

$$\delta \mathbf{D} = (\delta u - z\delta w_x)\mathbf{j}_1 + (\delta v - z\delta w_y)\mathbf{j}_2 + \delta w \mathbf{j}_3 \quad (7.2.6b)$$

Substituting (7.2.6a,b) into (7.2.2a) yields

$$\begin{aligned}
\delta T &= -\int_A \Big[(I_0 \ddot{u} - I_1 \ddot{w}_x)\,\delta u + (I_0 \ddot{v} - I_1 \ddot{w}_y)\,\delta v + I_0 \ddot{w}\,\delta w \\
&\qquad + (I_2 \ddot{w}_x - I_1 \ddot{u})\,\delta w_x + (I_2 \ddot{w}_y - I_1 \ddot{v})\,\delta w_y \Big] dA \\
&= -\int_A \Big\{ (I_0 \ddot{u} - I_1 \ddot{w}_x)\,\delta u + (I_0 \ddot{v} - I_1 \ddot{w}_y)\,\delta v \\
&\qquad + \Big[I_0 \ddot{w} - (I_2 \ddot{w}_x - I_1 \ddot{u})_x - (I_2 \ddot{w}_y - I_1 \ddot{v})_y \Big] \delta w \Big\} dA \\
&\quad - \int_y [I_2 \ddot{w}_x - I_1 \ddot{u}]_{x=0}^{x=a}\,\delta w\, dy - \int_x [I_2 \ddot{w}_y - I_1 \ddot{v}]_{y=0}^{y=b}\,\delta w\, dx \quad (7.2.7)
\end{aligned}$$

where

$$\{I_0, I_1, I_2\} \equiv \int_z \rho\{1, z, z^2\}dz \tag{7.2.8}$$

We note that $I_1 = 0$ if ρ is constant and the middle plane is chosen as the reference plane.

Substituting (7.2.4) into (7.2.2b) yields

$$\begin{aligned}
\delta\Pi &= \int_A \int_z [\sigma_{11}(\delta u_x - z\delta w_{xx}) + \sigma_{22}(\delta v_y - z\delta w_{yy}) \\
&\quad + \sigma_{12}(\delta u_y + \delta v_x - 2z\delta w_{xy})]dAdz \\
&= \int_A (N_1 \delta u_x + N_6 \delta u_y + N_2 \delta v_y + N_6 \delta v_x \\
&\quad - M_1 \delta w_{xx} - M_2 \delta w_{yy} - 2M_6 \delta w_{xy})dA \\
&= -\int_A [(N_{1x} + N_{6y})\delta u + (N_{2y} + N_{6x})\delta v \\
&\quad + (M_{1xx} + M_{2yy} + 2M_{6xy})\delta w]dA \\
&\quad + \int_y [N_1 \delta u + N_6 \delta v + (M_{1x} + 2M_{6y})\delta w - M_1 \delta w_x]_{x=0}^{x=a} dy \\
&\quad + \int_x [N_6 \delta u + N_2 \delta v + (M_{2y} + 2M_{6x})\delta w - M_2 \delta w_y]_{y=0}^{y=b} dx \\
&\quad - 2M_6 \delta w \Big|_{(x,y)=(a,0),(0,b)}^{(x,y)=(0,0),(a,b)}
\end{aligned} \tag{7.2.9}$$

where

$$\{N_1, N_2, N_6\} = \int_z \{\sigma_{11}, \sigma_{22}, \sigma_{12}\}dz$$
$$\{M_1, M_2, M_6\} = \int_z z\{\sigma_{11}, \sigma_{22}, \sigma_{12}\}dz \tag{7.2.10}$$

M_1, M_2, and M_6 represent the moment intensities acting on the edges of the plate element, as shown in Figure 7.2.2a. Figure 7.2.2b shows the inplane extension force intensities N_1 and N_2 and the inplane shear intensity N_6.

To relate the stress resultants and moments to the midplane strains $(u_x, v_y, u_y + v_x)$ and curvatures (w_{xx}, w_{yy}, w_{xy}), we first use $\epsilon_{13} = \epsilon_{23} = 0$ from (7.2.4) and $\sigma_{33} = 0$ (because $\sigma_{33} = 0$ is more appropriate than $\epsilon_{33} = 0$ for thin plates) to obtain the reduced stress-strain relation from (2.4.34) and (2.4.35) as

$$\left\{\begin{array}{c}\sigma_{11}\\\sigma_{22}\\\sigma_{12}\end{array}\right\} = \left[\begin{array}{ccc}\tilde{Q}_{11} & \tilde{Q}_{12} & \tilde{Q}_{16}\\\tilde{Q}_{12} & \tilde{Q}_{22} & \tilde{Q}_{26}\\\tilde{Q}_{16} & \tilde{Q}_{26} & \tilde{Q}_{66}\end{array}\right]\left(\left\{\begin{array}{c}u_x\\v_y\\u_y+v_x\end{array}\right\} - z\left\{\begin{array}{c}w_{xx}\\w_{yy}\\2w_{xy}\end{array}\right\}\right) \tag{7.2.11}$$

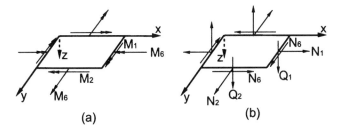

Fig. 7.2.2 (a) Moment resultants and (b) force resultants on a differential plate element.

Substituting (7.2.11) into (7.2.10), we obtain the internal forces and moments in terms of the midplane strains and curvatures as

$$\left\{\begin{array}{c} N_1 \\ N_2 \\ N_6 \\ M_1 \\ M_2 \\ M_6 \end{array}\right\} = \left[\begin{array}{cc} [A_{ij}] & [B_{ij}] \\ [B_{ij}] & [D_{ij}] \end{array}\right] \left\{\begin{array}{c} u_x \\ v_y \\ u_y + v_x \\ -w_{xx} \\ -w_{yy} \\ -2w_{xy} \end{array}\right\} \quad (7.2.12)$$

where

$$\{A_{ij}, B_{ij}, D_{ij}\} = \sum_{k=1}^{N} \int_{z_k}^{z_{k+1}} \tilde{Q}_{ij}^{(k)} \{1, z, z^2\} dz \quad \text{for } i,j = 1,2,6 \quad (7.2.13)$$

In (7.2.13), we assume that there are N laminae, the kth lamina is located between the two planes at $z = z_k$ and $z = z_{k+1}$, and $z_{N+1} - z_1 = h$.

Substituting (7.2.9) and (7.2.7) into (7.2.1) and setting each of the coefficients of δu, δv, and δw equal to zero, we obtain

$$N_{1x} + N_{6y} = I_0 \ddot{u} - I_1 \ddot{w}_x + \mu_1 \dot{u} \quad (7.2.14a)$$

$$N_{6x} + N_{2y} = I_0 \ddot{v} - I_1 \ddot{w}_y + \mu_2 \dot{v} \quad (7.2.14b)$$

$$M_{1xx} + 2M_{6xy} + M_{2yy} = I_0 \ddot{w} - (I_2 \ddot{w}_x - I_1 \ddot{u})_x - (I_2 \ddot{w}_y - I_1 \ddot{v})_y + \mu_3 \dot{w} \quad (7.2.14c)$$

where we added a linear viscous damping term to each of (7.2.14a-c) and the μ_i are the damping coefficients. The damping terms are due to the nonconservative energy W_{nc} in (7.2.1). The boundary conditions for the plate are of the form: specify

Along $x = 0, a$:

$\delta u = 0$ or N_1

$\delta v = 0$ or N_6

$\delta w = 0$ or $M_{1x} + 2M_{6y} - I_1 \ddot{u} + I_2 \ddot{w}_x$

$$\delta w_x = 0 \quad \text{or} \quad M_1$$

Along $y = 0, b$:
$$\delta u = 0 \quad \text{or} \quad N_6$$
$$\delta v = 0 \quad \text{or} \quad N_2$$
$$\delta w = 0 \quad \text{or} \quad M_{2y} + 2M_{6x} - I_1 \ddot{v} + I_2 \ddot{w}_y$$
$$\delta w_y = 0 \quad \text{or} \quad M_2$$

At $(x, y) = (0, 0), (a, b), (a, 0), (0, b)$:
$$\delta w = 0 \quad \text{or} \quad M_6 \qquad (7.2.15)$$

We note that (7.2.9) can be rewritten as

$$\delta \Pi = \int_A \Big(N_1 \delta u_x + N_6 \delta u_y + N_2 \delta v_y + N_6 \delta v_x$$
$$- M_1 \delta w_{xx} - M_2 \delta w_{yy} - M_6 \delta w_{xy} - M_6 \delta w_{yx}$$
$$+ Q_1 \delta w_x + Q_2 \delta w_y - Q_1 \delta w_x - Q_2 \delta w_y \Big) dA$$
$$= -\int_A \Big[(N_{1x} + N_{6y}) \delta u + (N_{2y} + N_{6x}) \delta v + (Q_{1x} + Q_{2y}) \delta w$$
$$- (M_{1x} + M_{6y} - Q_1) \delta w_x - (M_{2y} + M_{6x} - Q_2) \delta w_y \Big] dA$$
$$+ \int_y [N_1 \delta u + N_6 \delta v + (Q_1 + M_{6y}) \delta w - M_1 \delta w_x]_{x=0}^{x=a} dy$$
$$+ \int_x [N_6 \delta u + N_2 \delta v + (Q_2 + M_{6x}) \delta w - M_2 \delta w_y]_{y=0}^{y=b} dx$$
$$- 2 M_6 \delta w \Big|_{(x,y)=(a,0),(0,b)}^{(x,y)=(0,0),(a,b)} \qquad (7.2.16)$$

where the introduced variables Q_1 and Q_2 are shown below to be the transverse shear intensities. Substituting (7.2.16) and the first two lines of (7.2.7) into (7.2.1) and then setting each of the coefficients of δu, δv, δw, δw_y, and δw_x equal to zero, we obtain the following equations of motion:

$$N_{1x} + N_{6y} = I_0 \ddot{u} - I_1 \ddot{w}_x + \mu_1 \dot{u} \qquad (7.2.17a)$$
$$N_{6x} + N_{2y} = I_0 \ddot{v} - I_1 \ddot{w}_y + \mu_2 \dot{v} \qquad (7.2.17b)$$
$$Q_{1x} + Q_{2y} = I_0 \ddot{w} + \mu_3 \dot{w} \qquad (7.2.17c)$$
$$-M_{6x} - M_{2y} + Q_2 = I_2 \ddot{w}_y - I_1 \ddot{v} \qquad (7.2.17d)$$
$$M_{1x} + M_{6y} - Q_1 = -I_2 \ddot{w}_x + I_1 \ddot{u} \qquad (7.2.17e)$$

The boundary conditions for the plate are of the form: specify

Along $x = 0, a$:
$$\delta u = 0 \quad \text{or} \quad N_1$$

$$\delta v = 0 \quad \text{or} \quad N_6$$
$$\delta w = 0 \quad \text{or} \quad Q_1 + M_{6y}$$
$$\delta w_x = 0 \quad \text{or} \quad M_1$$

Along $y = 0, b$:
$$\delta u = 0 \quad \text{or} \quad N_6$$
$$\delta v = 0 \quad \text{or} \quad N_2$$
$$\delta w = 0 \quad \text{or} \quad Q_2 + M_{6x}$$
$$\delta w_y = 0 \quad \text{or} \quad M_2$$

At $(x, y) = (0,0), (a,b), (a,0), (0,b)$:
$$\delta w = 0 \quad \text{or} \quad M_6 \qquad (7.2.18)$$

Summing up $\mathbf{j}_1 \times$ (7.2.17a), $\mathbf{j}_2 \times$ (7.2.17b), and $\mathbf{j}_3 \times$ (7.2.17c), as well as $\mathbf{j}_1 \times$ (7.2.17d), $\mathbf{j}_2 \times$ (7.2.17e), and $\mathbf{j}_3 \times (N_6 - N_6)$, we put them in the following vector form:

$$\frac{\partial \mathbf{F}_\alpha}{\partial x} + \frac{\partial \mathbf{F}_\beta}{\partial y} = \mathbf{I}_F \qquad (7.2.19a)$$

$$\frac{\partial \mathbf{M}_\alpha}{\partial x} + \frac{\partial \mathbf{M}_\beta}{\partial y} + \mathbf{j}_1 \times \mathbf{F}_\alpha + \mathbf{j}_2 \times \mathbf{F}_\beta = \mathbf{I}_M \qquad (7.2.19b)$$

where

$$\mathbf{F}_\alpha = N_1 \mathbf{j}_1 + N_6 \mathbf{j}_2 + Q_1 \mathbf{j}_3, \quad \mathbf{F}_\beta = N_6 \mathbf{j}_1 + N_2 \mathbf{j}_2 + Q_2 \mathbf{j}_3$$
$$\mathbf{M}_\alpha = -M_6 \mathbf{j}_1 + M_1 \mathbf{j}_2, \quad \mathbf{M}_\beta = -M_2 \mathbf{j}_1 + M_6 \mathbf{j}_2$$
$$\mathbf{I}_F = (I_0 \ddot{u} - I_1 \ddot{w}_x + \mu_1 \dot{u})\mathbf{j}_1 + (I_0 \ddot{v} - I_1 \ddot{w}_y + \mu_2 \dot{v})\mathbf{j}_2 + (I_0 \ddot{w} + \mu_3 \dot{w})\mathbf{j}_3$$
$$\mathbf{I}_M = (I_2 \ddot{w}_y - I_1 \ddot{v})\mathbf{j}_1 + (-I_2 \ddot{w}_x + I_1 \ddot{u})\mathbf{j}_2 \qquad (7.2.20)$$

One can see from (7.2.20) that Q_1 and Q_2 are along the $\mathbf{j}_3$ direction and hence they represent the transverse shear intensities. The governing equations (7.2.19a,b) can be easily derived from Figure 7.2.2 using the Newtonian approach by considering the balancing of forces and moments. However, the boundary conditions cannot be easily figured out in the Newtonian approach, especially the corner conditions. Because transverse shear deformations are assumed to be zero in the classical plate theory, the transverse shear intensities Q_1 and Q_2 need to be obtained from the governing equations (7.2.17d,e) as

$$Q_2 = M_{2y} + M_{6x} + I_2 \ddot{w}_y - I_1 \ddot{v} \qquad (7.2.21a)$$

$$Q_1 = M_{1x} + M_{6y} + I_2 \ddot{w}_x - I_1 \ddot{u} \qquad (7.2.21b)$$

It follows from Section 2.4.4 that for isotropic materials

$$\left\{ \begin{array}{c} \sigma_{11} \\ \sigma_{22} \\ \sigma_{12} \end{array} \right\} = \frac{E}{1-\nu^2} \left[\begin{array}{ccc} 1 & \nu & 0 \\ \nu & 1 & 0 \\ 0 & 0 & (1-\nu)/2 \end{array} \right] \left\{ \begin{array}{c} u_x \\ v_y \\ u_y + v_x \end{array} \right\} - z \left\{ \begin{array}{c} w_{xx} \\ w_{yy} \\ 2w_{xy} \end{array} \right\}$$
$$(7.2.22)$$

Substituting (7.2.22) into (7.2.10) yields

$$\left\{\begin{array}{c}N_1\\N_2\\N_6\end{array}\right\} = \frac{Eh}{1-\nu^2}\begin{bmatrix}1 & \nu & 0\\\nu & 1 & 0\\0 & 0 & (1-\nu)/2\end{bmatrix}\left\{\begin{array}{c}u_x\\v_y\\u_y+v_x\end{array}\right\} \qquad (7.2.23)$$

$$\left\{\begin{array}{c}M_1\\M_2\\M_6\end{array}\right\} = -\frac{Eh^3}{12(1-\nu^2)}\begin{bmatrix}1 & \nu & 0\\\nu & 1 & 0\\0 & 0 & (1-\nu)/2\end{bmatrix}\left\{\begin{array}{c}w_{xx}\\w_{yy}\\2w_{xy}\end{array}\right\} \qquad (7.2.24)$$

Substituting (7.2.23) and (7.2.24) into (7.2.14) yields

$$\frac{Eh}{1-\nu^2}\left[u_{xx} + \frac{1}{2}(1+\nu)v_{xy} + \frac{1}{2}(1-\nu)u_{yy}\right] = \rho h\ddot{u} + \mu_1\dot{u} \qquad (7.2.25)$$

$$\frac{Eh}{1-\nu^2}\left[v_{yy} + \frac{1}{2}(1+\nu)u_{xy} + \frac{1}{2}(1-\nu)v_{xx}\right] = \rho h\ddot{v} + \mu_2\dot{v} \qquad (7.2.26)$$

$$\frac{Eh^3}{12(1-\nu^2)}\nabla^4 w = -\rho h\ddot{w} + \frac{1}{12}\rho h^3(\ddot{w}_{xx} + \ddot{w}_{yy}) - \mu_3\dot{w} \qquad (7.2.27)$$

where $I_1 = 0$ and $I_2 = \frac{1}{12}\rho h^3$ when ρ is a constant and the middle surface is chosen as the reference plane. We note that the equation governing the transverse displacement is uncoupled from those governing the inplane displacements.

7.2.2 Circular Plates

We consider a circular plate over the domain $0 \le r \le a$ and $0 \le \theta \le \theta_0$ and having thickness h. From Figure 7.2.3, we obtain

$$dx = dr, \quad dy = rd\theta \qquad (7.2.28a)$$

$$\frac{\partial \mathbf{j}_1}{\partial y} = \frac{1}{r}\mathbf{j}_2, \quad \frac{\partial \mathbf{j}_2}{\partial y} = -\frac{1}{r}\mathbf{j}_1 \qquad (7.2.28b)$$

and all other spatial derivatives of the base vectors $\mathbf{j}_1$, $\mathbf{j}_2$, and $\mathbf{j}_3$ of the xyz system are zero. Hence, it follows from (6.1.6) and (6.1.7) that the initial curvatures are

$$k_1^0 = k_2^0 = k_{61}^0 = k_{62}^0 = k_5^0 = 0, \quad [K_1^0] = [0], \quad k_4^0 = \frac{1}{r} \qquad (7.2.28c)$$

We point out here that k_1^0, k_2^0, k_{61}^0, and k_{62}^0 are always zero in plate analyses because the axes x and y are on the same plane. It follows from Figures 7.2.3 and 7.2.1 that the small displacements u_i are the same as those in (7.2.3). However, because the y axis is a curvilinear coordinate, we use vectors in the derivation of strains in order to account for the changes in the directions of the unit vectors. We write the displacement vector $\mathbf{D}$ as

$$\mathbf{D} = u_1\mathbf{j}_1 + u_2\mathbf{j}_2 + u_3\mathbf{j}_3 = (u - zw_r)\mathbf{j}_1 + \left(v - \frac{z}{r}w_\theta\right)\mathbf{j}_2 + w\mathbf{j}_3 \qquad (7.2.29)$$

LINEAR CLASSICAL PLATE THEORY   389

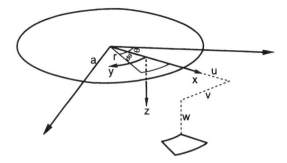

Fig. 7.2.3 A circular plate described by a polar coordinate system.

Taking the derivatives of (7.2.29) and using (7.2.28b) and (6.1.6) yields

$$\frac{\partial \mathbf{D}}{\partial x} = (u_r - zw_{rr})\mathbf{j}_1 + \left(v_r - \frac{z}{r}w_{r\theta} + \frac{z}{r^2}w_\theta\right)\mathbf{j}_2 + w_r\mathbf{j}_3$$

$$\frac{\partial \mathbf{D}}{\partial y} = \frac{1}{r}\left(u_\theta - zw_{r\theta} - v + \frac{z}{r}w_\theta\right)\mathbf{j}_1 + \frac{1}{r}\left(v_\theta - \frac{z}{r}w_{\theta\theta} + u - zw_r\right)\mathbf{j}_2 + \frac{1}{r}w_\theta\mathbf{j}_3$$

$$\frac{\partial \mathbf{D}}{\partial z} = -w_r\mathbf{j}_1 - \frac{1}{r}w_\theta\mathbf{j}_2 \qquad (7.2.30)$$

Using (7.2.30), we obtain the strains as

$$\epsilon_{11} = \frac{\partial \mathbf{D}}{\partial x} \cdot \mathbf{j}_1 = u_r - zw_{rr}$$

$$\epsilon_{22} = \frac{\partial \mathbf{D}}{\partial y} \cdot \mathbf{j}_2 = \frac{1}{r}\left[v_\theta + u - z\left(\frac{1}{r}w_{\theta\theta} + w_r\right)\right]$$

$$\epsilon_{12} = \frac{\partial \mathbf{D}}{\partial x} \cdot \mathbf{j}_2 + \frac{\partial \mathbf{D}}{\partial y} \cdot \mathbf{j}_1 = v_r + \frac{1}{r}\left[u_\theta - v - z\left(2w_{r\theta} - \frac{2}{r}w_\theta\right)\right]$$

$$\epsilon_{33} = \epsilon_{13} = \epsilon_{23} = 0 \qquad (7.2.31)$$

Taking the time derivatives and the variation of **D** in (7.2.29) and then substituting the results into (7.2.2a) yields (7.2.7). However, it is better to present the boundaries as $r = 0$, a and $\theta = 0$, θ_0 instead of $x = 0$, a and $y = 0$, $r\theta_0$ because a polar coordinate system is used here. Hence,

$$\delta T = -\int_A \left[(I_0\ddot{u} - I_1\ddot{w}_r)\,\delta u + \left(I_0\ddot{v} - \frac{1}{r}I_1\ddot{w}_\theta\right)\delta v + I_0\ddot{w}\delta w\right.$$

$$\left. + (I_2\ddot{w}_r - I_1\ddot{u})\,\delta w_r + \frac{1}{r}\left(\frac{1}{r}I_2\ddot{w}_\theta - I_1\ddot{v}\right)\delta w_\theta\right]rdrd\theta$$

$$= -\int_A \left\{(I_0\ddot{u} - I_1\ddot{w}_r)\,\delta u + \left(I_0\ddot{v} - \frac{1}{r}I_1\ddot{w}_\theta\right)\delta v\right.$$

$$\left. + \left[I_0\ddot{w} - \frac{1}{r}(I_2r\ddot{w}_r - I_1r\ddot{u})_r - \frac{1}{r}\left(\frac{1}{r}I_2\ddot{w}_\theta - I_1\ddot{v}\right)_\theta\right]\delta w\right\}rdrd\theta$$

$$-\int_\theta [I_2\ddot{w}_r - I_1\ddot{u}]_{r=0}^{r=a} \delta wr d\theta - \int_r \left[\frac{1}{r}I_2\ddot{w}_\theta - I_1\ddot{v}\right]_{\theta=0}^{\theta=\theta_0} \delta w dr \qquad (7.2.32)$$

Substituting (7.2.31) into (7.2.2b), we obtain

$$\begin{aligned}
\delta\Pi &= \int_A \bigg(N_1 \delta u_r + \frac{1}{r} N_6 \delta u_\theta + \frac{1}{r} N_2 \delta v_\theta + N_6 \delta v_r \\
&\quad - M_1 \delta w_{rr} - \frac{1}{r^2} M_2 \delta w_{\theta\theta} - \frac{1}{r} M_6 \delta w_{r\theta} - \frac{1}{r} M_6 \delta w_{\theta r} \\
&\quad + \frac{1}{r} N_2 \delta u - \frac{1}{r} N_6 \delta v - \frac{1}{r} M_2 \delta w_r + \frac{2}{r^2} M_6 \delta w_\theta \\
&\quad + Q_1 \delta w_r + \frac{1}{r} Q_2 \delta w_\theta - Q_1 \delta w_r - \frac{1}{r} Q_2 \delta w_\theta \bigg) r dr d\theta \\
&= -\int_A \bigg\{ [(rN_1)_r + N_{6\theta} - N_2] \delta u + [N_{2\theta} + (rN_6)_r + N_6] \delta v \\
&\quad + [(rQ_1)_r + Q_{2\theta}] \delta w - [(rM_1)_r + M_{6\theta} - rQ_1 - M_2] \delta w_r \\
&\quad - \frac{1}{r}[M_{2\theta} + rM_{6r} - rQ_2 + 2M_6] \delta w_\theta \bigg\} dr d\theta \\
&\quad + \int_\theta \bigg[N_1 \delta u + N_6 \delta v + (Q_1 + \frac{1}{r}M_{6\theta})\delta w - M_1 \delta w_r \bigg]_{r=0}^{r=a} r d\theta \\
&\quad + \int_r \bigg[N_6 \delta u + N_2 \delta v + (Q_2 + M_{6r})\delta w - \frac{1}{r}M_2 \delta w_\theta \bigg]_{\theta=0}^{\theta=\theta_0} dr \\
&\quad - 2 M_6 \delta w \bigg|_{(r,\theta)=(a,0),(0,\theta_0)}^{(r,\theta)=(0,0),(a,\theta_0)}
\end{aligned} \qquad (7.2.33)$$

where the stress resultants and moments are defined in (7.2.10) and the introduced variables Q_1 and Q_2 can be shown to be the transverse shear intensities.

Similar to (7.2.11) the reduced stress-strain relations take the form

$$\left\{\begin{array}{c} \sigma_{11} \\ \sigma_{22} \\ \sigma_{12} \end{array}\right\} = \left[\begin{array}{ccc} \tilde{Q}_{11} & \tilde{Q}_{12} & \tilde{Q}_{16} \\ \tilde{Q}_{12} & \tilde{Q}_{22} & \tilde{Q}_{26} \\ \tilde{Q}_{16} & \tilde{Q}_{26} & \tilde{Q}_{66} \end{array}\right]$$
$$\times \left(\left\{\begin{array}{c} u_r \\ (v_\theta + u)/r \\ (u_\theta + rv_r - v)/r \end{array}\right\} - z \left\{\begin{array}{c} w_{rr} \\ (w_{\theta\theta} + rw_r)/r^2 \\ (2rw_{r\theta} - 2w_\theta)/r^2 \end{array}\right\}\right) \qquad (7.2.34)$$

Substituting (7.2.34) into (7.2.10), we obtain the internal forces and moments in terms of the midplane strains and curvatures as

$$\left\{\begin{array}{c} N_1 \\ N_2 \\ N_6 \\ M_1 \\ M_2 \\ M_6 \end{array}\right\} = \left[\begin{array}{cc} [A_{ij}] & [B_{ij}] \\ [B_{ij}] & [D_{ij}] \end{array}\right] \left\{\begin{array}{c} u_r \\ (v_\theta + u)/r \\ (u_\theta + rv_r - v)/r \\ -w_{rr} \\ -(w_{\theta\theta} + rw_r)/r^2 \\ -(2rw_{r\theta} - 2w_\theta)/r^2 \end{array}\right\} \qquad (7.2.35)$$

where the A_{ij}, B_{ij}, and D_{ij} are defined in (7.2.13).

Substituting (7.2.32) and (7.2.33) into (7.2.1) and then setting each of the coefficients of δu, δv, δw, δw_θ, and δw_r equal to zero, we obtain the following equations of motion:

$$\frac{\partial N_1}{\partial r} + \frac{1}{r}\frac{\partial N_6}{\partial \theta} + \frac{N_1 - N_2}{r} = I_0 \ddot{u} - I_1 \ddot{w}_r + \mu_1 \dot{u} \qquad (7.2.36a)$$

$$\frac{\partial N_6}{\partial r} + \frac{1}{r}\frac{\partial N_2}{\partial \theta} + \frac{2N_6}{r} = I_0 \ddot{v} - \frac{I_1}{r}\ddot{w}_\theta + \mu_2 \dot{v} \qquad (7.2.36b)$$

$$\frac{\partial Q_1}{\partial r} + \frac{1}{r}\frac{\partial Q_2}{\partial \theta} + \frac{Q_1}{r} = I_0 \ddot{w} + \mu_3 \dot{w} \qquad (7.2.36c)$$

$$-\frac{\partial M_6}{\partial r} - \frac{1}{r}\frac{\partial M_2}{\partial \theta} - \frac{2M_6}{r} + Q_2 = \frac{I_2}{r}\ddot{w}_\theta - I_1 \ddot{v} \qquad (7.2.36d)$$

$$\frac{\partial M_1}{\partial r} + \frac{1}{r}\frac{\partial M_6}{\partial \theta} + \frac{M_1 - M_2}{r} - Q_1 = -I_2 \ddot{w}_r + I_1 \ddot{u} \qquad (7.2.36e)$$

The boundary conditions for the plate are of the form: specify

Along $r = 0, a$:

$\delta u = 0 \quad$ or $\quad N_1$

$\delta v = 0 \quad$ or $\quad N_6$

$\delta w = 0 \quad$ or $\quad Q_1 + \frac{1}{r}\frac{\partial M_6}{\partial \theta}$

$\delta w_r = 0 \quad$ or $\quad M_1$

Along $\theta = 0, \theta_0$:

$\delta u = 0 \quad$ or $\quad N_6$

$\delta v = 0 \quad$ or $\quad N_2$

$\delta w = 0 \quad$ or $\quad Q_2 + M_{6r}$

$\delta w_\theta = 0 \quad$ or $\quad M_2$

At $(r, \theta) = (0,0), (a, \theta_0), (a, 0), (0, \theta_0)$:

$\delta w = 0 \quad$ or $\quad M_6 \qquad (7.2.37)$

For isotropic plates, (7.2.35) reduces to

$$\left\{\begin{array}{c} N_1 \\ N_2 \\ N_6 \end{array}\right\} = \frac{Eh}{1-\nu^2}\left[\begin{array}{ccc} 1 & \nu & 0 \\ \nu & 1 & 0 \\ 0 & 0 & (1-\nu)/2 \end{array}\right]\left\{\begin{array}{c} u_r \\ (v_\theta + u)/r \\ (u_\theta + rv_r - v)/r \end{array}\right\} \qquad (7.2.38)$$

$$\left\{\begin{array}{c} M_1 \\ M_2 \\ M_6 \end{array}\right\} = -\frac{Eh^3}{12(1-\nu^2)}\left[\begin{array}{ccc} 1 & \nu & 0 \\ \nu & 1 & 0 \\ 0 & 0 & (1-\nu) \end{array}\right]\left\{\begin{array}{c} w_{rr} \\ (w_{\theta\theta} + rw_r)/r^2 \\ (rw_{r\theta} - w_\theta)/r^2 \end{array}\right\} \qquad (7.2.39)$$

Substituting (7.2.39) into (7.2.36c,d,e), setting $I_1 = 0$ and $I_2 = \frac{1}{12}\rho h^3$, and eliminating Q_1 and Q_2, we obtain

$$\frac{Eh^3}{12(1-\nu^2)}\nabla^4 w = -\rho h \ddot{w} + \frac{1}{12}\rho h^3 \nabla^2 \ddot{w} - \mu_3 \dot{w} \qquad (7.2.40)$$

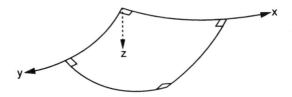

Fig. 7.2.4 A general plate described by an orthogonal curvilinear coordinate system.

which is the same as (7.2.27). Here,

$$\nabla^2 \equiv \frac{\partial^2}{\partial r^2} + \frac{1}{r}\frac{\partial}{\partial r} + \frac{1}{r^2}\frac{\partial^2}{\partial \theta^2}$$

7.2.3 General Plates

To describe the geometry of a general plate, one can use an orthogonal curvilinear coordinate system xyz with x and y being orthogonal curvilinear coordinates on the flat reference plane and z being a rectilinear coordinate, as shown in Figure 7.2.4. We consider a general plate over the domain $0 \leq x \leq X$ and $0 \leq y \leq Y$ and having thickness h. Because x and y are on the flat xy plane, it follows from (6.1.7) that the initial curvatures are given by

$$k_1^0 = k_2^0 = k_{61}^0 = k_{62}^0 = 0 \qquad (7.2.41\text{a})$$

$$\frac{\partial \mathbf{j}_1}{\partial x} = k_5^0 \mathbf{j}_2, \quad \frac{\partial \mathbf{j}_2}{\partial x} = -k_5^0 \mathbf{j}_1, \quad \frac{\partial \mathbf{j}_1}{\partial y} = k_4^0 \mathbf{j}_2, \quad \frac{\partial \mathbf{j}_2}{\partial y} = -k_4^0 \mathbf{j}_1 \qquad (7.2.41\text{b})$$

and all other spatial derivatives of the base vectors $\mathbf{j}_1$, $\mathbf{j}_2$, and $\mathbf{j}_3$ of the xyz system are zero. The displacement vector $\mathbf{D}$ of an arbitrary point is the same as (7.2.3) and is repeated here as

$$\mathbf{D} = (u - zw_x)\mathbf{j}_1 + (v - zw_y)\mathbf{j}_2 + w\mathbf{j}_3 \qquad (7.2.42)$$

where u, v, and w are the displacement components of the observed reference point with respect to the xyz system, and they are functions of x and y. Taking the derivatives of (7.2.42) and using (7.2.41a,b) and (6.1.6), we obtain

$$\frac{\partial \mathbf{D}}{\partial x} = \left(u_x - zw_{xx} - k_5^0 v + zk_5^0 w_y\right)\mathbf{j}_1 + \left(v_x - zw_{yx} + k_5^0 u - zk_5^0 w_x\right)\mathbf{j}_2 + w_x\mathbf{j}_3$$

$$\frac{\partial \mathbf{D}}{\partial y} = \left(u_y - zw_{xy} - k_4^0 v + zk_4^0 w_y\right)\mathbf{j}_1 + \left(v_y - zw_{yy} + k_4^0 u - zk_4^0 w_x\right)\mathbf{j}_2 + w_y\mathbf{j}_3$$

$$\frac{\partial \mathbf{D}}{\partial z} = -w_x\mathbf{j}_1 - w_y\mathbf{j}_2 \qquad (7.2.43)$$

Using (7.2.43), we obtain the following strains:

$$\epsilon_{11} = \frac{\partial \mathbf{D}}{\partial x} \cdot \mathbf{j}_1 = u_x - k_5^0 v - z\left(w_{xx} - k_5^0 w_y\right)$$

$$\epsilon_{22} = \frac{\partial \mathbf{D}}{\partial y} \cdot \mathbf{j}_2 = v_y + k_4^0 u - z\left(w_{yy} + k_4^0 w_x\right)$$

$$\epsilon_{12} = \frac{\partial \mathbf{D}}{\partial x} \cdot \mathbf{j}_2 + \frac{\partial \mathbf{D}}{\partial y} \cdot \mathbf{j}_1 = u_y + v_x - k_4^0 v + k_5^0 u$$
$$- z\left(w_{xy} + w_{yx} - k_4^0 w_y + k_5^0 w_x\right)$$

$$\epsilon_{33} = \epsilon_{13} = \epsilon_{23} = 0 \qquad (7.2.44)$$

Taking the time derivatives and the variation of $\mathbf{D}$ in (7.2.42) and then substituting the result into (7.2.2a) yields

$$\delta T = -\int_A \Big[(I_0 \ddot{u} - I_1 \ddot{w}_x)\,\delta u + (I_0 \ddot{v} - I_1 \ddot{w}_y)\,\delta v + I_0 \ddot{w}\,\delta w$$
$$+ (I_2 \ddot{w}_x - I_1 \ddot{u})\,\delta w_x + (I_2 \ddot{w}_y - I_1 \ddot{v})\,\delta w_y \Big] dA$$

$$= -\int_A \Big\{ (I_0 \ddot{u} - I_1 \ddot{w}_x)\,\delta u + (I_0 \ddot{v} - I_1 \ddot{w}_y)\,\delta v$$
$$+ \Big[I_0 \ddot{w} - (I_2 \ddot{w}_x - I_1 \ddot{u})_x - (I_2 \ddot{w}_y - I_1 \ddot{v})_y \Big]\,\delta w \Big\} dA$$

$$- \int_y [I_2 \ddot{w}_x - I_1 \ddot{u}]_{x=0}^{x=X}\,\delta w\, dy - \int_x [I_2 \ddot{w}_y - I_1 \ddot{v}]_{y=0}^{y=Y}\,\delta w\, dx \qquad (7.2.45)$$

Substituting (7.2.44) into (7.2.2b), we obtain

$$\delta \Pi = \int_A \Big(N_1 \delta u_x + N_6 \delta u_y + N_2 \delta v_y + N_6 \delta v_x$$
$$- M_1 \delta w_{xx} - M_2 \delta w_{yy} - M_6 \delta w_{xy} - M_6 \delta w_{yx}$$
$$+ k_4^0 N_2 \delta u - k_4^0 N_6 \delta v - k_4^0 M_2 \delta w_x + k_4^0 M_6 \delta w_y$$
$$+ k_5^0 N_6 \delta u - k_5^0 N_1 \delta v - k_5^0 M_6 \delta w_x + k_5^0 M_1 \delta w_y$$
$$+ Q_1 \delta w_x + Q_2 \delta w_y - Q_1 \delta w_x - Q_2 \delta w_y \Big) dA$$

$$= -\int_A \Big[\left(N_{1x} + N_{6y} - k_4^0 N_2 - k_5^0 N_6\right) \delta u$$
$$+ \left(N_{2y} + N_{6x} + k_4^0 N_6 + k_5^0 N_1\right) \delta v$$
$$+ (Q_{1x} + Q_{2y}) \delta w - \big(M_{1x} + M_{6y}$$
$$- Q_1 - k_4^0 M_2 - k_5^0 M_6\big) \delta w_x$$
$$- \left(M_{2y} + M_{6x} - Q_2 + k_4^0 M_6 + k_5^0 M_1\right) \delta w_y \Big] dA$$

$$+ \int_y [N_1 \delta u + N_6 \delta v + (Q_1 + M_{6y}) \delta w - M_1 \delta w_x]_{x=0}^{x=X}\, dy$$

$$+ \int_x [N_6 \delta u + N_2 \delta v + (Q_2 + M_{6x}) \delta w - M_2 \delta w_y]_{y=0}^{y=Y}\, dx$$

$$- 2 M_6 \delta w \Big|_{(x,y)=(X,0),(0,Y)}^{(x,y)=(0,0),(X,Y)} \qquad (7.2.46)$$

where the stress resultants and moments are defined in (7.2.10) and the introduced variables Q_1 and Q_2 are transverse shear intensities.

The reduced stress-strain relation is obtained using (2.4.34) and (2.4.35) and $\sigma_{33} = \epsilon_{13} = \epsilon_{23} = 0$ to be

$$\left\{ \begin{array}{c} \sigma_{11} \\ \sigma_{22} \\ \sigma_{12} \end{array} \right\} = \left[\begin{array}{ccc} \tilde{Q}_{11} & \tilde{Q}_{12} & \tilde{Q}_{16} \\ \tilde{Q}_{12} & \tilde{Q}_{22} & \tilde{Q}_{26} \\ \tilde{Q}_{16} & \tilde{Q}_{26} & \tilde{Q}_{66} \end{array} \right]$$
$$\times \left(\left\{ \begin{array}{c} u_x - k_5^0 v \\ v_y + k_4^0 u \\ u_y + v_x - k_4^0 v + k_5^0 u \end{array} \right\} - z \left\{ \begin{array}{c} w_{xx} - k_5^0 w_y \\ w_{yy} + k_4^0 w_x \\ w_{xy} + w_{yx} - k_4^0 w_y + k_5^0 w_x \end{array} \right\} \right) \quad (7.2.47)$$

Substituting (7.2.47) into (7.2.10), we obtain the internal forces and moments in terms of the midplane strains and curvatures as

$$\left\{ \begin{array}{c} N_1 \\ N_2 \\ N_6 \\ M_1 \\ M_2 \\ M_6 \end{array} \right\} = \left[\begin{array}{cc} [A_{ij}] & [B_{ij}] \\ [B_{ij}] & [D_{ij}] \end{array} \right] \left\{ \begin{array}{c} u_x - k_5^0 v \\ v_y + k_4^0 u \\ u_y + v_x - k_4^0 v + k_5^0 u \\ -w_{xx} + k_5^0 w_y \\ -w_{yy} - k_4^0 w_x \\ -w_{xy} - w_{yx} + k_4^0 w_y - k_5^0 w_x \end{array} \right\} \quad (7.2.48)$$

where the A_{ij}, B_{ij}, and D_{ij} are defined in (7.2.13).

Substituting (7.2.45) and (7.2.46) into (7.2.1) and then setting each of the coefficients of δu, δv, δw, δw_y, and δw_x equal to zero, we obtain the following equations of motion:

$$N_{1x} + N_{6y} - k_4^0 N_2 - k_5^0 N_6 = I_0 \ddot{u} - I_1 \ddot{w}_x + \mu_1 \dot{u} \quad (7.2.49a)$$

$$N_{6x} + N_{2y} + k_4^0 N_6 + k_5^0 N_1 = I_0 \ddot{v} - I_1 \ddot{w}_y + \mu_2 \dot{v} \quad (7.2.49b)$$

$$Q_{1x} + Q_{2y} = I_0 \ddot{w} + \mu_3 \dot{w} \quad (7.2.49c)$$

$$-M_{6x} - M_{2y} - k_4^0 M_6 - k_5^0 M_1 + Q_2 = I_2 \ddot{w}_y - I_1 \ddot{v} \quad (7.2.49d)$$

$$M_{1x} + M_{6y} - k_4^0 M_2 - k_5^0 M_6 - Q_1 = -I_2 \ddot{w}_x + I_1 \ddot{u} \quad (7.2.49e)$$

The boundary conditions for the plate are of the form: specify

Along $x = 0, X$:

$$\delta u = 0 \quad \text{or} \quad N_1$$
$$\delta v = 0 \quad \text{or} \quad N_6$$
$$\delta w = 0 \quad \text{or} \quad Q_1 + M_{6y}$$
$$\delta w_x = 0 \quad \text{or} \quad M_1$$

Along $y = 0, Y$:

$$\delta u = 0 \quad \text{or} \quad N_6$$

$$\delta v = 0 \quad \text{or} \quad N_2$$
$$\delta w = 0 \quad \text{or} \quad Q_2 + M_{6x}$$
$$\delta w_y = 0 \quad \text{or} \quad M_2$$

At $(x, y) = (0, 0), (X, Y), (X, 0), (0, Y)$:

$$\delta w = 0 \quad \text{or} \quad M_6 \qquad (7.2.50)$$

Summing up $\mathbf{j}_1 \times$ (7.2.49a), $\mathbf{j}_2 \times$ (7.2.49b), and $\mathbf{j}_3 \times$ (7.2.49c), as well as $\mathbf{j}_1 \times$ (7.2.49d), $\mathbf{j}_2 \times$ (7.2.49e), and $\mathbf{j}_3 \times (N_6 - N_6)$ and using (7.2.41b), we obtain the same equations as (7.2.19a,b) and (7.2.20). The transverse shear intensities Q_1 and Q_2 can be obtained from the governing equations (7.2.49d,e) as

$$Q_2 = M_{2y} + M_{6x} + k_4^0 M_6 + k_5^0 M_1 + I_2 \ddot{w}_y - I_1 \ddot{v} \qquad (7.2.51a)$$

$$Q_1 = M_{1x} + M_{6y} - k_4^0 M_2 - k_5^0 M_6 + I_2 \ddot{w}_x - I_1 \ddot{u} \qquad (7.2.51b)$$

Hence (7.2.49a,b,c) are the three equations governing u, v, and w.

We note that the governing equations and associated boundary conditions for rectangular and circular plates are special cases of the general plate theory given by (7.2.49) and (7.2.50). Equations (7.2.17) and (7.2.18) for rectangular plates can be obtained from (7.2.49) and (7.2.50) by setting $k_4^0 = k_5^0 = 0$.

For circular plates, we note that $k_5^0 = 0$, $k_4^0 = 1/r$, $dx = dr$, and $dy = rd\theta$. Hence, the differential area $dA = dxdy = rdrd\theta$ varies with r. We did not account for this effect when we integrated (7.2.46) by parts. To account for this effect, we note that

$$\frac{\partial(N_1 r) d\theta dr}{\partial r} \frac{1}{rd\theta dr} = \frac{1}{r} \frac{\partial(rN_1)}{\partial r}$$

Therefore, we replace N_{ix}, Q_{ix}, and M_{ix} as follows:

$$N_{ix} = \frac{1}{r} \frac{\partial(rN_i)}{\partial r}, \quad Q_{ix} = \frac{1}{r} \frac{\partial(rQ_i)}{\partial r}, \quad M_{ix} = \frac{1}{r} \frac{\partial(rM_i)}{\partial r} \qquad (7.2.52)$$

With these expressions, (7.2.49) reduces to (7.2.36) for circular plates.

We note that, for the general curvilinear coordinates x and y, one can show that

$$w_{yx} \equiv \frac{\partial w_y}{\partial x}\bigg|_{y=r\theta=\text{const}} = w_{xy} \equiv \frac{\partial w_x}{\partial y}\bigg|_{x=r=\text{const}} = \frac{1}{r} w_{\theta r} - \frac{1}{r^2} w_\theta - \frac{\theta}{r^2} w_{\theta\theta} \qquad (7.2.53)$$

However, for the polar coordinates r and θ, we have

$$w_{\theta r} \equiv \frac{\partial w_\theta}{\partial r}\bigg|_{\theta=\text{const}} = w_{r\theta} \equiv \frac{\partial w_r}{\partial \theta}\bigg|_{r=\text{const}}$$

$$w_{yx}|_{\theta=\text{const}} = \frac{\partial(w_\theta/r)}{\partial r}\bigg|_{\theta=\text{const}} = \frac{1}{r} w_{\theta r} - \frac{1}{r^2} w_\theta$$

$$w_{xy}|_{r=\text{const}} = \frac{\partial w_r}{r\partial\theta}\bigg|_{r=\text{const}} = \frac{1}{r} w_{r\theta} \qquad (7.2.54)$$

Substituting (7.2.54) into (7.2.48) yields (7.2.35).

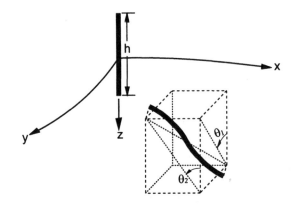

Fig. 7.3.1 A differential plate element with shear deformation.

7.3 LINEAR SHEAR-DEFORMABLE PLATE THEORIES

7.3.1 Formulation for Curvilinear Coordinate Systems

Figure 7.3.1 shows a differential plate element before and after deformation with transverse shear strains. We consider a general plate described by a curvilinear coordinate system xyz with $0 \leq x \leq X, 0 \leq y \leq Y$ and thickness h. The displacements u_i of an arbitrary point on the differential element are expressed as

$$u_1 = u + z\theta_2 + g(z)\gamma_5 = u - zw_x + g\gamma_5$$
$$u_2 = v - z\theta_1 + g(z)\gamma_4 = v - zw_y + g\gamma_4$$
$$u_3 = w \qquad (7.3.1)$$

where u, v, and w are the displacement components of the observed reference point with respect to the xyz coordinate system, which are independent of z. Moreover, $g(z)$ is the so-called shear-warping function and γ_4 and γ_5 are the shear rotation angles of the differential plate element, as shown in Figure 7.3.2. The displacement vector $\mathbf{D}$ can be written as

$$\mathbf{D} = u_1\mathbf{j}_1 + u_2\mathbf{j}_2 + u_3\mathbf{j}_3 = (u - zw_x + g\gamma_5)\mathbf{j}_1 + (v - zw_y + g\gamma_4)\mathbf{j}_2 + w\mathbf{j}_3 \qquad (7.3.2)$$

Taking the derivatives of (7.3.2) and using (7.2.41), we obtain

$$\frac{\partial \mathbf{D}}{\partial x} = \left(u_x - zw_{xx} - k_5^0 v + zk_5^0 w_y + g\gamma_{5x} - gk_5^0\gamma_4\right)\mathbf{j}_1$$
$$\quad + \left(v_x - zw_{yx} + k_5^0 u - zk_5^0 w_x + g\gamma_{4x} + gk_5^0\gamma_5\right)\mathbf{j}_2 + w_x\mathbf{j}_3$$
$$\frac{\partial \mathbf{D}}{\partial y} = \left(u_y - zw_{xy} - k_4^0 v + zk_4^0 w_y + g\gamma_{5y} - gk_4^0\gamma_4\right)\mathbf{j}_1$$

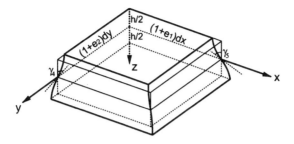

Fig. 7.3.2 The shear warpings and shear angles γ_4 and γ_5.

$$+ \left(v_y - zw_{yy} + k_4^0 u - zk_4^0 w_x + g\gamma_{4y} + gk_4^0 \gamma_5\right)\mathbf{j}_2 + w_y \mathbf{j}_3$$

$$\frac{\partial \mathbf{D}}{\partial z} = (g_z \gamma_5 - w_x)\mathbf{j}_1 + (g_z \gamma_4 - w_y)\mathbf{j}_2 \qquad (7.3.3)$$

Using (7.3.3), we obtain the following strains

$$\epsilon_{11} = \frac{\partial \mathbf{D}}{\partial x} \cdot \mathbf{j}_1 = u_x - k_5^0 v - z\left(w_{xx} - k_5^0 w_y\right) + g\left(\gamma_{5x} - k_5^0 \gamma_4\right)$$

$$\epsilon_{22} = \frac{\partial \mathbf{D}}{\partial y} \cdot \mathbf{j}_2 = v_y + k_4^0 u - z\left(w_{yy} + k_4^0 w_x\right) + g\left(\gamma_{4y} + k_4^0 \gamma_5\right)$$

$$\epsilon_{12} = \frac{\partial \mathbf{D}}{\partial x} \cdot \mathbf{j}_2 + \frac{\partial \mathbf{D}}{\partial y} \cdot \mathbf{j}_1 = u_y + v_x - k_4^0 v + k_5^0 u$$
$$\qquad - z\left(w_{xy} + w_{yx} - k_4^0 w_y + k_5^0 w_x\right)$$
$$\qquad + g\left(\gamma_{4x} + \gamma_{5y} + k_5^0 \gamma_5 - k_4^0 \gamma_4\right)$$

$$\epsilon_{13} = \frac{\partial \mathbf{D}}{\partial x} \cdot \mathbf{j}_3 + \frac{\partial \mathbf{D}}{\partial z} \cdot \mathbf{j}_1 = g_z \gamma_5$$

$$\epsilon_{23} = \frac{\partial \mathbf{D}}{\partial y} \cdot \mathbf{j}_3 + \frac{\partial \mathbf{D}}{\partial z} \cdot \mathbf{j}_2 = g_z \gamma_4$$

$$\epsilon_{33} = 0 \qquad (7.3.4)$$

Taking the time derivatives and variation of $\mathbf{D}$ in (7.3.2) and then substituting the results into (7.2.2a) yields

$$\delta T = -\int_A \Big[(I_0 \ddot{u} - I_1 \ddot{w}_x + I_3 \ddot{\gamma}_5)\delta u + (I_0 \ddot{v} - I_1 \ddot{w}_y + I_3 \ddot{\gamma}_4)\delta v + I_0 \ddot{w}\delta w$$
$$+ (I_2 \ddot{w}_x - I_1 \ddot{u} - I_4 \ddot{\gamma}_5)\delta w_x + (I_2 \ddot{w}_y - I_1 \ddot{v} - I_4 \ddot{\gamma}_4)\delta w_y$$
$$+ (I_3 \ddot{u} - I_4 \ddot{w}_x + I_5 \ddot{\gamma}_5)\delta \gamma_5 + (I_3 \ddot{v} - I_4 \ddot{w}_y + I_5 \ddot{\gamma}_4)\delta \gamma_4 \Big] dA \qquad (7.3.5)$$

where I_0, I_1, and I_2 are defined in (7.2.8) and

$$\{I_3, I_4, I_5\} \equiv \int_z \rho\{g, zg, g^2\}dz \qquad (7.3.6)$$

Substituting (7.3.4) into (7.2.2b), we obtain

$$\delta\Pi = \int_A \Big[N_1 \delta u_x + N_6 \delta u_y + N_2 \delta v_y + N_6 \delta v_x$$
$$- M_1 \delta w_{xx} - M_2 \delta w_{yy} - M_6 \delta w_{xy} - M_6 \delta w_{yx}$$
$$+ k_4^0 N_2 \delta u - k_4^0 N_6 \delta v - k_4^0 M_2 \delta w_x + k_4^0 M_6 \delta w_y$$
$$+ k_5^0 N_6 \delta u - k_5^0 N_1 \delta v - k_5^0 M_6 \delta w_x + k_5^0 M_1 \delta w_y$$
$$+ Q_1 \delta w_x + Q_2 \delta w_y - Q_1 \delta w_x - Q_2 \delta w_y$$
$$+ \left(q_2 - m_1 k_5^0 - m_6 k_4^0 \right) \delta \gamma_4 + m_6 \delta \gamma_{4x} + m_2 \delta \gamma_{4y}$$
$$+ \left(q_1 + m_2 k_4^0 + m_6 k_5^0 \right) \delta \gamma_5 + m_6 \delta \gamma_{5y} + m_1 \delta \gamma_{5x} \Big] dA$$

$$= - \int_A \Big[\left(N_{1x} + N_{6y} - k_4^0 N_2 - k_5^0 N_6 \right) \delta u + \left(N_{2y} + N_{6x} + k_4^0 N_6 + k_5^0 N_1 \right) \delta v$$
$$+ (Q_{1x} + Q_{2y}) \delta w - \left(M_{1x} + M_{6y} - Q_1 - k_4^0 M_2 - k_5^0 M_6 \right) \delta w_x$$
$$- \left(M_{2y} + M_{6x} - Q_2 + k_4^0 M_6 + k_5^0 M_1 \right) \delta w_y$$
$$+ \left(m_{6x} + m_{2y} + m_1 k_5^0 + m_6 k_4^0 - q_2 \right) \delta \gamma_4$$
$$+ \left(m_{6y} + m_{1x} - m_2 k_4^0 - m_6 k_5^0 - q_1 \right) \delta \gamma_5 \Big] dA$$

$$+ \int_y [N_1 \delta u + N_6 \delta v + (Q_1 + M_{6y}) \delta w - M_1 \delta w_x + m_1 \delta \gamma_5 + m_6 \delta \gamma_4]_{x=0}^{x=X} dy$$

$$+ \int_x [N_6 \delta u + N_2 \delta v + (Q_2 + M_{6x}) \delta w - M_2 \delta w_y + m_2 \delta \gamma_4 + m_6 \delta \gamma_5]_{y=0}^{y=Y} dx$$

$$- 2 M_6 \delta w \Big|_{(x,y)=(X,0),(0,Y)}^{(x,y)=(0,0),(X,Y)} \tag{7.3.7}$$

where Q_1 and Q_2 are the transverse shear intensities, the N_i and M_i are defined in (7.2.10), and

$$\{m_1, m_2, m_6\} \equiv \int_z g\{\sigma_{11}, \sigma_{22}, \sigma_{12}\} dz$$

$$\{q_1, q_2\} \equiv \int_z g_z \{\sigma_{13}, \sigma_{23}\} dz \tag{7.3.8}$$

which are higher-order stress resultants.

The reduced stress-strain relation is obtained using (2.4.34) and (2.4.35) and $\sigma_{33} = 0$ to be

$$\begin{Bmatrix} \sigma_{11} \\ \sigma_{22} \\ \sigma_{12} \end{Bmatrix} = \begin{bmatrix} \tilde{Q}_{11} & \tilde{Q}_{12} & \tilde{Q}_{16} \\ \tilde{Q}_{12} & \tilde{Q}_{22} & \tilde{Q}_{26} \\ \tilde{Q}_{16} & \tilde{Q}_{26} & \tilde{Q}_{66} \end{bmatrix} \left(\begin{Bmatrix} u_x - k_5^0 v \\ v_y + k_4^0 u \\ u_y + v_x - k_4^0 v + k_5^0 u \end{Bmatrix} \right.$$
$$\left. - z \begin{Bmatrix} w_{xx} - k_5^0 w_y \\ w_{yy} + k_4^0 w_x \\ w_{xy} + w_{yx} - k_4^0 w_y + k_5^0 w_x \end{Bmatrix} + g \begin{Bmatrix} \gamma_{5x} - k_5^0 \gamma_4 \\ \gamma_{4y} + k_4^0 \gamma_5 \\ \gamma_{4x} + \gamma_{5y} + k_5^0 \gamma_5 - k_4^0 \gamma_4 \end{Bmatrix} \right) \tag{7.3.9a}$$

$$\left\{ \begin{array}{c} \sigma_{23} \\ \sigma_{13} \end{array} \right\} = \left[\begin{array}{cc} \overline{Q}_{44} & \overline{Q}_{45} \\ \overline{Q}_{45} & \overline{Q}_{55} \end{array} \right] g_z \left\{ \begin{array}{c} \gamma_4 \\ \gamma_5 \end{array} \right\} \qquad (7.3.9b)$$

Substituting (7.3.9a,b) into (7.2.10) and (7.3.8), we obtain the internal forces and moments in terms of the midplane strains and curvatures as

$$\left\{ \begin{array}{c} N_1 \\ N_2 \\ N_6 \\ M_1 \\ M_2 \\ M_6 \\ m_1 \\ m_2 \\ m_6 \end{array} \right\} = \left[\begin{array}{ccc} [A_{ij}] & [B_{ij}] & [E_{ij}] \\ [B_{ij}] & [D_{ij}] & [F_{ij}] \\ [E_{ij}] & [F_{ij}] & [H_{ij}] \end{array} \right] \left\{ \begin{array}{c} u_x - k_5^0 v \\ v_y + k_4^0 u \\ u_y + v_x - k_4^0 v + k_5^0 u \\ -w_{xx} + k_5^0 w_y \\ -w_{yy} - k_4^0 w_x \\ -w_{xy} - w_{yx} + k_4^0 w_y - k_5^0 w_x \\ \gamma_{5x} - k_5^0 \gamma_4 \\ \gamma_{4y} + k_4^0 \gamma_5 \\ \gamma_{4x} + \gamma_{5y} + k_5^0 \gamma_5 - k_4^0 \gamma_4 \end{array} \right\}$$
(7.3.10a)

$$\left\{ \begin{array}{c} q_1 \\ q_2 \end{array} \right\} = [C_{ij}] \left\{ \begin{array}{c} \gamma_5 \\ \gamma_4 \end{array} \right\} \qquad (7.3.10b)$$

where the A_{ij}, B_{ij}, and D_{ij} are defined in (7.2.13) and

$$\{E_{ij}, F_{ij}, H_{ij}\} = \sum_{k=1}^{N} \int_{z_k}^{z_{k+1}} \tilde{Q}_{ij}^{(k)} \{g, gz, g^2\} dz \quad \text{for } i,j = 1,2,6 \quad (7.3.11a)$$

$$[C_{ij}] = \sum_{k=1}^{N} \int_{z_k}^{z_{k+1}} \left[\begin{array}{cc} \overline{Q}_{55} & \overline{Q}_{45} \\ \overline{Q}_{45} & \overline{Q}_{44} \end{array} \right] g_z^2 dz \qquad (7.3.11b)$$

Substituting (7.3.5) and (7.3.7) into (7.2.1) and then setting each of the coefficients of δu, δv, δw, $\delta \gamma_4$, $\delta \gamma_5$, δw_y, and δw_x equal to zero, we obtain the following equations of motion:

$$N_{1x} + N_{6y} - k_4^0 N_2 - k_5^0 N_6 = I_0 \ddot{u} - I_1 \ddot{w}_x + I_3 \ddot{\gamma}_5 + \mu_1 \dot{u} \qquad (7.3.12a)$$

$$N_{6x} + N_{2y} + k_4^0 N_6 + k_5^0 N_1 = I_0 \ddot{v} - I_1 \ddot{w}_y + I_3 \ddot{\gamma}_4 + \mu_2 \dot{v} \qquad (7.3.12b)$$

$$Q_{1x} + Q_{2y} = I_0 \ddot{w} + \mu_3 \dot{w} \qquad (7.3.12c)$$

$$m_{6x} + m_{2y} + m_1 k_5^0 + m_6 k_4^0 - q_2 = I_5 \ddot{\gamma}_4 + I_3 \ddot{v} - I_4 \ddot{w}_y + \mu_4 \dot{\gamma}_4 \qquad (7.3.12d)$$

$$m_{1x} + m_{6y} - m_2 k_4^0 - m_6 k_5^0 - q_1 = I_5 \ddot{\gamma}_5 + I_3 \ddot{u} - I_4 \ddot{w}_x + \mu_5 \dot{\gamma}_5 \qquad (7.3.12e)$$

$$-M_{6x} - M_{2y} - k_4^0 M_6 - k_5^0 M_1 + Q_2 = I_2 \ddot{w}_y - I_1 \ddot{v} - I_4 \ddot{\gamma}_4 \qquad (7.3.12f)$$

$$M_{1x} + M_{6y} - k_4^0 M_2 - k_5^0 M_6 - Q_1 = -I_2 \ddot{w}_x + I_1 \ddot{u} + I_4 \ddot{\gamma}_5 \qquad (7.3.12g)$$

where we added a linear viscous damping term to each of (7.3.12a-e) and the μ_i are damping coefficients. The boundary conditions for the plate are of the form: specify

Along $x = 0, X$:

$\delta u = 0$	or	N_1
$\delta v = 0$	or	N_6
$\delta w = 0$	or	$Q_1 + M_{6y}$
$\delta w_x = 0$	or	M_1
$\delta \gamma_4 = 0$	or	m_6
$\delta \gamma_5 = 0$	or	m_1

Along $y = 0, Y$:

$\delta u = 0$	or	N_6
$\delta v = 0$	or	N_2
$\delta w = 0$	or	$Q_2 + M_{6x}$
$\delta w_y = 0$	or	M_2
$\delta \gamma_4 = 0$	or	m_2
$\delta \gamma_5 = 0$	or	m_6

At $(x,y) = (0,0), (X,Y), (X,0), (0,Y)$:

$$\delta w = 0 \quad \text{or} \quad M_6 \qquad (7.3.13)$$

Summing up $\mathbf{j}_1 \times$ (7.3.12a), $\mathbf{j}_2 \times$ (7.3.12b), and $\mathbf{j}_3 \times$ (7.3.12c), as well as $\mathbf{j}_1 \times$ (7.3.12f), $\mathbf{j}_2 \times$ (7.3.12g), and $\mathbf{j}_3 \times (N_6 - N_6)$ and using (7.2.41b), we obtain the same equations as (7.2.19a,b) and (7.2.20), except that $\mathbf{I}_F$ and $\mathbf{I}_M$ need to be modified to account for the inertias due to the shear rotations as

$$\mathbf{I}_F = (I_0\ddot{u} - I_1\ddot{w}_x + I_3\ddot{\gamma}_5 + \mu_1\dot{u})\mathbf{j}_1 + (I_0\ddot{v} - I_1\ddot{w}_y + I_3\ddot{\gamma}_4 + \mu_2\dot{v})\mathbf{j}_2$$
$$+ (I_0\ddot{w} + \mu_3\dot{w})\mathbf{j}_3$$
$$\mathbf{I}_M = (I_2\ddot{w}_y - I_1\ddot{v} - I_4\ddot{\gamma}_4)\mathbf{j}_1 + (-I_2\ddot{w}_x + I_1\ddot{u} + I_4\ddot{\gamma}_5)\mathbf{j}_2 \qquad (7.3.14)$$

In other words, (7.3.12a,b,c,f,g) can be also derived using the Newtonian approach. However, the governing equations (7.3.12d,e) for shear deformations cannot be directly derived using the Newtonian approach.

One can solve for Q_1 and Q_2 from (7.3.12f,g) and then substitute the result into (7.3.12c). Then there are only five equations governing u, v, w, γ_4, and γ_5. We point out here that the transverse shear intensities Q_1 and Q_2 actually represent the geometric averages of the shear forces per unit length; that is,

$$\{Q_1, Q_2\} = \int_z \{\sigma_{13}, \sigma_{23}\} dz \qquad (7.3.15)$$

and the q_1 and q_2 in (7.3.8) represent the energy averages of the shear forces per unit length. If $g = z$ (the first-order shear theory), $Q_1 = q_1$ and $Q_2 = q_2$.

If the differential area $dA = dxdy$ changes with x and/or y, one needs to account for the area effect by replacing the N_{ix}, M_{ix}, N_{iy}, M_{iy}, Q_{ix}, Q_{iy}, m_{ix}, and m_{iy} in (7.3.12a-g) with appropriate terms, as in Section 7.2.3.

7.3.2 Rectangular and Circular Plates

For rectangular plates, we let $k_4^0 = k_5^0 = 0$ in (7.3.12a-g) and (7.3.13). For circular plates, we let $k_5^0 = 0$, $k_4^0 = 1/r$, $dx = dr$, $dy = rd\theta$, and

$$N_{ix} = \frac{1}{r}\frac{\partial(rN_i)}{\partial r}, \quad Q_{ix} = \frac{1}{r}\frac{\partial(rQ_i)}{\partial r}, \quad M_{ix} = \frac{1}{r}\frac{\partial(rM_i)}{\partial r}, \quad m_{ix} = \frac{1}{r}\frac{\partial(rm_i)}{\partial r}$$

in (7.3.12) and (7.3.13) and obtain

$$\frac{\partial N_1}{\partial r} + \frac{1}{r}\frac{\partial N_6}{\partial \theta} + \frac{N_1 - N_2}{r} = I_0\ddot{u} - I_1\ddot{w}_r + I_3\ddot{\gamma}_5 + \mu_1\dot{u} \quad (7.3.16a)$$

$$\frac{\partial N_6}{\partial r} + \frac{1}{r}\frac{\partial N_2}{\partial \theta} + \frac{2N_6}{r} = I_0\ddot{v} - \frac{I_1}{r}\ddot{w}_\theta + I_3\ddot{\gamma}_4 + \mu_2\dot{v} \quad (7.3.16b)$$

$$\frac{\partial Q_1}{\partial r} + \frac{1}{r}\frac{\partial Q_2}{\partial \theta} + \frac{Q_1}{r} = I_0\ddot{w} + \mu_3\dot{w} \quad (7.3.16c)$$

$$\frac{\partial m_6}{\partial r} + \frac{1}{r}\frac{\partial m_2}{\partial \theta} + \frac{2m_6}{r} - q_2 = I_5\ddot{\gamma}_4 - \frac{I_4}{r}\ddot{w}_\theta + I_3\ddot{v} + \mu_4\dot{\gamma}_4 \quad (7.3.16d)$$

$$\frac{\partial m_1}{\partial r} + \frac{1}{r}\frac{\partial m_6}{\partial \theta} + \frac{m_1 - m_2}{r} - q_1 = I_5\ddot{\gamma}_5 - I_4\ddot{w}_r + I_3\ddot{u} + \mu_5\dot{\gamma}_5 \quad (7.3.16e)$$

$$-\frac{\partial M_6}{\partial r} - \frac{1}{r}\frac{\partial M_2}{\partial \theta} - \frac{2M_6}{r} + Q_2 = \frac{I_2}{r}\ddot{w}_\theta - I_1\ddot{v} - I_4\ddot{\gamma}_4 \quad (7.3.16f)$$

$$\frac{\partial M_1}{\partial r} + \frac{1}{r}\frac{\partial M_6}{\partial \theta} + \frac{M_1 - M_2}{r} - Q_1 = -I_2\ddot{w}_r + I_1\ddot{u} + I_4\ddot{\gamma}_5 \quad (7.3.16g)$$

The boundary conditions for the plate are of the form: specify

Along $r = 0, a$:

$\delta u = 0$	or	N_1
$\delta v = 0$	or	N_6
$\delta w = 0$	or	$Q_1 + \frac{1}{r}\frac{\partial M_6}{\partial \theta}$
$\delta w_r = 0$	or	M_1
$\delta \gamma_4 = 0$	or	m_6
$\delta \gamma_5 = 0$	or	m_1

Along $\theta = 0, \theta_0$:

$\delta u = 0$	or	N_6
$\delta v = 0$	or	N_2
$\delta w = 0$	or	$Q_2 + M_{6r}$
$\delta w_\theta = 0$	or	M_2

$$\delta\gamma_4 = 0 \quad \text{or} \quad m_2$$
$$\delta\gamma_5 = 0 \quad \text{or} \quad m_6$$

At $(r,\theta) = (0,0), (a,\theta_0), (a,0), (0,\theta_0)$:
$$\delta w = 0 \quad \text{or} \quad M_6 \qquad (7.3.16h)$$

7.3.3 Different Shear-Warping Functions

In the third-order shear-deformation theory, the middle plane of a plate is chosen as the reference plane and it is assumed that

$$g(z) = z - \frac{4z^3}{3h^2} \qquad (7.3.17)$$

In the first-order shear-deformation theory (i.e., the Hencky-Mindlin theory), the middle plane is chosen as the reference plane and it is assumed that

$$g(z) = z \qquad (7.3.18)$$

Because (7.3.18) implies that

$$\epsilon_{23} = \gamma_4, \quad \epsilon_{13} = \gamma_5 \qquad (7.3.19)$$

shear-correction factors are needed in order to account for the non-uniform distribution of transverse shear stresses, as shown in Section 4.3.2. Substituting (7.3.18) into (7.3.6) and (7.3.8) and using (7.2.10) and (7.2.8), we find that

$$q_1 = Q_1, \quad q_2 = Q_2,$$
$$m_1 = M_1, \quad m_2 = M_2, \quad m_6 = M_6,$$
$$I_3 = I_1, \quad I_4 = I_5 = I_2 \qquad (7.3.20)$$

Then, we obtain from (7.3.12d,e) that

$$Q_2 = M_{2y} + M_{6x} + k_4^0 M_6 + k_5^0 M_1 + I_2 \ddot{w}_y - I_1 \ddot{v} - I_2 \ddot{\gamma}_4 - \mu_4 \dot{\gamma}_4 \qquad (7.3.21a)$$

$$Q_1 = M_{1x} + M_{6y} - k_4^0 M_2 - k_5^0 M_6 + I_2 \ddot{w}_x - I_1 \ddot{u} - I_2 \ddot{\gamma}_5 - \mu_5 \dot{\gamma}_5 \qquad (7.3.21b)$$

which are the same as (7.3.12f,g), except for the added damping terms.

For sandwich plates consisting of isotropic laminae, one can use the theory introduced in Section 4.3.1 to derive g. For example, for a plate made of three layers as shown in Figure 4.3.3,

$$g(z) = \left[U\left(z + \frac{1}{2}h\right) - U\left(z + \frac{1}{3}h\right)\right]\left(z + \frac{3}{h}z^2 + \frac{8}{3h^2}z^3\right)$$
$$+ \left[U\left(z + \frac{1}{3}h\right) - U\left(z - \frac{1}{3}h\right)\right]\left(z - \frac{19}{3h^2}z^3\right)$$
$$+ \left[U\left(z - \frac{1}{3}h\right) - U\left(z - \frac{1}{2}h\right)\right]\left(z - \frac{3}{h}z^2 + \frac{8}{3h^2}z^3\right) \qquad (7.3.22)$$

where $U(z)$ is the unit-step function. For laminated composites, shear coupling between γ_4 and γ_5 exists and it is derived in Section 7.8.1.

In the classical plate theory, it is assumed that

$$\gamma_4 = \gamma_5 = g = 0 \tag{7.3.23}$$

Hence, (7.3.12a,b,c) are the three governing equations.

7.4 NONLINEAR CLASSICAL PLATE THEORY

In this section, we develop the equations and boundary conditions governing the nonlinear static and dynamics of plates using a combination of the classical plate theory and the von Kármán-type nonlinearity. The classical plate theory is based on the Kirchhoff hypothesis (1850), which assumes that the cross-sections are flat and perpendicular to the reference plane before and after deformation.

7.4.1 Rectangular Plates

In Figure 7.2.1, we show the undeformed and deformed configurations of a differential plate element. Without shear-warpings, a plate differential element is a rigid straight filament with dimensions $dx \times dy \times h$, where h is the plate thickness. Thus, the small displacements u_i of an arbitrary point on the observed plate element are given by

$$u_1 = u - zw_x, \quad u_2 = v - zw_y, \quad u_3 = w \tag{7.4.1}$$

where u, v, and w are the displacement components of the observed reference point with respect to the x, y, z coordinate system and they are independent of z.

The displacement vector $\mathbf{D}$ is given by

$$\mathbf{D} = u_1 \mathbf{j}_1 + u_2 \mathbf{j}_2 + u_3 \mathbf{j}_3 \tag{7.4.2}$$

where $\mathbf{j}_1$, $\mathbf{j}_2$, and $\mathbf{j}_3$ denote the unit vectors along the axes x, y, and z, respectively. Taking the time derivatives and variation of the displacement vector $\mathbf{D}$, we obtain

$$\ddot{\mathbf{D}} = (\ddot{u} - z\ddot{w}_x)\mathbf{j}_1 + (\ddot{v} - z\ddot{w}_y)\mathbf{j}_2 + \ddot{w}\mathbf{j}_3 \tag{7.4.3a}$$

$$\delta\mathbf{D} = (\delta u - z\delta w_x)\mathbf{j}_1 + (\delta v - z\delta w_y)\mathbf{j}_2 + \delta w\mathbf{j}_3 \tag{7.4.3b}$$

Substituting (7.4.3a,b) into (7.2.2a) yields

$$\begin{aligned}\delta T = &- \int_A \big[(I_0\ddot{u} - I_1\ddot{w}_x)\delta u + (I_0\ddot{v} - I_1\ddot{w}_y)\delta v + I_0\ddot{w}\delta w \\ &+ (I_2\ddot{w}_x - I_1\ddot{u})\delta w_x + (I_2\ddot{w}_y - I_1\ddot{v})\delta w_y \big] dA \\ = &- \int_A \big\{ (I_0\ddot{u} - I_1\ddot{w}_x)\delta u + (I_0\ddot{v} - I_1\ddot{w}_y)\delta v \end{aligned}$$

$$+\left[I_0\ddot{w}-(I_2\ddot{w}_x-I_1\ddot{u})_x-(I_2\ddot{w}_y-I_1\ddot{v})_y\right]\delta w\bigg\}dA$$
$$-\int_y\left[I_2\ddot{w}_x-I_1\ddot{u}\right]_{x=0}^{x=a}\delta w\,dy-\int_x\left[I_2\ddot{w}_y-I_1\ddot{v}\right]_{y=0}^{y=b}\delta w\,dx \quad (7.4.4)$$

where the I_i are defined in (7.2.8).

The von Kármán nonlinear strains are given by

$$\epsilon_{11} = e_1 - zw_{xx}$$
$$\epsilon_{22} = e_2 - zw_{yy}$$
$$\epsilon_{12} = \gamma_6 - 2zw_{xy}$$
$$\epsilon_{33} = \epsilon_{13} = \epsilon_{23} = 0 \quad (7.4.5)$$

where

$$e_1 = u_x + \frac{1}{2}w_x^2$$
$$e_2 = v_y + \frac{1}{2}w_y^2$$
$$\gamma_6 = u_y + v_x + w_x w_y \quad (7.4.6)$$

Substituting (7.4.5) and (7.4.6) into (7.2.2b) yields

$$\begin{aligned}\delta\Pi &= \int_A\int_z\Big[\sigma_{11}\left(\delta u_x+w_x\delta w_x-z\delta w_{xx}\right)+\sigma_{22}\left(\delta v_y+w_y\delta w_y-z\delta w_{yy}\right) \\ &\quad +\sigma_{12}\left(\delta u_y+\delta v_x+w_x\delta w_y+w_y\delta w_x-2z\delta w_{xy}\right)\Big]dA\,dz \\ &= \int_A\Big[N_1\left(\delta u_x+w_x\delta w_x\right)+N_6\left(\delta u_y+\delta v_x+w_x\delta w_y+w_y\delta w_x\right) \\ &\quad +N_2\left(\delta v_y+w_y\delta w_y\right)-M_1\delta w_{xx}-M_2\delta w_{yy}-2M_6\delta w_{xy}\Big]dA \\ &= -\int_A\Big\{(N_{1x}+N_{6y})\delta u+(N_{2y}+N_{6x})\delta v \\ &\quad +\Big[M_{1xx}+M_{2yy}+2M_{6xy}+(N_1w_x)_x+(N_2w_y)_y \\ &\quad +(N_6w_x)_y+(N_6w_y)_x\Big]\delta w\Big\}dA \\ &\quad +\int_y\Big[N_1\delta u+N_6\delta v+\big(M_{1x}+2M_{6y} \\ &\quad +N_1w_x+N_6w_y\big)\delta w-M_1\delta w_x\Big]_{x=0}^{x=a}dy \\ &\quad +\int_x\Big[N_6\delta u+N_2\delta v+\big(M_{2y}+2M_{6x} \\ &\quad +N_2w_y+N_6w_x\big)\delta w-M_2\delta w_y\Big]_{y=0}^{y=b}dx \\ &\quad -2M_6\delta w\,\Big|_{(x,y)=(a,0),(0,b)}^{(x,y)=(0,0),(a,b)}\end{aligned}\quad(7.4.7)$$

where the N_i and M_i are defined in (7.2.10).

Substituting (7.4.4) and (7.4.7) into (7.2.1) and setting each of the coefficients of $\delta u, \delta v$, and δw equal to zero, we obtain the equations of motion as

$$N_{1x} + N_{6y} = I_0 \ddot{u} - I_1 \ddot{w}_x \qquad (7.4.8)$$

$$N_{6x} + N_{2y} = I_0 \ddot{v} - I_1 \ddot{w}_y \qquad (7.4.9)$$

$$M_{1xx} + 2M_{6xy} + M_{2yy} + (N_1 w_x + N_6 w_y)_x + (N_6 w_x + N_2 w_y)_y$$
$$= I_0 \ddot{w} - I_2 (\ddot{w}_{xx} + \ddot{w}_{yy}) + I_1 \ddot{u}_x + I_1 \ddot{v}_y \qquad (7.4.10)$$

The corresponding boundary conditions are

Along $x = 0, a$:

$\delta u = 0$	or	N_1
$\delta v = 0$	or	N_6
$\delta w = 0$	or	$M_{1x} + 2M_{6y} + N_1 w_x + N_6 w_y - I_1 \ddot{u} + I_2 \ddot{w}_x$
$\delta w_x = 0$	or	M_1

Along $y = 0, b$:

$\delta u = 0$	or	N_6
$\delta v = 0$	or	N_2
$\delta w = 0$	or	$M_{2y} + 2M_{6x} + N_6 w_x + N_2 w_y - I_1 \ddot{v} + I_2 \ddot{w}_y$
$\delta w_y = 0$	or	M_2

At $(x, y) = (0,0), (a,b), (a,0), (0,b)$:

$$\delta w = 0 \quad \text{or} \quad M_6 \qquad (7.4.11)$$

As in Section 7.2.1, the internal forces N_i and moments M_i are related to the midplane strains and curvatures as

$$\begin{Bmatrix} N_1 \\ N_2 \\ N_6 \\ M_1 \\ M_2 \\ M_6 \end{Bmatrix} = \begin{bmatrix} [A_{ij}] & [B_{ij}] \\ [B_{ij}] & [D_{ij}] \end{bmatrix} \begin{Bmatrix} e_1 \\ e_2 \\ \gamma_6 \\ -w_{xx} \\ -w_{yy} \\ -2w_{xy} \end{Bmatrix} \qquad (7.4.12)$$

where

$$\{A_{ij}, B_{ij}, D_{ij}\} = \sum_{k=1}^{N} \int_{z_k}^{z_{k+1}} \tilde{Q}_{ij}^{(k)} \{1, z, z^2\} dz \quad \text{for } i,j = 1, 2, 6 \qquad (7.4.13)$$

For a symmetric cross-ply laminate, $[B_{ij}] = 0$ and

$$[A_{ij}] = \begin{bmatrix} A_{11} & A_{12} & 0 \\ A_{12} & A_{22} & 0 \\ 0 & 0 & A_{66} \end{bmatrix} \text{ and } [D_{ij}] = \begin{bmatrix} D_{11} & D_{12} & 0 \\ D_{12} & D_{22} & 0 \\ 0 & 0 & D_{66} \end{bmatrix} \qquad (7.4.14)$$

Substituting (7.4.12) and (7.4.14) into (7.4.8)-(7.4.10), we obtain the following equations of motion expressed in terms of displacements:

$$A_{11}\frac{\partial^2 u}{\partial x^2} + (A_{12}+A_{66})\frac{\partial^2 v}{\partial x \partial y} + A_{66}\frac{\partial^2 u}{\partial y^2} + A_{11}\frac{\partial w}{\partial x}\frac{\partial^2 w}{\partial x^2} + A_{12}\frac{\partial w}{\partial y}\frac{\partial^2 w}{\partial x \partial y}$$

$$+A_{66}\left(\frac{\partial w}{\partial y}\frac{\partial^2 w}{\partial x \partial y} + \frac{\partial w}{\partial x}\frac{\partial^2 w}{\partial y^2}\right) = I_0 \frac{\partial^2 u}{\partial t^2} - I_1 \frac{\partial^3 w}{\partial x \partial t^2} \qquad (7.4.15)$$

$$(A_{12}+A_{66})\frac{\partial^2 u}{\partial x \partial y} + A_{66}\frac{\partial^2 v}{\partial x^2} + A_{22}\frac{\partial^2 v}{\partial y^2} + A_{12}\frac{\partial w}{\partial x}\frac{\partial^2 w}{\partial x \partial y} + A_{22}\frac{\partial w}{\partial y}\frac{\partial^2 w}{\partial y^2}$$

$$+A_{66}\left(\frac{\partial w}{\partial y}\frac{\partial^2 w}{\partial x^2} + \frac{\partial w}{\partial x}\frac{\partial^2 w}{\partial x \partial y}\right) = I_0 \frac{\partial^2 v}{\partial t^2} - I_1 \frac{\partial^3 w}{\partial y \partial t^2} \qquad (7.4.16)$$

$$D_{11}\frac{\partial^4 w}{\partial x^4} + 2(D_{12}+2D_{66})\frac{\partial^4 w}{\partial x^2 \partial y^2} + D_{22}\frac{\partial^4 w}{\partial y^4} + I_0 \frac{\partial^2 w}{\partial t^2}$$

$$= (A_{11}e_1 + A_{12}e_2)\frac{\partial^2 w}{\partial x^2} + 2A_{66}\gamma_6 \frac{\partial^2 w}{\partial x \partial y} + (A_{12}e_1 + A_{22}e_2)\frac{\partial^2 w}{\partial y^2}$$

$$+I_2 \frac{\partial^2}{\partial t^2}\left(\frac{\partial^2 w}{\partial x^2} + \frac{\partial^2 w}{\partial y^2}\right) + \frac{\partial w}{\partial x}\frac{\partial^2}{\partial t^2}\left(I_0 u - I_1 \frac{\partial w}{\partial x}\right)$$

$$+\frac{\partial w}{\partial y}\frac{\partial^2}{\partial t^2}\left(I_0 v - I_1 \frac{\partial w}{\partial y}\right) - I_1 \frac{\partial^2}{\partial t^2}\frac{\partial u}{\partial x} - I_1 \frac{\partial^2}{\partial t^2}\frac{\partial v}{\partial y} \qquad (7.4.17)$$

where e_1 and e_2 are defined in (7.4.6).

For a plate made of a single orthotropic material, $[B_{ij}] = 0$ and

$$[A_{ij}] = \begin{bmatrix} A_{11} & A_{12} & 0 \\ A_{12} & A_{22} & 0 \\ 0 & 0 & A_{66} \end{bmatrix}$$

$$= \frac{h}{1-\nu_{12}\nu_{21}}\begin{bmatrix} E_{11} & \nu_{21}E_{11} & 0 \\ \nu_{21}E_{11} & E_{22} & 0 \\ 0 & 0 & G_{12}(1-\nu_{12}\nu_{21}) \end{bmatrix} \qquad (7.4.18)$$

$$[D_{ij}] = \begin{bmatrix} D_{11} & D_{12} & 0 \\ D_{12} & D_{22} & 0 \\ 0 & 0 & D_{66} \end{bmatrix} = \frac{1}{12}h^2[A_{ij}] \qquad (7.4.19)$$

For a symmetric cross-ply laminate made of multiple orthotropic layers, $[B_{ij}] = 0$ and

$$\left\{\tilde{Q}_{11}^{(k)}, \tilde{Q}_{12}^{(k)}, \tilde{Q}_{22}^{(k)}\right\} = \frac{1}{1-\nu_{12}^{(k)}\nu_{21}^{(k)}}\left\{E_{11}^{(k)}, \nu_{21}^{(k)}E_{11}^{(k)}, E_{22}^{(k)}\right\}$$

$$\tilde{Q}_{16}^{(k)} = 0, \quad \tilde{Q}_{26}^{(k)} = 0, \quad \tilde{Q}_{66}^{(k)} = G_{12}^{(k)} \qquad (7.4.20)$$

For an isotropic plate, $E_{11} = E_{22} = E$, $\nu_{12} = \nu_{21} = \nu$, and (7.4.18) and (7.4.19) reduce to

$$[A_{ij}] = \frac{Eh}{1-\nu^2}\begin{bmatrix} 1 & \nu & 0 \\ \nu & 1 & 0 \\ 0 & 0 & (1-\nu)/2 \end{bmatrix} \qquad (7.4.21)$$

$$[D_{ij}] = \frac{1}{12}h^2[A_{ij}] \tag{7.4.22}$$

$$\tilde{Q}_{16}^{(k)} = 0, \quad \tilde{Q}_{26}^{(k)} = 0, \quad \tilde{Q}_{66}^{(k)} = G_{12}^{(k)} \tag{7.4.23}$$

Hence, (7.4.15)-(7.4.17) simplify to

$$\frac{\partial^2 u}{\partial x^2} + \frac{1}{2}(1+\nu)\frac{\partial^2 v}{\partial x \partial y} + \frac{1}{2}(1-\nu)\frac{\partial^2 u}{\partial y^2} + \frac{\partial w}{\partial x}\frac{\partial^2 w}{\partial x^2}$$
$$+ \frac{1}{2}(1+\nu)\frac{\partial w}{\partial y}\frac{\partial^2 w}{\partial x \partial y} + \frac{1}{2}(1-\nu)\frac{\partial w}{\partial x}\frac{\partial^2 w}{\partial y^2}$$
$$= \frac{1-\nu^2}{Eh}\left[I_0\frac{\partial^2 u}{\partial t^2} - I_1\frac{\partial^3 w}{\partial x \partial t^2}\right] \tag{7.4.24}$$

$$\frac{1}{2}(1+\nu)\frac{\partial^2 u}{\partial x \partial y} + \frac{1}{2}(1-\nu)\frac{\partial^2 v}{\partial x^2} + \frac{\partial^2 v}{\partial y^2} + \frac{1}{2}(1+\nu)\frac{\partial w}{\partial x}\frac{\partial^2 w}{\partial x \partial y} + \frac{\partial w}{\partial y}\frac{\partial^2 w}{\partial y^2}$$
$$+ \frac{1}{2}(1-\nu)\frac{\partial w}{\partial y}\frac{\partial^2 w}{\partial x^2} = \frac{1-\nu^2}{Eh}\left[I_0\frac{\partial^2 v}{\partial t^2} - I_1\frac{\partial^3 w}{\partial y \partial t^2}\right] \tag{7.4.25}$$

$$D\nabla^4 w + I_0\frac{\partial^2 w}{\partial t^2} = I_2\frac{\partial^2}{\partial t^2}(\nabla^2 w) + N_1\frac{\partial^2 w}{\partial x^2} + 2N_6\frac{\partial^2 w}{\partial x \partial y}$$
$$+ N_2\frac{\partial^2 w}{\partial y^2} - I_1\frac{\partial^2}{\partial t^2}\left(\frac{\partial u}{\partial x}\right) - I_1\frac{\partial^2}{\partial t^2}\left(\frac{\partial v}{\partial y}\right) \tag{7.4.26}$$

where $D \equiv \frac{Eh^3}{12(1-\nu^2)}$ and the nonlinear rotary inertia terms have been neglected.

When the inplane linear natural frequencies are very large compared to the natural frequencies of the transverse motion, the inplane inertia terms can be neglected from (7.4.8)-(7.4.10). The result is

$$N_{1x} + N_{6y} = 0 \tag{7.4.27}$$
$$N_{6x} + N_{2y} = 0 \tag{7.4.28}$$
$$M_{1xx} + 2M_{6xy} + M_{2yy} + N_1 w_{xx} + 2N_6 w_{xy} + N_2 w_{yy}$$
$$= I_0 \ddot{w} - I_2(\ddot{w}_{xx} + \ddot{w}_{yy}) \tag{7.4.29}$$

where I_1 is identically zero due to symmetry, as shown in (7.2.8). Equations (7.4.27) and (7.4.28) can be satisfied exactly by introducing the stress function $\Phi(x,y,t)$ defined by

$$N_1 = \Phi_{yy}, \quad N_2 = \Phi_{xx}, \quad N_6 = -\Phi_{xy} \tag{7.4.30}$$

Then, (7.4.29) becomes

$$M_{1xx} + 2M_{6xy} + M_{2yy} + w_{xx}\Phi_{yy} - 2w_{xy}\Phi_{xy} + w_{yy}\Phi_{xx} = I_0\ddot{w} - I_2(\ddot{w}_{xx} + \ddot{w}_{yy}) \tag{7.4.31}$$

Equation (7.4.31) needs to be augmented with a compatibility equation, as shown in (2.5.34). Eliminating u and v from (7.4.6) yields

$$\frac{\partial^2 e_1}{\partial y^2} + \frac{\partial^2 e_2}{\partial x^2} - 2\frac{\partial^2 e_6}{\partial x \partial y} = \left(\frac{\partial^2 w}{\partial x \partial y}\right)^2 - \frac{\partial^2 w}{\partial x^2}\frac{\partial^2 w}{\partial y^2} \tag{7.4.32}$$

where $e_6 = \gamma_6/2$. Next, we express the e_i in terms of $\Phi(x,y,t)$. To this end, we solve

$$\left\{\begin{array}{c} N_1 \\ N_2 \\ N_6 \end{array}\right\} = \left[\begin{array}{ccc} A_{11} & A_{12} & 0 \\ A_{12} & A_{22} & 0 \\ 0 & 0 & A_{66} \end{array}\right] \left\{\begin{array}{c} e_1 \\ e_2 \\ 2e_6 \end{array}\right\}$$

for the e_i and obtain

$$\left\{\begin{array}{c} e_1 \\ e_2 \\ 2e_6 \end{array}\right\} = \left[\begin{array}{ccc} S_{11} & S_{12} & 0 \\ S_{12} & S_{22} & 0 \\ 0 & 0 & S_{66} \end{array}\right] \left\{\begin{array}{c} N_1 \\ N_2 \\ N_6 \end{array}\right\} \quad (7.4.33)$$

where

$$(S_{11}, S_{12}, S_{22}) = \frac{(A_{22}, -A_{12}, A_{11})}{A_{11}A_{22} - A_{12}^2}$$

$$S_{66} = 1/A_{66}$$

Substituting for the N_i from (7.4.30) into (7.4.33) yields

$$e_1 = S_{11}\Phi_{yy} + S_{12}\Phi_{xx} \quad (7.4.34a)$$
$$e_2 = S_{12}\Phi_{yy} + S_{22}\Phi_{xx} \quad (7.4.34b)$$
$$2e_6 = -S_{66}\Phi_{xy} \quad (7.4.34c)$$

Substituting (7.4.34) into (7.4.32), we obtain the second equation

$$S_{22}\frac{\partial^4\Phi}{\partial x^4} + (2S_{12} + S_{66})\frac{\partial^4\Phi}{\partial x^2 \partial y^2} + S_{11}\frac{\partial^4\Phi}{\partial y^4} = \left(\frac{\partial^2 w}{\partial x \partial y}\right)^2 - \frac{\partial^2 w}{\partial x^2}\frac{\partial^2 w}{\partial y^2} \quad (7.4.35)$$

For an isotropic material,

$$[S_{ij}] = \frac{1}{Eh}\left[\begin{array}{ccc} 1 & -\nu & 0 \\ -\nu & 1 & 0 \\ 0 & 0 & 2(1+\nu) \end{array}\right]$$

and (7.4.35) becomes

$$\nabla^4 \Phi = Eh\left(w_{xy}^2 - w_{xx}w_{yy}\right) \quad (7.4.36)$$

Substituting for the M_i from (7.2.24) into (7.4.31), we obtain

$$D\nabla^4 w + \rho h \ddot{w} = w_{xx}\Phi_{yy} - 2w_{xy}\Phi_{xy} + w_{yy}\Phi_{xx} + I_2\nabla^2 \ddot{w} \quad (7.4.37)$$

7.4.2 von Kármán Plate Theory in Polar Coordinates

In this section, we extend the use of the von Kármán strains for rectangular plates to annular plates. Using (7.2.28a), we express the strains in (7.4.5) and (7.4.6) as

$$\epsilon_{11} = e_1 - zw_{rr}$$
$$\epsilon_{22} = e_2 - z\left(\frac{w_{\theta\theta}}{r^2} + \frac{w_r}{r}\right)$$
$$\epsilon_{12} = \gamma_6 - 2z\left(\frac{w_{r\theta}}{r} - \frac{w_\theta}{r^2}\right)$$
$$\epsilon_{33} = \epsilon_{13} = \epsilon_{23} = 0 \quad (7.4.38)$$

where

$$e_1 = u_r + \frac{1}{2}w_r^2$$
$$e_2 = \frac{v_\theta}{r} + \frac{u}{r} + \frac{w_\theta^2}{2r^2}$$
$$\gamma_6 = \frac{u_\theta}{r} - \frac{v}{r} + v_r + \frac{w_r w_\theta}{r} \qquad (7.4.39)$$

Therefore,

$$\begin{aligned}\delta\Pi &= \int_A \bigg[N_1 \delta u_r + \frac{1}{r} N_6 \delta u_\theta + \frac{1}{r} N_2 \delta v_\theta + N_6 \delta v_r + N_1 w_r \delta w_r \\ &+ \frac{1}{r^2} N_2 w_\theta \delta w_\theta + \frac{1}{r} N_6 w_r \delta w_\theta + \frac{1}{r} N_6 w_\theta \delta w_r - M_1 \delta w_{rr} - \frac{1}{r^2} M_2 \delta w_{\theta\theta} \\ &- \frac{1}{r} M_6 \delta w_{r\theta} - \frac{1}{r} M_6 \delta w_{\theta r} + \frac{1}{r} N_2 \delta u - \frac{1}{r} N_6 \delta v - \frac{1}{r} M_2 \delta w_r \\ &+ \frac{2}{r^2} M_6 \delta w_\theta + Q_1 \delta w_r + \frac{1}{r} Q_2 \delta w_\theta - Q_1 \delta w_r - \frac{1}{r} Q_2 \delta w_\theta \bigg] r\, dr\, d\theta \\ &= -\int_A \bigg\{ [(rN_1)_r + N_{6\theta} - N_2]\, \delta u + [N_{2\theta} + (rN_6)_r + N_6]\, \delta v \\ &+ \bigg[(rQ_1)_r + Q_{2\theta} + (rN_1 w_r + N_6 w_\theta)_r + \bigg(\frac{1}{r} N_2 w_\theta + N_6 w_r\bigg)_\theta \bigg]\, \delta w \\ &- [(rM_1)_r + M_{6\theta} - rQ_1 - M_2]\, \delta w_r \\ &- \frac{1}{r}[M_{2\theta} + rM_{6r} - rQ_2 + 2M_6]\, \delta w_\theta \bigg\}\, dr\, d\theta \\ &+ \int_\theta \bigg[N_1 \delta u + N_6 \delta v + \bigg(Q_1 + \frac{M_{6\theta}}{r} + N_1 w_r + \frac{N_6 w_\theta}{r}\bigg)\delta w \\ &- M_1 \delta w_r \bigg]_{r=a}^{r=b} r\, d\theta \\ &+ \int_r \bigg[N_6 \delta u + N_2 \delta v + \bigg(Q_2 + M_{6r} + \frac{N_2 w_\theta}{r} + N_6 w_r\bigg)\delta w \\ &- \frac{1}{r} M_2 \delta w_\theta \bigg]_{\theta=0}^{\theta=\theta_0} dr - 2 M_6 \delta w \bigg|_{(r,\theta)=(b,0),(a,\theta_0)}^{(r,\theta)=(a,0),(b,\theta_0)} \end{aligned} \qquad (7.4.40)$$

where b and a are the outer and inner radii. We note that $N_1 = N_{rr}$, $N_2 = N_{\theta\theta}$, $N_6 = N_{r\theta}$, $M_1 = M_{rr}$, $M_2 = M_{\theta\theta}$, and $M_6 = M_{r\theta}$.

Neglecting the rotary and inplane inertia, we obtain the following equations:

$$\frac{\partial N_1}{\partial r} + \frac{1}{r}\frac{\partial N_6}{\partial \theta} + \frac{N_1 - N_2}{r} = 0 \qquad (7.4.41)$$

$$\frac{\partial N_6}{\partial r} + \frac{1}{r}\frac{\partial N_2}{\partial \theta} + \frac{2 N_6}{r} = 0 \qquad (7.4.42)$$

$$\frac{\partial Q_1}{\partial r} + \frac{1}{r}\frac{\partial Q_2}{\partial \theta} + \frac{Q_1}{r} + \frac{1}{r}\frac{\partial}{\partial \theta}\bigg(\frac{N_2}{r}\frac{\partial w}{\partial \theta} + N_6 \frac{\partial w}{\partial r}\bigg)$$

$$+ \frac{1}{r}\frac{\partial}{\partial r}\left(rN_1\frac{\partial w}{\partial r} + N_6\frac{\partial w}{\partial \theta}\right) = I_0\ddot{w} \qquad (7.4.43)$$

$$\frac{\partial M_6}{\partial r} + \frac{1}{r}\frac{\partial M_2}{\partial \theta} + \frac{2M_6}{r} - Q_2 = 0 \qquad (7.4.44)$$

$$\frac{\partial M_1}{\partial r} + \frac{1}{r}\frac{\partial M_6}{\partial \theta} + \frac{M_1 - M_2}{r} - Q_1 = 0 \qquad (7.4.45)$$

The boundary conditions for the plate are of the form: specify

Along $r = b, a$:

$\delta u = 0$ or N_1

$\delta v = 0$ or N_6

$\delta w = 0$ or $Q_1 + \dfrac{M_{6\theta}}{r} + N_1 w_r + \dfrac{N_6 w_\theta}{r}$

$\delta w_r = 0$ or M_1

Along $\theta = 0, \theta_0$:

$\delta u = 0$ or N_6

$\delta v = 0$ or N_2

$\delta w = 0$ or $Q_2 + M_{6r} + \dfrac{N_2 w_\theta}{r} + N_6 w_r$

$\delta w_\theta = 0$ or M_2

At $(r, \theta) = (a, 0), (b, \theta_0), (b, 0), (a, \theta_0)$:

$\delta w = 0$ or M_6 \hfill (7.4.46)

When $a = 0$, the boundary conditions demand that all of the dependent variables be bounded at $r = 0$.

Using (7.4.44) and (7.4.45) to eliminate Q_1 and Q_2 from (7.4.43) and using (7.4.41) and (7.4.42) to simplify the result, we obtain

$$\frac{\partial^2 M_1}{\partial r^2} + \frac{2}{r}\frac{\partial M_1}{\partial r} + \frac{1}{r^2}\frac{\partial^2 M_2}{\partial \theta^2} - \frac{1}{r}\frac{\partial M_2}{\partial r} + \frac{2}{r}\frac{\partial^2 M_6}{\partial r \partial \theta} + \frac{2}{r^2}\frac{\partial M_6}{\partial \theta} + N_1\frac{\partial^2 w}{\partial r^2}$$
$$+ N_2\left(\frac{1}{r}\frac{\partial w}{\partial r} + \frac{1}{r^2}\frac{\partial^2 w}{\partial \theta^2}\right) + 2N_6\left(\frac{1}{r}\frac{\partial^2 w}{\partial r \partial \theta} - \frac{1}{r^2}\frac{\partial w}{\partial \theta}\right) = I_0\ddot{w} \quad (7.4.47)$$

For isotropic materials, the M_i are related to the displacements by (7.2.39), whereas the N_i are related to the displacements by

$$\left\{\begin{array}{c} N_1 \\ N_2 \\ N_6 \end{array}\right\} = \frac{Eh}{1-\nu^2}\left[\begin{array}{ccc} 1 & \nu & 0 \\ \nu & 1 & 0 \\ 0 & 0 & (1-\nu)/2 \end{array}\right]\left\{\begin{array}{c} e_1 \\ e_2 \\ \gamma_6 \end{array}\right\} \qquad (7.4.48)$$

Substituting for the M_i from (7.2.39) into (7.4.47) yields

$$D\nabla^4 w + \rho h\ddot{w} = N_1\frac{\partial^2 w}{\partial r^2} + N_2\left(\frac{1}{r}\frac{\partial w}{\partial r} + \frac{1}{r^2}\frac{\partial^2 w}{\partial \theta^2}\right) + 2N_6\left(\frac{1}{r}\frac{\partial^2 w}{\partial r \partial \theta} - \frac{1}{r^2}\frac{\partial w}{\partial \theta}\right) \qquad (7.4.49)$$

Substituting for the M_i from (7.2.39) into (7.4.44) and (7.4.45), we have

$$Q_1 = -D\frac{\partial}{\partial r}\left(\nabla^2 w\right) \quad \text{and} \quad Q_2 = -\frac{D}{r}\frac{\partial}{\partial \theta}\left(\nabla^2 w\right) \tag{7.4.50}$$

Most researchers combine (7.4.41) and (7.4.42) with (7.4.49) in solving for the nonlinear vibrations of circular plates. Moreover, they introduce a stress function $\Phi(r,\theta,t)$ that satisfies exactly (7.4.41) and (7.4.42) defined by

$$N_1 = \frac{1}{r}\frac{\partial \Phi}{\partial r} + \frac{1}{r^2}\frac{\partial^2 \Phi}{\partial \theta^2} \tag{7.4.51}$$

$$N_2 = \frac{\partial^2 \Phi}{\partial r^2} \tag{7.4.52}$$

$$N_6 = -\frac{1}{r}\frac{\partial^2 \Phi}{\partial r \partial \theta} + \frac{1}{r^2}\frac{\partial \Phi}{\partial \theta} \tag{7.4.53}$$

Substituting (7.4.51)–(7.4.53) into (7.4.49) and adding viscous damping with coefficient $2c$ and an external excitation $F(r,\theta,t)$, we obtain

$$\rho h \frac{\partial^2 w}{\partial t^2} + D\nabla^4 w = \frac{\partial^2 w}{\partial r^2}\left(\frac{1}{r}\frac{\partial \Phi}{\partial r} + \frac{1}{r^2}\frac{\partial^2 \Phi}{\partial \theta^2}\right) + \frac{\partial^2 \Phi}{\partial r^2}\left(\frac{1}{r}\frac{\partial w}{\partial r} + \frac{1}{r^2}\frac{\partial^2 w}{\partial \theta^2}\right)$$
$$- 2\left(\frac{1}{r}\frac{\partial^2 \Phi}{\partial r \partial \theta} - \frac{1}{r^2}\frac{\partial \Phi}{\partial \theta}\right)\left(\frac{1}{r}\frac{\partial^2 w}{\partial r \partial \theta} - \frac{1}{r^2}\frac{\partial w}{\partial \theta}\right)$$
$$- 2c\frac{\partial w}{\partial t} + F(r,\theta,t) \tag{7.4.54}$$

Equation (7.4.54) needs to be augmented with the compatibility equation. Eliminating u and v from (7.4.39) yields the following compatibility equation expressed in polar coordinates:

$$\frac{\partial^2 e_1}{\partial \theta^2} - \frac{\partial^2 (r\gamma_6)}{\partial r \partial \theta} + \frac{\partial}{\partial r}\left(r^2 \frac{\partial e_2}{\partial r}\right) - r\frac{\partial e_1}{\partial r} = \left(\frac{\partial^2 w}{\partial r \partial \theta} - \frac{1}{r}\frac{\partial w}{\partial \theta}\right)^2$$
$$- \frac{\partial^2 w}{\partial r^2}\left(r\frac{\partial w}{\partial r} + \frac{\partial^2 w}{\partial \theta^2}\right) \tag{7.4.55}$$

Solving (7.4.48) for the e_i in terms of the N_i and then using (7.4.51)–(7.4.53) to express the N_i in terms of Φ, we have

$$e_1 = \frac{1}{Eh}\left(\frac{1}{r}\frac{\partial \Phi}{\partial r} + \frac{1}{r^2}\frac{\partial^2 \Phi}{\partial \theta^2} - \nu\frac{\partial^2 \Phi}{\partial r^2}\right) \tag{7.4.56}$$

$$e_2 = \frac{1}{Eh}\left[\frac{\partial^2 \Phi}{\partial r^2} - \nu\left(\frac{1}{r}\frac{\partial \Phi}{\partial r} + \frac{1}{r^2}\frac{\partial^2 \Phi}{\partial \theta^2}\right)\right] \tag{7.4.57}$$

$$\gamma_6 = \frac{2(1+\nu)}{Eh}\left[\frac{1}{r^2}\frac{\partial \Phi}{\partial \theta} - \frac{1}{r}\frac{\partial^2 \Phi}{\partial r \partial \theta}\right] \tag{7.4.58}$$

Substituting (7.4.56)–(7.4.58) into (7.4.55) yields

$$\nabla^4 \Phi = Eh\left[\left(\frac{1}{r}\frac{\partial^2 w}{\partial r \partial \theta} - \frac{1}{r^2}\frac{\partial w}{\partial \theta}\right)^2 - \frac{\partial^2 w}{\partial r^2}\left(\frac{1}{r}\frac{\partial w}{\partial r} + \frac{1}{r^2}\frac{\partial^2 w}{\partial \theta^2}\right)\right] \tag{7.4.59}$$

Equations (7.4.54) and (7.4.59) are usually the starting equations for analyzing the nonlinear dynamics of isotropic circular and annular plates and are used in Chapter 8.

7.4.3 Thermoelastic Equations in Cartesian Coordinates

For a symmetric cross-ply laminate made of multiple orthotropic layers, it follows from (2.4.20b), (2.4.21), and (7.4.5) that

$$\left\{ \begin{array}{c} \sigma_{11} \\ \sigma_{22} \\ \sigma_{12} \end{array} \right\} = \left[\begin{array}{ccc} \tilde{Q}_{11} & \tilde{Q}_{12} & 0 \\ \tilde{Q}_{12} & \tilde{Q}_{22} & 0 \\ 0 & 0 & \tilde{Q}_{66} \end{array} \right] \left\{ \begin{array}{c} e_1 - zw_{xx} - \alpha_1(T - T_0) \\ e_2 - zw_{yy} - \alpha_2(T - T_0) \\ \gamma_6 - 2zw_{xy} \end{array} \right\} \quad (7.4.60)$$

where the $\tilde{Q}_{ij}$ are defined in (7.4.20), α_1 and α_2 are the coefficients of thermal expansion, and T_0 is the stress-free temperature. Hence, it follows from (7.2.10) that

$$\left\{ \begin{array}{c} N_1 \\ N_2 \\ N_6 \end{array} \right\} = [A_{ij}] \left\{ \begin{array}{c} e_1 \\ e_2 \\ \gamma_6 \end{array} \right\} - \left\{ \begin{array}{c} N_1^T \\ N_2^T \\ 0 \end{array} \right\} \quad (7.4.61)$$

$$\left\{ \begin{array}{c} M_1 \\ M_2 \\ M_6 \end{array} \right\} = -[D_{ij}] \left\{ \begin{array}{c} w_{xx} \\ w_{yy} \\ 2w_{xy} \end{array} \right\} - \left\{ \begin{array}{c} M_1^T \\ M_2^T \\ 0 \end{array} \right\} \quad (7.4.62)$$

where $[A_{ij}]$ and $[D_{ij}]$ are defined in (7.4.13) and

$$N_1^T = \sum_{k=1}^{N} \int_{z_k}^{z_{k+1}} \left(\alpha_1 \tilde{Q}_{11}^{(k)} + \alpha_2 \tilde{Q}_{12}^{(k)} \right) (T - T_0) dz \quad (7.4.63)$$

$$N_2^T = \sum_{k=1}^{N} \int_{z_k}^{z_{k+1}} \left(\alpha_1 \tilde{Q}_{12}^{(k)} + \alpha_2 \tilde{Q}_{22}^{(k)} \right) (T - T_0) dz \quad (7.4.64)$$

$$M_1^T = \sum_{k=1}^{N} \int_{z_k}^{z_{k+1}} \left(\alpha_1 \tilde{Q}_{11}^{(k)} + \alpha_2 \tilde{Q}_{12}^{(k)} \right) z(T - T_0) dz \quad (7.4.65)$$

$$M_2^T = \sum_{k=1}^{N} \int_{z_k}^{z_{k+1}} \left(\alpha_1 \tilde{Q}_{12}^{(k)} + \alpha_2 \tilde{Q}_{22}^{(k)} \right) z(T - T_0) dz \quad (7.4.66)$$

Consequently, (7.4.15)-(7.4.17) become

$$A_{11} \frac{\partial^2 u}{\partial x^2} + (A_{12} + A_{66}) \frac{\partial^2 v}{\partial x \partial y} + A_{66} \frac{\partial^2 u}{\partial y^2} + A_{11} \frac{\partial w}{\partial x} \frac{\partial^2 w}{\partial x^2} + A_{12} \frac{\partial w}{\partial y} \frac{\partial^2 w}{\partial x \partial y}$$
$$+ A_{66} \left(\frac{\partial w}{\partial y} \frac{\partial^2 w}{\partial x \partial y} + \frac{\partial w}{\partial x} \frac{\partial^2 w}{\partial y^2} \right) = I_0 \frac{\partial^2 u}{\partial t^2} - I_1 \frac{\partial^3 w}{\partial x \partial t^2} + \frac{\partial N_1^T}{\partial x} \quad (7.4.67)$$

$$(A_{12} + A_{66})\frac{\partial^2 u}{\partial x \partial y} + A_{66}\frac{\partial^2 v}{\partial x^2} + A_{22}\frac{\partial^2 v}{\partial y^2} + A_{12}\frac{\partial w}{\partial x}\frac{\partial^2 w}{\partial x \partial y} + A_{22}\frac{\partial w}{\partial y}\frac{\partial^2 w}{\partial y^2}$$
$$+ A_{66}\left(\frac{\partial w}{\partial y}\frac{\partial^2 w}{\partial x^2} + \frac{\partial w}{\partial x}\frac{\partial^2 w}{\partial x \partial y}\right) = I_0\frac{\partial^2 v}{\partial t^2} - I_1\frac{\partial^3 w}{\partial y \partial t^2} + \frac{\partial N_2^T}{\partial y} \quad (7.4.68)$$

$$D_{11}\frac{\partial^4 w}{\partial x^4} + 2(D_{12} + 2D_{66})\frac{\partial^4 w}{\partial x^2 \partial y^2} + D_{22}\frac{\partial^4 w}{\partial y^4} + N_1^T\frac{\partial^2 w}{\partial x^2} + N_2^T\frac{\partial^2 w}{\partial y^2} + I_0\frac{\partial^2 w}{\partial t^2}$$
$$= (A_{11}e_1 + A_{12}e_2)\frac{\partial^2 w}{\partial x^2} + 2A_{66}\gamma_6\frac{\partial^2 w}{\partial x \partial y} + (A_{12}e_1 + A_{22}e_2)\frac{\partial^2 w}{\partial y^2}$$
$$+ I_2\frac{\partial^2}{\partial t^2}\left(\frac{\partial^2 w}{\partial x^2} + \frac{\partial^2 w}{\partial y^2}\right) + \frac{\partial w}{\partial x}\frac{\partial^2}{\partial t^2}\left(I_0 u - I_1\frac{\partial w}{\partial x}\right)$$
$$+ \frac{\partial w}{\partial y}\frac{\partial^2}{\partial t^2}\left(I_0 v - I_1\frac{\partial w}{\partial y}\right) - I_1\frac{\partial^2}{\partial t^2}\frac{\partial u}{\partial x} - I_1\frac{\partial^2}{\partial t^2}\frac{\partial v}{\partial y}$$
$$- \frac{\partial^2 M_1^T}{\partial x^2} - \frac{\partial^2 M_2^T}{\partial y^2} \quad (7.4.69)$$

For isotropic plates, $\alpha_1 = \alpha_2 = \alpha$ and it follows from (7.4.63), (7.4.64), and (7.4.18) that

$$N_1^T = N_2^T = \frac{E\alpha}{1-\nu}\int_{-h/2}^{h/2}(T - T_0)dz \quad (7.4.70)$$

$$M_1^T = M_2^T = \frac{E\alpha}{1-\nu}\int_{-h/2}^{h/2}z(T - T_0)dz \quad (7.4.71)$$

And (7.4.24)-(7.4.26) become

$$\frac{\partial^2 u}{\partial x^2} + \frac{1}{2}(1+\nu)\frac{\partial^2 v}{\partial x \partial y} + \frac{1}{2}(1-\nu)\frac{\partial^2 u}{\partial y^2} + \frac{\partial w}{\partial x}\frac{\partial^2 w}{\partial x^2} + \nu\frac{\partial w}{\partial y}\frac{\partial^2 w}{\partial x \partial y}$$
$$+ \frac{1}{2}(1-\nu)\left(\frac{\partial w}{\partial y}\frac{\partial^2 w}{\partial x \partial y} + \frac{\partial w}{\partial x}\frac{\partial^2 w}{\partial y^2}\right)$$
$$= \frac{1-\nu^2}{Eh}\left[I_0\frac{\partial^2 u}{\partial t^2} - I_1\frac{\partial^3 w}{\partial x \partial t^2}\right] + \frac{(1+\nu)\alpha}{h}\int_{-h/2}^{h/2}\frac{\partial T}{\partial x}dz \quad (7.4.72)$$

$$\frac{1}{2}(1+\nu)\frac{\partial^2 u}{\partial x \partial y} + \frac{1}{2}(1-\nu)\frac{\partial^2 v}{\partial x^2} + \frac{\partial^2 v}{\partial y^2} + \nu\frac{\partial w}{\partial x}\frac{\partial^2 w}{\partial x \partial y} + \frac{\partial w}{\partial y}\frac{\partial^2 w}{\partial y^2}$$
$$+ \frac{1}{2}(1-\nu)\left(\frac{\partial w}{\partial y}\frac{\partial^2 w}{\partial x^2} + \frac{\partial w}{\partial x}\frac{\partial^2 w}{\partial x \partial y}\right)$$
$$= \frac{1-\nu^2}{Eh}\left[I_0\frac{\partial^2 v}{\partial t^2} - I_1\frac{\partial^3 w}{\partial y \partial t^2}\right] + \frac{(1+\nu)\alpha}{h}\int_{-h/2}^{h/2}\frac{\partial T}{\partial y}dz \quad (7.4.73)$$

$$D\nabla^4 w + I_0\frac{\partial^2 w}{\partial t^2} + \frac{E\alpha}{1-\nu}\nabla^2 w\int_{-h/2}^{h/2}(\Delta T)\,dz = I_2\frac{\partial^2}{\partial t^2}\left(\nabla^2 w\right)$$
$$- \frac{E\alpha}{1-\nu}\int_{-h/2}^{h/2}z\nabla^2 T dz + \frac{Eh}{1-\nu^2}\left[(e_1 + \nu e_2)\frac{\partial^2 w}{\partial x^2}\right.$$

$$+ (1-\nu)\gamma_6 \frac{\partial^2 w}{\partial x \partial y} + (\nu e_1 + e_2)\frac{\partial^2 w}{\partial y^2}\Bigg] - I_1 \frac{\partial^2}{\partial t^2}\frac{\partial u}{\partial x} - I_1 \frac{\partial^2}{\partial t^2}\frac{\partial v}{\partial y} \quad (7.4.74)$$

The temperature distribution is given by the heat conduction equation (Boley and Weiner, 1960; Hetnarski, 1987)

$$k\nabla^2 T + Q = \rho c_p \frac{\partial T}{\partial t} + \frac{E\alpha T}{1-2\nu}\frac{\partial e}{\partial t} \quad (7.4.75)$$

where $T(x,y,z,t)$ is the temperature distribution, e is the dilatational strain due to the thermal effect, ρ is the mass density, c_p is the heat capacity coefficient at constant pressure, E is the modulus of elasticity, α is the coefficient of thermal expansion, Q is the heat flux, and ν is Poisson's ratio. To express e in terms of the displacements, we need to account for ϵ_{33} because, although $\sigma_{33} \approx 0$, ϵ_{33} is not zero for the thermoelastic case. To this end, we note that

$$\epsilon_{11} = \frac{1}{E}(\sigma_{11} - \nu\sigma_{22}) + \alpha(T - T_0) \quad (7.4.76)$$

$$\epsilon_{22} = \frac{1}{E}(\sigma_{22} - \nu\sigma_{11}) + \alpha(T - T_0) \quad (7.4.77)$$

$$\epsilon_{33} = -\frac{\nu}{E}(\sigma_{11} + \sigma_{22}) + \alpha(T - T_0) \quad (7.4.78)$$

where T_0 is the stress-free temperature. Solving (7.4.76) and (7.4.77) for σ_{11} and σ_{22}, we obtain

$$\sigma_{11} = \frac{E}{1-\nu^2}[\epsilon_{11} - \nu\epsilon_{22} + (1+\nu)\alpha(T - T_0)] \quad (7.4.79)$$

$$\sigma_{22} = \frac{E}{1-\nu^2}[\epsilon_{22} + \nu\epsilon_{11} - (1+\nu)\alpha(T - T_0)] \quad (7.4.80)$$

Hence,

$$\epsilon_{33} = \frac{1}{1-\nu}[-\nu(\epsilon_{11} + \epsilon_{22}) + (1+\nu)\alpha(T - T_0)] \quad (7.4.81)$$

Substituting for ϵ_{11} and ϵ_{22} from (7.4.5) and (7.4.6) into

$$e = \epsilon_{11} + \epsilon_{22} + \epsilon_{33}$$

we obtain

$$e = \frac{1-2\nu}{1-\nu}\left[u_x + v_y - z(w_{xx} + w_{yy}) + \frac{1}{2}(w_x^2 + w_y^2)\right] + \frac{1+\nu}{1-\nu}\alpha(T - T_0) \quad (7.4.82)$$

Eliminating e from (7.4.75) and (7.4.82) yields

$$k\nabla^2 T + Q = \left[\rho c_p + \frac{E\alpha^2(1+\nu)T}{(1-\nu)(1-2\nu)}\right]\frac{\partial T}{\partial t}$$
$$+ \frac{E\alpha T}{1-\nu}\frac{\partial}{\partial t}\left[u_x + v_y - z(w_{xx} + w_{yy}) + \frac{1}{2}(w_x^2 + w_y^2)\right] \quad (7.4.83)$$

For most applications, (7.4.83) can be replaced with its linearized version; that is,

$$k\nabla^2 T + Q = \left[\rho c_p + \frac{E\alpha^2(1+\nu)T_0}{(1-\nu)(1-2\nu)}\right]\frac{\partial T}{\partial t}$$
$$+ \frac{E\alpha T_0}{1-\nu}\frac{\partial}{\partial t}[u_x + v_y - z(w_{xx} + w_{yy})] \quad (7.4.84)$$

7.4.4 Thermoelastic Equations in Polar Coordinates

For an orthotropic plate whose principal material axes coincide with the polar coordinates (r, θ), the constitutive relation can be expressed as

$$\left\{\begin{array}{c} \sigma_{11} \\ \sigma_{22} \\ \sigma_{12} \end{array}\right\} = \left[\begin{array}{ccc} Q_{11} & Q_{12} & 0 \\ Q_{12} & Q_{22} & 0 \\ 0 & 0 & Q_{66} \end{array}\right]\left\{\begin{array}{c} \epsilon_{11} - \alpha_1(T-T_0) \\ \epsilon_{22} - \alpha_2(T-T_0) \\ \epsilon_{12} \end{array}\right\} \quad (7.4.85)$$

where $\epsilon_{11}, \epsilon_{22}$, and ϵ_{12} are given by (7.4.38) and (7.4.39). Then, it follows from (7.2.10) that

$$\left\{\begin{array}{c} N_1 \\ N_2 \\ N_6 \end{array}\right\} = [A_{ij}]\left\{\begin{array}{c} e_1 \\ e_2 \\ \gamma_6 \end{array}\right\} - \left\{\begin{array}{c} N_1^T \\ N_2^T \\ 0 \end{array}\right\} \quad (7.4.86)$$

$$\left\{\begin{array}{c} M_1 \\ M_2 \\ M_6 \end{array}\right\} = -[D_{ij}]\left\{\begin{array}{c} \frac{\partial^2 w}{\partial r^2} \\ \frac{1}{r}\frac{\partial w}{\partial r} + \frac{1}{r^2}\frac{\partial^2 w}{\partial \theta^2} \\ \frac{2}{r}\frac{\partial^2 w}{\partial r \partial \theta} - \frac{2}{r^2}\frac{\partial w}{\partial \theta} \end{array}\right\} - \left\{\begin{array}{c} M_1^T \\ M_2^T \\ 0 \end{array}\right\} \quad (7.4.87)$$

where

$$(A_{ij}, D_{ij}) = \int_{-\frac{1}{2}h}^{\frac{1}{2}h} Q_{ij}(1, z^2)dz = \left(h, \frac{1}{12}h^3\right)Q_{ij} \quad (7.4.88)$$

$$N_1^T = \int_{-\frac{1}{2}h}^{\frac{1}{2}h}(\alpha_1 Q_{11} + \alpha_2 Q_{12})(T-T_0)dz \quad (7.4.89)$$

$$N_2^T = \int_{-\frac{1}{2}h}^{\frac{1}{2}h}(\alpha_1 Q_{12} + \alpha_2 Q_{22})(T-T_0)dz \quad (7.4.90)$$

$$M_1^T = \int_{-\frac{1}{2}h}^{\frac{1}{2}h}(\alpha_1 Q_{11} + \alpha_2 Q_{12})z(T-T_0)dz \quad (7.4.91)$$

$$M_2^T = \int_{-\frac{1}{2}h}^{\frac{1}{2}h}(\alpha_1 Q_{12} + \alpha_2 Q_{22})z(T-T_0)dz \quad (7.4.92)$$

For isotropic plates,

$$[A_{ij}] = \frac{Eh}{1-\nu^2}\left[\begin{array}{ccc} 1 & \nu & 0 \\ \nu & 1 & 0 \\ 0 & 0 & (1-\nu)/2 \end{array}\right] \quad (7.4.93)$$

$$[D_{ij}] = \frac{1}{12}h^2[A_{ij}]$$

Then, (7.4.86) and (7.4.87) become

$$\left\{\begin{array}{c} N_1 \\ N_2 \\ N_6 \end{array}\right\} = \frac{Eh}{1-\nu^2}\begin{bmatrix} 1 & \nu & 0 \\ \nu & 1 & 0 \\ 0 & 0 & (1-\nu)/2 \end{bmatrix}\left\{\begin{array}{c} e_1 \\ e_2 \\ \gamma_6 \end{array}\right\} - \frac{1}{1-\nu}\left\{\begin{array}{c} N^T \\ N^T \\ 0 \end{array}\right\}$$
(7.4.94)

$$\left\{\begin{array}{c} M_1 \\ M_2 \\ M_6 \end{array}\right\} = -D\begin{bmatrix} 1 & \nu & 0 \\ \nu & 1 & 0 \\ 0 & 0 & 1-\nu \end{bmatrix}\left\{\begin{array}{c} \frac{\partial^2 w}{\partial r^2} \\ \frac{1}{r}\frac{\partial w}{\partial r} + \frac{1}{r^2}\frac{\partial^2 w}{\partial \theta^2} \\ \frac{1}{r}\frac{\partial^2 w}{\partial r \partial \theta} - \frac{1}{r^2}\frac{\partial w}{\partial \theta} \end{array}\right\} - \frac{1}{1-\nu}\left\{\begin{array}{c} M^T \\ M^T \\ 0 \end{array}\right\}$$
(7.4.95)

where

$$N^T \equiv E\alpha \int_{-\frac{1}{2}h}^{\frac{1}{2}h}(T-T_0)dz \quad \text{and} \quad M^T \equiv E\alpha \int_{-\frac{1}{2}h}^{\frac{1}{2}h} z(T-T_0)dz \quad (7.4.96)$$

Substituting (7.4.95) into (7.4.47) yields

$$D\nabla^4 w + \rho h \ddot{w} = N_1\frac{\partial^2 w}{\partial r^2} + N_2\left(\frac{1}{r}\frac{\partial w}{\partial r} + \frac{1}{r^2}\frac{\partial^2 w}{\partial \theta^2}\right)$$
$$+ 2N_6\left(\frac{1}{r}\frac{\partial^2 w}{\partial r \partial \theta} - \frac{1}{r^2}\frac{\partial w}{\partial \theta}\right) - \frac{1}{1-\nu}\nabla^2 M^T \quad (7.4.97)$$

As in Section 7.4.2, (7.4.41) and (7.4.42) are satisfied exactly by introducing the stress function $\Phi(r,\theta,t)$ defined by (7.4.51)-(7.4.53). The compatibility equation is given by (7.4.55). Solving (7.4.94) for the e_i, we obtain

$$e_1 = \frac{1}{Eh}\left(\frac{1}{r}\frac{\partial \Phi}{\partial r} + \frac{1}{r^2}\frac{\partial^2 \Phi}{\partial \theta^2} - \nu\frac{\partial^2 \Phi}{\partial r^2}\right) + \frac{N^T}{Eh} \quad (7.4.98)$$

$$e_2 = \frac{1}{Eh}\left[\frac{\partial^2 \Phi}{\partial r^2} - \nu\left(\frac{1}{r}\frac{\partial \Phi}{\partial r} + \frac{1}{r^2}\frac{\partial^2 \Phi}{\partial \theta^2}\right)\right] + \frac{N^T}{Eh} \quad (7.4.99)$$

$$\gamma_6 = \frac{2(1+\nu)}{Eh}\left[\frac{1}{r^2}\frac{\partial \Phi}{\partial \theta} - \frac{1}{r}\frac{\partial^2 \Phi}{\partial r \partial \theta}\right] \quad (7.4.100)$$

Substituting (7.4.98)-(7.4.100) into (7.4.55), we obtain

$$\nabla^4 \Phi = Eh\left[\left(\frac{1}{r}\frac{\partial^2 w}{\partial r \partial \theta} - \frac{1}{r^2}\frac{\partial w}{\partial \theta}\right)^2 - \frac{\partial^2 w}{\partial r^2}\left(\frac{1}{r}\frac{\partial w}{\partial r} + \frac{1}{r^2}\frac{\partial^2 w}{\partial \theta^2}\right)\right] - \nabla^2 N^T$$
(7.4.101)

7.5 NONLINEAR MODELING OF RECTANGULAR SURFACES

Deformation of a plate consists of the deformation of its reference plane and shear-warping of its cross section with respect to the deformed reference plane. Different approximations for the deformation of the reference plane result in different geometrically nonlinear plate theories, and different approximations for the warping of cross section result in different shear-deformation theories. Rectangular plates are the most studied 2-D structures and are the simplest 2-D structures because they do not have initial curvatures. Hence, we use rectangular plates to show how to derive geometrically exact plate theories.

7.5.1 Coordinate Transformation, Inplane Strains, and Curvatures

We consider a rectangular plate over the domain $0 \leq x \leq a$ and $0 \leq y \leq b$ and having thickness h, as shown in Figure 7.5.1. The undeformed coordinate system xyz is a Cartesian coordinate system with the xy plane representing the undeformed reference plane, and the $\xi\eta\zeta$ is an orthogonal curvilinear coordinate system with the $\xi\eta$ plane representing the deformed reference plane. The middle plane is usually chosen as the reference plane in the analysis of isotropic plates. However, for anisotropic plates under pure bending, the middle plane may not be the neutral surface due to extension-bending coupling. Hence, it is not necessary to choose the middle plane as the reference plane.

As shown in Figure 7.5.2, the deformation of the reference surface is described by the displacements u, v, and w of the reference Point A, the transformation matrix $[T]$ that relates the two coordinate systems xyz and $\xi\eta\zeta$, the inplane extensional strains e_1 and e_2, the inplane shear strain γ_6, and the bending and twisting curvatures. $\mathbf{j}_1, \mathbf{j}_2$, and $\mathbf{j}_3$ represent unit vectors of the xyz system; $\mathbf{i}_1, \mathbf{i}_2$, and $\mathbf{i}_3$ represent unit vectors of the $\xi\eta\zeta$ system; and $\mathbf{i}_{\hat{1}}$ and $\mathbf{i}_{\hat{2}}$ are unit vectors along the convected axes $\hat{\xi}$ and $\hat{\eta}$,

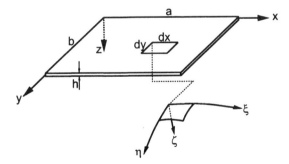

Fig. 7.5.1 A rectangular plate with a Cartesian coordinate system xyz to describe the undeformed geometry and an orthogonal curvilinear coordinate system $\xi\eta\zeta$ to describe the deformed geometry.

418 PLATES

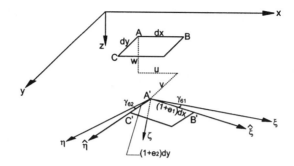

Fig. 7.5.2 The undeformed and deformed shapes of a differential rectangular reference surface.

respectively. We note that $\mathbf{i}_{\hat{1}}$ is not perpendicular to $\mathbf{i}_{\hat{2}}$ if the inplane shear strain $\gamma_6 (= \gamma_{61} + \gamma_{62}) \neq 0$. In the modeling of beams, the displacements u, v, and w of points on the reference axis cannot describe the twisting of a differential beam element with respect to the reference axis itself and hence a torsion variable ϕ is needed (see Figure 4.4.4). On the other hand, in the modeling of plates, u, v, and w are the displacement components of points on the reference plane and hence are functions of x and y and can describe any bending and twisting rotations. Therefore, there is no need of extra rotational variables in the modeling of plates, and hence only five equations of motion are expected for plates.

Because rectangular plates do not have initial curvatures, the coordinates of the corners of the reference plane shown in Figure 7.5.2 are

$$
\begin{aligned}
A &: (x, y, 0) & A' &: (x + u, y + v, w) \\
B &: (x + dx, y, 0) & B' &: (x + dx + u + u_x dx, y + v + v_x dx, w + w_x dx) \\
C &: (x, y + dy, 0) & C' &: (x + u + u_y dy, y + dy + v + v_y dy, w + w_y dy)
\end{aligned}
$$
(7.5.1)

It follows from Figure 7.5.2 and (7.5.1) that the axial strains along $\hat{\xi}$ and $\hat{\eta}$ are given by

$$e_1 = \frac{\overline{A'B'} - dx}{dx} = \sqrt{(1+u_x)^2 + v_x^2 + w_x^2} - 1 \quad (7.5.2)$$

$$e_2 = \frac{\overline{A'C'} - dy}{dy} = \sqrt{u_y^2 + (1+v_y)^2 + w_y^2} - 1 \quad (7.5.3)$$

and the unit vectors along $\hat{\xi}$ and $\hat{\eta}$ are given by

$$\mathbf{i}_{\hat{1}} = \frac{\overrightarrow{A'B'}}{(1+e_1)dx} = \hat{T}_{11}\mathbf{j}_1 + \hat{T}_{12}\mathbf{j}_2 + \hat{T}_{13}\mathbf{j}_3 \quad (7.5.4)$$

$$\mathbf{i}_{\hat{2}} = \frac{\overrightarrow{A'C'}}{(1+e_2)dy} = \hat{T}_{21}\mathbf{j}_1 + \hat{T}_{22}\mathbf{j}_2 + \hat{T}_{23}\mathbf{j}_3 \qquad (7.5.5)$$

where

$$\hat{T}_{11} = \frac{1+u_x}{1+e_1}, \quad \hat{T}_{12} = \frac{v_x}{1+e_1}, \quad \hat{T}_{13} = \frac{w_x}{1+e_1} \qquad (7.5.6)$$

$$\hat{T}_{21} = \frac{u_y}{1+e_2}, \quad \hat{T}_{22} = \frac{1+v_y}{1+e_2}, \quad \hat{T}_{23} = \frac{w_y}{1+e_2} \qquad (7.5.7)$$

Using (7.5.4) and (7.5.5), we represent the inplane shear deformation in terms of u, v, and w as

$$\gamma_6 = \gamma_{61} + \gamma_{62} = \sin^{-1}(\mathbf{i}_{\hat{1}} \cdot \mathbf{i}_{\hat{2}}) = \sin^{-1}\left(\hat{T}_{11}\hat{T}_{21} + \hat{T}_{12}\hat{T}_{22} + \hat{T}_{13}\hat{T}_{23}\right) \qquad (7.5.8)$$

Equations (7.5.2), (7.5.3), and (7.5.8) are fully nonlinear expressions of e_1, e_2, and γ_6 in terms of the displacements. The unit normal to the deformed reference plane is given by

$$\mathbf{i}_3 = \frac{\mathbf{i}_{\hat{1}} \times \mathbf{i}_{\hat{2}}}{|\mathbf{i}_{\hat{1}} \times \mathbf{i}_{\hat{2}}|} = T_{31}\mathbf{j}_1 + T_{32}\mathbf{j}_2 + T_{33}\mathbf{j}_3 \qquad (7.5.9)$$

where

$$T_{31} = \left(\hat{T}_{12}\hat{T}_{23} - \hat{T}_{13}\hat{T}_{22}\right)/R_0$$

$$T_{32} = \left(\hat{T}_{13}\hat{T}_{21} - \hat{T}_{11}\hat{T}_{23}\right)/R_0$$

$$T_{33} = \left(\hat{T}_{11}\hat{T}_{22} - \hat{T}_{12}\hat{T}_{21}\right)/R_0 \qquad (7.5.10)$$

and

$$R_0 \equiv \sqrt{\left(\hat{T}_{12}\hat{T}_{23} - \hat{T}_{13}\hat{T}_{22}\right)^2 + \left(\hat{T}_{13}\hat{T}_{21} - \hat{T}_{11}\hat{T}_{23}\right)^2 + \left(\hat{T}_{11}\hat{T}_{22} - \hat{T}_{12}\hat{T}_{21}\right)^2}$$

The influence of the inplane shear strain γ_6 on the surface analysis is discussed in Section 7.5.2. For thin plates and shells subjected to transverse loadings, the configuration change due to the inplane shear deformation γ_6 is negligible for most materials except rubber-like materials. If the influence of γ_6 on the deformed geometry is neglected, the axes ξ and η coincide with the axes $\hat{\xi}$ and $\hat{\eta}$, respectively, and

$$\mathbf{i}_1 = \mathbf{i}_{\hat{1}}\left(T_{1i} = \hat{T}_{1i}\right), \quad \mathbf{i}_2 = \mathbf{i}_{\hat{2}}\left(T_{2i} = \hat{T}_{2i}\right) \qquad (7.5.11)$$

It follows from (6.2.11a) that, whether the influence of γ_6 on the deformed geometry is negligible or not, the undeformed coordinate system xyz is always related to the deformed coordinate system $\xi\eta\zeta$ by

$$\{\mathbf{i}_{123}\} = [T]\{\mathbf{j}_{123}\}, \quad [T] \equiv \begin{bmatrix} T_{11} & T_{12} & T_{13} \\ T_{21} & T_{22} & T_{23} \\ T_{31} & T_{32} & T_{33} \end{bmatrix} \qquad (7.5.12)$$

Because the $\xi\eta\zeta$ is an orthogonal system, we have the following identities

$$\frac{\partial \mathbf{i}_m}{\partial x} \cdot \mathbf{i}_m = \frac{\partial \mathbf{i}_m}{\partial y} \cdot \mathbf{i}_m = 0, \quad \frac{\partial \mathbf{i}_m}{\partial x} \cdot \mathbf{i}_n = -\frac{\partial \mathbf{i}_n}{\partial x} \cdot \mathbf{i}_m, \quad \frac{\partial \mathbf{i}_m}{\partial y} \cdot \mathbf{i}_n = -\frac{\partial \mathbf{i}_n}{\partial y} \cdot \mathbf{i}_m \quad (7.5.13)$$

for $m, n = 1, 2, 3$. Using (7.5.13) we obtain

$$\frac{\partial}{\partial x}\{\mathbf{i}_{123}\} = [K_1]\{\mathbf{i}_{123}\}, \quad [K_1] \equiv \begin{bmatrix} 0 & k_5 & -k_1 \\ -k_5 & 0 & -k_{61} \\ k_1 & k_{61} & 0 \end{bmatrix} \quad (7.5.14)$$

$$\frac{\partial}{\partial y}\{\mathbf{i}_{123}\} = [K_2]\{\mathbf{i}_{123}\}, \quad [K_2] \equiv \begin{bmatrix} 0 & k_4 & -k_{62} \\ -k_4 & 0 & -k_2 \\ k_{62} & k_2 & 0 \end{bmatrix} \quad (7.5.15)$$

where $[K_1]$ and $[K_2]$ are the deformed curvature matrices and the deformed curvatures are given by

$$k_1 \equiv -\frac{\partial \mathbf{i}_1}{\partial x} \cdot \mathbf{i}_3 = -T_{11x}T_{31} - T_{12x}T_{32} - T_{13x}T_{33}$$

$$k_2 \equiv -\frac{\partial \mathbf{i}_2}{\partial y} \cdot \mathbf{i}_3 = -T_{21y}T_{31} - T_{22y}T_{32} - T_{23y}T_{33}$$

$$k_{61} \equiv -\frac{\partial \mathbf{i}_2}{\partial x} \cdot \mathbf{i}_3 = -T_{21x}T_{31} - T_{22x}T_{32} - T_{23x}T_{33}$$

$$k_{62} \equiv -\frac{\partial \mathbf{i}_1}{\partial y} \cdot \mathbf{i}_3 = -T_{11y}T_{31} - T_{12y}T_{32} - T_{13y}T_{33}$$

$$k_4 \equiv -\frac{\partial \mathbf{i}_2}{\partial y} \cdot \mathbf{i}_1 = -T_{21y}T_{11} - T_{22y}T_{12} - T_{23y}T_{13}$$

$$k_5 \equiv \frac{\partial \mathbf{i}_1}{\partial x} \cdot \mathbf{i}_2 = T_{11x}T_{21} + T_{12x}T_{22} + T_{13x}T_{23} \quad (7.5.16)$$

Here, k_1 and k_2 are the bending curvatures with respect to η and $-\xi$, respectively; k_{61} and k_{62} are the twisting curvatures with respect to $-\xi$ and η, respectively; and k_4 and k_5 are the spiral curvatures with respect to ζ of the axes η and ξ, respectively. We note that (7.5.16) is the same as (6.2.15) without initial curvatures.

7.5.2 Influence of the Inplane Shear Deformation

If the influence of γ_6 on the surface analysis is to be accounted for, the assumption (7.5.11) needs to be relaxed. In this section, we examine the influence of γ_6 by deriving exact nonlinear functions and expand each function up to cubic terms of the displacements. Substituting (7.5.2), (7.5.3), (7.5.6), and (7.5.7) into (7.5.8) and (7.5.10) and expanding the results in Taylor series, we obtain

$$\gamma_6 = u_y + v_x + w_x w_y - u_x v_x - u_y v_y$$
$$- u_y \left(\frac{1}{2}w_x^2 + \frac{1}{2}w_y^2 + \frac{1}{3}u_y^2 - v_y^2\right)$$
$$- v_x \left(\frac{1}{2}w_x^2 + \frac{1}{2}w_y^2 + \frac{1}{3}v_x^2 - u_x^2\right) - w_x w_y (u_x + v_y) \quad (7.5.17)$$

NONLINEAR MODELING OF RECTANGULAR SURFACES

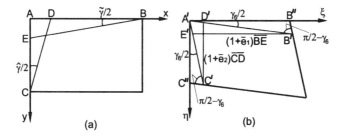

Fig. 7.5.3 The geometry of a differential reference surface: (a) before deformation and (b) after deformation.

$$T_{31} = -w_x + v_x w_y + u_x w_x + \frac{1}{2} w_x w_y^2 - v_x v_y w_y - u_x v_x w_y$$
$$+ \frac{1}{2} w_x^3 - u_y v_x w_x - u_x^2 w_x$$

$$T_{32} = -w_y + u_y w_x + v_y w_y + \frac{1}{2} w_y^3 + \frac{1}{2} w_x^2 w_y - v_y^2 w_y$$
$$- u_y v_x w_y - u_y v_y w_x - u_x u_y w_x$$

$$T_{33} = 1 - \frac{1}{2} w_x^2 - \frac{1}{2} w_y^2 + v_y w_y^2 + v_x w_x w_y + u_y w_x w_y + u_x w_x^2 \quad (7.5.18)$$

Next, we determine nonlinear expansions of the unit vectors $\mathbf{i}_1$ and $\mathbf{i}_2$. In order to simplify the derivation, we assume that

$$\gamma_{61} = \gamma_{62} = \frac{1}{2}\gamma_6 \quad (7.5.19)$$

as shown in Figure 7.5.3, instead of using the exact but implicit relation in (6.2.8b). It follows from Figures 7.5.2 and 7.5.3 and (7.5.1) that

$$\overrightarrow{A'C'} = [u_y \mathbf{j}_1 + (1 + v_y)\mathbf{j}_2 + w_y \mathbf{j}_3] \, dy \quad (7.5.20)$$

$$\overrightarrow{C''C'} = \overrightarrow{A'C'} \frac{\sin(\gamma_6/2)}{\sin[\pi + \gamma_6]/2} \mathbf{i}_{\hat{1}} = \overrightarrow{A'C'} \tan\left(\frac{1}{2}\gamma_6\right) \mathbf{i}_{\hat{1}} \quad (7.5.21)$$

Using (7.5.20), (7.5.21), (7.5.17), (7.5.4), and (7.5.6), we obtain

$$\mathbf{i}_2 = \frac{\overrightarrow{A'C''}}{\overrightarrow{A'C'''}} = \frac{\overrightarrow{A'C'} - \overrightarrow{C''C'}}{|\overrightarrow{A'C'} - \overrightarrow{C''C'}|} = T_{21}\mathbf{j}_1 + T_{22}\mathbf{j}_2 + T_{23}\mathbf{j}_3 \quad (7.5.22)$$

where

$$T_{21} = \frac{1}{2}u_y - \frac{1}{2}v_x - \frac{1}{2}w_x w_y + \frac{1}{2}u_x v_x - \frac{1}{2}u_y v_y + \frac{1}{4}v_x w_y^2 - \frac{1}{4}u_y w_y^2$$
$$+ \frac{1}{2}v_y w_x w_y + \frac{1}{2}u_x w_x w_y + \frac{1}{2}v_x w_x^2 + \frac{1}{2}u_y w_x^2 + \frac{1}{2}u_y v_y^2$$

$$+\frac{3}{16}v_x^3 - \frac{1}{16}u_y v_x^2 + \frac{1}{16}u_y^2 v_x - \frac{1}{2}u_x^2 v_x - \frac{3}{16}u_y^3$$

$$T_{22} = 1 - \frac{1}{8}v_x^2 - \frac{1}{2}w_y^2 - \frac{1}{8}u_y^2 + \frac{1}{4}u_y v_x + v_y w_y^2 + \frac{1}{4}v_x w_x w_y + \frac{3}{4}u_y w_x w_y$$

$$-\frac{1}{4}u_y v_x v_y + \frac{1}{4}u_y^2 v_y + \frac{1}{4}u_x v_x^2 - \frac{1}{4}u_x u_y v_x$$

$$T_{23} = w_y - \frac{1}{2}w_x u_y - \frac{1}{2}w_x v_x - v_y w_y - \frac{1}{2}w_y^3 - \frac{1}{2}w_y^2 w_y + v_y^2 w_y + \frac{3}{8}v_x^2 w_y$$

$$+\frac{3}{4}u_y v_x w_y - \frac{1}{8}u_y^2 w_y + \frac{1}{2}u_y v_y w_x + u_x v_x w_x + \frac{1}{2}u_x u_y w_x \qquad (7.5.23)$$

Using (7.5.18), (7.5.22), and (7.5.23), we obtain

$$\mathbf{i}_1 = \mathbf{i}_2 \times \mathbf{i}_3 = T_{11}\mathbf{j}_1 + T_{12}\mathbf{j}_2 + T_{13}\mathbf{j}_3 \qquad (7.5.24)$$

where

$$T_{11} = 1 - \frac{1}{8}u_y^2 - \frac{1}{2}w_x^2 - \frac{1}{8}v_x^2 + \frac{1}{4}v_x u_y + u_x w_x^2$$

$$+\frac{1}{4}u_y w_y w_x + \frac{3}{4}v_x w_y w_x - \frac{1}{4}v_x u_y u_x + \frac{1}{4}v_x^2 u_x + \frac{1}{4}v_y u_y^2 - \frac{1}{4}v_y v_x u_y$$

$$T_{12} = \frac{1}{2}v_x - \frac{1}{2}u_y - \frac{1}{2}w_y w_x + \frac{1}{2}v_y u_y - \frac{1}{2}v_x u_x$$

$$+\frac{1}{4}u_y w_x^2 - \frac{1}{4}v_x w_x^2 + \frac{1}{2}u_x w_y w_x + \frac{1}{2}v_y w_y w_x + \frac{1}{2}u_y w_y^2$$

$$+\frac{1}{2}v_x w_y^2 + \frac{1}{2}v_x u_x^2 + \frac{3}{16}u_y^3 - \frac{1}{16}v_x u_y^2 + \frac{1}{16}v_x^2 u_y - \frac{1}{2}v_y^2 u_y - \frac{3}{16}v_x^3$$

$$T_{13} = w_x - \frac{1}{2}w_y v_x - \frac{1}{2}w_y u_y - u_x w_x - \frac{1}{2}w_x^3 - \frac{1}{2}w_y^2 w_x + u_x^2 w_x + \frac{3}{8}u_y^2 w_x$$

$$+\frac{3}{4}v_x u_y w_x - \frac{1}{8}v_x^2 w_x + \frac{1}{2}v_x u_x w_y + v_y u_y w_y + \frac{1}{2}v_y v_x w_y \qquad (7.5.25)$$

Next, we determine the strains $\bar{e}_1$ and $\bar{e}_2$ along ξ and η. It follows from Figure 7.5.3 that

$$\tan \frac{1}{2}\hat{\gamma} = \frac{\overline{AD}}{dy} = \frac{\overline{C''C'}}{(1+e_1)dy} \qquad (7.5.26)$$

Using (7.5.2), (7.5.20), (7.5.21), (7.5.4), and (7.5.17), we obtain

$$\hat{\gamma} = 2\tan^{-1}\left(\tan \frac{1}{2}\hat{\gamma}\right)$$

$$= u_y + v_x + v_x v_y + w_x w_y - u_x u_y - 2u_x v_x + \frac{1}{6}u_y^3 + \frac{1}{2}v_x u_y^2 - v_x w_x^2$$

$$-\frac{5}{6}v_x^3 - 2u_x v_x v_y - 2u_x w_x w_y + 3u_x^2 v_x + u_x^2 u_y - \frac{1}{2}v_x^2 u_y - w_x^2 u_y \quad (7.5.27)$$

Using the identity

$$\frac{(1+\bar{e}_2)dy/\cos(\hat{\gamma}/2)}{\sin(90° - \gamma_6)} = \frac{(1+e_2)dy}{\sin(90° + \gamma_6/2)} \qquad (7.5.28)$$

we obtain the expanded form of the axial strain $\bar{e}_2$ along the η axis as

$$\bar{e}_2 = -1 + (1+e_2)\frac{\cos\gamma_6}{\cos(\gamma_6/2)}\cos(\hat{\gamma}/2)$$

$$= v_y - u_y v_x - \frac{1}{2}v_x^2 + \frac{1}{2}w_y^2 - u_y w_x w_y - v_x w_x w_y + \frac{3}{2}u_y u_x v_x$$

$$+\frac{5}{4}u_x v_x^2 + \frac{1}{4}u_y^2 u_x - \frac{1}{2}v_y w_y^2 - \frac{1}{4}u_y^2 v_y - \frac{1}{2}u_y v_x v_y - \frac{3}{4}v_x^2 v_y \quad (7.5.29)$$

Following a procedure similar to that used in obtaining $\hat{\gamma}$ and $\bar{e}_2$, we obtain

$$\tilde{\gamma} = u_y + v_x + u_x u_y + w_x w_y - v_y v_x - 2u_y v_y + \frac{1}{6}v_x^3 + \frac{1}{2}u_y v_x^2 - u_y w_y^2$$

$$-\frac{5}{6}u_y^3 - 2v_y u_x u_y - 2v_y w_x w_y + 3v_y^2 u_y + v_y^2 v_x - \frac{1}{2}u_y^2 v_x - w_y^2 v_x \quad (7.5.30)$$

and the expanded form of the axial strain along the ξ axis as

$$\bar{e}_1 = -1 + (1+e_1)\frac{\cos\gamma_6}{\cos(\gamma_6/2)}\cos(\tilde{\gamma}/2)$$

$$= u_x - v_x u_y - \frac{1}{2}u_y^2 + \frac{1}{2}w_x^2 - v_x w_x w_y - u_y w_x w_y + \frac{3}{2}v_x v_y u_y$$

$$+\frac{5}{4}v_y u_y^2 + \frac{1}{4}v_x^2 v_y - \frac{1}{2}u_x w_x^2 - \frac{1}{4}v_x^2 u_x - \frac{1}{2}v_x u_y u_x - \frac{3}{4}u_y^2 u_x \quad (7.5.31)$$

Equations (7.5.31), (7.5.29), and (7.5.17) represent the asymptotic expansions of the inplane strains $\bar{e}_1$, $\bar{e}_2$, and γ_6, respectively.

Using the definitions of the curvatures in (7.5.16), (7.5.18), (7.5.23), and (7.5.25), we expand the curvatures in terms of the global displacements u, v, and w as

$$k_1 = -\mathbf{i}_{1x} \cdot \mathbf{i}_3$$

$$= -w_{xx} + v_{xx}w_y + u_x w_{xx} + \frac{1}{2}v_x w_{xy} + \frac{1}{2}u_y w_{xy} + u_{xx} w_x$$

$$+\frac{1}{2}w_{xx}w_y^2 + \frac{1}{2}w_x w_y w_{xy} + w_x^2 w_{xx} - v_{xx}v_y w_y - u_x v_{xx} w_y$$

$$-\frac{1}{2}v_x v_{xy}w_y - \frac{1}{2}u_y v_{xy} w_y - u_{xx} v_x w_y + \frac{1}{8}v_x^2 w_{xx} - \frac{3}{4}u_y v_x w_{xx}$$

$$-\frac{3}{8}u_y^2 w_{xx} - u_x^2 w_{xx} - \frac{1}{2}v_x v_y w_{xy} - u_y v_y w_{xy} - \frac{1}{2}u_x v_x w_{xy}$$

$$-u_y v_{xx} w_x - \frac{1}{2}u_{xy} v_x w_x - \frac{1}{2}u_{xy} u_y w_x - 2u_x u_{xx} w_x \quad (7.5.32)$$

$$k_2 = -\mathbf{i}_{2y} \cdot \mathbf{i}_3$$

$$= -w_{yy} + u_{yy}w_x + v_y w_{yy} + \frac{1}{2}u_y w_{xy} + \frac{1}{2}v_x w_{xy} + v_{yy}w_y$$

$$+\frac{1}{2}w_{yy}w_x^2 + \frac{1}{2}w_y w_x w_{xy} + w_y^2 w_{yy} - u_{yy}u_x w_x - v_y u_{yy} w_x$$

$$-\frac{1}{2}u_y u_{xy} w_x - \frac{1}{2}v_x u_{xy} w_x - v_{yy} u_y w_x + \frac{1}{8}u_y^2 w_{yy} - \frac{3}{4}v_x u_y w_{yy}$$
$$-\frac{3}{8}v_x^2 w_{yy} - v_y^2 w_{yy} - \frac{1}{2}u_y u_x w_{xy} - v_x u_x w_{xy} - \frac{1}{2}v_y u_y w_{xy}$$
$$-v_x u_{yy} w_y - \frac{1}{2}v_{xy} u_y w_y - \frac{1}{2}v_{xy} v_x w_y - 2v_y v_{yy} w_y \qquad (7.5.33)$$

$$k_{61} = -\mathbf{i}_{2x} \cdot \mathbf{i}_3$$
$$= -w_{xy} + \frac{1}{2}u_y w_{xx} + \frac{1}{2}v_x w_{xx} + v_{xy} w_y + u_{xy} w_x + v_y w_{xy}$$
$$+ \frac{1}{2}w_x w_y w_{xx} + w_{xy} w_y^2 + \frac{1}{2}w_{xy} w_x^2$$
$$- \frac{1}{2}u_x u_y w_{xx} - u_x v_x w_{xx} - \frac{1}{2}u_y v_y w_{xx}$$
$$- \frac{1}{2}u_y u_{xx} w_x - \frac{1}{2}v_x u_{xx} w_x - 2v_{xy} v_y w_y - u_{xy} u_x w_x - \frac{1}{2}v_x v_{xx} w_y$$
$$- \frac{1}{2}u_y v_{xx} w_y - v_y u_{xy} w_x - u_{xy} v_x w_y$$
$$- v_{xy} u_y w_x - v_y^2 w_{xy} - \frac{3}{4}u_y v_x w_{xy} - \frac{3}{8}v_x^2 w_{xy} + \frac{1}{8}u_y^2 w_{xy} \qquad (7.5.34)$$

$$k_{62} = -\mathbf{i}_{1y} \cdot \mathbf{i}_3$$
$$= -w_{xy} + \frac{1}{2}v_x w_{yy} + \frac{1}{2}u_y w_{yy} + v_{xy} w_y + u_{xy} w_x$$
$$+ u_x w_{xy} + \frac{1}{2}w_x w_y w_{yy} + \frac{1}{2}w_{xy} w_y^2 + w_{xy} w_x^2 - \frac{1}{2}v_x v_y w_{yy}$$
$$- u_y v_y w_{yy} - \frac{1}{2}u_x v_x w_{yy} - \frac{1}{2}v_x v_{yy} w_y$$
$$- \frac{1}{2}u_y v_{yy} w_y - v_{xy} v_y w_y - 2u_{xy} u_x w_x$$
$$- \frac{1}{2}u_y u_{yy} w_x - \frac{1}{2}v_x u_{yy} w_x - u_x v_{xy} w_y - u_{xy} v_x w_y$$
$$- v_{xy} u_y w_x - u_x^2 w_{xy} - \frac{3}{4}u_y v_x w_{xy} + \frac{1}{8}v_x^2 w_{xy} - \frac{3}{8}u_y^2 w_{xy} \qquad (7.5.35)$$

$$k_4 = -\mathbf{i}_{2y} \cdot \mathbf{i}_1$$
$$= \frac{1}{2}\left(v_{xy} - u_{yy} - w_x w_{yy} + w_y w_{xy} + u_y v_{yy} + v_y u_{yy} - u_x v_{xy} - v_x u_{xy}\right) \qquad (7.5.36)$$

$$k_5 = \mathbf{i}_{1x} \cdot \mathbf{i}_2$$
$$= \frac{1}{2}\left(v_{xx} - u_{xy} + w_y w_{xx} - w_x w_{xy} + v_y u_{xy} - u_x v_{xx} + u_y v_{xy} - v_x u_{xx}\right) \qquad (7.5.37)$$

Here, k_4 and k_5 are expanded only up to quadratic terms because they only appear in nonlinear terms, as shown later. This completes the derivation of asymptotic expressions for the midplane strains ($\bar{e}_1, \bar{e}_2, \gamma_6$) and curvatures ($k_1, k_2, k_6$) and hence the surface analysis of the deformed reference plane.

We note that under the transformation

$$u \to v, \quad v \to u, \quad w \to -w, \quad x \to y, \quad y \to x, \quad z \to -z, \qquad (7.5.38)$$

which corresponds to the case in which the coordinate system xyz is rotated by 90° with respect to the z axis and then 180° with respect to the new x axis, the following pairs are interchangeable:

$$\bar{e}_1 \to \bar{e}_2, \quad \gamma_6 \to \gamma_6, \quad k_1 \to -k_2, \quad k_4 \to -k_5, \quad k_{61} \to k_{62}$$
$$k_{61} + k_{62} \to -(k_{61} + k_{62}) \qquad (7.5.39)$$

It shows that this mathematical model is symmetric.

The nonlinear von Kármán-type geometric nonlinearity is based on the following nonlinear inplane strains and linear curvatures:

$$\bar{e}_1 = u_x + \frac{1}{2}w_x^2, \quad \bar{e}_2 = v_y + \frac{1}{2}w_y^2, \quad \gamma_6 = u_y + v_x + w_x w_y \qquad (7.5.40a)$$

$$k_1 = -w_{xx}, \quad k_2 = -w_{yy}, \quad k_6 = 2k_{61} = 2k_{62} = -2w_{xy}, k_4 = k_5 = 0 \qquad (7.5.40b)$$

They are commonly used in nonlinear plate analyses. Comparing (7.5.40a) with (7.5.31), (7.5.29), and (7.5.17), we note that the von Kármán strains are correct to third order in w only if u and v are of second or higher order than w. Comparing (7.5.40b) with the asymptotic expansions in (7.5.32)-(7.5.35), we note that several quadratic and cubic terms may not be negligible if the inplane inertias are not negligible. Moreover, k_4 and k_5 are not used in most of the plate theories.

7.5.3 Variation of the Global Strains

To derive fully nonlinear governing equations, we need to derive variations of the global strains e_1, e_2, and γ_6. Using the assumption (7.5.11), the fact that $T_{11}^2 + T_{12}^2 + T_{13}^2 = T_{21}^2 + T_{22}^2 + T_{13}^2 = 1$, (7.5.6), and (7.5.7), we obtain

$$1 + e_1 = T_{11}(1 + u_x) + T_{12}v_x + T_{13}w_x \qquad (7.5.41a)$$

$$1 + e_2 = T_{21}u_y + T_{22}(1 + v_y) + T_{23}w_y \qquad (7.5.41b)$$

Taking the variations of (7.5.41a,b) and using (7.5.6) and (7.5.7), we obtain

$$\delta e_1 = T_{11}\delta u_x + T_{12}\delta v_x + T_{13}\delta w_x \qquad (7.5.42a)$$

$$\delta e_2 = T_{21}\delta u_y + T_{22}\delta v_y + T_{23}\delta w_y \qquad (7.5.42b)$$

Taking the variations of $\mathbf{i}_{\hat{1}}(= \mathbf{i}_1)$ and $\mathbf{i}_{\hat{2}}(= \mathbf{i}_2)$ in (7.5.4) and (7.5.5) and using (7.5.6) and (7.5.7), we obtain

$$\delta \mathbf{i}_{\hat{1}} = \frac{1}{1+e_1}(\delta u_x \mathbf{j}_1 + \delta v_x \mathbf{j}_2 + \delta w_x \mathbf{j}_3) - \frac{\delta e_1}{1+e_1}\mathbf{i}_1 \qquad (7.5.43a)$$

426 PLATES

$$\delta \mathbf{i}_2 = \frac{1}{1+e_2}[\delta u_y \mathbf{j}_1 + \delta v_y \mathbf{j}_2 + \delta w_y \mathbf{j}_3] - \frac{\delta e_2}{1+e_2}\mathbf{i}_2 \qquad (7.5.43\text{b})$$

Using (6.3.1a) and (7.5.43a,b), we obtain

$$(1+e_1)\delta\theta_2 + T_{31}\delta u_x + T_{32}\delta v_x + T_{33}\delta w_x = 0 \qquad (7.5.44\text{a})$$

$$-(1+e_2)\delta\theta_1 + T_{31}\delta u_y + T_{32}\delta v_y + T_{33}\delta w_y = 0 \qquad (7.5.44\text{b})$$

Also, it can be seen from Figure 7.5.2 that

$$\delta\gamma_6 = \delta\gamma_{61} + \delta\gamma_{62} = \delta\mathbf{i}_1 \cdot \mathbf{i}_{\hat{2}} + \delta\mathbf{i}_2 \cdot \mathbf{i}_{\hat{1}} \qquad (7.5.45)$$

Using (7.5.45), (7.5.43a,b), and (7.5.11), one can show that

$$\delta\gamma_6 = \frac{1}{1+e_1}(T_{21}\delta u_x + T_{22}\delta v_x + T_{23}\delta w_x)$$
$$+ \frac{1}{1+e_2}(T_{11}\delta u_y + T_{12}\delta v_y + T_{13}\delta w_y) \qquad (7.5.46)$$

We note that (7.5.42a), (7.5.42b), (7.5.44a), (7.5.44b), and (7.5.46) are the same as (6.3.3a), (6.3.3b), (6.3.5a), (6.3.5b), and (6.3.7a)+(6.3.7b) without initial curvatures.

7.6 GENERAL NONLINEAR CLASSICAL PLATE THEORY

Similar to the fact shown in Section 4.4.1 that a differential beam element without warpings is a perfectly rigid thin "plate", a differential plate element without shear warpings is a perfectly rigid straight filament having dimensions $dx \times dy \times h$ (see Figure 7.2.1). Classical plate theories are based on the assumption that an initially straight differential plate element is straight and perpendicular to the reference plane before and after deformation, and this assumption is the so-called Kirchhoff hypothesis.

We consider a rectangular plate, as shown in Figure 7.5.1. It follows from Figure 7.5.2 that, without shear warpings, the displacement vector **D** of an arbitrary point on the observed differential plate element (a straight line) is

$$\mathbf{D} = u\mathbf{j}_1 + v\mathbf{j}_2 + w\mathbf{j}_3 + z\mathbf{i}_3 - z\mathbf{j}_3 \qquad (7.6.1)$$

Taking the variation of **D** and using (6.3.1a), we obtain

$$\delta\mathbf{D} = \mathbf{j}_1\delta u + \mathbf{j}_2\delta v + \mathbf{j}_3\delta w + z(\mathbf{i}_1\delta\theta_2 - \mathbf{i}_2\delta\theta_1) \qquad (7.6.2\text{a})$$

Taking the time derivatives of **D** and using (7.5.12) yields

$$\ddot{\mathbf{D}} = \ddot{u}\mathbf{j}_1 + \ddot{v}\mathbf{j}_2 + \ddot{w}\mathbf{j}_3 + z(\ddot{T}_{31}\mathbf{j}_1 + \ddot{T}_{32}\mathbf{j}_2 + \ddot{T}_{33}\mathbf{j}_3) \qquad (7.6.2\text{b})$$

Substituting (7.6.2a,b) into (7.2.2a) and using (7.5.12), we obtain

$$\delta T = -\int_A (A_u \delta u + A_v \delta v + A_w \delta w + A_{\theta_1}\delta\theta_1 + A_{\theta_2}\delta\theta_2)dA \qquad (7.6.3)$$

where

$$A_u = I_0\ddot{u} + I_1\ddot{T}_{31}$$
$$A_v = I_0\ddot{v} + I_1\ddot{T}_{32}$$
$$A_w = I_0\ddot{w} + I_1\ddot{T}_{33}$$
$$A_{\theta_1} = -I_1(\ddot{u}T_{21} + \ddot{v}T_{22} + \ddot{w}T_{23}) - I_2(\ddot{T}_{31}T_{21} + \ddot{T}_{32}T_{22} + \ddot{T}_{33}T_{23})$$
$$A_{\theta_2} = I_1(\ddot{u}T_{11} + \ddot{v}T_{12} + \ddot{w}T_{13}) + I_2(\ddot{T}_{31}T_{11} + \ddot{T}_{32}T_{12} + \ddot{T}_{33}T_{13}) \quad (7.6.4a)$$

$$\{I_0, I_1, I_2\} \equiv \int_z \rho\{1, z, z^2\}dz \quad (7.6.4b)$$

and $I_1 = 0$ if ρ is constant and the middle surface is chosen as the reference plane. We note that (7.2.7) is a linear version of (7.6.3) when the displacements are small and hence

$$\theta_2 = T_{31} = -w_x, \quad \theta_1 = -T_{32} = w_y$$
$$T_{11} = T_{22} = T_{33} = 1, \quad T_{12} = T_{13} = T_{21} = T_{23} = 0 \quad (7.6.5)$$

To derive the variation of the strain energy $\delta\Pi$, we need to express the strains in terms of the displacements. We assume that the local displacements u_1, u_2, and u_3 (with respect to the local $\xi\eta\zeta$ coordinate system) of an arbitrary point on the plate element have the form

$$u_1(x, y, z, t) = u_1^0(x, y, t) + z\theta_2(x, y, t) \quad (7.6.6a)$$
$$u_2(x, y, z, t) = u_2^0(x, y, t) - z\theta_1(x, y, t) \quad (7.6.6b)$$
$$u_3(x, y, z, t) = u_3^0(x, y, t) \quad (7.6.6c)$$

where the u_i^0 ($i = 1, 2, 3$) are the local displacements (relative to the local coordinate system) of a point, which is located at $(x, y, 0)$ before deformation, and θ_1 and θ_2 are the rotation angles of the observed plate element (a straight line) with respect to the axes ξ and η, respectively. Because $\xi\eta\zeta$ is a local coordinate system attached to the observed plate element and the $\xi\eta$ plane is tangent to the deformed reference plane, we have

$$u_1^0 = u_2^0 = u_3^0 = \theta_1 = \theta_2 = \partial u_3^0/\partial x = \partial u_3^0/\partial y = 0 \quad (7.6.7)$$

If **U** denotes the local displacement vector; that is,

$$\mathbf{U} = u_1\mathbf{i}_1 + u_2\mathbf{i}_2 + u_3\mathbf{i}_3 \quad (7.6.8)$$

then it follows from (7.6.7) and (7.6.8) that

$$\frac{\partial \mathbf{U}}{\partial x} = \frac{\partial u_1}{\partial x}\mathbf{i}_1 + \frac{\partial u_2}{\partial x}\mathbf{i}_2 + \frac{\partial u_3}{\partial x}\mathbf{i}_3 + u_1\frac{\partial \mathbf{i}_1}{\partial x} + u_2\frac{\partial \mathbf{i}_2}{\partial x} + u_3\frac{\partial \mathbf{i}_3}{\partial x}$$
$$= (e_1 + zk_1)\mathbf{i}_1 + \left(\frac{\partial u_2^0}{\partial x} + zk_{61}\right)\mathbf{i}_2 \quad (7.6.9a)$$

$$\frac{\partial \mathbf{U}}{\partial y} = \frac{\partial u_1}{\partial y}\mathbf{i}_1 + \frac{\partial u_2}{\partial y}\mathbf{i}_2 + \frac{\partial u_3}{\partial y}\mathbf{i}_3 + u_1\frac{\partial \mathbf{i}_1}{\partial y} + u_2\frac{\partial \mathbf{i}_2}{\partial y} + u_3\frac{\partial \mathbf{i}_3}{\partial y}$$

$$= \left(\frac{\partial u_1^0}{\partial y} + zk_{62}\right)\mathbf{i}_1 + (e_2 + zk_2)\mathbf{i}_2 \qquad (7.6.9\mathrm{b})$$

$$\frac{\partial \mathbf{U}}{\partial z} = \frac{\partial u_1}{\partial z}\mathbf{i}_1 + \frac{\partial u_2}{\partial z}\mathbf{i}_2 + \frac{\partial u_3}{\partial z}\mathbf{i}_3 = 0 \qquad (7.6.9\mathrm{c})$$

and

$$\epsilon_{11} = \frac{\partial \mathbf{U}}{\partial x}\cdot\mathbf{i}_1 = e_1 + zk_1 \qquad (7.6.10\mathrm{a})$$

$$\epsilon_{22} = \frac{\partial \mathbf{U}}{\partial y}\cdot\mathbf{i}_2 = e_2 + zk_2 \qquad (7.6.10\mathrm{b})$$

$$\epsilon_{12} = \frac{\partial \mathbf{U}}{\partial x}\cdot\mathbf{i}_2 + \frac{\partial \mathbf{U}}{\partial y}\cdot\mathbf{i}_1 = \gamma_6 + zk_6 \qquad (7.6.10\mathrm{c})$$

$$\epsilon_{33} = \epsilon_{23} = \epsilon_{13} = 0 \qquad (7.6.10\mathrm{d})$$

where the ϵ_{ij} are the Jaumann strains and

$$e_1 \equiv \frac{\partial u_1^0}{\partial x},\quad e_2 \equiv \frac{\partial u_2^0}{\partial y},\quad \gamma_6 \equiv \frac{\partial u_1^0}{\partial y} + \frac{\partial u_2^0}{\partial x} \qquad (7.6.11\mathrm{a})$$

$$k_1 \equiv \frac{\partial \theta_2}{\partial x},\quad k_2 \equiv -\frac{\partial \theta_1}{\partial y},\quad k_{61} \equiv -\frac{\partial \theta_1}{\partial x},\quad k_{62} \equiv \frac{\partial \theta_2}{\partial y},\quad k_6 \equiv k_{61} + k_{62} \qquad (7.6.11\mathrm{b})$$

We note that (7.2.4) is a linear version of (7.6.10) with

$$e_1 = u_x,\quad e_2 = v_y,\quad \gamma_6 = u_y + v_x$$
$$k_1 = -w_{xx},\quad k_2 = -w_{yy},\quad k_6 = -2w_{xy} \qquad (7.6.12)$$

Substituting (7.6.10a-d) into (7.2.2b) and using (6.4.1a) (with $m = M_1$), (6.4.1b) (with $m = M_2$), and (6.4.1c,d) (with $m = M_6$), we obtain

$$\delta\Pi = \int_A (N_1\delta e_1 + N_2\delta e_2 + N_6\delta\gamma_6 + M_1\delta k_1 + M_2\delta k_2 + M_6\delta k_6)\,dxdy$$

$$= \int_A (N_1\delta e_1 + N_2\delta e_2 + N_6\delta\gamma_6 + \Theta_1\delta\theta_1 + \Theta_2\delta\theta_2 + \Theta_3\delta\theta_3)\,dxdy$$

$$+ \int_x [-M_2\delta\theta_1 + M_6\delta\theta_2]_{y=0}^{y=b}\,dx + \int_y [-M_6\delta\theta_1 + M_1\delta\theta_2]_{x=0}^{x=a}\,dy \qquad (7.6.13)$$

where

$$\{N_1, N_2, N_6\} = \int_z \{\sigma_{11}, \sigma_{22}, \sigma_{12}\}\,dz$$
$$\{M_1, M_2, M_6\} = \int_z z\{\sigma_{11}, \sigma_{22}, \sigma_{12}\}\,dz \qquad (7.6.14\mathrm{a})$$

$$\Theta_1 \equiv M_{6x} + M_{2y} + M_1 k_5 + M_6 k_4$$
$$\Theta_2 \equiv -M_{1x} - M_{6y} + M_2 k_4 + M_6 k_5$$
$$\Theta_3 \equiv M_1 k_{61} - M_2 k_{62} + M_6 k_2 - M_6 k_1 \qquad (7.6.14\mathrm{b})$$

GENERAL NONLINEAR CLASSICAL PLATE THEORY

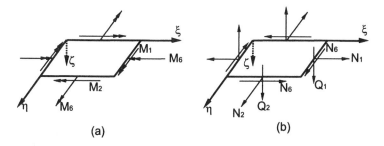

Fig. 7.6.1 (a) Moment resultants and (b) force resultants on a nonlinear classical plate element.

Here, M_1, M_2, and M_6 represent the moment intensities acting on the edges of the plate element, as shown in Figure 7.6.1a. Figure 7.6.1b shows the inplane extension force intensities N_1 and N_2 and the inplane shear intensity N_6. We note that substituting (7.6.12) into (7.6.13) yields (7.2.9).

To relate the stress resultants to the inplane strains, we first use $\epsilon_{13} = \epsilon_{23} = 0$ from (7.6.10d) and $\sigma_{33} = 0$ to obtain the reduced stress-strain relations from (2.4.34) and (2.4.35) as

$$\left\{ \begin{array}{c} \sigma_{11} \\ \sigma_{22} \\ \sigma_{12} \end{array} \right\} = \left[\begin{array}{ccc} \tilde{Q}_{11} & \tilde{Q}_{12} & \tilde{Q}_{16} \\ \tilde{Q}_{12} & \tilde{Q}_{22} & \tilde{Q}_{26} \\ \tilde{Q}_{16} & \tilde{Q}_{26} & \tilde{Q}_{66} \end{array} \right] \left(\left\{ \begin{array}{c} e_1 \\ e_2 \\ \gamma_6 \end{array} \right\} + z \left\{ \begin{array}{c} k_1 \\ k_2 \\ k_6 \end{array} \right\} \right) \quad (7.6.15)$$

where the σ_{ij} are the Jaumann stresses. Substituting (7.6.15) into (7.6.14a) yields the internal forces and moments in terms of the midplane strains and curvatures as

$$\left\{ \begin{array}{c} N_1 \\ N_2 \\ N_6 \\ M_1 \\ M_2 \\ M_6 \end{array} \right\} = \left[\begin{array}{cc} [A_{ij}] & [B_{ij}] \\ [B_{ij}] & [D_{ij}] \end{array} \right] \left\{ \begin{array}{c} e_1 \\ e_2 \\ \gamma_6 \\ k_1 \\ k_2 \\ k_6 \end{array} \right\} \quad (7.6.16)$$

where

$$\{A_{ij}, B_{ij}, D_{ij}\} = \sum_{k=1}^{N} \int_{z_k}^{z_{k+1}} \tilde{Q}_{ij}^{(k)} \{1, z, z^2\} dz \quad \text{for } i,j = 1,2,6 \quad (7.6.17)$$

In (7.6.17), we assume that there are N laminae, the kth lamina is located between the two planes at $z = z_k$ and $z = z_{k+1}$, and $z_{N+1} - z_1 = h$.

Substituting (7.5.42a,b) and (7.5.46) into (7.6.13) and adding the integrals of $Q_1\times$ (7.5.44a) and $Q_2\times$ (7.5.44b), we obtain

$$\delta\Pi = \int_A \left[\{N_1,\ N_6/(1+e_1),\ Q_1\}\,[T] \left\{\begin{array}{c} \delta u_x \\ \delta v_x \\ \delta w_x \end{array}\right\} \right.$$

$$+ \{N_6/(1+e_2),\ N_2,\ Q_2\}\,[T] \left\{\begin{array}{c} \delta u_y \\ \delta v_y \\ \delta w_y \end{array}\right\}$$

$$\left. + [\Theta_1 - Q_2(1+e_2)]\,\delta\theta_1 + [\Theta_2 + Q_1(1+e_1)]\,\delta\theta_2 + \Theta_3\delta\theta_3 \right] dxdy$$

$$+ \int_x [-M_2\delta\theta_1 + M_6\delta\theta_2]_{y=0}^{y=b}\,dx + \int_y [-M_6\delta\theta_1 + M_1\delta\theta_2]_{x=0}^{x=a}\,dy \quad (7.6.18)$$

We show later that the introduced Q_1 and Q_2 are transverse shear intensities. Substituting (7.6.18) and (7.6.3) into (7.2.1), using (7.6.14b) and (7.5.44a,b), and integrating the result by parts, we obtain

$$0 = \int_A \left[\left(\frac{\partial\left(\{F_\alpha\}^T[T]\right)}{\partial x} + \frac{\partial\left(\{F_\beta\}^T[T]\right)}{\partial y} - \{I_F\}^T \right) \left\{\begin{array}{c} \delta u \\ \delta v \\ \delta w \end{array}\right\} \right.$$

$$+ \left(\frac{\partial\{M_\alpha\}^T}{\partial x} + \{M_\alpha\}^T[K_1] + \frac{\partial\{M_\beta\}^T}{\partial y} + \{M_\beta\}^T[K_2] \right.$$

$$\left.\left. + \{Q_\alpha\}^T + \{Q_\beta\}^T - \{I_M\}^T \right) \left\{\begin{array}{c} \delta\theta_1 \\ \delta\theta_2 \\ \delta\theta_3 \end{array}\right\} \right] dxdy$$

$$- \int_x \left[\{F_\beta\}^T[T] \left\{\begin{array}{c} \delta u \\ \delta v \\ \delta w \end{array}\right\} + \left(\frac{M_6 T_{31}}{1+e_1}\right)_x \delta u + \left(\frac{M_6 T_{32}}{1+e_1}\right)_x \delta v \right.$$

$$\left. + \left(\frac{M_6 T_{33}}{1+e_1}\right)_x \delta w - M_2\delta\theta_1 \right]_{y=0}^{y=b} dx$$

$$- \int_y \left[\{F_\alpha\}^T[T] \left\{\begin{array}{c} \delta u \\ \delta v \\ \delta w \end{array}\right\} + \left(\frac{M_6 T_{31}}{1+e_2}\right)_y \delta u + \left(\frac{M_6 T_{32}}{1+e_2}\right)_y \delta v \right.$$

$$\left. + \left(\frac{M_6 T_{33}}{1+e_2}\right)_y \delta w + M_1\delta\theta_2 \right]_{x=0}^{x=a} dy$$

$$+ \left[\left(\frac{M_6}{1+e_1} + \frac{M_6}{1+e_2}\right)(T_{31}\delta u + T_{32}\delta v + T_{33}\delta w) \right]_{(x,y)=(a,0),(0,b)}^{(x,y)=(0,0),(a,b)} \quad (7.6.19)$$

where

$$\{F_\alpha\} \equiv \{N_1,\ N_6/(1+e_1),\ Q_1\}^T, \qquad \{F_\beta\} \equiv \{N_6/(1+e_2),\ N_2,\ Q_2\}^T$$

$$\{M_\alpha\} \equiv \{-M_6, M_1, 0\}^T, \qquad \{M_\beta\} \equiv \{-M_2, M_6, 0\}^T$$
$$\{Q_\alpha\} \equiv \{0, -Q_1(1+e_1), N_6\}^T, \qquad \{Q_\beta\} \equiv \{Q_2(1+e_2), 0, -N_6\}^T$$
$$\{I_F\} \equiv \{A_u, A_v, A_w\}^T, \qquad \{I_M\} \equiv \{A_{\theta_1}, A_{\theta_2}, 0\}^T \qquad (7.6.20)$$

Setting each of the coefficients of δu, δv, δw, $\delta\theta_1$, $\delta\theta_2$, and $\delta\theta_3$ in (7.6.19) equal to zero, we obtain the following equations of motion:

$$\frac{\partial}{\partial x}\left([T]^T\{F_\alpha\}\right) + \frac{\partial}{\partial y}\left([T]^T\{F_\beta\}\right) = \{I_F\} \qquad (7.6.21)$$

$$\frac{\partial\{M_\alpha\}}{\partial x} + [K_1]^T\{M_\alpha\} + \frac{\partial\{M_\beta\}}{\partial y} + [K_2]^T\{M_\beta\} + \{Q_\alpha\} + \{Q_\beta\} = \{I_M\} \qquad (7.6.22)$$

and boundary conditions: specify

Along $x = 0, a$:

$\delta u = 0$	or	$\mathbf{F}_\alpha \cdot \mathbf{j}_1 + [M_6 T_{31}/(1+e_2)]_y$
$\delta v = 0$	or	$\mathbf{F}_\alpha \cdot \mathbf{j}_2 + [M_6 T_{32}/(1+e_2)]_y$
$\delta w = 0$	or	$\mathbf{F}_\alpha \cdot \mathbf{j}_3 + [M_6 T_{33}/(1+e_2)]_y$
$\delta\theta_2 = 0$	or	M_1

Along $y = 0, b$:

$\delta u = 0$	or	$\mathbf{F}_\beta \cdot \mathbf{j}_1 + [M_6 T_{31}/(1+e_1)]_x$
$\delta v = 0$	or	$\mathbf{F}_\beta \cdot \mathbf{j}_2 + [M_6 T_{32}/(1+e_1)]_x$
$\delta w = 0$	or	$\mathbf{F}_\beta \cdot \mathbf{j}_3 + [M_6 T_{33}/(1+e_1)]_x$
$\delta\theta_1 = 0$	or	M_2

At $(x,y) = (0,0), (a,b), (a,0), (0,b)$:

$\delta u = 0$	or	$[M_6/(1+e_1) + M_6/(1+e_2)]T_{31}$
$\delta v = 0$	or	$[M_6/(1+e_1) + M_6/(1+e_2)]T_{32}$
$\delta w = 0$	or	$[M_6/(1+e_1) + M_6/(1+e_2)]T_{33}$ (7.6.23)

where

$$\mathbf{F}_\alpha = N_1\mathbf{i}_1 + N_6/(1+e_1)\mathbf{i}_2 + Q_1\mathbf{i}_3$$
$$\mathbf{F}_\beta = N_6/(1+e_2)\mathbf{i}_1 + N_2\mathbf{i}_2 + Q_2\mathbf{i}_3 \qquad (7.6.24a)$$

We also define

$$\mathbf{M}_\alpha = -M_6\mathbf{i}_1 + M_1\mathbf{i}_2$$
$$\mathbf{M}_\beta = -M_2\mathbf{i}_1 + M_6\mathbf{i}_2$$
$$\mathbf{I}_F = A_u\mathbf{j}_1 + A_v\mathbf{j}_2 + A_w\mathbf{j}_3$$
$$\mathbf{I}_M = A_{\theta_1}\mathbf{i}_1 + A_{\theta_2}\mathbf{i}_2 \qquad (7.6.24b)$$

432 PLATES

Using (7.6.24a,b), (7.5.12), (7.5.14), and (7.5.15), we put (7.6.21) and (7.6.22) in the following vector forms:

$$\frac{\partial \mathbf{F}_\alpha}{\partial x} + \frac{\partial \mathbf{F}_\beta}{\partial y} = \mathbf{I}_F \tag{7.6.25}$$

$$\frac{\partial \mathbf{M}_\alpha}{\partial x} + \frac{\partial \mathbf{M}_\beta}{\partial y} + (1+e_1)\mathbf{i}_1 \times \mathbf{F}_\alpha + (1+e_2)\mathbf{i}_2 \times \mathbf{F}_\beta = \mathbf{I}_M \tag{7.6.26}$$

One can see from (7.6.24a) that Q_1 and Q_2 are along the $\mathbf{i}_3$ direction and hence they represent the transverse shear intensities. However, it is assumed that there are no transverse shear deformations in the classical plate theory and hence (7.6.16) does not include structural constitutive equations for Q_1 and Q_2. Hence, Q_1 and Q_2 need to be obtained from the first two equations in (7.6.22) as

$$Q_1 = \frac{1}{1+e_1}(M_{1x} + M_{6y} - M_2 k_4 - M_6 k_5 - A_{\theta_2})$$

$$Q_2 = \frac{1}{1+e_2}(M_{2y} + M_{6x} + M_6 k_4 + M_1 k_5 + A_{\theta_1}) \tag{7.6.27}$$

Because the diameter of a plate element (i.e., $\sqrt{dx^2 + dy^2}$) is very small compared to the plate thickness h, its rotary inertia A_{θ_3} with respect to the ζ axis is very small and hence the third equation in (7.6.22) is a statement of the balance of internal moments about the ζ axis, which is given by

$$N_6 - N_6 + M_2 k_{62} - M_1 k_{61} + M_6 k_1 - M_6 k_2 = 0 \tag{7.6.28}$$

Hence, only the three partial-differential equations in (7.6.21) and the boundary conditions in (7.6.23) need to be solved for the vibration of classical plates.

Next, we show that the general nonlinear classical plate theory reduces to the nonlinear classical plate theory in Section 7.4 when the inplane dynamics are fast compared to the transverse dynamics. To this end, we note that, if the inplane natural frequencies are large compared with the transverse natural frequencies, the inplane inertia terms $\ddot{u}$ and $\ddot{v}$ can be neglected in low-frequency plate dynamics. Hence, if $w = O(\epsilon)$, where $\epsilon = h^2/a^2$ and h and a are representative plate thickness and length, then the inplane displacements u and v are $O(\epsilon^2)$. Using these orders of magnitude, we introduce nondimensional quantities as follows:

$$\tilde{x} = \frac{x}{a}, \quad \tilde{y} = \frac{y}{a}, \quad \tilde{w} = \frac{w}{\epsilon a}, \quad \tilde{u} = \frac{u}{\epsilon^2 a}, \quad \tilde{v} = \frac{v}{\epsilon^2 a}, \quad \tilde{t} = \frac{h\sqrt{E}}{a^2\sqrt{\rho}}t,$$

$$\tilde{N}_i = \frac{N_i}{Eh\epsilon^2}, \quad \frac{\partial \tilde{M}_i}{\partial \tilde{x}} = \frac{1}{Eh\epsilon^2}\frac{\partial M_i}{\partial x}, \quad \frac{\partial \tilde{M}_i}{\partial \tilde{y}} = \frac{1}{Eh\epsilon^2}\frac{\partial M_i}{\partial y} \tag{7.6.29}$$

Substituting (7.6.29) into (7.6.21), (7.6.24), and (7.6.4) and expanding the result for small ϵ, we obtain

$$\frac{\partial \tilde{N}_1}{\partial \tilde{x}} + \frac{\partial \tilde{N}_6}{\partial \tilde{y}} = O(\epsilon) \tag{7.6.30a}$$

$$\frac{\partial \tilde{N}_6}{\partial \tilde{x}} + \frac{\partial \tilde{N}_2}{\partial \tilde{y}} = O(\epsilon) \qquad (7.6.30\text{b})$$

$$\frac{\partial^2 \tilde{M}_1}{\partial \tilde{x}^2} + 2\frac{\partial^2 \tilde{M}_6}{\partial \tilde{x} \partial \tilde{y}} + \frac{\partial^2 \tilde{M}_2}{\partial \tilde{y}^2} - \tilde{I}_0 \frac{\partial^2 \tilde{w}}{\partial \tilde{t}^2} + \epsilon \frac{\partial}{\partial \tilde{x}}\left(\tilde{N}_1 \frac{\partial \tilde{w}}{\partial \tilde{x}} + \tilde{N}_6 \frac{\partial \tilde{w}}{\partial \tilde{y}}\right)$$
$$+ \epsilon \frac{\partial}{\partial \tilde{y}}\left(\tilde{N}_6 \frac{\partial \tilde{w}}{\partial \tilde{x}} + \tilde{N}_2 \frac{\partial \tilde{w}}{\partial \tilde{y}}\right) + \epsilon \tilde{I}_2 \frac{\partial}{\partial \tilde{t}}\left(\frac{\partial^2 \tilde{w}}{\partial \tilde{x}^2} + \frac{\partial^2 \tilde{w}}{\partial \tilde{y}^2}\right) = O(\epsilon^2) \quad (7.6.30\text{c})$$

where
$$\tilde{I}_0 = \frac{I_0}{\rho h}, \qquad \tilde{I}_2 = \frac{I_2}{\rho h^3}$$

Expressing (7.6.30) in terms of dimensional quantities and dropping the order of magnitude terms, we obtain

$$N_{1x} + N_{6y} = 0 \qquad (7.6.31\text{a})$$
$$N_{6x} + N_{2y} = 0 \qquad (7.6.31\text{b})$$
$$(N_1 w_x + N_6 w_y + M_{1x} + M_{6y})_x + (N_6 w_x + N_2 w_y + M_{6x} + M_{2y})_y$$
$$- I_0 \ddot{w} + I_2(\ddot{w}_{xx} + \ddot{w}_{yy}) = 0 \qquad (7.6.31\text{c})$$

which are the von Kármán equations (von Kármán, 1910). Equations (7.6.31) are the same as (7.4.8)-(7.4.10) if $I_1 = 0$ and $\ddot{u}$ and $\ddot{v}$ are neglected. The corresponding boundary conditions are of the form: specify

Along $x = 0, a$:

$\delta u = 0$ or N_1

$\delta v = 0$ or N_6

$\delta w = 0$ or $M_{1x} + 2M_{6y} + N_1 w_x + N_6 w_y + I_2 \ddot{w}_x$

$\delta w_x = 0$ or M_1

Along $y = 0, b$:

$\delta u = 0$ or N_6

$\delta v = 0$ or N_2

$\delta w = 0$ or $M_{2y} + 2M_{6x} + N_6 w_x + N_2 w_y + I_2 \ddot{w}_y$

$\delta w_y = 0$ or M_2

At $(x, y) = (0,0), (a,b), (a,0), (0,b)$:

$\delta w = 0$ or M_6 (7.6.32)

If the von Kármán-type geometric nonlinearity and the linear curvatures shown in (7.5.40) are used, we obtain the following strains:

$$\epsilon_{11} = u_x + \frac{1}{2}w_x^2 - z w_{xx}$$
$$\epsilon_{22} = v_y + \frac{1}{2}w_y^2 - z w_{yy}$$
$$\epsilon_{12} = u_y + v_x + w_x w_y - 2z w_{xy}$$
$$\epsilon_{33} = \epsilon_{13} = \epsilon_{23} = 0 \qquad (7.6.33)$$

Substituting (7.6.33) into (7.2.2b) yields

$$\delta\Pi = \int_A \int_z \left[\sigma_{11}(\delta u_x + w_x \delta w_x - z\delta w_{xx}) + \sigma_{22}(\delta v_y + w_y \delta w_y - z\delta w_{yy}) \right.$$
$$\left. + \sigma_{12}(\delta u_y + \delta v_x + w_x \delta w_y + w_y \delta w_x - 2z\delta w_{xy}) \right] dA dz$$
$$= \int_A \left[N_1(\delta u_x + w_x \delta w_x) + N_6(\delta u_y + \delta v_x + w_x \delta w_y + w_y \delta w_x) \right.$$
$$\left. + N_2(\delta v_y + w_y \delta w_y) - M_1 \delta w_{xx} - M_2 \delta w_{yy} - 2M_6 \delta w_{xy} \right] dA$$
$$= -\int_A \left\{ (N_{1x} + N_{6y}) \delta u + (N_{2y} + N_{6x}) \delta v \right.$$
$$+ \left[M_{1xx} + M_{2yy} + 2M_{6xy} + (N_1 w_x)_x + (N_2 w_y)_y \right.$$
$$\left. \left. + (N_6 w_x)_y + (N_6 w_y)_x \right] \delta w \right\} dA$$
$$+ \int_y \left[N_1 \delta u + N_6 \delta v + (M_{1x} + 2M_{6y} + N_1 w_x + N_6 w_y) \delta w - M_1 \delta w_x \right]_{x=0}^{x=a} dy$$
$$+ \int_x \left[N_6 \delta u + N_2 \delta v + (M_{2y} + 2M_{6x} + N_2 w_y + N_6 w_x) \delta w - M_2 \delta w_y \right]_{y=0}^{y=b} dx$$
$$- 2M_6 \delta w \Big|_{(x,y)=(a,0),(0,b)}^{(x,y)=(0,0),(a,b)} \tag{7.6.34}$$

Using (7.6.4) and (7.6.5) and neglecting $\ddot{u}$ and $\ddot{v}$, we obtain from (7.6.3) that

$$\delta T = -\int_A [I_0 \ddot{w} - I_2(\ddot{w}_{xx} + \ddot{w}_{yy})] \delta w dA - I_2 \ddot{w}_y \delta w \Big|_{y=0}^{y=b} - I_2 \ddot{w}_x \delta w \Big|_{x=0}^{x=a}$$
(7.6.35)

Consequently, (7.6.31) and (7.6.32) follow directly from (7.6.34) and (7.6.35).

Alternatively, the von Kármán equations and boundary conditions can be obtained from the general case by approximating the transformation matrix $[T]$ by

$$[T] = \begin{bmatrix} 1 & 0 & w_x \\ 0 & 1 & w_y \\ 0 & 0 & 1 \end{bmatrix} \tag{7.6.36}$$

Equation (7.6.36) shows that only linear terms in w are used to account for the configuration change. We note that the approximate matrix $[T]$ is not a unitary matrix (i.e., $[T]^{-1} = [T]^T$), as it should be. We also note that (7.6.36) is a special case of (7.5.18), (7.5.23), and (7.5.25). Using (7.6.36), (7.5.12), and (6.3.1a), we obtain

$$\delta \theta_1 = \mathbf{i}_3 \cdot \delta \mathbf{i}_2 = \delta w_y \tag{7.6.37a}$$

$$\delta\theta_2 = -\mathbf{i}_3 \cdot \delta\mathbf{i}_1 = -\delta w_x \qquad (7.6.37\text{b})$$

$$\delta\theta_3 = \frac{1}{2}(\mathbf{i}_2 \cdot \delta\mathbf{i}_1 - \mathbf{i}_1 \cdot \delta\mathbf{i}_2) = \frac{1}{2}w_y\delta w_x - \frac{1}{2}w_x\delta w_y \qquad (7.6.37\text{c})$$

In calculating $\delta\theta_3$, we used the average of the two expressions because the transformation (7.6.36) is not unitary and in particular $\mathbf{i}_1 \cdot \mathbf{i}_2 = w_x w_y \neq 0$. Using (7.6.36) in (7.5.16) yields the same k_1, k_2, and k_6 as those in (7.5.40b). However, k_4 and k_5 are non-zero. Substituting (7.6.36), (7.6.37a,b), (7.6.27), and (7.6.28) into (7.6.21) and assuming that $e_1 \approx e_2 \approx 0$, we obtain (7.6.31). Moreover, with these approximations, the boundary conditions (7.6.23) reduce to (7.6.32).

For isotropic materials, the M_i are related to the displacements by (7.2.24) and the N_i are related to the displacements by

$$\begin{Bmatrix} N_1 \\ N_2 \\ N_6 \end{Bmatrix} = \frac{Eh}{1-\nu^2} \begin{bmatrix} 1 & \nu & 0 \\ \nu & 1 & 0 \\ 0 & 0 & (1-\nu)/2 \end{bmatrix} \begin{Bmatrix} u_x + \frac{1}{2}w_x^2 \\ v_y + \frac{1}{2}w_y^2 \\ u_y + v_x + w_x w_y \end{Bmatrix} \qquad (7.6.38)$$

Substituting (7.2.24) and (7.6.38) into (7.6.31), we obtain the following equations:

$$u_{xx} + w_x w_{xx} + \nu(v_{xy} + w_y w_{xy})$$
$$+ \frac{1}{2}(1-\nu)(u_{yy} + v_{xy} + w_x w_{yy} + w_y w_{xy}) = 0 \quad (7.6.39)$$

$$v_{yy} + w_y w_{yy} + \nu(u_{xy} + w_x w_{xy})$$
$$+ \frac{1}{2}(1-\nu)(u_{xy} + v_{xx} + w_x w_{xy} + w_y w_{xx}) = 0 \quad (7.6.40)$$

$$D\nabla^4 w - \frac{Eh}{1-\nu^2}\left[u_x w_{xx} + \frac{1}{2}w_x^2 w_{xx} + v_y w_{yy} + \frac{1}{2}w_y^2 w_{yy} + \nu\left(v_y w_{xx}\right.\right.$$
$$\left.\left. + \frac{1}{2}w_y^2 w_{xx} + u_x w_{yy} + \frac{1}{2}w_x^2 w_{yy}\right)\right] - \frac{Eh}{1+\nu}(u_y + v_x + w_x w_y)w_{xy}$$
$$= -\rho h \ddot{w} + \frac{1}{12}\rho h^3 (\ddot{w}_{xx} + \ddot{w}_{yy}) \qquad (7.6.41)$$

7.7 NONLINEAR SHEAR-DEFORMABLE PLATE THEORY

We consider again the nonlinear motion of the rectangular plate shown in Figure 7.5.1, which has dimensions $a \times b \times h$. Here, we choose the middle plane as the reference plane, release the Kirchhoff hypothesis, and use the third-order shear function to describe the warping of cross sections. As shown in Figure 7.7.1, because of shear deformations, the deformed plate element is a crooked line segment and is not parallel to the ζ axis, which is perpendicular to the deformed middle plane (i.e., the $\xi\eta$ plane).

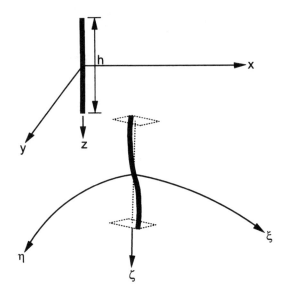

Fig. 7.7.1 A differential plate element with shear deformations and large displacements.

7.7.1 Equations of Motion

Considering the third-order shear-warping function (Bhimaraddi and Stevens, 1984; Reddy, 1984b) and (7.6.6a-c), we assume the local displacements u_1, u_2, and u_3 (with respect to the local $\xi\eta\zeta$ coordinate system) to have the form

$$u_1(x,y,z,t) = u_1^0(x,y,t) + z\theta_2(x,y,t) + g(z)\gamma_5(x,y,t) \tag{7.7.1a}$$

$$u_2(x,y,z,t) = u_2^0(x,y,t) - z\theta_1(x,y,t) + g(z)\gamma_4(x,y,t) \tag{7.7.1b}$$

$$u_3(x,y,z,t) = u_3^0(x,y,t) \tag{7.7.1c}$$

where

$$g(z) = z - \frac{4}{3h^2}z^3 \tag{7.7.1d}$$

is the so-called third-order shear-warping function, the u_i^0 ($i = 1, 2, 3$) are the local displacements (with respect to the local coordinate system) of a point which is located at $(x, y, 0)$ before deformation, γ_4 and γ_5 are the transverse shear-rotation angles at the midplane (i.e., $\gamma_5 \equiv \epsilon_{13}|_{\zeta=0}$ and $\gamma_4 \equiv \epsilon_{23}|_{\zeta=0}$, see Figure 7.3.2), and θ_1 and θ_2 are the rotation angles of the normal to the midplane with respect to the axes ξ and η, respectively. This displacement field accounts for a parabolic distribution of the transverse shear strains and satisfies the free shear-stress conditions $\sigma_{13}|_{\zeta=\pm h/2} = \sigma_{23}|_{\zeta=\pm h/2} = 0$.

NONLINEAR SHEAR-DEFORMABLE PLATE THEORY

Because $\xi\eta\zeta$ is a local coordinate system attached to the observed plate element and the $\xi\eta$ plane is tangent to the deformed reference-plane, we recall that

$$u_1^0 = u_2^0 = u_3^0 = \theta_1 = \theta_2 = \partial u_3^0/\partial x = \partial u_3^0/\partial y = 0 \quad (7.7.2)$$

and the local displacement vector is

$$\mathbf{U} = u_1 \mathbf{i}_1 + u_2 \mathbf{i}_2 + u_3 \mathbf{i}_3 \quad (7.7.3)$$

It follows from Figures 7.5.2 and 7.7.1 and (7.7.1a-c), (7.7.2), and (7.7.3) that the displacement vector $\mathbf{D}$ of an arbitrary point on the observed plate element is

$$\begin{aligned}\mathbf{D} &= u\mathbf{j}_1 + v\mathbf{j}_2 + w\mathbf{j}_3 + z\mathbf{i}_3 + \mathbf{U} - z\mathbf{j}_3 \\ &= u\mathbf{j}_1 + v\mathbf{j}_2 + w\mathbf{j}_3 + z\mathbf{i}_3 + g\gamma_5\mathbf{i}_1 + g\gamma_4\mathbf{i}_2 - z\mathbf{j}_3\end{aligned} \quad (7.7.4)$$

Taking the variation of $\mathbf{D}$ and using (6.3.1a) we obtain

$$\begin{aligned}\delta\mathbf{D} &= \mathbf{j}_1\delta u + \mathbf{j}_2\delta v + \mathbf{j}_3\delta w + z(\mathbf{i}_1\delta\theta_2 - \mathbf{i}_2\delta\theta_1) + g\gamma_5(\mathbf{i}_2\delta\theta_3 - \mathbf{i}_3\delta\theta_2) \\ &\quad + g\gamma_4(\mathbf{i}_3\delta\theta_1 - \mathbf{i}_1\delta\theta_3) + g\delta\gamma_5\mathbf{i}_1 + g\delta\gamma_4\mathbf{i}_2\end{aligned} \quad (7.7.5a)$$

Differentiating $\mathbf{D}$ twice with respect to t yields

$$\ddot{\mathbf{D}} = \ddot{u}\mathbf{j}_1 + \ddot{v}\mathbf{j}_2 + \ddot{w}\mathbf{j}_3 + z\ddot{\mathbf{i}}_3 + g\left(\ddot{\gamma}_5\mathbf{i}_1 + 2\dot{\gamma}_5\dot{\mathbf{i}}_1 + \gamma_5\ddot{\mathbf{i}}_1 + \ddot{\gamma}_4\mathbf{i}_2 + 2\dot{\gamma}_4\dot{\mathbf{i}}_2 + \gamma_4\ddot{\mathbf{i}}_2\right) \quad (7.7.5b)$$

Substituting (7.7.5a,b) into (7.2.2a) and using (7.5.12), we obtain

$$\delta T = -\int_A \left(A_u\delta u + A_v\delta v + A_w\delta w + A_{\gamma_4}\delta\gamma_4 + A_{\gamma_5}\delta\gamma_5 \right.\\ \left. + A_{\theta_1}\delta\theta_1 + A_{\theta_2}\delta\theta_2 + A_{\theta_3}\delta\theta_3\right)dA \quad (7.7.6)$$

where

$$\begin{aligned}A_u &= I_0\ddot{u} + I_4(\gamma_5 T_{11})\ddot{} + I_4(\gamma_4 T_{21})\ddot{} + I_1\ddot{T}_{31} \\ A_v &= I_0\ddot{v} + I_4(\gamma_5 T_{12})\ddot{} + I_4(\gamma_4 T_{22})\ddot{} + I_1\ddot{T}_{32} \\ A_w &= I_0\ddot{w} + I_4(\gamma_5 T_{13})\ddot{} + I_4(\gamma_4 T_{23})\ddot{} + I_1\ddot{T}_{33} \\ A_{\gamma_4} &= I_4(\ddot{u}T_{21} + \ddot{v}T_{22} + \ddot{w}T_{23}) + I_7 f_{41} + I_7 f_{42} + I_5 f_{43} \\ A_{\gamma_5} &= I_4(\ddot{u}T_{11} + \ddot{v}T_{12} + \ddot{w}T_{13}) + I_7 f_{51} + I_7 f_{52} + I_5 f_{53} \\ A_{\theta_1} &= I_4\gamma_4(\ddot{u}T_{31} + \ddot{v}T_{32} + \ddot{w}T_{33}) + I_7\gamma_4 f_{11} + I_7\gamma_4 f_{12} + I_5\gamma_4 f_{13} \\ &\quad - I_1(\ddot{u}T_{21} + \ddot{v}T_{22} + \ddot{w}T_{23}) - I_5 f_{41} - I_5 f_{42} - I_2 f_{43} \\ A_{\theta_2} &= I_1(\ddot{u}T_{11} + \ddot{v}T_{12} + \ddot{w}T_{13}) + I_5 f_{51} + I_5 f_{52} + I_2 f_{53} \\ &\quad - I_4\gamma_5(\ddot{u}T_{31} + \ddot{v}T_{32} + \ddot{w}T_{33}) - I_7\gamma_5 f_{11} - I_7\gamma_5 f_{12} - I_5\gamma_5 f_{13} \\ A_{\theta_3} &= I_4\gamma_5(\ddot{u}T_{21} + \ddot{v}T_{22} + \ddot{w}T_{23}) + I_7\gamma_5 f_{41} + I_7\gamma_5 f_{42} + I_5\gamma_5 f_{43} \\ &\quad - I_4\gamma_4(\ddot{u}T_{11} + \ddot{v}T_{12} + \ddot{w}T_{13}) - I_7\gamma_4 f_{51} - I_7\gamma_4 f_{52} - I_5\gamma_4 f_{53} \quad (7.7.7a)\end{aligned}$$

the mass inertias I_i are defined as

$$[I_0, I_1, I_2, I_4, I_5, I_7]$$
$$\equiv \int_{-h/2}^{h/2} \rho \left[1, z, z^2, z - \frac{4z^3}{3h^2}, z(z - \frac{4z^3}{3h^2}), \left(z - \frac{4z^3}{3h^2}\right)^2\right] dz \quad (7.7.7b)$$

and

$$f_{11} \equiv (\gamma_5 T_{11})\ddot{T}_{31} + (\gamma_5 T_{12})\ddot{T}_{32} + (\gamma_5 T_{13})\ddot{T}_{33}$$
$$f_{12} \equiv (\gamma_4 T_{21})\ddot{T}_{31} + (\gamma_4 T_{22})\ddot{T}_{32} + (\gamma_4 T_{23})\ddot{T}_{33}$$
$$f_{13} \equiv \ddot{T}_{31}T_{31} + \ddot{T}_{32}T_{32} + \ddot{T}_{33}T_{33}$$
$$f_{41} \equiv (\gamma_5 T_{11})\ddot{T}_{21} + (\gamma_5 T_{12})\ddot{T}_{22} + (\gamma_5 T_{13})\ddot{T}_{23}$$
$$f_{42} \equiv (\gamma_4 T_{21})\ddot{T}_{21} + (\gamma_4 T_{22})\ddot{T}_{22} + (\gamma_4 T_{23})\ddot{T}_{23}$$
$$f_{43} \equiv \ddot{T}_{31}T_{21} + \ddot{T}_{32}T_{22} + \ddot{T}_{33}T_{23}$$
$$f_{51} \equiv (\gamma_5 T_{11})\ddot{T}_{11} + (\gamma_5 T_{12})\ddot{T}_{12} + (\gamma_5 T_{13})\ddot{T}_{13}$$
$$f_{52} \equiv (\gamma_4 T_{21})\ddot{T}_{11} + (\gamma_4 T_{22})\ddot{T}_{12} + (\gamma_4 T_{23})\ddot{T}_{13}$$
$$f_{53} \equiv \ddot{T}_{31}T_{11} + \ddot{T}_{32}T_{12} + \ddot{T}_{33}T_{13} \quad (7.7.7c)$$

We note that $I_1 = I_4 = 0$ if ρ is constant.

For a thin plate, because the total thickness h is small, I_1, I_2, I_4, I_5, and I_7 are small and can be assumed to be of order ϵ^2. If u, v, w, γ_4, and γ_5 are assumed to be of order ϵ, one can use (7.5.18), (7.5.23), and (7.5.25) to expand the inertia terms in (7.7.7a) up to order ϵ^3 and obtain

$$A_u = I_0\ddot{u} - I_1\ddot{w}_x + I_4\ddot{\gamma}_5, \quad A_{\theta_1} = -I_1\ddot{v} + I_2\ddot{w}_y - I_5\ddot{\gamma}_4$$
$$A_v = I_0\ddot{v} - I_1\ddot{w}_y + I_4\ddot{\gamma}_4, \quad A_{\theta_2} = I_1\ddot{u} - I_2\ddot{w}_x + I_5\ddot{\gamma}_5$$
$$A_w = I_0\ddot{w}, \quad A_{\theta_3} = 0$$
$$A_{\gamma_4} = I_4\ddot{v} - I_5\ddot{w}_y + I_7\ddot{\gamma}_4$$
$$A_{\gamma_5} = I_4\ddot{u} - I_5\ddot{w}_x + I_7\ddot{\gamma}_5 \quad (7.7.8)$$

We note from (7.7.7a) that A_{θ_3}, the rotary inertia of the plate element with respect to the ζ axis, is a nonlinear term due to shear warpings. According to the ordering scheme, A_{θ_3} is small and of order ϵ^4 or higher and is negligible. Equation (7.7.8) shows that there are no nonlinear inertia terms of order ϵ^3 or lower.

Taking the derivatives of **U** in (7.7.3) and using (7.7.1a-c), (7.7.2), (7.6.11a,b), (7.5.14), and (7.5.15), we obtain

$$\frac{\partial \mathbf{U}}{\partial x} = \frac{\partial u_1}{\partial x}\mathbf{i}_1 + \frac{\partial u_2}{\partial x}\mathbf{i}_2 + \frac{\partial u_3}{\partial x}\mathbf{i}_3 + u_1\frac{\partial \mathbf{i}_1}{\partial x} + u_2\frac{\partial \mathbf{i}_2}{\partial x} + u_3\frac{\partial \mathbf{i}_3}{\partial x}$$
$$= (e_1 + zk_1 + g\gamma_{5x})\mathbf{i}_1 + \left(\frac{\partial u_2^0}{\partial x} + zk_{61} + g\gamma_{4x}\right)\mathbf{i}_2$$
$$+ g(-k_1\gamma_5\mathbf{i}_3 + k_5\gamma_5\mathbf{i}_2) - g(k_{61}\gamma_4\mathbf{i}_3 + k_5\gamma_4\mathbf{i}_1) \quad (7.7.9a)$$

$$\frac{\partial \mathbf{U}}{\partial y} = \frac{\partial u_1}{\partial y}\mathbf{i}_1 + \frac{\partial u_2}{\partial y}\mathbf{i}_2 + \frac{\partial u_3}{\partial y}\mathbf{i}_3 + u_1\frac{\partial \mathbf{i}_1}{\partial y} + u_2\frac{\partial \mathbf{i}_2}{\partial y} + u_3\frac{\partial \mathbf{i}_3}{\partial y}$$

$$= \left(\frac{\partial u_1^0}{\partial y} + zk_{62} + g\gamma_{5y}\right)\mathbf{i}_1 + (e_2 + zk_2 + g\gamma_{4y})\mathbf{i}_2$$

$$+ g(-k_{62}\gamma_5\mathbf{i}_3 + k_4\gamma_5\mathbf{i}_2) - g(k_2\gamma_4\mathbf{i}_3 + k_4\gamma_4\mathbf{i}_1) \quad (7.7.9\text{b})$$

$$\frac{\partial \mathbf{U}}{\partial z} = \frac{\partial u_1}{\partial z}\mathbf{i}_1 + \frac{\partial u_2}{\partial z}\mathbf{i}_2 + \frac{\partial u_3}{\partial z}\mathbf{i}_3$$

$$= \left(1 - \frac{4z^2}{h^2}\right)\gamma_5\mathbf{i}_1 + \left(1 - \frac{4z^2}{h^2}\right)\gamma_4\mathbf{i}_2 \quad (7.7.9\text{c})$$

Hence, the Jaumann strains are obtained as

$$\epsilon_{11} \equiv \frac{\partial \mathbf{U}}{\partial x} \cdot \mathbf{i}_1 = e_1 + zk_1 + g(\gamma_{5x} - k_5\gamma_4) \quad (7.7.10\text{a})$$

$$\epsilon_{22} \equiv \frac{\partial \mathbf{U}}{\partial y} \cdot \mathbf{i}_2 = e_2 + zk_2 + g(\gamma_{4y} + k_4\gamma_5) \quad (7.7.10\text{b})$$

$$\epsilon_{12} \equiv \frac{\partial \mathbf{U}}{\partial x} \cdot \mathbf{i}_2 + \frac{\partial \mathbf{U}}{\partial y} \cdot \mathbf{i}_1 = \gamma_6 + zk_6 + g(\gamma_{5y} + \gamma_{4x} + \gamma_5 k_5 - \gamma_4 k_4) \quad (7.7.10\text{c})$$

$$\epsilon_{13} \equiv \frac{\partial \mathbf{U}}{\partial x} \cdot \mathbf{i}_3 + \frac{\partial \mathbf{U}}{\partial z} \cdot \mathbf{i}_1 = \left(1 - \frac{4z^2}{h^2}\right)\gamma_5 - g(\gamma_5 k_1 + \gamma_4 k_{61}) \quad (7.7.10\text{d})$$

$$\epsilon_{23} \equiv \frac{\partial \mathbf{U}}{\partial y} \cdot \mathbf{i}_3 + \frac{\partial \mathbf{U}}{\partial z} \cdot \mathbf{i}_2 = \left(1 - \frac{4z^2}{h^2}\right)\gamma_4 - g(\gamma_4 k_2 + \gamma_5 k_{62}) \quad (7.7.10\text{e})$$

$$\epsilon_{33} \equiv \frac{\partial \mathbf{U}}{\partial z} \cdot \mathbf{i}_3 = 0 \quad (7.7.10\text{f})$$

To reduce the three-dimensional problem to a two-dimensional one, we introduce the following stress resultants:

$$(N_1, N_2, N_6, Q_1, Q_2) \equiv \int_{-h/2}^{h/2} (\sigma_{11}, \sigma_{22}, \sigma_{12}, \sigma_{13}, \sigma_{23})\, dz \quad (7.7.11\text{a})$$

$$\left(\hat{R}_1, \hat{R}_2\right) \equiv \int_{-h/2}^{h/2} z^2(\sigma_{13}, \sigma_{23})\, dz, \quad (R_1, R_2) \equiv \frac{4}{h^2}\left(\hat{R}_1, \hat{R}_2\right) \quad (7.7.11\text{b})$$

$$(M_1, M_2, M_6) \equiv \int_{-h/2}^{h/2} z(\sigma_{11}, \sigma_{22}, \sigma_{12})\, dz \quad (7.7.11\text{c})$$

$$\left(\hat{P}_1, \hat{P}_2, \hat{P}_6\right) \equiv \int_{-h/2}^{h/2} z^3(\sigma_{11}, \sigma_{22}, \sigma_{12})\, dz, \quad (P_1, P_2, P_6) \equiv \frac{4}{3h^2}\left(\hat{P}_1, \hat{P}_2, \hat{P}_6\right)$$

$$(7.7.11\text{d})$$

$$(S_1, S_2) \equiv \int_{-h/2}^{h/2} z(\sigma_{13}, \sigma_{23})\, dz \quad (7.7.11\text{e})$$

$$\left(\hat{G}_1, \hat{G}_2\right) \equiv \int_{-h/2}^{h/2} z^3 \left(\sigma_{13}, \sigma_{23}\right) dz, \quad (G_1, G_2) \equiv \frac{4}{3h^2}\left(\hat{G}_1, \hat{G}_2\right) \quad (7.7.11\text{f})$$

Here, N_1, N_2, and N_6 are membrane force intensities (i.e., inplane extension and shearing forces); Q_1 and Q_2 are transverse shearing force intensities; R_1 and R_2 are higher-order transverse shearing force intensities associated with Q_1 and Q_2, respectively; M_1 and M_2 are bending moment intensities; M_6 is a twisting moment intensity; P_6 is a higher-order twisting moment intensity associated with M_6; P_1 and P_2 are higher-order bending moment intensities associated with M_1 and M_2, respectively; S_1 and S_2 are moment intensities; G_1 and G_2 are higher-order moment intensities associated with S_1 and S_2, respectively; and the S_i and G_i are usually not mentioned in the literature and they only appear as nonlinear terms in the equations of motion, as shown later. Using $\sigma_{33}=0$ and (7.7.10a-e), (2.4.34), and (2.4.35), we obtain

$$\begin{Bmatrix} \sigma_{23} \\ \sigma_{13} \end{Bmatrix} = \begin{bmatrix} \overline{Q}_{44} & \overline{Q}_{45} \\ \overline{Q}_{45} & \overline{Q}_{55} \end{bmatrix}$$
$$\times \left(\begin{Bmatrix} \gamma_4 \\ \gamma_5 \end{Bmatrix} - \frac{4z^2}{h^2} \begin{Bmatrix} \gamma_4 \\ \gamma_5 \end{Bmatrix} - \left(z - \frac{4z^3}{3h^2}\right) \begin{Bmatrix} \gamma_4 k_2 + \gamma_5 k_{62} \\ \gamma_5 k_1 + \gamma_4 k_{61} \end{Bmatrix} \right) \quad (7.7.12\text{a})$$

$$\begin{Bmatrix} \sigma_{11} \\ \sigma_{22} \\ \sigma_{12} \end{Bmatrix} = \begin{bmatrix} \tilde{Q}_{11} & \tilde{Q}_{12} & \tilde{Q}_{16} \\ \tilde{Q}_{12} & \tilde{Q}_{22} & \tilde{Q}_{26} \\ \tilde{Q}_{16} & \tilde{Q}_{26} & \tilde{Q}_{66} \end{bmatrix} \left(\begin{Bmatrix} e_1 \\ e_2 \\ \gamma_6 \end{Bmatrix} + z \right.$$
$$\times \left. \begin{Bmatrix} k_1 + \gamma_{5x} - k_5\gamma_4 \\ k_2 + \gamma_{4y} + k_4\gamma_5 \\ k_6 + \gamma_{4x} + \gamma_{5y} + k_5\gamma_5 - k_4\gamma_4 \end{Bmatrix} - \frac{4z^3}{3h^2} \begin{Bmatrix} \gamma_{5x} - k_5\gamma_4 \\ \gamma_{4y} + k_4\gamma_5 \\ \gamma_{4x} + \gamma_{5y} + k_5\gamma_5 - k_4\gamma_4 \end{Bmatrix} \right) \quad (7.7.12\text{b})$$

Substituting (7.7.12a,b) into (7.7.11a-f), we obtain the internal force and moment resultants in terms of the midplane strains and curvatures as

$$\begin{Bmatrix} N_1 \\ N_2 \\ N_6 \\ M_1 \\ M_2 \\ M_6 \\ \hat{P}_1 \\ \hat{P}_2 \\ \hat{P}_6 \end{Bmatrix} = \begin{bmatrix} [A_{ij}] & [B_{ij}] & [E_{ij}] \\ [B_{ij}] & [D_{ij}] & [F_{ij}] \\ [E_{ij}] & [F_{ij}] & [H_{ij}] \end{bmatrix} \begin{Bmatrix} e_1 \\ e_2 \\ \gamma_6 \\ k_1 + \gamma_{5x} - k_5\gamma_4 \\ k_2 + \gamma_{4y} + k_4\gamma_5 \\ k_6 + \gamma_{5y} + \gamma_{4x} - k_4\gamma_4 + k_5\gamma_5 \\ -4(\gamma_{5x} - k_5\gamma_4)/3h^2 \\ -4(\gamma_{4y} + k_4\gamma_5)/3h^2 \\ -4(\gamma_{5y} + \gamma_{4x} - k_4\gamma_4 + k_5\gamma_5)/3h^2 \end{Bmatrix} \quad (7.7.13)$$

where $i, j = 1, 2, 6$; and

$$\begin{Bmatrix} Q_2 \\ Q_1 \\ \hat{R}_2 \\ \hat{R}_1 \end{Bmatrix} = \begin{bmatrix} [A_{ij}] & [D_{ij}] \\ [D_{ij}] & [F_{ij}] \end{bmatrix} \begin{Bmatrix} \gamma_4 \\ \gamma_5 \\ -4\gamma_4/h^2 \\ -4\gamma_5/h^2 \end{Bmatrix}$$

$$+ \begin{bmatrix} [B_{ij}] & [E_{ij}] \\ [E_{ij}] & [C_{ij}] \end{bmatrix} \begin{Bmatrix} -\gamma_4 k_2 - \gamma_5 k_{62} \\ -\gamma_5 k_1 - \gamma_4 k_{61} \\ 4(\gamma_4 k_2 + \gamma_5 k_{62})/3h^2 \\ 4(\gamma_5 k_1 + \gamma_4 k_{61})/3h^2 \end{Bmatrix} \quad (7.7.14)$$

$$\begin{Bmatrix} S_2 \\ S_1 \\ \hat{G}_2 \\ \hat{G}_1 \end{Bmatrix} = \begin{bmatrix} [B_{ij}] & [E_{ij}] \\ [E_{ij}] & [C_{ij}] \end{bmatrix} \begin{Bmatrix} \gamma_4 \\ \gamma_5 \\ -4\gamma_4/h^2 \\ -4\gamma_5/h^2 \end{Bmatrix}$$

$$+ \begin{bmatrix} [D_{ij}] & [F_{ij}] \\ [F_{ij}] & [H_{ij}] \end{bmatrix} \begin{Bmatrix} -\gamma_4 k_2 - \gamma_5 k_{62} \\ -\gamma_5 k_1 - \gamma_4 k_{61} \\ 4(\gamma_4 k_2 + \gamma_5 k_{62})/3h^2 \\ 4(\gamma_5 k_1 + \gamma_4 k_{61})/3h^2 \end{Bmatrix} \quad (7.7.15)$$

where $i, j = 4, 5$. The stiffness matrices are symmetric and their components are given by

$$(A_{ij}, B_{ij}, D_{ij}, E_{ij}, F_{ij}, C_{ij}, H_{ij})$$
$$= \sum_{k=1}^{N} \int_{z_k}^{z_{k+1}} \tilde{Q}_{ij}^{(k)} \left(1, z, z^2, z^3, z^4, z^5, z^6\right) dz \quad (7.7.16)$$

for $i, j = 1, 2, 4, 5, 6$. For $i, j = 4, 5$, use $\overline{Q}_{ij}^{(k)}$ instead of $\tilde{Q}_{ij}^{(k)}$ in (7.7.16).

Substituting (7.7.10a-e) into (7.2.2b), using (7.7.11a-f) and (6.4.2a,b), and integrating by parts, we obtain the variation of the elastic energy in terms of the stress resultants and midplane curvatures and strains as

$$\delta \Pi = \int_A \Big(N_1 \delta e_1 + N_2 \delta e_2 + N_6 \delta \gamma_6 + \Theta_1 \delta \theta_1 + \Theta_2 \delta \theta_2 + \Theta_3 \delta \theta_3$$
$$+ \Gamma_4 \delta \gamma_4 + \Gamma_5 \delta \gamma_5 \Big) dx dy$$
$$+ \int_x \bigg[-\hat{M}_2 \delta \theta_1 + \hat{M}_{62} \delta \theta_2 + m_{32} \delta \theta_3 + (M_2 - P_2) \delta \gamma_4$$
$$+ (M_6 - P_6) \delta \gamma_5 \bigg]_{y=0}^{y=b} dx$$
$$+ \int_y \bigg[-\hat{M}_{61} \delta \theta_1 + \hat{M}_1 \delta \theta_2 + m_{31} \delta \theta_3 + (M_6 - P_6) \delta \gamma_4$$
$$+ (M_1 - P_1) \delta \gamma_5 \bigg]_{x=0}^{x=a} dy \quad (7.7.17)$$

where

$$\hat{M}_1 \equiv M_1 - (S_1 - G_1)\gamma_5 \quad (7.7.18a)$$

$$\hat{M}_2 \equiv M_2 - (S_2 - G_2)\gamma_4 \quad (7.7.18b)$$

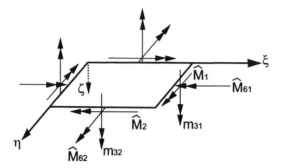

Fig. 7.7.2 Moment resultants on a differential shear-deformable plate element.

$$\hat{M}_{61} \equiv M_6 - (S_1 - G_1)\gamma_4 \qquad (7.7.18c)$$

$$\hat{M}_{62} \equiv M_6 - (S_2 - G_2)\gamma_5 \qquad (7.7.18d)$$

$$m_{31} \equiv (M_6 - P_6)\gamma_5 - (M_1 - P_1)\gamma_4 \qquad (7.7.18e)$$

$$m_{32} \equiv -(M_6 - P_6)\gamma_4 + (M_2 - P_2)\gamma_5 \qquad (7.7.18f)$$

$$\Theta_1 \equiv \hat{M}_{61x} + \hat{M}_{2y} - m_{31}k_1 - m_{32}k_{62} + \hat{M}_1 k_5 + \hat{M}_{62} k_4 \qquad (7.7.18g)$$

$$\Theta_2 \equiv -\hat{M}_{1x} - \hat{M}_{62y} - m_{31}k_{61} - m_{32}k_2 + \hat{M}_2 k_4 + \hat{M}_{61} k_5 \qquad (7.7.18h)$$

$$\Theta_3 \equiv -m_{31x} - m_{32y} + \hat{M}_1 k_{61} - \hat{M}_2 k_{62} + \hat{M}_{62} k_2 - \hat{M}_{61} k_1 \qquad (7.7.18i)$$

$$\Gamma_4 \equiv -(M_2 - P_2)_y - (M_6 - P_6)_x + Q_2 - R_2 - (S_1 - G_1)k_{61}$$
$$-(S_2 - G_2)k_2 - (M_1 - P_1)k_5 - (M_6 - P_6)k_4 \qquad (7.7.18j)$$

$$\Gamma_5 \equiv -(M_1 - P_1)_x - (M_6 - P_6)_y + Q_1 - R_1 - (S_1 - G_1)k_1$$
$$-(S_2 - G_2)k_{62} + (M_2 - P_2)k_4 + (M_6 - P_6)k_5 \qquad (7.7.18k)$$

Here, $\hat{M}_1$, $\hat{M}_2$, $\hat{M}_{61}$, $\hat{M}_{62}$, m_{31}, and m_{32} represent the total moment intensities acting on the edges of the plate element, as shown in Figure 7.7.2. The inplane extension force intensities N_1 and N_2, inplane shear intensity N_6, and transverse shear intensities Q_1 and Q_2 are the same as those in Figure 7.6.1b. Equations (7.7.18a-k) show that m_{31}, m_{32}, and Θ_3 are nonlinear and S_1, S_2, G_1, G_2, k_4, and k_5 only appear in nonlinear terms.

Next, we assume that $\gamma_6 = 0$ and hence (7.5.11) is valid. Substituting (7.5.42a,b) and (7.5.46) and adding the integrals of $Q_1 \times$ (7.5.44a) and $Q_2 \times$ (7.5.44b) into

(7.7.17), we obtain

$$\delta\Pi = \int_R \bigg\{ \{N_1,\ N_6/(1+e_1),\ Q_1\}[T] \left\{ \begin{array}{c} \delta u_x \\ \delta v_x \\ \delta w_x \end{array} \right\}$$

$$+ \{N_6/(1+e_2),\ N_2,\ Q_2\}[T] \left\{ \begin{array}{c} \delta u_y \\ \delta v_y \\ \delta w_y \end{array} \right\}$$

$$+ [\Theta_1 - Q_2(1+e_2)]\delta\theta_1 + [\Theta_2 + Q_1(1+e_1)]\delta\theta_2 + \Theta_3\delta\theta_3$$

$$+ \Gamma_4\delta\gamma_4 + \Gamma_5\delta\gamma_5 \bigg\} dxdy$$

$$+ \int_x \bigg[-\hat{M}_2\delta\theta_1 + \hat{M}_{62}\delta\theta_2 + m_{32}\delta\theta_3$$

$$+ (M_2 - P_2)\delta\gamma_4 + (M_6 - P_6)\delta\gamma_5 \bigg]_{y=0}^{y=b} dx$$

$$+ \int_y \bigg[-\hat{M}_{61}\delta\theta_1 + \hat{M}_1\delta\theta_2 + m_{31}\delta\theta_3$$

$$+ (M_6 - P_6)\delta\gamma_4 + (M_1 - P_1)\delta\gamma_5 \bigg]_{x=0}^{x=a} dy \qquad (7.7.19)$$

Substituting (7.7.19) and (7.7.6) into (7.2.1), using (7.7.18g-i) and (7.5.44a,b), and integrating by parts, we have

$$0 = \int_A \bigg[\left(\frac{\partial \left(\{F_\alpha\}^T[T]\right)}{\partial x} + \frac{\partial \left(\{F_\beta\}^T[T]\right)}{\partial y} - \{I_F\}^T \right) \left\{ \begin{array}{c} \delta u \\ \delta v \\ \delta w \end{array} \right\}$$

$$+ \left(\frac{\partial \{M_\alpha\}^T}{\partial x} + \{M_\alpha\}^T[K_1] + \frac{\partial \{M_\beta\}^T}{\partial y} + \{M_\beta\}^T[K_2] \right.$$

$$+ \{Q_\alpha\}^T + \{Q_\beta\}^T - \{I_M\}^T \bigg) \left\{ \begin{array}{c} \delta\theta_1 \\ \delta\theta_2 \\ \delta\theta_3 \end{array} \right\}$$

$$- (\Gamma_4 + A_{\gamma_4})\delta\gamma_4 - (\Gamma_5 + A_{\gamma_5})\delta\gamma_5 \bigg] dxdy$$

$$- \int_x \bigg[\{F_\beta\}^T[T] \left\{ \begin{array}{c} \delta u \\ \delta v \\ \delta w \end{array} \right\} + \left(\frac{\hat{M}_{62}T_{31}}{1+e_1} \right)_x \delta u + \left(\frac{\hat{M}_{62}T_{32}}{1+e_1} \right)_x \delta v$$

$$+ \left(\frac{\hat{M}_{62}T_{33}}{1+e_1} \right)_x \delta w - \hat{M}_2\delta\theta_1 + m_{32}\delta\theta_3$$

$$+ (M_2 - P_2)\delta\gamma_4 + (M_6 - P_6)\delta\gamma_5 \bigg]_{y=0}^{y=b} dx$$

$$-\int_y \left[\{F_\alpha\}^T [T] \begin{Bmatrix} \delta u \\ \delta v \\ \delta w \end{Bmatrix} + \left(\frac{\hat{M}_{61} T_{31}}{1+e_2} \right)_y \delta u + \left(\frac{\hat{M}_{61} T_{32}}{1+e_2} \right)_y \delta v \right.$$

$$+ \left(\frac{\hat{M}_{61} T_{33}}{1+e_2} \right)_y \delta w + \hat{M}_1 \delta \theta_2 + m_{31} \delta \theta_3$$

$$\left. + (M_6 - P_6)\delta\gamma_4 + (M_1 - P_1)\delta\gamma_5 \right]_{x=0}^{x=a} dy$$

$$+ \left[\left(\frac{\hat{M}_{62}}{1+e_1} + \frac{\hat{M}_{61}}{1+e_2} \right) (T_{31}\delta u + T_{32}\delta v + T_{33}\delta w) \right]_{(x,y)=(a,0),(0,b)}^{(x,y)=(0,0),(a,b)}$$

(7.7.20)

where

$$\{F_\alpha\} \equiv \{N_1, N_6/(1+e_1), Q_1\}^T, \quad \{F_\beta\} \equiv \{N_6/(1+e_2), N_2, Q_2\}^T$$
$$\{M_\alpha\} \equiv \{-\hat{M}_{61}, \hat{M}_1, m_{31}\}^T, \quad \{M_\beta\} \equiv \{-\hat{M}_2, \hat{M}_{62}, m_{32}\}^T$$
$$\{Q_\alpha\} \equiv \{0, -Q_1(1+e_1), N_6\}^T, \quad \{Q_\beta\} \equiv \{Q_2(1+e_2), 0, -N_6\}^T$$
$$\{I_F\} \equiv \{A_u, A_v, A_w\}^T, \quad \{I_M\} \equiv \{A_{\theta_1}, A_{\theta_2}, A_{\theta_3}\}^T \quad (7.7.21)$$

Setting each of the coefficients of δu, δv, δw, $\delta\gamma_4$, $\delta\gamma_5$, $\delta\theta_1$, $\delta\theta_2$, and $\delta\theta_3$ equal to zero and using (7.7.18j,k) yields

$$\frac{\partial}{\partial x} \left([T]^T \{F_\alpha\} \right) + \frac{\partial}{\partial y} \left([T]^T \{F_\beta\} \right) = \{I_F\} \quad (7.7.22)$$

$$(M_2 - P_2)_y + (M_6 - P_6)_x - Q_2 + R_2 + (S_1 - G_1)k_{61} + (S_2 - G_2)k_2$$
$$+ (M_1 - P_1)k_5 + (M_6 - P_6)k_4 = A_{\gamma_4} \quad (7.7.23)$$

$$(M_1 - P_1)_x + (M_6 - P_6)_y - Q_1 + R_1 + (S_1 - G_1)k_1 + (S_2 - G_2)k_{62}$$
$$- (M_2 - P_2)k_4 - (M_6 - P_6)k_5 = A_{\gamma_5} \quad (7.7.24)$$

$$\frac{\partial \{M_\alpha\}}{\partial x} + [K_1]^T \{M_\alpha\} + \frac{\partial \{M_\beta\}}{\partial y} + [K_2]^T \{M_\beta\} + \{Q_\alpha\} + \{Q_\beta\} = \{I_M\}$$
(7.7.25)

We note that (7.7.22) and (7.7.25) are the same as (7.6.21) and (7.6.22) except that M_1, M_2, and M_6 are replaced by $\hat{M}_1$, $\hat{M}_2$, and $\hat{M}_{61}$ (or $\hat{M}_{62}$), where the difference is due to transverse shear deformations. Moreover, there are no m_{31} and m_{32} in the classical plate theory. However, (7.7.23) and (7.7.24) are statements of the balance of moments when there are shearing rotations γ_4 and γ_5 with respect to the deformed midplane. These two equations of motion cannot be directly derived using the vector approach. The boundary conditions are obtained from (7.7.20) to be: specify

Along $x = 0, a$:

$$\delta u = 0 \quad \text{or} \quad \mathbf{F}_\alpha \cdot \mathbf{j}_1 + \left[\hat{M}_{61} T_{31}/(1+e_2) \right]_y$$

$\delta v = 0$	or	$\mathbf{F}_\alpha \cdot \mathbf{j}_2 + \left[\hat{M}_{61}T_{32}/(1+e_2)\right]_y$
$\delta w = 0$	or	$\mathbf{F}_\alpha \cdot \mathbf{j}_3 + \left[\hat{M}_{61}T_{33}/(1+e_2)\right]_y$
$\delta\theta_2 = 0$	or	$\hat{M}_1$
$\delta\gamma_4 = 0$	or	$M_6 - P_6$
$\delta\gamma_5 = 0$	or	$M_1 - P_1$
$\delta\theta_3 = 0$	or	m_{31}

Along $y = 0, b$:

$\delta u = 0$	or	$\mathbf{F}_\beta \cdot \mathbf{j}_1 + \left[\hat{M}_{62}T_{31}/(1+e_1)\right]_x$
$\delta v = 0$	or	$\mathbf{F}_\beta \cdot \mathbf{j}_2 + \left[\hat{M}_{62}T_{32}/(1+e_1)\right]_x$
$\delta w = 0$	or	$\mathbf{F}_\beta \cdot \mathbf{j}_3 + \left[\hat{M}_{62}T_{33}/(1+e_1)\right]_x$
$\delta\theta_1 = 0$	or	$\hat{M}_2$
$\delta\gamma_4 = 0$	or	$M_2 - P_2$
$\delta\gamma_5 = 0$	or	$M_6 - P_6$
$\delta\theta_3 = 0$	or	m_{32}

At $(x, y) = (0, 0), (a, b), (a, 0), (0, b)$:

$\delta u = 0$	or	$\left[\hat{M}_{62}/(1+e_1) + \hat{M}_{61}/(1+e_2)\right]T_{31}$
$\delta v = 0$	or	$\left[\hat{M}_{62}/(1+e_1) + \hat{M}_{61}/(1+e_2)\right]T_{32}$
$\delta w = 0$	or	$\left[\hat{M}_{62}/(1+e_1) + \hat{M}_{61}/(1+e_2)\right]T_{33}$ (7.7.26)

We note that (7.7.25) states that

$$Q_2 = \frac{1}{1+e_2}(\Theta_1 + A_{\theta_1}) \tag{7.7.27a}$$

$$Q_1 = \frac{-1}{1+e_1}(\Theta_2 + A_{\theta_2}) \tag{7.7.27b}$$

$$\Theta_3 + A_{\theta_3} = 0 \tag{7.7.27c}$$

Because A_{θ_3} is due to nonlinear effects as shown in (7.7.7a) and is usually negligible (see (7.7.8)), (7.7.27c) is a statement of the balance of internal moments about the ζ axis. Hence (7.7.27c) and the corresponding boundary conditions on m_{31}, m_{32}, and $\delta\theta_3$ are not needed in solving the nonlinear plate problem. Moreover, (7.7.27a,b) can be used to express Q_1 and Q_2 in terms of other internal stress resultants. Hence, (7.7.22)-(7.7.24) are the five governing equations need to be solved with the boundary conditions shown in (7.7.26). To expand the fully nonlinear governing equations and boundary conditions shown in (7.7.22)-(7.7.24), (7.7.26), (7.7.7a), and (7.7.13)-(7.7.15) into polynomials of displacement variables, one only needs to expand e_1, e_2, γ_6, $T_{1j}(=\hat{T}_{1j})$, $T_{2j}(=\hat{T}_{2j})$, T_{3j}, and k_{ij} using (7.5.2), (7.5.3), (7.5.8), (7.5.6), (7.5.7), (7.5.10), and (7.5.16), respectively.

We note that, if x and y are curvilinear coordinates and hence k_4^0 and k_5^0 are not zero, one can use (7.2.41b) and follow the steps shown in this section to derive the equations of motion.

7.7.2 Nonlinear First-Order Theory

In the Hencky-Mindlin type first-order shear-deformation theory, it is assumed that the shear strains are uniform across the cross section (i.e., $\epsilon_{13} = \gamma_5$, $\epsilon_{23} = \gamma_4$), which is equivalent to neglecting the cubic term $4z^3/3h^2$ in (7.7.1d). Hence, it follows from (7.7.11a-f) and (7.7.7b) that

$$P_1 = P_2 = P_6 = R_1 = R_2 = G_1 = G_2 = 0$$
$$I_1 = I_4, \qquad I_2 = I_5 = I_7 \qquad (7.7.28)$$

If, furthermore, the influence of the strain γ_6 on the deformed configuration is neglected, the equations of motion are the same as those shown in (7.7.22)-(7.7.25) except that

$$\hat{M}_1 = M_1 - S_1\gamma_5, \qquad \hat{M}_2 = M_2 - S_2\gamma_4$$
$$\hat{M}_{61} = M_6 - S_1\gamma_4, \qquad \hat{M}_{62} = M_6 - S_2\gamma_5$$
$$m_{31} = M_6\gamma_5 - M_1\gamma_4, \qquad m_{32} = -M_6\gamma_4 + M_2\gamma_5$$
$$\Gamma_4 = -M_{2y} - M_{6x} + Q_2 - S_1 k_{61} - S_2 k_2 - M_1 k_5 - M_6 k_4$$
$$\Gamma_5 = -M_{1x} - M_{6y} + Q_1 - S_1 k_1 - S_2 k_{62} + M_2 k_4 + M_6 k_5 \qquad (7.7.29)$$

However, this theory requires the use of shear-correction factors for the transverse stiffnesses to account for the nonuniform distribution of shear stresses (Pai, 1995).

7.7.3 Third-Order Theory with von Kármán Nonlinearity

If the von Kármán strains shown in (7.5.40a,b) are used to account for geometric nonlinearity, (7.6.36) and (7.6.37a-c) continue to hold. One can obtain the corresponding equations of motion and boundary conditions by substituting for $[T]$ from (7.6.36), the k_{ij} from (7.5.40b), the e_1, e_2, and γ_6 from (7.5.40a), and the $\delta\theta_i$ from (7.6.37a-c) into (7.6.22)-(7.6.25).

We note that the von Kármán nonlinearity appears only in the equation of motion governing w. The obtained governing equations are the same as those of Reddy (1984a) and Bhimaraddi (1987c).

7.8 NONLINEAR LAYERWISE SHEAR-DEFORMABLE PLATE THEORY

We consider a rectangular laminated piezoelectric plate over the domain $0 \leq x \leq a$, $0 \leq y \leq b$, and $z_1 \leq z \leq z_{N+1}$. In Figure 7.8.1, we show the construction of a typical laminated piezoelectric plate, where the ith lamina is located between the two planes at $z = z_i$ and $z = z_{i+1}$, there are N plies, and the thickness $h = z_{N+1} - z_1$.

NONLINEAR LAYERWISE SHEAR-DEFORMABLE PLATE THEORY

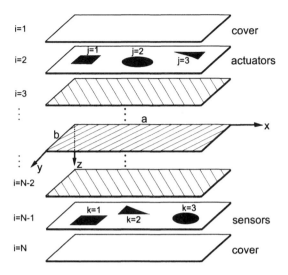

Fig. 7.8.1 A typical arrangement of laminated piezoelectric plates.

Every composite and piezoelectric lamina is assumed to be orthotropic, and the shapes of the actuators and sensors can be arbitrarily designed. The two coordinate systems xyz and $\xi\eta\zeta$ shown in Figure 7.5.1 are used to describe the undeformed and deformed geometries.

For a plate with imbedded piezoelectric actuators (see Figure 7.8.2a), the cross-section warping caused by external loadings is different from the warping caused by the actuation of the imbedded actuator. A typical deformation due to external loading is shown in Figure 7.8.2b, and a typical deformation due to piezoelectric actuation is shown in Figure 7.8.2c. An actual deformation may be a combination of both. Because the warping functions in Figures 7.8.2b and c are different, we need to obtain these warping functions separately.

7.8.1 Warpings Due to External Loads and Actuators

To include shear deformations in the mathematical model of a general anisotropic laminated plate subjected to external loads, see Figure 7.8.2b, we assume a displacement field for each layer. We assume that the local displacements $u_1^{(i)}$, $u_2^{(i)}$, and $u_3^{(i)}$ (with respect to the $\xi\eta\zeta$ coordinate system) of the ith lamina have the form

$$u_1^{(i)}(x,y,z,t) = u_1^0(x,y,t) + z\theta_2(x,y,t) + \gamma_5 z + \alpha_1^{(i)}(x,y,t)z^2 + \beta_1^{(i)}(x,y,t)z^3 \tag{7.8.1a}$$

$$u_2^{(i)}(x,y,z,t) = u_2^0(x,y,t) - z\theta_1(x,y,t) + \gamma_4 z + \alpha_2^{(i)}(x,y,t)z^2 + \beta_2^{(i)}(x,y,t)z^3 \tag{7.8.1b}$$

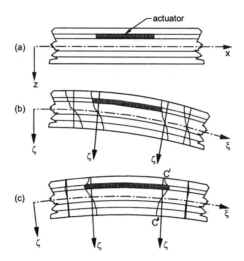

Fig. 7.8.2 (a) A cross section of an undeformed laminated piezoelectric plate, (b) the cross-section warping due to external loading, and (c) the cross-section warping due to actuation of an imbedded piezo-actuator.

$$u_3^{(i)}(x,y,z,t) = u_3^0(x,y,t) \tag{7.8.1c}$$

where the u_i^0 ($i = 1, 2, 3$) are the displacements (with respect to the local coordinate system) of a point which is located at $(x, y, 0)$ before deformation, γ_4 and γ_5 are the transverse shear rotations at the reference plane with respect to the ξ and η axes, respectively, and θ_1 and θ_2 are the rotation angles of the normal to the reference plane with respect to the ξ and η axes, respectively. Moreover, the $\alpha_j^{(i)}$ and $\beta_j^{(i)}$ are functions to be determined by imposing continuity conditions on the inplane displacements and interlaminar shear stresses and the free surface conditions. Because $\xi\eta\zeta$ is a local coordinate system and the $\xi\eta$ plane is tangent to the deformed reference-plane, we have

$$u_1^0 = u_2^0 = u_3^0 = \theta_1 = \theta_2 = \partial u_3^0/\partial x = \partial u_3^0/\partial y = 0 \tag{7.8.2}$$

Using (7.8.1a-c) and (7.8.2), we obtain the local transverse shear strains as

$$\epsilon_{23}^{(i)} = \frac{\partial u_2^{(i)}}{\partial z} + \frac{\partial u_3^{(i)}}{\partial y} = \gamma_4 + 2\alpha_2^{(i)} z + 3\beta_2^{(i)} z^2 \tag{7.8.3a}$$

$$\epsilon_{13}^{(i)} = \frac{\partial u_1^{(i)}}{\partial z} + \frac{\partial u_3^{(i)}}{\partial x} = \gamma_5 + 2\alpha_1^{(i)} z + 3\beta_1^{(i)} z^2 \tag{7.8.3b}$$

The stress-strain relations for the ith lamina can be obtained from (2.4.34) and (2.4.35) as

$$\left\{ \begin{array}{c} \sigma_{11}^{(i)} \\ \sigma_{22}^{(i)} \\ \sigma_{12}^{(i)} \end{array} \right\} = \left[\tilde{Q}^{(i)} \right] \left\{ \begin{array}{c} \epsilon_{11}^{(i)} \\ \epsilon_{22}^{(i)} \\ \epsilon_{12}^{(i)} \end{array} \right\}, \quad \left\{ \begin{array}{c} \sigma_{23}^{(i)} \\ \sigma_{13}^{(i)} \end{array} \right\} = \left[\hat{Q}^{(i)} \right] \left\{ \begin{array}{c} \epsilon_{23}^{(i)} \\ \epsilon_{13}^{(i)} \end{array} \right\} \quad (7.8.4a)$$

where

$$[\tilde{Q}^{(i)}] \equiv \begin{bmatrix} \tilde{Q}_{11}^{(i)} & \tilde{Q}_{12}^{(i)} & \tilde{Q}_{16}^{(i)} \\ \tilde{Q}_{12}^{(i)} & \tilde{Q}_{22}^{(i)} & \tilde{Q}_{26}^{(i)} \\ \tilde{Q}_{16}^{(i)} & \tilde{Q}_{26}^{(i)} & \tilde{Q}_{66}^{(i)} \end{bmatrix}, \quad [\hat{Q}^{(i)}] \equiv \begin{bmatrix} \overline{Q}_{44}^{(i)} & \overline{Q}_{45}^{(i)} \\ \overline{Q}_{45}^{(i)} & \overline{Q}_{55}^{(i)} \end{bmatrix} \quad (7.8.4b)$$

Here, $[\tilde{Q}^{(i)}]$ is the reduced stiffness matrix obtained by using the assumption $\sigma_{33}^{(i)} = 0$.

We assume that there is no delamination, and hence the inplane displacements u_1 and u_2 and interlaminar shear stresses σ_{13} and σ_{23} are continuous across the interface of two contiguous laminae. Moreover, we assume that there are no applied shear loads on the bounding surfaces and hence $\sigma_{13} = \sigma_{23} = 0$ at the $z = z_1$ and $z = z_{N+1}$ planes. Hence, we have

$$\epsilon_{13}^{(1)}(x, y, z_1, t) = 0$$
$$\epsilon_{23}^{(1)}(x, y, z_1, t) = 0$$
$$u_1^{(i)}(x, y, z_{i+1}, t) - u_1^{(i+1)}(x, y, z_{i+1}, t) = 0 \quad \text{for} \quad i = 1, 2, ..., N-1$$
$$u_2^{(i)}(x, y, z_{i+1}, t) - u_2^{(i+1)}(x, y, z_{i+1}, t) = 0 \quad \text{for} \quad i = 1, 2, ..., N-1$$
$$\sigma_{23}^{(i)}(x, y, z_{i+1}, t) - \sigma_{23}^{(i+1)}(x, y, z_{i+1}, t) = 0 \quad \text{for} \quad i = 1, 2, ..., N-1$$
$$\sigma_{13}^{(i)}(x, y, z_{i+1}, t) - \sigma_{13}^{(i+1)}(x, y, z_{i+1}, t) = 0 \quad \text{for} \quad i = 1, 2, ..., N-1$$
$$\epsilon_{13}^{(N)}(x, y, z_{N+1}, t) = 0$$
$$\epsilon_{23}^{(N)}(x, y, z_{N+1}, t) = 0 \quad (7.8.5)$$

These $4N$ algebraic equations can be used to determine the $4N$ unknowns (i.e., $\alpha_1^{(i)}$, $\alpha_2^{(i)}$, $\beta_1^{(i)}$, $\beta_2^{(i)}$ for $i = 1, 2, ..., N$) by using (7.8.3a,b), (7.8.1a,b), and (7.8.4a,b) in terms of γ_4 and γ_5 as

$$\alpha_1^{(i)} = a_{14}^{(i)} \gamma_4 + a_{15}^{(i)} \gamma_5$$
$$\alpha_2^{(i)} = a_{24}^{(i)} \gamma_4 + a_{25}^{(i)} \gamma_5$$
$$\beta_1^{(i)} = b_{14}^{(i)} \gamma_4 + b_{15}^{(i)} \gamma_5$$
$$\beta_2^{(i)} = b_{24}^{(i)} \gamma_4 + b_{25}^{(i)} \gamma_5 \quad (7.8.6)$$

for $i = 1, 2, ..., N$, where the $a_{kl}^{(i)}$ and $b_{kl}^{(i)}$ are functions of the z_j, $\overline{Q}_{44}^{(j)}$, $\overline{Q}_{45}^{(j)}$, and $\overline{Q}_{55}^{(j)}$.

It follows from (7.8.1a-c) and (7.8.6) that a general displacement field describing the deformation due to any load can be represented by

$$u_1^{(i)}(x,y,z,t) = u_1^0(x,y,t) + z\theta_2(x,y,t) + \gamma_5 g_{15}^{(i)} + \gamma_4 g_{14}^{(i)} \quad (7.8.7a)$$

$$u_2^{(i)}(x,y,z,t) = u_2^0(x,y,t) - z\theta_1(x,y,t) + \gamma_4 g_{24}^{(i)} + \gamma_5 g_{25}^{(i)} \quad (7.8.7b)$$

$$u_3^{(i)}(x,y,z,t) = u_3^0(x,y,t) \quad (7.8.7c)$$

where the $g_{14}^{(i)}$, $g_{15}^{(i)}$, $g_{24}^{(i)}$, and $g_{25}^{(i)}$ are polynomial functions of z. For example, substituting (7.8.6) into (7.8.1a-c) yields

$$g_{14}^{(i)} \equiv a_{14}^{(i)} z^2 + b_{14}^{(i)} z^3$$
$$g_{15}^{(i)} \equiv z + a_{15}^{(i)} z^2 + b_{15}^{(i)} z^3$$
$$g_{24}^{(i)} \equiv z + a_{24}^{(i)} z^2 + b_{24}^{(i)} z^3$$
$$g_{25}^{(i)} \equiv a_{25}^{(i)} z^2 + b_{25}^{(i)} z^3 \quad (7.8.8)$$

Using (7.8.7a-c) and (7.8.2), we obtain the local strains as

$$\left\{ \begin{array}{c} \epsilon_{11}^{(i)} \\ \epsilon_{22}^{(i)} \\ \epsilon_{12}^{(i)} \end{array} \right\} = \left\{ \begin{array}{c} \partial u_1^{(i)}/\partial x \\ \partial u_2^{(i)}/\partial y \\ \partial u_1^{(i)}/\partial y + \partial u_2^{(i)}/\partial x \end{array} \right\} = [S_1^{(i)}]\{\psi\} \quad (7.8.9a)$$

$$\left\{ \begin{array}{c} \epsilon_{23}^{(i)} \\ \epsilon_{13}^{(i)} \end{array} \right\} = \left\{ \begin{array}{c} \partial u_2^{(i)}/\partial z + \partial u_3^{(i)}/\partial y \\ \partial u_1^{(i)}/\partial z + \partial u_3^{(i)}/\partial x \end{array} \right\} = [S_2^{(i)}] \left\{ \begin{array}{c} \gamma_4 \\ \gamma_5 \end{array} \right\} \quad (7.8.9b)$$

where

$$[S_1^{(i)}] \equiv \begin{bmatrix} 1 & 0 & 0 & z & 0 & 0 & g_{14}^{(i)} & 0 & g_{15}^{(i)} & 0 \\ 0 & 1 & 0 & 0 & z & 0 & 0 & g_{24}^{(i)} & 0 & g_{25}^{(i)} \\ 0 & 0 & 1 & 0 & 0 & z & g_{24}^{(i)} & g_{14}^{(i)} & g_{25}^{(i)} & g_{15}^{(i)} \end{bmatrix} \quad (7.8.10a)$$

$$[S_2^{(i)}] \equiv \begin{bmatrix} g_{24z}^{(i)} & g_{25z}^{(i)} \\ g_{14z}^{(i)} & g_{15z}^{(i)} \end{bmatrix} \quad (7.8.10b)$$

$$\{\psi\} \equiv \{e_1, e_2, \gamma_6, k_1, k_2, k_6, \gamma_{4x}, \gamma_{4y}, \gamma_{5x}, \gamma_{5y}\}^T \quad (7.8.10c)$$

$$e_1 \equiv \frac{\partial u_1^0}{\partial x}, \quad e_2 \equiv \frac{\partial u_2^0}{\partial y}, \quad \gamma_6 \equiv \frac{\partial u_1^0}{\partial y} + \frac{\partial u_2^0}{\partial x} \quad (7.8.10d)$$

$$k_1 \equiv \frac{\partial \theta_2}{\partial x}, \quad k_2 \equiv -\frac{\partial \theta_1}{\partial y}, \quad k_{61} \equiv -\frac{\partial \theta_1}{\partial x}, \quad k_{62} \equiv \frac{\partial \theta_2}{\partial y}, \quad k_6 \equiv k_{61} + k_{62} \quad (7.8.10e)$$

Here, we used the fact that $\xi\eta\zeta$ is a local coordinate system and the $\xi\eta$ plane is tangent to the deformed reference plane.

Because the orthotropy of piezoelectric laminae is caused by manufacturing and poling processes, we assume that the principal mechanical or material directions are the same as the principal piezoelectric directions. For the ith piezoelectric layer, when a voltage is applied across its thickness, the stress-strain relations are given by

$$\left\{\begin{array}{c} \sigma_{11}^{(i)} \\ \sigma_{22}^{(i)} \\ \sigma_{12}^{(i)} \end{array}\right\} = \left[\tilde{Q}^{(i)}\right] \left\{\begin{array}{c} \epsilon_{11}^{(i)} - \bar{\Lambda}_1 R \\ \epsilon_{22}^{(i)} - \bar{\Lambda}_2 R \\ \epsilon_{12}^{(i)} - \bar{\Lambda}_{12} R \end{array}\right\}, \quad \left\{\begin{array}{c} \sigma_{23}^{(i)} \\ \sigma_{13}^{(i)} \end{array}\right\} = \left[\hat{Q}^{(i)}\right] \left\{\begin{array}{c} \epsilon_{23}^{(i)} \\ \epsilon_{13}^{(i)} \end{array}\right\} \quad (7.8.11)$$

where

$$\Lambda_1 = \frac{d_{31} V_3}{z_{I+1} - z_I}, \quad \Lambda_2 = \frac{d_{32} V_3}{z_{I+1} - z_I}$$
$$\bar{\Lambda}_1 = \Lambda_1 \cos^2 \alpha + \Lambda_2 \sin^2 \alpha$$
$$\bar{\Lambda}_2 = \Lambda_1 \sin^2 \alpha + \Lambda_2 \cos^2 \alpha$$
$$\bar{\Lambda}_{12} = 2(\Lambda_1 - \Lambda_2) \cos \alpha \sin \alpha \quad (7.8.12)$$

Here, V_3 is the applied voltage across the actuator, which is located between the two planes at $z = z_I$ and $z = z_{I+1}$; d_{31} and d_{32} are the piezoelectric strain constants with respect to the principal piezoelectric directions 1 and 2, which may be functions of the induced strains ϵ_{11} and ϵ_{22} (Lazarus and Crawley, 1989); Λ_1 and Λ_2 are the free actuating strains along the 1 and 2 directions, respectively; $\bar{\Lambda}_1$ and $\bar{\Lambda}_2$ are the free actuating strains along the deformed structural axes ξ and η; $\bar{\Lambda}_{12}$ is the inplane shear actuating strain (Jones, 1975; Lazarus and Crawley, 1989); the angle α is obtained by rotating the axis ξ with respect to the ζ axis to the axis 1; and R is a function that describes the location of the actuator, which is given by

$$R = [H(z - z_I) - H(z - z_{I+1})] L(x, y) \quad (7.8.13a)$$

where $H(z)$ is the Heavyside step function defined as

$$H(z) = 1, \ z > 0$$
$$= 0, \ z < 0 \quad (7.8.13b)$$

and $L(x, y)$ describes the location of the actuator on the reference plane and is defined as

$$L(x, y) = \pm 1, \quad \text{if the point } (x, y) \text{ is covered by the actuator}$$
$$= 0, \quad \text{if the point } (x, y) \text{ is not covered by the actuator} \quad (7.8.13c)$$

Here, the $\pm$ signs are used to account for the possibility that the piezoelectric material can have different poling directions on the same lamina (Lee, 1990). For example, if the actuator is a rectangular one, then

$$L(x, y) = \pm [H(x - x_{j1}) - H(x - x_{j2})] [H(y - y_{j1}) - H(y - y_{j2})] \quad (7.8.13d)$$

where $x = x_{j1}$, $x = x_{j2}$, $y = y_{j1}$, and $y = y_{j2}$ represent the border lines of the jth actuator on the Ith lamina.

It follows from (7.8.11) that the stresses are proportional to the mechanical strains, which are the differences between the total strains $\epsilon_{kl}^{(i)}$ and the free actuation strains $(\bar{\Lambda}_1, \bar{\Lambda}_2, \bar{\Lambda}_{12})$. Also, (7.8.11) and (7.8.12) show that an induced strain actuator can only introduce inplane extension or compression and/or inplane shear deformations.

To solve for the actuator induced warpings with free boundaries and no external or inertia loads, we assume that the displacement field for the ith composite lamina has the same form given in (7.8.7a-c) and (7.8.8), but the values of the $a_{kl}^{(i)}$ and $b_{kl}^{(i)}$ are different from those in (7.8.6). For the actuator located at the Ith lamina, we assume that its displacement field is the same as that given by (7.8.7a-c) but

$$\begin{aligned}
g_{14}^{(I)} &\equiv a_{14}^{(I)} z^2 + b_{14}^{(I)} z^3 + c_{14}^{(I)} z^4 \\
g_{15}^{(I)} &\equiv z + a_{15}^{(I)} z^2 + b_{15}^{(I)} z^3 + c_{15}^{(I)} z^4 + \bar{c}_{15}^{(I)} z^5 \\
g_{24}^{(I)} &\equiv z + a_{24}^{(I)} z^2 + b_{24}^{(I)} z^3 + c_{24}^{(I)} z^4 + \bar{c}_{24}^{(I)} z^5 \\
g_{25}^{(I)} &\equiv a_{25}^{(I)} z^2 + b_{25}^{(I)} z^3 + c_{25}^{(I)} z^4
\end{aligned} \qquad (7.8.14)$$

Because the peak of the warping function should be at the actuation layer and the signs of the shear angle should change within the actuation layer, we assume that

$$g_{15z}(z_I) + g_{15z}(z_{I+1}) = g_{24z}(z_I) + g_{24z}(z_{I+1}) = 0 \qquad (7.8.15a)$$

Because a thin piezoelectric actuator can only provide inplane strains ϵ_{11}, ϵ_{22}, and ϵ_{12} but no transverse shear actuating forces, we have

$$\sum_{i=1}^{N} \int_{z_i}^{z_{i+1}} \sigma_{23}^{(i)} dz = \sum_{i=1}^{N} \int_{z_i}^{z_{i+1}} \sigma_{13}^{(i)} dz = 0 \qquad (7.8.15b)$$

Substituting (7.8.8), (7.8.14), (7.8.7a-c), (7.9.9b), and (7.8.11) into (7.8.5) and (7.8.15) and then setting each of the coefficients of γ_4 and γ_5 equal to zero yields

$$\begin{aligned}
2a_{14}^{(1)} z_1 + 3b_{14}^{(1)} z_1^2 + \delta_{1I} 4c_{14}^{(1)} z_1^3 &= 0 \\
2a_{15}^{(1)} z_1 + 3b_{15}^{(1)} z_1^2 + \delta_{1I} 4c_{15}^{(1)} z_1^3 + \delta_{1I} 5\bar{c}_{15}^{(1)} z_1^4 &= -1 \\
2a_{24}^{(1)} z_1 + 3b_{24}^{(1)} z_1^2 + \delta_{1I} 4c_{24}^{(1)} z_1^3 + \delta_{1I} 5\bar{c}_{24}^{(1)} z_1^4 &= -1 \\
2a_{25}^{(1)} z_1 + 3b_{25}^{(1)} z_1^2 + \delta_{1I} 4c_{25}^{(1)} z_1^3 &= 0
\end{aligned} \qquad (7.8.16a)$$

$$a_{14}^{(i)} z_j^2 + b_{14}^{(i)} z_j^3 + \delta_{iI} c_{14}^{(i)} z_j^4 - a_{14}^{(j)} z_j^2 - b_{14}^{(j)} z_j^3 - \delta_{jI} c_{14}^{(j)} z_j^4 = 0$$

$$a_{15}^{(i)} z_j^2 + b_{15}^{(i)} z_j^3 + \delta_{iI} c_{15}^{(i)} z_j^4 + \delta_{iI} \bar{c}_{15}^{(i)} z_j^5 - a_{15}^{(j)} z_j^2 - b_{15}^{(j)} z_j^3$$
$$- \delta_{jI} c_{15}^{(j)} z_j^4 - \delta_{jI} \bar{c}_{15}^{(j)} z_j^5 = 0$$

$$a_{24}^{(i)} z_j^2 + b_{24}^{(i)} z_j^3 + \delta_{iI} c_{24}^{(i)} z_j^4 + \delta_{iI} \bar{c}_{24}^{(i)} z_j^5 - a_{24}^{(j)} z_j^2 - b_{24}^{(j)} z_j^3$$
$$- \delta_{jI} c_{24}^{(j)} z_j^4 - \delta_{jI} \bar{c}_{24}^{(j)} z_j^5 = 0$$

$$a_{25}^{(i)} z_j^2 + b_{25}^{(i)} z_j^3 + \delta_{iI} c_{25}^{(i)} z_j^4 - a_{25}^{(j)} z_j^2 - b_{25}^{(j)} z_j^3 - \delta_{jI} c_{25}^{(j)} z_j^4 = 0$$

$$\overline{Q}_{44}^{(i)}\left(2a_{24}^{(i)}z_j + 3b_{24}^{(i)}z_j^2 + \delta_{iI}4c_{24}^{(i)}z_j^3 + \delta_{iI}5\bar{c}_{24}^{(i)}z_j^4\right)$$
$$+\overline{Q}_{45}^{(i)}\left(2a_{14}^{(i)}z_j + 3b_{14}^{(i)}z_j^2 + \delta_{iI}4c_{14}^{(i)}z_j^3\right)$$
$$-\overline{Q}_{44}^{(j)}\left(2a_{24}^{(j)}z_j + 3b_{24}^{(j)}z_j^2 + \delta_{jI}4c_{24}^{(j)}z_j^3 + \delta_{jI}5\bar{c}_{24}^{(j)}z_j^4\right)$$
$$-\overline{Q}_{45}^{(j)}\left(2a_{14}^{(j)}z_j + 3b_{14}^{(j)}z_j^2 + \delta_{jI}4c_{14}^{(j)}z_j^3\right) = \overline{Q}_{44}^{(j)} - \overline{Q}_{44}^{(i)}$$

$$\overline{Q}_{44}^{(i)}\left(2a_{25}^{(i)}z_j + 3b_{25}^{(i)}z_j^2 + \delta_{iI}4c_{25}^{(i)}z_j^3\right)$$
$$+\overline{Q}_{45}^{(i)}\left(2a_{15}^{(i)}z_j + 3b_{15}^{(i)}z_j^2 + \delta_{iI}4c_{15}^{(i)}z_j^3 + \delta_{iI}5\bar{c}_{15}^{(i)}z_j^4\right)$$
$$-\overline{Q}_{44}^{(j)}\left(2a_{25}^{(j)}z_j + 3b_{25}^{(j)}z_j^2 + \delta_{jI}4c_{25}^{(j)}z_j^3\right)$$
$$-\overline{Q}_{45}^{(j)}\left(2a_{15}^{(j)}z_j + 3b_{15}^{(j)}z_j^2 + \delta_{jI}4c_{15}^{(j)}z_j^3 + \delta_{jI}5\bar{c}_{15}^{(j)}z_j^4\right) = \overline{Q}_{45}^{(j)} - \overline{Q}_{45}^{(i)}$$

$$\overline{Q}_{45}^{(i)}\left(2a_{24}^{(i)}z_j + 3b_{24}^{(i)}z_j^2 + \delta_{iI}4c_{24}^{(i)}z_j^3 + \delta_{iI}5\bar{c}_{24}^{(i)}z_j^4\right)$$
$$+\overline{Q}_{55}^{(i)}\left(2a_{14}^{(i)}z_j + 3b_{14}^{(i)}z_j^2 + \delta_{iI}4c_{14}^{(i)}z_j^3\right)$$
$$-\overline{Q}_{45}^{(j)}\left(2a_{24}^{(j)}z_j + 3b_{24}^{(j)}z_j^2 + \delta_{jI}4c_{24}^{(j)}z_j^3 + \delta_{jI}5\bar{c}_{24}^{(j)}z_j^4\right)$$
$$-\overline{Q}_{55}^{(j)}\left(2a_{14}^{(j)}z_j + 3b_{14}^{(j)}z_j^2 + \delta_{jI}4c_{14}^{(j)}z_j^3\right) = \overline{Q}_{45}^{(j)} - \overline{Q}_{45}^{(i)}$$

$$\overline{Q}_{45}^{(i)}\left(2a_{25}^{(i)}z_j + 3b_{25}^{(i)}z_j^2 + \delta_{iI}4c_{25}^{(i)}z_j^3\right)$$
$$+\overline{Q}_{55}^{(i)}\left(2a_{15}^{(i)}z_j + 3b_{15}^{(i)}z_j^2 + \delta_{iI}4c_{15}^{(i)}z_j^3 + \delta_{iI}5\bar{c}_{15}^{(i)}z_j^4\right)$$
$$-\overline{Q}_{55}^{(j)}\left(2a_{15}^{(j)}z_j + 3b_{15}^{(j)}z_j^2 + \delta_{jI}4c_{15}^{(j)}z_j^3 + \delta_{jI}5\bar{c}_{15}^{(j)}z_j^4\right)$$
$$-\overline{Q}_{45}^{(j)}\left(2a_{25}^{(j)}z_j + 3b_{25}^{(j)}z_j^2 + \delta_{jI}4c_{25}^{(j)}z_j^3\right) = \overline{Q}_{55}^{(j)} - \overline{Q}_{55}^{(i)} \qquad (7.8.16b)$$

for $i = 1, 2, ..., N - 1$, where $j \equiv i + 1$,

$$2a_{14}^{(N)}z_{N+1} + 3b_{14}^{(N)}z_{N+1}^2 + \delta_{NI}4c_{14}^{(N)}z_{N+1}^3 = 0$$
$$2a_{15}^{(N)}z_{N+1} + 3b_{15}^{(N)}z_{N+1}^2 + \delta_{NI}4c_{15}^{(N)}z_{N+1}^3 + \delta_{NI}5\bar{c}_{15}^{(N)}z_{N+1}^4 = -1$$
$$2a_{24}^{(N)}z_{N+1} + 3b_{24}^{(N)}z_{N+1}^2 + \delta_{NI}4c_{24}^{(N)}z_{N+1}^3 + \delta_{NI}5\bar{c}_{24}^{(N)}z_{N+1}^4 = -1$$
$$2a_{25}^{(N)}z_{N+1} + 3b_{25}^{(N)}z_{N+1}^2 + \delta_{NI}4c_{25}^{(N)}z_{N+1}^3 = 0 \qquad (7.8.16c)$$

$$2a_{15}^{(I)}(z_I + z_{I+1}) + 3b_{15}^{(I)}(z_I^2 + z_{I+1}^2) + 4c_{15}^{(I)}(z_I^3 + z_{I+1}^3)$$
$$+5\bar{c}_{15}^{(I)}(z_I^4 + z_{I+1}^4) = -2$$
$$2a_{24}^{(I)}(z_I + z_{I+1}) + 3b_{24}^{(I)}(z_I^2 + z_{I+1}^2) + 4c_{24}^{(I)}(z_I^3 + z_{I+1}^3)$$
$$+5\bar{c}_{24}^{(I)}(z_I^4 + z_{I+1}^4) = -2 \qquad (7.8.16d)$$

$$\sum_{i=1}^{N} \int_{z_i}^{z_{i+1}} \left[\overline{Q}_{44}^{(i)} \left(1 + 2a_{24}^{(i)} z + 3b_{24}^{(i)} z^2 + \delta_{iI} 4c_{24}^{(i)} z^3 + \delta_{iI} 5\bar{c}_{24}^{(i)} z^4 \right) \right.$$
$$\left. + \overline{Q}_{45}^{(i)} \left(2a_{14}^{(i)} z + 3b_{14}^{(i)} z^2 + \delta_{iI} 4c_{14}^{(i)} z^3 \right) \right] dz = 0$$

$$\sum_{i=1}^{N} \int_{z_i}^{z_{i+1}} \left[\overline{Q}_{44}^{(i)} \left(2a_{25}^{(i)} z + 3b_{25}^{(i)} z^2 + \delta_{iI} 4c_{25}^{(i)} z^3 \right) \right.$$
$$\left. + \overline{Q}_{45}^{(i)} \left(1 + 2a_{15}^{(i)} z + 3b_{15}^{(i)} z^2 + \delta_{iI} 4c_{15}^{(i)} z^3 + \delta_{iI} 5\bar{c}_{15}^{(i)} z^4 \right) \right] dz = 0$$

$$\sum_{i=1}^{N} \int_{z_i}^{z_{i+1}} \left[\overline{Q}_{45}^{(i)} \left(1 + 2a_{24}^{(i)} z + 3b_{24}^{(i)} z^2 + \delta_{iI} 4c_{24}^{(i)} z^3 + \delta_{iI} 5\bar{c}_{24}^{(i)} z^4 \right) \right.$$
$$\left. + \overline{Q}_{55}^{(i)} \left(2a_{14}^{(i)} z + 3b_{14}^{(i)} z^2 + \delta_{iI} 4c_{14}^{(i)} z^3 \right) \right] dz = 0$$

$$\sum_{i=1}^{N} \int_{z_i}^{z_{i+1}} \left[\overline{Q}_{45}^{(i)} \left(2a_{25}^{(i)} z + 3b_{25}^{(i)} z^2 + \delta_{iI} 4c_{25}^{(i)} z^3 \right) \right.$$
$$\left. + \overline{Q}_{55}^{(i)} \left(1 + 2a_{15}^{(i)} z + 3b_{15}^{(i)} z^2 + \delta_{iI} 4c_{15}^{(i)} z^3 + \delta_{iI} 5\bar{c}_{15}^{(i)} z^4 \right) \right] dz = 0$$

(7.8.16e)

where δ_{iI} is the Kronecker delta function. These $8N + 6$ algebraic equations can be solved for the $8N + 6$ unknowns — $a_{14}^{(i)}$, $a_{15}^{(i)}$, $a_{24}^{(i)}$, $a_{25}^{(i)}$, $b_{14}^{(i)}$, $b_{15}^{(i)}$, $b_{24}^{(i)}$, $b_{25}^{(i)}$, for $i = 1, 2, ..., N$, and $\bar{c}_{15}^{(I)}$, $\bar{c}_{24}^{(I)}$, $c_{14}^{(I)}$, $c_{15}^{(I)}$, $c_{24}^{(I)}$, $c_{25}^{(I)}$. We note that the reference plane cannot be chosen as the contacting surface of any two laminae because it will make the shear angles continuous at $z = 0$. Also, we note that (7.8.5) is a special case of (7.8.16a-e). If there are several actuators covering the same area of the reference plane, a superposition method can be applied to obtain the warping functions if the input voltages to these actuators are proportional to one another.

For isotropic plates or one-layer orthotropic plates with an arbitrary ply angle, the external-load-induced warping functions are obtained from (7.8.16a-e) as

$$g_{15}^{(i)} = g_{24}^{(i)} = z - \frac{4z^3}{3h^2} \text{ and } g_{14}^{(i)} = g_{25}^{(i)} = 0 \quad (7.8.17)$$

They are shown in Figure 7.8.3a. This is the so-called third-order shear-deformation theory (Bhimaraddi and Stevens, 1984; Reddy, 1984a,b). We note that there are no coupling between the two transverse shear rotations γ_4 and γ_5. However, for general laminated plates, $g_{14}^{(i)}$ and $g_{25}^{(i)}$ are non-trivial and hence γ_4 and γ_5 are coupled.

In Figures 7.8.3b-f, we show the external load-induced warping functions for a five-layer graphite-epoxy laminated composite plate with different lay-ups. The material properties of each lamina are

$$E_{11} = 1.92 \times 10^7 \; psi, \; E_{22} = E_{33} = 1.56 \times 10^6 \; psi$$

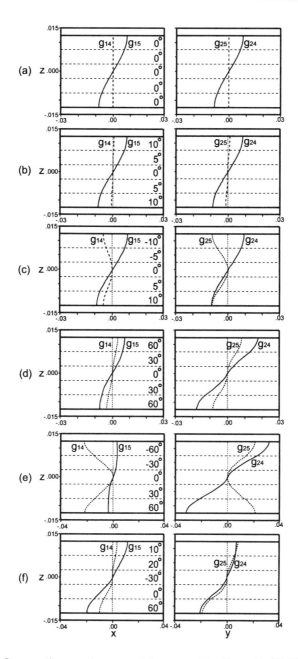

Fig. 7.8.3 Cross-section warpings caused by external loading: (a) $[0°/0°/0°/0°/0°]$, (b) $[10°/5°/0°/5°/10°]$, (c) $[10°/5°/0°/-5°/-10°]$, (d) $[60°/30°/0°/30°/60°]$, (e) $[60°/30°/0°/-30°/-60°]$, and (f) $[60°/0°/-30°/20°/10°]$.

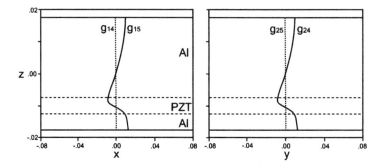

Fig. 7.8.4 Actuator-induced warping functions of an aluminum-piezoceramics-aluminum laminate.

$$G_{23} = 5.23 \times 10^5 \, psi, \; G_{12} = G_{13} = 8.20 \times 10^5 \, psi$$
$$\nu_{12} = \nu_{13} = 0.24, \; \nu_{23} = 0.49 \tag{7.8.18}$$

and the lamina thickness $t_k = 0.005 \, in.$ It follows from Figures 7.8.3b and c that anti-symmetric lamination results in even shear coupling functions $g_{14}^{(i)}$ and $g_{25}^{(i)}$ whereas symmetric lamination results in odd shear coupling functions. Moreover, anti-symmetric lamination results in more significant shear coupling effects. It follows from Figure 7.8.3d that, for a symmetric laminate with large ply angles, the warping functions $g_{15}^{(i)}$ and $g_{24}^{(i)}$ are quite different from those used in the third-order shear theory. Figure 7.8.3e shows again that anti-symmetric laminates have more significant shear coupling effects. Moreover, it follows from Figure 7.8.3f that, for a general laminated plate, the warping functions $g_{15}^{(i)}$ and $g_{24}^{(i)}$ are not odd functions and the shear coupling functions $g_{14}^{(i)}$ and $g_{25}^{(i)}$ are neither odd nor even functions.

In Figure 7.8.4, we show the actuator-induced warping functions of a three-layer laminate with the second layer being an actuator. The passive layers are aluminum with the properties

$$E = 1.03 \times 10^7 \, psi \; \text{and} \; \nu = 0.334 \tag{7.8.19a}$$

and the piezoelectric actuator is an isotropic $G1195$ piezoelectric patch (Piezo System, 1987) with the properties

$$E = 9.14 \times 10^6 \, psi \; \text{and} \; \nu = 0.28 \tag{7.8.19b}$$

Because both of the actuator and substrate are isotropic, there are no shear couplings (i.e., $g_{14}^{(i)} = g_{25}^{(i)} = 0$) and $g_{15}^{(i)} = g_{24}^{(i)}$.

In Figures 7.8.5a and b, we show the actuator-induced warping functions for a seven-layer laminate with the second layer being an actuator and the layup being $[10°/0°/20°/40°/-30°/90°/45°]$. The material properties of the composite laminae and the actuator are given in (7.8.18) and (7.8.19b), respectively. We note that, in this case, there are shear couplings and the warping functions are very different

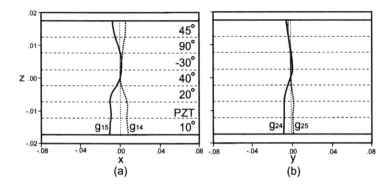

Fig. 7.8.5 Actuator-induced warping functions of a graphite-epoxy laminate where the layup is $[10°/0°/20°/40°/-30°/90°/45°]$: (a) g_{15} and g_{14} and (b) g_{24} and g_{25}.

from those of isotropic plates and depend on the stacking sequence. We point out that the actuator-induced extension of the reference plane and the bending rotations are not included in the warping functions because the warping functions represent displacements with respect to the deformed local coordinate system.

For areas not covered by actuators, one can use the shear-warping function shown in Figure 7.8.2b. For the area covered by an actuator, one can use the shear-warping function shown in Figure 7.8.2c if the actuator is actuated and there is no externally applied load. An appropriate combination of both warping functions needs to be used for areas covered by actuators if there are external loads.

7.8.2 Equations of Motion

The relative displacement vector **U** with respect to the deformed local coordinate system $\xi\eta\zeta$ is given by

$$\mathbf{U} = u_1^{(i)}\mathbf{i}_1 + u_2^{(i)}\mathbf{i}_2 + u_3^{(i)}\mathbf{i}_3 \qquad (7.8.20)$$

Taking the derivatives of (7.8.20) and using (7.8.7a-c), (7.5.14), (7.5.15), (7.8.2), and (7.8.10d,e), we obtain

$$\begin{aligned}\frac{\partial \mathbf{U}}{\partial x} &= \frac{\partial u_1^{(i)}}{\partial x}\mathbf{i}_1 + \frac{\partial u_2^{(i)}}{\partial x}\mathbf{i}_2 + \frac{\partial u_3^{(i)}}{\partial x}\mathbf{i}_3 + u_1^{(i)}\frac{\partial \mathbf{i}_1}{\partial x} + u_2^{(i)}\frac{\partial \mathbf{i}_2}{\partial x} + u_3^{(i)}\frac{\partial \mathbf{i}_3}{\partial x} \\ &= \left[e_1 + zk_1 + \gamma_{5x}g_{15}^{(i)} + \gamma_{4x}g_{14}^{(i)}\right]\mathbf{i}_1 \\ &\quad + \left[\frac{\partial u_2^0}{\partial x} + zk_{61} + \gamma_{4x}g_{24}^{(i)} + \gamma_{5x}g_{25}^{(i)}\right]\mathbf{i}_2 \\ &\quad + \left(\gamma_5 g_{15}^{(i)} + \gamma_4 g_{14}^{(i)}\right)(-k_1\mathbf{i}_3 + k_5\mathbf{i}_2) \\ &\quad - \left(\gamma_4 g_{24}^{(i)} + \gamma_5 g_{25}^{(i)}\right)(k_{61}\mathbf{i}_3 + k_5\mathbf{i}_1)\end{aligned} \qquad (7.8.21a)$$

$$\frac{\partial \mathbf{U}}{\partial y} = \frac{\partial u_1^{(i)}}{\partial y}\mathbf{i}_1 + \frac{\partial u_2^{(i)}}{\partial y}\mathbf{i}_2 + \frac{\partial u_3^{(i)}}{\partial y}\mathbf{i}_3 + u_1^{(i)}\frac{\partial \mathbf{i}_1}{\partial y} + u_2^{(i)}\frac{\partial \mathbf{i}_2}{\partial y} + u_3^{(i)}\frac{\partial \mathbf{i}_3}{\partial y}$$

$$= \left[\frac{\partial u_1^0}{\partial y} + zk_{62} + \gamma_{5y}g_{15}^{(i)} + \gamma_{4y}g_{14}^{(i)}\right]\mathbf{i}_1$$

$$+ \left[e_2 + zk_2 + \gamma_{4y}g_{24}^{(i)} + \gamma_{5y}g_{25}^{(i)}\right]\mathbf{i}_2$$

$$+ \left(\gamma_5 g_{15}^{(i)} + \gamma_4 g_{14}^{(i)}\right)(-k_{62}\mathbf{i}_3 + k_4\mathbf{i}_2)$$

$$- \left(\gamma_4 g_{24}^{(i)} + \gamma_5 g_{25}^{(i)}\right)(k_2\mathbf{i}_3 + k_4\mathbf{i}_1) \qquad (7.8.21b)$$

$$\frac{\partial \mathbf{U}}{\partial z} = \frac{\partial u_1^{(i)}}{\partial z}\mathbf{i}_1 + \frac{\partial u_2^{(i)}}{\partial z}\mathbf{i}_2 + \frac{\partial u_3^{(i)}}{\partial z}\mathbf{i}_3$$

$$= \left(\gamma_5 g_{15z}^{(i)} + \gamma_4 g_{14z}^{(i)}\right)\mathbf{i}_1 + \left(\gamma_4 g_{24z}^{(i)} + \gamma_5 g_{25z}^{(i)}\right)\mathbf{i}_2 \qquad (7.8.21c)$$

Hence, the Jaumann strains $\hat{\epsilon}_{ij}^{(i)}$ are obtained as

$$\hat{\epsilon}_{11}^{(i)} \equiv \frac{\partial \mathbf{U}}{\partial x} \cdot \mathbf{i}_1 = \epsilon_{11}^{(i)} - k_5\left(\gamma_4 g_{24}^{(i)} + \gamma_5 g_{25}^{(i)}\right) \qquad (7.8.22a)$$

$$\hat{\epsilon}_{22}^{(i)} \equiv \frac{\partial \mathbf{U}}{\partial y} \cdot \mathbf{i}_2 = \epsilon_{22}^{(i)} + k_4\left(\gamma_5 g_{15}^{(i)} + \gamma_4 g_{14}^{(i)}\right) \qquad (7.8.22b)$$

$$\hat{\epsilon}_{12}^{(i)} \equiv \frac{\partial \mathbf{U}}{\partial x} \cdot \mathbf{i}_2 + \frac{\partial \mathbf{U}}{\partial y} \cdot \mathbf{i}_1 = \epsilon_{12}^{(i)} - k_4\left(\gamma_4 g_{24}^{(i)} + \gamma_5 g_{25}^{(i)}\right) + k_5\left(\gamma_5 g_{15}^{(i)} + \gamma_4 g_{14}^{(i)}\right)$$
$$(7.8.22c)$$

$$\hat{\epsilon}_{23}^{(i)} \equiv \frac{\partial \mathbf{U}}{\partial z} \cdot \mathbf{i}_2 + \frac{\partial \mathbf{U}}{\partial y} \cdot \mathbf{i}_3 = \epsilon_{23}^{(i)} - k_2\left(\gamma_4 g_{24}^{(i)} + \gamma_5 g_{25}^{(i)}\right) - k_{62}\left(\gamma_5 g_{15}^{(i)} + \gamma_4 g_{14}^{(i)}\right)$$
$$(7.8.22d)$$

$$\hat{\epsilon}_{13}^{(i)} \equiv \frac{\partial \mathbf{U}}{\partial z} \cdot \mathbf{i}_1 + \frac{\partial \mathbf{U}}{\partial x} \cdot \mathbf{i}_3 = \epsilon_{13}^{(i)} - k_{61}\left(\gamma_4 g_{24}^{(i)} + \gamma_5 g_{25}^{(i)}\right) - k_1\left(\gamma_5 g_{15}^{(i)} + \gamma_4 g_{14}^{(i)}\right)$$
$$(7.8.22e)$$

Comparing the two strain expressions shown in (7.8.9a,b) and (7.8.22a-e), we note that the nonlinear terms in (7.8.22a-e) are due to the shear rotations γ_4 and γ_5 and the rotation of the coordinate system.

Using (7.8.22a-e), (7.8.9a,b), (7.8.10a-c), and (6.4.2a,b), we obtain the variation of the elastic energy as

$$\delta \Pi = \sum_{i=1}^{N}\int_{z_i}^{z_{i+1}}\int_A \left(\sigma_{11}^{(i)}\delta\hat{\epsilon}_{11}^{(i)} + \sigma_{22}^{(i)}\delta\hat{\epsilon}_{22}^{(i)} + \sigma_{12}^{(i)}\delta\hat{\epsilon}_{12}^{(i)}\right.$$

$$\left. + \sigma_{23}^{(i)}\delta\hat{\epsilon}_{23}^{(i)} + \sigma_{13}^{(i)}\delta\hat{\epsilon}_{13}^{(i)}\right)dxdydz$$

$$= \int_A \left(N_1\delta e_1 + N_2\delta e_2 + N_6\delta\gamma_6 + \Theta_1\delta\theta_1 + \Theta_2\delta\theta_2 + \Theta_3\delta\theta_3\right.$$

NONLINEAR LAYERWISE SHEAR-DEFORMABLE PLATE THEORY

$$+ \Gamma_4 \delta\gamma_4 + \Gamma_5 \delta\gamma_5 \Big) dx dy$$
$$+ \int_x \Big[-\hat{M}_2 \delta\theta_1 + \hat{M}_{62}\delta\theta_2 + m_{32}\delta\theta_3 + m_2\delta\gamma_4 + m_{62}\delta\gamma_5 \Big]_{y=0}^{y=b} dx$$
$$+ \int_y \Big[-\hat{M}_{61}\delta\theta_1 + \hat{M}_1\delta\theta_2 + m_{31}\delta\theta_3 + m_{61}\delta\gamma_4 + m_1\delta\gamma_5 \Big]_{x=0}^{x=a} dy$$
(7.8.23)

where

$$\hat{M}_1 \equiv M_1 - \bar{s}_{11}\gamma_5 - \tilde{s}_{11}\gamma_4 \tag{7.8.24a}$$

$$\hat{M}_2 \equiv M_2 - \tilde{s}_{22}\gamma_5 - \bar{s}_{22}\gamma_4 \tag{7.8.24b}$$

$$\hat{M}_{61} \equiv M_6 - \tilde{s}_{12}\gamma_5 - \bar{s}_{12}\gamma_4 \tag{7.8.24c}$$

$$\hat{M}_{62} \equiv M_6 - \bar{s}_{21}\gamma_5 - \tilde{s}_{21}\gamma_4 \tag{7.8.24d}$$

$$m_{31} \equiv (\overline{m}_{62} - \tilde{m}_1)\gamma_5 + (\tilde{m}_{62} - \overline{m}_1)\gamma_4 \tag{7.8.24e}$$

$$m_{32} \equiv (\overline{m}_2 - \tilde{m}_{61})\gamma_5 + (\tilde{m}_2 - \overline{m}_{61})\gamma_4 \tag{7.8.24f}$$

$$\Theta_1 \equiv \hat{M}_{61x} + \hat{M}_{2y} - m_{31}k_1 - m_{32}k_{62} + \hat{M}_1 k_5 + \hat{M}_{62}k_4 \tag{7.8.24g}$$

$$\Theta_2 \equiv -\hat{M}_{1x} - \hat{M}_{62y} - m_{31}k_{61} - m_{32}k_2 + \hat{M}_2 k_4 + \hat{M}_{61}k_5 \tag{7.8.24h}$$

$$\Theta_3 \equiv -m_{31x} - m_{32y} + \hat{M}_1 k_{61} - \hat{M}_2 k_{62} + \hat{M}_{62}k_2 - \hat{M}_{61}k_1 \tag{7.8.24i}$$

$$\Gamma_4 \equiv -m_{61x} - m_{2y} + q_2 - \tilde{s}_{11}k_1 - \bar{s}_{22}k_2 - \bar{s}_{12}k_{61} - \tilde{s}_{21}k_{62}$$
$$+ (\tilde{m}_2 - \overline{m}_{61})k_4 + (\tilde{m}_{62} - \overline{m}_1)k_5 \tag{7.8.24j}$$

$$\Gamma_5 \equiv -m_{1x} - m_{62y} + q_1 - \bar{s}_{11}k_1 - \tilde{s}_{22}k_2 - \tilde{s}_{12}k_{61} - \bar{s}_{21}k_{62}$$
$$+ (\overline{m}_2 - \tilde{m}_{61})k_4 + (\overline{m}_{62} - \tilde{m}_1)k_5 \tag{7.8.24k}$$

The stress resultants and moments are defined as

$$\{N_1, N_2, N_6, M_1, M_2, M_6, m_{61}, m_2, m_1, m_{62}\}$$
$$\equiv \sum_{i=1}^{N} \int_{z_i}^{z_{i+1}} \left\{ \sigma_{11}^{(i)}, \sigma_{22}^{(i)}, \sigma_{12}^{(i)} \right\} \left[S_1^{(i)} \right] dz$$
$$= \sum_{i=1}^{N} \int_{z_i}^{z_{i+1}} \Big\{ \sigma_{11}^{(i)}, \sigma_{22}^{(i)}, \sigma_{12}^{(i)}, z\sigma_{11}^{(i)}, z\sigma_{22}^{(i)}, z\sigma_{12}^{(i)}, \sigma_{11}^{(i)}g_{14}^{(i)} + \sigma_{12}^{(i)}g_{24}^{(i)},$$
$$\sigma_{22}^{(i)}g_{24}^{(i)} + \sigma_{12}^{(i)}g_{14}^{(i)}, \sigma_{11}^{(i)}g_{15}^{(i)} + \sigma_{12}^{(i)}g_{25}^{(i)}, \sigma_{22}^{(i)}g_{25}^{(i)} + \sigma_{12}^{(i)}g_{15}^{(i)} \Big\} dz \tag{7.8.25a}$$

$$\{q_2, q_1\} = \sum_{i=1}^{N} \int_{z_i}^{z_{i+1}} \left\{\sigma_{23}^{(i)}, \sigma_{13}^{(i)}\right\} \left[S_2^{(i)}\right] dz$$

$$= \sum_{i=1}^{N} \int_{z_i}^{z_{i+1}} \left\{\sigma_{23}^{(i)} g_{24z}^{(i)} + \sigma_{13}^{(i)} g_{14z}^{(i)}, \sigma_{23}^{(i)} g_{25z}^{(i)} + \sigma_{13}^{(i)} g_{15z}^{(i)}\right\} dz \quad (7.8.25b)$$

$$\{\overline{m}_1, \tilde{m}_1, \tilde{m}_2, \overline{m}_2, \tilde{m}_{62}, \overline{m}_{62}, \overline{m}_{61}, \tilde{m}_{61}\}$$

$$= \sum_{i=1}^{N} \int_{z_i}^{z_{i+1}} \left\{\sigma_{11}^{(i)}, \sigma_{22}^{(i)}, \sigma_{12}^{(i)}\right\} \left[S_3^{(i)}\right] dz$$

$$= \sum_{i=1}^{N} \int_{z_i}^{z_{i+1}} \left\{\sigma_{11}^{(i)} g_{24}^{(i)}, \sigma_{11}^{(i)} g_{25}^{(i)}, \sigma_{22}^{(i)} g_{14}^{(i)}, \sigma_{22}^{(i)} g_{15}^{(i)}, \sigma_{12}^{(i)} g_{14}^{(i)}, \right.$$

$$\left. \sigma_{12}^{(i)} g_{15}^{(i)}, \sigma_{12}^{(i)} g_{24}^{(i)}, \sigma_{12}^{(i)} g_{25}^{(i)}\right\} dz \quad (7.8.25c)$$

$$\{\tilde{s}_{21}, \overline{s}_{21}, \overline{s}_{22}, \tilde{s}_{22}\} = \sum_{i=1}^{N} \int_{z_i}^{z_{i+1}} \left\{\sigma_{23}^{(i)} g_{14}^{(i)}, \sigma_{23}^{(i)} g_{15}^{(i)}, \sigma_{23}^{(i)} g_{24}^{(i)}, \sigma_{23}^{(i)} g_{25}^{(i)}\right\} dz$$

$$(7.8.25d)$$

$$\{\tilde{s}_{11}, \overline{s}_{11}, \overline{s}_{12}, \tilde{s}_{12}\} = \sum_{i=1}^{N} \int_{z_i}^{z_{i+1}} \left\{\sigma_{13}^{(i)} g_{14}^{(i)}, \sigma_{13}^{(i)} g_{15}^{(i)}, \sigma_{13}^{(i)} g_{24}^{(i)}, \sigma_{13}^{(i)} g_{25}^{(i)}\right\} dz$$

$$(7.8.25e)$$

where

$$\left[S_3^{(i)}\right] \equiv \begin{bmatrix} g_{24}^{(i)} & g_{25}^{(i)} & 0 & 0 & 0 & 0 & 0 & 0 \\ 0 & 0 & g_{14}^{(i)} & g_{15}^{(i)} & 0 & 0 & 0 & 0 \\ 0 & 0 & 0 & 0 & g_{14}^{(i)} & g_{15}^{(i)} & g_{24}^{(i)} & g_{25}^{(i)} \end{bmatrix} \quad (7.8.25f)$$

We note that the stress resultants defined in (7.8.25c-e) are due to nonlinear effects and they only appear in nonlinear terms, as shown in (7.8.24a-k). Moreover, $\hat{M}_1$, $\hat{M}_2$, $\hat{M}_{61}$, $\hat{M}_{62}$, m_{31}, and m_{32} represent the total moment intensities acting on the edges of the plate element. Equations (7.8.24a-k) show that m_{31}, m_{32}, and Θ_3 are nonlinear terms. We note that the effects of k_4 and k_5 are nonlinear. It can be seen from (7.8.22a-e) and (7.8.23) that, if changes in the elastic energy due to changes in the configuration are not included (i.e., the ϵ_{ij} are used instead of the $\hat{\epsilon}_{ij}$) and linear curvature expressions are used, then there are no nonlinear terms in (7.8.24a-k) and $m_{31} = m_{32} = \Theta_3 = 0$. Substituting (7.5.42a,b), (7.5.44a,b), and (7.5.46) into (7.8.23), we obtain

$$\delta\Pi = \int_A \left[\{N_1, N_6/(1+e_1), -\Theta_2/(1+e_1)\}[T] \begin{Bmatrix} \delta u_x \\ \delta v_x \\ \delta w_x \end{Bmatrix} \right.$$

$$\left. + \{N_6/(1+e_2), N_2, \Theta_1/(1+e_2)\}[T] \begin{Bmatrix} \delta u_y \\ \delta v_y \\ \delta w_y \end{Bmatrix} \right.$$

$$+\Theta_3\delta\theta_3 + \Gamma_4\delta\gamma_4 + \Gamma_5\delta\gamma_5\bigg]dxdy$$

$$+\int_x \left[-\hat{M}_2\delta\theta_1 + \hat{M}_{62}\delta\theta_2 + m_{32}\delta\theta_3 + m_2\delta\gamma_4 + m_{62}\delta\gamma_5\right]_{y=0}^{y=b}dx$$

$$+\int_y \left[-\hat{M}_{61}\delta\theta_1 + \hat{M}_1\delta\theta_2 + m_{31}\delta\theta_3 + m_{61}\delta\gamma_4 + m_1\delta\gamma_5\right]_{x=0}^{x=a}dy \quad (7.8.26)$$

Using (7.8.7a-c) and (7.8.2), we obtain the absolute displacement vector $\mathbf{D}$ of an arbitrary point within the ith layer as

$$\begin{aligned}\mathbf{D} &= u\mathbf{j}_1 + v\mathbf{j}_2 + w\mathbf{j}_3 + z\mathbf{i}_3 - z\mathbf{j}_3 + \mathbf{U} \\ &= u\mathbf{j}_1 + v\mathbf{j}_2 + w\mathbf{j}_3 + z\mathbf{i}_3 - z\mathbf{j}_3 + \left(\gamma_5 g_{15}^{(i)} + \gamma_4 g_{14}^{(i)}\right)\mathbf{i}_1 + \left(\gamma_4 g_{24}^{(i)} + \gamma_5 g_{25}^{(i)}\right)\mathbf{i}_2\end{aligned}$$
$$(7.8.27)$$

Taking the variation and the time derivative of (7.8.27), we obtain

$$\begin{aligned}\delta\mathbf{D} &= \mathbf{j}_1\delta u + \mathbf{j}_2\delta v + \mathbf{j}_3\delta w + \left(g_{14}^{(i)}\mathbf{i}_1 + g_{24}^{(i)}\mathbf{i}_2\right)\delta\gamma_4 + \left(g_{15}^{(i)}\mathbf{i}_1 + g_{25}^{(i)}\mathbf{i}_2\right)\delta\gamma_5 \\ &+ \left[\left(\gamma_4 g_{24}^{(i)} + \gamma_5 g_{25}^{(i)}\right)\mathbf{i}_3 - z\mathbf{i}_2\right]\delta\theta_1 + \left[z\mathbf{i}_1 - \left(\gamma_5 g_{15}^{(i)} + \gamma_4 g_{14}^{(i)}\right)\mathbf{i}_3\right]\delta\theta_2 \\ &+ \left[\left(\gamma_5 g_{15}^{(i)} + \gamma_4 g_{14}^{(i)}\right)\mathbf{i}_2 - \left(\gamma_4 g_{24}^{(i)} + \gamma_5 g_{25}^{(i)}\right)\mathbf{i}_1\right]\delta\theta_3 \end{aligned} \quad (7.8.28a)$$

and

$$\begin{aligned}\ddot{\mathbf{D}} &= \ddot{u}\mathbf{j}_1 + \ddot{v}\mathbf{j}_2 + \ddot{w}\mathbf{j}_3 + z\ddot{\mathbf{i}}_3 + \left(\ddot{\gamma}_5 g_{15}^{(i)} + \ddot{\gamma}_4 g_{14}^{(i)}\right)\mathbf{i}_1 \\ &+ 2\left(\dot{\gamma}_5 g_{15}^{(i)} + \dot{\gamma}_4 g_{14}^{(i)}\right)\dot{\mathbf{i}}_1 + \left(\gamma_5 g_{15}^{(i)} + \gamma_4 g_{14}^{(i)}\right)\ddot{\mathbf{i}}_1 + \left(\ddot{\gamma}_4 g_{24}^{(i)} + \ddot{\gamma}_5 g_{25}^{(i)}\right)\mathbf{i}_2 \\ &+ 2\left(\dot{\gamma}_4 g_{24}^{(i)} + \dot{\gamma}_5 g_{25}^{(i)}\right)\dot{\mathbf{i}}_2 + \left(\gamma_4 g_{24}^{(i)} + \gamma_5 g_{25}^{(i)}\right)\ddot{\mathbf{i}}_2 \end{aligned} \quad (7.8.28b)$$

where

$$\begin{aligned}\mathbf{i}_k &= T_{k1}\mathbf{j}_1 + T_{k2}\mathbf{j}_2 + T_{k3}\mathbf{j}_3 \\ \dot{\mathbf{i}}_k &= \dot{T}_{k1}\mathbf{j}_1 + \dot{T}_{k2}\mathbf{j}_2 + \dot{T}_{k3}\mathbf{j}_3 \\ \ddot{\mathbf{i}}_k &= \ddot{T}_{k1}\mathbf{j}_1 + \ddot{T}_{k2}\mathbf{j}_2 + \ddot{T}_{k3}\mathbf{j}_3 \end{aligned} \quad (7.8.28c)$$

for $k = 1, 2, 3$. Using (7.8.28a-c) and the identity $\dot{\mathbf{i}}_k \cdot \mathbf{i}_k = 0$ in (7.2.2a), we obtain

$$\begin{aligned}\delta T = -\int_A \Big(&A_u\delta u + A_v\delta v + A_w\delta w + A_{\gamma_4}\delta\gamma_4 + A_{\gamma_5}\delta\gamma_5 \\ &+ A_{\theta_1}\delta\theta_1 + A_{\theta_2}\delta\theta_2 + A_{\theta_3}\delta\theta_3\Big)dxdy \end{aligned} \quad (7.8.29)$$

where the inertia terms $A_u, A_v, A_w, A_{\gamma_4}, A_{\gamma_5}, A_{\theta_1}, A_{\theta_2}$, and A_{θ_3} are given by

$$A_u \equiv I_0\ddot{u} + I_1\ddot{T}_{31} + \left[(\gamma_5 I_6 + \gamma_4 I_5)T_{11}\right]^{\cdot\cdot} + \left[(\gamma_4 I_7 + \gamma_5 I_8)T_{21}\right]^{\cdot\cdot} \quad (7.8.30a)$$

$$A_v \equiv I_0\ddot{v} + I_1\ddot{T}_{32} + [(\gamma_5 I_6 + \gamma_4 I_5) T_{12}\ddot{]} + [(\gamma_4 I_7 + \gamma_5 I_8) T_{22}\ddot{]} \quad (7.8.30\text{b})$$

$$A_w \equiv I_0\ddot{w} + I_1\ddot{T}_{33} + [(\gamma_5 I_6 + \gamma_4 I_5) T_{13}\ddot{]} + [(\gamma_4 I_7 + \gamma_5 I_8) T_{23}\ddot{]} \quad (7.8.30\text{c})$$

$$\begin{aligned}
A_{\gamma_4} \equiv\ & I_5\left(\ddot{u}T_{11} + \ddot{v}T_{12} + \ddot{w}T_{13}\right) + I_{51}\ddot{\mathbf{i}}_3 \cdot \mathbf{i}_1 \\
& + \ddot{\gamma}_5 I_{56} + \ddot{\gamma}_4 I_{55} + (\gamma_5 I_{56} + \gamma_4 I_{55})\ddot{\mathbf{i}}_1 \cdot \mathbf{i}_1 \\
& + 2\left(\dot{\gamma}_4 I_{57} + \dot{\gamma}_5 I_{58}\right)\dot{\mathbf{i}}_2 \cdot \mathbf{i}_1 + (\gamma_4 I_{57} + \gamma_5 I_{58})\ddot{\mathbf{i}}_2 \cdot \mathbf{i}_1 \\
& + I_7\left(\ddot{u}T_{21} + \ddot{v}T_{22} + \ddot{w}T_{23}\right) + I_{71}\ddot{\mathbf{i}}_3 \cdot \mathbf{i}_2 \\
& + \ddot{\gamma}_4 I_{77} + \ddot{\gamma}_5 I_{78} + (\gamma_4 I_{77} + \gamma_5 I_{78})\ddot{\mathbf{i}}_2 \cdot \mathbf{i}_2 \\
& + 2\left(\dot{\gamma}_5 I_{67} + \dot{\gamma}_4 I_{57}\right)\dot{\mathbf{i}}_1 \cdot \mathbf{i}_2 + (\gamma_5 I_{67} + \gamma_4 I_{57})\ddot{\mathbf{i}}_1 \cdot \mathbf{i}_2 \quad (7.8.30\text{d})
\end{aligned}$$

$$\begin{aligned}
A_{\gamma_5} \equiv\ & I_6\left(\ddot{u}T_{11} + \ddot{v}T_{12} + \ddot{w}T_{13}\right) + I_{61}\ddot{\mathbf{i}}_3 \cdot \mathbf{i}_1 \\
& + \ddot{\gamma}_5 I_{66} + \ddot{\gamma}_4 I_{56} + (\gamma_5 I_{66} + \gamma_4 I_{56})\ddot{\mathbf{i}}_1 \cdot \mathbf{i}_1 \\
& + 2\left(\dot{\gamma}_4 I_{67} + \dot{\gamma}_5 I_{68}\right)\dot{\mathbf{i}}_2 \cdot \mathbf{i}_1 + (\gamma_4 I_{67} + \gamma_5 I_{68})\ddot{\mathbf{i}}_2 \cdot \mathbf{i}_1 \\
& + I_8\left(\ddot{u}T_{21} + \ddot{v}T_{22} + \ddot{w}T_{23}\right) + I_{81}\ddot{\mathbf{i}}_3 \cdot \mathbf{i}_2 \\
& + \ddot{\gamma}_4 I_{78} + \ddot{\gamma}_5 I_{88} + (\gamma_4 I_{78} + \gamma_5 I_{88})\ddot{\mathbf{i}}_2 \cdot \mathbf{i}_2 \\
& + 2\left(\dot{\gamma}_5 I_{68} + \dot{\gamma}_4 I_{58}\right)\dot{\mathbf{i}}_1 \cdot \mathbf{i}_2 + (\gamma_5 I_{68} + \gamma_4 I_{58})\ddot{\mathbf{i}}_1 \cdot \mathbf{i}_2 \quad (7.8.30\text{e})
\end{aligned}$$

$$\begin{aligned}
A_{\theta_1} \equiv\ & -I_1\left(\ddot{u}T_{21} + \ddot{v}T_{22} + \ddot{w}T_{23}\right) - I_2\ddot{\mathbf{i}}_3 \cdot \mathbf{i}_2 - 2\left(\dot{\gamma}_5 I_{61} + \dot{\gamma}_4 I_{51}\right)\dot{\mathbf{i}}_1 \cdot \mathbf{i}_2 \\
& -(\gamma_5 I_{61} + \gamma_4 I_{51})\ddot{\mathbf{i}}_1 \cdot \mathbf{i}_2 - \ddot{\gamma}_4 I_{71} - \ddot{\gamma}_5 I_{81} - (\gamma_4 I_{71} + \gamma_5 I_{81})\ddot{\mathbf{i}}_2 \cdot \mathbf{i}_2 \\
& + (\gamma_4 I_7 + \gamma_5 I_8)\left(\ddot{u}T_{31} + \ddot{v}T_{32} + \ddot{w}T_{33}\right) + (\gamma_4 I_{71} + \gamma_5 I_{81})\ddot{\mathbf{i}}_3 \cdot \mathbf{i}_3 \\
& + 2\left(\dot{\gamma}_5\gamma_4 I_{67} + \dot{\gamma}_5\gamma_5 I_{68} + \dot{\gamma}_4\gamma_4 I_{57} + \dot{\gamma}_4\gamma_5 I_{58}\right)\dot{\mathbf{i}}_1 \cdot \mathbf{i}_3 \\
& + (\gamma_5\gamma_4 I_{67} + \gamma_5\gamma_5 I_{68} + \gamma_4\gamma_4 I_{57} + \gamma_4\gamma_5 I_{58})\ddot{\mathbf{i}}_1 \cdot \mathbf{i}_3 \\
& + 2\left(\dot{\gamma}_4\gamma_4 I_{77} + \dot{\gamma}_4\gamma_5 I_{78} + \dot{\gamma}_5\gamma_4 I_{78} + \dot{\gamma}_5\gamma_5 I_{88}\right)\dot{\mathbf{i}}_2 \cdot \mathbf{i}_3 \\
& + (\gamma_4\gamma_4 I_{77} + 2\gamma_4\gamma_5 I_{78} + \gamma_5\gamma_5 I_{88})\ddot{\mathbf{i}}_2 \cdot \mathbf{i}_3 \quad (7.8.30\text{f})
\end{aligned}$$

$$\begin{aligned}
A_{\theta_2} \equiv\ & I_1\left(\ddot{u}T_{11} + \ddot{v}T_{12} + \ddot{w}T_{13}\right) + I_2\ddot{\mathbf{i}}_3 \cdot \mathbf{i}_1 + \ddot{\gamma}_5 I_{61} + \ddot{\gamma}_4 I_{51} \\
& + (\gamma_5 I_{61} + \gamma_4 I_{51})\ddot{\mathbf{i}}_1 \cdot \mathbf{i}_1 + 2\left(\dot{\gamma}_4 I_{71} + \dot{\gamma}_5 I_{81}\right)\dot{\mathbf{i}}_2 \cdot \mathbf{i}_1 + (\gamma_4 I_{71} + \gamma_5 I_{81})\ddot{\mathbf{i}}_2 \cdot \mathbf{i}_1 \\
& - (\gamma_5 I_6 + \gamma_4 I_5)\left(\ddot{u}T_{31} + \ddot{v}T_{32} + \ddot{w}T_{33}\right) - (\gamma_5 I_{61} + \gamma_4 I_{51})\ddot{\mathbf{i}}_3 \cdot \mathbf{i}_3 \\
& - 2\left(\dot{\gamma}_5\gamma_5 I_{66} + \dot{\gamma}_5\gamma_4 I_{56} + \dot{\gamma}_4\gamma_5 I_{56} + \dot{\gamma}_4\gamma_4 I_{55}\right)\dot{\mathbf{i}}_1 \cdot \mathbf{i}_3 \\
& - (\gamma_5\gamma_5 I_{66} + 2\gamma_5\gamma_4 I_{56} + \gamma_4\gamma_4 I_{55})\ddot{\mathbf{i}}_1 \cdot \mathbf{i}_3 \\
& - 2\left(\dot{\gamma}_4\gamma_5 I_{67} + \dot{\gamma}_4\gamma_4 I_{57} + \dot{\gamma}_5\gamma_5 I_{68} + \dot{\gamma}_5\gamma_4 I_{58}\right)\dot{\mathbf{i}}_2 \cdot \mathbf{i}_3 \\
& - (\gamma_4\gamma_5 I_{67} + \gamma_4\gamma_4 I_{57} + \gamma_5\gamma_5 I_{68} + \gamma_5\gamma_4 I_{58})\ddot{\mathbf{i}}_2 \cdot \mathbf{i}_3 \quad (7.8.30\text{g})
\end{aligned}$$

$$\begin{aligned}
A_{\theta_3} \equiv\ & (\gamma_5 I_6 + \gamma_4 I_5)\left(\ddot{u}T_{21} + \ddot{v}T_{22} + \ddot{w}T_{23}\right) + (\gamma_5 I_{61} + \gamma_4 I_{51})\ddot{\mathbf{i}}_3 \cdot \mathbf{i}_2 \\
& + 2\left(\dot{\gamma}_5\gamma_5 I_{66} + \dot{\gamma}_5\gamma_4 I_{56} + \dot{\gamma}_4\gamma_5 I_{56} + \dot{\gamma}_4\gamma_4 I_{55}\right)\dot{\mathbf{i}}_1 \cdot \mathbf{i}_2 \\
& + (\gamma_5\gamma_5 I_{66} + 2\gamma_5\gamma_4 I_{56} + \gamma_4\gamma_4 I_{55})\ddot{\mathbf{i}}_1 \cdot \mathbf{i}_2 \\
& + \ddot{\gamma}_4\gamma_5 I_{67} + \ddot{\gamma}_4\gamma_4 I_{57} + \ddot{\gamma}_5\gamma_5 I_{68} + \ddot{\gamma}_5\gamma_4 I_{58} \\
& + (\gamma_4\gamma_5 I_{67} + \gamma_4\gamma_4 I_{57} + \gamma_5\gamma_5 I_{68} + \gamma_5\gamma_4 I_{58})\ddot{\mathbf{i}}_2 \cdot \mathbf{i}_2
\end{aligned}$$

$$- (\gamma_4 I_7 + \gamma_5 I_8)(\ddot{u}T_{11} + \ddot{v}T_{12} + \ddot{w}T_{13}) - (\gamma_4 I_{71} + \gamma_5 I_{81})\ddot{\mathbf{i}}_3 \cdot \mathbf{i}_1$$
$$-\ddot{\gamma}_5\gamma_4 I_{67} - \ddot{\gamma}_5\gamma_5 I_{68} - \ddot{\gamma}_4\gamma_4 I_{57} - \ddot{\gamma}_4\gamma_5 I_{58}$$
$$- (\gamma_5\gamma_4 I_{67} + \gamma_5\gamma_5 I_{68} + \gamma_4\gamma_4 I_{57} + \gamma_4\gamma_5 I_{58})\ddot{\mathbf{i}}_1 \cdot \mathbf{i}_1$$
$$-2(\dot{\gamma}_4\gamma_4 I_{77} + \dot{\gamma}_4\gamma_5 I_{78} + \dot{\gamma}_5\gamma_4 I_{78} + \dot{\gamma}_5\gamma_5 I_{88})\dot{\mathbf{i}}_2 \cdot \mathbf{i}_1$$
$$- (\gamma_4\gamma_4 I_{77} + 2\gamma_4\gamma_5 I_{78} + \gamma_5\gamma_5 I_{88})\ddot{\mathbf{i}}_2 \cdot \mathbf{i}_1 \qquad (7.8.30\text{h})$$

The inertias are defined as

$$\{I_0, I_1, I_2, I_5, I_6, I_7, I_8\}$$
$$\equiv \sum_{i=1}^{N} \int_{z_i}^{z_{i+1}} \rho^{(i)} \left\{1, z, z^2, g_{14}^{(i)}, g_{15}^{(i)}, g_{24}^{(i)}, g_{25}^{(i)}\right\} dz$$

$$\{I_{51}, I_{61}, I_{71}, I_{81}\} \equiv \sum_{i=1}^{N} \int_{z_i}^{z_{i+1}} \rho^{(i)} \left\{g_{14}^{(i)}z, g_{15}^{(i)}z, g_{24}^{(i)}z, g_{25}^{(i)}z\right\} dz$$

$$\{I_{55}, I_{56}, I_{57}, I_{58}\}$$
$$\equiv \sum_{i=1}^{N} \int_{z_i}^{z_{i+1}} \rho^{(i)} \left\{g_{14}^{(i)}g_{14}^{(i)}, g_{14}^{(i)}g_{15}^{(i)}, g_{14}^{(i)}g_{24}^{(i)}, g_{14}^{(i)}g_{25}^{(i)}\right\} dz$$

$$\{I_{66}, I_{67}, I_{68}, I_{77}, I_{78}, I_{88}\}$$
$$\equiv \sum_{i=1}^{N} \int_{z_i}^{z_{i+1}} \rho^{(i)} \left\{g_{15}^{(i)}g_{15}^{(i)}, g_{15}^{(i)}g_{24}^{(i)}, g_{15}^{(i)}g_{25}^{(i)}, g_{24}^{(i)}g_{24}^{(i)}, g_{24}^{(i)}g_{25}^{(i)}, g_{25}^{(i)}g_{25}^{(i)}\right\} dz$$

$$(7.8.31)$$

where

$$\dot{\mathbf{i}}_j \cdot \mathbf{i}_k = \dot{T}_{j1}T_{k1} + \dot{T}_{j2}T_{k2} + \dot{T}_{j3}T_{k3}$$
$$\ddot{\mathbf{i}}_j \cdot \mathbf{i}_k = \ddot{T}_{j1}T_{k1} + \ddot{T}_{j2}T_{k2} + \ddot{T}_{j3}T_{k3} \qquad (7.8.32)$$

for $j, k = 1, 2, 3$. We note that, when there are no shear couplings, $g_{14}^{(i)} = g_{25}^{(i)} = 0$, $I_5 = I_8 = I_{51} = I_{81} = I_{55} = I_{56} = I_{57} = I_{58} = I_{68} = I_{78} = I_{88} = 0$.

Substituting (7.8.29) and (7.8.26) into (7.2.1), using (7.5.44a,b), integrating the results by parts, and then setting each of the coefficients of δu, δv, δw, $\delta \gamma_4$, and $\delta \gamma_5$ equal to zero, we obtain the following equations of motion:

$$F_{11x} + F_{12y} = A_u + \mu_1 \dot{u} \qquad (7.8.33\text{a})$$

$$F_{21x} + F_{22y} = A_v + \mu_2 \dot{v} \qquad (7.8.33\text{b})$$

$$F_{31x} + F_{32y} = A_w + \mu_3 \dot{w} \qquad (7.8.33\text{c})$$

$$m_{61x} + m_{2y} - q_2 + \tilde{s}_{11}k_1 + \bar{s}_{22}k_2 + \bar{s}_{12}k_{61} + \tilde{s}_{21}k_{62} - (\tilde{m}_2 - \overline{m}_{61})k_4$$
$$- (\tilde{m}_{62} - \overline{m}_1)k_5 = A_{\gamma_4} + \mu_4 \dot{\gamma}_4 \qquad (7.8.33\text{d})$$

$$m_{1x} + m_{62y} - q_1 + \bar{s}_{11}k_1 + \bar{s}_{22}k_2 + \tilde{s}_{12}k_{61} + \bar{s}_{21}k_{62} - (\overline{m}_2 - \tilde{m}_{61})k_4$$
$$- (\overline{m}_{62} - \tilde{m}_1)k_5 = A\dot{\gamma}_5 + \mu_5\dot{\gamma}_5 \qquad (7.8.33e)$$

where

$$\{F_{11}, F_{21}, F_{31}\} = \{N_1, N_6/(1+e_1), -\bar{\Theta}_2/(1+e_1)\}\,[T]$$
$$\{F_{12}, F_{22}, F_{32}\} = \{N_6/(1+e_2), N_2, \bar{\Theta}_1/(1+e_2)\}\,[T]$$
$$\bar{\Theta}_1 \equiv \Theta_1 + A_{\theta_1}$$
$$\bar{\Theta}_2 \equiv \Theta_2 + A_{\theta_2} \qquad (7.8.34)$$

We added a linear viscous damping term to each of (7.8.33a-e), where the μ_i are the damping coefficients. The boundary conditions for the plate are of the form: specify

Along $x = 0, a$:

$\delta u = 0$	or	$F_{11} + \left[\hat{M}_{61}T_{31}/(1+e_2)\right]_y$
$\delta v = 0$	or	$F_{21} + \left[\hat{M}_{61}T_{32}/(1+e_2)\right]_y$
$\delta w = 0$	or	$F_{31} + \left[\hat{M}_{61}T_{33}/(1+e_2)\right]_y$
$\delta\theta_2 = 0$	or	$\hat{M}_1$
$\delta\gamma_4 = 0$	or	m_{61}
$\delta\gamma_5 = 0$	or	m_1

Along $y = 0, b$:

$\delta u = 0$	or	$F_{12} + \left[\hat{M}_{62}T_{31}/(1+e_1)\right]_x$
$\delta v = 0$	or	$F_{22} + \left[\hat{M}_{62}T_{32}/(1+e_1)\right]_x$
$\delta w = 0$	or	$F_{32} + \left[\hat{M}_{62}T_{33}/(1+e_1)\right]_x$
$\delta\theta_1 = 0$	or	$\hat{M}_2$
$\delta\gamma_4 = 0$	or	m_2
$\delta\gamma_5 = 0$	or	m_{62}

At $(x,y) = (0,0), (a,b), (a,0), (0,b)$:

$\delta u = 0$	or	$-T_{31}\left[\hat{M}_{62}/(1+e_1) + \hat{M}_{61}/(1+e_2)\right]$
$\delta v = 0$	or	$-T_{32}\left[\hat{M}_{62}/(1+e_1) + \hat{M}_{61}/(1+e_2)\right]$
$\delta w = 0$	or	$-T_{33}\left[\hat{M}_{62}/(1+e_1) + \hat{M}_{61}/(1+e_2)\right]$ (7.8.35)

The corner conditions for u and v are nonlinear as can be seen from (7.8.35). The stress resultants and moments, which include piezoelectric actuating forces are separated into two parts: one is due to deformations (indicated by the superscript d) and the other is due to the actuating forces (indicated by superscript f). The result is

$$\{N_1, N_2, N_6, M_1, M_2, M_6, m_{61}, m_2, m_1, m_{62}\}$$
$$\equiv \{N_1^d, N_2^d, N_6^d, M_1^d, M_2^d, M_6^d, m_{61}^d, m_2^d, m_1^d, m_{62}^d\}$$
$$- H\left\{N_1^f, N_2^f, N_6^f, M_1^f, M_2^f, M_6^f, m_{61}^f, m_2^f, m_1^f, m_{62}^f\right\} \quad (7.8.36a)$$

$$\{\overline{m}_1, \tilde{m}_1, \tilde{m}_2, \overline{m}_2, \tilde{m}_{62}, \overline{m}_{62}, \overline{m}_{61}, \tilde{m}_{61}\}$$
$$\equiv \{\overline{m}_1^d, \tilde{m}_1^d, \tilde{m}_2^d, \overline{m}_2^d, \tilde{m}_{62}^d, \overline{m}_{62}^d, \overline{m}_{61}^d, \tilde{m}_{61}^d\}$$
$$- H\left\{\overline{m}_1^f, \tilde{m}_1^f, \tilde{m}_2^f, \overline{m}_2^f, \tilde{m}_{62}^f, \overline{m}_{62}^f, \overline{m}_{61}^f, \tilde{m}_{61}^f\right\} \quad (7.8.36b)$$

where $H(x,y)$ is the Heavyside function and

$$\{N_1^d, N_2^d, N_6^d, M_1^d, M_2^d, M_6^d, m_{61}^d, m_2^d, m_1^d, m_{62}^d\}$$
$$\equiv \{\psi\}^T \sum_{i=1}^{N} \int_{z_i}^{z_{i+1}} \left[S_1^{(i)}\right]^T \left[\tilde{Q}^{(i)}\right] \left[S_1^{(i)}\right] dz$$
$$\left\{N_1^f, N_2^f, N_6^f, M_1^f, M_2^f, M_6^f, m_{61}^f, m_2^f, m_1^f, m_{62}^f\right\}$$
$$\equiv \{\bar{\Lambda}_1, \bar{\Lambda}_2, \bar{\Lambda}_{12}\} \int_{z_I}^{z_{I+1}} \left[\tilde{Q}^{(I)}\right] \left[S_1^{(I)}\right] dz \quad (7.8.37a)$$

$$\{\overline{m}_1^d, \tilde{m}_1^d, \tilde{m}_2^d, \overline{m}_2^d, \tilde{m}_{62}^d, \overline{m}_{62}^d, \overline{m}_{61}^d, \tilde{m}_{61}^d\}$$
$$\equiv \{\psi\}^T \sum_{i=1}^{N} \int_{z_i}^{z_{i+1}} \left[S_1^{(i)}\right]^T \left[\tilde{Q}^{(i)}\right] \left[S_3^{(i)}\right] dz$$
$$\left\{\overline{m}_1^f, \tilde{m}_1^f, \tilde{m}_2^f, \overline{m}_2^f, \tilde{m}_{62}^f, \overline{m}_{62}^f, \overline{m}_{61}^f, \tilde{m}_{61}^f\right\}$$
$$\equiv \{\bar{\Lambda}_1, \bar{\Lambda}_2, \bar{\Lambda}_{12}\} \int_{z_I}^{z_{I+1}} \left[\tilde{Q}^{(I)}\right] \left[S_3^{(I)}\right] dz \quad (7.8.37b)$$

Because the rotary inertia A_{θ_3} about the ζ axis is due to nonlinear effects and is relatively small, we have

$$\Theta_3 = 0 \quad (7.8.38)$$

which is a statement of the balance of the internal moments with respect to the ζ axis and is not needed in the analysis.

7.8.3 Linear Piezoelectric Plate Theory

To develop a linear piezoelectric plate theory, we first use (7.5.6), (7.5.7), (7.5.10), and (7.5.11) to expand the transformation matrix $[T]$ as

$$[T] = \begin{bmatrix} 1 & v_x & w_x \\ u_y & 1 & w_y \\ -w_x & -w_y & 1 \end{bmatrix} \quad (7.8.39)$$

Substituting (7.8.39) and (7.5.11) into (7.8.30a-h) yields the linear inertia terms as

$$A_u = I_0\ddot{u} - I_1\ddot{w}_x + I_6\ddot{\gamma}_5 + I_5\ddot{\gamma}_4$$
$$A_v = I_0\ddot{v} - I_1\ddot{w}_y + I_7\ddot{\gamma}_4 + I_8\ddot{\gamma}_5$$
$$A_w = I_0\ddot{w}$$
$$A_{\gamma_4} = (I_{55} + I_{77})\ddot{\gamma}_4 + (I_{56} + I_{78})\ddot{\gamma}_5 + I_5\ddot{u} + I_7\ddot{v} - I_{51}\ddot{w}_x - I_{71}\ddot{w}_y$$
$$A_{\gamma_5} = (I_{66} + I_{88})\ddot{\gamma}_5 + (I_{56} + I_{78})\ddot{\gamma}_4 + I_6\ddot{u} + I_8\ddot{v} - I_{61}\ddot{w}_x - I_{81}\ddot{w}_y$$
$$A_{\theta_1} = I_2\ddot{w}_y - I_1\ddot{v} - I_{71}\ddot{\gamma}_4 - I_{81}\ddot{\gamma}_5$$
$$A_{\theta_2} = -I_2\ddot{w}_x + I_1\ddot{u} + I_{61}\ddot{\gamma}_5 + I_{51}\ddot{\gamma}_4$$
$$A_{\theta_3} = 0 \qquad (7.8.40)$$

Linearizing (7.8.33a-e), we obtain the following the linear equations of motion for piezoelectric plates:

$$N^d_{1x} + N^d_{6y} - N^f_1 L_x - N^f_6 L_y = I_0\ddot{u} - I_1\ddot{w}_x + I_6\ddot{\gamma}_5 + I_5\ddot{\gamma}_4 + \mu_1\dot{u} \quad (7.8.41a)$$

$$N^d_{6x} + N^d_{2y} - N^f_6 L_x - N^f_2 L_y = I_0\ddot{v} - I_1\ddot{w}_y + I_7\ddot{\gamma}_4 + I_8\ddot{\gamma}_5 + \mu_2\dot{v} \quad (7.8.41b)$$

$$M^d_{1xx} + 2M^d_{6xy} + M^d_{2yy} - M^f_1 L_{xx} - 2M^f_6 L_{xy} - M^f_2 L_{yy} = I_0\ddot{w} + \mu_3\dot{w}$$
$$+ (I_1\ddot{u} - I_2\ddot{w}_x + I_{61}\ddot{\gamma}_5 + I_{51}\ddot{\gamma}_4)_x + (I_1\ddot{v} - I_2\ddot{w}_y + I_{71}\ddot{\gamma}_4 + I_{81}\ddot{\gamma}_5)_y$$
$$(7.8.41c)$$

$$m^d_{61x} + m^d_{2y} - q_2 - m^f_{61} L_x - m^f_2 L_y = (I_{55} + I_{77})\ddot{\gamma}_4 + (I_{56} + I_{78})\ddot{\gamma}_5$$
$$+ I_5\ddot{u} + I_7\ddot{v} - I_{51}\ddot{w}_x - I_{71}\ddot{w}_y + \mu_4\dot{\gamma}_4 \quad (7.8.41d)$$

$$m^d_{1x} + m^d_{62y} - q_1 - m^f_1 L_x - m^f_{62} L_y = (I_{66} + I_{88})\ddot{\gamma}_5 + (I_{56} + I_{78})\ddot{\gamma}_4$$
$$+ I_6\ddot{u} + I_8\ddot{v} - I_{61}\ddot{w}_x - I_{81}\ddot{w}_y + \mu_5\dot{\gamma}_5 \quad (7.8.41e)$$

where the stress resultants are defined in (7.8.37a) with

$$\{\psi\} = \{u_x, v_y, u_y + v_x, -w_{xx}, -w_{yy}, -2w_{xy}, \gamma_{4x}, \gamma_{4y}, \gamma_{5x}, \gamma_{5y}\}^T \qquad (7.8.42)$$

and

$$\{q_2, q_1\} \equiv \{\gamma_4, \gamma_5\} \sum_{i=1}^{N} \int_{z_i}^{z_{i+1}} \left[S_2^{(i)}\right]^T \left[\hat{Q}^{(i)}\right] \left[S_2^{(i)}\right] dz \qquad (7.8.43)$$

The corresponding linear boundary conditions are of the form: specify

Along $x = 0, a$:

$\delta u = 0$ or N_1

$\delta v = 0$ or N_6

$\delta w = 0$ or $M_{1x} + 2M_{6y} + I_2\ddot{w}_x - I_1\ddot{u} - I_{61}\ddot{\gamma}_5 - I_{51}\ddot{\gamma}_4$

$\delta w_x = 0$ or M_1

$\delta \gamma_4 = 0$ or m_{61}

$\delta\gamma_5 = 0$ or m_1

Along $y = 0, b$:

     $\delta u = 0$ or N_6

     $\delta v = 0$ or N_2

     $\delta w = 0$ or $M_{2y} + 2M_{6x} + I_2 \ddot{w}_y - I_1 \ddot{v} - I_{71}\ddot{\gamma}_4 - I_{81}\ddot{\gamma}_5$

     $\delta w_y = 0$ or M_2

     $\delta\gamma_4 = 0$ or m_2

     $\delta\gamma_5 = 0$ or m_{62}

At $(x, y) = (0, 0), (0, b), (a, 0), (a, b)$:

     $\delta w = 0$ or $2M_6$ (7.8.44)

7.8.4 Actuator-Induced Loads

It follows from (7.8.25a-e) that all stress resultants and moments are defined with respect to the deformed coordinate system $\xi\eta\zeta$ because the Jaumann stresses are defined with respect to the deformed coordinate system. Moreover, to solve the governing equations (7.8.33a-e) subject to the boundary conditions (7.8.35), one needs to evaluate the integral

$$\int_A m^f \left(L_x, L_y, L_{xx}, L_{xy}, L_{yy} \right) dx dy \qquad (7.8.45)$$

where m^f denotes either a stress resultant or a moment due to piezoelectric actuation. Because $L(x, y)$ is a Heaviside function (see (7.8.13c)), L_x and L_y are Kronecker delta functions δ, and L_{xx}, L_{xy}, and L_{yy} are derivatives of δ. Hence, the induced actuating forces and moments act only along the boundaries of the actuators (Lee, 1990). Because of this discontinuity in the actuating loads, the resulting structural strains are discontinuous, especially around the boundaries of the actuators. Obtaining analytical solutions for such problems by assuming displacement functions that account for this discontinuity in the strains is almost an impossible task. On the other hand, a numerical approach based on the finite-element method can be used to account for this discontinuity in the strains as well as arbitrarily designed and distributed actuators.

7.8.5 Thermal and Moisture Effects

Induced strains due to thermal and moisture expansion or contraction are of the same form as those due to piezoelectric actuation except that the location function $R(x, y, z)$ is not needed because thermal and moisture effects are usually continuously distributed over the whole structure. Hence, the stress-strain relations are the same as those shown in (7.8.11) and (7.8.12) except that $R = 1$ and

$$\Lambda_1 = \alpha_1(T - T_0), \ \Lambda_2 = \alpha_2(T - T_0) \qquad (7.8.46a)$$

in the case of a thermal effect, and

$$\Lambda_1 = \beta_1 \delta m, \quad \Lambda_2 = \beta_2 \delta m \qquad (7.8.46b)$$

in the case of a moisture effect. Here, the α_i are the coefficients of thermal expansion, T_0 is the stress-free temperature change, the β_i are the coefficients of hygro-thermal expansion, and δm is the percentage weight increase due to moisture. Moreover, the definitions of the stress resultants and moments due to induced-strain actuation (see (7.8.37a,b)) need to be redefined as

$$\left\{ N_1^f, N_2^f, N_6^f, M_1^f, M_2^f, M_6^f, m_{61}^f, m_2^f, m_1^f, m_{62}^f \right\}$$

$$\equiv \sum_{i=1}^{N} \int_{z_i}^{z_{i+1}} \left\{ \begin{array}{c} \bar{\Lambda}_1^{(i)} \\ \bar{\Lambda}_2^{(i)} \\ \bar{\Lambda}_{12}^{(i)} \end{array} \right\}^T \left[\tilde{Q}^{(i)} \right] \left[S_1^{(i)} \right] dz$$

$$\left\{ \overline{m}_1^f, \tilde{m}_1^f, \tilde{m}_2^f, \overline{m}_2^f, \tilde{m}_{62}^f, \overline{m}_{62}^f, \overline{m}_{61}^f, \tilde{m}_{61}^f \right\}$$

$$\equiv \sum_{i=1}^{N} \int_{z_i}^{z_{i+1}} \left\{ \begin{array}{c} \bar{\Lambda}_1^{(i)} \\ \bar{\Lambda}_2^{(i)} \\ \bar{\Lambda}_{12}^{(i)} \end{array} \right\}^T \left[\tilde{Q}^{(i)} \right] \left[S_3^{(i)} \right] dz \qquad (7.8.47)$$

because all layers are subjected to thermal and/or moisture effects.

We point out that, in this section, the distributions of electric charges, temperature, and moisture are assumed to be known. If these distributions are unknown, equations and boundary conditions describing their variations in space and time need to be derived.

8
DYNAMICS OF PLATES

The linear and nonlinear free and forced vibrations of metallic and composite plates have received considerable attention. Comprehensive accounts of linear vibrations can be found in the books of Lekhnitski (1968), Leissa (1993a), Ambartsumian (1970), Timoshenko (1974), Whitney (1987), Soedel (1993), Reddy (1997, 2003), Ye (2003), Sathyamoorthy (1997), and Qatu (2004), and the review articles of Leissa (1978, 1981a, 1981b, 1981c, 1987a, 1987b). Comprehensive accounts of the nonlinear vibrations of plates can be found in the books of Nayfeh and Mook (1979), Chia (1980), Sathyamoorthy (1997), and Nayfeh (2000), the PhD Dissertations of J. Nayfeh (1990), Hadian (1991), Oh (1994), and Alhazza (2002), and the review article of Kapania and Raciti (1989c). In this chapter, we discuss examples of linear and nonlinear vibrations of plates, including microplates and thermally excited plates.

8.1 LINEAR VIBRATIONS OF RECTANGULAR PLATES

We employ a Cartesian coordinate system xyz, as shown in Figure 7.5.1. The lengths of the plate along the x and y axes are taken to be a and b, respectively. It follows from (7.4.26) that the equation governing the natural frequencies and mode shapes of a rectangular plate subject to an inplane axial load N_1 in the x direction can be expressed as

$$\rho h \frac{\partial^2 w}{\partial t^2} + D\nabla^4 w - N_1 \frac{\partial^2 w}{\partial x^2} = 0 \qquad (8.1.1)$$

470   DYNAMICS OF PLATES

where D, ρ, and h are the flexural rigidity, density, and thickness of the plate. In Section 8.1.1, we consider the case of hinged edges, and in Section 8.1.2 we consider the case of two hinged opposite edges.

8.1.1 Hinged Edges

In this section, we consider the case of simply supported edges (Navier solution). The boundary conditions demand that the displacements and moments vanish at all edges. It follows from (7.2.24) and (7.2.15) that

$$w = 0 \text{ and } w_{xx} + \nu w_{yy} = 0 \text{ at } x = 0 \text{ and } x = a \qquad (8.1.2)$$
$$w = 0 \text{ and } w_{yy} + \nu w_{xx} = 0 \text{ at } y = 0 \text{ and } y = b \qquad (8.1.3)$$

To determine the natural frequencies and mode shapes of the plate, we let

$$w = \phi(x, y) e^{i\omega t} \qquad (8.1.4)$$

in (8.1.1)-(8.1.3) and obtain the eigenvalue problem

$$D\nabla^4 \phi - N_1 \phi_{xx} - \rho h \omega^2 \phi = 0 \qquad (8.1.5)$$

subject to the boundary conditions

$$\phi = 0 \text{ and } \phi_{xx} + \nu \phi_{yy} = 0 \text{ at } x = 0 \text{ and } x = a \qquad (8.1.6)$$
$$\phi = 0 \text{ and } \phi_{yy} + \nu \phi_{xx} = 0 \text{ at } y = 0 \text{ and } y = b \qquad (8.1.7)$$

We seek a solution of (8.1.5)-(8.1.7) in the form

$$\phi = \sin\left(\frac{m\pi x}{a}\right) \sin\left(\frac{n\pi y}{b}\right) \qquad (8.1.8)$$

that satisfies (8.1.6) and (8.1.7) and obtain from (8.1.5) that

$$\omega = \left[\frac{D\pi^4}{\rho h}\left(\frac{m^2}{a^2} + \frac{n^2}{b^2}\right)^2 + \frac{N_1 m^2 \pi^2}{\rho h a^2}\right]^{\frac{1}{2}} \qquad (8.1.9)$$

For a compressive load N_1, ω might vanish or become complex. This condition corresponds to buckling of the plate.

8.1.2 Two Hinged Opposite Edges

We consider the case of Levy solutions in which the edges along the x axis are hinged and the edges along the y axis take on different conditions: clamped-clamped, free-free, hinged-clamped, clamped-free, and hinged-free conditions. Thus,

$$w = 0 \text{ and } w_{yy} + \nu w_{xx} = 0 \text{ at } y = 0 \text{ and } y = b \qquad (8.1.10)$$

and either
$$w = 0 \text{ and } w_x = 0 \text{ at } x = 0 \text{ and } x = a \tag{8.1.11}$$
for clamped-clamped conditions or
$$w_{xx} + \nu w_{yy} = 0 \text{ and } w_{xxx} + (2-\nu)w_{xyy} = 0 \text{ at } x = 0 \text{ and } x = a \tag{8.1.12}$$
for free-free conditions or
$$w = 0 \text{ and } w_{xx} + \nu w_{yy} = 0 \text{ at } x = 0 \tag{8.1.13}$$
$$w = 0 \text{ and } w_x = 0 \text{ at } x = a \tag{8.1.14}$$
for hinged-clamped conditions or
$$w = 0 \text{ and } w_x = 0 \text{ at } x = 0 \tag{8.1.15}$$
$$w_{xx} + \nu w_{yy} = 0 \text{ and } w_{xxx} + (2-\nu)w_{xyy} = 0 \text{ at } x = a \tag{8.1.16}$$
for clamped-free conditions or
$$w = 0 \text{ and } w_{xx} + \nu w_{yy} = 0 \text{ at } x = 0 \tag{8.1.17}$$
$$w_{xx} + \nu w_{yy} = 0 \text{ and } w_{xxx} + (2-\nu)w_{xyy} = 0 \text{ at } x = a \tag{8.1.18}$$
for hinged-free conditions.

In all cases, we seek a solution for w that satisfies the boundary conditions (8.1.10) in the form
$$w = X(x)\sin\left(\frac{n\pi y}{b}\right)e^{i\omega t} \tag{8.1.19}$$
Substituting (8.1.19) into (8.1.1) yields
$$X^{iv} - \left(\frac{2n^2\pi^2}{b^2} + \frac{N_1}{D}\right)X'' + \left(\frac{n^4\pi^4}{b^4} - \frac{\rho h \omega^2}{D}\right)X = 0 \tag{8.1.20}$$
The general solution of (8.1.20) is given by
$$X = c_1 \sin(s_1 x) + c_2 \cos(s_1 x) + c_3 \sinh(s_2 x) + c_4 \cosh(s_2 x) \tag{8.1.21}$$
where the c_i are constants and
$$s_{2,1} = \pm\left[\left(\frac{n^2\pi^2}{b^2} + \frac{N_1}{2D}\right) \pm \sqrt{\frac{\rho h \omega^2}{D} + \left(\frac{n^2\pi^2}{b^2} + \frac{N_1}{2D}\right)^2 - \frac{n^4\pi^4}{b^4}}\right]^{\frac{1}{2}} \tag{8.1.22}$$

Next, we consider the five cases separately.

Clamped-Clamped Case

Substituting (8.1.19) into (8.1.11) yields

$$X = 0 \text{ and } X' = 0 \text{ at } x = 0 \text{ and } x = a \qquad (8.1.23)$$

Substituting (8.1.21) into (8.1.23), we obtain

$$c_2 + c_4 = 0 \qquad (8.1.24)$$
$$s_1 c_1 + s_2 c_3 = 0 \qquad (8.1.25)$$
$$c_1 \sin(s_1 a) + c_2 \cos(s_1 a) + c_3 \sinh(s_2 a) + c_4 \cosh(s_2 a) = 0 \qquad (8.1.26)$$
$$c_1 s_1 \cos(s_1 a) - c_2 s_1 \sin(s_1 a) + c_3 s_2 \cosh(s_2 a) + c_4 s_2 \sinh(s_2 a) = 0 \qquad (8.1.27)$$

Equations (8.1.24)-(8.1.27) represent a system of four algebraic equations in the c_i and the natural frequency ω. For a nontrivial solution, the determinant of the coefficient matrix of (8.1.24)-(8.1.27) must vanish; that is,

$$\begin{vmatrix} 0 & 1 & 0 & 1 \\ s_1 & 0 & s_2 & 0 \\ \sin(s_1 a) & \cos(s_1 a) & \sinh(s_2 a) & \cosh(s_2 a) \\ s_1 \cos(s_1 a) & -s_1 \sin(s_1 a) & s_2 \cosh(s_2 a) & s_2 \sinh(s_2 a) \end{vmatrix} = 0 \qquad (8.1.28)$$

which yields the characteristic equation

$$2 s_1 s_2 \left[\cos(s_1 a) \cosh(s_2 a) - 1\right] + \left(s_1^2 - s_2^2\right) \sin(s_1 a) \sinh(s_2 a) = 0 \qquad (8.1.29)$$

Solving this equation for ω, we obtain the natural frequencies and corresponding mode shapes of the plate.

Free-Free Case

Substituting (8.1.19) into (8.1.12) yields

$$X'' - \nu \beta^2 X = 0 \text{ and } X''' - (2-\nu)\beta^2 X' = 0 \text{ at } x = 0 \text{ and } x = a \qquad (8.1.30)$$

where

$$\beta \equiv \frac{n\pi}{b} \qquad (8.1.31)$$

Substituting (8.1.21) into (8.1.30), we obtain

$$\lambda_1 c_2 - \lambda_2 c_4 = 0 \qquad (8.1.32)$$
$$\lambda_3 c_1 - \lambda_4 c_3 = 0 \qquad (8.1.33)$$
$$\lambda_1 \sin(s_1 a) c_1 + \lambda_1 \cos(s_1 a) c_2 - \lambda_2 \sinh(s_2 a) c_3 - \lambda_2 \cosh(s_2 a) c_4 = 0 \qquad (8.1.34)$$
$$\lambda_3 \cos(s_1 a) c_1 - \lambda_3 \sin(s_1 a) c_2 - \lambda_4 \cosh(s_2 a) c_3 - \lambda_4 \sinh(s_2 a) c_4 = 0 \qquad (8.1.35)$$

where

$$\lambda_1 \equiv s_1^2 + \nu \beta^2, \quad \lambda_2 \equiv s_2^2 - \nu \beta^2,$$
$$\lambda_3 \equiv \left[s_1^2 + (2-\nu)\beta^2\right] s_1, \text{ and } \lambda_4 \equiv \left[s_2^2 - (2-\nu)\beta^2\right] s_2 \qquad (8.1.36)$$

LINEAR VIBRATIONS OF RECTANGULAR PLATES

For a nontrivial solution, the determinant of the coefficient matrix of (8.1.32)-(8.1.35) must vanish; that is,

$$\begin{vmatrix} 0 & \lambda_1 & 0 & -\lambda_2 \\ \lambda_3 & 0 & -\lambda_4 & 0 \\ \lambda_1 \sin(s_1 a) & \lambda_1 \cos(s_1 a) & -\lambda_2 \sinh(s_2 a) & -\lambda_2 \cosh(s_2 a) \\ \lambda_3 \cos(s_1 a) & -\lambda_3 \sin(s_1 a) & -\lambda_4 \cosh(s_2 a) & -\lambda_4 \sinh(s_2 a) \end{vmatrix} = 0 \quad (8.1.37)$$

which yields the characteristic equation

$$2\lambda_1 \lambda_2 \lambda_3 \lambda_4 \left[-1 + \cos(s_1 a) \cosh(s_2 a)\right] = \left(\lambda_1^2 \lambda_4^2 - \lambda_2^2 \lambda_3^2\right) \sin(s_1 a) \sinh(s_2 a) \quad (8.1.38)$$

Hinged-Clamped Case

Substituting (8.1.19) into (8.1.13) and (8.1.14) yields

$$X = 0 \text{ and } X'' - \nu \beta^2 X = 0 \text{ at } x = 0 \quad (8.1.39)$$
$$X = 0 \text{ and } X' = 0 \text{ at } x = a \quad (8.1.40)$$

Substituting (8.1.21) into (8.1.39) and (8.1.40), we obtain

$$c_2 + c_4 = 0 \quad (8.1.41)$$
$$\lambda_1 c_2 - \lambda_2 c_4 = 0 \quad (8.1.42)$$
$$\sin(s_1 a) c_1 + \cos(s_1 a) c_2 + \sinh(s_2 a) c_3 + \cosh(s_2 a) c_4 = 0 \quad (8.1.43)$$
$$s_1 \cos(s_1 a) c_1 - s_1 \sin(s_1 a) c_2 + s_2 \cosh(s_2 a) c_3 + s_2 \sinh(s_2 a) c_4 = 0 \quad (8.1.44)$$

Setting the determinant of the coefficient matrix of (8.1.41)-(8.1.44) equal to zero yields the characteristic equation

$$s_2 \sin(s_1 a) \cosh(s_2 a) = s_1 \cos(s_1 a) \sinh(s_2 a) \quad (8.1.45)$$

Clamped-Free Case

Substituting (8.1.19) into (8.1.15) and (8.1.16) gives

$$X = 0 \text{ and } X' = 0 \text{ at } x = 0 \quad (8.1.46)$$
$$X'' - \nu \beta^2 X = 0 \text{ and } X''' - (2-\nu)\beta^2 X' = 0 \text{ at } x = a \quad (8.1.47)$$

Substituting (8.1.21) into (8.1.46) and (8.1.47), we obtain

$$c_2 + c_4 = 0 \quad (8.1.48)$$
$$s_1 c_1 + s_2 c_3 = 0 \quad (8.1.49)$$
$$\lambda_1 \sin(s_1 a) c_1 + \lambda_1 \cos(s_1 a) c_2 - \lambda_2 \sinh(s_2 a) c_3 - \lambda_2 \cosh(s_2 a) c_4 = 0 \quad (8.1.50)$$
$$\lambda_3 \cos(s_1 a) c_1 - \lambda_3 \sin(s_1 a) c_2 - \lambda_4 \cosh(s_2 a) c_3 - \lambda_4 \sinh(s_2 a) c_4 = 0 \quad (8.1.51)$$

474 DYNAMICS OF PLATES

Setting the determinant of the coefficient matrix of (8.1.48)-(8.1.51) equal to zero yields the characteristic equation

$$(s_2\lambda_1\lambda_4 - s_1\lambda_2\lambda_3)\sin(s_1 a)\sinh(s_2 a) - (s_1\lambda_1\lambda_4 + s_2\lambda_2\lambda_3)\cos(s_1 a)\cosh(s_1 a)$$
$$= s_1\lambda_2\lambda_4 + s_2\lambda_1\lambda_3 \quad (8.1.52)$$

Hinged-Free Case

Substituting (8.1.19) into (8.1.17) and (8.1.18) yields

$$X = 0 \text{ and } X'' - \nu\beta^2 X = 0 \text{ at } x = 0 \quad (8.1.53)$$
$$X'' - \nu\beta^2 X = 0 \text{ and } X''' - (2-\nu)\beta^2 X' = 0 \text{ at } x = a \quad (8.1.54)$$

Substituting (8.1.21) into (8.1.53) and (8.1.54), we obtain

$$c_2 + c_4 = 0 \quad (8.1.55)$$
$$\lambda_1 c_2 - \lambda_2 c_4 = 0 \quad (8.1.56)$$
$$\lambda_1 \sin(s_1 a) c_1 + \lambda_1 \cos(s_1 a) c_2 - \lambda_2 \sinh(s_2 a) c_3 - \lambda_2 \cosh(s_2 a) c_4 = 0 \quad (8.1.57)$$
$$\lambda_3 \cos(s_1 a) c_1 - \lambda_3 \sin(s_1 a) c_2 - \lambda_4 \cosh(s_2 a) c_3 - \lambda_4 \sinh(s_2 a) c_4 = 0 \quad (8.1.58)$$

Setting the determinant of the coefficient matrix of (8.1.55)-(8.1.58) equal to zero yields the characteristic equation

$$\lambda_2\lambda_3 \cos(s_1 a)\sinh(s_2 a) = \lambda_1\lambda_4 \sin(s_1 a)\cosh(s_2 a) \quad (8.1.59)$$

8.2 LINEAR VIBRATIONS OF MEMBRANES

We consider the mode shapes and corresponding natural frequencies of a uniform circular membrane with radius a, density ρ, thickness h, and uniform tension T. We employ a polar coordinate system (r, θ) with its origin at the center of the membrane. Because membranes are pretensioned plates without bending stiffness, it follows from (7.4.49), with $N_1 = N_2 = T$ and $N_6 = D = 0$, that the equation of motion governing the linear vibration frequencies and mode shapes can be expressed as

$$T\nabla^2 w - \rho h \ddot{w} = 0 \quad (8.2.1)$$

In the following section, we consider the case of a circular membrane; in Section 8.2.2, we consider the case of a nearly circular membrane; and in Section 8.2.3, we consider the case of an elliptic membrane.

8.2.1 Circular Membranes

In this case, the boundary conditions are

$$w = 0 \text{ at } r = a \text{ and } w < \infty \text{ at } r = 0 \quad (8.2.2)$$

For time harmonic variation, we let

$$w(r, \theta, t) = \phi(r, \theta) \cos(\omega t + \tau) \tag{8.2.3}$$

in (8.2.1) and (8.2.2) and obtain the eigenvalue problem

$$\nabla^2 \phi + \kappa^2 \phi = 0 \tag{8.2.4}$$

$$\phi = 0 \text{ at } r = a \text{ and } \phi < \infty \text{ at } r = 0 \tag{8.2.5}$$

where $\kappa^2 \equiv \rho h \omega^2 / T$.

We use the method of separation of variables and express $\phi(r, \theta)$ as

$$\phi(r, \theta) = R(r) e^{in\theta} \tag{8.2.6}$$

where n is an integer. Substituting (8.2.6) into (8.2.4) and (8.2.5) yields

$$\frac{d^2 R}{dr^2} + \frac{1}{r}\frac{dR}{dr} + \left(\kappa^2 - \frac{n^2}{r^2}\right) R = 0 \tag{8.2.7}$$

$$R = 0 \text{ at } r = a \text{ and } R < \infty \text{ at } r = 0 \tag{8.2.8}$$

The solution of (8.2.7) that is bounded at the origin is

$$R = c J_n(\kappa r) \tag{8.2.9}$$

where J_n is Bessel's function of the first kind. Imposing the first boundary condition in (8.2.8) yields the characteristic equation

$$J_n(\kappa a) = 0 \tag{8.2.10}$$

Using the following asymptotic expansion of $J_n(\alpha)$ for large α:

$$J_n(\alpha) = \sqrt{\frac{2}{\pi \alpha}} \left[\cos\left(\alpha - \frac{1}{2} n\pi - \frac{1}{4}\pi\right) - \frac{4n^2 - 1}{8\alpha} \sin\left(\alpha - \frac{1}{2} n\pi - \frac{1}{4}\pi\right) \right]$$
$$+ O\left(\frac{1}{\alpha^3}\right) \tag{8.2.11}$$

where $\alpha = \kappa a$, we rewrite (8.2.10) as

$$\cos\left(\alpha - \frac{1}{2} n\pi - \frac{1}{4}\pi\right) = \frac{4n^2 - 1}{8\alpha} \sin\left(\alpha - \frac{1}{2} n\pi - \frac{1}{4}\pi\right) + \cdots \tag{8.2.12}$$

or

$$\cot\left(\alpha - \frac{1}{2} n\pi - \frac{1}{4}\pi\right) = \frac{4n^2 - 1}{8\alpha} + \cdots \tag{8.2.13}$$

To the first approximation,

$$\cos\left(\alpha - \frac{1}{2} n\pi - \frac{1}{4}\pi\right) = 0$$

Hence,
$$\alpha = \left(m + \frac{1}{2}n + \frac{3}{4}\right)\pi$$
where n is the integer. Then, we let
$$\alpha = \left(m + \frac{1}{2}n + \frac{3}{4}\right)\pi + \delta \tag{8.2.14}$$
in (8.2.11) and obtain
$$\frac{\cot\left[\left(m + \frac{1}{2}n + \frac{3}{4}\right)\pi\right]\cot\delta - 1}{\cot\left[\left(m + \frac{1}{2}n + \frac{3}{4}\right)\pi\right] + \cot\delta} = \frac{4n^2 - 1}{8\left(m + \frac{1}{2}n + \frac{3}{4}\right)\pi + 8\delta} + \cdots$$
or
$$-\tan\delta = \frac{4n^2 - 1}{8\left(m + \frac{1}{2}n + \frac{3}{4}\right)\pi + 8\delta} + \cdots$$
Hence,
$$\delta \approx -\frac{4n^2 - 1}{8\left(m + \frac{1}{2}n + \frac{3}{4}\right)\pi}$$
and
$$\alpha_{nm} \approx \left(m + \frac{1}{2}n + \frac{3}{4}\right)\pi - \frac{4n^2 - 1}{8\left(m + \frac{1}{2}n + \frac{3}{4}\right)\pi} \tag{8.2.15}$$

Using the values of α_{nm} calculated from (8.2.15) as initial guesses, we solved (8.2.10) numerically. These results are compared with the asymptotic values in Table 8.1. Clearly, the larger the root is and the smaller n is, the closer the asymptotic value to the numerically calculated value is. To improve the asymptotic values for large n, we need to use the asymptotic expansion of $J_n(\alpha)$ for large n.

Table 8.1 Values of the eigenvalues α_{nm} of a circular membrane for different combinations of $n \geq 0$ and $m \geq 1$.

n	Method	α_{n1}	α_{n2}	α_{n3}	α_{n4}	α_{n5}
0	Numerical	2.405	5.200	8.654	11.792	14.931
	Asymptotic	2.409	5.201	8.654	11.792	14.931
1	Numerical	3.832	7.016	10.174	13.324	16.471
	Asymptotic	3.832	7.016	10.173	13.324	16.471
2	Numerical	5.136	8.417	11.620	14.796	17.960
	Asymptotic	5.157	8.422	11.622	14.797	17.960
3	Numerical	6.380	9.761	13.015	16.224	19.410
	Asymptotic	6.450	9.782	13.024	16.228	19.412

Therefore, the natural frequencies of the membrane are given by

$$\omega_{nm} = \sqrt{\frac{T}{\rho h}} \frac{\alpha_{nm}}{a} \qquad (8.2.16)$$

Combining (8.2.6) and (8.2.9), we find that

$$\phi_{nm} = J_n\left(\frac{\alpha_{nm} r}{a}\right)[c_1 \cos n\theta + c_2 \sin n\theta] \qquad (8.2.17)$$

We note that, for every frequency ω_{nm}, there are two possible modes of oscillation, namely

$$J_n\left(\frac{\alpha_{nm} r}{a}\right) \cos n\theta \quad \text{and} \quad J_n\left(\frac{\alpha_{nm} r}{a}\right) \sin n\theta \qquad (8.2.18)$$

Consequently, when $n \neq 0$, we speak of the circular membrane as a degenerate system because there are more than one eigenfunction (mode shape) corresponding to a given frequency. The degeneracy is the result of symmetry and it can be removed by introducing asymmetries as shown in the next section.

8.2.2 Near Circular Membranes

In this case, the boundary conditions are

$$w = 0 \text{ at } r = a\left[1 + \epsilon f(\theta)\right] \quad \text{and} \quad w < \infty \text{ at } r = 0 \qquad (8.2.19)$$

where $r = a + \epsilon a f(\theta)$ describes the boundary of the membrane and ϵ is a small nondimensional parameter. We assume that $r = a$ is the mean radius of the membrane so that

$$\int_0^{2\pi} f(\theta) d\theta = 0 \qquad (8.2.20)$$

Substituting (8.2.3) into (8.2.1) and (8.2.19), we obtain an eigenvalue problem consisting of (8.2.4) subject to the boundary conditions

$$\phi = 0 \text{ at } r = a + \epsilon a f(\theta) \qquad (8.2.21)$$
$$\phi < \infty \text{ at } r = 0 \qquad (8.2.22)$$

We use the method of strained parameters to determine a first-order uniform expansion of the solutions of (8.2.4), (8.2.21), and (8.2.22). To this end, we let

$$\phi(r, \theta; \epsilon) = \phi_0(r, \theta) + \epsilon \phi_1(r, \theta) + \cdots \qquad (8.2.23)$$
$$\kappa = \kappa_0 + \epsilon \kappa_1 + \cdots \qquad (8.2.24)$$

As a first step, we transfer the boundary condition (8.2.21) from $r = a + \epsilon a f(\theta)$ to $r = a$ by expanding $\phi\left[a + \epsilon a f(\theta), \theta\right]$ in a Taylor series. The result is

$$\phi(a, \theta) + \epsilon a \frac{\partial \phi}{\partial r}(a, \theta) f(\theta) + \cdots = 0 \qquad (8.2.25)$$

478   DYNAMICS OF PLATES

Substituting (8.2.23) and (8.2.24) into (8.2.4), (8.2.22), and (8.2.25) and equating coefficients of like powers of ϵ, we obtain

Order ϵ^0:

$$\nabla^2 \phi_0 + \kappa_0^2 \phi_0 = 0 \quad (8.2.26)$$

$$\phi_0 = 0 \text{ at } r = a \quad (8.2.27)$$

$$\phi_0 < \infty \text{ at } r = 0 \quad (8.2.28)$$

Order ϵ:

$$\nabla^2 \phi_1 + \kappa_0^2 \phi_1 = -2\kappa_0 \kappa_1 \phi_0 \quad (8.2.29)$$

$$\phi_1 = -a \frac{\partial \phi_0}{\partial r} f(\theta) \text{ at } r = a \quad (8.2.30)$$

$$\phi_1 < \infty \text{ at } r = 0 \quad (8.2.31)$$

Equations (8.2.26)-(8.2.28) are the same as (8.2.4) and (8.2.5), and hence we express their solution as

$$\phi_0 = J_n \left(\frac{\alpha_{nm} r}{a}\right) \left(A_{nm} e^{in\theta} + \bar{A}_{nm} e^{-in\theta}\right) \quad (8.2.32)$$

where A_{nm} is a complex-valued constant, $\alpha_{nm} = \kappa_{nm} a$, and $\kappa_{nm} = \kappa_0$. Substituting (8.2.32) into (8.2.29) and (8.2.30), we have

$$\nabla^2 \phi_1 + \kappa_{nm}^2 \phi_1 = -2\kappa_{nm} \kappa_1 J_n \left(\frac{\alpha_{nm} r}{a}\right) \left(A_{nm} e^{in\theta} + \bar{A}_{nm} e^{-in\theta}\right) \quad (8.2.33)$$

$$\phi_1 = -\alpha_{nm} J_n' (\alpha_{nm}) f(\theta) \left(A_{nm} e^{in\theta} + \bar{A}_{nm} e^{-in\theta}\right) \text{ at } r = a \quad (8.2.34)$$

Because the homogeneous first-order problem, (8.2.33), (8.2.34), and (8.2.31), is the same as (8.2.26)-(8.2.28) and because the latter has a nontrivial solution, the first-order problem has a solution only if a solvability condition is satisfied.

To determine the solvability condition, we use the general Green identity

$$\oiint_S (u \nabla^2 \phi - \phi \nabla^2 u) \, dS = \oint_\Gamma \left(u \frac{\partial \phi}{\partial n} - \phi \frac{\partial u}{\partial n}\right) ds \quad (8.2.35)$$

where **n** is the unit vector normal to the boundary curve. Thus, multiplying (8.2.33) with the adjoint $u(r, \theta)$, integrating the result over the domain $r = 0$ to $r = a$ and $\theta = 0$ to $\theta = 2\pi$, and using the Green identity, we obtain

$$\int_0^a \int_0^{2\pi} \phi_1 \left(\nabla^2 u + \kappa_{nm}^2 u\right) r \, dr \, d\theta + \int_0^{2\pi} \left(u \frac{\partial \phi_1}{\partial r} - \phi_1 \frac{\partial u}{\partial r}\right) a \, d\theta$$

$$= -2\kappa_{nm} \kappa_1 \int_0^a \int_0^{2\pi} u J_n \left(\frac{\alpha_{nm} r}{a}\right) \left(A_{nm} e^{in\theta} + \bar{A}_{nm} e^{-in\theta}\right) r \, dr \, d\theta \quad (8.2.36)$$

LINEAR VIBRATIONS OF MEMBRANES

We define the adjoint equation by setting the coefficient of ϕ_1 in the area integrand in (8.2.36) equal to zero; that is,

$$\nabla^2 u + \kappa_{nm}^2 u = 0 \tag{8.2.37}$$

To define the adjoint boundary conditions, we consider the homogeneous problem. Then, (8.2.36) becomes

$$\int_0^{2\pi} u \frac{\partial \phi_1}{\partial r} d\theta = 0 \tag{8.2.38}$$

Because $\partial \phi_1/\partial r$ is arbitrary at $r = a$, we require that

$$u = 0 \quad \text{at} \quad r = a \tag{8.2.39}$$

Moreover, we demand that $u < \infty$ at $r = 0$. Consequently, the adjoint problem is the same as the homogeneous first-order problem. Hence, the latter is self-adjoint and its solutions are given by

$$u = J_n\left(\frac{\alpha_{nm} r}{a}\right) e^{-in\theta} \tag{8.2.40}$$

where k is an integer.

Using the adjoint, we reduce (8.2.36) to

$$\alpha_{nm}\alpha_{km} J'_n(\alpha_{nm}) J'_k(\alpha_{km}) \int_0^{2\pi} f(\theta) \left[A_{nm} e^{i(n-k)\theta} + \bar{A}_{nm} e^{-i(k+n)\theta}\right] d\theta$$

$$= -2\kappa_{nm}\kappa_1 \int_0^a \int_0^{2\pi} r J_k\left(\frac{\alpha_{km} r}{a}\right) J_n\left(\frac{\alpha_{nm} r}{a}\right)$$

$$\times \left[A_{nm} e^{i(n-k)\theta} + \bar{A}_{nm} e^{-i(k+n)\theta}\right] dr d\theta \tag{8.2.41}$$

on account of (8.2.34). But

$$\int_0^a r J_k\left(\frac{\alpha_{km} r}{a}\right) J_n\left(\frac{\alpha_{nm} r}{a}\right) dr = \frac{1}{2} J'^2_n(\alpha_{nm}) \delta_{kn} \tag{8.2.42}$$

and $\int_0^{2\pi} f(\theta) d\theta = 0$, therefore (8.2.41) reduces to

$$\kappa_1 A_{nm} = -a\alpha_{nm} f_{2n} \bar{A}_{nm} \tag{8.2.43}$$

where

$$f_{2n} = \frac{1}{2\pi} \int_0^{2\pi} f(\theta) e^{-2ik\theta} d\theta \tag{8.2.44}$$

To analyze (8.2.43), we express A_{nm} and f_{2n} in polar form as

$$A_{nm} = \frac{1}{2} a_{nm} e^{i\beta_{nm}} \quad \text{and} \quad f_{2n} = F_{2n} e^{i\nu_{2n}} \tag{8.2.45}$$

and obtain

$$\kappa_1 = -a\alpha_{nm} F_{2n} e^{i(\nu_{2n} - 2\beta_{nm})} \tag{8.2.46}$$

Because κ_1 is real, it follows from (8.2.46) that

$$\nu_{2n} - 2\beta_{nm} = 0 \text{ or } \pi$$

Hence,

$$\beta_{nm} = \frac{1}{2}\nu_{2n} \text{ or } \frac{1}{2}(\nu_{2n} - \pi) \tag{8.2.47}$$

and it follows from (8.2.46) that

$$\kappa_1 = -a\alpha_{nm}F_{2n} \text{ or } a\alpha_{nm}F_{2n} \tag{8.2.48}$$

Substituting (8.2.45), (8.2.47), and (8.2.48) into (8.2.23), (8.2.24), and (8.2.3), we obtain to the first approximation

$$w^{(1)} = a_{nm} J_n\left(\frac{\alpha_{nm}}{a}r\right) \cos\left(n\theta + \frac{1}{2}\nu_{2n}\right) \cos\left(\omega_{nm}^{(1)} t + \tau\right) + \cdots \tag{8.2.49}$$

$$w^{(2)} = a_{nm} J_n\left(\frac{\alpha_{nm}}{a}r\right) \sin\left(n\theta + \frac{1}{2}\nu_{2n}\right) \cos\left(\omega_{nm}^{(2)} t + \tau\right) + \cdots \tag{8.2.50}$$

where

$$\omega_{nm}^{(1)} = \sqrt{\frac{T}{\rho h}} \frac{\alpha_{nm}}{a} (1 - \epsilon F_{2n}) + \cdots \tag{8.2.51}$$

$$\omega_{nm}^{(2)} = \sqrt{\frac{T}{\rho h}} \frac{\alpha_{nm}}{a} (1 + \epsilon F_{2n}) + \cdots \tag{8.2.52}$$

We note that, in the circular case, β_{nm} is arbitrary, and hence, there are two eigenfunctions $\sin n\theta$ and $\cos n\theta$ corresponding to the same eigenvalue ω_{nm}. However, in the near circular case, there are two different eigenfunctions corresponding to the two different eigenvalues $\omega_{nm}^{(1)}$ and $\omega_{nm}^{(2)}$ if $F_{2n} \neq 0$. If $F_{2n} = 0$, one needs to continue the expansion to higher order in order that the degeneracy may be removed.

8.2.3 Elliptic Membranes

We consider the elliptic boundary to be a deviation from a circle whose radius is determined in the course of analysis. The equation of an ellipse with a semi-major axis a^* and a semi-minor axis b^* is given by

$$\frac{x^2}{a^{*2}} + \frac{y^2}{b^{*2}} = 1 \tag{8.2.53}$$

Introducing the transformation

$$x = r\cos\theta \text{ and } y = r\sin\theta \tag{8.2.54}$$

where θ is measured counter-clockwise from the positive major axis, we rewrite (8.2.53) in polar form as

$$r = b^* \left(1 - \mu \cos^2\theta\right)^{-1/2} \tag{8.2.55}$$

where $\mu = 1 - b^{*2}/a^{*2}$. We rewrite (8.2.55) as

$$r = a\,[1 + f(\theta)] \tag{8.2.56}$$

where a is the average radius given by

$$a \equiv \frac{b^*}{2\pi} \int_0^{2\pi} \frac{1}{\sqrt{1 - \mu \cos^2 \theta}} d\theta = b^* f_0(\mu) \tag{8.2.57}$$

and

$$f(\theta) = \frac{1}{f_0(\mu)\sqrt{1 - \mu \cos^2 \theta}} - 1 \tag{8.2.58}$$

Therefore, it follows from (8.2.44) that

$$f_{2n} = F_{2n} = \frac{1}{2\pi f_0(\mu)} \int_0^{2\pi} \frac{\cos 2n\theta}{\sqrt{1 - \mu \cos^2 \theta}} d\theta \tag{8.2.59}$$

and the frequencies of the elliptic membrane are given by

$$\Omega_{nm}^{(2,1)} \equiv \frac{\omega_{nm}^{(2,1)}}{\omega_{nm}} = 1 \pm \frac{1}{2\pi f_0(\mu)} \int_0^{2\pi} \frac{\cos 2n\theta}{\sqrt{1 - \mu \cos^2 \theta}} d\theta \tag{8.2.60}$$

where the ω_{nm} are the natural frequencies of the circular membrane.
When $b^*/a^* = 0.9$, $\mu = 0.19$, and the $\Omega_{nm}^{(2,1)}$ are

$$\begin{array}{lll} n = 1: & 1.02632 & 0.97368 \\ n = 2: & 1.00104 & 0.99896 \\ n = 3: & 1.00005 & 0.99995 \end{array}$$

For $n \geq 4$, the $\Omega_{nm}^{(2,1)}$ are equal to 1 to within five significant figures.
When $b^*/a^* = 0.8$, $\mu = 0.36$, and the $\Omega_{nm}^{(2,1)}$ are

$$\begin{array}{lll} n = 1: & 1.05564 & 0.94436 \\ n = 2: & 1.00464 & 0.99536 \\ n = 3: & 1.00043 & 0.99957 \\ n = 4: & 1.00004 & 0.99996 \end{array}$$

For $n \geq 5$, the $\Omega_{nm}^{(2,1)}$ are equal to 1 to within five significant figures.

8.3 LINEAR VIBRATIONS OF CIRCULAR AND ANNULAR PLATES

For free undamped linear vibrations of circular plates without pretension, it follows from (7.4.49) with $N_1 = N_2 = N_6 = 0$ that

$$D\nabla^4 w + \rho h \ddot{w} = 0 \tag{8.3.1}$$

For time harmonic vibrations,

$$w(r,\theta,t) = \phi(r,\theta)e^{i\omega t} \tag{8.3.2}$$

and (8.3.1) becomes

$$\nabla^4 \phi - \kappa^4 \phi = 0 \tag{8.3.3}$$

$$\kappa^4 = \frac{\rho h \omega^2}{D}$$

The operator ∇^4 can be factored and hence (8.3.3) can be rewritten as

$$\left(\nabla^2 - \kappa^2\right)\left(\nabla^2 + \kappa^2\right)\phi = 0 \tag{8.3.4}$$

Hence, the general solution of (8.3.4) can be obtained by the superposition of the solutions of the two following Bessel equations:

$$\left(\nabla^2 - \kappa^2\right)\phi = 0 \text{ and } \left(\nabla^2 + \kappa^2\right)\phi = 0 \tag{8.3.5}$$

Next, we separate the θ variation by letting

$$\phi(r,\theta) = R(r)e^{in\theta} \tag{8.3.6}$$

where n is an integer and obtain from (8.3.5) that

$$\left(\nabla^2 - \kappa^2\right)R = 0 \text{ and } \left(\nabla^2 + \kappa^2\right)R = 0 \tag{8.3.7}$$

where

$$\nabla^2 = \frac{d^2}{dr^2} + \frac{1}{r}\frac{d}{dr} - \frac{n^2}{r^2} \tag{8.3.8}$$

In Section 8.3.1, we consider the case of circular plates; in Section 8.3.2, we consider the case of near circular and elliptic plates; and in Section 8.3.3, we consider the case of annular plates.

8.3.1 Circular Plates

We consider the mode shapes and corresponding natural frequencies of a circular plate with radius a. In this case, $\phi < \infty$ at $r = 0$ and hence $R < \infty$ at $r = 0$. Consequently, the general solution of (8.3.7) and (8.3.8) that is bounded at the origin can be expressed as

$$R(r) = c_1 J_n(\kappa r) + c_2 I_n(\kappa r) \tag{8.3.9}$$

Next, we consider three cases: clamped, simply supported, and free circular plates.

Clamped Case

In this case, the boundary conditions are

$$\phi = 0 \quad \text{and} \quad \phi_r = 0 \quad \text{at} \quad r = a \tag{8.3.10}$$

or

$$R = 0 \quad \text{and} \quad R' = 0 \quad \text{at} \quad r = a \tag{8.3.11}$$

Substituting (8.3.9) into (8.3.11) yields

$$c_1 J_n(\alpha) + c_2 I_n(\alpha) = 0 \tag{8.3.12}$$
$$c_1 J'_n(\alpha) + c_2 I'_n(\alpha) = 0 \tag{8.3.13}$$

where $\alpha \equiv \kappa a$ and the prime indicates the derivative with respect to α. For a nontrivial solution, the determinant of the coefficient matrix in (8.3.12) and (8.3.13) must vanish, resulting in the characteristic equation

$$J_n(\alpha) I'_n(\alpha) - J'_n(\alpha) I_n(\alpha) = 0 \tag{8.3.14}$$

Using the recursion relations

$$\alpha J'_n(\alpha) = n J_n(\alpha) - \alpha J_{n+1}(\alpha) \tag{8.3.15}$$
$$\alpha I'_n(\alpha) = n I_n(\alpha) + \alpha I_{n+1}(\alpha) \tag{8.3.16}$$

we rewrite (8.3.14) as

$$J_n(\alpha) I_{n+1}(\alpha) + J_{n+1}(\alpha) I_n(\alpha) = 0 \tag{8.3.17}$$

Using the following asymptotic expansions of $J_k(\alpha)$ and $I_k(\alpha)$ for large α:

$$J_k(\alpha) \sim \sqrt{\frac{2}{\pi\alpha}} \left\{ \cos\left(\alpha - \frac{1}{2}k\pi - \frac{1}{4}\pi\right)\left[1 - \frac{(4k^2-1)(4k^2-9)}{128\alpha^2}\right] \right.$$
$$\left. - \frac{4k^2-1}{8\alpha} \sin\left(\alpha - \frac{1}{2}k\pi - \frac{1}{4}\pi\right) \right\}$$

$$I_k(\alpha) \sim \frac{e^\alpha}{\sqrt{2\pi\alpha}} \left\{ 1 - \frac{4k^2-1}{8\alpha} + \frac{(4k^2-1)(4k^2-9)}{128\alpha^2} \right\}$$

we rewrite (8.3.17) as

$$\tan\left(\alpha - \frac{1}{2}n\pi\right) = \frac{1 - 4n^2}{8}\left(\frac{1}{\alpha} + \frac{1}{\alpha^2}\right) + \cdots \tag{8.3.18}$$

For large α, the right-hand side of (8.3.18) can be neglected and hence

$$\alpha = \left(m + \frac{1}{2}n\right)\pi$$

where m is an integer. To obtain a better approximation, we let

$$\alpha = \left(m + \frac{1}{2}n\right)\pi + \delta \tag{8.3.19}$$

where δ is a correction to $\left(m + \frac{1}{2}n\right)\pi$.

Substituting (8.3.19) into (8.3.18) yields

$$\tan\delta = \frac{1 - 4n^2}{4(2m+n)\pi + 8\delta} + \frac{1 - 4n^2}{2[(2m+n)\pi + 2\delta]^2} + \cdots$$

Hence,

$$\delta = \frac{1 - 4n^2}{4(2m+n)\pi} + \frac{1 - 4n^2}{2(2m+n)^2\pi^2} + \cdots$$

Therefore,

$$\alpha \approx \left(m + \frac{1}{2}n\right)\pi + \frac{(1 - 4n^2)(2 + 2m\pi + n\pi)}{4(2m+n)^2\pi^2} \tag{8.3.20}$$

Using the values of α_{nm} calculated from (8.3.20) as initial guesses, we solved (8.3.17) numerically. These results are compared with the asymptotic values in Table 8.2. Clearly, the larger the root is and the smaller n is, the closer is the asymptotic value to the numerically calculated value.

Table 8.2 Values of the eigenvalues α_{nm} of a clamped circular plate for different combinations of $n \geq 0$ and $m \geq 1$.

n	Method	α_{n1}	α_{n2}	α_{n3}	α_{n4}	α_{n5}
0	Numerical	3.1962	6.3064	9.4395	12.5771	15.7164
	Asymptotic	3.1941	6.3063	9.4395	12.5771	15.7164
	% Error	0.0680	0.0030	0.0005	0.0002	0.0001
1	Numerical	4.6109	7.7993	10.9581	14.1086	17.2557
	Asymptotic	4.6159	7.8002	10.9584	14.1088	17.2558
	% Error	0.1090	0.0113	0.0027	0.0010	0.0004
2	Numerical	5.9057	9.1969	12.4022	15.5795	18.7440
	Asymptotic	5.9373	9.2047	12.4053	15.5810	18.7448
	% Error	0.5350	0.0853	0.0247	0.0097	0.0045
3	Numerical	7.1435	10.5367	13.7951	17.0053	20.1923
	Asymptotic	7.2260	10.5615	13.8058	17.0109	20.1956
	% Error	1.1547	0.2357	0.0779	0.0330	0.0163

Simply Supported Case

In this case, the deflection and moment must vanish at the boundary leading to

$$w = 0 \text{ and } \frac{\partial^2 w}{\partial r^2} + \nu\left(\frac{1}{r}\frac{\partial w}{\partial r} + \frac{1}{r^2}\frac{\partial^2 w}{\partial \theta^2}\right) = 0 \text{ at } r = a \tag{8.3.21}$$

or

$$R = 0 \text{ and } \frac{d^2R}{dr^2} + \frac{\nu}{r}\frac{dR}{dr} - \frac{\nu n^2}{r^2}R = 0 \text{ at } r = a \qquad (8.3.22)$$

Substituting (8.3.9) into (8.3.22) yields

$$c_1 J_n(\alpha) + c_2 I_n(\alpha) = 0 \qquad (8.3.23)$$

$$c_1 \left[J_n''(\alpha) + \frac{\nu}{\alpha} J_n'(\alpha) \right] + c_2 \left[I_n''(\alpha) + \frac{\nu}{\alpha} I_n'(\alpha) \right] = 0 \qquad (8.3.24)$$

where the prime in (8.3.24) indicates the derivative with respect to α. Setting the determinant of the coefficient matrix in (8.3.23) and (8.3.24) equal to zero yields the characteristic equation

$$J_n(\alpha) \left[\alpha I_n''(\alpha) + \nu I_n'(\alpha) \right] - I_n(\alpha) \left[\alpha J_n''(\alpha) + \nu J_n'(\alpha) \right] = 0 \qquad (8.3.25)$$

Using the recursion relations (8.3.15) and (8.3.16), we rewrite (8.3.25) as

$$\frac{J_{n+1}(\alpha)}{J_n(\alpha)} + \frac{I_{n+1}(\alpha)}{I_n(\alpha)} = \frac{2\alpha}{1-\nu} \qquad (8.3.26)$$

Using a method similar to that used earlier, we find that, to the second approximation, the large eigenvalues are given by

$$\alpha_{nm} \approx (4m + 2n - 1)\frac{\pi}{4} - \frac{\left[8(1-\nu)^2 + (4m + 2n - 1)(4n^2 + 3 - 4\nu)\pi\right]}{2(4m + 2n - 1)^2 \pi^2} \qquad (8.3.27)$$

When $\nu = 0.3$, we compare in Table 8.3 the eigenvalues obtained with the asymptotic expansion with those obtained numerically using (8.3.26).

Free-Boundary Case

In this case, the boundary conditions demand the vanishing of the moment and shear force at the edge; that is,

$$\frac{\partial^2 w}{\partial r^2} + \nu \left(\frac{1}{r}\frac{\partial w}{\partial r} + \frac{1}{r^2}\frac{\partial^2 w}{\partial \theta^2} \right) = 0 \text{ at } r = a \qquad (8.3.28)$$

$$\frac{\partial}{\partial r}(\nabla^2 w) + \frac{1-\nu}{r}\left(\frac{1}{r}\frac{\partial^3 w}{\partial r \partial \theta^2} - \frac{1}{r^2}\frac{\partial^2 w}{\partial \theta^2} \right) = 0 \text{ at } r = a \qquad (8.3.29)$$

Substituting (8.3.2) and (8.3.6) into (8.3.28) and (8.3.29) yields

$$R'' + \frac{\nu}{r}R' - \frac{\nu n^2}{r^2}R = 0 \text{ at } r = a \qquad (8.3.30)$$

$$R''' + \frac{1}{r}R'' - \frac{1 + 2n^2 - \nu n^2}{r^2}R' + \frac{(3-\nu)n^2}{r^3}R = 0 \text{ at } r = a \qquad (8.3.31)$$

Table 8.3 Values of the eigenvalues α_{nm} of a simply supported circular plate for different combinations of $n \geq 0$ and $m \geq 1$.

n	Method	α_{n1}	α_{n2}	α_{n3}	α_{n4}	α_{n5}
0	Numerical	2.2215	5.4516	8.6114	11.7609	14.9069
	Asymptotic	2.2386	5.4528	8.6117	11.7610	14.9069
	% Error	0.7705	0.0221	0.0035	0.0010	0.0004
1	Numerical	3.7280	6.9627	10.1377	13.2967	16.4489
	Asymptotic	3.7344	6.9636	10.1380	13.2968	16.4490
	% Error	0.1718	0.0127	0.0027	0.0009	0.0004
2	Numerical	5.0610	8.3736	11.5887	14.7717	17.9399
	Asymptotic	5.0890	8.3802	11.5912	14.7729	17.9406
	% Error	0.5546	0.0789	0.0219	0.0083	0.0038
3	Numerical	6.3212	9.7236	12.9875	16.2014	19.3910
	Asymptotic	6.3977	9.7462	12.9972	16.2064	19.3940
	% Error	1.2102	0.2324	0.0747	0.0312	0.0153

Substituting (8.3.9) into (8.3.30) and (8.3.31) yields

$$[\alpha^2 J_n''(\alpha) + \nu \alpha J_n'(\alpha) - \nu n^2 J_n(\alpha)] c_1$$
$$+ [\alpha^2 I_n''(\alpha) + \nu \alpha I_n'(\alpha) - \nu n^2 I_n(\alpha)] c_2 = 0 \qquad (8.3.32)$$

$$[\alpha^3 J_n'''(\alpha) + \alpha^2 J_n''(\alpha) - (1 + 2n^2 - \nu n^2) \alpha J_n'(\alpha) + (3 - \nu) n^2 J_n(\alpha)] c_1$$
$$+ [\alpha^3 I_n'''(\alpha) + \alpha^2 I_n''(\alpha) - (1 + 2n^2 - \nu n^2) \alpha I_n'(\alpha) + (3 - \nu) n^2 I_n(\alpha)] c_2 = 0$$
$$(8.3.33)$$

where $\alpha = \kappa a$. Setting the determinant of the coefficient matrix in (8.3.32) and (8.3.33) equal to zero, using the Bessel equations to express the second- and third-order derivatives in terms of J_n, I_n, and the recursion relations (8.3.15) and (8.3.16), we obtain the characteristic equation

$$[\alpha^4 + (1 - \nu)^2 n^2 (n^2 - 1)] [J_{n-1}(\alpha) I_{n+1}(\alpha) + J_{n+1}(\alpha) I_{n-1}(\alpha)]$$
$$- 2\alpha^2 (1 - \nu) n [(n - 1) I_{n-1}(\alpha) J_{n-1}(\alpha) + (n + 1) I_{n+1}(\alpha) J_{n+1}(\alpha)] = 0$$
$$(8.3.34)$$

where $n \neq 0$. When $n = 0$, the vanishing of the determinant of the coefficient matrix in (8.3.32) and (8.3.33) yields

$$\alpha J_0(\alpha) I_1(\alpha) + \alpha I_0(\alpha) J_1(\alpha) - 2(1 - \nu) I_1(\alpha) J_1(\alpha) = 0 \qquad (8.3.35)$$

Again, to the second approximation, the large eigenvalues of (8.3.34) are given by

$$\alpha_{nm} \approx (2m + n - 2) \frac{\pi}{2} - \frac{7 + 4n^2 - 8\nu}{4\pi (2m + n - 2)} + \frac{1 + 8(1 - \nu)\nu + 4n^2(3 - 4\nu)}{2\pi^2 (2m + n - 2)^2}$$
$$(8.3.36)$$

Table 8.4 Values of the eigenvalues α_{nm} of a free circular plate for different combinations of $n \geq 0$ and $m \geq 1$. NA denotes not applicable.

n	Method	α_{n1}	α_{n2}	α_{n3}	α_{n4}	α_{n5}
0	Numerical	0.0000	3.0005	6.2003	9.3675	12.5227
	Asymptotic	NA	2.9925	6.2002	9.3675	12.5227
	% Error	NA	0.2671	0.0016	0.0003	0.0002
1	Numerical	0.0000	4.5249	7.7338	10.9068	14.0667
	Asymptotic	NA	4.5399	7.7371	10.9080	14.0673
	% Error	NA	0.3315	0.0431	0.0116	0.0044
2	Numerical	2.3148	5.9380	9.1851	12.3817	15.5575
	Asymptotic	NA	5.9730	9.1959	12.3864	15.5600
	% Error	NA	0.5897	0.1171	0.0380	0.0159
3	Numerical	3.5269	7.2806	10.5804	13.8091	17.0070
	Asymptotic	NA	7.3446	10.6038	13.8204	17.0133
	% Error	NA	0.8778	0.2210	0.0816	0.0371

When $\nu = 0.3$, we compare in Table 8.4 the eigenvalues obtained using asymptotic expansion with those obtained numerically using (8.3.34).

8.3.2 Near Circular and Elliptic Plates

As in Section 8.2.2, we represent the boundary of the plate by $r = a + \epsilon a f(\theta)$, where a is the mean radius of the plate so that the mean of $f(\theta)$ is zero as in (8.2.20). We consider the case of a clamped edge so that the boundary conditions are

$$w = 0 \text{ and } \frac{\partial w}{\partial n} = 0 \text{ at } r = a + \epsilon a f(\theta)$$

or

$$\phi = 0 \text{ and } \frac{\partial \phi}{\partial n} = 0 \text{ at } r = a + \epsilon a f(\theta) \qquad (8.3.37)$$

where n is the normal to the boundary and w and ϕ are bounded at $r = 0$. To express the second boundary condition in (8.3.37) in terms of r and θ, we write the equation describing the boundary as

$$F(r, \theta) = r - a - \epsilon a f(\theta) = 0$$

Thus,

$$\nabla F = \mathbf{e}_r - \epsilon a f'(\theta) \mathbf{e}_\theta$$

is normal to the boundary. Here, $\mathbf{e}_r$ and $\mathbf{e}_\theta$ are unit vectors along the r and θ axes, respectively. Consequently,

$$\frac{\partial \phi}{\partial n} \propto \nabla F \cdot \nabla \phi = 0 \text{ at } r = a + \epsilon a f(\theta)$$

or
$$\frac{\partial \phi}{\partial r} - \epsilon a f' \frac{\partial \phi}{\partial \theta} = 0 \quad \text{at} \quad r = a + \epsilon a f(\theta) \tag{8.3.38}$$

To proceed further, we transfer the boundary conditions from $r = a + \epsilon a f(\theta)$ to $r = a$ using Taylor-series expansions. The result is

$$\phi + \epsilon a \frac{\partial \phi}{\partial r} f + \cdots = 0 \quad \text{at} \quad r = a \tag{8.3.39}$$

$$\frac{\partial \phi}{\partial r} + \epsilon a \frac{\partial^2 \phi}{\partial r^2} f - \epsilon a f' \frac{\partial \phi}{\partial \theta} + \cdots = 0 \quad \text{at} \quad r = a \tag{8.3.40}$$

Next, we use the method of strained parameters to determine a first-order uniform expansion of the solution of (8.3.3), (8.3.39), and (8.3.40) in the form (8.2.23) and (8.2.24). Substituting these expansions into (8.3.3), (8.3.39), and (8.3.40) and equating coefficients of like powers of ϵ, we obtain

Order ϵ^0:

$$\nabla^4 \phi_0 - \kappa_0^4 \phi_0 = 0 \tag{8.3.41}$$

$$\phi_0 = 0 \quad \text{and} \quad \phi_{0r} = 0 \quad \text{at} \quad r = a \tag{8.3.42}$$

Order ϵ^1:

$$\nabla^4 \phi_1 - \kappa_0^4 \phi_1 = 4\kappa_0^3 \kappa_1 \phi_0 \tag{8.3.43}$$

$$\phi_1 = 0 \quad \text{at} \quad r = a \tag{8.3.44}$$

$$\phi_{1r} = -a \phi_{0rr} f + a f' \phi_{0\theta} \quad \text{at} \quad r = a \tag{8.3.45}$$

At the origin,

$$\phi_0 < \infty \quad \text{and} \quad \phi_1 < \infty \quad \text{at} \quad r = 0 \tag{8.3.46}$$

The solution of (8.3.41), (8.3.42), and (8.3.46) was obtained in the preceding section, which we express as

$$\phi_0 = R_{nm}(r) \left(A_{nm} e^{in\theta} + \bar{A}_{nm} e^{-in\theta} \right) \tag{8.3.47}$$

where

$$R_{nm}(r) = \frac{1}{a} \left[\frac{J_n\left(\frac{\alpha_{nm} r}{a}\right)}{J_n(\alpha_{nm})} - \frac{I_n\left(\frac{\alpha_{nm} r}{a}\right)}{I_n(\alpha_{nm})} \right] \tag{8.3.48}$$

We note that

$$\int_0^a r R_{nm}(r) R_{kl}(r) dr = \delta_{nk} \delta_{ml} \tag{8.3.49}$$

on account of the characteristic equation (8.3.17).

Substituting (8.3.47) into (8.3.43)-(8.3.45), we have

$$\nabla^4 \phi_1 - \frac{\alpha_{nm}^4}{a^4}\phi_1 = \frac{4\alpha_{nm}^3 \kappa_1}{a^3} R_{nm}\left(A_{nm}e^{in\theta} + \bar{A}_{nm}e^{-in\theta}\right) \qquad (8.3.50)$$

$$\phi_1 = 0 \quad \text{at} \quad r = a \qquad (8.3.51)$$

$$\phi_{1r} = -aR''_{nm}f\left(A_{nm}e^{in\theta} + \bar{A}_{nm}e^{-in\theta}\right) \quad \text{at} \quad r = a \; (8.3.52)$$

Because the homogeneous part of the problem governing ϕ_1 is the same as the problem governing ϕ_0 and since the latter has a nontrivial solution, the nonhomogeneous problem governing ϕ_1 has a solution only if a solvability condition is satisfied. To determine this solvability condition, we multiply (8.3.50) by the adjoint $u(r,\theta)$ of the homogeneous problem, integrate the result over the surface of a circle of radius a, and obtain

$$\int_0^a \int_0^{2\pi} u\left(\nabla^4 \phi_1 - \frac{\alpha_{nm}^4}{a^4}\phi_1\right) r\,dr\,d\theta$$

$$= \frac{4\alpha_{nm}^3 \kappa_1}{a^3}\int_0^a \int_0^{2\pi} uR_{nm}\left(A_{nm}e^{in\theta} + \bar{A}_{nm}e^{-in\theta}\right) r\,dr\,d\theta \quad (8.3.53)$$

Next, we use Green's identify (8.2.35) twice to rewrite (8.3.53) as

$$\int_0^a \int_0^{2\pi} \phi_1\left(\nabla^4 u - \frac{\alpha_{nm}^4}{a^4}u\right) r\,dr\,d\theta$$

$$+ a\int_0^{2\pi} \left[u\frac{\partial}{\partial r}(\nabla^2 \phi_1) - \nabla^2 \phi_1 \frac{\partial u}{\partial r} + \frac{\partial \phi_1}{\partial r}\nabla^2 u - \phi_1 \frac{\partial}{\partial r}(\nabla^2 u)\right] d\theta$$

$$= \frac{4\alpha_{nm}^3 \kappa_1}{a^3}\int_0^a \int_0^{2\pi} uR_{nm}\left(A_{nm}e^{in\theta} + \bar{A}_{nm}e^{-in\theta}\right) r\,dr\,d\theta \qquad (8.3.54)$$

Following a procedure similar to that used in the preceding section, we find that u is defined by

$$\nabla^4 u - \frac{\alpha_{nm}^4}{a^4}u = 0 \qquad (8.3.55)$$

$$u = 0 \quad \text{and} \quad u_r = 0 \quad \text{at} \quad r = a \qquad (8.3.56)$$

$$u < \infty \quad \text{at} \quad r = 0 \qquad (8.3.57)$$

Hence, the problem is self-adjoint.

Letting $u = R_{nm}e^{-in\theta}$ and the boundary conditions (8.3.51) and (8.3.52), we obtain from (8.3.54) the solvability condition

$$-a^2 R''^2_{nm}(\alpha_{nm})\int_0^{2\pi} f(\theta)\left[A_{nm}e^{i(n-k)\theta} + \bar{A}_{nm}e^{-i(n+k)\theta}\right] d\theta$$

$$= \frac{4\alpha_{nm}^3 \kappa_1}{a^3}\int_0^a \int_0^{2\pi} R^2_{nm}\left[A_{nm}e^{i(n-k)\theta} + \bar{A}_{nm}e^{-i(n+k)\theta}\right] r\,dr\,d\theta \quad (8.3.58)$$

The integral on the right-hand side of (8.3.58) vanishes unless $k = n$ or $-n$. Choosing $k = n$ and using (8.3.49), we rewrite (8.3.58) as

$$\kappa_1 = -\frac{a^5 R''^2_{nm}(\alpha_{nm})}{4\alpha^3_{nm}} f_{2n} \frac{\bar{A}_{nm}}{A_{nm}} \tag{8.3.59}$$

Using (8.2.45), we rewrite (8.3.59) as

$$\kappa_1 = -\frac{a^5 R''^2_{nm}(\alpha_{nm})}{4\alpha^3_{nm}} F_{2n} e^{i(\nu_{2n}-2\beta_{nm})} \tag{8.3.60}$$

It follows from (8.3.48) that

$$R''_{nm}(\alpha_{nm}) = \frac{\alpha^2_{nm}}{a^3} \left[\frac{J''_n(\alpha_{nm})}{J_n(\alpha_{nm})} - \frac{I''_n(\alpha_{nm})}{I_n(\alpha_{nm})} \right] \tag{8.3.61}$$

But $J_n(x)$ and $I_n(x)$ are governed by

$$J''_n + \frac{1}{x} J'_n + \left(1 - \frac{n^2}{x^2}\right) J_n = 0$$

$$I''_n + \frac{1}{x} I'_n - \left(1 - \frac{n^2}{x^2}\right) I_n = 0$$

hence (8.3.61) can be rewritten as

$$R''_{nm}(\alpha_{nm}) = \frac{\alpha^2_{nm}}{a^3} \left[-\frac{J'_n(\alpha_{nm})}{J_n(\alpha_{nm})} + \frac{I'_n(\alpha_{nm})}{I_n(\alpha_{nm})} - 2 \right] \tag{8.3.62}$$

Using the recursion relations (8.3.15) and (8.3.16), we obtain

$$R''_{nm}(\alpha_{nm}) = \frac{\alpha_{nm}}{a^3} \left[\frac{J_{n+1}(\alpha_{nm})}{J_n(\alpha_{nm})} + \frac{I_{n+1}(\alpha_{nm})}{I_n(\alpha_{nm})} - 2 \right] = -\frac{2\alpha^2_{nm}}{a^3} \tag{8.3.63}$$

according to the characteristic equation (8.3.17).

Substituting (8.3.63) into (8.3.60), we have

$$\kappa_1 = -\frac{\alpha_{nm}}{a} F_{2n} e^{i(\nu_{2n}-2\beta_{nm})} \tag{8.3.64}$$

Because κ_1 is real,

$$\beta_{nm} = \frac{1}{2}\nu_{2n} \quad \text{or} \quad \frac{1}{2}(\nu_{2n} - \pi) \tag{8.3.65}$$

and it follows from (8.3.64) that

$$\kappa_1 = -\frac{\alpha_{nm} F_{2n}}{a} \quad \text{or} \quad \frac{\alpha_{nm} F_{2n}}{a} \tag{8.3.66}$$

Consequently, there are two eigenfunctions

$$w^{(1)}(r) = R_{nm}\left(\frac{\alpha_{nm} r}{a}\right) \cos\left(n\theta + \frac{1}{2}\nu_{2n}\right) \cos\left[\omega^{(1)}_{nm} t + \tau\right] \tag{8.3.67}$$

$$w^{(2)}(r) = R_{nm}\left(\frac{\alpha_{nm} r}{a}\right) \sin\left(n\theta + \frac{1}{2}\nu_{2n}\right) \cos\left[\omega^{(2)}_{nm} t + \tau\right] \tag{8.3.68}$$

corresponding to the natural frequencies

$$\omega_{nm}^{(1)} = \sqrt{\frac{D}{\rho h} \frac{\alpha_{nm}^2}{a^2}} [1 - 2F_{2n}] + \cdots \quad (8.3.69)$$

$$\omega_{nm}^{(2)} = \sqrt{\frac{D}{\rho h} \frac{\alpha_{nm}^2}{a^2}} [1 + 2F_{2n}] + \cdots \quad (8.3.70)$$

For an elliptic plate with semi-major axis a^* and semi-minor axis b^*, a and $f(\theta)$ are given by (8.2.57) and (8.2.58). Hence, $\nu_{2n} = 0$ and F_{2n} is given by (8.2.59). Therefore, the frequencies of the elliptic plate are given by

$$\Omega_{nm}^{(2,1)} = \frac{\omega_{nm}^{(2,1)}}{\omega_{nm}} = 1 \pm \frac{1}{\pi f_0(\mu)} \int_0^{2\pi} \frac{\cos 2n\theta}{\sqrt{1 - \mu \cos^2 \theta}} d\theta \quad (8.3.71)$$

where $\omega_{nm} = \sqrt{\frac{D}{\rho} \frac{\alpha_{nm}^2}{a^2}}$.

8.3.3 Annular Plates

We consider a plate that consists of a circular outer boundary of radius a and a concentric circular inner boundary of radius b. There exists nine possible combinations of simple boundary conditions (i.e., clamped, simply supported, or free). In this book, we consider the case of clamped inner and outer boundaries and refer the reader to Leissa (1993a) for solutions of the other cases. The problem is governed by (8.3.1) subject to the boundary conditions

$$w = 0 \quad \text{and} \quad w_r = 0 \quad \text{at} \quad r = a \text{ and } b \quad (8.3.72)$$

Substituting (8.3.2) and (8.3.6) into (8.3.1) and (8.3.72) yields (8.3.7) and the boundary conditions

$$R = 0 \quad \text{and} \quad R' = 0 \quad \text{at} \quad r = a \text{ and } b \quad (8.3.73)$$

The general solution of (8.3.7) can be expressed as

$$R(r) = c_1 J_n(\kappa r) + c_2 I_n(\kappa r) + c_3 Y_n(\kappa r) + c_4 K_n(\kappa r) \quad (8.3.74)$$

where the Y_n are Bessel functions of the second kind and the K_n are the modified Bessel functions of the second kind. Substituting (8.3.74) into (8.3.73) yields

$$c_1 J_n(\alpha) + c_2 I_n(\alpha) + c_3 Y_n(\alpha) + c_4 K_n(\alpha) = 0 \quad (8.3.75)$$
$$c_1 J_n'(\alpha) + c_2 I_n'(\alpha) + c_3 Y_n'(\alpha) + c_4 K_n'(\alpha) = 0 \quad (8.3.76)$$
$$c_1 J_n(\alpha\beta) + c_2 I_n(\alpha\beta) + c_3 Y_n(\alpha\beta) + c_4 K_n(\alpha\beta) = 0 \quad (8.3.77)$$
$$c_1 J_n'(\alpha\beta) + c_2 I_n'(\alpha\beta) + c_3 Y_n'(\alpha\beta) + c_4 K_n'(\alpha\beta) = 0 \quad (8.3.78)$$

where $\alpha = \kappa a$ and $\beta = \frac{b}{a}$. Setting the determinant of the coefficient matrix in (8.3.75)-(8.3.78) equal to zero yields the characteristic equation governing the α_{nm}.

Clamped/Clamped Case

For a clamped/clamped annular plate, the boundary conditions are $w = 0$ and $w_r = 0$ at $r = a$ and $r = b$. The second-order asymptotic expansion is

$$\alpha_{nm}^{(2)} \approx \frac{(2m+1)\pi}{2(1-\beta)} + \frac{(4n^2-1)(1-\beta)}{4(2m+1)\pi\beta} - \frac{(4n^2-1)(1+\beta^2)(1-\beta)}{2(2m+1)^2\pi^2\beta^2} \quad (8.3.79)$$

In Table 8.5, we show some of the results obtained from the numerical (NUM) and second-order asymptotic (ASY) solutions and present the relative errors (%ERR) for the cases $\beta = 0.2$ and $\beta = 0.8$. For $\beta = 0.2$, we note that as n increases, significant errors occur in predicting the first eigenvalue. However, using the order-three approximation

$$\alpha_{nm}^{(3)} \approx \alpha_{nm}^{(2)} + \frac{(4n^2-1)\left[(4n^2+23)\beta^2 + (29-20n^2)\beta + 4n^2+23\right](1-\beta)^3}{48(2m+1)^3\pi^3\beta^3} \quad (8.3.80)$$

we obtain $\alpha_{11} = 6.0003$, $\alpha_{21} = 6.4511$, and $\alpha_{31} = 7.2432$, yielding percent errors of 0.1381, 0.2428, and 0.8691, respectively.

Clamped/Simply Supported Case

For a clamped/simply supported annular plate, the boundary conditions are $w = 0$ and $w_r = 0$ at $r = a$ and $w = 0$ and $M_1 = 0$ at $r = b$. The second-order approximation of the eigenvalues is

$$\alpha_{nm} \approx \frac{(4m+1)\pi}{4(1-\beta)} - \frac{\left[(4n^2-1)\beta - 4n^2 - 3 + 4\nu\right]}{2(4m+1)\pi\beta} \\ - \frac{2\left[(4n^2-1)\beta^2 + 2(1-\nu)^2\right](1-\beta)}{(4m+1)^2\pi^2\beta^2} \quad (8.3.81)$$

In Table 8.6, we present values for the first few eigenvalues determined numerically and asymptotically for the cases $\beta = 0.2$ and $\beta = 0.8$.

Clamped/Free Case

For a clamped/free annular plate, the boundary conditions are $w = 0$ and $w_r = 0$ at $r = a$ and $M_1 = 0$ and $Q_1 + M_{6\theta}/r = 0$ at $r = b$. The second-order approximation

Table 8.5 Values of the eigenvalues α_{nm} of a clamped/clamped annular plate, obtained from numerical (NUM) and second-order asymptotic (ASY) approaches, and the percent relative error (%ERR) for different combinations of $n \geq 0$ and $m \geq 1$.

n	Method	α_{n1}	α_{n2}	α_{n3}	α_{n4}	α_{n5}
$\beta = b/a = 0.2$						
0	NUM	5.8830	9.7847	13.7167	17.6473	21.5772
	ASY	5.9015	9.7960	13.7205	17.6491	21.5782
	%ERR	0.3145	0.1149	0.0275	0.0102	0.0046
1	NUM	6.0086	9.9135	13.8285	17.7443	21.6623
	ASY	5.8576	9.8820	13.8164	17.7385	21.6591
	%ERR	2.5138	0.3179	0.0877	0.0322	0.0144
2	NUM	6.4668	10.3207	14.1726	18.0395	21.9198
	ASY	5.7258	10.1402	14.1040	18.0068	21.9019
	%ERR	11.4586	1.7493	0.4842	0.1809	0.0815
3	NUM	7.3067	11.0312	14.7647	18.5424	22.3556
	ASY	5.5062	10.5704	14.5833	18.4540	22.3065
	%ERR	24.6412	4.1768	1.2283	0.4768	0.2197
$\beta = b/a = 0.8$						
0	NUM	23.6466	39.2631	54.9757	70.6839	86.3922
	ASY	23.5582	39.2670	54.9756	70.6839	86.3922
	%ERR	0.3739	0.0099	0.0003	0.0000	0.0000
1	NUM	23.6610	39.2749	54.9850	70.6915	86.3986
	ASY	23.5732	39.2787	54.9848	70.6915	86.3986
	%ERR	0.3710	0.0098	0.0003	0.0000	0.0000
2	NUM	23.7042	39.3103	55.0127	70.7142	86.4177
	ASY	23.6181	39.3140	55.0126	70.7142	86.4177
	%ERR	0.3631	0.0095	0.0003	0.0000	0.0000
3	NUM	23.7769	39.3694	55.0590	70.7520	86.4496
	ASY	23.6931	39.3728	55.0588	70.7520	86.4496
	%ERR	0.3525	0.0088	0.0004	0.0000	0.0000

of the eigenvalues is

$$\alpha_{nm} \approx \frac{(2m-1)\pi}{2(1-\beta)} - \frac{\left[(4n^2-1)\beta - 4n^2 + 8\nu - 7\right]}{4(2m-1)\pi\beta} \\ - \frac{\left[(4n^2-1)\beta^2 + 4(4\nu-3)n^2 + 8\nu^2 - 8\nu - 1\right](1-\beta)}{2(2m-1)^2\pi^2\beta^2} \quad (8.3.82)$$

In Table 8.7, we present values for the first few eigenvalues determined numerically and asymptotically for the cases $\beta = 0.2$ and $\beta = 0.8$. We note that for a given n, the asymptotic solution fails to predict the first, and in some cases the second, eigenvalues when β is small (i.e., < 0.5). However, when $\beta \to 1$, the asymptotic solution yields

Table 8.6 Values of the eigenvalues α_{nm} of a clamped/simply supported annular plate, obtained from numerical (NUM) and asymptotic (ASY) approaches, and the percent relative error (%ERR) for different combinations of $n \geq 0$ and $m \geq 1$.

n	Method	α_{n1}	α_{n2}	α_{n3}	α_{n4}	α_{n5}
$\beta = b/a = 0.2$						
0	NUM	5.1708	8.9846	12.8701	16.7738	20.6858
	ASY	5.0747	8.9655	12.8626	16.7701	20.6839
	%ERR	1.8592	0.2127	0.0583	0.0215	0.0096
1	NUM	5.4082	9.1912	13.0344	16.9076	20.7979
	ASY	5.5580	9.2405	13.0547	16.9177	20.8036
	%ERR	2.7701	0.5366	0.1556	0.0597	0.0274
2	NUM	6.1335	9.7986	13.5223	17.3072	21.1335
	ASY	7.0081	10.0653	13.6308	17.3603	21.1630
	%ERR	14.2587	2.7217	0.8024	0.3071	0.1399
3	NUM	7.1900	10.7420	14.3114	17.9646	21.6894
	ASY	9.4249	11.4400	14.5910	18.0981	21.7620
	%ERR	31.0826	6.4974	1.9542	0.7430	0.3347
$\beta = b/a = 0.8$						
0	NUM	19.7413	35.3999	51.0905	66.7892	82.4914
	ASY	19.7375	35.4001	51.0905	66.7892	82.4914
	%ERR	0.0193	0.0006	0.0000	0.0000	0.0000
1	NUM	19.7657	35.4154	51.1018	66.7980	82.4986
	ASY	19.7629	35.4158	51.1018	66.7980	82.4986
	%ERR	0.0143	0.0011	0.0001	0.0000	0.0000
2	NUM	19.8389	35.4619	51.1354	66.8243	82.5202
	ASY	19.8389	35.4629	51.1357	66.8244	82.5202
	%ERR	0.0001	0.0026	0.0005	0.0002	0.0001
3	NUM	19.9611	35.5394	51.1915	66.8681	82.5561
	ASY	19.9657	35.5413	51.1921	66.8684	82.5563
	%ERR	0.0231	0.0052	0.0012	0.0004	0.0002

reasonably good results for the first eigenvalues and excellent results for the second and higher ones.

Simply Supported/Clamped Case

For a simply supported/clamped annular plate, the boundary conditions are $w = 0$ and $M_1 = 0$ at $r = a$ and $w = 0$ and $w_r = 0$ at $r = b$. The second-order approximation of the eigenvalues is

$$\alpha_{nm} \approx \frac{(4m+1)\pi}{4(1-\beta)} - \frac{\left[1 - 4n^2 + \left(3 + 4n^2 - 4\nu\right)\beta\right]}{2(4m+1)\pi\beta}$$

LINEAR VIBRATIONS OF CIRCULAR AND ANNULAR PLATES   495

$$+ \frac{2(1-\beta)\left[1 - 4n^2 - 2(1-\nu)^2 \beta^2\right]}{(4m+1)^2 \pi^2 \beta^2} \quad (8.3.83)$$

where $\beta \equiv b/a$. In Table 8.8, we present values for the first few eigenvalues determined numerically and asymptotically for the cases $\beta = 0.2$ and $\beta = 0.8$.

Simply Supported/Simply Supported Case

For a simply supported/simply supported annular plate, the boundary conditions are $w = 0$ and $M_1 = 0$ at $r = a$ and $r = b$. The second-order approximation of the

Table 8.7 Values of the eigenvalues α_{nm} of a clamped/free annular plate, obtained from numerical (NUM) and asymptotic (ASY) approaches, and the percent relative error (%ERR) for different combinations of $n \geq 0$ and $m \geq 1$.

n	Method	α_{n1}	α_{n2}	α_{n3}	α_{n4}	α_{n5}
$\beta = b/a = 0.2$						
0	NUM	3.2261	6.5590	10.2475	14.0508	17.9072
	ASY	NA	6.8333	10.3097	14.0735	17.9177
	%ERR	NA	4.1831	0.6068	0.1618	0.0586
1	NUM	4.5333	7.5430	10.8201	14.4074	18.1542
	ASY	NA	8.0503	10.8497	14.4010	18.1472
	%ERR	NA	6.7254	0.2733	0.0445	0.0382
2	NUM	5.8082	8.9909	12.0988	15.3656	18.8611
	ASY	NA	11.7012	12.4696	15.3834	18.8358
	%ERR	NA	30.1449	3.0647	0.1157	0.1338
3	NUM	7.1067	10.4121	13.5362	16.6549	19.9198
	ASY	NA	NA	15.1694	17.0207	19.9835
	%ERR	NA	NA	12.0651	2.1962	0.3197
$\beta = b/a = 0.8$						
0	NUM	9.6455	23.6609	39.3828	55.0554	70.7461
	ASY	8.4437	23.7468	39.3794	55.0557	70.7462
	%ERR	12.4596	0.3632	0.0085	0.0005	0.0001
1	NUM	9.7270	23.6968	39.4017	55.0683	70.7559
	ASY	8.5967	23.7815	39.3983	55.0685	70.7559
	%ERR	11.6197	0.3577	0.0085	0.0005	0.0001
2	NUM	9.9646	23.8037	39.4581	55.1069	70.7851
	ASY	9.0558	23.8856	39.4549	55.1071	70.7852
	%ERR	9.1198	0.3442	0.0082	0.0005	0.0001
3	NUM	10.3406	23.9796	39.5519	55.1711	70.8338
	ASY	9.8210	24.0590	39.5491	55.1715	70.8339
	%ERR	5.0247	0.3310	0.0071	0.0006	0.0001

eigenvalues is

$$\alpha_{nm} \approx \frac{m\pi}{(1-\beta)} + \frac{(1-\beta)\left(3+4n^2-4\nu\right)}{8m\pi\beta} - \frac{(1-\beta)\left(1+\beta^2\right)(1-\nu)^2}{4m^2\pi^2\beta^2} \quad (8.3.84)$$

where $\beta \equiv b/a$. In Table 8.9, we present values for the first few eigenvalues determined numerically and asymptotically for the cases $\beta = 0.2$ and $\beta = 0.8$.

Table 8.8 Values of the eigenvalues α_{nm} of a simply supported/clamped annular plate, obtained from numerical (NUM) and asymptotic (ASY) approaches, and the percent relative error (%ERR) for different combinations of $n \geq 0$ and $m \geq 1$.

n	Method	α_{n1}	α_{n2}	α_{n3}	α_{n4}	α_{n5}
$\beta = b/a = 0.2$						
0	NUM	4.7660	8.7489	12.6974	16.6372	20.5728
	ASY	4.8481	8.7636	12.7025	16.6395	20.5740
	%ERR	1.7222	0.1672	0.0402	0.0139	0.0060
1	NUM	4.9267	8.8913	12.8180	16.7403	20.6622
	ASY	4.7089	8.8464	12.8025	16.7332	20.6585
	%ERR	4.4208	0.5058	0.1214	0.0421	0.0182
2	NUM	5.4853	9.3419	13.1895	17.0541	20.9332
	ASY	4.2914	9.0948	13.1023	17.0143	20.9120
	%ERR	21.7647	2.6459	0.6605	0.2334	0.1015
3	NUM	6.4288	10.1182	13.8260	17.5881	21.3917
	ASY	3.5956	9.5088	13.6021	17.4828	21.3345
	%ERR	44.0702	6.0229	1.6193	0.5986	0.2673
$\beta = b/a = 0.8$						
0	NUM	19.5308	35.2893	51.0136	66.7304	82.4437
	ASY	19.5388	35.2893	51.0137	66.7304	82.4437
	%ERR	0.0409	0.0000	0.0001	0.0000	0.0000
1	NUM	19.5534	35.3040	51.0244	66.4389	82.4508
	ASY	19.5605	35.3038	51.0244	66.7389	82.4507
	%ERR	0.0363	0.0004	0.0000	0.0000	0.0000
2	NUM	19.6213	35.3481	51.0568	66.7644	82.4718
	ASY	19.6256	35.3475	51.0567	66.7643	82.4718
	%ERR	0.0220	0.0016	0.0003	0.0001	0.0000
3	NUM	19.7347	35.4215	51.1107	66.8069	82.5069
	ASY	19.7341	35.4203	51.1104	66.8067	82.5068
	%ERR	0.0030	0.0035	0.0007	0.0002	0.0001

Table 8.9 Values of the eigenvalues α_{nm} of a simply supported/simply supported annular plate, obtained from numerical (NUM) and asymptotic (ASY) approaches, and the percent relative error (%ERR) for different combinations of $n \geq 0$ and $m \geq 1$.

n	Method	α_{n1}	α_{n2}	α_{n3}	α_{n4}	α_{n5}
$\beta = b/a = 0.2$						
0	NUM	4.0963	7.9606	11.8575	15.7679	19.6843
	ASY	3.9553	7.9327	11.8478	15.7634	19.6819
	%ERR	3.4421	0.3504	0.0821	0.0283	0.0122
1	NUM	4.3843	8.1918	12.0369	15.9115	19.8032
	ASY	4.5919	8.2510	12.0600	15.9226	19.8092
	%ERR	4.7351	0.7220	0.1920	0.0696	0.0307
2	NUM	5.2192	8.8630	12.5664	16.3391	20.1584
	ASY	6.5018	9.2059	12.6966	16.4001	20.1912
	%ERR	24.5735	3.8696	1.0365	0.3731	0.1630
3	NUM	6.3489	9.8750	13.4113	17.0382	20.7451
	ASY	9.6849	10.7975	13.7576	17.1958	20.8278
	%ERR	52.5438	9.3416	2.5826	0.9250	0.3988
$\beta = b/a = 0.8$						
0	NUM	15.7191	31.4233	47.1292	62.8359	78.5431
	ASY	15.7195	31.4233	47.1292	62.8359	78.5431
	%ERR	0.0025	0.0001	0.0000	0.0000	0.0000
1	NUM	15.7583	31.4431	47.1424	62.8459	78.5511
	ASY	15.7593	31.4432	47.1424	62.8459	78.5511
	%ERR	0.0064	0.0002	0.0000	0.0000	0.0000
2	NUM	15.8752	31.5025	47.1821	62.8757	78.5749
	ASY	15.8787	31.5029	47.1822	62.8757	78.5750
	%ERR	0.0215	0.0012	0.0002	0.0001	0.0000
3	NUM	16.0683	31.6012	47.2482	62.9253	78.6147
	ASY	16.0776	31.6023	47.2485	62.9255	78.6148
	%ERR	0.0578	0.0037	0.0007	0.0002	0.0001

Free/Clamped Case

For a free/clamped annular plate, the boundary conditions are $M_1 = 0$ and $Q_1 + M_{6\theta}/r = 0$ at $r = a$ and $w = 0$ and $w_r = 0$ at $r = b$. The second-order approximation of the eigenvalues is

$$\alpha_{nm} \approx \frac{(2m-1)\pi}{2(1-\beta)} - \frac{\left[1 - 4n^2 + \beta\left(4n^2 - 8\nu + 7\right)\right]}{4(2m-1)\pi\beta}$$
$$- \frac{\left\{\beta^2\left[4(4\nu - 3)n^2 + 8\nu^2 - 8\nu - 1\right] + (4n^2 - 1)\right\}(1-\beta)}{2(2m-1)^2 \pi^2 \beta^2} \quad (8.3.85)$$

Table 8.10 Values of the eigenvalues α_{nm} of a free/clamped annular plate, obtained from numerical (NUM) and asymptotic (ASY) approaches, and the percent relative error (%ERR) for different combinations of $n \geq 0$ and $m \geq 1$.

n	Method	α_{n1}	α_{n2}	α_{n3}	α_{n4}	α_{n5}
$\beta = b/a = 0.2$						
0	NUM	2.2762	5.6825	9.6997	13.6540	17.5984
	ASY	2.3214	5.7605	9.7096	13.6582	17.6004
	%ERR	1.9849	1.3717	0.1015	0.0311	0.0114
1	NUM	2.1939	5.8759	9.8465	13.7754	17.7012
	ASY	N/A	5.7670	9.8138	13.7634	17.6955
	%ERR	N/A	1.8525	0.3324	0.0873	0.0325
2	NUM	2.5391	6.4776	10.3017	14.1461	18.0130
	ASY	N/A	5.7866	10.1264	14.0788	17.9806
	%ERR	N/A	10.6680	1.7020	0.4763	0.1800
3	NUM	3.5517	7.4438	11.0733	14.7755	18.5405
	ASY	N/A	5.8192	10.6474	14.6044	18.4558
	%ERR	N/A	21.8253	3.8464	1.1576	0.4569
$\beta = b/a = 0.8$						
0	NUM	9.1924	23.3126	39.1824	54.9120	70.6346
	ASY	7.4314	23.4115	39.1785	54.9122	70.6346
	%ERR	19.1565	0.4244	0.0098	0.0004	0.0000
1	NUM	9.2283	23.3415	39.1987	54.9236	70.6436
	ASY	7.5206	23.4391	39.1948	54.9238	70.6436
	%ERR	18.5048	0.4182	0.0098	0.0004	0.0000
2	NUM	9.3377	23.4280	39.2475	54.9583	70.6705
	ASY	7.7883	23.5219	39.2437	54.9585	70.6705
	%ERR	16.5937	0.4009	0.0096	0.0004	0.0000
3	NUM	9.5246	23.5710	39.3288	55.0161	70.7153
	ASY	8.2343	23.6599	39.3252	55.0163	70.7153
	%ERR	13.5475	0.3772	0.0091	0.0004	0.0000

where $\beta \equiv b/a$. In Table 8.10, we present values for the first few eigenvalues determined numerically and asymptotically for the cases $\beta = 0.2$ and $\beta = 0.8$.

8.4 NONLINEAR VIBRATIONS OF CIRCULAR PLATES

The free and forced nonlinear vibrations of plates were studied extensively in the literature (Chu and Herrmann, 1956; Chia, 1980; Ostiguy and Evan-Iwanowski, 1993; Sathyamoorthy, 1997). Modal interactions in the response of isotropic thin circular membranes, disks, and plates were considered by Williams and Tobias (1963), Nowinski (1964), Efstathiades (1971), Sridhar, Mook, and Nayfeh (1975, 1978), Nayfeh and

Mook (1979), Hadian and Nayfeh (1990), T. Nayfeh and Vakakis (1994), Lee and Kim (1995), Yeo and Lee (2002), Lee and Yeo (2003), Lee, Yeo, and Samoilenko (2003), and Arafat and Nayfeh (2004a,b); in the response of near square plates by Yang and Sethna (1991, 1992); in the response of rectangular plates by Chang, Bajaj, and Krousgrill (1993), Oh (1994), Abe, Kobayashi, and Yamada (1998a,b), and Oh and Nayfeh (1998); in the response of rectangular membranes by Yasuda and Asano (1986); and in the response of circular and square membranes by Yasuda and Torii (1986, 1987). Hui (1983, 1985a) investigated the effects of geometric imperfections on the large-amplitude vibration of circular plates. He found out that geometric imperfections may significantly raise the linear vibration frequencies and may cause a qualitative change in the plate behavior.

Tobias (1957, 1958), Williams and Tobias (1963), and Williams (1966) studied the vibrations of imperfect disks and demonstrated the phenomenon of preferential modes; that is, corresponding to each asymmetric mode of a perfect disk there are two modes having slightly different frequencies in the imperfect disk. Moreover, they found that large-amplitude oscillations of a circular disk can include a traveling wave component. The existence of the traveling wave component in the response was attributed (and confirmed by experiments) to the nonlinear coupling between preferential modes; that is, the one-to-one internal resonance. Yasuda and Uno (1983) investigated the multimode response of a circular membrane.

We follow Sridhar, Mook, and Nayfeh (1978) and consider the nonlinear response of a uniform circular plate to a transverse harmonic excitation. Efstathiadis (1971) derived the equations governing the free undamped oscillations of nonuniform circular plates. The governing equations are given by (7.4.54) and (7.4.59).

It is convenient to rewrite these equations in terms of nondimensional variables, denoted by asterisks, which are defined as follows:

$$r = ar^*, \qquad t = a^2 \left(\frac{\rho h}{D}\right)^{1/2} t^*, \qquad w = \frac{h^2}{a} w^*,$$

$$u = \frac{h^4}{a^3} u^*, \qquad v = \frac{h^4}{a^3} v^*, \qquad \Phi = \frac{Eh^5}{a^2} \Phi^*$$

$$c = \frac{24(1-\nu^2)}{a^4} (\rho h^5 D)^{1/2} c^*, \qquad F = \frac{12(1-\nu^2)Dh^4}{a^7} F^*,$$

where a is the radius of the plate. We are concerned with generating an approximate solution which is valid as h/a approaches zero; each of the nondimensional variables defined above is presumed to be $O(1)$ in this limit. In addition, we define e_r^*, e_θ^*, $e_{r\theta}^*$, N_r^*, N_θ^*, $N_{r\theta}^*$, M_r^*, M_θ^*, $M_{r\theta}^*$, Q_r^*, and Q_θ^*, which are also presumed to be $O(1)$ as h/a approaches zero, as follows:

$$(e_1, e_2, e_6) = \frac{h^4}{a^4} (e_1^*, e_2^*, e_6^*),$$

$$(N_1, N_2, N_6) = \frac{Eh^5}{a^4} (N_1^*, N_1^*, N_6^*),$$

and

$$(M_1, M_2, M_6, Q_1, Q_2) = \frac{Eh^5}{12(1-\nu^2)a^3}\left(M_1^*, M_2^*, M_6^*, \frac{Q_1^*}{a}, \frac{Q_2^*}{a}\right)$$

Substituting these definitions into (7.4.54), (7.4.56)-(7.4.59), and (7.4.39) and dropping the asterisks in the result, one obtains

$$\frac{\partial^2 w}{\partial t^2} + \nabla^4 w = \epsilon \Bigg[\frac{\partial^2 w}{\partial r^2}\left(\frac{1}{r}\frac{\partial \Phi}{\partial r} + \frac{1}{r^2}\frac{\partial^2 \Phi}{\partial \theta^2}\right) + \frac{\partial^2 \Phi}{\partial r^2}\left(\frac{1}{r}\frac{\partial w}{\partial r} + \frac{1}{r^2}\frac{\partial^2 w}{\partial \theta^2}\right)$$
$$-2\left(\frac{1}{r}\frac{\partial^2 \Phi}{\partial r \partial \theta} - \frac{1}{r^2}\frac{\partial \Phi}{\partial \theta}\right)\left(\frac{1}{r}\frac{\partial^2 w}{\partial r \partial \theta} - \frac{1}{r^2}\frac{\partial w}{\partial \theta}\right)$$
$$-2c\frac{\partial w}{\partial t} + F\Bigg] \qquad (8.4.1)$$

$$\nabla^4 \Phi = \left(\frac{1}{r}\frac{\partial^2 w}{\partial r \partial \theta} - \frac{1}{r^2}\frac{\partial w}{\partial \theta}\right)^2 - \frac{\partial^2 w}{\partial r^2}\left(\frac{1}{r}\frac{\partial w}{\partial r} + \frac{1}{r^2}\frac{\partial^2 w}{\partial \theta^2}\right) \qquad (8.4.2)$$

where $\epsilon = 12\left(1-\nu^2\right)h^2/a^2$ and

$$e_1 = \frac{1}{r}\frac{\partial \Phi}{\partial r} + \frac{1}{r^2}\frac{\partial^2 \Phi}{\partial \theta^2} - \nu\frac{\partial^2 \Phi}{\partial r^2} \qquad (8.4.3a)$$

$$e_2 = \frac{\partial^2 \Phi}{\partial r^2} - \nu\left(\frac{1}{r}\frac{\partial \Phi}{\partial r} + \frac{1}{r^2}\frac{\partial^2 \Phi}{\partial \theta^2}\right) \qquad (8.4.3b)$$

$$\frac{1}{2}\gamma_6 = (1+\nu)\left(\frac{1}{r^2}\frac{\partial \Phi}{\partial \theta} - \frac{1}{r}\frac{\partial^2 \Phi}{\partial r \partial \theta}\right) \qquad (8.4.3c)$$

$$e_1 = \frac{\partial u}{\partial r} + \frac{1}{2}\left(\frac{\partial w}{\partial r}\right)^2 \qquad (8.4.4a)$$

$$e_2 = \frac{u}{r} + \frac{1}{r}\frac{\partial v}{\partial \theta} + \frac{1}{2r^2}\left(\frac{\partial w}{\partial \theta}\right)^2 \qquad (8.4.4b)$$

$$\gamma_6 = \frac{1}{r}\frac{\partial u}{\partial \theta} + \frac{\partial v}{\partial r} - \frac{v}{r} + \frac{1}{r}\frac{\partial w}{\partial r}\frac{\partial w}{\partial \theta} \qquad (8.4.4c)$$

At a clamped immovable edge,

$$w = 0, \quad \frac{\partial w}{\partial r} = 0, \quad u = 0, \text{ and } v = 0 \qquad (8.4.5)$$

It follows from (8.4.4b) and (8.4.5) that

$$e_2 = 0 \text{ at } r = 1 \qquad (8.4.6)$$

and it follows from (8.4.4a), (8.4.4c), and (8.4.5) that

$$e_1 = \frac{\partial u}{\partial r}, \quad \gamma_6 = \frac{\partial v}{\partial r} \quad \text{at} \quad r = 1 \tag{8.4.7}$$

To eliminate u and v from (8.4.7), we multiply (8.4.4b) with r, differentiate the result with respect to r, and obtain

$$\frac{\partial}{\partial r}(re_2) = \frac{\partial u}{\partial r} + \frac{\partial^2 v}{\partial r \partial \theta} - \frac{1}{2r^2}\left(\frac{\partial w}{\partial \theta}\right)^2 + \frac{1}{r}\frac{\partial w}{\partial \theta}\frac{\partial^2 w}{\partial r \partial \theta}$$

which upon using (8.4.5) becomes

$$\frac{\partial}{\partial r}(re_2) = \frac{\partial u}{\partial r} + \frac{\partial^2 v}{\partial r \partial \theta} \tag{8.4.8}$$

Eliminating u and v from (8.4.7) and (8.4.8) yields

$$\frac{\partial}{\partial r}(re_2) - e_1 - \frac{\partial}{\partial \theta}(\gamma_6) = 0 \tag{8.4.9}$$

Substituting (8.4.3) into (8.4.9) and (8.4.6) yields the following two conditions on Φ:

$$\frac{\partial^2 \Phi}{\partial r^2} - \nu\left(\frac{1}{r}\frac{\partial \Phi}{\partial r} + \frac{1}{r^2}\frac{\partial^2 \Phi}{\partial \theta^2}\right) = 0 \tag{8.4.10}$$

$$\frac{\partial^3 \Phi}{\partial r^3} + \frac{1}{r}\frac{\partial^2 \Phi}{\partial r^2} - \frac{1}{r^2}\frac{\partial \Phi}{\partial r} + \frac{2+\nu}{r^2}\frac{\partial^3 \Phi}{\partial r \partial \theta^2} - \frac{3+\nu}{r^3}\frac{\partial^2 \Phi}{\partial \theta^2} = 0 \tag{8.4.11}$$

At a free edge,

$$N_1 = 0, \; N_6 = 0, \; M_1 = 0, \; \text{and} \; Q_1 + \frac{1}{r}\frac{\partial M_6}{\partial \theta} = 0 \tag{8.4.12}$$

Combining the nondimensionalized forms of (7.2.39), (7.4.50), (7.4.51), and (7.4.53) yields the following nondimensional boundary conditions at a free edge:

$$\frac{\partial^2 w}{\partial r^2} + \frac{\nu}{r}\frac{\partial w}{\partial r} + \frac{\nu}{r^2}\frac{\partial^2 w}{\partial \theta^2} = 0 \tag{8.4.13}$$

$$\frac{\partial}{\partial r}(\nabla^2 w) + \frac{1-\nu}{r^2}\frac{\partial^2}{\partial \theta^2}\left(\frac{\partial w}{\partial r} - \frac{w}{r}\right) = 0 \tag{8.4.14}$$

$$\frac{\partial \Phi}{\partial r} + \frac{1}{r}\frac{\partial^2 \Phi}{\partial \theta^2} = 0 \tag{8.4.15}$$

$$\frac{\partial^2 \Phi}{\partial r \partial \theta} - \frac{1}{r}\frac{\partial \Phi}{\partial \theta} = 0 \tag{8.4.16}$$

For the axisymmetric case, the governing equations can be obtained from (8.4.1) and (8.4.2) by dropping the derivatives with respect to θ. The result is

$$\frac{\partial^2 w}{\partial t^2} + \nabla^4 w = \epsilon\left[\frac{1}{r}\frac{\partial^2 w}{\partial r^2}\frac{\partial \Phi}{\partial r} + \frac{1}{r}\frac{\partial w}{\partial r}\frac{\partial^2 \Phi}{\partial r^2} - 2c\frac{\partial w}{\partial t} + F\right] \tag{8.4.17}$$

$$\nabla^4 \Phi = -\frac{1}{r}\frac{\partial w}{\partial r}\frac{\partial^2 w}{\partial r^2} \tag{8.4.18}$$

Also for the axisymmetric case, the compatibility equation and boundary conditions can be simplified considerably. It follows from (8.4.3c) that $\gamma_6 = 0$ and hence, it follows from (8.4.4c) that

$$\frac{\partial v}{\partial r} - \frac{v}{r} = 0$$

whose solution is

$$v(r,t) = c(t)r$$

Dropping the derivatives with respect to θ from (8.4.3a), (8.4.3b), (8.4.4a), and (8.4.4b), we obtain

$$e_1 = \frac{1}{r}\frac{\partial \Phi}{\partial r} - \nu \frac{\partial^2 \Phi}{\partial r^2} \qquad (8.4.19)$$

$$e_2 = \frac{\partial^2 \Phi}{\partial r^2} - \frac{\nu}{r}\frac{\partial \Phi}{\partial r} \qquad (8.4.20)$$

$$e_1 = \frac{\partial u}{\partial r} + \frac{1}{2}\left(\frac{\partial w}{\partial r}\right)^2 \qquad (8.4.21)$$

$$e_2 = \frac{u}{r} \qquad (8.4.22)$$

Eliminating u from (8.4.21) and (8.4.22) yields

$$e_1 - \frac{\partial (re_2)}{\partial r} = \frac{1}{2}\left(\frac{\partial w}{\partial r}\right)^2 \qquad (8.4.23)$$

Substituting (8.4.19) and (8.4.20) into (8.4.23) yields the simplified compatibility equation

$$r\frac{\partial^3 \Phi}{\partial r^3} + \frac{\partial^2 \Phi}{\partial r^2} - \frac{1}{r}\frac{\partial \Phi}{\partial r} = -\frac{1}{2}\left(\frac{\partial w}{\partial r}\right)^2 \qquad (8.4.24)$$

At a clamped edge, $v = 0$ at $r = 1$ and therefore $c(t) = 0$ and hence, $v(r,t) \equiv 0$. Consequently, this boundary condition does not lead to a boundary condition on $\Phi(r,t)$, and therefore we have only one instead of two boundary conditions on $\Phi(r,t)$. This condition corresponds to $u = re_2 = 0$ or

$$\frac{\partial^2 \Phi}{\partial r^2} - \frac{\nu}{r}\frac{\partial \Phi}{\partial r} = 0 \qquad (8.4.25)$$

Hence, $\Phi(r,t)$ is unique to within an arbitrary function of time, the latter does not influence $w(r,t)$ according to (8.4.17).

At a free edge, dropping the derivatives with respect to θ from (8.4.13)-(8.4.16) leads to

$$\frac{\partial^2 w}{\partial r^2} + \frac{\nu}{r}\frac{\partial w}{\partial r} = 0 \qquad (8.4.26)$$

$$\frac{\partial}{\partial r}(\nabla^2 w) = 0 \qquad (8.4.27)$$

$$\frac{\partial \Phi}{\partial r} = 0 \qquad (8.4.28)$$

and the boundary condition (8.4.16) has disappeared. Again, we have only one instead of two boundary conditions on $\Phi(r,t)$, and hence $\Phi(r,t)$ is unique to within an arbitrary function of time, the latter does not influence $w(r,t)$ according to (8.4.17).

In this section, we consider the case of a circular isotropic plate clamped along its edge so that the motion is described by (8.4.1) and (8.4.2) subject to the boundary conditions (8.4.5), (8.4.10), and (8.4.11) at $r = 1$. In addition, we require the solution to be bounded at the origin.

The linear undamped natural frequencies and mode shapes are governed by

$$\frac{\partial^2 w}{\partial t^2} + \nabla^4 w = 0 \tag{8.4.29}$$

$$w = 0 \text{ and } \frac{\partial w}{\partial r} = 0 \text{ at } r = 1 \tag{8.4.30}$$

$$w(r,\theta,t) < \infty \text{ at } r = 0 \tag{8.4.31}$$

As in Section 8.3.1, the solution of (8.4.29)-(8.4.31) can be expressed as

$$w(r,\theta,t) = \sum_{m=1}^{\infty} \sum_{n=1}^{\infty} (c_{nm} \cos n\theta + b_{nm} \sin n\theta) \phi_{nm}(r) e^{i\omega_{nm}t} + cc \tag{8.4.32}$$

where the $\phi_{nm}(r)$ are the linear free-oscillation modes given by

$$\phi_{nm}(r) = \frac{J_n\left(\sqrt{\omega_{nm}}r\right)}{J_n\left(\sqrt{\omega_{nm}}\right)} - \frac{I_n\left(\sqrt{\omega_{nm}}r\right)}{I_n\left(\sqrt{\omega_{nm}}\right)} \tag{8.4.33}$$

and the ω_{nm} are the corresponding linear undamped natural frequencies, which are the roots of

$$I_n\left(\sqrt{\omega}\right) J'_n\left(\sqrt{\omega}\right) - I'_n\left(\sqrt{\omega}\right) J_n(\sqrt{\omega}) = 0 \tag{8.4.34}$$

The J_n and I_n are the Bessel and modified Bessel functions of order n. We note that there is a one-to-one internal resonance between the modes $\phi_{nm}(r) \cos n\theta$ and $\phi_{nm}(r) \sin n\theta$ when $n \geq 1$. Moreover, we note that the mode shapes corresponding to different frequencies are orthonormal; that is,

$$\int_0^1 r\phi_{nm}(r)\phi_{k\ell}(r)dr = \delta_{nk}\delta_{m\ell}$$

where δ_{nk} is the Kronecker delta. Next, we consider axisymmetric vibrations and then consider asymmetric vibrations.

8.4.1 Axisymmetric Vibrations

In this case, the motion is described by (8.4.17) and (8.4.24) subject to the boundary conditions (8.4.5) and (8.4.25) at $r = 1$ and that $w(r,t)$ and $\Phi(r,t)$ are bounded at the origin. We consider an axisymmetric primary resonance excitation of the mth axisymmetric mode having the form

$$F(r,t) = \hat{f}(r) \cos \Omega t \tag{8.4.35}$$

where
$$\Omega = \omega_m + \epsilon\sigma \qquad (8.4.36)$$

We assume that the mth axisymmetric mode is not involved in any internal or combination resonance with any other mode. Sridhar, Mook, and Nayfeh (1978), Hadian and Nayfeh (1990), and Lee and Kim (1995) treated the case of internal resonance among three axisymmetric modes, and the results are summarized by Nayfeh (2000).

In the presence of damping, all modes that are not directly excited or indirectly excited via an internal or combination resonance decay with time. Because the nonlinearity is cubic and because we are considering the case of primary resonance, we can use the Galerkin procedure to reduce the distributed parameter problem into a single nonlinear second-order ordinary-differential equation. To this end, we approximate $w(r,t)$ by

$$w(r,t) \approx \eta_m(t)\phi_m(r) \qquad (8.4.37)$$

where ϕ_m is the mth axisymmetric mode shape. Substituting (8.4.37) into (8.4.24) yields

$$r\frac{\partial^3 \Phi}{\partial r^3} + \frac{\partial^2 \Phi}{\partial r^2} - \frac{1}{r}\frac{\partial \Phi}{\partial r} = -\frac{1}{2}\eta_m^2 \phi_m'^2 \qquad (8.4.38)$$

We seek the solution of (8.4.38) that satisfies the boundary conditions as

$$\Phi(r,t) = \psi_m(r)\eta_m^2 \qquad (8.4.39)$$

where ψ_m is given by the boundary-value problem

$$r\frac{d}{dr}\left(\frac{d^2\psi_m}{dr^2} + \frac{1}{r}\frac{d\psi_m}{dr}\right) = -\frac{1}{2}\phi_m'^2 \qquad (8.4.40)$$

$$\psi_m'' - \nu\psi_m' = 0 \text{ at } r = 1 \qquad (8.4.41)$$

$$\psi_m < \infty \text{ at } r = 0 \qquad (8.4.42)$$

Equation (8.4.40) can be rewritten as

$$r\frac{d^2v}{dr^2} + \frac{dv}{dr} - \frac{1}{r}v = -\frac{1}{2}\phi_m'^2 \qquad (8.4.43)$$

where $v = \psi_m'$. Using the method of variation of parameters and noting that two linear independent solutions of the homogeneous equation (8.4.40) are r and r^{-1}, we express the general solution of (8.4.40) as

$$v = c_2(r)r + c_3(r)r^{-1} \qquad (8.4.44)$$

Differentiating (8.4.44) once with respect to r yields

$$v' = c_2'r + c_3'r^{-1} + c_2 - c_3 r^{-2}$$

Setting
$$c_2'r + c_3'r^{-1} = 0 \qquad (8.4.45)$$

we obtain
$$v' = c_2 - c_3 r^{-2} \tag{8.4.46}$$

Substituting (8.4.44) and (8.4.46) into (8.4.43), we have
$$rc_2' - c_3' r^{-1} = -\frac{1}{2}\phi_m'^2(r) \tag{8.4.47}$$

Solving (8.4.45) and (8.4.47) for c_2' and c_3', we obtain
$$c_2' = -\frac{1}{4r}\phi_m'^2(r) \text{ and } c_3' = \frac{1}{4}r\phi_m'^2(r)$$

and hence
$$c_2 = -\frac{1}{4}\int_0^r \frac{\phi_m'^2(\xi)}{\xi}d\xi + c_{20} \text{ and } c_3 = \frac{1}{4}\int_0^r \xi\phi_m'^2(\xi)d\xi + c_{30} \tag{8.4.48}$$

Substituting (8.4.48) into (8.4.44) and noting that $v(r) = \psi_m'(r)$, we obtain
$$\psi_m'(r) = -\frac{1}{4}r\int_0^r \frac{\phi_m'^2(\xi)}{\xi}d\xi + \frac{1}{4r}\int_0^r \xi\phi_m'^2(\xi)d\xi + c_{20}r + \frac{c_{30}}{r} \tag{8.4.49}$$

The boundary condition (8.4.42) demands that $c_{30} = 0$ and the boundary condition (8.4.41) demands that
$$c_{20} = \frac{1}{4}\int_0^1 \frac{\phi_m'^2(\xi)}{\xi}d\xi + \frac{1+\nu}{4(1-\nu)}\int_0^1 \xi\phi_m'^2(\xi)d\xi \tag{8.4.50}$$

Alternatively, we express ψ_m' in terms of the $J_1(\xi_m r)$ as follows. Because
$$\left(r^2\frac{d^2}{dr^2} + r\frac{d}{dr} - 1\right)J_1(\xi r) = -\xi^2 r^2 J_1(\xi r) \tag{8.4.51}$$

we express ψ_m' as
$$\psi_m'(r) = \sum_{k=1}^{\infty} b_{km} J_1(\xi_k r) \tag{8.4.52}$$

Using the boundary condition (8.4.41), we find that the ξ_k are the roots of
$$\xi_k J_1(\xi_k) - (1+\nu)J_0(\xi_k) = 0 \tag{8.4.53}$$

The lowest ten roots are 1.57883, 5.27284, 8.5006, 11.6802, 14.8433, 17.9988, 21.1502, 24.299, 27.4461, and 30.5921. Substituting (8.4.52) into (8.4.43) and using (8.4.51), we obtain
$$\sum_{k=1}^{\infty} b_{km}\xi_k^2 r J_1(\xi_k r) = \frac{1}{2}\phi_m'^2(r) \tag{8.4.54}$$

Multiplying (8.4.54) with $J_1(\xi_s r)$ and integrating the result with respect to r from $r = 0$ to 1, we have

$$b_{sm} = \frac{\int_0^1 J_1(\xi_s r) \phi_m'^2(r) dr}{2\xi_s^2 \int_0^1 r J_1^2(\xi_s r) dr} \tag{8.4.55}$$

Substituting (8.4.37) and (8.4.39) into (8.4.17) yields

$$\ddot{\eta}_m \phi_m + \eta_m \nabla^4 \phi_m \approx \epsilon \left[(\phi_m'' \psi_m' + \phi_m' \psi_m'') \eta_m^3 - 2c\dot{\eta}_m \phi_m + F \right] \tag{8.4.56}$$

Multiplying (8.4.56) by $r\phi_m$, noting that $\nabla^4 \phi_m = \omega_m^2 \phi_m$, and integrating the result over the domain, we obtain

$$\ddot{\eta}_m + \omega_m^2 \eta_m = \epsilon \left[-2\mu_m \dot{\eta}_m - \alpha_{em} \eta_m^3 + f_m \cos \Omega t \right] \tag{8.4.57}$$

where

$$\alpha_{em} = -\int_0^1 \phi_m (\phi_m' \psi_m')' dr = \int_0^1 \phi_m'^2 \psi_m' dr \tag{8.4.58}$$

and

$$\mu_m = \int_0^1 cr\phi_m^2 dr \quad \text{and} \quad f_m = \int_0^1 r\hat{f}\phi_m dr$$

To calculate α_{em}, we first calculate the roots of (8.4.34) when $n = 0$. The first three roots are 3.1962, 6.3064, and 9.4395. Then, we calculate the corresponding c_{20} from (8.4.50); the results are 11.0376, 102.429, and 341.797. Finally, we calculate the corresponding α_{em} from (8.4.58); the results for $\nu = 0.3$ are 52.3598, 1,810.8, and 11,240.6. Alternatively, we calculate the b_{sm} from (8.4.55), then calculate ψ_m' from (8.4.52), and finally calculate α_{em} from (8.4.58). We note that the effective nonlinearities α_{em} are positive, and hence the nonlinearity is of the hardening type.

Equation (8.4.57) is a standard Duffing equation whose solutions can be found in many texts on perturbation methods, nonlinear vibrations, and dynamics. Here, we develop an approximate solution by using the method of multiple scales in the form

$$\eta_m(t; \epsilon) = \eta_{m0}(T_0, T_1) + \epsilon \eta_{m1}(T_0, T_1) + \cdots \tag{8.4.59}$$

Substituting (8.4.59) into (8.4.57) and equating coefficients of like powers of ϵ, we obtain

Order ϵ^0:

$$D_0^2 \eta_{m0} + \omega_m^2 \eta_{m0} = 0 \tag{8.4.60}$$

Order ϵ:

$$D_0^2 \eta_{m1} + \omega_m^2 \eta_{m1} = -2D_0 D_1 \eta_{m0} - 2\mu_m D_0 \eta_{m0} - \alpha_{em} \eta_{m0}^3 + f_m \cos \Omega T_0 \tag{8.4.61}$$

The solution of (8.4.60) can be expressed as

$$\eta_{m0} = A_m(T_1) e^{i\omega_m T_0} + cc \tag{8.4.62}$$

Substituting (8.4.62) into (8.4.61) yields

$$D_0^2 \eta_{m1} + \omega_m^2 \eta_{m1} = - \left[2i\omega_m (A_m' + \mu_m A_m) + 3\alpha_{em} A_m^2 \bar{A}_m \right] e^{i\omega_m T_0}$$
$$- \alpha_{em} A_m^3 e^{3i\omega_m T_0} + \frac{1}{2} f_m e^{i\Omega T_0} + cc \qquad (8.4.63)$$

Using (8.4.36) in eliminating the terms that produce secular terms in (8.4.63), we obtain

$$2i\omega_m (A_m' + \mu_m A_m) + 3\alpha_{em} A_m^2 \bar{A}_m - \frac{1}{2} f_m e^{i\sigma T_1} = 0 \qquad (8.4.64)$$

Substituting the polar form

$$A_m = \frac{1}{2} a_m e^{i(\sigma T_1 - \gamma_m)} \qquad (8.4.65)$$

into (8.4.64), multiplying the result by $\exp[i(\gamma_m - \sigma T_1)]$, and separating real and imaginary parts, we obtain

$$a_m' = -\mu_m a_m + \frac{f_m}{2\omega_m} \sin \gamma_m \qquad (8.4.66)$$

$$a_m \gamma_m' = \sigma a_m - \frac{3\alpha_{em}}{8\omega_m} a_m^3 + \frac{f_m}{2\omega_m} \cos \gamma_m \qquad (8.4.67)$$

Using (8.4.37), (8.4.33), (8.4.62), and (8.4.65), we find that the deflection, to the first approximation, is given by

$$w(r,t) \approx a_m \left[\frac{J_0(\sqrt{\omega_m} r)}{J_0(\sqrt{\omega_m})} - \frac{I_0(\sqrt{\omega_m} r)}{I_0(\sqrt{\omega_m})} \right] \cos(\Omega t - \gamma_m) \qquad (8.4.68)$$

where a_m and γ_m are given by (8.4.66) and (8.4.67).

It follows from (8.4.68) that periodic responses of the plate correspond to constant a_m and γ_m. Setting a_m' and γ_m' equal to zero in (8.4.66) and (8.4.67) and eliminating γ_m from the resulting equations yields the frequency-response equation

$$\mu_m^2 + \left(\sigma - \frac{3\alpha_{em}}{8\omega_m} a_m^2 \right)^2 = \frac{f_m^2}{4\omega_m^2 a_m^2}$$

In Figures 8.4.1 and 8.4.2, we show typical frequency- and force-response curves. The frequency-response curve is bent to the right reflecting the hardening nature of the effective nonlinearity.

8.4.2 Asymmetric Vibrations

We consider a primary resonant excitation of the nmth mode having the form

$$F(r, \theta, t) = \hat{f}(r) \cos n\theta e^{i\Omega t} + cc$$

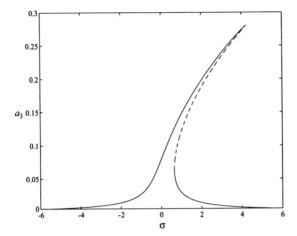

Fig. 8.4.1 The frequency-response curve obtained for the third mode when $\mu_3 = 0.1$ and $f_3 = 5$.

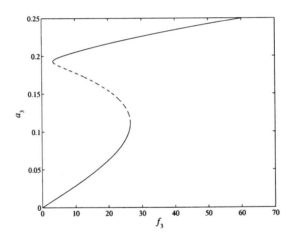

Fig. 8.4.2 The force-response curve obtained for the third mode when $\mu_3 = 0.1$ and $\sigma = 2$.

where $n \geq 1$ and
$$\Omega = \omega_{nm} + \epsilon\sigma$$

We refer to the mode $\phi_{nm}(r)\cos n\theta$ as the directly excited mode and refer to the mode $\phi_{nm}(r)\sin n\theta$ as the companion mode. If neither of these modes is involved in an internal resonance with any other mode, then in the presence of damping all of the other modes will decay with time and the dynamics of the plate can be described by the directly excited mode and its companion mode.

Because the nonlinearity is cubic, we use the Galerkin procedure to condense the partial-differential equations and their boundary conditions into two second-order nonlinear ordinary-differential equations. To this end, we approximate $w(r,\theta,t)$ by

$$w(r,\theta,t) \approx [\eta(t)\cos n\theta + \zeta(t)\sin n\theta]\,\phi(r) \qquad (8.4.69)$$

where the subscript nm has been eliminated for ease of notation. Substituting (8.4.69) into (8.4.2) yields

$$\nabla^4 \Phi = \frac{1}{2}(\eta^2 + \zeta^2)\chi_1(r) + \left[\frac{1}{2}(\zeta^2 - \eta^2)\cos 2n\theta - \eta\zeta\sin 2n\theta\right]\chi_2(r) \qquad (8.4.70)$$

where

$$\chi_1 = \frac{n^2}{r^4}(r\phi' - \phi)^2 - \frac{\phi''}{r^2}(r\phi' - n^2\phi)$$

$$\chi_2 = \frac{n^2}{r^4}(r\phi' - \phi)^2 + \frac{\phi''}{r^2}(r\phi' - n^2\phi)$$

The boundary conditions are (8.4.10) and (8.4.11) evaluated at $r = 1$ and

$$\Phi(r,\theta,t) < \infty \quad \text{at} \quad r = 0$$

We seek the solution of (8.4.70) that satisfies the boundary conditions as

$$\Phi = \frac{1}{2}\psi_1(r)(\eta^2 + \zeta^2) + \left[\frac{1}{2}(\zeta^2 - \eta^2)\cos 2n\theta - \eta\zeta\sin 2n\theta\right]\psi_2(r) \qquad (8.4.71)$$

where ψ_1 and ψ_2 are given by the two-point boundary-value problems

$$\left(\frac{d^2}{dr^2} + \frac{1}{r}\frac{d}{dr}\right)^2 \psi_1 = \chi_1(r) \qquad (8.4.72)$$

$$\psi_1'' - \nu\psi_1' = 0 \quad \text{at} \quad r = 1 \qquad (8.4.73)$$

$$\psi_1''' + \psi_1'' - \psi_1' = 0 \quad \text{at} \quad r = 1 \qquad (8.4.74)$$

$$\psi_1(0) < \infty \qquad (8.4.75)$$

and

$$\left(\frac{d^2}{dr^2} + \frac{1}{r}\frac{d}{dr} - \frac{4n^2}{r^2}\right)^2 \psi_2 = \chi_2(r) \qquad (8.4.76)$$

$$\psi_2'' - \nu\psi_2' + 4\nu n^2\psi_2 = 0 \quad \text{at} \quad r = 1 \qquad (8.4.77)$$

$$\psi_2''' + \psi_2'' - (1 + 8n^2 + 4n^2\nu)\psi_2' + 4n^2(3 + \nu)\psi_2 = 0 \quad \text{at} \quad r = 1 \qquad (8.4.78)$$

$$\psi_2(0) < \infty \qquad (8.4.79)$$

Substituting (8.4.69) and (8.4.71) into (8.4.1), multiplying the outcome with $r\phi(r)\cos n\theta$ and $r\phi(r)\sin n\theta$, respectively, and integrating the results from $\theta = 0$ to $\theta = 2\pi$ and $r = 0$ to $r = 1$, we obtain

$$\ddot{\eta} + \omega^2\eta = -2\epsilon\mu\dot{\eta} - \epsilon\alpha\eta^3 - \epsilon\alpha\eta\zeta^2 + \epsilon f\cos\Omega t \qquad (8.4.80)$$

$$\ddot{\zeta} + \omega^2\zeta = -2\epsilon\mu\dot{\zeta} - \epsilon\alpha\eta^2\zeta - \epsilon\alpha\zeta^3 \qquad (8.4.81)$$

where

$$\alpha = -\int_0^1 \phi\left[\frac{1}{2}(\phi'\psi_1'' + \phi''\psi_1') - \frac{1}{4}(\phi'\psi_2'' + \phi''\psi_2') - \frac{n^2}{r^2}(\phi'\psi_2 + \phi\psi_2')\right.$$
$$\left. + \frac{n^2}{r}\left(\phi'\psi_2' + \phi''\psi_2 - \frac{1}{2}\phi\psi_1'' + \frac{1}{4}\phi\psi_2''\right) + \frac{n^2}{r^3}\phi\psi_2\right]dr \quad (8.4.82)$$

$$\mu = \int_0^1 rc\phi^2 dr \quad \text{and} \quad f = \int_0^1 r\hat{f}\phi\, dr \quad (8.4.83)$$

To calculate α, we need to determine ψ_1 and ψ_2. Multiplying (8.4.72) with r, integrating the result over $0 \leq r \leq 1$, and noting that

$$r\chi_1 = \frac{d}{dr}\left[-\frac{1}{2}\phi'^2 + \frac{n^2}{2r^2}(2r\phi\phi' - \phi^2)\right]$$

we obtain

$$r\frac{d}{dr}\left[\frac{d^2\psi_1}{dr^2} + \frac{1}{r}\frac{d\psi_1}{dr}\right] = -\frac{1}{2}\phi'^2 + \frac{n^2}{2r^2}(2r\phi\phi' - \phi^2) + c_{10} \quad (8.4.84)$$

Imposing the boundary condition (8.4.74) and using the boundary conditions $\phi(1) = 0$ and $\phi'(1) = 0$, we find that $c_{10} = 0$. Following a procedure similar to that used in the preceding section, we find that

$$\psi_1' = -\frac{1}{4}r\int_0^r \left\{\frac{\phi'^2(\xi)}{\xi} - \frac{n^2}{\xi^3}\left[2\xi\phi(\xi)\phi'(\xi) - \phi^2(\xi)\right]\right\}d\xi$$
$$+ \frac{1}{4r}\int_0^r \left\{\xi\phi'^2(\xi) - \frac{n^2}{\xi}\left[2\xi\phi(\xi)\phi'(\xi) - \phi^2(\xi)\right]\right\}d\xi + c_{20}r \quad (8.4.85)$$

where

$$c_{20} = \frac{1}{4}\int_0^1 \left\{\frac{\phi'^2(\xi)}{\xi} - \frac{n^2}{\xi^3}\left[2\xi\phi(\xi)\phi'(\xi) - \phi^2(\xi)\right]\right\}d\xi$$
$$+ \frac{1+\nu}{4(1-\nu)}\int_0^1 \left\{\xi\phi'^2(\xi) - \frac{n^2}{\xi}\left[2\xi\phi(\xi)\phi'(\xi) - \phi^2(\xi)\right]\right\}d\xi \quad (8.4.86)$$

Although (8.4.85) is an analytical expression, which is attractive, one has to be careful in the numerical evaluation of ψ_1. An alternate scheme is described later in this section.

To determine ψ_2, we use the method of variation of parameters. We note that, when $n > 0$, r^{2n}, r^{2+2n}, r^{2-2n}, and r^{-2n} are four linearly independent solutions of the homogeneous equation (8.4.76). Therefore, we seek the solution of (8.4.76)-(8.4.79) in the form

$$\psi_2 = c_1(r)r^{2n} + c_2(r)r^{2+2n} + c_3(r)r^{2-2n} + c_4(r)r^{-2n} \quad (8.4.87)$$

Differentiating (8.4.87) once with respect to r and letting

$$c_1'r^{2n} + c_2'r^{2+2n} + c_3'r^{2-2n} + c_4'r^{-2n} = 0 \quad (8.4.88)$$

we obtain

$$\psi_2' = 2nc_1 r^{2n-1} + (2+2n)c_2 r^{1+2n} + (2-2n)c_3 r^{1-2n} - 2nc_4 r^{-1-2n} \quad (8.4.89)$$

Differentiating (8.4.89) once with respect to r and letting

$$2nc_1' r^{2n-1} + (2+2n)c_2' r^{1+2n} + (2-2n)c_3' r^{1-2n} - 2nc_4' r^{-1-2n} = 0 \quad (8.4.90)$$

we obtain

$$\psi_2'' = 2n(2n-1)c_1 r^{2n-2} + (2+2n)(1+2n)c_2 r^{2n} + (2-2n)(1-2n)c_3 r^{-2n}$$
$$+ 2n(1+2n)c_4 r^{-2-2n} \quad (8.4.91)$$

Differentiating (8.4.91) once with respect to r and letting

$$2n(2n-1)c_1' r^{2n-2} + (2+2n)(1+2n)c_2' r^{2n} + (2-2n)(1-2n)c_3' r^{-2n}$$
$$+ 2n(1+2n)c_4' r^{-2-2n} = 0 \quad (8.4.92)$$

we obtain

$$\psi_2''' = 2n(2n-1)(2n-2)c_1 r^{2n-3} + 2n(2+2n)(1+2n)c_2 r^{2n-1}$$
$$- 2n(2-2n)(1-2n)c_3 r^{-2n-1} - 2n(1+2n)(2+2n)c_4 r^{-3-2n} \quad (8.4.93)$$

Substituting (8.4.87), (8.4.89), (8.4.91), and (8.4.93) into (8.4.76), we have

$$4n(1-3n+2n^2)\left[c_1' r^{-3+2n} - c_3' r^{-1-2n}\right]$$
$$+4n(1+3n+2n^2)\left[c_2' r^{-1+2n} - c_4' r^{-3-2n}\right] = \chi_2 \quad (8.4.94)$$

Solving (8.4.88), (8.4.90), (8.4.92), and (8.4.94) yields

$$c_1' = \frac{r^{3-2n}\chi_2}{16n(1-2n)}, \quad c_2' = \frac{r^{1-2n}\chi_2}{16n(2n+1)}, \quad c_3' = \frac{r^{1+2n}\chi_2}{16n(2n-1)},$$
$$c_4' = -\frac{r^{3+2n}\chi_2}{16n(2n+1)} \quad (8.4.95)$$

Therefore,

$$\psi_2 = \frac{r^{2n}}{16n(1-2n)}\int_0^r r^{3-2n}\chi_2 dr + \frac{r^{2+2n}}{16n(2n+1)}\int_0^r r^{1-2n}\chi_2 dr$$
$$+ \frac{r^{2-2n}}{16n(2n-1)}\int_0^r r^{1+2n}\chi_2 dr - \frac{r^{-2n}}{16n(2n+1)}\int_0^r r^{3+2n}\chi_2 dr$$
$$+ c_{10}r^{2n} + c_{20}r^{2+2n} + c_{30}r^{2-2n} + c_{40}r^{-2n} \quad (8.4.96)$$

When $n \geq 2$, the boundary conditions (8.4.79) demand that $c_{30} = 0$ and $c_{40} = 0$ because ψ_2 and its derivatives must be bounded at $r = 0$ and the boundary conditions

(8.4.77) and (8.4.78) demand that

$$c_{10} = \frac{1}{16n(2n-1)} \int_0^1 r^{3-2n} \chi_2 dr - \frac{1+\nu}{16n(3-\nu)} \int_0^1 r^{3+2n} \chi_2 dr$$
$$+ \frac{2 - 2\nu + n^2(1+\nu)^2}{8n^2(2n-1)(1+\nu)(3-\nu)} \int_0^1 r^{1+2n} \chi_2 dr \qquad (8.4.97)$$

$$c_{20} = -\frac{1}{16n(2n+1)} \int_0^1 r^{1-2n} \chi_2 dr - \frac{1+\nu}{16n(3-\nu)} \int_0^1 r^{1+2n} \chi_2 dr$$
$$+ \frac{1+\nu}{8(2n+1)(3-\nu)} \int_0^1 r^{3+2n} \chi_2 dr \qquad (8.4.98)$$

The above solution does not apply to the case $n = 1$. Moreover, one has to be careful in the numerical evaluation of ψ_2 by either replacing 0 with a small value δ or finding an analytical solution for small δ and letting $\delta \to 0$.

Alternatively, we expand ψ_2 in terms of the eigenfunctions of

$$\left(\frac{d^2}{dr^2} + \frac{1}{r}\frac{d}{dr} - \frac{4n^2}{r^2} \right)^2 \Psi = \lambda^4 \Psi \qquad (8.4.99)$$

$$\Psi'' - \nu\Psi + 4\nu n^2 \Psi = 0 \quad \text{at} \quad r = 1 \qquad (8.4.100)$$

$$\Psi''' + \Psi'' - \left(1 + 8n^2 + 4n^2\nu\right)\Psi' + 4n^2(3+\nu)\Psi = 0 \quad \text{at} \quad r = 1 \qquad (8.4.101)$$

$$\Psi(0) < \infty \qquad (8.4.102)$$

The solution of (8.4.99), which is bounded at the origin, can be expressed as

$$\Psi = c_1 J_{2n}(\lambda r) + c_2 I_{2n}(\lambda r) \qquad (8.4.103)$$

Imposing the boundary conditions (8.4.100) and (8.4.101) yields

$$c_1 \left[\lambda^2 J''_{2n} - \nu\lambda J'_{2n} + 4\nu n^2 J_{2n} \right] + c_2 \left[\lambda^2 I''_{2n} - \nu\lambda I'_{2n} + 4n^2 \nu I_{2n} \right] = 0 \quad (8.4.104)$$

$$c_1 \left[\lambda^3 J'''_{2n} + \lambda^2 J''_{2n} - \left(1 + 8n^2 + 4n^2\nu\right) J'_{2n} + 4n^2(3+\nu) J_{2n} \right]$$
$$+ c_2 \left[\lambda^3 I'''_{2n} + \lambda^2 I''_{2n} - \left(1 + 8n^2 + 4n^2\nu\right) \lambda I'_{2n} + 4n^2(3+\nu) I_{2n} \right] = 0 \quad (8.4.105)$$

The values of the effective nonlinearity coefficient α for $\nu = 0.3$ are presented in Table 8.11. They were determined by retaining 15 terms in the series expansions of $\psi_1(r)$ and $\psi_2(r)$. The coefficients for $n > 1$ were also calculated from the closed-form solution discussed earlier by taking the lower limit of integration to be 0.001 instead of 0. The results are in excellent agreement with those presented in Table 8.11.

For detailed analysis of the solutions of (8.4.80) and (8.4.81), we refer the reader to Nayfeh (2000) and the references cited therein. We note that the modulation equations derived by Sridhar, Mook, and Nayfeh (1978) were corrected by Yeo and Lee (2002).

Table 8.11 Values of the effective nonlinearity coefficient α_{nm} for different combinations of $n \geq 1$ and $m \geq 1$ when taking $\nu = 0.3$ and using 15 terms in the series expansions for $\psi_1(r)$ and $\psi_2(r)$.

n	α_{n1}	α_{n2}	α_{n3}	α_{n4}	α_{n5}
1	188.066	2546.61	12158.62	37809.12	92238.52
2	486.619	4168.05	16701.17	47084.60	107923.56
3	1152.590	6966.53	24061.67	62179.36	134553.46
4	2382.940	11187.37	34222.61	82116.44	168944.28
5	4438.137	17228.35	47676.23	107363.78	211311.16

8.5 NONLINEAR VIBRATIONS OF ROTATING DISKS

In this section, we consider the nonlinear transverse vibrations of rotating circular disks. Such disks are principal components of many engineering systems, such as turbines, circular saws, and memory disk drives. The importance of flexural waves in causing spinning disk and membrane failures was recognized a long time ago by Richards (1872). Such waves have been observed in practice and studied experimentally either as free or forced motions in disks that have been spun in air (Campbell, 1924; von Freudenreich, 1925; Tobias and Arnold, 1957; Tobias, 1958; Williams and Tobias, 1963; Krauter and Bulkeley, 1970; MacBain, Horner, Stange, and Ogg, 1979; Yasuda, Torii, and Shimizu, 1992; Torii, Yasuda, and Yamuda, 1995a,b; and Torii, Yasuda, and Toyada, 1998). Nowinski (1964) and Nowinski and Woodall (1964) studied the nonlinear vibrations of spinning disks and membranes, respectively. Using experimental results and a dynamic impact model, Kumar, Khanna, and Jayantha (1994) revealed the complex nature of the head-disk shock-failure mechanism. Chen and Wong (1998) used the method of finite elements to investigate the vibration and stability of a spinning disk in contact with evenly spaced stationary load systems when two modes are almost degenerate; that is, two different modes having approximately equal frequencies.

Maganty and Bickford (1987, 1988) studied the influence of a one-to-one internal resonance on the free and primary resonance response of thin circular rings. Natsiavas (1994a) studied free oscillations of thin rotating rings and found that one-to-one internal resonances involving two bending modes are activated only for slowly rotating rings.

Sueoka, Yoshitake, Tamura, and Horita (1986) and Yoshitake, Sueoka, Ryn, and Tamura (1987) used the method of averaging and direct numerical integration to investigate self-excited oscillations in a rotating circular plate induced by a frictional force concentrated at its outer circumference.

The equations governing the motions of rotating disks are given by (7.4.41)-(7.4.45) modified by adding the centrifugal force $\rho h \Omega^2 r$ in the radial direction. Eliminating the Q_i and substituting for the M_i from (7.2.39), we obtain the following

514 DYNAMICS OF PLATES

equations of motion:

$$\frac{\partial N_1}{\partial r} + \frac{1}{r}\frac{\partial N_6}{\partial \theta} + \frac{N_1 - N_2}{r} + \rho h \Omega^2 r = 0 \qquad (8.5.1)$$

$$\frac{1}{r}\frac{\partial N_2}{\partial \theta} + \frac{\partial N_6}{\partial r} + \frac{2N_6}{r} = 0 \qquad (8.5.2)$$

$$\rho h \frac{\partial^2 w}{\partial t^2} + D\nabla^4 w = \frac{1}{r}\frac{\partial}{\partial r}\left(rN_1\frac{\partial w}{\partial r}\right) + \frac{1}{r^2}\frac{\partial}{\partial \theta}\left(N_2\frac{\partial w}{\partial \theta}\right)$$

$$+ \frac{1}{r}\frac{\partial}{\partial \theta}\left(N_6\frac{\partial w}{\partial r}\right) + \frac{1}{r}\frac{\partial}{\partial r}\left(N_6\frac{\partial w}{\partial \theta}\right) - c\frac{\partial w}{\partial t} + F(r,\theta,t) \quad (8.5.3)$$

Equations (8.5.1) and (8.5.2) can be satisfied exactly by introducing the stress function $\Phi(r,\theta,t)$ defined by

$$N_1 = \frac{1}{r}\frac{\partial \Phi}{\partial r} + \frac{1}{r^2}\frac{\partial^2 \Phi}{\partial \theta^2} - \frac{1}{2}\rho h \Omega^2 r^2 \qquad (8.5.4)$$

$$N_2 = \frac{\partial^2 \Phi}{\partial r^2} - \frac{1}{2}\rho h \Omega^2 r^2 \qquad (8.5.5)$$

$$N_6 = -\frac{1}{r}\frac{\partial^2 \Phi}{\partial r \partial \theta} + \frac{1}{r^2}\frac{\partial \Phi}{\partial \theta} \qquad (8.5.6)$$

Substituting (8.5.4)-(8.5.6) into (8.5.3) yields

$$\rho h \frac{\partial^2 w}{\partial t^2} + D\nabla^4 w + \frac{1}{2}\rho h \Omega^2 r^2 \nabla^2 w + \rho h \Omega^2 r \frac{\partial w}{\partial r}$$

$$= \frac{\partial^2 w}{\partial r^2}\left(\frac{1}{r}\frac{\partial \Phi}{\partial r} + \frac{1}{r^2}\frac{\partial^2 \Phi}{\partial \theta^2}\right) + \frac{\partial^2 \Phi}{\partial r^2}\left(\frac{1}{r}\frac{\partial w}{\partial r} + \frac{1}{r^2}\frac{\partial^2 w}{\partial \theta^2}\right)$$

$$- 2\left(\frac{1}{r}\frac{\partial^2 \Phi}{\partial r \partial \theta} - \frac{1}{r^2}\frac{\partial \Phi}{\partial \theta}\right)\left(\frac{1}{r}\frac{\partial^2 w}{\partial r \partial \theta} - \frac{1}{r^2}\frac{\partial w}{\partial \theta}\right)$$

$$- c\frac{\partial w}{\partial t} + F(r,\theta,t) \qquad (8.5.7)$$

Solving (7.4.48) for e_1, e_2, and γ_6 and using (8.5.4)-(8.5.6), we obtain

$$e_1 = \frac{1}{Eh}\left[\frac{1}{r}\frac{\partial \Phi}{\partial r} + \frac{1}{r^2}\frac{\partial^2 \Phi}{\partial \theta^2} - \nu \frac{\partial^2 \Phi}{\partial r^2} - \frac{1}{2}(1-\nu)\rho h \Omega^2 r^2\right] \qquad (8.5.8)$$

$$e_2 = \frac{1}{Eh}\left[\frac{\partial^2 \Phi}{\partial r^2} - \nu\left(\frac{1}{r}\frac{\partial \Phi}{\partial r} + \frac{1}{r^2}\frac{\partial^2 \Phi}{\partial \theta^2}\right) - \frac{1}{2}(1-\nu)\rho h \Omega^2 r^2\right] \qquad (8.5.9)$$

$$\gamma_6 = \frac{2(1+\nu)}{Eh}\left[\frac{1}{r^2}\frac{\partial \Phi}{\partial \theta} - \frac{1}{r}\frac{\partial^2 \Phi}{\partial r \partial \theta}\right] \qquad (8.5.10)$$

Substituting (8.5.8)-(8.5.10) into the compatibility equation (7.4.55) yields

$$\nabla^4 \Phi - 2(1-\nu)\rho h \Omega^2 = Eh\left[\left(\frac{1}{r}\frac{\partial^2 w}{\partial r \partial \theta} - \frac{1}{r^2}\frac{\partial w}{\partial \theta}\right)^2 \right.$$

$$\left. - \frac{\partial^2 w}{\partial r^2}\left(\frac{1}{r}\frac{\partial w}{\partial r} + \frac{1}{r^2}\frac{\partial^2 w}{\partial \theta^2}\right)\right] \qquad (8.5.11)$$

Next, we introduce nondimensional variables as in the preceding section and $\Omega = R^2\sqrt{\rho h/D}\Omega^*$, drop the asterisks, and rewrite (8.5.7) and (8.5.11) as

$$\frac{\partial^2 w}{\partial t^2} + \nabla^4 w + \frac{1}{2}\Omega^2 r^2 \nabla^2 w + \Omega^2 r \frac{\partial w}{\partial r}$$
$$= \epsilon\left[\frac{\partial^2 w}{\partial r^2}\left(\frac{1}{r}\frac{\partial \Phi}{\partial r} + \frac{1}{r^2}\frac{\partial^2 \Phi}{\partial \theta^2}\right) + \frac{\partial^2 \Phi}{\partial r^2}\left(\frac{1}{r}\frac{\partial w}{\partial r} + \frac{1}{r^2}\frac{\partial^2 w}{\partial \theta^2}\right)\right.$$
$$\left. - 2\left(\frac{1}{r}\frac{\partial^2 \Phi}{\partial r \partial \theta} - \frac{1}{r^2}\frac{\partial \Phi}{\partial \theta}\right)\left(\frac{1}{r}\frac{\partial^2 w}{\partial r \partial \theta} - \frac{1}{r^2}\frac{\partial w}{\partial \theta}\right) - 2c\frac{\partial w}{\partial t} + F\right] \quad (8.5.12)$$

$$\nabla^4 \Phi - \frac{2(1-\nu)\Omega^2}{\epsilon} = \left(\frac{1}{r}\frac{\partial^2 w}{\partial r \partial \theta} - \frac{1}{r^2}\frac{\partial w}{\partial \theta}\right)^2 - \frac{\partial^2 w}{\partial r^2}\left(\frac{1}{r}\frac{\partial w}{\partial r} + \frac{1}{r^2}\frac{\partial^2 w}{\partial \theta^2}\right) \quad (8.5.13)$$

where $\epsilon = 12\left(1 - \nu^2\right) h^2/a^2$. Equations (8.4.5)-(8.4.8) remain unchanged and (8.5.8) and (8.5.9) become

$$e_1 = \frac{1}{r}\frac{\partial \Phi}{\partial r} + \frac{1}{r^2}\frac{\partial^2 \Phi}{\partial \theta^2} - \nu\frac{\partial^2 \Phi}{\partial r^2} - \frac{1-\nu}{2\epsilon}\Omega^2 r^2 \quad (8.5.14)$$

$$e_2 = \frac{\partial^2 \Phi}{\partial r^2} - \nu\left(\frac{1}{r}\frac{\partial \Phi}{\partial r} + \frac{1}{r^2}\frac{\partial^2 \Phi}{\partial \theta^2}\right) - \frac{1-\nu}{2\epsilon}\Omega^2 r^2 \quad (8.5.15)$$

We consider a memory disk of outer radius a and inner radius βa, which is clamped at the inner radius and free at the outer radius. Hence, the boundary conditions at the inner radius are

$$w = 0 \text{ and } \frac{\partial w}{\partial r} = 0 \text{ at } r = \beta \quad (8.5.16)$$

$$\frac{\partial^2 \Phi}{\partial r^2} - \nu\left(\frac{1}{r}\frac{\partial \Phi}{\partial r} + \frac{1}{r^2}\frac{\partial^2 \Phi}{\partial \theta^2}\right) - \frac{1-\nu}{2\epsilon}\Omega^2 r^2 = 0 \text{ at } r = \beta \quad (8.5.17)$$

$$\frac{\partial^3 \Phi}{\partial r^3} + \frac{1}{r}\frac{\partial^2 \Phi}{\partial r^2} - \frac{1}{r^2}\frac{\partial \Phi}{\partial r} + \frac{2+\nu}{r^2}\frac{\partial^3 \Phi}{\partial r \partial \theta^2} - \frac{3+\nu}{r^3}\frac{\partial^2 \Phi}{\partial \theta^2}$$
$$- \frac{1-\nu}{\epsilon}\Omega^2 r = 0 \text{ at } r = \beta \quad (8.5.18)$$

At the outer radius, the boundary conditions are

$$\frac{\partial^2 w}{\partial r^2} + \frac{\nu}{r}\frac{\partial w}{\partial r} + \frac{\nu}{r^2}\frac{\partial^2 w}{\partial \theta^2} = 0 \text{ at } r = 1 \quad (8.5.19)$$

$$\frac{\partial}{\partial r}\left(\nabla^2 w\right) + \frac{1-\nu}{r^2}\frac{\partial^2}{\partial \theta^2}\left(\frac{\partial w}{\partial r} - \frac{w}{r}\right) = 0 \text{ at } r = 1 \quad (8.5.20)$$

$$\frac{\partial \Phi}{\partial r} + \frac{1}{r}\frac{\partial^2 \Phi}{\partial \theta^2} - \frac{1}{2\epsilon}\Omega^2 r^3 = 0 \text{ at } r = 1 \quad (8.5.21)$$

$$\frac{\partial^2 \Phi}{\partial r \partial \theta} - \frac{1}{r}\frac{\partial \Phi}{\partial \theta} = 0 \text{ at } r = 1 \quad (8.5.22)$$

For a memory disk, the excitation is a rotating point force having the form

$$F(r,\theta,t) = \hat{f}\delta(r-r_0)\delta(\theta-\Omega t) \tag{8.5.23}$$

where δ is the Dirac delta function. The forcing is scaled to be $O(\epsilon)$ to investigate primary resonances.

8.5.1 Static Problem

In the absence of the external excitation and the time variations, we seek the solution of (8.5.12) and (8.5.13) subject to the boundary conditions as

$$w = 0 \quad \text{and} \quad \Phi = \frac{g(r)\Omega^2}{\epsilon} \tag{8.5.24}$$

Substituting (8.5.24) into (8.5.13) and (8.5.17), (8.5.18), (8.5.21), and (8.5.22), we obtain the following problem governing $g(r)$:

$$\left(\frac{d^2}{dr^2} + \frac{1}{r}\frac{d}{dr}\right)^2 g = 2(1-\nu) \tag{8.5.25}$$

$$g'' - \frac{\nu}{\beta}g' = \frac{1}{2}(1-\nu)\beta^2 \quad \text{at} \quad r = \beta \tag{8.5.26}$$

$$g''' + \frac{1}{\beta}g'' - \frac{1}{\beta^2}g' = (1-\nu)\beta \quad \text{at} \quad r = \beta \tag{8.5.27}$$

$$g' = \frac{1}{2} \quad \text{at} \quad r = 1 \tag{8.5.28}$$

We note that the boundary condition (8.5.22) does not yield a boundary condition for g, and hence it can be determined only to within an arbitrary constant. However, as shown below, we do not need g but its derivatives.

Multiplying (8.5.25) with r and integrating the result once yields

$$r\frac{d}{dr}\left(\frac{d^2g}{dr^2} + \frac{1}{r}\frac{dg}{dr}\right) = (1-\nu)r^2 + c_1 \tag{8.5.29}$$

Evaluating (8.5.29) at $r = \beta$ and comparing the result with (8.5.27), we conclude that $c_1 = 0$. Then, we rewrite (8.5.29) as

$$\frac{d}{dr}\left(\frac{d^2g}{dr^2} + \frac{1}{r}\frac{dg}{dr}\right) = (1-\nu)r$$

which, upon integrating once, yields

$$\frac{d^2g}{dr^2} + \frac{1}{r}\frac{dg}{dr} = \frac{1}{2}(1-\nu)r^2 + c_2 \tag{8.5.30}$$

Multiplying (8.5.30) with r and integrating the result once, we have

$$g' = \frac{1}{8}(1-\nu)r^3 + \frac{1}{2}c_2 r + \frac{c_3}{r} \tag{8.5.31}$$

Imposing the boundary conditions (8.5.26) and (8.5.28) yields

$$\frac{1}{2}c_2 + c_3 = \frac{1}{8}(3+\nu)$$

$$\frac{1}{2}(1-\nu)c_2 - \frac{1+\nu}{\beta^2}c_3 = \frac{1}{8}(1-\nu^2)\beta^2$$

whose solution is

$$c_2 = \frac{(1+\nu)[3+\nu+(1-\nu)\beta^4]}{4[1+\nu+(1-\nu)\beta^2]}$$

$$c_3 = \frac{\beta^2(1-\nu)[3+\nu-(1+\nu)\beta^2]}{8[1+\nu+(1-\nu)\beta^2]}$$

8.5.2 Natural Frequencies and Mode Shapes

The first linear analysis of rotating disks was performed by Lamb and Southwell (1921). They calculated the natural frequencies and mode shapes of the transverse vibrations of a disk accounting for the effects of the centrifugal and bending stresses. In a subsequent paper, Southwell (1922) extended the previous results to the case of a centrally clamped annulus. Eversman and Dodson (1969) analyzed free vibrations of a centrally clamped spinning circular disk by using a power-series solution of the eigenfunctions. Barash and Chen (1972) used a numerical method to reduce and integrate the equations of motion of a freely rotating disk. The influence of moving massive loads on stationary circular disks was first studied by Iwan and Stahl (1973) and later extended by Iwan and Moeller (1976) to the case of spinning elastic disks. Using a series solution, Adams (1987) calculated the critical speeds at which a uniform flexible elastic disk is unable to support arbitrary spatially fixed transverse loads. He also accounted for the effect of an elastic foundation; he found that it raises the critical speed of rotation. Cole and Benson (1988) proposed a technique for determining the forced response to spatially fixed loads using an eigenfunction approach that predicts, in advance, the modes that dominate disk deflections. Shen and Mote (1991) analyzed the stability mechanisms of a stationary circular plate subjected to rotating forces. Chen and Bogy (1992) obtained the derivatives of the eigenvalues of a flexible spinning disk with a stationary load with respect to certain parameters in the system. Chen (1994, 1997) emphasized the effect of the relative motion between the disk and the edge load on the stability and natural frequencies of the loaded disk. They later analyzed the parametric resonance of a spinning disk under space-fixed pulsating edge loads. Yang and Hutton (1998) analyzed the vibrations of an elastically constrained rotating disk using a power-series expansion for the eigenfunctions. Nayfeh, Jilani, and Manzione (2001) calculated the natural frequencies and mode shapes of spinning hard and floppy disks for various rotating speeds. In this section, we present some of their calculations.

To determine the equations and boundary conditions governing the dynamics of rotating disks, we let

$$\Phi = \frac{g(r)}{\epsilon}\Omega^2 + \Phi_0(r,\theta,t) \tag{8.5.32}$$

in (8.5.12) and (8.5.13), use the definition of $g(r)$, and obtain

$$\frac{\partial^2 w}{\partial t^2} + \nabla^4 w + \Omega^2 \left[\frac{1}{2} r^2 \nabla^2 w + r \frac{\partial w}{\partial r} - \frac{g'}{r} \frac{\partial^2 w}{\partial r^2} - \frac{g''}{r} \frac{\partial w}{\partial r} - \frac{g''}{r^2} \frac{\partial^2 w}{\partial \theta^2} \right]$$

$$= \epsilon \left[\frac{\partial^2 w}{\partial r^2} \left(\frac{1}{r} \frac{\partial \Phi_0}{\partial r} + \frac{1}{r^2} \frac{\partial^2 \Phi_0}{\partial \theta^2} \right) + \frac{\partial^2 \Phi_0}{\partial r^2} \left(\frac{1}{r} \frac{\partial w}{\partial r} + \frac{1}{r^2} \frac{\partial^2 w}{\partial \theta^2} \right) \right.$$

$$\left. - 2 \left(\frac{1}{r} \frac{\partial^2 \Phi_0}{\partial r \partial \theta} - \frac{1}{r^2} \frac{\partial \Phi_0}{\partial \theta} \right) \left(\frac{1}{r} \frac{\partial^2 w}{\partial r \partial \theta} - \frac{1}{r^2} \frac{\partial w}{\partial \theta} \right) - 2c \frac{\partial w}{\partial t} + F \right] \quad (8.5.33)$$

$$\nabla^4 \Phi_0 = \left(\frac{1}{r} \frac{\partial^2 w}{\partial r \partial \theta} - \frac{1}{r^2} \frac{\partial w}{\partial \theta} \right)^2 - \frac{\partial^2 w}{\partial r^2} \left(\frac{1}{r} \frac{\partial w}{\partial r} + \frac{1}{r^2} \frac{\partial^2 w}{\partial \theta^2} \right) \quad (8.5.34)$$

The boundary conditions (8.5.16), (8.5.19), and (8.5.20) remain unchanged and the boundary conditions (8.5.17), (8.5.18), (8.5.21), and (8.5.22) become

$$\frac{\partial^2 \Phi_0}{\partial r^2} - \nu \left(\frac{1}{r} \frac{\partial \Phi_0}{\partial r} + \frac{1}{r^2} \frac{\partial^2 \Phi_0}{\partial \theta^2} \right) = 0 \text{ at } r = \beta \quad (8.5.35)$$

$$\frac{\partial^3 \Phi_0}{\partial r^3} + \frac{1}{r} \frac{\partial^2 \Phi_0}{\partial r^2} - \frac{1}{r^2} \frac{\partial \Phi_0}{\partial r} + \frac{2+\nu}{r^2} \frac{\partial^3 \Phi_0}{\partial r \partial \theta^2}$$

$$- \frac{3+\nu}{r^3} \frac{\partial^2 \Phi_0}{\partial \theta^2} = 0 \text{ at } r = \beta \quad (8.5.36)$$

$$\frac{\partial \Phi_0}{\partial r} + \frac{1}{r} \frac{\partial^2 \Phi_0}{\partial \theta^2} = 0 \text{ at } r = 1 \quad (8.5.37)$$

$$\frac{\partial^2 \Phi_0}{\partial r \partial \theta} - \frac{1}{r} \frac{\partial \Phi_0}{\partial \theta} = 0 \text{ at } r = 1 \quad (8.5.38)$$

The equation governing the mode shapes and corresponding natural frequencies can be obtained by dropping the nonlinear, damping, and forcing terms from (8.5.33). The result is

$$\frac{\partial^2 w}{\partial t^2} + \nabla^4 w + \Omega^2 \left[\frac{1}{2} r^2 \nabla^2 w + r \frac{\partial w}{\partial r} - \frac{g'}{r} \frac{\partial^2 w}{\partial r^2} - \frac{g''}{r} \frac{\partial w}{\partial r} - \frac{g''}{r^2} \frac{\partial^2 w}{\partial \theta^2} \right] = 0$$

(8.5.39)

The boundary conditions are (8.5.16), (8.5.19), and (8.5.20). Next, we let

$$w = \phi(r) e^{i\omega t + in\theta} \quad (8.5.40)$$

in (8.5.39), (8.5.16), (8.5.19), and (8.5.20) and obtain the eigenvalue problem

$$\nabla^4 \phi + \Omega^2 \left[\frac{1}{2} r^2 \nabla^2 \phi + r \phi' - \frac{g'}{r} \phi'' - \frac{g''}{r^2} (r\phi' - n^2 \phi) \right] - \omega^2 \phi = 0 \quad (8.5.41)$$

$$\phi = 0 \text{ and } \phi' = 0 \text{ at } r = \beta \quad (8.5.42)$$

$$\phi''' + \phi'' - (1 + 2n^2 - n^2 \nu) \phi' + n^2 (3 - \nu) \phi = 0 \text{ at } r = 1 \quad (8.5.43)$$

$$\phi'' + \nu \left(\phi' - n^2 \phi \right) = 0 \text{ at } r = 1 \quad (8.5.44)$$

Following a procedure similar to that used in Section 8.3.3, one can show that this eigenvalue problem is self-adjoint.

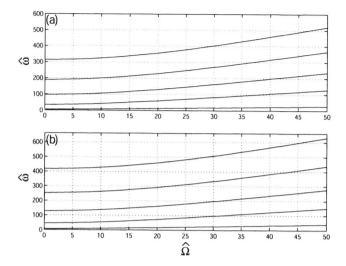

Fig. 8.5.1 Variation of the natural frequency $\hat{\omega}$ with $\hat{\Omega}$ for $n = 0$ nodal diameters for (a) floppy disk and (b) hard disk.

Nayfeh, Jilani, and Manzione (2001) obtained numerically the free vibration characteristics of two commercially available disk storage devices. Representative specifications for these disks are provided in Table 8.12. To this end, they first reduced the fourth-order problem, (8.5.41)-(8.5.44), into a system of four first-order differential equations and boundary conditions. The general solution of $\phi(r)$ consists of a linear combination of four solutions, obtained using a fifth-order Runge-Kutta method, corresponding to four sets of initial conditions. Each set of initial conditions chosen corresponds to a column in the (4×4) identity matrix. The boundary conditions (8.5.42)-(8.5.44) lead to a homogeneous system of four equations for the unknown coefficients in the linear combination determining $\phi(r)$. Setting the determinant of this system equal to zero, one can determine the eigenvalues and eigenfunctions.

Table 8.12 Representative properties of the studied memory disks (Cole and Benson, 1988).

Properties	Floppy Disk	Hard Disk
$a\ (mm)$	66.7	65
$\beta a\ (mm)$	13.34	20
$h\ (mm)$	0.076	2.03
$\rho\ (Ns^2/m^4)$	1400	2700
$E\ (MN/m^2)$	3800	70000
ν	0.23	0.35

520 DYNAMICS OF PLATES

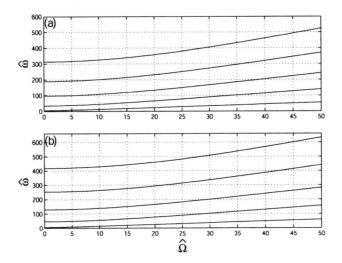

Fig. 8.5.2 Variation of the natural frequency $\hat{\omega}$ with $\hat{\Omega}$ for $n = 1$ nodal diameters for (a) floppy disk and (b) hard disk.

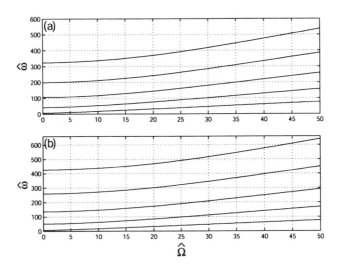

Fig. 8.5.3 Variation of the natural frequency $\hat{\omega}$ with $\hat{\Omega}$ for $n = 2$ nodal diameters for (a) floppy disk and (b) hard disk.

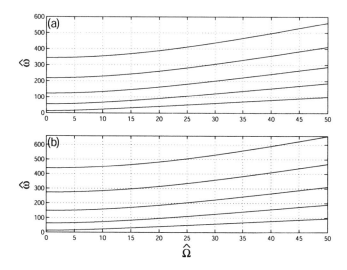

Fig. 8.5.4 Variation of the natural frequency $\hat{\omega}$ with $\hat{\Omega}$ for $n = 3$ nodal diameters for (a) floppy disk and (b) hard disk.

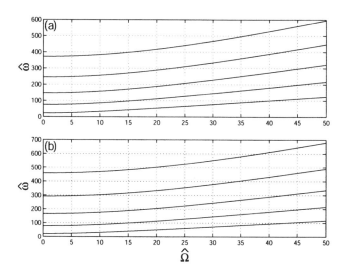

Fig. 8.5.5 Variation of the natural frequency $\hat{\omega}$ with $\hat{\Omega}$ for $n = 4$ nodal diameters for (a) floppy disk and (b) hard disk.

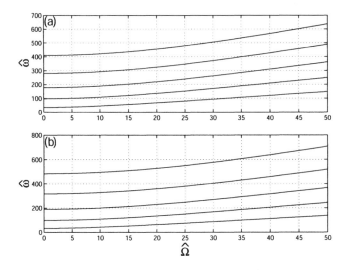

Fig. 8.5.6 Variation of the natural frequency $\hat{\omega}$ with $\hat{\Omega}$ for $n = 5$ nodal diameters for (a) floppy disk and (b) hard disk.

Nayfeh, Jilani, and Manzione (2001) calculated the dimensional natural frequencies $\hat{\omega}$ as a function of the angular dimensional velocity $\hat{\Omega}$ for $n = 0$ to $n = 5$ nodal diameters and for $m = 0$ to $m = 4$ nodal circles for the two memory disks specified in Table 8.12. In Figures 8.5.1-8.5.6, we show their results. The positive natural frequency ω_{nm} is traced; this frequency coexists with its negative value and describes an oscillatory behavior of the system. We note that the natural frequencies of the disk grow with the rotational speed $\hat{\Omega}$ due to the phenomenon known as "rotational stiffening" and the solution converges to a membrane-type solution (Southwell, 1922). Moreover the critical speed at which one of the eigenvalues is equal to zero ($\hat{\Omega} = n\hat{\omega}_{nm}$) in a nonrotating reference frame occurs for modes with $m = 0$ only when $n \geq 2$.

8.5.3 Response to a Primary-Resonance Excitation

We follow Nayfeh, Jilani, and Manzione (2001) and consider the case of primary resonance of two orthogonal modes having n nodal diameters where $n \geq 1$. We assume that the directly excited modes are not involved in an internal resonance with any other mode. Because in the presence of damping all modes that are not directly excited or indirectly excited by an internal resonance decay with time, the long-time response consists, to the second approximation, of the two directly excited modes. Moreover, because the nonlinearity is cubic, we can find a two-degree-of-freedom reduced-order model by using the Galerkin approximation. Thus, we approximate the transverse deflection $w(r,\theta,t)$ by

$$w(r,\theta,t) \approx [\eta_{nm}(t)\cos n\theta + \zeta_{nm}(t)\sin n\theta]\phi_{nm}(r) \qquad (8.5.45)$$

where the $\phi_{nm}(r)$ are the linear free transverse vibrations modes and the ω_{nm} are the linear undamped natural frequencies. We assume that the mode shapes have been normalized so that

$$\int_\beta^1 r\phi_{nm}(r)\phi_{nk}(r)dr = \delta_{mk} \qquad (8.5.46)$$

where δ_{mk} is the Kronecker delta function. For ease of notation, we drop the subscript nm.

Substituting (8.5.45) into (8.5.34) yields

$$\nabla^4 \Phi_0 = \left[\frac{1}{2}(\zeta^2-\eta^2)\cos 2n\theta - \eta\zeta\sin 2n\theta\right]\chi_2(r) + \frac{1}{2}(\eta^2+\zeta^2)\chi_1(r) \qquad (8.5.47)$$

where

$$\chi_1 = \left[\frac{n^2}{r^4}(r\phi'-\phi)^2 - \frac{\phi''}{r^2}(r\phi'-n^2\phi)\right]$$

$$\chi_2 = \left[\frac{n^2}{r^4}(r\phi'-\phi)^2 + \frac{\phi''}{r^2}(r\phi'-n^2\phi)\right]$$

We express the solution of (8.5.47) and (8.5.35)-(8.5.38) as

$$\Phi_0 = \frac{1}{2}\psi_1(r)(\eta^2+\zeta^2) + \frac{1}{2}(\zeta^2-\eta^2)\psi_2(r)\cos 2n\theta$$
$$- \eta\zeta\psi_2(r)\sin 2n\theta \qquad (8.5.48)$$

where the functions $\psi_1(r)$ and $\psi_2(r)$ are governed by the following two boundary-value problems:

$$\left(\frac{d^2}{dr^2} + \frac{1}{r}\frac{d}{dr}\right)^2 \psi_1 = \chi_1$$

$$\psi_1'' - \frac{\nu}{\beta}\psi_1' = 0 \quad \text{at} \quad r=\beta$$

$$\psi_1''' + \frac{1}{\beta}\psi_1'' - \frac{1}{\beta^2}\psi_1' = 0 \quad \text{at} \quad r=\beta$$

$$\psi_1' = 0 \quad \text{at} \quad r=1$$

and

$$\left(\frac{d^2}{dr^2} + \frac{1}{r}\frac{d}{dr} - \frac{4n^2}{r^2}\right)^2 \psi_2 = \chi_2$$

$$\psi_2'' - \frac{\nu}{\beta}\left(\psi_2' - \frac{n^2}{\beta}\psi_2\right) = 0 \quad \text{at} \quad r=\beta$$

$$\psi_2''' + \frac{1}{\beta}\psi_2'' - \frac{1}{\beta^2}\psi_2' - \frac{4(2+\nu)n^2}{\beta^2}\psi_2' + \frac{4(3+\nu)n^2}{\beta^3}\psi_2 = 0 \quad \text{at} \quad r = \beta$$

$$\psi_2' - 4n^2\psi_2 = 0 \quad \text{at} \quad r = 1$$

$$\psi_2' - \psi_2 = 0 \quad \text{at} \quad r = 1$$

Again, the boundary condition (8.5.38) does not yield a boundary condition for ψ_1 and hence it can be determined only to within an arbitrary constant. However, as shown below, we do not need ψ_1 but its derivatives.

Substituting (8.5.45) and (8.5.48) into (8.5.33), multiplying the result with $r\phi\cos n\theta$ and $r\phi\sin n\theta$, respectively, integrating the outcomes from $\theta = 0$ to $\theta = 2\pi$ and $r = \beta$ to $r = 1$, and using (8.5.46), we obtain

$$\ddot{\eta} + \eta = \epsilon(-2\mu\dot{\eta} - \alpha\eta^3 - \alpha\eta\zeta^2 + f\cos\tilde{\Omega}t) \tag{8.5.49}$$

$$\ddot{\zeta} + \zeta = \epsilon(-2\mu\dot{\zeta} - \alpha\zeta^3 - \alpha\zeta\eta^2 + f\sin\tilde{\Omega}t) \tag{8.5.50}$$

where

$$\alpha = -\int_\beta^1 \phi\left[\frac{1}{2}(\phi'\psi_1'' + \phi''\psi_1') - \frac{1}{4}(\phi'\psi_2'' + \phi''\psi_2') - \frac{n^2}{r^2}(\phi'\psi_2 + \phi\psi_2')\right.$$
$$\left. + \frac{n^2}{r}\left(\phi'\psi_2 + \phi''\psi_2 - \frac{1}{2}\phi\psi_1'' + \frac{1}{4}\phi\psi_2''\right) + \frac{n^2}{r^3}\phi\psi_2\right]dr \tag{8.5.51}$$

$$\mu = \int_\beta^1 c\phi^2(r)dr, \quad f = \frac{1}{\pi r_0}\int_\beta^1 \hat{f}\delta(r-r_0)\phi r dr = \frac{\hat{f}\phi(r_0)}{\pi}, \quad \tilde{\Omega} = \frac{n\Omega}{\omega} \tag{8.5.52}$$

Equation (8.5.51) is the same as (8.4.82) obtained for the case of nonrotating circular plates.

To investigate the nonlinear dynamics of the disk, Nayfeh, Jilani, and Manzione (2001) calculated the parameter α in (8.5.51) for the floppy disk for different values of the rotational speed $\hat{\Omega}$ and for $n = 0, 1, 2, 3, 4$ nodal diameters and $m = 0, 1, 2, 4$ nodal circles. The results are listed in Table 8.13.

We use the method of multiple scales to determine a first-order approximate solution of (8.5.49) and (8.5.50) for the case of primary resonance for small- but finite-amplitude motions. To this end, we introduce the detuning parameter σ defined according to

$$\tilde{\Omega} = 1 + \epsilon\sigma \tag{8.5.53}$$

where σ is a detuning parameter. We seek a first-order uniform approximation to the solution of (8.5.49) and (8.5.50) in the form

$$\eta(t;\epsilon) \simeq \eta_0(T_0, T_1) + \epsilon\eta_1(T_0, T_1) \tag{8.5.54}$$

$$\zeta(t;\epsilon) \simeq \zeta_0(T_0, T_1) + \epsilon\zeta_1(T_0, T_1) \tag{8.5.55}$$

Substituting (8.5.54) and (8.5.55) into (8.5.49) and (8.5.50) and equating coefficients of like powers of ϵ leads to the following problems:

Table 8.13 Values of α for $n = 1, 2, 3, 4$ nodal diameters and $m = 0, 1, 2, 3, 4$ nodal circles for the floppy disk.

n	$m = 0$	$m = 1$	$m = 2$	$m = 3$	$m = 4$
$\hat{\Omega} = 0$					
0	$2.7983E + 0$	$1.1209E + 2$	$1.4961E + 3$	$6.7663E + 3$	$2.0451E + 4$
1	$2.7690E - 1$	$1.4532E + 2$	$1.4917E + 3$	$6.9843E + 3$	$2.1647E + 4$
2	$6.0810E + 0$	$2.3224E + 2$	$2.0940E + 3$	$9.3409E + 3$	$2.8969E + 4$
3	$5.5225E + 1$	$5.0060E + 2$	$3.1184E + 3$	$1.1656E + 4$	$3.4411E + 4$
4	$2.2264E + 2$	$1.0918E + 3$	$5.1687E + 3$	$1.5742E + 4$	$4.1568E + 4$
$\hat{\Omega} = 10$					
0	$2.4914E + 0$	$1.5630E + 2$	$1.4347E + 3$	$6.5691E + 3$	$2.0059E + 4$
1	$2.2742E - 1$	$1.4370E + 2$	$1.4325E + 3$	$6.7922E + 3$	$2.1235E + 4$
2	$6.1110E + 0$	$2.2502E + 2$	$2.0171E + 3$	$9.0950E + 3$	$2.8436E + 4$
3	$5.5785E + 1$	$4.8242E + 2$	$3.0339E + 3$	$1.1420E + 4$	$3.3851E + 4$
4	$2.2677E + 2$	$1.0531E + 3$	$5.0598E + 3$	$1.5519E + 4$	$4.1021E + 4$
$\hat{\Omega} = 20$					
0	$2.3760E + 0$	$1.5212E + 2$	$1.3066E + 3$	$6.0883E + 3$	$1.8982E + 4$
1	$2.0112E - 1$	$1.3982E + 2$	$1.3065E + 3$	$6.2965E + 3$	$2.0098E + 4$
2	$6.1516E + 0$	$2.1314E + 2$	$1.8464E + 3$	$8.4775E + 3$	$2.6984E + 4$
3	$5.6598E + 1$	$4.4934E + 2$	$2.8254E + 3$	$1.0793E + 4$	$3.2283E + 4$
4	$2.3383E + 2$	$9.7715E + 2$	$4.7620E + 3$	$1.4890E + 4$	$3.9452E + 4$
$\hat{\Omega} = 30$					
0	$2.3587E + 0$	$1.4894E + 2$	$1.1902E + 3$	$5.5309E + 3$	$1.7549E + 4$
1	$1.9378E - 1$	$1.3720E + 2$	$1.1910E + 3$	$5.7096E + 3$	$1.8553E + 4$
2	$6.1750E + 0$	$2.0495E + 2$	$1.6800E + 3$	$7.7283E + 3$	$2.4949E + 4$
3	$5.7133E + 1$	$4.2404E + 2$	$2.5984E + 3$	$1.0003E + 4$	$3.0141E + 4$
4	$2.3938E + 2$	$9.1461E + 2$	$4.3985E + 3$	$1.4026E + 4$	$3.7243E + 4$

Order ϵ^0:

$$D_0^2 \eta_0 + \eta_0 = 0 \tag{8.5.56}$$

$$D_0^2 \zeta_0 + \zeta_0 = 0 \tag{8.5.57}$$

Order ϵ^1:

$$D_0^2 \eta_1 + \eta_1 = -2\mu D_0 \eta_0 - 2D_0 D_1 \eta_0 - \alpha \eta_0^3 - \alpha \eta_0 \zeta_0^2 + f \cos(\tilde{\Omega} T_0) \tag{8.5.58}$$

$$D_0^2 \zeta_1 + \zeta_1 = -2\mu D_0 \zeta_0 - 2D_0 D_1 \zeta_0 - \alpha \zeta_0^3 - \alpha \eta_0^2 \zeta_0 + f \sin(\tilde{\Omega} T_0) \tag{8.5.59}$$

The solutions of (8.5.56) and (8.5.57) can be expressed as

$$\eta_0 = A_1(T_1)e^{iT_0} + cc \text{ and } \zeta_0 = A_2(T_1)e^{iT_0} + cc \tag{8.5.60}$$

Substituting (8.5.60) into (8.5.58) and (8.5.59) and eliminating the terms that produce secular terms, we obtain the following modulation equations:

$$2iA_1' = -2i\mu A_1 - 3\alpha A_1^2 \bar{A}_1 - 2\alpha A_1 A_2 \bar{A}_2 - \alpha A_2^2 \bar{A}_1 + \frac{1}{2} f e^{i\sigma T_1} \quad (8.5.61)$$

$$2iA_2' = -2i\mu A_2 - 3\alpha A_2^2 \bar{A}_2 - 2\alpha A_1 A_2 \bar{A}_1 - \alpha A_1^2 \bar{A}_2 - \frac{1}{2} i f e^{i\sigma T_1} \quad (8.5.62)$$

where the prime indicates the derivative with respect to T_1.

We note that (8.5.61) and (8.5.62) do not admit single-mode equilibrium solutions because the excitation is spatially fixed, and hence it appears in both equations. If the excitation were rotating, then single-mode solutions would be possible. Using the transformation

$$B_1 = \frac{1}{2}(A_1 + iA_2); \quad B_2 = \frac{1}{2}(A_1 - iA_2) \quad (8.5.63)$$

we obtain the equivalent modulation equations

$$2iB_1' = -2i\mu B_1 - 4\alpha B_1(B_1 \bar{B}_1 + 2B_2 \bar{B}_2) + \frac{1}{2} f e^{i\sigma T_1} \quad (8.5.64)$$

$$2iB_2' = -2i\mu B_2 - 4\alpha B_2(2B_1 \bar{B}_1 + B_2 \bar{B}_2) \quad (8.5.65)$$

where the forcing appears in one equation only. Clearly, $B_2 \equiv 0$ is a possible equilibrium solution.

To analyze the solutions of (8.5.64) and (8.5.65), we introduce the polar transformation

$$B_m = \frac{1}{2} b_m(T_1) e^{i\beta_m(T_1)} \quad (8.5.66)$$

into (8.5.64) and (8.5.65), separate real and imaginary parts, and obtain

$$b_1' = -\mu b_1 + \frac{1}{2} f \sin(\sigma T_1 - \beta_1) \quad (8.5.67)$$

$$b_1 \beta_1' = \frac{1}{2} \alpha b_1^3 + \alpha b_2^2 b_1 - \frac{1}{2} f \cos(\sigma T_1 - \beta_1) \quad (8.5.68)$$

$$b_2' = -\mu b_2 \quad (8.5.69)$$

$$b_2 \beta_2' = \alpha b_1^2 b_2 + \frac{1}{2} \alpha b_2^3 \quad (8.5.70)$$

It follows from (8.5.69) that $b_2 \to 0$ and hence $B_2 \to 0$ as $T_1 \to \infty$ and hence $t \to \infty$. Therefore, the long-time response of the disk is given by

$$b' = -\mu b + \frac{1}{2} f \sin\gamma \quad (8.5.71)$$

$$b\gamma' = \sigma b - \frac{1}{2} \alpha b^3 + \frac{1}{2} f \cos\gamma \quad (8.5.72)$$

where $b = b_1$ and $\gamma = \sigma T_1 - \beta_1$. Equations (8.5.71) and (8.5.72) have the same form as the modulation equations governing the amplitude and phase of the Duffing oscillator in the case of primary resonance excitation (Nayfeh, 1973b, 1981; Nayfeh

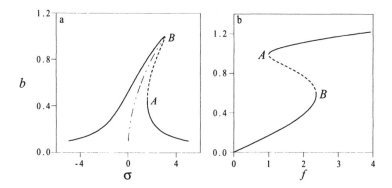

Fig. 8.5.7 Bifurcation diagrams for the mode with $m = 0$ nodal circles, $n = 2$ nodal diameters, $\Omega = 5.14265$, $\mu = 0.5$: (a) the control parameter used is σ and $f = 1$ and (b) the control parameter used is f and $\sigma = 3$.

and Mook, 1979). We note that α is always negative for any combination of m, n, Ω, and hence the effective nonlinearity is of the hardening type. It follows from (8.5.71) and (8.5.72) that the frequency-response equation is given by

$$\sigma = -\frac{1}{2}\alpha b^2 \pm \sqrt{\frac{f^2}{4b^2} - \mu^2} \qquad (8.5.73)$$

where the subscript 1 has been dropped. In Figure 8.5.7, we display typical frequency-response and force-response curves. The dashed lines indicate unstable solutions and solid lines indicate stable ones. The Points A and B are saddle-node bifurcations.

Because $B_2 \to 0$ as $t \to \infty$, in the long-time response,

$$A_2 = -iA_1 \qquad (8.5.74)$$

Then, it follows from (8.5.63) that $B_1 = A_1$. Moreover, it follows from (8.5.60), (8.5.63), and (8.5.66) that

$$\eta \approx \frac{1}{2}be^{i(\tilde{\Omega}t-\gamma)} + cc \text{ and } \zeta \approx -\frac{1}{2}ibe^{i(\tilde{\Omega}t-\gamma)} + cc \qquad (8.5.75)$$

Substituting (8.5.75) into (8.5.45), we obtain

$$w(r,\theta,t) \approx b\phi_{nm}(r)\cos(\tilde{\Omega}t - n\theta - \gamma) \qquad (8.5.76)$$

Therefore, the long-time asymmetric behavior is always a traveling wave.

8.6 NONLINEAR VIBRATIONS OF NEAR-SQUARE PLATES

The nonlinear free and forced vibrations of metallic and composite plates have been extensively studied by many investigators. Large-amplitude vibrations of rectangular plates were studied by Cummings (1964), Chia (1980), and Prathap and Varadan

(1978). The most comprehensive review of geometrically nonlinear laminated composite plates can be found in the book by Chia (1980). Other reviews on the nonlinear vibration of plates have been conducted by Nayfeh and Mook (1979), Nayfeh (2000), and Sathyamoorthy (1997). In this section, we consider the nonlinear vibrations of a plate that is initially flat and use the von Kármán equations (7.4.24)-(7.4.26). We follow Yang and Sethna (1991, 1992) and treat the case of a near-square simply supported plate for which the boundary conditions are

$$u = w = 0 \text{ and } w_{xx} + \nu w_{yy} = 0 \text{ at } x = 0 \text{ and } x = a \quad (8.6.1)$$
$$v = w = 0 \text{ and } w_{yy} + \nu w_{xx} = 0 \text{ at } y = 0 \text{ and } y = b \quad (8.6.2)$$

where a and b are the edge lengths, with $b \approx a$.

For thin plates, h/a is very small. To describe nonlinear motions of the plate, we introduce a small parameter ϵ and scale the various quantities in terms of ϵ as

$$\epsilon = \frac{h^2}{a^2}, \quad \tilde{x} = \frac{x}{a}, \quad \tilde{y} = \frac{y}{b}, \quad \tilde{w} = \frac{w}{\epsilon a}, \quad \tilde{u} = \frac{u}{\epsilon^2 a}, \quad \tilde{v} = \frac{v}{\epsilon^2 a},$$
$$\tilde{t} = t \frac{c_p}{a}\sqrt{\frac{\epsilon}{12}}, \quad \tilde{F} = \frac{12a(1-\nu^2)F}{\epsilon^3 E h}, \quad \frac{a^2}{b^2} = 1 + \frac{1}{2}\epsilon\sigma$$

Substituting these variables into (7.4.24)-(7.4.26), dropping the tilde for convenience, and considering the case of no inplane edge loads, we obtain

$$u_{xx} + \frac{1}{2}(1+\nu)v_{xy} + \frac{1}{2}(1-\nu)u_{yy} = -w_x w_{xx}$$
$$- \frac{1}{2}(1+\nu)w_y w_{xy} - \frac{1}{2}(1-\nu)w_x w_{yy} + O(\epsilon^2) \quad (8.6.3)$$

$$v_{yy} + \frac{1}{2}(1+\nu)u_{xy} + \frac{1}{2}(1-\nu)v_{xx} = -w_y w_{yy}$$
$$- \frac{1}{2}(1+\nu)w_x w_{xy} - \frac{1}{2}(1-\nu)w_y w_{xx} + O(\epsilon^2) \quad (8.6.4)$$

$$\ddot{w} + w_{xxxx} + 2w_{xxyy} + w_{yyyy} = -\epsilon\Big\{\sigma(w_{xxyy} + w_{yyyy}) - 12u_x w_{xx}$$
$$- 6w_x^2 w_{xx} - 12\nu\left(v_y w_{xx} + \frac{1}{2}w_y^2 w_{xx}\right) - 12v_y w_{yy} - 6w_y^2 w_{yy}$$
$$- 12\nu\left(u_x w_{yy} + \frac{1}{2}w_x^2 w_{yy}\right) - 12(1-\nu)\big(u_y w_{xy} + v_x w_{xy}$$
$$+ w_x w_y w_{xy}\big)\Big\} + \epsilon F + O(\epsilon^2) \quad (8.6.5)$$

For simply supported edges, the boundary conditions are

$$u = w = 0 \text{ and } w_{xx} + \nu\left(1 + \frac{1}{2}\epsilon\sigma\right)w_{yy} = 0 \text{ at } x = 0 \text{ and } x = 1 \quad (8.6.6)$$
$$v = w = 0 \text{ and } \left(1 + \frac{1}{2}\epsilon\sigma\right)w_{yy} + \nu w_{xx} = 0 \text{ at } y = 0 \text{ and } y = 1 \quad (8.6.7)$$

The above ordering implies that the inplane dynamics is fast compared to the transverse dynamics. Georgiou and Schwartz (1996), Georgiou, Bajaj, and Corless (1998), and Georgiou, Corless, and Bajaj (1999) developed a systematic analytic/computational technique based on singular perturbations and invariant manifolds of motion to separate the fast and slow dynamics. Georgiou, Schwartz, Emaci, and Vakakis (1999) carried out experiments on a beam/pendulum system to validate the theoretical findings.

To investigate the case of one-to-one internal resonance between the mnth and nmth modes, we let

$$w = \eta(t) \sin m\pi x \sin n\pi y + \zeta(t) \sin n\pi x \sin m\pi y \qquad (8.6.8)$$

in (8.6.3) and (8.6.4), neglect $\ddot{u}$ and $\ddot{v}$, and obtain

$$u_{xx} + \frac{1}{2}(1+\nu)v_{xy} + \frac{1}{2}(1-\nu)u_{yy}$$
$$= \frac{1}{4}\pi^3 \eta^2 m(m^2 - \nu n^2)\sin 2m\pi x + \frac{1}{4}\pi^3 \zeta^2 n(n^2 - \nu m^2)\sin 2n\pi x$$
$$- \frac{1}{4}\pi^3 \eta^2 m(m^2 + n^2)\sin 2m\pi x \cos 2n\pi y$$
$$- \frac{1}{4}\pi^3 \zeta^2 n(m^2 + n^2)\sin 2n\pi x \cos 2m\pi y$$
$$+ \pi^3 \eta \zeta n \left[m^2 + \frac{1}{2}(1-\nu)n^2\right]\sin m\pi x \cos n\pi x \sin m\pi y \sin n\pi y$$
$$+ \pi^3 \eta \zeta m \left[n^2 + \frac{1}{2}(1-\nu)m^2\right]\cos m\pi x \sin n\pi x \sin m\pi y \sin n\pi y$$
$$- \frac{1}{2}(1+\nu)mn^2\pi^3 \eta \zeta \sin m\pi x \cos n\pi x \cos m\pi y \cos n\pi y$$
$$- \frac{1}{2}(1+\nu)m^2 n\pi^3 \eta \zeta \cos m\pi x \sin n\pi x \cos m\pi y \cos n\pi y \qquad (8.6.9)$$
$$v_{yy} + \frac{1}{2}(1+\nu)u_{xy} + \frac{1}{2}(1-\nu)v_{xx}$$
$$= \frac{1}{4}\pi^3 \eta^2 n(n^2 - \nu m^2)\sin 2n\pi y + \frac{1}{4}\pi^3 \zeta^2 m(m^2 - \nu n^2)\sin 2m\pi y$$
$$- \frac{1}{4}\pi^3 \eta^2 n(m^2 + n^2)\sin 2n\pi y \cos 2m\pi x$$
$$- \frac{1}{4}\pi^3 \zeta^2 m(m^2 + n^2)\sin 2m\pi y \cos 2n\pi x$$
$$+ \pi^3 \eta \zeta n \left[m^2 + \frac{1}{2}(1-\nu)n^2\right]\sin m\pi x \sin n\pi x \sin m\pi y \cos n\pi y$$
$$+ \pi^3 \eta \zeta m \left[n^2 + \frac{1}{2}(1-\nu)m^2\right]\sin m\pi x \sin n\pi x \cos m\pi y \sin n\pi y$$
$$- \frac{1}{2}(1+\nu)m^2 n\pi^3 \eta \zeta \cos m\pi x \cos n\pi x \cos m\pi y \sin n\pi y$$

$$-\frac{1}{2}(1+\nu)mn^2\pi^3\eta\zeta\cos m\pi x\cos n\pi x\sin m\pi y\cos n\pi y \qquad (8.6.10)$$

The solution of (8.6.9) and (8.6.10) that satisfies (8.6.6) and (8.6.7) is

$$\begin{aligned} u = &\; \frac{m\pi}{16}\eta^2\left(\cos 2n\pi y - 1 + \frac{\nu n^2}{m^2}\right)\sin 2m\pi x \\ &+ \frac{n\pi}{16}\zeta^2\left(\cos 2m\pi y - 1 + \frac{\nu m^2}{n^2}\right)\sin 2n\pi x \\ &+ \pi\eta\zeta\Big[c_1\sin m\pi x\cos n\pi x\sin m\pi y\sin n\pi y \\ &\quad + c_2\sin m\pi x\cos n\pi x\cos m\pi y\cos n\pi y \\ &\quad + c_3\cos m\pi x\sin n\pi x\sin m\pi y\sin n\pi y \\ &\quad + c_4\cos m\pi x\sin n\pi x\cos m\pi y\cos n\pi y\Big] \end{aligned} \qquad (8.6.11)$$

$$\begin{aligned} v = &\; \frac{n\pi}{16}\eta^2\left(\cos 2m\pi x - 1 + \frac{\nu m^2}{n^2}\right)\sin 2n\pi y \\ &+ \frac{m\pi}{16}\zeta^2\left(\cos 2n\pi x - 1 + \frac{\nu n^2}{m^2}\right)\sin 2m\pi y \\ &+ \pi\eta\zeta\Big[c_1\sin m\pi x\sin n\pi x\sin m\pi y\cos n\pi y \\ &\quad + c_2\cos m\pi x\cos n\pi x\sin m\pi y\cos n\pi y \\ &\quad + c_3\sin m\pi x\sin n\pi x\cos m\pi y\sin n\pi y \\ &\quad + c_4\cos m\pi x\cos n\pi x\cos m\pi y\sin n\pi y\Big] \end{aligned} \qquad (8.6.12)$$

where

$$c_1 = \frac{n\pi[m^2 - n^2 + \nu(m^2 + n^2)]}{4(n^2 - m^2)}, \quad c_2 = \frac{mn^2\nu}{2(n^2 - m^2)},$$

$$c_3 = \frac{m\pi[m^2 - n^2 - \nu(m^2 + n^2)]}{4(n^2 - m^2)}, \quad c_4 = \frac{m^2 n\nu}{2(m^2 - n^2)} \qquad (8.6.13)$$

Substituting (8.6.8), (8.6.11), and (8.6.12) into (8.6.5), using the Galerkin procedure because the nonlinearity is only cubic, and adding linear viscous damping, we obtain (Yang and Sethna, 1991)

$$\ddot{\eta} + 2\epsilon\mu\dot{\eta} + \eta + \epsilon[\sigma\delta_1\eta + \alpha_1\eta^3 + \alpha_2\eta\zeta^2] = \epsilon f_1 \qquad (8.6.14)$$
$$\ddot{\zeta} + 2\epsilon\mu\dot{\zeta} + \zeta + \epsilon[\sigma\delta_2\zeta + \alpha_1\zeta^3 + \alpha_2\zeta\eta^2] = \epsilon f_2 \qquad (8.6.15)$$

where

$$\alpha_1 = \frac{3}{4}\left[(3-\nu^2)\frac{r^4+1}{(1+r^2)^2} + 4\nu\frac{r^2}{(1+r^2)^2}\right]$$

$$\alpha_2 = \frac{3}{4}\left[(1+2\nu-\nu^2)\frac{r^4+1}{(1+r^2)^2} + 4(2-\nu^2)\frac{r^2}{(1+r^2)^2}\right]$$

$$\delta_1 = \frac{1}{1+r^2}, \quad \delta_2 = \frac{r^2}{1+r^2}$$
$$f_1 = f_2 = \frac{(1 - \cos m\pi)(1 - \cos n\pi)}{mn\pi^2} \tag{8.6.16}$$

and $r = m/n$. For a detailed analysis of the solutions of (8.6.14) and (8.6.15), we refer the reader to Nayfeh (2000) and the references cited therein.

8.7 MICROPUMPS

Micropumps are designed to handle small and precise volumes in medical, biomedical, and chemical applications. Usually, micropumps are integrated with other microflow devices and sensors to allow precise sensing of flow and control on the order of microliters. Reciprocating micropumps are widely used. Their actuation principles include piezoelectric, pneumatic, electrostatic, and thermopneumatic. Electrostatic actuation is gaining popularity because of its simplicity and high-flow output pressures (Tay and Phoon, 1997). Different shapes of electrostatic actuating elements are reported in the literature, such as rectangular, square, circular, and annular (Francais and Dufour, 1999, 2000; Saif, Alaca, and Sehitoglu, 1999; Jiang, Ng, and Lam, 2000). Annular actuating elements have also been proposed for use in micropumps by some companies. They are also used in other MEMS devices, such as sensors, valves, deformable micromirrors, and micro-turbo generators (Wiegele, 1996).

The electrostatic load applied to a plate has an upper limit beyond which the mechanical restoring force can no longer resist the opposing electrostatic force, thereby leading to the collapse of the structure. This structural instability phenomenon is known as 'pull in', and the critical voltage associated with it is called the 'pull in voltage.'

Saif, Alaca, and Sehitoglu (1999) treated an annular actuating element in a micropump as a membrane, therefore neglecting the bending stiffness of the element. Malhaire, Le Berre, Febvre, Barbier, and Pinard (1999) found experimentally that, in the case of clamped actuating elements in MEMS devices, the bending stiffness cannot be neglected. Faris, Abdel-Rahman, and Nayfeh (2002) studied the mechanical behavior of an electrostatic micropump employing an annular actuation member. They adopted the von Kármán plate model under axisymmetric loading conditions to account for the bending stiffness and large deflections. They considered two cases of boundary conditions to simulate practical boundary conditions in MEMS devices. Vogl and Nayfeh (2003) presented a reduced-order model for an electrically actuated clamped circular plate by using the Galerkin approach, thereby reducing the distributed-parameter equations into a finite system of ordinary-differential equations in time. The reduced-order model accounts for residual stresses and geometric and electric nonlinearities. They solved the reduced-order model for the equilibrium positions due to an electrostatic force and determined the natural frequencies of the axisymmetric modes. They validated the model using data from experiments performed on silicon-based MEMS systems.

In Section 8.7.1, we follow Faris, Abdel-Rahman, and Nayfeh (2002) and consider an electrostatically actuated micropump employing an annular actuating member. The boundary-value problem describing the static deflection of the plate under the electrostatic force is solved numerically. Then, the eigenvalue problem describing the vibration of the plate around the equilibrium position is also solved numerically for the natural frequencies and mode shapes. In Section 8.7.2, we follow Vogl and Nayfeh (2003) and consider a micropump employing a clamped circular member. We use the Galerkin procedure to discretize the governing partial-differential equations and associated boundary conditions into a finite set of ordinary-differential equations. Then, we solve for the equilibrium positions under an electrostatic force and calculate the natural frequencies of vibrations around these equilibrium positions.

The general equations governing the axisymmetric statics and dynamics of an isotropic circular or annular plate with zero residual stress are given in (8.4.17) and (8.4.18). Adding the effect of a residual stress σ and an electric force (Griffiths, 1981) to these equations, we obtain

$$D\nabla^4 w + \rho h \frac{\partial^2 w}{\partial t^2} = \frac{1}{r}\frac{\partial^2 w}{\partial r^2}\frac{\partial \Phi}{\partial r} + \frac{1}{r}\frac{\partial w}{\partial r}\frac{\partial^2 \Phi}{\partial r^2} + \frac{\sigma h}{r}\frac{\partial}{\partial r}\left(r\frac{\partial w}{\partial r}\right)$$
$$+ \frac{\epsilon_0 V^2}{2(d-w)^2} \quad (8.7.1)$$

$$\nabla^4 \Phi = -\frac{Eh}{r}\frac{\partial^2 w}{\partial r^2}\frac{\partial w}{\partial r} \quad (8.7.2)$$

where t is time, ρ is the mass density, h is the plate thickness, d is the capacitor gap distance, ϵ_0 is the dielectric constant of the medium, Φ is the stress function, and $D = Eh^3/[12(1-\nu^2)]$ is the bending rigidity. In the next section, we consider the case of annular plates, and in Section 8.7.2 we consider the case of circular plates.

8.7.1 Annular Plates

We follow Faris, Abdel-Rahman, and Nayfeh (2002) and model the actuating member of the micropump as a clamped-clamped annular plate bending under an applied electrostatic load and no residual stresses. This configuration was proposed as a micropump design by Saif, Alaca, and Sehitoglu (1999). We introduce nondimensional variables, denoted by asterisks and defined as follows:

$$r = Rr^*, \quad w = dw^*, \quad t = R^2\left(\frac{\rho h}{D}\right)^{1/2} t^*, \quad \Phi = Ehd^2\Phi^* \quad (8.7.3)$$

where R is the outer radius of the annular plate. Substituting (8.7.3) into (8.7.1) and (8.7.2) and dropping the asterisks, we obtain

$$\frac{\partial^2 w}{\partial t^2} + \nabla^4 w = \frac{\alpha_1}{r}\frac{\partial}{\partial r}\left(\frac{\partial w}{\partial r}\frac{\partial \phi}{\partial r}\right) + \frac{\alpha_2 V^2}{(1-w)^2} \quad (8.7.4)$$

$$\nabla^4 \Phi = -\frac{1}{2r}\frac{\partial}{\partial r}\left(\frac{\partial w}{\partial r}\right)^2 \quad (8.7.5)$$

where

$$\alpha_1 = \frac{12(1-\nu^2)d^2}{h^2} \quad \text{and} \quad \alpha_2 = \frac{\epsilon_0 R^4}{2Dd^3}$$

It follows from (8.4.5), (8.4.10), and (8.4.11) that the nondimensional boundary conditions are

$$w = 0 \text{ and } \frac{\partial w}{\partial r} = 0 \text{ at } r = b \text{ and } 1 \tag{8.7.6}$$

$$\frac{\partial^2 \Phi}{\partial r^2} - \frac{\nu}{r} \frac{\partial \Phi}{\partial r} = 0 \text{ at } r = b \text{ and } 1 \tag{8.7.7}$$

$$\frac{\partial^3 \Phi}{\partial r^3} + \frac{1}{r} \frac{\partial^2 \Phi}{\partial r^2} - \frac{1}{r^2} \frac{\partial \Phi}{\partial r} = 0 \text{ at } r = b \text{ and } 1 \tag{8.7.8}$$

where b is the nondimensional inner radius of the plate.

We express the plate deflection and the stress function as the summation of static components, denoted by $w_s(r)$ and $\Phi_s(r)$, and dynamic components, denoted by $u(r,t)$ and $\Phi_d(r,t)$; that is,

$$w(r,t) = w_s(r) + u(r,t) \tag{8.7.9}$$
$$\Phi(r,t) = \Phi_s(r) + \Phi_d(r,t) \tag{8.7.10}$$

To calculate the static deflection and stress function, we substitute (8.7.9) and (8.7.10) into (8.7.4) and (8.7.5), set $u(r,t)$ and $\Phi_d(r,t)$ equal to zero, and obtain

$$\nabla^4 w_s = \frac{\alpha_1}{r} \frac{d}{dr}\left(\frac{dw_s}{dr} \frac{d\Phi_s}{dr}\right) + \frac{\alpha_2 V^2}{(1-w_s)^2} \tag{8.7.11}$$

$$\nabla^4 \Phi_s = -\frac{1}{2r} \frac{d}{dr}\left(\frac{dw_s}{dr}\right)^2 \tag{8.7.12}$$

We generate the equations of motion governing the dynamic behavior of the plate around the deflected shape by substituting (8.7.9) and (8.7.10) into (8.7.4) and (8.7.5), expanding the electrostatic forcing term in a Taylor series around $w_s(r)$ up to third-order, and subtracting (8.7.11) and (8.7.12). The result is

$$\frac{\partial^2 u}{\partial t^2} + \nabla^4 u = \frac{\alpha_1}{r} \frac{\partial}{\partial r}\left(\frac{\partial u}{\partial r} \frac{d\Phi_s}{dr} + \frac{dw_s}{dr} \frac{\partial \Phi_d}{\partial r} + \frac{\partial u}{\partial r} \frac{\partial \Phi_d}{\partial r}\right)$$
$$+ \frac{2\alpha_2 V^2 u}{(1-w_s)^3} + \frac{3\alpha_2 V^2 u^2}{(1-w_s)^4} + \frac{4\alpha_2 V^2 u^3}{(1-w_s)^5} \tag{8.7.13}$$

$$\nabla^4 \Phi_d = -\frac{1}{2r} \frac{\partial}{\partial r}\left[2\frac{\partial u}{\partial r} \frac{dw_s}{dr} + \left(\frac{\partial u}{\partial r}\right)^2\right] \tag{8.7.14}$$

Faris, Abdel-Rahman, and Nayfeh (2002) calculated the static deflection and stress function by solving numerically (8.7.11) and (8.7.12) subject to the boundary conditions

$$w_s = 0 \text{ and } w_s' = 0 \text{ at } r = b \text{ and } 1 \tag{8.7.15}$$

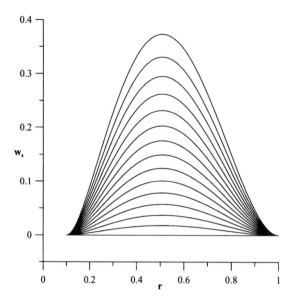

Fig. 8.7.1 Nondimensional deflection $w_s(r)$ of the annular plate under an electrostatic force ranging from $\alpha_2 V^2 = 0$ to 140.6.

$$\Phi_s'' - \frac{\nu}{r}\Phi_s' = 0 \quad \text{at} \quad r = b \text{ and } 1 \tag{8.7.16}$$

$$\Phi_s''' + \frac{1}{r}\Phi_s'' - \frac{1}{r^2}\Phi_s' = 0 \quad \text{at} \quad r = b \text{ and } 1 \tag{8.7.17}$$

They considered a polysilicon annular plate with outer radius $50\mu m$, inner radius $5\mu m$, thickness $1\mu m$, capacitor gap $2\mu m$, and a range of electrostatic forces varying from zero to the forcing level where the numerical solver diverges. Figure 8.7.1 shows an axisymmetric cross-section of the plate as it deflects under an upward electrostatic force increasing from zero to a forcing level corresponding to $\alpha_2 V^2 = 140.6$. The evolution of the static deflection is most pronounced in the mid-region of the cross-section. Figure 8.7.2 shows variation of the maximum deflection W_{Max} for the same range of $\alpha_2 V^2$ as that in Figure 8.7.1. As the electrostatic force, represented by $\alpha_2 V^2$, increases, the maximum deflection increases. These results are in qualitative agreement with the experimental results of Francais and Dufour (1999, 2000) for a square plate.

Having determined $w_s(r)$ and $\Phi_s(r)$, Faris, Abdel-Rahman, and Nayfeh (2002) calculated the mode shapes and corresponding natural frequencies by dropping the nonlinear terms from (8.7.13) and (8.7.14) and obtaining

$$\frac{\partial^2 u}{\partial t^2} + \nabla^4 u = \frac{\alpha_1}{r}\frac{\partial}{\partial r}\left(\frac{\partial u}{\partial r}\frac{d\Phi_s}{dr} + \frac{dw_s}{dr}\frac{\partial \Phi_d}{\partial r}\right) + \frac{2\alpha_2 V^2}{(1-w_s)^3}u \tag{8.7.18}$$

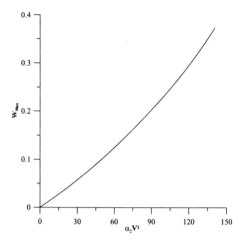

Fig. 8.7.2 Variation of the maximum deflection W_{Max} of the annular plate with $\alpha_2 V^2$.

$$\nabla^4 \Phi_d = -\frac{1}{r}\frac{\partial}{\partial r}\left(\frac{\partial u}{\partial r}\frac{dw_s}{dr}\right) \tag{8.7.19}$$

$$u = 0 \text{ and } u' = 0 \quad \text{at} \quad r = b \text{ and } 1 \tag{8.7.20}$$

$$\Phi_d'' - \frac{\nu}{r}\Phi_d' = 0 \quad \text{at} \quad r = b \text{ and } 1 \tag{8.7.21}$$

$$\Phi_d''' + \frac{1}{r}\Phi_d'' - \frac{1}{r^2}\Phi_d' = 0 \quad \text{at} \quad r = b \text{ and } 1 \tag{8.7.22}$$

To solve these equations for the undamped mode shapes and natural frequencies of the plate, they assumed a harmonic motion in the form

$$u(r,t) = \phi(r)e^{i\omega t} \quad \text{and} \quad \Phi_d(r,t) = \chi(r)e^{i\omega t} \tag{8.7.23}$$

Substituting (8.7.23) into (8.7.18)-(8.7.22) yields

$$\nabla^4 \phi + \frac{\alpha_1}{r}\left(\phi''\Phi_s' + \phi'\Phi_s'' + w_s''\chi' + w_s'\chi''\right) = \left[\frac{2\alpha_2 V^2}{(1-w_s)^3} + \omega^2\right]\phi \tag{8.7.24}$$

$$\nabla^4 \chi = -\frac{1}{r}\frac{d}{dr}(w_s'\phi') \tag{8.7.25}$$

$$\phi = 0 \text{ and } \phi' = 0 \quad \text{at} \quad r = b \text{ and } 1 \tag{8.7.26}$$

$$\chi'' - \frac{\nu}{r}\chi' = 0 \quad \text{at} \quad r = b \text{ and } 1 \tag{8.7.27}$$

$$\chi''' + \frac{1}{r}\chi'' - \frac{1}{r^2}\chi' = 0 \quad \text{at} \quad r = b \text{ and } 1 \tag{8.7.28}$$

To solve this eigenvalue problem, Faris, Abdel-Rahman, and Nayfeh (2002) released one of the boundary conditions at $r = 1$ and solved the boundary-value problem

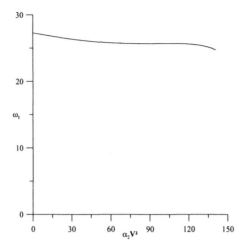

Fig. 8.7.3 Variation of the fundamental natural frequency of the annular plate with $\alpha_2 V^2$.

iteratively using a shooting method until ω, ϕ, and χ converge to within predefined tolerances. This procedure was repeated for each mode shape, associated natural frequency, and stress function, using various sets of initial guesses.

Figure 8.7.3 shows variation of the fundamental natural frequency ω_1 of the clamped-clamped plate with $\alpha_2 V^2$. The fundamental natural frequency of the plate decreases at low and high levels of the electrostatic force. In the mid-range, the fundamental natural frequency remains nearly constant due to the counter-balance between the electrostatic force and inplane stresses in the plate. As pull-in develops, the fundamental natural frequency is expected to approach zero. However, the results shown in Figure 8.7.3 do not reflect this fact because the numerical routine is sensitive to the stiffness of the system of differential equations (8.7.24)-(8.7.28). The stiffness of these equations increases as $\alpha_2 V^2$ increases. As a result, it becomes harder to generate adequate initial guesses as pull-in is approached.

Figures 8.7.4 and 8.7.5 show the first and second mode shapes of the plate as the electrostatic force is increased from zero and approaches the pull-in level. In both figures, the mode shape is normalized with respect to the absolute maximum of the mode shape for a straight plate. The natural frequencies and mode shapes are in qualitative agreement with those reported by Amabili, Frosali, and Kwak (1996). These results indicate that the mode shapes of the flat plate form an adequate basis set for Galerkin reduction, which we pursue in the next section.

8.7.2 Circular Plates

In this section, we follow Vogl and Nayfeh (2003) and consider a clamped circular plate actuated by an axisymmetric electrostatic force. As an alternative to the nondi-

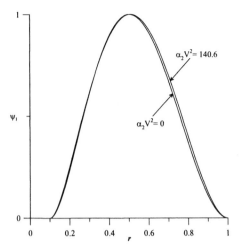

Fig. 8.7.4 The first mode shape of the annular plate as the electrostatic force increases from zero and approaches pull-in level.

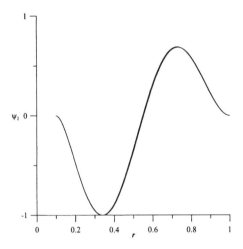

Fig. 8.7.5 The second mode shape of the annular plate as the electrostatic force increases from zero and approaches pull-in level.

mensionalization in the preceding section, we nondimensionalize the equations and boundary conditions according to

$$r = Rr^*, \; t = R^2 \left(\frac{\rho h}{D}\right)^{1/2} t^*, \; w = \frac{h^2}{R} w^*,$$

$$V^2 = \frac{2Dh^6}{\epsilon_0 R^7} V^{*2}, \quad \sigma = \frac{Eh^4}{R^4} \sigma^*, \quad \text{and} \quad \Phi = \frac{Eh^5}{R^2} \Phi^* \qquad (8.7.29)$$

where the nondimensional variables are denoted by asterisks. By using (8.7.29) and dropping the asterisks, we transform (8.7.1) and (8.7.2) into

$$\frac{\partial^2 w}{\partial t^2} + \nabla^4 w = \epsilon \left[\frac{1}{r} \frac{\partial}{\partial r} \left(\frac{\partial w}{\partial r} \frac{\partial \Phi}{\partial r} \right) + \frac{\sigma}{r} \frac{\partial}{\partial r} \left(r \frac{\partial w}{\partial r} \right) \right] + \frac{V^2}{(w_{max} - w)^2} \qquad (8.7.30)$$

$$\nabla^4 \Phi = -\frac{1}{2r} \frac{\partial}{\partial r} \left(\frac{\partial w}{\partial r} \right)^2 \qquad (8.7.31)$$

where

$$\epsilon = 12(1 - \nu^2) h^2 / R^2 \quad \text{and} \quad w_{max} = dR/h^2 \qquad (8.7.32)$$

Furthermore, the nondimensional boundary conditions become

$$w = 0 \quad \text{and} \quad \frac{\partial w}{\partial r} = 0 \quad \text{at} \quad r = 1 \qquad (8.7.33)$$

$$\Phi < \infty \quad \text{and} \quad w < \infty \quad \text{at} \quad r = 0 \qquad (8.7.34)$$

$$\frac{\partial^2 \Phi}{\partial r^2} - \frac{\nu}{r} \frac{\partial \Phi}{\partial r} = 0 \quad \text{at} \quad r = 1 \qquad (8.7.35)$$

$$\frac{\partial^3 \Phi}{\partial r^3} + \frac{1}{r} \frac{\partial^2 \Phi}{\partial r^2} - \frac{1}{r^2} \frac{\partial \Phi}{\partial r} = 0 \quad \text{at} \quad r = 1 \qquad (8.7.36)$$

Next, we use the Galerkin procedure to reduce (8.7.30), (8.7.31), and (8.7.33)-(8.7.36) into a system of finite number of ordinary-differential equations by letting

$$w(r, t) = \sum_{m=1}^{N} \eta_m(t) \phi_m(r) \qquad (8.7.37)$$

where $\phi_m(r)$ is the mth shape function, $\eta_m(t)$ is the mth generalized coordinate, and N is the number of chosen shape functions. We choose the shape functions $\phi_m(r)$ in (8.7.37) to be the first N linear undamped axisymmetric modes with no residual stress and no electrostatic forcing. For this case,

$$\phi_m(r) = \frac{J_0(r\sqrt{\omega_m})}{J_0(\sqrt{\omega_m})} - \frac{I_0(r\sqrt{\omega_m})}{I_0(\sqrt{\omega_m})} \qquad (8.7.38)$$

according to (8.4.33), where $J_0(r)$ and $I_0(r)$ are Bessel functions of the first and second kind, respectively. These mode shapes are orthonormalized so that

$$\int_0^1 r \phi_m(r) \phi_n(r) dr = \delta_{mn} \qquad (8.7.39)$$

where δ_{mn} is the Kronecker delta.

Substituting (8.7.37) into (8.7.31) yields

$$\nabla^4 \Phi = -\frac{1}{2r} \sum_{m,n=1}^{N} \eta_m(t)\eta_n(t) \left[\phi'_m(r)\phi'_n(r)\right]' \qquad (8.7.40)$$

The solution of (8.7.40) subject to the boundary conditions (8.7.34)-(8.7.36) can be expressed as

$$\Phi(r,t) = \sum_{m,n=1}^{N} \eta_m(t)\eta_n(t)\psi_{mn}(r) \qquad (8.7.41)$$

where the functions $\psi_{mn}(r)$ are given by the two-point boundary-value problems

$$\left(\frac{d^2}{dr^2} + \frac{1}{r}\frac{d}{dr}\right)^2 \psi_{mn} = -\frac{1}{2r}\left[\phi'_m(r)\phi'_n(r)\right]' \qquad (8.7.42)$$

$$\psi''_{mn} - \nu\psi'_{mn} = 0 \text{ at } r = 1 \qquad (8.7.43)$$

$$\psi'''_{mn} + \psi''_{mn} - \psi'_{mn} = 0 \text{ at } r = 1 \qquad (8.7.44)$$

$$\psi_{mn}(0) < \infty \qquad (8.7.45)$$

The solution of (8.7.42)-(8.4.45) can be obtained as in Section 8.4.1 to be

$$\psi'_{mn}(r) = -\frac{r}{4}\int_0^r \frac{\phi'_m \phi'_n}{\xi} d\xi + \frac{1}{4r}\int_0^r \xi \phi'_m \phi'_n d\xi$$

$$+ \frac{r}{4}\int_0^1 \frac{\phi'_m \phi'_n}{\xi} d\xi + \frac{r}{4}\frac{1+\nu}{1-\nu}\int_0^1 \xi \phi'_m \phi'_n d\xi \qquad (8.7.46)$$

We multiply (8.7.30) through with $r\phi_q(r)(w_{max} - w)^2$, substitute (8.7.37) and (8.7.41) into the resulting equation, integrate the outcome over $r \in [0,1]$, use the orthogonality condition (8.7.39), and obtain

$$w_{max}^2 (\ddot{\eta}_q + \omega_q^2 \eta_q) - 2w_{max} \sum_{i,m=1}^{N} (\ddot{\eta}_m + \omega_m^2 \eta_m) \eta_i \int_0^1 r\phi_i \phi_m \phi_q \, dr$$

$$+ \sum_{i,j,m=1}^{N} (\ddot{\eta}_m + \omega_m^2 \eta_m) \eta_i \eta_j \int_0^1 r\phi_i \phi_j \phi_m \phi_q \, dr$$

$$= \epsilon \Bigg[-\omega_{max}^2 \sum_{m,n,p=1}^{N} \eta_m \eta_n \eta_p \int_0^1 \phi'_q \phi'_m \psi'_{np} \, dr$$

$$+ 2\omega_{max} \sum_{i,m,n,p=1}^{N} \eta_i \eta_m \eta_n \eta_p \int_0^1 (\phi_i \phi_q)' \phi'_m \psi'_{np} \, dr$$

$$- \sum_{i,j,m,n,p=1}^{N} \eta_i \eta_j \eta_m \eta_n \eta_p \int_0^1 (\phi_i \phi_j \phi_q)' \phi'_m \psi'_{np} \, dr$$

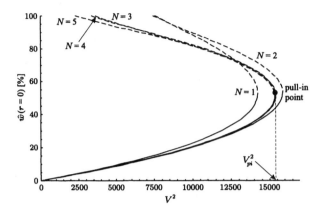

Fig. 8.7.6 Variation of the normalized deflection at the center of the circular plate with the electric forcing level.

$$-\sigma \sum_{i,j,m=1}^{N} \eta_i \eta_j \eta_m \int_0^1 r\phi'_m (\phi_i \phi_j \phi_q)' \, dr$$

$$+2\sigma w_{max} \sum_{i,m=1}^{N} \eta_i \eta_m \int_0^1 r\phi'_m (\phi_i \phi_q)' \, dr$$

$$-\sigma w_{max}^2 \sum_{m=1}^{N} \eta_m \int_0^1 r\phi'_m \phi'_q dr \bigg] + V^2 \int_0^1 r\phi_q dr \quad (8.7.47)$$

where $q = 1, 2, \ldots, N$.

To determine the static deflection, Vogl and Nayfeh (2003) let $\eta_m(t)$ be independent of time, set all of the time derivatives equal to zero and reduced the system of coupled ordinary-differential equations (8.7.47) to a system of coupled algebraic equations. Choosing values for the parameters ϵ, w_{max}, ν, σ, N, and the DC electric voltage V, they solved (8.4.37) for all fixed points η_m.

Figure 8.7.6 shows variation of the static normalized deflection $\hat{w}$ at the plate center ($r = 0$) with the electric forcing v^2 for $\epsilon = 0.1$, $w_{max} = 10$, $\nu = 0.3$, $\sigma = 0$, and various values of N, where $\hat{w}$ is defined as

$$\hat{w}(r) = w(r)/w_{max} \quad (8.7.48)$$

The solid branches in Figure 8.7.6 are stable solutions, whereas the dashed branches are unstable solutions. The maximum percentage of $\hat{w}$ used for plotting is 100% because the center of the plate contacts the electrode when it deflects 100% of $\hat{w} = 1$. The solution converges as the number N of terms retained in the discretization increases. In fact, the curve obtained with six modes ($N = 6$) is not plotted in Figure 8.7.6 because it cannot be distinguished from the curve for five modes ($N = 5$). The unstable solution approaches the physical deflection limit of 100% for $v = 0$ as

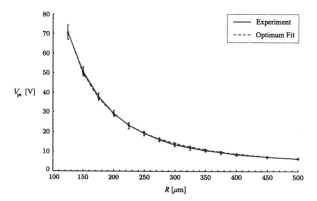

Fig. 8.7.7 Variation of the pull-in voltage with the plate radius.

N increases. Also, the slopes of all curves become infinite at the pull-in point, as expected, when the voltage V approaches the critical pull-in voltage V_{pi}.

Upon comparison of the curves for various N in Figure 8.7.6, we conclude that using five modes ($N = 5$) for discretizing $w(r)$ is sufficient to model adequately the static deflection over most of its physical range $\hat{w} \in [0, 1]$. Consequently, Vogl and Nayfeh (2003) used five modes to describe the dynamics of the plate.

Vogl and Nayfeh (2003) validated the model by comparing it with the experimental data of Osterberg et al. (1994) and Osterberg (1995) who measured the pull-in voltage v_{pi} for multiple radii R of clamped circular plates made of silicon. These plates had a thickness h of about 3 μm with an initial gap width d of about 1 μm. Osterberg developed a statistics-based model to approximate V_{pi} and solved for the optimal statistical coefficients by fitting his model to the experimental data. Vogl and Nayfeh fit the above reduced-order model to the experimental data by solving for the values of E, σ, ν, d, and h that minimize the difference between the theoretical and experimental pull-in values for the 14 different experimental radii. They found that $d = 1.014$ μm, $h = 3.01$ μm, $E = 150.6$ GPa, $\nu = 0.0436$, and $\sigma = 7.82$ MPa. The pull-in voltages from this optimum model are displayed in Figure 8.7.7 along with the experimental data. Standard deviation bars for the experimental data are also shown in Figure 8.7.7. The optimum fit (dashed curve) from the reduced-order model matches fairly well the experimental data curve (solid curve) in Figure 8.7.7 for all radii. Except for the right-most data point in Figure 8.7.7, the optimum values for V_{pi} are within the standard deviation bars for each data point.

The optimum geometric and material properties of Vogl and Nayfeh (2003) also match fairly well the values obtained by Osterberg (1995). They found that $h = 2.94 \pm 0.16$ μm, $d = 1.05 \pm 0.01$ μm, $\frac{E}{1-\nu^2} = 152 \pm 12$ GPa, and $\sigma = 6.8 \pm 0.7$ MPa. Consequently, they concluded that the reduced-order model is adequate for determining the material properties (Young's modulus E, Poisson's ratio ν, and

Table 8.14 Comparison between the pull-in voltages predicted by the statistics-based model of Osterberg (1995), the reduced-order model of Vogl and Nayfeh (2003), and the 3-D CoSolve-EM FEM simulations of Osterberg.

σ [MPa]	V_{pi} Osterberg (O)	V_{pi} Vogl and Nayfeh (V)	V_{pi} CoSolve-EM (C)	$\Delta\%$ between O and C	$\Delta\%$ between V and C
0	313	315	319	2.0	1.2
500	362	364	369	1.8	1.2

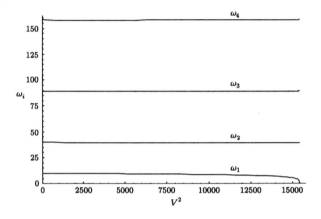

Fig. 8.7.8 Variation of the first four natural frequencies with the square of the voltage.

residual stress σ) and geometric properties (thickness h, gap width d) of the system through use of experimental pull-in data.

The reduced-order model pull-in values are also comparable to those from full 3-D electromechanical simulations. For a clamped circular plate with $d = 1\mu m$, $h = 20\mu m$, $E = 169\ GPa$, and $\nu = 0.3$, Vogl and Nayfeh (2003) used their reduced-order model and Osterberg's model (1995) to calculate the pull-in voltage V_{pi} for both $\sigma = 0$ and $\sigma = 500\ MPa$. The results are given in Table 8.14 along with the pull-in voltages from the simulations obtained by Osterberg with the FEM software CoSolve-EM. The pull-in values obtained using the reduced-order model deviate by 1.2% from those obtained with the full 3-D CoSolve-EM.

To determine the natural frequencies of the axisymmetric modes of the plate around its deflected position, Vogl and Nayfeh (2003) perturbed the reduced-order model (8.7.47) around its equilibrium $\{\eta_m^{eq}\}$ with a small harmonic perturbation $\{\zeta_m\}$ and obtained the linear system

$$[Z(\omega)]\{\zeta_m\} = \{0\} \qquad (8.7.49)$$

where $[Z(\omega)]$ is an $N \times N$ matrix, which is a function of ω^2. For nontrivial ζ_m, the determinant of $[Z(\omega)]$ must be equal to zero, which yields an algebraic equation of order N in ω^2. Positive roots correspond to stable static deflections with ω being the frequency and negative or complex roots correspond to unstable static deflections. The modes of vibration for the stable deformed state can be determined from (8.7.49).

Figure 8.7.8 shows variation of the first four nondimensional axisymmetric natural frequencies associated with the stable deflected equilibrium for the system of Figure 8.7.6 with V^2. Figure 8.7.9 displays variation of the fundamental natural frequency ω_1, normalized with respect to its value Ω_1 at zero V, with V^2. The fundamental natural frequency decreases as the electric voltage V increases, reaching a value of zero at pull-in, as seen in Figure 8.7.9. The fundamental natural frequency is zero at pull-in because the mechanical restoring force of the clamped plate is negated by the equal and opposite electrostatic force from the electrode at pull-in.

We note that reduced-order models have a computational advantage over other models in calculating the natural frequencies. As with the analysis of Younis, Abdel-Rahman, and Nayfeh (2003) for a clamped microbeam, the natural frequencies near pull-in can be calculated from such models as easily as those well before pull-in. Consequently, reduced-order models are robust because they do not suffer from numerical stiffness when calculating the natural frequencies. The computational time is then relatively low compared to other methods, such as the shooting method.

8.8 THERMALLY LOADED PLATES

Thermal loads developed in structural elements due to different forms of heat transfer may influence significantly their static as well as dynamic behavior (Thornton, 1993),

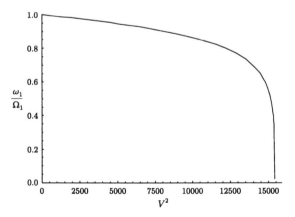

Fig. 8.7.9 Variation of the normalized fundamental natural frequency with the electrostatic forcing level.

such as buckling, frequency shifts, and noise generation, especially in MEMS devices. These effects have favorable impact in some devices and adverse impact in others. Therefore, understanding these effects is important to designers and structural analysts. Examples of MEMS applications include sensors working in high-temperature environments, such as in automotive applications (Dunbar and Sager, 2000), and thermoelastic dampers in micromechanical and nanomechanical systems (Lifshitz and Roukes, 2000).

Fedorov (1976) examined the linear thermostability problem of elastically clamped variable-stiffness annular plates under axisymmetric radial nonuniform thermal loads. He found that, in certain cases, variation of Young's modulus and Poisson's ratios with temperature cannot be neglected. Irie and Yamada (1978a) investigated the linear free vibrations of elastically supported circular and annular plates, with one edge exposed to an axisymmetric sinusoidal heat flux and the other edge thermally insulated. Majima and Hayashi (1984) analyzed the vibrations of elastically buckled annular plates under axisymmetric inplane mechanical and thermal loads. Arafat, Nayfeh, and Faris (2004) analyzed the free vibrations and buckling of circular and annular plates with clamped immovable boundary conditions subjected to axisymmetric inplane thermal loads. They used the dynamic analogue of the Saint Venant plate theory (Timoshenko, 1959) to model the plates and explored the influence of key thermal and geometric design parameters on their responses.

The study of large-amplitude vibrations of plates induced by thermal loads started by Sunakawa (1964) who investigated the influence of temperature changes on the large-amplitude free vibrations of rectangular plates using the method of successive approximations. He showed that these changes are not negligible. In a series of publications, Pal (1969, 1970, 1973) examined large deformations of thermally loaded circular plates using Berger's approximation. Jones, Mazumdar, and Cheung (1980) investigated large-amplitude vibrations and buckling of a rectangular plate at elevated temperatures using also Berger's approximation; they included viscoelastic effects. Chang and Wan (1986) solved the thermo-coupled Berger equation for flat plate vibrations. Bhimaraddi and Chandrashekhara (1993) investigated nonlinear vibrations of heated antisymmetric angle-ply laminated plates. The nonlinear free vibrations of isotropic and axisymmetric orthotropic circular and annular plates subjected to thermal loads were investigated by Li and Zhou (2001, 2003) and Li, Zhou, and Song (2002). They assumed the nonlinear responses to be harmonic and applied the Kantorovich averaging method to the von Kármán equations. Arafat and Nayfeh (2004a,b) investigated modal interactions involving axisymmetric modes in heated clamped-clamped annular plates due to three-to-one and subcombination internal resonances. Nayfeh and Faris (2002, 2003) investigated large-amplitude vibrations of a simply supported plate with movable edge and a clamped circular plate with immovable edge under a thermal load using the method of multiple scales. In this section, we consider the latter case for a thermal load, which produces a temperature consisting of a constant component and a harmonically varying component.

We consider the coupled heat equation and the dynamic version of the von Kármán equations for a thermally excited circular plate. The temperature distribution is given by the heat conduction equation (Boley and Weiner, 1960; Hetnarski, 1987)

$$k\nabla^2 T + Q = \rho c_p \frac{\partial T}{\partial t} + \frac{E\alpha T_0}{1-2\nu}\frac{\partial e}{\partial t} \tag{8.8.1}$$

where $T(r,\theta,z,t)$ is the temperature distribution, T_0 is the initial stress-free temperature, e is the dilatational strain due to the thermal effect, ρ is the mass density, c_p is the heat capacity coefficient at constant pressure, E is the modulus of elasticity, α is the coefficient of thermal expansion, Q is the heat flux, ν is Poisson's ratio, and k is the thermal conductivity. The equations governing the plate deflection are given by (7.4.97) and (7.4.101).

For boundary conditions, we consider a clamped circular plate of radius R. This case is important in most MEMS devices, such as sensors and micropumps, as it is a more realistic representation of the actual boundary conditions. The thermal boundary condition may involve specification of T, $\partial T/\partial r$, or $T + \lambda \partial T/\partial r$, where λ may be constant, spatially varying, or a combination of both. In this section, we consider the case of specifying T. Therefore, the boundary conditions are

$$w = 0, \quad \frac{\partial w}{\partial r} = 0, \quad u = 0, \quad v = 0, \quad \text{and } T = T_b \text{ at } r = R \tag{8.8.2}$$

Because all of the dependent variables must be bounded at the center of the plate,

$$\Phi < \infty, \quad w < \infty, \quad \text{and } T < \infty \text{ at } r = 0 \tag{8.8.3}$$

As in Section 8.4, it follows from (8.8.2) and (8.4.4) that these boundary conditions lead to the following alternate boundary conditions:

$$e_2 = 0 \tag{8.8.4}$$

and

$$\frac{\partial}{\partial r}(re_2) - e_1 - \frac{\partial}{\partial \theta}(\gamma_6) = 0 \tag{8.8.5}$$

as shown in (8.4.6) and (8.4.9). Substituting for the e_i from (7.4.98)-(7.4.100) into (8.8.4) and (8.8.5), we obtain the following boundary conditions in place of $u = 0$ and $v = 0$:

$$\frac{\partial^2 \Phi}{\partial r^2} - \nu\left(\frac{1}{r}\frac{\partial \Phi}{\partial r} + \frac{1}{r^2}\frac{\partial^2 \Phi}{\partial \theta^2}\right) = -N^T \text{ at } r = R \tag{8.8.6}$$

$$\frac{\partial^3 \Phi}{\partial r^3} + \frac{1}{r}\frac{\partial^2 \Phi}{\partial r^2} - \frac{1}{r^2}\frac{\partial \Phi}{\partial r} + \frac{2+\nu}{r^2}\frac{\partial^3 \Phi}{\partial r \partial \theta^2} - \frac{3+\nu}{r^3}\frac{\partial^2 \Phi}{\partial \theta^2}$$
$$= -\frac{\partial N^T}{\partial r} \text{ at } r = R \tag{8.8.7}$$

where

$$N^T = E\alpha \int_{-\frac{1}{2}h}^{\frac{1}{2}h} [T(r,\theta,z,t) - T_0]\, dz \tag{8.8.8}$$

We introduce nondimensional variables, denoted by asterisks, defined as follows:

$$r = Rr^*, \quad t = R^2 \left(\frac{\rho h}{D}\right)^{1/2} t^*, \quad w = \frac{h^2}{R} w^*, \quad z = hz^*,$$

$$c = \frac{12(1-\nu^2)}{R^4} (\rho h^5 D)^{1/2} c^*, \quad \Phi = \frac{Eh^5}{R^2} \Phi^*, \quad T = T_0 T^*,$$

$$F = \frac{12(1-\nu^2)h^4 D}{R^7} F^*, \quad Q = \frac{kh^4}{\alpha R^6} Q^*, \quad T_b = T_0 T_b^* \tag{8.8.9}$$

Substituting (8.8.9) into (8.8.1)-(8.8.3), (7.4.97), (7.4.101), (8.8.6), and (8.8.7) and dropping the asterisks, we obtain

$$\nabla^2 T + \frac{h^4}{\alpha T_0 R^4} Q = \Gamma_1 \frac{\partial T}{\partial t} + \Gamma_2 \frac{\partial e}{\partial t} \tag{8.8.10}$$

$$\frac{\partial^2 w}{\partial t^2} + \nabla^4 w = \epsilon \left[\frac{\partial^2 w}{\partial r^2} \left(\frac{1}{r} \frac{\partial \Phi}{\partial r} + \frac{1}{r^2} \frac{\partial^2 \Phi}{\partial \theta^2} \right) + \frac{\partial^2 \Phi}{\partial r^2} \left(\frac{1}{r} \frac{\partial w}{\partial r} + \frac{1}{r^2} \frac{\partial^2 w}{\partial \theta^2} \right) \right.$$
$$\left. - 2 \left(\frac{1}{r} \frac{\partial^2 \Phi}{\partial r \partial \theta} - \frac{1}{r^2} \frac{\partial \Phi}{\partial \theta} \right) \left(\frac{1}{r} \frac{\partial^2 w}{\partial r \partial \theta} - \frac{1}{r^2} \frac{\partial w}{\partial \theta} \right) - 2c \frac{\partial w}{\partial t} + F \right]$$
$$- \frac{12\alpha T_0 (1+\nu) R^3}{h^3} \int_{-\frac{1}{2}}^{\frac{1}{2}} z \nabla^2 T \, dz \tag{8.8.11}$$

$$\nabla^4 \Phi = \left(\frac{1}{r} \frac{\partial^2 w}{\partial r \partial \theta} - \frac{1}{r^2} \frac{\partial w}{\partial \theta} \right)^2 - \frac{\partial^2 w}{\partial r^2} \left(\frac{1}{r} \frac{\partial w}{\partial r} + \frac{1}{r^2} \frac{\partial^2 w}{\partial \theta^2} \right)$$
$$- \frac{\alpha T_0 R^4}{h^4} \int_{-\frac{1}{2}}^{\frac{1}{2}} \nabla^2 T \, dz \tag{8.8.12}$$

$$w = 0, \quad \frac{\partial w}{\partial r} = 0, \quad \text{and} \quad T = T_b \quad \text{at} \quad r = 1 \tag{8.8.13}$$

$$\Phi < \infty, \quad w < \infty, \quad \text{and} \quad T < \infty \quad \text{at} \quad r = 0 \tag{8.8.14}$$

$$\frac{\partial^2 \Phi}{\partial r^2} - \nu \left(\frac{1}{r} \frac{\partial \Phi}{\partial r} + \frac{1}{r^2} \frac{\partial^2 \Phi}{\partial \theta^2} \right) = -\frac{\alpha T_0 R^4}{h^4} \int_{-\frac{1}{2}}^{\frac{1}{2}} (T-1) dz \quad \text{at} \quad r = 1 \tag{8.8.15}$$

$$\frac{\partial^3 \Phi}{\partial r^3} + \frac{1}{r} \frac{\partial^2 \Phi}{\partial r^2} - \frac{1}{r^2} \frac{\partial \Phi}{\partial r} + \frac{2+\nu}{r^2} \frac{\partial^3 \Phi}{\partial r \partial \theta^2} - \frac{3+\nu}{r^3} \frac{\partial^2 \Phi}{\partial \theta^2}$$
$$= -\frac{\alpha T_0 R^4}{h^4} \int_{-\frac{1}{2}}^{\frac{1}{2}} \frac{\partial T}{\partial r} dz \quad \text{at} \quad r = 1 \tag{8.8.16}$$

where

$$\Gamma_1 = \frac{\rho c_p}{k} \left(\frac{D}{\rho h}\right)^{1/2}, \quad \Gamma_2 = \frac{E\alpha}{(1-2\nu)k} \left(\frac{D}{\rho h}\right)^{1/2}, \quad \text{and} \quad \epsilon = \frac{12(1-\nu^2)h^2}{R^2}$$

The last two terms on the right-hand side of (8.8.10) represent the diffusion of heat and thermoelastic coupling (Hetnarski, 1987).

Usually, the thermal diffusion and thermoelastic coupling terms are negligible because Γ_1 and Γ_2 are very large as shown in Table 8.15 and also because we are not interested in dissipation (Hetnarski, 1987) due to heat transfer. Hence, (8.8.10) is reduced to

$$\nabla^2 T + \frac{h^4}{\alpha T_0 R^4} Q = 0 \qquad (8.8.17)$$

Table 8.15 Thermal and mechanical properties of used materials.

Material	Al	Steel	SiNi
$\rho\,(kg/m^3)$	2702	7800	2330
$E\,(GPa)$	70	207	100
$k\,(W/m.K)$	237	60	148
$c_p\,(J/kg.K)$	903	430	712
$\alpha(1/K)$	24.0×10^{-6}	12.0×10^{-6}	2.9×10^{-6}
Γ_1	$15.86h \times 10^6$	$97.43h \times 10^6$	$22.24h \times 10^6$
Γ_2	$27.3h \times 10^6$	$161.34h \times 10^6$	$9.72h \times 10^6$

Next, we consider the case of a time-harmonic uniform heat flux so that $Q = 4q\cos(\Omega t)$. Then, solving (8.8.17) subject to the boundary conditions $T = T_b$ at $r = 1$ and $T < \infty$ at $r = 0$, we find that the nondimensional temperature distribution is given by

$$T(r,t) = T_b + \frac{h^4}{\alpha T_0 R^4}\left(1 - r^2\right) q\cos(\Omega t) \qquad (8.8.18)$$

Substituting (8.8.18) into (8.8.11), (8.8.12), (8.8.15), and (8.8.16), we obtain

$$\frac{\partial^2 w}{\partial t^2} + \nabla^4 w = \epsilon\left[\frac{\partial^2 w}{\partial r^2}\left(\frac{1}{r}\frac{\partial \Phi}{\partial r} + \frac{1}{r^2}\frac{\partial^2 \Phi}{\partial \theta^2}\right) + \frac{\partial^2 \Phi}{\partial r^2}\left(\frac{1}{r}\frac{\partial w}{\partial r} + \frac{1}{r^2}\frac{\partial^2 w}{\partial \theta^2}\right) \right.$$
$$\left. -2\left(\frac{1}{r}\frac{\partial^2 \Phi}{\partial r\partial\theta} - \frac{1}{r^2}\frac{\partial \Phi}{\partial \theta}\right)\left(\frac{1}{r}\frac{\partial^2 w}{\partial r\partial\theta} - \frac{1}{r^2}\frac{\partial w}{\partial \theta}\right) - 2c\frac{\partial w}{\partial t} + F\right]$$
$$\qquad (8.8.19)$$

$$\nabla^4 \Phi = \left(\frac{1}{r}\frac{\partial^2 w}{\partial r\partial\theta} - \frac{1}{r^2}\frac{\partial w}{\partial \theta}\right)^2 - \frac{\partial^2 w}{\partial r^2}\left(\frac{1}{r}\frac{\partial w}{\partial r} + \frac{1}{r^2}\frac{\partial^2 w}{\partial \theta^2}\right) + 4q\cos(\Omega t)$$
$$\qquad (8.8.20)$$

$$\frac{\partial^2 \Phi}{\partial r^2} - \nu\left(\frac{1}{r}\frac{\partial \Phi}{\partial r} + \frac{1}{r^2}\frac{\partial^2 \Phi}{\partial \theta^2}\right) = -\frac{\alpha T_0 R^4 (T_b - 1)}{h^4} \quad \text{at } r = 1 \qquad (8.8.21)$$

$$\frac{\partial^3 \Phi}{\partial r^3} + \frac{1}{r}\frac{\partial^2 \Phi}{\partial r^2} - \frac{1}{r^2}\frac{\partial \Phi}{\partial r} + \frac{2+\nu}{r^2}\frac{\partial^3 \Phi}{\partial r\partial\theta^2} - \frac{3+\nu}{r^3}\frac{\partial^2 \Phi}{\partial \theta^2}$$
$$= 2q\cos(\Omega t) \text{ at } r = 1 \qquad (8.8.22)$$

Exact solutions of the above system of equations with its nonhomogeneous boundary conditions are not available yet. In what follows, we introduce a transformation to eliminate the nonhomogeneous terms. To this end, we find a solution to the linear problem consisting of (8.8.14) and (8.8.20)-(8.8.22) in the form

$$\Phi_l = c_1 - \frac{\alpha T_0 R^4 (T_b - 1)}{2(1-\nu)h^4} r^2 + \frac{1}{16}\left[r^4 - \frac{2(3-\nu)}{1-\nu}r^2\right] q\cos\Omega t \qquad (8.8.23)$$

where c_1 is an arbitrary constant. This constant is not needed because we need only the derivatives of Φ_l. Next, we let

$$\Phi = c_1 - \frac{\alpha T_0 R^4 (T_b - 1)}{2(1-\nu)h^4} r^2 + \frac{1}{16}\left[r^4 - \frac{2(3-\nu)}{1-\nu}r^2\right] q\cos\Omega t + \Phi_h \qquad (8.8.24)$$

in (8.8.19)-(8.8.22) and obtain

$$\frac{\partial^2 w}{\partial t^2} + p\nabla^2 w + \nabla^4 w =$$

$$+ \frac{1}{4}\epsilon\left\{\frac{\partial^2 w}{\partial r^2}\left[r^2 - \frac{3-\nu}{1-\nu}\right] + \frac{\partial w}{\partial r}\left[3r - \frac{3-\nu}{(1-\nu)r}\right]\right\} q\cos(\Omega t)$$

$$+ \epsilon\left[\frac{\partial^2 w}{\partial r^2}\left(\frac{1}{r}\frac{\partial \Phi_h}{\partial r} + \frac{1}{r^2}\frac{\partial^2 \Phi_h}{\partial \theta^2}\right) + \frac{\partial^2 \Phi_h}{\partial r^2}\left(\frac{1}{r}\frac{\partial w}{\partial r} + \frac{1}{r^2}\frac{\partial^2 w}{\partial \theta^2}\right)\right.$$

$$\left. - 2\left(\frac{1}{r}\frac{\partial^2 \Phi_h}{\partial r \partial \theta} - \frac{1}{r^2}\frac{\partial \Phi_h}{\partial \theta}\right)\left(\frac{1}{r}\frac{\partial^2 w}{\partial r \partial \theta} - \frac{1}{r^2}\frac{\partial w}{\partial \theta}\right) - 2c\frac{\partial w}{\partial t} + F\right] \qquad (8.8.25)$$

$$\nabla^4 \Phi_h = \left(\frac{1}{r}\frac{\partial^2 w}{\partial r \partial \theta} - \frac{1}{r^2}\frac{\partial w}{\partial \theta}\right)^2 - \frac{\partial^2 w}{\partial r^2}\left(\frac{1}{r}\frac{\partial w}{\partial r} + \frac{1}{r^2}\frac{\partial^2 w}{\partial \theta^2}\right) \qquad (8.8.26)$$

$$\frac{\partial^2 \Phi_h}{\partial r^2} - \nu\left(\frac{1}{r}\frac{\partial \Phi_h}{\partial r} + \frac{1}{r^2}\frac{\partial^2 \Phi_h}{\partial \theta^2}\right) = 0 \text{ at } r = 1 \qquad (8.8.27)$$

$$\frac{\partial^3 \Phi_h}{\partial r^3} + \frac{1}{r}\frac{\partial^2 \Phi_h}{\partial r^2} - \frac{1}{r^2}\frac{\partial \Phi_h}{\partial r} + \frac{2+\nu}{r^2}\frac{\partial^3 \Phi_h}{\partial r \partial \theta^2} - \frac{3+\nu}{r^3}\frac{\partial^2 \Phi_h}{\partial \theta^2} = 0 \text{ at } r = 1 \qquad (8.8.28)$$

where

$$p = \frac{12\alpha T_0(1+\nu)R^2(T_b - 1)}{h^2}$$

In the next section, we investigate the linear natural frequencies and their corresponding mode shapes, and in Section 8.8.2, we consider combination parametric resonance of two axisymmetric modes.

8.8.1 Linear Natural Frequencies and Mode Shapes

The linear natural frequencies and mode shapes can be obtained from (8.8.25), (8.8.13), and (8.8.14) by letting $\epsilon = 0$, seeking solutions in the form

$$w(r,\theta,t) = \phi(r)e^{i\omega t + in\theta}$$

and obtaining

$$\nabla^4 \phi + p\nabla^2 \phi - \omega^2 \phi = 0 \qquad (8.8.29)$$

$$\phi = 0 \text{ and } \phi' = 0 \text{ at } r = 1 \qquad (8.8.30)$$

$$\phi < \infty \text{ at } r = 0 \qquad (8.8.31)$$

where
$$\nabla^2 = \frac{d^2}{dr^2} + \frac{1}{r}\frac{d}{dr} - \frac{n^2}{r^2}$$

To determine the solution of (8.8.29)-(8.8.31), we let

$$\nabla^2 \phi = -\kappa^2 \phi \tag{8.8.32}$$

and obtain

$$\kappa^4 - p\kappa^2 - \omega^2 = 0 \tag{8.8.33}$$

whose solutions are

$$\kappa^2 = \frac{1}{2}\left[\sqrt{p^2 + 4\omega^2} \pm p\right] \tag{8.8.34}$$

Therefore, it follows from (8.8.32) and (8.8.34) that the general solution of (8.8.29) can be expressed as

$$\phi(r) = c_1 J_n(\kappa_1 r) + c_2 Y_n(\kappa_1 r) + c_3 I_n(\kappa_2 r) + c_4 K_n(\kappa_2 r) \tag{8.8.35}$$

where

$$\kappa_1^2 = \frac{1}{2}\left[\sqrt{p^2 + 4\omega^2} + p\right] \quad \text{and} \quad \kappa_2^2 = \frac{1}{2}\left[\sqrt{p^2 + 4\omega^2} - p\right] \tag{8.8.36}$$

The boundary condition (8.8.14) demands that $c_2 = 0$ and $c_4 = 0$, whereas the boundary condition $\phi(1) = 0$ demands that $c_3 = -c_1 J_n(\kappa_1)/I_n(\kappa_2)$. Hence, (8.8.35) can be rewritten as

$$\phi(r) = c\left[\frac{J_n(\kappa_1 r)}{J_n(\kappa_1)} - \frac{I_n(\kappa_2 r)}{I_n(\kappa_2)}\right] \tag{8.8.37}$$

where c is an arbitrary constant. Imposing the boundary condition $\phi'(1) = 0$, we obtain the transcendental characteristic equation

$$\kappa_1 J_n'(\kappa_1) I_n(\kappa_2) - \kappa_2 J_n(\kappa_1) I_n'(\kappa_2) = 0 \tag{8.8.38}$$

which can be solved numerically for ω as a function of p and hence as a function of T_b. In Figure 8.8.1, we show variation of the lowest three axisymmetric natural frequencies ($n = 0$) with p. As p increases, the first natural frequency decreases and approaches zero when $p \approx 14.682$, corresponding to the first buckling condition. This value can be determined directly as follows. Setting $\omega = 0$ in (8.8.36) yields $\kappa_1 = \sqrt{p}$ and $\kappa_2 = 0$. Consequently, the characteristic equation is reduced to $J_0'(\sqrt{p}) = 0$, whose first root corresponds to $p \approx 14.862$.

8.8.2 Combination Parametric Resonance of Two Axisymmetric Modes

We consider the case of combination parametric resonance of the additive type in which the frequency Ω of the thermal load is approximately equal to $\omega_m + \omega_n$,

550 DYNAMICS OF PLATES

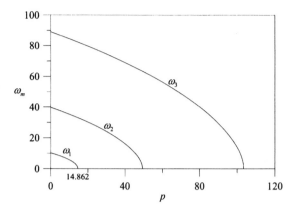

Fig. 8.8.1 Variation of the lowest three nondimensional natural frequencies with the nondimensional temperature p.

where ω_m and ω_n are the mth and nth axisymmetric natural frequencies of the plate. Moreover, we assume that neither of these modes is involved in an internal resonance with any other axisymmetric or asymmetric mode.

We use the method of multiple scales and seek a first-order expansion of the response of the plate in the form

$$w(r, t; \epsilon) = w_0(r, t_0, t_1) + \epsilon w_1(r, t_0, t_1) + \cdots \qquad (8.8.39)$$

$$\Phi_h(r, t; \epsilon) = \Phi_0(r, t_0, t_1) + \cdots \qquad (8.8.40)$$

where $t_0 = t$ and $t_1 = \epsilon t$. Substituting (8.8.39) and (8.8.40) into (8.8.25)-(8.8.28), (8.8.13), and (8.8.14), dropping F, neglecting the derivatives with respect to θ, and equating coefficients of like powers of ϵ, we obtain

Order ϵ^0

$$D_0^2 w_0 + p\nabla^2 w_0 + \nabla^4 w_0 = 0 \qquad (8.8.41)$$

$$\nabla^4 \Phi_0 = -\frac{1}{2r}\frac{\partial}{\partial r}\left(\frac{\partial w_0}{\partial r}\right)^2 \qquad (8.8.42)$$

$$w_0 = 0 \text{ and } w_0' = 0 \text{ at } r = 1 \qquad (8.8.43)$$

$$w_0 < \infty \text{ and } \Phi_0 < \infty \text{ at } r = 0 \qquad (8.8.44)$$

$$\Phi_0'' - \nu\Phi_0' = 0 \text{ at } r = 1 \qquad (8.8.45)$$

$$\Phi_0''' + \Phi_0'' - \Phi_0' = 0 \text{ at } r = 1 \qquad (8.8.46)$$

Order ϵ

$$D_0^2 w_1 + p\nabla^2 w_1 + \nabla^4 w_1 = -2D_0 D_1 w_0 + \frac{1}{r}\frac{\partial^2 w_0}{\partial r^2}\frac{\partial \Phi_0}{\partial r} + \frac{1}{r}\frac{\partial w_0}{\partial r}\frac{\partial^2 \Phi_0}{\partial r^2}$$

$$-2cD_0w_0 + \frac{1}{4r}\left[\frac{\partial w_0}{\partial r}\left(r^3 - \frac{3-\nu}{1-\nu}r\right)\right]' q\cos(\Omega t_0) \quad (8.8.47)$$

$$w_1 < \infty \text{ at } r = 0 \quad (8.8.48)$$

$$w_1 = 0 \text{ and } w_1' = 0 \text{ at } r = 1 \quad (8.8.49)$$

where the prime indicates the partial derivative with respect to r.

Because in the presence of damping all modes that are not directly excited by the thermal loads or indirectly excited by internal resonances decay with time, the solution of (8.8.41), (8.8.43), and (8.8.44) can be taken to consist of the two directly excited mth and nth modes if they are not involved in an internal resonance with any other mode; that is,

$$w_0 = A_m(t_1)\phi_m(r)e^{i\omega_m t_0} + A_n(t_1)\phi_n(r)e^{i\omega_n t_0} + cc \quad (8.8.50)$$

where $\phi_m(r)$ and $\phi_n(r)$ are the mth and nth mode shapes corresponding to the natural frequencies ω_m and ω_n, respectively. Substituting (8.8.50) into (8.8.42) yields

$$\nabla^4 \Phi_0 = -\frac{1}{2r}\sum_{j,k=m,n}\left[A_k A_j e^{i(\omega_k+\omega_j)t_0} + A_k \bar{A}_j e^{i(\omega_k-\omega_j)t_0}\right](\phi_k'\phi_j')' + cc$$

$$(8.8.51)$$

The solution of (8.8.51) subject to the boundary conditions (8.8.44)-(8.8.46) can be expressed as

$$\Phi_0 = \sum_{j,k=m,n}\left[A_k A_j e^{i(\omega_k+\omega_j)t_0} + A_k \bar{A}_j e^{i(\omega_k-\omega_j)t_0}\right]\psi_{kj}(r) + cc \quad (8.8.52)$$

where

$$\nabla^4 \psi_{kj} = -\frac{1}{2r}(\phi_k'\phi_j')' \quad (8.8.53)$$

$$\psi_{kj}(0) < \infty \quad (8.8.54)$$

$$\psi_{kj}'' - \nu\psi_{kj}' = 0 \text{ at } r = 1 \quad (8.8.55)$$

$$\psi_{kj}''' + \psi_{kj}'' - \psi_{kj}' = 0 \text{ at } r = 1 \quad (8.8.56)$$

Using the method of variation of parameters and following a procedure similar to that used in Section 8.4.1, we obtain the solution of (8.8.53)-(8.8.56) as

$$\psi_{kj}'(r) = -\frac{1}{4}r\int_0^r \frac{\phi_k'(\xi)\phi_j'(\xi)}{\xi}d\xi + \frac{1}{4r}\int_0^r \xi\phi_k'(\xi)\phi_j'(\xi)d\xi + c_{20}r \quad (8.8.57)$$

where

$$c_{20} = \frac{1}{4}\int_0^1 \frac{\phi_k'(\xi)\phi_j'(\xi)}{\xi}d\xi + \frac{1+\nu}{4(1-\nu)}\int_0^1 \xi\phi_k'(\xi)\phi_j'(\xi)d\xi \quad (8.8.58)$$

Alternatively, we use the method of series expansions. Multiplying (8.8.53) with r and integrating the result, we have

$$\left(r\frac{d^2}{dr^2} + \frac{d}{dr} - \frac{1}{r}\right)\frac{d\psi_{kj}}{dr} = -\frac{1}{2}\phi'_k\phi'_j \qquad (8.8.59)$$

where the constant of integration is zero on account of the boundary condition (8.8.56). Because

$$\left(r^2\frac{d^2}{dr^2} + r\frac{d}{dr} - 1\right)J_1(\xi r) = -\xi^2 r^2 J_1(\xi r) \qquad (8.8.60)$$

we express ψ'_{kj} as

$$\psi'_{kj}(r) = \sum_{m=1}^{\infty} b_{mkj} J_1(\xi_m r) \qquad (8.8.61)$$

Using the boundary condition (8.8.55), we find that the ξ_m are the roots of

$$\xi_m J_1(\xi_m) - (1+\nu)J_0(\xi_m) = 0 \qquad (8.8.62)$$

The lowest ten roots are 1.57883, 5.27284, 8.5006, 11.6802, 14.8433, 17.9988, 21.1502, 24.299, 27.4461, and 30.5921. Substituting (8.8.61) into (8.8.59) and using (8.8.60), we obtain

$$\sum_{m=1}^{\infty} b_{mkj}\xi_m^2 r J_1(\xi_m r) = \frac{1}{2}\phi'_k\phi'_j \qquad (8.8.63)$$

Multiplying (8.8.63) with $J_1(\xi_s r)$ and integrating the result with respect to r from $r=0$ to 1, we have

$$b_{skj} = \frac{\int_0^1 J_1(\xi_s r)\phi'_k(r)\phi'_j(r)dr}{2\xi_s^2 \int_0^1 rJ_1^2(\xi_s r)\,dr} \qquad (8.8.64)$$

Substituting (8.8.50) and (8.8.52) into (8.8.47) yields

$$D_0^2 w_1 + p\nabla^2 w_1 + \nabla^4 w_1 = -2i\sum_{k=m,n}\omega_k(A'_k + cA_k)\phi_k e^{i\omega_k t_0}$$

$$+ \frac{1}{8r}qe^{i\Omega t_0}\sum_{k=m,n}\left[\left(r^3 - \frac{3-\nu}{1-\nu}r\right)\phi'_k\right]'\left[A_k e^{i\omega_k t_0} + \bar{A}_k e^{-i\omega_k t_0}\right]$$

$$+ \frac{1}{r}\sum_{k,j,s}(\phi'_s\psi'_{kj})'\left[A_s A_k A_j e^{i(\omega_k+\omega_j+\omega_s)t_0} + A_s A_k \bar{A}_j e^{i(\omega_s+\omega_k-\omega_j)t_0}\right.$$

$$\left. + A_k A_j \bar{A}_s e^{i(\omega_k+\omega_j-\omega_s)t_0} + A_k \bar{A}_j \bar{A}_s e^{i(\omega_k-\omega_j-\omega_s)t_0}\right] + cc \qquad (8.8.65)$$

To describe the nearness of the combination resonance, we introduce the detuning parameter σ defined by

$$\Omega = \omega_n + \omega_m + \epsilon\sigma \qquad (8.8.66)$$

Substituting (8.8.66) into (8.8.65) and keeping the terms that lead to secular terms yields

$$D_0^2 w_1 + p\nabla^2 w_1 + \nabla^4 w_1 = -2i \sum_{k=m,n} \omega_k \left(A_k' + cA_k\right) \phi_k e^{i\omega_k t_0}$$

$$+ \frac{3}{r} \left(\phi_n' \psi_{nn}'\right)' A_n^2 \bar{A}_n e^{i\omega_n t_0} + \frac{2}{r} \left[2 \left(\phi_m' \psi_{mn}'\right)' + \left(\phi_n' \psi_{mm}'\right)'\right] A_m \bar{A}_m A_n e^{i\omega_n t_0}$$

$$+ \frac{3}{r} \left(\phi_m' \psi_{mm}'\right)' A_m^2 \bar{A}_m e^{i\omega_m t_0} + \frac{2}{r} \left[2 \left(\phi_n' \psi_{mn}'\right)' + \left(\phi_m' \psi_{nn}'\right)'\right] A_n \bar{A}_n A_m e^{i\omega_m t_0}$$

$$+ \frac{1}{8r} q \left[\left(r^3 - \frac{3-\nu}{1-\nu} r\right) \phi_m'\right]' \bar{A}_m e^{i\omega_n t_0 + i\sigma t_1}$$

$$+ \frac{1}{8r} q \left[\left(r^3 - \frac{3-\nu}{1-\nu} r\right) \phi_n'\right]' \bar{A}_n e^{i\omega_m t_0 + i\sigma t_1} + NST + cc \qquad (8.8.67)$$

Because the homogeneous problem consisting of (8.8.67), (8.8.48), and (8.8.49) has nontrivial solutions, the corresponding nonhomogeneous problem has solutions only if the nonhomogeneous part in (8.8.67) is orthogonal to every solution of the adjoint homogeneous problem. Because the problem is self-adjoint, the solvability condition demands that the right-hand side of (8.8.67) is orthogonal to every $r\phi_j(r)e^{-i\omega_j t_0}$. Multiplying the right-hand side of (8.8.67) once with $r\phi_n(r)e^{-i\omega_n t_0}$ and once with $r\phi_m(r)e^{-i\omega_m t_0}$, integrating each result from $r = 0$ to $r = 1$, and setting each outcome equal to zero, we obtain

$$2i\omega_n \left(A_n' + \mu_n A_n\right) + \alpha_{nn} A_n^2 \bar{A}_n + \alpha_{nm} A_m \bar{A}_m A_n + q_{nm} \bar{A}_m e^{i\sigma t_1} = 0 \qquad (8.8.68)$$

$$2i\omega_m \left(A_m' + \mu_m A_m\right) + \alpha_{mm} A_m^2 \bar{A}_m + \alpha_{mn} A_n \bar{A}_n A_m + q_{mn} \bar{A}_n e^{i\sigma t_1} = 0 \qquad (8.8.69)$$

where

$$\alpha_{jj} = -3 \int_0^1 \phi_j \left(\phi_j' \psi_{jj}'\right)' dr = 3 \int_0^1 \phi_j'^2 \psi_{jj}' dr \qquad (8.8.70)$$

$$q_{nm} = q_{mn} = -\frac{1}{8} q \int_0^1 \phi_n \left[\left(r^3 - \frac{3-\nu}{1-\nu} r\right) \phi_m'\right]' dr$$

$$= \frac{1}{8} q \int_0^1 \left(r^3 - \frac{3-\nu}{1-\nu} r\right) \phi_m' \phi_n' dr \qquad (8.8.71)$$

$$\alpha_{nm} = -\int_0^1 \phi_n \left(4\phi_m' \psi_{mn}' + 2\phi_n' \psi_{mm}'\right)' dr$$

$$= \int_0^1 \phi_n' \left(4\phi_m' \psi_{mn}' + 2\phi_n' \psi_{mm}'\right) dr \qquad (8.8.72)$$

$$\alpha_{mn} = -\int_0^1 \phi_m \left(4\phi_n' \psi_{mn}' + 2\phi_m' \psi_{nn}'\right)' dr$$

$$= \int_0^1 \phi_m' \left(4\phi_n'\psi_{mn}' + 2\phi_m'\psi_{nn}'\right) dr \tag{8.8.73}$$

In order that (8.8.68) and (8.8.69) be derivable from a Lagrangian in the absence of damping, α_{mn} must be equal to α_{nm}. We note that the first two terms in the integrands in (8.8.72) and (8.8.73) are the same. To show that the last two integral terms are equal, we use (8.8.57) and express the last term in (8.8.72) as

$$2\int_0^1 \phi_n'^2 \psi_{mm}' dr$$

$$= -\frac{1}{2}\int_0^1 r\phi_n'^2 \left[\int_0^r \frac{\phi_m'^2(\xi)}{\xi} d\xi\right] dr + \frac{1}{2}\int_0^1 \frac{\phi_n'^2}{r}\left[\int_0^r \xi\phi_m'^2(\xi) d\xi\right] dr$$

$$+ 2c_{20}\int_0^1 r\phi_n'^2 dr$$

$$= \frac{1}{2}\int_0^1 \frac{\phi_m'^2}{r}\left[\int_0^r \xi\phi_n'^2(\xi) d\xi\right] dr - \frac{1}{2}\int_0^1 r\phi_m'^2\left[\int_0^r \frac{\phi_n'^2(\xi)}{\xi} d\xi\right] dr$$

$$-\frac{1}{2}\left(\int_0^1 \frac{\phi_m'^2}{r} dr\right)\left[\int_0^r r\phi_n'^2(r) dr\right] + 2c_{20}\int_0^1 r\phi_n'^2 dr$$

$$= \frac{1}{2}\int_0^1 \frac{\phi_m'^2}{r}\left[\int_0^r \xi\phi_n'^2(\xi) d\xi\right] dr - \frac{1}{2}\int_0^1 r\phi_m'^2\left[\int_0^r \frac{\phi_n'^2(\xi)}{\xi} d\xi\right] dr$$

$$+ 2c_{20}\int_0^1 r\phi_m'^2 dr = 2\int_0^1 \phi_m'^2 \psi_{nn}' dr \tag{8.8.74}$$

upon using integration by parts and (8.8.58).

To investigate the solutions of (8.8.68) and (8.8.69), we introduce the polar transformations

$$A_m = \frac{1}{2}a_m e^{i(\beta_m + \frac{1}{2}\sigma t_1)} \quad \text{and} \quad A_n = \frac{1}{2}a_n e^{i(\beta_n + \frac{1}{2}\sigma t_1)} \tag{8.8.75}$$

separate real and imaginary parts and obtain

$$a_m' = -\mu_m a_m + \frac{q_{mn}}{2\omega_m}a_n \sin\gamma \tag{8.8.76}$$

$$a_m\beta_m' = -\frac{1}{2}\sigma a_m + \frac{\alpha_{mm}}{8\omega_m}a_m^3 + \frac{\alpha_{mn}}{8\omega_m}a_m a_n^2 + \frac{q_{mn}}{2\omega_m}a_n \cos\gamma \tag{8.8.77}$$

$$a_n' = -\mu_n a_n + \frac{q_{mn}}{2\omega_n}a_m \sin\gamma \tag{8.8.78}$$

$$a_n\beta_n' = -\frac{1}{2}\sigma a_n + \frac{\alpha_{nn}}{8\omega_n}a_n^3 + \frac{\alpha_{mn}}{8\omega_n}a_n a_m^2 + \frac{q_{mn}}{2\omega_n}a_m \cos\gamma \tag{8.8.79}$$

where

$$\gamma = \beta_m + \beta_n \tag{8.8.80}$$

The equilibrium solutions or fixed points of (8.8.76)-(8.8.79) correspond to $a_m' = 0$, $a_n' = 0$, and $\gamma' = 0$. There are two possible equilibrium solutions: (a) $a_m = 0$

and $a_n = 0$ corresponding to nonactivation of the combination resonance and (b) $a_m \neq 0$ and $a_n \neq 0$. In the latter case, (8.8.77) and (8.8.79) can be used to eliminate β_m and β_n from (8.8.80) to obtain the following equation for γ':

$$\gamma' = -\sigma + \left(\frac{\alpha_{mm}}{8\omega_m} + \frac{\alpha_{mn}}{8\omega_n}\right) a_m^2 + \left(\frac{\alpha_{mn}}{8\omega_m} + \frac{\alpha_{nn}}{8\omega_n}\right) a_n^2$$
$$+ \left(\frac{a_n}{2a_m\omega_m} + \frac{a_m}{2a_n\omega_n}\right) q_{mn} \cos\gamma \qquad (8.8.81)$$

Consequently, for nontrivial solutions, the modulation equations are reduced from four to three first-order differential equations. For equilibrium solutions, we set $a_m' = 0$, $a_n' = 0$, and $\gamma' = 0$ in (8.8.76), (8.8.78), and (8.8.81), and obtain

$$\alpha_e a_m^2 = \sigma \pm \frac{\mu_m + \mu_n}{\sqrt{\mu_m \mu_n}} \sqrt{\frac{q_{mn}^2}{4\omega_m \omega_n} - \mu_m \mu_n} \qquad (8.8.82)$$

$$a_n^2 = \frac{\mu_m \omega_m}{\mu_n \omega_n} a_m^2 \qquad (8.8.83)$$

$$\sin\gamma = \frac{2\mu_m \omega_m a_m}{q_{mn} a_n} = \frac{2\mu_n \omega_n a_n}{q_{mn} a_m} = \pm 2\sqrt{\frac{\mu_m \mu_n \omega_m \omega_n}{q_{mn}^2}} \qquad (8.8.84)$$

where the effective nonlinearity α_e is given by

$$\alpha_e = \frac{1}{8}\left[\frac{\alpha_{mm}}{\omega_m} + \frac{\alpha_{mn}}{\omega_n} + \left(\frac{\alpha_{mn}}{\omega_m} + \frac{\alpha_{nn}}{\omega_n}\right)\frac{\mu_m \omega_m}{\mu_n \omega_n}\right] \qquad (8.8.85)$$

In Figure 8.8.2, a frequency-response curve is shown for $q = 3$ and nondimensional temperature $p = 7$. The curve is bent to the right, indicating a hardening-type nonlinearity. The bending of the frequency-response curve leads to multivalued amplitudes and hence jumps. To illustrate the jump phenomenon, let us suppose that an experiment is conducted where the flux amplitude is kept constant (i.e., $q =$ constant) while the flux frequency is varied very slowly. We start from a frequency far above the sum of the two natural frequencies involved in the combination resonance (i.e., positive detuning) and decrease it. We choose the initial conditions so that the response amplitude is trivial. As σ is decreased, the response amplitude remains trivial until Point B is reached, where the response amplitude experiences a sudden jump up to Point C. Point B is a subcritical or reverse pitchfork bifurcation. The response amplitude decreases as the flux frequency is further decreased until Point D is reached, where it becomes trivial again. Point D is a supercritical pitchfork bifurcation. Starting at Point C and sweeping σ upward, we note that, to the first-order approximation, the response amplitude increases without bound as the flux frequency is increased. In reality, the stable and unstable branches merge to produce a saddle-node bifurcation. Increasing the flux frequency above this critical value results in a jump down of the response amplitude to the trivial solution. Between Point B and the saddle-node bifurcation, there are three possible solutions: the trivial solution, which is stable, and two nontrivial solutions, the larger of which is stable and the smaller is unstable.

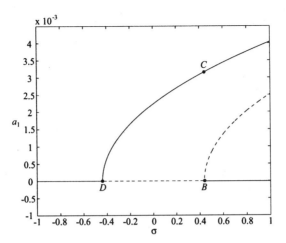

Fig. 8.8.2 Variation of the equilibrium solution a_1 with the frequency detuning parameter σ when $q = 3$ and $p = 7$.

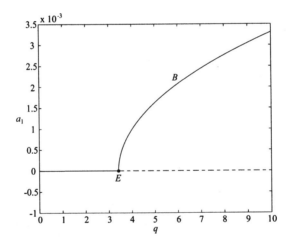

Fig. 8.8.3 Variation of the equilibrium solution a_1 with the flux amplitude q for the case of a negatively detuning system ($\sigma = -0.5$).

In this interval, the response may be trivial or nontrivial, depending on the initial conditions.

In Figure 8.8.3, a force-response curve is generated when the system is negatively detuned by 0.5. Starting from a flux amplitude corresponding to Point E and increasing it results in a monotonic increase in the response amplitude. Point E

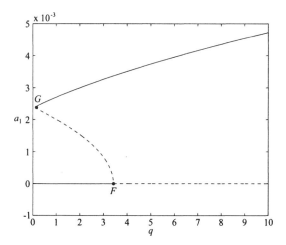

Fig. 8.8.4 Variation of the equilibrium solution a_1 with the flux amplitude q for the case of a positively detuning system ($\sigma = 0.5$).

is a supercritical pitchfork bifurcation. For any flux amplitude above Point E, the response amplitude settles on the branch EB, irrespective of the initial conditions. Sweeping q down, one finds that the response amplitude follows the branch BE. Decreasing q below Point E results in a smooth transition of the response from a nontrivial to a trivial response. Increasing the magnitude of the detuning parameter results in a shift of the critical Point E to the right.

We show in Figure 8.8.4 a force-response curve when the system is positively detuned by 0.5. In contrast with the preceding case, there are multivalued responses. Point F is a subcritical pitchfork bifurcation and Point G is a saddle-node bifurcation. Between these bifurcations, the response may be trivial or nontrivial depending on the initial conditions. Whereas the linear theory predicts stable trivial responses for flux amplitudes below Point F and unbounded responses above it, the nonlinear analysis predicts instabilities of the trivial response between Points F and G, a subcritical instability. Moreover, the nonlinearity puts a cap on the response amplitude. Sweeping q in the forward direction beyond Point F, one finds that the response amplitude jumps up. On the other hand, sweeping q in the reverse direction beyond Point G results in a jump to the trivial response.

9
SHELLS

A shell is a pre-curved structure with one of its dimensions being much smaller than the other two characteristic dimensions. The modeling and dynamics of shells are more complicated than those of plates just because of their initial curvatures. Two important degenerate classes of shells are plates (shells without initial curvatures) and membranes (shells whose walls offer no resistance to bending moments and transverse shear forces). For accounts of the statics and dynamics of shells as well as assessments of the computational methods, we refer the reader to the books of Novozhilov (1958), Timoshenko (1959), Gol'denveizer (1961), Ambartsumian (1964), Kraus (1967), Koiter (1969), Flügge (1973), Leissa (1993b), Nayfeh and Mook (1979), Soedel (1993), Liew, Kitipornchai, and Wang (1993), Sathyamoorthy (1997), Nayfeh (2000), and Qatu (2004); and to the review articles of Kapania (1989), Noor and Burton (1990), Mirza (1991), Qatu (1992, 2002a, 2002b), Noor, Hadian, and Peters (1994), Soldatos (1994, 1999), Noor, Burton, and Peters (1995), Noor, Burton, and Bert (1996), Liew, Lim, and Kitipornchai (1997), Carrera (1999a, 1999b, 2002), Wu and Wu (2000), Basar and Omurtag (2000), Kadivar and Samani (2000), Birman, Griffin, and Knowle (2000), and Moussaoui and Benamar (2002).

9.1 INTRODUCTION

Because of their initial curvatures, curvilinear coordinates are needed for the modeling of shells, quadratic terms are introduced into the nonlinear governing equations of shells, and the symmetry of the governing equations, which exists in plate equations, is broken. Furthermore, loads on a shell are carried by a combination of bending and stretching actions because the extension and bending motions are coupled due to its initially curved shape. On the other hand, loads on a plate can be carried only by bending for some special loading and boundary conditions. Consequently, the strength of a shell depends on its shape as well as its material.

Thin-shell theories are engineering approximations that model deformations of a shell with a small value of h/a (thickness/radius of curvature) in terms of the kinematics of its reference plane. For a very thin shell, stretching and inplane shear deformations contribute most of the elastic energy, and hence it can be modeled as a membrane. However, for general anisotropic shells, the extension-bending coupling makes it necessary to include bending effects in the modeling of very thin shells. As the ratio h/a increases, the bending energy and even the transverse shear-strain energy may become significant, especially for high-frequency vibration modes.

9.1.1 Shell Theories

Motion of a shell is described by the deformation of its reference surface and shear-warpings of its cross-sections with respect to the deformed reference surface. Different approximations for the deformation of the reference surface result in different geometrically nonlinear shell theories, and different approximations for the warpings of cross-sections result in different shear-deformation theories.

Love (1944) established the foundation of the linear classical theory of thin shells by applying the Kirchhoff hypothesis and hence neglecting the transverse shear strains. In Love's shell theory, it is assumed that (1) the shell is thin, (2) the displacements and rotations are small, (3) the normals to the reference surface before deformation remain normal after deformation, and (4) the transverse normal stress is negligible. Hence, Love's shell theory is essentially an extension of Kirchhoff flat-plate theory and is often called the Kirchhoff-Love shell theory. In contrast to plate theories, there is no universally accepted shell formulation, even for small-deformation cases, because neglecting or simplifying the trapezoidal-edge effect and other secondary effects at different stages in the formulation results in different theories (Donnell, 1938; Love, 1944; Reissner, 1956; Vlasov, 1949; Sanders, 1959; Timoshenko, 1959; Naghdi and Berry, 1964; Flügge, 1973). Sanders (1959), Budiansky and Radkowski (1963), and Budiansky and Sanders (1963) developed a linear shell theory by using modified symmetric measures of the stress resultants, stress couples, and strains in a way that results in a reduction in the number of unknowns and satisfies both the basic conditions of classical mechanics and some considerations of symmetry. This theory is said to be the "best" in the sense that it fulfills some of the requirements of classical shell theory, including symmetry, the vanishing of all strains for any arbitrary rigid-body motion, establishment of the static-geometry analogy, non-violation of the constitutive equations of the sixth non-differential equation of equilibrium, and the derivation of the equations of motion in general tensor form for arbitrary shells. However, this theory is not the best in the sense of producing better accuracy than other shell theories. Koiter (1960) pointed out that, from the point of view of accuracy, there is little difference among these classical linear shell theories. To improve analytical solutions in the dynamic case, Seide (1975) included rotary inertias. But, Koiter (1960) pointed out, based on an order-of-magnitude analysis, that refinements of Love's first approximation theory of thin elastic shells are meaningless unless the effects of transverse shear and normal stresses are taken into account. For more analyses and comparisons among these theories, the reader is referred to the works of Koiter (1960), Dym (1973), Markus (1988), and Soldatos (1984, 1985).

Shear effects are quite significant for thick shells and high-frequency vibration modes of thin shells. Moreover, it is a well-known fact that shear deformations are significant in composite structures. Analytical studies (e.g., Bert and Reddy, 1982; Soldatos, 1984, 1985) and experimental results show that classical shell theories underpredict the deflections and overpredict the natural frequencies and critical buckling loads because they neglect the transverse shear strains; the effect of transverse shears increases with the vibration mode number. Hence, an adequate shell theory must account for accurate distributions of the transverse shear stresses.

To include transverse shear deformations in the modeling of a general anisotropic laminated shell, one needs to assume a displacement field for each layer because the material properties are not uniform through the thickness. But, the large number of governing equations needed in such a three-dimensional analysis makes the problem intractable. For thin laminated shells, lamination theories that employ a single expansion for the displacement field throughout the entire thickness are commonly used in the dynamic analysis of composite shells. There are several refined shear-deformable shell theories, such as the first-order (Reddy, 1983), third-order (Bhimaraddi, 1984; Reddy and Liu, 1985; Soldatos, 1986, 1987, 1991; Dennis and Palazotto, 1989, 1990), and other higher-order theories (Mirsky and Herrmann, 1957; Zukas and Vinson, 1971; Whitney and Sun, 1974; Librescu and Schmidt, 1988; Voyiadjis and Shi, 1991). The first-order shear theory (Reddy, 1983) and some higher-order shear theories (Mirsky and Herrmann, 1957; Whitney and Sun, 1974; Voyiadjis and Shi, 1991) require the use of shear-correction factors, which depend on the shear angles, complex stress state, and shape of the shell, and cannot be easily determined (Soldatos, 1986). On the other hand, the third-order shear theories (Bhimaraddi, 1984; Reddy and Liu, 1985; Soldatos, 1986, 1987, 1991; Dennis and Palazotto, 1989, 1990; Pai and Nayfeh, 1992b) do not need shear-correction factors. The numerical studies of Bhimaraddi (1984), Reddy and Liu (1985), Soldatos (1986), and Khdeir and Reddy (1990) showed that the third-order shear theory gives more accurate global responses than classical shell theories. However, the third-order theory does not account for the continuity of interlaminar shear stresses and the elastic coupling between two transverse shear deformations, and hence it cannot provide accurate predictions of interlaminar shear stresses, which play the key role in the failure analysis of laminated shells.

For moderately thick composite shells, Di Sciuva (1987) and Librescu and Schmidt (1991) used a piecewise linear displacement field to fulfill the interlaminar continuity of shear stresses, thereby keeping the number of equations of motion independent of the number of layers. Pai et al. (1993) extended the piecewise linear displacement field used by Di Sciuva (1987) by using quadratic and cubic interpolation functions to satisfy continuity of the interlaminar shear stresses, to accommodate free shear-stress conditions on the bonding surfaces, and to account for non-uniform distributions of the transverse shear stresses in each layer. Also, Pai and Nayfeh (1994b) extended the higher-order layerwise shear-deformable plate theory of Pai et al. for general surface structures by including the effect of initial curvatures. Moreover, Pai and Palazotto (2001) introduced a new concept of sublaminates into the higher-order layerwise shear-deformable plate theory (Pai et al., 1993) to account for the non-uniform distribution of the transverse normal stress and the change of shear-warping functions with spatial locations, vibration frequencies, and loading and boundary conditions.

The postbuckling strength of thin panels plays an important role in the design of aircraft structures because conventional aircraft structural elements are often designed to operate in the postbuckling range. Conventional nonlinear problems of elastic shells were mostly those dealing with the predictions of stability and postbuckling behavior and nonlinear analysis of panel flutter. However, highly flexible structures have recently received considerable attention because of their wide application in large space structures and advanced aerospace vehicles. Investigating the nonlinear dynamics (e.g., bifurcations and chaos) of such flexible structures undergoing large deformations but small strains requires the use of nonlinear theories, where geometric nonlinearities play an important role in accounting for the load-carrying capability of membrane forces and couple the extension, bending, torsion, and shearing deformations.

Nonlinear theories of shells were first derived for special shell geometries. Donnell (1933, 1938) and Mushtari (1938) proposed, independently, one of the most widely used simplifications in shell theories. They neglected the influence of inplane displacements on the bending strain energy, and they also neglected the inplane inertia and out-of-plane shear. They used these assumptions to derive the equations of motion for circular cylindrical shells. Vlasov (1951) generalized this approach to shells of any geometry and noted that these assumptions are particularly accurate for shallow shells. This became known as the shallow shell theory or the Donnell-Mushtari-Vlasov equations. Some nonlinear shell theories were derived by Reissner (1950, 1963), Novozhilov (1953), Sanders (1959, 1963), Leonard (1961), Naghdi and Nordgren (1963), Koiter (1966), Budiansky (1968), Marlowe (1968), Simmonds and Danielson (1972), Simmonds (1979), Pietraszkiewicz and Szwabowicz (1981), Schmidt (1984, 1985), and Libai and Simmonds (1988). The main differences among these nonlinear theories are the approximations used in the derivation of the kinematic relationships. Naghdi (1963a,b) pointed out that many of these nonlinear shell theories maintain some inconsistencies. Koiter and Simmonds (1972) discussed the problem of obtaining error bounds in shell theories.

In the derivation of a geometrically nonlinear shell theory, if a vector approach is adopted in the formulation, one applies Newton's second law to obtain the equilibrium equations in terms of stress resultants and moments and accounts for the geometric nonlinearities by considering the deformation of the reference surface. On the other hand, if an energy approach is adopted in the formulation, one uses the extended Hamilton's principle and calculus of variation to derive the governing equations, where nonlinear strain-displacement relations are used to account for the geometric nonlinearities. It is very often that the von Kármán strain-displacement relations are used to account for the geometric nonlinearities (e.g., Habip, 1965; Ebcioglu, 1970; Oshima, Seguchi, and Shindo, 1970; Brush and Almroth, 1975; Librescu, 1976; Reddy and Chandrashekhara, 1985; Librescu and Schmidt, 1988), and such nonlinearities are called the von Kármán nonlinearities. Although the von Kármán nonlinearities produce some quadratic and cubic terms in the governing equations of surface structures, they do not fully account for the geometric nonlinearities under moderate rotations. In the literature, the geometric nonlinearities have not received the attention they deserve in shell analyses. Pai and Nayfeh (1992b) extended the new concepts of local displacements, local stress and strain measures, and orthogonal virtual rotations underlying the development of their nonlinear beam theory (1992a) and nonlinear plate theory (1991c) to derive an asymptotic nonlinear theory for circular cylindrical shells, where a vector approach was used to derive fully nonlinear curvatures and mid-plane strains.

In the modeling of shells, Lamé parameters are usually used because of the use of curvilinear coordinates, and either the Newtonian approach or the energy approach is used in the formulation. However, the use of Lamé parameters makes the formulation of each shell problem very individual. Pai and Nayfeh (1994b) pointed out that both plates and shells are two-dimensional structures described by two orthogonal curvilinear (or rectilinear if rectangular plates) coordinates and one rectilinear coordinate perpendicular to the reference surface. Hence, the only difference in the modeling of plates and shells is the initial curvatures of the two coordinates on the reference surface. Therefore, the formulations of plates and shells can be simplified by using initial curvatures (instead of Lamé parameters) and can be unified into one with all possible curvatures included. Moreover, using the concepts of local displacements, local stress and strain measures, and orthogonal virtual rotations, Pai and Nayfeh (1994b) showed that the Newtonian and energy approaches are fully correlated even when large displacements and rotations are involved.

9.1.2 Nonlinear Vibrations of Shells

Methods for solving geometrically nonlinear shell problems can be classified into three types: purely numerical methods (e.g., Stricklin et al., 1971; Mente, 1973), purely analytical methods such as perturbation methods (e.g., Maewal, 1978; Simmonds, 1979), and a combination of analytical and numerical methods in which the partial-differential equations are discretized into a set of temporal nonlinear ordinary-differential equations (e.g., Nayfeh and Mook, 1979; Nayfeh, 1988, 2000). A popular approach is the Galerkin procedure in which the assumed spatial modes are used as weighting functions to minimize the residual over the spatial domain. A common practice is to truncate these equations to a finite number of modes and integrate the remaining equations numerically in the time domain. Mente (1973) studied the nonlinear response of cylindrical shells by integrating up to fifteen circumferential harmonics and three axial harmonics. However, many researchers retained only one or two modes (Lakis, Selmane, and Toledano, 1998; Abe, Kobayashi, and Yamada, 2000; J. Nayfeh and Rivieccio, 2000).

Chin and Nayfeh (1996b) studied the nonlinear vibrations of infinitely long circular cylindrical shells by directly attacking the governing partial-differential equations of motion and by attacking a two-mode Galerkin discretization of these equations. They found that attacking the two-mode discretized equations might produce erroneous results. Abe, Kobayashi, and Yamada (2001) studied the responses of cross-ply laminated shallow shells using an improved displacement function to overcome the short comings of the Galerkin approximation. Alhazza and Nayfeh (2001) studied the nonlinear response of a doubly-curved cross-ply laminated shallow shell with simply supported boundary conditions to a primary resonance of its fundamental mode using a finite-order Galerkin discretization. They investigated how the predicted response depends on the number of modes retained in the discretization. They found out that single-mode or two-mode approximations might lead to qualitative as well as quantitative errors. In another paper, Nayfeh and Alhazza (2001) investigated the dynamic instability and nonlinear vibrations of simply supported cross-ply doubly-curved shallow shells. They used a combination of a shooting technique and Floquet theory to calculate the periodic responses and their stability. They found out that a single-mode approximation may miss some important dynamics, such as period-doubling bifurcations.

The phenomenon of modal interaction might occur when the linear natural frequencies of some modes are commensurate or nearly commensurate. The first studies of modal interactions in the response of shells were initiated by McIvor (1962, 1966), Goodier and McIvor (1964), McIvor and Sonstegard (1966), and McIvor and Lovell (1968). They analyzed the responses of cylindrical and spherical shells to radial and nearly radial impulses, taking into account the coupling of a breathing mode and a flexural mode when their frequencies are in the ratio of two-to-one. They transformed their equations into a Mathieu-type equation and studied its stability regions. They numerically integrated the governing equations of motion and found that the energy is continuously exchanged between the resonant modes. Yasuda and Kushida (1984) studied theoretically and experimentally the axisymmetric response of shallow spherical shells. They studied the case of primary resonance of a higher flexural mode in the presence and absence of a two-to-one internal resonance. They analyzed the stability of periodic solutions and verified their analysis experimentally. Their theoretical analysis showed regions of almost periodic motions for some values of the excitation frequency. They experimentally observed periodic responses of the shell but did not report any experimental observations of almost periodic motions. Nayfeh and Raouf (1987a,b) used the method of multiple time scales to study the forced nonlinear response of infinitely long circular cylindrical shells to primary resonant excitations of the breathing mode and a flexural mode in the presence of a two-to-one internal resonance. They showed that, above a certain threshold of excitation, the amplitude of the excited mode saturates and spills over the extra input energy into the coupled flexural mode, which responds nonlinearly. They showed that the response can exhibit a Hopf bifurcation, resulting in amplitude- and phase-modulated motions. Raouf and Nayfeh (1990a) used the method of multiple scales to determine the response of a closed spherical shell to a primary-resonance radial load in the case of two-to-one internal resonance. They found periodic and periodically and chaotically modulated motions. Nayfeh, Raouf, and J. Nayfeh (1991) analyzed the response of a circular cylindrical shell to a harmonic radial load whose frequency is twice that of the breathing mode, which in turn is twice the frequencies of two orthogonal flexural modes.

The first study of the interaction between the driven and companion modes in thin structures was done by Evensen (1966). He used the method of averaging to study the influence of a one-to-one internal resonance on the inextensional forced flexural response of thin circular rings. He found that the response involves either a single bending mode or two coupled bending modes. He reported experimental observations of unsteady vibrations and verified them by using analog-computer simulations. Chen and Babcock (1975) used the Lindstédt-Poincaré technique to study the nonlinear response of finite cylindrical shells to harmonic excitations. They studied the interaction of the driven and companion modes. They also reported experimental observations of "nonstationary" responses "in which the amplitude drifts from one value to another." A similar approach was used by Bieniek, Fan, and Lackman (1966) to study the symmetric response of cylindrical shells.

Maewal (1981) studied the interaction between backward and forward traveling waves in rotating rings. Natsiavas (1995b) used the method of multiple scales to study the nonlinear response of a rotating ring to a parametric excitation originating from small periodic perturbations of the spin speed. Using a nonlinear inextensionality constraint, Maewal (1986b) studied the modal interaction between the driven and companion modes of axisymmetric shells. He used the method of multiple scales to derive a set of evolution equations for the amplitudes and phases of the modes. He noted that these equations have the same form as those derived earlier by Miles (1984a,b) for internally resonant surface waves in a circular cylinder. He integrated numerically these evolution equations and showed that, for certain ranges of the excitation frequency, the response is chaotically modulated.

Maganty and Bickford (1987) used the method of multiple scales to study nonlinear free vibrations of circular rings in the presence of a one-to-one internal resonance between an inplane mode and out-of-plane mode. They found a continuous exchange of energy between the coupled modes. In another investigation, they (1988) studied the forced response of the same system to a harmonic inplane or out-of-plane excitation. They showed that, for certain values of the excitation frequency, an inplane excitation can produce nonplanar oscillations and vice versa.

Raouf and Nayfeh (1990b) analyzed the nonlinear dynamic response of infinitely long circular cylindrical shells to a primary resonant excitation of the nth flexural mode. They used the method of multiple scales to attack directly the governing partial-differential equations, taking into account the interaction of the excited nth flexural mode with it companion mode and assuming modal damping. Chin and Nayfeh (1996b) corrected the solution of Raouf and Nayfeh and demonstrated the symmetry of the modulation equations. Also, Chin and Nayfeh (1996b) compared the results derived by attacking the governing partial-differential equations with those derived from a two-mode discretization. They showed that the discretization approach fails to account for the spatial variation at second order and completely misses the influence of the quadratic nonlinearity on the interaction coefficients. Chin and Nayfeh (2001)

566 SHELLS

considered the response of infinitely long circular cylindrical shells to a primary resonance of the breathing mode, taking into account a two-to-one internal resonance between the breathing mode and two orthogonal flexural modes.

9.2 LINEAR CLASSICAL SHELL THEORY

The main difference between plates and shells is that shells have initial curvatures, which make membrane forces play an important role in resisting external loads. Similar to plates, an infinitesimal shell element without warpings is a perfectly rigid straight parallelpiped having a finite length h (the thickness) and two infinitesimal side lengths dx and dy, where x and y are curvilinear coordinates. Classical shell theories are based on the Kirchhoff hypothesis.

9.2.1 Different Shell Geometries

Figures 9.2.1-9.2.6 show several different shells. For any shell, the undeformed geometry of its reference surface can be described by an orthogonal curvilinear coordinate system xyz with x and y being curvilinear axes on the undeformed reference surface and z being a rectilinear axis perpendicular to the undeformed reference surface. An inertial Cartesian coordinate system abc is also needed for reference. Moreover, $\mathbf{j}_1$, $\mathbf{j}_2$, and $\mathbf{j}_3$ are unit vectors along the x, y, and z axes; and $\mathbf{i}_a$, $\mathbf{i}_b$, and $\mathbf{i}_c$ are unit vectors along the a, b, and c axes. The base vectors of the curvilinear coordinate system xyz are related to the base vectors of the reference coordinate system abc by the transformation matrix $[T^x]$ as shown in (6.1.3).

Cylindrical Shells

As a first example, we consider the cylindrical shell shown in Figure 9.2.1. Cylindrical shells are also called translational shells because they can be formed by translating its generator. Because the position vector $\mathbf{R}$ of an arbitrary point on the generator is

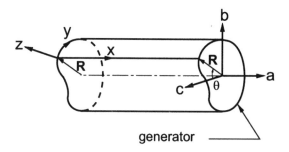

Fig. 9.2.1 A cylindrical shell.

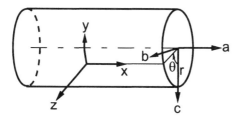

Fig. 9.2.2 A circular cylindrical shell.

given by
$$\mathbf{R} = B(\theta)\mathbf{i}_b + C(\theta)\mathbf{i}_c \quad (9.2.1a)$$
we have
$$dx = dx,\ dy = |d\mathbf{R}| = \sqrt{(B_\theta d\theta)^2 + (C_\theta d\theta)^2} = d\theta\sqrt{B_\theta^2 + C_\theta^2} \quad (9.2.1b)$$
where $B_\theta \equiv dB/d\theta$ and $C_\theta \equiv dC/d\theta$. Hence,
$$\mathbf{j}_1 = \mathbf{i}_a$$
$$\mathbf{j}_2 = \frac{\partial \mathbf{R}}{\partial y} = \frac{B_\theta}{\sqrt{B_\theta^2 + C_\theta^2}}\mathbf{i}_b + \frac{C_\theta}{\sqrt{B_\theta^2 + C_\theta^2}}\mathbf{i}_c \quad (9.2.2a)$$
and
$$\mathbf{j}_3 = \mathbf{j}_1 \times \mathbf{j}_2 = -\frac{C_\theta}{\sqrt{B_\theta^2 + C_\theta^2}}\mathbf{i}_b + \frac{B_\theta}{\sqrt{B_\theta^2 + C_\theta^2}}\mathbf{i}_c \quad (9.2.2b)$$

Then, one can substitute (9.2.2a) and (9.2.2b) into (6.1.3) to obtain $[T^x]$ and then use (6.1.7) to calculate the initial curvatures.

For the circular cylindrical shell shown in Figure 9.2.2,
$$\mathbf{R} = r\sin\theta\,\mathbf{i}_b + r\cos\theta\,\mathbf{i}_c \quad (9.2.3a)$$
where r is the radius of the shell. Then, it follows from (9.2.1b) and (9.2.2) that
$$dx = dx,\ dy = r\,d\theta \quad (9.2.3b)$$
$$\mathbf{j}_1 = \mathbf{i}_a,\ \mathbf{j}_2 = \cos\theta\,\mathbf{i}_b - \sin\theta\,\mathbf{i}_c,\ \mathbf{j}_3 = \sin\theta\,\mathbf{i}_b + \cos\theta\,\mathbf{i}_c \quad (9.2.3c)$$

Substituting (9.2.3c) into (6.1.3) to obtain $[T^x]$ and then using (6.1.7), we obtain the initial curvatures as
$$k_1^0 = k_{61}^0 = k_5^0 = k_{62}^0 = k_4^0 = 0,\ k_2^0 = \frac{1}{r} \quad (9.2.4)$$

Next, we use helical coordinates to describe the reference surface of the circular cylindrical shell shown in Figure 9.2.3a. It follows from Figure 9.2.3b that
$$\mathbf{R} = r\cos(\theta - \phi)\,\mathbf{i}_a + r\sin(\theta - \phi)\,\mathbf{i}_b + (r\theta\tan\psi + r\phi\cot\psi)\,\mathbf{i}_c \quad (9.2.5a)$$

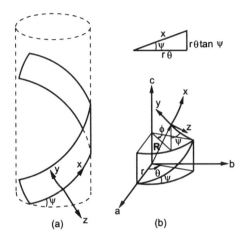

Fig. 9.2.3 A twisted circular cylindrical shell.

$$dx = \sqrt{(rd\theta)^2 + (rd\theta \tan \psi)^2} = \frac{r}{\cos \psi} d\theta$$
$$dy = \sqrt{(rd\phi)^2 + (rd\phi \cot \psi)^2} = \frac{r}{\sin \psi} d\phi \qquad (9.2.5b)$$

Hence,

$$\mathbf{j}_1 = \frac{\partial \mathbf{R}}{\partial x} = -\sin(\theta - \phi) \cos \psi \, \mathbf{i}_a + \cos(\theta - \phi) \cos \psi \, \mathbf{i}_b + \sin \psi \, \mathbf{i}_c$$
$$\mathbf{j}_2 = \frac{\partial \mathbf{R}}{\partial y} = \sin(\theta - \phi) \sin \psi \, \mathbf{i}_a - \cos(\theta - \phi) \sin \psi \, \mathbf{i}_b + \cos \psi \, \mathbf{i}_c$$
$$\mathbf{j}_3 = \cos(\theta - \phi) \, \mathbf{i}_a + \sin(\theta - \phi) \, \mathbf{i}_b \qquad (9.2.6)$$

Substituting (9.2.6) into (6.1.3) to obtain $[T^x]$ and then using (6.1.7), we obtain the initial curvatures as

$$k_4^0 = k_5^0 = 0, \; k_1^0 = \frac{\cos^2 \psi}{r}, \; k_2^0 = \frac{\sin^2 \psi}{r}, \; k_{61}^0 = k_{62}^0 = -\frac{\sin 2\psi}{2r} \qquad (9.2.7)$$

We note that k_{61}^0 and k_{62}^0 are non-zero, and they are equal because the surface is smooth.

Conical Shells

Next, we consider the conical shell shown in Figure 9.2.4. In this case,

$$\mathbf{R} = x \sin \alpha \cos \theta \, \mathbf{i}_a + x \sin \alpha \sin \theta \, \mathbf{i}_b + (C_0 - x \cos \alpha) \mathbf{i}_c \qquad (9.2.8a)$$

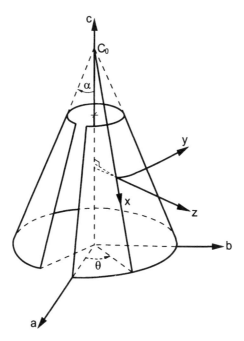

Fig. 9.2.4 A conical shell.

$$dx = dx, \quad dy = x \sin\alpha\, d\theta \tag{9.2.8b}$$

$$\mathbf{j}_1 = \frac{\partial \mathbf{R}}{\partial x} = \sin\alpha\cos\theta\, \mathbf{i}_a + \sin\alpha\sin\theta\, \mathbf{i}_b - \cos\alpha\, \mathbf{i}_c$$
$$\mathbf{j}_2 = \frac{\partial \mathbf{R}}{\partial y} = -\sin\theta\, \mathbf{i}_a + \cos\theta\, \mathbf{i}_b$$
$$\mathbf{j}_3 = \mathbf{j}_1 \times \mathbf{j}_2 = \cos\alpha\cos\theta\, \mathbf{i}_a + \cos\alpha\sin\theta\, \mathbf{i}_b + \sin\alpha\, \mathbf{i}_c \tag{9.2.8c}$$

Substituting (9.2.8c) into (6.1.3) to obtain $[T^x]$ and then using (6.1.7), we obtain the initial curvatures as

$$k_1^0 = k_{61}^0 = k_5^0 = k_{62}^0 = 0, \quad k_2^0 = \frac{1}{x\tan\alpha}, \quad k_4^0 = \frac{1}{x} \tag{9.2.9}$$

Shells of Revolution

Next, we consider the shell of revolution shown in Figure 9.2.5. Shells of revolution are also called rotational or axisymmetric shells because they are formed by rotating its generator. The position vector $\mathbf{R}$ of an arbitrary point on the reference surface is given by

$$\mathbf{R} = a\mathbf{i}_a + r\sin\theta\, \mathbf{i}_b + r\cos\theta\, \mathbf{i}_c \tag{9.2.10a}$$

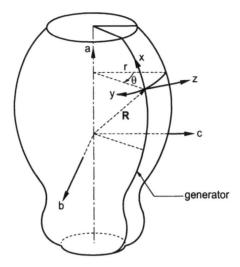

Fig. 9.2.5 A revolutional shell.

where $r = r(a)$. Moreover,

$$dx = \sqrt{da^2 + dr^2} = da\sqrt{1 + r_a^2}, \quad dy = rd\theta \qquad (9.2.10b)$$

where $r_a \equiv dr/da$. Consequently,

$$\mathbf{j}_1 = \frac{\partial \mathbf{R}}{\partial x} = \frac{1}{\sqrt{1 + r_a^2}}(\mathbf{i}_a + r_a \sin\theta\, \mathbf{i}_b + r_a \cos\theta\, \mathbf{i}_c)$$

$$\mathbf{j}_2 = \frac{\partial \mathbf{R}}{\partial y} = \cos\theta\, \mathbf{i}_b - \sin\theta\, \mathbf{i}_c$$

$$\mathbf{j}_3 = \mathbf{j}_1 \times \mathbf{j}_2 = \frac{1}{\sqrt{1 + r_a^2}}(-r_a \mathbf{i}_a + \sin\theta\, \mathbf{i}_b + \cos\theta\, \mathbf{i}_c) \qquad (9.2.11)$$

Substituting (9.2.11) into (6.1.3) to obtain $[T^x]$ and then using (6.1.7), one can obtain the initial curvatures.

Spherical Shells

Next, we consider the spherical shell shown in Figure 9.2.6. In this case,

$$\mathbf{R} = r \sin\theta \cos\phi\, \mathbf{i}_a + r \sin\theta \sin\phi\, \mathbf{i}_b + r \cos\theta\, \mathbf{i}_c \qquad (9.2.12a)$$

where r is the radius of the shell. Hence,

$$dx = rd\theta, \quad dy = r \sin\theta d\phi \qquad (9.2.12b)$$

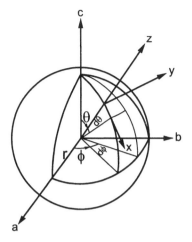

Fig. 9.2.6 A spherical shell.

Therefore,

$$\mathbf{j}_1 = \frac{\partial \mathbf{R}}{\partial x} = \cos\theta\cos\phi\,\mathbf{i}_a + \cos\theta\sin\phi\,\mathbf{i}_b - \sin\theta\,\mathbf{i}_c$$

$$\mathbf{j}_2 = \frac{\partial \mathbf{R}}{\partial y} = -\sin\phi\,\mathbf{i}_a + \cos\phi\,\mathbf{i}_b$$

$$\mathbf{j}_3 = \mathbf{j}_1 \times \mathbf{j}_2 = \sin\theta\cos\phi\,\mathbf{i}_a + \sin\theta\sin\phi\,\mathbf{i}_b + \cos\theta\,\mathbf{i}_c \quad (9.2.12c)$$

Substituting (9.2.12c) into (6.1.3) to obtain $[T^x]$ and then using (6.1.7), we obtain the initial curvatures as

$$k_{61}^0 = k_5^0 = k_{62}^0 = 0, \quad k_1^0 = \frac{1}{r}, \quad k_2^0 = \frac{1}{r}, \quad k_4^0 = \frac{1}{r\tan\theta} \quad (9.2.13)$$

We point out here that cylindrical and conical shells can be made of initially flat plates and hence are called developable structures. The Gaussian curvature k_G^0, which is defined as

$$k_G^0 = k_\alpha^0 k_\beta^0 \quad (9.2.14)$$

with k_α^0 and k_β^0 being the principal curvatures of these developable structures equal to zero.

9.2.2 Doubly-Curved Shell Theory

For general shells, the undeformed coordinates x and y are orthogonal curvilinear, as shown in Figure 6.1.1. We consider a general shell over the domain $0 \le x \le X$ and $0 \le y \le Y$ having a thickness h and the initial curvatures shown in (6.1.7). The

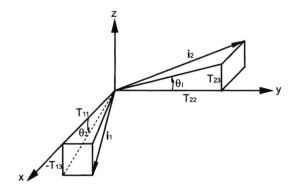

Fig. 9.2.7 The relation between small rotation angles and the T_{ij}.

displacement vector **D** of an arbitrary point on the observed shell element is

$$\mathbf{D} = u_1\mathbf{j}_1 + u_2\mathbf{j}_2 + u_3\mathbf{j}_3 = (u + z\theta_2)\mathbf{j}_1 + (v - z\theta_1)\mathbf{j}_2 + w\mathbf{j}_3 \qquad (9.2.15)$$

where u, v, and w are the displacement components of the observed reference point with respect to the xyz coordinate system and θ_1 and θ_2 are rotation angles of the deformed shell element with respect to the x and y axes, respectively. Because of non-zero initial curvatures, $\theta_1 \neq w_y$ and $\theta_2 \neq -w_x$. Moreover, because rotations are not vector quantities, the expressions for θ_1 and θ_2 depend on the sequence of rotations. However, in linear structural theories, θ_1 and θ_2 are assumed to be infinitesimal. Hence, it follows from Figure 9.2.7 and (6.2.7) that

$$\theta_1 = \tan^{-1}\frac{T_{23}}{T_{22}} = \tan^{-1}\frac{\hat{T}_{23}}{\hat{T}_{22}} = w_y - uk_{62}^0 - vk_2^0$$

$$\theta_2 = -\tan^{-1}\frac{T_{13}}{T_{11}} = -\tan^{-1}\frac{\hat{T}_{13}}{\hat{T}_{11}} = -w_x + uk_1^0 + vk_{61}^0 \qquad (9.2.16)$$

where only linear terms are retained. Differentiating (9.2.15) and using (9.2.16) and (6.1.6), we obtain

$$\frac{\partial \mathbf{D}}{\partial x} = (u_x + z\theta_{2x} - vk_5^0 + z\theta_1 k_5^0 + wk_1^0)\mathbf{j}_1$$

$$+ (v_x - z\theta_{1x} + uk_5^0 + z\theta_2 k_5^0 + wk_{61}^0)\mathbf{j}_2$$

$$+ (w_x - uk_1^0 - z\theta_2 k_1^0 - vk_{61}^0 + z\theta_1 k_{61}^0)\mathbf{j}_3$$

$$\frac{\partial \mathbf{D}}{\partial y} = (u_y + z\theta_{2y} - vk_4^0 + z\theta_1 k_4^0 + wk_{62}^0)\mathbf{j}_1$$

$$+ (v_y - z\theta_{1y} + uk_4^0 + z\theta_2 k_4^0 + wk_2^0)\mathbf{j}_2$$

$$+ (w_y - uk_{62}^0 - z\theta_2 k_{62}^0 - vk_2^0 + z\theta_1 k_2^0)\mathbf{j}_3$$

$$\frac{\partial \mathbf{D}}{\partial z} = \theta_2 \mathbf{j}_1 - \theta_1 \mathbf{j}_2 \qquad (9.2.17)$$

LINEAR CLASSICAL SHELL THEORY

It follows from (9.2.17) that the engineering strains ϵ_{ij} are

$$\epsilon_{11} = \frac{\partial \mathbf{D}}{\partial x} \cdot \mathbf{j}_1 = u_x - k_5^0 v + k_1^0 w + z(\theta_{2x} + k_5^0 \theta_1)$$

$$\epsilon_{22} = \frac{\partial \mathbf{D}}{\partial y} \cdot \mathbf{j}_2 = v_y + k_4^0 u + k_2^0 w + z(-\theta_{1y} + k_4^0 \theta_2)$$

$$\epsilon_{12} = \frac{\partial \mathbf{D}}{\partial x} \cdot \mathbf{j}_2 + \frac{\partial \mathbf{D}}{\partial y} \cdot \mathbf{j}_1$$

$$= u_y + v_x - k_4^0 v + k_5^0 u + k_6^0 w + z(\theta_{2y} - \theta_{1x} + k_4^0 \theta_1 + k_5^0 \theta_2)$$

$$\epsilon_{33} = 0, \quad \epsilon_{13} = z(k_{61}^0 \theta_1 - k_1^0 \theta_2), \quad \epsilon_{23} = z(k_2^0 \theta_1 - k_{62}^0 \theta_2) \qquad (9.2.18)$$

where it is assumed that $k_{61}^0 = k_{62}^0 = k_6^0/2$, which is true for smooth undeformed shells. Because the transverse shear deformations are assumed to be zero in (9.2.15), ϵ_{13} and ϵ_{23} should be zero; that is,

$$z(k_{61}^0 \theta_1 - k_1^0 \theta_2) = z(k_2^0 \theta_1 - k_{62}^0 \theta_2) = 0 \qquad (9.2.19)$$

Apparently (9.2.19) may not be satisfied at $z \neq 0$. Because of the initial curvatures, the undeformed side lengths at $z \neq 0$ are different from dx and dy, but dx and dy are used as the undeformed lengths in (9.2.18). This is the so-called trapezoidal-edge effect, which is discussed more later in this chapter.

Taking the time derivatives and variations of the displacement vector $\mathbf{D}$ in (9.2.15), we obtain variation of the kinetic energy δT as

$$\delta T = -\int_A \int_z \rho \ddot{\mathbf{D}} \cdot \delta \mathbf{D} \, dz \, dA$$

$$= -\int_A [(I_0 \ddot{u} + I_1 \ddot{\theta}_2)\delta u + (I_0 \ddot{v} - I_1 \ddot{\theta}_1)\delta v + I_0 \ddot{w}\delta w$$

$$+ (I_2 \ddot{\theta}_2 + I_1 \ddot{u})\delta\theta_2 + (I_2 \ddot{\theta}_1 - I_1 \ddot{v})\delta\theta_1] dA \qquad (9.2.20)$$

where A is the area of the reference surface and I_0, I_1, and I_2 are defined in (7.2.8).

It follows from (9.2.18), (9.2.19), and (9.2.16) that variation of the elastic energy $\delta \Pi$ is given by

$$\delta \Pi = \int_A \int_z (\sigma_{11} \delta\epsilon_{11} + \sigma_{22}\delta\epsilon_{22} + \sigma_{12}\delta\epsilon_{12}) dz \, dA$$

$$= \int_A \bigg(N_1 \delta u_x + N_6 \delta u_y + N_2 \delta v_y + N_6 \delta v_x$$

$$+ M_1 \delta\theta_{2x} - M_2 \delta\theta_{1y} + M_6 \delta\theta_{2y} - M_6 \delta\theta_{1x}$$

$$+ k_4^0 N_2 \delta u - k_4^0 N_6 \delta v + k_4^0 M_2 \delta\theta_2 + k_4^0 M_6 \delta\theta_1$$

$$+ k_5^0 N_6 \delta u - k_5^0 N_1 \delta v + k_5^0 M_6 \delta\theta_2 + k_5^0 M_1 \delta\theta_1$$

$$+ k_1^0 N_1 \delta w + k_2^0 N_2 \delta w + k_6^0 N_6 \delta w$$

$$+ Q_1 \delta\theta_2 - Q_2 \delta\theta_1 - Q_1 \delta\theta_2 + Q_2 \delta\theta_1 \bigg) dA$$

$$
\begin{aligned}
= -\int_A &\bigg[\left(N_{1x} + N_{6y} - k_4^0 N_2 - k_5^0 N_6 + k_1^0 Q_1 + k_{62}^0 Q_2\right) \delta u \\
&+ \left(N_{2y} + N_{6x} + k_4^0 N_6 + k_5^0 N_1 + k_{61}^0 Q_1 + k_2^0 Q_2\right) \delta v \\
&+ \left(Q_{1x} + Q_{2y} - k_1^0 N_1 - k_2^0 N_2 - k_6^0 N_6\right) \delta w \\
&+ \left(M_{1x} + M_{6y} - Q_1 - k_4^0 M_2 - k_5^0 M_6\right) \delta\theta_2 \\
&- \left(M_{2y} + M_{6x} - Q_2 + k_4^0 M_6 + k_5^0 M_1\right) \delta\theta_1 \bigg] dA \\
+ \int_y &\bigg[\left(N_1 + k_{62}^0 M_6\right) \delta u + \left(N_6 + k_2^0 M_6\right) \delta v \\
&+ \left(Q_1 + M_{6y}\right) \delta w + M_1 \delta\theta_2 \bigg]_{x=0}^{x=X} dy \\
+ \int_x &\bigg[\left(N_6 + k_1^0 M_6\right) \delta u + \left(N_2 + k_{61}^0 M_6\right) \delta v \\
&+ \left(Q_2 + M_{6x}\right) \delta w - M_2 \delta\theta_1 \bigg]_{y=0}^{y=Y} dx \\
- 2M_6 \delta w &\bigg|_{(x,y)=(X,0),(0,Y)}^{(x,y)=(0,0),(X,Y)}
\end{aligned} \tag{9.2.21}
$$

where the stress resultants and moments are the same as those in (7.2.10) and the introduced variables Q_1 and Q_2 are transverse shear intensities.

The reduced stress-strain relation can be obtained from (2.4.34), (2.4.35), and (9.2.18) with $\sigma_{33} = \epsilon_{13} = \epsilon_{23} = 0$ to be

$$
\left\{ \begin{array}{c} \sigma_{11} \\ \sigma_{22} \\ \sigma_{12} \end{array} \right\} = \left[\begin{array}{ccc} \tilde{Q}_{11} & \tilde{Q}_{12} & \tilde{Q}_{16} \\ \tilde{Q}_{12} & \tilde{Q}_{22} & \tilde{Q}_{26} \\ \tilde{Q}_{16} & \tilde{Q}_{26} & \tilde{Q}_{66} \end{array} \right] \left(\left\{ \begin{array}{c} u_x - k_5^0 v + k_1^0 w \\ v_y + k_4^0 u + k_2^0 w \\ u_y + v_x - k_4^0 v + k_5^0 u + k_6^0 w \end{array} \right\} + z \left\{ \begin{array}{c} \theta_{2x} + k_5^0 \theta_1 \\ -\theta_{1y} + k_4^0 \theta_2 \\ \theta_{2y} - \theta_{1x} + k_4^0 \theta_1 + k_5^0 \theta_2 \end{array} \right\} \right) \tag{9.2.22}
$$

Substituting (9.2.22) into (7.2.10), we obtain the internal forces and moments in terms of the mid-plane strains and curvatures as

$$
\left\{ \begin{array}{c} N_1 \\ N_2 \\ N_6 \\ M_1 \\ M_2 \\ M_6 \end{array} \right\} = \left[\begin{array}{cc} [A_{ij}] & [B_{ij}] \\ [B_{ij}] & [D_{ij}] \end{array} \right] \left\{ \begin{array}{c} u_x - k_5^0 v + k_1^0 w \\ v_y + k_4^0 u + k_2^0 w \\ u_y + v_x - k_4^0 v + k_5^0 u + k_6^0 w \\ \theta_{2x} + k_5^0 \theta_1 \\ -\theta_{1y} + k_4^0 \theta_2 \\ \theta_{2y} - \theta_{1x} + k_4^0 \theta_1 + k_5^0 \theta_2 \end{array} \right\} \tag{9.2.23}
$$

where the A_{ij}, B_{ij}, and D_{ij} are defined in (7.2.13).

Substituting (9.2.20), (9.2.21), and $\delta W_{nc} = 0$ into the extended Hamilton's principle (i.e., (7.2.1)) and then setting each of the coefficients of δu, δv, δw, $\delta\theta_1$, and $\delta\theta_2$ equal to zero, we obtain the following equations of motion:

$$N_{1x} + N_{6y} - k_4^0 N_2 - k_5^0 N_6 + k_1^0 Q_1 + k_{62}^0 Q_2 = I_0 \ddot{u} + I_1 \ddot{\theta}_2 \quad (9.2.24a)$$
$$N_{6x} + N_{2y} + k_4^0 N_6 + k_5^0 N_1 + k_{61}^0 Q_1 + k_2^0 Q_2 = I_0 \ddot{v} - I_1 \ddot{\theta}_1 \quad (9.2.24b)$$
$$Q_{1x} + Q_{2y} - k_1^0 N_1 - k_2^0 N_2 - k_6^0 N_6 = I_0 \ddot{w} \quad (9.2.24c)$$
$$-M_{2y} - M_{6x} - k_4^0 M_6 - k_5^0 M_1 + Q_2 = I_2 \ddot{\theta}_1 - I_1 \ddot{v} \quad (9.2.24d)$$
$$M_{1x} + M_{6y} - k_4^0 M_2 - k_5^0 M_6 - Q_1 = I_2 \ddot{\theta}_2 + I_1 \ddot{u} \quad (9.2.24e)$$

The boundary conditions are of the form: specify

Along $x = 0, X$:

$\delta u = 0$	or	$N_1 + k_{62}^0 M_6$
$\delta v = 0$	or	$N_6 + k_2^0 M_6$
$\delta w = 0$	or	$Q_1 + M_{6y}$
$\delta\theta_2 = 0$	or	M_1

Along $y = 0, Y$:

$\delta u = 0$	or	$N_6 + k_1^0 M_6$
$\delta v = 0$	or	$N_2 + k_{61}^0 M_6$
$\delta w = 0$	or	$Q_2 + M_{6x}$
$\delta\theta_1 = 0$	or	M_2

At $(x, y) = (0, 0), (0, Y), (X, 0), (X, Y)$:

$$\delta w = 0 \quad \text{or} \quad M_6 \quad (9.2.25)$$

Summing up $\mathbf{j}_1\times$ (9.2.24a), $\mathbf{j}_2\times$ (9.2.24b), $\mathbf{j}_3\times$ (9.2.24c), $\mathbf{j}_1\times$ (9.2.24d), and $\mathbf{j}_2\times$ (9.2.24e) and using (6.1.6), we obtain

$$\frac{\partial \mathbf{F}_\alpha}{\partial x} + \frac{\partial \mathbf{F}_\beta}{\partial y} = \mathbf{I}_F \quad (9.2.26a)$$

$$\frac{\partial \mathbf{M}_\alpha}{\partial x} + \frac{\partial \mathbf{M}_\beta}{\partial y} + \mathbf{j}_1 \times \mathbf{F}_\alpha + \mathbf{j}_2 \times \mathbf{F}_\beta = \mathbf{I}_M \quad (9.2.26b)$$

where

$$\mathbf{F}_\alpha = N_1\mathbf{j}_1 + N_6\mathbf{j}_2 + Q_1\mathbf{j}_3, \quad \mathbf{F}_\beta = N_6\mathbf{j}_1 + N_2\mathbf{j}_2 + Q_2\mathbf{j}_3$$
$$\mathbf{M}_\alpha = -M_6\mathbf{j}_1 + M_1\mathbf{j}_2, \quad \mathbf{M}_\beta = -M_2\mathbf{j}_1 + M_6\mathbf{j}_2$$
$$\mathbf{I}_F = (I_0\ddot{u} + I_1\ddot{\theta}_2)\mathbf{j}_1 + (I_0\ddot{v} - I_1\ddot{\theta}_1)\mathbf{j}_2 + I_0\ddot{w}\mathbf{j}_3$$
$$\mathbf{I}_M = (I_2\ddot{\theta}_1 - I_1\ddot{v})\mathbf{j}_1 + (I_2\ddot{\theta}_2 + I_1\ddot{u})\mathbf{j}_2 \quad (9.2.27)$$

One can see from (9.2.27) that Q_1 and Q_2 are along the $\mathbf{j}_3$-direction and hence represent the transverse shear intensities. Equations (9.2.26a,b) can be directly obtained using the Newtonian approach and the stress resultants and moments. However, the boundary conditions cannot be easily obtained using the Newtonian approach, especially the corner condition. We note that the coefficient of $\mathbf{j}_3$ in (9.2.26b) is

$$N_6 - N_6 + k_1^0 M_6 - k_2^0 M_6 + k_{62}^0 M_2 - k_{61}^0 M_1 = 0 \qquad (9.2.28)$$

which is not obtained in this linear formulation. In Section 9.4, we show that (9.2.28) is a statement of the balance of moments about the z axis.

Because the transverse shear deformations and stresses are assumed to be zero in such a classical shell theory, the transverse shear intensities Q_1 and Q_2 cannot be defined as stress resultants and need to be obtained from the governing equations (9.2.24d,e) as

$$Q_2 = M_{2y} + M_{6x} + k_4^0 M_6 + k_5^0 M_1 + I_2 \ddot{\theta}_1 - I_1 \ddot{v} \qquad (9.2.29a)$$

$$Q_1 = M_{1x} + M_{6y} - k_4^0 M_2 - k_5^0 M_6 - I_2 \ddot{\theta}_2 - I_1 \ddot{u} \qquad (9.2.29b)$$

Hence, only the three governing equations (9.2.24a-c) need to be solved with the boundary conditions (9.2.25).

9.2.3 Circular Cylindrical Shell Theory

For circular cylindrical shells, we recall from (9.2.3b) and (9.2.4) that

$$dy = a d\theta, \; k_1^0 = k_{61}^0 = k_5^0 = k_{62}^0 = k_4^0 = 0, \; k_2^0 = \frac{1}{a} \qquad (9.2.30a)$$

where a denotes the radius of the cylinder. Hence, (9.2.16) and (9.2.29a,b) become

$$\theta_1 = w_y - \frac{1}{a} v, \; \theta_2 = -w_x \qquad (9.2.30b)$$

$$Q_2 = M_{2y} + M_{6x} + I_2 \left(\ddot{w}_y - \frac{1}{a} \ddot{v} \right) - I_1 \ddot{v} \qquad (9.2.30c)$$

$$Q_1 = M_{1x} + M_{6y} + I_2 \ddot{w}_x - I_1 \ddot{u} \qquad (9.2.30d)$$

Also, the strains in (9.2.18) reduce to

$$\epsilon_{11} = u_x - z w_{xx}$$

$$\epsilon_{22} = v_y + \frac{1}{a} w - z \left(w_{yy} - \frac{1}{a} v_y \right)$$

$$\epsilon_{12} = u_y + v_x - z \left(2 w_{xy} - \frac{1}{a} v_x \right) \qquad (9.2.31)$$

We note that $w_{xy} = w_{yx} = w_{x\theta}/a = w_{\theta x}/a$ for this case. Also, (9.2.23) reduces to

$$\begin{Bmatrix} N_1 \\ N_2 \\ N_6 \\ M_1 \\ M_2 \\ M_6 \end{Bmatrix} = \begin{bmatrix} [A_{ij}] & [B_{ij}] \\ [B_{ij}] & [D_{ij}] \end{bmatrix} \begin{Bmatrix} u_x \\ v_y + w/a \\ u_y + v_x \\ -w_{xx} \\ -w_{yy} + v_y/a \\ -2w_{xy} + v_x/a \end{Bmatrix} \qquad (9.2.32)$$

Substituting (9.2.30) into (9.2.24) yields the following equations of motion:

$$N_{1x} + N_{6y} = I_0 \ddot{u} - I_1 \ddot{w}_x \tag{9.2.33a}$$

$$N_{6x} + N_{2y} + \frac{1}{a}(M_{6x} + M_{2y}) = \left(I_0 + \frac{2}{a}I_1 + \frac{1}{a^2}I_2\right)\ddot{v} - \left(I_1 + \frac{1}{a}I_2\right)\ddot{w}_y \tag{9.2.33b}$$

$$M_{1xx} + 2M_{6xy} + M_{2yy} - \frac{1}{a}N_2 = I_0 \ddot{w} + (I_1\ddot{u} - I_2\ddot{w}_x)_x$$
$$+ \left(I_1\ddot{v} - I_2\ddot{w}_y + \frac{1}{a}I_2\ddot{v}\right)_y \tag{9.2.33c}$$

where the dependence of the N_i and M_i on u, v, and w is given in (9.2.32). The boundary conditions are of the form: specify

Along $x = 0, X$:

$\delta u = 0$	or	N_1
$\delta v = 0$	or	$N_6 + M_6/a$
$\delta w = 0$	or	$Q_1 + M_{6y}$
$\delta w_x = 0$	or	M_1

Along $y = 0, Y$:

$\delta u = 0$	or	N_6
$\delta v = 0$	or	N_2
$\delta w = 0$	or	$Q_2 + M_{6x}$
$\delta(w_y - v/a) = 0$	or	M_2

At $(x, y) = (0, 0), (0, Y), (X, 0), (X, Y)$:

$$\delta w = 0 \quad \text{or} \quad M_6 \tag{9.2.34}$$

We point out here again that the trapezoidal-edge effect is neglected, which is discussed in Section 9.5.4.

9.3 LINEAR SHEAR-DEFORMABLE SHELL THEORIES

9.3.1 Formulation for General Shells

We add shear deformations to the displacement vector **D** in (9.2.15) as

$$\mathbf{D} = u_1 \mathbf{j}_1 + u_2 \mathbf{j}_2 + u_3 \mathbf{j}_3 = (u + z\theta_2 + g\gamma_5)\mathbf{j}_1 + (v - z\theta_1 + g\gamma_4)\mathbf{j}_2 + w\mathbf{j}_3 \tag{9.3.1}$$

where g is the shear-warping function, u, v, and w are the displacement components of the observed reference point with respect to the xyz coordinate system, θ_1 and θ_2 are rotation angles of the normal of the deformed reference plane with respect to the *x* and *y* axes, respectively, and γ_4 and γ_5 are extra rotations of the shell element due to shear deformations. Because θ_1 and θ_2 are the rotations of the normal, (9.2.16) is still valid. Differentiating (9.3.1) and using (9.2.16) and (6.1.6), we obtain

$$\frac{\partial \mathbf{D}}{\partial x} = (u_x + z\theta_{2x} - vk_5^0 + z\theta_1 k_5^0 + wk_1^0 + g\gamma_{5x} - gk_5^0 \gamma_4)\mathbf{j}_1$$
$$+ (v_x - z\theta_{1x} + uk_5^0 + z\theta_2 k_5^0 + wk_{61}^0 + g\gamma_{4x} + gk_5^0 \gamma_5)\mathbf{j}_2$$
$$+ (w_x - uk_1^0 - z\theta_2 k_1^0 - vk_{61}^0 + z\theta_1 k_{61}^0 - gk_1^0 \gamma_5 - gk_{61}^0 \gamma_4)\mathbf{j}_3$$

$$\frac{\partial \mathbf{D}}{\partial y} = (u_y + z\theta_{2y} - vk_4^0 + z\theta_1 k_4^0 + wk_{62}^0 + g\gamma_{5y} - gk_4^0 \gamma_4)\mathbf{j}_1$$
$$+ (v_y - z\theta_{1y} + uk_4^0 + z\theta_2 k_4^0 + wk_2^0 + g\gamma_{4y} + gk_4^0 \gamma_5)\mathbf{j}_2$$
$$+ (w_y - uk_{62}^0 - z\theta_2 k_{62}^0 - vk_2^0 + z\theta_1 k_2^0 - gk_{62}^0 \gamma_5 - gk_2^0 \gamma_4)\mathbf{j}_3$$

$$\frac{\partial \mathbf{D}}{\partial z} = (\theta_2 + g_z \gamma_5)\mathbf{j}_1 + (-\theta_1 + g_z \gamma_4)\mathbf{j}_2 \quad (9.3.2)$$

Using (9.3.2) and (9.2.16) and (9.2.19), we obtain the engineering strains ϵ_{ij} as

$$\epsilon_{11} = \frac{\partial \mathbf{D}}{\partial x} \cdot \mathbf{j}_1 = u_x - k_5^0 v + k_1^0 w + z(\theta_{2x} + k_5^0 \theta_1) + g(\gamma_{5x} - k_5^0 \gamma_4)$$

$$\epsilon_{22} = \frac{\partial \mathbf{D}}{\partial y} \cdot \mathbf{j}_2 = v_y + k_4^0 u + k_2^0 w + z(-\theta_{1y} + k_4^0 \theta_2) + g(\gamma_{4y} + k_4^0 \gamma_5)$$

$$\epsilon_{12} = \frac{\partial \mathbf{D}}{\partial x} \cdot \mathbf{j}_2 + \frac{\partial \mathbf{D}}{\partial y} \cdot \mathbf{j}_1$$
$$= u_y + v_x - k_4^0 v + k_5^0 u + k_6^0 w + z(\theta_{2y} - \theta_{1x} + k_4^0 \theta_1 + k_5^0 \theta_2)$$
$$+ g(\gamma_{4x} + \gamma_{5y} + k_5^0 \gamma_5 - k_4^0 \gamma_4)$$

$$\epsilon_{13} = g_z \gamma_5 - g(k_1^0 \gamma_5 + k_{61}^0 \gamma_4)$$
$$\epsilon_{23} = g_z \gamma_4 - g(k_{62}^0 \gamma_5 + k_2^0 \gamma_4)$$
$$\epsilon_{33} = 0 \quad (9.3.3)$$

Taking the time derivatives and variation of **D** in (9.3.1), we obtain variation of the kinetic energy as

$$\delta T = -\int_A \int_z \rho \ddot{\mathbf{D}} \cdot \delta \mathbf{D} \, dz \, dA$$
$$= -\int_A \Big[(I_0 \ddot{u} + I_1 \ddot{\theta}_2 + I_3 \ddot{\gamma}_5) \delta u + (I_0 \ddot{v} - I_1 \ddot{\theta}_1 + I_3 \ddot{\gamma}_4) \delta v + I_0 \ddot{w} \delta w$$
$$+ (I_2 \ddot{\theta}_2 + I_1 \ddot{u} + I_4 \ddot{\gamma}_5) \delta \theta_2 + (I_2 \ddot{\theta}_1 - I_1 \ddot{v} - I_4 \ddot{\gamma}_4) \delta \theta_1$$
$$+ (I_3 \ddot{u} + I_4 \ddot{\theta}_2 + I_5 \ddot{\gamma}_5) \delta \gamma_5 + (I_3 \ddot{v} - I_4 \ddot{\theta}_1 + I_5 \ddot{\gamma}_4) \delta \gamma_4 \Big] dA \quad (9.3.4)$$

where I_0, I_1, and I_2 are defined in (7.2.8) and I_3, I_4, and I_5 are defined in (7.3.6).

Using (9.3.3), we obtain variation of the elastic energy as

$$\delta \Pi = \int_A \int_z (\sigma_{11} \delta \epsilon_{11} + \sigma_{22} \delta \epsilon_{22} + \sigma_{12} \delta \epsilon_{12} + \sigma_{13} \delta \epsilon_{13} + \sigma_{23} \delta \epsilon_{23}) dz \, dA$$

$$= \int_A \bigg(N_1 \delta u_x + N_6 \delta u_y + N_2 \delta v_y + N_6 \delta v_x$$
$$+ M_1 \delta \theta_{2x} - M_2 \delta \theta_{1y} + M_6 \delta \theta_{2y} - M_6 \delta \theta_{1x}$$
$$+ k_4^0 N_2 \delta u - k_4^0 N_6 \delta v + k_4^0 M_2 \delta \theta_2 + k_4^0 M_6 \delta \theta_1$$
$$+ k_5^0 N_6 \delta u - k_5^0 N_1 \delta v + k_5^0 M_6 \delta \theta_2 + k_5^0 M_1 \delta \theta_1$$
$$+ k_1^0 N_1 \delta w + k_2^0 N_2 \delta w + k_6^0 N_6 \delta w$$
$$+ (q_2 - m_1 k_5^0 - m_6 k_4^0 - s_1 k_{61}^0 - s_2 k_2^0) \delta \gamma_4 + m_6 \delta \gamma_{4x} + m_2 \delta \gamma_{4y}$$
$$+ (q_1 + m_2 k_4^0 + m_6 k_5^0 - s_1 k_1^0 - s_2 k_{62}^0) \delta \gamma_5 + m_6 \delta \gamma_{5y} + m_1 \delta \gamma_{5x}$$
$$+ Q_1 \delta \theta_2 - Q_2 \delta \theta_1 - Q_1 \delta \theta_2 + Q_2 \delta \theta_1 \bigg) dA$$

$$= -\int_A \bigg[(N_{1x} + N_{6y} - k_4^0 N_2 - k_5^0 N_6 + k_1^0 Q_1 + k_{62}^0 Q_2) \delta u$$
$$+ (N_{2y} + N_{6x} + k_4^0 N_6 + k_5^0 N_1 + k_{61}^0 Q_1 + k_2^0 Q_2) \delta v$$
$$+ (Q_{1x} + Q_{2y} - k_1^0 N_1 - k_2^0 N_2 - k_6^0 N_6) \delta w$$
$$+ (M_{1x} + M_{6y} - Q_1 - k_4^0 M_2 - k_5^0 M_6) \delta \theta_2$$
$$- (M_{2y} + M_{6x} - Q_2 + k_4^0 M_6 + k_5^0 M_1) \delta \theta_1$$
$$- (m_{6x} + m_{2y} + m_1 k_5^0 + m_6 k_4^0 - q_2 + s_1 k_{61}^0 + s_2 k_2^0) \delta \gamma_4$$
$$- (m_{1x} + m_{6y} - m_2 k_4^0 - m_6 k_5^0 - q_1 + s_1 k_1^0 + s_2 k_{62}^0) \delta \gamma_5 \bigg] dA$$

$$+ \int_y \bigg[(N_1 + k_{62}^0 M_6) \delta u + (N_6 + k_2^0 M_6) \delta v + (Q_1 + M_{6y}) \delta w$$
$$+ M_1 \delta \theta_2 + m_1 \delta \gamma_5 + m_6 \delta \gamma_4 \bigg]_{x=0}^{x=X} dy$$

$$+ \int_x \bigg[(N_6 + k_1^0 M_6) \delta u + (N_2 + k_{61}^0 M_6) \delta v + (Q_2 + M_{6x}) \delta w$$
$$- M_2 \delta \theta_1 + m_2 \delta \gamma_4 + m_6 \delta \gamma_5 \bigg]_{y=0}^{y=Y} dx$$

$$- 2 M_6 \delta w \bigg|_{(x,y)=(X,0),(0,Y)}^{(x,y)=(0,0),(X,Y)} \qquad (9.3.5)$$

where the stress resultants and moments are defined in (7.2.10) and (7.3.8) and

$$s_1 \equiv \int_z g \sigma_{13} dz$$
$$s_2 \equiv \int_z g \sigma_{23} dz \qquad (9.3.6)$$

The introduced variables Q_1 and Q_2 are the transverse shear intensities.

The reduced stress-strain relations are obtained from (2.4.34), (2.4.35), and (9.3.3) with $\sigma_{33} = 0$ to be

$$\left\{\begin{array}{c}\sigma_{11}\\ \sigma_{22}\\ \sigma_{12}\end{array}\right\} = \left[\begin{array}{ccc}\tilde{Q}_{11} & \tilde{Q}_{12} & \tilde{Q}_{16}\\ \tilde{Q}_{12} & \tilde{Q}_{22} & \tilde{Q}_{26}\\ \tilde{Q}_{16} & \tilde{Q}_{26} & \tilde{Q}_{66}\end{array}\right]\left(\left\{\begin{array}{c}u_x - k_5^0 v + k_1^0 w\\ v_y + k_4^0 u + k_2^0 w\\ u_y + v_x - k_4^0 v + k_5^0 u + k_6^0 w\end{array}\right\}\right.$$

$$+z\left\{\begin{array}{c}\theta_{2x} + k_5^0\theta_1\\ -\theta_{1y} + k_4^0\theta_2\\ \theta_{2y} - \theta_{1x} + k_4^0\theta_1 + k_5^0\theta_2\end{array}\right\} + g\left.\left\{\begin{array}{c}\gamma_{5x} - k_5^0\gamma_4\\ \gamma_{4y} + k_4^0\gamma_5\\ \gamma_{4x} + \gamma_{5y} + k_5^0\gamma_5 - k_4^0\gamma_4\end{array}\right\}\right) \quad (9.3.7a)$$

$$\left\{\begin{array}{c}\sigma_{23}\\ \sigma_{13}\end{array}\right\} = \left[\begin{array}{cc}\overline{Q}_{44} & \overline{Q}_{45}\\ \overline{Q}_{45} & \overline{Q}_{55}\end{array}\right]\left(g_z\left\{\begin{array}{c}\gamma_4\\ \gamma_5\end{array}\right\} - g\left\{\begin{array}{c}k_{62}^0\gamma_5 + k_2^0\gamma_4\\ k_1^0\gamma_5 + k_{61}^0\gamma_4\end{array}\right\}\right) \quad (9.3.7b)$$

Substituting (9.3.7) into (7.2.10), (7.3.8), and (9.3.6), we obtain the internal forces and moments in terms of the mid-plane strains and curvatures as

$$\left\{\begin{array}{c}N_1\\ N_2\\ N_6\\ M_1\\ M_2\\ M_6\\ m_1\\ m_2\\ m_6\end{array}\right\} = \left[\begin{array}{ccc}[A_{ij}] & [B_{ij}] & [E_{ij}]\\ [B_{ij}] & [D_{ij}] & [F_{ij}]\\ [E_{ij}] & [F_{ij}] & [H_{ij}]\end{array}\right]\left\{\begin{array}{c}u_x - k_5^0 v + k_1^0 w\\ v_y + k_4^0 u + k_2^0 w\\ u_y + v_x - k_4^0 v + k_5^0 u + k_6^0 w\\ \theta_{2x} + k_5^0\theta_1\\ -\theta_{1y} + k_4^0\theta_2\\ \theta_{2y} - \theta_{1x} + k_4^0\theta_1 + k_5^0\theta_2\\ \gamma_{5x} - k_5^0\gamma_4\\ \gamma_{4y} + k_4^0\gamma_5\\ \gamma_{4x} + \gamma_{5y} + k_5^0\gamma_5 - k_4^0\gamma_4\end{array}\right\}$$

(9.3.8a)

$$\left\{\begin{array}{c}q_2\\ q_1\\ s_2\\ s_1\end{array}\right\} = \left[\begin{array}{cc}[C_{ij}] & [G_{ij}]\\ [G_{ij}] & [H_{ij}]\end{array}\right]\left\{\begin{array}{c}\gamma_4\\ \gamma_5\\ k_{62}^0\gamma_5 + k_2^0\gamma_4\\ k_1^0\gamma_5 + k_{61}^0\gamma_4\end{array}\right\} \quad (9.3.8b)$$

where the A_{ij}, B_{ij}, and D_{ij} are defined in (7.2.13), the E_{ij}, F_{ij}, and H_{ij}, $i, j = 1, 2, 6$, are defined in (7.3.11a), the C_{ij} are defined in (7.3.11b), and

$$G_{ij} = \sum_{k=1}^{N}\int_{z_k}^{z_{k+1}}\overline{Q}_{ij}^{(k)}g_z g\,dz, \quad H_{ij} = \sum_{k=1}^{N}\int_{z_k}^{z_{k+1}}\overline{Q}_{ij}^{(k)}g^2\,dz, \quad i, j = 4, 5 \quad (9.3.9)$$

Substituting (9.3.4), (9.3.5), and $\delta W_{nc} = 0$ into the extended Hamilton's principle (i.e., (7.2.1)) and setting each of the coefficients of δu, δv, δw, $\delta\gamma_4$, $\delta\gamma_5$, $\delta\theta_1$, and $\delta\theta_2$ equal to zero, we obtain the following equations of motion:

$$N_{1x} + N_{6y} - k_4^0 N_2 - k_5^0 N_6 + k_1^0 Q_1 + k_{62}^0 Q_2 = I_0\ddot{u} + I_1\ddot{\theta}_2 \quad (9.3.10a)$$

$$N_{6x} + N_{2y} + k_4^0 N_6 + k_5^0 N_1 + k_{61}^0 Q_1 + k_2^0 Q_2 = I_0\ddot{v} - I_1\ddot{\theta}_1 \quad (9.3.10b)$$

$$Q_{1x} + Q_{2y} - k_1^0 N_1 - k_2^0 N_2 - k_6^0 N_6 = I_0\ddot{w} \quad (9.3.10c)$$

$$m_{6x} + m_{2y} + m_1 k_5^0 + m_6 k_4^0 - q_2 + s_1 k_{61}^0 + s_2 k_2^0 = I_5\ddot{\gamma}_4 + I_3\ddot{v} - I_4\ddot{\theta}_1 \quad (9.3.10d)$$

$$m_{1x} + m_{6y} - m_2 k_4^0 - m_6 k_5^0 - q_1 + s_1 k_1^0 + s_2 k_{62}^0 = I_5\ddot{\gamma}_5 + I_3\ddot{u} + I_4\ddot{\theta}_2 \quad (9.3.10e)$$

$$-M_{2y} - M_{6x} - k_4^0 M_6 - k_5^0 M_1 + Q_2 = I_2 \ddot{\theta}_1 - I_1 \ddot{v} \qquad (9.3.10\text{f})$$

$$M_{1x} + M_{6y} - k_4^0 M_2 - k_5^0 M_6 - Q_1 = I_2 \ddot{\theta}_2 + I_1 \ddot{u} \qquad (9.3.10\text{g})$$

The boundary conditions are of the form: specify

Along $x = 0, X$:

$$\begin{array}{lll} \delta u = 0 & \text{or} & N_1 + k_{62}^0 M_6 \\ \delta v = 0 & \text{or} & N_6 + k_2^0 M_6 \\ \delta w = 0 & \text{or} & Q_1 + M_{6y} \\ \delta \theta_2 = 0 & \text{or} & M_1 \\ \delta \gamma_4 = 0 & \text{or} & m_6 \\ \delta \gamma_5 = 0 & \text{or} & m_1 \end{array}$$

Along $y = 0, Y$:

$$\begin{array}{lll} \delta u = 0 & \text{or} & N_6 + k_1^0 M_6 \\ \delta v = 0 & \text{or} & N_2 + k_{61}^0 M_6 \\ \delta w = 0 & \text{or} & Q_2 + M_{6x} \\ \delta \theta_1 = 0 & \text{or} & M_2 \\ \delta \gamma_4 = 0 & \text{or} & m_2 \\ \delta \gamma_5 = 0 & \text{or} & m_6 \end{array}$$

At $(x, y) = (0, 0), (0, Y), (X, 0), (X, Y)$:

$$\delta w = 0 \qquad \text{or} \qquad M_6 \qquad (9.3.11)$$

Because (9.3.10a-c,f,g) are the same as (9.2.24a-e), (9.2.26), (9.2.28), and (9.2.29) are also valid here. Hence, (9.3.10a-e) are the five governing equations.

9.3.2 Equations of Motion for Different Shells

Equations (9.3.10) and (9.3.11) are valid for any shell geometry, but different shell geometries have different initial curvatures. To obtain equations of motion for a specific shell, one only needs to substitute the initial curvatures of the specific shell into (9.3.10) and (9.3.11) and notes that w_{xy} may be different from w_{yx}, as shown in Section 7.2.3. If the differential area $dA = dxdy$ changes with x and/or y, one needs to account for the area effect as shown in (7.2.52).

9.3.3 Shear-Warping Functions

Shear-warping functions are the same as those discussed in Section 7.3.3. Shear-warping functions that account for shear coupling effects are discussed in Section 9.6.

9.4 NONLINEAR CLASSICAL THEORY FOR DOUBLY-CURVED SHELLS

We consider the general shell shown in Figure 6.1.1. Because of the use of Kirchhoff hypothesis in the classical shell theory, there is no shear-warping and the displacement vector **D** of an arbitrary point on an infinitesimal shell element (a straight line having a length h) is given by

$$\mathbf{D} = u\mathbf{j}_1 + v\mathbf{j}_2 + w\mathbf{j}_3 + z\mathbf{i}_3 - z\mathbf{j}_3 \qquad (9.4.1)$$

Taking the variation of **D** and using (6.3.1a) we obtain

$$\delta\mathbf{D} = \mathbf{j}_1\delta u + \mathbf{j}_2\delta v + \mathbf{j}_3\delta w + z(\mathbf{i}_1\delta\theta_2 - \mathbf{i}_2\delta\theta_1) \qquad (9.4.2a)$$

Taking the time derivatives of **D** and using (6.2.11a) yields

$$\ddot{\mathbf{D}} = \ddot{u}\mathbf{j}_1 + \ddot{v}\mathbf{j}_2 + \ddot{w}\mathbf{j}_3 + z(\ddot{T}_{31}\mathbf{j}_1 + \ddot{T}_{32}\mathbf{j}_2 + \ddot{T}_{33}\mathbf{j}_3) \qquad (9.4.2b)$$

Using (9.4.2a,b) we obtain variation of the kinetic energy as

$$\delta T = -\int_z \int_A \rho \ddot{\mathbf{D}} \cdot \delta \mathbf{D} \, dA \, dz$$
$$= -\int_A (A_u\delta u + A_v\delta v + A_w\delta w + A_{\theta_1}\delta\theta_1 + A_{\theta_2}\delta\theta_2) dA \qquad (9.4.3)$$

where

$$A_u = I_0\ddot{u} + I_1\ddot{T}_{31}$$
$$A_v = I_0\ddot{v} + I_1\ddot{T}_{32}$$
$$A_w = I_0\ddot{w} + I_1\ddot{T}_{33}$$
$$A_{\theta_1} = -I_1(\ddot{u}T_{21} + \ddot{v}T_{22} + \ddot{w}T_{23}) - I_2(\ddot{T}_{31}T_{21} + \ddot{T}_{32}T_{22} + \ddot{T}_{33}T_{23})$$
$$A_{\theta_2} = I_1(\ddot{u}T_{11} + \ddot{v}T_{12} + \ddot{w}T_{13}) + I_2(\ddot{T}_{31}T_{11} + \ddot{T}_{32}T_{12} + \ddot{T}_{33}T_{13}) \quad (9.4.4a)$$

$$\{I_0, I_1, I_2\} \equiv \int_z \rho\{1, z, z^2\} dz \qquad (9.4.4b)$$

We note that (9.4.4a,b) are the same as (7.6.4a,b), but the T_{ij} here include the influence of the initial curvatures through (6.2.7) and (6.2.11a).

To calculate variation of the strain energy, we need to obtain the strain-displacement relations. We assume that the local displacements u_1, u_2, and u_3 (with respect to the $\xi\eta\zeta$ coordinate system) of an arbitrary point on the shell element have the form

$$u_1(x, y, z, t) = u_1^0(x, y, t) + z[\theta_2(x, y, t) - \theta_{20}(x, y)] \qquad (9.4.5a)$$

$$u_2(x, y, z, t) = u_2^0(x, y, t) - z[\theta_1(x, y, t) - \theta_{10}(x, y)] \qquad (9.4.5b)$$

$$u_3(x, y, z, t) = u_3^0(x, y, t) \qquad (9.4.5c)$$

where the u_i^0 ($i = 1, 2, 3$) are the displacements (with respect to the local coordinate system $\xi\eta\zeta$) of the observed reference point (i.e., the point where the observed shell element crosses the reference surface), θ_1 and θ_2 are the rotation angles of the observed shell element with respect to the ξ and η axes, and θ_{10} and θ_{20} are the initial rotation angles of the observed shell element with respect to the ξ and η axes, respectively. Because $\xi\eta\zeta$ is a local coordinate system attached to the observed shell element and the $\xi\eta$ plane is tangent to the deformed reference surface, we have

$$u_1^0 = u_2^0 = u_3^0 = \theta_1 = \theta_2 = \theta_{10} = \theta_{20} = \partial u_3^0/\partial x = \partial u_3^0/\partial y = 0 \qquad (9.4.6)$$

Using **U** to denote the local displacement vector; that is,

$$\mathbf{U} = u_1 \mathbf{i}_1 + u_2 \mathbf{i}_2 + u_3 \mathbf{i}_3 \qquad (9.4.7)$$

taking the derivatives of (9.4.7), and using (9.4.6), we obtain

$$\frac{\partial \mathbf{U}}{\partial x} = \frac{\partial u_1}{\partial x}\mathbf{i}_1 + \frac{\partial u_2}{\partial x}\mathbf{i}_2 + \frac{\partial u_3}{\partial x}\mathbf{i}_3 + u_1\frac{\partial \mathbf{i}_1}{\partial x} + u_2\frac{\partial \mathbf{i}_2}{\partial x} + u_3\frac{\partial \mathbf{i}_3}{\partial x}$$
$$= \left[e_1 + z\left(k_1 - k_1^0\right)\right]\mathbf{i}_1 + \left[\frac{\partial u_2^0}{\partial x} + z\left(k_{61} - k_{61}^0\right)\right]\mathbf{i}_2 \qquad (9.4.8a)$$

$$\frac{\partial \mathbf{U}}{\partial y} = \frac{\partial u_1}{\partial y}\mathbf{i}_1 + \frac{\partial u_2}{\partial y}\mathbf{i}_2 + \frac{\partial u_3}{\partial y}\mathbf{i}_3 + u_1\frac{\partial \mathbf{i}_1}{\partial y} + u_2\frac{\partial \mathbf{i}_2}{\partial y} + u_3\frac{\partial \mathbf{i}_3}{\partial y}$$
$$= \left[\frac{\partial u_1^0}{\partial y} + z\left(k_{62} - k_{62}^0\right)\right]\mathbf{i}_1 + \left[e_2 + z\left(k_2 - k_2^0\right)\right]\mathbf{i}_2 \qquad (9.4.8b)$$

$$\frac{\partial \mathbf{U}}{\partial z} = \frac{\partial u_1}{\partial z}\mathbf{i}_1 + \frac{\partial u_2}{\partial z}\mathbf{i}_2 + \frac{\partial u_3}{\partial z}\mathbf{i}_3 = 0 \qquad (9.4.8c)$$

It follows from (9.4.8) that the Jaumann strains ϵ_{ij} are

$$\epsilon_{11} = \frac{\partial \mathbf{U}}{\partial x} \cdot \mathbf{i}_1 = e_1 + z(k_1 - k_1^0) \qquad (9.4.9a)$$

$$\epsilon_{22} = \frac{\partial \mathbf{U}}{\partial y} \cdot \mathbf{i}_2 = e_2 + z(k_2 - k_2^0) \qquad (9.4.9b)$$

$$\epsilon_{12} = \frac{\partial \mathbf{U}}{\partial x} \cdot \mathbf{i}_2 + \frac{\partial \mathbf{U}}{\partial y} \cdot \mathbf{i}_1 = \gamma_6 + z(k_6 - k_6^0) \qquad (9.4.9c)$$

$$\epsilon_{33} = \epsilon_{23} = \epsilon_{13} = 0 \qquad (9.4.9d)$$

where

$$e_1 \equiv \frac{\partial u_1^0}{\partial x}, \quad e_2 \equiv \frac{\partial u_2^0}{\partial y}, \quad \gamma_6 \equiv \frac{\partial u_1^0}{\partial y} + \frac{\partial u_2^0}{\partial x} \qquad (9.4.10a)$$

$$k_1 \equiv \frac{\partial \theta_2}{\partial x}, \quad k_2 \equiv -\frac{\partial \theta_1}{\partial y}, \quad k_{61} \equiv -\frac{\partial \theta_1}{\partial x}, \quad k_{62} \equiv \frac{\partial \theta_2}{\partial y}, \quad k_6 \equiv k_{61} + k_{62} \qquad (9.4.10b)$$

$$k_1^0 \equiv \frac{\partial \theta_{20}}{\partial x}, \quad k_2^0 \equiv -\frac{\partial \theta_{10}}{\partial y}, \quad k_{61}^0 \equiv -\frac{\partial \theta_{10}}{\partial x}, \quad k_{62}^0 \equiv \frac{\partial \theta_{20}}{\partial y}, \quad k_6^0 \equiv k_{61}^0 + k_{62}^0 \qquad (9.4.10c)$$

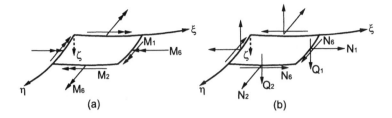

Fig. 9.4.1 Loadings on a nonlinear classical doubly-curved shell element: (a) stress couples and (b) stress resultants.

Using (9.4.9) and (6.4.1a-d) and the fact that variations of the known initial curvatures are zero, we obtain

$$\delta\Pi = \int_A \int_z (\sigma_{11}\delta\epsilon_{11} + \sigma_{22}\delta\epsilon_{22} + \sigma_{12}\delta\epsilon_{12})dzdA$$
$$= \int_A (N_1\delta e_1 + N_2\delta e_2 + N_6\delta\gamma_6 + M_1\delta k_1 + M_2\delta k_2 + M_6\delta k_6)dxdy$$
$$= \int_A (N_1\delta e_1 + N_2\delta e_2 + N_6\delta\gamma_6 + \Theta_1\delta\theta_1 + \Theta_2\delta\theta_2 + \Theta_3\delta\theta_3)dxdy$$
$$+ \int_x [-M_2\delta\theta_1 + M_6\delta\theta_2]_{y=0}^{y=Y} dx + \int_y [-M_6\delta\theta_1 + M_1\delta\theta_2]_{x=0}^{x=X} dy \quad (9.4.11)$$

where

$$(N_1, N_2, N_6) = \int_z (\sigma_{11}, \sigma_{22}, \sigma_{12})dz$$
$$(M_1, M_2, M_6) = \int_z z(\sigma_{11}, \sigma_{22}, \sigma_{12})dz \quad (9.4.12a)$$

$$\Theta_1 \equiv M_{6x} + M_{2y} + M_1 k_5 + M_6 k_4$$
$$\Theta_2 \equiv -M_{1x} - M_{6y} + M_2 k_4 + M_6 k_5$$
$$\Theta_3 \equiv M_1 k_{61} - M_2 k_{62} + M_6 k_2 - M_6 k_1 \quad (9.4.12b)$$

Here, M_1, M_2, and M_6 represent the moment intensities acting on the edges of the shell element, as shown in Figure 9.4.1a. Figure 9.4.1b shows the inplane extension force intensities N_1 and N_2 and inplane shear intensity N_6.

To relate the stress resultants and moments to the inplane strains, we first use (9.4.9a-d) with $\sigma_{33} = 0$ to obtain the reduced stress-strain relations from (2.4.34) and (2.4.35) as

$$\left\{\begin{array}{c}\sigma_{11}\\\sigma_{22}\\\sigma_{12}\end{array}\right\} = \left[\begin{array}{ccc}\tilde{Q}_{11} & \tilde{Q}_{12} & \tilde{Q}_{16}\\\tilde{Q}_{12} & \tilde{Q}_{22} & \tilde{Q}_{26}\\\tilde{Q}_{16} & \tilde{Q}_{26} & \tilde{Q}_{66}\end{array}\right]\left(\left\{\begin{array}{c}e_1\\e_2\\\gamma_6\end{array}\right\} + z\left\{\begin{array}{c}k_1 - k_1^0\\k_2 - k_2^0\\k_6 - k_6^0\end{array}\right\}\right) \quad (9.4.13)$$

Substituting (9.4.13) into (9.4.12a), we obtain the internal forces and moments in terms of the mid-plane strains and curvatures as

$$\left\{\begin{array}{c} N_1 \\ N_2 \\ N_6 \\ M_1 \\ M_2 \\ M_6 \end{array}\right\} = \left[\begin{array}{cc} [A_{ij}] & [B_{ij}] \\ [B_{ij}] & [D_{ij}] \end{array}\right] \left\{\begin{array}{c} e_1 \\ e_2 \\ \gamma_6 \\ k_1 - k_1^0 \\ k_2 - k_2^0 \\ k_6 - k_6^0 \end{array}\right\} \quad (9.4.14)$$

where

$$(A_{ij},\ B_{ij},\ D_{ij}) = \sum_{k=1}^{N} \int_{z_k}^{z_{k+1}} \tilde{Q}_{ij}^{(k)}(1,\ z,\ z^2)\,dz \quad \text{for } i,j = 1,2,6 \quad (9.4.15)$$

In (9.4.15), we assume that there are N laminae and the kth lamina is located between the $z = z_k$ and $z = z_{k+1}$ planes and $z_{N+1} - z_1 = h$.

Neglecting the influence of γ_6 on the deformed geometry, substituting (6.3.3a,b) and (6.3.7a,b) ($\gamma_6 = \gamma_{61} + \gamma_{62}$) into (9.4.11), and adding the integrals of $Q_1 \times$ (6.3.5a) and $Q_2 \times$ (6.3.5b), we obtain

$$\delta\Pi = \int_A \left\{ \{N_1,\ N_6/(1+e_1),\ Q_1\}[T] \left\{\begin{array}{c} \delta u_x \\ \delta v_x \\ \delta w_x \end{array}\right\}\right.$$

$$+ \{N_1,\ N_6/(1+e_1),\ Q_1\}[T][K_1^0]^T \left\{\begin{array}{c} \delta u \\ \delta v \\ \delta w \end{array}\right\}$$

$$+ \{N_6/(1+e_2),\ N_2,\ Q_2\}[T] \left\{\begin{array}{c} \delta u_y \\ \delta v_y \\ \delta w_y \end{array}\right\}$$

$$+ \{N_6/(1+e_2),\ N_2,\ Q_2\}[T][K_2^0]^T \left\{\begin{array}{c} \delta u \\ \delta v \\ \delta w \end{array}\right\}$$

$$+ [\Theta_1 - Q_2(1+e_2)]\delta\theta_1 + [\Theta_2 + Q_1(1+e_1)]\delta\theta_2 + \Theta_3\delta\theta_3 \Bigg\} dxdy$$

$$+ \int_x [-M_2\delta\theta_1 + M_6\delta\theta_2]_{y=0}^{y=Y}\,dx + \int_y [-M_6\delta\theta_1 + M_1\delta\theta_2]_{x=0}^{x=X}\,dy$$

(9.4.16)

It is shown later that the introduced Q_1 and Q_2 are transverse shear intensities. Substituting (9.4.16), (9.4.3), and $\delta W_{nc} = 0$ into the extended Hamilton's principle (i.e., (7.2.1)), integrating by parts, and using (9.4.12b), we obtain

$$0 = \int_A \left\{ \left(\frac{\partial \{F_\alpha\}^T [T]}{\partial x} + \frac{\partial \{F_\beta\}^T [T]}{\partial y} - \{F_\alpha\}^T [T][K_1^0]^T \right.\right.$$

$$
\begin{aligned}
&\left.-\{F_\beta\}^T[T][K_2^0]^T - \{I_F\}^T\right)\begin{Bmatrix}\delta u \\ \delta v \\ \delta w\end{Bmatrix} \\
&+\left(\frac{\partial\{M_\alpha\}^T}{\partial x} + \{M_\alpha\}^T[K_1] + \frac{\partial\{M_\beta\}^T}{\partial y} + \{M_\beta\}^T[K_2]\right. \\
&\left.+\{Q_\alpha\}^T + \{Q_\beta\}^T - \{I_M\}^T\right)\begin{Bmatrix}\delta\theta_1 \\ \delta\theta_2 \\ \delta\theta_3\end{Bmatrix}\bigg\}dxdy \\
&-\int_x\bigg\{\{F_\beta\}^T[T]\begin{Bmatrix}\delta u\\ \delta v\\ \delta w\end{Bmatrix} + \left[\left(\frac{M_6 T_{31}}{1+e_1}\right)_x + T_{33}k_1^0 - T_{32}k_5^0\right]\delta u \\
&+\left[\left(\frac{M_6 T_{32}}{1+e_1}\right)_x + T_{33}k_{61}^0 + T_{31}k_5^0\right]\delta v \\
&+\left[\left(\frac{M_6 T_{33}}{1+e_1}\right)_x - T_{32}k_{61}^0 - T_{31}k_1^0\right]\delta w - M_2\delta\theta_1\bigg\}_{y=0}^{y=Y}dx \\
&-\int_y\bigg\{\{F_\alpha\}^T[T]\begin{Bmatrix}\delta u\\ \delta v\\ \delta w\end{Bmatrix} + \left[\left(\frac{M_6 T_{31}}{1+e_2}\right)_y + T_{33}k_{62}^0 - T_{32}k_4^0\right]\delta u \\
&+\left[\left(\frac{M_6 T_{32}}{1+e_2}\right)_y + T_{33}k_2^0 + T_{31}k_4^0\right]\delta v \\
&+\left[\left(\frac{M_6 T_{33}}{1+e_2}\right)_y - T_{32}k_2^0 - T_{31}k_{62}^0\right]\delta w + M_1\delta\theta_2\bigg\}_{x=0}^{x=X}dx \\
&+\left[\left(\frac{M_6}{1+e_1} + \frac{M_6}{1+e_2}\right)(T_{31}\delta u + T_{32}\delta v + T_{33}\delta w)\right]_{(x,y)=(X,0),(0,Y)}^{(x,y)=(0,0),(X,Y)}
\end{aligned}
$$
(9.4.17)

where

$$
\begin{aligned}
\{F_\alpha\} &\equiv \{N_1,\ N_6/(1+e_1),\ Q_1\}^T, & \{F_\beta\} &\equiv \{N_6/(1+e_2),\ N_2,\ Q_2\}^T \\
\{M_\alpha\} &\equiv \{-M_6,\ M_1,\ 0\}^T, & \{M_\beta\} &\equiv \{-M_2,\ M_6,\ 0\}^T \\
\{Q_\alpha\} &\equiv \{0,\ -Q_1(1+e_1),\ N_6\}^T, & \{Q_\beta\} &\equiv \{Q_2(1+e_2),\ 0,\ -N_6\}^T \\
\{I_F\} &\equiv \{A_u,\ A_v,\ A_w\}^T, & \{I_M\} &\equiv \{A_{\theta_1},\ A_{\theta_2},\ 0\}^T
\end{aligned}
$$
(9.4.18)

Setting each of the coefficients of δu, δv, δw, $\delta\theta_1$, and $\delta\theta_2$ in (9.4.17) equal to zero, we obtain the following equations of motion:

$$\frac{\partial}{\partial x}\left([T]^T\{F_\alpha\}\right) - [K_1^0][T]^T\{F_\alpha\} + \frac{\partial}{\partial y}\left([T]^T\{F_\beta\}\right) - [K_2^0][T]^T\{F_\beta\} = \{I_F\}$$
(9.4.19)

$$\frac{\partial \{M_\alpha\}}{\partial x} + [K_1]^T \{M_\alpha\} + \frac{\partial \{M_\beta\}}{\partial y} + [K_2]^T \{M_\beta\} + \{Q_\alpha\} + \{Q_\beta\} = \{I_M\} \tag{9.4.20}$$

The boundary conditions are of the form: specify

Along $x = 0, X$:

$\delta u = 0$	or	$\mathbf{F}_\alpha \cdot \mathbf{j}_1 + [M_6 T_{31}/(1+e_2)]_y + T_{33}k_{62}^0 - T_{32}k_4^0$
$\delta v = 0$	or	$\mathbf{F}_\alpha \cdot \mathbf{j}_2 + [M_6 T_{32}/(1+e_2)]_y + T_{33}k_2^0 + T_{31}k_4^0$
$\delta w = 0$	or	$\mathbf{F}_\alpha \cdot \mathbf{j}_3 + [M_6 T_{33}/(1+e_2)]_y - T_{32}k_2^0 - T_{31}k_{62}^0$
$\delta\theta_2 = 0$	or	M_1

Along $y = 0, Y$:

$\delta u = 0$	or	$\mathbf{F}_\beta \cdot \mathbf{j}_1 + [M_6 T_{31}/(1+e_1)]_x + T_{33}k_1^0 - T_{32}k_5^0$
$\delta v = 0$	or	$\mathbf{F}_\beta \cdot \mathbf{j}_2 + [M_6 T_{32}/(1+e_1)]_x + T_{33}k_{61}^0 + T_{31}k_5^0$
$\delta w = 0$	or	$\mathbf{F}_\beta \cdot \mathbf{j}_3 + [M_6 T_{33}/(1+e_1)]_x - T_{31}k_1^0 - T_{32}k_{61}^0$
$\delta\theta_1 = 0$	or	M_2

At $(x, y) = (0, 0), (0, Y), (X, 0), (X, Y)$:

$$\begin{aligned}
\delta u &= 0 \quad \text{or} \quad [M_6/(1+e_1) + M_6/(1+e_2)]T_{31} \\
\delta v &= 0 \quad \text{or} \quad [M_6/(1+e_1) + M_6/(1+e_2)]T_{32} \\
\delta w &= 0 \quad \text{or} \quad [M_6/(1+e_1) + M_6/(1+e_2)]T_{33}
\end{aligned} \tag{9.4.21}$$

where

$$\begin{aligned}
\mathbf{F}_\alpha &= N_1 \mathbf{i}_1 + N_6/(1+e_1)\mathbf{i}_2 + Q_1 \mathbf{i}_3 \\
\mathbf{F}_\beta &= N_6/(1+e_2)\mathbf{i}_1 + N_2 \mathbf{i}_2 + Q_2 \mathbf{i}_3
\end{aligned} \tag{9.4.22a}$$

We also define

$$\begin{aligned}
\mathbf{M}_\alpha &= -M_6 \mathbf{i}_1 + M_1 \mathbf{i}_2 \\
\mathbf{M}_\beta &= -M_2 \mathbf{i}_1 + M_6 \mathbf{i}_2 \\
\mathbf{I}_F &= A_u \mathbf{j}_1 + A_v \mathbf{j}_2 + A_w \mathbf{j}_3 \\
\mathbf{I}_M &= A_{\theta_1} \mathbf{i}_1 + A_{\theta_2} \mathbf{i}_2
\end{aligned} \tag{9.4.22b}$$

Using (6.2.13) and (6.2.14) and the identity $[T]^T = [T]^{-1}$, we rewrite (9.4.19) as

$$[T]^T \left(\frac{\partial \{F_\alpha\}}{\partial x} + [K_1]^T \{F_\alpha\} + \frac{\partial \{F_\beta\}}{\partial y} + [K_2]^T \{F_\beta\} \right) = \{I_F\} \tag{9.4.23}$$

Using (9.4.22), (6.2.13), and (6.2.14), we put (9.4.23) and (9.4.20) in the following vector forms:

$$\frac{\partial \mathbf{F}_\alpha}{\partial x} + \frac{\partial \mathbf{F}_\beta}{\partial y} = \mathbf{I}_F \tag{9.4.24}$$

$$\frac{\partial \mathbf{M}_\alpha}{\partial x} + \frac{\partial \mathbf{M}_\beta}{\partial y} + (1+e_1)\mathbf{i}_1 \times \mathbf{F}_\alpha + (1+e_2)\mathbf{i}_2 \times \mathbf{F}_\beta = \mathbf{I}_M \qquad (9.4.25)$$

One can see from (9.4.22a) that Q_1 and Q_2 are along the i_3-direction and hence represent the transverse shear intensities. However, it is assumed that there are no transverse shear deformations in the classical shell theory and hence Q_1 and Q_2 do not appear in (9.4.14). Hence, Q_1 and Q_2 need to be obtained from the first two equations in (9.4.20) as

$$Q_1 = \frac{1}{1+e_1}(M_{1x} + M_{6y} - M_2 k_4 - M_6 k_5 - A_{\theta_2})$$
$$Q_2 = \frac{1}{1+e_2}(M_{2y} + M_{6x} + M_6 k_4 + M_1 k_5 + A_{\theta_1}) \qquad (9.4.26a)$$

Because the diameter of an infinitesimal shell element is zero, its rotary inertia A_{θ_3} with respect to the ζ axis is zero and hence the third equation in (9.4.20) is a statement of the balance of internal moments about the ζ axis, which is given by

$$N_6 - N_6 + M_2 k_{62} - M_1 k_{61} + M_6 k_1 - M_6 k_2 = 0 \qquad (9.4.26b)$$

Hence, only the three equations in (9.4.23) and the boundary conditions in (9.4.21) need to be solved in the analysis of nonlinear classical doubly-curved shells. We note that (9.4.24) and (9.4.25) are the same as (7.6.25) and (7.6.26) except that x and y here are curvilinear coordinates.

9.5 NONLINEAR SHEAR-DEFORMABLE THEORIES FOR CIRCULAR CYLINDRICAL SHELLS

We consider a thin-walled circular cylindrical shell panel of length L, wall thickness h, undeformed middle-surface radius a, and $0 \leq \theta \leq \theta_0$, as shown in Figure 9.5.1a. Figure 9.5.1b shows a midsurface element before and after deformation. Two orthogonal coordinate systems are used in the derivation; the xyz is an inertial cylindrical coordinate system with the y and z axes being on the undeformed midsurface and the x axis being parallel to the center line of the shell, and the $\xi\eta\zeta$ is an orthogonal curvilinear coordinate system with the ξ and η axes being on the deformed midsurface. Here, z and ζ are rectilinear axes, $\hat{\xi}$ and $\hat{\eta}$ represent the deformed x and y, and ξ and η coincide with $\hat{\xi}$ and $\hat{\eta}$ if the inplane shear deformation $\gamma_6 (= \gamma_{61} + \gamma_{62})$ is zero. Moreover, $\mathbf{j}_1, \mathbf{j}_2$, and $\mathbf{j}_3$ are unit vectors along the x, y, and z axes; $\mathbf{i}_1, \mathbf{i}_2$, and $\mathbf{i}_3$ are unit vectors along the ξ, η, and ζ axes; and $\mathbf{i}_{\hat{1}}$ and $\mathbf{i}_{\hat{2}}$ are unit vectors along the $\hat{\xi}$ and $\hat{\eta}$ axes, respectively. We let u, v, and w be the longitudinal (along x axis), circumferential (along y axis), and radial (along z axis) displacements of the observed reference point A. Because the xyz is a cylindrical coordinate system, we recall from (6.1.6), (9.2.3b), and (9.2.4) that

$$\partial y = a\partial\theta, \quad k_1^0 = k_{61}^0 = k_5^0 = k_{62}^0 = k_4^0 = 0, \quad k_2^0 = \frac{1}{a} \qquad (9.5.1a)$$

NONLINEAR SHEAR-DEFORMABLE THEORIES FOR CIRCULAR CYLINDRICAL SHELLS 589

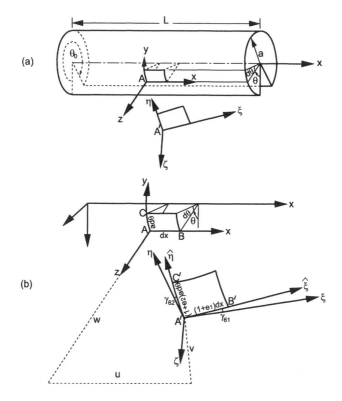

Fig. 9.5.1 A circular cylindrical shell: (a) coordinate systems and (b) the undeformed and deformed geometries of a midsurface element.

$$\frac{\partial}{\partial x}\{\mathbf{j}_{123}\} = [K_1^0]\{\mathbf{j}_{123}\}, \ [K_1^0] = [0] \tag{9.5.1b}$$

$$\frac{\partial}{\partial y}\{\mathbf{j}_{123}\} = [K_2^0]\{\mathbf{j}_{123}\}, \ [K_2^0] = \begin{bmatrix} 0 & 0 & 0 \\ 0 & 0 & -1/a \\ 0 & 1/a & 0 \end{bmatrix} \tag{9.5.1c}$$

9.5.1 Equations of Motion

We consider u, v, and w to be small but finite and assume that u, v, w are $O(\epsilon)$, where ϵ is a small dimensionless bookkeeping parameter. Also, we neglect the influence of the inplane shear deformation γ_6 on the deformed system configuration and adopt (6.3.2) in the following derivation. The goal here is to derive third-order nonlinear equations of motion. Substituting (9.5.1a) into (6.2.3) and (6.2.4) and expanding the

590 SHELLS

results, we obtain the axial strains along the ξ and η directions as

$$e_1 = \frac{\overline{A'B'} - dx}{dx} = \sqrt{(1+u_x)^2 + v_x^2 + w_x^2} - 1$$

$$= u_x + \frac{1}{2}(v_x^2 + w_x^2) - \frac{1}{2}u_x(v_x^2 + w_x^2) \qquad (9.5.2)$$

$$e_2 = \frac{\overline{A'C'} - dy}{dy} = \sqrt{u_y^2 + \left(1 + \frac{w}{a} + v_y\right)^2 + \left(w_y - \frac{v}{a}\right)^2} - 1$$

$$= v_y + \frac{1}{a}\left(w + \frac{1}{2a}v^2 - vw_y\right) + \frac{1}{2}\left(u_y^2 + w_y^2\right) - \frac{1}{2}v_y\left(u_y^2 + w_y^2\right)$$

$$+ \frac{1}{a}\left(vv_y w_y - \frac{1}{2}ww_y^2 - \frac{1}{2}wu_y^2 + \frac{1}{a}vww_y - \frac{1}{2a}v^2 v_y - \frac{1}{2a^2}v^2 w\right) \quad (9.5.3)$$

where fourth- and higher-order terms have been neglected.

Substituting (6.3.2) and (9.5.1a) into (6.2.7), (6.2.9), and (6.2.11a), we obtain

$$\{\mathbf{i}_{123}\} = [T]\{\mathbf{j}_{123}\}, \quad [T] = \begin{bmatrix} T_{11} & T_{12} & T_{13} \\ T_{21} & T_{22} & T_{23} \\ T_{31} & T_{32} & T_{33} \end{bmatrix} \qquad (9.5.4a)$$

where

$$T_{11} = \frac{1+u_x}{1+e_1} = 1 - \frac{1}{2}w_x^2 - \frac{1}{2}v_x^2 + u_x w_x^2 + u_x v_x^2$$

$$T_{12} = \frac{v_x}{1+e_1} = v_x - u_x v_x + u_x^2 v_x - \frac{1}{2}v_x w_x^2 - \frac{1}{2}v_x^3$$

$$T_{13} = \frac{w_x}{1+e_1} = w_x - u_x w_x + u_x^2 w_x - \frac{1}{2}v_x^2 w_x - \frac{1}{2}w_x^3 \qquad (9.5.4b)$$

$$T_{21} = \frac{u_y}{1+e_2}$$

$$= u_y - \frac{1}{a}u_y w - u_y v_y + u_y v_y^2 - \frac{1}{2}u_y w_y^2 - \frac{1}{2}u_y^3$$

$$+ \frac{1}{a}\left(u_y vw_y + \frac{1}{a}u_y w^2 + 2u_y v_y w - \frac{1}{2a}u_y v^2\right)$$

$$T_{22} = \frac{1 + w/a + v_y}{1+e_2}$$

$$= 1 - \frac{1}{2}w_y^2 - \frac{1}{2}u_y^2 + \frac{1}{a}vw_y - \frac{1}{2a^2}v^2 + v_y w_y^2 + u_y^2 v_y$$

$$+ \frac{1}{a}\left(ww_y^2 - 2vv_y w_y + u_y^2 w - \frac{2}{a}vww_y + \frac{1}{a}v^2 v_y + \frac{1}{a^2}v^2 w\right)$$

$$T_{23} = \frac{w_y - v/a}{1+e_2}$$

$$= w_y - \frac{1}{a}\left(v + ww_y - vv_y - \frac{1}{a}vw\right) - v_y w_y + v_y^2 w_y - \frac{1}{2}u_y^2 w_y - \frac{1}{2}w_y^3$$

NONLINEAR SHEAR-DEFORMABLE THEORIES FOR CIRCULAR CYLINDRICAL SHELLS

$$+\frac{1}{a}\left(\frac{3}{2}vw_y^2 + 2v_y ww_y + \frac{1}{2}u_y^2 v - vv_y^2 + \frac{1}{a}w^2 w_y - \frac{3}{2a}v^2 w_y - \frac{2}{a}vv_y w\right)$$

$$+\frac{1}{a^3}\left(\frac{1}{2}v^3 - vw^2\right) \tag{9.5.4c}$$

$$T_{31} = -w_x + v_x w_y + u_x w_x - \frac{1}{a}vv_x + \frac{1}{2}w_x w_y^2 - v_x v_y w_y$$

$$-u_x v_x w_y + \frac{1}{2}w_x^3 - u_y v_x w_x - u_x^2 w_x$$

$$+\frac{1}{a}\left(vv_x v_y + u_x vv_x - vw_x w_y - v_x ww_y + \frac{1}{2a}v^2 w_x + \frac{1}{a}vv_x w\right)$$

$$T_{32} = -w_y + \frac{1}{a}\left(v + ww_y - vv_y - \frac{1}{a}vw\right) + u_y w_x + v_y w_y + \frac{1}{2}w_y^3$$

$$+\frac{1}{2}w_x^2 w_y - v_y^2 w_y - u_y v_x w_y - u_y v_y w_x$$

$$-u_x u_y w_x + \frac{1}{a}\left(vv_y^2 - \frac{3}{2}vw_y^2 - 2v_y ww_y - \frac{1}{2}vw_x^2 - u_y ww_x + u_y vv_x\right)$$

$$-\frac{1}{a}w^2 w_y + \frac{3}{2a}v^2 w_y + \frac{2}{a}vv_y w + \frac{1}{a^2}vw^2 - \frac{1}{2a^2}v^3$$

$$T_{33} = 1 - \frac{1}{2}w_x^2 - \frac{1}{2}w_y^2 + \frac{1}{a}vw_y - \frac{1}{2a^2}v^2$$

$$+\frac{1}{a}\left(ww_y^2 - 2vv_y w_y - vv_x w_x - u_y vw_x - \frac{2}{a}vww_y + \frac{1}{a}v^2 v_y + \frac{1}{a^2}v^2 w\right)$$

$$+v_y w_y^2 + v_x w_x w_y + u_y w_x w_y + u_x w_x^2 \tag{9.5.4d}$$

Substituting (9.5.4b,c) into (6.2.8a) yields the expanded form of the inplane shear strain as

$$\gamma_6 = u_y + v_x + w_x w_y - u_x v_x - u_y v_y - \frac{1}{a}(vw_x + u_y w) - w_x w_y(u_x + v_y)$$

$$-u_y\left(\frac{1}{2}w_x^2 + \frac{1}{2}w_y^2 + \frac{1}{3}u_y^2 - v_y^2\right) - v_x\left(\frac{1}{2}w_x^2 + \frac{1}{2}w_y^2 + \frac{1}{3}v_x^2 - u_x^2\right)$$

$$+\frac{1}{a}\left(vv_x w_y - ww_x w_y + vu_y w_y + vv_y w_x + u_x vw_x\right.$$

$$\left.+2u_y v_y w + \frac{1}{a}vww_x + \frac{1}{a}u_y w^2 - \frac{1}{2a}v^2 v_x - \frac{1}{2a}u_y v^2\right) \tag{9.5.5}$$

Because e_1 and e_2 of flexible shells are usually negligible, it follows from (6.2.8b) that $\gamma_{61} = \gamma_{62}$. Hence, without loss of accuracy it is reasonable to assume that

$$\gamma_{61} = \sin^{-1}(\sin \gamma_{61}) = \sin^{-1}(\mathbf{i}_{\hat{1}} \cdot \mathbf{i}_2) = \frac{1}{2}\gamma_6 \tag{9.5.6a}$$

$$\gamma_{62} = \sin^{-1}(\sin \gamma_{62}) = \sin^{-1}(\mathbf{i}_{\hat{2}} \cdot \mathbf{i}_1) = \frac{1}{2}\gamma_6 \tag{9.5.6b}$$

592 SHELLS

Substituting (9.5.1a) and (9.5.4b-d) into (6.2.15), we obtain the deformed curvatures as

$$k_1 = -w_{xx} + v_{xx}w_y + u_xw_{xx} + u_{xx}w_x - \frac{1}{a}vv_{xx} + \frac{1}{2}w_{xx}w_y^2 + w_x^2w_{xx}$$

$$-v_{xx}v_yw_y - u_xv_{xx}w_y - u_{xx}v_xw_y + \frac{1}{2}v_x^2w_{xx} - u_x^2w_{xx}$$

$$-u_yv_{xx}w_x - 2u_xu_{xx}w_x + \frac{1}{a}(vv_{xx}v_y + u_xvv_{xx} + u_{xx}vv_x$$

$$-vw_{xx}w_y - v_{xx}ww_y + \frac{1}{2a}v^2w_{xx} + \frac{1}{a}vv_{xx}w) \qquad (9.5.7)$$

$$k_2 = \frac{1}{a} - w_{yy} + \frac{1}{a}v_y + u_{yy}w_x + v_yw_{yy} + v_{yy}w_y + \frac{1}{a}\left(ww_{yy} + w_y^2 - \frac{1}{2}w_x^2\right)$$

$$-vv_{yy} - v_y^2 - \frac{1}{2}v_x^2 - \frac{1}{a}vw_y - \frac{1}{a}v_yw\right) + \frac{1}{2}w_{yy}w_x^2 + w_y^2w_{yy}$$

$$-u_{yy}u_xw_x - v_yu_{yy}w_x - v_{yy}u_yw_x + \frac{1}{2}u_y^2w_{yy} - v_y^2w_{yy} - v_xu_{yy}w_y$$

$$-2v_yv_{yy}w_y + \frac{1}{a}\left(u_{yy}vv_x - 2vw_yw_{yy} - 2v_yww_{yy} - 3v_yw_y^2 - u_yw_xw_y\right.$$

$$-2v_{yy}ww_y - \frac{1}{2}v_yw_x^2 - u_{yy}ww_x + 2vv_yv_{yy} + v_y^3 - \frac{1}{2}u_y^2v_y + u_xv_x^2$$

$$+u_xw_x^2\right) + \frac{1}{a^2}\left(-w^2w_{yy} + v^2w_{yy} - 2ww_y^2 + 4vv_yw_y + 2vv_{yy}w + 2v_y^2w\right)$$

$$+\frac{1}{a^3}\left(2vww_y + v_yw^2 - v^2v_y\right) \qquad (9.5.8)$$

$$k_{61} = -w_{xy} + \frac{1}{a}v_x + v_{xy}w_y + u_{xy}w_x + v_yw_{xy}$$

$$+\frac{1}{a}\left(w_xw_y + ww_{xy} - v_xv_y - vv_{xy} - \frac{1}{a}vw_x - \frac{1}{a}v_xw\right) + w_{xy}w_y^2$$

$$+\frac{1}{2}w_{xy}w_x^2 - 2v_{xy}v_yw_y - u_{xy}u_xw_x - v_yu_{xy}w_x - u_{xy}v_xw_y - v_{xy}u_yw_x$$

$$-v_y^2w_{xy} + \frac{1}{2}u_y^2w_{xy} + \frac{1}{a}\left(u_{xy}vv_x - v_xw_y^2 - 2vw_{xy}w_y - 2v_yw_xw_y\right.$$

$$-2v_{xy}ww_y - 2v_yww_{xy} - \frac{1}{2}v_xw_x^2 - u_yw_x^2 - u_{xy}ww_x + v_xv_y^2$$

$$+2vv_{xy}v_y - \frac{1}{2}u_y^2v_x\right) + \frac{1}{a^2}\left(-2ww_xw_y + 2vv_xw_y - w^2w_{xy}\right.$$

$$+v^2w_{xy} + 2vv_yw_x + 2v_xv_yw + 2vv_{xy}w\right)$$

$$+\frac{1}{a^3}\left(2vww_x + v_xw^2 - v^2v_x\right) \qquad (9.5.9)$$

$$k_{62} = -w_{xy} - \frac{1}{a}u_y + v_{xy}w_y + u_{xy}w_x + u_xw_{xy} + \frac{1}{a}\left(u_yv_y - vv_{xy} + \frac{1}{a}u_yw\right)$$

$$+\frac{1}{2}w_{xy}w_y^2 + w_{xy}w_x^2 - v_{xy}v_yw_y - 2u_{xy}u_xw_x - u_xv_{xy}w_y - u_{xy}v_xw_y$$

$$-v_{xy}u_yw_x - u_x^2w_{xy} + \frac{1}{2}v_x^2w_{xy} + \frac{1}{a}\left(u_{xy}vv_x - vw_{xy}w_y - v_{xy}ww_y\right.$$

$$+vv_{xy}v_y + u_xvv_{xy} + \frac{1}{2}u_yw_y^2 + \frac{1}{2}u_y^3 - u_yv_y^2\bigg) + \frac{1}{a^2}\bigg(\frac{1}{2}v^2w_{xy}$$

$$+vv_{xy}w - u_yvw_y - 2u_yv_yw - \frac{1}{a}u_yw^2 + \frac{1}{2a}u_yv^2\bigg) \tag{9.5.10}$$

$$k_4 = -u_{yy} + \frac{1}{a}w_x - w_xw_{yy} + u_yv_{yy} + u_{yy}v_y$$

$$+ \frac{1}{a}\left(u_yw_y + v_yw_x + u_{yy}w - v_xw_y - u_xw_x + \frac{1}{a}vv_x\right) \tag{9.5.11}$$

$$k_5 = v_{xx} + w_{xx}w_y - u_xv_{xx} - u_{xx}v_x - \frac{1}{a}vw_{xx} \tag{9.5.12}$$

Only up to quadratic terms are needed in the expansions of k_4 and k_5, as shown later. The curvature k_{62} in (9.5.10) is obtained using (6.2.15); that is,

$$k_{62} = -(T_{11y}T_{31} + T_{12y}T_{32} + T_{13y}T_{33}) - \frac{1}{a}T_{21}$$

Using the orthogonality of the $\xi\eta\zeta$ system, one can express k_{62} as

$$k_{62} = -(T_{11y}T_{31} + T_{12y}T_{32} + T_{13y}T_{33}) + \frac{1}{a}(T_{12}T_{33} - T_{13}T_{32})$$

of which the linear part is $k_{62} = -w_{xy} + v_x/a$. But the linear part of (9.5.10) is $k_{62} = -w_{xy} - u_y/a$. In other words, $u_y = -v_x$ because the influence of $\gamma_6(= u_y + v_x)$ on the deformed configuration has been neglected.

Next, we derive nonlinear strain-displacement relations using the Jaumann strains and local displacements. The movement of a shell element consists of two parts: a rigid-body motion, which translates the vertex A (see Figure 9.5.1b) of the element to A' and rotates the sides dx and dy of the element to be parallel to the ξ and η axes, respectively; and local strainable displacements with respect to the $\xi\eta\zeta$ system, which result in strain values and the deformed geometry $A'B'C'$. Because the rigid-body motion does not result in any strain values, to obtain the elastic energy we only need to deal with the local strainable displacements. In view of the initial curvature $1/a$ and the displacement fields used by Pai and Nayfeh (1991c) for nonlinear plate theories and by Bhimaraddi (1984), Reddy and Liu (1985), and Kovarik (1980) for linear shear-deformable shell theories, we assume that the local displacements u_1, u_2, and u_3 (with respect to the $\xi\eta\zeta$ coordinate system) have the following form:

$$u_1(x,\theta,z,t) = u_1^0(x,\theta,t) + z\theta_2(x,\theta,t) + \left(z - \frac{4z^3}{3h^2}\right)\gamma_5(x,\theta,t) \tag{9.5.13a}$$

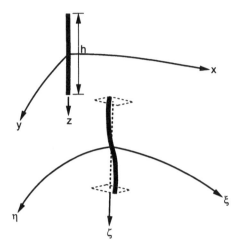

Fig. 9.5.2 A differential shell element with third-order shear-warpings.

$$u_2(x, \theta, z, t) = u_2^0(x, \theta, t) - z[\theta_1(x, \theta, t) - \theta_{10}(x, \theta)]$$
$$+ \left(1 + \frac{z}{a}\right)\left(z - \frac{4z^3}{3h^2}\right)\gamma_4(x, \theta, t) \qquad (9.5.13b)$$

$$u_3(x, \theta, z, t) = u_3^0(x, \theta, t) \qquad (9.5.13c)$$

The first terms on the right sides of (9.5.13a,b) are due to inplane stretching (membrane deformations), the second terms are due to bending, and the third terms are due to transverse shear deformations. Here, the u_i^0 ($i = 1, 2, 3$) are the displacements (with respect to the $\xi\eta\zeta$ system) of the observed reference point A, γ_4 and γ_5 are the transverse shear strains at the midsurface (i.e., $\gamma_5 \equiv \epsilon_{13}|_{z=0}$ and $\gamma_4 \equiv \epsilon_{23}|_{z=0}$), θ_1 and θ_2 are the rotation angles of the normal to the deformed midsurface with respect to the ξ and η axes, respectively, θ_{10} accounts for the initial angle with respect to the ξ axis, and h is the total thickness of the shell. Equation (9.5.13c) implies that the transverse normal strain is neglected in the modeling. This displacement field accounts for a parabolic distribution of the transverse shear strains and satisfies the free shear stress and strain conditions $\sigma_{13} = \sigma_{23} = \epsilon_{13} = \epsilon_{23} = 0$ at $z = \pm h/2$, as shown later. Because an undeformed infinitesimal shell element is a straight line with a finite length h and two infinitesimal side lengths dx and $ad\theta$, the deformed infinitesimal shell element with third-order shear deformations is a curved line, as shown in Figure 9.5.2.

Because the Jaumann strains are referred to the undeformed length, we use the Lagrangian coordinates x, θ, and z in the derivation, thereby satisfying conservation of mass identically and avoiding the calculation of co-moving derivatives. Taking the derivatives of the local displacement vector $\mathbf{U} = u_1\mathbf{i}_1 + u_2\mathbf{i}_2 + u_3\mathbf{i}_3$, we obtain

$$\frac{\partial \mathbf{U}}{\partial x} = \frac{\partial u_1}{\partial x}\mathbf{i}_1 + \frac{\partial u_2}{\partial x}\mathbf{i}_2 + \frac{\partial u_3}{\partial x}\mathbf{i}_3 + u_1\frac{\partial \mathbf{i}_1}{\partial x} + u_2\frac{\partial \mathbf{i}_2}{\partial x} + u_3\frac{\partial \mathbf{i}_3}{\partial x}$$

$$= \left[e_1 + zk_1 + \hat{z}\gamma_{5x} - k_5\gamma_4\hat{z}\left(1+\frac{z}{a}\right)\right]\mathbf{i}_1$$

$$+ \left[\gamma_{61} - \theta_3 + zk_{61} + \left(1+\frac{z}{a}\right)\hat{z}\gamma_{4x} + k_5\gamma_5\hat{z}\right]\mathbf{i}_2$$

$$- \left[k_1\gamma_5 + k_{61}\gamma_4\left(1+\frac{z}{a}\right)\right]\hat{z}\mathbf{i}_3$$

$$\frac{\partial \mathbf{U}}{(a+z)\partial\theta} = \frac{a}{a+z}\left[\frac{\partial u_1}{\partial y}\mathbf{i}_1 + \frac{\partial u_2}{\partial y}\mathbf{i}_2 + \frac{\partial u_3}{\partial y}\mathbf{i}_3 + u_1\frac{\partial \mathbf{i}_1}{\partial y} + u_2\frac{\partial \mathbf{i}_2}{\partial y} + u_3\frac{\partial \mathbf{i}_3}{\partial y}\right]$$

$$= \left[\frac{a}{a+z}(\gamma_{62} + \theta_3) + \frac{az}{a+z}k_{62} + \frac{a}{a+z}\hat{z}\gamma_{5y} - k_4\gamma_4\hat{z}\right]\mathbf{i}_1$$

$$+ \left[\frac{a}{a+z}e_2 + \frac{az}{a+z}\left(k_2 - \frac{1}{a}\right) + \hat{z}\gamma_{4y} + k_4\gamma_5\frac{\hat{z}a}{a+z}\right]\mathbf{i}_2$$

$$- \left(k_{62}\gamma_5\frac{a}{a+z} + k_2\gamma_4\right)\hat{z}\mathbf{i}_3$$

$$\frac{\partial \mathbf{U}}{\partial z} = \frac{\partial u_1}{\partial z}\mathbf{i}_1 + \frac{\partial u_2}{\partial z}\mathbf{i}_2 + \frac{\partial u_3}{\partial z}\mathbf{i}_3$$

$$= \left(1 - \frac{4z^2}{h^2}\right)\gamma_5\mathbf{i}_1 + \left[\left(1+\frac{z}{a}\right)\left(1 - \frac{4z^2}{h^2}\right)\gamma_4 + \gamma_4\frac{\hat{z}}{a}\right]\mathbf{i}_2$$

and

$$\epsilon_{11} = \frac{\partial \mathbf{U}}{\partial x}\cdot\mathbf{i}_1 = e_1 + zk_1 + \hat{z}\left[\gamma_{5x} - k_5\gamma_4\left(1+\frac{z}{a}\right)\right] \quad (9.5.14a)$$

$$\epsilon_{22} = \frac{\partial \mathbf{U}}{(a+z)\partial\theta}\cdot\mathbf{i}_2 = \frac{a}{a+z}\left[e_2 + z\bar{k}_2 + \hat{z}\left(\frac{a+z}{a}\gamma_{4y} + k_4\gamma_5\right)\right] \quad (9.5.14b)$$

$$\epsilon_{33} = \frac{\partial \mathbf{U}}{\partial z}\cdot\mathbf{i}_3 = 0 \quad (9.5.14c)$$

$$e_{12} = \frac{\partial \mathbf{U}}{\partial x}\cdot\mathbf{i}_2 = \gamma_{61} - \theta_3 + zk_{61} + \hat{z}\left[\left(1+\frac{z}{a}\right)\gamma_{4x} + k_5\gamma_5\right] \quad (9.5.14d)$$

$$e_{21} = \frac{\partial \mathbf{U}}{(a+z)\partial\theta}\cdot\mathbf{i}_1 = \frac{a}{a+z}\left[\gamma_{62} + \theta_3 + zk_{62} + \hat{z}\left(\gamma_{5y} - k_4\gamma_4\frac{a+z}{a}\right)\right] \quad (9.5.14e)$$

$$\epsilon_{13} = \frac{\partial \mathbf{U}}{\partial z}\cdot\mathbf{i}_1 + \frac{\partial \mathbf{U}}{\partial x}\cdot\mathbf{i}_3 = \left(1 - \frac{4z^2}{h^2}\right)\gamma_5 - \hat{z}\left[k_1\gamma_5 + \left(1+\frac{z}{a}\right)k_{61}\gamma_4\right] \quad (9.5.14f)$$

$$\epsilon_{23} = \frac{\partial \mathbf{U}}{\partial z}\cdot\mathbf{i}_2 + \frac{\partial \mathbf{U}}{(a+z)\partial\theta}\cdot\mathbf{i}_3 = \left(1+\frac{z}{a}\right)\left(1 - \frac{4z^2}{h^2}\right)\gamma_4 - \hat{z}\left(\bar{k}_2\gamma_4 + \frac{a}{a+z}k_{62}\gamma_5\right) \quad (9.5.14g)$$

where θ_3 is the rotation angle of the normal to the deformed midsurface with respect to the ζ axis and

$$\frac{\partial u_1^0}{\partial x} \equiv e_1, \quad \frac{\partial u_2^0}{\partial y} \equiv e_2, \quad \frac{\partial \theta_{10}}{\partial \theta} = -1, \quad \frac{\partial \theta_{10}}{\partial x} = 0 \quad (9.5.15a)$$

596   SHELLS

$$\frac{\partial \theta_2}{\partial x} \equiv k_1, \quad \frac{\partial \theta_1}{\partial y} \equiv -k_2, \quad \frac{\partial \theta_1}{\partial x} \equiv -k_{61}, \quad \frac{\partial \theta_2}{\partial y} \equiv k_{62} \qquad (9.5.15b)$$

$$\frac{\partial u_2^0}{\partial x} \equiv \gamma_{61} - \theta_3, \quad \frac{\partial u_1^0}{\partial y} \equiv \gamma_{62} + \theta_3, \quad \bar{k}_2 \equiv k_2 - \frac{1}{a}, \quad \hat{z} \equiv z - \frac{4z^3}{3h^2} \qquad (9.5.15c)$$

We note that the nonlinear terms in (9.5.14a-g) are due to the shear rotations γ_4 and γ_5 and the rotation of the coordinate system. Because the $\xi\eta\zeta$ coordinate system is a local system and the $\xi\eta$ plane is tangent to the deformed midsurface,

$$u_1^0 = u_2^0 = u_3^0 = \theta_1 = \theta_{10} = \theta_2 = \theta_3 = \partial u_3^0/\partial x = \partial u_3^0/\partial y = 0 \qquad (9.5.15d)$$

Although $\theta_3 = 0$, we keep it in the strain expressions because $\delta\theta_3(\neq 0)$ is needed in deriving nonlinear governing equations using the extended Hamilton's principle. The constitutive equations which relate the Jaumann stresses to the Jaumann strains are given in (2.4.34) and (2.4.35).

The displacement vector $\mathbf{D}$ of an arbitrary point of the observed element is given by

$$\mathbf{D} = u\mathbf{j}_1 + v\mathbf{j}_2 + w\mathbf{j}_3 + z\mathbf{i}_3 - z\mathbf{j}_3 + u_1\mathbf{i}_1 + u_2\mathbf{i}_2 + u_3\mathbf{i}_3 \qquad (9.5.16)$$

Variation of the kinetic energy of the element consists of three parts. The first part is the translational kinetic energy due to the virtual displacements δu, δv, and δw. The second part is the rotational kinetic energy due to the virtual shear rotations $\delta\gamma_4$ and $\delta\gamma_5$. And the third part is the rotational kinetic energy due to the virtual rotations $\delta\theta_i$ of the local coordinate system. Because the rotations of the $\xi\eta\zeta$ coordinate system correspond to rigid-body rotations of the element, they result in the main part of the rotational kinetic energy (the other part of the rotational kinetic energy is due to shear rotations). Using (9.5.16), (9.5.13a-c), (9.5.15d), (9.5.4a), and (6.3.1a,b), we obtain variation of the kinetic energy δT as

$$\delta T = -\int_{-h/2}^{h/2}\int_A \rho\ddot{\mathbf{D}}\cdot\delta\mathbf{D}(1+z/a)dxdydz$$

$$= -\int_A \Big\{\delta u\Big[I_1\ddot{u} + I_4(\gamma_5 T_{11})\ddot{} + \bar{I}_4(\gamma_4 T_{21})\ddot{} + I_2\ddot{T}_{31}\Big]$$

$$+\delta v\Big[I_1\ddot{v} + I_4(\gamma_5 T_{12})\ddot{} + \bar{I}_4(\gamma_4 T_{22})\ddot{} + I_2\ddot{T}_{32}\Big]$$

$$+\delta w\Big[I_1\ddot{w} + I_4(\gamma_5 T_{13})\ddot{} + \bar{I}_4(\gamma_4 T_{23})\ddot{} + I_2\ddot{T}_{33}\Big]$$

$$+\delta\gamma_4\Big[\bar{I}_4(\ddot{u}T_{21} + \ddot{v}T_{22} + \ddot{w}T_{23}) + \bar{I}_7 f_{41} + \tilde{I}_7 f_{42} + \bar{I}_5 f_{43}\Big]$$

$$+\delta\gamma_5\Big[I_4(\ddot{u}T_{11} + \ddot{v}T_{12} + \ddot{w}T_{13}) + I_7 f_{51} + \bar{I}_7 f_{52} + I_5 f_{53}\Big]$$

$$+\delta\theta_1\Big[\bar{I}_4\gamma_4(\ddot{u}T_{31} + \ddot{v}T_{32} + \ddot{w}T_{33}) + \bar{I}_7\gamma_4 f_{11} + \tilde{I}_7\gamma_4 f_{12} + \bar{I}_5\gamma_4 f_{13}$$

$$-I_2(\ddot{u}T_{21} + \ddot{v}T_{22} + \ddot{w}T_{23}) - I_5 f_{41} - \bar{I}_5 f_{42} - I_3 f_{43}\Big]$$

$$+\delta\theta_2\Big[I_2(\ddot{u}T_{11} + \ddot{v}T_{12} + \ddot{w}T_{13}) + I_5 f_{51} + \bar{I}_5 f_{52} + I_3 f_{53}$$

$$-I_4\gamma_5(\ddot{u}T_{31} + \ddot{v}T_{32} + \ddot{w}T_{33}) - I_7\gamma_5 f_{11} - \bar{I}_7\gamma_5 f_{12} - I_5\gamma_5 f_{13}\Big]$$

$$+\delta\theta_3\Big[I_4\gamma_5(\ddot{u}T_{21} + \ddot{v}T_{22} + \ddot{w}T_{23}) + I_7\gamma_5 f_{41} + \bar{I}_7\gamma_5 f_{42} + I_5\gamma_5 f_{43}$$

$$-\bar{I}_4\gamma_4(\ddot{u}T_{11} + \ddot{v}T_{12} + \ddot{w}T_{13}) - \bar{I}_7\gamma_4 f_{51}$$

$$-\tilde{I}_7\gamma_4 f_{52} - \bar{I}_5\gamma_4 f_{53}\Big]\bigg\}dxdy \quad (9.5.17)$$

where the factor $(1 + z/a)$ in (9.5.17) is due to the fact that the undeformed area of a surface element at a distance z from the middle surface is $(a + z)d\theta dx$ instead of $ad\theta dx(= dydx)$, which is the so-called trapezoidal-edge effect. Moreover,

$$f_{11} \equiv (\gamma_5 T_{11})\ddot{T}_{31} + (\gamma_5 T_{12})\ddot{T}_{32} + (\gamma_5 T_{13})\ddot{T}_{33}$$
$$f_{12} \equiv (\gamma_4 T_{21})\ddot{T}_{31} + (\gamma_4 T_{22})\ddot{T}_{32} + (\gamma_4 T_{23})\ddot{T}_{33}$$
$$f_{13} \equiv \ddot{T}_{31}T_{31} + \ddot{T}_{32}T_{32} + \ddot{T}_{33}T_{33}$$
$$f_{41} \equiv (\gamma_5 T_{11})\ddot{T}_{21} + (\gamma_5 T_{12})\ddot{T}_{22} + (\gamma_5 T_{13})\ddot{T}_{23}$$
$$f_{42} \equiv (\gamma_4 T_{21})\ddot{T}_{21} + (\gamma_4 T_{22})\ddot{T}_{22} + (\gamma_4 T_{23})\ddot{T}_{23}$$
$$f_{43} \equiv \ddot{T}_{31}T_{21} + \ddot{T}_{32}T_{22} + \ddot{T}_{33}T_{23}$$
$$f_{51} \equiv (\gamma_5 T_{11})\ddot{T}_{11} + (\gamma_5 T_{12})\ddot{T}_{12} + (\gamma_5 T_{13})\ddot{T}_{13}$$
$$f_{52} \equiv (\gamma_4 T_{21})\ddot{T}_{11} + (\gamma_4 T_{22})\ddot{T}_{12} + (\gamma_4 T_{23})\ddot{T}_{13}$$
$$f_{53} \equiv \ddot{T}_{31}T_{11} + \ddot{T}_{32}T_{12} + \ddot{T}_{33}T_{13}$$

and the I_i are the inertias, which are defined as

$$[I_1, I_2, I_3, I_4, I_5, I_7] \equiv \int_{-h/2}^{h/2} \rho\left[1, z, z^2, \hat{z}, z\hat{z}, \hat{z}^2\right]\left(1 + \frac{z}{a}\right)dz \quad (9.5.18a)$$

$$[\bar{I}_4, \bar{I}_5, \bar{I}_7, \tilde{I}_7] \equiv \int_{-h/2}^{h/2} \rho\left[\hat{z}, z\hat{z}, \hat{z}^2, \hat{z}^2\left(1 + \frac{z}{a}\right)\right]\left(1 + \frac{z}{a}\right)^2 dz \quad (9.5.18b)$$

Because the thickness h of flexible shells is usually small and hence the inertias are small, we assume that I_2, I_3, I_4, I_5, I_7, $\bar{I}_4$, $\bar{I}_5$, $\bar{I}_7$, and $\tilde{I}_7$ are $O(\epsilon^2)$. We note that $I_2 = I_4 = \bar{I}_4 = 0$ when ρ is constant and $z/a \to 0$.

Next, we expand the coefficients of δu, δv, δw, $\delta\gamma_4$, $\delta\gamma_5$, $\delta\theta_1$, $\delta\theta_2$, and $\delta\theta_3$ in (9.5.17) to $O(\epsilon^3)$ and obtain

$$\delta T = -\int_A \bigg\{(I_1\ddot{u} - I_2\ddot{w}_x + I_4\ddot{\gamma}_5)\,\delta u + (I_1\ddot{v} - I_2\ddot{w}_y + I_2\ddot{v}/a + \bar{I}_4\ddot{\gamma}_4)\delta v$$
$$+I_1\ddot{w}\delta w + (I_2\ddot{u} - I_3\ddot{w}_x + I_5\ddot{\gamma}_5)\delta\theta_2 - (I_2\ddot{v} - I_3\ddot{w}_y + I_3\ddot{v}/a + \bar{I}_5\ddot{\gamma}_4)\delta\theta_1$$
$$+(I_4\ddot{u} - I_5\ddot{w}_x + I_7\ddot{\gamma}_5)\delta\gamma_5 + (\bar{I}_4\ddot{v} - \bar{I}_5\ddot{w}_y + \bar{I}_5\ddot{v}/a + \tilde{I}_7\ddot{\gamma}_4)\delta\gamma_4\bigg\}dxdy$$

$$(9.5.19)$$

In (9.5.19), the coefficients of $\delta\theta_1$ and $\delta\theta_2$ are the rotary inertias of the shell element with respect to the ξ and η axes, respectively. We note that the coefficient of $\delta\theta_3$ is zero because the rotary inertia of the shell element with respect to the ζ axis is small and of order ϵ^4 or higher. Equation (9.5.19) shows that there are no nonlinear inertia terms of order ϵ^3 or lower. Because the coefficients of $\delta\theta_1$ and $\delta\theta_2$ are all of $O(\epsilon^3)$, we only need to expand $\delta\theta_1$ and $\delta\theta_2$ in (6.3.1b) as

$$\delta\theta_1 = \delta w_y - \delta v/a, \quad \delta\theta_2 = -\delta w_x \tag{9.5.20}$$

on account of (9.5.4b-d). Substituting (9.5.20) into (9.5.19) and integrating the result by parts yields

$$\begin{aligned}\delta T = -\int_A \Big\{ & (I_1\ddot{u} - I_2\ddot{w}_x + I_4\ddot{\gamma}_5)\delta u \\
& + (I_1\ddot{v} - I_2\ddot{w}_y + \bar{I}_4\ddot{\gamma}_4 + 2I_2\ddot{v}/a - I_3\ddot{w}_y/a + I_3\ddot{v}/a^2 + \bar{I}_5\ddot{\gamma}_4/a)\delta v \\
& + [I_1\ddot{w} + (I_2\ddot{u} - I_3\ddot{w}_x + I_5\ddot{\gamma}_5)_x + (I_2\ddot{v} - I_3\ddot{w}_y + I_3\ddot{v}/a + \bar{I}_5\ddot{\gamma}_4)_y]\delta w \\
& + (I_4\ddot{u} - I_5\ddot{w}_x + I_7\ddot{\gamma}_5)\delta\gamma_5 + (\bar{I}_4\ddot{v} - \bar{I}_5\ddot{w}_y + \bar{I}_5\ddot{v}/a + \tilde{I}_7\ddot{\gamma}_4)\delta\gamma_4 \Big\} dxdy \\
& + \int_y (I_2\ddot{u} - I_3\ddot{w}_x + I_5\ddot{\gamma}_5)\delta w \Big|_{x=0}^{x=L} dy \\
& + \int_x (I_2\ddot{v} - I_3\ddot{w}_y + I_3\ddot{v}/a + \bar{I}_5\ddot{\gamma}_4)\delta w \Big|_{\theta=0}^{\theta=\theta_0} dx \end{aligned} \tag{9.5.21}$$

Using (9.5.14a-g), we obtain variation of the elastic energy $\delta\Pi$ as

$$\begin{aligned}\delta\Pi = & \int_{-h/2}^{h/2}\int_A \Big\{\sigma_{11}\delta\epsilon_{11} + \sigma_{12}\delta\epsilon_{12} + \sigma_{21}\delta\epsilon_{21} + \sigma_{13}\delta\epsilon_{13} \\
& + \sigma_{22}\delta\epsilon_{22} + \sigma_{23}\delta\epsilon_{23}\Big\}\left(1+\frac{z}{a}\right)dxdydz \\
= & \int_{-h/2}^{h/2}\int_A \Big\{\sigma_{11}\left(1+\frac{z}{a}\right)\delta\left[e_1 + zk_1 + \hat{z}\left[\gamma_{5x} - k_5\gamma_4\left(1+\frac{z}{a}\right)\right]\right] \\
& + \sigma_{12}\left(1+\frac{z}{a}\right)\delta\left[\gamma_{61} - \theta_3 + zk_{61} + \hat{z}\left[\left(1+\frac{z}{a}\right)\gamma_{4x} + k_5\gamma_5\right]\right] \\
& + \sigma_{13}\left(1+\frac{z}{a}\right)\delta\left[\left(1-\frac{4z^2}{h^2}\right)\gamma_5 - \hat{z}\left[k_1\gamma_5 + \left(1+\frac{z}{a}\right)k_{61}\gamma_4\right]\right] \\
& + \sigma_{23}\delta\left[\left(1-\frac{4z^2}{h^2}\right)\gamma_4\left(1+\frac{2}{a}z+\frac{z^2}{a^2}\right) - \hat{z}\left[\left(1+\frac{z}{a}\right)\bar{k}_2\gamma_4\right.\right. \\
& \left.\left.+ k_{62}\gamma_5\right]\right] + \sigma_{22}\delta\left[e_2 + z\bar{k}_2 + \hat{z}\left[\left(1+\frac{z}{a}\right)\gamma_{4y} + k_4\gamma_5\right]\right] \\
& + \sigma_{21}\delta\left[\gamma_{62} + \theta_3 + zk_{62} + \hat{z}\left[\gamma_{5y} - k_4\gamma_4\left(1+\frac{z}{a}\right)\right]\right]\Big\}dxdydz \end{aligned}$$

$$= \int_A \Big[N_1\delta e_1 + M_1\delta k_1 + (M_1 - P_1)\delta(\gamma_{5x} - k_5\gamma_4)$$
$$+ N_{61}\delta\gamma_{61} - N_{61}\delta\theta_3 + M_{61}\delta k_{61} + (M_{61} - P_{61})\delta(\gamma_{4x} + k_5\gamma_5)$$
$$+ (Q_1 - R_1)\delta\gamma_5 - (S_1 - G_1)\delta(k_1\gamma_5 + k_{61}\gamma_4)$$
$$+ N_{62}\delta\gamma_{62} + N_{62}\delta\theta_3 + M_{62}\delta k_{62} + (M_{62} - P_{62})\delta(\gamma_{5y} - k_4\gamma_4)$$
$$+ N_2\delta e_2 + M_2\delta \bar{k}_2 + (M_2 - P_2)\delta(\gamma_{4y} + k_4\gamma_5)$$
$$+ (Q_2 - R_2 + 2S_2/a - 6G_2/a)\delta\gamma_4$$
$$- (S_2 - G_2)\delta(\bar{k}_2\gamma_4 + k_{62}\gamma_5)\Big]dxdy \qquad (9.5.22)$$

Since the theory developed here is for thin shells undergoing small strains but moderate rotations, the geometric nonlinearity is the most important effect and it is represented by the transformation matrix $[T]$ and the nonlinear curvatures k_j. On the other hand, because the strains are small, the term $(1 + z/a)$ in the coefficients of the nonlinear terms $k_i\gamma_j$ and the derivatives of γ_4 and γ_5 is approximated by unity and z^2/a^2 is neglected. Moreover, the stress resultants, stress couples, and modified stresses are defined as

$$(N_1, N_2, N_{61}, N_{62}) = \int_{-h/2}^{h/2}(\tilde{\sigma}_{11}, \sigma_{22}, \tilde{\sigma}_{12}, \sigma_{21})dz \qquad (9.5.23a)$$

$$(Q_1, Q_2) = \int_{-h/2}^{h/2}(\tilde{\sigma}_{13}, \sigma_{23})dz \qquad (9.5.23b)$$

$$\left(\hat{R}_1, \hat{R}_2\right) = \int_{-h/2}^{h/2}(\tilde{\sigma}_{13}, \sigma_{23})z^2 dz, \quad (R_1, R_2) = \frac{4}{h^2}\left(\hat{R}_1, \hat{R}_2\right) \qquad (9.5.23c)$$

$$(M_1, M_2, M_{61}, M_{62}) = \int_{-h/2}^{h/2}(\tilde{\sigma}_{11}, \sigma_{22}, \tilde{\sigma}_{12}, \sigma_{21})z\,dz \qquad (9.5.23d)$$

$$\left(\hat{P}_1, \hat{P}_2, \hat{P}_{61}, \hat{P}_{62}\right) = \int_{-h/2}^{h/2}(\tilde{\sigma}_{11}, \sigma_{22}, \tilde{\sigma}_{12}, \sigma_{21})z^3 dz$$
$$(P_1, P_2, P_{61}, P_{62}) = \frac{4}{3h^2}\left(\hat{P}_1, \hat{P}_2, \hat{P}_{61}, \hat{P}_{62}\right) \qquad (9.5.23e)$$

$$(S_1, S_2) = \int_{-h/2}^{h/2}(\tilde{\sigma}_{13}, \sigma_{23})z\,dz \qquad (9.5.23f)$$

$$\left(\hat{G}_1, \hat{G}_2\right) = \int_{-h/2}^{h/2}(\tilde{\sigma}_{13}, \sigma_{23})z^3 dz, \quad (G_1, G_2) = \frac{4}{3h^2}\left(\hat{G}_1, \hat{G}_2\right) \qquad (9.5.23g)$$

$$\tilde{\sigma}_{11} \equiv \sigma_{11}\left(1 + \frac{z}{a}\right), \quad \tilde{\sigma}_{12} \equiv \sigma_{12}\left(1 + \frac{z}{a}\right), \quad \tilde{\sigma}_{13} \equiv \sigma_{13}\left(1 + \frac{z}{a}\right) \qquad (9.5.23h)$$

We note that
$$N_{61} = N_{62} + \frac{1}{a}M_{62} \qquad (9.5.23i)$$

Here, N_1, N_2, N_{61}, and N_{62} are membrane force intensities; Q_1 and Q_2 are transverse shear intensities; R_1 and R_2 are higher-order transverse shearing force intensities associated with Q_1 and Q_2, respectively; M_1 and M_2 are bending moment intensities; M_{61} and M_{62} are twisting moment intensities; P_{61} and P_{62} are higher-order twisting moment intensities associated with M_{61} and M_{62}; P_1 and P_2 are higher-order bending moment intensities associated with M_1 and M_2, respectively; S_1 and S_2 are moment intensities; and G_1 and G_2 are higher-order moment intensities associated with S_1 and S_2, respectively. We note that the S_i and G_i only appear as nonlinear terms in the equations of motion, as shown later. One can define different artificial stress resultants and couples, but the conjugate strain expressions will be different (Budiansky and Sanders, 1963; Atluri, 1984).

Substituting (6.4.2a,b) into (9.5.22), we obtain the variation of elastic energy in terms of the stress resultants, stress couples, and midsurface curvatures and strains as

$$\delta\Pi = \int_A \{N_1\delta e_1 + N_2\delta e_2 + N_{61}\delta\gamma_{61} + N_{62}\delta\gamma_{62} + \Theta_1\delta\theta_1 + \Theta_2\delta\theta_2 + \Theta_3\delta\theta_3$$

$$+\Gamma_4\delta\gamma_4 + \Gamma_5\delta\gamma_5\}dxdy + \int_x \Big[-\hat{M}_2\delta\theta_1 + \hat{M}_{62}\delta\theta_2 + m_{32}\delta\theta_3$$

$$+ (M_2 - P_2)\delta\gamma_4 + (M_{62} - P_{62})\delta\gamma_5 \Big]_{\theta=0}^{\theta=\theta_0} dx + \int_y \Big[-\hat{M}_{61}\delta\theta_1 + \hat{M}_1\delta\theta_2$$

$$+ m_{31}\delta\theta_3 + (M_{61} - P_{61})\delta\gamma_4 + (M_1 - P_1)\delta\gamma_5 \Big]_{x=0}^{x=L} dy \qquad (9.5.24)$$

where

$$\hat{M}_1 \equiv M_1 - (S_1 - G_1)\gamma_5 \qquad (9.5.25a)$$
$$\hat{M}_2 \equiv M_2 - (S_2 - G_2)\gamma_4 \qquad (9.5.25b)$$
$$\hat{M}_{61} \equiv M_{61} - (S_1 - G_1)\gamma_4 \qquad (9.5.25c)$$
$$\hat{M}_{62} \equiv M_{62} - (S_2 - G_2)\gamma_5 \qquad (9.5.25d)$$
$$m_{31} \equiv (M_{61} - P_{61})\gamma_5 - (M_1 - P_1)\gamma_4 \qquad (9.5.25e)$$
$$m_{32} \equiv -(M_{62} - P_{62})\gamma_4 + (M_2 - P_2)\gamma_5 \qquad (9.5.25f)$$
$$\Theta_1 \equiv \hat{M}_{61x} + \hat{M}_{2y} - m_{31}k_1 - m_{32}k_{62} + \hat{M}_1k_5 + \hat{M}_{62}k_4 \qquad (9.5.25g)$$
$$\Theta_2 \equiv -\hat{M}_{1x} - \hat{M}_{62y} - m_{31}k_{61} - m_{32}k_2 + \hat{M}_2k_4 + \hat{M}_{61}k_5 \qquad (9.5.25h)$$
$$\Theta_3 \equiv -m_{31x} - m_{32y} + \hat{M}_1k_{61} - \hat{M}_2k_{62} - \hat{M}_{61}k_1 + \hat{M}_{62}k_2 + N_{62} - N_{61} \qquad (9.5.25i)$$

$$\Gamma_4 \equiv -(M_2 - P_2)_y - (M_{61} - P_{61})_x + Q_2 - R_2 + 2S_2/a - 6G_2/a$$
$$-(S_1 - G_1)k_{61} - (S_2 - G_2)\bar{k}_2 - (M_1 - P_1)k_5 - (M_{62} - P_{62})k_4$$
$$\qquad (9.5.25j)$$
$$\Gamma_5 \equiv -(M_1 - P_1)_x - (M_{62} - P_{62})_y + Q_1 - R_1$$
$$-(S_1 - G_1)k_1 - (S_2 - G_2)k_{62} + (M_2 - P_2)k_4 + (M_{61} - P_{61})k_5$$
$$\qquad (9.5.25k)$$

Here, $\hat{M}_1$, $\hat{M}_2$, $\hat{M}_{61}$, $\hat{M}_{62}$, m_{31}, and m_{32} represent the stress couples acting on the edges of the shell element, as shown in Figure 9.5.3a. In Figure 9.5.3b, we show the inplane extension force intensities N_1 and N_2, inplane shear intensities N_{61} and

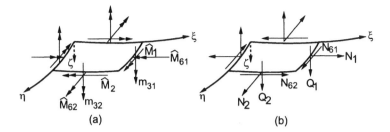

Fig. 9.5.3 Loadings on a nonlinear shear-deformable shell element: (a) stress couples and (b) stress resultants.

N_{62}, and transverse shear intensities Q_1 and Q_2. One can see from (9.5.25e,f,i) that m_{31}, m_{32}, and Θ_3 are nonlinear. Moreover, the effects of S_1, S_2, G_1, G_2, k_4, and k_5 are nonlinear. It can be seen from (9.5.22) and (9.5.14a-g) that if changes in the elastic energy due to changes in the configuration are not included (i.e., the nonlinear terms in (9.5.14a-g) are neglected) and linear curvature expressions are used, then there are no nonlinear terms in (9.5.25a-k) and $m_{31} = m_{32} = 0$ and $\Theta_3 = M_{62}/a + N_{62} - N_{61} = 0$ on account of (9.5.23i).

Substituting (9.5.1a), (6.3.3a,b), (6.3.5a,b), and (6.3.7a,b) into (9.5.24), we obtain

$$\delta\Pi = \int_A \left\{ [N_1, N_{61}/(1+e_1), -\Theta_2/(1+e_1)][T] \begin{Bmatrix} \delta u_x \\ \delta v_x \\ \delta w_x \end{Bmatrix} \right.$$

$$+ [N_{62}/(1+e_2), N_2, \Theta_1/(1+e_2)][T] \begin{bmatrix} 0 & 0 & 0 \\ 0 & 0 & 1/a \\ 0 & -1/a & 0 \end{bmatrix} \begin{Bmatrix} \delta u \\ \delta v \\ \delta w \end{Bmatrix}$$

$$+ [N_{62}/(1+e_2), N_2, \Theta_1/(1+e_2)][T] \begin{Bmatrix} \delta u_y \\ \delta v_y \\ \delta w_y \end{Bmatrix}$$

$$\left. + \Theta_3\delta\theta_3 + \Gamma_4\delta\gamma_4 + \Gamma_5\delta\gamma_5 \right\} dxdy + \int_x \left[-\hat{M}_2\delta\theta_1 + \hat{M}_{62}\delta\theta_2 + m_{32}\delta\theta_3 \right.$$

$$+ (M_2 - P_2)\delta\gamma_4 + (M_{62} - P_{62})\delta\gamma_5 \bigg]_{\theta=0}^{\theta=\theta_0} dx + \int_y \left[-\hat{M}_{61}\delta\theta_1 + \hat{M}_1\delta\theta_2 \right.$$

$$\left. + m_{31}\delta\theta_3 + (M_{61} - P_{61})\delta\gamma_4 + (M_1 - P_1)\delta\gamma_5 \right]_{x=0}^{x=L} dy \quad (9.5.26)$$

Substituting for $\delta\theta_1$ and $\delta\theta_2$ from (6.3.5a,b) into (9.5.26), substituting (9.5.26) and (9.5.21) into the extended Hamilton's principle (see (7.2.1)), integrating the result by parts, and then setting each of the coefficients of δu, δv, δw, $\delta\gamma_4$, and $\delta\gamma_5$ equal to zero, we obtain the following equations of motion:

$$F_{11x} + F_{12y} = I_1\ddot{u} - I_2\ddot{w}_x + I_4\ddot{\gamma}_5 + \mu_1\dot{u} \qquad (9.5.27)$$

$$F_{21x} + F_{22y} + \frac{1}{a}F_{32} = I_1\ddot{v} - I_2\ddot{w}_y + \bar{I}_4\ddot{\gamma}_4 + \frac{2}{a}I_2\ddot{v} - \frac{1}{a}I_3\ddot{w}_y$$
$$+ \frac{1}{a^2}I_3\ddot{v} + \frac{1}{a}\bar{I}_5\ddot{\gamma}_4 + \mu_2\dot{v} \qquad (9.5.28)$$

$$F_{31x} + F_{32y} - \frac{1}{a}F_{22} + q = I_1\ddot{w} + \mu_3\dot{w} + (I_2\ddot{u} - I_3\ddot{w}_x + I_5\ddot{\gamma}_5)_x$$
$$+ \left(I_2\ddot{v} - I_3\ddot{w}_y + \frac{1}{a}I_3\ddot{v} + \bar{I}_5\ddot{\gamma}_4\right)_y \qquad (9.5.29)$$

$$(M_2 - P_2)_y + (M_{61} - P_{61})_x - Q_2 + R_2 - 2S_2/a + 6G_2/a + (S_1 - G_1)k_{61}$$
$$+ (S_2 - G_2)\bar{k}_2 + (M_1 - P_1)k_5 + (M_{62} - P_{62})k_4$$
$$= \bar{I}_4\ddot{v} - \bar{I}_5\ddot{w}_y + \frac{1}{a}\bar{I}_5\ddot{v} + \tilde{I}_7\ddot{\gamma}_4 + \mu_4\dot{\gamma}_4 \qquad (9.5.30)$$

$$(M_1 - P_1)_x + (M_{62} - P_{62})_y - Q_1 + R_1 + (S_1 - G_1)k_1 + (S_2 - G_2)k_{62}$$
$$- (M_2 - P_2)k_4 - (M_{61} - P_{61})k_5 = I_4\ddot{u} - I_5\ddot{w}_x + I_7\ddot{\gamma}_5 + \mu_5\dot{\gamma}_5$$
$$(9.5.31)$$

where

$$\{F_{11}, F_{21}, F_{31}\} = \{\tilde{F}_\alpha\}^T[T], \quad \{F_{12}, F_{22}, F_{32}\} = \{\tilde{F}_\beta\}^T[T]$$
$$\{\tilde{F}_\alpha\} = \{N_1, N_{61}/(1+e_1), -\Theta_2/(1+e_1)\}$$
$$\{\tilde{F}_\beta\} = \{N_{62}/(1+e_2), N_2, \Theta_1/(1+e_2)\} \qquad (9.5.32)$$

We added a linear viscous damping term to each of (9.5.27)-(9.5.31), where the μ_i are the damping coefficients. The boundary conditions for the shell are of the form: specify

Along $x = 0, L$:

$\delta u = 0$	or	$F_{11} + [\hat{M}_{61}T_{31}/(1+e_2)]_y$
$\delta v = 0$	or	$F_{21} + [\hat{M}_{61}T_{32}/(1+e_2)]_y + \hat{M}_{61}T_{33}/(a+ae_2)$
$\delta w = 0$	or	$F_{31} + [\hat{M}_{61}T_{33}/(1+e_2)]_y - \hat{M}_{61}T_{32}/(a+ae_2)$
		$- I_2\ddot{u} + I_3\ddot{w}_x - I_5\ddot{\gamma}_5$
$\delta w_x = 0$	or	$\hat{M}_1$
$\delta \gamma_4 = 0$	or	$M_{61} - P_{61}$
$\delta \gamma_5 = 0$	or	$M_1 - P_1$

Along $\theta = 0, \theta_0$:

$\delta u = 0$	or	$F_{12} + [\hat{M}_{62}T_{31}/(1+e_1)]_x$
$\delta v = 0$	or	$F_{22} + [\hat{M}_{62}T_{32}/(1+e_1)]_x$

$$\delta w = 0 \quad \text{or} \quad F_{32} + [\hat{M}_{62}T_{33}/(1+e_1)]_x$$
$$- I_2\ddot{v} + I_3\ddot{w}_y - I_3\ddot{v}/a - \bar{I}_5\ddot{\gamma}_4$$
$$\delta w_y = 0 \quad \text{or} \quad \hat{M}_2$$
$$\delta\gamma_4 = 0 \quad \text{or} \quad M_2 - P_2$$
$$\delta\gamma_5 = 0 \quad \text{or} \quad M_{62} - P_{62}$$

At $(x,\theta) = (0,0), (L,0), (0,\theta_0), (L,\theta_0)$:

$$\delta u = 0 \quad \text{or} \quad -T_{31}[\hat{M}_{62}/(1+e_1) + \hat{M}_{61}/(1+e_2)]$$
$$\delta v = 0 \quad \text{or} \quad -T_{32}[\hat{M}_{62}/(1+e_1) + \hat{M}_{61}/(1+e_2)]$$
$$\delta w = 0 \quad \text{or} \quad -T_{33}[\hat{M}_{62}/(1+e_1) + \hat{M}_{61}/(1+e_2)] \quad (9.5.33)$$

The fourth geometric boundary condition along the edges $x = 0, L$ should be $\delta\theta_2 = 0$, which is equivalent to $\delta w_x = 0$. For example, at a free or hinged edge, the boundary condition is to specify $\hat{M}_1$. At a fixed edge, $w = w_x = w_y = v = 0$ and hence $T_{31}/(1+e_1) = T_{32}/(1+e_1) = 0$ in (6.3.5a) and in turn $\delta\theta_2 = -T_{33}\delta w_x/(1+e_1)$. Similarly, the fourth kinematic boundary condition along the edges $\theta = 0, \theta_0$ should be $\delta\theta_1 = 0$, which is equivalent to $\delta w_y = 0$. For a closed circular cylinder (i.e., $\theta_0 = 2\pi$), there are no boundary conditions for edges where θ is constant and there is no $\hat{M}_{62}$ in the natural corner conditions. The natural corner conditions for u and v are nonlinear effects as can be seen from (9.5.33). Because the rotary inertia about the ζ axis is zero (see (9.5.19)), it follows from (9.5.26) that

$$\Theta_3 = 0 \qquad (9.5.34a)$$
$$m_{32} = 0 \quad \text{at} \quad \theta = 0, \ \theta_0 \qquad (9.5.34b)$$
$$m_{31} = 0 \quad \text{at} \quad x = 0, \ L \qquad (9.5.34c)$$

Equations (9.5.34a-c) represent a statement of the balance of the internal moments with respect to the ζ axis, which do not need to be solved in vibration analyses.

9.5.2 Simplified Shell Theories

Equations of Motion From a Vector Approach

Substituting (9.5.1a), (6.3.3), and (6.3.7) into (9.5.24) and adding the integrals of $Q_1 \times$ (6.3.5a) and $Q_2 \times$ (6.3.5b) to the result, we obtain

$$\delta\Pi = \int_A \left\{ [N_1, \ N_{61}/(1+e_1), \ Q_1][T] \begin{Bmatrix} \delta u_x \\ \delta v_x \\ \delta w_x \end{Bmatrix} \right.$$
$$\left. + [N_{62}/(1+e_2), \ N_2, \ Q_2][T] \begin{Bmatrix} \delta u_y \\ \delta v_y \\ \delta w_y \end{Bmatrix} + \Gamma_4\delta\gamma_4 + \Gamma_5\delta\gamma_5 \right.$$

$$+ [N_{62}/(1+e_2), N_2, Q_2] [T] \begin{bmatrix} 0 & 0 & 0 \\ 0 & 0 & 1/a \\ 0 & -1/a & 0 \end{bmatrix} \begin{Bmatrix} \delta u \\ \delta v \\ \delta w \end{Bmatrix}$$

$$+ [\Theta_1 - Q_2(1+e_2)] \delta\theta_1 + [\Theta_2 + Q_1(1+e_1)] \delta\theta_2 + \Theta_3 \delta\theta_3 \Bigg\} dxdy$$

$$+ \int_x \Big[-\hat{M}_2 \delta\theta_1 + \hat{M}_{62}\delta\theta_2 + m_{32}\delta\theta_3$$

$$+ (M_2 - P_2)\delta\gamma_4 + (M_{62} - P_{62})\delta\gamma_5 \Big]_{\theta=0}^{\theta=\theta_0} dx$$

$$+ \int_y \Big[-\hat{M}_{61}\delta\theta_1 + \hat{M}_1 \delta\theta_2 + m_{31}\delta\theta_3$$

$$+ (M_{61} - P_{61})\delta\gamma_4 + (M_1 - P_1)\delta\gamma_5 \Big]_{x=0}^{x=L} dy \qquad (9.5.35)$$

If we integrate the first two terms in (9.5.35) by parts, substitute the result and (9.5.21) into the extended Hamilton's principle, use (9.5.1b,c), set the coefficients of δu, δv, and δw equal to zero, and put them in vector form, we obtain

$$\frac{\partial \mathbf{F}_\alpha}{\partial x} + \frac{\partial \mathbf{F}_\beta}{\partial y} = \mathbf{I}_F \qquad (9.5.36)$$

where

$$\mathbf{F}_\alpha \equiv N_1 \mathbf{i}_1 + N_{61}/(1+e_1)\mathbf{i}_2 + Q_1 \mathbf{i}_3, \quad \mathbf{F}_\beta \equiv N_{62}/(1+e_2)\mathbf{i}_1 + N_2 \mathbf{i}_2 + Q_2 \mathbf{i}_3$$
$$\mathbf{I}_F \equiv (I_1 \ddot{u} - I_2 \ddot{w}_x + I_4 \ddot{\gamma}_5)\mathbf{j}_1 + (I_1 \ddot{v} - I_2 \ddot{w}_y + I_2 \ddot{v}/a + \bar{I}_4 \ddot{\gamma}_4)\mathbf{j}_2 + I_1 \ddot{w} \mathbf{j}_3$$
$$(9.5.37)$$

And, setting the coefficients of $\delta\theta_1$, $\delta\theta_2$, and $\delta\theta_3$ equal to zero and using (6.2.13) and (6.2.14) yields

$$\frac{\partial \mathbf{M}_\alpha}{\partial x} + \frac{\partial \mathbf{M}_\beta}{\partial y} + (1+e_1)\mathbf{i}_1 \times \mathbf{F}_\alpha + (1+e_2)\mathbf{i}_2 \times \mathbf{F}_\beta = \mathbf{I}_M \qquad (9.5.38)$$

where

$$\mathbf{M}_\alpha \equiv -\hat{M}_{61}\mathbf{i}_1 + \hat{M}_1 \mathbf{i}_2 + m_{31}\mathbf{i}_3, \quad \mathbf{M}_\beta \equiv -\hat{M}_2 \mathbf{i}_1 + \hat{M}_{62}\mathbf{i}_2 + m_{32}\mathbf{i}_3$$
$$\mathbf{I}_M \equiv (-I_2 \ddot{v} + I_3 \ddot{w}_y - I_3 \ddot{v}/a - \bar{I}_5 \ddot{\gamma}_4)\mathbf{i}_1 + (I_2 \ddot{u} - I_3 \ddot{w}_x + I_5 \ddot{\gamma}_5)\mathbf{i}_2 \quad (9.5.39)$$

Equation (9.5.36) is a statement of the balance of forces and (9.5.38) is a statement of the balance of moments for a shell element, which can be directly obtained using the vector (Newtonian) approach and stress resultants and moments. Thus, the nonlinear equations of motion can be obtained by using either an energy formulation (i.e., the extended Hamilton's principle) or a vector approach. However, with shear deformations it is difficult to obtain the geometric relationship between the deformed and undeformed geometries using the vector approach. Moreover, (9.5.30) and (9.5.31) are statements of the balance of moments when there are shearing rotations γ_4 and γ_5 with respect to the deformed midsurface. These two equations of motion may not be easily visualized and derived by using a vector approach. Moreover, the boundary conditions in (9.5.33) are very difficult to derive using a vector approach.

Shell Theory with von Kármán Nonlinearity

To account for geometric nonlinearities in an energy formulation of shells, the von Kármán strains are usually used (e.g., Habip, 1965; Ebcioglu, 1970; Oshima, Seguchi, and Shindo, 1970; Brush and Almroth, 1975; Librescu, 1976; Reddy and Chandrashekhara, 1985; Librescu and Schmidt, 1988). For circular cylindrical shells, the von Kármán strains are given by

$$e_1 = u_x + \frac{1}{2}w_x^2, \; e_2 = v_y + \frac{w}{a} + \frac{1}{2}w_y^2, \; \gamma_6 = u_y + v_x + w_x w_y \qquad (9.5.40)$$

which are a special case of (9.5.2), (9.5.3), and (9.5.5). Because the undeformed coordinate system is used,

$$\delta\gamma_6 = \delta v_x + w_y \delta w_x + \delta u_y + w_x \delta w_y$$
$$\delta\gamma_{61} = \delta v_x + w_y \delta w_x, \; \delta\gamma_{62} = \delta u_y + w_x \delta w_y \qquad (9.5.41)$$

It follows from (9.5.24) and (9.5.41) that

$$N_1 \delta e_1 + N_2 \delta e_2 + N_{61} \delta\gamma_{61} + N_{62} \delta\gamma_{62}$$
$$= N_1(\delta u_x + w_x \delta w_x) + N_2(\delta v_y + w_y \delta w_y + \frac{1}{a}\delta w)$$
$$+ N_{61}(\delta v_x + w_y \delta w_x) + N_{62}(\delta u_y + w_x \delta w_y)$$
$$= (N_1, \; N_{61}, \; 0) \begin{bmatrix} 1 & 0 & w_x \\ 0 & 1 & w_y \\ 0 & 0 & 1 \end{bmatrix} \begin{Bmatrix} \delta u_x \\ \delta v_x \\ \delta w_x \end{Bmatrix}$$
$$+ (N_{62}, \; N_2, \; 0) \begin{bmatrix} 1 & 0 & w_x \\ 0 & 1 & w_y \\ 0 & 0 & 1 \end{bmatrix} \begin{Bmatrix} \delta u_y \\ \delta v_y \\ \delta w_y \end{Bmatrix}$$
$$+ (N_{62}, \; N_2, \; 0) \begin{bmatrix} 1 & 0 & 0 \\ 0 & 1 & 0 \\ 0 & 0 & 1 \end{bmatrix} \begin{bmatrix} 0 & 0 & 0 \\ 0 & 0 & 1/a \\ 0 & -1/a & 0 \end{bmatrix} \begin{Bmatrix} \delta u \\ \delta v \\ \delta w \end{Bmatrix} \qquad (9.5.42)$$

Comparing (9.5.42) and (9.5.26), we note that, in the von Kármán approximation, e_1 and e_2 are neglected and the transformation matrix $[T]$ is approximated by

$$[T] = \begin{bmatrix} 1 & 0 & w_x \\ 0 & 1 & w_y \\ 0 & 0 & 1 \end{bmatrix} \qquad (9.5.43)$$

Moreover, $[T]$ is approximated in the third term in (9.5.42) by the identity matrix. We note that (9.5.43) is a special case of (9.5.4a). It follows from (9.5.43) that only linear terms in w are used to account for the configuration change. We note that the matrix (9.5.43) is not a unitary matrix (i.e., $[T]^{-1} = [T]^T$), as it should be.

It follows from (6.3.1b), (9.5.4b-d), and (9.5.43) that

$$\delta\theta_1 = \mathbf{i}_3 \cdot \delta \mathbf{i}_2 = \delta T_{23} = \delta w_y \qquad (9.5.44a)$$

$$\delta\theta_2 = -\mathbf{i}_3 \cdot \delta\mathbf{i}_1 = -\delta T_{13} = -\delta w_x \qquad (9.5.44\text{b})$$

$$\delta\theta_3 = \frac{1}{2}(\mathbf{i}_2\cdot\delta\mathbf{i}_1 - \mathbf{i}_1\cdot\delta\mathbf{i}_2) = \frac{1}{2}(w_y\delta T_{13} - w_x\delta T_{23}) = \frac{1}{2}w_y\delta w_x - \frac{1}{2}w_x\delta w_y \quad (9.5.44\text{c})$$

In calculating $\delta\theta_3$, we used the average of the two expressions in (6.3.1b) because the transformation (9.5.43) is not unitary and in particular $\mathbf{i}_1 \cdot \mathbf{i}_2 = w_x w_y \neq 0$. Moreover, substituting the elements of $[T]$ in (9.5.43) into (6.2.15), we obtain

$$k_1 = -w_{xx}, \quad k_2 = -w_{yy}, \quad k_{61} + k_{62} = -2w_{xy} \qquad (9.5.45)$$

Equations (9.5.45) show that linearization of the curvatures in the von Kármán nonlinearities is consistent with linearization of the transformation matrix $[T]$ shown in (9.5.43).

Using the linear curvature expressions (9.5.45) and neglecting the nonlinear terms in (9.5.25a-k) (i.e., neglecting the changes in the configuration or neglecting the nonlinear terms in (9.5.14a-g)), we obtain from (9.5.25a-k) that

$$m_{31} = m_{32} = \Theta_3 = 0, \quad \hat{M}_1 = M_1, \quad \hat{M}_2 = M_2, \quad \hat{M}_{61} = M_{61}, \quad \hat{M}_{62} = M_{62},$$
$$\Theta_1 = M_{61x} + M_{2y}, \quad \Theta_2 = -M_{1x} - M_{62y}$$
$$\Gamma_4 = -(M_2 - P_2)_y - (M_{61} - P_{61})_x + Q_2 - R_2 + 2S_2/a - 6G_2/a$$
$$\Gamma_5 = -(M_1 - P_1)_x - (M_{62} - P_{62})_y + Q_1 - R_1 \qquad (9.5.46)$$

Substituting (9.5.43), (9.5.44a-c), and (9.5.46) into (9.5.26), letting $1 + e_1 \approx 1 + e_2 \approx 1$, and integrating by parts yields

$$\delta\Pi = \int_A \left\{ -\left[N_{1x}, N_{61x}, (N_1 w_x + N_{61} w_y + M_{1x} + M_{62y})_x\right] \left\{ \begin{array}{c} \delta u \\ \delta v \\ \delta w \end{array} \right\} \right.$$
$$+ \frac{1}{a}N_2\delta w - \frac{1}{a}(M_{61x} + M_{2y})\delta v + \Gamma_4\delta\gamma_4 + \Gamma_5\delta\gamma_5$$
$$\left. - \left[N_{62y}, N_{2y}, (N_{62} w_x + N_2 w_y + M_{61x} + M_{2y})_y\right] \left\{ \begin{array}{c} \delta u \\ \delta v \\ \delta w \end{array} \right\} \right\} dxdy$$
$$+ \int_x \left[N_{62}\delta u + N_2\delta v + (N_{62} w_x + N_2 w_y + M_{61x} + M_{62x} + M_{2y})\delta w \right.$$
$$\left. -M_2\delta w_y + (M_2 - P_2)\delta\gamma_4 + (M_{62} - P_{62})\delta\gamma_5\right]_{\theta=0}^{\theta=\theta_0} dx$$
$$+ \int_y \left[N_1\delta u + N_{61}\delta v + (N_1 w_x + N_{61} w_y + M_{1x} + M_{61y} + M_{62y})\delta w \right.$$
$$\left. -M_1\delta w_x + (M_{61} - P_{61})\delta\gamma_4 + (M_1 - P_1)\delta\gamma_5\right]_{x=0}^{x=L} dy$$
$$-(M_{61} + M_{62})\delta w \left|_{(L,0),(0,\theta_0)}^{(0,0),(L,\theta_0)} \right. \qquad (9.5.47)$$

Substituting (9.5.47) and (9.5.21) into the extended Hamilton's principle and setting each of the coefficients of δu, δv, δw, $\delta\gamma_4$, and $\delta\gamma_5$ equal to zero, we obtain

$$N_{1x} + N_{62y} = I_1\ddot{u} - I_2\ddot{w}_x + I_4\ddot{\gamma}_5 + \mu_1\dot{u} \qquad (9.5.48)$$

$$N_{61x} + N_{2y} + \frac{1}{a}(M_{61x} + M_{2y}) = I_1\ddot{v} - I_2\ddot{w}_y + \bar{I}_4\ddot{\gamma}_4 + \frac{2}{a}I_2\ddot{v}$$
$$-\frac{1}{a}I_3\ddot{w}_y + \frac{1}{a^2}I_3\ddot{v} + \frac{1}{a}\bar{I}_5\ddot{\gamma}_4 + \mu_2\dot{v} \tag{9.5.49}$$
$$\left(N_1 w_x + N_{61} w_y + M_{1x} + M_{62y}\right)_x + (N_{62} w_x + N_2 w_y + M_{61x} + M_{2y})_y$$
$$-\frac{1}{a}N_2 + q = I_1\ddot{w} + (I_2\ddot{u} - I_3\ddot{w}_x + I_5\ddot{\gamma}_5)_x$$
$$+(I_2\ddot{v} - I_3\ddot{w}_y + \frac{1}{a}I_3\ddot{v} + \bar{I}_5\ddot{\gamma}_4)_y + \mu_3\dot{w} \tag{9.5.50}$$
$$\left(M_2 - P_2\right)_y + (M_{61} - P_{61})_x - Q_2 + R_2 - 2S_2/a + 6G_2/a$$
$$= \bar{I}_4\ddot{v} - \bar{I}_5\ddot{w}_y + \frac{1}{a}\bar{I}_5\ddot{v} + \tilde{I}_7\ddot{\gamma}_4 + \mu_4\dot{\gamma}_4 \tag{9.5.51}$$
$$\left(M_1 - P_1\right)_x + (M_{62} - P_{62})_y - Q_1 + R_1 = I_4\ddot{u} - I_5\ddot{w}_x + I_7\ddot{\gamma}_5 + \mu_5\dot{\gamma}_5 \tag{9.5.52}$$

The boundary conditions for the shell are of the form: specify

Along $x = 0, L$:

$\delta u = 0$	or	N_1
$\delta v = 0$	or	N_{61}
$\delta w = 0$	or	$N_1 w_x + N_{61} w_y + M_{1x} + M_{61y} + M_{62y}$
		$\quad - I_2\ddot{u} + I_3\ddot{w}_x - I_5\ddot{\gamma}_5$
$\delta w_x = 0$	or	M_1
$\delta \gamma_4 = 0$	or	$M_{61} - P_{61}$
$\delta \gamma_5 = 0$	or	$M_1 - P_1$

Along $\theta = 0, \theta_0$:

$\delta u = 0$	or	N_{62}
$\delta v = 0$	or	N_2
$\delta w = 0$	or	$N_{62} w_x + N_2 w_y + M_{61x} + M_{62x} + M_{2y}$
		$\quad - I_2\ddot{v} + I_3\ddot{w}_y - I_3\ddot{v}/a - \bar{I}_5\ddot{\gamma}_4$
$\delta w_y = 0$	or	M_2
$\delta \gamma_4 = 0$	or	$M_2 - P_2$
$\delta \gamma_5 = 0$	or	$M_{62} - P_{62}$

At $(x, \theta) = (0, 0), (L, 0), (0, \theta_0), (L, \theta_0)$

$$\delta w = 0 \quad \text{or} \quad M_{61} + M_{62} \tag{9.5.53}$$

We note that the von Kármán nonlinearity appears only in the equation governing w.

Nonlinear First-Order Shear-Deformation Theory

In the first-order shear-deformation theory, it is assumed that the shear strains are uniform across the cross-section (i.e., $\epsilon_{13} = \gamma_5$, $\epsilon_{23} = \gamma_4$), which are equivalent to neglecting the cubic terms $4z^3\gamma_5/3h^2$ and $4z^3\gamma_4/3h^2$ in (9.5.13a,b). Hence, it follows from (9.5.23c,e,g) and (9.5.18a,b) that

$$P_1 = P_2 = P_{61} = P_{62} = R_1 = R_2 = G_1 = G_2 = 0$$
$$I_2 = I_4, \quad I_3 = I_5 = I_7, \quad \bar{I}_5 = \bar{I}_7 \qquad (9.5.54)$$

The equations of motion are the same as (9.5.27)-(9.5.31) except that (9.5.25a-f,j-k) are reduced to

$$\hat{M}_1 = M_1 - S_1\gamma_5$$
$$\hat{M}_2 = M_2 - S_2\gamma_4$$
$$\hat{M}_{61} = M_{61} - S_1\gamma_4$$
$$\hat{M}_{62} = M_{62} - S_2\gamma_5$$
$$m_{31} = M_{61}\gamma_5 - M_1\gamma_4$$
$$m_{32} = -M_{62}\gamma_4 + M_2\gamma_5$$
$$\Gamma_4 = -M_{2y} - M_{61x} + Q_2 + 2S_2/a - S_1k_{61} - S_2\bar{k}_2 - M_1k_5 - M_{62}k_4$$
$$\Gamma_5 = -M_{1x} - M_{62y} + Q_1 - S_1k_1 - S_2k_{62} + M_2k_4 + M_{61}k_5 \qquad (9.5.55)$$

This theory requires the use of shear-correction factors for the transverse stiffnesses to account for the nonuniform distribution of shear stresses.

Nonlinear Plate Theory

Taking the limit $a \to \infty$ in (9.5.27)-(9.5.33), one obtains the equations of motion and boundary conditions for plates, see Section 7.7.

9.5.3 Stiffness Matrices

Substituting (9.5.14a-g), (2.4.34), and (2.4.35) into (9.5.23a-g), one can express the stress resultants and couples in terms of the midsurface strains, curvatures, and stiffness matrices for any material, but the stiffness matrices are not symmetric. As pointed out by Calladine (1983), in most practical problems the magnitudes of M_{61} and M_{62} are small in comparison with the magnitude of either M_1 or M_2 and play only a minor part in the resistance to loadings. Moreover, because the theory developed here is for thin shells undergoing small strains but moderate rotations, the coefficient z/a in the nonlinear terms $k_i\gamma_j$ and the derivatives of γ_4 and γ_5 are negligible in the strain-displacement relations. Hence, to obtain symmetric stiffness matrices, to simplify these expressions, and to reduce the computational effort without significant loss of accuracy, we assume that z/a is small for the case of thin shells. Expanding the term $a/(a+z)$ for large a and neglecting the terms z^n/a and z^n/a^n for $n \geq 2$

in the strain-displacement relations (9.5.14a-g), we obtain

$$\epsilon_{11} = e_1 + zk_1 + \left(z - \frac{4z^3}{3h^2}\right)(\gamma_{5x} - k_5\gamma_4) \qquad (9.5.56a)$$

$$\epsilon_{22} = \left(1 - \frac{z}{a}\right)e_2 + z\bar{k}_2 + \left(z - \frac{4z^3}{3h^2}\right)(\gamma_{4y} + k_4\gamma_5) \qquad (9.5.56b)$$

$$\epsilon_{12} = \hat{e}_{12} + \hat{e}_{21} = \gamma_{61} + \left(1 - \frac{z}{a}\right)\gamma_{62} + zk_6$$
$$+ \left(z - \frac{4z^3}{3h^2}\right)(\gamma_{4x} + \gamma_{5y} + k_5\gamma_5 - k_4\gamma_4) \qquad (9.5.56c)$$

$$\epsilon_{13} = \left(1 - \frac{4z^2}{h^2}\right)\gamma_5 - \left(z - \frac{4z^3}{3h^2}\right)(k_1\gamma_5 + k_{61}\gamma_4) \qquad (9.5.56d)$$

$$\epsilon_{23} = \left(1 - \frac{4z^2}{h^2}\right)\gamma_4 - z\left(\tilde{k}_2\gamma_4 + k_{62}\gamma_5\right) + \frac{4z^3}{3h^2}\left(\bar{k}_2\gamma_4 + k_{62}\gamma_5\right) \qquad (9.5.56e)$$

where

$$\tilde{k}_2 \equiv k_2 - \frac{2}{a}, \quad k_6 \equiv k_{61} + k_{62} \qquad (9.5.56f)$$

Also, we define the stress resultants and couples as

$$\left(\tilde{N}_1, N_2, N_{62}\right) = \int_{-h/2}^{h/2} (\sigma_{11}, \sigma_{22}, \sigma_{21})\, dz \qquad (9.5.57a)$$

$$\left(\tilde{Q}_1, Q_2\right) = \int_{-h/2}^{h/2} (\sigma_{13}, \sigma_{23})\, dz \qquad (9.5.57b)$$

$$\left(\hat{R}_1, \hat{R}_2\right) = \int_{-h/2}^{h/2} (\sigma_{13}, \sigma_{23})\, z^2 dz, \quad (R_1, R_2) = \frac{4}{h^2}\left(\hat{R}_1, \hat{R}_2\right) \qquad (9.5.57c)$$

$$(M_1, M_2, M_{62}) = \int_{-h/2}^{h/2} (\sigma_{11}, \sigma_{22}, \sigma_{21})\, z\, dz \qquad (9.5.57d)$$

$$\left(\hat{P}_1, \hat{P}_2, \hat{P}_{62}\right) = \int_{-h/2}^{h/2} (\sigma_{11}, \sigma_{22}, \sigma_{21})\, z^3 dz$$
$$(P_1, P_2, P_{62}) = \frac{4}{3h^2}\left(\hat{P}_1, \hat{P}_2, \hat{P}_{62}\right) \qquad (9.5.57e)$$

$$(S_1, S_2) = \int_{-h/2}^{h/2} (\sigma_{13}, \sigma_{23})\, z\, dz \qquad (9.5.57f)$$

$$\left(\hat{G}_1, \hat{G}_2\right) = \int_{-h/2}^{h/2} (\sigma_{13}, \sigma_{23})\, z^3 dz, \quad (G_1, G_2) = \frac{4}{3h^2}\left(\hat{G}_1, \hat{G}_2\right) \qquad (9.5.57g)$$

$$N_1 = \int_{-h/2}^{h/2} \sigma_{11}\left(1 + \frac{z}{a}\right) dz = \tilde{N}_1 + \frac{1}{a}M_1 \qquad (9.5.57h)$$

$$N_{61} = \int_{-h/2}^{h/2} \sigma_{12}(1 + \frac{z}{a}) dz = N_{62} + \frac{1}{a}M_{62} \qquad (9.5.57i)$$

$$Q_1 = \int_{-h/2}^{h/2} \sigma_{13}\left(1 + \frac{z}{a}\right) dz = \tilde{Q}_1 + \frac{1}{a}S_1 \qquad (9.5.57\text{j})$$

$$M_{61} = M_{62} \qquad (9.5.57\text{k})$$

$$P_{61} = P_{62} \qquad (9.5.57\text{l})$$

It follows from (9.5.56a-e), (2.4.34), and (2.4.35) that

$$\left\{ \begin{array}{c} \sigma_{23} \\ \sigma_{13} \end{array} \right\} = \left[\begin{array}{cc} \overline{Q}_{44} & \overline{Q}_{45} \\ \overline{Q}_{45} & \overline{Q}_{55} \end{array} \right] \left(\left\{ \begin{array}{c} \gamma_4 \\ \gamma_5 \end{array} \right\} - \frac{4z^2}{h^2} \left\{ \begin{array}{c} \gamma_4 \\ \gamma_5 \end{array} \right\} - z \left\{ \begin{array}{c} \gamma_4 \bar{k}_2 + \gamma_5 k_{62} \\ \gamma_5 k_1 + \gamma_4 k_{61} \end{array} \right\} \right.$$

$$\left. + \frac{4z^3}{3h^2} \left\{ \begin{array}{c} \gamma_4 \bar{k}_2 + \gamma_5 k_{62} \\ \gamma_5 k_1 + \gamma_4 k_{61} \end{array} \right\} \right) \qquad (9.5.58\text{a})$$

$$\left\{ \begin{array}{c} \sigma_{11} \\ \sigma_{22} \\ \sigma_{12} \end{array} \right\} = \left[\begin{array}{ccc} \tilde{Q}_{11} & \tilde{Q}_{12} & \tilde{Q}_{16} \\ \tilde{Q}_{12} & \tilde{Q}_{22} & \tilde{Q}_{26} \\ \tilde{Q}_{16} & \tilde{Q}_{26} & \tilde{Q}_{66} \end{array} \right] \left(-\frac{4z^3}{3h^2} \left\{ \begin{array}{c} \gamma_{5x} - k_5\gamma_4 \\ \gamma_{4y} + k_4\gamma_5 \\ \gamma_{5y} + \gamma_{4x} - k_4\gamma_4 + k_5\gamma_5 \end{array} \right\} \right.$$

$$\left. + \left\{ \begin{array}{c} e_1 \\ e_2 \\ \gamma_6 \end{array} \right\} + z \left\{ \begin{array}{c} k_1 + \gamma_{5x} - k_5\gamma_4 \\ \bar{k}_2 + \gamma_{4y} + k_4\gamma_5 - e_2/a \\ k_6 + \gamma_{5y} + \gamma_{4x} - k_4\gamma_4 + k_5\gamma_5 - \gamma_{62}/a \end{array} \right\} \right) \qquad (9.5.58\text{b})$$

Substituting (9.5.58a,b) into (9.5.57) yields the internal force and moment resultants in terms of the midsurface strains and curvatures as

$$\left\{ \begin{array}{c} \tilde{N}_1 \\ N_2 \\ N_{62} \\ M_1 \\ M_2 \\ M_{62} \\ \hat{P}_1 \\ \hat{P}_2 \\ \hat{P}_{62} \end{array} \right\} = \left[\begin{array}{ccc} [A_{ij}] & [B_{ij}] & [E_{ij}] \\ [B_{ij}] & [D_{ij}] & [F_{ij}] \\ [E_{ij}] & [F_{ij}] & [H_{ij}] \end{array} \right]$$

$$\times \left\{ \begin{array}{c} e_1 \\ e_2 \\ \gamma_6 \\ k_1 + \gamma_{5x} - k_5\gamma_4 \\ \bar{k}_2 + \gamma_{4y} + k_4\gamma_5 - e_2/a \\ k_6 + \gamma_{5y} + \gamma_{4x} - k_4\gamma_4 + k_5\gamma_5 - \gamma_{62}/a \\ -4(\gamma_{5x} - k_5\gamma_4)/3h^2 \\ -4(\gamma_{4y} + k_4\gamma_5)/3h^2 \\ -4(\gamma_{5y} + \gamma_{4x} - k_4\gamma_4 + k_5\gamma_5)/3h^2 \end{array} \right\} \qquad (9.5.59)$$

where $i, j = 1, 2, 6$; and

$$\left\{\begin{array}{c} Q_2 \\ \tilde{Q}_1 \\ \hat{R}_2 \\ \hat{R}_1 \end{array}\right\} = \left[\begin{array}{cc} [A_{ij}] & [D_{ij}] \\ [D_{ij}] & [F_{ij}] \end{array}\right] \left\{\begin{array}{c} \gamma_4 \\ \gamma_5 \\ -4\gamma_4/h^2 \\ -4\gamma_5/h^2 \end{array}\right\}$$

$$+ \left[\begin{array}{cc} [B_{ij}] & [E_{ij}] \\ [E_{ij}] & [C_{ij}] \end{array}\right] \left\{\begin{array}{c} -\gamma_4 \tilde{k}_2 - \gamma_5 k_{62} \\ -\gamma_5 k_1 - \gamma_4 k_{61} \\ 4(\gamma_4 \tilde{k}_2 + \gamma_5 k_{62})/3h^2 \\ 4(\gamma_5 k_1 + \gamma_4 k_{61})/3h^2 \end{array}\right\} \quad (9.5.60)$$

$$\left\{\begin{array}{c} S_2 \\ S_1 \\ \hat{G}_2 \\ \hat{G}_1 \end{array}\right\} = \left[\begin{array}{cc} [B_{ij}] & [E_{ij}] \\ [E_{ij}] & [C_{ij}] \end{array}\right] \left\{\begin{array}{c} \gamma_4 \\ \gamma_5 \\ -4\gamma_4/h^2 \\ -4\gamma_5/h^2 \end{array}\right\}$$

$$+ \left[\begin{array}{cc} [D_{ij}] & [F_{ij}] \\ [F_{ij}] & [H_{ij}] \end{array}\right] \left\{\begin{array}{c} -\gamma_4 \tilde{k}_2 - \gamma_5 k_{62} \\ -\gamma_5 k_1 - \gamma_4 k_{61} \\ 4(\gamma_4 \tilde{k}_2 + \gamma_5 k_{62})/3h^2 \\ 4(\gamma_5 k_1 + \gamma_4 k_{61})/3h^2 \end{array}\right\} \quad (9.5.61)$$

where $i, j = 4, 5$. We note that the stiffness matrices are symmetric and their components are defined in (7.7.16). Substituting (9.5.2), (9.5.3), (9.5.5), (9.5.4b-d), (9.5.7)-(9.5.12), and (9.5.59)-(9.5.61) into (9.5.27)-(9.5.31) and expanding the results up to $O(\epsilon^3)$, one can obtain five third-order nonlinear equations of motion.

9.5.4 Classical Linear Theories of Circular Cylindrical Shells

Setting γ_4 and γ_5 equal to zero and keeping only the linear parts of (9.5.27)-(9.5.29) yields the following linear equations of motion of classical shells:

$$N_{1x} + N_{62y} = I_1 \ddot{u} - I_2 \ddot{w}_x + \mu_1 \dot{u} \quad (9.5.62a)$$

$$N_{61x} + N_{2y} + \frac{1}{a}(M_{61x} + M_{2y}) = I_1 \ddot{v} - I_2 \ddot{w}_y + \frac{2}{a} I_2 \ddot{v} - \frac{1}{a} I_3 \ddot{w}_y + \frac{1}{a^2} I_3 \ddot{v} + \mu_2 \dot{v} \quad (9.5.62b)$$

$$M_{1xx} + M_{61xy} + M_{62xy} + M_{2yy} - \frac{1}{a} N_2 + q = I_1 \ddot{w}$$

$$+ (I_2 \ddot{u} - I_3 \ddot{w}_x)_x + \left(I_2 \ddot{v} - I_3 \ddot{w}_y + \frac{1}{a} I_3 \ddot{v}\right)_y + \mu_3 \dot{w} \quad (9.5.62c)$$

Setting γ_4 and γ_5 equal to zero and substituting the linear parts of $e_1, e_2, \gamma_{61}, \gamma_{62}$, k_1, k_2, k_{61}, and k_{62} into (9.5.14a,b,d,e), we obtain

$$\epsilon_{11} = e_1 + zk_1 = u_x - zw_{xx} \quad (9.5.63a)$$

$$\epsilon_{22} = \frac{a}{a+z}(e_2 + z\bar{k}_2) = \frac{a}{a+z}\left[v_y + \frac{w}{a} - z\left(w_{yy} - \frac{v_y}{a}\right)\right] \quad (9.5.63b)$$

$$e_{12} = \gamma_{61} + zk_{61} = \frac{1}{2}(u_y + v_x) - z\left(w_{xy} - \frac{v_x}{a}\right) \quad (9.5.63c)$$

$$e_{21} = \frac{a}{a+z}(\gamma_{62} + zk_{62}) = \frac{a}{a+z}\left[\frac{1}{2}(u_y + v_x) - z\left(w_{xy} + \frac{u_y}{a}\right)\right] \quad (9.5.63d)$$

The stress resultants and moments, which act on the element edges, become

$$[N_1, N_{61}, M_1, M_{61}] \equiv \int_{-h/2}^{h/2} [\sigma_{11}, \sigma_{12}, \sigma_{11}z, \sigma_{12}z]\left(1 + \frac{z}{a}\right)dz \quad (9.5.64a)$$

$$[N_2, N_{62}, M_2, M_{62}] \equiv \int_{-h/2}^{h/2} [\sigma_{22}, \sigma_{21}, \sigma_{22}z, \sigma_{21}z]\,dz \quad (9.5.64b)$$

Moreover, variation of the elastic energy becomes

$$\delta\Pi = \int_A \int_{-h/2}^{h/2} (\sigma_{11}\delta\epsilon_{11} + \sigma_{22}\delta\epsilon_{22} + \sigma_{12}\delta e_{12} + \sigma_{21}\delta e_{21})\left(1 + \frac{z}{a}\right)dzdxdy \quad (9.5.65)$$

Next, we show that the strain-displacement relations (9.5.63a-d) include most of those used in existing linear theories as special cases.

The "Best" Theory of Sanders

Next, we discuss the so-called "best" theory of Sanders (Sanders, 1959; Budiansky and Sanders, 1963). Substituting (9.5.63a-d) into (9.5.65) and using (9.5.64a,b), we obtain

$$\delta\Pi = \int_A \int_{-h/2}^{h/2} \left\{\sigma_{11}\left(1 + \frac{z}{a}\right)\delta(u_x - zw_{xx}) + \sigma_{22}\delta\left[v_y + \frac{w}{a} - z\left(w_{yy} - \frac{v_y}{a}\right)\right]\right.$$

$$+ \sigma_{12}\left(1 + \frac{z}{a}\right)\delta\left[\left(\frac{1}{2} + \frac{z}{4a}\right)(u_y + v_x) - z\left(w_{xy} - \frac{3v_x}{4a} + \frac{u_y}{4a}\right)\right]$$

$$+ \sigma_{21}\delta\left[\left(\frac{1}{2} - \frac{z}{4a}\right)(u_y + v_x) - z\left(w_{xy} + \frac{3u_y}{4a} - \frac{v_x}{4a}\right)\right]\bigg\}dzdxdy$$

$$= \int_A \left[N_1\delta u_x - M_1\delta w_{xx} + N_2\delta\left(v_y + \frac{w}{a}\right) - M_2\delta\left(w_{yy} - \frac{v_y}{a}\right)\right.$$

$$+ \tilde{N}_6\delta(u_y + v_x) - 2\tilde{M}_6\delta\left(w_{xy} - \frac{3v_x}{4a} + \frac{u_y}{4a}\right)\bigg]dxdy \quad (9.5.66)$$

where

$$\tilde{M}_6 \equiv \frac{1}{2}(M_{61} + M_{62})$$

$$\tilde{N}_6 \equiv \frac{1}{2}(N_{61} + N_{62}) + \frac{1}{4a}(M_{61} - M_{62}) \quad (9.5.67)$$

In obtaining (9.5.66), we replaced $-w_{xy} - 3u_y/4a + v_x/4a$ with $-w_{xy} + 3v_x/4a - u_y/4a$ because $u_y = -v_x$, as discussed right after (9.5.12). Now if there exists a modified stress $\tilde{\sigma}_{12}$ such that

$$\delta \Pi = \int_A \int_{-h/2}^{h/2} \left\{ \tilde{\sigma}_{11} \delta \left(u_x - z w_{xx}\right) + \sigma_{22} \delta \left[v_y + \frac{w}{a} - z \left(w_{yy} - \frac{v_y}{a}\right)\right] \right.$$
$$\left. + \tilde{\sigma}_{12} \delta \left[u_y + v_x - 2z \left(w_{xy} - \frac{3v_x}{4a} + \frac{u_y}{4a}\right)\right] \right\} dz dx dy \quad (9.5.68)$$

where $\tilde{\sigma}_{11} \equiv \sigma_{11}(1+z/a)$, then the conjugate strains (Sanders, 1959) of the modified stresses are

$$\epsilon_{11} = u_x - z w_{xx} \quad (9.5.69a)$$

$$\epsilon_{22} = v_y + \frac{w}{a} - z \left(w_{yy} - \frac{v_y}{a}\right) \quad (9.5.69b)$$

$$\epsilon_{12} = u_y + v_x - 2z \left(w_{xy} - \frac{3v_x}{4a} + \frac{u_y}{4a}\right) \quad (9.5.69c)$$

To determine the expression of $\tilde{\sigma}_{12}$, we rearrange (9.5.66) as

$$\delta \Pi = \int_A \int_{-h/2}^{h/2} \left\{ \tilde{\sigma}_{11} \delta \left(u_x - z w_{xx}\right) + \sigma_{22} \delta \left[v_y + \frac{w}{a} - z \left(w_{yy} - \frac{v_y}{a}\right)\right] \right.$$
$$\left. + \bar{\sigma}_{12} \delta \left(u_y + v_x\right) - 2z \hat{\sigma}_{12} \delta \left(w_{xy} - \frac{3v_x}{4a} + \frac{u_y}{4a}\right) \right\} dz dx dy \quad (9.5.70)$$

where

$$\bar{\sigma}_{12} = \sigma_{12} \left(1 + \frac{z}{2a} + \frac{z^2}{4a^2}\right), \quad \hat{\sigma}_{12} = \sigma_{12} \left(1 + \frac{z}{2a}\right) \quad (9.5.71)$$

Then, if we neglect the term $z^2/4a^2$ in (9.5.71), we obtain from (9.5.68) and (9.5.70) that $\tilde{\sigma}_{12} = \bar{\sigma}_{12} = \hat{\sigma}_{12} = \sigma_{12}(1 + z/2a)$. We note that (9.5.68) is called the "best" because the number of stress resultants and moments is less than that in (9.5.66).

Next, we examine other linear theories, where the term z/a is neglected at different stages.

Theory of Love (1944) and Timoshenko (1959)

Substituting (9.5.63a-d) into (9.5.65) yields

$$\delta \Pi = \int_A \int_{-h/2}^{h/2} \left\{ \tilde{\sigma}_{11} \delta \left(u_x - z w_{xx}\right) + \sigma_{22} \delta \left[v_y + \frac{w}{a} - z \left(w_{yy} - \frac{v_y}{a}\right)\right] \right.$$
$$\left. + \tilde{\sigma}_{12} \delta \left[u_y + v_x - 2z \left(w_{xy} - \frac{v_x}{a}\right)\right] \right\} dz dx dy$$
$$= \int_A \left[N_1 \delta u_x - M_1 \delta w_{xx} + N_2 \delta \left(v_y + \frac{w}{a}\right) - M_2 \delta \left(w_{yy} - \frac{v_y}{a}\right) \right.$$
$$\left. + \tilde{N}_6 \delta \left(u_y + v_x\right) - 2 \tilde{M}_6 \delta \left(w_{xy} - \frac{v_x}{a}\right) \right] dx dy \quad (9.5.72)$$

where the modified stresses are

$$\tilde{\sigma}_{11} = \sigma_{11}\left(1 + \frac{z}{a}\right), \quad \tilde{\sigma}_{12} = \sigma_{12}\left(1 + \frac{z}{2a}\right) \quad (9.5.73)$$

The conjugate strains (Love, 1944; Timoshenko, 1959) are

$$\epsilon_{11} = u_x - zw_{xx} \quad (9.5.74a)$$

$$\epsilon_{22} = v_y + \frac{w}{a} - z\left(w_{yy} - \frac{v_y}{a}\right) \quad (9.5.74b)$$

$$\epsilon_{12} = u_y + v_x - 2z\left(w_{xy} - \frac{v_x}{a}\right) \quad (9.5.74c)$$

and the modified couple and stress resultant are

$$\tilde{N}_6 \equiv \int_{-h/2}^{h/2} \tilde{\sigma}_{12} dz = \int_{-h/2}^{h/2} \left[\frac{1}{2}\left(1 + \frac{z}{a}\right)\sigma_{12} + \frac{1}{2}\sigma_{21}\right] dz$$

$$= \frac{1}{2}(N_{61} + N_{62})$$

$$\tilde{M}_6 \equiv \int_{-h/2}^{h/2} \tilde{\sigma}_{12} z\, dz = \frac{1}{2}(M_{61} + M_{62}) \quad (9.5.75)$$

In obtaining (9.5.72), we put $k_{62} = -w_{xy} + v_x/a$ because the change in configuration due to γ_6 is neglected and hence $u_y = -v_x$. Similar to the theory of Sanders, the theory of Love and Timoshenko does not include any expansion or truncation of the term z/a. However, Love and Timoshenko did not include the influence of the longitudinal displacement u on k_{62}. But, the difference between $k_{62} = -w_{xy} + v_x/a$ and $k_{62} = -w_{xy} - u_y/a$ is $(u_y + v_x)/a$, which is negligible when a is large and/or when the inplane shear strain $\gamma_6(= u_y + v_x)$ is small. Hence, the theory of Love and Timoshenko is very close to that of Sanders (1959).

Theory of Donnell (1938)

Neglecting v_y/a and v_x/a in (9.5.74a-c) results in the Donnell strain-displacement expressions

$$\epsilon_{11} = u_x - zw_{xx} \quad (9.5.76a)$$

$$\epsilon_{22} = v_y + \frac{w}{a} - zw_{yy} \quad (9.5.76b)$$

$$\epsilon_{12} = u_y + v_x - 2zw_{xy} \quad (9.5.76c)$$

Theory of Flügge (1973)

Neglecting $z(u_y+v_x)/2a(= z\gamma_6/2a)$ in (9.5.63a-d) yields the Flügge strain-displacement expressions

$$\epsilon_{11} = u_x - zw_{xx} \quad (9.5.77a)$$

$$\epsilon_{22} = \frac{a}{a+z}\left[v_y + \frac{w}{a} - z\left(w_{yy} - \frac{v_y}{a}\right)\right] \quad (9.5.77b)$$

$$\epsilon_{12} = \frac{a}{a+z}\left[u_y + v_x - 2z\left(1 + \frac{z}{2a}\right)\left(w_{xy} - \frac{v_x}{a}\right)\right] \quad (9.5.77c)$$

9.6 NONLINEAR LAYERWISE SHEAR-DEFORMABLE SHELL THEORY

An infinitesimal shell element without shear deformation is a straight line. With shear deformations assumed to be of third-order (see (9.5.13a,b)), the deformed shell element is a crooked line, as shown in Figure 9.5.2. However, the third-order shear deformation theory shown in (9.5.13a,b) is only accurate for isotropic structures. For anisotropic laminated composite shells, shear deformations are more complex than those shown in (9.5.13a,b) because of the nonuniform distribution of the material properties through the thickness. Next, we derive a layerwise higher-order shear-deformation theory for flexible doubly-curved laminated composite shells.

9.6.1 Strains and Shear-Warping Functions

Because the material properties are not uniform through the thickness, we assume that the local displacements $u_1^{(i)}$, $u_2^{(i)}$, and $u_3^{(i)}$ (with respect to the $\xi\eta\zeta$ coordinate system) of the ith lamina have the form

$$u_1^{(i)}(x,y,z,t) = u_1^0(x,y,t) + z\left[\theta_2(x,y,t) - \theta_{20}(x,y)\right]$$
$$+ \gamma_5 z + \alpha_1^{(i)}(x,y,t) z^2 + \beta_1^{(i)}(x,y,t) z^3 \quad (9.6.1a)$$

$$u_2^{(i)}(x,y,z,t) = u_2^0(x,y,t) - z\left[\theta_1(x,y,t) - \theta_{10}(x,y)\right]$$
$$+ \gamma_4 z + \alpha_2^{(i)}(x,y,t) z^2 + \beta_2^{(i)}(x,y,t) z^3 \quad (9.6.1b)$$

$$u_3^{(i)}(x,y,z,t) = u_3^0(x,y,t) \quad (9.6.1c)$$

where the u_j^0 ($j = 1, 2, 3$) are the displacements (with respect to the $\xi\eta\zeta$ system) of the observed reference point A in Figure 6.1.2, γ_4 and γ_5 are the transverse shear-rotation angles of the infinitesimal shell element at the reference point A with respect to the $-\xi$ and η axes, respectively, θ_1 and θ_2 are the rotation angles of the normal to the deformed reference surface with respect to the ξ and η axes, respectively, and θ_{10} and θ_{20} are the initial rotation angles of the normal with respect to the ξ and η axes, respectively. Moreover, the $\alpha_j^{(i)}$ and $\beta_j^{(i)}$ are unknown functions for the ith layer to be determined by imposing continuity conditions of the inplane displacements and interlaminar shear stresses and the free surface conditions. Because $\xi\eta\zeta$ is a local coordinate system and the $\xi\eta$ plane is tangent to the deformed reference surface, (6.5.11a-d) are also valid here. Letting $\mathbf{U}$ denote the local displacement vector, that is,

$$\mathbf{U} = u_1^{(i)}\mathbf{i}_1 + u_2^{(i)}\mathbf{i}_2 + u_3^{(i)}\mathbf{i}_3 \quad (9.6.2)$$

taking the derivatives of (9.6.2), and using (6.5.11a-d), we obtain

$$\frac{\partial \mathbf{U}}{\partial x} = \frac{\partial u_1^{(i)}}{\partial x}\mathbf{i}_1 + \frac{\partial u_2^{(i)}}{\partial x}\mathbf{i}_2 + \frac{\partial u_3^{(i)}}{\partial x}\mathbf{i}_3 + u_1^{(i)}\frac{\partial \mathbf{i}_1}{\partial x} + u_2^{(i)}\frac{\partial \mathbf{i}_2}{\partial x} + u_3^{(i)}\frac{\partial \mathbf{i}_3}{\partial x}$$
$$= \left[e_1 + z\left(k_1 - k_1^0\right) + G_{1x} - k_5 G_2\right]\mathbf{i}_1 + \left[\frac{\partial u_2^0}{\partial x} + z\left(k_{61} - k_{61}^0\right)\right.$$
$$\left. + G_{2x} + k_5 G_1\right]\mathbf{i}_2 - \left(k_1 G_1 + k_{61} G_2\right)\mathbf{i}_3 \quad (9.6.3a)$$

$$\frac{\partial \mathbf{U}}{\partial y} = \frac{\partial u_1^{(i)}}{\partial y}\mathbf{i}_1 + \frac{\partial u_2^{(i)}}{\partial y}\mathbf{i}_2 + \frac{\partial u_3^{(i)}}{\partial y}\mathbf{i}_3 + u_1^{(i)}\frac{\partial \mathbf{i}_1}{\partial y} + u_2^{(i)}\frac{\partial \mathbf{i}_2}{\partial y} + u_3^{(i)}\frac{\partial \mathbf{i}_3}{\partial y}$$

$$= \left[\frac{\partial u_1^0}{\partial y} + z\left(k_{62} - k_{62}^0\right) + G_{1y} - k_4 G_2\right]\mathbf{i}_1 + \left[e_2 + z\left(k_2 - k_2^0\right)\right.$$
$$\left. + G_{2y} + k_4 G_1\right]\mathbf{i}_2 - \left(k_{62} G_1 + k_2 G_2\right)\mathbf{i}_3 \qquad (9.6.3b)$$

$$\frac{\partial \mathbf{U}}{\partial z} = \frac{\partial u_1^{(i)}}{\partial z}\mathbf{i}_1 + \frac{\partial u_2^{(i)}}{\partial z}\mathbf{i}_2 + \frac{\partial u_3^{(i)}}{\partial z}\mathbf{i}_3 = G_{1z}\mathbf{i}_1 + G_{2z}\mathbf{i}_2 \qquad (9.6.3c)$$

where

$$G_1 \equiv \gamma_5 z + \alpha_1^{(i)} z^2 + \beta_1^{(i)} z^3, \quad G_2 \equiv \gamma_4 z + \alpha_2^{(i)} z^2 + \beta_2^{(i)} z^3 \qquad (9.6.4)$$

Hence, the fully nonlinear Jaumann strains are

$$\epsilon_{11}^{(i)} = \frac{\partial \mathbf{U}}{\partial x} \cdot \mathbf{i}_1 = e_1 + z\left(k_1 - k_1^0\right) + G_{1x} - k_5 G_2 \qquad (9.6.5a)$$

$$\epsilon_{22}^{(i)} = \frac{\partial \mathbf{U}}{\partial y} \cdot \mathbf{i}_2 = e_2 + z\left(k_2 - k_2^0\right) + G_{2y} + k_4 G_1 \qquad (9.6.5b)$$

$$\epsilon_{12}^{(i)} = \frac{\partial \mathbf{U}}{\partial x} \cdot \mathbf{i}_2 + \frac{\partial \mathbf{U}}{\partial y} \cdot \mathbf{i}_1 = \gamma_6 + z\left(k_6 - k_6^0\right) + G_{1y} + G_{2x} + k_5 G_1 - k_4 G_2 \qquad (9.6.5c)$$

$$\epsilon_{13}^{(i)} = \frac{\partial \mathbf{U}}{\partial x} \cdot \mathbf{i}_3 + \frac{\partial \mathbf{U}}{\partial z} \cdot \mathbf{i}_1 = G_{1z} - k_1 G_1 - k_{61} G_2 \qquad (9.6.5d)$$

$$\epsilon_{23}^{(i)} = \frac{\partial \mathbf{U}}{\partial y} \cdot \mathbf{i}_3 + \frac{\partial \mathbf{U}}{\partial z} \cdot \mathbf{i}_2 = G_{2z} - k_{62} G_1 - k_2 G_2 \qquad (9.6.5e)$$

$$\epsilon_{33}^{(i)} = \frac{\partial \mathbf{U}}{\partial z} \cdot \mathbf{i}_3 = 0 \qquad (9.6.5f)$$

We point out here that we neglect the trapezoidal-edge effect in deriving the strains in order to simplify the following derivations. However, it is straightforward to include the trapezoidal-edge effect in all of the derivations presented in this section, and the only thing that will be changed is the definitions of stress resultants and moments (see Section 9.5). Because the shear deformations of a flexible shell are usually small, one can substitute the deformed curvatures with the initial curvatures in (9.6.5d,e) without significant loss of accuracy and rewrite (9.6.5d,e) as

$$\epsilon_{13}^{(i)} = -\gamma_4 \left(k_{61}^0 z\right) + \gamma_5 \left(1 - k_1^0 z\right) + \alpha_1^{(i)} \left(2z - k_1^0 z^2\right) + \beta_1^{(i)} \left(3z^2 - k_1^0 z^3\right)$$
$$- \alpha_2^{(i)} \left(k_{61}^0 z^2\right) - \beta_2^{(i)} \left(k_{61}^0 z^3\right)$$

$$\epsilon_{23}^{(i)} = \gamma_4 \left(1 - k_2^0 z\right) - \gamma_5 \left(k_{62}^0 z\right) - \alpha_1^{(i)} \left(k_{62}^0 z^2\right) - \beta_1^{(i)} \left(k_{62}^0 z^3\right)$$
$$+ \alpha_2^{(i)} \left(2z - k_2^0 z^2\right) + \beta_2^{(i)} \left(3z^2 - k_2^0 z^3\right) \qquad (9.6.6a,b)$$

It follows from (2.4.34) and (2.4.35) that the stress-strain relations for the ith lamina is

$$\begin{Bmatrix} \sigma_{11}^{(i)} \\ \sigma_{22}^{(i)} \\ \sigma_{33}^{(i)} \\ \sigma_{23}^{(i)} \\ \sigma_{13}^{(i)} \\ \sigma_{12}^{(i)} \end{Bmatrix} = \begin{bmatrix} \overline{Q}_{11}^{(i)} & \overline{Q}_{12}^{(i)} & \overline{Q}_{13}^{(i)} & 0 & 0 & \overline{Q}_{16}^{(i)} \\ \overline{Q}_{12}^{(i)} & \overline{Q}_{22}^{(i)} & \overline{Q}_{23}^{(i)} & 0 & 0 & \overline{Q}_{26}^{(i)} \\ \overline{Q}_{13}^{(i)} & \overline{Q}_{23}^{(i)} & \overline{Q}_{33}^{(i)} & 0 & 0 & \overline{Q}_{36}^{(i)} \\ 0 & 0 & 0 & \overline{Q}_{44}^{(i)} & \overline{Q}_{45}^{(i)} & 0 \\ 0 & 0 & 0 & \overline{Q}_{45}^{(i)} & \overline{Q}_{55}^{(i)} & 0 \\ \overline{Q}_{16}^{(i)} & \overline{Q}_{26}^{(i)} & \overline{Q}_{36}^{(i)} & 0 & 0 & \overline{Q}_{66}^{(i)} \end{bmatrix} \begin{Bmatrix} \epsilon_{11}^{(i)} \\ \epsilon_{22}^{(i)} \\ \epsilon_{33}^{(i)} \\ \epsilon_{23}^{(i)} \\ \epsilon_{13}^{(i)} \\ \epsilon_{12}^{(i)} \end{Bmatrix} \quad (9.6.7)$$

where $[\overline{Q}^{(i)}]$ is the transformed stiffness matrix of the ith lamina. We note that transverse shear stresses $\sigma_{13}^{(i)}$ and $\sigma_{23}^{(i)}$ are only related to the transverse shear strains $\epsilon_{13}^{(i)}$ and $\epsilon_{23}^{(i)}$. We assume that there is no delamination and hence the inplane displacements u_1 and u_2 and interlaminar shear stresses σ_{13} and σ_{23} are continuous across the interface of two contiguous laminae. Moreover, we assume that there are no applied shear loads on the bonding surfaces and hence $\sigma_{13} = \sigma_{23} = \epsilon_{13} = \epsilon_{23} = 0$ at the $z = z_1, z_{N+1}$ planes, where N is the total number of layers. Hence, we have

$$\epsilon_{13}^{(1)}(x, y, z_1, t) = 0$$
$$\epsilon_{23}^{(1)}(x, y, z_1, t) = 0$$
$$u_1^{(i)}(x, y, z_{i+1}, t) - u_1^{(i+1)}(x, y, z_{i+1}, t) = 0 \text{ for } i = 1, 2, ..., N-1$$
$$u_2^{(i)}(x, y, z_{i+1}, t) - u_2^{(i+1)}(x, y, z_{i+1}, t) = 0 \text{ for } i = 1, 2, ..., N-1$$
$$\sigma_{23}^{(i)}(x, y, z_{i+1}, t) - \sigma_{23}^{(i+1)}(x, y, z_{i+1}, t) = 0 \text{ for } i = 1, 2, ..., N-1$$
$$\sigma_{13}^{(i)}(x, y, z_{i+1}, t) - \sigma_{13}^{(i+1)}(x, y, z_{i+1}, t) = 0 \text{ for } i = 1, 2, ..., N-1$$
$$\epsilon_{13}^{(N)}(x, y, z_{N+1}, t) = 0$$
$$\epsilon_{23}^{(N)}(x, y, z_{N+1}, t) = 0 \quad (9.6.8)$$

These $4N$ algebraic equations can be used to determine the $4N$ unknowns (i.e., $\alpha_1^{(i)}$, $\alpha_2^{(i)}$, $\beta_1^{(i)}$, $\beta_2^{(i)}$ for $i = 1, 2, ..., N$). Using (9.6.8), (9.6.6a,b), (9.6.7), (9.6.1a,b), and (6.5.11a), we obtain that

$$\alpha_1^{(i)} = a_{14}^{(i)} \gamma_4 + a_{15}^{(i)} \gamma_5, \quad \alpha_2^{(i)} = a_{24}^{(i)} \gamma_4 + a_{25}^{(i)} \gamma_5$$
$$\beta_1^{(i)} = b_{14}^{(i)} \gamma_4 + b_{15}^{(i)} \gamma_5, \quad \beta_2^{(i)} = b_{24}^{(i)} \gamma_4 + b_{25}^{(i)} \gamma_5 \quad (9.6.9)$$

for $i = 1, 2, ..., N$, where the $a_{kl}^{(i)}$ and $b_{kl}^{(i)}$ are functions of the z_j, k_1^0, k_2^0, k_{61}^0, k_{62}^0, $\overline{Q}_{44}^{(j)}$, $\overline{Q}_{45}^{(j)}$, and $\overline{Q}_{55}^{(j)}$.

Hence, it follows from (9.6.1a-c) and (9.6.9) that a general displacement field can be represented by

$$u_1^{(i)}(x, y, z, t) = u_1^0(x, y, t) + z\left[\theta_2(x, y, t) - \theta_{20}(x, y)\right] + \gamma_5 g_{15}^{(i)} + \gamma_4 g_{14}^{(i)} \quad (9.6.10a)$$

$$u_2^{(i)}(x,y,z,t) = u_2^0(x,y,t) - z\left[\theta_1(x,y,t) - \theta_{10}(x,y)\right] + \gamma_4 g_{24}^{(i)} + \gamma_5 g_{25}^{(i)} \quad (9.6.10b)$$

$$u_3^{(i)}(x,y,z,t) = u_3^0(x,y,t) \quad (9.6.10c)$$

where

$$g_{15}^{(i)} \equiv z + a_{15}^{(i)} z^2 + b_{15}^{(i)} z^3, \quad g_{14}^{(i)} \equiv a_{14}^{(i)} z^2 + b_{14}^{(i)} z^3$$
$$g_{24}^{(i)} \equiv z + a_{24}^{(i)} z^2 + b_{24}^{(i)} z^3, \quad g_{25}^{(i)} \equiv a_{25}^{(i)} z^2 + b_{25}^{(i)} z^3 \quad (9.6.11)$$

and

$$G_1 = \gamma_5 g_{15}^{(i)} + \gamma_4 g_{14}^{(i)}, \quad G_2 = \gamma_4 g_{24}^{(i)} + \gamma_5 g_{25}^{(i)} \quad (9.6.12)$$

We rewrite the strains in (9.6.5a-f) as

$$\epsilon_{11}^{(i)} = e_1 + z\left(k_1 - k_1^0\right) + \gamma_{5x} g_{15}^{(i)} + \gamma_{4x} g_{14}^{(i)} - k_5\left(\gamma_4 g_{24}^{(i)} + \gamma_5 g_{25}^{(i)}\right) \quad (9.6.13a)$$

$$\epsilon_{22}^{(i)} = e_2 + z\left(k_2 - k_2^0\right) + \gamma_{4y} g_{24}^{(i)} + \gamma_{5y} g_{25}^{(i)} + k_4\left(\gamma_5 g_{15}^{(i)} + \gamma_4 g_{14}^{(i)}\right) \quad (9.6.13b)$$

$$\epsilon_{12}^{(i)} = \gamma_6 + z\left(k_6 - k_6^0\right) + \gamma_{5y} g_{15}^{(i)} + \gamma_{4y} g_{14}^{(i)} + \gamma_{4x} g_{24}^{(i)} + \gamma_{5x} g_{25}^{(i)}$$
$$+ k_5\left(\gamma_5 g_{15}^{(i)} + \gamma_4 g_{14}^{(i)}\right) - k_4\left(\gamma_4 g_{24}^{(i)} + \gamma_5 g_{25}^{(i)}\right) \quad (9.6.13c)$$

$$\epsilon_{13}^{(i)} = \gamma_5 g_{15z}^{(i)} + \gamma_4 g_{14z}^{(i)} - k_1\left(\gamma_5 g_{15}^{(i)} + \gamma_4 g_{14}^{(i)}\right) - k_{61}\left(\gamma_4 g_{24}^{(i)} + \gamma_5 g_{25}^{(i)}\right) \quad (9.6.13d)$$

$$\epsilon_{23}^{(i)} = \gamma_4 g_{24z}^{(i)} + \gamma_5 g_{25z}^{(i)} - k_{62}\left(\gamma_5 g_{15}^{(i)} + \gamma_4 g_{14}^{(i)}\right) - k_2\left(\gamma_4 g_{24}^{(i)} + \gamma_5 g_{25}^{(i)}\right) \quad (9.6.13e)$$

$$\epsilon_{33}^{(i)} = 0 \quad (9.6.13f)$$

Or, one can put them in the following matrix form:

$$\left\{\begin{array}{c} \epsilon_{11}^{(i)} \\ \epsilon_{22}^{(i)} \\ \epsilon_{12}^{(i)} \end{array}\right\} = \left[S_1^{(i)}\right]\{\psi\} + \left[\begin{array}{cc} -k_5 & 0 \\ 0 & k_4 \\ -k_4 & k_5 \end{array}\right]\left[S_4^{(i)}\right]\left\{\begin{array}{c} \gamma_4 \\ \gamma_5 \end{array}\right\} \quad (9.6.14a)$$

$$\left\{\begin{array}{c} \epsilon_{23}^{(i)} \\ \epsilon_{13}^{(i)} \end{array}\right\} = \left(\left[S_2^{(i)}\right] - \left[\begin{array}{cc} k_2 & k_{62} \\ k_{61} & k_1 \end{array}\right]\left[S_4^{(i)}\right]\right)\left\{\begin{array}{c} \gamma_4 \\ \gamma_5 \end{array}\right\} \quad (9.6.14b)$$

where

$$\left[S_1^{(i)}\right] \equiv \left[\begin{array}{ccccccccc} 1 & 0 & 0 & z & 0 & 0 & g_{14}^{(i)} & 0 & g_{15}^{(i)} & 0 \\ 0 & 1 & 0 & 0 & z & 0 & 0 & g_{24}^{(i)} & 0 & g_{25}^{(i)} \\ 0 & 0 & 1 & 0 & 0 & z & g_{24}^{(i)} & g_{14}^{(i)} & g_{25}^{(i)} & g_{15}^{(i)} \end{array}\right] \quad (9.6.15a)$$

$$\left[S_2^{(i)}\right] \equiv \left[\begin{array}{cc} g_{24z}^{(i)} & g_{25z}^{(i)} \\ g_{14z}^{(i)} & g_{15z}^{(i)} \end{array}\right], \quad \left[S_4^{(i)}\right] \equiv \left[\begin{array}{cc} g_{24}^{(i)} & g_{25}^{(i)} \\ g_{14}^{(i)} & g_{15}^{(i)} \end{array}\right] \quad (9.6.15b,c)$$

$$\{\psi\} \equiv \left\{e_1, e_2, \gamma_6, k_1 - k_1^0, k_2 - k_2^0, k_6 - k_6^0, \gamma_{4x}, \gamma_{4y}, \gamma_{5x}, \gamma_{5y}\right\}^T \quad (9.6.15d)$$

NONLINEAR LAYERWISE SHEAR-DEFORMABLE SHELL THEORY 619

Substituting (9.6.11), (9.6.10a-c), (6.5.11a), (9.6.13d,e), and (9.6.7) into (9.6.8) and then setting each of the coefficients of γ_4 and γ_5 equal to zero yields

$$a_{14}^{(1)}\left(2z_1 - k_1^0 z_1^2\right) + b_{14}^{(1)}\left(3z_1^2 - k_1^0 z_1^3\right) - a_{24}^{(1)} k_{61}^0 z_1^2 - b_{24}^{(1)} k_{61}^0 z_1^3 = k_{61}^0 z_1$$
$$a_{15}^{(1)}\left(2z_1 - k_1^0 z_1^2\right) + b_{15}^{(1)}\left(3z_1^2 - k_1^0 z_1^3\right) - a_{25}^{(1)} k_{61}^0 z_1^2 - b_{25}^{(1)} k_{61}^0 z_1^3 = -1 + k_1^0 z_1$$
$$-a_{14}^{(1)} k_{62}^0 z_1^2 - b_{14}^{(1)} k_{62}^0 z_1^3 + a_{24}^{(1)}\left(2z_1 - k_2^0 z_1^2\right) + b_{24}^{(1)}\left(3z_1^2 - k_2^0 z_1^3\right) = -1 + k_2^0 z_1$$
$$-a_{15}^{(1)} k_{62}^0 z_1^2 - b_{15}^{(1)} k_{62}^0 z_1^3 + a_{25}^{(1)}\left(2z_1 - k_2^0 z_1^2\right) + b_{25}^{(1)}\left(3z_1^2 - k_2^0 z_1^3\right) = k_{62}^0 z_1$$

(9.6.16a)

$$a_{14}^{(i)} z_j^2 + b_{14}^{(i)} z_j^3 - a_{14}^{(j)} z_j^2 - b_{14}^{(j)} z_j^3 = 0$$
$$a_{15}^{(i)} z_j^2 + b_{15}^{(i)} z_j^3 - a_{15}^{(j)} z_j^2 - b_{15}^{(j)} z_j^3 = 0$$
$$a_{24}^{(i)} z_j^2 + b_{24}^{(i)} z_j^3 - a_{24}^{(j)} z_j^2 - b_{24}^{(j)} z_j^3 = 0$$
$$a_{25}^{(i)} z_j^2 + b_{25}^{(i)} z_j^3 - a_{25}^{(j)} z_j^2 - b_{25}^{(j)} z_j^3 = 0$$

$$\overline{Q}_{44}^{(i)}\left[-a_{14}^{(i)} k_{62}^0 z_j^2 - b_{14}^{(i)} k_{62}^0 z_j^3 + a_{24}^{(i)}\left(2z_j - k_2^0 z_j^2\right) + b_{24}^{(i)}\left(3z_j^2 - k_2^0 z_j^3\right)\right]$$
$$+\overline{Q}_{45}^{(i)}\left[a_{14}^{(i)}\left(2z_j - k_1^0 z_j^2\right) + b_{14}^{(i)}\left(3z_j^2 - k_1^0 z_j^3\right) - a_{24}^{(i)} k_{61}^0 z_j^2 - b_{24}^{(i)} k_{61}^0 z_j^3\right]$$
$$-\overline{Q}_{44}^{(j)}\left[-a_{14}^{(j)} k_{62}^0 z_j^2 - b_{14}^{(j)} k_{62}^0 z_j^3 + a_{24}^{(j)}\left(2z_j - k_2^0 z_j^2\right) + b_{24}^{(j)}\left(3z_j^2 - k_2^0 z_j^3\right)\right]$$
$$-\overline{Q}_{45}^{(j)}\left[a_{14}^{(j)}\left(2z_j - k_1^0 z_j^2\right) + b_{14}^{(j)}\left(3z_j^2 - k_1^0 z_j^3\right) - a_{24}^{(j)} k_{61}^0 z_j^2 - b_{24}^{(j)} k_{61}^0 z_j^3\right]$$
$$= \left(1 - k_2^0 z_j\right)\left(\overline{Q}_{44}^{(j)} - \overline{Q}_{44}^{(i)}\right) - k_{61}^0 z_j\left(\overline{Q}_{45}^{(j)} - \overline{Q}_{45}^{(i)}\right)$$

$$\overline{Q}_{44}^{(i)}\left[-a_{15}^{(i)} k_{62}^0 z_j^2 - b_{15}^{(i)} k_{62}^0 z_j^3 + a_{25}^{(i)}\left(2z_j - k_2^0 z_j^2\right) + b_{25}^{(i)}\left(3z_j^2 - k_2^0 z_j^3\right)\right]$$
$$+\overline{Q}_{45}^{(i)}\left[a_{15}^{(i)}\left(2z_j - k_1^0 z_j^2\right) + b_{15}^{(i)}\left(3z_j^2 - k_1^0 z_j^3\right) - a_{25}^{(i)} k_{61}^0 z_j^2 - b_{25}^{(i)} k_{61}^0 z_j^3\right]$$
$$-\overline{Q}_{44}^{(j)}\left[-a_{15}^{(j)} k_{62}^0 z_j^2 - b_{15}^{(j)} k_{62}^0 z_j^3 + a_{25}^{(j)}\left(2z_j - k_2^0 z_j^2\right) + b_{25}^{(j)}\left(3z_j^2 - k_2^0 z_j^3\right)\right]$$
$$-\overline{Q}_{45}^{(j)}\left[a_{15}^{(j)}\left(2z_j - k_1^0 z_j^2\right) + b_{15}^{(j)}\left(3z_j^2 - k_1^0 z_j^3\right) - a_{25}^{(j)} k_{61}^0 z_j^2 - b_{25}^{(j)} k_{61}^0 z_j^3\right]$$
$$= \left(1 - k_1^0 z_j\right)\left(\overline{Q}_{45}^{(j)} - \overline{Q}_{45}^{(i)}\right) - k_{62}^0 z_j\left(\overline{Q}_{44}^{(j)} - \overline{Q}_{44}^{(i)}\right)$$

$$\overline{Q}_{45}^{(i)}\left[-a_{14}^{(i)} k_{62}^0 z_j^2 - b_{14}^{(i)} k_{62}^0 z_j^3 + a_{24}^{(i)}\left(2z_j - k_2^0 z_j^2\right) + b_{24}^{(i)}\left(3z_j^2 - k_2^0 z_j^3\right)\right]$$
$$+\overline{Q}_{55}^{(i)}\left[a_{14}^{(i)}\left(2z_j - k_1^0 z_j^2\right) + b_{14}^{(i)}\left(3z_j^2 - k_1^0 z_j^3\right) - a_{24}^{(i)} k_{61}^0 z_j^2 - b_{24}^{(i)} k_{61}^0 z_j^3\right]$$
$$-\overline{Q}_{45}^{(j)}\left[-a_{14}^{(j)} k_{62}^0 z_j^2 - b_{14}^{(j)} k_{62}^0 z_j^3 + a_{24}^{(j)}\left(2z_j - k_2^0 z_j^2\right) + b_{24}^{(j)}\left(3z_j^2 - k_2^0 z_j^3\right)\right]$$
$$-\overline{Q}_{55}^{(j)}\left[a_{14}^{(j)}\left(2z_j - k_1^0 z_j^2\right) + b_{14}^{(j)}\left(3z_j^2 - k_1^0 z_j^3\right) - a_{24}^{(j)} k_{61}^0 z_j^2 - b_{24}^{(j)} k_{61}^0 z_j^3\right]$$
$$= \left(1 - k_2^0 z_j\right)\left(\overline{Q}_{45}^{(j)} - \overline{Q}_{45}^{(i)}\right) - k_{61}^0 z_j\left(\overline{Q}_{55}^{(j)} - \overline{Q}_{55}^{(i)}\right)$$

$$\overline{Q}_{45}^{(i)}\left[-a_{15}^{(i)} k_{62}^0 z_j^2 - b_{15}^{(i)} k_{62}^0 z_j^3 + a_{25}^{(i)}\left(2z_j - k_2^0 z_j^2\right) + b_{25}^{(i)}\left(3z_j^2 - k_2^0 z_j^3\right)\right]$$

$$+\overline{Q}_{55}^{(i)}\left[a_{15}^{(i)}\left(2z_j - k_1^0 z_j^2\right) + b_{15}^{(i)}\left(3z_j^2 - k_1^0 z_j^3\right) - a_{25}^{(i)} k_{61}^0 z_j^2 - b_{25}^{(i)} k_{61}^0 z_j^3\right]$$
$$-\overline{Q}_{45}^{(j)}\left[-a_{15}^{(j)} k_{62}^0 z_j^2 - b_{15}^{(j)} k_{62}^0 z_j^3 + a_{25}^{(j)}\left(2z_j - k_2^0 z_j^2\right) + b_{25}^{(j)}\left(3z_j^2 - k_2^0 z_j^3\right)\right]$$
$$-\overline{Q}_{55}^{(j)}\left[a_{15}^{(j)}\left(2z_j - k_1^0 z_j^2\right) + b_{15}^{(j)}\left(3z_j^2 - k_1^0 z_j^3\right) - a_{25}^{(j)} k_{61}^0 z_j^2 - b_{25}^{(j)} k_{61}^0 z_j^3\right]$$
$$= \left(1 - k_1^0 z_j\right)\left(\overline{Q}_{55}^{(j)} - \overline{Q}_{55}^{(i)}\right) - k_{62}^0 z_j\left(\overline{Q}_{45}^{(j)} - \overline{Q}_{45}^{(i)}\right) \qquad (9.6.16b)$$

for $i = 1, 2, ..., N-1$, where $j \equiv i+1$, and

$$a_{14}^{(N)}\left(2z_{N+1} - k_1^0 z_{N+1}^2\right) + b_{14}^{(N)}\left(3z_{N+1}^2 - k_1^0 z_{N+1}^3\right)$$
$$-a_{24}^{(N)} k_{61}^0 z_{N+1}^2 - b_{24}^{(N)} k_{61}^0 z_{N+1}^3 = k_{61}^0 z_{N+1}$$
$$a_{15}^{(N)}\left(2z_{N+1} - k_1^0 z_{N+1}^2\right) + b_{15}^{(N)}\left(3z_{N+1}^2 - k_1^0 z_{N+1}^3\right)$$
$$-a_{25}^{(N)} k_{61}^0 z_{N+1}^2 - b_{25}^{(N)} k_{61}^0 z_{N+1}^3 = -1 + k_1^0 z_{N+1}$$
$$-a_{14}^{(N)} k_{62}^0 z_{N+1}^2 - b_{14}^{(N)} k_{62}^0 z_{N+1}^3 + a_{24}^{(N)}\left(2z_{N+1} - k_2^0 z_{N+1}^2\right)$$
$$+b_{24}^{(N)}\left(3z_{N+1}^2 - k_2^0 z_{N+1}^3\right) = -1 + k_2^0 z_{N+1}$$
$$-a_{15}^{(N)} k_{62}^0 z_{N+1}^2 - b_{15}^{(N)} k_{62}^0 z_{N+1}^3 + a_{25}^{(N)}\left(2z_{N+1} - k_2^0 z_{N+1}^2\right)$$
$$+b_{25}^{(N)}\left(3z_{N+1}^2 - k_2^0 z_{N+1}^3\right) = k_{62}^0 z_{N+1} \qquad (9.6.16c)$$

Here, we neglect the effect of the local rotations due to strainable displacements and replace the deformed curvatures with the undeformed curvatures in (9.6.13d,e). These $8N$ algebraic equations can be solved for the $8N$ unknowns — $a_{14}^{(i)}, a_{15}^{(i)}, a_{24}^{(i)}, a_{25}^{(i)}, b_{14}^{(i)}, b_{15}^{(i)}, b_{24}^{(i)}, b_{25}^{(i)}$, for $i = 1, 2, ..., N$. We note that the reference surface cannot be chosen as the contacting surface of any two laminae because it will make the transverse shear strains continuous at $z = 0$.

9.6.2 Inertia Terms

It follows from Figures 6.1.2 and 9.5.2 and (9.6.10a-c) and (6.5.11a) that the absolute displacement vector **D** of an arbitrary point on the observed shell element is

$$\mathbf{D} = u\mathbf{j}_1 + v\mathbf{j}_2 + w\mathbf{j}_3 + z\mathbf{i}_3 - z\mathbf{j}_3 + u_1^{(i)}\mathbf{i}_1 + u_2^{(i)}\mathbf{i}_2 + u_3^{(i)}\mathbf{i}_3$$
$$= u\mathbf{j}_1 + v\mathbf{j}_2 + w\mathbf{j}_3 + z\mathbf{i}_3 - z\mathbf{j}_3 + \left(\gamma_5 g_{15}^{(i)} + \gamma_4 g_{14}^{(i)}\right)\mathbf{i}_1 + \left(\gamma_4 g_{24}^{(i)} + \gamma_5 g_{25}^{(i)}\right)\mathbf{i}_2$$
$$(9.6.17)$$

Taking the variation and time derivatives of (9.6.17) and using (6.3.1a) and (6.2.11a), we obtain

$$\delta\mathbf{D} = \mathbf{j}_1 \delta u + \mathbf{j}_2 \delta v + \mathbf{j}_3 \delta w + \left(g_{14}^{(i)}\mathbf{i}_1 + g_{24}^{(i)}\mathbf{i}_2\right)\delta\gamma_4 + \left(g_{15}^{(i)}\mathbf{i}_1 + g_{25}^{(i)}\mathbf{i}_2\right)\delta\gamma_5$$
$$+ \left[\left(\gamma_4 g_{24}^{(i)} + \gamma_5 g_{25}^{(i)}\right)\mathbf{i}_3 - z\mathbf{i}_2\right]\delta\theta_1 + \left[z\mathbf{i}_1 - \left(\gamma_5 g_{15}^{(i)} + \gamma_4 g_{14}^{(i)}\right)\mathbf{i}_3\right]\delta\theta_2$$
$$+ \left[\left(\gamma_5 g_{15}^{(i)} + \gamma_4 g_{14}^{(i)}\right)\mathbf{i}_2 - \left(\gamma_4 g_{24}^{(i)} + \gamma_5 g_{25}^{(i)}\right)\mathbf{i}_1\right]\delta\theta_3 \qquad (9.6.18a)$$

and

$$\ddot{\mathbf{D}} = \ddot{u}\mathbf{j}_1 + \ddot{v}\mathbf{j}_2 + \ddot{w}\mathbf{j}_3 + \ddot{z}\mathbf{i}_3 + \left(\ddot{\gamma}_5 g_{15}^{(i)} + \ddot{\gamma}_4 g_{14}^{(i)}\right)\mathbf{i}_1 + 2\left(\dot{\gamma}_5 g_{15}^{(i)} + \dot{\gamma}_4 g_{14}^{(i)}\right)\dot{\mathbf{i}}_1$$
$$+ \left(\gamma_5 g_{15}^{(i)} + \gamma_4 g_{14}^{(i)}\right)\ddot{\mathbf{i}}_1 + \left(\ddot{\gamma}_4 g_{24}^{(i)} + \ddot{\gamma}_5 g_{25}^{(i)}\right)\mathbf{i}_2$$
$$+ 2\left(\dot{\gamma}_4 g_{24}^{(i)} + \dot{\gamma}_5 g_{25}^{(i)}\right)\dot{\mathbf{i}}_2 + \left(\gamma_4 g_{24}^{(i)} + \gamma_5 g_{25}^{(i)}\right)\ddot{\mathbf{i}}_2 \qquad (9.6.18\text{b})$$

where

$$\mathbf{i}_k = T_{k1}\mathbf{j}_1 + T_{k2}\mathbf{j}_2 + T_{k3}\mathbf{j}_3, \quad \dot{\mathbf{i}}_k = \dot{T}_{k1}\mathbf{j}_1 + \dot{T}_{k2}\mathbf{j}_2 + \dot{T}_{k3}\mathbf{j}_3$$
$$\ddot{\mathbf{i}}_k = \ddot{T}_{k1}\mathbf{j}_1 + \ddot{T}_{k2}\mathbf{j}_2 + \ddot{T}_{k3}\mathbf{j}_3 \qquad (9.6.18\text{c})$$

for $k = 1, 2, 3$. Using (9.6.18a-c) and the identity $\mathbf{i}_k \cdot \mathbf{i}_k = 0$ in (7.2.2a), we obtain variation of the kinetic energy δT as

$$\delta T = -\int_A \left(A_u \delta u + A_v \delta v + A_w \delta w + A_{\gamma_4}\delta\gamma_4 + A_{\gamma_5}\delta\gamma_5 \right.$$
$$\left. + A_{\theta_1}\delta\theta_1 + A_{\theta_2}\delta\theta_2 + A_{\theta_3}\delta\theta_3\right)dxdy \qquad (9.6.19)$$

where the inertia terms A_u, A_v, A_w, A_{γ_4}, A_{γ_5}, A_{θ_1}, A_{θ_2}, and A_{θ_3} are the same as those defined in (7.8.30a-h) and (7.8.31), but the entries of the transformation matrix $[T]$ are different because of initial curvatures. To obtain linear expressions for the inertia terms, one can expand $[T]$ as

$$[T] = \begin{bmatrix} 1 & v_x + uk_5^0 + wk_{61}^0 & w_x - uk_1^0 - vk_{61}^0 \\ u_y - vk_4^0 + wk_{62}^0 & 1 & w_y - uk_{62}^0 - vk_2^0 \\ -w_x + uk_1^0 + vk_{61}^0 & -w_y + uk_{62}^0 + vk_2^0 & 1 \end{bmatrix}$$
$$(9.6.20)$$

Substituting (9.6.20) and (9.6.18c) into (7.8.30) yields the linear inertia terms

$$A_u = I_0\ddot{u} + I_1\left(-\ddot{w}_x + \ddot{u}k_1^0 + \ddot{v}k_{61}^0\right) + I_6\ddot{\gamma}_5 + I_5\ddot{\gamma}_4$$
$$A_v = I_0\ddot{v} + I_1\left(-\ddot{w}_y + \ddot{u}k_{62}^0 + \ddot{v}k_2^0\right) + I_7\ddot{\gamma}_4 + I_8\ddot{\gamma}_5$$
$$A_w = I_0\ddot{w}$$
$$A_{\theta_1} = I_2\left(\ddot{w}_y - \ddot{u}k_{62}^0 - \ddot{v}k_2^0\right) - I_1\ddot{v} - I_{71}\ddot{\gamma}_4 - I_{81}\ddot{\gamma}_5$$
$$A_{\theta_2} = -I_2\left(\ddot{w}_x - \ddot{u}k_1^0 - \ddot{v}k_{61}^0\right) + I_1\ddot{u} + I_{61}\ddot{\gamma}_5 + I_{51}\ddot{\gamma}_4$$
$$A_{\theta_3} = 0$$
$$A_{\gamma_4} = (I_{55} + I_{77})\ddot{\gamma}_4 + (I_{56} + I_{78})\ddot{\gamma}_5 + I_5\ddot{u} + I_7\ddot{v}$$
$$\quad - I_{51}\left(\ddot{w}_x - \ddot{u}k_1^0 - \ddot{v}k_{61}^0\right) - I_{71}\left(\ddot{w}_y - \ddot{u}k_{62}^0 - \ddot{v}k_2^0\right)$$
$$A_{\gamma_5} = (I_{66} + I_{88})\ddot{\gamma}_5 + (I_{56} + I_{78})\ddot{\gamma}_4 + I_6\ddot{u} + I_8\ddot{v}$$
$$\quad - I_{61}\left(\ddot{w}_x - \ddot{u}k_1^0 - \ddot{v}k_{61}^0\right) - I_{81}\left(\ddot{w}_y - \ddot{u}k_{62}^0 - \ddot{v}k_2^0\right) \qquad (9.6.21)$$

We note that the differences between (9.6.20) and (7.8.39) and (9.6.21) and (7.8.40) are due to the initial curvatures.

9.6.3 Structural Terms

Using the assumption that $\sigma_{33}^{(i)} = 0$, we separate (9.6.7) into

$$\left\{\begin{array}{c}\sigma_{11}^{(i)}\\ \sigma_{22}^{(i)}\\ \sigma_{12}^{(i)}\end{array}\right\} = \left[\tilde{Q}^{(i)}\right]\left\{\begin{array}{c}\epsilon_{11}^{(i)}\\ \epsilon_{22}^{(i)}\\ \epsilon_{12}^{(i)}\end{array}\right\},\ \left[\tilde{Q}^{(i)}\right] \equiv \left[\begin{array}{ccc}\tilde{Q}_{11}^{(i)} & \tilde{Q}_{12}^{(i)} & \tilde{Q}_{16}^{(i)}\\ \tilde{Q}_{12}^{(i)} & \tilde{Q}_{22}^{(i)} & \tilde{Q}_{26}^{(i)}\\ \tilde{Q}_{16}^{(i)} & \tilde{Q}_{26}^{(i)} & \tilde{Q}_{66}^{(i)}\end{array}\right] \quad (9.6.22)$$

$$\left\{\begin{array}{c}\sigma_{23}^{(i)}\\ \sigma_{13}^{(i)}\end{array}\right\} = \left[\hat{Q}^{(i)}\right]\left\{\begin{array}{c}\epsilon_{23}^{(i)}\\ \epsilon_{13}^{(i)}\end{array}\right\},\ \left[\hat{Q}^{(i)}\right] \equiv \left[\begin{array}{cc}\overline{Q}_{44}^{(i)} & \overline{Q}_{45}^{(i)}\\ \overline{Q}_{45}^{(i)} & \overline{Q}_{55}^{(i)}\end{array}\right] \quad (9.6.23)$$

For flexible structures undergoing large rotation and small strain vibrations, the important geometric nonlinearities are included in the transformation matrix $[T]$. On the other hand, because the strains are small, one can replace the deformed curvatures with the initial curvatures in the strain-displacement relations (9.6.14a,b) without significant loss of accuracy. Hence, we rewrite (9.6.14a,b) as

$$\left\{\begin{array}{c}\epsilon_{11}^{(i)}\\ \epsilon_{22}^{(i)}\\ \epsilon_{12}^{(i)}\end{array}\right\} = \left[\bar{S}_1^{(i)}\right]\{\bar{\psi}\},\ \left\{\begin{array}{c}\epsilon_{23}^{(i)}\\ \epsilon_{13}^{(i)}\end{array}\right\} = \left[\bar{S}_2^{(i)}\right]\left\{\begin{array}{c}\gamma_4\\ \gamma_5\end{array}\right\} \quad (9.6.24a,b)$$

where

$$\left[\bar{S}_1^{(i)}\right] \equiv \left[\begin{array}{cccccccccc}1 & 0 & 0 & z & 0 & 0 & g_{14}^{(i)} & 0 & g_{15}^{(i)} & 0\\ 0 & 1 & 0 & 0 & z & 0 & 0 & g_{24}^{(i)} & 0 & g_{25}^{(i)}\\ 0 & 0 & 1 & 0 & 0 & z & g_{24}^{(i)} & g_{14}^{(i)} & g_{25}^{(i)} & g_{15}^{(i)}\end{array}\right.$$

$$\left.\begin{array}{cc}-k_5^0 g_{24}^{(i)} & -k_5^0 g_{25}^{(i)}\\ k_4^0 g_{14}^{(i)} & k_4^0 g_{15}^{(i)}\\ k_5^0 g_{14}^{(i)} - k_4^0 g_{24}^{(i)} & k_5^0 g_{15}^{(i)} - k_4^0 g_{25}^{(i)}\end{array}\right] \quad (9.6.25a)$$

$$\left[\bar{S}_2^{(i)}\right] \equiv \left[\begin{array}{cc}g_{24z}^{(i)} - k_2^0 g_{24}^{(i)} - k_{62}^0 g_{14}^{(i)} & g_{25z}^{(i)} - k_2^0 g_{25}^{(i)} - k_{62}^0 g_{15}^{(i)}\\ g_{14z}^{(i)} - k_{61}^0 g_{24}^{(i)} - k_1^0 g_{14}^{(i)} & g_{15z}^{(i)} - k_{61}^0 g_{25}^{(i)} - k_1^0 g_{15}^{(i)}\end{array}\right] \quad (9.6.25b)$$

$$\{\bar{\psi}\} \equiv \{e_1, e_2, \gamma_6, k_1 - k_1^0, k_2 - k_2^0, k_6 - k_6^0, \gamma_{4x}, \gamma_{4y}, \gamma_{5x}, \gamma_{5y}, \gamma_4, \gamma_5\}^T \quad (9.6.25c)$$

Next, we define the following stress resultants and moments:

$$\{N_1, N_2, N_6, M_1, M_2, M_6, m_{61}, m_2, m_1, m_{62}\}$$

$$\equiv \sum_{i=1}^{N}\int_{z_i}^{z_{i+1}}\left\{\sigma_{11}^{(i)}, \sigma_{22}^{(i)}, \sigma_{12}^{(i)}\right\}\left[S_1^{(i)}\right]dz$$

$$= \{\bar{\psi}\}^T \sum_{i=1}^{N}\int_{z_i}^{z_{i+1}}\left[\bar{S}_1^{(i)}\right]^T\left[\tilde{Q}^{(i)}\right]\left[S_1^{(i)}\right]dz \quad (9.6.26)$$

$$\{q_2, q_1\} \equiv \sum_{i=1}^{N} \int_{z_i}^{z_{i+1}} \left\{\sigma_{23}^{(i)}, \sigma_{13}^{(i)}\right\} \left[S_2^{(i)}\right] dz$$

$$= \{\gamma_4, \gamma_5\} \sum_{i=1}^{N} \int_{z_i}^{z_{i+1}} \left[\bar{S}_2^{(i)}\right]^T \left[\hat{Q}^{(i)}\right] \left[S_2^{(i)}\right] dz \qquad (9.6.27)$$

$$\{\overline{m}_1, \tilde{m}_1, \tilde{m}_2, \overline{m}_2, \tilde{m}_{62}, \overline{m}_{62}, \overline{m}_{61}, \tilde{m}_{61}\}$$

$$\equiv \sum_{i=1}^{N} \int_{z_i}^{z_{i+1}} \left\{\sigma_{11}^{(i)}, \sigma_{22}^{(i)}, \sigma_{12}^{(i)}\right\} \left[S_3^{(i)}\right] dz$$

$$= \{\bar{\psi}\}^T \sum_{i=1}^{N} \int_{z_i}^{z_{i+1}} \left[\bar{S}_1^{(i)}\right]^T \left[\tilde{Q}^{(i)}\right] \left[S_3^{(i)}\right] dz \qquad (9.6.28)$$

$$\{\tilde{s}_{21}, \bar{s}_{21}, \bar{s}_{22}, \tilde{s}_{22}\} \equiv \sum_{i=1}^{N} \int_{z_i}^{z_{i+1}} \sigma_{23}^{(i)} \{g_{14}^{(i)}, g_{15}^{(i)}, g_{24}^{(i)}, g_{25}^{(i)}\} dz$$

$$= \{\gamma_4, \gamma_5\} \sum_{i=1}^{N} \int_{z_i}^{z_{i+1}} \left[\bar{S}_2^{(i)}\right]^T \left\{\begin{array}{c} \overline{Q}_{44}^{(i)} \\ \overline{Q}_{45}^{(i)} \end{array}\right\} \{g_{14}^{(i)}, g_{15}^{(i)}, g_{24}^{(i)}, g_{25}^{(i)}\} dz \quad (9.6.29)$$

$$\{\tilde{s}_{11}, \bar{s}_{11}, \bar{s}_{12}, \tilde{s}_{12}\} \equiv \sum_{i=1}^{N} \int_{z_i}^{z_{i+1}} \sigma_{13}^{(i)} \left\{g_{14}^{(i)}, g_{15}^{(i)}, g_{24}^{(i)}, g_{25}^{(i)}\right\} dz$$

$$= \{\gamma_4, \gamma_5\} \sum_{i=1}^{N} \int_{z_i}^{z_{i+1}} \left[\bar{S}_2^{(i)}\right]^T \left\{\begin{array}{c} \overline{Q}_{45}^{(i)} \\ \overline{Q}_{55}^{(i)} \end{array}\right\} \{g_{14}^{(i)}, g_{15}^{(i)}, g_{24}^{(i)}, g_{25}^{(i)}\} dz \quad (9.6.30)$$

where

$$\left[S_3^{(i)}\right] \equiv \begin{bmatrix} g_{24}^{(i)} & g_{25}^{(i)} & 0 & 0 & 0 & 0 & 0 & 0 \\ 0 & 0 & g_{14}^{(i)} & g_{15}^{(i)} & 0 & 0 & 0 & 0 \\ 0 & 0 & 0 & 0 & g_{14}^{(i)} & g_{15}^{(i)} & g_{24}^{(i)} & g_{25}^{(i)} \end{bmatrix} \qquad (9.6.31)$$

Substituting (9.6.13a-e) and $\sigma_{33} = 0$ into (7.2.2b) and using (6.4.2a,b) and the fact that variations of the initial curvatures are zero, we obtain variation of the elastic energy in terms of the stress resultants, stress moments, and inplane strains as

$$\delta\Pi = \int_A \left(N_1 \delta e_1 + N_2 \delta e_2 + N_6 \delta\gamma_6 + \Theta_1 \delta\theta_1 + \Theta_2 \delta\theta_2 + \Theta_3 \delta\theta_3 \right.$$
$$\left. + \Gamma_4 \delta\gamma_4 + \Gamma_5 \delta\gamma_5\right) dx dy$$
$$+ \int_x \left[-\hat{M}_2 \delta\theta_1 + \hat{M}_{62} \delta\theta_2 + m_{32} \delta\theta_3 + m_2 \delta\gamma_4 + m_{62} \delta\gamma_5\right]_{y=0}^{y=Y} dx$$
$$+ \int_y \left[-\hat{M}_{61} \delta\theta_1 + \hat{M}_1 \delta\theta_2 + m_{31} \delta\theta_3 + m_{61} \delta\gamma_4 + m_1 \delta\gamma_5\right]_{x=0}^{x=X} dy \quad (9.6.32)$$

where the modified stress resultants and moments are defined as

$$\hat{M}_1 \equiv M_1 - \bar{s}_{11}\gamma_5 - \tilde{s}_{11}\gamma_4 \tag{9.6.33a}$$

$$\hat{M}_2 \equiv M_2 - \tilde{s}_{22}\gamma_5 - \bar{s}_{22}\gamma_4 \tag{9.6.33b}$$

$$\hat{M}_{61} \equiv M_6 - \tilde{s}_{12}\gamma_5 - \bar{s}_{12}\gamma_4 \tag{9.6.33c}$$

$$\hat{M}_{62} \equiv M_6 - \bar{s}_{21}\gamma_5 - \tilde{s}_{21}\gamma_4 \tag{9.6.33d}$$

$$m_{31} \equiv (\overline{m}_{62} - \tilde{m}_1)\gamma_5 + (\tilde{m}_{62} - \overline{m}_1)\gamma_4 \tag{9.6.33e}$$

$$m_{32} \equiv (\overline{m}_2 - \tilde{m}_{61})\gamma_5 + (\tilde{m}_2 - \overline{m}_{61})\gamma_4 \tag{9.6.33f}$$

$$\Theta_1 \equiv \hat{M}_{61x} + \hat{M}_{2y} - m_{31}k_1 - m_{32}k_{62} + \hat{M}_1 k_5 + \hat{M}_{62} k_4 \tag{9.6.33g}$$

$$\Theta_2 \equiv -\hat{M}_{1x} - \hat{M}_{62y} - m_{31}k_{61} - m_{32}k_2 + \hat{M}_2 k_4 + \hat{M}_{61} k_5 \tag{9.6.33h}$$

$$\Theta_3 \equiv -m_{31x} - m_{32y} + \hat{M}_1 k_{61} - \hat{M}_2 k_{62} + \hat{M}_{62} k_2 - \hat{M}_{61} k_1 \tag{9.6.33i}$$

$$\Gamma_4 \equiv -m_{61x} - m_{2y} + q_2 - \tilde{s}_{11}k_1 - \bar{s}_{22}k_2 - \bar{s}_{12}k_{61} - \tilde{s}_{21}k_{62}$$
$$+ (\tilde{m}_2 - \overline{m}_{61})k_4 + (\tilde{m}_{62} - \overline{m}_1)k_5 \tag{9.6.33j}$$

$$\Gamma_5 \equiv -m_{1x} - m_{62y} + q_1 - \bar{s}_{11}k_1 - \tilde{s}_{22}k_2 - \tilde{s}_{12}k_{61} - \bar{s}_{21}k_{62}$$
$$+ (\overline{m}_2 - \tilde{m}_{61})k_4 + (\overline{m}_{62} - \tilde{m}_1)k_5 \tag{9.6.33k}$$

Here, $\hat{M}_1$, $\hat{M}_2$, $\hat{M}_{61}$, $\hat{M}_{62}$, m_{31}, and m_{32} represent the total moment intensities acting on the edges of the shell element, as shown in Figure 9.5.3a; and N_1, N_2, and N_6 ($=N_{61}=N_{62}$) are the force intensities, as shown in Figure 9.5.3b. Equations (9.6.33e,f) show that m_{31} and m_{32} are nonlinear terms. Also, we note that (9.6.32) and (9.6.33a-k) are the same as those obtained in the nonlinear plate theory (i.e., (7.8.23) and (7.8.24a-k)), except that the deformed curvatures are functions of the undeformed curvatures as well as u, v, and w.

Substituting (6.3.3a,b), (6.3.5a,b), and (6.3.7a,b) into (9.6.32), we obtain

$$\delta\Pi = \int_A \Biggl\{ \{N_1, N_6/(1+e_1), -\Theta_2/(1+e_1)\}$$
$$\times \left([T] \begin{Bmatrix} \delta u_x \\ \delta v_x \\ \delta w_x \end{Bmatrix} + [T][K_1^0]^T \begin{Bmatrix} \delta u \\ \delta v \\ \delta w \end{Bmatrix} \right) + \Theta_3 \delta\theta_3 + \Gamma_4 \delta\gamma_4 + \Gamma_5 \delta\gamma_5$$
$$+ \{N_6/(1+e_2), N_2, \Theta_1/(1+e_2)\}$$
$$\times \left([T] \begin{Bmatrix} \delta u_y \\ \delta v_y \\ \delta w_y \end{Bmatrix} + [T][K_2^0]^T \begin{Bmatrix} \delta u \\ \delta v \\ \delta w \end{Bmatrix} \right) \Biggr\} dxdy$$
$$+ \int_x \left[-\hat{M}_2 \delta\theta_1 + \hat{M}_{62} \delta\theta_2 + m_{32}\delta\theta_3 + m_2\delta\gamma_4 + m_{62}\delta\gamma_5 \right]_{y=0}^{y=Y} dx$$
$$+ \int_y \left[-\hat{M}_{61}\delta\theta_1 + \hat{M}_1\delta\theta_2 + m_{31}\delta\theta_3 + m_{61}\delta\gamma_4 + m_1\delta\gamma_5 \right]_{x=0}^{x=X} dy \tag{9.6.34}$$

on account of $\gamma_6 = \gamma_{61} + \gamma_{62}$.

9.6.4 Equations of Motion

Substituting (6.3.5a,b), (9.6.19), and (9.6.34) into the extended Hamilton's principle, (7.2.1), integrating the result by parts, and then setting each of the coefficients of δu, δv, δw, $\delta \gamma_4$, and $\delta \gamma_5$ equal to zero, we obtain the following equations of motion:

$$F_{11x} + F_{12y} - F_1 = A_u + \mu_1 \ddot{u} \qquad (9.6.35a)$$

$$F_{21x} + F_{22y} - F_2 = A_v + \mu_2 \ddot{v} \qquad (9.6.35b)$$

$$F_{31x} + F_{32y} - F_3 = A_w + \mu_3 \ddot{w} \qquad (9.6.35c)$$

$$m_{61x} + m_{2y} - q_2 + \tilde{s}_{11}k_1 + \bar{s}_{22}k_2 + \tilde{s}_{12}k_{61} + \bar{s}_{21}k_{62}$$
$$-(\tilde{m}_2 - \overline{m}_{61})k_4 - (\tilde{m}_{62} - \overline{m}_1)k_5 = A_{\gamma_4} + \mu_4 \ddot{\gamma}_4 \qquad (9.6.35d)$$

$$m_{1x} + m_{62y} - q_1 + \bar{s}_{11}k_1 + \bar{s}_{22}k_2 + \bar{s}_{12}k_{61} + \bar{s}_{21}k_{62}$$
$$-(\overline{m}_2 - \tilde{m}_{61})k_4 - (\overline{m}_{62} - \tilde{m}_1)k_5 = A_{\gamma_5} + \mu_5 \ddot{\gamma}_5 \qquad (9.6.35e)$$

where

$$\{F_{11}, F_{21}, F_{31}\} = \{N_1, N_6/(1+e_1), -(\Theta_2 + A_{\theta_2})/(1+e_1)\}[T]$$
$$\{F_{12}, F_{22}, F_{32}\} = \{N_6/(1+e_2), N_2, (\Theta_1 + A_{\theta_1})/(1+e_2)\}[T]$$
$$\{F_1, F_2, F_3\} = \{N_1, N_6/(1+e_1), -(\Theta_2 + A_{\theta_2})/(1+e_1)\}[T][K_1^0]^T$$
$$+ \{N_6/(1+e_2), N_2, (\Theta_1 + A_{\theta_1})/(1+e_2)\}[T][K_2^0]^T \qquad (9.6.36)$$

We added a linear viscous damping term to each of (9.6.35a-e), where the μ_i are the damping coefficients. The boundary conditions are of the form: specify

Along $x = 0, X$:

$\delta u = 0$	or	$F_{11} + [\hat{M}_{61}T_{31}/(1+e_2)]_y$
		$- (k_4^0 T_{32} - k_{62}^0 T_{33})\hat{M}_{61}/(1+e_2)$
$\delta v = 0$	or	$F_{21} + [\hat{M}_{61}T_{32}/(1+e_2)]_y$
		$+ (k_4^0 T_{31} + k_2^0 T_{33})\hat{M}_{61}/(1+e_2)$
$\delta w = 0$	or	$F_{31} + [\hat{M}_{61}T_{33}/(1+e_2)]_y$
		$- (k_2^0 T_{32} + k_{62}^0 T_{31})\hat{M}_{61}/(1+e_2)$
$\delta \theta_2 = 0$	or	$\hat{M}_1$
$\delta \gamma_4 = 0$	or	m_{61}
$\delta \gamma_5 = 0$	or	m_1

Along $y = 0, Y$:

$\delta u = 0$	or	$F_{12} + [\hat{M}_{62}T_{31}/(1+e_1)]_x$
		$- (k_5^0 T_{32} - k_1^0 T_{33})\hat{M}_{62}/(1+e_1)$
$\delta v = 0$	or	$F_{22} + [\hat{M}_{62}T_{32}/(1+e_1)]_x$

$\delta w = 0$	or	$+ (k_5^0 T_{31} + k_{61}^0 T_{33})\hat{M}_{62}/(1+e_1)$ $F_{32} + [\hat{M}_{62} T_{33}/(1+e_1)]_x$ $- (k_1^0 T_{31} + k_{61}^0 T_{32})\hat{M}_{62}/(1+e_1)$
$\delta\theta_1 = 0$	or	$\hat{M}_2$
$\delta\gamma_4 = 0$	or	m_2
$\delta\gamma_5 = 0$	or	m_{62}

At $x, y) = (0,0), (X,0), (0,Y), (X,Y)$:

$\delta u = 0$	or	$-T_{31}[\hat{M}_{62}/(1+e_1) + \hat{M}_{61}/(1+e_2)]$
$\delta v = 0$	or	$-T_{32}[\hat{M}_{62}/(1+e_1) + \hat{M}_{61}/(1+e_2)]$
$\delta w = 0$	or	$-T_{33}[\hat{M}_{62}/(1+e_1) + \hat{M}_{61}/(1+e_2)]$ (9.6.37)

Because the rotary inertia A_{θ_3} about the ζ axis is due to nonlinear effects (see (9.6.21)) and is negligibly small,

$$\Theta_3 + A_{\theta_3} = 0 \quad (9.6.38a)$$
$$m_{31} = 0 \quad \text{along } x = 0, X \quad (9.6.38b)$$
$$m_{32} = 0 \quad \text{along } y = 0, Y \quad (9.6.38c)$$

is a statement of the balance of the internal moments with respect to the ζ axis, which have no significant influence on the shell dynamics.

Adding $Q_1 \times$ (6.3.5a) and $Q_2 \times$ (6.3.5b) to (9.6.34) yields

$$\delta\Pi = \int_A \left\{ \{N_1, N_6/(1+e_1), Q_1\} \left([T] \left\{ \begin{array}{c} \delta u_x \\ \delta v_x \\ \delta w_x \end{array} \right\} + [T][K_1^0]^T \left\{ \begin{array}{c} \delta u \\ \delta v \\ \delta w \end{array} \right\} \right) \right.$$
$$+ [\Theta_1 - Q_2(1+e_2)]\delta\theta_1 + [\Theta_2 + Q_1(1+e_1)]\delta\theta_2 + \Theta_3\delta\theta_3$$
$$+ \Gamma_4\delta\gamma_4 + \Gamma_5\delta\gamma_5 + \{N_6/(1+e_2), N_2, Q_2\}$$
$$\left. \times \left([T] \left\{ \begin{array}{c} \delta u_y \\ \delta v_y \\ \delta w_y \end{array} \right\} + [T][K_2^0]^T \left\{ \begin{array}{c} \delta u \\ \delta v \\ \delta w \end{array} \right\} \right) \right\} dxdy$$
$$+ \int_x \left[-\hat{M}_2\delta\theta_1 + \hat{M}_{62}\delta\theta_2 + m_{32}\delta\theta_3 + m_2\delta\gamma_4 + m_{62}\delta\gamma_5 \right]_{y=0}^{y=Y} dx$$
$$+ \int_y \left[-\hat{M}_{61}\delta\theta_1 + \hat{M}_1\delta\theta_2 + m_{31}\delta\theta_3 + m_{61}\delta\gamma_4 + m_1\delta\gamma_5 \right]_{x=0}^{x=X} dy \quad (9.6.39)$$

where the introduced variables Q_1 and Q_2 are shear intensities. Substituting (9.6.39) and (9.6.19) into (7.2.1), integrating by parts, and setting each of the coefficients of $\delta u, \delta v, \delta w, \delta\theta_1, \delta\theta_2$ and $\delta\theta_3$ equal to zero, we obtain the equations of motion

$$\frac{\partial}{\partial x}\left([T]^T\{F_\alpha\}\right) - [K_1^0][T]^T\{F_\alpha\} + \frac{\partial}{\partial y}\left([T]^T\{F_\beta\}\right) - [K_2^0][T]^T\{F_\beta\} = \{I_F\}$$
$$(9.6.40)$$

$$\frac{\partial \{M_\alpha\}}{\partial x} + [K_1]^T \{M_\alpha\} + \frac{\partial \{M_\beta\}}{\partial y} + [K_2]^T \{M_\beta\} + \{Q_\alpha\} + \{Q_\beta\} = \{I_M\}$$

(9.6.41)

where

$$\{F_\alpha\} \equiv \{N_1, \ N_6/(1+e_1), \ Q_1\}^T, \ \{F_\beta\} \equiv \{N_6/(1+e_2), \ N_2, \ Q_2\}^T$$
$$\{M_\alpha\} \equiv \{-\hat{M}_{61}, \ \hat{M}_1, \ m_{31}\}^T, \ \{M_\beta\} \equiv \{-\hat{M}_2, \ \hat{M}_{62}, \ m_{32}\}^T$$
$$\{Q_\alpha\} \equiv \{0, \ -Q_1(1+e_1), \ N_6\}^T, \ \{Q_\beta\} \equiv \{Q_2(1+e_2), \ 0, \ -N_6\}^T$$
$$\{I_F\} \equiv \{A_u, \ A_v, \ A_w\}^T, \ \{I_M\} \equiv \{A_{\theta_1}, \ A_{\theta_2}, \ A_{\theta_3}\}^T \quad (9.6.42)$$

Using (6.2.13) and (6.2.14) and the identity $[T]^T = [T]^{-1}$, we rewrite (9.6.40) as

$$[T]^T \left(\frac{\partial \{F_\alpha\}}{\partial x} + [K_1]^T \{F_\alpha\} + \frac{\partial \{F_\beta\}}{\partial y} + [K_2]^T \{F_\beta\} \right) = \{I_F\} \quad (9.6.43)$$

Using (6.2.13) and (6.2.14), we put (9.6.43) and (9.6.41) in the following vector forms:

$$\frac{\partial \mathbf{F}_\alpha}{\partial x} + \frac{\partial \mathbf{F}_\beta}{\partial y} = \mathbf{I}_F \quad (9.6.44)$$

$$\frac{\partial \mathbf{M}_\alpha}{\partial x} + \frac{\partial \mathbf{M}_\beta}{\partial y} + (1+e_1)\mathbf{i}_1 \times \mathbf{F}_\alpha + (1+e_2)\mathbf{i}_2 \times \mathbf{F}_\beta = \mathbf{I}_M \quad (9.6.45)$$

where

$$\mathbf{F}_\alpha = N_1 \mathbf{i}_1 + N_6/(1+e_1)\mathbf{i}_2 + Q_1 \mathbf{i}_3, \ \mathbf{F}_\beta = N_6/(1+e_2)\mathbf{i}_1 + N_2 \mathbf{i}_2 + Q_2 \mathbf{i}_3$$
$$\mathbf{M}_\alpha = -\hat{M}_{61}\mathbf{i}_1 + \hat{M}_1 \mathbf{i}_2 + m_{31}\mathbf{i}_3, \ \mathbf{M}_\beta = -\hat{M}_2 \mathbf{i}_1 + \hat{M}_{62}\mathbf{i}_2 + m_{32}\mathbf{i}_3$$
$$\mathbf{I}_F = A_u \mathbf{j}_1 + A_v \mathbf{j}_2 + A_w \mathbf{j}_3, \ \mathbf{I}_M = A_{\theta_1}\mathbf{i}_1 + A_{\theta_2}\mathbf{i}_2 + A_{\theta_3}\mathbf{i}_3 \quad (9.6.46)$$

One can see from (9.6.46) and Figure 9.5.3b that Q_1 and Q_2 are along the $\mathbf{i}_3$-direction and hence represent the transverse shear intensities. However, because Q_1 and Q_2 are artificially introduced, they need to be obtained from the first two equations of (9.6.41) as

$$Q_1 = \frac{1}{1+e_1}\left(\hat{M}_{1x} + \hat{M}_{62y} + m_{31}k_{61} + m_{32}k_2 - \hat{M}_2 k_4 - \hat{M}_{61}k_5 - A_{\theta_2}\right)$$
$$Q_2 = \frac{1}{1+e_2}\left(\hat{M}_{61x} + \hat{M}_{2y} - m_{31}k_1 - m_{32}k_{62} + \hat{M}_1 k_5 + \hat{M}_{62}k_4 + A_{\theta_1}\right)$$

(9.6.47)

9.6.5 Shear-Warping Functions

From (9.6.16a-c) one can see that the shear-warping functions of shells depend on the initial curvatures k_1^0, k_2^0, k_{61}^0, and k_{62}^0. For isotropic or one-layer orthotropic cylindrical shells with an arbitrary ply angle θ (see Figures 2.4.2 and 9.5.1a), the warping functions are obtained to be

$$g_{15}^{(i)} = z - \frac{4z^3}{3h^2}, \quad g_{14}^{(i)} = g_{25}^{(i)} = 0 \qquad (9.6.48a)$$

Because the shear coupling functions $g_{14}^{(i)}$ and $g_{25}^{(i)}$ are zero, the transverse shear deformations γ_4 and γ_5 are independent. Moreover, the $g_{15}^{(i)}$ are the so-called third-order shear functions (Bhimaraddi and Stevens, 1984; Reddy, 1984b). We also note that

$$g_{24}^{(i)} = z - \frac{4z^3}{3h^2} \qquad (9.6.48b)$$

only if the initial curvature $k_2^0 = 1/a = 0$ (i.e., $a \to \infty$), which corresponds to plates. Neglecting the effect of the local rotations and replacing the deformed curvatures with the initial curvatures in (9.6.13e) and using (9.6.48a) and (9.2.4), we obtain

$$\epsilon_{23}^{(i)} = \gamma_4 g_{24z}^{(i)} - k_2^0 \gamma_4 g_{24}^{(i)} \qquad (9.6.49)$$

Equation (9.6.49) shows that, if $k_2^0 = 0$ (i.e., plates), $g_{24z}^{(i)} = 0$ on the bonding surfaces where $\epsilon_{23}^{(i)} = 0$. However, for cylindrical shells, $g_{24z}^{(i)} \neq 0$ on the bonding surfaces because $k_2^0 \neq 0$, as shown in Figure 9.6.1. In Figure 9.6.1, we also note that $g_{24}^{(i)}$ deviates from the third-order shear function and is not an odd function when $k_2^0 \neq 0$.

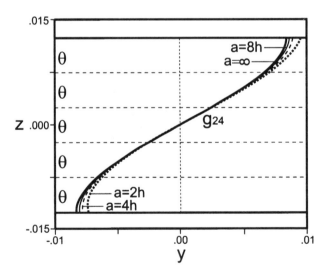

Fig. 9.6.1 The shear-warping function g_{24} of circular cylindrical orthotropic shells having $h = 0.025\ in$ and $a = \infty$ (i.e., plates), $8h, 4h, 2h$.

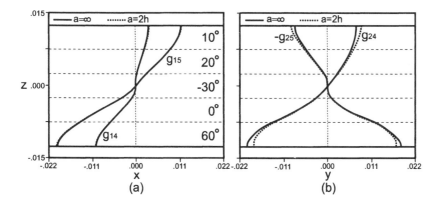

Fig. 9.6.2 Shear-warping functions of plates (solid lines) and circular cylindrical shells (dashed lines) having $a = 2h, h = 0.025\ in$, and the layup $[60°/0°/-30°/20°/10°]$: (a) g_{14} and g_{15} and (b) g_{24} and $-g_{25}$.

For general laminated plates and shells, $g_{14}^{(i)}$ and $g_{25}^{(i)}$ are nontrivial and hence γ_4 and γ_5 are coupled. In Figures 9.6.2a and b, we show the shear-warping functions of five-layer graphite-epoxy laminated composite shells with the following material properties of each lamina:

$$E_{11} = 1.92 \times 10^7\ psi,\quad E_{22} = E_{33} = 1.56 \times 10^6\ psi$$
$$G_{23} = 5.23 \times 10^5\ psi,\quad G_{12} = G_{13} = 8.20 \times 10^5\ psi$$
$$\nu_{12} = \nu_{13} = 0.24,\quad \nu_{23} = 0.49 \qquad (9.6.50)$$

We note that the changes in $g_{14}^{(i)}$ and $g_{15}^{(i)}$ are caused by shear coupling and are very small. However, the changes in $g_{24}^{(i)}$ and $g_{25}^{(i)}$ due to the initial curvature k_2^0 are significant. In Figures 9.6.3a and b, we compare the shear-warping functions of a twisted circular cylindrical shell having $a = 4h$ and $\psi = 30°$ (see Figures 9.2.3 and 9.2.7) with those of a circular cylindrical shell (i.e., $\psi = 90°$) having $a = 4h$. We note that the initial twisting curvatures k_{61}^0 and k_{62}^0 have a significant influence on the shear-warping functions. In Figures 9.6.4a and b, we compare the shear-warping functions of a circular cylindrical shell having $a = 4h$ ($k_1^0 = 0, k_2^0 = 1/a$) with those of a spherical shell having $a = 4h$ ($k_1^0 = k_2^0 = 1/a$). We note that the initial bending curvature k_1^0 of the x axis only affects $g_{14}^{(i)}$ and $g_{15}^{(i)}$, and the small changes in $g_{24}^{(i)}$ and $g_{25}^{(i)}$ are due to shear coupling.

One can see from Figures 9.6.1-9.6.4 that, if a/h is 8 or larger, the influence of the initial curvatures on the shear-warping functions is small. In other words, the influence of the initial curvatures can be neglected in deriving shear-warping functions for very thin shells.

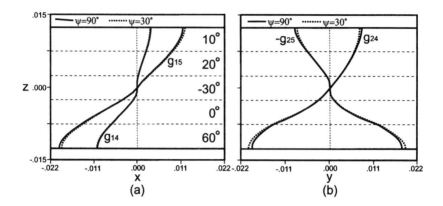

Fig. 9.6.3 Shear-warping functions of a circular cylindrical shells (i.e., $\psi = 90°$, solid lines) and a twisted circular cylindrical shell ($\psi = 30°$, dashed lines) having $h = 0.025\ in$, $a = 4h$, and the layup $[60°/0°/-30°/20°/10°]$: (a) g_{14} and g_{15} and (b) g_{24} and $-g_{25}$.

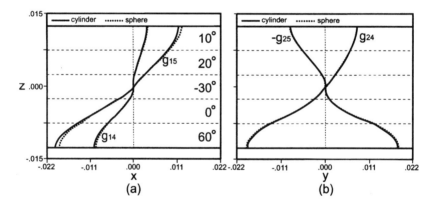

Fig. 9.6.4 Shear-warping functions of a circular cylindrical shell (solid lines) and a spherical shells (dashed lines) having $h = 0.025\ in$, $a = 4h$, and the layup $[60°/0°/-30°/20°/10°]$: (a) g_{14} and g_{15} and (b) g_{24} and $-g_{25}$.

9.7 NONLINEAR DYNAMICS OF INFINITELY LONG CIRCULAR CYLINDRICAL SHELLS

We consider the interaction between the breathing mode and a flexural mode in an infinitely long isotropic cylindrical shell in the case of two-to-one internal resonance when the breathing mode is excited by a primary resonant internal pressure.

9.7.1 Governing Equations

We consider a long circular cylindrical shell subjected to loadings that make its deformation at points away from its two ends being independent of the axial coordinate (i.e., the b axis of the rectangular system abc shown in Figure 9.7.1).

The undeformed and deformed position vectors $\mathbf{R}$ and $\mathbf{r}$ of an arbitrary point on the reference surface of such a shell are given by

$$\mathbf{R} = a\sin\theta \mathbf{i}_a + a\cos\theta \mathbf{i}_c \tag{9.7.1}$$

$$\mathbf{r} = r\sin\phi \mathbf{i}_a + r\cos\phi \mathbf{i}_c \tag{9.7.2}$$

where a is the radius of the undeformed cross section. Moreover, the undeformed and deformed differential circumferential lengths ds and $d\tilde{s}$ of a midsurface element are given by

$$ds = ad\theta \tag{9.7.3}$$

$$d\tilde{s} = \sqrt{dr^2 + r^2 d\phi^2} = \sqrt{r'^2 + r^2\phi'^2}\,d\theta \tag{9.7.4}$$

where the prime indicates the partial derivative with respect to θ. Hence, the circumferential Jaumann strain e_1 of the deformed reference surface is given by

$$e_1 = \frac{d\tilde{s} - ds}{ds} = \frac{\sqrt{r'^2 + r^2\phi'^2} - a}{a} \tag{9.7.5}$$

It follows from (9.7.1) and (9.7.2) and Figure 9.7.1 that the unit vectors along the undeformed and deformed coordinate systems are given by

$$\mathbf{j}_1 = \frac{d\mathbf{R}}{ds} = \cos\theta \mathbf{i}_a - \sin\theta \mathbf{i}_c, \quad \mathbf{j}_3 = \sin\theta \mathbf{i}_a + \cos\theta \mathbf{i}_c \tag{9.7.6}$$

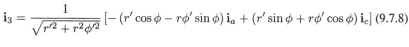

$$\mathbf{i}_1 = \frac{d\mathbf{r}}{d\tilde{s}} = \frac{1}{\sqrt{r'^2 + r^2\phi'^2}}\big[(r'\sin\phi + r\phi'\cos\phi)\mathbf{i}_a + (r'\cos\phi - r\phi'\sin\phi)\mathbf{i}_c\big] \tag{9.7.7}$$

$$\mathbf{i}_3 = \frac{1}{\sqrt{r'^2 + r^2\phi'^2}}\big[-(r'\cos\phi - r\phi'\sin\phi)\mathbf{i}_a + (r'\sin\phi + r\phi'\cos\phi)\mathbf{i}_c\big] \tag{9.7.8}$$

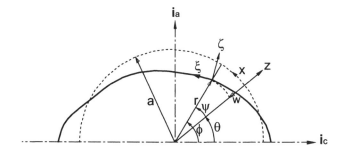

Fig. 9.7.1 A cross section of a long circular cylindrical shell.

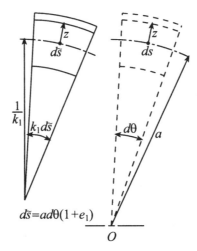

Fig. 9.7.2 A schematic of the deformed and undeformed fiber element of the shell.

where $\mathbf{j}_1 \cdot \mathbf{j}_3 = 0$ and $\mathbf{i}_1 \cdot \mathbf{i}_3 = 0$ were used. It follows from (6.1.7) that the undeformed curvature k_1^0 can be obtained as

$$k_1^0 = -\frac{d\mathbf{j}_1}{ds} \cdot \mathbf{j}_3 = \frac{1}{a} \qquad (9.7.9)$$

And, it follows from (6.2.15) that the actual deformed curvature k_1 can be obtained as

$$k_1 = -\frac{d\mathbf{i}_1}{d\tilde{s}} \cdot \mathbf{i}_3 = \frac{\phi'(r^2\phi'^2 - rr'' + 2r'^2) + \phi''rr'}{(r'^2 + r^2\phi'^2)^{3/2}} \qquad (9.7.10)$$

Next, we express the Jaumann circumferential strain ϵ_{11} of a fiber located at a distance z from the midsurface (Figure 9.7.2) in terms of the strain e_1 and curvature k_1 of the midsurface. We neglect shear deformations and assume that normals to the midsurface before deformation remain normal to the midsurface after deformation. Moreover, we assume that the distance z of the element is not affected by deformation; that is, we neglect the strain in the z direction. The initial lengths of the midsurface and the fiber are $ds = ad\theta$ and $d\hat{s} = (a + z)d\theta$. After deformation, it follows from Figure 9.7.2 that the length of the midsurface element is $d\tilde{s} = a(1 + e_1)d\theta$ and the angle subtended by this element at the origin is $a(1 + e_1)k_1 d\theta$. Hence, the length of the deformed fiber is $a(1 + e_1)(1 + k_1 z)d\theta$. Therefore, the Jaumann strain ϵ_{11} is

$$\epsilon_{11} = \frac{a(1 + e_1)(1 + k_1 z)d\theta - (a + z)d\theta}{(a + z)d\theta} \qquad (9.7.11a)$$

Expanding (9.7.11a) in powers of z/a gives

$$\epsilon_{11} = e_1 + (k_1 - k_1^0)(1 + e_1)\left[z - k_1^0 z^2 + (k_1^0)^2 z^3 + \cdots\right] \qquad (9.7.11b)$$

where $k_1^0 = 1/a$ is the initial curvature.

For a long cylinder, $\epsilon_{22} = 0$ and hence the Jaumann stress $\sigma_{11} = \epsilon_{11}E/(1-\nu^2)$. Hence, variation of the elastic energy $\delta \Pi$ is given by

$$\delta\Pi = aL \int_0^{2\pi} \int_{-h/2}^{h/2} \sigma_{11}\delta\epsilon_{11}\left(1+\frac{z}{a}\right)d\theta dz$$

$$= aL \int_0^{2\pi} (Ce_1\delta e_1 + Dk_{1m}\delta k_{1m})\,d\theta \qquad (9.7.12)$$

where L is the cylinder length and

$$k_{1m} \equiv (k_1 - k_1^0)(1+e_1),\ C \equiv \frac{Eh}{1-\nu^2},\ D \equiv \frac{Eh^3}{12(1-\nu^2)} \qquad (9.7.13)$$

For an arbitrary point on the reference plane, the displacement vector $\mathbf{D} = \mathbf{r} - \mathbf{R}$. Hence, if the rotary inertia is assumed to be negligible, variation of the kinetic energy δT is given by

$$\delta T = -aL \int_0^{2\pi}\int_{-h/2}^{h/2} \rho \ddot{\mathbf{D}}\cdot\delta\mathbf{D}\left(1+\frac{z}{a}\right)d\theta dz$$

$$= -aL \int_0^{2\pi}\int_{-h/2}^{h/2} \rho \ddot{\mathbf{r}}\cdot\delta\mathbf{r}\left(1+\frac{z}{a}\right)d\theta dz$$

$$= -ahL \int_0^{2\pi} \rho\left[\left(\ddot{r}-r\dot{\phi}^2\right)\delta r + \left(2r\dot{r}\dot{\phi}+r^2\ddot{\phi}\right)\delta\phi\right]d\theta \qquad (9.7.14)$$

where the overdot indicates the derivative with respect to the dimensional time $\hat{t}$.

Variation of the non-conservative work δW_{nc} due to the internal pressure P is given by

$$\delta W_{nc} = \int_0^{2\pi} P\mathbf{i}_3\cdot\delta\mathbf{D}L d\tilde{s} = \int_0^{2\pi} P\mathbf{i}_3\cdot\delta\mathbf{r}L d\tilde{s} \qquad (9.7.15)$$

where $L d\tilde{s}$ accounts for the actual deformed reference area and P is assumed to be constant. It follows from (9.7.2) and (9.7.8) that

$$\delta\mathbf{r} = (\sin\phi\delta r + r\cos\phi\delta\phi)\mathbf{i}_a + (\cos\phi\delta r - r\sin\phi\delta\phi)\mathbf{i}_b \qquad (9.7.16)$$

$$\mathbf{i}_3\cdot\delta\mathbf{r} = \frac{1}{\sqrt{r'^2+r^2\phi'^2}}(r\phi'\delta r - rr'\delta\phi) \qquad (9.7.17)$$

Substituting (9.7.17) and (9.7.4) into (9.7.15) yields

$$\delta W_{nc} = PL \int_0^{2\pi} (r\phi'\delta r - rr'\delta\phi)\,d\theta \qquad (9.7.18)$$

Next, we introduce the nondimensional variables w and ψ and time t as

$$w \equiv \frac{W}{a} = \frac{a-r}{a},\ \psi \equiv \phi - \theta,\ t \equiv \frac{\hat{t}}{a}\sqrt{\frac{E}{\rho(1-\nu^2)}} \qquad (9.7.19)$$

634 SHELLS

where W is the radial displacement shown in Figure 9.7.1. Hence, we have

$$r' = -aw', \quad \phi' = \psi' + 1, \quad \delta r = -a\delta w$$
$$\delta \phi = \delta \psi, \quad \dot{r} = -a\dot{w}, \quad \dot{\phi} = \dot{\psi} \quad (9.7.20)$$

where $\dot{\phi} = \partial \phi / \partial t$. To express the governing equations in polynomial form, we expand e_1 and k_1 in Taylor series up to cubic terms and obtain

$$e_1 = \psi' - w + \frac{1}{2}w'^2 - w\psi' - \frac{1}{2}\psi'w'^2 + \frac{1}{2}ww'^2 + \ldots \quad (9.7.21)$$

$$k_1 - k_1^0 = \frac{1}{a}\left[w + w'' + w^2 + 2ww'' + \frac{1}{2}w'^2 - 2\psi'w'' - w'\psi''\right.$$
$$+ w^3 + 3w^2 w'' - 2w\psi''w' - 4w\psi'w'' + \frac{3}{2}ww'^2$$
$$\left.+ 3\psi''\psi'w' + 3\psi'^2 w'' - \psi'w'^2 - \frac{3}{2}w''w'^2\right] + \ldots \quad (9.7.22)$$

The governing equations up to cubic nonlinearities are then obtained by substituting (9.7.19)-(9.7.22) into (9.7.12)-(9.7.14) and (9.7.18) and then into the extended Hamilton principle (2.5.37) and performing integration by parts. The result is

$$\ddot{w} + w - \psi' + \alpha^2\left(w^{iv} + 2w'' + w\right) = -\dot{\psi}^2 - \frac{1}{2}w'^2 - 2w\psi' + \psi'^2 + \psi'w''$$
$$- ww'' + w'\psi'' + w\dot{\psi}^2 - 2w'\psi''\psi' + \frac{1}{2}w'^2\psi' - w''\psi'^2 + \frac{3}{2}w''w'^2 - w\psi'^2$$
$$+ ww'\psi'' - ww'^2 + ww''\psi' - w^2 w'' + \alpha^2\left(4w''\psi' - 2w\psi' - \frac{3}{2}w'^2\right.$$
$$+ 3w''\psi''' - 3w''^2 + 4w'''\psi'' - 4w'''w' + 2w^{iv}\psi' - w\psi'''$$
$$\left.- 3ww'' - 2ww^{iv} + w'\psi^{iv}\right) - \frac{a(1-\nu^2)}{Eh}P(1+\psi')(1-w) \quad (9.7.23)$$

$$\ddot{\psi} - \psi'' + w' = 2w\ddot{\psi} + 2\dot{w}\dot{\psi} + w'w'' - 2w'\psi' - 2w\psi'' + 2ww' + ww''w'$$
$$+ w^2\psi'' - w'^2\psi'' + \frac{1}{2}w'^3 + 2ww'\psi' - 2w''w'\psi' - 2w\dot{w}\dot{\psi} - w^2\ddot{\psi}$$
$$+ \alpha^2\left(2ww' - 2w''w'''\right) + \frac{a(1-\nu^2)}{Eh}(1-w)w'P \quad (9.7.24)$$

where

$$\alpha^2 \equiv h^2/(12a^2)$$

To determine approximate solutions to the dynamics of the shell, one can either attack the governing equations (9.7.23) and (9.7.24) or attack the Lagrangian. In the latter case, the extended Hamilton's principle can be expressed as

$$\delta \int_{t_1}^{t_2} CaL dt = \delta \int_{t_1}^{t_2} [T - \Pi + W_{nc}] dt = 0 \quad (9.7.25)$$

where T is the kinetic energy, Π is the total strain energy, which is the sum of the membrane strain energy Π_m and the bending strain energy Π_b, and W_{nc} is the work done by the internal pressure. To fourth order, they are given by

$$\Pi_m = \frac{1}{2}CaL\int_0^{2\pi} e_1^2 d\theta = \frac{EahL}{2(1-\nu^2)}\int_0^{2\pi}\left[\psi'^2 - 2\psi'w + w^2 + \psi'w'^2\right.$$
$$\left. +2\psi'w^2 - 2w\psi'^2 - ww'^2 - \psi'^2 w'^2 + w\psi'w'^2 - w^2w'^2 + \frac{1}{4}w'^4 + w^2\psi'^2\right]d\theta$$
(9.7.26)

$$\Pi_b = \frac{1}{2}DaL\int_0^{2\pi} k_{1m}^2 d\theta = \frac{Eah^3L}{24(1-\nu^2)}\int_0^{2\pi}\left[2w''w + w^2 + w''^2 + w'^2w\right.$$
$$\left. +2w^2\psi - 2ww'\psi'' + w'^2w'' - 2w''^2\psi' - 2w'w''\psi'' + 2w^2w'' + 2ww''^2\right]d\theta$$
(9.7.27)

$$T = \frac{EhL}{2a(1-\nu^2)}\int_0^{2\pi}\left(\dot{r}^2 + r^2\dot{\phi}^2\right)d\theta$$
$$= \frac{EahL}{2(1-\nu^2)}\int_0^{2\pi}\left[\dot{w}^2 + (1-w)^2\dot{\psi}^2\right]d\theta \qquad (9.7.28)$$

$$W_{nc} = PL\int_0^{2\pi}\frac{1}{2}r^2\phi' d\theta - PLa^2\pi$$
$$= \frac{1}{2}PLa^2\int_0^{2\pi}\left[(1-w)^2\psi' - 2w + w^2\right]d\theta \qquad (9.7.29)$$

Using calculus of variation, one finds that the equations of motion are given by the following Euler-Lagrange equations:

$$\frac{\partial}{\partial t}\left(\frac{\partial \mathcal{L}}{\partial \dot{w}}\right) - \frac{\partial^2}{\partial \theta^2}\left(\frac{\partial \mathcal{L}}{\partial w''}\right) + \frac{\partial}{\partial \theta}\left(\frac{\partial \mathcal{L}}{\partial w'}\right) - \frac{\partial \mathcal{L}}{\partial w} = 0 \qquad (9.7.30)$$

$$\frac{\partial}{\partial t}\left(\frac{\partial \mathcal{L}}{\partial \dot{\psi}}\right) + \frac{\partial}{\partial \theta}\left(\frac{\partial \mathcal{L}}{\partial \psi'}\right) - \frac{\partial \mathcal{L}}{\partial \psi} = 0 \qquad (9.7.31)$$

Using (9.7.25)-(9.7.29) in (9.7.30) and (9.7.31), we obtain (9.7.23) and (9.7.24).

9.7.2 Natural Frequencies and Mode Shapes

The natural frequencies and mode shapes of the shell are governed by

$$\ddot{w} + w - \psi' + \alpha^2\left(w^{iv} + 2w'' + w\right) = 0 \qquad (9.7.32)$$
$$\ddot{\psi} - \psi'' + w' = 0 \qquad (9.7.33)$$

We let

$$w = c_1 e^{i(\omega t + n\theta)} \text{ and } \psi = c_2 e^{i(\omega t + n\theta)} \qquad (9.7.34)$$

in (9.7.32) and (9.7.33) and obtain

$$[1 - \omega^2 + \alpha^2 (n^2 - 1)^2] c_1 - inc_2 = 0 \tag{9.7.35}$$
$$inc_1 - (\omega^2 - n^2) c_2 = 0 \tag{9.7.36}$$

where ω is the natural frequency. For a nontrivial solution,

$$-(\omega^2 - n^2)\left[1 - \omega^2 + \alpha^2 (n^2 - 1)^2\right] - n^2 = 0$$

or

$$\omega^4 - \left[n^2 + 1 + \alpha^2 (n^2 - 1)^2\right]\omega^2 + \alpha^2 n^2 (n^2 - 1)^2 = 0 \tag{9.7.37}$$

Then, it follows from (9.7.35) that

$$\frac{c_2}{c_1} = -i\Gamma_n \tag{9.7.38}$$

where

$$\Gamma_n = \frac{1}{n}\left[1 + \alpha^2 (n^2 - 1)^2 - \omega_n^2\right] \tag{9.7.39}$$

when $n \neq 0$. When $n = 0$, it follows from (9.7.35) and (9.7.36) that $c_2 = 0$, c_1 is arbitrary, and

$$\omega_0^2 = 1 + \alpha^2 \tag{9.7.40}$$

In summary, when $n = 0$, the mode shape is given by

$$w = \cos \omega_0 t \text{ and } \psi = 0 \tag{9.7.41}$$

It is the so-called breathing mode because the shell radius oscillates uniformly around the undeflected configuration. When $n \neq 0$, there are two modes of oscillation corresponding to the same frequency ω_n:

$$w = \cos n\theta \cos \omega_n t \text{ and } \psi = \Gamma_n \sin n\theta \cos \omega_n t \tag{9.7.42}$$
$$w = \sin n\theta \cos \omega_n t \text{ and } \psi = -\Gamma_n \cos n\theta \cos \omega_n t \tag{9.7.43}$$

These modes are orthogonal and are called flexural modes. In this case, there is one-to-one internal resonance between these two degenerate modes. The directly excited mode of these two is called the driven mode, whereas the other orthogonal mode is called the companion mode. In the next section, we need solutions of the adjoint of (9.7.32) and (9.7.33).

To determine the adjoint equations, we multiply (9.7.32) and (9.7.33) with w^* and ψ^*, respectively, add the results, integrate the outcome over the space and time domain, and obtain

$$\int_0^{2\pi} \int_0^{2\pi/\omega} \left\{ w^* \left[\ddot{w} + w - \psi' + \alpha^2 (w^{iv} + 2w'' + w)\right] \right.$$
$$\left. + \psi^* \left(\ddot{\psi} - \psi'' + w'\right) \right\} dt d\theta = 0 \tag{9.7.44}$$

Integrating (9.7.44) by parts to transfer the derivatives from w and ψ to w^* and ψ^* and using the fact that $w, \psi, w^*,$ and ψ^* are periodic in θ and t, we obtain

$$\int_0^{2\pi} \int_0^{2\pi/\omega} \left\{ w \left[\ddot{w}^* + w^* - \psi^{*\prime} + \alpha^2 \left(w^{*iv} + 2w^{*\prime\prime} + w^* \right) \right] \right.$$
$$\left. + \psi \left(\ddot{\psi}^* - \psi^{*\prime\prime} + w^{*\prime} \right) \right\} dt\, d\theta = 0 \qquad (9.7.45)$$

Setting each of the coefficients of w and ψ in the integrand in (9.7.45) equal to zero, we obtain the adjoint equations

$$\ddot{w}^* + w^* - \psi^{*\prime} + \alpha^2 \left(w^{*iv} + 2w^{*\prime\prime} + w^* \right) = 0 \qquad (9.7.46)$$

$$\ddot{\psi}^* - \psi^{*\prime\prime} + w^{*\prime} = 0 \qquad (9.7.47)$$

The adjoint equations are the same as the original equations, and hence these equations are self-adjoint. Consequently, solutions of the adjoint equations are

$$(w^*, \psi^*) = \Big(\{1, 0\}, \{\cos n\theta, \Gamma_n \sin n\theta\}, \{\sin n\theta, -\Gamma_n \cos n\theta\} \Big) \qquad (9.7.48)$$

The first study of the interaction between the driven and companion modes in thin structures was done by Evensen (1966). He used the method of averaging to study the influence of a one-to-one internal resonance on the inextensional forced flexural response of thin circular rings. He found that the response involves either a single bending mode or two coupled bending modes. He reported experimental observations of unsteady vibrations and verified them by using analog-computer simulations. Chen and Babcock (1975) used the Lindstédt-Poincaré technique to study the nonlinear response of finite cylindrical shells to harmonic excitations. They studied the interaction of the driven and companion modes. They also reported experimental observations of "nonstationary" responses "in which the amplitude drifts from one value to another."

Maewal (1981) studied the interaction between backward and forward traveling waves in rotating rings. Natsiavas (1995b) used the method of multiple scales to study the nonlinear response of a rotating ring to a parametric excitation originating from small periodic perturbations of the spin speed. Using a nonlinear inextensionality constraint, Maewal (1986b) studied the modal interaction between the driven and companion modes of axisymmetric shells. He used the method of multiple scales to derive a set of evolution equations for the amplitudes and phases of the modes. He noted that these equations have the same form as those derived earlier by Miles (1984a,b) for internally resonant surface waves in a circular cylinder. He integrated numerically these evolution equations and showed that, for certain ranges of the excitation frequency, the response is chaotically modulated.

Maganty and Bickford (1987) used the method of multiple scales to study nonlinear free vibrations of circular rings in the presence of a one-to-one internal resonance between an inplane mode and out-of-plane mode. They found a continuous exchange of energy between the coupled modes. In another investigation, they (1988) studied the forced response of the same system to a harmonic inplane or out-of-plane excitation. They showed that, for certain values of the excitation frequency, an inplane excitation can produce nonplanar oscillations and vice versa.

Raouf and Nayfeh (1990a) analyzed the nonlinear dynamic response of infinitely long circular cylindrical shells to a primary resonant excitation of the nth flexural mode of the form

$$\frac{a(1-\nu^2)}{Eh}P = \frac{1}{2}\epsilon^3 f \cos n\theta e^{i\Omega t} + cc$$

$$\Omega = \omega_n + \epsilon^2 \sigma$$

where σ is a detuning parameter. They used the method of multiple scales to attack directly (9.7.23) and (9.7.24), taking into account the interaction of the excited nth flexural mode with it companion mode and assuming modal damping. Chin and Nayfeh (1996b) corrected the solution of Raouf and Nayfeh (1990a) and demonstrated the symmetry of the modulation equations. Also, Chin and Nayfeh (1996b) compared the results derived by attacking the governing partial-differential equations with those derived from a two-mode discretization. They showed that the discretization approach fails to account for the spatial variation at second order and completely misses the influence of the quadratic nonlinearity on the interaction coefficients. Chin and Nayfeh (2001) considered the response of infinitely long circular cylindrical shell to a primary resonance of the breathing mode, taking into account a two-to-one internal resonance between the breathing mode and two orthogonal flexural modes.

When $a/h = 0.028$, $\alpha^2 = 0.00065333$, and $\omega_0 \approx 1.00033$ and $\omega_8 \approx 0.50528$. Because $\omega_0 \approx 2\omega_8$, a two-to-one internal resonance might be activated between the breathing mode and the two orthogonal eighth flexural modes. This case is considered in the next section.

9.7.3 Primary Resonance of the Breathing Mode

We consider the general case of two-to-one internal resonance $\omega_0 \approx 2\omega_n$ when the breathing mode is excited by a harmonic radial load with frequency $\Omega \approx \omega_0$. We use the method of multiple scales to determine a second-order uniform expansion in the form

$$w(\theta, t; \epsilon) = \epsilon w_1(\theta, T_0, T_1) + \epsilon^2 w_2(\theta, T_0, T_1) + \ldots \quad (9.7.49)$$

$$\psi(\theta, t; \epsilon) = \epsilon \psi_1(\theta, T_0, T_1) + \epsilon^2 \psi_2(\theta, T_0, T_1) + \ldots \quad (9.7.50)$$

and scale the radial load as

$$\frac{a(1-\nu^2)}{Eh}P = \epsilon^2 F \cos \Omega T_0 \quad (9.7.51)$$

where ϵ is a small nondimensional bookkeeping parameter. Substituting (9.7.49)-(9.7.51) into (9.7.23) and (9.7.24) and separating coefficients of like powers of ϵ, we obtain

Order ϵ:

$$D_0^2 w_1 + w_1 - \psi_1' + \alpha^2 \left(w_1^{iv} + 2w_1'' + w_1\right) = 0 \qquad (9.7.52)$$

$$D_0^2 \psi_1 - \psi_1'' + w_1' = 0 \qquad (9.7.53)$$

Order ϵ^2:

$$\begin{aligned}
D_0^2 w_2 + w_2 - \psi_2' + \alpha^2 &\left(w_2^{iv} + 2w_2'' + w_2\right) = -2D_0 D_1 w_1 \\
&- (D_0 \psi_1)^2 - \frac{1}{2} w_1'^2 - 2w_1 \psi_1' + \psi_1'^2 + \psi_1' w_1'' \\
&- w_1 w_1'' + w_1' \psi_1'' + \alpha^2 \left(4 w_1'' \psi_1' + 3 w_1'' \psi_1''' - 2 w_1 \psi_1' - \frac{3}{2} w_1'^2 \right. \\
&- 3 w_1''^2 + 4 w_1''' \psi_1'' - 4 w_1' w_1''' + 2 w_1^{iv} \psi_1' \\
&\left. - w_1 \psi_1''' - 3 w_1 w_1'' - 2 w_1 w_1^{iv} + w_1' \psi_1^{iv} \right) - F \cos \Omega T_0 \qquad (9.7.54)
\end{aligned}$$

$$\begin{aligned}
D_0^2 \psi_2 - \psi_2'' + w_2' &= -2 D_0 D_1 \psi_1 + 2 w_1 D_0^2 \psi_1 + 2 D_0 w_1 D_0 \psi_1 \\
&+ w_1' w_1'' - 2 w_1' \psi_1' - 2 w_1 \psi_1'' + 2 w_1 w_1' \\
&+ \alpha^2 \left(2 w_1 w_1' - 2 w_1'' w_1''' \right) \qquad (9.7.55)
\end{aligned}$$

In the presence of damping, all modes that are not directly or indirectly excited by an internal resonance decay with time. Therefore, the solution of (9.7.52) and (9.7.53) can be expressed as

$$w_1 = A_0(T_1) e^{i\omega_0 T_0} + [A_{1n}(T_1) \cos n\theta + A_{2n}(T_1) \sin n\theta] e^{i\omega_n T_0} + cc \qquad (9.7.56)$$

$$\psi_1 = [A_{1n}(T_1) \sin n\theta - A_{2n}(T_1) \cos n\theta] \Gamma_n e^{i\omega_n T_0} + cc \qquad (9.7.57)$$

Substituting (9.7.56) and (9.7.57) into (9.7.54) and (9.7.55), we obtain

$$\begin{aligned}
D_0^2 w_2 + w_2 - \psi_2' + \alpha^2 &\left(w_2^{iv} + 2 w_2'' + w_2\right) = -2i\omega_0 A_0' e^{i\omega_0 T_0} \\
&- 2i\omega_n \left(\cos n\theta A_{1n}' - \sin n\theta A_{2n}'\right) e^{i\omega_n T_0} \\
&+ \chi_1 \left(\cos n\theta \bar{A}_{1n} + \sin n\theta \bar{A}_{2n}\right) A_0 e^{i(\omega_0 - \omega_n) T_0} \\
&+ \Lambda \left(A_{1n}^2 + A_{2n}^2\right) e^{2i\omega_n T_0} - \frac{1}{2} F e^{i\Omega T_0} + cc + NST \qquad (9.7.58)
\end{aligned}$$

$$\begin{aligned}
D_0^2 \psi_2 - \psi_2'' + w_2' &= -2i\omega_n \Gamma_n \left(\sin n\theta A_{1n}' + \cos n\theta A_{2n}'\right) e^{i\omega_n T_0} \\
&+ \chi_2 \left(\cos n\theta \bar{A}_{2n} - \sin n\theta \bar{A}_{1n}\right) A_0 e^{i(\omega_0 - \omega_n) T_0} \\
&+ cc + NST \qquad (9.7.59)
\end{aligned}$$

where

$$\begin{aligned}
\Lambda &= \frac{1}{4} \left(n^2 - 4n\Gamma_n + 2n^2 \Gamma_n^2 + 2\omega_n^2 \Gamma_n^2 \right) \\
&+ \frac{1}{4} \alpha^2 \left(3n^2 - 2n^4 - 4n\Gamma_n + 2n^3 \Gamma_n \right) \qquad (9.7.60)
\end{aligned}$$

$$\chi_1 = n^2 - 2n\Gamma_n + \alpha^2 \left(3n^2 - 2n^4 - 2n\Gamma_n + n^3 \Gamma_n \right) \qquad (9.7.61)$$

$$\chi_2 = 2n - 2n^2 \Gamma_n - 2\omega_0 \omega_n \Gamma_n + 2\omega_n^2 \Gamma_n + 2n\alpha^2 \qquad (9.7.62)$$

640 SHELLS

To describe quantitatively the nearness of the internal and external resonances, we introduce the detuning parameters σ_1 and σ_2 defined by

$$\omega_0 = 2\omega_n + \epsilon\sigma_1 \quad \text{and} \quad \Omega = \omega_0 + \epsilon\sigma_2 \qquad (9.7.63)$$

Because the homogeneous equations (9.7.58) and (9.7.59) have nontrivial solutions, the nonhomogeneous equations (9.7.58) and (9.7.59) have a solution only if their right-hand sides are orthogonal to every solution of the adjoint homogeneous problem.

Using (9.7.63) and requiring the right-hand sides of (9.7.58) and (9.7.59) be orthogonal to each of the adjoints in (9.7.48), we obtain the solvability conditions

$$2i\omega_0 A_0' = \Lambda\left(A_{1n}^2 + A_{2n}^2\right)e^{-i\sigma_1 T_1} + \frac{1}{2}Fe^{i\sigma_2 T_0} \qquad (9.7.64)$$

$$2i\omega_n\left(1+\Gamma_n^2\right)A_{1n}' = 2\Lambda A_0 \bar{A}_{1n}e^{i\sigma_1 T_1} \qquad (9.7.65)$$

$$2i\omega_n\left(1+\Gamma_n^2\right)A_{2n}' = 2\Lambda A_0 \bar{A}_{2n}e^{i\sigma_1 T_1} \qquad (9.7.66)$$

When $n=8$ and $h/a = 0.028$, $\omega_0 \approx 1.00033$, $\omega_8 \approx 0.50528$, $\Gamma_8 \approx 0.125501$, $\Lambda_1 \approx 15.2437$, and $\Lambda_2 \approx 60.9747$.

The modulation equations (9.7.64)-(9.7.66) can be obtained alternatively by averaging the Lagrangian

$$\mathcal{L} = T - \Pi_m - \Pi_b + W_{nc} \qquad (9.7.67)$$

where T, Π_m, Π_b, W_{nc}, and P are defined in (9.7.26)-(9.7.29) and (9.7.51). To this end, we seek an approximate solution in the form

$$w \approx \epsilon A_0(T_1)e^{i\omega_0 T_0} + \epsilon\left[A_{1n}(T_1)\cos n\theta + A_{2n}(T_1)\sin n\theta\right]e^{i\omega_n T_0} + cc \qquad (9.7.68)$$

$$\psi \approx \epsilon\left[A_{1n}(T_1)\sin n\theta - A_{2n}(T_1)\cos n\theta\right]\Gamma_n e^{i\omega_n T_0} + cc \qquad (9.7.69)$$

Substituting (9.7.68) and (9.7.69) into (9.7.67) and (9.7.26)-(9.7.29), integrating the result from $\theta = 0$ to $\theta = 2\pi$, and keeping the slowly varying terms only, we obtain

$$\frac{1-\nu^2}{\epsilon^3 EahL}<\mathcal{L}> = i\omega_n\left(1+\Gamma_n^2\right)\left(A_{1n}\bar{A}_{1n}' - A_{1n}'\bar{A}_{1n} + A_{2n}\bar{A}_{2n}' - A_{2n}'\bar{A}_{2n}\right)$$

$$+i\omega_0\left(A_0\bar{A}_0' - A_0'\bar{A}_0\right) + \Lambda\left(A_0\bar{A}_{1n}^2 + A_0\bar{A}_{2n}^2\right)e^{i\sigma_1 T_1}$$

$$+\Lambda\left(\bar{A}_0 A_{1n}^2 + \bar{A}_0 A_{2n}^2\right)e^{-i\sigma_1 T_1} + \frac{1}{2}F\bar{A}_0 e^{i\sigma_2 T_1}$$

$$+\frac{1}{2}FA_0 e^{-i\sigma_2 T_1} \qquad (9.7.70)$$

where Λ is defined in (9.7.60). Substituting (9.7.70) into the Euler-Lagrange equations

$$\frac{d}{dT_1}\left(\frac{\partial <\mathcal{L}>}{\partial \bar{A}_0'}\right) - \frac{\partial <\mathcal{L}>}{\partial \bar{A}_0} = 0 \qquad (9.7.71)$$

$$\frac{d}{dT_1}\left(\frac{\partial <\mathcal{L}>}{\partial \bar{A}_{in}'}\right) - \frac{\partial <\mathcal{L}>}{\partial \bar{A}_{in}} = 0 \qquad (9.7.72)$$

we obtain (9.7.64)-(9.7.66).

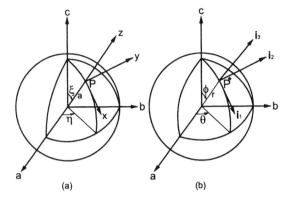

Fig. 9.8.1 A spherical shell under axisymmetric loading: (a) undeformed geometry and (b) deformed geometry.

It follows from (9.7.64)-(9.7.66) that, up to second order, the degeneracy in the two orthogonal flexural modes has not been removed and one needs to extend the analysis to third order. An extensive analysis of the solutions of (9.7.64) and (9.7.65) when $A_{2n} = 0$ can be found in Nayfeh and Raouf (1987a,b) and Nayfeh (2000).

9.8 NONLINEAR DYNAMICS OF AXISYMMETRIC MOTION OF CLOSED SPHERICAL SHELLS

We consider the case of two-to-one internal resonance between two axisymmetric modes of a closed spherical shell of thickness h and radius a when one of them is excited by a primary resonant internal pressure.

9.8.1 Equations of Motion

Let P be a point on the undeformed middle surface at time $\hat{t}$ (Figure 9.8.1.a) and P^* be the same point on the deformed middle surface at time t^* (Figure 9.8.1.b). For axisymmetric deformations, $\theta = \eta$, $\partial()/\partial\eta = 0$, and the inplane shear strain $\epsilon_{12} = 0$. The undeformed and deformed position vectors $\mathbf{R}$ and $\mathbf{r}$ of an arbitrary point on the reference surface of such a shell are given by

$$\mathbf{R} = a\sin\xi\cos\eta\mathbf{i}_a + a\sin\xi\sin\eta\mathbf{i}_b + a\cos\xi\mathbf{i}_c \quad (9.8.1)$$

$$\mathbf{r} = r\sin\phi\cos\eta\mathbf{i}_a + r\sin\phi\sin\eta\mathbf{i}_b + r\cos\phi\mathbf{i}_c \quad (9.8.2)$$

where a is the radius of the undeformed middle surface. Moreover, the undeformed differential circumferential lengths dx and dy and their deformed ones dx^* and dy^*

are given by

$$dx = ad\xi, \quad dy = a\sin\xi \, d\eta \tag{9.8.3}$$
$$dx^{*2} = (rd\phi)^2 + (dr)^2 = \left(r'^2 + r^2\phi'^2\right) d\xi^2$$
$$dy^* = r\sin\phi \, d\eta \tag{9.8.4}$$

where the prime indicates the derivative with respect to ξ. Hence, the circumferential strains e_1 and e_2 of the deformed reference surface are given by

$$e_1 = \frac{dx^* - dx}{dx} = \frac{\sqrt{r'^2 + r^2\phi'^2} - a}{a} \tag{9.8.5}$$
$$e_2 = \frac{dy^* - dy}{dy} = \frac{r\sin\phi - a\sin\xi}{a\sin\xi} \tag{9.8.6}$$

It follows from (9.8.1) and Figure 9.8.1 that the unit vectors along the undeformed coordinate system are given by

$$\mathbf{j}_1 = \frac{1}{a}\frac{\partial \mathbf{R}}{\partial \xi} = \cos\xi\cos\eta \, \mathbf{i}_a + \cos\xi\sin\eta \, \mathbf{i}_b - \sin\xi \, \mathbf{i}_c \tag{9.8.7}$$
$$\mathbf{j}_2 = \frac{1}{a\sin\xi}\frac{\partial \mathbf{R}}{\partial \eta} = -\sin\eta \, \mathbf{i}_a + \cos\eta \, \mathbf{i}_b \tag{9.8.8}$$

Therefore,

$$\mathbf{j}_3 = \mathbf{j}_1 \times \mathbf{j}_2 = \sin\xi\cos\eta \, \mathbf{i}_a + \sin\xi\sin\eta \, \mathbf{i}_b + \cos\xi \, \mathbf{i}_c \tag{9.8.9}$$

Then, it follows from (6.1.7) that the initial curvatures are

$$k_1^0 = -\frac{1}{a}\frac{\partial \mathbf{j}_1}{\partial \xi} \cdot \mathbf{j}_3 = \frac{1}{a} \tag{9.8.10}$$
$$k_2^0 = -\frac{1}{a\sin\xi}\frac{\partial \mathbf{j}_2}{\partial \eta} \cdot \mathbf{j}_3 = \frac{1}{a} \tag{9.8.11}$$
$$k_4^0 = \frac{1}{a\sin\xi}\frac{\partial \mathbf{j}_1}{\partial \eta} \cdot \mathbf{j}_2 = \frac{1}{a\tan\xi} \tag{9.8.12}$$
$$k_{61}^0 = 0, \quad k_{62}^0 = 0, \quad k_5^0 = 0 \tag{9.8.13}$$

It follows from (9.8.2) and Figure 9.8.1 that the unit vectors along the deformed coordinate system are given by

$$\mathbf{i}_1 = \frac{d\mathbf{r}}{dx^*} = \frac{1}{\sqrt{r'^2 + r^2\phi'^2}}\frac{\partial \mathbf{r}}{\partial \xi} = \frac{\cos\eta(r'\sin\phi + r\phi'\cos\phi)}{\sqrt{r'^2 + r^2\phi'^2}}\mathbf{i}_a$$
$$+ \frac{\sin\eta(r'\sin\phi + r\phi'\cos\phi)}{\sqrt{r'^2 + r^2\phi'^2}}\mathbf{i}_b + \frac{r'\cos\phi - r\phi'\sin\phi}{\sqrt{r'^2 + r^2\phi'^2}}\mathbf{i}_c \tag{9.8.14}$$
$$\mathbf{i}_2 = \frac{d\mathbf{r}}{dy^*} = \frac{1}{r\sin\phi}\frac{\partial \mathbf{r}}{\partial \eta} = -\sin\eta \, \mathbf{i}_a + \cos\eta \, \mathbf{i}_b \tag{9.8.15}$$

Therefore,

$$\mathbf{i}_3 = \mathbf{i}_1 \times \mathbf{i}_2 = \frac{\cos\eta(-r'\cos\phi + r\phi'\sin\phi)}{\sqrt{r'^2 + r^2\phi'^2}}\mathbf{i}_a$$

$$+ \frac{\sin\eta(-r'\cos\phi + r\phi'\sin\phi)}{\sqrt{r'^2 + r^2\phi'^2}}\mathbf{i}_b + \frac{r'\sin\phi + r\phi'\cos\phi}{\sqrt{r'^2 + r^2\phi'^2}}\mathbf{i}_c \quad (9.8.16)$$

Then, it follows from (6.2.15) that the curvatures of the deformed middle surface are given by

$$k_1 = -\frac{1}{\sqrt{r'^2 + r^2\phi'^2}} \frac{\partial \mathbf{i}_1}{\partial \xi} \cdot \mathbf{i}_3 = \frac{rr'\phi'' + r^2\phi'^3 + 2r'^2\phi' - rr''\phi'}{(r'^2 + r^2\phi'^2)^{3/2}} \quad (9.8.17)$$

$$k_2 = -\frac{1}{r\sin\phi} \frac{\partial \mathbf{i}_2}{\partial \eta} \cdot \mathbf{i}_3 = \frac{-r'\cot\phi + r\phi'}{r\sqrt{r'^2 + r^2\phi'^2}} \quad (9.8.18)$$

$$k_4 = \frac{1}{r\sin\phi} \frac{\partial \mathbf{i}_1}{\partial \eta} \cdot \mathbf{i}_2 = \frac{r' + r\phi'\cot\phi}{r\sqrt{r'^2 + r^2\phi'^2}} \quad (9.8.19)$$

$$k_{61} = 0, \quad k_{62} = 0, \quad k_5 = 0 \quad (9.8.20)$$

It follows from (9.4.9a,b) that the circumferential Jaumann strains of an arbitrary point at $\zeta = z$ are given by

$$\epsilon_{11} = e_1 + (k_1 - k_1^0)(1 + e_1)(z - k_1^0 z^2) \quad (9.8.21)$$

$$\epsilon_{22} = e_2 + (k_2 - k_2^0)(1 + e_2)(z - k_2^0 z^2) \quad (9.8.22)$$

and $\epsilon_{12} = 0$ in the case of axisymmetric deformations. Hence, the Jaumann stresses are

$$\sigma_{11} = \frac{E}{1 - \nu^2}(\epsilon_{11} + \nu\epsilon_{22}) \quad (9.8.23)$$

$$\sigma_{22} = \frac{E}{1 - \nu^2}(\epsilon_{22} + \nu\epsilon_{11}) \quad (9.8.24)$$

Therefore, the elastic energy is given by

$$\Pi = \frac{1}{2}\int_0^\pi \int_0^{2\pi} \int_{-h/2}^{h/2} (\sigma_{11}\epsilon_{11} + \sigma_{22}\epsilon_{22})(a+z)^2 \sin\xi \, d\xi \, d\eta \, dz$$

$$= \frac{Eha^2\pi}{1-\nu^2}\int_0^\pi \bigg[(1+\alpha^2)(e_1^2 + e_2^2 + 2\nu e_1 e_2) + \alpha^2(k_{1m}^2 + k_{2m}^2 + 2\nu k_{1m}k_{2m}$$

$$+ 2e_1 k_{1m} + 2e_2 k_{2m} + 2\nu e_1 k_{2m} + 2\nu e_2 k_{1m})\bigg]\sin\xi \, d\xi \quad (9.8.25)$$

where

$$k_{im} \equiv (k_i - k_i^0)(1 + e_i) \quad \text{and} \quad \alpha^2 \equiv \frac{h^2}{12a^2}$$

For an arbitrary point on the reference plane, the displacement vector $\mathbf{D} = \mathbf{r} - \mathbf{R}$. Hence, if the rotary inertia is assumed to be negligible, the kinetic energy T is given by

$$T = \frac{1}{2}\int_{vol} \rho \dot{\mathbf{D}} \cdot \dot{\mathbf{D}} dV = \frac{1}{2}\int_{vol} \rho\left[\left(\frac{\partial r}{\partial t}\right)^2 + r^2\left(\frac{\partial \phi}{\partial t}\right)^2\right] dV \qquad (9.8.26)$$

The work done by the internal pressure P is given by

$$W_{nc} = \int_{\delta vol} P dV = \int_0^\pi P\pi (r\sin\phi)^2 \frac{d(-r\cos\phi)}{d\xi} d\xi - \frac{4}{3}\pi a^3 P$$

$$= \int_0^\pi P\pi r^2 \sin^2\phi (r\phi' \sin\phi - r'\cos\phi) d\xi - \frac{4}{3}\pi a^3 P \qquad (9.8.27)$$

We introduce the nondimensional displacements w and u and the nondimensional time t defined by

$$\epsilon w = \frac{r}{a} - 1, \quad \epsilon u = \phi - \xi, \quad \text{and} \quad t \equiv \frac{\hat{t}}{a}\sqrt{\frac{E}{\rho(1-\nu^2)}} \qquad (9.8.28)$$

where $\epsilon = w_m/a \ll 1$ is a nondimensional perturbation parameter and w_m is the maximum radial displacement. Substituting (9.8.28) into (9.8.5) and (9.8.6), expanding the result for small ϵ, and keeping terms up to $O(\epsilon^2)$, we obtain

$$e_1 = \epsilon(u' + w) + \epsilon^2\left(\frac{1}{2}w'^2 + wu'\right) + O(\epsilon^3) \qquad (9.8.29)$$

$$e_2 = \epsilon(u\cot\xi + w) + \epsilon^2\left(uw\cot\xi - \frac{1}{2}u^2\right) + O(\epsilon^3) \qquad (9.8.30)$$

Substituting (9.8.28) into (9.8.17) and (9.8.18), expanding the result for small ϵ, and keeping terms up to $O(\epsilon)$, we obtain

$$k_1 - k_1^0 = \frac{1}{a}\epsilon(w'' + w) + O(\epsilon^2) \qquad (9.8.31)$$

$$k_2 - k_2^0 = \frac{1}{a}\epsilon(w'\cot\xi + w) + O(\epsilon^2) \qquad (9.8.32)$$

Substituting (9.8.29)-(9.8.32) into (9.8.21) and (9.8.22) and substituting the result into (9.8.25), we obtain

$$\Pi = \frac{\pi a^2 Eh}{1-\nu^2}\int_0^\pi \left\{\epsilon^2\left[u'^2 + 2(1+\nu)(wu' + w^2 + uw\cot\xi) + 2\nu uu'\cot\xi\right.\right.$$
$$\left.+ u^2\cot^2\xi\right] - \epsilon^3\left[(1+\nu)(u^2w - 2uw^2\cot\xi - 2w^2u' - ww'^2) - u^3\cot\xi\right.$$
$$\left.\left.+ 2u^2w\cot^2\xi - \nu u^2 u' + 4\nu uwu'\cot\xi + 2wu'^2 + \nu uw'^2\cot\xi + u'w'^2\right]\right.$$

$$+\epsilon^2\alpha^2\left[u'^2 + u^2\cot^2\xi - 2w''u' - 2w'u\cot^2\xi + w''^2 + w'^2\cot^2\xi\right.$$
$$\left.\left.+2\nu\left(uu'\cot\xi - w'u'\cot\xi - w''u\cot\xi + ww''\cot\xi\right)\right]\right\}\sin\xi d\xi \quad (9.8.33)$$

Substituting (9.8.28) into (9.8.26) yields

$$T = \frac{\pi a^2 E h}{1-\nu^2}\int_0^\pi \left[\epsilon^2(\dot{w}^2 + \dot{u}^2) + 2\epsilon^3 w\dot{u}^2\right]\sin\xi d\xi \quad (9.8.34)$$

Substituting (9.8.33) and (9.8.34) into (9.7.30) and (9.7.31) with ψ replaced with u, we obtain the following equations of motion (McIvor and Sonstegard, 1966):

$$\ddot{u} + L_1(w,u) + \epsilon\left\{2w\ddot{u} + 2\dot{w}\dot{u} - 2w'u' - 2wu'' - w'w'' - \frac{1}{2}(1-\nu)w'^2\cot\xi\right.$$
$$-2wu'\cot\xi - \frac{1}{2}(3-\nu)u^2\cot\xi - wu\left(1 - \nu - 2\cot^2\xi\right)$$
$$\left.-2(1+\nu)ww' - 2\nu w'u\cot\xi\right\} = 0 \quad (9.8.35)$$

$$\ddot{w} + L_2(w,u) + \epsilon\left\{(1+\nu)\left[2wu' - \frac{1}{2}w'^2 - \frac{1}{2}u^2 + 2wu\cot\xi - ww''\right.\right.$$
$$\left.-ww'\cot\xi - w'u'\cot\xi\right] + u'^2 - w''u' + u^2\cot^2\xi$$
$$\left.-u''w' + 2\nu uu'\cot\xi - \nu w''u\cot\xi + \nu w'u - \dot{u}^2\right\}$$
$$= \epsilon^2 F_r(\xi,\tau) \quad (9.8.36)$$

where $F_r(\xi,\tau)$ is the generalized radial force and L_1 and L_2 are spatial linear differential operators defined as

$$L_1(w,u) \equiv -u'' - u'\cot\xi + u(\nu + \cot^2\xi) - (1+\nu)w'$$
$$+\alpha^2\left[(\nu + \cot^2\xi)(u - w') - u'' - u'\cot\xi + w''' + w''\cot\xi\right] \quad (9.8.37)$$

$$L_2(w,u) \equiv (1+\nu)(u' + u\cot\xi + 2w) + \alpha^2\left\{-u''' - 2u''\cot\xi\right.$$
$$+u'(1 + \nu + \cot^2\xi) + (\nu - 2 - \cot^2\xi)u\cot\xi + w^{iv}$$
$$+2w'''\cot\xi - (1 + \nu + \cot^2\xi)w'' + 2w'\cot\xi$$
$$\left.+(\cot^2\xi - \nu)w'\cot\xi\right\} \quad (9.8.38)$$

We note that the excitation is ordered as ϵ^2 for the case of primary resonance.

9.8.2 Natural Frequencies and Mode Shapes

The natural frequencies and modes shapes are given by

$$-\omega^2 u + L_1(w, u) = 0 \qquad (9.8.39)$$
$$-\omega^2 w + L_2(w, u) = 0 \qquad (9.8.40)$$

where L_1 and L_2 are given by (9.8.37) and (9.8.38).

To simplify the problem, we introduce the variable ψ such that

$$u = \psi' \qquad (9.8.41)$$

and the operator H defined as

$$H(y) = \frac{d^2 y}{d\xi^2} + \cot \xi \frac{dy}{d\xi} + 2y \qquad (9.8.42)$$

Then, we rewrite (9.8.39) as

$$\frac{d}{d\xi}\left[-\omega^2 \psi - H(\psi) + (1+\alpha^2)(1+\nu)(\psi - w) - \alpha^2 H(\psi - w)\right] = 0 \quad (9.8.43)$$

Integrating (9.8.43) once yields

$$-\omega^2 \psi - H(\psi) + (1+\alpha^2)(1+\nu)(\psi - w) - \alpha^2 H(\psi - w) = 0 \qquad (9.8.44)$$

where the constant of integration is chosen to be zero because an additive constant in ψ does not change u. Using the operator H, we rewrite (9.8.40) as

$$-\omega^2 w + (1+\nu)H(\psi) - 2(1+\nu)(1+\alpha^2)(\psi - w) + (3+\nu)\alpha^2 H(\psi - w)$$
$$- \alpha^2 H^2(\psi - w) = 0 \qquad (9.8.45)$$

Equation (9.8.42) suggests seeking the solutions of (9.8.44) and (9.8.45) in terms of the Legendre polynomials $P_n(\cos \xi)$; that is,

$$w = c_1 P_n(\cos \xi) \quad \text{and} \quad \psi = c_2 P_n(\cos \xi) \qquad (9.8.46)$$

where

$$\frac{d^2 P_n}{d\xi^2} + \cot \xi \frac{dP_n}{d\xi} + n(n+1)P_n = 0 \qquad (9.8.47)$$

Substituting (9.8.46) into (9.8.44) and (9.8.45) and using the fact that $H(P_n) = -\lambda_n P_n$, where $\lambda_n = n(n+1) - 2$, yields

$$\begin{bmatrix} a_{11} & a_{12} \\ a_{21} & a_{22} \end{bmatrix} \begin{Bmatrix} c_1 \\ c_2 \end{Bmatrix} = 0 \qquad (9.8.48)$$

where

$$a_{11} = -\alpha^2 \lambda_n - (1+\alpha^2)(1+\nu) \qquad (9.8.49a)$$

$$a_{12} = (1+\alpha^2)(1+\nu+\lambda_n) - \omega_n^2 \qquad (9.8.49b)$$

$$a_{21} = 2(1+\alpha^2)(1+\nu) + \alpha^2\lambda_n(\lambda_n + 3 + \nu) - \omega_n^2 \qquad (9.8.49c)$$

$$a_{22} = -(1+\nu)(2+\lambda_n) - \alpha^2\left[\lambda_n^2 + (3+\nu)\lambda_n + 2 + 2\nu\right] \qquad (9.8.49d)$$

For a nontrivial solution, the determinant of the coefficient matrix in (9.8.48) must be equal to zero. Hence, ω_n is given by the characteristic equation

$$\omega_n^4 - \left[\lambda_n + 3 + 3\nu + \alpha^2\left(\lambda_n^2 + \nu\lambda_n + 4\lambda_n + 3 + 3\nu\right)\right]\omega_n^2$$
$$+ \lambda_n\left[(1-\nu^2) + \alpha^2\left(\lambda_n^2 + 2\lambda_n + 1 - \nu^2\right)\right] = 0 \qquad (9.8.50)$$

Moreover, it follows from (9.8.48) and (9.8.49) that

$$c_{2n} = \Gamma_n c_{1n} \qquad (9.8.51a)$$

where

$$\Gamma_n = \frac{\alpha^2\lambda_n + (1+\alpha^2)(1+\nu)}{(1+\alpha^2)(1+\nu+\lambda_n) - \omega_n^2} \qquad (9.8.51b)$$

The $n=0$ mode corresponds to a pure radial motion and it is called the breathing mode. Its corresponding frequency when $\nu = 0.3$ is given by

$$\omega_0 = 1.61245 - 3.55024 \times 10^{-18}\alpha^4$$

which is almost independent of α. For each $n \geq 1$, there are two modes corresponding to the frequencies ω_{nc} and ω_{nb}, respectively. In Tables 9.1 and 9.2, we list the lowest ten ω_{nc} and ω_{nb} frequencies, respectively. The frequencies ω_{nc} are relatively independent of a/h, whereas the frequencies ω_{nb} are sensitive to changes in a/h. This observation suggests a classification of the modes with frequencies ω_{nc} as membrane modes and those with frequencies ω_{nb} as bending modes.

Table 9.1 Variation of the membrane frequencies ω_{nc} with the mode number and a/h when $\nu = 0.3$.

n	$a/h = 20$	$a/h = 60$	$a/h = 100$	$a/h = 150$	$a/h = 200$
1	1.97505	1.97486	1.97485	1.97485	1.97484
2	2.72286	2.72200	2.72193	2.72191	2.72190
3	3.63640	3.63491	3.63479	3.63475	3.63474
4	4.59846	4.59641	4.59624	4.59619	4.59617
5	5.57787	5.57528	5.57508	5.57501	5.57499
6	6.56509	6.56198	6.56174	6.56166	6.56163
7	7.55646	7.55283	7.55254	7.55245	7.55242
8	8.55029	8.54614	8.54581	8.54571	8.54567
9	9.54570	9.54103	9.54066	9.54054	9.54050
10	10.54220	10.53700	10.53660	10.53650	10.53640

Table 9.2 Variation of the bending frequencies ω_{nb} with the mode number and a/h when $\nu = 0.3$.

n	$a/h = 20$	$a/h = 60$	$a/h = 100$	$a/h = 150$	$a/h = 200$
1	0	0	0	0	0
2	0.70268	0.70113	0.70101	0.70097	0.70096
3	0.84097	0.83118	0.83039	0.83014	0.83006
4	0.91577	0.88455	0.88200	0.88121	0.88093
5	0.98824	0.91502	0.90890	0.90698	0.90630
6	1.08145	0.93887	0.92651	0.92261	0.92124
7	1.20688	0.96317	0.94093	0.93387	0.93139
8	1.36999	0.99197	0.95525	0.94348	0.93932
9	1.57242	1.02808	0.97138	0.95297	0.94644
10	1.81373	1.07364	0.99074	0.96337	0.95360

Inspection of Table 9.2 suggests that a two-to-one internal resonance may be activated between the breathing mode with frequency 1.61245 and the third bending mode with frequency 0.84097 when $a/h = 20$, with frequency 0.83118 when $a/h = 60$, and with frequency 0.83039 when $a/h = 100$. As shown in Section 9.8.3, these modes interact.

Moreover, inspection of Tables 9.1 and 9.2 suggests that a two-to-one internal resonance might be activated between the first membrane mode and either the fifth or the sixth bending modes with frequencies 0.98824 and 1.08145 when $a/h = 20$, with either the eighth or the ninth bending modes with frequencies 0.99197 and 1.02808 when $a/h = 60$, and with the tenth mode with frequency 0.99074 when $a/h = 100$. However, it turns out that these modes are orthogonal and hence they do not interact.

Furthermore, inspection of Tables 9.1 and 9.2 suggests that a two-to-one internal resonance might be activated when $a/h = 20$ between the second membrane mode with frequency 2.72286 and the eighth bending mode with frequency 1.36999 or between the third membrane mode with frequency 3.6364 and the tenth bending mode with frequency 1.81373. As shown in Section 9.8.3, the first resonance can be activated whereas the second resonance cannot be activated.

In the next section, we need solutions of the adjoint of (9.8.39) and (9.8.40). To determine the adjoint, we multiply (9.8.39) and (9.8.40) by $u^* \sin \xi$ and $w^* \sin \xi$, respectively, add the results, and rewrite them as

$$\int_0^\pi \left\{ u^* \sin \xi \left[-(1+\alpha^2)(u'' + u' \cot \xi) + (1+\alpha^2)(\nu + \cot^2 \xi)u - (1+\nu)w' \right. \right.$$
$$\left. + \alpha^2 \frac{d}{d\xi}(w'' + w' \cot \xi) + \alpha^2(1-\nu)w' \right] + w^* \sin \xi \left[(1+\nu)(u' + u \cot \xi + 2w) \right.$$
$$-\alpha^2 \left(\frac{d^2}{d\xi^2} + \frac{d}{d\xi} \cot \xi \right)(u' + u \cot \xi) + \alpha^2 \left(\frac{d^2}{d\xi^2} + \frac{d}{d\xi} \cot \xi \right)(w'' + w' \cot \xi)$$
$$\left. \left. -\alpha^2(1-\nu)(u' + u \cot \xi) + \alpha^2(1-\nu)(w'' + w' \cot \xi) \right] \right\} d\xi = 0 \qquad (9.8.52)$$

Integrating (9.8.52) by parts to transfer the derivatives from u and w to u^* and w^* and rearranging the result, we obtain

$$\int_0^\pi \Bigg\{ u\sin\xi \bigg[-(1+\alpha^2)\left(u^{*\prime\prime} + u^{*\prime}\cot\xi\right) + (1+\alpha^2)(\nu + \cot^2\xi)u^*$$

$$+\alpha^2 \frac{d}{d\xi}\left(w^{*\prime\prime} + w^{*\prime}\cot\xi\right) + \alpha^2(1-\nu)w^{*\prime}\bigg]$$

$$+w\sin\xi\bigg[(1+\nu)\left(u^{*\prime} + u^*\cot\xi + 2w^*\right)$$

$$-\alpha^2\left(\frac{d^2}{d\xi^2} + \frac{d}{d\xi}\cot\xi\right)\left(u^{*\prime} + u^*\cot\xi\right)$$

$$-\alpha^2(1-\nu)\left(u^{*\prime} + u^*\cot\xi\right)$$

$$+\alpha^2\left(\frac{d^2}{d\xi^2} + \frac{d}{d\xi}\cot\xi\right)\left(w^{*\prime\prime} + w^{*\prime}\cot\xi\right)$$

$$+\alpha^2(1-\nu)\left(w^{*\prime\prime} + w^{*\prime}\cot\xi\right)\bigg]\Bigg\}d\xi = 0 \qquad (9.8.53)$$

where use has been made of the fact that the boundary terms vanish. Setting each of the coefficients of u and w in (9.8.53) equal to zero and expanding the results, we obtain (9.8.39) and (9.8.40) with u and w being replaced with u^* and w^*. Therefore, (9.8.39) and (9.8.40) are self-adjoint and hence solutions of the adjoint equations are given by

$$w^* = P_n(\cos\xi) \quad \text{and} \quad u^* = \Gamma_n P_n'(\cos\xi) \qquad (9.8.54)$$

for $n = 0, 1, 2, \ldots$.

9.8.3 Two-to-One Internal Resonance

Again, we use the method of multiple time scales to determine a first-order uniformly valid expansion of the solution of (9.8.35) and (9.8.36) for small but finite amplitudes. We expand the excitation F_r in terms of Legendre polynomials as

$$F_r = \sum_{n=0}^{\infty} f_n P_n(\cos\xi)\cos(\Omega\tau) \qquad (9.8.55)$$

We seek an asymptotic expansion of the displacement field in the form

$$w(\xi,\tau;\epsilon) = \epsilon w_1(\xi, T_0, T_1) + \epsilon^2 w_2(\xi, T_0, T_1) + \cdots \qquad (9.8.56)$$
$$u(\xi,\tau;\epsilon) = \epsilon u_1(\xi, T_0, T_1) + \epsilon^2 u_2(\xi, T_0, T_1) + \cdots \qquad (9.8.57)$$

Substituting (9.8.55)-(9.8.57) into (9.8.35) and (9.8.36) and equating coefficients of like powers of ϵ, we obtain

Order ϵ:

$$D_0^2 u_1 + L_1(w_1, u_1) = 0 \qquad (9.8.58)$$
$$D_0^2 w_1 + L_2(w_1, u_1) = 0 \qquad (9.8.59)$$

Order ϵ^2:

$$D_0^2 u_2 + L_1(w_2, u_2) = -2D_0 D_1 u_1 + 2w_1' u_1' + 2w_1 u_1'' + w_1' w_1'' + 2w_1 u_1' \cot \xi$$
$$+ (1-\nu)\left(\frac{1}{2}w_1'^2 \cot \xi + w_1 u_1\right) + \frac{1}{2}(3-\nu) u_1^2 \cot \xi - 2u_1 w_1 \cot^2 \xi$$
$$+ 2\nu u_1 w_1' \cot \xi + 2(1+\nu) w_1 w_1' - 2D_0 w_1 D_0 u_1 - 2w_1 D_0^2 u_1 \quad (9.8.60)$$

$$D_0^2 w_2 + L_2(w_2, u_2) = -2D_0 D_1 w_1 + (D_0 u_1)^2 - (1+\nu)\left(2w_1 u_1' - \frac{1}{2}w_1'^2\right.$$
$$\left. - \frac{1}{2}u_1^2 + 2w_1 u_1 \cot \xi - w_1 w_1' \cot \xi - w_1' u_1' \cot \xi - w_1 w_1''\right) - u_1^2 \cot^2 \xi$$
$$+ u_1'' w_1' - 2\nu u_1 u_1' \cot \xi + \nu w_1'' u_1 \cot \xi - u_1'^2 + w_1'' u_1' - \nu w_1' u_1$$
$$+ \sum_{n=0}^{\infty} f_n P_n \cos(\Omega T_0) \quad (9.8.61)$$

In the presence of damping, only the directly and indirectly excited modes are involved in the long-term response. Thus, to proceed further, we need to define the resonance conditions. We consider the case of a primary-resonance excitation of the mth mode and introduce the detuning parameter σ_1 defined by

$$\Omega = \omega_m + \epsilon \sigma_1 \quad (9.8.62)$$

Moreover, we consider the case of two-to-one internal resonance between the mth and kth modes and introduce the detuning parameter σ_2 defined by

$$\omega_m = 2\omega_k + \epsilon \sigma_2, \quad m > k \quad (9.8.63)$$

Thus, we can express w_1 and u_1 as

$$w_1(\xi, T_0, T_1) = A_m(T_1) P_m(\cos \xi) e^{i\omega_m T_0} + A_n(T_1) P_k(\cos \xi) e^{i\omega_k T_0} + cc \quad (9.8.64)$$

$$u_1(\xi, T_0, T_1) = \Gamma_m A_m(T_1) P_m'(\cos \xi) e^{i\omega_m T_0} + \Gamma_k A_k(T_1) P_k'(\cos \xi) e^{i\omega_k T_0} + cc \quad (9.8.65)$$

Substituting (9.8.64) and (9.8.65) into (9.8.60) and (9.8.61) and using (9.8.62) and (9.8.63) yields

$$D_0^2 u_2 + L_1(w_2, u_2) = -2i\omega_m \Gamma_m A_m' P_m' e^{i\omega_m T_0} - 2i\omega_k \Gamma_k A_k' P_k' e^{i\omega_k T_0}$$
$$+ \chi_1 A_k^2 e^{i(\omega_m T_0 - \sigma_2 T_1)} + \chi_3 A_m \bar{A}_k e^{i(\omega_k T_0 + \sigma_2 T_1)}$$
$$+ cc + NST \quad (9.8.66)$$

$$D_0^2 w_2 + L_2(w_2, u_2) = -2i\omega_m A_m' P_m e^{i\omega_m T_0} - 2i\omega_k A_k' P_k e^{i\omega_k T_0}$$
$$+ \chi_2 A_k^2 e^{i(\omega_m T_0 - \sigma_2 T_1)} + \chi_4 A_m \bar{A}_k e^{i(\omega_k T_0 + \sigma_2 T_1)}$$
$$+ \frac{1}{2}\sum_0^{\infty} f_n P_n e^{i(\omega_m T_0 + \sigma_1 T_1)} + cc + NST \quad (9.8.67)$$

where

$$\chi_1 = \left[(1 - \nu - 2\cot^2\xi + 4\omega_k^2)\Gamma_k + 2 + 2\nu - k^2 - k\right] P_k P_k'$$
$$+ \left[\frac{1}{2}(3-\nu)\Gamma_k^2 - \frac{1}{2}(1+\nu) + 2\nu\Gamma_k\right] P_k'^2 \cot\xi \qquad (9.8.68)$$

$$\chi_3 = \left[\Gamma_k\Gamma_m(3-\nu) - \nu - 1 + 2\nu(\Gamma_k + \Gamma_m)\right] P_k' P_m' \cot\xi$$
$$+ \left[\Gamma_k\left(1 - \nu - 2\cot^2\xi + 2\omega_k^2 - 2\omega_k\omega_w\right) - m^2 - m + 2(1+\nu)\right] P_k' P_m$$
$$+ \left[\Gamma_m\left(1 - \nu - 2\cot^2\xi - 2\omega_k\omega_m + 2\omega_m^2\right) + 2 + 2\nu - k^2 - k\right] P_m' P_k$$
$$(9.8.69)$$

$$\chi_2 = -(1+\nu)(k+k^2)P_k^2 - \Gamma_k(1+\nu)(2+k+k^2)P_k'P_k \cot\xi$$
$$+ \left[\frac{1}{2}(1+\nu) - \nu\Gamma_k\left(1+\cot^2\xi\right) + \frac{1}{2}\Gamma_k^2\left(1+\nu - 2\cot^2\xi\right)\right] P_k'^2 \qquad (9.8.70)$$

$$\chi_4 = -(1+\nu)\left(k+k^2+m+m^2\right)P_m P_k$$
$$-\Gamma_k\left(2+2\nu+\nu m+\nu m^2\right)P_m P_k' \cot\xi$$
$$+ \left[2-k-k^2+2\nu+\Gamma_m\left(1-\nu-2\cot^2\xi-2\omega_k\omega_m+2\omega_m^2\right)\right]P_m' P_k \cot\xi$$
$$- \left[1+\nu-2\nu(\Gamma_k+\Gamma_m)-(3-\nu)\Gamma_k\Gamma_m\right]P_m' P_k' \qquad (9.8.71)$$

Because the homogeneous parts of (9.8.66) and (9.8.67) have a nontrivial solution, the nonhomogeneous equations have a solution only if the nonhomogeneous parts are orthogonal to every solution of the adjoint homogeneous problem (Nayfeh, 1981). Requiring the right-hand sides of (9.8.66) and (9.8.67) be orthogonal to

$$(\Gamma_m P_m', P_m) e^{-i\omega_k T_0} \quad \text{and} \quad (\Gamma_k P_k', P_k) e^{-i\omega_k T_0}$$

we obtain the modulation equations

$$\frac{4i\omega_m}{2m+1}\left[1+m(m+1)\Gamma_m^2\right]\left(\frac{dA_m}{dT_1} + \mu_m A_m\right) - \Lambda_1 A_k^2 e^{-i\sigma_2 T_1} + fe^{i\sigma_1 T_1} = 0$$
$$(9.8.72)$$

$$\frac{4i\omega_k}{2k+1}\left[1+k(k+1)\Gamma_k^2\right]\left(\frac{dA_k}{dT_1} + \mu_k A_k\right) - \Lambda_2 A_m \bar{A}_k e^{i\sigma_2 T_1} = 0 \qquad (9.8.73)$$

where f is proportional to f_n and

$$\Lambda_1 = \int_0^\pi (\Gamma_m \chi_1 P_m' + \chi_2 P_m) \sin\xi d\xi \qquad (9.8.74)$$

$$\Lambda_2 = \int_0^\pi (\Gamma_k \chi_3 P_k' + \chi_4 P_k) \sin\xi d\xi \qquad (9.8.75)$$

Alternatively, we use the time-averaged Lagrangian method. To this end, we let

$$u = \epsilon A_k(T_1)\Gamma_k P_k'(\xi)e^{i\omega_k T_0} + \epsilon A_m(T_1)\Gamma_m P_m'(\xi)e^{i\omega_m T_0} + cc \quad (9.8.76)$$
$$w = \epsilon A_k(T_1)P_k(\xi)e^{i\omega_k T_0} + \epsilon A_m(T_1)P_m(\xi)e^{i\omega_m T_0} + cc \qquad (9.8.77)$$

652   SHELLS

in $\mathcal{L} = T - \Pi$, where Π and T are given by (9.8.33) and (9.8.34), keep the slowly varying terms only, integrate the result from $\xi = 0$ to π, and obtain

$$\frac{1-\nu^2}{\epsilon^3 \pi a^2 Eh} <\mathcal{L}> = \frac{2i\omega_k}{2k+1}\left[1+k(k+1)\Gamma_k^2\right]\left(A_k \bar{A}_k' - A_k' \bar{A}_k\right)$$
$$+ \frac{2i\omega_m}{2m+1}\left[1+m(m+1)\Gamma_m^2\right]\left(A_m \bar{A}_m' - A_m' \bar{A}_m\right)$$
$$+ \Lambda\left(A_m \bar{A}_k^2 e^{i\sigma_2 T_1} + \bar{A}_m A_k^2 e^{-i\sigma_2 T_1}\right)$$
$$+ \text{fast-varying terms} \quad (9.8.78)$$

where

$$\Lambda = \int_0^\pi \left\{ -2(1+\nu)\Gamma_k \cos\xi P_k P_k' P_m - (1+\nu)\Gamma_m \cos\xi P_k^2 P_m' \right.$$
$$- \left[\frac{\cos^2\xi}{\sin\xi}\Gamma_k^2 + \frac{1}{2}\sin\xi(1+\nu)\left(1-\Gamma_k^2\right) - \Gamma_k^2 \omega_k^2 \sin\xi\right] P_k'^2 P_m$$
$$- \left[\frac{2\cos^2\xi}{\sin\xi}\Gamma_k\Gamma_m + \sin\xi(1+\nu)\left(1-\Gamma_k\Gamma_m\right) - 2\Gamma_k\Gamma_m\omega_k\omega_m \sin\xi\right] P_k P_k' P_m'$$
$$\left. - \frac{1}{2}\left(2\nu\Gamma_k + \nu\Gamma_m - 3\Gamma_k^3\Gamma_m\right)\cos\xi P_k'^2 P_m' \right\} d\xi \quad (9.8.79)$$

Substituting (9.8.78) into the Euler-Lagrange equation

$$\frac{d}{dT_1}\left(\frac{\partial <\mathcal{L}>}{\partial \bar{A}_n'}\right) - \frac{\partial <\mathcal{L}>}{\partial \bar{A}_n} = -f \quad (9.8.80)$$

where f stands for the nonconservative forces, we obtain the modulation equations

$$\frac{4i\omega_m}{2m+1}\left[1+m(m+1)\Gamma_m^2\right]\left(\frac{dA_m}{dT_1} + \mu_m A_m\right) - \Lambda A_k^2 e^{-i\sigma_2 T_1} + fe^{i\sigma_1 T_1} = 0$$
$$(9.8.81)$$

$$\frac{4i\omega_k}{2k+1}\left[1+k(k+1)\Gamma_k^2\right]\left(\frac{dA_k}{dT_1} + \mu_k A_k\right) - 2\Lambda A_m \bar{A}_k e^{i\sigma_2 T_1} = 0 \quad (9.8.82)$$

Comparing (9.8.81) and (9.8.82) with (9.8.72) and (9.8.73), we note that they are the same and that the following conditions must hold:

$$\Lambda_1 = \Lambda \quad \text{and} \quad \Lambda_2 = 2\Lambda \quad (9.8.83)$$

Next, we give the values of Γ_m, Γ_k, and Λ for $\nu = 0.3$ and $a/h = 20$ and few cases in which one frequency is approximately twice another frequency.

When $a/h = 20$, $\alpha = 1.44338 \times 10^{-2}$ and the lowest membrane and bending frequencies are listed in Tables 9.1 and 9.2, respectively. The frequency of the breathing mode $\omega_{0c} \approx 1.61245$. Inspecting Tables 9.1 and 9.2, we find that few pairs of natural frequencies are in the ratio of two-to-one. Next, we present $\Gamma_m, \Gamma_k, \sigma_2$, and Λ for some of these pairs.

In this case, $\omega_{0c} \approx 2\omega_{3b}$
$\omega_m = \omega_{0c} = 1.61245, \omega_k = \omega_{3b} = 0.84097, \sigma_2 = -0.06949, \Gamma_m = -0.39388$, $\Gamma_k = 0.12292, \Lambda_1 = \Lambda = -1.72117$, and $\Lambda_2 = 2\Lambda = -3.44234$. Hence, a two-to-one internal resonance between the breathing mode and the third bending mode can be activated.

In this case, $\omega_{1c} \approx 2\omega_{5b}$
$\omega_{1c} = 1.97505, \omega_{5b} = 0.98824, \sigma_2 = -0.00144, \Gamma_m = -0.5, \Gamma_k = 0.04610$, $\Lambda_1 = \Lambda = 0$, and $\Lambda_2 = 0$. Consequently, the first membrane mode is orthogonal to the fifth bending mode, and a two-to-one internal resonance between them cannot be activated.

In this case, $\omega_{2c} \approx 2\omega_{8b}$
$\omega_{2c} = 2.72286, \omega_{8b} = 1.36999, \sigma_2 = -0.01712, \Gamma_m = -0.61580, \Gamma_k = 0.01894, \Lambda_1 = \Lambda = -2.57548, \Lambda_2 = 2\Lambda = -5.15095$. Hence, a two-to-one internal resonance between the second membrane mode and the eighth bending mode can be activated.

In this case, $\omega_{3c} \approx 2\omega_{10b}$
$\omega_{3c} = 3.63640, \omega_{10b} = 1.81373, \sigma_2 = 0.00894, \Gamma_m = -0.67795, \Gamma_k = 0.01248$, $\Lambda_1 = \Lambda = 0$, and $\Lambda_2 = 0$. Therefore, the third membrane mode is orthogonal to the tenth bending mode.

In this case, $\omega_{5c} \approx 2\omega_{2c}$
$\omega_{5c} = 5.57787, \omega_{2c} = 2.72286, \sigma_2 = 0.13215, \Gamma_m = -0.72300, \Gamma_k = -0.61580$, $\Lambda_1 = \Lambda = 0$, and $\Lambda_2 = 0$. Consequently, the fifth membrane mode is orthogonal to the second membrane mode.

BIBLIOGRAPHY

Abarcar, R. B., and Cunniff, P. F. (1972). The vibration of cantilever beams of fiber reinforced material, *Journal of Composite Materials* 6, 504-517. *180*

Abdel-Rahman, E. M. See Faris, W. F., Abdel-Rahman, E. M., and Nayfeh, A. H.; Younis, M. I., Abdel-Rahman, E. M., and Nayfeh, A. H.

Abe, A., Kobayashi, Y., and Yamada, G. (1998a). Analysis of subharmonic resonance of moderately thick antisymmetric angle-ply laminated plates by using method of multiple scales, *Journal of Sound and Vibration* 217, 467-484. *381, 499*

Abe, A., Kobayashi, Y., and Yamada, G. (1998b). Two-mode response of simply supported rectangular laminated plates, *Nonlinear Dynamics* 33, 675-690. *499*

Abe, A., Kobayashi, Y., and Yamada, G. (2000). Nonlinear vibration characteristics of clamped laminated shallow shells, *Journal of Sound and Vibration* 234, 405-426. *563*

Abe, A., Kobayashi, Y., and Yamada, G. (2001). One-to-one internal resonance of symmetric crossply laminated shallow shells, *Journal of Applied Mechanics* 68, 640-649. *564*

Abdel-Rahman, E. M., Younis, M. I., and Nayfeh, A. H. (2002). Characterization of the mechanical behavior of an electrically actuated microbeam, *Journal of Micromechanics and Microengineering* 12, 759-766. *342, 344-346*

Abhyankar, N. S., Hall, E. K., and Hanagud, S. V. (1993). Chaotic vibrations of beams: Numerical solution of partial differential equations, *Journal of Applied Mechanics* 60, 167-174. *318*

Abou-Rayan, A. M., Nayfeh, A. H., and Mook, D. T. (1993). Nonlinear response of a parametrically excited buckled beam, *Nonlinear Dynamics* 4, 499-525. *318*

Adams, G. G. (1987). Critical speeds for a flexible spinning disk, *International Journal of Mechanical Sciences* 29, 525-531. *517*

Adams, R. D., and Bacon, D. G. C. (1973). Measurement of the flexural damping capacity and dynamic Young's modulus of metals and reinforced plastics, *Journal of Physics D: Applied Physics* 6, 27-41. *8, 179*

Adams, R. D., Fox, M. A. O., Flood, R. J. L., Friend, R. J., and Hewitt, R. L. (1969). The dynamic properties of unidirectional carbon and glass fiber reinforced plastics in torsion and flexure, *Journal of Composite Materials* 3, 594-603. *8*

Adams, R. D. See Ni, R. G., and Adams, R. D.

Adan, M. See Sheinman, I., and Adan, M.

Afaneh, A. A., and Ibrahim, R. A. (1992). Nonlinear response of an initially buckled beam with 1:1 internal resonance to sinusoidal excitation, *Nonlinear Dynamics* 4, 547-571. *319*

Afaneh, A. A. See Chang, W. K., Ibrahim, R. A., and Afaneh, A. A.; Chang, W. K., Ibrahim, R. A., and Afaneh, A. A.

Ahn, Y., Guckel, H., and Zook, J. D. (2001). Capacitive microbeam resonator design, *Journal of Micromechanics and Microengineering* 11, 70-80. *342*

Alaca, B. E. See Saif, M. T. A., Alaca, B. E., and Sehitoglu, H.

Alaggio, R., and Rega, G. (2000). Characterizing bifurcations and classes of motion in the transition to chaos through 3D-tori of a continuous experimental system in solid mechanics, *Physica D* 137, 70-93. *137*

Alaggio, R. See Benedettini, F., Rega, G., and Alaggio, R.; Rega, G., and Alaggio, R.; Rega, G., Alaggio, R., and Benedettini, F.

Alhazza, K. See Nayfeh, A. H., and Alhazza, K.

Alhazza, K. A. (2002). *Nonlinear Vibrations of Doubly Curved Cross-Ply Shallow Shells*, PhD Dissertation, Virginia Polytechnic Institute and State University, Blacksburg, VA. *469*

Alhazza, K. A., and Nayfeh, A. H. (2001). Nonlinear vibrations of doubly-curved cross-ply shallow shells, in *Proceedings of the 42nd AIAA/ASME/ASCE/AHS/*

ASC Structures, Structural Dynamics, and Materials Conference, AIAA Paper No. 2001-1661, Seattle, Washington, April 16-19. *53, 564*

Ali, S. A. (1986). Dynamic responses of sagged cables, *Computers & Structures* 23, 51-57. *136*

Ali, S. A. See Al-Noury, S. I., and Ali, S. A.

Al-Khattat, I. M. See Huang, C. L., and Al-Khattat, I. M.

Alkire, K. (1984). An analysis of rotor blade twist variables associated with different Euler sequences and pretwist treatments, NASA TM 84394. *177, 202, 238, 376*

Almroth, B. O. See Brush, D. O., and Almroth, B. O.

Al-Noury, S. I., and Ali, S. A. (1985). Large-amplitude vibrations of parabolic cables, *Journal of Sound and Vibration* 101, 451-462. *136, 137*

Amabili, M., Frosali, G., and Kwak, M. K. (1996). Free vibrations of annular plates coupled with fluids, *Journal of Sound and Vibration* 191, 825-846. *536*

Ambartsumian, S. A. (1964). *Theory of Anisotropic Shells*, Fizmargiz, Moskva, English translation, NASA TTF-118. *559*

Ambartsumian, S. A. (1970). *Theory of Anisotropic Plates*, J. E. Ashton ed., translated from Russian by T. Cheron, Technomic Publishing Co., Lancaster, Pennsylvania. *469*

Ames, W. F. See Lee, S. Y., and Ames, W. F.

Anand, G. V. (1973). Negative resistance mode of forced oscillations of a string, *Journal of the Acoustical Society of America* 54, 692-698. *112*

Ananthasuresh, G. K. See Grétillat, M.-A., Yang, Y.-J., Hung, E. S., Rabinovich, V., Ananthasuresh, G. K., deRooij, N. F., and Senturia, S. D.

Anderson, T. J., Balachandran, B., and Nayfeh, A. H. (1992). Observations of nonlinear interactions in a flexible cantilever beam, in *Proceedings of the 33rd AIAA/ASME/ASCE/AHS/ASC Structures, Structural Dynamics, and Materials Conference*, AIAA Paper No. 92-2332-CP, Dallas, Texas, April 13-15, pp. 1678-1686. *267, 270*

Anderson, T. J., Balachandran, B., and Nayfeh, A. H. (1994). Nonlinear resonances in a flexible cantilever beam, *Journal of Vibration and Acoustics* 116, 480-484. *267, 270, 278*

Anderson, T. J., Nayfeh, A. H., and Balachandran, B. (1996). Experimental verification of the importance of the nonlinear curvature in the response of a cantilever beam, *Journal of Vibration and Acoustics* 118, 21-27. *7, 223, 270, 277, 278*

Anderson, T. J., and Nayfeh A. H. (1996). Natural frequencies and mode shapes of laminated composite plates: Experiments and finite element analysis, *Journal of Vibration and Control* 2, 381-414. *378*

Anderson, T. J. See Nayfeh, A. H., Kreider, W., and Anderson, T. J.

Andrews, M. K. See Turner, G. C., and Andrews, M. K.

Arafat, H. N. (1999). *Nonlinear Response of Cantilever Beams*, PhD Dissertation, Virginia Polytechnic Institute and State University, Blacksburg, Virginia. *309, 310*

Arafat, H. N., and Nayfeh, A. H. (1999). Nonlinear bending-torsion oscillations of cantilever beams to combination parametric excitations, in *Proceedings of the 17th International Modal Analysis Conference, IMAC XVII*, Kissimmee, Florida, February 8-11, pp. 1203-1209. *53, 138, 278*

Arafat, H. N., and Nayfeh, A. H. (2000). Nonlinear response of symmetrically laminated composite cantilever beams, in *Proceedings of the 41st AIAA/ASME/ASCE/AHS/ASC Structures, Structural Dynamics, and Materials Conference*, AIAA Paper No. 2000-1471, Atlanta, Georgia, April 3-6. *138*

Arafat, H. N., and Nayfeh, A. H. (2003). Nonlinear responses of suspended cables to primary resonance excitations, *Journal of Sound and Vibration* 266, 325-354. *53, 136, 164-166*

Arafat, H. N., and Nayfeh, A. H. (2004a). Modal interactions in the vibrations of a heated annular plate, *International Journal of Non-Linear Mechanics*, in press. *499, 544*

Arafat, H. N., and Nayfeh, A. H. (2004b). Nonlinear interactions in the vibrations of heated annular plates, *Nonlinear Dynamics*, in press. *499, 544*

Arafat, H. N., Nayfeh, A. H., and Chin, C-M. (1998). Nonlinear nonplanar dynamics of parametrically excited cantilever beams, *Nonlinear Dynamics* 15, 31-61. *138, 284, 285, 287, 304*

Arafat, H. N., Nayfeh, A. H., and Faris, W. (2004). Natural frequencies of heated annular and circular plates, *International Journal of Solids and Structures* 41, 3031-3051. *544*

Arafat, H. N. See Nayfeh A. H., and Arafat, H. N.; Nayfeh, A. H., Arafat, H. N., Chin, C-M., and Lacarbonara, W.

Arnold, R. N. See Tobias, S. A., and Arnold, R. N.

Asano, T. See Yasuda, K., and Asano, T.

Asfar, O. R., Masad, J. A., and Nayfeh, A. H. (1990). A method for calculating the stability of boundary layers, *Computers & Fluids* 18, 305-315. *379*

Ashton, J. E. (1969). Approximate solutions for unsymmetrically laminated plates, *Journal of Composite Materials* 3, 189-191. *377, 378*

Atilgan, A. R., and Hodges, D. H. (1991). Unified nonlinear analysis for nonhomogeneous anisotropic beams with closed cross sections, *AIAA Journal* 29, 1990-1999. *176*

Atilgan, A. R. See Hodges, D. H., Atilgan, A. R., Cesnik, C. E. S., and Fulton, M. V.

Atluri, S. N. (1984). Alternate stress and conjugate strain measures, and mixed variational formulations involving rigid rotations, for computational analyses of finitely deformed solids, with application to plates and shells-I Theory, *Computers & Structures* 18, 93-116. *600*

Atluri, S. N. See Borri, M., Mello, F., Iura, M., and Atluri, S. N.; Im, S., and Atluri, S. N.

Awrejcewicz, A. (1989). A route to chaos in a nonlinear oscillator with delay, *Acta Mechanica* 77, 111-120. *25*

Ayela, F., and Fournier, T. (1998). An experimental study of an harmonic micromachined silicon resonators, *Measurement, Science and Technology* 9, 1821-1830. *342*

Ayyappan, M. See Manoj, T., Ayyappan, M., Krishnan, K. S., and Rao, B. N.

Babcock, C. D. See Chen, J. C., and Babcock, C. D.

Bacon, D. G. C. See Adams, R. D., and Bacon, D. G. C.

Baharlou, B., and Leissa, A. W. (1987). Vibration and buckling of generally laminated composite plates with arbitrary edge conditions, *International Journal of Mechanical Sciences* 29, 545-555. *377*

Bailey, C. D. (1975a). A new look at Hamilton's principle, *Foundation Physics* 5, 433-451. *71*

Bailey, C. D. (1975b). Application of Hamilton's law of varying action, *AIAA Journal* 13, 1154-1157. *71*

Bajaj, A. K., and Johnson, J. M. (1990). Asymptotic techniques and complex dynamics in weakly non-linear forced mechanical systems, *International Journal of Non-Linear Mechanics* 25, 211-226. *132*

Bajaj, A. K., and Johnson, J. M. (1992). On the amplitude dynamics and crisis in resonant motions of stretched strings, *Philosophical Transactions of the Royal Society of London* 338, 1-41. *112, 132, 135*

Bajaj, A. K. See Chang, S. I., Bajaj, A. K., and Krousgrill, C. M.; Georgiou, I. T., Bajaj, A. K., and Corless, M.; Georgiou, I. T., Corless, M. J., and Bajaj, A. K.;

Johnson, J. M., and Bajaj, A. K.; Streit, D. A., Bajaj, A. K., and Krousgrill, C. M.; Tousi, S., and Bajaj, A. K.

Balachandran, B., and Nayfeh, A. H. (1990). Nonlinear motions of a beam-mass structure, *Nonlinear Dynamics* 1, 36-59. *25*

Balachandran, B., Nayfeh, A. H., Smith, S. W., and Pappa, R. S. (1994). Identification of nonlinear interactions in structures, *Journal of Guidance, Control, and Dynamics* 17, 257-262. *376*

Balachandran, B. See Anderson, T. J., Balachandran, B., and Nayfeh, A. H.; Anderson, T. J., Balachandran, B., and Nayfeh, A. H.; Anderson, T. J., Nayfeh, A. H., and Balachandran, B.; Nayfeh, A. H., and Balachandran, B.; Nayfeh, A. H., and Balachandran, B.

Banan, M. R., Karami, G., and Farshad, M. (1991). Nonlinear theory for elastic spatial rods, *International Journal of Solids and Structures* 27, 713-724. *177*

Banerjee, A. K. See Kane, T. R., Ryan, R. R., and Banerjee, A. K.

Barash, S., and Chen, Y. (1972). On the vibration of a rotating disk, *Journal of Applied Mechanics* 29, 1143-1144. *517*

Barbier, D. See Malhaire, C., Le Berre, M., Febvre, M., Barbier, D., and Pinard, P.

Barr, A. D. S. See Haddow, A. G., Barr, A. D. S., and Mook, D. T.

Baruch, M., and Riff, R. (1982). Hamilton's principle, Hamilton's law -6^n correct formulations, *AIAA Journal* 20, 687-692. *71*

Basar, Y., and Omurtag, M. H. (2000). Free-vibration analysis of thin-thick laminated structures by layer-wise shell models, *Computers & Structures* 74(4), 409-427. *559*

Basset, A. B. (1890). On the extension and flexure of cylindrical and spherical thin elastic shells, *Philosophical Transactions of the Royal Society of London Series A* 181, 433-480. *372*

Bauchau, O. A., and Hong, C. H. (1987). Large displacement analysis of naturally curved and twisted composite beams, *AIAA Journal* 25, 1469-1475. *174, 177, 180*

Bauchau, O. A., and Hong, C. H. (1988). Nonlinear composite beam theory, *Journal of Applied Mechanics* 55, 156-163. *174, 180*

Baz, A., and Poh, S. (1989). Independent modal space control with positive position feedback, in *Proceedings of the Seventh VPI&SU Symposium on Dynamics and Control of Large Structures*, Blacksburg, Virginia, May 8-10. *8*

Beatty, M. F., and Chow, A. C. (1984). Free vibrations of a loaded rubber string, *International Journal of Non-Linear Mechanics* 19, 69-82. *136*

Beaulieu, G. See Pastorel, H., and Beaulieu, G.

Bellman, R. E., and Kalaba, R. E. (1965). *Quasilinearization and Nonlinear Boundary-Value Problems,* American Elsevier, New York. *50-52*

Benamar, R. See Moussaoui, F., and Benamar, R.

Benedettini, F., and Rega, G. (1987). Non-linear dynamics of an elastic cable under planar excitation, *International Journal of Non-Linear Mechanics* 22, 497-509. *136, 137, 161*

Benedettini, F., and Rega, G. (1990). Numerical simulations of chaotic dynamics in a model of an elastic cable, *Nonlinear Dynamics* 1, 23-38. *136*

Benedettini, F., and Rega, G. (1994). Analysis of finite oscillations of elastic cables under internal/external resonance conditions, in *Proceedings of the International Mechanical Engineering Congress and Exposition*, Chicago, Illinois, November 6-11, AMD-Vol. 192/DE-Vol. 78, pp. 39-46. *136, 138*

Benedettini, F., and Rega, G. (1997). Experimental investigation of the nonlinear response of a hanging cable. Part II: Global analysis, *Nonlinear Dynamics* 14, 119-138. *136, 137*

Benedettini, F., Rega, G., and Alaggio, R. (1995). Non-linear oscillations of a four-degree-of-freedom model of a suspended cable under multiple internal resonance conditions, *Journal of Sound and Vibration* 182, 775-798. *138*

Benedettini, F., Rega, G., and Vestroni, F. (1986). Modal coupling in the free nonplanar finite motion of an elastic cable, *Meccanica* 21, 38-46. *136, 137*

Benedettini, F. See Rega, G., Alaggio, R., and Benedettini, F.; Rega, G., Vestroni, F., and Benedettini, F.

Bennett, J. A. (1971). Nonlinear vibration of simply supported angle ply laminated plates, *AIAA Journal* 9, 1997-2003. *380*

Bennett, J. A. See Eisley, J. G., and Bennett, J. A.

Benson, R. C. See Cole, K. A., and Benson, R. C.

Berdichevskii, V. L. (1981). On the energy of an elastic rod, *Journal of Applied Mathematics and Mechanics (PMM U.S.S.R.)* 45, 518-529. *175*

Berger, H. M. (1955). A new approach to the analysis of large deflection of plates, *Journal of Applied Mechanics* 22, 465-472. *373, 375, 380, 544*

Berman, J. H. See Oyibo, G. A., and Berman, J. H.

Berry, J. G. See Naghdi, P. M., and Berry, J. G.

Bert, C. W. (1973). Nonlinear vibration of rectangular plates arbitrarily laminated of anisotropic materials, *Journal of Applied Mechanics* 40, 452-458. *380*

Bert, C. W. (1985). Research on dynamic behavior of composite and sandwich plates, *Shock and Vibration Digest* 17, 3-15. *378*

Bert, C. W., and Chen, T. L. C. (1978). Effect of shear deformation on vibration of antisymmetric angle-ply laminated rectangular plates, *International Journal of Solids and Structures* 14, 465-473. *378*

Bert, C. W., and Mayberry, B. L. (1969). Free vibration of unsymmetrically laminated anisotropic plates with clamped edges, *Journal of Composite Materials* 3, 282-293. *8, 377, 378*

Bert, C. W., and Reddy, V. S. (1982). Cylindrical shells of bimodulus composite material, *ASCE Journal of the Engineering Mechanics Division* 108, 675-688. *561*

Bert, C. W. See Noor, A. K., Burton, W. S., and Bert, C. W.; Ray, J. D., and Bert, C. W.; Siu, C. C., and Bert, C. W.

Bhat, R. B. See Rajalingham, C., Bhat, R. B., and Xistris, G. D.

Bhimaraddi, A. (1984). A higher order theory for free vibration analysis of circular cylindrical shells, *International Journal of Solids and Structures* 20, 623-630. *561, 593*

Bhimaraddi, A. (1987a). Nonlinear flexural vibrations of rectangular plates subjected to in-plane forces using a new shear deformation theory, *Thin-Walled Structures* 5, 309-327.

Bhimaraddi, A. (1987b). Static and transient response of cylindrical shells, *Thin-Walled Structures* 5, 157-179.

Bhimaraddi, A. (1987c). Static and transient response of rectangular plates, *Thin-Walled Structures* 5, 125-143. *375, 446*

Bhimaraddi, A. (1990). Nonlinear vibrations of in-plane loaded, imperfect, orthotropic plates using the perturbation technique, *International Journal of Solids and Structures* 25, 563-575. *381*

Bhimaraddi, A. (1993). Large amplitude vibrations of imperfect antisymmetric angle-ply laminated plates, *Journal of Sound and Vibration* 162, 457-470. *381*

Bhimaraddi, A., and Chandrashekhara, K. (1993). Nonlinear vibrations of heated antisymmetric angle-ply laminated plates, *International Journal of Solids and Structures* 30, 1255-1268. *544*

Bhimaraddi, A., and Stevens, L. K. (1984). A higher order theory for free vibration of orthotropic, homogeneous, and laminated rectangular plates, *Journal of Applied Mechanics* 51, 195-198. *374, 376, 377, 436, 454, 628*

Bickford, W. B. See Maganty, S. P., and Bickford, W. B.; Maganty, S. P., and Bickford, W. B.

Bieniek, M. P., Fan, T. C., and Lackman, L. M. (1966). Dynamic stability of cylindrical shells, *AIAA Journal* 4, 495-500. *565*

Birman, V., Griffin, S., and Knowle, G. (2000). Axisymmetric dynamics of composite spherical shells with active piezoelectric and composite stiffeners, *Acta Mechanica* 141, 71-83. *559*

Birman, V., and Zahed, H. (1989). Nonlinear problems of parametric vibrations of imperfect laminated plates, *Computers & Structures* 12, 181-191. *381*

Bishop, R. E. D. See Pendered, J. W., and Bishop, R. E. D.

Bleich, F. (1952). *Buckling Strength of Metal Structures*, McGraw-Hill, New York. *379*

Bogy, D. B. See Chen, J. S., and Bogy, D. B.

Boley, B. A., and Weiner, J. H. (1960). *Theory of Thermal Stresses*, Wiley, New York. *414, 545*

Bolotin, V. V. (1964). *The Dynamic Stability of Elastic Systems*, Translated by V. I. Weingarten et al., Holden-Day, San Francisco, California. *24, 177, 223*

Borri, M. (1986). Helicopter rotor dynamics by finite element time approximation, *Special Issue of Computers and Mathematics with Applications* 12A, 149-160. *71*

Borri, M., Ghiringhelli, G. L., Lanz, M., Mantegazza, P., and Merlini, T. (1985). Dynamic response of mechanical systems by a weak Hamiltonian formulation, *Computers & Structures* 20, 495-508. *71*

Borri, M., Mello, F., Iura, M., and Atluri, S. N. (1988). Primal and mixed forms of Hamilton's principle for constrained rigid and flexible dynamical systems: Numerical studies, in *Second Conference on Nonlinear Vibrations, Stability, and Dynamics of Structures and Mechanisms*, Blacksburg, Virginia, June 1-3. *71*

Borri, M., and Merlini, T. (1986). A large displacement formulation for anisotropic beam analysis, *Meccanica* 21, 30-37. *175, 176*

Borri, M. See Giavotto, V., Borri, M., Mantegazza, P., Ghiringhelli, G., Carmaschi, V., Maffioli, G. C., and Mussi, F.

Bouguerra, H. See Nayfeh, A. H., and Bouguerra, H.

Bowlns, J. A., Palazotto, A., N., and Whitney, J. M. (1987). Vibration of symmetrically laminated rectangular plates considering deformation and rotary inertia, *AIAA Journal* 25, 1500-1511. *377*

Brooks, G. W. See Houbolt, J. C., and Brooks, G. W.

Brunelle, E. J., and Robertson, S. R. (1976). Vibrations of an initially stressed thick plate, *Journal of Sound and Vibration* 45, 405-416. *373*

Brush, D. O., and Almroth, B. O. (1975). *Buckling of Bars, Plates, and Shells,* McGraw-Hill, New York. *379, 382, 563, 605*

Bryan, G. H. (1891). On the stability of a plane plate under thrusts in its own plane, with applications to the buckling of the sides of a ship, *Proceedings of the London Mathematical Society* 22, 54-67. *378*

Budiansky, B. (1968). Notes on nonlinear shell theory, *Journal of Applied Mechanics* 35, 392-401. *562*

Budiansky, B., and Radkowski, P. P. (1963). Numerical analysis of unsymmetrical bending of shells of revolution, *AIAA Journal* 1, 1833-1842. *560*

Budiansky, B., and Sanders, J. L. (1963). On the "best" first-order linear shell theory, in *Progress in Applied Mechanics,* The Prager Anniversary Volume, Macmillan, New York, pp. 129-140. *560, 600, 612*

Bulkeley, P. Z. See Krauter, A. L., and Bulkeley, P. Z.

Bulson, P. S. (1969). *The Stability of Flat Plates,* American Elsevier, New York. *379*

Burgreen, D. (1951). Free vibrations of a pin-ended column with constant distance between pin ends, *Journal of Applied Mechanics* 18, 135-139. *317*

Burns, D. W. See Zook, J. D., Burns, D. W., Guckel, H., Sniegowski, J. J., Engelstad, R. L., and Feng, Z.

Burton, T. D., and Kolowith, M. (1988). Nonlinear resonances and chaotic motion in a flexible parametrically excited beam, in *Second Conference on Nonlinear Vibrations, Stability, and Dynamics of Structures and Mechanisms,* Blacksburg, Virginia, June 1-3. *182*

Burton, W. S. See Noor, A. K., and Burton, W. S.; Noor, A. K., Burton, W. S., and Bert, C. W.; Noor, A. K., Burton, W. S., and Peters, J. M.

Busby, H. R. Jr., and Weingarten, V. I. (1972). Nonlinear response of a beam to periodic loading, *International Journal of Non-Linear Mechanics* 7, 289-303. *178*

Cai, X. See Osterberg, P., Yie, H., Cai, X., White, J., and Senturia, S.

Calladine, C. R. (1983). *Theory of Shell Structures,* Cambridge University Press, New York. *608*

Campbell, W. (1924). The protection of stream-turbine disk wheels from axial vibration, *Transactions of ASME* 46, 31-160. *513*

Canosa, J., and Penafiel, H. R. (1973). Parallel shooting solution of the neutron transport equation in spherical geometry, *Journal of Computational Physics* 13, 380-397. *52*

Carmaschi, V. See Giavotto, V., Borri, M., Mantegazza, P., Ghiringhelli, G., Carmaschi, V., Maffioli, G. C., and Mussi, F.

Carrera, E. (1999a). Multilayered shell theories accounting for layerwise mixed description. Part 1: Governing equations, *AIAA Journal* 37, 1107-1116. *559*

Carrera, E. (1999b). Multilayered shell theories accounting for layerwise mixed description. Part 2: Numerical evaluation, *AIAA Journal* 37, 1117-1124. *559*

Carrera, E. (2002). Theories and finite elements for multilayered, anisotropic, composite plates and shells, *Archives of Computational Methods in Engineering* 9, 87-140. *559*

Cartmell, M. P., and Roberts, J. W. (1987). Simultaneous combination resonances in a parametrically excited cantilever beam, *Strain* 23, 117-126. *278*

Castañer, L. M., and Senturia, S. D. (1999). Speed-energy optimization of electrostatic actuators based on pull-in, *Journal of Microelectromechanical Systems* 8, 290-298. *342*

Caughey, T. K. See Irvine, H. M., and Caughey, T. K.

Cesnik, C. E. S. See Hodges, D. H., Atilgan, A. R., Cesnik, C. E. S., and Fulton, M. V.

Chai, G. (1996). Free vibration of laminated composite plates with a central circular hole, *Journal of Composite Structures* 35, 357-368. *378*

Chamis, C. C. See Saravanos, D. A., and Chamis, C. C.; Saravanos, D. A., and Chamis, C. C.; Saravanos, D. A., and Chamis, C. C.

Chan, E. K., Garikipati, K., and Dutton, R. W. (1999). Characterization of contact electromechanics through capacitance-voltage measurements and simulations, *Journal of Microelectromechanical Systems* 8, 208-217. *341*

Chandra, R. (1976). Large deflection vibration of cross-ply laminated plates with certain edge conditions, *Journal of Sound and Vibration* 47, 509-514. *378, 380*

Chandra, R., and Raju, B. B. (1975a). Large deflection vibration of angle-ply laminated plates, *Journal of Sound and Vibration* 40, 393-408. *380*

Chandra, R., and Raju, B. B. (1975b). Large amplitude flexural vibration of cross-ply laminated composite plates, *Fiber Science and Technology* 8, 243-263. *378, 380*

Chandrashekhara, K., and Kolli, M. (1997). Free vibration of eccentrically stiffened laminated plates, *Journal of Reinforced Plastics and Composites* 16, 884-902. *378*

Chandrashekhara, K. See Bhimaraddi, A., and Chandrashekhara, K.; Reddy, J. N., and Chandrashekhara, K.; Tenneti, R., and Chandrashekhara, K.

Chang, S. I., Bajaj, A. K., and Krousgrill, C. M. (1993). Non-linear vibrations and chaos in harmonically excited rectangular plates with one-to-one internal resonance, *Nonlinear Dynamics* 4, 433-460. *499*

Chang, W. K., and Ibrahim, R. A. (1997). Multiple internal resonance in suspended cables under random in-plane loading, *Nonlinear Dynamics* 12, 275-303. *138*

Chang, W. K., Ibrahim, R. A., and Afaneh, A. A. (1994). Planar and non-planar nonlinear dynamics of suspended cables under random in-plane loading, in *Proceedings of the International Mechanical Engineering Congress and Exposition*, Chicago, Illinois, November 6-11, AMD-Vol. 192/DE-Vol. 78, pp. 201-214. *137*

Chang, W. K., Ibrahim R. A., and Afaneh, A. A. (1996). Planar and non-planar non-linear dynamics of suspended cables under random in-plane loading - I. Single internal resonance, *International Journal of Non-Linear Mechanics* 31, 837-859. *137*

Chang, W. K. See Ibrahim, R. A., and Chang, W. K.

Chang, W. P., and Wan, S. M. (1986). Thermomechanically coupled non-Linear vibration of plates, *International Journal of Non-Linear Mechanics* 21, 375-389. *544*

Chao, W. C. See Reddy, J. N., and Chao, W. C.; Reddy, J. N., and Chao, W. C.

Chen, F. Y. See Young, T. H., and Chen, F. Y.

Chen, J. C., and Babcock, C. D. (1975). Nonlinear vibrations of cylindrical shells, *AIAA Journal* 13, 868-876. *637*

Chen, J. S. (1994). Stability analysis of a spinning elastic disk under a stationary concentrated edge load, *Journal of Applied Mechanics* 61, 788-792. *517, 565*

Chen, J. S. (1997). Parametric resonance of a spinning disk under space-fixed pulsating edge loads, *Journal of Applied Mechanics* 64, 139-143. *517*

Chen, J. S. (2000). Nonlinear transient analysis of rectangular composite laminated plates, *Computers & Structures* 49, 129-139. *381*

Chen, J. S., and Bogy, D. B. (1992). Effects of load parameters on the natural frequencies and stability of a flexible spinning disk with a stationary load system, *Journal of Applied Mechanics* 59, S230-S235. *517*

Chen, J. S., and Wong, C. C. (1998). Vibration and stability of a spinning disk in contact with evenly-spaced stationary load systems, *Journal of Vibration and Acoustics* 120, 301-302. *513*

Chen, S. See Yao, Z. J., Chen, S., Eshelman, S., Denniston, D., and Goldsmith, C.

Chen, T. J. See Doong, J. L., and Chen, T. J.

Chen, T. L. C. See Bert, C. W., and Chen, T. L. C.

Chen, Y. See Barash, S., and Chen, Y.

Cheung, Y. K. See Jones, R., Mazumdar, J., and Cheung, Y. K.

Chia, C. Y. (1980). *Nonlinear Analysis of Plates,* McGraw-Hill, New York. *382, 469, 498, 527, 528*

Chia, C. Y. (1982). Large amplitude vibrations of laminated rectangular plates, *Fibre Science and Technology* 17, 123-131. *380*

Chia, C. Y. (1985). Nonlinear oscillation of unsymmetric angle-ply plate on elastic foundation having nonuniform edge supports, *Journal of Composite Structures* 4, 161-178. *380*

Chia, C. Y., and Prabhakara, M. K. (1976). Large deflection of unsymmetric crossply and angle-ply plates, *Journal of Mechanics and Engineering Science* 18, 179-183. *380*

Chia, C. Y., and Prabhakara, M. K. (1978). A general mode approach to nonlinear flexural vibration of laminated rectangular plates, *Journal of Applied Mechanics* 45, 623-628. *380, 381*

Chia, C. Y. See Sathyamoorthy, M., and Chia, C. Y.; Sivakumaran, K. S., and Chia, C. Y.; Sivakumaran, K. S., and Chia, C. Y.

Chin, C., and Nayfeh, A. H. (1996a). Nonlinear dynamics of crane operation at sea, in *Proceedings of the 37th AIAA/ASME/ASCE/AHS/ASC Structures, Structural Dynamics, and Materials Conference*, AIAA Paper No. 96-1485, Salt Lake City, Utah, April 15-17. *53*

Chin, C., Nayfeh, A. H., and Lacarbonara, W. (1997). Two-to-one internal resonances in parametrically excited buckled beams, in *Proceedings of the 38th AIAA/ASME/ASCE/AHS/ASC Structures, Structural Dynamics, and Materials Conference*, AIAA Paper No. 97-1081, Kissimmee, Florida, April 7-10. *323*

Chin, C-M., and Nayfeh, A. H. (1996b). Bifurcation and chaos in externally excited circular cylindrical shells, *Journal of Applied Mechanics* 63, 565-574. *53, 564, 565, 638*

Chin, C-M., and Nayfeh, A. H. (1997). Three-to-one internal resonances in hinged-clamped beams, *Nonlinear Dynamics* 12, 129-154. *181*

Chin, C-M., and Nayfeh, A. H. (1999). Three-to-one internal resonances in parametrically excited hinged-clamped beams, *Nonlinear Dynamics* 20, 131-158. *181*

Chin, C-M., and Nayfeh, A. H. (2001). A second-order approximation of multimodal interactions in externally excited circular cylindrical shells, *Nonlinear Dynamics* 26, 45-66. *565, 638*

Chin, C-M. See Arafat, H. N., Nayfeh, A. H., and Chin, C-M.; Nayfeh, A. H., Arafat, H. N., Chin, C-M., and Lacarbonara, W.; Nayfeh, A. H., and Chin, C-M.; Nayfeh, A. H., and Chin, C-M.; Nayfeh, A. H., and Chin, C-M.; Nayfeh, A. H., Lacarbonara, W., and Chin, C-M.; Rega, G., Lacarbonara, W., Nayfeh, A. H., and Chin, C-M.; Rega, G., Lacarbonara, W., Nayfeh, A. H., and Chin, C-M.

Chow, A. C. See Beatty, M. F., and Chow, A. C.

Chow, S. N., and Hale, J. K. (1982). *Methods of Bifurcation Theory*, Springer-Verlag, New York. *24*

Chree, C. (1889). The equations of an isotropic elastic solid in polar and cylindrical coordinates, their solution and application, *Transactions of the Cambridge Philosophical Society* 14, 250. *172*

Christensen, R. M. (1979). *Mechanics of Composite Materials*, Wiley-Interscience, New York. *7*

Christensen, R. M. See Lo, K. H., Christensen, R. M., and Wu, E. M.; Lo, K. H., Christensen, R. M., and Wu, E. M.

Chu, H-N., and Herrmann, G. (1956). Influence of large amplitudes on free flexural vibrations of rectangular elastic plates, *Journal of Applied Mechanics* 23, 532-540. *498*

Chua, L. O., and Lin, P. M. (1975). *Computer-Aided Analysis of Electronic Circuits*, Prentice-Hall, New Jersey. *25*

Chung, J. H. See Lee, J. M., Chung, J. H., and Chung, T. Y.

Chung, T. Y. See Lee, J. M., Chung, J. H., and Chung, T. Y.

Ciliberto, S., and Gollub, J. P. (1985). *Chaotic mode competition in parametrically forced surface waves, Journal of Fluid Mechanics* 158, 381-398. *26*

Clary, R. R. (1972). Vibration characteristics of unidirectional filamentary composite material panels, in *Composite Materials: Testing and Design,* 2nd Conference American Society for Testing and Materials, Special Technical Publication, Vol. 497, pp. 415-438. *8*

Cole, K. A., and Benson, R. C. (1988). A fast eigenfunction approach for computing spinning disk deflections, *Journal of Applied Mechanics* 55, 453-457. *517, 519*

Conte, S. D. (1966). The numerical solution of linear boundary value problems, *Society for Industrial and Applied Mathematics Review* 8, 309-321. *52*

Corless, M. See Georgiou, I. T., Bajaj, A. K., and Corless, M.

Corless, M. J. See Georgiou, I. T., Corless, M. J., and Bajaj, A. K.

Cowper, G. R. (1966). The shear coefficient in Timoshenko's beam theory, *Journal of Applied Mechanics* 33, 335-340. *173*

Cowper, G. R. (1968). On the accuracy of Timoshenko's beam theory, *Journal of the Engineering Mechanics Division* 94, 1447-1453. *172*

Cox, H. L. (1963). *The Buckling of Plates and Shells,* Macmillan, New York. *379*

Crary, S., and Zhang, Y. (1990). CAEMENS: An integrated computer-aided engineering workbench for micro-electro-mechanical systems, in *Proceedings of the IEEE Workshop on Microelectromechanical Systems, MEMS'90*, Napa Valley, California, pp. 113-114. *341*

Crary, S. B. See Seeger, J. I., and Crary, S. B.

Crawley, E. F. See Lazarus, K. B., and Crawley, E. F.

Crespo da Silva, M. R. M. (1988). Nonlinear flexural-flexural-torsional-extensional dynamics of beams–I. Formulation, *International Journal of Solids and Structures* 24, 1225-1234. *177*

Crespo da Silva, M. R. M., and Glynn, C. C. (1978a). Nonlinear flexural-flexural-torsional dynamics of inextensional beams–I. Equations of motion, *Journal of Structural Mechanics* 6, 437-448. *177, 182, 223, 243-245, 376*

Crespo da Silva, M. R. M., and Glynn, C. C. (1978b). Nonlinear flexural-flexural-torsional dynamics of inextensional beams–II. Forced motions, *Journal of Structural Mechanics* 6, 449-461. *177, 223, 307*

Crespo da Silva, M. R. M., and Glynn, C. C. (1979a). Out-of-plane vibrations of a beam including nonlinear inertia and nonlinear curvature effects, *International Journal of Non-Linear Mechanics* 13, 261-270. *181*

Crespo da Silva, M. R. M., and Glynn, C. C. (1979b). Nonlinear nonplanar resonant oscillations in fixed-free beams with support asymmetry, *International Journal of Solids and Structures* 15, 209-219. *182*

Crespo da Silva, M. R. M., and Hodges, D. H. (1986a). Nonlinear flexure and torsion of rotating beams, with application to helicopter rotor blades–I. Formulation, *Vertica* 10, 151-169. *178*

Crespo da Silva, M. R. M., and Hodges, D. H. (1986b). Nonlinear flexure and torsion of rotating beams, with application to helicopter rotor blades–II. Response and stability results, *Vertica* 10, 171-186. *178*

Crespo da Silva, M. R. M., and Zaretzky, C. L. (1988). Nonlinear modal coupling in the response of inextensional beams, in *Second Conference on Nonlinear Vibrations, Stability, and Dynamics of Structures and Mechanisms*, Blacksburg, Virginia, June 1-3. *182*

Crespo da Silva, M. R. M., and Zaretzky, C. L. (1991). Effects of approximations on the static and dynamic responses of a cantilever with a tip mass, *International Journal of Solids and Structures* 27, 565-583. *179*

Crespo da Silva, M. R. M. See Hodges, D. H., Crespo da Silva, M. R. M., and Peters, D. A.; Zaretzky, C. L., and Crespo da Silva, M. R. M.

Croquette, V., and Pointou, C. (1981). Cascade of period doubling bifurcations and large stochasticity in the motion of a compass, *Journal of Physique-Letters* 42, L537-L539. *23*

Cummings, B. E. (1964). Large-amplitude vibration and response of curved panels, *AIAA Journal* 2, 709-716. *527*

Cunha, J. (1997). Free vibration of composite laminated plates by using finite element technique with a higher-order theory, *Science and Engineering Journal* 6, 73-81. *378*

Cunniff, P. F. See Abarcar, R. B., and Cunniff, P. F.

Daniel, J. W. See Noble, B., and Daniel, J. W.

Danielson, D. A. See Simmonds, J. G., and Danielson, D. A.

Das, Y. C. See Singh, P. N., Sundararajan, V., and Das, Y. C.

Davey, A. (1973). A simple numerical method for solving Orr-Sommerfeld problems, *Quarterly Journal of Mechanics and Applied Mathematics* 26, 401-411. *52*

Dawe, D. J. See Roufaeil, O. L., and Dawe, D. J.

De Almeida, S. F., and Hansen, J. S. (1997). Free vibration analysis of composite plates with tailored thermal residual stresses, in *Proceedings of the ASME International Mechanical Engineering Congress and Exposition* AD Vol. 55, pp. 183-190. *378*

Dennis, S. T., and Palazotto, A. N. (1989). Transverse shear deformation in orthotropic cylindrical pressure vessels using a higher order shear theory, *AIAA Journal* 27, 1441-1447. *561*

Dennis, S. T., and Palazotto, A. N. (1990). Large displacement and rotational formulation for laminated shells including parabolic transverse shear, *International Journal of Non-Linear Mechanics* 25, 67-85. *561*

Denniston, D. See Yao, Z. J., Chen, S., Eshelman, S., Denniston, D., and Goldsmith, C.

deRooij, N. F. See Grétillat, M.-A., Yang, Y.-J., Hung, E. S., Rabinovich, V., Ananthasuresh, G. K., deRooij, N. F., and Senturia, S. D.

Di Sciuva, M. (1986). Bending, vibration, and buckling of simply supported thick multilayered orthotropic plates: An evaluation of a new displacement model, *Journal of Sound and Vibration* 105, 425-442. *377*

Di Sciuva, M. (1987). An improved shear-deformation theory for moderately thick multilayered anisotropic shells and plates, *Journal of Applied Mechanics* 54, 589-596. *374, 561*

Dodson, R. O. See Eversman, W., and Dodson, R. O.

Donnell, L. H. (1933). Stability of thin walled tubes under torsion, NACA Report No. 479. *562*

Donnell, L. H. (1938). A discussion of thin shell theory, in *Proceedings of the 5th International Congress of Applied Mechanics,* Wiley-Interscience, New York, pp. 66-70. *560, 562, 614*

Doong, J. L., and Chen, T. J. (1987). Vibration and stability of an initially stressed laminated plate based on a higher-order deformation theory, *Composite Structures* 7, 285-310. *377*

Dowell, E. H. (1975). *Aeroelasticity of Plates and Shells,* Noordhoff, Leyden, The Netherlands. *31*

Dowell, E. H. (1989). *A Modern Course in Aeroelasticity,* Kluwer, Dordrecht, The Netherlands. *31*

Dowell, E. H., and Pezeshki, C. (1986). On the understanding of chaos in Duffing's equation including a comparison with experiment, *Journal of Applied Mechanics* 53, 5-9. *24*

Dowell, E. H., Traybar, J., and Hodges, D. H. (1977). An experimental-theoretical correlation study of nonlinear bending and torsion deformations of a cantilever beam, *Journal of Sound and Vibration* 50, 533-544. *177, 178, 376*

Dowell, E. H. See Hodges, D. H., and Dowell, E. H.; Smelova-Reynolds, T., and Dowell, E. H.; Smelova-Reynolds, T., and Dowell, E. H.; Tang, D. M., and Dowell, E. H.

Dufour, I. See Francais, O., and Dufour, I.; Francais, O., and Dufour, I.

Dugundji, J., and Mukhopadhyay, V. (1973). Lateral bending-torsion vibrations of a thin beam under parametric excitation, *Journal of Applied Mechanics* 40, 693-698. *278*

Dugundji, J. See Minguet, P., and Dugundji, J.; Minguet, P., and Dugundji, J.; Tseng, W. Y., and Dugundji, J.

Dunbar, M. L., and Sager, K. (2000). A novel, media-compatible pressure sensor for automotive applications, *Sensors* 17, 28-36. *544*

Durvasula, S. See Kamal, K., and Durvasula, S.

Dutton, R. W. See Chan, E. K., Garikipati, K., and Dutton, R. W.

Dym, C. L. (1973). Some new results for the vibrations of circular cylinders, *Journal of Sound and Vibration* 29, 189-205. *561*

Dym, C. L. See Shames, I. H., and Dym, C. L.

Ebcioglu, I. K. (1970). Nonlinear theory of shells, *International Journal of Non-Linear Mechanics* 6, 469-478. *563, 605*

Efstathiades, G. J. (1971). A new approach to the large-deflection vibrations of imperfect circular disks using Galerkin's procedure, *Journal of Sound and Vibration* 16, 231-253. *498, 499*

Eisley, J. G. (1964a). Large amplitude vibration of buckled beams and rectangular plates, *AIAA Journal* 2, 2207-2209. *317, 379*

Eisley, J. G. (1964b). Nonlinear vibration of beams and rectangular plates, *Zeitschrift fur Angewandte Mathematik und Physik (ZAMP)* 15, 167-175. *317*

Eisley, J. G., and Bennett, J. A. (1970). Stability of large amplitude forced motion of a simply supported beam, *International Journal of Non-Linear Mechanics* 5, 645-657. *318*

Eisley, J. G. See Ho, C.-H., Scott, R. A., and Eisley, J. G.; Ho, C.-H., Scott, R. A., and Eisley, J. G.; Min, G. B., and Eisley, J. G.

Eller, A. I. (1972). Driven non-linear oscillations of a string, *Journal of the Acoustical Society of America* 51, 960-966. *112*

Elwespoek, M. See Tilmans, H. A., Elwespoek, M., and Fluitman, J. H.

Emaci, E. See Georgiou, I. T., Schwartz, I. B., Emaci, E., and Vakakis, A.

Emam, S. (2002). *A Theoretical and Experimental Study of Nonlinear Dynamics of Buckled Beams*, PhD Dissertation, Virginia Polytechnic Institute and State University, Blacksburg, Virginia. *319, 331*

Emam, S. A., and Nayfeh, A. H. (2002). Nonlinear dynamics of a clamped-clamped buckled beam, in *Proceedings of the 43rd AIAA/ASME/ASCE/AHS/ASC Structures, Structural Dynamics, and Materials Conference*, AIAA Paper No. 2002-1751, Denver, Colorado, April 22-25. *319, 336, 338-340*

Emam, S. A., and Nayfeh, A. H. (2004a). On the nonlinear dynamics of a buckled beam subjected to a primary-resonance excitation, *Nonlinear Dynamics* 35, 1-17. *319, 331, 336, 338, 339*

Emam, S. A., and Nayfeh, A. H. (2004b). Nonlinear responses of buckled beams to subharmonic-resonance excitations, *Nonlinear Dynamics* 35, 105-122. *319*

Engelstad, R. L. See Zook, J. D., Burns, D. W., Guckel, H., Sniegowski, J. J., Engelstad, R. L., and Feng, Z.

Eshelman, S. See Yao, Z. J., Chen, S., Eshelman, S., Denniston, D., and Goldsmith, C.

Eslami, H., and Kandil, O. A. (1989a). Nonlinear forced vibration of orthotropic rectangular plates using the method of multiple scales, *AIAA Journal* 27, 955-960. *381*

Eslami, H., and Kandil, O. A. (1989b). Two-mode nonlinear vibration of orthotropic plates using method of multiple scales, *AIAA Journal* 27, 961-967. *381*

Evan-Iwanowski, R. M. (1976). *Resonance Oscillations in Mechanical Systems*, Elsevier, New York. *24*

Evan-Iwanowski, R. M., Sanford, W. F., and Kehagioglou, T. (1970). Nonstationary parametric response of a nonlinear column, *Developments in Theoretical and Applied Mechanics* 5, 715-743. *178, 223*

Evan-Iwanowski, R. M. See Evensen, D. A., and Evan-Iwanowski, R. M.; Ostiguy, G. L., and Evan-Iwanowski, R. M.

Evensen, D. A. (1966). Nonlinear flexural vibrations of thin circular rings, *Journal of Applied Mechanics* 33, 553-560. *565, 637*

Evensen, D. A., and Evan-Iwanowski, R. M. (1966). Effects of longitudinal inertia upon the parametric response of elastic columns, *Journal of Applied Mechanics* 33, 141-148. *223*

Eversman, W., and Dodson, R. O. (1969). Free vibration of a centrally clamped spinning circular disk, *AIAA Journal* 7, 2010-2012. *517*

Ewins, D. J. (1984). *Modal Testing: Theory and Practice*, Wiley, New York. *8*

Falkenberg, J. C. (1968). A method for integration of unstable systems of ordinary differential equations subject to two-point boundary conditions, *BIT* 8, 86-103. *52*

Fan, J., and Ye, J. (1990). An exact solution for the statics and dynamics of laminated thick plates with orthotropic layers, *International Journal of Solids and Structures* 26, 655-662. *378*

Fan, T. C. See Bieniek, M. P., Fan, T. C., and Lackman, L. M.

Faris, W. See Arafat, H. N., Nayfeh, A. H., and Faris, W.; Nayfeh, A. H., and Faris, W.; Nayfeh, A. H., and Faris, W.

Faris, W. F., Abdel-Rahman, E. M., and Nayfeh, A. H. (2002). Mechanical behavior of an electrostatically actuated micropump, in *Proceedings of the 43rd AIAA/ASME/ASCE/AHS/ASC Structures, Structural Dynamics, and Materials Conference*, AIAA Paper No. 2002-1303, Denver, Colorado, April 22-25. *531-535*

Farshad, M. See Banan, M. R., Karami, G., and Farshad, M.

Febvre, M. See Malhaire, C., Le Berre, M., Febvre, M., Barbier, D., and Pinard, P.

Fedorov, V. A. (1976). Thermostability of elastically clamped variable-stiffness annular plates, *Soviet Aeronautics* 19, 105-109. *544*

Feigenbaum, M. J. (1979). The universal metric properties of nonlinear transformations, *Journal of Statistical Physics* 21, 669-706. *25*

Feng, Z. See Zook, J. D., Burns, D. W., Guckel, H., Sniegowski, J. J., Engelstad, R. L., and Feng, Z.

Feng, Z. C., and Leal, L. G. (1995). Symmetries of the amplitude equations of an inextensional beam with internal resonance, *Journal of Applied Mechanics* 62, 235-238. *138, 284, 285*

Feng, Z. C., and Sethna, P. R. (1989). Symmetry-breaking bifurcations in resonant surface waves, *Journal of Fluid Mechanics* 199, 495-518. *26*

Flood, R. J. L. See Adams, R. D., Fox, M. A. O., Flood, R. J. L., Friend, R. J., and Hewitt, R. L.

Flügge, W. (1973). *Stresses in Shells,* Springer-Verlag, Berlin. *559, 560, 562, 614*

Fluitman, J. H. See Tilmans, H. A., Elwespoek, M., and Fluitman, J. H.

Forsberg, K. (1964). Influence of boundary conditions on the modal characteristics of thin cylindrical shells, *AIAA Journal* 2, 2150-2167.

Fortier, R. C., and Rossettos, J. N. (1973). On the vibration of shear deformable curved anisotropic composite plates, *Journal of Applied Mechanics* 40, 299-301. *378*

Fournier, T. See Ayela, F., and Fournier, T.

Fox, L., and Parker, I. B. (1968). *Chebyshev Polynomials in Numerical Analysis*, Oxford University Press, New York. *47*

Fox, M. A. O. See Adams, R. D., Fox, M. A. O., Flood, R. J. L., Friend, R. J., and Hewitt, R. L.

Francais, O., and Dufour, I. (1999). Normalized abacus for the global behavior of diaphragms: Pneumatic, electrostatic, piezoelectric or electromagnetic actuation, *Journal of Modeling and Simulation of Microsystems* 1, 149-160. *531, 534*

Francais, O., and Dufour, I. (2000). Enhancement of elementary displaced volume with electrostatically actuated diaphragms: Application to electrostatic micropumps, *Journal of Micromechanical Microengineering* 10, 282-286. *531, 534*

Friedmann, P. See Rosen, A., and Friedmann, P.

Friedmann, P. P. (1977). Recent developments in rotary-wing aeroelasticity, *Journal of Aircraft* 14, 1027-1041. *178*

Friend, R. J. See Adams, R. D., Fox, M. A. O., Flood, R. J. L., Friend, R. J., and Hewitt, R. L.

Frosali, G. See Amabili, M., Frosali, G., and Kwak, M. K.

Fujino, Y., Warnitchai, P., and Pacheco, B. M. (1993). An experimental and analytical study of autoparametric resonance in a 3dof model of cable-stayed-beam, *Nonlinear Dynamics* 4, 111-138. *136, 376*

Fulton, M. V. See Hodges, D. H., Atilgan, A. R., Cesnik, C. E. S., and Fulton, M. V.

Fung, Y. C. (1969). *The Theory of Aeroelasticity*, Dover, New York. *6, 31*

Funk, J. See Korvink, J. G., Funk, J., Roos, M., and Wachutka, G.

Gabbay, L. D., Mehner, J. E., and Senturia, S. D. (2000). Computer-aided generation of nonlinear reduced-order dynamic macromodels - I: Non-stress-stiffened case, *Journal of Microelectromechanical Systems* 9, 262-269. *342*

Gabbay, L. D. See Mehner, J. E., Gabbay, L. D., and Senturia, S. D.

Gadre, M. See Tzou, H. S., and Gadre, M.

Ganapathi, M., Varadan, T. K., and Sarma, B. S. (1991). Nonlinear flexural vibrations of laminated orthotropic plates, *Computers & Structures* 39, 685-688. *380*

Ganesan, R. See Ramu, S. A., Sankar, T. S., and Ganesan, R.

Garikipati, K. See Chan, E. K., Garikipati, K., and Dutton, R. W.

Gattulli, V., Pasca, M., and Vestroni, F. (1997). Nonlinear oscillations of a nonresonant cable under in-plane excitation with a longitudinal control, *Nonlinear Dynamics* 14, 139-156. *137*

Georgiou, I. T., Bajaj, A. K., and Corless, M. (1998). Slow and fast invariant manifolds, and normal modes in a two-degree-of-freedom structural dynamical system with multiple equilibrium states, *International Journal of Non-Linear Mechanics* 33, 275-300. *529*

Georgiou, I. T., Corless, M. J., and Bajaj, A. K. (1999). Dynamics of nonlinear structures with multiple equilibria: A singular perturbation-invariant manifold approach, *Zeitschrift fur Angewandte Mathematik und Physik (ZAMP)*, 50, 892-924. *529*

Georgiou, I. T., and Schwartz, I. B. (1996). The slow invariant manifold of a conservative pendulum-oscillator system, *International Journal of Bifurcation and Chaos* 6, 673-692. *529*

Georgiou, I. T., Schwartz, I. B., Emaci, E., and Vakakis, A. (1999). Interaction between slow and fast oscillations in an infinite-degree-of-freedom linear system coupled to a nonlinear subsystem: Theory and experiment, *Journal of Applied Mechanics* 66, 448-459. *529*

Gere, J. M. See Timoshenko, S. P., and Gere, J. M.

Gersting, J. M., and Jankowski, D. F. (1972). Numerical methods for Orr-Sommerfeld problems, *International Journal for Numerical Methods in Engineering* 4, 195-206. *52*

Ghiringhelli, G. See Giavotto, V., Borri, M., Mantegazza, P., Ghiringhelli, G., Carmaschi, V., Maffioli, G. C., and Mussi, F.

Ghiringhelli, G. L. See Borri, M., Ghiringhelli, G. L., Lanz, M., Mantegazza, P., and Merlini, T.

Ghosh, A. See Hatwal, H., Mallik, A. K., and Ghosh, A.; Hatwal, H., Mallik, A. K., and Ghosh, A.

Giavotto, V., Borri, M., Mantegazza, P., Ghiringhelli, G., Carmaschi, V., Maffioli, G. C., and Mussi, F. (1983). Anisotropic beam theory and applications, *Computers & Structures* 16, 403-413. *175, 176, 251*

Glynn, C. C. See Crespo da Silva, M. R. M., and Glynn, C. C.; Crespo da Silva, M. R. M., and Glynn, C. C.; Crespo da Silva, M. R. M., and Glynn, C. C.; Crespo da Silva, M. R. M., and Glynn, C. C.

Godunov, S. K. (1961). On the numerical solution of boundary-value problems for systems of linear ordinary differential equations, *Uspekhi Matematicheskikh Nauk* 16, 171-174. *52*

Gol'denveizer, A. L. (1961). *Theory of Elastic Thin Shells,* Pergamon Press, New York. *559*

Goldsmith, C. See Yao, Z. J., Chen, S., Eshelman, S., Denniston, D., and Goldsmith, C.

Gollub, J. P. See Simonelli, F., and Gollub, J. P.

Goodier, J. N., and McIvor, I. K. (1964). The elastic cylindrical shell under nearly uniform radial impulse, *Journal of Applied Mechanics* 31, 259-266. *564*

Goodier, J. N. See Timoshenko, S. P., and Goodier, J. N.

Gottlieb, H. P. W. (1990). Nonlinear vibration of a constant-tension string, *Journal of Sound and Vibration* 143, 455-460. *112*

Green, A. E., and Naghdi, P. M. (1981). A theory of laminated composite plates and rods, Report UCB/AM-81-3, University of California, Berkeley, California. *373*

Green, G. (1837). On the laws of reflexion and refraction of light at the common surface of two non-crystallized media, *Transactions of the Cambridge Philosophical Society* 7, 245-269. *86*

Green, G. (1839). On the propagation of light in crystallized media, *Transactions of the Cambridge Philosophical Society* 7, 293-311. *86*

Greer, J. M. Jr. See Pai, P. F., Palazotto, A. N., and Greer, J. M. Jr.

Grétillat, M.-A., Yang, Y.-J., Hung, E. S., Rabinovich, V., Ananthasuresh, G. K., deRooij, N. F., and Senturia, S. D. (1997). Nonlinear electromechanical behavior of an electrostatic microrelay, in *TRANSDUCERS'97: International Conference on Solid-State Sensors and Actuators,* Chicago, Illinois, pp. 1141-1144. *342*

Griffin, S. See Birman, V., Griffin, S., and Knowle, G.

Griffiths, D. J. (1981). *Introduction to Electrodynamics,* Prentice-Hall, Englewood Cliffs, New Jersey. *343, 532*

Guckel, H. See Ahn, Y., Guckel, H., and Zook, J. D.; Zook, J. D., Burns, D. W., Guckel, H., Sniegowski, J. J., Engelstad, R. L., and Feng, Z.

Guckenheimer, J., and Holmes, P. (1983). *Nonlinear Oscillations, Dynamical Systems and Bifurcations of Vector Fields,* Springer-Verlag, New York. *24*

Gurgoze, M. (1986). Parametric vibrations of a restrained beam with an end mass under displacement excitation, *Journal of Sound and Vibration* 108, 73-84. *24*

Habip, L. M. (1965). Theory of elastic shells in the reference state, *Ingenieur Archiv* 34, 228-237. *563, 605*

Haddow, A. G., Barr, A. D. S., and Mook, D. T. (1984). Theoretical and experimental study of modal interaction in a two-degree-of-freedom structure, *Journal of Sound and Vibration* 97, 451-473. *25*

Haddow, A. G., and Hasan, S. M. (1988). Nonlinear oscillation of a flexible cantilever: Experimental results, in *Second Conference on Nonlinear Vibrations, Stability, and Dynamics of Structures and Mechanisms*, Blacksburg, Virginia, June 1-3. *182*

Hadian, J. (1991). *Modal Interactions in the Dynamic Response of Isotropic and Composite Plates*, PhD Dissertation, Virginia Polytechnic Institute and State University, Blacksburg, VA. *469*

Hadian, J., and Nayfeh, A. H. (1990). Modal interaction in circular plates, *Journal of Sound and Vibration* 142, 279-292. *52, 499, 504*

Hadian, J., and Nayfeh, A. H. (1993). Free vibration and buckling of shear-deformable cross-ply laminated plates using the state-space concept, *Computers & Structures* 48, 677-693. *379*

Hadian, M. J. See Noor, A. K., Hadian, M. J., and Peters, J. M.

Hagedorn, P., and Schäfer, B. (1980). On non-linear free vibrations of an elastic cable, *International Journal of Non-Linear Mechanics* 15, 333-340. *137*

Hahn, H. T. See Tsai, S. W., and Hahn, H. T.

Haight, E. C., and King, W. W. (1971). Stability of parametrically excited vibrations of an elastic rod, *Developments in Theoretical and Applied Mechanics* 5, 677-714. *180, 223*

Haight, E. C., and King, W. W. (1972). Stability of nonlinear oscillations of an elastic rod, *Journal of the Acoustical Society of America* 52, 899-911. *180, 181*

Haisler, W. E. See Stricklin, J. A., Martinez, J. E., Tillerson, J. R., Hong, J. H., and Haisler, W. E.

Hale, J. K. (1963). *Oscillations in Nonlinear Systems,* McGraw-Hill, New York. *24*

Hale, J. K. See Chow, S. N., and Hale, J. K.

Hall, E. K. See Abhyankar, N. S., Hall, E. K., and Hanagud, S. V.

Hamilton, W. R. (1834). On a general method in dynamics, *Philosophical Transactions of the Royal Society of London* 125, 247-308. *71*

Hanagud, S. V. See Abhyankar, N. S., Hall, E. K., and Hanagud, S. V.; Lestari, W., and Hanagud, S.

Hansen, C. H. See Ji, J.-C., and Hansen, C. H.

Hansen, J. S. See De Almeida, S. F., and Hansen, J. S.

HaQuang, N. (1986). *The Response of Multidegree-of-Freedom Systems with Quadratic and Cubic Nonlinearities Subjected to Parametric and External Excitation*, Ph.D. Dissertation, Virginia Polytechnic Institute and State University, Blacksburg, Virginia. *24*

Harichandran, R. S., and Naja, M. K. (1997). Random vibration of laminated composite plates with material nonlinearity, *International Journal of Non-Linear Mechanics* 32, 707-720. *381*

Harris, R. See Yie, H., Harris, R., Napadenski, G., and Maseeh, F.

Harris, R. M. See Senturia, S. D., Harris, R. M., Johnson, B. P., Kim, S., Nabors, K., Shulman, M. A., and White, J. K.

Harrison, H. (1948). Plane and circular motion of a string, *Journal of the Acoustical Society of America* 20, 874-875. *111, 112, 121, 132*

Hartley, S. J. See Noor, A. K., and Hartley, S. J.

Hasan, S. M. See Haddow, A. G., and Hasan, S. M.

Hassard, B. D., Kazarinoff, N. D., and Wan, Y. H. (1981). *Theory and Application of Hopf Bifurcation*, Cambridge University Press, Cambridge, Massachusetts. *24*

Hatfield, S. J. See Pagano, N. J., and Hatfield, S. J.

Hatwal, H., Mallik, A. K., and Ghosh, A. (1983a). Forced nonlinear oscillations of an autoparametric system-part I: Periodic responses, *Journal of Applied Mechanics* 50, 657-662. *25*

Hatwal, H., Mallik, A. K., and Ghosh, A. (1983b). Forced nonlinear oscillations of an autoparametric system-part II: Chaotic responses, *Journal of Applied Mechanics* 50, 663-668. *25*

Hayashi, K. See Majima, O., and Hayashi, K.

Hencky, H. (1947). Uber die berucksichtigung der schubverzerrung in ebenen platten, *Ingenieur Archiv* 16, 72-76. *372*

Herrmann, G. See Chu, H-N., and Herrmann, G.; Mirsky, I., and Herrmann, G.

Hetnarski, R. B. (1987). *Thermal Stresses II*, North-Holland, Amsterdam. *414, 545-547*

Hewitt, R. L. See Adams, R. D., Fox, M. A. O., Flood, R. J. L., Friend, R. J., and Hewitt, R. L.

Heyliger, P. R., and Reddy, J. N. (1988). A higher order beam finite element for bending and vibration problems, *Journal of Sound and Vibration* 126, 309-326. *174, 177*

Hildebrand, F. B., Reissner, E., and Thomas, G. B. (1949). Notes on the foundations of the theory of small displacements of orthotropic shells, NASA Technical Note No. 1833. *372*

Hinnant, H. E., and Hodges, D. H. (1988). Nonlinear analysis of a cantilever beam, *AIAA Journal* 26, 1521-1527. *178*

Hitzl, D. L. (1980). Implementing Hamilton's law of varying action with shifted Legendre polynominals, *Journal of Computational Physics* 38, 185-211. *71*

Ho, C.-H., Scott, R. A., and Eisley, J. G. (1975). Nonplanar, nonlinear oscillations of a beam–I. Forced motions, *International Journal of Non-Linear Mechanics* 10, 113-127. *177, 180*

Ho, C.-H., Scott, R. A., and Eisley, J. G. (1976). Nonplanar, nonlinear oscillations of a beam–II. Free motions, *Journal of Sound and Vibration* 47, 333-339. *181*

Hodges, D. H. (1976). Nonlinear equations of motion for cantilever rotor blades in hover with pitch link flexibility, twist, precone, droop, sweep, torque offset, and blade root offset, NASA TM X-73,112. *177, 263*

Hodges, D. H. (1990). Review of composite rotor blade modeling, *AIAA Journal Technical Note* 28, 561-565. *175, 176*

Hodges, D. H., Atilgan, A. R., Cesnik, C. E. S., and Fulton, M. V. (1992). On a simplified strain energy function for geometrically nonlinear behavior of anisotropic beams, *Composites Engineering* 2, 513-526. *176*

Hodges, D. H., Crespo da Silva, M. R. M., and Peters, D. A. (1988). Nonlinear effects in the static and dynamic behavior of beams and rotor blades, *Vertica* 12, 243-256. *177, 178, 376*

Hodges, D. H., and Dowell, E. H. (1974). Nonlinear equations of motion for the elastic bending and torsion of twisted nonuniform rotor blades, NASA TN D-7818. *177, 178, 376*

Hodges, D. H., and Peters, D. A. (1975). On the lateral buckling of uniform slender cantilever beams, *International Journal of Solids and Structures* 11, 1269-1280. *178*

Hodges, D. H. See Atilgan, A. R., and Hodges, D. H.; Crespo da Silva, M. R. M., and Hodges, D. H.; Crespo da Silva, M. R. M., and Hodges, D. H.; Dowell, E. H., Traybar, J., and Hodges, D. H.; Hinnant, H. E., and Hodges, D. H.

Holmes, P. See Guckenheimer, J., and Holmes, P.

Holmes, P. J. (1979). A nonlinear oscillator with a strange attractor, *Philosophical Transactions of the Royal Society of London A* 292, 419-448. *318*

Holmes, P. J., and Moon, F. C. (1983). Strange attractors and chaos in nonlinear mechanics, *Journal of Applied Mechanics* 50, 1021-1032. *23, 318*

Holmes, P. J. See O'Reilly, O., and Holmes, P. J.

Hong, C. H. See Bauchau, O. A., and Hong, C. H.; Bauchau, O. A., and Hong, C. H.

Hong, J. H. See Stricklin, J. A., Martinez, J. E., Tillerson, J. R., Hong, J. H., and Haisler, W. E.

Horita, K. See Sueoka, A., Yoshitake, Y., Tamura, H., and Horita, K.

Horner, J. E. See MacBain, J. C., Horner, J. E., Stange, W. A., and Ogg, J. S.

Houbolt, J. C., and Brooks, G. W. (1956). Differential equations of motion for combined flapwise bending, chordwise bending, and torsion of twisted nonuniform rotor blades, NACA Report 1346. (Supersedes NACA TN 3905, 1957). *178*

Huang, C. (1973). Finite amplitude vibrations of an orthotropic circular plate with an isotropic core, *AIAA Journal* 12, 388-390. *379*

Huang, C., and Sandman, B. E. (1971). Large amplitude vibrations of a rigidly clamped circular plate, *International Journal of Non-Linear Mechanics* 6, 451-468. *379*

Huang, C. L., and Al-Khattat, I. M. (1977). Finite amplitude vibrations of a circular plate, *International Journal of Non-Linear Mechanics* 12, 297-306. *375*

Huang, T. C. (1961). The effect of rotatory inertia and of shear deformation on the frequency and normal mode equations of uniform beams with simple end conditions, *Journal of Applied Mechanics* 28, 579-584. *179*

Hui, D. (1983). Large-amplitude axisymmetric vibrations of geometrically imperfect circular plates, *Journal of Sound and Vibration* 91, 239-246. *499*

Hui, D. (1985a). Effects of geometric imperfections on frequency-load interaction of biaxially compressed antisymmetric angle-ply rectangular plates, *Journal of Applied Mechanics* 52, 155-162. *381, 499*

Hui, D. (1985b). Soft-spring nonlinear vibrations of antisymmetrically laminated rectangular plates, *International Journal of Mechanical Science* 27, 397-408. *381*

Hung, E. S., and Senturia, S. D. (1999). Generating efficient dynamical models for microelectromechanical systems from a few finite-element simulation runs, *Journal of Microelectromechanical Systems* 8, 280-289. *342*

Hung, E. S. See Grétillat, M.-A., Yang, Y.-J., Hung, E. S., Rabinovich, V., Ananthasuresh, G. K., deRooij, N. F., and Senturia, S. D.

Hutton, S. G. See Yang, L., and Hutton, S. G.

Hyer, M. W. (1979). Whirling of a base-excited cantilever beam, *Journal of the Acoustical Society of America* 65, 931-939. *181*

Ibrahim, R. A. (1985). *Parametric Random Vibration*, Wiley-Interscience, New York. *24*

Ibrahim, R. A., and Chang, W. K. (1999). Stochastic excitation of suspended cables involving three simultaneous internal resonances using Monte Carlo simulation, *Computer Methods in Applied Mechanics and Engineering* 168, 285-304. *138*

Ibrahim, R. A. See Afaneh, A. A., and Ibrahim, R. A.; Chang, W. K., and Ibrahim, R. A.; Chang, W. K., Ibrahim, R. A., and Afaneh, A. A.; Chang, W. K., Ibrahim, R. A., and Afaneh, A. A.; Pilipchuk, V. N., and Ibrahim, R. A.

Iesan, D. (1987). St. Venant's problem, in *Lecture Notes in Mathematics*, No. 1279, A. Dold and B. Eckmann, eds., Springer-Verlag, Berlin, Germany. *7, 175*

Ijntema, D. J., and Tilmans, H. A. (1992). Static and dynamic aspects of an air-gap capacitor, *Sensors and Actuators A* 35, 121-128. *342*

Im, S., and Atluri, S. N. (1989). Effects of a piezo-actuator on a finitely deformed beams subjected to general loading, *AIAA Journal* 27, 1801-1807. *376*

Iooss, G., and Joseph, D. D. (1980). *Elementary Stability and Bifurcation Theory*, Springer-Verlag, New York. *24*

Irie, T., and Yamada, G. (1978). Thermally induced vibration of circular plates, *Bulletin of JSME* 21, 1703-1709. *544*

Irvine, H. M. (1981). *Cable Structures*, MIT Press, Cambridge, Massachusetts. *137*

Irvine, H. M., and Caughey, T. K. (1974). The linear theory of free vibrations of a suspended cable, *Proceedings of the Royal Society of London A* 341, 299-315. *137, 164*

Ito, M. See Shimizu, T., Yamaguchi, H., and Ito, M.; Yamaguchi, H., Miyata, T., and Ito, M.

Iura, M. See Borri, M., Mello, F., Iura, M., and Atluri, S. N.

Ivanov, Y. A. See Khoroshun, L. P., and Ivanov, Y. A.

Iwan, W. D., and Moeller T. L. (1976). The stability of a spinning elastic disk with a transverse load system, *Journal of Applied Mechanics* 43, 485-490. *517*

Iwan, W. D., and Stahl, K. J. (1973). The response of an elastic disk with a moving mass system, *Journal of Applied Mechanics* 40, 445-450. *517*

Iyengar, N. G. R. See Singh, G., Rao, V. G., and Iyengar, N. G. R.

Iyengar, R. N. See Rao, G. V., and Iyengar, R. N.; Visweswara Rao, G., and Iyengar, R. N.

Izadpanah, A. P. See Peters, D. A., and Izadpanah, A. P.

Jaakkola, O. See Veijola, T., Mattila, T., Jaakkola, O., Kiihamaki, J., Lamminmaki, T., Oja, A., Ruokonen, K., Seppala, H., and Tittonen, I.

Jankowski, D. F. See Gersting, J. M., and Jankowski, D. F.

Jegley, D. C. See Stein, M., and Jegley, D. C.

Ji, J.-C., and Hansen, C. H. (2000). Non-linear response of a post-buckled beam subjected to a harmonic axial excitation, *Journal of Sound and Vibration* 237, 303-318. *319*

Jiang, T. Y., Ng, T. Y., and Lam, K. Y. (2000). Dynamic analysis of an electrostatic micropump, in *Proceedings of the International Conference on Modeling and Simulation of Microsystems, MSM 2000*, San Diego, California, March 27-29, pp. 632-635. *342, 531*

Jilani, A. See Nayfeh, A. H., Jilani, A., and Manzione, P.

Joga Rao, C. V. See Srinivas, S., Joga Rao, C. V., and Rao, A. K.

John, F. (1965). Estimates for the derivatives of the stresses in a thin shell and interior shell equations, *Communications on Pure and Applied Mathematics* 18, 235-267. *372*

Johnson, B. P. See Senturia, S. D., Harris, R. M., Johnson, B. P., Kim, S., Nabors, K., Shulman, M. A., and White, J. K.

Johnson, J. M., and Bajaj, A. K. (1989). Amplitude modulated and chaotic dynamics in resonant motion of strings, *Journal of Sound and Vibration* 128, 87-107. *112*

Johnson, J. M. See Bajaj, A. K., and Johnson, J. M.; Bajaj, A. K., and Johnson, J. M.

Jones, R., Mazumdar, J., and Cheung, Y. K. (1980). Vibration and buckling of plates at elevated temperatures, *International Journal of Solids and Structures* 16, 61-70. *544*

Jones, R. M. (1973). Buckling and vibration of unsymmetrically laminated cross-ply rectangular plates, *AIAA Journal* 11, 1626-1632. *377, 378*

Jones, R. M. (1975). *Mechanics of Composite Materials,* Hemisphere, New York. *7, 372, 451*

Jordan, D. W., and Smith, P. (1977). *Nonlinear Ordinary Differential Equations,* Clarendon Press, Oxford, England. *24*

Joseph, D. D. See Iooss, G., and Joseph, D. D.

Kadivar, M. H., and Samani, K. (2000). Free vibration of rotating thick composite cylindrical shells using layerwise laminate theory, *Mechanics Research Communication* 27, 679-684. *559*

Kalaba, R. E. See Bellman, R. E., and Kalaba, R. E.

Kamal, K., and Durvasula, S. (1986). Some studies on free vibration of composite laminates, *Composite Structures* 5, 177-202. *378*

Kandil, O. A. See Eslami, H., and Kandil, O. A.; Eslami, H., and Kandil, O. A.

Kane, T. R., Ryan, R. R., and Banerjee, A. K. (1987). Dynamics of a cantilever beam attached to a moving base, *Journal of Guidance, Control, and Dynamics* 10, 139-151. *174*

Kapania, P. K. (1989). Review on the analysis of laminated shells, *Journal of Pressure Vessel Technology* 111, 88-96. *559*

Kapania, R. K., and Raciti, S. (1989a). Recent advances in analysis of laminated beams and plates, Part I: Shear effects and buckling, *AIAA Journal* 27, 923-934. *179, 381*

Kapania, R. K., and Raciti, S. (1989b). Recent advances in analysis of laminated beams and plates, Part II: Vibrations and wave propagation, *AIAA Journal* 27, 935-946. *179, 381*

Kapania, R. K., and Raciti, S. (1989c). Nonlinear vibrations of unsymmetrically laminated beams, *AIAA Journal* 27, 201-210. *179, 180, 469*

Kar, R. C., and Sujata, T. (1990). Parametric instability of an elastically restrained cantilever beam, *Computers & Structures* 34, 469-475. *278*

Kar, R. C., and Sujata, T. (1992). Dynamic stability of a rotating, pretwisted, and preconed cantilever beam including Coriolis effects, *Computers & Structures* 42, 741-750. *278*

Karami, G. See Banan, M. R., Karami, G., and Farshad, M.

Kazarinoff, N. D. See Hassard, B. D., Kazarinoff, N. D., and Wan, Y. H.

Kehagioglou, T. See Evan-Iwanowski, R. M., Sanford, W. F., and Kehagioglou, T.

Khanna, V. D. See Kumar, S., and Khanna, V. D.

Khdeir, A. A. (1988a). Free vibration and buckling of symmetric cross-ply laminated plates by an exact method, *Journal of Sound and Vibration* 126, 447-461. *379*

Khdeir, A. A. (1988b). Free vibration of antisymmetrically angle-ply laminated plates including various boundary conditions, in *Proceedings of the 7th ASCE, EMD Conference on Engineering Mechanics*, R. A. Heller and M. P. Singh, eds., ASME, New York. *377*

Khdeir, A. A. (1989). Free vibration and buckling of unsymmetric cross-ply laminated plates using a refined theory, *Journal of Sound and Vibration* 128, 377-395. *379*

Khdeir, A. A., and Librescu, L. (1988). Analysis of symmetric cross-ply laminated elastic plates using a higher-order theory: Part II - Buckling and free vibration, *Composite Structures* 9, 259-277. *374*

Khdeir, A. A., and Reddy, J. N. (1988). Dynamic response of antisymmetric angle-ply laminated plates subjected to arbitrary loading, *Journal of Sound and Vibration* 126, 437-445.

Khdeir, A. A., and Reddy, J. N. (1989). Exact solutions for the transient response of symmetric cross-ply laminates using a higher-order plate theory, *Composites Science and Technology* 34, 205-224.

Khdeir, A. A., and Reddy, J. N. (1990). Influence of edge conditions on the modal characteristics of cross-ply laminated shells, *Computers & Structures* 34, 817-826. *561*

Khdeir, A. A. See Librescu, L., and Khdeir, A. A.; Librescu, L., Khdeir, A. A., and Reddy, J. N.; Nayfeh, A. H., and Khdeir, A. A.; Nayfeh, A. H., and Khdeir, A. A.; Reddy, J. N., and Khdeir, A. A.; Reddy, J. N., Khdeir, A. A., and Librescu, L.

Khoroshun, L. P. (1978). Derivation equations for laminated plates and shells, *Soviet Applied Mechanics* 14, 1021-1034. *374*

Khoroshun, L. P., and Ivanov, Y. A. (1981). Solving the problem of a harmonic wave in a double-layer plate on the basis of the generalized shear model, *Soviet Applied Mechanics* 17, 147-153. *374*

Kiihamaki, J. See Veijola, T., Mattila, T., Jaakkola, O., Kiihamaki, J., Lamminmaki, T., Oja, A., Ruokonen, K., Seppala, H., and Tittonen, I.

Kim, C. H. See Lee, W. K., and Kim, C. H.

Kim, J. H. See Kim,T., W., and Kim, J. H.

Kim, S. See Senturia, S. D., Harris, R. M., Johnson, B. P., Kim, S., Nabors, K., Shulman, M. A., and White, J. K.

Kim, T. W., and Kim, J. H. (2002). Nonlinear vibration of viscoelastic laminated composite plates, *International Journal of Solids and structures* 39, 2857-2870. *381*

King, A. L. (1974). Oscillations of a loaded rubber band, *American Journal of Physics* 42, 699-701. *136*

King, A. L. See Lawton, R. W., and King, A. L.

King, W. W. See Haight, E. C., and King, W. W.; Haight, E. C., and King, W. W.

Kingsbury, H. B. See Mohan, D., and Kingsbury, H. B.

Kirchhoff, G. R. (1850). Über das Gleichgewicht und die Bewegung einer Elastischen Scheibe, *Journal Mathematik (Crelle)* 40, 51-88. *382, 403*

Kitipornchai, S. See Liew, K. M., Kitipornchai, S., and Wang, C. M.; Liew, K. M., Lim, C. W., and Kitipornchai, S.

Knowle, G. See Birman, V., Griffin, S., and Knowle, G.

Kobayashi, Y. See Abe, A., Kobayashi, Y., and Yamada, G.; Abe, A., Kobayashi, Y., and Yamada, G.; Abe, A., Kobayashi, Y., and Yamada, G.; Abe, A., Kobayashi, Y., and Yamada, G.

Koiter, W. T. (1960). A consistent first approximation in the general theory of thin elastic shells, in *Proceedings of IUTAM Symposium on the Theory of Thin Elastic Shells*, Delft, The Netherlands, August 24-28, 1959, W. T. Koiter, ed., North-Holland, Amsterdam, pp. 12-33. *372, 560, 561*

Koiter, W. T. (1966). On the nonlinear theory of thin elastic shells, in *Proceedings of the Koninklijke Nederlandse Akademie van Wetenschappen* Series B69, pp. 1-54. *562*

Koiter, W. T. (1967). Foundations and basic equations of shell theory–A survey of recent progress, in *Theory of Thin Shells, Proceedings of IUTAM Symposium*, F. I. Niordson, ed., Copenhagen, Denmark, pp. 93-105. *372*

Koiter, W. T. (1969). *Theory of Thin Shells*, F. L. Niordson, ed., Springer-Verlag, New York, pp. 93-105. *559*

Koiter, W. T., and Simmonds, J. G. (1972). Foundations of shell theory, WTHD 40, Delft University of Technology, Delft, The Netherlands. *562*

Kojima, H., Nagaya, J., Shiraishi, H., and Yamashita, A. (1985). Nonlinear vibrations of a beam with a mass subjected to alternating electromagnetic force, *Bulletin of JSME* 18, 468-475. *24*

Kolli, M. See Chandrashekhara, K., and Kolli, M.

Kolowith, M. See Burton, T. D., and Kolowith, M.

Konishi, Y. See Takahashi, K., and Konishi, Y.; Takahashi, K., and Konishi, Y.

Korvink, J. G., Funk, J., Roos, M., and Wachutka, G. (1994). SESES: A comprehensive MEMS modelling system, in *Proceedings of the IEEE Workshop on Micro Electro Mechanical Systems, MEMS'94*, Oiso, Japan, January 25-28, pp. 22-27. *341*

Kovarik, V. (1980). Cylindrical orthotropic sandwich-type shells under general loading– I. Formulation of the problem, *Acta Technica (CSAV)* 4, 499-518. *593*

Kraus, H. (1967). *Thin Elastic Shells*, John Wiley and Sons, New York. *559*

Krauter, A. L., and Bulkeley, P. Z. (1970). The effect of central clamping on transverse vibrations of spinning membrane disks, *Journal of Applied Mechanics* 37, 1037-1042. *513*

Kreider, W. (1995). *Linear and Nonlinear Vibrations of Buckled Beams*, MSc Thesis, Virginia Polytechnic Institute and State University, Blacksburg, Virginia. *319*

Kreider, W., and Nayfeh, A. H. (1998). Experimental investigation of single-mode responses in a fixed-fixed buckled beam, *Nonlinear Dynamics* 15, 155-177. *319, 334-337, 340*

Kreider, W. See Lacarbonara, W., Nayfeh, A. H., and Kreider, W.; Nayfeh, A. H., Kreider, W., and Anderson, T. J.

Krenk, S. (1983). A linear theory for pretwisted elastic beams, *Journal of Applied Mechanics* 50, 137-142. *180*

Krishna Murty, A. V. (1970). Vibrations of short beams, *AIAA Journal* 8, 34-38. *173, 174*

Krishnamoorthy, C. S. See Ramesh, G., and Krishnamoorthy, C. S.

Krishnan, K. S. See Manoj, T., Ayyappan, M., Krishnan, K. S., and Rao, B. N.

Krousgrill, C. M. See Chang, S. I., Bajaj, A. K., and Krousgrill, C. M.; Streit, D. A., Bajaj, A. K., and Krousgrill, C. M.

Kumar, S., Khanna, V. D., and Jayantha, M. S. (1994). Study of the head disk interface shock failure mechanism, *IEEE Transactions of Magnetics* 30, 4155-4157. *513*

Kung, G. C., and Pao, Y. H. (1972). Nonlinear flexural vibration of a clamped circular plate, *Journal of Applied Mechanics* 39, 1050-1054. *379*

Kuo, W. S., and Yang, I. H. (1989). Generic nonlinear behavior of antisymmetric angle-ply laminated plates, *International Journal of Mechanical Sciences* 31, 131-143. *380*

Kushida, G. See Yasuda, K., and Kushida, G.

Kwak, M. K. See Amabili, M., Frosali, G., and Kwak, M. K.

Lacarbonara, W. (1998). *Approximate Methods and Models in Nonlinear Vibrations of One-Dimensional Systems with Initial Curvature*, PhD Dissertation, University of Rome La Sapienza, Rome, Italy. *53, 324, 330, 331*

Lacarbonara, W., Nayfeh, A. H., and Kreider, W. (1998). Experimental validation of reduction methods for nonlinear vibrations of distributed-parameter systems: Analysis of a buckled beam, *Nonlinear Dynamics* 17, 95-117. *53, 138, 324, 331-333*

Lacarbonara, W. See Chin, C., Nayfeh, A. H., and Lacarbonara, W.; Nayfeh, A. H., Arafat, H. N., Chin, C-M., and Lacarbonara, W.; Nayfeh, A. H., and Lacarbonara, W.; Nayfeh, A. H., and Lacarbonara, W.; Nayfeh, A. H., Lacarbonara, W., and Chin, C-M.; Rega, G., Lacarbonara, W., Nayfeh, A. H., and Chin, C-M.; Rega, G., Lacarbonara, W., Nayfeh, A. H., and Chin, C-M.

Lackman, L. M. See Bieniek, M. P., Fan, T. C., and Lackman, L. M.

Lakis, A. A., Selmane, A., and Toledano, A. (1998). Nonlinear free vibration of laminated orthotropic cylindrical shells, *International Journal of Mechanical Sciences* 40, 27-94. *563*

Lam, K. Y., Ng, T. Y., and Qian, W. (2000). Vibration analysis of think laminated composite cylindrical shells, *AIAA Journal* 38, 1102-1107.

Lamb, H., and Southwell, R. V. (1921). The vibrations of a spinning disc, *Proceedings of the Royal Society A* 99, 272-280. *517*

Lamminmaki, T. See Veijola, T., Mattila, T., Jaakkola, O., Kiihamaki, J., Lamminmaki, T., Oja, A., Ruokonen, K., Seppala, H., and Tittonen, I.

Lanczos, C. (1986). *The Variational Principles of Mechanics,* University of Toronto Press, Toronto, Canada. *47*

Lanz, M. See Borri, M., Ghiringhelli, G. L., Lanz, M., Mantegazza, P., and Merlini, T.

Lawton, R. W., and King, A. L. (1951). Free longitudinal vibrations of rubber and tissue strips, *Journal of Applied Physics* 22, 1340-1343. *136*

Lazarus, K. B., and Crawley, E. F. (1989). Induced strain actuation of composite plates, Massachusetts Institute of Technology, Cambridge, Massachusetts, GTL Report No. 197. *376, 451*

Le Berre, M. See Malhaire, C., Le Berre, M., Febvre, M., Barbier, D., and Pinard, P.

Leal, L. G. See Feng, Z. C., and Leal, L. G.

Lee, C. K. (1990). Theory of laminated piezoelectric plates for the design of distributed sensors/actuators. Part I: Governing equations and reciprocal relationships, *Journal of the Acoustical Society of America* 87, 1144-1158. *376, 451, 467*

Lee, C. K., and Moon, F. C. (1989). Laminated piezopolymer plates for torsion and bending sensors and actuators, *Journal of the Acoustical Society of America* 85, 2432-2439. *376*

Lee, C. L., and Perkins, N. C. (1992). Nonlinear oscillations of suspended cables containing a two-to-one internal resonance, *Nonlinear Dynamics* 3, 465-490. *136, 137*

Lee, C. L., and Perkins, N. C. (1995a). Experimental investigation of isolated and simultaneous internal resonances in suspended cables, *Journal of Vibration and Acoustics* 117, 385-391. *136, 137*

Lee, C. L., and Perkins, N. C. (1995b). Three-dimensional oscillations of suspended cables involving simultaneous internal resonances, *Nonlinear Dynamics* 8, 45-63. *136, 138*

Lee, J. M., Chung, J. H., and Chung, T. Y. (1997). Free vibration analysis of symmetrically laminated composite rectangular plates, *Journal of Sound and Vibration* 199, 71-85. *378*

Lee, S. W. See Stemple, A. D., and Lee, S. W.

Lee, S. Y., and Ames, W. F. (1973). A class of general solutions to the nonlinear dynamic equations of elastic strings. *Journal of Applied Mechanics* 40, 1035-1039. *112*

Lee, W. K., and Kim, C. H. (1995). Combination resonances of a circular plate with three-mode interaction, *Journal of Applied Mechanics* 62, 1015–1022. *499, 504*

Lee, W. K., and Yeo, M. H. (2003). Non-linear interactions in asymmetric vibrations of a circular plate, *Journal of Sound and Vibration* 263, 1017-1030. *499*

Lee, W. K., Yeo, M. H., and Samoilenko, S. B. (2003). The effect of the number of nodal diameters on non-linear interactions in two asymmetric vibration modes of a circular plate, *Journal of Sound and Vibration* 268, 1013-1023. *499*

Lee, W. K. See Yeo, M. H., and Lee, W. K.

Legtenberg, R., and Tilmans, H. A. (1994). Electrostatically driven vacuum-encapsulated polysilicon resonators. Part I. Design and fabrication, *Sensors and Actuators A* 45, 57-66. *350*

Legtenberg, R. See Tilmans, H. A., and Legtenberg, R.

Leissa, A. W. (1978). Recent research in plate vibrations, 1973-1976: Complicating effects, *Shock and Vibration Digest* 10, 21-35. *469*

Leissa, A. W. (1981a). Advances in vibrations, buckling, and postbuckling studies on composite plates, in *Composite Structures,* T. H. Marshall, ed., Applied Science Publishers, London, pp. 312-334. *378, 469*

Leissa, A. W. (1981b). Recent research in plate vibrations, 1975-1980: Complicating effects, *Shock and Vibration Digest* 13, 19-36. *469*

Leissa, A. W. (1981c). Vibrational aspects of rotating turbomachinery blades, *Applied Mechanics Reviews* 34, 629-635. *469*

Leissa, A. W. (1987a). A review of laminated composite plate buckling, *Applied Mechanics Reviews* 40, 575-591. *469*

Leissa, A. W. (1987b). Recent studies in plate vibrations 1981-1985: Complicating effects, *Shock and Vibration Digest* 19, 10-24. *469*

Leissa, A. W. (1993a). *Vibration of Plates*, published for the Acoustical Society of America through the American Institute of Physics, (originally issued as NASA SP-160, 1969). *469, 491*

Leissa, A. W. (1993b). *Vibration of Shells*, published for the Acoustical Society of America through the American Institute of Physics, (originally issued as NASA SP-288, 1973). *559*

Leissa, A. W. See Baharlou, B., and Leissa, A. W.; Qatu, M. S., and Leissa, A. W.; Whitney, J. M., and Leissa, A. W.

Lekhnitski, S. T. (1968). *Anisotropic Plates*, Gordon and Breach, New York. *372, 378, 469*

Leonard, R. W. (1961). *Nonlinear First Approximation Thin Shell and Membrane Theory*, MSc Thesis, Virginia Polytechnic Institute and State University, Blacksburg, Virginia. *562*

Lestari, W., and Hanagud, S. (2001). Nonlinear vibration of buckled beams: Some exact solutions, *International Journal of Solids and Structures* 38, 4741-4757. *318*

Levinson, M. (1980). An accurate, simple theory of the statics and dynamics of elastic plates, *Mechanics Research Communications* 7, 343-350. *374*

Li, S.-R., and Zhou, Y.-H. (2001). Shooting method for non-linear vibration and buckling of heated orthotropic circular plates, *Journal of Sound and Vibration* 248, 379-386. *544*

Li, S.-R., and Zhou, Y.-H. (2003). Nonlinear vibration of heated orthotropic annular plates with immovably hinged edges, *Journal of Thermal Stresses* 26, 691-700. *544*

Li, S.-R., Zhou, Y.-H., and Song, X. (2002). Non-linear vibration and thermal buckling of an orthotropic annular plate with a centric rigid mass, *Journal of Sound and Vibration* 251, 141-152. *544*

Libai, A., and Simmonds, J. G. (1988). *The Nonlinear Theory of Elastic Shells, One Spatial Dimension,* Academic Press, Orlando, Florida. *562*

Librescu, L. (1966). The thermoelastic problem of shells and plates approached by eliminating the Love-Kirchhoff hypothesis, *Studii si Cercetari de Macanica Aplicata* 21, 351-365. *374*

Librescu, L. (1969). The elasto-kinetic problems in the theory of anisotropic shells and plates–II. Theory of plates, *Revue Roumaine des Sciences Techniques. Serie de Mecanique Appliquee* 14, 637-654. *374*

Librescu, L. (1976). *Elastostatics and Kinetics of Anisotropic and Heterogeneous Shell-Type Structures,* Noordhoff, Leyden. *563, 605*

Librescu, L., and Khdeir, A. A. (1988). Analysis of symmetric cross-ply laminated elastic plates using a higher-order theory: Part I - Stress and displacement, *Composite Structures* 9, 189-213. *374*

Librescu, L., Khdeir, A. A., and Reddy, J. N. (1987). A comprehensive analysis of the state of stress of elastic anisotropic flat plates using refined theories, *Acta Mechanica* 70, 57-81. *374, 378*

Librescu, L., and Reddy, J. N. (1989). A few remarks concerning several refined theories of anisotropic composite laminated plates, *International Journal of Engineering Science* 27, 515-527. *374*

Librescu, L., and Schmidt, R. (1988). Refined theories of elastic anisotropic shells accounting for small strains and moderate rotations, *International Journal of Non-Linear Mechanics* 23, 217-229. *561, 563, 605*

Librescu, L., and Schmidt, R. (1991). Substantiation of a shear-deformable theory of anisotropic composite laminated shells accounting for the interlaminae continuity conditions, *International Journal of Engineering Science* 29, 669-683. *374, 561*

Librescu, L., and Simovich, J. (1988). A general formulation for the aeroelastic divergence of composite sweptforward wing structures, *Journal of Aircraft* 4, 364-371. *175*

Librescu, L. See Khdeir, A. A., and Librescu, L.; Reddy, J. N., Khdeir, A. A., and Librescu, L.

Liew, K. M., Kitipornchai, S., and Wang, C. M. (1993). Research development in the analysis of plates and shells, *Journal of Construction Steel Research* 26, 231-248. *559*

Liew, K. M., Lim, C. W., and Kitipornchai, S. (1997). Vibration of shallow shells: A review with bibliography, *Applied Mechanics Reviews* 50, 431-444. *559*

Lifshitz, R., and Roukes, M. L. (2000). Thermoelastic damping in micro- and nanomechanical systems, *Physical Review B* 61, 5600-5609. *544*

Lim, C. W. See Liew, K. M., Lim, C. W., and Kitipornchai, S.

Lin, P. M. See Chua, L. O., and Lin, P. M.

Liu, C. F. See Reddy, J. N., and Liu, C. F.

Lo, K. H., Christensen, R. M., and Wu, E. M. (1977a). A high-order theory of plate deformation-Part 1: Homogeneous plates, *Journal of Applied Mechanics* 44, 663-668. *374*

Lo, K. H., Christensen, R. M., and Wu, E. M. (1977b). A high-order theory of plate deformation, Part 2: Laminated plates, *Journal of Applied Mechanics* 44, 669-676. *374*

Lobitz, D. W., Nayfeh, A. H., and Mook, D. T. (1977). Nonlinear analysis of vibrations of irregular plates, *Journal of Sound and Vibration* 50, 203-217. *381*

Lobitz, D. W. See Nayfeh, A. H., Mook, D. T., and Lobitz, D. W.

Loewy, R. G. See Rosen, A., Loewy, R. G., and Mathew, M. B.; Rosen, A., Loewy, R. G., and Mathew, M. B.; Rosen, A., Loewy, R. G., and Mathew, M. B.

Lorch, D. R. See Nelson, R. B., and Lorch, D. R.

Love, A. E. H. (1944). *A Treatise on the Mathematical Theory of Elasticity*, 4th ed., Dover, New York. *175, 199, 249, 560, 613, 614*

Lovell, E. G. See McIvor, I. K., and Lovell, E. G.

Luongo, A., Paolone, A., and Piccardo, G. (1998). Postcritical behavior of cables undergoing two simultaneous galloping modes, *Meccanica* 33, 229-242. *136*

Luongo, A., Rega, G., and Vestroni, F. (1984). Planar non-linear free vibrations of an elastic cable, *International Journal of Non-Linear Mechanics* 19, 39-52. *136, 137*

Luongo, A., and Vestroni, F. (1994). Non-linear free periodic oscillations of a tethered satellite system, *Journal of Sound and Vibration* 175, 299-315. *136*

Luongo, A., and Vestroni, F. (1996). Bifurcations and stability of amplitude modulated planar oscillations of an orbiting string with internal resonances, *Nonlinear Dynamics* 9, 305-325. *136*

MacBain, J. C., Horner, J. E., Stange, W. A., and Ogg, J. S. (1979). Vibration analysis of a spinning disk using image-derotated holographic interferometry, *Experimental Mechanics* 19, 17-22. *513*

Maewal, A. (1978). Nonlinear flexural vibration of an elastic ring, *Journal of Applied Mechanics* 45, 428-428. *563*

Maewal, A. (1981). Nonlinear harmonic oscillations of gyroscopic structural systems and the case of a rotating ring, *Journal of Applied Mechanics* 48, 627-633. *565, 637*

Maewal, A. (1986a). Chaos in a harmonically excited elastic beam, *Journal of Applied Mechanics* 53, 625-632. *26*

Maewal, A. (1986b). Finite element analysis of steady nonlinear harmonic oscillations of axisymmetric shells, *Computer Methods in Applied Mechanics and Engineering* 68, 37-50. *565, 637*

Maffioli, G. C. See Giavotto, V., Borri, M., Mantegazza, P., Ghiringhelli, G., Carmaschi, V., Maffioli, G. C., and Mussi, F.

Maganty, S. P., and Bickford, W. B. (1987). Large amplitude oscillations of thin circular rings, *Journal of Applied Mechanics* 54, 315-322. *181, 513, 565, 637*

Maganty, S. P., and Bickford, W. B. (1988). Influence of internal resonance on the non-linear oscillations of a circular ring under primary resonance conditions, *Journal of Sound and Vibration* 122, 507-521. *513, 565, 637*

Majima, O., and Hayashi, K. (1984). Elastic buckling and flexural vibration of annular plates under axisymmetric in-plane forces, *Bulletin of the JSME* 27, 2088-2094. *544*

Majumder, S. See Zavracky, P. M., Majumder, S., and McGruer, N. E.

Malatkar, P. (2003). *Nonlinear Vibrations of Cantilever Beams and Plates*, PhD Dissertation, Virginia Polytechnic Institute and State University, Blacksburg, Virginia. *294*

Malatkar, P., and Nayfeh, A. H. (2003). A parametric identification technique for SDOF weakly nonlinear systems with cubic nonlinearities, *Journal of Vibration and Control* 9, 317-336. *294, 296*

Malhaire, C., Le Berre, M., Febvre, M., Barbier, D., and Pinard, P. (1999). Effect of clamping conditions and built-in stresses on the thermopneumatic deflection of SiO_2/Si membranes with various geometries, *Sensors and Actuators* 74, 174-177. *531*

Mallik, A. K. See Hatwal, H., Mallik, A. K., and Ghosh, A.; Hatwal, H., Mallik, A. K., and Ghosh, A.

Malvern, L. E. (1969). *Introduction to the Mechanics of a Continuous Medium*, Prentice-Hall, New Jersey. *85, 107, 110, 215, 252, 376*

Manoj, T., Ayyappan, M., Krishnan, K. S., and Rao, B. N. (2000). Nonlinear vibration analysis of thin laminated rectangular plates on elastic foundations, *Zeitschrift fur Angewandte Mathematik und Mechanik (ZAMM)* 80, 183-192. *380*

Mantegazza, P. See Borri, M., Ghiringhelli, G. L., Lanz, M., Mantegazza, P., and Merlini, T.; Giavotto, V., Borri, M., Mantegazza, P., Ghiringhelli, G., Carmaschi, V., Maffioli, G. C., and Mussi, F.

Manzione, P. See Nayfeh, A. H., Jilani, A., and Manzione, P.

Markus, S. (1988). *The Mechanics of Vibrations of Cylindrical Shells*, Elsevier, New York. *561*

Marlowe, M. B. (1968). *Some New Developments in the Foundations of Shell Theory*. Thesis, Stanford University, Stanford, California. *562*

Marsden, J. E., and McCracken, M. (1976). *Hopf Bifurcation and Its Application*, Springer-Verlag, New York. *24*

Marshall, L. R. See Mook, D. T., Marshall, L. R., and Nayfeh, A. H.; Nayfeh, A. H., Mook, D. T., and Marshall, L. R.

Martinez, J. E. See Stricklin, J. A., Martinez, J. E., Tillerson, J. R., Hong, J. H., and Haisler, W. E.

Masad, J. A. See Asfar, O. R., Masad, J. A., and Nayfeh, A. H.

Maseeh, F. See Yie, H., Harris, R., Napadenski, G., and Maseeh, F.

Mathew, M. B. See Rosen, A., Loewy, R. G., and Mathew, M. B.; Rosen, A., Loewy, R. G., and Mathew, M. B.; Rosen, A., Loewy, R. G., and Mathew, M. B.

Mattila, T. See Veijola, T., Mattila, T., Jaakkola, O., Kiihamaki, J., Lamminmaki, T., Oja, A., Ruokonen, K., Seppala, H., and Tittonen, I.

Mau, S. T. (1973). A refined laminated plate theory, *Journal of Applied Mechanics* 40, 606-607. *373*

Mavrakos, S. A. See Papazoglou, V. J., and Mavrakos, S. A.

Mayberry, B. L. See Bert, C. W., and Mayberry, B. L.

Mazumdar, J. See Jones, R., Mazumdar, J., and Cheung, Y. K.

McCracken, M. See Marsden, J. E., and McCracken, M.

McDonald, P. H. (1955). Nonlinear dynamic coupling in a beam vibration, *Journal of Applied Mechanics* 22, 573-578. *317*

McGruer, N. E. See Zavracky, P. M., Majumder, S., and McGruer, N. E.

McIvor, I. K. (1962). *Dynamic Stability and Nonlinear Oscillations of Cylindrical Shells (Plane Strain) Subjected to Impulsive Pressure*, PhD Dissertation, Stanford University, Stanford, California. *564*

McIvor, I. K. (1966). The elastic cylindrical shell under radial impulse, *Journal of Applied Mechanics* 33, 831-837. *564*

McIvor, I. K., and Lovell, E. G. (1968). Dynamic response of finite-length cylindrical shells to nearly uniform radial impulse, *AIAA Journal* 6, 2346-2351. *564*

McIvor, I. K., and Sonstegard, D. A. (1966). Axisymmetric response of a closed spherical shell to a nearly uniform radial impulse, *Journal of the Acoustical Society of America* 40, 1540-1547. *564, 645*

McIvor, I. K. See Goodier, J. N., and McIvor, I. K.

Medwadowski, S. J. (1958). A refined theory of elastic, orthotropic plates, *Journal of Applied Mechanics* 25, 437-443. *372*

Mehner, J. E., Gabbay, L. D., and Senturia, S. D. (2000). Computer-aided generation of nonlinear reduced-order dynamic macromodels - II. Non-stress-stiffened case, *Journal of Microelectromechanical Systems* 9, 270-278. *342*

Mehner, J. E. See Gabbay, L. D., Mehner, J. E., and Senturia, S. D.

Meirovitch, L. (1980). *Computational Methods in Structural Dynamics,* Sijthoff and Noordhoff, Rockville, Maryland. *19, 27*

Melde, F. (1859). Über erregung stehender wellen eines faden formigen körpers, *Annalen der Physik und Chemie* 109, 193-215. *112*

Mello, F. See Borri, M., Mello, F., Iura, M., and Atluri, S. N.

Mente, L. J. (1973). Dynamic nonlinear response of cylindrical shells to axisymmetric pressure loading, *AIAA Journal* 11, 793-800. *563*

Merlini, T. See Borri, M., Ghiringhelli, G. L., Lanz, M., Mantegazza, P., and Merlini, T.; Borri, M., and Merlini, T.

Miles, J. W. (1984a). Nonlinear Faraday resonance, *Journal of Fluid Mechanics* 146, 285-302. *26, 565, 637*

Miles, J. W. (1984b). Internally resonant surface waves in a circular cylinder, *Journal of Fluid Mechanics* 149, 1-14. *26, 565, 637*

Miles, J. W. (1984c). Resonant motion of a spherical pendulum, *Physica D* 11, 309-323. *26*

Miles, J. W. (1984d). Resonant, nonplanar motion of a stretched string, *Journal of the Acoustical Society of America* 75, 1505-1510. *112*

Miles, J. W. (1985). Parametric excitation of an internally resonant double pendulum, *Zeitschrift fur Angewandte Mathematik und Physik (ZAMP)* 36, 337-345. *26*

Min, G. B., and Eisley, J. G. (1972). Nonlinear vibrations of buckled beams, *Journal of Engineering for Industry* 94, 637-646. *318*

Mindlin, R. D. (1951). Influence of rotatory inertia and shear on flexural motions of isotropic, elastic plates, *Journal of Applied Mechanics* 18, A31-A38. *372*

Minguet, P., and Dugundji, J. (1990a). Experiments and analysis for composite blades under large deflections, Part I. Static behavior, *AIAA Journal* 28, 1573-1579. *177, 178, 180*

Minguet, P., and Dugundji, J. (1990b). Experiments and analysis for composite blades under large deflections, Part II. Dynamic behavior, *AIAA Journal* 28, 1580-1588. *178, 180*

Mirsky, I., and Herrmann, G. (1957). Nonaxially symmetric motions of cylindrical shells, *Journal of the Acoustical Society of America* 29, 1116-1124. *561*

Mirza, S. (1991). Recent research in vibration of layered shells, *Journal of Pressure Vessel Technology* 113, 321-325. *559*

Mitsugi, J., and Yasaka, T. (1991). Nonlinear static and dynamic analysis method of cable structures, *AIAA Journal* 29, 150-152. *136*

Miyata, T. See Yamaguchi, H., Miyata, T., and Ito, M.

Modeer, J. R. See Nash, W. A., and Modeer, J. R.

Moeller, T. L. See Iwan, W. D., and Moeller, T. L.

Mohan, D., and Kingsbury, H. B. (1971). Free vibrations of generally laminated orthotropic plates, *Journal of the Acoustical Society of America* 106, 266-269. *378*

Molteno, T. C. A. (1994). *Chaos and Crises in Strings*, PhD Dissertation, Otago Polytechnic, Dunedin, New Zealand. *135*

Molteno, T. C. A., and Tufillaro, N. B. (1990). Torus doubling and chaotic string vibrations: Experimental results, *Journal of Sound and Vibration* 137, 327-330. *112, 121, 132*

Moody, M. L. (1967). The parametric response of imperfect columns, in *Developments in Mechanics, Proceedings of the Tenth Midwestern Mechanics Conference*, Fort Collins, Colorado, August 21-23, Vol. 4, pp. 329-346. *177, 223*

Mook, D. T., Marshall, L. R., and Nayfeh, A. H. (1974). Subharmonic and superharmonic resonances in the pitch and roll modes of ship motions, *Journal of Hydrodynamics* 8, 32-40. *25*

Mook, D. T. See Abou-Rayan, A. M., Nayfeh, A. H., and Mook, D. T.; Haddow, A. G., Barr, A. D. S., and Mook, D. T.; Lobitz, D. W., Nayfeh, A. H., and Mook, D. T.; Nayfeh, A. H., and Mook, D. T.; Nayfeh, A. H., Mook, D. T., and Lobitz, D. W.; Nayfeh, A. H., Mook, D. T., and Marshall, L. R.; Nayfeh, A. H., Mook, D. T., and Nayfeh, J. F.; Nayfeh, A. H., Mook, D. T., and Sridhar, S.; Nayfeh, A. H., Nayfeh, J. F., and Mook, D. T.; Nayfeh, S. A., Nayfeh, A. H., and Mook, D. T.; Pai, P. F., Nayfeh, A. H., Oh, K., and Mook, D. T.; Shyu, I.-M. K., Mook, D. T., and Plaut, R. H.; Shyu, I.-M. K., Mook, D. T., and Plaut, R. H.; Shyu, I.-M. K., Mook, D. T., and Plaut, R. H.; Sridhar, S., Mook, D. T., and Nayfeh, A. H.; Sridhar, S., Mook, D. T., and Nayfeh, A. H.; Tezak, E. G., Mook, D. T., and Nayfeh, A. H.

Moon, F. C. (1980). Experiments on chaotic motions of a forced nonlinear oscillator: Strange attractors, *Journal of Applied Mechanics* 47, 638-644. *24, 318*

Moon, F. C. (1987). *Chaotic Vibrations, An Introduction for Applied Scientists and Engineers*, Wiley-Interscience, New York. *24, 268*

Moon, F. C. (1992). *Chaotic and Fractal Dynamics: An Introduction for Applied Scientists and Engineers*, Wiley-Interscience, New York. *24, 268*

Moon, F. C. See Holmes, P. J., and Moon, F. C.; Lee, C. K., and Moon, F. C.

Mote, Jr. C. D. See Shen, I. Y., and Mote, Jr. C. D.

Moussaoui, F., and Benamar, R. (2002). Non-linear vibrations of shell-type structures: A review with bibliography, *Journal of Sound and Vibration* 255, 161-184. *559*

Mukhopadhyay, V. See Dugundji, J., and Mukhopadhyay, V.

Murthy, G. S. S., and Ramakrishna, B. S. (1965). Nonlinear character of resonance in stretched strings, *Journal of the Acoustical Society of America* 38, 461-471. *112, 121*

Murthy, M. V. V. (1981). An improved transverse shear deformation theory for laminated anisotropic plates, NASA Technical Paper 1903. *374*

Mushtari, K. M. (1938). Certain generalizations of the theory of thin shells, *Izvestiya na Fizicheskiya* 11. *562*

Mussi, F. See Giavotto, V., Borri, M., Mantegazza, P., Ghiringhelli, G., Carmaschi, V., Maffioli, G. C., and Mussi, F.

Nabors, K. See Senturia, S. D., Harris, R. M., Johnson, B. P., Kim, S., Nabors, K., Shulman, M. A., and White, J. K.

Nagaya, J. See Kojima, H., Nagaya, J., Shiraishi, H., and Yamashita, A.

Naghdi, P. M. (1963a). Foundations of elastic shell theory, in *Progress in Solid Mechanics*, Vol. 4, pp. 1-90, I. N. Sneddon and R. Hill, eds., North-Holland, Amsterdam. *562*

Naghdi, P. M. (1963b). A new derivation of the general equations of elastic shells, *International Journal of Engineering Science* 1, 509-522. *562*

Naghdi, P. M., and Berry, J. G. (1964). On the equations of motion of cylindrical shells, *Journal of Applied Mechanics* 21, 160-166. *560*

Naghdi, P. M., and Nordgren, R. P. (1963). Nonlinear theory of elastic shells, *Quarterly of Applied Mathematics* 21, 19-59. *562*

Naghdi, P. M. See Green, A. E., and Naghdi, P. M.

Naja, M. See Harichandran, R. S., and Naja, M.

Napadenski, G. See Yie, H., Harris, R., Napadenski, G., and Maseeh, F.

Narasimha, R. (1968). Nonlinear vibration of an elastic string, *Journal of Sound and Vibration* 8, 134-146. *112, 120, 125*

Narayanan, S. See Rajgopal, S. V., Singh, G., Sadasiva Rao, Y. V. K., and Narayanan, S.

Nash, W. A., and Modeer, J. R. (1959). Certain approximate analysis of the nonlinear behavior of plates and shallow shells, in *Proceedings of the Symposium on Theory of Thin Elastic Shells*, Wiley-Interscience, New York, pp. 331-359. *375*

Natsiavas, S. (1994a). Dynamics and stability of non-linear free vibration of thin rotating rings, *International Journal of Non-Linear Mechanics* 29, 31-48. *513*

Natsiavas, S. (1994b). Free vibration of two coupled nonlinear oscillators, *Nonlinear Dynamics* 6, 69-86.

Natsiavas, S. (1995a). Modal interactions in self-excited oscillators under external primary resonance, *Journal of Sound and Vibration* 184, 261-280.

Natsiavas, S. (1995b). Non-linear parametric resonance of spinning rings, *Journal of Sound and Vibration* 184, 93-109. *565, 637*

Nayfeh, A. See Oh, K., and Nayfeh, A.

Nayfeh, Adnan H. (1995). *Wave Propagation in Layered Anisotropic Media*, North-Holland, Amsterdam. *373*

Nayfeh, Adnan H. See Taylor, T. W., and Nayfeh, Adnan H.

Nayfeh, A. H. (1973a). Nonlinear transverse vibrations of beams with properties that vary along the length, *Journal of the Acoustical Society of America* 53, 766-770. *179*

Nayfeh, A. H. (1973b). *Perturbation Methods*, Wiley-Interscience, New York. *57, 121, 138, 139, 271, 526*

Nayfeh, A. H. (1981). *Introduction to Perturbation Techniques*, Wiley, New York. *57-59, 121, 138, 139, 181, 271, 273, 526, 651*

Nayfeh, A. H. (1983). The response of single-degree-of-freedom systems with quadratic and cubic nonlinearities to a subharmonic excitation, *Journal of Sound and Vibration* 89, 457-470. *61*

Nayfeh, A. H. (1985). Perturbation methods in nonlinear dynamics, in *Nonlinear Dynamics Aspects of Particle Accelerators: Proceedings of the Joint US-CERN School on Particle Accelerators*, Santa Margherita di Pula, Sardinia, Italy, January 31-February 5, J. M. Jowett, M. Month, and S. Turner, eds., Springer-Verlag, Berlin, pp. 238-314. *139*

Nayfeh, A. H. (1987a). Parametric excitation of two internally resonant oscillators, *Journal of Sound and Vibration* 119, 95-109. *26*

Nayfeh, A. H. (1987b). Surface waves in closed basins under parametric and internal resonances, *Physics of Fluids* 30, 2976-2983. *26*

Nayfeh A. H. (1988). Numerical-perturbation methods in mechanics, *Computers & Structures* 30, 185-204. *24, 25, 563*

Nayfeh, A. H. (1997). On the discretization of weakly nonlinear spatially continuous systems, in *Differential Equations and Chaos Lectures on Selected Topics*, N. H. Ibragimov, F. M. Mahomed, D. P. Mason, and D. Sherwell, eds., New Age International Limited, New Delhi, pp. 3-39. *53, 138*

Nayfeh, A. H. (1998). Reduced-order models of weakly nonlinear spatially continuous systems, *Nonlinear Dynamics* 16, 105-125. *53*

Nayfeh, A. H. (2000). *Nonlinear Interactions,* Wiley, New York. *23, 138, 139, 181, 183, 316, 324, 376, 382, 469, 504, 512, 528, 531, 559, 563, 641*

Nayfeh, A. H., and Alhazza, K. (2001). Dynamic instability and nonlinear vibration of doubly-curves cross-play shallows shells, in *Proceedings of the ASME 2001 Design Engineering Technical Conferences*, ASME Paper No. DETC01/VIB-21410, Pittsburgh, PA, September 9-13. *53, 564*

Nayfeh, A. H., and Arafat, H. N. (1998). Nonlinear response of cantilever beams to combination and subcombination resonances, *Shock and Vibration* 5, 277-288. *278, 298*

Nayfeh, A. H., Arafat, H. N., Chin, C-M., and Lacarbonara, W. (2002). Multimode interactions in suspended cables, *Journal of Vibration and Control* 8, 337-387. *53, 138, 150*

Nayfeh, A. H., and Balachandran, B. (1989). Modal interactions in dynamical and structural systems, *Applied Mechanics Reviews* 42, 175-201. *183*

Nayfeh, A. H., and Balachandran, B. (1995). *Applied Nonlinear Dynamics,* Wiley-Interscience, New York. *24, 268, 277, 288, 289*

Nayfeh, A. H., and Bouguerra, H. (1990). Non-linear response of a fluid valve, *International Journal of Non-Linear Mechanics* 25, 433-449. *57*

Nayfeh, A. H., and Chin, C-M. (1995). Nonlinear interactions in a parametrically excited system with widely spaced frequencies, *Nonlinear Dynamics* 7, 195-216. *270*

Nayfeh, A. H., and Chin, C-M. (1999a). *Perturbation Methods with Maple*, Dynamic Press, Blacksburg, Virginia.

Nayfeh, A. H., and Chin, C-M. (1999b). *Perturbation Methods with Mathematica*, Dynamic Press, Blacksburg, Virginia. *290*

Nayfeh, A. H., and Faris, W. (2002). Thermally induced principal parametric resonance in circular plates, *Shock and Vibration* 9, 143-150. *544*

Nayfeh, A. H., and Faris, W. (2003). Dynamic behavior of circular structural elements under thermal loads, in *Proceedings of the 44th AIAA/ASME/ASCE/AHS/ ASC Structures, Structural Dynamics, and Materials Conference*, AIAA Paper No. 2003-1618, Norfolk, Virginia, April 7-10, 2003. *544*

Nayfeh, A. H., Jilani, A., and Manzione, P. (2001). Transverse vibrations of a centrally clamped rotating circular disk, *Nonlinear Dynamics* 26, 163-178. *517, 519, 522, 524*

Nayfeh, A. H., and Khdeir, A. A. (1986a). Nonlinear rolling of ships in regular beams seas, *International Shipbuilding Progress* 33, 40-49. *24*

Nayfeh, A. H., and Khdeir, A. A. (1986b). Nonlinear rolling of biased ships in regular beam waves, *International Shipbuilding Progress* 33, 84-93. *24*

Nayfeh, A. H., Kreider, W., and Anderson, T. J. (1995). Investigation of natural frequencies and mode shapes of buckled beams, *AIAA Journal* 33, 1121-1126. *321, 323*

Nayfeh, A. H., and Lacarbonara, W. (1997). On the discretization of distributed-parameter systems with quadratic and cubic nonlinearities, *Nonlinear Dynamics* 13, 203-220. *53, 138*

Nayfeh, A. H., and Lacarbonara, W. (1998). On the discretization of spatially continuous systems with quadratic and cubic nonlinearities, *JSME International Journal, Series C* 41, 510-531. *53, 138*

Nayfeh, A. H., Lacarbonara, W., and Chin, C-M. (1999). Nonlinear normal modes of buckled beams: Three-to-one and one-to-one internal resonances, *Nonlinear Dynamics* 18, 253-273. *323*

Nayfeh, A. H., and Mook, D. T. (1979). *Nonlinear Oscillations*, Wiley-Interscience, New York. *23, 112, 116, 117, 181, 223, 272, 292, 296, 309, 316, 325, 382, 469, 499, 527, 528, 559, 563*

Nayfeh, A. H., Mook, D. T., and Lobitz, D. W. (1974). Numerical-perturbation method for the nonlinear analysis of structural vibrations, *AIAA Journal* 12, 1222-1228.

Nayfeh, A. H., Mook, D. T., and Marshall, L. R. (1973). Nonlinear coupling of pitch and roll modes in ship motions, *Journal of Hydrodynamics* 7, 145-152. *25*

Nayfeh, A. H., Mook, D. T., and Nayfeh, J. F. (1987). Some aspects of modal interactions in the response of beams, in *Proceedings of the 28th AIAA/ASME/ASCE/AHS Structural Dynamics and Materials Conference*, AIAA Paper No. 87-0777, Monterey, California, April 6-8, pp. 206-216. *181*

Nayfeh, A. H., Mook, D. T., and Sridhar, S. (1974). Nonlinear analysis of the forced response of structural elements, *Journal of the Acoustical Society of America* 55, 281-291. *181*

Nayfeh, A. H., Nayfeh, J. F., and Mook, D. T. (1992). On methods for continuous systems with quadratic and cubic nonlinearities, *Nonlinear Dynamics* 3, 145-162. *53, 138*

Nayfeh, A. H., Nayfeh, S. A., and Pakdemirli, M. (1995). On the discretization of weakly nonlinear spatially continuous systems, in *Nonlinear Dynamics and*

Stochastic Mechanics, W. Kliemann and N. S. Namachchivaya, eds., CRC Press, Boca Raton, Florida, pp. 175-200. *53, 138*

Nayfeh, A. H., and Oh, I. G. (1995). Nonlinearly coupled pitch and roll motions in the presence of internal resonance: Part I, Theory, *International Shipbuilding Progress* 42, 295-324. *25*

Nayfeh, A. H., and Pai, P. F. (1989). Nonlinear nonplanar parametric responses of an inextensional beam, *International Journal of Non-Linear Mechanics* 24, 139-158. *177, 182, 223, 284, 289*

Nayfeh, A. H., and Raouf, R. A. (1987a). Nonlinear forced response of infinitely long circular cylindrical shells, *Journal of Applied Mechanics* 54, 571-577. *564, 641*

Nayfeh, A. H., and Raouf, R. A. (1987b). Non-linear oscillation of circular cylindrical shells, *International Journal of Solids and Structures* 23, 1625-1638. *564, 641*

Nayfeh, A. H., Raouf, R. A., and Nayfeh, J. F. (1991). Nonlinear response of infinitely long circular cylindrical shells to subharmonic radial loads, *Journal of Applied Mechanics* 58, 1033-1041. *565*

Nayfeh, A. H., and Zavodney, L. D. (1986). The response of two-degree-of-freedom system with quadratic non-linearities to a combination parametric resonance, *Journal of Sound and Vibration* 107, 329-350. *25*

Nayfeh, A. H. See Abdel-Rahman, E. M., Younis, M. I., and Nayfeh, A. H.; Abou-Rayan, A. M., Nayfeh, A. H., and Mook, D. T.; Alhazza, K. A., and Nayfeh, A. H.; Anderson, T. J., Balachandran, B., and Nayfeh, A. H.; Balachandran, B., Nayfeh, A. H., Smith, S. W., and Pappa, R. S.; Anderson, T. J., Balachandran, B., and Nayfeh, A. H.; Anderson, T. J., Nayfeh, A. H., and Balachandran, B.; Anderson, T. J., and Nayfeh, A. H.; Arafat, H. N., and Nayfeh, A. H.; Arafat, H. N., and Nayfeh, A. H.; Arafat, H. N., and Nayfeh, A. H.; Arafat, H. N., and Nayfeh, A. H.; Arafat, H., and Nayfeh, A. H.; Arafat, H. N., Nayfeh, A. H., and Chin, C.; Arafat, H. N., Nayfeh, A. H., and Faris, W.; Asfar, O. R., Masad, J. A., and Nayfeh, A. H.; Balachandran, B., and Nayfeh, A. H.; Chin, C-M., and Nayfeh, A. H.; Chin, C-M., and Nayfeh, A. H.; Chin, C-M., and Nayfeh, A. H.; Chin, C-M., and Nayfeh, A. H.; Chin, C., Nayfeh, A. H., and Lacarbonara, W.; Emam, S. A., and Nayfeh, A. H.; Emam, S. A., and Nayfeh, A. H.; Emam, S. A., and Nayfeh, A. H.; Faris, W. F., Abdel-Rahman, E. M., and Nayfeh, A. H.; Hadian, J., and Nayfeh, A. H.; Hadian, J., and Nayfeh, A. H.; Kreider, W., and Nayfeh, A. H.; Lacarbonara, W., Nayfeh, A. H., and Kreider, W.; Lobitz, D. W., Nayfeh, A. H., and Mook, D. T.; Malatkar, P., and Nayfeh, A. H.; Mook, D. T., Marshall, L. R., and Nayfeh, A. H.; Nayfeh, S. A., and Nayfeh, A. H.; Nayfeh, S. A., Nayfeh, A. H., and Mook, D. T.; Pai, P. F., and Nayfeh, A. H.; Pai, P. F., and Nayfeh, A. H.; Pai, P. F., and Nayfeh,

A. H.; Pai, P. F., and Nayfeh, A. H.; Pai, P. F., and Nayfeh, A. H.; Pai, P. F., and Nayfeh, A. H.; Pai, P. F., and Nayfeh, A. H.; Pai, P. F., and Nayfeh, A. H.; Pai, P. F., and Nayfeh, A. H.; Pai, P. F., and Nayfeh, A. H.; Pai, P. F., and Nayfeh, A. H.; Pai, P. F., Nayfeh, A. H., Oh, K., and Mook, D. T.; Pakdemirli, M., Nayfeh, S. A., and Nayfeh, A. H.; Raouf, R. A., and Nayfeh, A. H.; Raouf, R. A., and Nayfeh, A. H.; Rega, G., Lacarbonara, W., Nayfeh, A. H., and Chin, C-M.; Rega, G., Lacarbonara, W., Nayfeh, A. H., and Chin, C-M.; Sridhar, S., Mook, D. T., and Nayfeh, A. H.; Sridhar, S., Mook, D. T., and Nayfeh, A. H.; Tezak, E. G., Mook, D. T., and Nayfeh, A. H.; Vogl, G. W., and Nayfeh, A. H.; Younis, M. I., Abdel-Rahman, E. M., and Nayfeh, A. H.; Younis, M. I., and Nayfeh, A. H.; Zavodney, L. D., and Nayfeh, A. H.; Zavodney, L. D., and Nayfeh, A. H.; Zavodney, L. D., Nayfeh, A. H., and Sanchez, N. E.; Zavodney, L. D., Nayfeh, A. H., and Sanchez, N. E.

Nayfeh, J. F. (1990). *Dynamics of Composite Plates Using Higher-Order Shear Deformation Theory*, PhD Dissertation, Virginia Polytechnic Institute and State University, Blacksburg, Virginia. *469*

Nayfeh, J. F., and Rivieccio, N. J. (2000). Nonlinear vibration of composite shell subjected to resonant excitations, *Journal of Aerospace Engineering* 13, 59-68. *563*

Nayfeh, J. F. See Nayfeh, A. H., Mook, D. T., and Nayfeh, J. F.; Nayfeh, A. H., Nayfeh, J. F., and Mook, D. T.; Nayfeh, A. H., Raouf, R. A., and Nayfeh, J. F.

Nayfeh, S. A., and Nayfeh, A. H. (1993). Nonlinear interactions between two widely spaced modes – External excitation, *International Journal of Bifurcation and Chaos* 3, 417-427. *270*

Nayfeh, S. A., Nayfeh, A. H., and Mook, D. T. (1995). Nonlinear response of a taut string to longitudinal and transverse end excitation, *Journal of Vibration and Control* 1, 307-334. *113, 121, 124, 129, 130, 132, 133, 135*

Nayfeh, S. A. See Nayfeh, A. H., Nayfeh, S. A., and Pakdemirli, M.; Pakdemirli, M., Nayfeh, S. A., and Nayfeh, A. H.

Nayfeh, T. A., and Vakakis, A. F. (1994). Subharmonic travelling waves in a geometrically nonlinear circular plate, *International Journal of Non-Linear Mechanics* 29, 233-245. *499*

Nelson, R. B., and Lorch, D. R. (1974). A refined theory for laminated orthotropic plates, *Journal of Applied Mechanics* 41, 177-183. *374*

Ng, T. Y. See Lam, K. Y., Ng, T. Y., and Qian, W.

Ni, R. G., and Adams, R. D. (1984). The damping and dynamic moduli of symmetric laminated composite beams–Theoretical and experimental results, *Journal of Composite Materials* 18, 104-121. *8*

Noble, B., and Daniel, J. W. (1977). *Applied Linear Algebra,* Prentice-Hall, New Jersey. *52*

Noor, A. K. (1973a). Free vibration of multilayered composite plates, *AIAA Journal* 11, 1038-1039. *378*

Noor, A. K. (1973b). Mixed finite-difference scheme for simply supported thick plates, *Computers & Structures* 3, 967-982. *379*

Noor, A. K. (1975). Stability of multilayered composite plates, *Fibre Science and Technology* 8, 81-89. *379*

Noor, A. K., and Burton, W. S. (1990). Assessment of computational models for multilayered composite shells, *Applied Mechanics Reviews* 43, 67-97. *559*

Noor, A. K., Burton, W. S., and Bert, C. W. (1996). Computational models for sandwich panels and shells, *Applied Mechanics Reviews* 49, 155-199. *559*

Noor, A. K., Burton, W. S., and Peters, J. M. (1995). Hierarchical adaptive modeling of structural sandwiches and multilayered composite panels, *Engineering Fracture Mechanics* 50, 801-817. *559*

Noor, A. K., Hadian, M. J., and Peters, J. M. (1994). Reduced basic technique for evaluating the sensitivity of the nonlinear vibrational response of composite plates, *Computers & Structures* 52, 1097-1105. *559*

Noor, A. K., and Hartley, S. J. (1977). Effects of shear deformation and anisotropy on the non-linear response of composite plates, in *Development in Composite Materials*, G. Holister, ed., Applied Science Publishers, Barking, Essex, Vol. 1, pp. 55-65. *373*

Nordgren, R. P. See Naghdi, P. M., and Nordgren, R. P.

Norris, C. H. See Yang, P. C., Norris, C. H., and Stavsky, Y.

Novozhilov, V. V. (1953). *Foundations of the Nonlinear Theory of Elasticity,* Greylock Press, Rochester, New York. *562*

Novozhilov, V. V. (1958). *Thin Elastic Shells*, translated from 2nd Russian edition by Lowe, P. G., London. *559*

Nowinski, J. L. (1964). Nonlinear transverse vibrations of a spinning disk, *Journal of Applied Mechanics* 31, 72-78. *498, 513*

Nowinski, J. L., and Ohnabe, H. (1972). On certain inconsistencies in Berger equations for large deflection of plastic plates, *International Journal of Mechanical Science* 14, 165-170. *375*

Nowinski, J. L., and Woodall, S. R. (1964). Finite vibrations of free rotating anisotropic membrane, *Journal of the Acoustical Society of America* 36, 2113-2118. *513*

Ogg, J. S. See MacBain, J. C., Horner, J. E., Stange, W. A., and Ogg, J. S.

Oh, I. G. See Nayfeh, A. H., and Oh, I. G.

Oh, K. (1994). *A Theoretical and Experimental Study of Modal Interactions in Metallic and Laminated Composite Plates*, PhD Dissertation, Virginia Polytechnic Institute and State University, Blacksburg, VA. *469, 499*

Oh, K., and Nayfeh, A. H. (1998). High- to low-frequency modal interactions in a cantilever composite plate, *Journal of Vibration and Acoustics* 120, 579-587. *499*

Oh, K. See Pai, P. F., Nayfeh, A. H., Oh, K., and Mook, D. T.

Ohnabe, H. See Nowinski, J. L., and Ohnabe, H.

Oja, A. See Veijola, T., Mattila, T., Jaakkola, O., Kiihamaki, J., Lamminmaki, T., Oja, A., Ruokonen, K., Seppala, H., and Tittonen, I.

Omurtag, M. H. See Basar, Y., and Omurtag, M. H.

O'Reilly, O., and Holmes, P. J. (1992). Nonlinear, non-planar, and non-periodic vibrations of a string, *Journal of Sound and Vibration* 153, 413-435. *112, 132*

Oshima, I., Seguchi, Y., and Shindo, A. (1970). On nonlinear shell theories, *Bulletin of the JSME* 13, 1155-1164. *563, 605*

Osterberg, P. M. (1995). *Electrostatically Actuated Microelectromechanical Test Structures for Material Property Measurement*, PhD Dissertation, Massachusetts Institute of Technology, Cambridge, Massachusetts. *541, 542*

Osterberg, P., Yie, H., Cai, X., White, J., and Senturia, S. (1994). Self-consistent simulation and modeling of electrostatically deformed diaphragms, in *Proceedings of the IEEE Workshop on Micro Electro Mechanical Systems, MEMS'94*, Oiso, Japan, January 25-28, pp. 28-32. *541*

Ostiguy, G. L., and Evan-Iwanowski, R. M. (1993). On dynamic stability and nonlinear modal interaction of parametrically excited rectangular plates, in *Proceedings of the ASME 1993 Design Engineering Technical Conferences*, ASME DE-Vol. 56, Albuquerque, New Mexico, September 19-22, pp. 465-474. *498*

Otomi, K. See Sato, K., Saito, H., and Otomi, K.

Oyibo, G. A., and Berman, J. H. (1985). Anisotropic wing aeroelastic theories with warping effects, in *Second International Symposium on Aeroelasticity and Structural Dynamics*, DGLR Paper No. 85-57, Technical University of Aachen, West Germany, April 1-3, pp. 434-440. *175*

Pacheco, B. M. See Fujino, Y., Warnitchai, P., and Pacheco, B. M.

Pagano, N. J. (1969). Exact solutions for composite laminates in cylindrical bending, *Journal of Composite Materials* 3, 398-411. *372, 373*

Pagano, N. J. (1970). Exact solutions for rectangular bidirectional composites and sandwich plates, *Journal of Composite Materials* 4, 20-34. *372, 373*

Pagano, N. J., and Hatfield, S. J. (1972). Elastic behavior of multilayer bidirectional composites, *AIAA Journal* 10, 931-933. *372, 373*

Pagano, N. J. See Tsai, S. W., and Pagano, N. J.; Whitney, J. M., and Pagano, N. J.

Pai, P. F. (1995). A new look at shear correction factors and warping functions of anisotropic laminates. *International Journal of Solids and Structures* 32, 2295-2313. *373, 374, 446*

Pai, P. F., and Nayfeh, A. H. (1990a). Nonlinear nonplanar oscillations of a cantilever beam under lateral base excitations, *International Journal of Non-Linear Mechanics* 25, 455-474. *176, 182, 307-309*

Pai, P. F., and Nayfeh, A. H. (1990b). Three-dimensional nonlinear vibrations of composite beams – I. Equations of motion, *Nonlinear Dynamics* 1, 477-502. *177, 233, 376*

Pai, P. F., and Nayfeh, A. H. (1991a). Three-dimensional nonlinear vibrations of composite beams – II. Flapwise excitations, *Nonlinear Dynamics* 2, 1-34. *136, 180, 182, 379*

Pai, P. F., and Nayfeh, A. H. (1991b). Three-dimensional nonlinear vibrations of composite beams – III. Chordwise excitations, *Nonlinear Dynamics* 2, 137-156. *136, 180, 182, 379*

Pai, P. F., and Nayfeh, A. H. (1991c). A nonlinear composite plate theory, *Nonlinear Dynamics* 2, 445-477. *177, 375-377, 563, 593*

Pai, P. F., and Nayfeh, A. H. (1992a). A nonlinear composite beam theory, *Nonlinear Dynamics* 3, 273-303. *174, 176, 177, 563*

Pai, P. F., and Nayfeh, A. H. (1992b). A nonlinear composite shell theory, *Nonlinear Dynamics* 3, 431-463. *561, 563*

Pai, P. F., and Nayfeh, A. H. (1992c). Fully nonlinear model of cables, *AIAA Journal* 30, 2993-2996. *137*

Pai, P. F., and Nayfeh, A. H. (1994a). A fully nonlinear theory of curved and twisted composite rotor blades accounting for warpings and three-dimensional stress effects, *International Journal of Solids and Structures* 31, 1309-1340. *176, 177*

Pai, P. F., and Nayfeh, A. H. (1994b). A unified nonlinear formulation for plate and shell theories, *Nonlinear Dynamics* 6, 459-500. *562, 563*

Pai, P. F., and Nayfeh, A. H. (1994c). A new method for the modeling of geometric nonlinearities in structures, *Computers & Structures* 53, 877-895. *215*

Pai, P. F., Nayfeh, A. H., Oh, K., and Mook, D. T. (1993). A refined nonlinear model of composite plates with integrated piezoelectric actuators and sensors, *International Journal of Solids and Structures* 30, 1603-1630. *374, 376, 377, 561, 562*

Pai, P. F., and Palazotto, A. N. (1995). Polar decomposition theory in nonlinear analyses of solids and structures, *Journal of Engineering Mechanics* 121, 568-581. *376*

Pai, P. F., and Palazotto, A. N. (2001). A higher-order sandwich plate theory accounting for 3-D stresses, *International Journal of Solids and Structures* 38, 5045-5062. *562*

Pai, P. F., Palazotto, A. N., and Greer, J. M., Jr. (1998). Polar decomposition and appropriate strains and stresses for nonlinear structural analyses, *Computers & Structures* 66, 823-840. *136, 177*

Pai, P. F., and Schulz, M. J. (1999). Shear correction factors and an energy-consistent beam theory, *International Journal of Solids and Structures* 36, 1523-1540. *174, 176, 191, 247*

Pai, P. F. See Nayfeh, A. H., and Pai, P. F.

Pakdemirli, M., Nayfeh, S. A., and Nayfeh, A. H. (1995). Analysis of one-to-one autoparametric resonances in cables – Discretization vs. direct treatment, *Nonlinear Dynamics* 8, 65-83. *53, 136-138*

Pakdemirli, M. See Nayfeh, A. H., Nayfeh, S. A., and Pakdemirli, M.

Pal, M. C. (1969). Large deflections of heated circular plates, *Acta Mechanica* 8, 82-103. *544*

Pal, M. C. (1970). Large amplitude free vibration of circular plates subjected to aerodynamic heating, *International Journal of Solids and Structures* 6, 301-313. *544*

Pal, M. C. (1973). Static and dynamic non-linear behaviour of heated orthotropic circular plates, *International Journal of Non-Linear Mechanics* 8, 489-504. *544*

Palazotto, A. N. See Bowlns, J. A., Palazotto, A. N., and Whitney, J. M.; Dennis, S. T., and Palazotto, A. N.; Dennis, S. T., and Palazotto, A. N.; Pai, P. F., and Palazotto, A. N.; Pai, P. F., and Palazotto, A. N.; Pai, P. F., Palazotto, A. N., and Greer, J. M. Jr.

Pandalai, K. A. V. See Prathap, G., and Pandalai, K. A. V.

Pao, Y. H. See Kung, G. C., and Pao, Y. H.

Paolone, A. See Luongo, A., Paolone, A., and Piccardo, G.

Papazoglou, V. J., and Mavrakos, S. A. (1990). Nonlinear cable response and model testing in water, *Journal of Sound and Vibration* 140, 103-115. *136*

Pappa, R. S. See Balachandran, B., Nayfeh, A. H., Smith, S. W., and Pappa, R. S.

Parker, D. F. (1979a). An asymptotic analysis of large deflections and rotations of elastic rods, *International Journal of Solids and Structures* 15, 361-377. *174, 175*

Parker, D. F. (1979b). The role of Saint Venant's solutions in rod and beam theories, *Journal of Applied Mechanics* 46, 861-866. *174, 175*

Parker, I. B. See Fox, L., and Parker, I. B.

Pasca, M. See Gattulli, V., Pasca, M., and Vestroni, F.

Pastorel, H., and Beaulieu, G. (1985). Nonlinear vibrations of guy cable systems, *Computers & Structures* 21, 33-50. *136*

Penafiel, H. R. See Canosa, J., and Penafiel, H. R.

Pendered, J. W., and Bishop, R. E. D. (1963). A critical introduction to some industrial resonance testing techniques, *Journal of Mechanical Engineering Science* 5, 345-367. *8*

Perkins, N. C. (1992). Modal interactions in the non-linear response of elastic cables under parametric/external excitation, *International Journal of Non-Linear Mechanics* 27, 233-250. *137*

Perkins, N. C. See Lee, C. L., and Perkins, N. C.; Lee, C. L., and Perkins, N. C.; Lee, C. L., and Perkins, N. C.

Peters, D. A. (1977). The effect of principal bending curvature on the lateral buckling of uniform slender beams, *Journal of Applied Mechanics* 44, 311-316. *178*

Peters, D. A., and Izadpanah, A. P. (1988). HP-version finite elements for the space-time domain, *Computational Mechanics* 3, 73-88. *71*

Peters, D. A. See Hodges, D. H., Crespo da Silva, M. R. M., and Peters, D. A.; Hodges, D. H., and Peters, D. A.

Peters, J. M. See Noor, A. K., Burton, W. S., and Peters, J. M.; Noor, A. K., Hadian, M. J., and Peters, J. M.

Petyt, M. See Ribeiro, P., and Petyt, M.

Pezeshki, C. See Dowell, E. H., and Pezeshki, C.

Phan, N. D. See Reddy, J. N., and Phan, N. D.

Phoon, K. H. See Tay, E. H., and Phoon, K. H.

Piccardo, G. See Luongo, A., Paolone, A., and Piccardo, G.

Pietraszkiewicz, W., and Szwabowicz, M. L. (1981). Entirely Lagrangian nonlinear theory of thin shells, *Archives of Mechanics* 33, 273-288. *562*

Piezo Systems (1987). *Piezoelectric Motor/Actuator Kit Manual,* Piezo Electric Products, Inc., Advanced Technology Group, Cambridge, Maryland. *456*

Pilipchuk, V. N., and Ibrahim, R. A. (1997). Strong nonlinear modal interaction in shallow suspended cables with oscillating ends, *Chaos, Solitons & Fractals* 8, 637-657. *137*

Pinard, P. See Malhaire, C., Le Berre, M., Febvre, M., Barbier, D., and Pinard, P.

Plaut, R. H. See Shyu, I.-M. K., Mook, D. T., and Plaut, R. H.; Shyu, I.-M. K., Mook, D. T., and Plaut, R. H.; Shyu, I.-M. K., Mook, D. T., and Plaut, R. H.

Poh, S. See Baz, A., and Poh, S.

Pointou, C. See Croquette, V., and Pointou, C.

Prabhakara, M. K. See Chia, C. Y., and Prabhakara, M. K.; Chia, C. Y., and Prabhakara, M. K.

Prasad, M. E. See Sathyamoorthy, M., and Prasad, M. E.

Prathap, G., and Pandalai, K. A. V. (1979). Non-linear vibrations of transversely isotropic rectangular plates, *International Journal of Non-Linear Mechanics* 13, 285-294. *373*

Prathap, G., and Varadan, T. K. (1978). On the nonlinear vibrations of rectangular plates, *Journal of Sound and Vibration* 56, 521-530. *527*

Qatu, M. S. (1992). Review of shallow shell vibration research, *Shock and Vibration Digest* 24, 3-15. *559*

Qatu, M. S. (2002a). Recent research advances in the dynamic behavior of shells: 1989-2000. Part 1: Laminated composite shells, *Applied Mechanics Reviews* 55, 325-351. *559*

Qatu, M. S. (2002b). Recent research advances in the dynamic behavior of shells: 1989-2000. Part 2: Homogeneous shells, *Applied Mechanics Reviews* 55, 415-435. *559*

Qatu, M. S. (2004). *Vibration of Laminated Shells and Plates,* Elsevier, Oxford, UK. *372, 378, 469, 559*

Qatu, M. S., and Leissa, A. W. (1991). Natural frequency for cantilevered doubly-curved laminated composite shallow shells, *Journal of Composite Structures* 17, 227-255. *378*

Qian, W. See Lam, K. Y., Ng, T. Y., and Qian, W.

Rabinovich, V. See Grétillat, M.-A., Yang, Y.-J., Hung, E. S., Rabinovich, V., Ananthasuresh, G. K., deRooij, N. F., and Senturia, S. D.

Raciti, S. See Kapania, R. K., and Raciti, S.; Kapania, R. K., and Raciti, S.; Kapania, R. K., and Raciti, S.

Radkowski, P. P. See Budiansky, B., and Radkowski, P. P.

Rajalingham, C., Bhat, R. B., and Xistris, G. D. (1996). Vibration of rectangular plates using plate characteristic functions as shape functions in the Rayleigh-Ritz method, *Journal of Sound and Vibration* 193, 497-509. *378*

Rajgopal, S. V., Singh, G., Sadasiva Rao, Y. V. K., and Narayanan, S. (1986). Nonlinear vibration of sandwich plates, *Journal of Sound and Vibration* 110, 261-269. *373*

Raju, B. B. See Chandra, R., and Raju, B. B.; Chandra, R., and Raju, B. B.

Raju, I. S. See Rao, G. V., Raju, I. S., and Raju, K. K.

Raju, K. K. See Rao, G. V., Raju, I. S., and Raju, K. K.

Ramakrishna, B. S. See Murthy, G. S. S., and Ramakrishna, B. S.

Ramesh, G., and Krishnamoorthy, C. S. (1995). Geometrically non-linear analysis of plates and shallow shells by dynamic relaxation, *Computer Methods in Applied Mechanics and Engineering* 123, 15-32. *380*

Ramu, S. A., Sankar, T. S., and Ganesan, R. (1994). Bifurcations, catastrophes and chaos in a pre-buckled beam, *International Journal of Non-Linear Mechanics* 29, 449-462. *318*

Rand, O. See Rosen, A., and Rand, O.

Rao, A. K. See Srinivas, S., Joga Rao, C. V., and Rao, A. K.; Srinivas, S., and Rao, A. K.

Rao, B. N. See Manoj, T., Ayyappan, M., Krishnan, K. S., and Rao, B. N.

Rao, G. V., and Iyengar, R. N. (1991). Internal resonance and nonlinear response of a cable under periodic excitation, *Journal of Sound and Vibration* 149, 25-41. *136, 137*

Rao, G. V., Raju, I. S., and Raju, K. K. (1976). Nonlinear vibrations of beams considering shear deformation and rotary inertia, *AIAA Journal, Technical Notes* 14, 685-687. *179*

Rao, V. G. See Singh, G., Rao, V. G., and Iyengar, N. G. R.

Raouf, R. A., and Nayfeh, A. H. (1990a). One-to-one autoparametric resonances in infinitely long cylindrical shells, *Computers & Structures* 35, 163-173. *564, 638*

Raouf, R. A., and Nayfeh, A. H. (1990b). Nonlinear axisymmetric response of closed spherical shells to a radial harmonic excitation, *International Journal of Non-Linear Mechanics* 25, 475-492. *565*

Raouf, R. A. See Nayfeh, A. H., and Raouf, R. A.; Nayfeh, A. H., and Raouf, R. A.; Nayfeh, A. H., Raouf, R. A., and Nayfeh, J. F.

Ray, J. D., and Bert, C. W. (1969). Nonlinear vibrations of a beam with pinned ends, *Journal of Engineering for Industry* 91, 997-1004. *8, 182*

Reddy, J. N. (1980). A penalty plate-bending element for the analysis of laminated anisotropic composite plates, *International Journal for Numerical Methods in Engineering* 15, 1187-1206. *373*

Reddy, J. N. (1982). Large amplitude flexural vibrations of layered composite plates with cutouts, *Journal of Sound and Vibration* 83, 1-10. *380*

Reddy, J. N. (1983). Exact solutions of moderately thick laminated shells, *Journal of Engineering Mechanics* 110, 794-809. *377, 561*

Reddy, J. N. (1984a). A refined nonlinear theory of plates with transverse shear deformation, *International Journal of Solids and Structures* 20, 881-896. *374-378, 446, 454, 561*

Reddy, J. N. (1984b). A simple higher-order theory for laminated composite plates, *Journal of Applied Mechanics* 51, 745-752. *374, 376, 378, 379, 436, 454, 628*

Reddy, J. N. (1984c). *Energy and Variational Methods in Applied Mechanics*, Wiley-Interscience, New York. *374*

Reddy, J. N. (1985). A review of the literature on finite-element modeling of laminated composite plates, *Shock and Vibration Digest* 17, 3-8. *374, 375, 378*

Reddy, J. N. (1990). A general non-linear third-order theory of plates with moderate thickness, *International Journal of Non-Linear Mechanics* 25, 677-686. *374*

Reddy, J. N. (1997). *Mechanics of Laminated Composite Plates*, CRC Press, Boca Raton, Florida. *7, 372, 469*

Reddy, J. N. (2003). *Mechanics of Laminated Composite Plates and Shells: Theory and Analysis*, CRC Press, Boca Raton, Florida. *7, 372, 469*

Reddy, J. N., and Chandrashekhara, K. (1985). Geometrically nonlinear transient analysis of laminated, doubly curved shells, *International Journal of Non-Linear Mechanics* 20, 79-90. *563, 605*

Reddy, J. N., and Chao, W. C. (1981). Large deflection and large amplitude free vibrations of laminated composite material plates, *Computers & Structures* 13, 341-347. *373, 380*

Reddy, J. N., and Chao, W. C. (1982). Nonlinear oscillations of laminated anisotropic rectangular plates, *Journal of Applied Mechanics* 49, 396-402. *380*

Reddy, J. N., and Khdeir, A. A. (1989). Buckling and vibration of laminated composite plates using various plate theories, *AIAA Journal* 27, 1808-1817. *379*

Reddy, J. N., Khdeir, A. A., and Librescu, L. (1987). Levy type solutions for symmetrically laminated rectangular plates using first-order shear deformation theory, *Journal of Applied Mechanics* 54, 740-742. *377, 378*

Reddy, J. N., and Liu, C. F. (1985). A higher-order shear deformation theory of laminated elastic shells, *International Journal of Engineering Science* 23, 319-330. *561, 593*

Reddy, J. N., and Phan, N. D. (1985). Stability and vibration of isotropic, orthotropic, and laminated plates according to a higher-order deformation theory, *Journal of Sound and Vibration* 98, 157-170. *379*

Reddy, J. N. See Heyliger, P. R., and Reddy, J. N.; Khdeir, A. A., and Reddy, J. N.; Khdeir, A. A., and Reddy, J. N.; Khdeir, A. A., and Reddy, J. N.; Librescu, L., Khdeir, A. A., and Reddy, J. N.; Librescu, L., and Reddy, J. N.

Reddy, V. S. See Bert, C. W., and Reddy, V. S.

Redheffer, R. M. See Sokolnikoff, I. S., and Redheffer, R. M.

Rega, G. (1996). Non-linearity, bifurcation and chaos in the finite dynamics of different cable models, *Chaos, Solitons & Fractals* 7, 1507-1536.

Rega, G., and Alaggio, R. (2001). Spatio-temporal dimensionality in the overall complex dynamics of an experimental cable/mass system, *International Journal of Solids and Structures*, 38, 2049-2068. *136*

Rega, G., Alaggio, R., and Benedettini, F. (1997). Experimental investigation of the nonlinear response of a hanging cable. Part I: Local analysis, *Nonlinear Dynamics* 14, 89-117. *136, 137*

Rega, G., Lacarbonara, W., Nayfeh, A. H., and Chin, C-M. (1997). Multimodal resonances in suspended cables via a direct perturbation approach, in *Proceedings of the ASME 1997 Design Engineering Technical Conferences*, ASME Paper No. DETC97/VIB-4101, Sacramento, California, September 14-17. *53, 138, 150*

Rega, G., Lacarbonara, W., Nayfeh, A. H., and Chin, C-M. (1999). Multiple resonances in suspended cables: Direct versus reduced-order models, *International Journal of Non-Linear Mechanics* 34, 901-924. *53, 138, 150*

Rega, G., Vestroni, F., and Benedettini, F. (1984). Parametric analysis of large amplitude free vibrations of a suspended cable, *International Journal of Solids and Structures* 20, 95-105. *136, 150*

Rega, G. See Alaggio, R., and Rega, G.; Benedettini, F., and Rega, G.; Benedettini, F., and Rega, G.; Benedettini, F., and Rega, G.; Benedettini, F., and Rega, G.; Benedettini, F., Rega, G., and Alaggio, R.; Benedettini, F., Rega, G., and Vestroni, F.; Luongo, A., Rega, G., and Vestroni, F.

Rehfield, L. W. (1974). Large amplitude forced vibration of elastic structures, *AIAA Journal* 12, 388-390. *379*

Reissner, E. (1944). On the theory of bending of elastic plates, *Journal of Mathematical Physics* 23, 184-191. *372*

Reissner, E. (1945). The effect of transverse shear deformation on the bending of elastic plates, *Journal of Applied Mechanics* 12, A69-A77. *372*

Reissner, E. (1947). On bending of elastic plates, *Quarterly of Applied Mathematics* 5, 55-68. *372*

Reissner, E. (1948). Finite deflections of sandwich plates, *Journal of the Aeronautical Sciences* 15, 435-440. *375*

Reissner, E. (1950). An axisymmetric deformation theory of thin shells of revolution, *Proceedings of Symposia in Applied Mathematics* 3, 27-52. *562*

Reissner, E. (1953). On variational theory for finite elastic deformation, *Journal of Mathematical Physics* 32, 129-135. *375*

Reissner, E. (1956). Some aspects of the theory of thin elastic shells, *Journal of the Boston Society of Engineers* 42, 100-133. *560*

Reissner, E. (1963). On the equations for finite symmetrical deflections of thin shells of revolution, in *Progress in Applied Mechanics*, Prager Anniversary Volume. *562*

Reissner, E. See Hildebrand, F. B., Reissner, E., and Thomas, G. B.

Ribeiro, P., and Petyt, M. (1999). Nonlinear vibration of composite laminated plates by the hierarchical FEM, *Computers & Structures* 46, 197-208. *380*

Richards, J. B. (1872). *A treatise on the Construction and Operation of Wood-Working Machines*, E. & F. N. Spon, London. *513*

Riff, R. See Baruch, M., and Riff, R.

Rivieccio, N. J. See Nayfeh, J. F., and Rivieccio, N. J.

Robbins, K. A. (1977). A new approach to subcritical instability and turbulent transitions in a simple dynamo, *Mathematical Proceedings of the Cambridge Philosophical Society* 82, 309-325. *23*

Roberts, J. W. See Cartmell, M. P., and Roberts, J. W.

Roberts, S. M., and Shipman, J. S. (1972). *Two-Point Boundary Value Problem: Shooting Methods,* American Elsevier, New York. *52*

Robertson, S. R. See Brunelle, E. J., and Robertson, S. R.

Roos, M. See Korvink, J. G., Funk, J., Roos, M., and Wachutka, G.

Roseau, M. (1987). *Vibrations in Mechanical Systems: Analytical Methods and Applications,* Springer-Verlag, Berlin. *66*

Rosen, A., and Friedmann, P. (1979). The nonlinear behavior of elastic slender straight beams undergoing small strains and moderate rotations, *Journal of Applied Mechanics* 46, 161-168. *178*

Rosen, A., Loewy, R. G., and Mathew, M. B. (1987a). Nonlinear analysis of pretwisted rods using principal curvature transformation, Part I: Theoretical derivation, *AIAA Journal* 25, 470-478. *177, 178*

Rosen, A., Loewy, R. G., and Mathew, M. B. (1987b). Nonlinear analysis of pretwisted rods using principal curvature transformation, Part II: Numerical results, *AIAA Journal* 25, 598-604. *177-179, 238*

Rosen, A., Loewy, R. G., and Mathew, M. B. (1987c). Nonlinear dynamics of slender rods, *AIAA Journal* 25, 611-619. *178*

Rosen, A., and Rand, O. (1986). Numerical model of the nonlinear behavior of curved rods, *Computers & Structures* 22, 785-799. *177*

Rossettos, J. N. See Fortier, R. C., and Rossettos, J. N.

Rossler, O. E. (1976). Chemical turbulence: Chaos in a small reaction-diffusion system, *Zeitschrift fur Naturforschung* 31a, 1168-1172. *23*

Roufaeil, O. L., and Dawe, D. J. (1982). Raleigh-Ritz vibration analysis of rectangular Mindlin plates subjected to membrane stresses, *Journal of Sound and Vibration* 85, 263-275. *373*

Ruokonen, K. See Veijola, T., Mattila, T., Jaakkola, O., Kiihamaki, J., Lamminmaki, T., Oja, A., Ruokonen, K., Seppala, H., and Tittonen, I.

Ryan, R. R. See Kane, T. R., Ryan, R. R., and Banerjee, A. K.

Ryu, T. See Yoshitake, Y., Sueoka, A., Ryu, T., and Tamura, H.

Sadasiva Rao, Y. V. K. See Rajgopal, S. V., Singh, G., Sadasiva Rao, Y. V. K., and Narayanan, S.

Sager, K. See Dunbar, M. L., and Sager, K.

Saif, M. T. A., Alaca, B. E., and Sehitoglu, H. (1999). Analytical modeling of electrostatic membrane actuator for micropumps, *Journal of Microelectromechanical Systems* 8, 335-345. *531, 532*

Saito, H. See Sato, K., Saito, H., and Otomi, K.

Samani, K. See Kadivar, M. H., and Samani, K.

Samoilenko, S. B. See Lee, W. K., Yeo, M. H., and Samoilenko, S. B.

Sanchez, N. E. See Zavodney, L. D., Nayfeh, A. H., and Sanchez, N. E.; Zavodney, L. D., Nayfeh, A. H., and Sanchez, N. E.

Sanders, J. L. (1959). An improved first approximation theory of thin shells, NASA TR-24. *560, 562, 612-614*

Sanders, J. L. (1963). Nonlinear theories of thin shells, *Quarterly of Applied Mathematics* 21, 21-36. *562*

Sanders, J. L. See Budiansky, B., and Sanders, J. L.

Sandman, B. E. See Huang, C., and Sandman, B. E.

Sanford, W. F. See Evan-Iwanowski, R. M., Sanford, W. F., and Kehagioglou, T.

Sankar, T. S. See Ramu, S. A., Sankar, T. S., and Ganesan, R.

Saravanos, D. A., and Chamis, C. C. (1990a). Unified micromechanics of damping for unidirectional and off-axis fiber composites, *Journal of Composites Technology and Research* 12, 31-40. *8*

Saravanos, D. A., and Chamis, C. C. (1990b). Mechanics of damping for fiber composite laminates including hygro-thermal effects, *AIAA Journal* 28, 1813-1819. *8*

Saravanos, D. A., and Chamis, C. C. (1991). Computational simulation of damping in composite structures, *Journal of Reinforced Plastics and Composites* 10, 256-278. *8*

Sarma, B. S. See Ganapathi, M., Varadan, T. K., and Sarma, B. S.

Sathyamoorthy, M. (1978). Vibration of plates considering shear and rotary inertia, *AIAA Journal* 16, 285-286. *373*

Sathyamoorthy, M. (1979). Effects of large amplitude, shear and rotatory inertia on vibration of rectangular plates, *Journal of Sound and Vibration* 63, 161-167. *373*

Sathyamoorthy, M. (1981). Transverse shear and rotary inertia effects on nonlinear vibration of orthotropic circular plates, *Computers & Structures* 14, 129-134. *373*

Sathyamoorthy, M. (1982a). Nonlinear analysis of beams Part I: A survey of recent advances, *Shock and Vibration Digest* 14, 19-35. *179*

Sathyamoorthy, M. (1982b). Nonlinear analysis of beams Part II: Finite element methods, *Shock and Vibration Digest* 14, 7-18. *179*

Sathyamoorthy, M. (1984a). Multiple mode non-linear dynamic analysis of thick orthotropic elliptic plates, *Journal of Sound and Vibration* 96, 353-361. *373*

Sathyamoorthy, M. (1984b). Vibration of orthotropic thick plates, *AIAA Journal* 22, 851-854. *373*

Sathyamoorthy, M. (1997). *Nonlinear Analysis of Structures*, CRC Press, Boca Raton, Florida. *382, 469, 498, 528, 559*

Sathyamoorthy, M., and Chia, C. Y. (1980). Non-linear vibration of anisotropic rectangular plates including shear and rotatory inertia, *Fibre Science and Technology* 13, 337-361. *373*

Sathyamoorthy, M., and Prasad, M. E. (1983). Multiple mode nonlinear analysis of circular plates, *Journal of Engineering Mechanics* 109, 1114-1123. *373*

Sato, K., Saito, H., and Otomi, K. (1978). The parametric response of a horizontal beam carrying a concentrated mass under gravity, *Journal of Applied Mechanics* 45, 643-648. *179*

Schäfer, B. See Hagedorn, P., and Schäfer, B.

Schmidt, G., and Tondl, A. (1986). *Non-Linear Vibrations,* Akademie-Verlag, Berlin. *24*

Schmidt, R. (1977). Refined non-linear theory of plates with transverse shear deformation, *Industrial Mathematics* 27, 23-38. *373-375*

Schmidt, R. (1984). Thin elastic shells undergoing small strains and large rotations – A simple consistent theory and variational principle, in *Numerical Methods for Nonlinear Problems,* C. Taylor, E. Hinton, and D. Owen, eds., Pineridge Press, Swansea, Great Britain, Vol. 2, pp. 170-181. *562*

Schmidt, R. (1985). On the entirely Lagrangian first approximation theory of thin elastic shells undergoing small strains and arbitrary rotations, *Zeitschrift fur Angewandte Mathematik und Mechanik (ZAMM)* 65, 119-121. *562*

Schmidt, R. See Librescu, L., and Schmidt, R.; Librescu, L., and Schmidt, R.

Schultz, A. B., and Tsai, S. W. (1968). Dynamic moduli and damping ratios in fiber-reinforced composites, *Journal of Composite Materials* 2, 368-379. *8*

Schulz, M. J. See Pai, P. F., and Schulz, M. J.

Schwartz, I. B. See Georgiou, I. T., and Schwartz, I. B.; Georgiou, I. T., Schwartz, I. B., Emaci, E., and Vakakis, A.

Scott, M. R., and Watts, H. A. (1975). SUPORT - A computer code for two-point boundary-value problems via orthonormalization, SAND75-0198, Sandia Laboratories, Albuquerque, New Mexico. *50, 52*

Scott, M. R., and Watts, H. A. (1977). Computational solution of linear two-point boundary value problems and orthonormalization, *SIAM Journal on Numerical Analysis* 14, 40-70. *52*

Scott, R. A. See Ho, C.-H., Scott, R. A., and Eisley, J. G.; Ho, C.-H., Scott, R. A., and Eisley, J. G.

Seeger, J. I., and Crary, S. B. (1997). Stabilization of electrostatically actuated mechanical devices, in *TRANSDUCERS'97: International Conference on Solid-State Sensors and Actuators*, Chicago, Illinois, pp. 1133-1136. *342*

Seguchi, Y. See Oshima, I., Seguchi, Y., and Shindo, A.

Sehitoglu, H. See Saif, M. T. A., Alaca, B. E., and Sehitoglu, H.

Seide, P. (1975). *Small Elastic Deformations of Thin Shells,* Noordhoff, Leyden, The Netherlands. *372, 560*

Selmane, A. See Lakis, A. A., Selmane, A., and Toledano, A.

Senturia, S. See Osterberg, P., Yie, H., Cai, X., White, J., and Senturia, S.

Senturia, S. D., Harris, R. M., Johnson, B. P., Kim, S., Nabors, K., Shulman, M. A., and White, J. K. (1992). A computer-aided design system for microelectromechanical systems (MEMCAD), *Journal of Microelectromechanical Systems* 1, 3-13. *341*

Senturia, S. D. See Castañer, L. M., and Senturia, S. D.; Gabbay, L. D., Mehner, J. E., and Senturia, S. D.; Grétillat, M.-A., Yang, Y.-J., Hung, E. S., Rabinovich, V., Ananthasuresh, G. K., deRooij, N. F., and Senturia, S. D.; Hung, E. S., and Senturia, S. D.; Mehner, J. E., Gabbay, L. D., and Senturia, S. D.

Seppala, H. See Veijola, T., Mattila, T., Jaakkola, O., Kiihamaki, J., Lamminmaki, T., Oja, A., Ruokonen, K., Seppala, H., and Tittonen, I.

Sethna, P. R. (1965). Vibrations of dynamical systems with quadratic nonlinearities, *Journal of Applied Mechanics* 32, 576-582. *25*

Sethna, P. R. See Feng, Z. C., and Sethna, P. R.; Yang, X. L., and Sethna, P. R.; Yang, X. L., and Sethna, P. R.

Seydel, R. (1988). *From Equilibrium to Chaos: Practical Bifurcation and Stability Analysis,* Elsevier, New York. *24*

Seydel, R. (1994). *Practical Bifurcation and Stability Analysis,* Springer-Verlag, New York. *288, 289*

Shames, I. H., and Dym, C. L. (1985). *Energy and Finite Element Methods in Structural Mechanics,* McGraw-Hill, New York, pp. 197-204. *173*

Shaw, J., and Shaw, S. W. (1989). The onset of chaos in a two-degree-of-freedom impacting system, *Journal of Applied Mechanics* 56, 168-174. *26*

Shaw, S. W. See Shaw, J., and Shaw, S. W.

Sheinman, I., and Adan, M. (1987). The effect of shear deformation on postbuckling behavior of laminated beams, *Journal of Applied Mechanics* 54, 558-562. *174, 177*

Shen, I. Y., and Mote, Jr., C. D. (1991). On the mechanism of instability of a circular plate under a rotating spring-mass-dashpot system, *Journal of Sound and Vibration* 178, 307-318. *517*

Shi, G. See Voyiadjis, G. Z., and Shi, G.

Shimizu, T., Yamaguchi, H., and Ito, M. (1981). Experimental study of nonlinear dynamic response of a cable in three dimensions, in *Proceedings of the 36th Annual Conference of the Japan Society of Civil Engineers,* Vol. 1, pp. 371-372. *136*

Shimizu, T. See Yasuda, K., Torii, T., and Shimizu, T.

Shindo, A. See Oshima, I., Seguchi, Y., and Shindo, A.

Shipman, J. S. See Roberts, S. M., and Shipman, J. S.

Shiraishi, H. See Kojima, H., Nagaya, J., Shiraishi, H., and Yamashita, A.

Shulman, M. A. See Senturia, S. D., Harris, R. M., Johnson, B. P., Kim, S., Nabors, K., Shulman, M. A., and White, J. K.

Shyu, I.-M. K., Mook, D. T., and Plaut, R. H. (1993a). Whirling of a forced cantilevered beam with static deflection. I: Primary resonance, *Nonlinear Dynamics* 4, 227-249. *179*

Shyu, I.-M. K., Mook, D. T., and Plaut, R. H. (1993b). Whirling of a forced cantilevered beam with static deflection. II: Superharmonic and subharmonic resonances, *Nonlinear Dynamics* 4, 337-356. *179*

Shyu, I.-M. K., Mook, D. T., and Plaut, R. H. (1993c). Whirling of a forced cantilevered beam with static deflection. III: Passage through resonance, *Nonlinear Dynamics* 4, 461-481. *179*

Sierakowski, R. L. See Vinson, J. R., and Sierakowski, R. L.

Simmonds, J. G. (1979). Accurate nonlinear equations and a perturbation solution for the free vibrations of circular elastic rings, *Journal of Applied Mechanics* 46, 156-160. *562, 563*

Simmonds, J. G., and Danielson, D. A. (1972). Nonlinear shell theory with finite rotation and stress-function vectors, *Journal of Applied Mechanics* 39, 1085-1090. *562*

Simmonds, J. G. See Koiter, W. T., and Simmonds, J. G.; Libai, A., and Simmonds, J. G.

Simo, J. C., and Vu-Quoc, L. (1991). A geometrically-exact rod model incorporating shear and torsion-warping deformation, *International Journal of Solids and Structures* 27, 371-393. *177*

Simonelli, F., and Gollub, J. P. (1989). Surface wave mode interactions: Effects of symmetry and degeneracy, *Journal of Fluid Mechanics* 199, 471-494. *27*

Simovich, J. See Librescu, L., and Simovich, J.

Sinclair, G. B. (1979). The nonlinear bending of a cantilever beam with shear and longitudinal deformations, *International Journal of Non-Linear Mechanics* 14, 111-122. *179*

Singh, G. See Rajgopal, S. V., Singh, G., Sadasiva Rao, Y. V. K., and Narayanan, S.

Singh, P. N., Sundararajan, V., and Das, Y. C. (1974). Large amplitude vibration of some moderately thick structural elements, *Journal of Sound and Vibration* 36, 375-387. *373*

Singh, G., Rao, V. G., and Iyengar, N. G. R. (1995). Finite element analysis of the nonlinear vibrations of moderately thick unsymmetrically laminated composite plates, *Journal of Sound and Vibration* 181, 315-329. *380*

Siu, C. C., and Bert, C. W. (1974). Sinusoidal response of composite-material plates with material damping, *Journal of Engineering for Industry* 96, 603-610. *8*

Sivakumaran, K. S. (1987). Natural frequencies of symmetrically laminated rectangular plates with free edges, *Composite Structures* 7, 191-204. *377*

Sivakumaran, K. S., and Chia, C. Y. (1984). Nonlinear vibration of generally laminated anisotropic thick plates, *Ingenieur Archiv* 54, 220-231. *380*

Sivakumaran, K. S., and Chia, C. Y. (1985). Large-amplitude oscillations of unsymmetrically laminated anisotropic rectangular plates including shear, rotary inertia, and transverse normal stresses, *Journal of Applied Mechanics* 52, 536-542. *380*

Smelova-Reynolds, T., and Dowell, E. H. (1996a). The role of higher modes in the chaotic motion of the buckled beam-I, *International Journal of Non-Linear Mechanics* 31, 931-939. *318*

Smelova-Reynolds, T., and Dowell, E. H. (1996b). The role of higher modes in the chaotic motion of the buckled beam-II, *International Journal of Non-Linear Mechanics* 31, 941-950. *318*

Smith, P. See Jordan, D. W., and Smith, P.

Smith, S. W. See Balachandran, B., Nayfeh, A. H., Smith, S. W., and Pappa, R. S.

Sniegowski, J. J. See Zook, J. D., Burns, D. W., Guckel, H., Sniegowski, J. J., Engelstad, R. L., and Feng, Z.

Soedel, W. (1993). *Vibrations of Shells and Plates*, Marcel Dekker, New York. *469, 559*

Sokolnikoff, I. S. (1956). *Mathematical Theory of Elasticity*, 2nd ed., McGraw-Hill, New York. *175*

Sokolnikoff, I. S., and Redheffer, R. M. (1966). *Mathematics of Physics and Modern Engineering*, McGraw-Hill, New York. *382*

Soldatos, K. P. (1984). A comparison of some theories used for the dynamic analysis of cross-ply laminated circular cylindrical panels, *Journal of Sound and Vibration* 97, 305-319. *561*

Soldatos, K. P. (1985). On the theories used for the wave propagation in laminated composite thin elastic shells, *Zeitschrift fur Angewandte Mathematik und Physik (ZAMP)* 36, 120-133. *561*

Soldatos, K. P. (1986). On thickness shear deformation theories for the dynamic analysis of non-circular cylindrical shells, *International Journal of Solids and Structures* 22, 625-641. *561*

Soldatos, K. P. (1987). Buckling of axially compressed antisymmetric angle-ply laminated circular cylindrical panels according to a refined shear deformable shell theory, in *Recent Advances in Structural Dynamics*, H. Chung and H. D. Fisher, eds., ASME PVP-124, pp. 63-71. *561*

Soldatos, K. P. (1991). A refined laminated plate and shell theory with applications, *Journal of Sound and Vibration* 144, 109-129. *561*

Soldatos, K. P. (1994). Review of three dimensional dynamic analyses of circular cylinders and cylindrical shells, *Applied Mechanics Reviews* 47, 501-516. *559*

Soldatos, K. P. (1999). Mechanics of cylindrical shells with non-circular cross-section, *Applied Mechanics Reviews* 52, 237-274. *559*

Song, X. See Li, S.-R., Zhou, Y.-H., and Song, X.

Sonstegard, D. A. See McIvor, I. K., and Sonstegard, D. A.

Southwell, R. V. (1922). On the free transverse vibration of a uniform circular disk clamped at its center; and on the effects of rotation, *Proceedings of the Royal Society of London* 101, 133-153. *517, 522*

Southwell, R. V. See Lamb, H., and Southwell, R. V.

Sridhar, S., Mook, D. T., and Nayfeh, A. H. (1975). Non-linear resonances in the forced response of plates, Part I: Symmetric responses of circular plates, *Journal of Sound and Vibration* 41, 359-373. *381, 498*

Sridhar, S., Mook, D. T., and Nayfeh, A. H., (1978). Non-linear resonances in the forced responses of plates, Part II: Asymmetric responses of circular plates, *Journal of Sound and Vibration* 59, 159-170. *381, 498, 499, 504, 512*

Sridhar, S. See Nayfeh, A. H., Mook, D. T., and Sridhar, S.

Srinivas, S., Joga Rao, C. V., and Rao, A. K. (1970). An exact analysis for vibration of simply-supported and laminated thick rectangular plates, *Journal of Sound and Vibration* 12, 187-199. *372*

Srinivas, S., and Rao, A. K. (1970). Bending, vibration and buckling of simply supported thick orthotropic rectangular plates and laminates, *International Journal of Solids and Structures* 6, 1463-1481. *372, 373, 378*

Stahl, K. J. See Iwan, W. D., and Stahl, K. J.

Stange, W. A. See MacBain, J. C., Horner, J. E., Stange, W. A., and Ogg, J. S.

Stavsky, Y. (1965). On the theory of symmetrically heterogeneous plates having the same thickness variation of the elastic moduli, in *Topics in Applied Mechanics*, D. Abir, D. F. Ollendorff, and M. Reiner, eds., American Elsevier, New York. *372*

Stavsky, Y. See Yang, P. C., Norris, C. H., and Stavsky, Y.

Stein, M., and Jegley, D. C. (1987). Effects of transverse shearing on cylindrical bending, vibration, and buckling of laminated plates, *AIAA Journal* 25, 123-129. *377*

Stemple, A. D., and Lee, S. W. (1988). Finite-element model for composite beams with arbitrary cross-sectional warping, *AIAA Journal* 26, 1512-1520. *174*

Stevens, L. K. See Bhimaraddi, A., and Stevens, L. K.

Stewart, H. B. See Thompson, J. M. T., and Stewart, H. B.

Streit, D. A., Bajaj, A. K., and Krousgrill, C. M. (1988). Combination parametric resonances leading to periodic and chaotic response in two-degree-of-freedom system with quadratic non-linearities, *Journal of Sound and Vibration* 124, 297-314. *24*

Stricklin, J. A., Martinez, J. E., Tillerson, J. R., Hong, J. H., and Haisler, W. E. (1971). Nonlinear dynamic analysis of shells of revolution by matrix displacement method, *AIAA Journal* 9, 629-636. *563*

Sueoka, A., Yoshitake, Y., Tamura, H., and Horita, K. (1986). Self-excited vibration of a circular plate subjected to frictional forces exerted in two regions on its outer circumference (Part 2, the case of a circular plates with effect of internal resonances), *Bulletin of the JSME* 29, 3058-3065. *513*

Sueoka, A. See Yoshitake, Y., Sueoka, A., Ryu, T., and Tamura, H.

Sujata, T. See Kar, R. C., and Sujata, T.; Kar, R. C., and Sujata, T.

Sun, C. T., and Tan, T. M. (1981). Wave propagation in a graphite/epoxy laminate, in *NCKU/AAS International Symposium on Engineering Science and Mechanics*, Taiwan, Vol. 2, pp. 1320-1337. *374*

Sun, C. T., and Whitney, J. M. (1973). Theories for the dynamic response of laminated plates, *AIAA Journal* 11, 178-183. *373*

Sun, C. T. See Whitney, J. M., and Sun, C. T.; Whitney, J. M., and Sun, C. T.

Sunakawa, M. (1964). Influence of temperature changes and large amplitude on free flexural vibration of rectangular elastic plates, *Transactions of the JSME* 30, 558-566. *544*

Sundararajan, V. See Singh, P. N., Sundararajan, V., and Das, Y. C.

Szemplinska-Stupnicka, W. (1978). The generalized harmonic balance method for determining the combination resonance in the parametric dynamic systems, *Journal of Sound and Vibration* 58, 347-361. *25*

Szwabowicz, M. L. See Pietraszkiewicz, W., and Szwabowicz, M. L.

Tabaddor, M. (2000). Influence of nonlinear boundary conditions on the single-mode response of a cantilever beam, *International Journal of Solids and Structures* 37, 4915-4931. *294*

Tadjbakhsh, I. G., and Wang, Y. M. (1990). Wind-driven nonlinear oscillations of cables, *Nonlinear Dynamics* 1, 265-291. *136, 138*

Takahashi, K. (1991). Dynamic stability of cables subjected to an axial periodic load, *Journal of Sound and Vibration* 144, 323-330. *136*

Takahashi, K., and Konishi, Y. (1987a). Non-linear vibrations of cables in three dimensions, Part I: Non-linear free vibrations, *Journal of Sound and Vibration* 118, 69-84. *136, 137*

Takahashi, K., and Konishi, Y. (1987b). Non-linear vibrations of cables in three dimensions, Part II: Out-of-plane vibrations under in-plane sinusoidally time-varying load, *Journal of Sound and Vibration* 118, 85-97. *136, 137*

Tamura, H. See Sueoka, A., Yoshitake, Y., Tamura, H., and Horita, K.; Yoshitake, Y., Sueoka, A., Ryu, T., and Tamura, H.

Tan, T. M. See Sun, C. T., and Tan, T. M.

Tang, D. M., and Dowell, E. H. (1988). On the threshold force for chaotic motions for a forced buckled beam, *Journal of Applied Mechanics* 55, 190-196. *318*

Tay, E. H., and Phoon, K. H. (1997). Review of the technology of micropumps, *Journal of the Institution of Engineers, Singapore* 37, 68-72. *531*

Taylor, T. W., and Nayfeh, Adnan H. (1997). Damping characteristics of laminated thick plates. *Journal of Applied Mechanics* 64, 132-138. *374*

Tenneti, R., and Chandrashekhara, K. (1994). Nonlinear vibration of laminated plates using a refined shear flexible element, *Advanced Composite Materials* 4, 145-160. *380*

Tezak, E. G., Mook, D. T., and Nayfeh, A. H. (1978). Nonlinear analysis of the lateral response of columns to periodic loads, *Journal of Mechanical Design* 100, 651-659. *181*

Thomas, G. B. See Hildebrand, F. B., Reissner, E., and Thomas, G. B.

Thompson, J. M. T., and Stewart, H. B. (1986). *Nonlinear Dynamics and Chaos*, Wiley, Chichester, England. *24*

Thomson, W. T. (1981). *Theory of Vibration with Applications*, Prentice-Hall, New Jersey. *7, 173*

Thornton, E. A. (1993). Thermal buckling of plates and shells, *Applied Mechanics Reviews* 46, 485-506. *543*

Tillerson, J. R. See Stricklin, J. A., Martinez, J. E., Tillerson, J. R., Hong, J. H., and Haisler, W. E.

Tilmans, H. A., Elwespoek, M., and Fluitman, J. H. (1992). Micro resonant force gauges, *Sensors and Actuators A* 30, 35-53. *342*

Tilmans, H. A., and Legtenberg, R. (1994). Electrostatically driven vacuum-encapsulated polysilicon resonators. Part II. Theory and performance, *Sensors and Actuators A* 45, 67-84. *342, 345, 346, 350, 351, 353*

Tilmans, H. A. See Ijntema, D. J., and Tilmans, H. A.; Legtenberg, R., and Tilmans, H. A.

Timoshenko, S. P. (1921). On the correction for shear of the differential equation for transverse vibrations of prismatic bars, *Philosophical Magazine* 41, 744-746. *173, 191*

Timoshenko, S. P. (1922). On the transverse vibrations of bars of uniform cross sections, *Philosophical Magazine* Series 6, 43, 125-131. *173*

Timoshenko, S. (1959). *Theory of Plates and Shells,* McGraw-Hill, New York. *382, 544, 559, 560, 613, 614*

Timoshenko, S. (1974). *Vibration Problems in Engineering*, Wiley, New York. *469*

Timoshenko, S. P., and Gere, J. M. (1961). *Theory of Elastic Stability,* McGraw-Hill, New York. *379*

Timoshenko, S. P., and Goodier, J. N. (1970). *Theory of Elasticity,* 3rd ed., McGraw-Hill, New York. *172, 174, 175, 186, 234, 251, 309*

Tittonen, I. See Veijola, T., Mattila, T., Jaakkola, O., Kiihamaki, J., Lamminmaki, T., Oja, A., Ruokonen, K., Seppala, H., and Tittonen, I.

Tobias, S. A. (1957). Free undamped non-linear vibrations of imperfect circular disks, *Proceedings of the Institute of Mechanical Engineering* 171, 691-701. *499*

Tobias, S. A. (1958). Non-linear forced vibrations of circular disks. An experimental investigation, *Engineering* 186, 51-56. *379, 499, 513*

Tobias, S. A., and Arnold, R. N. (1957). The influence of dynamical imperfection on the vibration of rotating disks, *Proceedings of the Institute of Mechanical Engineering* 171, 669-690. *513*

Tobias, S. A. See Williams, C. J. H., and Tobias, S. A.

Toledano, A. See Lakis, A. A., Selmane, A., and Toledano, A.

Tondl, A. See Schmidt, G., and Tondl, A.

Torii, T., Yasuda, K., and Toyada, T. (1998). Nonlinear forced oscillations of a rotating disk excited at a point fixed in space (combination tones near primary resonance points), *JSME International Journal* 41, 84-90. *513*

Torii, T., Yasuda, K., and Yamada, Y. (1995a). Nonlinear forced oscillations of rotating disk excited at fixed point in space (combination tones occurring under condition of combination resonance), *Transactions of the JSME* 61, 3805-3811. *513*

Torii, T., Yasuda, K., and Yamada, Y. (1995b). Nonlinear forced oscillations of rotating disk excited at fixed point in space (occurrence of combination tones of super-combination resonance type), *Transactions of the JSME* 61, 1286-1292. *513*

Torii, T. See Yasuda, K., and Torii, T.; Yasuda, K., and Torii, T.; Yasuda, K., Torii, T., and Shimizu, T.

Tousi, S., and Bajaj, A. K. (1985). Period-doubling bifurcations and modulated motions in forced mechanical systems, *Journal of Applied Mechanics* 52, 446-452. *25*

Toyada, T. See Torii, T., Yasuda, K., and Toyada, T.

Traybar, J. See Dowell, E. H., Traybar, J., and Hodges, D. H.

Triantafyllou, M. S. (1984). Linear dynamics of cables and chains, *Shock and Vibration Digest* 16, 9-17. *137*

Triantafyllou, M. S. (1987). Dynamics of cables and chains, *Shock and Vibration Digest* 19, 3-5. *137*

Triantafyllou, M. S. (1991). Dynamics of cables, towing cables and mooring systems, *Shock and Vibration Digest* 23, 3-8. *137*

Tsai, S. W., and Hahn, H. T. (1980). *Introduction to Composite Materials,* Technomic, Westport, Connecticut. *7*

Tsai, S. W., and Pagano, N. J. (1968). Invariant properties of composite materials, in *Composite Materials Workshop,* S. W. Tsai, J. C. Halpin, and N. J. Pagono, eds., Technomic, Westport, Connecticut, pp. 233-253. *97*

Tsai, S. W. See Schultz, A. B., and Tsai, S. W.

Tseng, W. Y., and Dugundji, J. (1971). Nonlinear vibrations of a buckled beam under harmonic excitation, *Journal of Applied Mechanics* 38, 467-476. *318*

Tso, W. K. (1968). Parametric torsional stability of a bar under axial excitation, *Journal of Applied Mechanics* 35, 13-19. *182*

Tsuiji, T. (1985). Free vibrations of thin-walled pretwisted beams under axial loadings, *Bulletin of the JSME* 28(239), 894-898. *175*

Tufillaro, N. B. (1989). Nonlinear and chaotic string vibrations, *American Journal of Physics* 57, 408-414. *112, 132*

Tufillaro, N. B. See Molteno, T. C. A., and Tufillaro, N. B.

Turner, G. C., and Andrews, M. K. (1995). Frequency stabilization of electrostatic oscillators, in *Proceedings of the 8th International Conference on Solid-State*

Sensors and Actuators, TRANSDUCERS'95-EUROSENSORS IX, Stockholm, Sweden, June 25-29, pp. 624-626. *342*

Tzou, H. S., and Gadre, M. (1989). Theoretical analysis of a multi-layered thin shell coupled with piezoelectric shell actuators for distributed vibration control, *Journal of Sound and Vibration* 132, 433-450. *376*

Ueda, Y. (1979). Randomly transitional phenomena in the system governed by Duffing's equation, *Journal of Statistical Physics* 20, 181-196. *23*

Uno, H. See Yasuda, K., and Uno, H.

Urabe, M. (1967). *Nonlinear Autonomous Oscillations,* Academic Press, New York. *24*

Vakakis, A. See Georgiou, I. T., Schwartz, I. B., Emaci, E., and Vakakis, A.

Vakakis, A. F. See Nayfeh, T. A., and Vakakis, A. F.

Vanderbauwhede, A. (1982). *Local Bifurcations and Symmetry,* Pitman, Great Britain. *24*

Varadan, T. K. See Ganapathi, M., Varadan, T. K., and Sarma, B. S.; Prathap, G., and Varadan, T. K.

Veijola, T., Mattila, T., Jaakkola, O., Kiihamaki, J., Lamminmaki, T., Oja, A., Ruokonen, K., Seppala, H., and Tittonen, I. (2000). Large-displacement modelling and simulation of micromechanical electrostatically driven resonators using the harmonic balance method, *Microwave Symposium Digest* 1, 11-16. *342*

Vestroni, F. See Benedettini, F., Rega, G., and Vestroni, F.; Gattulli, V., Pasca, M., and Vestroni, F.; Luongo, A., Rega, G., and Vestroni, F.; Luongo, A., and Vestroni, F.; Luongo, A., and Vestroni, F.; Rega, G., Vestroni, F., and Benedettini, F.

Vinson, J. R., and Sierakowski, R. L. (1986). *The Behavior of Structures Composed of Composite Materials*, Nijhoff, Boston, Massachusetts. *7, 372*

Vinson, J. R. See Wu, C.-I., and Vinson, J. R.; Wu, C.-I., and Vinson, J. R.; Wu, C.-I., and Vinson, J. R.; Zukas, J. A., and Vinson, J. R.

Visweswara Rao, G., and Iyengar, R. N. (1991). Internal resonance and non-linear response of a cable under periodic excitation, *Journal of Sound and Vibration* 149, 25-41. *137*

Vlasov, A. S. (1949). *General Theory of Shells and its Application in Engineering,* Gostekhizdat, Moscow (in Russian). *560*

Vlasov, V. Z. (1951). Basic differential equations in the general theory of elastic shells, NACA TM 1241 (Translated from 1944 Russian version). *562*

Vogl, G. W., and Nayfeh, A. H. (2003). A reduced-order model for electrically actuated clamped circular plates, in *Proceedings of the ASME 2003 Design Engineering Technical Conferences*, ASME Paper No. DETC2003/VIB-48530, Chicago, Illinois, September 2-6. *531, 532, 536, 540-542*

von Freudenreich, J. (1925). Vibration of steam turbine discs, *Engineering* 119, 2-4 and 31-34. *513*

von Kármán, T. (1910). Festigkeitsprobleme in maschinenbau, in *Encyklopädie der Mathematischen Wissenschaften*, P. R. Halmos, ed., American Mathematical Society, Vol. 3, pp. 211-385. *375, 433*

Voyiadjis, G. Z., and Shi, G. (1991). A refined two-dimensional theory for thick cylindrical shells, *International Journal of Solids and Structures* 27, 261-282. *561*

Vu-Quoc, L. See Simo, J. C., and Vu-Quoc, L.

Wachutka, G. See Korvink, J. G., Funk, J., Roos, M., and Wachutka, G.

Wah, T. (1963). Large amplitude flexural vibrations of rectangular plates, *International Journal of Mechanical Sciences* 5, 425-438. *375*

Wan, S. M. See Chang, W. P., and Wan, S. M.

Wan, Y. H. See Hassard, B. D., Kazarinoff, N. D., and Wan, Y. H.

Wang, C. M. See Liew, K. M., Kitipornchai, S., and Wang, C. M.

Wang, Y. M. See Tadjbakhsh, I. G., and Wang, Y. M.

Warnitchai, P. See Fujino, Y., Warnitchai, P., and Pacheco, B. M.

Watts, H. A. See Scott, M. R., and Watts, H. A.; Scott, M. R., and Watts, H. A.

Weiner, J. H. See Boley, B. A., and Weiner, J. H.

Weingarten, V. I. See Busby, H. R. Jr., and Weingarten, V. I.

White, J. See Osterberg, P., Yie, H., Cai, X., White, J., and Senturia, S.

White, J. K. See Senturia, S. D., Harris, R. M., Johnson, B. P., Kim, S., Nabors, K., Shulman, M. A., and White, J. K.

Whitney, J. M. (1969a). Bending-extensional coupling in laminated plates under transverse loading, *Journal of Composite Materials* 3, 20-28. *373*

Whitney, J. M. (1969b). The effect of transverse shear deformation on the bending of laminated plates, *Journal of Composite Materials* 3, 534-547. *373, 377, 378*

Whitney, J. M. (1970). The effect of boundary condition on the response of laminated composites, *Journal of Composite Materials* 4, 192-203. *377*

Whitney, J. M. (1987). *Structural Analysis of Laminated Anisotropic Plates*, Technomic, Lancaster, Pennsylvania. *7, 180, 372, 469*

Whitney, J. M., and Leissa, A. W. (1969). Analysis of heterogeneous anisotropic plates, *Journal of Applied Mechanics* 36, 261-266. *378, 380*

Whitney, J. M., and Pagano, N. J. (1970). Shear deformation in heterogeneous anisotropic plates, *Journal of Applied Mechanics* 37, 1031-1036. *378*

Whitney, J. M., and Sun, C. T. (1973). A higher order theory for extensional motion of laminated composites, *Journal of Sound and Vibration* 30, 85-97. *374*

Whitney, J. M., and Sun, C. T. (1974). A refined theory for laminated anisotropic cylindrical shells, *Journal of Applied Mechanics* 41, 471-476. *561*

Whitney, J. M. See Bowlns, J. A., Palazotto, A. N., and Whitney, J. M.; Sun, C. T., and Whitney, J. M.

Wiegele, T. G. (1996). Micro-turbo-generator design and fabrication: A preliminary study, in *Proceedings of the 31st Intersociety Energy Conversion Engineering Conference, IECEC'96*, Washington, District of Columbia, August 11-16, Vol. 4, pp. 2308-2313. *531*

Williams, C. J. H. (1966). The stability of nodal patterns in disk vibration, *International Journal of Mechanical Science* 8, 421-432. *499*

Williams, C. J. H., and Tobias, S. A. (1963). Forced undamped nonlinear vibrations of imperfect circular discs, *Journal of Mechanical Engineering Science* 5, 325-335. *498, 499, 513*

Woias, P. (2001). Micropumps: Summarizing the first two decades, in *Proceedings of the SPIE, Microfluidics and BioMEMS*, Vol. 4560, C. H. Mastrangelo, and H. Becker, eds., pp. 39-52.

Wong, C. C. See Chen, J. S., and Wong, C. C.

Woodall, S. R. See Nowinski, J. L., and Woodall, S. R.

Wu, C. H. See Wu, C. P., and Wu, C. H.

Wu, C.-I., and Vinson, J. R. (1969a). Influences of large amplitude, transverse shear deformation, and rotatory inertia on lateral vibrations of transversely isotropic plates, *Journal of Applied Mechanics* 36, 254-260. *373*

Wu, C.-I., and Vinson, J. R. (1969b). On the nonlinear oscillations of plates composed of composite materials, *Journal of Composite Materials* 3, 548-561. *373*

Wu, C.-I., and Vinson, J. R. (1971). Nonlinear oscillations of laminated specially orthotropic plates with clamped and simply supported edges, *Journal of the Acoustical Society of America* 49, 1561-1567. *373, 380*

Wu, C. P., and Wu, C. H. (2000). Asymptotic differential quadrature solutions for the free vibration of laminated conical shells, *Computational Mechanics* 25, 346-357. *559*

Wu, E. M. See Lo, K. H., Christensen, R. M., and Wu, E. M.; Lo, K. H., Christensen, R. M., and Wu, E. M.

Xistris, G. D. See Rajalingham, C., Bhat, R. B., and Xistris, G. D.

Yamada, G. See Abe, A., Kobayashi, Y., and Yamada, G.; Abe, A., Kobayashi, Y., and Yamada, G.; Abe, A., Kobayashi, Y., and Yamada, G.; Abe, A., Kobayashi, Y., and Yamada, G.; Irie, T., and Yamada, G.

Yamada, Y. See Torii, T., Yasuda, K., and Yamada, Y.; Torii, T., Yasuda, K., and Yamada, Y.

Yamaguchi, H., Miyata, T., and Ito, M. (1981). Time response analysis of a cable under harmonic excitations, *Proceedings of Japan Society of Civil Engineers* 308, 37-45. *136*

Yamaguchi, H. See Shimizu, T., Yamaguchi, H., and Ito, M.

Yamaki, N. (1961). Influence of large amplitudes on flexural vibrations of elastic plates, *Zeitschrift fur Angewandte Mathematik und Mechanik (ZAMM)* 41, 501-510. *379*

Yamamoto, T., and Yasuda, K. (1977). On the internal resonance in a nonlinear two-degree-of-freedom system – forced vibrations near the lower resonance point when the natural frequencies are in the ratio 1:2, *Bulletin of the JSME* 20, 169-175. *25*

Yamashita, A. See Kojima, H., Nagaya, J., Shiraishi, H., and Yamashita, A.

Yang, I. H. See Kuo, W. S., and Yang, I. H.

Yang, L., and Hutton, S. G. (1998). Nonlinear vibrations of elastically-constrained rotating disks, *Journal of Vibration and Acoustics* 120, 475-483. *517*

Yang, P. C., Norris, C. H., and Stavsky, Y. (1966). Elastic wave propagation in heterogeneous plates, *International Journal of Solids and Structures* 2, 665-684. *373*

Yang, X. L., and Sethna, P. R. (1991). Local and global bifurcations in parametrically excited vibrations of nearly square plates, *International Journal of Non-Linear Mechanics* 26, 199-220. *499, 528, 530*

Yang, X. L., and Sethna, P. R. (1992). Non-linear phenomena in forced vibrations of a nearly square plate: Antisymmetric case, *Journal of Sound and Vibration* 155, 413-441. *499, 528*

Yang, Y.-J. See Grétillat, M.-A., Yang, Y.-J., Hung, E. S., Rabinovich, V., Ananthasuresh, G. K., deRooij, N. F., and Senturia, S. D.

Yao, Z. J., Chen, S., Eshelman, S., Denniston, D., and Goldsmith, C. (1999). Micromachined low-loss microwave switches, *Journal of Microelectromechanical Systems* 8, 129-134. *341*

Yasaka, T. See Mitsugi, J., and Yasaka, T.

Yasuda, K., and Asano, T. (1986). Nonlinear forced oscillations of a rectangular membrane with degenerate modes, *Bulletin of the JSME* 29, 3090-3095. *499*

Yasuda, K., and Kushida, G. (1984). Nonlinear forced oscillations of a shallow spherical shell, *Bulletin of the JSME* 27, 2233-2240. *564*

Yasuda, K., and Torii, T. (1986). Nonlinear forced oscillations of a string, 2nd report: Various types of responses near resonance points, *Bulletin of the JSME* 29, 1253-1260. *112, 499*

Yasuda, K., and Torii, T. (1987). Multi-mode response of a square membrane, *JSME International Journal* 30, 963-969. *499*

Yasuda, K., Torii, T., and Shimizu, T. (1992). Self-excited oscillations of a circular disk rotating in air, *JSME International Journal* 35, 347-352. *513*

Yasuda, K., and Uno, H. (1983). Multimode response of a circular membrane, *Bulletin of the JSME* 26, 1050-1058. *499*

Yasuda, K. See Torii, T., Yasuda, K., and Toyada, T.; Torii, T., Yasuda, K., and Yamada, Y.; Torii, T., Yasuda, K., and Yamada, Y.; Yamamoto, T., and Yasuda, K.

Ye, J. (2003). *Laminated Composite Plates and Shells: 3D Modeling*, Springer, Berlin. *372, 469*

Ye, J. See Fan, J., and Ye, J.

Yeo, M. H., and Lee, W. K. (2002). Corrected solvability conditions for non-linear asymmetric vibrations of a circular plate, *Journal of Sound and Vibration* 257, 653-665. *499, 512*

Yeo, M. H. See Lee, W. K., and Yeo, M. H.; Lee, W. K., Yeo, M. H., and Samoilenko, S. B.

Yie, H., Harris, R., Napadenski, G., and Maseeh, F. (1996). A virtual prototype manufacturing software system for MEMS, in *Proceedings of the IEEE Workshop on Micro Electro Mechanical Systems, MEMS'96*, San Diego, California, February 11-15, pp. 122-126. *341*

Yie, H. See Osterberg, P., Yie, H., Cai, X., White, J., and Senturia, S.

Yoshitake, Y., Sueoka, A., Ryu, T., and Tamura, H. (1987). Self-excited vibrations of a rotating circular plate (2nd report, the case of a circular plate with internal resonance), *Transactions of the JSME* 53, 364-369. *513*

Yoshitake, Y. See Sueoka, A., Yoshitake, Y., Tamura, H., and Horita, K.

Young, T. H., and Chen, F. Y. (1995). Non-linear vibration of cantilever skew plates subjected to aerodynamic and in-plane exciting forces, *Journal of Sound and Vibration* 182, 427-440. *379*

Younis, M. I., Abdel-Rahman, E. M., and Nayfeh, A. H. (2003). A reduced-order model for electrically actuated microbeam-based MEMS, *IEEE Journal of Microelectromechanical Systems*, 12, 672-680. *351, 352, 543*

Younis, M. I., and Nayfeh, A. H. (2003). A study of the nonlinear response of a resonant microbeam to an electric actuation, *Nonlinear Dynamics* 31, 91-117. *350, 351*

Younis, M. I. See Abdel-Rahman, E. M., Younis, M. I., and Nayfeh, A. H.

Zahed, H. See Birman, V., and Zahed, H.

Zaretzky, C. L., and Crespo da Silva, M. R. M. (1994). Experimental investigation of non-linear modal coupling in the response of cantilever beams, *Journal of Sound and Vibration* 174, 145-167. *312-316*

Zaretzky, C. L. See Crespo da Silva, M. R. M., and Zaretzky, C. L.; Crespo da Silva, M. R. M., and Zaretzky, C. L.

Zavodney, L. D. (1987). *A Theoretical and Experimental Investigation of Parametrically Excited Nonlinear Mechanical Systems*, PhD Dissertation, Virginia Polytechnic Institute and State University, Blacksburg, Virginia. *179, 183, 267*

Zavodney, L. D., and Nayfeh, A. H. (1988). The response of a single-degree-of-freedom system with quadratic and cubic non-linearities to a fundamental parametric resonance, *Journal of Sound and Vibration* 120, 63-93. *24*

Zavodney, L. D., and Nayfeh, A. H. (1989). The nonlinear response of a slender beam carrying a lumped mass to a principal parametric excitation: Theory and experiment, *International Journal of Non-Linear Mechanics* 24, 105-125. *24*

Zavodney, L. D., Nayfeh, A. H., and Sanchez, N. E. (1989). The response of a single-degree-of-freedom system with quadratic and cubic nonlinearities to a principal parametric resonance, *Journal of Sound and Vibration* 129, 417-442. *24*

Zavodney, L. D., Nayfeh, A. H., and Sanchez, N. E. (1990). Bifurcations and chaos in parametrically excited single-degree-of-freedom systems, *Nonlinear Dynamics* 1, 1-21. *24*

Zavodney, L. D. See Nayfeh, A. H., and Zavodney, L. D.

Zavracky, P. M., Majumder, S., and McGruer, N. E. (1997). Micromechanical switches fabricated using nickel surface micromachining, *Journal of Microelectromechanical Systems* 6, 3-9. *342*

Zhang, Y. See Crary, S., and Zhang, Y.

Zhou, Y.-H. See Li, S.-R., and Zhou, Y.-H.; Li, S.-R., and Zhou, Y.-H.; Li, S.-R., Zhou, Y.-H., and Song, X.

Zook, J. D., Burns, D. W., Guckel, H., Sniegowski, J. J., Engelstad, R. L., and Feng, Z. (1992). Characteristics of polysilicon resonant microbeams, *Sensors and Actuators A* 35, 290-294. *341*

Zukas, J. A., and Vinson, J. R. (1971). Laminated transversely isotropic cylindrical shells, *Journal of Applied Mechanics* 38, 400-407. *561*

SUBJECT INDEX

Absorber, vibration, 26
Accelerometer, 132, 268, 294, 312-314
Adjoint,
    buckled beam, 326
    cable, 154, 155, 158-160
    cantilever beams, 273, 274, 293
    continuous systems, 36, 37
    cylindrical shell, 636, 637, 640
    hinged-free beam, 30-32, 35
    near-circular membranes, 478, 479
    near-circular plates, 489-501
    non-self-, 35
    not self-, 31, 32
    self-, 30-32, 35-37, 47, 60, 155, 158, 274, 326, 348, 479, 489, 518, 553, 637, 649
    spherical shell, 648, 649, 651
    thermally loaded plate, 553
    valve, 60
Aeroelastic, 27
    stability, 178, 263
    tailoring, 6
Aerospace, 6, 171, 562

Anisotropic, 78
    beam, 172, 174, 175, 183, 226, 235
    material, 85, 86
    plate, 372-374, 378, 380, 417, 447
    shell, 560, 561, 615
Annular, 408-412, 481-498, 510, 544
    microplates, 531-537
Arclength continuation, 113
Autocorrelation function, 268
Autoparametric, *see* Internal resonance
Averaging, method of, 25, 26, 178, 180, 513, 565, 637

Backlash, 4
Bar, 2, 168, 169
Beam, 171
    bending modes, 304-306
    box, 174
    cantilever, 223, 240, 267-316
    equations of motion, 186, 189
    Euler-Bernoulli, 4, 172-174, 183-186, 215-245
    laminated, 174, 192-194

SUBJECT INDEX 733

parametrically excited, 267-291
pretwisted, 180, 290
stubby, 179
subjected to a follower force, 278
thick, 253
Timoshenko, 191, 192
theories, 171-263
torsional modes, 182, 278, 285, 379
whirl, 177, 180-182
Beam theory,
    Euler-Bernoulli, 4, 172-174, 183-186, 215-245
    layerwise shear-deformable, 192-194
    linear shear-deformable, 186-194
    nonlinear 3-D, 226-245
    nonlinear 3-D curved, 245-265
    nonlinear 2-D, 215-226
    polynomial equations, 235-238
Bending, cylindrical, 372, 377
Berger's theory, 372, 373, 375, 380
Bernoulli, Euler, beam theory, 4, 172-174, 183-186, 215-245
Bessel functions, 475-498, 515, 550
Bifurcation,
    Hopf, 23, 26, 112, 124, 130, 131, 138, 145, 146, 194, 289-301, 310, 339, 340, 564
    normal form of a Hopf, 289, 290
    pitchfork, 130, 131, 277, 282, 283, 289-291, 302, 303, 309, 310, 555, 557
    saddle-node, 112, 277, 282, 289, 290, 303, 310, 527, 555, 557,
    secondary Hopf, 23, 25, 319, 336
    torus-doubling, 112
    see also Hopf bifurcation, Period
Bonding, 7, 562, 617, 628
Boron-epoxy plates, 6,8
Boundary conditions,
    nonlinear, 4, 186, 344, 659
Boundary,
    crisis, 112, 135
    layer, 6, 7, 24, 175, 176, 379

-value problem, 30, 31, 107, 298, 310, 344, 348, 352, 379, 504, 509, 523, 532, 535, 539
Breathing mode,
    cylindrical shell, 564-566, 636-641
    spherical shell, 647-653
Buckled beam, 316-341
    clamps, 334, 335
    experiment, 331-338
    global dynamics, 334-341
    linear vibration problem, 321-324
    nonlinear vibration, 324-331
Buckled plates, 379
Buckling, 4
    level, 319, 322, 323, 330-333
    load, 48, 52, 178
    mode shape, 317
    post, 4, 5, 171, 176, 318
    problem, 320, 321

Cable, 136-167
    compressibility, 136, 137, 142
    discretization, 150-152, 161-167
    Lagrangian strain, 136
    mode shapes, 148-150
    multiple resonance, 137, 138
    natural frequencies, 148-150
    neo-Hookean material, 136
    one-to-one resonance, 137, 138, 150
    Poisson's effect, 136, 137, 140
    single-mode response, 152-167
    static deflection, 141-145
    two-to-one resonance, 137, 138, 150, 163, 167
    von Kármán strain, 136
Calculus of variations, 27, 64, 66, 562, 635
Cantilever beams, 6, 178, 179, 225
    autoparametric resonance, 182
    combination parametric resonance, 278-284
    Coriolis effect, 278
    curvature nonlinearity, 275, 277

734 SUBJECT INDEX

discrete mode of, 66
energy transfer, 270
equations of motion, 240-245
experiment, 267-270, 312-316
follower force, 278
Galerkin procedure, 318, 319
in a supersonic stream, 31
inertia nonlinearity, 275, 277
modulated motion, 270, 285, 310
natural frequencies, 294, 299, 303, 305, 313
nonlinear dynamics, 223
nonplanar dynamics, 284-291, 304-312
parameter estimation, 296, 297
preconed, 278
primary resonance, 291-295
principal parametric resonance, 182, 223, 270-278
quadratic damping, 270, 277
slow dynamics, 270
subcombination resonance, 298-304
time-averaged Lagrangian, 285, 306
tiploaded, 249, 250
torsional mode, 278, 285, 315
Capacitive microswitches, 342
Cauchy stresses, 99, 100, 177
Chaotic attractor, 5, 39, 124
Chordwise, 182
Circular,
    cross-section, 234, 235, 237, 240, 249, 251
    cylinder, 26
    cylindrical shell, 562-568, 576, 588-614
    membrane, 474-480
    plate, 373, 375, 379, 381, 388-392, 395, 401, 402, 411, 481-513, 531-557
    ring, 181, 265
    see also Disk
Circulatory effect, 19
Clamps, 334, 335

Classical, 382-395
    laminate theory, 378
    linear plate theory, 89, 96, 371, 372, 375, 379
    linear shell theory, 560, 561, 566-577, 611-614
    non-, 1, 2, 7, 174
    nonlinear plate theory, 403-416, 425-435
    nonlinear shell theory, 582-588
Clock frequency, 268
Codimension, 285
Combination resonance, 181, 183, 267, 504
    in cantilever beams, 269, 278-284
    in circular plates, 544, 549-557
    simultaneous, 137, 181
    sub, 298-304, 544
    see also Parametric combination resonance
Commensurate, 122, 181, 182, 301, 319, 336, 340, 564
Companion mode, 508, 565, 636-638
Compatibility, 65, 107, 108, 411, 416, 502, 514
Compliance matrix, 89-98
Concentrated,
    mass, 4, 54, 179
    load, 177, 513
Concentration, stress, 7
Conduction, heat equation, 414, 533, 544, 545, 547
Constitutive equations, 4, 254, 560
    anisotropic, 85, 86
    hygro-thermal, 98, 468
    isotropic, 90-94
    orthotropic, 87-90
    thermoelastic, 88, 93, 415
Constraint, 3, 4, 65, 66, 70-72, 176, 222, 231, 242, 318, 372, 565
Continuity-based method, 379
Coordinate,
    curvilinear, 80-82, 99, 100, 107, 195-200, 216, 246, 355
    cylindrical, 77, 78, 103, 104, 588

helical, 567, 568
polar, 389, 395, 408-412
spherical, 78-81, 105, 106
Coriolis effect, 4, 278
Correction factor, 174, 191, 373, 374, 402, 446, 561, 608
Coulomb friction, 7
Coupling,
elastic, 7, 172, 174, 180, 226, 374, 376, 561
extension-bending, 168, 417, 560
thermoelastic, 546, 547
Crisis, 113, 135
Crossover, 137, 138, 150, 164, 165, 318, 330
Cross-ply, 373, 378-381, 405, 406, 412, 564
Curvature,
bending, 198, 358, 361, 420
deformed, 206, 247, 252, 253, 358-361, 365, 420, 616, 620-628, 632
initial, 2, 201, 238, 252-255, 262-265, 355-357, 392, 417-420, 426, 559-566
normalized, 199, 219
real, 199, 214, 219, 361, 368
twisting, 177, 212, 238, 263, 358, 361, 366, 417, 420, 629
variation of, 210, 355, 365, 366
Curved, 2, 355, 559
beam, 2, 172, 176, 200, 207-212, 245, 264,
Curvilinear coordinates, 76, 80, 99, 108, 195, 198, 200, 216, 248, 355, 356, 358, 388, 392, 395, 396, 417, 446, 559, 563, 566, 588
Cutouts, 380
Cylindrical bending, 372, 377

Damping, 7
estimation of, 296, 297
quadratic, 270, 277
Damper, 9, 66-68, 70, 72, 98, 544

Degenerate,
classes of shells, 559
mode, 27, 513, 636
Delamination, 4, 7, 193, 381, 449, 617
Dielectric constant, 342, 532
Diffusion of heat, 546
Dislocation, 7, 8
Discontinuities, 374, 467
Discretization, *see* Galerkin procedure
Disk,
circular, 498, 513-527
imperfect, 499
rotating, 23, 513, 517
rotating, equations of motion, 517
vibration of, 499
Displacement,
concept of local, 211, 376, 563
global, 44, 75, 174, 423
local, 214, 215, 218-220, 229, 251, 252, 258, 366, 427, 436, 437, 447, 582, 583, 593, 595, 615
warping, 176, 194, 215, 369
Driven mode, 565, 636, 637
Duffing equation, 381, 506

Effect,
circulatory, 19
Coriolis, 278
gyroscopic, 19
hygro-thermal, 98
nonclassical, 7, 174
piezoelectric, 98
Poisson's, 115, 117, 136, 137, 185
shear, 4, 173, 179, 376, 381, 561
shortening, 182, 221
stretching, 181, 186, 224
thermal, 414, 467, 545
Effective nonlinearity, 125, 158, 160, 163-167, 223, 275, 289, 294, 296, 297, 329, 330, 331, 349-351, 506, 507, 512, 513, 517, 555
Elastic,
constraint, 318
coupling, 6, 7, 172, 180, 226, 561

foundation, 380, 517
Elasticity, 65-110, 172, 174-177, 223, 234, 247, 248, 251, 309, 372, 378, 379 414, 545
Elastic-plastic, 7
Elliptic function, 318
Elliptical path, 180
Equations of motion for,
- buckled beam, 316
- cantilever beam, 66, 242
- circular plate, 481, 498
- cylindrical shell, 631, 589
- general plate, 372, 377
- membrane, 474-477
- microbeam, 342
- microplate, 532-537
- rectangular plate, 469
- rotating disk, 516, 517
- shell, 563, 566, 577, 581, 589, 625
- spherical shell, 641
- thermoelastic plate, 412-415

Estimation, parameter, 296
Euler,
- angles, 177, 195, 196, 197, 200, 202, 204, 208-210, 226, 236, 238, 240, 243, 244, 257
- -Bernoulli beam, 4, 172, 174, 183-186, 217-245
- buckling load, 178
- -Lagrange equations, 61, 64, 72, 73, 635, 640, 652
- -like rotations, 177, 178, 240
- rotations, 177, 179, 376

Fiber-reinforced lamina, 96
Finite-difference method, 318
Finite-element method, 41-43, 64, 174, 178, 179, 373, 378, 380, 381, 467
Flapwise motion, 178, 182, 379
Flexural, -flexural vibration, 244
- mode, 285, 564-566, 630, 638, 641
Floppy disk, 517, 519-522, 524, 525
Floquet,
- multiplier, 339
- theory, 310, 564

Flutter, 1, 5, 6, 23, 176, 375, 376, 562
Follower force, 278
Foundation, elastic, 380, 517
Fourier transform, 374
Fundamental matrix method, 379

Galerkin procedure, 53, 61
- circular plates, 379, 504, 509
- microbeam, 319, 324, 327, 331
- microplate, 531, 532, 536, 538
- rectangular plates, 379-381, 530
- rotating disk, 522
- shells, 563, 564
- shortcomings of, 53
- strings, 126, 138, 150, 161
Gram-Schmidt procedure, 52
Graphite-epoxy, 6, 182, 374, 379, 454, 629
Green identity, 478, 489
Green-Lagrange strain, 176, 177
Gudonov procedure, 52
Gyration, radius of, 179, 224, 225, 317
Gyroscopic effect, 19

Hamilton, 72, 73, 112
- equations, 74
- extended principle, 28, 54, 55, 66, 71, 109, 110, 140, 177, 178, 187, 226, 231, 254, 260, 284, 285, 287, 298, 300, 307, 377, 379, 380, 382, 575, 580, 585, 596, 601, 604, 606, 625, 634
- theories, 70
Hamiltonian, 74, 112
Hardening nonlinearity, 296, 309
Hardening response, behavior, 137, 161, 330, 331, 350, 381
Harmonic balance, method of, 25, 318, 380, 381
Helical spring, 53, 54, 55, 264
Helicopter, 6, 171, 178, 180, 240, 245, 246, 263

Heterogeneity, 174, 176
Heterogeneous, 7, 85, 175, 176
Hierarchical finite element, 380
Hencky-Mindlin theory, 402, 446
Hopf bifurcation, 23, 25, 26, 112, 130, 131, 133, 134, 138, 182, 289, 290, 291, 310, 339, 564
    generic, 289
    normal form of, 289, 290
    secondary, 23, 319, 336
Hygro-thermal
    effect, 98
    expansion coefficient, 98, 468

Identification, 377
Imperfect, 7, 381, 499
Imperfection, 312, 381, 499
Incommensurate, 302, 319, 336, 340
Inconsistency, 186, 220
Inextensionality, 222, 225, 240, 565, 638
Initial-value methods, 35, 48, 52, 142
Integrability, 107
Interaction, 313
    coefficient, 122, 323, 565, 638
    modal, 5, 23, 25, 181-183, 270, 373, 376, 377, 381, 498, 544, 564, 565, 569, 630, 637, 638
    nonlinear, 183
    stretching-bending, 137
    structure-fluid, 8
    wave-mode, 27
Interlaminar shear stress, 4, 193, 374, 375, 377, 448, 449, 561, 615, 617
Internal resonance, 58, 122, 152, 153, 158, 161, 180-183, 272, 292, 298, 318, 323, 324, 331, 504, 508, 522, 550, 551, 639
    buckled beams, 316
    cables, 137, 138
    circular plates, 504
    combination, 183
    cylindrical shells, 564, 566, 630, 638, 639
    multiple, 137, 138
    near-square plates, 524
    one-to-one, 26, 112, 137, 138, 150, 181-183, 286, 319, 323, 499, 503, 513, 529, 565, 636, 637
    spherical shells, 564, 641, 648, 649, 653
    strings, 112, 122
    three-to-one, 183, 323, 324, 544
    two-to-one, 25, 26, 137, 138, 150, 163, 167, 183, 323, 564, 566, 630, 638, 641, 648-650, 653
Invariant, 84, 90, 288
    manifold, 524
    transformation, 83, 87, 288

Jaumann,
    strains, 215, 251, 252, 366, 369, 376, 382, 428, 439, 458, 583, 593, 594, 596, 616, 631, 632, 643
    stresses, 254, 382, 429, 467, 596, 633

Kantorovich method, 378, 379
Kirchhoff hypothesis, 371, 376, 382, 403, 426, 435, 560, 566, 582
Kirchhoff's kinetic analogy, 199, 207

Lagrange multipliers, 231
Lagrangian, 4, 25, 54, 61, 63, 64, 71, 76, 136, 138, 139, 225, 226, 252, 284-286, 298, 300, 304, 306, 554, 594, 634, 640
    see Time-averaged Lagrangian
Lamé parameters, 355, 563
Laminated,
    beams, 8, 174, 179, 180
    plates, 4, 8, 372-374, 376-381, 446-448, 454, 456, 528, 544, 624
    shells, 561, 564, 615, 629
Lamination,

angle-ply, 180, 182, 377, 378, 380, 381, 544
anisotropic, 172, 175, 183, 226, 235, 254, 372-374, 380, 417, 447, 561, 615
cross-ply, 373, 378-380, 405, 406, 412, 564
layerwise, 192, 446, 615
Laplace transform, 13, 22, 374
Layerwise,
    shear-deformable beam theory, 172, 192-194
    shear-deformable plate theory, 376, 377, 446-467, 562
    shear-deformable shell theory, 615-630
Levy solution, 377-379, 470, 471
Limit cycle, 5, 23, 25, 26, 112, 289, 290, 310, 319
    period-five, 319, 336, 337, 340
    period-one, 335-337, 339-341
    period-three, 319, 336, 337, 340
    period-two, 335-337, 339, 340
    period-six, 319
Lindstédt-Poincaré method, 379, 565, 637
Linear continuous systems, analysis of, 27
Locking, frequency, 336, 340
Logarithmic decrement, 133, 313

Macromechanics, 7
Magnetic, 4, 24, 111, 171
Magnetoelastic, 8
Material stiffness matrices, 94
Mathematica, 290
Mathieu-type equation, 24, 564
Membrane, 1-3, 474, 522, 531, 559, 560, 594
    circular, 474-477, 499
    elliptic, 474, 480, 481
    equations, 474
    force, 562, 566
    force intensity, 440, 599
    modes and frequencies of spherical shell, 647, 648, 652, 653
    natural frequencies, 474-481
    near-circular, 474, 477-480
    rectangular, 499
    square, 499
    strain energy, 635
    stress, 2
    spinning, 513
Memory disk, 513, 516
MEMS, 341, 342, 351, 531, 544, 545
Microbeam, 341-353
    applications of, 342, 345, 346, 351
    microswitch, 342
    natural frequencies of, 345
    nonlinear resonance frequency, 349-351
    pull-in voltage, 345, 352, 353, 541, 542
    reduced-order model, 351
    resonator, 342
    response to primary resonance, 346
    static deflection, 342
Microcracking, 8
Microelectromechanical, 341
Microplate, 469
Micropump, 531-543, 545
Midplane stretching, 4, 181
    in beams, 181
    in columns, 181
    in microbeams, 342, 345
Modal, 17, 34, 35, 180
    amplitudes, 72, 132
    coordinates, 17, 21, 34
    coupling, 5, 26, 181, 182
    damping, 8, 9, 18, 123, 127, 638
    excitation, 21
    interactions, 5, 23, 25, 181-183, 373, 376, 377, 498, 544, 564, 565, 637
    matrix, 17, 20, 21
    shaker, 267
    structure, 180
    testing, 8

Moment, 6, 75, 85, 100, 440
   intensity, 175, 384, 385, 440, 442, 460, 584, 600, 624
   method, 27, 48, 217, 429
Multiple scales, method of, 26, 57, 61, 152
   applied to axisymmetric shells, 565, 637
   applied to buckled beams, 319, 324, 328
   applied to cables, 138, 139, 152, 161
   applied to cantilever beams, 271, 278, 284
   applied to circular plates, 506
   applied to cylindrical shells, 565, 638
   applied to heated plates, 544, 550
   applied to microbeams, 346
   applied to orthotropic plates, 381
   applied to rings, 565, 637
   applied to rotating disks, 524
   applied to spherical shells, 564, 565
   applied to strings, 112, 113, 121, 127

Natural frequencies for, 5, 9-13, 15, 23, 30, 52, 58, 181
   annular plates, 491-498
   buckled beams, 317, 321-323, 330-332, 335
   cables, 146, 148-150, 164
   cantilever beams, 179-182, 244, 267, 270, 278, 286, 294, 296, 299, 303-305, 308, 313, 379
   circular plates, 482-487, 503
   classical plate theory, 372, 407, 432
   classical shell theory, 561
   composite beams, 179, 379
   composite plates, 373
   cylindrical shells, 635, 636
   elliptic membranes, 480, 481
   elliptic plates, 487, 491
   heated plates, 543, 544, 548-551, 555
   inhomogeneous beams, 179
   membranes, 474, 477
   microbeams, 342, 345-348, 350, 353
   microplates, 531, 532, 534-536, 542, 543
   near-circular membranes, 481
   near-circular plates, 491
   prismatic beams, 181
   rectangular plates, 469-474
   shear-deformable beams, 179
   spherical shells, 646, 652
   strings, 119, 126, 130, 133
Navier solution, 378, 470
Newton, 9, 14, 65-68, 114, 178, 184, 217, 310, 562,
Newton-Raphson, 142
Newtonian, 604
   approach, 192, 382, 387, 400, 563, 576, 604
   coordinate, 72, 75
   formulation, 189, 233, 262
Nonclassical effects, 1, 6, 7, 174
Nonlinear, 1, 2, 4
   analysis of continuous systems, 53, 373, 562
   boundary conditions, 4, 186, 344
   discrete systems, 22
   elastic characteristics, 53
   Ludwick foundation, 179
   Ramberg-Osgood foundation, 179
Nonlinear dynamics of,
   annular plates, 412
   beams, 267
   buckled beams, 318
   cables, 137
   cantilever beams, 177, 223
   circular plates, 412
   cylindrical shells, 565, 630-641
   plates, 376
   panels, 562
   rotating disks, 524
   shells, 562

spherical shells, 641-653
strings, 112, 126
Nonlinear response of,
    annular plates, 544
    buckled beams, 319, 324, 328, 334
    cantilever beams, 24, 182, 267, 278, 291, 304
    circular plates, 499
    composite plates, 373, 380
    cylindrical shells, 563-565, 637
    doubly-curved shells, 564
    heated plates, 544
    microbeams, 346
    rotating rings, 565, 637
    strings, 113, 119, 125, 129-135
    systems, 24, 183
    valves, 53
Nonlinear vibrations of,
    buckled beams, 317-319, 321, 338
    cables, 111
    cantilever beams, 178
    cantilever plates, 379
    circular plates, 411, 498-513
    composite plates, 380, 381
    cylindrical shells, 564
    doubly-curved shells, 564
    heated plates, 469, 544
    microplates, 469
    near-square plates, 527-531
    panels, 381
    plates, 469, 528
    rotating disks, 513-527
    sandwich plates, 373
    shells, 563, 564
    strings, 112
    valves, 56
Nonlinearities, 4, 24-26, 53, 61, 65, 121, 163, 178, 223, 278, 299, 309, 317, 319, 327, 330, 332, 506, 634
    classification of, 4
    curvature, 223
    electric, 4, 531
    geometric, 4, 7, 65, 136, 168, 169, 174, 176-179, 194, 223, 229, 251, 278, 298, 355, 375-377, 531, 562, 563, 605, 622
    inertia, 4, 177, 178, 180, 223, 275, 277, 278, 289, 298, 310, 438, 598
    material, 4, 65-169, 177, 179
    von Kármán, 563, 606
Nonplanar dynamics, 267
    in the response of cables, 137
    in the response of cantilever beams, 180-182, 284-291, 304-312
    in the response of rings, 565, 638
    in the response of taut strings, 112
Nonstationary,
    parametric, 178, 223
    responses, 278, 565, 637
Normal form, 298, 290

Orthonormalization, 52, 379
Orthotropic, 87-89, 94, 95, 97, 372, 373, 377-381, 406, 412, 415, 447, 454, 544, 627, 628

Parametric excitation, 24, 112, 181, 223, 268-270, 284, 319, 565, 637
Parametric resonance, 179, 267, 517
    combination, 25, 137, 267, 269, 278, 279, 281-284, 303, 548, 549
    principal, 24, 26, 137, 267, 269, 270, 273, 285, 286
Peeling stress, 1, 4, 6
Period,
    -demultiplying bifurcation, 319
    -doubling bifurcation, 23-26, 112, 181, 318, 319, 332, 335, 336, 339, 564
    -halving bifurcation, 112
    -multiplying bifurcations, 23, 24, 26, 182
Piecewise linear, 374, 561
Piezoactuator, 448

Piezoceramic, 456
Piezoelectric, 98, 376, 377, 446-448, 450-452, 456, 464-467, 531
    accelerometer, 312
    effect, 98
Piola-Kirchhoff stresses, 177
Plates,
    circular, 355, 373, 375, 379, 381, 388, 389, 395, 401, 411, 412, 481-488, 498-513, 517, 524, 532, 536-543
    composite, 372-381, 403, 447, 454, 456, 469, 527, 528
    cylindrical bending of, 372, 377
    governing equations, 372, 375, 382, 387, 395, 400, 403, 425, 445-447, 498, 501
    heated, 544
    internal resonance, 499, 503, 504, 508, 513, 519, 529, 550, 551
    micro, 531-543
    near circular, 487
    near square, 499, 527-529
    rectangular, 372, 377-380, 382-387, 395, 401, 403-408, 417, 418, 426, 435, 469-473, 499, 527, 544, 563
    spinning, 513, 517
    von Kármán's equations, 375, 379-381, 433, 434, 528, 544
    see also Disk
Poincaré,
    asymptotic expansion, 379, 565, 637
    section, 182, 268, 269, 340, 341
Polar,
    coordinates, 335, 389, 395, 408, 411, 415, 474
    decomposition, 110, 215, 252, 366, 376
Polarization voltage, 342
Postbuckling, 1, 4, 5, 23, 171, 176, 318, 319, 375, 381, 562
Preconed, 263, 278
Pretwisted beams, 180, 278

Primary resonance of, 12, 25, 57, 61, 62
    buckled beams, 319, 324, 328, 331
    cables, 137, 138, 152, 153, 158, 161, 164
    cantilever beams, 181, 268, 269, 291, 292, 306, 312
    circular plates, 503, 504
    cylindrical shells, 566, 638
    doubly-curved shells, 564
    microbeams, 346
    panels, 381
    rings, 513
    rotating disks, 516, 522, 524, 526
    spherical shells, 564, 645, 650
    taut strings, 112, 121, 124-127, 129
Principle of virtual work, 69, 70, 110
Prismatic, 169, 175, 181, 189-191, 223, 246, 247, 249
Pseudo arclength, 288
Pseudo-state plane, 268
Pull-in voltage, 344, 345, 352, 353, 536, 537, 541-543
Pump, micro, 531-543, 545
PVDF, 376
PZT, 376, 456, 457

Quasiperiodic, 12, 23, 25, 112, 135, 137, 138, 182, 269, 281, 301, 319, 336, 337, 339, 340

Radius of gyration, 179, 224, 225, 317
Random vibration, 8, 137, 138, 268, 381
Rayleigh-Ritz method, 378
Reconstitution, method of, 139
Rectilinear coordinates, 355, 392, 563
Reduced-order model, 53
    buckled beam, 330, 331
    microbeam, 342, 345, 351, 353
    microplate, 531, 541-543
    rotating disk, 523

Refined shear-deformable theories, 376, 378, 561
Reissner's variational principle, 28, 380
Reissner-Mindlin theory, 372, 373, 379
Resonator, 342, 346, 353
Rigid-body, 65, 66, 69, 72, 109, 195, 229, 246, 251, 252, 256, 257, 263, 593, 560
    displacement, 107, 251
    rotation, 65, 98, 110, 207, 208, 214, 256, 362, 596,
    transformation, 256
    translation, 110, 214, 259, 368
Rigidity, 6
    bending, 28, 178, 470, 532
    shear, 98, 111
    torsional, 168, 174, 234, 235, 244
Ring,
    circular, 181, 265, 513, 565, 637
    rotating, 513, 565, 637
Ritz method, 27, 35, 36, 38-42, 45, 47, 61, 180, 377, 378
Rocking, 315
Rope, jump, 112
Rotary inertia, 14, 15, 28, 31, 67, 98, 179, 181, 184, 189, 217, 221, 223, 224, 229, 236, 241, 243, 270, 372, 373, 377, 380, 407, 409, 432, 438, 465, 561, 588, 598, 603, 626, 633, 644
Rotating, 23, 90, 97, 246, 517, 526, 569
    beam, 178, 246, 278
    disk, 23, 513, 516, 517
    ring, 513, 565, 637
    see also Spinning
Rotations,
    Euler, 177, 179, 376
    Euler-like, 177, 178, 240
    finite, 177, 237
    virtual, 177, 207, 210, 226, 355, 362, 376, 563, 596
Rotor, 6, 171, 176, 178, 180, 240, 245, 246, 263, 264

Saint Venant's
    plate theory, 544
    principle, 175
    semi-inverse method, 172
    solution, 175
    warping solution, 174-176
Sandwich, 192-194, 373, 402
Saturation phenomenon, 24, 25, 138
Saw, 513
Secondary resonances, 25, 181
Self-equilibrating, 6, 7, 175
Self-excited oscillations, 6, 23, 513
Semiconductor industry, 341
Shaker, 130, 132, 267, 294, 312, 313, 334
Shear, 2, 4, 8, 94, 95, 174, 179, 181, 186, 191, 252, 270, 375, 382, 485, 559, 578, 596, 604, 615, 616, 632
    center, 246
    -correction factor, 173, 174, 373, 374, 402, 446, 561, 608
    coupling, 269, 377, 403, 456, 463, 581, 628
    -deformable beam theory, 172, 186, 187, 189
    effects, 4, 173, 174, 376, 381, 456, 561, 581
    energy average, 191, 400
    first-order, 395, 576, 579, 584, 585, 588, 599-601, 626, 627
    geometric average, 191, 400
    higher-order, 440, 499
    inertia, 261
    inplane, 2, 174, 358, 360, 362, 363, 366, 367, 417-420, 429, 440, 442, 451, 452, 560, 584, 588, 589, 591, 600, 614, 641
    intensity, 384, 386, 387, 390, 394, 398, 400, 429, 442, 574, 575
    interlaminar, 4, 374-376, 448, 449, 561, 615, 617
    load, 190, 193, 249, 251, 449, 617
    modulus, 95, 249, 376

SUBJECT INDEX 743

out-of-plane, 174, 562
resultant, 220, 249
rigidity, 98, 111
strength, 372
strain, 1, 172-174, 184, 191, 222, 251, 253, 355, 360, 362, 363, 371-374, 396, 417, 418, 617, 620, 641
stress, 100, 172-174, 186, 191, 193, 218, 249, 373-377, 402, 446, 561, 562, 588, 608, 617
third-order, 436, 594, 628
transverse, 2, 4, 6, 173, 174, 179, 184, 263, 372, 373, 375-377, 380, 430, 432, 436, 440, 442, 444, 448, 452, 454, 560-562, 573-576, 585, 588, 594, 599, 601, 615, 617, 620, 627, 628
warping, 173, 175, 176, 186, 192-194, 246, 247, 249, 250, 372, 382, 396, 397, 402, 403, 417, 426, 436, 438, 457, 560, 562, 577, 581, 582, 594, 615, 627, 629, 630
Shear-deformation theories, 4, 173, 186, 372, 373, 376, 378, 379, 381, 383, 387, 396, 417, 435, 446, 560, 577, 588, 593, 601
first-order, 173, 174, 191, 372, 375, 378, 400, 402, 446, 561, 608
higher-order, 4, 172, 174, 373-377, 379, 561, 615
layerwise, 192, 446, 562, 615
plates, 396, 435, 442
refined, 378, 380, 561
shells, 577, 588, 593, 601, 608
third-order, 172, 189, 190, 374, 376, 379, 402, 435, 454, 456, 561, 615
Shell, 1, 2, 24, 171, 355-357, 367, 371, 372, 378, 380, 559-649
axisymmetric, 565, 641
circular cylindrical, 564, 566, 576, 588, 611

classical, theories, 560, 566
composite, 561, 615
conical, 568
cylindrical, 24, 566
doubly-curved, 571, 582
equations of motion for cylindrical shells, 581, 589, 631
equations of motion, Donnell's, 614
laminated, panels, 561, 615
layerwise shear-deformable, 615
linear shear-deformable, 561, 577
nonlinear classical theory of, 582
of revolution, 569
simplified, theories, 603
spherical, 570
theories, 560, 605, 611, 613, 614
twisted circular cylindrical, 568, 629
Sidebands, 270, 319, 335, 336, 340
unexplained, 319, 336
Sine-dwell sweep, 13
Singular perturbation, 529
Skew plate, 379
Slenderness ratio, 174, 224
Slipping, 335
Slow dynamics, 270, 529
Snapping through, 278, 318, 319, 338, 339
Softening nonlinearity, 163, 165, 167, 178, 223, 275, 277, 282, 289, 296, 303, 310, 330-332, 350, 381
Softening response, behavior, 137, 161, 282, 330-332, 350, 381
Solvability condition, 58, 59, 158, 160, 272-274, 279, 293, 326, 348, 478, 489, 553, 640
Spherical shells, 564, 570, 571, 629, 630, 641-653
Spinning,
disk, 513, 517
membrane, 513
Spring, helical, 53-55, 121, 264
State-space, 23, 48, 377-379

744 SUBJECT INDEX

Strain, 6, 75, 76, 84-86, 95, 98, 107, 110, 113, 117, 136, 139, 141, 168, 172-175, 184, 185, 191, 197, 205, 214-216, 222, 237, 239, 242, 247, 251, 294, 295, 360, 367, 368, 374, 414, 423, 427, 446, 451, 452, 458, 468, 545, 582, 593, 594, 596, 614, 622, 632, 635
  conjugate, 600
  engineering, 110, 248, 376, 573, 578
  Green-Lagrange, 177
  inplane, 363, 417-419, 591
  Jaumann, 369, 631, 632
  local, measures, 177, 194, 213, 220, 229, 251, 377, 563
  nonlinear, 239
Strain-displacement, 4, 65, 75, 76, 107, 185, 186, 192, 214, 215, 218, 377, 383, 562, 563, 582, 593, 608, 612, 614, 622
  Cartesian coordinates, 76
  curvilinear coordinates, 80
  cylindrical coordinates, 77
  spherical coordinates, 78
Stress, 3, 4, 6, 7, 28, 76, 94, 110, 114, 172-175, 191, 193, 218, 230, 233, 235, 242, 247, 249, 254, 262, 371, 372, 374-376, 407, 411, 416, 514, 532, 533, 536, 560-562, 579, 584, 599, 600, 609, 612-614, 616, 617
  engineering, 248, 376
  Jaumann, 467, 633
  local, measures, 7, 177, 184, 220, 229, 377, 563
  residual, 532, 538, 542
Stress resultants for,
  circular plates, 390
  heated plates, 467, 468
  laminated plates, 429, 439, 441, 445
  rectangular plates, 384

Stress-strain relations, 65, 75, 85, 96, 97, 107, 168, 189, 384, 390, 394, 398, 429, 449, 451, 467, 574, 617
  anisotropic, 86
  orthotropic, 87
  thermoelastic, 412
Stretching, *see* Midplane stretching
String, 2, 111, 112, 134
  approximate equations, 114-117
  attacking three equations, 121
  discretization model, 126
  evaluation of two-equation model, 124
  exact equations, 113
  experiments, 132
  mode shapes, 118, 119, 126
  modeling of, 113
  modulation equations for, 113
  multiple resonances, 138
  natural frequencies, 133
  nonlinear response, 112, 113, 119, 125, 128
  Poisson's effect, 114-117
  response to combined parametric and external resonances, 137
  response to primary resonance, 121, 124-127, 129
  two-equation model, 118-121, 124, 125
  whirl, 112, 125 ,130 ,132-134
Subcombination resonance, 278, 291, 298, 302-304, 544
Subdomain method, 27, 48
Supports, 7
  horizontal, 148
  moving, 137
Surface analysis, 355, 377, 419, 420, 425
Switch, micro, 342

Temperature, 8, 88, 98, 412, 414, 468, 544, 545, 547, 550, 555

Thermal, 88, 94, 98, 412, 468, 544, 545, 547
  effect, 98, 414, 467, 468, 545
  load, 543, 544, 549, 551
Thermoelastic, 7, 96, 414
  constitutive relations, 88, 89, 93, 96
  coupling, 544, 547
  damper, 544
  equations, 412, 415
Thermopneumatic, 531
Thin-walled structures, 4, 75, 362
Time-averaged Lagrangian, 63, 64, 286, 306, 651
Timoshenko beam, 173, 174, 179
Torsional, 2, 6, 8, 23, 180, 182, 202, 226, 244, 264, 278, 285, 315, 379
  rigidity, 111, 168, 174, 234, 235, 244
  variable, 240, 244
  warping, 174, 176, 234, 235, 246, 251, 253
Transfer, energy, 5, 23, 181, 270, 319, 637
Transformation of,
  coordinates, 17, 21, 83, 102, 195, 202, 360, 376, 377, 417
  matrix, 83, 101, 103, 105, 177, 198, 201, 202, 203, 208, 209, 217, 237, 240, 256, 263, 355, 357, 417, 435, 466, 566, 599, 605, 606, 612, 622
  stresses and strains, 82
Transverse shear, 1, 2, 4, 6, 172-174, 179, 184, 249, 371-377, 380, 386, 387, 390, 394-396, 398, 400, 402, 430, 432, 436, 442, 444, 448, 452, 454, 559, 560-562, 573-576, 579, 585, 588, 594, 599, 601, 617, 620, 627, 628
Trapezoidal-edge effect, 560, 573, 577, 597, 616
Tuning fork, 112

Turbine, 6, 171
Twist, 6, 177, 180, 237, 240, 245
Twisted, 278, 568, 629, 630
  beam, 172, 180, 200, 209, 211, 212, 264
  blade, 176, 178

Valve, 53-55, 57, 531
Variation of curvatures, 210, 365
Variational, 41, 66, 71, 218, 375
  methods, 27, 28, 42, 53, 186
  principle, 28, 374, 375, 380
Virtual,
  displacement, 55, 69, 70, 109, 596
  principle of, 69, 110
  rotations, 177, 207, 208, 210, 226, 355, 362, 376, 563, 596
  work, 55, 69, 70, 109, 110, 140, 284-286, 298, 300, 304, 306
Viscoelastic, 381, 544
von Kármán, 531, 605
  equations, 375, 379, 380, 381 433, 434, 528, 544
  nonlinearity, 372, 373, 379, 380, 403, 425, 433, 446, 563, 605-607
  strains, 136, 176, 177, 375, 377, 404, 408, 425, 446, 563, 605

Warping, 1, 6, 172-176, 194, 200, 211, 213, 215, 220, 233, 245, 247, 248, 253, 256, 261, 262, 369, 377, 417, 426, 435, 447, 448, 452, 454-456, 560, 566
  bending-induced, 173
  displacement, 174, 176, 194, 215, 369
  extension-reduced, 173
  function, 175, 176, 186, 192-194, 247-251, 396, 402, 436, 447, 452, 454, 456, 457, 562, 577, 581, 615, 627-630
  shear-induced, 173-176, 249, 372, 382, 397, 403, 417, 426, 438, 560, 582, 594

torsion-induced, 173-175, 234, 235, 246, 251, 253
Wave propagation, 373, 374, 381
Waves, 6, 513, 565, 637
    surface, 24, 26, 637
Weighted residuals, method of, 27, 35, 46, 47, 53, 61, 342
Whipping, 278
Whirling motion, 112, 121, 125, 130, 132-134, 177, 180-182
Winkler-Pasternak foundation, 380